Common Graphs

Linear Function
$y = mx + b$

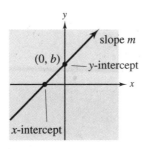

slope m

$(0, b)$ — y-intercept

x

x-intercept

Parabolic Function
$y = x^2$

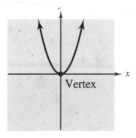

Vertex

Absolute Value Function
$y = |x|$

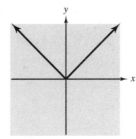

Square Root Function
$y = \sqrt{x}$

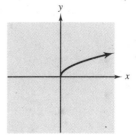

Exponential Function
$y = b^x, b > 1$

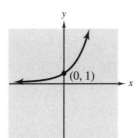

$(0, 1)$

Logarithmic Function
$y = \log_b x, b > 1$

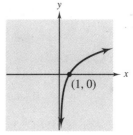

$(1, 0)$

Cubic Function
$y = x^3$

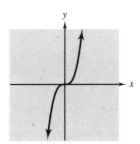

Study Smarter!

Get the best grade possible with a little help from these specially designed study aids.

The **Student Solutions Guide** (0-669-33321-2) includes solutions to odd-numbered text exercises.

Here's how the **Graphing Calculator Keystroke Guide** (0-669-33322-0) can help you. It provides:
- step-by-step keystroke instructions for the TI-81, TI-82, TI-85, Casio fx-7700G, Casio fx-9700GE and fx-7700GE, Sharp EL-9200/9300, and the HP 48G series
- references for the key word icons in the text which allow you to locate the information you need quickly

The **Tutorial Software** for IBM 3 $\frac{1}{2}$", 5 $\frac{1}{4}$ ", and Macintosh provides you with practice and reinforcement of important concepts.

The **Algebra: A Graphing Approach Videotape** series offers clear coverage of algebra topics in an engaging presentation.

Look for these supplements in your bookstore.

If you don't find them, check with your bookstore manager or call D. C. Heath toll free at 1-800-334-3284. In Canada, call toll free at 1-800-268-2472. Shipping, handling, and state tax may be added where applicable (tell the operator you are placing a #1-PREFER order).

Algebra

A Graphing Approach

Elaine Hubbard
Ronald D. Robinson

Kennesaw State College

D. C. HEATH AND COMPANY

Lexington, Massachusetts   Toronto

Address editorial correspondence to:
D. C. Heath and Company
125 Spring Street
Lexington, MA 02173

Acquisitions Editor: Charles Hartford
Developmental Editor: Kathleen Sessa-Federico
Production Editor: Kathleen A. Deselle
Designer: Cornelia Boynton
Art Editor: Gary Crespo
Production Coordinator: Lisa Merrill
Cover: "Color Blocks #19," 1992 © Nancy Crow. Photo by J. Kevin Fitzsimons.

Published simultaneously in Canada.

Printed in the United States of America.

International Standard Book Number: 0-669-33318-2

Library of Congress Catalog Number: 94-76112

10   9   8   7   6   5   4   3   2   1

Preface

The Approach

Every learning theory emphasizes the value of widening the sensory spectrum. To accomplish this in teaching mathematics, we believe one must give students the opportunity to progress beyond mere symbol manipulation by providing the visual connection that allows students to "see" mathematics in context.

In *Algebra: A Graphing Approach,* we use our extensive teaching experience to create a balance between traditional algebra and the use of the graphing calculator. We have developed an approach to teaching with a calculator that works successfully for us, for our colleagues, and for students. In fact, we have not hesitated to include ideas that are based on our students' observations.

We have learned how to use the graphing calculator to provide an efficient and manageable way for students to begin at a visual, concrete level before moving to the more formal, abstract level of mathematics. Our students are surprised and pleased with their ability to control their own environment and to make things happen. We have found that collaboration evolves naturally.

Students can create a visual illustration of a concept, see relationships, and estimate outcomes. By asking "What if?" questions, they can experiment, explore, and discover. In short, the graphing calculator allows our students to be active participants in the learning process.

Our classroom teaching experience has been that a graphing calculator, far from replacing traditional mathematics, actually motivates it. Students benefit from the graphics overview that they can create for themselves, but they invariably come to understand the need for the refinements and precision that formal algebraic methods provide. The power in this pedagogical approach lies in the student's eventual motivation to formulate definitions and rules and to develop methods for accomplishing mathematical tasks.

These positive outcomes are nurtured by the protocols that lie at the heart of this textbook. Whenever possible, we *explore, estimate,* and *discover* graphically; we *verify, generalize,* and *determine* algebraically. This process preserves mathematics as the authority.

In this book we strive to maintain a proper perspective. The focus is on mathematics, with the graphing calculator serving as a tool for better understanding. We remind students that a calculator is designed to execute the rules of mathematics, not the other way around. And we encourage students to view the calculator as a means to an end rather than an end itself.

This text is written for students who need to begin at the elementary level and progress to the intermediate level. The first two chapters provide a solid review of the fundamentals. The graphing calculator begins to play a prominent role with the introduction of the coordinate plane in Chapter 3. In all of the succeeding chapters, a calculator-based approach is used in the exposition of all topics appropriate to the elementary and intermediate level of study.

In the following pages, the special features of the text are highlighted and discussed in detail.

iii

Key Features

Chapter Opener

Each chapter begins with a short introduction to a real-data application that is
covered later in the chapter as well as an
accompanying graph of the data to moti-
vate students to want to learn more about
the related mathematical ideas. This
introduction also includes a helpful
overview of the topics that will be covered in the chapter and a list of the section titles.

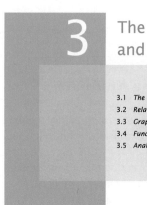

3   The Coordinate Plane and Functions

3.1   *The Coordinate Plane*
3.2   *Relations and Functions*
3.3   *Graphing on a Calculator*
3.4   *Functions: Notation and Evaluation*
3.5   *Analysis of Functions*

Since the Endangered Species Act of 1973,
the gray wolf population in the continental United States has
increased. The bar graph shows the estimated population of
gray wolves between 1960 and 1990.
Superimposed on the bar graph is the graph of a
function that models the data. Does the shape of the graph suggest
that the gray wolf population will increase indefinitely? What
change the shape of the graph in the future?
ow this model function
ises 75–78 at the end of

r 3 we introduce the
ons, and functions. We also
antages of a graphing
the graphs of functions.
with notation, evaluation
ns.

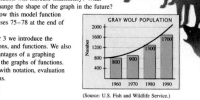

GRAY WOLF POPULATION

(Source: U.S. Fish and Wildlife Service.)

117

---

700   CHAPTER 10   *Quadratic Equations*

10.6   Quadratic Inequalities

Graphing and Algebraic Methods • Special Cases • Applications

Graphing and Algebraic Methods

The standard form of a quadratic equation is $ax^2 + bx + c = 0$. When the equal
sign is replaced by an inequality symbol, we obtain a **quadratic inequality.**

> **Definition of a Quadratic Inequality**
>
> A **quadratic inequality** is an inequality that can be written in the form
> $ax^2 + bx + c > 0$, where a, b, and c are real numbers and $a \neq 0$. We call
> this the *standard form* of a quadratic inequality. The symbols $<$, $\leq$, and $\geq$
> can also be used.

The **solution set** of a quadratic inequality is the set of all replacements for the
variable that make the inequality true.

 EXPLORATION 1

Using a Graph to Solve a Quadratic Inequality

(a) Produce the graph of $y = x^2 + 2x - 8$ and estimate the x-intercepts.

(b) Verify the x-intercepts algebraically.

(c) For what x-values is $y < 0$?

(d) What is the solution set of the quadratic inequality $x^2 + 2x - 8 < 0$?

Discovery

Figure 10.14

$y = x^2 + 2x - 8$

(a) Figure 10.14 shows the graph of $y = x^2 + 2x - 8$. The x-intercepts appear to
be $(-4, 0)$ and $(2, 0)$.

(b) We can verify that the x-intercepts are $(-4, 0)$ and $(2, 0)$ by substitution, or
we can algebraically determine the x-intercepts by solving the quadratic
equation.

$$x^2 + 2x - 8 = 0$$
$$(x + 4)(x - 2) = 0 \qquad \text{Factor.}$$
$$x + 4 = 0 \quad \text{or} \quad x - 2 = 0 \qquad \text{Zero Factor Property}$$
$$x = -4 \quad \text{or} \qquad x = 2 \qquad \text{Solve each case for } x.$$

(c) For $y < 0$, we look for that portion of the graph that is *below* the x-axis. From
Fig. 10.14 we see that $y < 0$ in the interval between the two x-intercepts, that
is, for all values of x such that $-4 < x < 2$.

Figure 10.15

(d) Solving $x^2 + 2x - 8 < 0$ is equivalent to solving $y < 0$. In interval notation,
the solution set of the quadratic inequality is $(-4, 2)$. (See Fig. 10.15.)

Section Opener

Each section begins with a list of subsection titles that provides a brief outline of
the material that follows.

Exploration / Discovery Examples

Whenever possible, we introduce topics via Exploration/Discovery examples. These titled examples help students to discover the rules and properties of algebra themselves. Usually these examples encourage students to experiment with their graphing calculators and answer "What if?" questions about the relationships between equations, graphs, and specific data points. Asking guiding questions in the Exploration and then showing how one looks for patterns and makes generalizations in the Discovery helps students obtain a deeper understanding of the mathematical relationships as well as develop problem-solving skills, key goals of the NCTM Standards.

---

344   CHAPTER 6 *Systems of Linear Equations*

6.1   Graphing and Substitution Methods

Systems of Equations • Solving by Graphing • Substitution Method • Special Cases • Applications

Systems of Equations

We write equations to describe the conditions and relationships given in applied problems. To this point, applications have led to a single equation in one or two variables. However, applications often lead to more than one equation.

EXPLORATION I   *Two Equations Involving Numbers*

(a) Write an equation to describe two numbers whose sum is 6. Produce the graph that represents all the solutions of the equation.

(b) Write an equation to describe two numbers whose difference is 22. Produce the graph that represents all the solutions of this second equation.

(c) Produce the graphs in parts (a) and (b) on the same coordinate system. Trace to the point of intersection of the two lines. What is the significance of this point?

Discovery

(a) The equation is $x + y = 6$. We write $y = -x + 6$ and produce the graph. (See Fig. 6.1.) The equation has infinitely many solutions.

Figure 6.1   Figure 6.2

$y = -x + 6$ $y = x - 22$

(b) The equation is $x - y = 22$. We write $y = x - 22$ and produce the graph. (See Fig. 6.2.) This equation also has infinitely many solutions.

(c) Figure 6.3 shows both graphs on the same coordinate system. The point of intersection is $(14, -8)$.

By substitution, we can easily verify that the point of intersection has coordinates that satisfy the conjunction $x + y = 6$ *and* $x - y = 22$.

SECTION 10.4

Now use the Quadratic Formula to solve

$$t = \frac{-12 \pm \sqrt{12^2 - 4(-6)(-4)}}{2(-6)}$$

$$t = \frac{-12 \pm \sqrt{144 - 96}}{-12}$$

$$t = \frac{-12 \pm \sqrt{48}}{-12}$$

$$t \approx 1.58 \quad \text{or} \quad t \approx 0.42$$

Note that these solutions are consistent with
10.6.

Figure 6.3

$y = -x + 6$ $y = x - 22$

X=14   Y=-8

Radical Equations

In Section 9.8 we developed methods for solving equations involving radicals. An early step in the routine was to raise both sides of the equation to a power equal to the index of the radical. This step can produce a quadratic equation that we can solve with our present methods. Recall, however, that raising both sides of an equation to a power can produce extraneous solutions. We must check all solutions.

EXAMPLE 2   *Solving Equations with Radical Expressions*

Solve the equation $\sqrt{x + 1} = x - 2$.

Solution

Figure 10.7

$y_1 = \sqrt{x+1}$ $y_2 = x - 2$

X=4.3473684
Y=2.3124378

Figure 10.7 shows the graphs of $y_1 = \sqrt{x + 1}$ and $y_2 = x - 2$. The one point of intersection represents the one real number solution of the equation. The solution is approximately 4.3.

$$\sqrt{x + 1} = x - 2$$
$$\left(\sqrt{x + 1}\right)^2 = (x - 2)^2 \qquad \text{Square both sides of the equation.}$$
$$x + 1 = x^2 - 4x + 4 \qquad \text{Square the binomial.}$$
$$0 = x^2 - 5x + 3 \qquad \text{Write the equation in standard form.}$$

Use the Quadratic Formula to solve for x.

$$x = \frac{5 \pm \sqrt{(-5)^2 - 4(1)(3)}}{2(1)} = \frac{5 \pm \sqrt{13}}{2}$$

$$x \approx 4.30 \quad \text{or} \quad x \approx 0.70$$

The solution 4.30 is consistent with the solution that we estimated from the graph. However, when we check the value 0.70, we find that it is an extraneous solution.

Standard Examples

All sections contain numerous standard, titled examples, many with multiple parts graded by difficulty, that serve to reinforce the concepts and techniques introduced in the Exploration/Discovery examples. Helpful comments appear to the right of the detailed solution steps.

Key Word Icons

Whenever we introduce a mathematical technique that can be performed on a graphing calculator, we also include a graphing calculator–related word, usually in the margin. Each word references a discussion in the accompanying *Graphing Calculator Keystroke Guide,* where specific keystroke information for several popular calculator models can be found. Students need only look up the key word in the guide to obtain the location of the appropriate keystroke discussion. We repeat key word icons whenever we feel that students would benefit from a review of that particular calculator technique.

The Graphing Calculator

We can use a graphing calculator to display the rectangular coordinate system or some particular portion of it.

 ERASE GRAPH

Each time you begin a new graphing problem, you will probably want to erase old graphs from the *graph screen*.

The *default viewing rectangle* for most calculators is oriented so that the origin is centered on the graphing screen. (See Fig. 3.2.) The tick marks along the *x*-axis are evenly distributed on both sides of the origin. The same is true for the tick marks along the *y*-axis.

DEFAULT

GRAPH

Figure 3.2

We may need to customize the viewing rectangle to meet the needs of a particular graph. Important features may be more apparent if the graph is drawn over a wider range. You may use a smaller range to see the behavior of the graph near particular points.

EXPLORATION 1

Changing the Range on a Calculator

To represent the first quadrant with 20 units along the *x*-axis, 12 units along the *y*-axis, and each tick mark indicating one unit, we use the following *range-and-scale* notation.

[X: 0, 20, 1]   [Y: 0, 12, 1]

The X and the Y indicate the *x*- and *y*-axes. The first two numbers in each block are the *range values*. The first numbers are the minimum *x*- and *y*-values, and the second numbers are the maximum *x*- and *y*-values. The third numbers represent the ___ance between the tick marks on each axis. We call these numbers the *scale* ___es.

___ enter the given range values and display the viewing rectangle.

___covery

___e that the *y*-axis is along the left edge of the screen and the *x*-axis is along the ___om edge of the screen. (See Fig. 3.3.) In this case, the 20 tick marks along the ___is and the 12 tick marks along the *y*-axis represent one unit each.

___ddition to adjusting the minimum and maximum values along either axis, we ___ adjust the scale along either axis so that each tick mark represents something ___r than one unit.

The Quadratic Formula

For real numbers *a*, *b*, and *c*, with $a \neq 0$, the solutions of the quadratic equation $ax^2 + bx + c = 0$ are given by the **Quadratic Formula**

$$x = \frac{-b \pm \sqrt{b^2 - 4ac}}{2a}.$$

NOTE: A quadratic equation must be written in standard form before the Quadratic Formula can be used to solve it. ∎

EXAMPLE 1   *Using the Quadratic Formula: Two Rational Number Solutions*

Use the Quadratic Formula to solve $3x^2 - 13x = 30$.

Solution

$$3x^2 - 13x = 30$$
$$3x^2 - 13x - 30 = 0 \quad \text{Write the equation in standard form.}$$

Substitute $a = 3$, $b = -13$, and $c = -30$ into the Quadratic Formula.

$$x = \frac{-b \pm \sqrt{b^2 - 4ac}}{2a}$$

$$x = \frac{(-13) \pm \sqrt{(-13)^2 - 4(3)(-30)}}{2(3)} \quad \text{Note that the opposite of } b \text{ is 13.}$$

$$x = \frac{13 \pm \sqrt{169 + 360}}{6} \quad \text{Simplify the radicand first.}$$

$$x = \frac{13 \pm \sqrt{529}}{6} \quad \text{Note that 529 is a perfect square.}$$

$$x = \frac{13 \pm 23}{6} \quad \begin{array}{l}\text{Use the + sign for one solution}\\\text{and the − sign for the other}\\\text{solution.}\end{array}$$

$$x = 6 \quad \text{or} \quad x = -\frac{5}{3}$$ ∎

The equation in Example 1 has two real number solutions, and the equation could have been solved by factoring. Note that after substituting into the Quadratic Formula and simplifying, the radicand is a perfect square.

EXAMPLE 2   *Using the Quadratic Formula: Double Root*

Use the Quadratic Formula to solve $16t^2 + 9 = 24t$.

Definitions, Properties, and Procedures

Important definitions, properties, and procedures are shaded and titled for easy reference.

Notes

Special remarks and cautionary notes that offer additional insight appear throughout the text.

Graphs

Numerous graphs throughout the exposition and exercises help students develop important visualization skills. Graphing calculator displays are proportioned like actual calculator screens to resemble what students obtain on their own calculators. Traditional coordinate plane graphs are also included where appropriate.

All curves have been computer-generated for accuracy. In addition, the axes and the curves appear in the same color, eliminating inaccuracies caused by color registration problems during printing.

A list of important functions and their graphs is provided on the inside front cover.

---

136   CHAPTER 3   *The Coordinate Plane and Functions*

EXAMPLE 5   *Using the Vertical Line Test*

Determine whether each graph represents a function.

(a) Figure 3.21

(b) Figure 3.22

Solution

(a) There are many locations where a vertical line would intersect the graph twice. Figure 3.23 shows one example. The graph does not pass the Vertical Line Test and is not the graph of a function.

(b) Figure 3.24 shows that no vertical line can be drawn to intersect the graph more than once. The graph passes the Vertical Line Test and is the graph of a function.

Figure 3.23

Figure 3.24

If a relation is expressed in the form of a rule, such as an equation, determining whether the relation is a function requires more careful study.

EXPLORATION 1   *Equations That Describe Functions*

Consider the following two relations.

$$G = \{(x, y) \mid x = y^2\} \qquad H = \{(x, y) \mid y = x^2\}$$

(a) Is it possible to list two pairs in set G for which the first coordinate is associated with more than one second coordinate?

---

154   CHAPTER 3   *The Coordinate Plane and Functions*

Figure 3.43

$y = \sqrt{x - 5}$

(b) To evaluate $f(20)$, store 20 in x and eval 3.87.

(c) Your calculator reports an error message

(d) Because function f is already entered, we (See Fig. 3.43.)

From the graph, we see that the domain calculator displayed an error when we not in the domain of the function. Note which is not a real number.

EXAMPLE 4   *Evaluating a Function with a Calculator*

Let

$$f(x) = \frac{x^2 + x^{-2}}{5 - 2(x - 4)^3}.$$

Use your calculator to determine $f(4)$.

Solution   Figure 3.44 shows the entry of function f. Note the parentheses around the numerator and denominator. The calculator will not treat the expression as a fraction without them.

Figure 3.44

```
:Y₁=(X²+X^(-2))/
(5-2(X-4)^3)
:Y₂=
:Y₃=
:Y₄=
```

Figure 3.45

```
4→X
                    4
Y₁
            3.2125
```

In Fig. 3.45, we store 4 in x and evaluate y. From the display, we see that $f(4) = 3.2125$.

The hardest part of using a calculator to evaluate a function is keying in the function. But using the calculator avoids the arithmetic that would be needed to evaluate the function by hand.

As the next two examples illustrate, a function can be evaluated for a variable expression instead of a specific value.

EXAMPLE 5   *Evaluating a Function for a Variable Expression*

Let $g(x) = \sqrt{x^2 - 4}$. Find $g(t)$ and $g(3t)$.

Solution

$$g(t) = \sqrt{(t)^2 - 4} = \sqrt{t^2 - 4} \qquad \text{Replace } x \text{ with } t.$$
$$g(3t) = \sqrt{(3t)^2 - 4} = \sqrt{9t^2 - 4} \qquad \text{Replace } x \text{ with } 3t.$$

Screen Displays of Calculations

In selected locations, especially in the early chapters when students might need guidance on how to use their calculators, we have included screen displays of calculations from a graphing calculator. These show students how a calculation will appear on their screens if it has been entered correctly.

Exercises

This text contains over 9,000 exercises. At the beginning of the problem sets, exercises are usually graded and paired and are intended to follow up on the examples in the text. Later in the set, we provide mixed exercises for the synthesis of concepts and methods. The answers to the odd-numbered section exercises are included at the back of the text.

Writing Exercises

Most exercise sets begin with a pair of simple writing exercises to help students gain confidence in their ability to write about mathematics. Additional writing exercises are scattered throughout the sets and provide a way for students to develop a more complete understanding of the important ideas.

 3.1 *Exercises*

 1. Explain in your own words how to plot the point associated with the ordered pair $A(-2, 4)$.

2. Give an example of a point that is not in any quadrant and describe its location.

In Exercises 3–6, plot the points associated with the given ordered pairs.

3. $A(4, -6)$, $B(-6, 4)$, $C(3, 5)$

4. $A(-2, -3)$, $B(-3, -2)$, $C(-5, 6)$

5. $A(0, -3)$, $B(-5, 0)$, $C(-2, -1)$

6. $A(3, 1)$, $B(3, 7)$, $C(4, -5)$

In Exercises 7–12, plot the points associated with the ordered pairs in the given set. From the resulting pattern, what do you think is the next ordered pair?

7. $\{(-4, -3), (-2, -2), (0, -1), (2, 0), (4, 1), \ldots\}$

8. $\{(-5, 6), (-2, 4), (1, 2), (4, 0), (7, -2), \ldots\}$

9. $\{(-3, 5), (-2, 5), (-1, 5), (0, 5), (1, 5), \ldots\}$

10. $\{(4, 2), (4, 1), (4, 0), (4, -1), (4, -2), \ldots\}$

11. $\{(-4, 0), (-2, 3), (0, 0), (2, 3), (4, 0), \ldots\}$

12. $\{(-1, 1), (0, 0), (1, -1), (2, 0), (3, 1), \ldots\}$

Figure for 19–24

25. Give an example of an ordered pair for which (x, y) is the same as (y, x). Explain why, in general, (x, y) is *not* the same as (y, x).

26. Determine the number of points in the graph of each of the following sets. If the number cannot be determined, explain why.

$A = \{(3, -2), (-4, -5), (-2, 7), (0, 9)\}$
$B = \{(x, y) \mid x = 3 \text{ and } -3 \le y \le 5\}$
$C = \{(x, y) \mid y = x\}$
$D = \{(x, y) \mid x = 3 \text{ and } -3 \le y \le 5 \text{ and } y \text{ is an integer}\}$

In Exercises 27–32, draw the graph of each set.

27. $A = \{(x, y) \mid x \text{ is an integer}, -3 \le x \le 3, y = 2\}$

28. $B = \{(x, y) \mid y \text{ is an integer}, -2 \le y \le 5, x = -2\}$

29. $C = \{(x, y) \mid y \text{ is an integer}, -4 \le y \le 2, x = -y\}$

30. $D = \{(x, y) \mid x \text{ is an integer}, -3 \le x \le 4, y = -x + 1\}$

31. $C = \{(x, y) \mid x \text{ and } y \text{ are integers}, -2 \le x \le 1, 1 \le y \le 3\}$

32. $D = \{(x, y) \mid x \text{ and } y \text{ are integers}, 1 \le x \le 4, -3 \le y \le 2\}$

---

61. When we say, "Find the quotient of $x + 2$ and $x^2 + 5x - 3$," do we divide $x + 2$ by $x^2 + 5x - 3$ or do we divide $x^2 + 5x - 3$ by $x + 2$?

62. Letting $P(x) = x^2 + 2x - 15$, $D(x) = x - 3$, and $Q(x) = x + 5$, show that

$$\frac{P(x)}{D(x)} = Q(x).$$

Now try to verify the following.

(a) $\dfrac{P(2)}{D(2)} = Q(2)$ (b) $\dfrac{P(3)}{D(3)} = Q(3)$

In which part were you unsuccessful? Why?

63. Find the quotient of $x^3 + 2x - 3$ and the product of x and $x - 3$.

64. Divide the product of $x^2 + 2$ and $x - 1$ by $x + 2$.

65. Subtract $4x - 1$ from the quotient of $6x^2 + 11x - 7$ and $2x - 1$.

66. If the sum of $x^2 + 3x - 2$ and $-2x^2 + x - 1$ is divided by $x - 1$, what is the quotient?

67. What polynomial divided by $3x^2$ results in $3x^3 + 4x^2 + x - 4$?

68. When $x^2 - 2x - 3$ is divided by $P(x)$, the quotient is $x + 2$ and the remainder is 5. What is $P(x)$?

69. The width of a rectangle is $x - 3$. The area of the rectangle is $x^2 + 5x - 24$. What expression represents the length of the rectangle? By how much does the length exceed the width?

70. A box of width w and length $w + 8$ has a volume of $w^3 + 10w^2 + 16w$. What is the height of the box? By how much does the height of the box exceed the width?

71. The length in feet of a roll of fencing is represented by $x^3 - 65x - 6$. The fencing is to be stretched across $x + 9$ posts, which are evenly spaced.

(a) Write an expression that represents the distance between the posts. (Hint: The number of spans between the posts is 1 less than the number of posts.)

(b) How much fencing will be left over?

72. The total cost of producing n items is $n^2 + 52n$ dollars. If 50 additional items are produced, the total cost is increased by $100.

(a) What is the unit cost of the original n items?

(b) Does producing 50 additional items increase or decrease the unit cost? By how much?

Exploring with Real Data

In Exercises 73–76, refer to the following background information and data.

The following table shows an increase in the number of cable television systems in operation from 1983 through 1991. Also shown in the table is the number (in thousands) of cable TV subscribers for the same period.

Year	Cable TV Systems	Subscribers (in thousands)
1983	5,600	25,000
1985	6,600	32,000
1987	7,900	41,100
1989	9,050	47,500
1991	10,704	51,000

(Source: *Television and Cable Factbook*, Warren Publishing Co., Washington, D.C.)

The number $S(t)$ of cable TV systems can be modeled by the function $S(t) = 612.0t + 3677.2$, where t is the number of years since 1980.

The number $N(t)$ of cable TV subscribers can be modeled by $N(t) = -143.6t^2 + 5333.6t + 10{,}491.7$, where t is the number of years since 1980.

73. Calculate $N(t) \div S(t)$. Use your calculator to perform the necessary arithmetic, and round all results to the nearest tenth. Let $Q(t)$ represent the quotient and ignore the remainder.

74. Interpret the meaning of $Q(t)$.

75. Note that $Q(t)$ is in slope–intercept form. By inspecting the slope, judge whether $Q(t)$ is increasing rapidly, decreasing rapidly, or remaining relatively constant.

76. Of the three quantities $S(t)$, $N(t)$, and $Q(t)$, which do you expect will increase significantly in the future? Why?

Exploring with Real Data

Most sections contain real-data problems with source acknowledgments that show the relevancy of mathematics to a wide variety of subject areas. Typically, the student is given a mathematical model of the data and is asked to calculate, predict, and interpret. A unique feature is the frequent use of a concluding discussion question of an interdisciplinary nature. An index of the real-data exercises is on the inside back cover.

Other Applications

There are numerous other interesting real-world applications, with some sections devoted entirely to them. A complete list of applications, organized by subject matter, appears in the Index of Applications.

Concept Extension Exercises

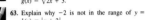

Groups of problems that are not specifically illustrated by examples are listed in the *Instructor's Resource Guide with Tests*. (For example, see problems 55 and 56 on page 169.) These Concept Extension exercises offer slight deviations from the kinds of examples presented in the exposition, and thus require students to extend the ideas and techniques they have learned to solve them.

SECTION 3.5 *Analysis of Functions* **169**

In Exercises 55 and 56, sketch an example of a graph that meets the given conditions.

55. The graph has three x-intercepts: $(-3, 0)$, $(1, 0)$, and $(5, 0)$.

The graph has local maximums at $(-1, 4)$ and at $(3, 7)$.

The graph has a local minimum at $(1, 0)$.

56. The graph has an absolute minimum at $(-3, -5)$.

The graph has one x-intercept: the origin.

The graph has a local maximum at $(2, 4)$.

The graph has a local minimum at $(5, 3)$.

57. For function f, $f(1) = 5$ and $f(3) = -2$. Explain how you can conclude that there is at least one x-intercept $(a, 0)$ such that $1 \le a \le 3$. (Assume that the graph of function f is a smooth curve with no breaks in it.)

58. Suppose function g has a local maximum at $(-1, 7)$ and another local maximum at $(1, 5)$. If the y-intercept is $(0, b)$, can we conclude that $5 \le b \le 7$? Sketch an example of a graph that supports your answer.

59. At the local hardware store, the cash register is programmed to add a 5% sales tax to the price x of all goods purchased. The total cost C is then computed as follows.

$$C(x) = x + 0.05x = 1.05x$$

In Exercises 60 and 61, produce a complete graph of the given function on your calculator. Then answer the following questions.

(a) How many x- and y-intercepts are there?

(b) How many local minimum and maximum points are there?

(c) What are your estimates of the domain and range?

(d) What are the approximate coordinates of the x-intercept that is the farthest to the right?

60. $f(x) = x^4 + 3x^3 - 2x^2 - 5x + 3$

61. $g(x) = 4x^2 - 2x^3 - x^4 + 3x - 5$

62. Explain why -5 is not in the domain of $y = g(x) = \sqrt{2x + 3}$.

63. Explain why -2 is not in the range of $y = h(x) = |x + 2|$.

In Exercises 64–67, use the integer setting to graph the given function on your calculator. Then answer the following questions.

(a) What are the x- and y-intercepts?

(b) What is the absolute minimum and the absolute maximum?

(c) What are your estimates of the domain and range?

64. $f(x) = |x + 2| + 3$ **65.** $g(x) = 5 - |x - 2|$

66. $h(x) = \sqrt{x - 1}$ **67.** $h(x) = 5 - \sqrt{x + 4}$

68. Explain why function f has an absolute minimum but function g does not. What is the absolute minimum of function f?

$f(x) = x + 1$ Domain: $\{x \mid x \ge 0\}$

$g(x) = x + 1$ Domain: $\{$all real numbers$\}$

69. Explain why function f has an absolute maximum but function g does not. What is the absolute maximum of function f?

$f(x) = x^2$ Domain: $\{x \mid 0 \le x \le 1\}$

$g(x) = x^2$ Domain: $\{$all real numbers$\}$

SECTION 7.2 *Multiplication and Special Products* **435**

The number $B(t)$ of textbooks sold can be approximated by $B(t) = 103.96 + 5.91t$, where t is the number of years since 1984 (for 1985, $t = 1$).

The average book price $P(t)$ for textbooks can be approximated by $P(t) = 0.16t^2 - 0.61t + 14.75$, where t is the number of years since 1984.

117. Use the data in the table to calculate the total expenditure for college textbooks in 1990.

118. The total expenditure $E(t)$ for college textbooks for any year t is the product of the number sold and the average price. Write a function for $E(t) = B(t) \cdot P(t)$.

119. Use your function $E(t)$ to approximate the total expenditures for college textbooks in 1990.

120. Use the expression

$$\frac{[E(6) - \text{Actual Total Expenditures in 1990}]}{\text{Actual Total Expenditures in 1990}}$$

to show that the model is in error by only about 1%.

Challenge

In Exercises 121 and 128, multiply.

121. $(x^n + 2)(x^n + 4)$

122. $(x^a + y^b)(2x^a - y^b)$

123. $(x^n - 2)^2$ **124.** $(2x^n + y^n)^2$

125. $(x^n + 4)(x^n - 4)$

126. $(2x^n - 3y^n)(2x^n + 3y^n)$

127. $x^{n-3}(x^{n+6} - x^3)$ **128.** $(x + 1)(x^{2n} - x^n + 1)$

129. In the accompanying figure, the total area of the largest rectangle is the sum of the areas of the four inner rectangles. Use this fact to show the following.

$$(a + b)^2 = a^2 + 2ab + b^2$$

130. In the accompanying figure, the total area of the largest rectangle is the sum of the areas of the three inner rectangles. Use this fact to show the following.

$$(a - b)(a + b) = a^2 - b^2$$

131. Explain why the following two products are the same. Which is easier to multiply?

(i) $(x - a)[(x + a)(x + a)]$

(ii) $[(x - a)(x + a)](x + a)$

132. Explain how to use a special product to find the product of the trinomials

$$(x + 3y - 4)(x + 3y + 4).$$

In Exercises 133 and 134, decide the best way to group the factors and then perform the multiplication.

133. $(2x + 1)(4x^2 + 1)(2x - 1)$

134. $(1 + 3a)(9a^2 - 1)(3a - 1)$

In Exercises 135–138, decide how to rearrange terms and to group within the factors to obtain a special product. Then perform the multiplication.

135. $(x + y + 2)(2 + x + y)$

136. $(x + y + 2)(-2 + x - y)$

137. $(x^2 + 5x + 3)(x^2 + 5x - 3)$

138. $(x^2 - 5x + 3)(x^2 + 5x - 3)$

In Exercises 139–142, multiply.

139. $[2y - (x - 3)]^2$

140. $[(2a + 3) - b]^2$

141. $[(x - 3) + 5y][(x - 3) - 5y]$

142. $[2 + (a + 3b)][2 - (a + 3b)]$

Geometric Connections Exercises

Problems that specifically reference topics and formulas in geometry appear throughout the exercise sets. A useful list of common geometry formulas is printed on the inside back cover.

Challenge Exercises

These problems appear at the end of most exercise sets and offer more challenging work than the standard and Concept Extension problems.

Quick Reference Summary

Quick References appear at the end of all sections except those that are devoted exclusively to applications. These detailed summaries of the important definitions, rules, properties, and procedures are grouped by subsection for a handy reference and review tool.

Chapter Review Exercises

Each chapter ends with a set of review exercises. These exercises include helpful section references that direct students to the appropriate sections for review. The answers to the odd-numbered review exercises are included at the back of the text.

Chapter Tests

A chapter test follows each chapter review. The answers to all the test questions, with the appropriate section references, are included at the back of the text.

Cumulative Tests

A cumulative test appears at the end of Chapters 2, 4, 6, 8, and 11. The answers to all the test questions, with section references, are included at the back of the text.

---

126   CHAPTER 3 *The Coordinate Plane and Functions*

3.1   Quick Reference

Rectangular Coordinate System

- An **ordered pair** consists of two numbers written in parentheses with the order in which the numbers are written being significant.

- Ordered pairs of real numbers are represented by points plotted in a **rectangular** (or **Cartesian**) **coordinate system.** This system consists of two perpendicular number lines called **axes** intersecting at the **origin.** The horizontal number line is called the *x*-axis; the vertical number line is called the *y*-axis.

- The axes divide the plane into four regions called **quadrants.** The quadrants are numbered counterclockwise from I to IV beginning with the upper right-hand

---

Chapter Review Exercises   **407**

6   Chapter Review Exercises

Section 6.1

1. Suppose you draw the graphs of $y_1 = ax + b$ and $y_2 = cx + d$ on the same coordinate axes. If the graphs intersect only at point Q, what is the significance of that point?

2. Suppose you use the substitution method to solve the following system.

   $2x + 3y = 3$

In Exer ordered equatio

3. (2

4. (

In Exe the giv

5. y

7. x

13. $y = -3x + 5$
   $y + 3x = 7$

14. $0.25x + 0.25y = -0.5$
   $-2x = 4 + 2y$

15. $y = -x + 4$
   $y = x - 3$

Chapter Test   **411**

6   Chapter Test

1. Use the graphing method to estimate the solution of the following system of equations. Show your verification of the solution by substitution.

   $y = -2x + 3$
   $x - y = 3$

In Questions 2 and 3, use the substitution method to solve the system of equations.

2. $x + 2y = 10$
   $3x + 4y = 8$

3. $2x + y = 7$
   $x - y = 8$

5-6   Cumulative Test

1. Find the values of a and b so that the ordered pairs are solutions of the given equation.
   $2x + 3y = 7$ $(3, a)$ $(b, 1)$

2. Determine the intercepts of the graph of $5x + y = -10$.

3. For each of the given equations, state whether the slope of the graph is positive, negative, zero, or undefined.
   (a) $x + 2y = 3$ (b) $x + 2 = 3$ (c) $0 \cdot x + 2y = 3$ (d) $x = 2y + 3$

4. Write $x - 2y + 4 = 0$ in slope–intercept form. Then, use what you know about the slope and *y*-intercept to draw a graph of the equation.

5. Consider the equation $y = \frac{2}{3}x - 1$. Describe the rate of change of y with respect to x.

6. Line L_1 contains point $P(-4, 3)$. Line L_2 is perpendicular to L_1. If L_1 and L_2 intersect at point $Q(3, -2)$, what is the slope of L_2?

7. Write an equation of a line that contains points $A(-2, -5)$ and $B(6, 3)$. Express the equation in the form $Ax + By = C$.

8. The equation of line L_1 is $2x + 3y = 6$. Line L_2 is parallel to L_1 and contains the point $P(1, -4)$. Write the equation of L_2 in slope–intercept form.

9. The graph of $x - y < 7$ is a shaded half-plane.
   (a) Describe the boundary line.
   (b) Describe two methods for determining which half-plane to shade.

10. Draw the graph of $|x - y| \leq 5$.

11. Each of the following is a graph of a system of two linear equations in two variables. From the graph, describe the solution set of the system.

   (a)
   (b)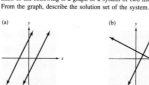
   (c)

12. Use the substitution method to solve the following system of equations.
   $x + 7y = 40$
   $3x - 2y = -18$

13. Use the addition method to solve each of the following systems. Indicate if a system is inconsistent or if the equations are dependent.

Supplements

Graphing Calculator Keystroke Guide   This guide provides helpful keystroke instruction for various graphing calculator models and is referenced by the key word icons in the text.

Instructor's Resource Guide with Tests   This item contains worked-out solutions to the even-numbered exercises in the text as well as formatted chapter and cumulative tests and their answers.

Student Solutions Guide   Worked-out solutions to the odd-numbered text exercises are included in this guide.

Computerized Testing   A computerized test bank of multiple-choice and free response questions for the IBM PC (in DOS and Windows versions) and the Apple Macintosh is available to instructors free of charge.

Tutorial Software   This supplement offers students additional problem-solving practice and concept reinforcement. This software is available for the IBM PC (in DOS and Windows versions) and the Apple Macintosh.

Videotapes   A series of videotapes provide concept review and worked-out examples to reinforce the text presentations.

Acknowledgments

Writing a textbook is a time-consuming and difficult task. Such a project cannot succeed without the understanding and assistance of many other people.

We would like to thank the following colleagues who reviewed the manuscript and made many helpful suggestions:

Donald Clayton, Madisonville Community College; Barbara Edwards, Portland State University; Doris Edwards, Motlow State Community College; Nancy Forrester, Northeast State Technical Community College; Roberta Gehrmann, California State University; Ruth Ann Henke, Southern Illinois University; Marijo LeVan, Eastern Kentucky University; Julia Polk, Okaloosa Walton Junior College; Allan Skillman, Casper College; Betsy Whitman, Framingham State College; Mary Jane Wolfe, University of Rio Grande; and Karl Zilm, Lewis and Clark Community College.

Helen Medley exhibited great patience and care in reviewing the text and answers to ensure accuracy.

We thank Kennesaw State College for the support that we received to make it possible for us to complete this book. We express our appreciation to our colleagues at Kennesaw State College for their patience, cooperation, and helpful suggestions. In particular, we thank Joanne Fowler for assisting with editing the early versions of the manuscript, for arranging release time for the project, and for providing moral support. We thank Harriet Gustafson for her input on the initial version of the exercises as well as class-testing the manuscript. We sincerely appreciate Ken Kiesler, Carla Moldavan, Cathy Nuse, and Ann Powell for their complete cooperation in field-testing the manuscript.

Finally, we thank the staff at D. C. Heath for their assistance and contributions to the project. We are particularly appreciative to Charlie Hartford, Acquisitions Editor; Kathy Sessa-Federico, Developmental Editor; Kathy Deselle, Production Editor; Cornelia Boynton, Designer; Gary Crespo, Art Editor; Carolyn Johnson, Editorial Associate; and Lisa Merrill, Production Coordinator.

Elaine Hubbard

Ronald D. Robinson

Contents

1 The Real Number System

The mean elevation of the continental United States is 2500 feet. The table shows the highest points in six selected states.

The difference between the altitude of Mount Rainier and the mean elevation is the **positive number** +11,908, which indicates that Mount Rainier is considerably higher than the mean elevation. The difference between the altitude of Iron Mountain and the mean elevation is the **negative number** −2175, which indicates that Iron Mountain is lower than the mean elevation. (See Exercises 115–118 at the end of Section 1.3 for more on this topic.)

Chapter 1 begins with a discussion of the structure and order of the real numbers. We develop rules for the four basic operations with real numbers and for working with exponents and square roots. We discuss the order in which all these operations are performed and conclude with some of the basic properties of the real number system.

State	Highest Point	Altitude (in feet)
Colorado	Mount Elbert	14,431
Florida	Iron Mountain	325
Illinois	Charles Mound	1,241
Rhode Island	Durfee Hill	805
Tennessee	Clingmans Dome	6,642
Washington	Mount Rainier	14,408

(Source: U.S. Geological Survey.)

1.1   The Real Numbers

Structure of the Real Numbers   ▪   The Number Line   ▪   Order of the Real Numbers   ▪   Opposites and Absolute Value

Structure of the Real Numbers

It is convenient to organize the numbers of mathematics into sets. A **set** is *any* collection of objects, but our focus will be on sets of numbers.

We use braces to enclose the **members** or **elements** of a set, and we usually assign a letter name to the set. For example, the set *N* of **natural,** or **counting numbers,** is written as follows.

$$N = \{1, 2, 3, 4, \ldots\}$$

The dots ... mean that the list continues without end. Because *N* has infinitely many elements, it is an example of an **infinite set.**

A set with no elements is called the **empty set** and is named ∅.

NOTE: The set {0}, which contains *one element,* 0, is not the same as the empty set ∅, which contains *no elements.*   ∎

The number of elements in *A* = {2, 4, 6, 8} is the counting number 4. If the number of elements of a set is 0 (empty set) or a counting number, the set is a **finite set.** To indicate that 8 is an element of set *A,* we write 8 ∈ *A.*

We wrote sets *N* and *A* by listing all of the elements of the set. This method of describing a set is called the **roster method.**

When the number 0 is included with the natural numbers, the collection of numbers is called the set of **whole numbers,** which we denote by

$$W = \{0, 1, 2, 3, \ldots\}.$$

A **variable** is a symbol, usually a letter, that represents a number. In the following, we use the variable *n* to describe set *A* in **set-builder notation.**

$$A = \{n \mid n \text{ is an even number between 2 and 8, inclusive}\}$$

In this context, the symbol | means "such that." We read the set as "the set of all numbers *n* such that *n* is an even number between 2 and 8, inclusive."

 NEGATIVE
When the temperature falls 7° below zero, we use a negative number to describe the temperature: −7°. The whole numbers and their negatives form the set of **integers.**

$$J = \{\ldots, -3, -2, -1, 0, 1, 2, 3, \ldots\}$$

There are infinitely many numbers that are not integers. For example, fractions such as $\frac{1}{2}$, $\frac{2}{3}$, and $-\frac{3}{8}$ are called **rational numbers.**

$$Q = \left\{ \frac{p}{q} \mid p \text{ and } q \text{ are integers}, q \neq 0 \right\}$$

Note that the integers are also rational numbers because we can write any integer n as $\frac{n}{1}$.

DECIMAL

DIVIDE

It can be shown that the decimal representation of any rational number is either a terminating or a repeating decimal. For example, $\frac{3}{4} = 0.75$, which is a terminating decimal, and $\frac{3}{11} = 0.272727\ldots$, which is a repeating decimal, usually written $0.\overline{27}$.

SQUARE ROOT

The set I of decimal numbers that neither terminate nor repeat is called the set of **irrational numbers.** Examples of irrational numbers are π and most square roots, such as $\sqrt{6}$, $\sqrt{15}$, and $-\sqrt{7}$. Note that $\sqrt{25} = 5$, and so $\sqrt{25}$ is a rational number.

The rational numbers and the irrational numbers are two distinct sets of numbers. Taken together, the two sets form the set R of **real numbers.** We can think of the real numbers as all numbers with decimal representations. Figure 1.1 shows the relationships of the various sets of numbers we have discussed.

Figure 1.1   The Real Numbers

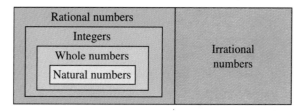

EXAMPLE 1   *Classifying the Real Numbers*

Name every set to which the following real numbers belong.

$$3.2, \quad 0, \quad -4, \quad 3\frac{5}{6}, \quad -\frac{3}{4}, \quad \sqrt{9}, \quad \sqrt{6}$$

Solution

	3.2	0	−4	$3\frac{5}{6}$	$-\frac{3}{4}$	$\sqrt{9}$	$\sqrt{6}$
Irrational							✓
Rational	✓	✓	✓	✓	✓	✓	
Integer		✓	✓			✓	
Whole		✓				✓	
Natural						✓	

The Number Line

We use a **number line** to visualize real numbers and their relation to each other.

To construct a number line, we choose a point corresponding to the number 0. Points at equally spaced intervals are then associated with the integers. The positive integers are to the right of 0, and negative integers are to the left of 0. All other real numbers are associated with intervening points.

The number associated with a point is called the **coordinate** of the point. The point associated with zero is called the **origin.**

Using a number line to highlight points corresponding to the numbers in a set is called **plotting** the points. The resulting set of plotted points is called the **graph** of the set of numbers.

EXAMPLE 2   *Graphing a Set of Real Numbers*

Graph the set $\left\{-4,\ -1.5,\ 2,\ \pi,\ 4\frac{1}{2},\ \sqrt{36}\right\}$.

Solution

Figure 1.2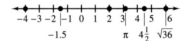

Order of the Real Numbers

The equality symbol is used to indicate that the number to its left has the same value as the number to its right. The following are four basic properties of equality.

> ### Basic Properties of Equality
>
> The following properties hold for real numbers *a, b,* and *c.*
>
> 1. $a = a$.   Reflexive Property
>
> 2. If $a = b$, then $b = a$.   Symmetric Property
>
> 3. If $a = b$ and $b = c$, then $a = c$.   Transitive Property
>
> 4. If $a = b$, then we can replace *a*   Substitution Property
>   with *b* in any expression.

EXAMPLE 3   *Examples of the Properties of Equality*

(a) $3x = 3x$.   Reflexive Property

(b) If $6 = x$, then $x = 6$.   Symmetric Property

(c) If $x + 2 = y$ and $y = 2x - 3$, then $x + 2 = 2x - 3$.   Transitive Property

(d) If $x = y - 3$ and $x + 6 = 10$, then $(y - 3) + 6 = 10$.   Substitution Property

■

The number line emphasizes the relative values or **order** of the real numbers. For any two real numbers a and b, we say that "a is less than b" if a is to the left of b on the number line. We say that "a is greater than b" if a is to the right of b on the number line.

 TEST

We use the symbol $<$ for the phrase "is less than" and the symbol $>$ for the phrase "is greater than." Thus, $\frac{37}{38} < \frac{38}{39}$ and $-15.23 > -16$.

EXAMPLE 4   *Writing Inequalities*

Insert an inequality symbol to make the statement true.

(a) 0 ____ -5 (b) $\frac{1}{3}$ ____ $\frac{1}{4}$ (c) -7 ____ -6

Solution

(a) $0 > -5$ (b) $\frac{1}{3} > \frac{1}{4}$ (c) $-7 < -6$ ■

Inequality symbols can be combined with the equality symbol. The symbol $\leq$ means "less than or equal to" and the symbol $\geq$ means "greater than or equal to."

To describe all of the numbers in the set that are less than or equal to 2, we write $\{x \mid x \leq 2\}$. Figure 1.3 shows the graph of this set.

Figure 1.3

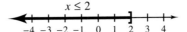
$$x \leq 2$$

We use a bracket to indicate that 2 is included in the set. The same notation is used to write the set in **interval notation:** $(-\infty, 2]$. The infinity symbol $-\infty$ indicates that the interval extends to the left without end.

NOTE: The symbol ∞ does not represent a number. It is used in interval notation to indicate that the interval extends without end.   ■

Note the difference between the graph in Fig. 1.3 and the graph of $\{x \mid x > 5\}$ in Fig. 1.4.

Figure 1.4

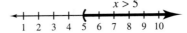
$$x > 5$$

We use a parenthesis to indicate that 5 is not included in the set. The corresponding interval notation is $(5, \infty)$. Here, the infinity symbol ∞ indicates that the interval extends to the right without end.

EXAMPLE 5   *Graphs and Interval Notation*

Graph the given set and write the set in interval notation.

(a) $\{x \mid 0 < x < 7\}$ (b) $\{x \mid -3 < x \le 5\}$

Solution

(a) We use parentheses to indicate that neither 0 nor 7 is included in the set. In interval notation the set is (0, 7). (See Fig. 1.5.)

(b) We use a left parenthesis at -3 and a bracket at 5. The interval notation is $(-3, 5]$. (See Fig. 1.6.)

Figure 1.5   Figure 1.6

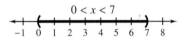

There are two basic properties of inequality.

Basic Properties of Inequality

The following properties hold for real numbers a, b, and c.

1. Exactly one of the following is true:

   $a < b, \quad a = b, \quad \text{or} \quad a > b.$ Trichotomy Property

2. If $a < b$ and $b < c$, then $a < c$.   Transitive Property

---

EXAMPLE 6   *Examples of the Properties of Inequality*

(a) For a real number x, $x < -1$, $x = -1$, or $x > -1$.   Trichotomy Property

(b) If $x < 0$ and $0 < y$, then $x < y$.   Transitive Property

Opposites and Absolute Value

On the number line, two different numbers that are on opposite sides of zero and that are the same distance from zero are called **opposites** (or **additive inverses**).

The numbers 4 and -4 are examples of opposites. (See Fig. 1.7.)

Figure 1.7

So that every real number has an opposite, we shall agree that the opposite of 0 is 0. For any real number x, we use the symbol $-x$ to represent "the opposite of x."

NOTE: It is incorrect to read $-x$ as "minus x" because minus refers to subtraction. It is also incorrect to read $-x$ as "negative x" because $-x$ does not necessarily represent a negative number. The symbol $-x$ refers to the *opposite* of the number represented by x. ∎

Figure 1.7 also shows that the opposite of -4 is 4. We write this fact in the following way.

$$-(-4) = 4$$

We read the statement, "The opposite of negative 4 is 4."

The Opposite of the Opposite of a Number

For any real number a, $-(-a) = a$.

ABSOLUTE VALUE

On the number line, the distance between any number and 0 is called the **absolute value** of the number. For a real number a, we denote its absolute value by $|a|$.

We can also state an algebraic definition of absolute value.

Algebraic Definition of Absolute Value

For any real number a, $|a| = \begin{cases} a & \text{if } a \geq 0. \\ -a & \text{if } a < 0. \end{cases}$

NOTE: In the definition of absolute value, if $a < 0$, then $-a$ represents a *positive* number. The absolute value of a number is never negative. ∎

A **numerical expression** is any combination of numbers and arithmetic operations. When we **evaluate** a numerical expression, we determine the value of the expression by performing the indicated operations.

EXAMPLE 7   *Evaluating Absolute Value Expressions*

Evaluate each expression.

(a) $|10|$ (b) $|-5|$ (c) $-|8|$

(d) $-|-7|$ (e) $|8| + |-7|$ (f) $|-12| - |-9|$

Solution

(a) $|10| = 10$ (b) $|-5| = 5$

(c) $-|8| = -8$ (d) $-|-7| = -7$

(e) $|8| + |-7| = 8 + 7 = 15$

(f) $|-12| - |-9| = 12 - 9 = 3$

1.1   Quick Reference

Structure of the Real Numbers

- A **set** is any collection of objects, such as numbers. The objects are called **members** or **elements** of the set.

- The set $N = \{1, 2, 3, \ldots\}$ is called the set of **natural,** or **counting numbers.**

- A set with no elements is called the **empty set.** If the number of elements in a set is 0 or a counting number, the set is a **finite set.** If a set has infinitely many elements, it is an **infinite set.**

- The set of **whole numbers** is $W = \{0, 1, 2, 3, \ldots\}$.

- Listing the elements of a set is called the **roster method.**

- A **variable** is a symbol that represents a number. A variable is used to write a set in **set-builder notation.**

- The set of **integers** is

$$J = \{\ldots, -3, -2, -1, 0, 1, 2, 3, \ldots\}.$$

- The set of **rational numbers** is

$$Q = \left\{ \frac{p}{q} \mid p \text{ and } q \text{ are integers, } q \neq 0 \right\}.$$

The decimal representations of rational numbers are either terminating or repeating decimals.

- The set of decimal numbers that neither terminate nor repeat is called the set of **irrational numbers.**

- The set of **real numbers** is the set of rational numbers together with the set of irrational numbers.

The Number Line

- A **number line** is a set of points with each point associated with a real number called the **coordinate** of the point. The point whose coordinate is 0 is the **origin.**

- Using a number line to highlight points corresponding to the numbers in a set is called **plotting** the points. The resulting set of plotted points is called the **graph** of the set of numbers.

Order of the Real Numbers

- The following are true for any real numbers a, b, and c.

1. $a = a$.   Reflexive Property
2. If $a = b$, then $b = a$.   Symmetric Property
3. If $a = b$ and $b = c$, then $a = c$.   Transitive Property
4. If $a = b$, then we can replace a Substitution Property
   with b in any expression.

- The **order** of two real numbers can be interpreted as their relative positions on a number line. If a is to the left of b, then $a < b$; if a is to the right of b, then $a > b$.

- The real numbers that satisfy a given inequality can be described with set-builder notation, with **interval notation,** or with a number line graph.

- The following are true for any real numbers a, b, and c.

  1. Exactly one of the following is true:

     $$a < b, \quad a = b, \quad a > b.$$ Trichotomy Property

  2. If $a < b$ and $b < c$, then $a < c$.   Transitive Property

---

Opposites and Absolute Value

- On a number line, two different numbers that are on opposite sides of 0 and the same distance from 0 are called **opposites** (or **additive inverses**).

- The opposite of any real number x is denoted by $-x$.

- For any real number a, $-(-a) = a$.

- On the number line, the distance between any number and 0 is called the **absolute value** of the number. For a real number a, we denote its absolute value by $|a|$.

- For any real number a, $|a| = \begin{cases} a & \text{if} \quad a \geq 0. \\ -a & \text{if} \quad a < 0. \end{cases}$

- A **numerical expression** is any combination of numbers and arithmetic operations. When we **evaluate** a numerical expression, we determine the value of the expression by performing the indicated operations.

1.1 *Exercises*

Use the following for all of the exercises in this section.

$N = \{\text{natural numbers}\}$ $Q = \{\text{rational numbers}\}$

$W = \{\text{whole numbers}\}$ $I = \{\text{irrational numbers}\}$

$J = \{\text{integers}\}$ $R = \{\text{real numbers}\}$

 1. Write a sentence to describe the decimal name of a rational number.

 2. Write a sentence to describe the decimal name of an irrational number.

In Exercises 3–6, state whether the given number has a terminating or repeating decimal name.

3. $\dfrac{3}{8}$ **4.** $-2\dfrac{23}{100}$

5. $\dfrac{35}{99}$ **6.** $\dfrac{5}{11}$

In Exercises 7–16, determine if the statement is true or false.

7. $-7 \in N$ **8.** $-12 \in J$

9. $\dfrac{2}{3} \in R$ **10.** $-\dfrac{7}{4} \in Q$

11. $\sqrt{5} \in Q$ **12.** $\sqrt{7} \in I$

13. $0 \in W$ **14.** $0 \in I$

15. $\dfrac{\pi}{2} \in I$ **16.** $\pi \in R$

In Exercises 17–20, use the roster method to write the given set.

17. $\{w \mid w \in W \text{ and } w \text{ is less than } 7\}$

18. $\{i \mid i \in J \text{ and } i \text{ is between } -3 \text{ and } 3, \text{ inclusive}\}$

19. $\{x \mid x \in W \text{ and } 0 < x < 1\}$

20. $\{n \mid n \in N \text{ and } n \leq 1\}$

In Exercises 21–24, describe, in words, numbers that satisfy the given conditions.

21. Integers that are not whole numbers

22. Rational numbers that are not integers

23. Rational numbers that are integers

24. Integers that are neither negative nor natural numbers

In Exercises 25–32, determine if the statement is true or false.

25. Every integer is also a whole number.

26. Every integer is a rational number.

27. Every rational number is also an irrational number.

28. Every whole number is also an integer.

29. Every natural number is also a whole number.

30. Every nonterminating decimal repeats.

31. The number $\frac{2}{3}$ is a rational number but $-\frac{2}{3}$ is not.

32. The number 0 is a whole number but not an integer.

 33. Explain the difference between 1.75 and $1.\overline{75}$.

 34. Is $0.\overline{438}$ a terminating or nonterminating decimal? Explain why $0.\overline{438}$ is not an irrational number.

In Exercises 35–42, use your calculator to write the given number in decimal form. What does the decimal name suggest as to whether the number is rational or irrational?

35. $\frac{35}{37}$

36. $\frac{8}{11}$

37. $-\frac{\pi}{3}$

38. $\frac{\pi}{4}$

39. $\frac{\sqrt{2}}{\sqrt{8}}$

40. $-\frac{\sqrt{3}}{\sqrt{27}}$

41. $-\frac{\sqrt{5}}{2}$

42. $\frac{\sqrt{7}}{5}$

In Exercises 43 and 44, list all the numbers in the given set that are members of the following sets.

(a) N (b) W (c) J

(d) Q (e) I

43. $\left\{ -4, 3\frac{5}{8}, 0, \frac{3}{5}, \sqrt{7}, 498, 0.25, -17, 0.\overline{63}, \pi, \sqrt{16} \right\}$

44. $\left\{ -\frac{4}{9}, -24, -\sqrt{5}, -4\frac{7}{8}, 32, 0.\overline{36}, \frac{\sqrt{3}}{2}, \frac{\pi}{7}, 0.2863, -\pi, \sqrt{9} \right\}$

In Exercises 45–52, use the given property of equality to fill in the blank.

45. Symmetric: If $2x + 3 = y$, then $y =$ _____.

46. Symmetric: If $2\pi r = A$, then $A =$ _____.

47. Reflexive: $x^2 - 9 =$ _____.

48. Reflexive: $6 + 3t =$ _____.

49. Transitive: If $2L + 2W = P$ and $P = 50$, then _____.

50. Transitive: If $x - 2y = 7$ and $7 = x - y$, then _____.

51. Substitution: If $x + 7 = 10$ and $x = 2t$, then _____.

52. Substitution: If $d = rt$ and $r = 3s$, then _____.

In Exercises 53–56, identify the property of inequality that justifies the given statement.

53. If $x < y$ and $y < 10$, then $x < 10$.

54. If $x < \sqrt{7}$ and $\sqrt{7} < y$, then $x < y$.

55. If x is a real number, then $x = 0$, $x > 0$, or $x < 0$.

56. If x is a real number, then $-2x = 10$, $-2x < 10$, or $-2x > 10$.

In Exercises 57–62, draw the graph of the given set.

57. $\{1, 2, 3, -2, -4\}$ **58.** $\{-1, -3, 5, 3\}$

59. $\left\{\dfrac{1}{2}, 0.75, -\dfrac{4}{3}, 0, -1.2\right\}$

60. $\left\{-2\dfrac{2}{3}, -\dfrac{3}{2}, -1, 3\dfrac{4}{5}\right\}$

61. $\left\{-3, -\dfrac{1}{2}, 3, -\pi, 4.3, \sqrt{4}, -\sqrt{4}, -2\dfrac{1}{3}\right\}$

62. $\left\{5, -4, 2.5, \dfrac{\pi}{2}, -\sqrt{9}, -\sqrt{6}, 3\dfrac{2}{3}, \dfrac{7}{2}\right\}$

 63. Interpret $a < b$ and $a > b$ in terms of "left" and "right" on the number line.

 64. Suppose a and b represent nonzero real numbers. Write a condition under which a is definitely to the left of b on the number line.

In Exercises 65–74, insert $<$ or $>$ to make the statement true.

65. -12 ____ -5 **66.** 11 ____ 5

67. $-\dfrac{15}{7}$ ____ $-\dfrac{7}{15}$ **68.** $\dfrac{22}{23}$ ____ $\dfrac{23}{24}$

69. $\dfrac{22}{7}$ ____ π **70.** 3.14 ____ π

71. $\sqrt{7}$ ____ $2.\overline{645}$

72. $-\sqrt{30}$ ____ $-5.4\overline{7}$

73. $|-4|$ ____ -4

74. $-|-3|$ ____ $-(-3)$

 75. On the number line, -4.1 is to the left of -4. Which is the correct way to describe the order of these numbers, $-4.1 < -4$ or $-4 > -4.1$? Explain.

 76. Describe and compare the graphs of $x < 3$ and $3 > x$.

In Exercises 77–80, write an inequality with the same meaning, but with the inequality symbol reversed.

77. $x < 10$ **78.** $5 \geq -1$

79. $-14 > -14.2$ **80.** $y \leq 0$

In Exercises 81–84, four numbers are given. Without using your calculator, estimate their order from least to greatest. Then use your calculator to verify your estimates.

81. $3.1416, \pi, \sqrt{10}, \dfrac{22}{7}$

82. $-2.6137, -\dfrac{9}{4}, -2\dfrac{2}{3}, -\sqrt{7}$

83. $\dfrac{\pi}{2}, \sqrt{3}, \sqrt{2}, \dfrac{11}{7}$

84. $-3.\overline{3}, -\sqrt{10}, -\sqrt{11}, -\dfrac{16}{5}$

In Exercises 85–94, translate the given information into an inequality.

85. 3 is less than y.

86. a is greater than -3.

87. x is between -2 and 3, inclusive.

88. b is at least 2.

89. y is at most -1.

90. n is between -4 and 3, including 3 but not -4.

91. a is negative.

92. b is nonnegative.

93. x is between -2 and 2.

94. y is greater than or equal to 0.

 95. Explain the difference between the graphs of $x > 2$ and $x \geq 2$.

 96. Explain the difference between the graphs of $x > 2$ and $x < 2$.

In Exercises 97–106, graph the given set on a real number line. Then describe the set with interval notation.

97. $\{x \mid x < -4\}$ **98.** $\{x \mid x \leq 6\}$

99. $\{x \mid x \geq -5\}$ **100.** $\{x \mid x > 4\}$

101. $\{x \mid -3 < x < -1\}$ **102.** $\{x \mid -2 < x < 8\}$

103. $\{x \mid 2 \leq x \leq 5\}$ **104.** $\{x \mid -2 \leq x \leq 6\}$

105. $\{x \mid -3 < x \leq 7\}$ **106.** $\{x \mid 3 \leq x < 8\}$

In Exercises 107–110, a set of numbers is described with interval notation. Write the set in set-builder notation.

107. $[-3, 7)$ **108.** $[0, 5]$

109. $(-\infty, 0]$ **110.** $(-1, \infty)$

 111. Using the word *distance*, explain why $|7| = |-7|$.

 112. Give an example to show that $|-a| = a$ is not necessarily true.

In Exercises 113–116, a number n is described in reference to a number line. State all possible values of n.

113. The number n is 5 units from 0.

114. The number n is 3 units from -2.

115. When increased by 2, n is 4 units from 0.

116. The number n has an absolute value of 8.

In Exercises 117–130, evaluate the given quantity.

117. $|-6|$ **118.** $|-4|$ **119.** $-|7|$

120. $-|-10|$ **121.** $|0|$ **122.** $-|12|$

123. $\left|\dfrac{3}{5}\right|$ **124.** $-\left|-\dfrac{9}{7}\right|$

125. $|-2| - |2|$ **126.** $|-4| + |5|$

127. $|-6| - |-5|$ **128.** $|-12| - |-10|$

129. $3 \cdot |-7|$ **130.** $\left|\dfrac{-8}{-2}\right|$

131. Explain why it is misleading to read $-a$ as "negative a."

132. On a number line, what is the location of a number n relative to the number $-n$? What is the location of $-(-n)$ relative to $-n$? Based on your answers, what can you conclude about n and $-(-n)$?

In Exercises 133–138, write the opposite of each number.

133. -6 **134.** 13 **135.** 0

136. $|-4|$ **137.** $-\dfrac{1}{2}$ **138.** $-\dfrac{1}{-9}$

Exploring with Real Data

In Exercises 139–142, refer to the following background information and data.

In recent years, Americans have consumed an average of about 5 gallons of frozen desserts per person per year. The accompanying figure indicates the popularity of the four basic categories. (Source: U.S. Department of Agriculture.)

Figure for 139–142

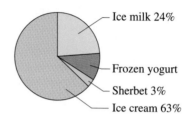

139. On the average how many gallons of ice cream does each American consume annually?

140. What percent of the total consumption is frozen yogurt?

141. On the average how many ounces of sherbet does each American consume annually? (There are 128 ounces in a gallon.)

142. The cost of compiling statistics such as these is borne by the taxpayer. How, if at all, does business and the consumer benefit?

Challenge

In Exercises 143–152, fill in the blanks with $<$, $>$, or $=$.

143. If $a < 0$, then $-a$ ____ 0.

144. If $a > 0$, then $-a$ ____ 0.

145. If $a > 0$, then $|a|$ ____ $-a$.

146. If $a < 0$, then $|a|$ ____ $-a$.

147. If $a < 0$, then $|a|$ ____ a.

148. If $a > 0$, then $|a|$ ____ a.

149. If $a = 0$, then $|a|$ ____ a.

150. If $a = 0$, then $|a|$ ____ $-a$.

151. If $a < 0$, then $-(-a)$ ____ 0.

152. If $a > 0$, then $-(-a)$ _____ 0.

153. For what value(s) of x are the following statements true?

   (a) $|x| = |-x|$ (b) $|-x| = -|x|$

154. For what value(s) of x are the following statements true?

   (a) $\dfrac{|-x|}{x} = 1$ (b) $\dfrac{|-x|}{-x} = -1$

   (c) $\dfrac{|x|}{x} = -1$

155. For what value(s) of x are the following statements true?

   (a) $|x| + 1 = |x + 1|$

   (b) $|1 - x| = 1 - |x|$

In Exercises 156–159, write the given number as the simplified quotient of two integers.

156. 2.7   **157.** 0.352

158. $0.\overline{3}$ **159.** $1.\overline{6}$

1.2 Addition

Addends with Like Signs ▪ *Addends with Unlike Signs* ▪ *Applications*

Addends with Like Signs

The basic operations of arithmetic are addition, subtraction, multiplication, and division. In this section we consider addition.

In an expression such as $7 + 5$, the numbers 7 and 5 are called **addends.** The expression and the result of performing the addition are both called a **sum.** As we will see, the methods for adding two real numbers will differ depending on the signs of the addends. We begin with addends that have like signs.

 EXPLORATION 1

Addends with Like Signs

Use your calculator to determine each sum in column A. Determine the sums in column B without your calculator. Then, answer the questions that follow.

 ADD

Column A	Column B
(a) $3 + 7$	$|3| + |7|$
(b) $-3 + (-7)$	$|-3| + |-7|$
(c) $10 + 2$	$|10| + |2|$
(d) $-10 + (-2)$	$|-10| + |-2|$

   (e) For which sums in column A are both addends positive? negative? In each case, what is the sign of the sum?

   (f) From your observations in part (e), propose a rule for determining the sign of a sum when the addends have like signs.

   (g) Now find the absolute value of each sum in column A and compare it to the corresponding sum in column B.

(h) From your comparisons in part (g), propose a rule for determining the absolute value of a sum when the addends have like signs.

Discovery

Column A	Column B
(a) $3 + 7 = 10$	$|3| + |7| = 10$
(b) $-3 + (-7) = -10$	$|-3| + |-7| = 10$
(c) $10 + 2 = 12$	$|10| + |2| = 12$
(d) $-10 + (-2) = -12$	$|-10| + |-2| = 12$

(e) The addends in parts (a) and (c) are positive, and the sums are positive. The addends in parts (b) and (d) are negative, and the sums are negative.

(f) These results suggest the following rule: If the addends have like signs, then the sign of the sum is the same as the sign of the addends.

(g) In each part the absolute value of the sum in column A is the same as the sum of the absolute values of the addends in column B.

(h) These results suggest the following rule: If the addends have like signs, then the absolute value of the sum is the sum of the absolute values of the addends.

Although the rules suggested by Exploration 1 can be proven to be true, we accept them as given.

Rules for Adding Addends with Like Signs

If the addends of a sum have like signs, then

1. the sign of the sum is the same as the sign of the addends, and

2. the absolute value of the sum is the sum of the absolute values of the addends.

EXAMPLE 1   *Adding Addends with Like Signs*

Determine each sum without your calculator. Then use your calculator to verify your results.

(a) $5 + 12$

(b) $-9 + (-7)$

(c) $-3.7 + (-2.5)$

(d) $-\dfrac{1}{2} + \left(-\dfrac{3}{4}\right)$

Solution

(a) $5 + 12 = 17$

(b) $-9 + (-7) = -16$

(c) $-3.7 + (-2.5) = -6.2$

(d) $-\dfrac{1}{2} + \left(-\dfrac{3}{4}\right) = -\dfrac{2}{4} + \left(-\dfrac{3}{4}\right) = -\dfrac{5}{4} = -1.25$

Addends with Unlike Signs

In Exploration 2 we investigate methods for adding numbers with unlike signs.

 **EXPLORATION 2**

Addends with Unlike Signs

Use your calculator to determine each sum in column A. Determine the differences in column B without your calculator. Then, answer the questions that follow.

Column A	*Column B*
(a) $-4 + 7$	$\lvert 7 \rvert - \lvert -4 \rvert$
(b) $4 + (-7)$	$\lvert -7 \rvert - \lvert 4 \rvert$
(c) $-9 + 3$	$\lvert -9 \rvert - \lvert 3 \rvert$
(d) $9 + (-3)$	$\lvert 9 \rvert - \lvert -3 \rvert$

(e) For each sum in column A, look carefully at the addends. Which addend seems to dictate the sign of the sum?

(f) From your observations in part (e), propose a rule for determining the sign of a sum when the addends have unlike signs.

(g) Now find the absolute value of each sum in column A and compare it to the corresponding difference in column B.

(h) From your comparisons in part (g), propose a rule for determining the absolute value of a sum when the addends have unlike signs.

Discovery

Column A	*Column B*
(a) $-4 + 7 = 3$	$\lvert 7 \rvert - \lvert -4 \rvert = 3$
(b) $4 + (-7) = -3$	$\lvert -7 \rvert - \lvert 4 \rvert = 3$
(c) $-9 + 3 = -6$	$\lvert -9 \rvert - \lvert 3 \rvert = 6$
(d) $9 + (-3) = 6$	$\lvert 9 \rvert - \lvert -3 \rvert = 6$

(e) In parts (a) and (d), the addend with the larger absolute value is positive, and the sum is positive. In parts (b) and (c), the addend with the larger absolute value is negative and the sum is negative. Thus, it appears that the addend with the larger absolute value dictates the sign of the sum.

(f) These results suggest the following rule: If the addends have unlike signs, then the sign of the sum is the same as the sign of the addend with the larger absolute value.

(g) In each part, the absolute value of the sum in column A is the same as the difference of the absolute values of the addends in column B.

(h) These results suggest the following rule: If the addends have unlike signs, then to determine the absolute value of the sum, take the absolute values of the addends and subtract the smaller absolute value from the larger absolute value.

Again, the rules suggested by Exploration 2 can be proven, but we accept them as given and summarize them as follows.

> ### Rules for Adding Addends with Unlike Signs
>
> If the addends of a sum have unlike signs, then
>
> 1. the sign of the sum is the same as the sign of the addend with the larger absolute value, and
>
> 2. to determine the absolute value of the sum, take the absolute values of the addends and subtract the smaller absolute value from the larger absolute value.

EXAMPLE 2 *Adding Addends with Unlike Signs*

Determine each sum without your calculator. Then, use your calculator to verify your results.

(a) $15 + (-5)$

(b) $-20 + 14$

(c) $4.3 + (-7.5)$

(d) $-\dfrac{2}{3} + \dfrac{1}{2}$

Solution

(a) $15 + (-5) = 10$

(b) $-20 + 14 = -6$

(c) $4.3 + (-7.5) = -3.2$

(d) $-\dfrac{2}{3} + \dfrac{1}{2} = -\dfrac{4}{6} + \dfrac{3}{6} = -\dfrac{1}{6} = -0.16666\ldots$ ∎

While the calculator is a valuable tool, do not become too dependent on it. Remember, too, that the calculator reports sums in decimal form, which may or may not be what you want when you perform addition involving fractions.

EXAMPLE 3 *Using the Addition Rules to Calculate Sums*

(a) $-8 + 5 = -3$

(b) $-3 + (-12) = -15$

(c) $15 + (-6) = 9$

(d) $\dfrac{5}{6} + \left(-\dfrac{4}{9}\right) = \dfrac{15}{18} + \left(-\dfrac{8}{18}\right) = \dfrac{15 + (-8)}{18} = \dfrac{7}{18} \approx 0.39$

(e) $-3 + \dfrac{2}{7} = \dfrac{-3}{1} + \dfrac{2}{7} = \dfrac{-21}{7} + \dfrac{2}{7} = \dfrac{-21 + 2}{7} = \dfrac{-19}{7} \approx -2.71$

(f) $-584 + 379 = -205$

(g) $-18.97 + (-34.5) = -53.47$ ∎

When a sum contains three or more addends, grouping symbols, such as parentheses or brackets, may be used to indicate the addends that are to be added first. If there are no grouping symbols, we simply add from left to right.

EXAMPLE 4   *Sums with More Than Two Addends*

(a) $-4 + 7 + 9 = 3 + 9 = 12$ There are no parentheses.
                                        Add from left to right.

(b) $(-2 + 5) + [6 + (-9)] = 3 + (-3) = 0$ Add inside the grouping
                                               symbols first.

Applications

We need to be able to add signed numbers in order to solve applications such as the following.

EXAMPLE 5   *Calculating a Depth Below Sea Level*

A marine biologist descended 20 feet below the ocean surface to gather data on ocean plant life. After gathering the data, she descended another 50 feet to collect more information. Finally, she descended another 70 feet for the deepest sample. From there, she ascended 30 feet for a final sample before returning to the surface. The depth changes are given by -20, -50, -70, and $+30$ feet. How deep was the biologist when she collected her deepest sample? How deep was she when she collected her last sample?

Solution

The deepest descent is the sum of her three descents.

$$-20 + (-50) + (-70) = -140$$

We interpret the result as 140 feet below sea level.

The location of the final sample is the sum of all of the biologist's depth changes.

$$-20 + (-50) + (-70) + 30 = -110$$

We interpret the result as 110 feet below sea level.

1.2   *Quick Reference*

Addends with Like Signs
- In the expression $a + b$, a and b are the **addends.** Both the expression and the value of the expression are called a **sum.**

- If a and b are real numbers with like signs, then $a + b$ is determined as follows.

   1. Add the absolute values of the addends.

   2. The sign of the sum is the same as the sign of the addends.

Addends with **Unlike Signs**	▪ If a and b are real numbers with unlike signs, then $a + b$ is determined as follows.

                1. Take the absolute values of the addends and subtract the smaller absolute value from the larger absolute value.

                2. The sign of the sum is the same as the sign of the addend with the larger absolute value.

      ▪ When a sum consists of three or more addends, grouping symbols, such as parentheses or brackets, may be used to indicate the operations that are to be performed first. Otherwise, add from left to right.

1.2 Exercises

1. Explain how to determine the sum of two numbers that have like signs.

2. Explain how to determine the sum of two numbers that have unlike signs.

In Exercises 3–34, determine the sum.

3. $8 + (-5)$	4. $9 + (-4)$
5. $-6 + 2$	6. $-7 + 4$
7. $-3 + (-7)$	8. $-5 + (-10)$
9. $-6 + (-2)$	10. $-9 + (-5)$
11. $-16 + 16$	12. $-14 + 14$
13. $-3 + 5$	14. $-7 + 5$
15. $12 + (-4)$	16. $13 + (-7)$
17. $-6 + 14$	18. $-14 + 8$
19. $-7 + 0$	20. $-9 + 0$
21. $4 + 16$	22. $5 + 18$
23. $0 + (-55)$	24. $0 + (-22)$
25. $-9 + 17$	26. $14 + (-20)$
27. $-12 + 12$	28. $17 + (-10)$
29. $-11 + (-11)$	30. $-13 + (-8)$
31. $100 + (-101)$	32. $-50 + (-5)$
33. $-14 + 3$	34. $36 + (-9)$

In Exercises 35–44, determine the sum.

35. $-14 + (-2) + (-1)$

36. $5 + (-8) + (-12)$

37. $-3 + [6 + (-9)]$

38. $-16 + [3 + (-7)]$

39. $6 + (-7) + 2$

40. $-20 + 6 + (-1)$

41. $-12 + (-2) + (-5) + 10$

42. $15 + (-1) + 2 + (-6)$

43. $-6 + 8 + [(-5) + 3]$

44. $(-2) + [(-5) + (-3)] + 9$

In Exercises 45–52, predict the sign of the sum and use your calculator to find the sum.

45. $-2.3 + (-4.5)$	46. $2.8 + (-4.6)$
47. $-6.7 + 9.5$	48. $-8.6 + 7.4$
49. $-45.84 + (-5.73)$	50. $-0.35 + 78.3$
51. $24.47 + (-34.78)$	52. $-975.3 + (-537.37)$

In Exercises 53–60, write a numerical expression and then evaluate it.

53. What is the sum of -964 and 351?

54. What is 11 more than -6?

55. What is -9 increased by 5?

56. If 26, -73, and 40 are the addends, what is the sum?

57. What is the sum of -7, 3, and -10?

58. If 12 is added to the sum of -5 and 2, what is the result?

59. What is the sum of 7 and -2 increased by 3?

60. If -10 is increased by the sum of -5 and 2, what is the result?

In Exercises 61–70, perform the indicated operations.

61. $-\dfrac{1}{3} + \dfrac{5}{3}$ **62.** $-\dfrac{3}{4} + \dfrac{1}{4}$

63. $-\dfrac{2}{5} + \left(-\dfrac{4}{5}\right)$ **64.** $-\dfrac{3}{7} + \left(-\dfrac{6}{7}\right)$

65. $-\dfrac{2}{3} + \dfrac{3}{5}$ **66.** $\dfrac{4}{7} + \left(-\dfrac{1}{3}\right)$

67. $\dfrac{5}{16} + \left(-\dfrac{3}{8}\right)$ **68.** $-\dfrac{6}{7} + \dfrac{5}{14}$

69. $-5 + \left(-\dfrac{2}{3}\right)$ **70.** $-7 + \left(-\dfrac{3}{4}\right)$

71. The sum of what number and 2 is -5?

72. The sum of -3 and what number is 7?

73. Two of three addends are -1 and 4. If the sum of the three numbers is -9, what is the third addend?

74. If the sum of three numbers is zero, and two of the numbers are 2 and -2, what is the third number?

In Exercises 75–78, perform the indicated operations.

75. $-|3| + |-10|$ **76.** $|7| + \left(-|12|\right)$

77. $|-5| + |-8|$ **78.** $-|-2| + \left(-|-9|\right)$

In Exercises 79–82, estimate the order of the two given sums and fill in the blank with a less than or greater than symbol. Verify your answer by calculating the sums with your calculator.

79. $8 + (-5)$ ___$\geq$___ $8 + (-5.1)$

80. $-3 + 3$ ___$\leq$___ $-3 + 3.1$

81. $6 + |-6|$ _____ $-6 + |6|$

82. $-4 + (-9)$ _____ $-4 + (-9.5)$

In Exercises 83–108, perform the indicated operations.

83. $17 + (-8)$

84. $-36 + (-9)$

85. $6 + (-8) + (-2)$

86. $-5 + 10 + (-7)$

87. $-6 + |-8|$

88. $|4| + |-3|$

89. $187.24 + (-187.24)$

90. $72.72 + (-73.73)$

91. $-6 + \dfrac{3}{5}$

92. $-\dfrac{3}{4} + 3$

93. $-50 + 5$

94. $-8 + (-9)$

95. $-10 + (-2) + 15$

96. $18 + 3 + (-10)$

97. $0 + (-18.3)$

98. $17.459 + 0$

99. $-\dfrac{3}{7} + \dfrac{1}{3}$

100. $-\dfrac{35}{6} + \left(-\dfrac{23}{4}\right)$

101. $-12 + (-3) + 10 + (-2)$

102. $3 + (-8) + (-6) + 11$

103. $|-7| + (-7)$

104. $|-5| + 2$

105. $-0.003 + (-0.219)$

106. $-0.5 + 0.3$

107. $-\dfrac{3}{4} + \left(-\dfrac{21}{8}\right)$

108. $-\dfrac{2}{3} + \left(-\dfrac{5}{8}\right)$

109. In their first possession of the football, a team had a loss of 9 yards on the first down, a gain of 4 yards on the second down, and a gain of 7 yards on the third down. What was the total gain or loss of yardage in these three downs?

110. A football team took possession of the ball on the 50 yard line. The team ran seven plays with the following results. Determine whether the team scored a touchdown.

Play	Yards
1	32 (Gain)
2	5 (Loss)
3	9 (Gain)
4	10 (Gain)
5	11 (Loss)
6	3 (Loss)
7	19 (Gain)

111. Suppose that you write an electric bill check for $76.27, and you notice that your new checkbook balance is −$26.84. You decide not to mail the electric bill until you make a deposit of $35.25. What is the checkbook balance after the deposit and the payment of the electric bill?

112. A wrestler's weight was 216 pounds when he went on a diet. His change in weight for each of 7 weeks was as follows. What was the wrestler's weight after the 7 weeks?

Week	Weight (lbs.)
1	5 (Loss)
2	1 (Gain)
3	2 (Loss)
4	1 (Loss)
5	4 (Gain)
6	3 (Loss)
7	2 (Gain)

For Exercises 113 and 114, the **average** or **mean** of n numbers is found by adding the n numbers and dividing the sum by n.

113. During a 1-week period in International Falls, Minnesota, the daily high temperatures (in Fahr-

enheit) were as follows: −10°, 17°, 5°, −9°, 0°, 6°, and 4°. What was the average high temperature for the week?

114. During a particularly active week, the stock prices for a utility company changed by the following amounts: $-\frac{3}{8}$, $+\frac{1}{2}$, $+\frac{3}{4}$, $-\frac{5}{16}$, and $+\frac{1}{8}$. What was the average change in price for the week?

Exploring with Real Data

In Exercises 115–118, refer to the following background information and data.

Research indicates that about 60% of American adults had automatic teller machine (ATM) cards in 1993. The pie chart shows the percentages of these cardholders who used their cards a certain number of times per month. (Source: Research Partnership, 1993.)

Figure for 115–118

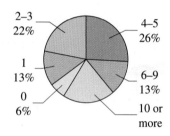

115. What percent of the cardholders used their cards fewer than four times per month?

116. What percent of the cardholders used their cards at least ten times per month?

117. Why do the given data not allow you to determine the actual number of cardholders in any of the categories?

118. To use your ATM card, you must enter a personal identification number (PIN), typically a four-digit number. If your card is stolen, what are the thief's odds of guessing your PIN number on the first try?

Challenge

In Exercises 119–122, let a represent a negative number and b represent a positive number. Determine whether the given sum is positive or negative.

119. $|a| + b$

120. $(-a) + b$

121. $a + (-b)$

122. $-b + b + a$

123. If a person takes one step backward and then two steps forward, how many steps must the person take to be 27 steps from the starting point?

124. (a) Calculate $1 + 2 + (-3) + (-4)$.

   (b) Calculate
$1 + 2 + (-3) + (-4) + 5 + 6 + (-7) + (-8)$.

   (c) Consider a sum that follows the same pattern as in part (b) with the last two addends being -47 and -48. What is your conjecture about the value of the expression?

1.3 Subtraction

Definition ▪ *Performing Subtraction* ▪ *Distance Between Points* ▪
Applications

Definition

In this section we consider subtraction of signed numbers.

In the expression $8 - 5$, the number 8 is called the **minuend** and the number 5 is called the **subtrahend.** Both the expression and the value of the expression are called the **difference.**

EXPLORATION 1 *Comparing Sums and Differences*

Use a calculator to determine the differences in parts (a)–(d). For each part, calculate the sum according to the rules for addition. Then answer the questions that follow.

 SUBTRACT

  (a) $9 - 4$ $9 + (-4)$

  (b) $-8 - 3$ $-8 + (-3)$

  (c) $5 - (-2)$ $5 + 2$

  (d) $-3 - (-7)$ $-3 + 7$

  (e) What do you observe about the results in each pair?

  (f) How do the expressions differ in each pair?

  (g) Using your observations in part (f), predict how the following differences can be written as sums.

$$35 - (-17) \qquad \underline{\hspace{0.6cm}} + \underline{\hspace{0.6cm}}$$
$$-10 - 19 \qquad \underline{\hspace{0.6cm}} + \underline{\hspace{0.6cm}}$$

  (h) Use your calculator to verify that each difference and sum pair in part (g) have the same value.

Discovery

(a) $9 - 4 = 5$ $9 + (-4) = 5$

(b) $-8 - 3 = -11$ $-8 + (-3) = -11$

(c) $5 - (-2) = 7$ $5 + 2 = 7$

(d) $-3 - (-7) = 4$ $-3 + 7 = 4$

(e) In each pair, the difference and the sum are the same.

(f) Comparing the difference to the sum, we see that the minus sign is changed to a plus sign and the subtrahend is changed to its opposite.

(g)
$$35 - (-17) \qquad \frac{35 \;+\; 17}{-10 \;-\; 19} \quad \frac{}{-10 \;+\; (-19)}$$

(h) $35 - (-17) = 35 + 17 = 52$

$-10 - 19 = -10 + (-19) = -29$

Exploration 1 suggests the following definition of subtraction.

Definition of Subtraction

For any real numbers a and b, $a - b = a + (-b)$.

According to this definition, we can write any difference as an equivalent sum by doing the following:

1. Change the minus sign to a plus sign.

2. Change the subtrahend to its opposite.

An obvious benefit of converting a difference into a sum is that we already know the rules of addition. Therefore, no additional rules are needed for subtraction.

Performing Subtraction

For simple subtraction problems, you should be able to apply the definition mentally and not have to rely on a calculator. (When you do use a calculator, you do not need to apply the definition because the calculator does it for you.)

EXAMPLE 1   *Performing Subtraction with Real Numbers*

Perform the indicated operations. Use your calculator in parts (f) and (g) and to verify the results in all other parts.

(a) $5 - 12$ (b) $-20 - 4$ (c) $16 - (-5)$

(d) $-7 - (-3)$ (e) $\dfrac{1}{3} - \dfrac{5}{2}$ (f) $-2.47 - (-6.91)$

(g) $-538 - 297$

Solution

(a) $5 - 12 = 5 + (-12) = -7$

(b) $-20 - 4 = -20 + (-4) = -24$

(c) $16 - (-5) = 16 + 5 = 21$

(d) $-7 - (-3) = -7 + 3 = -4$

(e) $\dfrac{1}{3} - \dfrac{5}{2} = \dfrac{2}{6} - \dfrac{15}{6} = \dfrac{2 - 15}{6} = \dfrac{2 + (-15)}{6} = \dfrac{-13}{6} \approx -2.167$

(f) $-2.47 - (-6.91) = 4.44$

(g) $-538 - 297 = -835$ ■

We now investigate the effect of reversing the order of the numbers in a difference.

EXPLORATION 2

Order of Subtraction

In parts (a)–(d), perform each pair of indicated operations and compare the results. Then, answer the questions that follow.

(a) $9 - 5$ $5 - 9$

(b) $3 - 12$ $12 - 3$

(c) $-3 - (-8)$ $-8 - (-3)$

(d) $-7 - 10$ $10 - (-7)$

(e) What word would you use to describe the two results in each part?

(f) In words, what general relationship between the expressions $a - b$ and $b - a$ is suggested by parts (a) through (d)?

(g) Write the relationship in part (f) symbolically.

Discovery

(a) $9 - 5 = 4$ $5 - 9 = -4$

(b) $3 - 12 = -9$ $12 - 3 = 9$

(c) $-3 - (-8) = 5$ $-8 - (-3) = -5$

(d) $-7 - 10 = -17$ $10 - (-7) = 17$

(e) In each pair, the two results are *opposites*.

(f) The expressions $a - b$ and $b - a$ are opposites.

(g) Symbolically, $a - b = -(b - a)$ and $-(a - b) = b - a$.

Exploration 2 suggests the following generalization.

Property of the Opposite of a Difference

For any real numbers a and b, $-(a - b) = b - a$.

As we will see in our later work, this is an important rule. We will use it to simplify expressions and to reverse the order of subtraction when we want to do so.

In our next example, we use the fact that operations inside grouping symbols are to be performed first. Otherwise, we add and subtract from left to right.

EXAMPLE 2   *Performing Mixed Operations*

(a) $12 - (3 - 7) = 12 - [3 + (-7)] = 12 - (-4) = 12 + 4 = 16$

(b) $4 - (5 + 9) - 7 = 4 - 14 - 7 = 4 + (-14) + (-7) = -17$

(c) $(6 - 9) - (-8 + 2) = -3 - (-6) = -3 + 6 = 3$

(d) $5 - [4 - (1 - 2)] = 5 - [4 - (-1)] = 5 - [4 + 1] = 5 - 5 = 0$ ■

EXAMPLE 3   *Translating Phrases into Numerical Expressions*

Translate each of the following phrases into a numerical expression and evaluate the expression.

(a) Four less than -3 (b) Four decreased by -3

(c) The difference between 4 and -3

Solution

(a) $-3 - 4 = -7$

(b) $4 - (-3) = 7$

(c) $4 - (-3) = 7$ The first given number is the minuend, and the second given number is the subtrahend.   ■

Distance Between Points

Figure 1.8

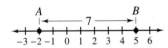

The distance between the two points A and B on the number line in Fig. 1.8 can be found by counting the units from -2 to 5. The distance is 7 units.

We can also calculate the distance between points A and B by subtracting the coordinates of the points: $5 - (-2) = 7$.

If we subtract the coordinates in reverse order, the result is $-2 - 5 = -7$. This result is not valid because distance is not negative. However, we can subtract in either order if we agree to take the absolute value of the result.

> ### Distance Between Points of the Number Line
>
> If the coordinates of points A and B are a and b, respectively, then the distance d between the points is determined as follows.
>
> $$d = |a - b| = |b - a|$$

NOTE: When you use your calculator to determine $|a - b|$, you will need to enclose $a - b$ in parentheses.   ■

EXAMPLE 4   *Finding the Distance Between Points on a Number Line*

The coordinates of points P and Q are given. Use your calculator to determine the distance d between the points.

(a) -26 and -5 (b) 7 and $-\dfrac{5}{3}$ (c) -4 and 3.9

Figure 1.9

```
abs(-26-(-5))
                 21
abs(7-(-5/3))
        8.666666667
abs(-4-3.9)
                7.9
```

Solution

(a) $d = |-26 - (-5)| = 21$ (b) $d = \left|7 - \left(-\dfrac{5}{3}\right)\right| = 8.\overline{6}$

(c) $d = |-4 - 3.9| = 7.9$ ■

Applications

Application problems, particularly those involving a calculation of distance, often require a subtraction of signed numbers.

EXAMPLE 5   *Distance Between Points Above and Below Sea Level*

The altitude gauge of an airplane indicates that the plane is 760 feet above sea level. The plane is to drop relief supplies to a remote desert location that is 42 feet below sea level. When the plane is directly above the drop site, what is the distance between the plane and the drop site?

Solution

Consider a vertical number line whose origin is at sea level. The plane is at a point whose coordinate is $+760$ and the drop site is at a point whose coordinate is -42. (See Fig. 1.10.)

Figure 1.10

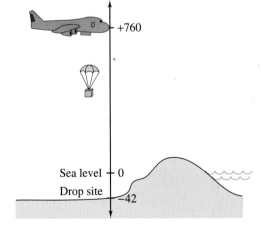

We calculate the distance d between the points as follows.

$$d = |760 - (-42)| = |760 + 42| = |802| = 802$$

The distance from the plane to the drop site is 802 feet.   ■

1.3   Quick Reference

Definition	• In the expression $a - b$, a is the **minuend** and b is the **subtrahend.** Both the expression and the value of the expression are called a **difference.**
	• For any real numbers a and b, $a - b$ is defined as $a + (-b)$.
Performing Ssubtraction	• The definition of subtraction gives us a way to perform the operation by applying the rules for addition. We convert a difference to a sum as follows.
	1. Change the minus sign to a plus sign.
	2. Change the subtrahend to its opposite.
	• If we reverse the order of the numbers in a difference, the value of the result is the opposite of the original value. Symbolically, $-(a - b) = b - a$.
Distance Between Points	• We determine the distance between two points of a number line by subtracting the coordinates of the points and taking the absolute value of the result.

1.3   Exercises

1. What two changes must be made to $-5 - 8$ to convert the expression into a sum with the same value?

2. In the expression $7 + 5$, the 7 and the 5 are both called addends. But in the expression $7 - 5$, the 7 and the 5 have two different names. What are they? Why is it necessary for 7 and 5 to have different names in the expression $7 - 5$, but not in the expression $7 + 5$?

In Exercises 3–26, determine the difference.

3. $4 - 15$

4. $3 - 17$

5. $-7 - 3$

6. $-6 - 5$

7. $9 - (-4)$

8. $8 - (-7)$

9. $-2 - (-8)$

10. $-4 - (-12)$

11. $-5 - 8$

12. $-12 - 9$

13. $0 - (-52)$

14. $0 - 25$

15. $13 - (-5)$

16. $15 - (-8)$

17. $-12 - 12$

18. $12 - (-12)$

19. $-6 - (-3)$

20. $-10 - (-5)$

21. $24 - (-24)$

22. $5 - (-24)$

23. $-4 - 16$

24. $-17 - 9$

25. $-8 - (-5)$

26. $-17 - 24$

In Exercises 27–38, perform the indicated operations.

27. $15 - 20 - 5$

28. $-3 - 5 - 7$

29. $-20 - (-14) - (-1)$

30. $6 - (-5) - 8$

31. $10 - (6 - 9)$

32. $-5 - (-2 + 12)$

33. $-8 - 10 - (-3) + 7$

34. $18 - (-2) - 25 + 8$

35. $-3 + (-2) - (-1 + 10)$

36. $6 + 2 - (5 - 12)$

37. $16 - [(-3) + 10 - (-9)]$

38. $-7 + [3 - (-11) - 13]$

39. What is the difference between -17 and -18?

40. What is -3 decreased by 4?

41. What is 9 less than -1?

42. If the minuend of an expression is -6 and the subtrahend is -8, what is the difference?

43. Subtract -3 from 10. What is the result?

44. What is -10 less -5?

In Exercises 45–54, perform the indicated operations.

45. $-\dfrac{1}{3} - \dfrac{4}{3}$ **46.** $-\dfrac{3}{4} - \dfrac{1}{4}$

47. $-\dfrac{1}{5} - \left(-\dfrac{3}{5}\right)$ **48.** $-\dfrac{2}{7} - \left(-\dfrac{5}{7}\right)$

49. $-\dfrac{3}{5} - \left(-\dfrac{1}{2}\right)$ **50.** $-\dfrac{3}{4} - \left(-\dfrac{2}{3}\right)$

51. $-5 - \left(-\dfrac{1}{3}\right)$ **52.** $-8 - \left(-\dfrac{3}{4}\right)$

53. $\dfrac{7}{16} - \left(-\dfrac{5}{8}\right)$ **54.** $-\dfrac{5}{7} - \dfrac{9}{14}$

 55. In your own words, interpret $-(a - b) = b - a$.

56. Suppose you enter X − Y in your calculator and the displayed result is -4. What will you obtain if you enter Y − X? Why?

In Exercises 57–62, use your calculator to perform the indicated operations.

57. $-22 - 14 - (-13) - 4 + (-7)$

58. $-35 - (-14) + (-23) - (-13) + 5 - (-8)$

59. $32 + (-35) - 12 - (-36) - 14 + 26$

60. $22 + (-27) - 16 - 8 - (-7)$

61. $13.84 - (-26.19) - (-27.5) - 25 - (-31.7)$

62. $-17.2 - (-82) - 45.23 + (-23.75) - (-54.1)$

63. What number must be subtracted from 3 to obtain -2?

64. The difference between what number and -6 is 4?

65. If decreasing a number by 7 results in -1, what is the number?

66. If 9 less than a number is -2, what is the number?

In Exercises 67–70, calculate the distances AB, BC, and AC. Then show that $AB + BC = AC$.

67.

A	B	C
-7	-1	9

68.

A	B	C
-12	-4	6

69.

A	B	C
$-\frac{7}{2}$	$\frac{3}{4}$	5

70.

A	B	C
-4	$-\frac{1}{5}$	$\frac{2}{3}$

In Exercises 71–74, perform the indicated operations. Use your calculator to verify the results.

71. $|-7| - |-3|$

72. $|5| - |-11|$

73. $|-2| - |2| - 2$

74. $|-3| - \left(4 - |-1|\right)$

In Exercises 75–80, determine whether the statement is true or false.

75. $|5 + 7| = |5| + |7|$

76. $|9 + 2| = |9| + |2|$

77. $|7 - 5| = |7| - |5|$

78. $|5 - 7| = |5| - |7|$

79. $|4 - 9| \geq |4| - |9|$

80. $|9 - 4| \leq |9| - |4|$

In Exercises 81–110, perform the indicated operations.

81. $-7 - (-2)$

82. $-12 - (-12)$

83. $-47.87 - (-6.74)$ 84. $-0.46 - 76.5$

85. $7 - (-36)$ 86. $21 - (-35)$

87. $|-8| - |-3|$ 88. $|-10| - (-6)$

89. $-\dfrac{2}{3} - \dfrac{4}{5}$ 90. $-\dfrac{31}{6} - \left(-\dfrac{25}{4}\right)$

91. $-6 - 10 + (-5)$ 92. $12 - 8 - (-2)$

93. $-16 - 16$ 94. $-22 - 37$

95. $-\dfrac{3}{4} - 4$ 96. $-\dfrac{7}{8} - 6$

97. $0 - (-32)$ 98. $-72 - (-72)$

99. $23.89 - (-35.87)$

100. $-974.3 - (-678.39)$

101. $-45 - 0$ 102. $23 - (-23)$

103. $4 - 9 - (-5)$ 104. $14 - 17 - 7$

105. $|4| - |-12|$ 106. $|-3| - 9$

107. $-24 - 56$ 108. $35 - 47$

109. $9 - 27 + 2 - (-1)$

110. $-12 + 3 - (-8) - (-5)$

111. In the state of California, the highest elevation is Mount Whitney with an altitude of 14,495 feet. The lowest altitude is Death Valley at 280 feet below sea level. What is the difference between the two altitudes?

112. Mauna Loa on the island of Hawaii has an elevation of 4169 feet. This peak is 32,024 feet above the ocean floor near Hawaii. How deep is the ocean floor?

113. The balance in a savings account at the beginning of June was $523.46. During June a deposit of $445 and two withdrawals of $310.99 and $246.48 were made. Disregarding the interest for the month, what was the savings account balance on June 30?

114. A person had a balance of $564.28 in his checkbook when he wrote a check for $834.75 to purchase a new stereo system. What was his resulting checkbook balance? Will a quick deposit of $300 rescue his credit rating?

Exploring with Real Data

In Exercises 115–118, refer to the following background information and data.

The mean elevation of the continental United States is 2500 feet. The following table shows the highest points in six selected states.

State	Highest Point	Altitude (in feet)	Difference
Colorado	Mount Elbert	14,431	
Florida	Iron Mountain	325	
Illinois	Charles Mound	1,241	
Rhode Island	Durfee Hill	805	
Tennessee	Clingmans Dome	6,642	
Washington	Mount Rainier	14,408	

(Source: U.S. Geological Survey.)

115. For each state in the table, record the difference between the altitude of the highest point and the mean elevation of the continental United States. Use positive numbers for altitudes above the mean and negative numbers for altitudes below the mean.

116. Calculate the mean of the differences for the six entries in the table. Why is the mean difference so large?

117. How many miles high is Mount Rainier?

118. As the words are conventionally used, we would rank *mound, hill,* and *mountain* from least to greatest in height. Look at the entries for Illinois, Rhode Island, and Florida, and note the altitudes. What do you observe?

Challenge

In Exercises 119–122, let a represent a negative number and b represent a positive number. Determine the sign of the expression.

119. $a - b$

120. $b - a$

121. $a - |a|$

122. $|a| - a$

123. Exercises 75 and 76 suggest that

$$|a + b| = |a| + |b|$$

is true for any real numbers a and b. Is this a valid rule? If you don't think so, give an example to show that the relationship is not always true.

 124. What is your opinion of the statement

$$|a - b| = |b - a|$$

for all real numbers a and b? Explain your reasoning.

 125. Are $|a| + |b|$ and $|a| - |b|$ always nonnegative? Explain your reasoning.

126. The value of the following expression is 5.

$$1 - 2 + 3 - 4 + 5 - 6 + 7 - 8 + 9$$

Insert parentheses (as many as are needed) so that the value of the expression is 7.

127. Is it possible to insert plus and minus signs in the blanks of the following expression in order for the expression to have a value of zero? Why?

1 _____ 2 _____ 3 _____ 4 _____ 5 _____
6 _____ 7 _____ 8 _____ 9

1.4 Multiplication

Sign Rules ▪ *Applications*

Sign Rules

Our third basic operation with real numbers is multiplication.

In algebra, multiplication is usually indicated by a dot · or by parentheses. For example, the product of 3 and 7 could be written in any of the following ways.

$3 \cdot 7$ $3(7)$ $(3)(7)$ $(3)7$

A calculator may display 3*7.

The numbers that are multiplied are called **factors.** Both the expression and the result are called the **product.**

In our first exploration, we investigate the sign rules for multiplication when at least one factor is negative.

 **EXPLORATION 1**   *Sign Rules*

Use your calculator to determine each product in column A. Determine the products in column B without your calculator. Then, answer the questions that follow.

 MULTIPLY

Column A	Column B
(a) $-4 \cdot 10$	$\lvert -4 \rvert \cdot \lvert 10 \rvert$
(b) $5(-3)$	$\lvert 5 \rvert \cdot \lvert -3 \rvert$
(c) $(-2)(-10)$	$\lvert -2 \rvert \cdot \lvert -10 \rvert$
(d) $-8(-6)$	$\lvert -8 \rvert \cdot \lvert -6 \rvert$

(e) For which products in column A do the factors have unlike signs? like signs?

(f) What is the sign of the result when the factors have unlike signs? like signs?

(g) Now find the absolute value of each product in column A and compare it to the corresponding product in column B.

(h) Based on your observations, propose a rule for multiplying signed numbers.

Discovery

Column A	Column B
(a) $-4 \cdot 10 = -40$	$\lvert-4\rvert \cdot \lvert10\rvert = 40$
(b) $5(-3) = -15$	$\lvert5\rvert \cdot \lvert-3\rvert = 15$
(c) $(-2)(-10) = 20$	$\lvert-2\rvert \cdot \lvert-10\rvert = 20$
(d) $-8(-6) = 48$	$\lvert-8\rvert \cdot \lvert-6\rvert = 48$

(e) The factors in parts (a) and (b) have unlike signs; the factors in parts (c) and (d) have like signs.

(f) When the factors have unlike signs, the result is negative; when the factors have like signs, the result is positive.

(g) In each part, the absolute value of the product in column A is the same as the corresponding product in column B.

(h) These results suggest the following rule: If two factors have unlike signs, the product is negative; if two factors have like signs, the product is positive. The absolute value of the product is the product of the absolute values of the factors.

One detail remains. What if at least one factor of a product is 0? Try the following operations on your calculator.

$$7 \cdot 0 \qquad 0 \cdot (-9) \qquad 0 \cdot 0$$

The results suggest that a product is 0 whenever at least one factor is 0.

We can now summarize the rules for multiplying two real numbers.

Rules for Multiplying Two Real Numbers

1. If at least one factor is 0, then the product is 0.

2. If neither factor is zero, the product is found as follows.

    (a) If the factors have like signs, the product is positive; if the factors have unlike signs, the product is negative.

    (b) The absolute value of the product is the product of the absolute values of the factors.

NOTE: As is true for addition, the sign rules for multiplication can be proven to be true. It is important to understand that a calculator does not *determine* the sign rules. Rather, a calculator is designed to *obey* the sign rules. ∎

EXAMPLE 1   *The Signs of Products*

Predict the sign, if any, of each product. Use your calculator to perform the indicated operation and verify your conclusions.

(a) $-56 \cdot 14$

(b) $-3.78(-6.3)$

(c) $(-67)(0)$

Solution

(a) Because the two factors have unlike signs, the product is negative:
$$-56 \cdot 14 = -784.$$

(b) Because the two factors have like signs, the product is positive:
$$-3.78(-6.3) = 23.814.$$

(c) The product is zero because one factor is zero.   ■

NOTE: Because we can easily determine the sign of a product in advance, it is not necessary to enter signs in your calculator when you use it to multiply. For example, in part (b) of Example 1, we know that the product will be positive, so we can simply enter 3.78 * 6.3.   ■

If a product consists of three or more factors, grouping symbols may be used to indicate the factors that are to be multiplied first. If a product does not contain grouping symbols, we multiply from left to right.

 **EXPLORATION 2**   *Multiplying More Than Two Factors*

Use your calculator to determine the products in parts (a) and (b). Then answer the questions that follow.

(a) $5(-3) \cdot 2 \cdot 3(-6)(-1)(-2)$

(b) $5(-3) \cdot 2 \cdot 3(-6) \cdot 1 \cdot (-2)$

(c) The absolute values of the factors are the same in both expressions. Compare the two results.

(d) In part (a), how many factors are negative? What is the sign of the product?

(e) In part (b), how many factors are negative? What is the sign of the product?

(f) Propose a method for determining the sign of a product involving more than two factors.

Discovery

(a) $5(-3) \cdot 2 \cdot 3(-6)(-1)(-2) = 1080$

(b) $5(-3) \cdot 2 \cdot 3(-6) \cdot 1 \cdot (-2) = -1080$

(c) Although the absolute values of the factors are the same, the results are opposites.

(d) There are four negative factors, and the product is positive.

(e) There are three negative factors, and the product is negative.

(f) If there is an even number of negative factors, the product is positive; if there is an odd number of negative factors, the product is negative.

EXAMPLE 2   *Determining the Sign of a Product of Three or More Factors*

Predict the sign of each product.

(a) $-2(-1)(-3)(-2)(-1)$ 　　　　　　(b) $(-1)(-5)(-2)(-3)$

Solution

(a) There is an odd number of negative factors, so the product is negative.

(b) There is an even number of negative factors, so the product is positive. ■

EXAMPLE 3   *Evaluating Products*

Evaluate the following products. Use your calculator as needed.

(a) $(-7)(-3)$ 　　　　　　　　(b) $8(-4)$

(c) $-8(4)$ 　　　　　　　　(d) $\dfrac{2}{3} \cdot \left(-\dfrac{5}{7}\right)$

(e) $(-1)(-2)(-3)(-4)(-5)(-84)(-93)(0)$

(f) $8 \cdot \left(-\dfrac{1}{2}\right)$ 　　　　　(g) $\left(-\dfrac{3}{2}\right)\left(-\dfrac{4}{3}\right)\left(-\dfrac{5}{4}\right)\left(-\dfrac{6}{5}\right)$

(h) $-274(-19)$ 　　　　　(i) $(-5)(-4)(-3)(-2)(-1)$

(j) $(-0.08)(126)(-0.47)$

Solution

(a) $(-7)(-3) = 21$ 　　　　　　(b) $8(-4) = -32$

(c) $-8(4) = -32$ 　　　　　(d) $\dfrac{2}{3} \cdot \left(-\dfrac{5}{7}\right) = -\dfrac{10}{21}$

(e) Because one factor is 0, the product is 0.

(f) $8 \cdot \left(-\dfrac{1}{2}\right) = -4$

(g) There is an even number of negative factors, so the product is positive.

$$\left(-\dfrac{3}{2}\right)\left(-\dfrac{4}{3}\right)\left(-\dfrac{5}{4}\right)\left(-\dfrac{6}{5}\right) = 3$$

(h) $-274(-19) = 5206$

(i) There is an odd number of negative factors, so the product is negative: $(-5)(-4)(-3)(-2)(-1) = -120$.

(j) $(-0.08)(126)(-0.47) = 4.7376$ ■

Applications

The Dow-Jones Average is used as an indicator of the performance of the stock market. The evening news usually includes a report on the number of *points* by which the Dow-Jones Average increased or decreased.

EXAMPLE 4   *Changes in the Dow-Jones Average*

The Dow-Jones Average declined by 7 points each day for 4 consecutive days. What was the total change in the average for the 4-day period? If the opening average was 3246, what was the average at the end of the 4-day period?

Solution   The change each day was -7. Therefore, the change for the 4-day period was $4(-7)$ or -28 points.

The average at the end of the period was $3246 + (-28) = 3218$.

1.4 *Quick Reference*

Sign Rules
- In the expression $a \cdot b$, a and b are called **factors.** Both the expression and the value of the expression are called the **product.**

- The following are the rules for multiplying two real numbers.

  1. If at least one factor is 0, then the product is 0.

  2. If neither factor is 0, the product is found as follows.

     (a) If the factors have like signs, the product is positive; if the factors have unlike signs, the product is negative.

     (b) The absolute value of the product is the product of the absolute values of the factors.

- If a product has an even number of negative factors, then the product is positive; if a product has an odd number of negative factors, then the product is negative.

1.4 *Exercises*

 1. Explain how to determine the product of two numbers with like signs.

 2. Explain how to determine the product of two numbers with unlike signs.

In Exercises 3–20, determine the product.

3. $7 \cdot (-4)$

4. $8 \cdot (-5)$

5. $-5 \cdot 2$

6. $-7 \cdot 6$

7. $-3 \cdot (-4)$

8. $-6 \cdot (-10)$

9. $-6 \cdot (-2)$

10. $-8 \cdot (-5)$

11. $-10 \cdot 10$

12. $-9 \cdot 9$

13. $11 \cdot (-4)$

14. $8 \cdot (-7)$

15. $-7 \cdot 1$

16. $-1 \cdot 0$

17. $4 \cdot (-6)$

18. $5 \cdot (-1)$

19. $0 \cdot (-45)$

20. $-1 \cdot (-92)$

In Exercises 21–28, predict the sign of the product. Then, use your calculator to multiply.

21. $-2.3 \cdot (-2.7)$

22. $-4.5 \cdot (-3.4)$

23. $2.1 \cdot (-2.3)$

24. $0.5 \cdot (-4.3)$

25. $2.4 \cdot (-13.5)$

26. $-25.7 \cdot (-21.2)$

27. $-26.4 \cdot (-57.8)$

28. $-45.1 \cdot 32.3$

In Exercises 29–36, write a numerical expression that models the problem and then evaluate it.

29. What is the product of 29 and 0.63?

30. What is -8 multiplied by -7?

31. If the two factors are -1.8 and 1.8, what is the product?

32. What is the product of -8 and -3 times -2?

33. What is $-\frac{3}{4}$ of 32?

34. What is 20% of -19?

35. What is twice the product of 5 and -3?

36. Triple the product of -7 and 2. What is the result?

In Exercises 37–48, multiply the fractions. Use your calculator to verify your results.

37. $-\frac{1}{3} \cdot \frac{3}{5}$

38. $-\frac{2}{5} \cdot \left(-\frac{5}{4}\right)$

39. $-\frac{4}{5} \cdot \left(-\frac{1}{2}\right)$

40. $\frac{5}{16} \cdot \left(-\frac{8}{3}\right)$

41. $-\frac{2}{5} \cdot \left(-\frac{5}{8}\right)$

42. $-\frac{3}{4} \cdot \left(-\frac{8}{9}\right)$

43. $-5 \cdot \left(-\frac{2}{3}\right)$

44. $-7 \cdot \left(-\frac{3}{4}\right)$

45. $-\frac{3}{4} \cdot 8$

46. $-\frac{7}{8} \cdot 16$

47. $-\frac{34}{5} \cdot \left(-\frac{25}{4}\right)$

48. $-\frac{21}{4} \cdot \left(-\frac{16}{3}\right)$

49. Suppose a product consists of nine nonzero factors and six of the factors are negative. Explain how to determine the sign of the product.

50. Suppose a product consists of 17 factors. Eight factors are positive, eight factors are negative, and one factor is zero. Explain how to evaluate the product.

In Exercises 51–62, determine the product. Use your calculator to verify your results.

51. $(-2)(-3)(-4)$

52. $-4(-2)(5)$

53. $-5(7)(-2)$

54. $3(-6)(2)$

55. $(-2)(-1)(3)(2)$

56. $(-1)(-4)(3)(-2)$

57. $6(-1)(4)(-2)$

58. $-2(-2)(-3)(10)$

59. $(-1)(-2)(3)(1)(-6)$

60. $(-1)(-1)(2)(-3)(-5)$

61. $(-1)(-2)(3)(-1)(6)(23)(0)$

62. $(-1)(-1)(2)(3)(-5)(-35)(0)$

In Exercises 63–66, without evaluating, determine the order of the two given products and fill in the blank with a less than or greater than symbol. Then verify by calculating the products.

63. $2(-3)$ _____ $-2(-3)$

64. $(-1)(-1)$ _____ $0 \cdot (-1)$

65. $(-1)(-1)(-1)(-1)(-1)$ _____ $(-1)(-1)(-1)(-1)$

66. $8965(-213)$ _____ $1 \cdot 2$

In Exercises 67–70, determine the unknown number.

67. The product of what number and -8 is 24?

68. Five times what number is -20?

69. What number multiplied by -3 is -18?

70. If the product of three factors is 0 and two of the factors are 120 and -813, what is the third factor?

In Exercises 71–74, perform the indicated operations. Use your calculator to verify the results.

71. $|-6| \cdot |-3|$

72. $-|2| \cdot |-8|$

73. $|-9| \cdot \left(-|11|\right)$

74. $\left(-|-5|\right) \cdot \left(-|-10|\right)$

In Exercises 75–100, perform the indicated operation.

75. $\dfrac{11}{4} \cdot \left(-\dfrac{4}{3}\right)$ **76.** $-\dfrac{6}{11} \cdot \dfrac{11}{3}$

77. $-5.76 \cdot (-4.95)$ **78.** $-0.34 \cdot 25.3$

79. $-5|4|$ **80.** $|-9| \cdot |-10|$

81. $5(-2)(4)$ **82.** $2(-2)(-3)$

83. $-\dfrac{3}{7} \cdot \left(-\dfrac{4}{9}\right)$ **84.** $-9 \cdot \left(-\dfrac{2}{7}\right)$

85. $36.47 \cdot (-4.47)$

86. $-95.3 \cdot (-57.37)$

87. $(-2)(-1)(-2)(-3)$

88. $-6(-1)(3)(-2)$

89. $|9| \cdot |-8|$

90. $|-7| \cdot (-6)$

91. $\dfrac{3}{5} \cdot \dfrac{5}{7} \cdot \left(-\dfrac{7}{9}\right)$ **92.** $-\dfrac{2}{7} \cdot \dfrac{4}{5} \cdot \left(-\dfrac{3}{8}\right)$

93. $-4 \cdot (-24)$ **94.** $-7 \cdot (-9)$

95. $(-7)(-2)(0)(2)(-3)$

96. $(-1)(-1)(-4)(-1)(-3)$

97. $-\dfrac{3}{5} \cdot \left(-\dfrac{4}{7}\right) \cdot \left(-\dfrac{7}{9}\right) \cdot \left(-\dfrac{3}{4}\right)$

98. $-\dfrac{8}{9} \cdot \left(-\dfrac{4}{5}\right) \cdot \left(-\dfrac{5}{12}\right) \cdot \left(-\dfrac{3}{8}\right)$

99. $-14 \cdot 10$ **100.** $-4 \cdot 80$

In Exercises 101–108, find two numbers whose sum is the first number and product is the second number.

101. $5, 6$ **102.** $-6, 8$

103. $1, -12$ **104.** $2, -15$

105. $-8, 12$ **106.** $-10, 24$

107. $-7, -8$ **108.** $-4, -21$

109. An employee has obtained a car loan for \$1075 from her credit union. She has an automatic payroll deduction of \$225.39 each month for the next 5 months to pay off the loan. At the end of the loan period, how much interest will she have paid?

110. A dog owner buys a 50-pound bag of dog food for his golden retriever. If the dog eats 12 ounces of food each day, how many pounds of dog food are left after 3 weeks? (There are 16 ounces in 1 pound.)

111. Instead of recording the actual number of strokes, a golfer used the following to record her scores on each hole.

eagle: -2 (2 below par)
birdie: -1 (1 below par)
par: 0 (standard score)
bogey: $+1$ (1 above par)
double bogey: $+2$ (2 above par)

At the end of the round, the golfer found that her performance on the 18 holes was as follows.

eagle: 1
birdie: 3
par: 9
bogey: 2
double bogey: 3

If par for the course was 72, what was the golfer's score?

112. To discourage guessing, a test grading scheme is as follows.

Correct answer $+1$
Answer left blank   0
Incorrect answer -1

On a 100-question test, a student left 13 answers blank and answered 50 questions correctly. What was the test score?

113. Consider the patterns in the following columns of products.

Column A		Column B	
$3 \cdot 3$	$= 9$	$-3 \cdot 3$	$= -9$
$3 \cdot 2$	$= 6$	$-3 \cdot 2$	$= -6$
$3 \cdot 1$	$= 3$	$-3 \cdot 1$	$= -3$
$3 \cdot 0$	$= 0$	$-3 \cdot 0$	$= 0$
$3 \cdot (-1) =$		$-3 \cdot (-1) =$	
$3 \cdot (-2) =$		$-3 \cdot (-2) =$	
$3 \cdot (-3) =$		$-3 \cdot (-3) =$	

Note that in column A, the first factor 3 remains the same in each product; in column B, the first factor -3 remains the same in each product.

(a) In both columns, what pattern is the second factor following from one product to the next?

(b) What pattern is the result following from one product to the next in column A? in column B?

(c) Assuming the patterns continue, what would be the last three results in each column?

(d) What known rules do these experiments support?

114. The product $5 \cdot 3$ can be interpreted as repeated addition.

$$5 \cdot 3 = 3 + 3 + 3 + 3 + 3$$

Interpret $7(-2)$ in a similar way and verify that the product and the sum have the same value.

115. Which of the following products is easier to calculate without a calculator? Why?

(i) $(-1)(-1)(-1)(-1)(-1)(-1)(-1)(-1)$

(ii) $(1.37)(6.42)$

Exploring with Real Data

In Exercises 116–119, refer to the following background information and data.

The federal government has not balanced its budget in over 130 years. In 1993, 22% of the federal budget was used just to pay interest on the public debt. To pay off this debt, it would have cost every American citizen $16,500. (Source: U.S. Treasury Department.)

116. Using $-\$16,500$ as the average cost per citizen to pay off the debt, and assuming a population of 250 million people, what was the national debt in 1993?

117. Use the result in Exercise 116 to calculate the interest paid on the national debt in 1993 at an average interest rate of 7%.

118. Use the result in Exercise 117 to calculate the 1993 cost per citizen to pay just the interest on the national debt.

119. Even if the national debt were cut in half, what would be the average cost for a family of four to pay off the debt?

Challenge

In Exercises 120–125, let a and b represent negative numbers and let c represent a positive number. What is the sign of the given expression?

120. ab

121. bc

122. $a \cdot |b| \cdot c$

123. abc

124. $a \cdot a$

125. $a \cdot b \cdot b$

1.5 Division

Definition and Rules ▪ *Division Involving 0* ▪ *Multiple Operations and Fractions* ▪ *Applications*

Definition and Rules

The last of our four basic operations is division of real numbers. Division is written with the division symbol ÷ or with a fraction bar. For example, we write 19 divided by 7 as $19 \div 7$ or $\frac{19}{7}$. Your calculator may display 19/7.

The number 19 is called the **dividend,** and the number 7 is called the **divisor.** The expression and the result of the division are both called the **quotient.**

Division is defined in terms of multiplication.

> ### Definition of Division
> For real numbers a, b, and c, with $b \neq 0$,
> $$a \div b = c \quad \text{if} \quad c \cdot b = a.$$

In the following exploration, we use the definition of division to derive the sign rules for division.

EXPLORATION 1

DIVIDE

Sign Rules for Division

Use your calculator to determine each quotient. Then, answer the questions that follow.

(a) $-8 \div (-2)$ (b) $24.71 \div 5.38$

(c) $56 \div (-7)$ (d) $-66 \div 6$

(e) Verify your results in parts (a)–(d) by using the definition of division. That is, if you found that $a \div b = c$, then check that $c \cdot b = a$.

(f) How do the sign rules for division seem to compare to the sign rules for multiplication?

Discovery

(a) $-8 \div (-2) = 4$ (b) $24.71 \div 5.38 \approx 4.593$

(c) $56 \div (-7) = -8$ (d) $-66 \div 6 = -11$

(e) (a) $4(-2) = -8$ (b) $4.593(5.38) \approx 24.71$
    (c) $-8(-7) = 56$ (d) $-11(6) = -66$

(f) When the dividend and divisor have like signs, as in parts (a) and (b), the quotient is positive. When the dividend and divisor have unlike signs, as in parts (c) and (d), the quotient is negative. The sign rules for division appear to be the same as the sign rules for multiplication.

Exploration 1 suggests the following generalizations.

1. The sign rules for division are the same as the sign rules for multiplication.

2. The absolute value of a quotient is the quotient of the absolute values of the dividend and divisor.

Division Involving 0

We have seen that if any factor of a product is 0, then the product is 0. In our next exploration, we investigate the role of 0 in division.

 EXPLORATION 2   *Division Involving* 0

Use your calculator to perform each operation, if possible. Then answer the questions that follow.

(a) $\dfrac{0}{2}$ (b) $\dfrac{0}{-4}$ (c) $\dfrac{5}{0}$ (d) $\dfrac{0}{0}$

(e) In parts (a) and (b), show that the results satisfy the definition of division.

(f) How did your calculator respond to parts (c) and (d)? Does the definition of division apply to these problems? Why?

Discovery

(a) $\dfrac{0}{2} = 0$ (b) $\dfrac{0}{-4} = 0$

(c) Error message.   (d) Error message.

(e) $\dfrac{0}{2} = 0$ and $(0)(2) = 0$

$\dfrac{0}{-4} = 0$ and $(0)(-4) = 0$

(f) Division by 0 causes a calculator to display an error message because the definition prohibits division by 0.

Exploration 2 suggests that 0 can be divided by any nonzero number and the result is 0. However, division *by* 0 is prohibited by the definition.

Suppose there were no such restriction and we attempt to use the definition to perform $5 \div 0$. Then the quotient c would have the property that $c \cdot 0 = 5$. But this is not possible because $c \cdot 0 = 0$ for any number c.

Also, suppose we attempt to use the definition to perform $0 \div 0$. Then the quotient c would have the property that $c \cdot 0 = 0$. But we cannot determine a specific value of c in this case because $c \cdot 0 = 0$ is true for any number c. We say that $0 \div 0$ is *indeterminant*.

In short, *division by 0 is undefined.*

EXAMPLE 1   *Determining the Sign of a Quotient*

For each of the following, state whether the quotient is defined. If it is, determine the sign, if any, of the quotient.

(a) $\dfrac{-4.78}{-5.2}$ (b) $\dfrac{0}{-3.2}$ (c) $\dfrac{2}{3} \div 0$ (d) $-\dfrac{3}{4} \div \dfrac{5}{6}$

Solution

(a) The quotient is defined and is positive.

(b) The quotient is defined and is 0.

(c) Because the divisor is 0, the expression is undefined.

(d) The quotient is defined and is negative.   ■

Multiple Operations and Fractions

Operations in grouping symbols are performed first. If there are no grouping symbols present, we multiply and divide from left to right.

Also, when either the dividend or the divisor is a fraction, we multiply the dividend by the reciprocal of the divisor.

EXAMPLE 2   *Performing Division*

Determine the quotient, if possible.

(a) $36 \div (-4)$ (b) $-13 \div (-9)$

(c) $36 \div (-12 \div 3)$ (d) $-4 \cdot 6 \div (-3)$

(e) $-\dfrac{1}{3} \div 5$ (f) $\dfrac{5}{7} \div \left(-\dfrac{3}{4}\right)$

Solution

(a) $36 \div (-4) = \dfrac{36}{-4} = -9$

(b) $-13 \div (-9) = \dfrac{-13}{-9} = \dfrac{13}{9} \approx 1.44$

(c) $36 \div (-12 \div 3) = 36 \div (-4) = -9$ Perform operations within grouping symbols first.

(d) $-4 \cdot 6 \div (-3) = -24 \div (-3) = 8$ Operate from left to right.

(e) $-\dfrac{1}{3} \div 5 = -\dfrac{1}{3} \cdot \dfrac{1}{5}$ The reciprocal of a positive number is positive.

$= -\dfrac{1}{15}$

≈ -0.067

(f) $\dfrac{5}{7} \div \left(-\dfrac{3}{4}\right) = \dfrac{5}{7} \cdot \left(-\dfrac{4}{3}\right)$ The reciprocal of a negative number is negative.

$= -\dfrac{20}{21}$

≈ -0.95 ∎

In part (a) of Example 2, we found that $\dfrac{36}{-4} = -9$. The same result is obtained for $\dfrac{-36}{4}$ and $-\dfrac{36}{4}$. Thus, $\dfrac{36}{-4} = \dfrac{-36}{4} = -\dfrac{36}{4}$. If a fraction has a single opposite sign, it can be placed in the numerator, in the denominator, or in front of the fraction.

In part (b) of Example 2, we found that $\dfrac{-13}{-9} = \dfrac{13}{9}$. This suggests that when a fraction has two opposite signs, they can both be removed.

If a fraction has three opposite signs, any two of them can be removed:

$$-\dfrac{-2}{-3} = \dfrac{-2}{3} = \dfrac{2}{-3} = -\dfrac{2}{3}.$$

Opposite Signs in Quotients

For real numbers a and b ($b \neq 0$), each of the following is true.

$$-\dfrac{a}{b} = \dfrac{-a}{b} = \dfrac{a}{-b}$$

$$\dfrac{-a}{-b} = \dfrac{a}{b}$$

$$-\dfrac{-a}{-b} = \dfrac{-a}{b} = \dfrac{a}{-b} = -\dfrac{a}{b}$$

Applications

If a quantity has decreased over a period of time, calculating the average decrease may require dividing a negative number by a positive number.

EXAMPLE 3   *Average Change in Temperature*

The temperature dropped 32°F in a 4-hour period. What was the average hourly change in temperature?

Solution   The average change was $\dfrac{-32}{4}$ or -8°F per hour. ∎

1.5 *Quick Reference*

Definition and Rules	■ In the expression $a \div b$, a is called the **dividend** and b is called the **divisor.** Both the expression and the value of the expression are called the **quotient.**

■ For real numbers a, b, and c, with $b \neq 0$,

$$a \div b = c \text{ if } c \cdot b = a.$$

■ The sign rules for division are the same as the sign rules for multiplication.

■ The absolute value of a quotient is the quotient of the absolute values of the dividend and divisor. |

Division Involving 0	■ If a is any nonzero real number, $0 \div a = 0$.

■ Division by 0 is undefined. |

Multiple Operations and Fractions	■ Operations in grouping symbols are performed first. If there are no grouping symbols present, we multiply and divide from left to right.

■ When the dividend or divisor is a fraction, multiply the dividend by the reciprocal of the divisor.

■ For real numbers a and b, $b \neq 0$, the following are true.

1. $-\dfrac{a}{b} = \dfrac{-a}{b} = \dfrac{a}{-b}$

2. $\dfrac{-a}{-b} = \dfrac{a}{b}$

3. $-\dfrac{-a}{-b} = \dfrac{-a}{b} = \dfrac{a}{-b} = -\dfrac{a}{b}$ |

1.5 *Exercises*

1. Explain how to determine the quotient of two numbers with like signs.

2. Explain how to determine the quotient of two numbers with unlike signs.

In Exercises 3–22, divide, if possible.

3. $8 \div (-4)$

4. $-14 \div 2$

5. $\dfrac{15}{-5}$

6. $\dfrac{-24}{6}$

7. $-40 \div (-5)$

8. $-60 \div (-10)$

9. $-16 \div (-2)$

10. $-10 \div 10$

11. $\dfrac{-80}{-5}$

12. $\dfrac{-8}{8}$

13. $-21 \div 7$

14. $55 \div (-55)$

15. $\dfrac{-20}{5}$

16. $\dfrac{56}{-7}$

17. $-7 \div 1$

18. $-4 \div 0$

19. $\dfrac{-2}{0}$

20. $\dfrac{0}{-1}$

21. $0 \div (-35)$

22. $-92 \div (-92)$

In Exercises 23–30, perform the indicated operations.

23. $-6 \div 3 \div (-4)$

24. $-30 \div (-5) \div (-2)$

25. $24 \div [-6 \div (-2)]$

26. $-12 \div [-6 \div (-3)]$

27. $-16 \cdot 2 \div 8$

28. $20 \div 5 \cdot 3$

29. $18 \div (-6 \cdot 3)$

30. $4 \cdot [-8 \div (-16)]$

  31. In your own words, interpret $n/n = 1,\, n \neq 0$.

32. In your own words, interpret $n/1 = n$.

In Exercises 33–40, predict the sign of the quotient. Then, use your calculator to perform the operation.

33. $-5.9 \div (-2.7)$

34. $-4.5 \div (-3.4)$

35. $22.1 \div (-2.3)$

36. $50.5 \div (-4.3)$

37. $-4.7 \div 9.3$

38. $-5.4 \div 7.3$

39. $42.4 \div (-13.5)$

40. $-25.7 \div (-21.2)$

In Exercises 41–48, write a numerical expression that models the problem and then evaluate it.

41. What is the quotient of -16.4 and -4.1?

42. What is 500 divided by -1000?

43. If the dividend is -11 and the divisor is 9, what is the quotient?

44. If both the dividend and divisor are -3, what is the quotient?

45. What is 0 divided by -10?

46. What is 0 divided into 7?

47. What is -20 divided into -10?

48. What is 12 divided by $\frac{1}{2}$?

In Exercises 49–58, perform the indicated division. Use your calculator only to verify your results.

49. $-\dfrac{1}{3} \div \dfrac{5}{9}$

50. $-\dfrac{3}{4} \div \dfrac{3}{16}$

51. $-\dfrac{2}{5} \div \left(-\dfrac{3}{10}\right)$

52. $-\dfrac{3}{14} \div \left(-\dfrac{6}{7}\right)$

53. $\dfrac{5}{16} \div \left(-\dfrac{3}{8}\right)$

54. $-\dfrac{3}{4} \div \left(-\dfrac{3}{8}\right)$

55. $-5 \div \left(-\dfrac{2}{3}\right)$

56. $-7 \div \left(-\dfrac{3}{4}\right)$

57. $\left(-\dfrac{3}{4}\right) \div 8$

58. $\left(-\dfrac{7}{8}\right) \div 16$

In Exercises 59–66, determine

   (a) the opposite of the given number and

   (b) the reciprocal of the given number.

59. 5

60. -8

61. $-\dfrac{3}{4}$

62. $\dfrac{2}{5}$

63. 0

64. $-\dfrac{4}{5}$

65. 4.5

66. 0.2

In Exercises 67–70, estimate the order of the two given quotients and fill in the blank with a less than or greater than symbol. Verify your answer by calculating the quotients with your calculator.

67. $\dfrac{3}{7}$ _____ $\dfrac{3}{8}$

68. $-200 \div 27.68$ _____ $0 \div (-17)$

69. $-8 \div 9$ _____ $-\dfrac{7}{9}$

70. $(-24 \div 8) \div (-3)$ _____ $(24 \div 8) \div (-3)$

In Exercises 71–74, determine the unknown number.

71. The quotient of what number and -7 is -2?

72. Ten divided by what number is -5?

73. If the quotient is 10 and the divisor is -4, what is the dividend?

74. What is the divisor if the dividend is -8 and the quotient is 8?

In Exercises 75–106, perform the indicated operations.

75. $-26.4 \div (-57.8)$

76. $-5.76 \div (-4.95)$

77. $-\dfrac{2}{3} \div \dfrac{8}{27}$

78. $\dfrac{6}{7} \div \left(-\dfrac{3}{28}\right)$

79. $-|-5| \div |-5|$

80. $|-63| \div \left(-|9|\right)$

81. $-56 \div 4 \cdot 2$

82. $-8 \cdot 3 \div 12$

83. $-7 \div \left(-\dfrac{2}{5}\right)$

84. $-\dfrac{1}{6} \div 6$

85. $\dfrac{9}{-3}$

86. $\dfrac{-48}{-6}$

87. $18 \div (-2) \div (-3)$

88. $-32 \div 8 \div (-2)$

89. $12 \div (-2 \cdot 3)$

90. $-50 \div [5 \cdot (-2)]$

91. $0 \div (-1.274)$

92. $-43.56 \div 43.56$

93. $86.47 \div (-4.47)$

94. $-95.3 \div (-57.37)$

95. $-20 \div (-10) \div (-6)$

96. $-7 \div 7 \div (-4)$

97. $\dfrac{|-48|}{|-12|}$

98. $\dfrac{|-42|}{|7|}$

99. $27 \div (-3)$

100. $-27 \div (-9)$

101. $72 \div (-6 \div 3)$

102. $-60 \div (12 \div 4)$

103. $\dfrac{0}{0}$

104. $\dfrac{-64}{8}$

105. $-11 \div 0$

106. $-37 \div 10$

 107. A nonzero number divided by itself always results in 1. For example, $5 \div 5 = 1$ and $-8 \div (-8) = 1$. Explain, then, why $0 \div 0$ does not equal 1.

108. The definition of division states that $a \div b = c$ if $c \cdot b = a$. Using the definition, we might claim that $0 \div 0 = 0$ because $0 \cdot 0 = 0$. What is your opinion of this?

In Exercises 109–114, rewrite the given fraction with fewer opposite signs.

109. $\dfrac{-5}{-8}$

110. $-\dfrac{-6}{-11}$

111. $-\dfrac{-1}{-6}$

112. $-\dfrac{3}{-5}$

113. $-\dfrac{-4}{7}$

114. $\dfrac{-7}{-9}$

115. The total change in the stock market over an 8-day period was -39 points. What was the average daily change?

116. An instructor grades a set of 20-answer quizzes by indicating the total number of incorrect answers. Here are the scores for six of the students: $-5, -1, -7, -4, -5, -2$.

(a) What was the average number of incorrect answers?

(b) What was the average percentage of correct answers?

Exploring with Real Data

In Exercises 117–120, refer to the following background information and data.

Harley-Davidson, Inc., a manufacturer of motorcycles, succeeded in reducing its long-term debt during the period 1986–1991.

	1986	1987	1988	1989	1990	1991
Long-Term Debt (millions)	191.6	178.8	135.2	74.8	48.3	46.9

(Source: Harley-Davidson, Inc.)

117. Calculate the difference between the debt in 1991 and the debt in 1986.

118. How do you interpret the fact that the result in Exercise 117 is a negative number?

119. What was the average annual reduction in long-term debt during the given period?

120. During this same period, Harley-Davidson's working capital increased. If you had been an investor during this time, why would you have felt more confident about buying stock in Harley-Davidson?

Challenge

In Exercises 121–126, let *a* represent a negative number and *b* represent a positive number. Determine the sign of the expression.

121. $\dfrac{a}{b}$

122. $\dfrac{0}{b}$

123. $\dfrac{a}{0}$

124. $\dfrac{a}{-b}$

125. $\dfrac{-a}{b}$

126. $\dfrac{|a|}{-b}$

In Exercises 127–130, evaluate the given quotient.

127. $\dfrac{\frac{5}{8}}{\frac{3}{4}}$

128. $\dfrac{\frac{14}{3}}{\frac{-7}{6}}$

129. $\dfrac{\frac{-20}{9}}{\frac{10}{-3}}$

130. $\dfrac{\frac{8}{-5}}{\frac{-4}{15}}$

1.6   Exponents, Square Roots, Order of Operations

Exponents ▪ *Square Roots* ▪ *Order of Operations*

Exponents

To indicate repeated multiplication, we use an exponent.

$$3^4 = 3 \cdot 3 \cdot 3 \cdot 3 = 81$$

The number 3 is called the **base,** and 4 is called the **exponent.** The expression 3^4 is read "three to the fourth power." For exponents of 2 and 3, the words *squared* and *cubed* are sometimes used rather than "to the second power" and "to the third power."

The expression 3^4 is in **exponential form,** and the expression $3 \cdot 3 \cdot 3 \cdot 3$ is in **product** (or **factored**) **form.** When the exponent is a natural number, the base can be any real number.

EXAMPLE 1   *Exponential and Product Forms*

Write each product form in exponential form and each exponential form in product form. Use your calculator to determine the value of both forms.

EXPONENT

(a) $5 \cdot 5 \cdot 5 \cdot 5 \cdot 5 \cdot 5 \cdot 5$

(b) $(-3) \cdot (-3) \cdot (-3) \cdot (-3)$

(c) 2^4

(d) $(-4)^5$

Solution

(a) $5 \cdot 5 \cdot 5 \cdot 5 \cdot 5 \cdot 5 \cdot 5 = 5^7 = 78{,}125$

(b) $(-3) \cdot (-3) \cdot (-3) \cdot (-3) = (-3)^4 = 81$

(c) $2^4 = 2 \cdot 2 \cdot 2 \cdot 2 = 16$

(d) $(-4)^5 = (-4) \cdot (-4) \cdot (-4) \cdot (-4) \cdot (-4) = -1024$ ∎

 **EXPLORATION 1** *Evaluating an Exponential Expression When the Base Is Negative*

Use your calculator to evaluate each expression. Then answer the questions that follow.

(a) $(-2)^3$ (b) $(-1)^7$

(c) $(-4)^6$ (d) $(-2)^{10}$

(e) In each part the base is negative. Under what condition is the result negative? positive?

(f) Write a sentence summarizing your conclusions in parts (a)–(e).

Discovery

(a)–(b)   Figure 1.11   (c)–(d)   Figure 1.12

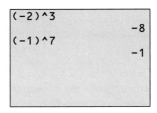   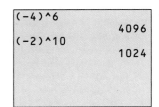

(e) In parts (a) and (b), the exponents are odd, and the result is negative. (See Fig. 1.11.) In parts (c) and (d), the exponents are even, and the result is positive. (See Fig. 1.12.)

(f) If the base is negative and the exponent is a natural number, then the expression is positive if the exponent is even and negative if the exponent is odd.

EXAMPLE 2 *Evaluating Exponential Expressions*

Evaluate each exponential expression. Use your calculator for parts (j)–(l).

(a) 3^2 (b) 2^3

(c) $(-4)^2$ (d) $(-4)^3$

(e) 0^{157} (f) 1^{157}

(g) $(-1)^{157}$ (h) $\left(\dfrac{3}{2}\right)^2$

(i) $\left(-\dfrac{2}{3}\right)^3$ (j) 9^5

(k) $(-3)^{10}$ (l) 2.94^6

Solution

(a) $3^2 = 3 \cdot 3 = 9$

(b) $2^3 = 2 \cdot 2 \cdot 2 = 8$

(c) $(-4)^2 = (-4) \cdot (-4) = 16$ Negative base, even exponent

(d) $(-4)^3 = (-4) \cdot (-4) \cdot (-4) = -64$ Negative base, odd exponent

(e) $0^{157} = 0$ If any factor is 0, the product is 0.

(f) $1^{157} = 1$ If all factors are 1, the product is 1.

(g) $(-1)^{157} = -1$ Negative base, odd exponent

(h) $\left(\dfrac{3}{2}\right)^2 = \dfrac{3}{2} \cdot \dfrac{3}{2} = \dfrac{3^2}{2^2} = \dfrac{9}{4}$

(i) $\left(-\dfrac{2}{3}\right)^3 = \dfrac{-2}{3} \cdot \dfrac{-2}{3} \cdot \dfrac{-2}{3} = \dfrac{(-2)^3}{3^3} = -\dfrac{8}{27}$ Recall that $-\dfrac{2}{3} = \dfrac{-2}{3}$.

(j)–(l)   Figure 1.13

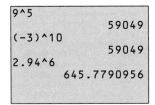

```
9^5
                    59049
(-3)^10
                    59049
2.94^6
          645.7790956
```

NOTE: Parts (h) and (i) of Example 2 suggest that raising a fraction to a power is the same as raising both the numerator and denominator to the power. We will formalize this rule in Chapter 2.

Square Roots

Because 16 can be written as 4^2, 4 is called a **square root** of 16. Similarly, 16 can be written as $(-4)^2$, so -4 is also a square root of 16.

> *Definition of Square Root*
>
> If $b^2 = a$, then b is a **square root** of a.

Note that in the system of real numbers, b^2 cannot be negative. Therefore, a negative number cannot have a square root. Also, because 0 is the only number that can be squared to obtain 0, 0 has only one square root. All positive numbers have two square roots, one positive and one negative.

The positive (or **principal**) square root of a number is indicated by a **radical sign** $\sqrt{}$. If $b^2 = a$, where $b \geq 0$, then $\sqrt{a} = b$. The number or expression under the radical sign is called the **radicand.**

EXAMPLE 3 *Determining Square Roots*

(a) What are the square roots of 81?

(b) What is $\sqrt{81}$?

(c) What is $-\sqrt{81}$?

(d) What is $\sqrt{-81}$?

(e) What is $\sqrt{0}$?

Solution

(a) Because $9^2 = 81$ and $(-9)^2 = 81$, the square roots of 81 are 9 and -9.

(b) $\sqrt{81} = 9$ The symbol $\sqrt{81}$ refers to the *principal* square root of 81

(c) $-\sqrt{81} = -1 \cdot \sqrt{81} = -1 \cdot 9 = -9$

(d) Because no real number can be squared to obtain -81, $\sqrt{-81}$ is not a real number.

(e) $\sqrt{0} = 0$ ■

It is easy to evaluate square roots like $\sqrt{9}$ or $\sqrt{100}$ without a calculator because the radicands are familiar perfect squares. However, we can make good use of a calculator for evaluating square roots like $\sqrt{57,121}$ and $\sqrt{43}$.

EXAMPLE 4 *Evaluating Square Roots with a Calculator*

Use your calculator to evaluate the following.

 SQUARE ROOT

(a) $\sqrt{57,121}$

(b) $\sqrt{43}$

(c) $\sqrt{36.27}$

(d) $\sqrt{0.0082}$

Solution

(a)–(b)   Figure 1.14   (c)–(d)   Figure 1.15

```
√57121
              239
√43
       6.557438524
```

```
√36.27
         6.02245797
√.0082
          .0905538514
```

Because $2^3 = 8$, 2 is called a **cube root** of 8. Symbolically, $\sqrt[3]{8} = 2$. Because $3^4 = 81$, 3 is a **fourth root** of 81. Symbolically, $\sqrt[4]{81} = 3$. The concept can be extended to higher and higher roots. We will discuss roots in greater detail in Chapter 9.

Order of Operations

When more than one operation is required in an expression, the order in which the operations are performed can produce different results.

EXPLORATION 2   *A Calculator's Order of Operations*

Use your calculator to evaluate each expression. In each case, which operation does the calculator perform first?

(a) $2 + 3 \cdot 5$ (b) $2 \cdot 3^2$

Discovery

See Fig. 1.16.

Figure 1.16

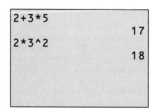

```
2+3*5
              17
2*3^2
              18
```

(a) Multiplication is done first, and then the addition is performed.

(b) The exponential factor is evaluated first, and then the multiplication is performed.

The following is the order in which operations in an expression are performed.

The Order of Operations

1. If grouping symbols are present, perform all operations inside of them according to the following order. For grouping symbols within grouping symbols, start with the innermost group and work outward.

2. Evaluate exponents and roots.

3. Perform multiplication and division from left to right.

4. Perform addition and subtraction from left to right.

NOTE: We must understand that this Order of Operations is not based on the order in which calculators perform operations. Rather, calculators are designed *in accordance with* the Order of Operations.   ■

By grouping symbols, we usually mean parentheses, brackets, or braces as in $2(5 + 4)$ or $3 + \{2[5 - 2(1 - 8)]\}$. However, fraction bars and radical signs also serve as grouping symbols. For example, $\frac{5+3}{7}$ means $\frac{(5+3)}{7}$ and $\sqrt{16-9}$ means $\sqrt{(16-9)}$.

EXAMPLE 5   *The Order of Operations*

Perform the indicated operations. Then, verify the results using your calculator. (Remember to supply implied grouping symbols when you use the calculator.)

(a) $5 + 4 \cdot 2$ (b) $5 - 5(3^2 - 7)$

(c) $2 - [4 - (9 - 12)]$ (d) $24 \div 6 \cdot 3 \div 2$

(e) $(-4)^2$ (f) -4^2

(g) $\dfrac{-3 + 5 \cdot 2}{14}$ (h) $7 + 2 \div 2 - 8 \cdot 3^2$

(i) $7 - \sqrt{34 - 6 \cdot 3}$

Solution

(a) $5 + 4 \cdot 2 = 5 + 8 = 13$ Multiplication ranks higher than addition.

(b) $5 - 5(3^2 - 7) = 5 - 5(9 - 7)$ Grouping symbols have the highest priority.
$$= 5 - 5(2)$$ Within the parentheses, the exponent ranks first.
$$= 5 - 10$$ Multiplication ranks higher than subtraction.
$$= -5$$

(c) $2 - [4 - (9 - 12)] = 2 - [4 - (-3)]$ Start with the innermost grouping symbols.
$$= 2 - [7]$$
$$= -5$$

(d) $24 \div 6 \cdot 3 \div 2 = 4 \cdot 3 \div 2$ Multiplication and division rank equally, so proceed from left to right.
$$= 12 \div 2$$
$$= 6$$

(e) $(-4)^2 = (-4)(-4) = 16$

(f) $-4^2 = -1 \cdot 4^2 = -1 \cdot 16 = -16$ Note that -4^2 does *not* mean $(-4)(-4)$.

(g) The fraction bar acts as a grouping symbol. We evaluate the numerator before dividing by 14. Enclose the numerator in parentheses when you use your calculator.
$$\frac{-3 + 5 \cdot 2}{14} = \frac{-3 + 10}{14} = \frac{7}{14} = \frac{1}{2}$$

(h) $7 + 2 \div 2 - 8 \cdot 3^2 = 7 + 2 \div 2 - 8 \cdot 9$ Exponent ranks first.
$$= 7 + 1 - 72$$ Multiplication and division rank next.
$$= -64$$ Addition and subtraction rank last.

(i) The radical sign serves as a grouping symbol, so the radicand is evaluated first. Enclose the radicand in parentheses when you use your calculator.
$$7 - \sqrt{34 - 6 \cdot 3} = 7 - \sqrt{34 - 18} = 7 - \sqrt{16} = 7 - 4 = 3$$ ∎

1.6   *Quick Reference*

Exponents

- The expression b^n is in **exponential form.** The number b is called the **base,** and the number n is called the **exponent.**

- If the base b is a negative number, then b^n is

  1. positive if n is an even natural number, and

  2. negative if n is an odd natural number.

- Raising a fraction to a power is the same as raising both the numerator and the denominator to the power.

Square Roots

- If $b^2 = a$, then b is a **square root** of a.

- A negative number cannot have a square root. All positive numbers have two square roots, one positive and one negative. The square root of 0 is 0.

- The **principal** (positive) square root of a number is indicated by a **radical sign** $\sqrt{}$. The expression under the radical sign is called the **radicand.**

Order of Operations

- The following is the order in which operations are performed.

  1. If grouping symbols are present, perform all operations inside of them according to the following order. For grouping symbols within grouping symbols, start with the innermost group and work outward.

  2. Evaluate exponents and roots.

  3. Perform multiplication and division from left to right.

  4. Perform addition and subtraction from left to right.

- In addition to parentheses, brackets, and braces, fraction bars and radical signs serve as grouping symbols.

1.6   *Exercises*

1. What are the meanings of the words *base* and *radicand*? Illustrate the use of these words with examples.

2. Name five kinds of symbols that are used as grouping symbols. For each kind, give an example of an expression in which the grouping symbols are used.

In Exercises 3–10, write the given product in exponential form.

3. $7 \cdot 7 \cdot 7 \cdot 7 \cdot 7 \cdot 7$

4. $(-11) \cdot (-11) \cdot (-11) \cdot (-11)$

5. $-[(-5) \cdot (-5) \cdot (-5)]$

6. $-(2 \cdot 2 \cdot 2 \cdot 2 \cdot 2)$

7. $a \cdot a \cdot a$

8. $x \cdot x \cdot x \cdot y \cdot y$

9. $(2b)(2b)(2b)(2b)$

10. $(xy)(xy)(xy)$

In Exercises 11–18, write the given exponential form in factored form.

11. $\left(\dfrac{7}{9}\right)^4$

12. 11^5

13. $2x^2$

14. $-x^4$

15. $5^2 y^3$

16. $3^4 x^2$

17. $-4xy^3$

18. $2x^3 y^2$

In Exercises 19–32, evaluate the given exponential expression.

19. 5^2

20. 3^4

21. $(-2)^3$

22. $(-4)^2$

23. -7^2

24. -6^2

25. 2^5

26. 10^2

27. 7^1

28. -5^1

29. $\left(\dfrac{2}{3}\right)^2$

30. $\left(-\dfrac{3}{4}\right)^3$

31. $(-1)^6$

32. $(-1)^9$

In Exercises 33–38, use your calculator to perform the indicated operations. Round decimal answers to the nearest hundredth.

33. 12^5

34. 11^4

35. $(-3)^{11}$

36. $(-4)^9$

37. $(3.47)^7$

38. $(2.58)^8$

 39. Explain why 36 has two square roots, but it is *incorrect* to say $\sqrt{36} = 6$ or -6.

 40. Explain how to verify that $\sqrt{2809} = 53$.

In Exercises 41–50, determine the square roots of the given number, if possible. Use a calculator as needed.

41. 4

42. 16

43. 36

44. 100

45. -25

46. 0

47. $\dfrac{9}{16}$

48. $\dfrac{1}{25}$

49. 529

50. 1764

In Exercises 51–56, evaluate each square root.

51. $\sqrt{4}$

52. $\sqrt{16}$

53. $-\sqrt{36}$

54. $-\sqrt{100}$

55. $\sqrt{\dfrac{4}{9}}$

56. $-\sqrt{\dfrac{25}{36}}$

In Exercises 57–66, use your calculator to determine the square root. Round results to the nearest hundredth.

57. $\sqrt{676}$

58. $\sqrt{1849}$

59. $\sqrt{8.41}$

60. $\sqrt{9.61}$

61. $\sqrt{534}$

62. $\sqrt{747}$

63. $\sqrt{63{,}876}$

64. $\sqrt{78{,}432}$

65. $\sqrt{0.0076}$

66. $\sqrt{0.0054}$

 67. Describe the order in which operations are performed for $(-6)^2$ and -6^2. Show that the two expressions have different values.

 68. For the following two expressions, describe the difference in the order in which the operations are to be performed.

    (i) $5 - 3^2$ (ii) $(5 - 3)^2$

In Exercises 69–90, perform the indicated operations. Use your calculator to verify results.

69. $6 + 4(-3)$

70. $-3 - 2^3$

71. $\dfrac{240}{6 \cdot 4}$

72. $30 \div 5 \cdot 6$

73. $1.3 \cdot 10^3$

74. $9.8 \cdot 10^5$

75. $3 \cdot 6 - 4 \cdot 2$

76. $-3^2 + 6(-1)^5$

77. $-2|-4| + |5|$

78. $-2\sqrt{16} - \sqrt{6+3}$

79. $6 - 4 \cdot 2 + 20 \div 5$

80. $6 - 5(5 - 3)^2$

81. $\dfrac{5^2 + 7}{11 - 3^2}$　　　　**82.** $\dfrac{4 + 6 \cdot 3}{-5 - 3 \cdot 2}$

83. $\left(\dfrac{3}{4} - \dfrac{1}{2}\right) \cdot \left(\dfrac{3}{4} + \dfrac{1}{2}\right)$

84. $\left(3 + \dfrac{1}{3}\right) \cdot \left(\dfrac{1}{2} - \dfrac{5}{6}\right)$

85. $\dfrac{9 - 3}{3 - 9}$　　　　**86.** $\dfrac{5 \cdot 6 - 2(-3)}{6 - (-6)}$

87. $12 - (5 + 3) \div 2 + 6$

88. $1 - (2 + 7) \div (5 + 4)$

89. $5 - [3(2 - 1) - 4(3 - 5)]$

90. $6 - 4[6 - 4(6 - 4)]$

In Exercises 91–96, translate the verbal description into a numerical expression. Then, evaluate the expression.

91. Four less than the product of -2 and 7

92. Six more than the square of -8, with the result divided by 10

93. Add -3 to 7 and divide the sum by 2. Then, subtract -4 from the quotient.

94. Subtract the sum of 3 and -5 from 4. Multiply the result by -2. Subtract this result from 7.

95. The square of the difference of 3 and -2

96. One more than three-fourths of the sum of the squares of -4 and 6

In Exercises 97–100, insert grouping symbols so that the numerical expression has the values given in parts (a), (b), and (c).

97. $-3 \cdot 4 - 5$　　　　**98.** $7 - 5^2 + 1$

   (a) 3　　　　　　　　(a) 5

   (b) -17　　　　　　(b) -19

   (c) -7　　　　　　　(c) -17

99. $8 + 12 \div 4 - 1$　　**100.** $4 - 2 \cdot 3 - 4 - 6$

   (a) 4　　　　　　　　(a) 8

   (b) 12　　　　　　　(b) 0

   (c) 10　　　　　　　(c) -4

In Exercises 101–118, use your calculator to perform the operations. Round decimal answers to the nearest hundredth.

101. $-4 \cdot 3^2$　　　　**102.** $2^3 - 6^1$

103. $-2^4 + 5^1 \cdot 2^3$　　**104.** $-(-2)^4 + 3\sqrt{16}$

105. $2 \cdot 3^2 - 4 \cdot 3 + 1$　**106.** $-(-1)^2 - (-1) - 1$

107. $\left(\sqrt{9}\right)^2 - \sqrt{9} + 1$

108. $5 - 4\sqrt{3 \cdot 7} - 5$

109. $\dfrac{6(4 - 2) - 3 \cdot 2^2}{5 - 3 \cdot 7}$

110. $\dfrac{2^2 - 3^2}{10 - 5(9 - 7)}$　　**111.** $\sqrt{16 + 9}$

112. $\dfrac{5^2 - 3^2}{\sqrt{5^2 - 3^2}}$

113. $(5 - 2)^2 + (-3 - 1)^2$

114. $\sqrt{(-7 - 1)^2 + (2 - 5)^2}$

115. $(-3)^2 - 4(2)(-1)$

116. $\sqrt{6^2 - 4(3)(1)}$

117. $\dfrac{5 - \sqrt{5^2 - 4(2)(1)}}{2(2)}$

118. $\dfrac{-3 + \sqrt{(-3)^2 - 4(1)(2)}}{2(1)}$

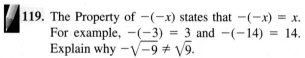
119. The Property of $-(-x)$ states that $-(-x) = x$. For example, $-(-3) = 3$ and $-(-14) = 14$. Explain why $-\sqrt{-9} \neq \sqrt{9}$.

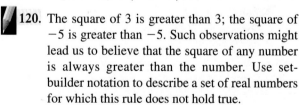
120. The square of 3 is greater than 3; the square of -5 is greater than -5. Such observations might lead us to believe that the square of any number is always greater than the number. Use set-builder notation to describe a set of real numbers for which this rule does not hold true.

121. Complete the table by entering the values of $\sqrt{x}$ and x^2 for the given values of x. Cross out any entries in which the value is not a real number.

x	-4	-1	0	1	4	9	16	25
$\sqrt{x}$								
x^2								

122. Complete the table by entering the values of $\sqrt{x}$ and $\sqrt{-x}$ for the given values of x. Cross out any entries in which the value is not a real number.

x	-4	-1	0	1	4
$\sqrt{x}$					
$\sqrt{-x}$					

Exploring with Real Data

In Exercises 123–126, refer to the following background information and data.

Archer Daniels Midland Corporation (ADM) declared a 5% stock dividend every year from 1977 through 1993. (Source: Archer Daniels Midland Corporation.) For the purposes of these exercises, assume that you bought one share of stock in 1977 and that you received your first stock dividend at the end of 1978. For any given year, the total number of shares owned is equal to the number of shares owned at the end of the previous year plus 5% of the number of shares owned at the end of the previous year.

123. Complete the middle column of the following table.

Year	Total Shares of ADM Stock	Column 3
1977	1	
1978		$(1.05)^1$
1979		$(1.05)^2$
1980		$(1.05)^3$
1981		$(1.05)^4$

124. Evaluate the exponential expressions in column 3 and compare the values with your results in the middle column.

125. Based on your observations in Exercise 124, write an exponential expression for the total number of shares owned at the end of year N, where N is the number of years since 1977.

126. Use the expression in Exercise 125 to calculate the total number of shares you would own at the end of the year 2000.

Challenge

127. For what value(s) of x is $-x^2$
   (a) positive?
   (b) 0?
   (c) negative?

128. For what value(s) of x is $(-x)^2$
   (a) positive?
   (b) 0?
   (c) negative?

129. For the following expressions, $a > 0$, $b > 0$, and $c < 0$. Determine whether the given expression represents a positive number or a negative number or either.
   (a) $a - c$ (b) $a + c$
   (c) ab (d) ac
   (e) $bc - a$ (f) a^2c
   (g) ac^2 (h) $\dfrac{a + b}{c}$

130. Insert $+$, $-$, $\cdot$, and $\div$ in the blanks so that the resulting expression has a value of 100.

$$1 \underline{\quad} 2 \underline{\quad} 3 \underline{\quad} 4 \underline{\quad} 5 \underline{\quad}$$
$$6 \underline{\quad} 7 \underline{\quad} 8 \underline{\quad} 9$$

131. One way to write 3 with three 2's is $2 \div 2 + 2$. Write 30 with three 3's.

132. Using four 4's, we can write $1 = \frac{44}{44}$ and $2 = \frac{4}{4} + \frac{4}{4}$. Write the numbers 3 through 10 with four 4's.

1.7 Properties of the Real Numbers

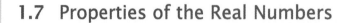

Commutative Properties ▪ *Associative Properties* ▪ *Identity Properties* ▪ *Inverse Properties* ▪ *Special Multiplication Properties* ▪ *Distributive Property*

In previous sections we informally used certain rules to perform operations with real numbers. In this section, we summarize and formally state these rules and other important properties of real numbers.

Commutative Properties

In addition, the order of the addends does not affect the sum. In multiplication, the order of the factors does not affect the product.

> ### Commutative Properties of Addition and Multiplication
> The following are true for all real numbers a and b.
>
> $a + b = b + a$ Commutative Property of Addition
>
> $ab = ba$ Commutative Property of Multiplication

Note that subtraction and division are *not* commutative operations.

$$5 - 3 \neq 3 - 5$$
$$12 \div 6 \neq 6 \div 12$$

Associative Properties

In addition, the grouping of three or more addends does not affect the sum. In multiplication, the grouping of three or more factors does not affect the product.

> ### Associative Properties of Addition and Multiplication
> The following are true for all real numbers a, b, and c.
>
> $(a + b) + c = a + (b + c)$ Associative Property of Addition
>
> $(ab)c = a(bc)$ Associative Property of Multiplication

Subtraction and division are *not* associative operations.

$$(12 - 8) - 2 = 2, \quad \text{but} \quad 12 - (8 - 2) = 6$$
$$(24 \div 4) \div 2 = 3, \quad \text{but} \quad 24 \div (4 \div 2) = 12$$

The associative properties are used to *regroup* addends and factors while the commutative properties are used to *reorder* addends and factors.

Although subtraction is neither commutative nor associative, we frequently wish to reorder and regroup an expression involving subtraction.

EXAMPLE 1   *Reordering and Regrouping a Numerical Expression*

Simplify $5 - 3 + (-7) - (-8) + 2$.

Solution

By first converting subtractions into additions, we can take advantage of the commutative and associative properties of addition to group the positive numbers and the negative numbers.

$$5 - 3 + (-7) - (-8) + 2$$

$= 5 + (-3) + (-7) + 8 + 2$	Definition of Subtraction
$= 5 + 8 + 2 + (-3) + (-7)$	Commutative Property of Addition
$= [5 + 8 + 2] + [(-3) + (-7)]$	Associative Property of Addition
$= 15 + (-10)$	
$= 5$	

Identity Properties

The sum of a number and 0 is the original number, and the product of a number and 1 is the original number. We call 0 the **additive identity,** and we call 1 the **multiplicative identity.**

Identity Properties

The following are true for any real number a.

$a + 0 = 0 + a = a$	Additive Identity Property
$a \cdot 1 = 1 \cdot a = a$	Multiplicative Identity Property

Inverse Properties

The sum of a number and its opposite is 0, which is the additive identity. The formal statement of this property usually uses **additive inverse** rather than **opposite.**

> ### Property of Additive Inverses
>
> For every real number a, there is a unique number $-a$ such that
>
> $$a + (-a) = -a + a = 0.$$

RECIPROCAL

A corresponding rule exists for multiplication. Each *nonzero* real number a has a **multiplicative inverse** or **reciprocal** represented by the symbol $\frac{1}{a}$. The product of a nonzero number and its reciprocal is 1, which is the multiplicative identity.

> ### Property of Multiplicative Inverses
>
> For every nonzero real number a, there is a unique number $\frac{1}{a}$ such that
>
> $$a \cdot \frac{1}{a} = \frac{1}{a} \cdot a = 1.$$

From the sign rules for multiplication, we can see that a number and its reciprocal always have the same sign.

For a nonzero rational number $\frac{p}{q}$, the reciprocal is $\frac{q}{p}$. For example, because $\frac{3}{4} \cdot \frac{4}{3} = 1$, the reciprocal of $\frac{3}{4}$ is $\frac{4}{3}$.

EXAMPLE 2   *Inverses of Real Numbers*

Write the additive inverse (opposite) and the multiplicative inverse (reciprocal) for each number.

(a) -12 (b) $\dfrac{9}{5}$ (c) 4.2

Solution

	Additive Inverse	*Multiplicative Inverse*
(a) -12	$-(-12)$ or 12	$\dfrac{1}{-12}$ or $-\dfrac{1}{12}$
(b) $\dfrac{9}{5}$	$-\dfrac{9}{5}$	$\dfrac{5}{9}$
(c) 4.2	-4.2	$\dfrac{1}{4.2}$ or $0.\overline{238095}$

In Section 1.5, we defined the quotient $a \div b$ $(b \neq 0)$ as the number c such that $c \cdot b = a$. Division is sometimes defined as follows.

> ### Alternate Definition of Division
>
> For real numbers a and b, $b \neq 0$,
>
> $$a \div b = a \cdot \frac{1}{b}.$$

In words, to divide a by b, multiply a by the reciprocal of b.

Special Multiplication Properties

There are two multiplication properties that involve 0 and -1.

> ### Multiplication Properties of 0 and -1
>
> The following are true for any real number a.
>
> | | |
> |---|---|
> | $a \cdot 0 = 0 \cdot a = 0$ | Multiplication Property of 0 |
> | $a \cdot (-1) = -1 \cdot a = -a$ | Multiplication Property of -1 |

Distributive Property

All the properties summarized so far have involved either addition or multiplication. The Distributive Property involves both operations.

> ### Distributive Property
>
> For all real numbers a, b, and c,
>
> $$a(b + c) = ab + ac.$$

More precisely, the foregoing property is the Distributive Property of Multiplication Over Addition. There is also a Distributive Property of Multiplication Over Subtraction.

$$a(b - c) = ab - ac$$

The proof is left as an exercise. We will refer to both versions simply as the Distributive Property.

The Distributive Property can be read in either direction.

1. $a(b + c) = ab + ac$

2. $ab + ac = a(b + c)$

In (1), we change an indicated product into a sum. In this case, we are using the Distributive Property to *multiply*. In (2) we change an indicated sum into a product. In this case, we are using the Distributive Property to *factor*.

EXAMPLE 3   *Using the Distributive Property to Multiply and Factor*

Use the Distributive Property to rewrite each indicated product as a sum and each indicated sum as a product.

(a) $-7(x - 3)$

(b) $3x + 3y$

(c) $a(y - z)$

(d) $-4a - 4b$

Solution

(a) $-7(x - 3) = -7 \cdot x - (-7) \cdot 3 = -7x + 21$

(b) $3x + 3y = 3 \cdot x + 3 \cdot y = 3(x + y)$

(c) $a(y - z) = ay - az$

(d) $-4a - 4b = -4a + (-4)b = -4(a + b)$

Some properties of real numbers can be deduced from previously stated properties. The following is an example.

$$-(a - b) = (-1)(a - b)$$ Multiplication Property of -1

$$= (-1)a - (-1)b$$ Distributive Property

$$= -a - (-b)$$ Multiplication Property of -1

$$= -a + b$$ Definition of Subtraction

$$= b + (-a)$$ Commutative Property of Addition

$$= b - a$$ Definition of Subtraction

We have proven the following important and useful property.

Property of the Opposite of a Difference

For any real numbers a and b, $-(a - b) = b - a$.

EXAMPLE 4   *Identifying the Properties of Real Numbers*

Identify the property that justifies the given statement.

(a) $x + (2 + x) = x + (x + 2)$ Commutative Property of Addition

(b) $1 \cdot (y - 2) = y - 2$ Multiplicative Identity Property

(c) $(x + 5) - 1 = x + (5 - 1)$ Associative Property of Addition

(d) $-3(2 - y) = -6 + 3y$ Distributive Property

(e) $(0.2)5 = 1$ Property of Multiplicative Inverses

(f) $0 \cdot (5x) = 0$ Multiplication Property of 0

(g) $-(3b) + (3b) = 0$ Property of Additive Inverses

(h) $-(3 - w) = w - 3$ Property of the Opposite of a Difference

EXAMPLE 5 *Using the Properties to Rewrite Expressions*

Use the given property to rewrite the expression.

(a) Commutative Property of Multiplication: $x \cdot 3 = $ _____.

(b) Distributive Property: $2a + 3a = $ _____.

(c) Associative Property of Multiplication: $\frac{1}{2}(6x) = $ _____.

(d) Additive Identity Property: $0 + y = $ _____.

Solution

(a) $x \cdot 3 = 3x$

(b) $2a + 3a = (2 + 3)a$

(c) $\frac{1}{2}(6x) = \left(\frac{1}{2} \cdot 6 \right) \cdot x$

(d) $0 + y = y$

1.7 Quick Reference

The following is a summary of the properties of real numbers, where a, b, and c represent any real number.

Commutative Properties	▪ $a + b = b + a$	Commutative Property of Addition
	▪ $ab = ba$	Commutative Property of Multiplication
	▪ Subtraction and division are not commutative.	
	▪ The commutative properties are used to reorder addends and factors.	

Associative Properties	▪ $(a + b) + c = a + (b + c)$	Associative Property of Addition
	▪ $(ab)c = a(bc)$	Associative Property of Multiplication
	▪ Subtraction and division are not associative.	
	▪ The associative properties are used to regroup addends and factors.	

Identity Properties	▪ $a + 0 = 0 + a = a$	Additive Identity Property
	▪ $a \cdot 1 = 1 \cdot a = a$	Multiplicative Identity Property

Inverse Properties	▪ $a + (-a) = -a + a = 0$	Property of Additive Inverses
	▪ $a \cdot \dfrac{1}{a} = \dfrac{1}{a} \cdot a = 1 \ (a \neq 0)$	Property of Multiplicative Inverses

- For any real number a, $-a$ is unique; for any nonzero real number a, $1/a$ is unique.

- An alternate definition of division is $a \div b = a \cdot (1/b)$, where $b \neq 0$.

Special Multiplication Properties	- $a \cdot 0 = 0 \cdot a = 0$	Multiplication Property of 0
	- $a \cdot (-1) = -1 \cdot a = -a$	Multiplication Property of -1

Distributive Property	- $a(b + c) = ab + ac$ and $a(b - c) = ab - ac$
	- We use $a(b + c) = ab + ac$ to *multiply;* we use $ab + ac = a(b + c)$ to *factor.*
	- The Distributive Property can be used to prove the Property of the Opposite of a Difference:

$$-(a - b) = b - a.$$

1.7   Exercises

1. Although subtraction is not commutative, it is correct to rewrite $-9 - 4x + x^2$ as $x^2 - 4x - 9$. Why?

2. (a) Name a number that is its own opposite. Show how your answer satisfies the Property of Additive Inverses.

   (b) Name two numbers that are their own reciprocals. Show how your answers satisfy the Property of Multiplicative Inverses.

In Exercises 3–22, name the property that is illustrated.

3. $3 + (2 + b) = (3 + 2) + b$

4. $-x + 2 = 2 - x$

5. $4x + 0 = 4x$

6. $-6x + 6x = 0$

7. $4(x - 5) = 4x - 20$

8. $(x^2 + 1) \cdot \dfrac{1}{x^2 + 1} = 1$

9. $-\dfrac{1}{5} \cdot (-5) = 1$

10. $\dfrac{2}{3}x - \dfrac{2}{3} = \dfrac{2}{3}(x - 1)$

11. $10 \cdot \left(\dfrac{1}{2}x\right) = \left(10 \cdot \dfrac{1}{2}\right)x$

12. $5(x - 7) = 5x - 5 \cdot 7$

13. $5 \cdot 1 = 5$

14. $a \cdot 9 = 9a$

15. $(2x - 7) \cdot 0 = 0$

16. $-1 \cdot (2y) = -2y$

17. $-3 - x = -1 \cdot (3 + x)$

18. $5x - 3x = (5 - 3)x$

19. $6 \cdot \left[\dfrac{1}{2}(x + 1)\right] = \left(6 \cdot \dfrac{1}{2}\right)(x + 1)$

20. $(y + 5)(y - 3) = (y - 3)(y + 5)$

21. $(a + 2) \cdot (3 + b) = (3 + b) \cdot (a + 2)$

22. $(x - 5) \cdot (5 + x) = (x - 5) \cdot (x + 5)$

In Exercises 23–32, match the expression in column A with the corresponding expression in column B.

Column A	*Column B*
23. $x \cdot (-7)$	(a) $b - a$
24. $-a + b$	(b) $(y + 2) + 9$
25. $y + (2 + 9)$	(c) $2x - y$
26. $3(x + 7)$	(d) 1
27. $1(2x - y)$	(e) 0
28. $-(c - 2)$	(f) $(2 \cdot 3)x$
29. $2(3x)$	(g) $3x + 21$
30. $0 \cdot (4 + z)$	(h) $-1(c - 2)$
31. $2x \cdot \dfrac{1}{2x}$ $(x \neq 0)$	(i) $-7x$
32. $-(a - b)$	

In Exercises 33–46, use the given property to write the expression in an alternate form.

33. Associative Property of Addition:
$(x + 4) + 3 = $ _____

34. Commutative Property of Addition:
$3x + 8 = $ _____

35. Commutative Property of Multiplication:
$(x + 6) \cdot 3 = $ _____

36. Associative Property of Multiplication:
$6(2x) = $ _____

37. Commutative Property of Addition:
$8 - x = $ _____

38. Commutative Property of Addition:
$4(x + 7) = $ _____

39. Additive Identity Property: $(x + 0) + 6 = $ _____

40. Property of Additive Inverses:
$-(y + 3) + (y + 3) = $ _____

41. Property of Multiplicative Inverses:
$\left(3 \cdot \dfrac{1}{3}\right)x = $ _____

42. Multiplicative Identity Property:
$1(n - 4) = $ _____

43. Distributive Property: $12z + 15 = $ _____

44. Distributive Property: $-3[a + (-6)] = $ _____

45. Multiplication Property of 0: $(x + y) \cdot 0 = $ _____

46. Multiplication Property of -1:
$-1(x - 4) = $ _____

47. Explain why 0 does not have a reciprocal.

48. Use the definition of reciprocal to show why the reciprocal of p/q is q/p, where $p \neq 0$, $q \neq 0$.

In Exercises 49–56, fill in the blank and state the property that justifies your answer.

49. $5 \cdot $ _____ $= 1$

50. $-7 + $ _____ $= 0$

51. $-\dfrac{2}{3} \cdot $ _____ $= -\dfrac{2}{3}$

52. $8 + (-x + x) = 8 + $ _____

53. _____ $(2x - 5) = -2x + 5$

54. $-x + 5 = -1 \cdot $ _____

55. $7 \cdot \left[\dfrac{1}{7}(x - 5)\right] = $ _____ $(x - 5)$

56. $-\dfrac{2}{3}\left[\dfrac{3}{2}(4 + x)\right] = $ _____ $(4 + x)$

In Exercises 57–64, name the property or definition that justifies each step.

57. $(y + 0) + (-y)$
$= (0 + y) + (-y)$ (a) _____
$= 0 + [y + (-y)]$ (b) _____
$= 0 + 0$ (c) _____
$= 0$ (d) _____

58. $(6 \cdot 1) \cdot \dfrac{1}{6}$

$= (1 \cdot 6) \cdot \dfrac{1}{6}$ (a) _____

$= 1 \cdot \left(6 \cdot \dfrac{1}{6}\right)$ (b) _____

$= 1 \cdot 1$ (c) _____

$= 1$ (d) _____

59. $-(2x - 3)$ (a) _____

    $= -1(2x - 3)$ (b) _____

    $= -2x + 3$ (c) _____

    $= 3 - 2x$ (d) _____

60. $(a - b) + b$

    $= [a + (-b)] + b$ (a) _____

    $= a + [(-b) + b]$ (b) _____

    $= a + 0$ (c) _____

    $= a$ (d) _____

61. $(x + 5) - 1$

    $= x + (5 - 1)$ (a) _____

    $= x + 4$

62. $5x - 7x$

    $= (5 - 7)x$ (a) _____

    $= -2x$

63. $5\left(\dfrac{1}{5}x\right)$

    $= \left(5 \cdot \dfrac{1}{5}\right)x$ (a) _____

    $= 1x$ (b) _____

    $= x$ (c) _____

64. $-3x \cdot 2x$

    $= -3 \cdot 2 \cdot x \cdot x$ (a) _____

    $= -6x^2$

In Exercises 65–76, use the associative properties to rewrite the given expression in simpler or exponential form.

65. $(x + 3) + 2$ **66.** $-4 + (8 + y)$

67. $(y + 1) - 6$ **68.** $(2z + 7) - 2$

69. $(3x - 5) + 7$ **70.** $(y - 3) + 1$

71. $-5(3x)$ **72.** $(2y)y$

73. $-\dfrac{3}{5}\left(-\dfrac{5}{3}a\right)$ **74.** $-6\left(\dfrac{5}{6}z\right)$

75. $-\dfrac{5}{6} \cdot \left(-\dfrac{9}{10}x\right)$ **76.** $\dfrac{3}{4}(2xy)$

In Exercises 77–98, use the Distributive Property to rewrite each expression.

77. $5(4 + 3b)$ **78.** $-2(3a - 7)$

79. $-\dfrac{3}{5}(5x + 20)$ **80.** $\dfrac{1}{3}(12x - 3)$

81. $-3(5 - 2x)$ **82.** $-4(-3x + 5)$

83. $x(y + z)$ **84.** $s(t + 3)$

85. $\dfrac{3}{4}\left(y + \dfrac{4}{3}\right)$ **86.** $\dfrac{5}{8}(x + 8)$

87. $5x + 5y$ **88.** $11a - 11b$

89. $7(3x) - 7(2y)$ **90.** $4(5a) + 4(2b)$

91. $\dfrac{7}{4}x + \dfrac{7}{4}(3)$ **92.** $-\dfrac{1}{6}y + \left(-\dfrac{1}{6}\right) \cdot 5$

93. $3x + 12$ **94.** $7y - 14$

95. $5y + 5$ **96.** $2 + 2b$

97. $6 - 3x$ **98.** $2x - 2$

In Exercises 99–102, write the given expression without grouping symbols.

99. $-(x - 4)$ **100.** $-(3 + 5x)$

101. $-(2x + 5y - 3)$ **102.** $-(-4x - y + 7)$

In Exercises 103–114, use the commutative and associative properties to rewrite the given expression in simpler or exponential form.

103. $(3x) \cdot 4$ **104.** $y(3y)$

105. $(x \cdot x) \cdot 5x$ **106.** $(-6x)(2x)$

107. $3x \cdot 4y$ **108.** $\dfrac{1}{3}x \cdot 3y$

109. $(6y)\left(\dfrac{4}{9}\right)$ **110.** $\left(\dfrac{3}{4}x\right) \cdot \left(\dfrac{2}{3}y\right)$

111. $(5 + x) + 3$ **112.** $x + [7 + (-x)]$

113. $-5x + (3 + 5x)$ **114.** $-7 + (2x - 3)$

In Exercises 115–118, use the commutative and associative properties to write the given numerical expression so that the operations can be easily performed. Then, evaluate the expression.

115. $\dfrac{2}{9} + \dfrac{3}{7} + \dfrac{5}{9} + \dfrac{4}{7} + \dfrac{2}{9}$

116. $77 + 40 + 23 + 30 + 70 + 60$

117. $2 \cdot (-87) \cdot 5 \cdot (-1)$

118. $5 \cdot \dfrac{1}{3} \cdot \dfrac{1}{5} \cdot 3$

119. The following is a proof of the Distributive Property of Multiplication Over Subtraction. Fill in the blanks with the property or definition that justifies each step.

For real numbers a, b, and c,

$a(b - c)$

$= a[b + (-c)]$ (a) ____

$= a \cdot b + a \cdot (-c)$ (b) ____

$= ab + (-c)a$ (c) ____

$= ab + (-1c)a$ (d) ____

$= ab + (-1)(ca)$ (e) ____

$= ab + (-1)(ac)$ (f) ____

$= ab + (-ac)$ (g) ____

$= ab - ac$ (h) ____

120. Show how the Distributive Property can be used to prove that

$$\dfrac{a + b}{3} = \dfrac{a}{3} + \dfrac{b}{3}$$

The Distributive Property can be useful in performing multiplication mentally. The following are examples.

$25 \cdot 103 = 25 \cdot (100 + 3)$

$= 25 \cdot 100 + 25 \cdot 3$

$= 2500 + 75$

$= 2575$

$30 \cdot 19 = 30(20 - 1)$

$= 30 \cdot 20 - 30 \cdot 1$

$= 600 - 30$

$= 570$

In Exercises 121–126, show how you would use the Distributive Property to determine the given product mentally.

121. $7 \cdot 108$ **122.** $30 \cdot 109$

123. $15 \cdot 98$ **124.** $4 \cdot 49$

125. $18(1.5)$ **126.** $24(1.75)$

127. As a baseball statistician, you find that you must frequently multiply a number by 9. Use the Distributive Property to explain why multiplying the number by 10 and then subtracting the number produces the desired result.

128. Explain why the opposite of a nonzero number can never be the same as the reciprocal of the number.

In Exercises 129–132, give the additive inverse (opposite) and the multiplicative inverse (reciprocal), if any, of the given number. Round decimal answers to the nearest hundredth.

129. $\dfrac{5}{7}$ **130.** 1.57

131. 0 **132.** -8

In Exercises 133–136, write the multiplicative inverse. Assume all variables represent positive numbers.

133. $3x$ **134.** $x + 6$

135. $\dfrac{2}{3}x$ **136.** $\dfrac{2}{y}$

In Exercises 137–140, write the additive inverse of the given expression.

137. $x + 3$ **138.** $y - 5$

139. $4 - 3x$ **140.** $2y$

In Exercises 141–148, for nonzero numbers a and b, indicate whether the given statement is always true, always false, or sometimes true. If sometimes, give examples of a and b for which the statement is true and for which the statement is false.

141. $a + b = b + a$

142. $a - b = b - a$

143. $\dfrac{a}{b} = \dfrac{b}{a}$

144. $ab = ba$

145. $|a + b| = |a| + |b|$

146. $|a - b| = |a| - |b|$

147. $(a + b)^2 = a^2 + b^2$

148. $\sqrt{a + b} = \sqrt{a} + \sqrt{b}$

Challenge

149. Is addition distributive with respect to multiplication? Show whether the following is true.

$$a + (b \cdot c) = (a + b)(a + c)$$

150. Suppose $a - b < 0$. Show that $|a - b| = b - a$.

151. Show whether the following is true:

$$\sqrt{a + b} + c = a + \sqrt{b + c}.$$

152. Suppose $*$ is an operation such that $n * a = a * n = n$. What name would you give the number a for this operation?

---

1 Chapter Review Exercises

Section 1.1

In Exercises 1–7, determine whether the statement is true or false. Use your calculator to determine decimal forms if necessary.

1. The number -8 is a natural number.

2. The number $-\pi$ is a rational number.

3. Every whole number is also an integer.

4. Because the decimal form of $\sqrt{11}$ is nonterminating and nonrepeating, $\sqrt{11}$ is an irrational number.

5. The value of the expression $\sqrt{4.8} / \sqrt{2.7}$ is an irrational number.

6. The statement $n > 0$ means that n represents a positive number.

7. For any real number n, $-n$ represents a negative real number.

8. Let $A = \left\{ -2, \frac{5}{9}, \pi/2, -\sqrt{10}, -\sqrt{9}, 2.34, 5 \right\}$. List all the members of set A that are members of the following sets.

(a) J (Integers)

(b) Q (Rational Numbers)

(c) I (Irrational Numbers)

 9. Explain the difference between the sets of numbers represented by $(-3, 5]$ and $[-3, 5)$.

 10. The absolute value of a number is nonnegative, but there are numbers n such that $|n| = -n$. Why?

11. Graph the following sets on the real number line.

(a) $\{x \mid -2 < x \le 7\}$

(b) $\{x \mid x > -7\}$

(c) $\{x \mid x \le 3\}$

12. Insert $<$ or $>$ to make the statement true.

(a) $-\dfrac{28}{29}$ _____ $-\dfrac{27}{28}$

(b) 3.1416 _____ π

(c) $|-3|$ _____ -3

(d) $-\sqrt{3}$ _____ $-\dfrac{\pi}{2}$

Section 1.2

 13. Explain the meaning of the word *addends*. Give an example.

 14. Under what circumstances is subtraction used to determine a sum?

In Exercises 15–20, determine the sum. Use your calculator only to verify the results.

15. $8 + (-11)$

16. $(-12) + (-8)$

17. $-\dfrac{2}{3} + \dfrac{3}{4}$

18. $-9 + \left(-\dfrac{4}{5}\right)$

19. $-|-2| + |-5|$

20. $-10 + (-11) + 5 + (-7) + 9$

21. The sum of what number and -5 is 6?

22. What is 3 increased by -5?

23. On a certain day, the high temperature was 26°F and the low was -4°F. What was the average temperature for the day?

24. After writing a check for $325.50, a person discovered that his bank account was overdrawn by $6.38. What was the checking account balance before the check was written?

Section 1.3

25. What are the meanings of the words *minuend* and *subtrahend*? Give an example.

26. Compare $a - b$ with $b - a$ for any two real numbers a and b.

In Exercises 27–32, perform the indicated operations. Use your calculator only to verify the results.

27. $7 - (-13)$

28. $(-15) - (-6)$

29. $-\dfrac{2}{5} - \dfrac{3}{4}$

30. $-7 - \left(-\dfrac{2}{3}\right)$

31. $-|-3| - |-7|$

32. $-10 - (-11) - 5 - (-7) - 9$

33. What number must be subtracted from -5 to obtain -9?

34. On the number line, what is the distance from point A at -5 to point B at 8?

35. The mean (average) elevation of the continental United States is 2500 feet. If Montana's highest point is Granite Peak at 12,850 feet, what is the difference between the highest point in Montana and the mean elevation?

36. The temperature range for any day is the difference between the high and low temperatures. What was the temperature range for a day when the high was 19°F and the low was -8°F?

Section 1.4

37. What do we mean by the *factors* of a product?

38. Suppose a product consists of 11 nonzero factors. Describe a quick way to determine the sign of the product.

In Exercises 39–46, multiply. Use your calculator to verify your results.

39. $5 \cdot (-4)$

40. $-7 \cdot (-5)$

41. $-6 \cdot \left(-\dfrac{5}{12}\right)$

42. $\dfrac{7}{8} \cdot \left(-\dfrac{4}{3}\right)$

43. $-\dfrac{20}{33} \cdot \left(-\dfrac{22}{15}\right) \cdot \left(-\dfrac{3}{16}\right)$

44. $(-1)(-2)(-1)(-3)(-3)$

45. $(-1)(-3)(0)(4)$

46. $-|-4| \cdot \left(-|-5|\right)$

47. What number must be multiplied by -2 to obtain 9?

48. The product of three factors is 0. If two of the factors are 5 and -5, what is the third factor?

49. A water tank is being drained at a rate of 100 gallons per minute. At the same time, the tank is being replenished at a rate of 60 gallons per minute. At 1:15 P.M., how much water is in the tank compared to the amount that was in the tank at noon?

50. Each of 30 students is given either a $+1$ or a -1. If the sum of their numbers is $+2$, what is the product of their numbers?

Section 1.5

51. What are the meanings of the words *divisor* and *dividend*?

52. Using the definition of division, explain why division by 0 is undefined.

In Exercises 53–58, divide, if possible. Use your calculator only to verify your results.

53. $\dfrac{10}{-2}$

54. $\dfrac{-45}{-9}$

55. $0 \div (-9)$

56. $-\dfrac{3}{5} \div \dfrac{5}{6}$

57. $-\dfrac{4}{9} \div (-9)$

58. $\dfrac{3}{16} \div \left(-\dfrac{6}{7}\right)$

59. If the quotient is -12 and the dividend is 6, what is the divisor?

60. If 9.6 inches of snow fell during the previous night, and if it then melts at an average rate of 0.4 inches per hour, how long will it take for the snow to melt completely?

Section 1.6

 61. Explain the order of operations used in evaluating -5^2 and $(-5)^2$. Are the values the same?

 62. Is $\sqrt{-25} = -\sqrt{25}$? Explain your answer.

63. Write $(-3)(-3)(-3)(-3)(-3)$ in exponential form. Then, use your calculator to evaluate the exponential expression.

64. Translate the following verbal description into a numerical expression. Then, evaluate the expression.

Multiply the sum of -7 and the square of 3 by -5. Then add -21 to the product.

In Exercises 65–72, perform the indicated operations. Use your calculator only to verify the results.

65. 2^4

66. 4^2

67. 0^{33}

68. $(-1)^{33}$

69. $(-1)^{36}$

70. $\left(-\dfrac{2}{3}\right)^3$

71. $\sqrt{49}$

72. $-\sqrt{49}$

In Exercises 73–78, perform the indicated operations. Use your calculator only to verify the results.

73. $5 + 7 \cdot 2 - 6 \div 3 + 8$

74. $8 + 2[8 + 2(8 + 2)]$

75. $-6^2 - 4(5 - 2^3)$

76. $18 - 7 + 2 - 8 \div 2$

77. $18 - (7 + 2) - 8 \div 2$

78. $\dfrac{2 \cdot (-7) - 4}{9 + 6 \cdot (-3)}$

Section 1.7

 79. Explain why we can write $-x$ as $-1x$.

 80. Consider the task of putting on your shoes and socks. Is this task commutative? Explain your answer.

In Exercises 81–86, name the property that is illustrated.

81. $3 + (5 + c) = 3 + (c + 5)$

82. $1 \cdot 6 = 6$

83. $4 \cdot (3 \cdot x) = (4 \cdot 3) \cdot x$

84. $7(c + d) = 7c + 7d$

85. $6 + 0 = 6$

86. $8 \cdot (-1) = -8$

In Exercises 87–92, use the stated property to write the expression in an alternate form.

87. Commutative Property of Multiplication:

$(c + 5) \cdot 7 = $ _____

88. Associative Property of Addition:

$x + (-2 + 2) = $ _____

89. Distributive Property:

$5 \cdot x + 5 \cdot 8 = $ _____

90. Property of Additive Inverses:

$x + (-2 + 2) = $ _____

91. Multiplicative Identity Property:

$-5\left(-\tfrac{1}{5}\right) \cdot 1 = $ _____

92. Multiplication Property of -1:

$-1 \cdot 7 = $ _____

Chapter Test

In Questions 1–3, state which elements of $\left\{-3.7,\ -2,\ -0.\overline{14},\ 0,\ \frac{2}{3},\ \pi,\ \sqrt{7},\ 6,\ \sqrt{9}\right\}$ also belong to the given sets.

1. Integers

2. Irrational numbers

3. Rational numbers

4. Use the roster method to write the set $\{n \mid n \text{ is an integer and } -3 < n \leq 5\}$.

In Questions 5 and 6, graph the given set on a number line.

5. $\{x \mid x \leq -3\}$

6. $\left\{-5,\ -\sqrt{2},\ \dfrac{1}{2},\ 2.7\right\}$

In Questions 7–16, evaluate the given expression.

7. $-7 - (-3)$

8. $\left(-\dfrac{3}{10}\right)\left(\dfrac{2}{9}\right)$

9. $\dfrac{2 - (-3)}{3(7) - 2(-8)}$

10. $3 - 2[4 - (2 - 3)]$

11. $\dfrac{-3 + \sqrt{3^2 - 4(2)}}{2}$

12. $|6 - 2(5)|$

13. $-\dfrac{3}{4} + \left(-\dfrac{1}{2}\right)$

14. $-2.5 - 6.7$

15. $(-2)(5)(-4)(0)(6)$

16. $-32 \div (-4)$

17. If possible, perform the indicated operation. If the operation is not defined, explain why.

   (a) $\dfrac{0}{-3}$ (b) $\dfrac{-3}{0}$

In Questions 18–20, use the indicated property to complete the statement.

18. Distributive Property: $3(x - 4) = $ _____

19. Commutative Property of Addition: $4 - x = $ _____

20. Associative Property of Addition: $(3 + t) - 3r = $ _____

21. What is the opposite of $-|-4|$?

22. What is the reciprocal of -2?

23. Use your calculator to evaluate the following expression.

$$\frac{-3.7 + (-4.13)}{4 - (3.1)^2}$$

24. The sum of the numbers 4 and -9 is multiplied by 2. The result is squared and then divided by 3 less than -17. Write the numerical expression and evaluate it.

25. A certain pharmaceutical is manufactured under very controlled temperature conditions. If the temperature in the lab is $-2.45°C$ and is decreased by $1.04°$, what is the new temperature?

2 Algebra Basics

As more and more states enacted helmet laws for bicycle riders, the sales of helmets increased dramatically. The bar graph shows the sales (in millions) of bicycle helmets for selected years in the period 1985–1993.

We can write an **algebraic expression** to describe the trend in helmet sales during this time. Using this expression, we can, for example, estimate the helmet sales in 1991 and predict sales for 1996. (To learn more about how this model expression can be used, see Exercises 99–102 at the end of Section 2.1.)

Chapter 2 begins with a discussion of algebraic expressions and how to evaluate and simplify them. Then, we turn to the topic of exponents and their properties, and we define and use the zero exponent and negative exponents. Finally, we consider the representation of numbers in scientific notation.

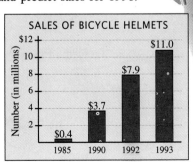

SALES OF BICYCLE HELMETS

Number (in millions)

Year	1985	1990	1992	1993
Sales	$0.4	$3.7	$7.9	$11.0

(Source: Bell Sports, U.S. Bicycle Federation.)

2.1 Algebraic Expressions and Formulas

Evaluating Algebraic Expressions   ▪   *Equations and Formulas*

Evaluating Algebraic Expressions

In the previous chapter, we described a **numerical expression** as any combination of numbers and operations. We also described a **variable** as a symbol (usually a letter) used to represent a number.

We can use variables to represent numbers in expressions. An **algebraic expression** is any combination of numbers, variables, grouping symbols, and operation symbols. The following are examples of algebraic expressions.

$$\frac{a^2 + 4}{6} \qquad 5(x + 6y) - 4 \qquad r - \sqrt{2t}$$

When we replace the variables in an algebraic expression with specific values, the expression becomes a numerical expression. Performing all indicated operations to determine the value of the expression is called **evaluating the expression.**

Example 1 illustrates several methods for using a calculator to evaluate expressions.

EXAMPLE 1 *Evaluating Algebraic Expressions*

Evaluate each expression.

(a) $5x + 2(1 - x)$, $x = -3$ (b) $-x^2 + 3x - 4$, $x = -2$

(c) $3x - |y - 2x|$, $x = 3$, $y = 1$ (d) $\dfrac{a + 2b}{2a^2b}$, $a = -1$, $b = 3$

Solution

(a) A calculator method for evaluating $5x + 2(1 - x)$ for $x = -3$ is to replace x with -3 and then calculate the resulting *numerical* expression:

$5x + 2(1 - x) = 5(-3) + 2[1 - (-3)]$.

See Fig. 2.1.

Figure 2.1

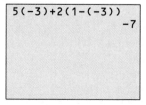

Figure 2.2

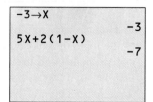

An alternate method begins with storing -3 in the variable x. Then, we enter $5x + 2(1 - x)$ and evaluate the *algebraic* expression. (See Fig. 2.2.) These

EVALUATE

two calculator methods work well if the expression is to be evaluated for just one value of the variable. If an expression is to be evaluated for multiple values of the variable, we are better served by another technique, which we will introduce later.

(b) $-x^2 + 3x - 4 = -(-2)^2 + 3(-2) - 4 = -4 - 6 - 4 = -14$

(c) $3x - |y - 2x| = 3(3) - |1 - 2(3)| = 9 - |-5| = 4$

ALPHA

If you evaluate the *algebraic* expression with your calculator, you will need to store values for both x and y.

(d) $\dfrac{a + 2b}{2a^2b} = \dfrac{-1 + 2 \cdot 3}{2(-1)^2 \cdot 3} = \dfrac{5}{6}$

FRACTION

Calculator evaluation will give a decimal result, but you can request the fraction form on some calculators. ■

Equations and Formulas

An **equation** is a statement that two expressions have the same value. If both expressions are numerical, then the truth of the equation can be determined simply by evaluating the expressions.

If one or both sides of an equation are algebraic expressions, then the equation may be true or false, depending on the replacement for the variable. A value of the variable that makes the equation *true* is called a **solution** of the equation.

EXAMPLE 2   *Testing Solutions of an Equation*

TEST

For each equation, determine whether the given number is a solution.

(a) $5 - x^2 = 3x - (2x + 1)$; 2   (b) $\sqrt{4 - x} - \sqrt{x + 6} = 2$; 3

Solution

(a) $5 - x^2 = 3x - (2x + 1)$

    $5 - 2^2 = 3 \cdot 2 - (2 \cdot 2 + 1)$ Replace x with 2.

        $1 = 1$ True

The last equation is true, so 2 is a solution.

(b) $\sqrt{4 - x} - \sqrt{x + 6} = 2$

    $\sqrt{4 - 3} - \sqrt{3 + 6} = 2$ Replace x with 3.

        $1 - 3 = 2$

          $-2 = 2$ False

The last equation is false, so 3 is not a solution. ■

A **formula** is an equation that uses an expression to represent some specific quantity. For example, if the base of a triangle is b and the height is h, then the formula $A = \frac{1}{2}bh$ tells us how to determine the area A of the triangle.

An important formula is contained in the **Pythagorean Theorem.**

> ### The Pythagorean Theorem
>
>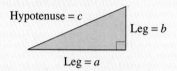
>
> A triangle is a right triangle if and only if the sum of the squares of the **legs** is equal to the square of the **hypotenuse.** Symbolically, $a^2 + b^2 = c^2$.

We can use the formula in the Pythagorean Theorem to determine whether a triangle is a right triangle.

EXAMPLE 3   *Determining Whether a Triangle Is a Right Triangle*

The three numbers given in each part are the lengths of the sides of a triangle. Determine if the triangle is a right triangle.

(a) 2, 4, 5   (b) 5, 12, 13

Solution

(a) We must determine whether the Pythagorean Theorem formula is true for the given lengths. The longest side is 5, so $c = 5$. It does not matter how we assign the values for a and b.

$$a^2 + b^2 = c^2$$
$$2^2 + 4^2 = 5^2$$
$$4 + 16 = 25$$
$$20 = 25 \qquad \text{False}$$

The triangle is not a right triangle.

(b) The longest side is 13, so $c = 13$.

$$a^2 + b^2 = c^2$$
$$5^2 + 12^2 = 13^2$$
$$25 + 144 = 169$$
$$169 = 169 \qquad \text{True}$$

The triangle is a right triangle.

2.1   Quick Reference

Evaluating Algebraic Expressions

- An **algebraic expression** is any combination of numbers, variables, grouping symbols, and operation symbols.

- To **evaluate** an algebraic expression, we replace each variable with a specific value and then perform all indicated operations.

Equations and Formulas

- An **equation** is a statement that two expressions have the same value.

- If an equation contains a variable, a value of the variable that makes the equation true is called a **solution.**

- A **formula** is an equation that uses an expression to represent some specific quantity.

- In a right triangle, the sides forming the right angle are called **legs;** the side opposite the right angle is called the **hypotenuse.**

- For a right triangle with legs a and b and hypotenuse c, the **Pythagorean Theorem** states that $a^2 + b^2 = c^2$.

- If the lengths of the three sides of a triangle satisfy the formula $a^2 + b^2 = c^2$, where c is the length of the longest side, then the triangle is a right triangle.

2.1 Exercises

 1. Explain why the expression $\dfrac{x + 1}{x - 2}$ cannot be evaluated for $x = 2$.

 2. The expression $\dfrac{x - 7}{7 - x}$ has a value of -1 for any replacement of x except 7. Why?

In Exercises 3–18, evaluate each expression for the given value of the variable. Use your calculator only to verify your results.

3. $5 - x$ for $x = -3$

4. $4 - 5x$ for $x = -2$

5. $-4x$ for $x = -1$

6. $3x - 2$ for $x = 4$

7. $5 - 3(x + 4)$ for $x = 2$

8. $x - 2(x - 1)$ for $x = -3$

9. $2x^2 - 4x - 9$ for $x = 3$

10. $x^2 - 5x$ for $x = -2$

11. $5 - |t + 3|$ for $t = -5$

12. $|3 - t| - 4$ for $t = 7$

13. $\dfrac{x + 3}{5 - x}$ for $x = 5$

14. $\dfrac{6}{x} + \dfrac{7}{x - 2}$ for $x = 2$

15. $\dfrac{2x - 3}{3 - 2x}$ for $x = 1$

16. $\dfrac{x^2 - 3}{x - 4}$ for $x = -1$

17. $\sqrt{11 - 2x}$ for $x = 1$

18. $\sqrt{7x + 4}$ for $x = 3$

In Exercises 19–30, evaluate each expression for the given values of the variables. Use your calculator only to verify your results.

19. $2x + 4y$ for $x = 3$ and $y = -2$

20. $x - y$ for $x = 4$ and $y = -3$

21. $2x - (x - y)$ for $x = 0$ and $y = -5$

22. $5 + x(x + y)$ for $x = -4$ and $y = 3$

23. $|2s - t|$ for $s = -4$ and $t = 2$

24. $5t - 2|t - s|$ for $s = 5$ and $t = 2$

25. $\dfrac{2x + y}{2x - y}$ for $x = 3$ and $y = 6$

26. $\dfrac{xy}{xy - 12}$ for $x = 4$ and $y = 3$

27. $\dfrac{2x + 3y}{x^2 + y^2}$ for $x = -6$ and $y = 4$

28. $\dfrac{xy^2}{(x + y)^2}$ for $x = 0$ and $y = -5$

29. $\dfrac{(x - 4)^2}{16} + \dfrac{(y + 2)^2}{25}$ for $x = 2$ and $y = 3$

30. $\dfrac{4y^2}{9} + \dfrac{x^2}{4}$ for $x = -2$ and $y = 6$

In Exercises 31–34, evaluate $\dfrac{-b + \sqrt{b^2 - 4ac}}{2a}$ for the given values.

31. $a = 2$, $b = -3$, and $c = -2$

32. $a = 3$, $b = -5$, and $c = 2$

33. $a = 9$, $b = 12$, and $c = 4$

34. $a = 2$, $b = 3$, and $c = -14$

In Exercises 35–38, evaluate $\dfrac{y_2 - y_1}{x_2 - x_1}$ for the given values.

35. $x_1 = -2$, $x_2 = -4$, $y_1 = 3$, $y_2 = -1$

36. $x_1 = -12$, $x_2 = 4$, $y_1 = 5$, $y_2 = 5$

37. $x_1 = 0$, $x_2 = 5$, $y_1 = 2$, $y_2 = 0$

38. $x_1 = -3$, $x_2 = -3$, $y_1 = 4$, $y_2 = -2$

In Exercises 39–48, use your calculator to evaluate each expression for the given value(s) of the variable(s). Round decimal answers to the nearest hundredth.

39. $3x - 2$ for $x = -2.47$

40. $t^2 - 3t$ for $t = -4.56$

41. $-y + 3z$ for $y = 2.1$, $z = -3.7$

42. $m^2 - 2n^2$ for $m = -2.14$ and $n = 7.12$

43. $\sqrt{13y} + \sqrt{15x}$ for $x = 8$ and $y = 5$

44. $\sqrt{5a} - |-25b|$ for $a = 2$ and $b = 10$

45. $\sqrt{x^2 - 3x + 2}$ for $x = 2$

46. $\sqrt{x^2 - 3x}$ for $x = 2$

47. $|3 - 5x| - |4 - 7y|$ for $x = 4$ and $y = 2$

48. $\sqrt{1 - x}$ for $x = -5$

 49. Suppose that you store 5 in t and then you evaluate $\sqrt{4 - t}$. Explain the result.

50. Suppose that someone turns on your calculator and presses the X key. How does the calculator respond? Why?

In Exercises 51–56, suppose that you store 5 for x. What value would you need to store for y in order for each of the following expressions to have a value of 9? Use your calculator to verify your answers.

51. $x + y$

52. $x - y$

53. xy

54. $\dfrac{y}{x}$

55. $|y - x|$

56. y

 57. Explain the difference between evaluating an expression and verifying that a number is a solution of an equation.

 58. Explain how to determine which member of the set $A = \{-8, -1, 0, 2.3, 8, 37.92\}$ is a solution of the equation $3 \cdot (2x) = (3 \cdot 2)x$ without substituting any of the numbers in set A.

In Exercises 59–64, determine which member of the following set is a solution of the given equation.

$$\left\{-5, -3, -\dfrac{1}{2}, 2, \dfrac{5}{2}, 4\right\}$$

59. $7 - 3x = 1$

60. $2x + 3 = 2$

61. $3x - 7 = 5$

62. $5 - x = 8$

63. $13 - 4x = 3$

64. $3x + 17 = 2$

In Exercises 65–72, for each equation, determine whether the given number is a solution.

65. $3x - 12 = -3(4 - x)$, $x = 0$

66. $x - 7 = 7 - x$, $x = 2$

67. $x(x - 3) = 2$, $x = 2$

68. $(x - 5)(x + 2) = 0$, $x = -2$

69. $x^2 - x - 6 = 0$, $x = -2$

70. $2x^2 - 7x + 3 = 0, x = 1$

71. $\frac{x}{2} - 3 = 5, x = 4$

72. $\frac{x}{3} + \frac{x}{6} = \frac{3}{2}, x = 3$

In Exercises 73–78, translate each sentence into an equation.

73. Five less than a number is 3.

74. One more than twice a number is 7.

75. The product of 8 and a number is 11.

76. The quotient of a number and 4 is 3 times the number.

77. Twice a number is 3 less than the number.

78. If 25 is added to the square of a number, the result is 10 times the number.

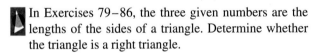

In Exercises 79–86, the three given numbers are the lengths of the sides of a triangle. Determine whether the triangle is a right triangle.

79. 5, 7, 4

80. 12, 13, 17

81. 17, 8, 15

82. 24, 25, 7

83. 9, 12, 15

84. 25, 24, 10

85. $1, 1, \sqrt{2}$

86. $1, 2, \sqrt{3}$

87. Some community volunteers have laid out a softball field at the local park. The distance between the bases is 60 feet. The distance from home plate to second base is 90 feet. (See figure.) Is the infield perfectly square? If not, what should be the distance from home plate to second base (to the nearest tenth)?

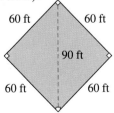

60 ft   60 ft

90 ft

60 ft   60 ft

88. Standard letter paper is 8.5 inches wide and 11 inches long. To the nearest tenth, what is the diagonal distance d across the paper? (See figure.)

Figure for 88

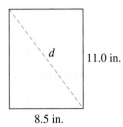

d 11.0 in.

8.5 in.

Figure for 89

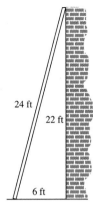

24 ft   22 ft

6 ft

89. A building inspector is checking on an apartment building that is under construction. She finds that a window ledge, exactly 22 feet from the level ground, can be reached with a ladder that is 24 feet long. (See figure.) The bottom of the ladder is 6 feet away from the base of the wall and the top of the ladder just reaches the window ledge. Is the wall perfectly vertical? Explain.

90. **Perpendicular lines** are lines that intersect to form right angles. A surveyor places pins at three corners of a property. (See figure.) The distance along one side of the property is 80 feet, and the distance along the other side is 150 feet. Explain what measurement the surveyor can use to determine if the two sides of the property are perpendicular. What should the measurement be?

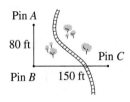

Pin A

80 ft

Pin B   150 ft

Pin C

91. A tourist is planning a January trip from his home in London to San Francisco. To plan his wardrobe, he learns that the average January temperature is 50°F. To convert to Celsius, the temperature scale in England, he uses the formula $C = \frac{5}{9}(F - 32)$. What is the corresponding Celsius temperature?

92. A radio station in Toronto, Canada, reports that the current temperature is $-10°$. Listeners who live in Buffalo, New York, know that this is a Celsius temperature. To convert to Fahrenheit, they use the formula $F = \frac{9}{5}C + 32$. What is the Fahrenheit temperature?

93. A driver traveled 50 kilometers in 45 minutes. Determine the average speed (rate) in kilometers per hour. (Hint: You will need to change minutes to hours and use: rate = distance/time.)

94. A stock car driver warmed up a car by making ten laps around a 5-mile race track. If the total time was 35 minutes, what was the average speed (rate)? See the hint in Exercise 93.

95. At a simple interest rate r, an investment of P dollars will grow to an amount A in t years according to the formula $A = P + Prt$. If $1000 is invested at 6% ($r = 0.06$), what will be the value of the investment after 5 years?

96. Using the formula in Exercise 95, determine the value of an investment of $2500 made at 7% ($r = 0.07$) after 4 years.

97. An archer shoots an arrow upward at an initial rate of 100 feet per second. The height h (in feet) of the arrow after t seconds is given by the formula $h = 100t - 16t^2$. Determine the height of the arrow after the following elapsed times.

  (a) 1 second

  (b) 3 seconds

  (c) 6.25 seconds

98. A major league pitcher throws a ball upward at an initial rate of 130 feet per second. The height h (in feet) of the ball after t seconds is given by the formula $h = 130t - 16t^2$. Determine the height of the ball after the following elapsed times.

  (a) 2 seconds

  (b) 4 seconds

  (c) 5 seconds

Exploring with Real Data

In Exercises 99–102, refer to the following background information and data.

As more and more states enacted helmet laws for bicycle riders, the sales of helmets increased dramatically. The bar graph shows sales (in millions) of bicycle helmets for selected years in the period 1985–1993. (Source: Bell Sports, U.S. Bicycle Federation.)

Figure for 99–102

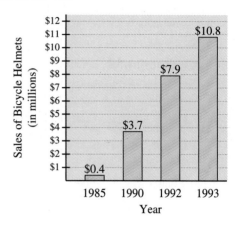

Letting t represent the number of years since 1980, we can model the data in the figure with the equation

$$S = 0.21t^2 - 2.53t + 7.72$$

where S is the sales (in millions) of helmets.

99. What values should be substituted for t in order to obtain approximations of the sales in the figure?

100. If you substitute 15 for t and evaluate the expression $0.21t^2 - 2.53t + 7.72$, what is your interpretation of the result?

101. Use the model to estimate helmet sales in 1980. What does the result suggest about the validity of the model prior to 1985?

102. Sunshine laws prevent legislators from holding closed hearings on certain matters of public interest. How might legislators have profited from enacting helmet laws if there were no sunshine laws?

Challenge

103. The following equations illustrate sets of three integers that satisfy the Pythagorean Theorem formula $a^2 + b^2 = c^2$.

$$3^2 + 4^2 = 5^2$$

$$5^2 + 12^2 = 13^2$$

$$7^2 + 24^2 = 25^2$$

$$9^2 + 40^2 = \;?$$

What pattern is revealed by these equations? Write the next two equations in the list and use your calculator to verify that they satisfy the Pythagorean Theorem formula.

104. In the accompanying figure, ΔABC is a right triangle. Squares I, II, and III are drawn so that one side of each square is also a side of the triangle. Explain how the area of Square I is related to the areas of the other two squares. (Hint: Consider the Pythagorean Theorem.)

Figure for 104

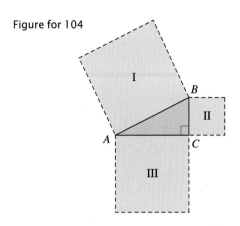

2.2 Simplifying Expressions

Terms and Coefficients ▪ *Simplifying Algebraic Expressions* ▪ *Translations*

Terms and Coefficients

We usually refer to the addends of an algebraic expression as **terms.** A term with no variable is called a **constant term.**

The expression $3x^2 + 8 - 5a^3b^5 + x$ can be written

$$3x^2 + 8 + (-5a^3b^5) + x,$$

and so the terms are $3x^2$, 8, $-5a^3b^5$, and x. In this expression, 8 is a constant term.

Grouping symbols can affect the number of terms of an expression.

Expression	Terms
$x^2 - 5x + 3$	Three terms: x^2, $-5x$, 3
$x^2 - (5x + 3)$	Two terms: x^2, $-(5x + 3)$

NOTE: An easy way to identify the terms of an expression is to determine the parts of the expression that are separated by plus or minus signs that are not inside grouping symbols. ▪

The numerical factor in a term is called the **numerical coefficient** or simply **coefficient.**

Expression	Coefficients
$4a^2 - 3ab + 5$	4, −3, 5
$2(x + 1) - 6$	2, −6

When the coefficient of a term is not written, it is understood to be 1 or -1.

Term	Coefficient
$x^2 = 1x^2$	1
$-xy = -1xy$	-1

EXAMPLE 1   *Identifying Terms and Coefficients*

Identify the terms and coefficients in each expression.

(a) $y - (x - 3) - 2$

(b) $3y - x + \dfrac{y - 5}{2}$

Solution

	Terms	Coefficients
(a) $y - (x - 3) - 2$	$y,\ -(x - 3),\ -2$	$1,\ -1,\ -2$
(b) $3y - x + \dfrac{y - 5}{2}$	$3y,\ -x,\ \dfrac{y - 5}{2}$	$3,\ -1,\ \dfrac{1}{2}$

Note that $\dfrac{y - 5}{2} = \dfrac{1}{2}(y - 5)$. Therefore, the coefficient is $\dfrac{1}{2}$. ∎

Two terms are called **like terms** if they are both constant terms or if they have the same variable factors with the same exponents.

Like Terms	Unlike Terms
$2x,\ 7x$	$2x,\ 7$
$5y,\ 3y$	$5y,\ 3y^2$
$3a^2b,\ 6a^2b$	$3a^2b,\ 6ab^2$

Simplifying Algebraic Expressions

Equivalent expressions are expressions that have the same defined value regardless of the replacements for the variables.

To **simplify** an algebraic expression, we remove grouping symbols and combine like terms. The result is an expression that is equivalent to the original expression.

Combining like terms is accomplished by applying the Distributive Property. Stated in reverse, the Distributive Property is $ab + ac = a(b + c)$. To simplify $3a + 5a$, we use this form of the Distributive Property.

$$3a + 5a = a(3 + 5) = a \cdot 8 = 8a$$

Because the Commutative Property of Multiplication allows us to write factors in any order, we can save a step when we combine like terms by using the form $ab + ac = (b + c)a$. For example,

$$6n^2 - 7n^2 = (6 - 7)n^2 = -1n^2 = -n^2.$$

EXAMPLE 2 *Combining Like Terms*

(a) $2y - 3 - 5y + 1 = (2 - 5)y + (-3 + 1) = -3y - 2$

(b) $2a^2 + b - 2b - a^2 = (2 - 1)a^2 + (1 - 2)b = a^2 - b$

(c) $a^2b - ab^2 + 3a^2b - 5ab^2 = 4a^2b - 6ab^2$ ◼

We can also use the Distributive Property to remove certain grouping symbols. An expression of the form $a(b + c)$ is equivalent to the expression $ab + ac$.

An expression of the form $-(b + c)$ is equivalent to the expression $-1(b + c)$, which is equivalent to $-b - c$. In effect, the opposite sign in front of the group has the effect of changing the sign of each term inside the group.

EXAMPLE 3 *Simplifying Algebraic Expressions*

Simplify each algebraic expression.

(a) $(3 + 4x) - (5x - 1)$ (b) $2(3x^2 - 2) - 3(x + 5)$

(c) $3x + x(5 - x) - 7x$ (d) $7 - 3[4 - (x - 2)]$

Solution

(a) $(3 + 4x) - (5x - 1) = 3 + 4x - 5x + 1$ Remove parentheses.
$$= -x + 4$$ Combine like terms.

(b) $2(3x^2 - 2) - 3(x + 5) = 6x^2 - 4 - 3x - 15$ Distributive Property
$$= 6x^2 - 3x - 19$$ Combine like terms.

(c) $3x + x(5 - x) - 7x = 3x + 5x - x^2 - 7x$ Distributive Property
$$= -x^2 + x$$ Combine like terms.

(d) $7 - 3[4 - (x - 2)] = 7 - 3[4 - x + 2]$ Remove parentheses.
$$= 7 - 3[6 - x]$$ Combine like terms.
$$= 7 - 18 + 3x$$ Remove brackets.
$$= -11 + 3x$$ Combine like terms. ◼

Translations

When we use algebra to solve problems, it is often necessary to write the given information as an algebraic expression that symbolically models the conditions of the problem and the operations to be performed.

The following are some typical translations involving the four basic operations.

Addition

Phrase	*Expression*
5 added to a number	$5 + x$
a number increased by 8	$x + 8$
7 more than a number	$x + 7$
the sum of two numbers	$x + y$

Subtraction

Phrase	**Expression**
-3 subtracted from a number	$x - (-3)$
a number decreased by 7	$x - 7$
6 less than a number	$x - 6$
the difference between 8 and a number	$8 - x$
a number less 4	$x - 4$

Multiplication

Phrase	**Expression**
a number multiplied by -4	$-4x$
two-thirds of a number	$\dfrac{2}{3}x$
the product of two numbers	xy
twice a number	$2x$
6% of a number	$0.06x$

Division

Phrase	**Expression**
the quotient of -3 and a number	$\dfrac{-3}{x}$
the ratio of two numbers	$\dfrac{x}{y}$
a number divided by 5	$\dfrac{x}{5}$

In addition to knowing how to translate such typical phrases, some common knowledge is sometimes needed to write expressions.

EXAMPLE 4   *Translating Word Phrases into Algebraic Expressions*

Translate the given verbal expressions into algebraic expressions.

(a) The value in cents of n nickels and d dimes.

(b) The difference between two numbers divided by the sum of their squares.

(c) The reciprocal of a number x increased by two-thirds of the number.

(d) The year-end value of y dollars invested at a 7% simple interest rate.

Solution

(a) $5n + 10d$

(b) $\dfrac{x - y}{x^2 + y^2}$

(c) $\dfrac{1}{x} + \dfrac{2}{3}x$

(d) $y + 0.07y$ or $1.07y$

■

2.2 Quick Reference

Terms and
Coefficients

- The addends of an algebraic expression are called **terms.** A term with no variable is a **constant term.**

- The numerical factor in a term is called the **coefficient.**

- Two terms are called **like terms** if they are both constant terms or if they have the same variable factors with the same exponents.

Simplifying
Algebraic
Expressions

- **Equivalent expressions** are expressions that have the same defined value regardless of the replacements for the variables.

- To **simplify** an algebraic expression, remove grouping symbols and combine like terms.

- We use the Distributive Property to combine like terms. The procedure is to combine the coefficients of the terms and retain the common variable factors.

- We also use the Distributive Property to remove grouping symbols.

  1. $a(b + c) = ab + ac$

  2. $-(b + c) = -b - c$

2.2 Exercises

 1. Use the Additive Identity Property to explain why every algebraic expression has a constant term.

 2. Use the Multiplicative Identity Property to explain why every term has a numerical coefficient.

In Exercises 3–10, identify the terms and coefficients in each expression.

3. $2x - y + 5$

4. $6a - 5b + c - 2d$

5. $3a^3 - 4b^2 + 5c - 6d$

6. $3x^2 + 5x - 6$

7. $7y^2 - 2(3x - 4) + \dfrac{2}{3}$

8. $\dfrac{5x}{3} - 2(y - 3)$ **9.** $\dfrac{x + 4}{5} - 5x + 2y^3$

10. $\dfrac{2x^2}{5} - y^2 - 8x + 2$

 11. Explain why $2x + 3$ has two terms, but $2(x + 3)$ is one term.

 12. Explain why $5x^3y^2$ and $5x^2y^3$ are not like terms.

In Exercises 13–28, simplify the expressions by combining like terms.

13. $2x + 1 - 3x + 2$

14. $3 - x + 3x - 5$

15. $2x - 3y + x - 5$

16. $x - y - 3y + 4$

17. $2x + 4x - 7y$

18. $5y - 8y + 3x$

19. $-x + 3 - 4x + 1 + 5x$

20. $4 + 2x + x - 3 - x - 7$

21. $\dfrac{3}{2}x + \dfrac{2}{3}y - \left(-\dfrac{1}{3}\right)y - \dfrac{5}{2}x$

22. $\dfrac{3}{7}m - \dfrac{3}{8}n + \dfrac{4}{7}m + \dfrac{5}{8}n$

23. $8x^2 + 9x^2 - 7x + 10x$

24. $17x^3 - 4x^2 + 5x^3 - 7x^2$

25. $3ac^2 - 7ac + 7ac^2 - 6ac$

26. $10ax^2 + 10a^2x - 8ax^2 - 2a^2x$

27. $5.23x^2 - 5.23x^3 + 7.98x^3 - 9.63x^2$

28. $7.36a^2 - 8.97b^2 + 4.86a^2$

29. Explain why we can remove the parentheses in $-(2x - 5)$ and in $3 - (4a + 1)$ by changing the sign of each term inside the grouping symbols.

30. Do the expressions $2x^2z$ and $2xyz$ have the same number of factors? Why?

In Exercises 31–38, simplify each expression by removing the grouping symbols.

31. $-(a - b)$ **32.** $-(3 + n)$

33. $5(2x - 3)$ **34.** $-3(4 - x)$

35. $-(2a - 3b + 7d)$ **36.** $-(3h - 7k - 8g)$

37. $3(x - 2y) - (z - 2)$

38. $-3(2x + 4y) - 7(z - 2)$

In Exercises 39–44, match the given expression in column A with its equivalent simplified expression in column B.

Column A	Column B
39. $2(x - 3) - 5x$	(a) $-(x - 1)$
40. $-(x - a) + x$	(b) $x^2y + xy^2$
41. $-x^2 + (-x)^2$	(c) $4x + 3$
42. $\dfrac{1}{2}(8x + 6)$	(d) $-3x - 6$
	(e) a
43. $-[-(1 - x)]$	(f) 0
44. $xy^2 + x^2y$	

In Exercises 45–68, simplify the expressions by removing grouping symbols and combining like terms.

45. $2x - (3x + 1)$ **46.** $5 + (3x - 4)$

47. $3(x - 4) - 2x$ **48.** $-2(3 - x) - 4$

49. $-(2a + 4b) - (5a - 2b)$

50. $-(3a - 7b) - (8b - a)$

51. $3x + 2(x - 4) - 5x$

52. $4 - 2(2x - 5) - x + 3$

53. $3(x + 5) + 2(1 - 3x)$

54. $5 - (x + 3) - 4$

55. $-(x - 1) + 2x + 1$

56. $5x - 1 + (x - 5)$

57. $(x + 1) + (2x - 1) + 5$

58. $-x - y - (x + y) + x$

59. $5(2t + 1) - 2(t - 4)$

60. $-(t + 1) - (2t - 1)$

61. $3x - 5y + 2x - (3x - 5y + 4)$

62. $7y - 6x + 9y - (7y - 3x - 8y)$

63. $6c - 5[6c - 5(c - 5)]$

64. $4a - 3[4a - 3(2a - 7)]$

65. $2(3x - 5) - 3(4x - 7) - 8(x + 6)$

66. $3(4x - 7) - 10(2x - 3) - 9(4 - x)$

67. $7 + 3[-(x - 2) - (3 - x)]$

68. $6 + 4[-2(5 - 3x) - (5x - 3)]$

In Exercises 69–72, simplify the expressions by removing grouping symbols and combining like terms.

69. $2x^2y^3 - 5x^3y^2 + 8x^3y^2 - x^2y^3$

70. $3x^4y - (x^2y^2 + x^4y)$

71. $-(x^5y^2 - x^2y^5) - (x^2y^5 - x^5y^2)$

72. $xyz + xy^2 - y(xz - xy)$

In Exercises 73–76, translate the given verbal expressions into algebraic expressions. Let n represent the number. Simplify the expression, if possible.

73. The quotient of twice the number and 12

74. The product of the number and the number increased by 4

75. Three times the number plus the number itself

76. Eight more than the number less twice the number

In Exercises 77–80, column A contains a word phrase describing a number n. Match the word phrase with the algebraic expression in column B.

Column A	*Column B*
77. Five less than a number	(a) $5 - n$
78. The difference between 5 and a number	(b) $n - 5$
79. The difference between a number and 5	
80. Five more than the opposite of a number	

In Exercises 81–94, translate the verbal description into an algebraic expression.

81. The perimeter of a rectangle whose width is half its length L

82. The area of a triangle whose base is twice its height h

83. The value (in cents) of q quarters and d dimes

84. The distance (in miles) traveled in h hours at an average speed of 55 mph

85. A person's age in 7 years if she is x years old today

86. The portion of a job done in 1 hour if it takes h hours to do the entire job

87. The sum of three consecutive integers, the smallest being n

88. The quotient of two consecutive odd integers, the smallest being n

89. The annual interest earned on d dollars if the simple interest rate is 8%

90. The amount needed to pay off a 1-year loan of d dollars if the simple interest rate is 9%

91. The hypotenuse of a right triangle if one leg is 2 units less than the other leg whose length is a

92. The selling price of a television set purchased at a 35% discount off the original price c

93. A person's salary after a 5% raise if the old salary was s

94. The amount of acid in L liters of a 20% solution

95. Which is the correct translation of "the product of 3 and a number increased by 7": $3x + 7$ or $3(x + 7)$? Explain.

96. Evaluate the expression $\sqrt{a + b}$ and the expression $\sqrt{a} + \sqrt{b}$ for $a = 1$ and $b = 0$. Are these equivalent expressions? Use the definition of equivalent expressions to explain your answer.

Exploring with Real Data

In Exercises 97–100, refer to the following background information and data.

In 1991 it was estimated that a total of 24 billion dollars was spent on sport (noncommercial) fishing. (Source: Fish and Wildlife Service, Bureau of Census.)

97. If a total of d dollars was spent on bait, what algebraic expression represents the percent of the total spent on bait?

98. If x percent of the total dollars spent was for fishing equipment, what algebraic expression represents the total dollars spent for this fishing equipment?

99. It was estimated that 49% of the total spent on sport fishing in 1991 was for travel, lodging, and food. What does this represent in dollars?

100. Environmentalists claim that keeping our rivers, lakes, and streams ecologically sound is also good for the economy. How does your answer in Exercise 99 support this claim?

Challenge

In Exercises 101–109, determine if the given expression is equivalent to (a) x^2? (b) $-x^2$?

101. $(-x)^2$ **102.** $-(-x)^2$ **103.** $|x^2|$

104. $|-x^2|$ **105.** $-|x^2|$ **106.** $|x|^2$

107. $|-x|^2$ **108.** $-|x|^2$ **109.** $-|-x|^2$

In Exercises 110–113, use the given information about x and y to simplify the given expression.

110. $x \cdot |x| + x^2$ where $x > 0$

111. $x \cdot |x| + x^2$ where $x < 0$

112. $|xy| + x(x - y)$ where $x < 0$ and $y < 0$

113. $|x(x - y)|$ where $x > 0$, $y > 0$, and $x > y$

2.3 Properties of Exponents

Bases ▪ Product Rule for Exponents ▪ Power to a Power Rule ▪ Products and Quotients to a Power ▪ Quotient Rule for Exponents

Bases

In Section 1.6 we saw how a positive integer exponent can be used to represent repeated multiplication. The **base** is the number being multiplied, and the **exponent** indicates the number of factors of that number.

$$\overset{\text{exponent}}{\underset{\text{base}}{3^4}} = 3 \cdot 3 \cdot 3 \cdot 3 \qquad \overset{\text{exponent}}{\underset{\text{base}}{b^5}} = b \cdot b \cdot b \cdot b \cdot b$$

When no exponent is indicated, we assume it to be 1: $b = b^1$. We refer to expressions involving exponents as **exponential expressions.**

EXAMPLE 1   *Identifying Bases in Exponential Expressions*

In each pair of exponential expressions, identify the bases.

(a) $(-3)^2$ -3^2

(b) $2x^4$ $(2x)^4$

(c) $x + y^2$ $(x + y)^2$

Solution

(a) $(-3)^2$ means $(-3)(-3)$. The base is -3.
   -3^2 means $-1 \cdot 3^2 = -1 \cdot 3 \cdot 3$. The base is 3.

(b) $2x^4$ means $2 \cdot x \cdot x \cdot x \cdot x$. The base is x.
   $(2x)^4$ means $2x \cdot 2x \cdot 2x \cdot 2x$. The base is $2x$.

(c) $x + y^2$ means $x + y \cdot y$. The base for the exponent 2 is y.
   $(x + y)^2$ means $(x + y)(x + y)$. The base is $x + y$.

Product Rule for Exponents

In the following exploration, we consider products of exponential expressions.

EXPLORATION 1   *Multiplying Exponential Expressions*

Use your calculator to evaluate each pair of expressions. Compare the results and answer the questions that follow.

(a) $3^2 \cdot 3^5$ $\qquad\qquad$ 3^7

(b) $(-2)^4 \cdot (-2)^2$ $\qquad$ $(-2)^6$

(c) $\left(\dfrac{3}{4}\right)^2 \cdot \dfrac{3}{4}$ $\qquad$ $\left(\dfrac{3}{4}\right)^3$

(d) $(5.2)^2 \cdot (5.2)^3$ $\qquad$ $(5.2)^5$

(e) Note that the bases are the same in each product. Propose a multiplication rule based on your observations.

(f) Use the rule to find the product $x^3 \cdot x^5$.

Discovery

(a) $3^2 \cdot 3^5 = 2187$ $\qquad\qquad$ $3^7 = 2187$

(b) $(-2)^4 \cdot (-2)^2 = 64$ $\qquad$ $(-2)^6 = 64$

(c) $\left(\dfrac{3}{4}\right)^2 \cdot \dfrac{3}{4} = 0.421875$ $\qquad$ $\left(\dfrac{3}{4}\right)^3 = 0.421875$

(d) $(5.2)^2 \cdot (5.2)^3 = 3802.04032$ $\qquad$ $(5.2)^5 = 3802.04032$

(e) Each pair of values is the same. In each pair the sum of the exponents in the first expression is the exponent in the second expression. Thus, when multiplying expressions for which the bases are the same, add the exponents.

(f) $x^3 \cdot x^5 = x^{3+5} = x^8$

We can justify our conclusion in Exploration 1 by returning to the original purpose of an exponent. Because a positive integer exponent indicates repeated multiplication, the product in part (f) is as follows.

$$x^3 \cdot x^5 = \underbrace{x \cdot x \cdot x}_{x^3} \cdot \underbrace{x \cdot x \cdot x \cdot x \cdot x}_{x^5} = x^8$$

(A total of eight factors of *x*.)

Product Rule for Exponents

For any real number b, and for any positive integers m and n,

$$b^m \cdot b^n = b^{m+n}.$$

NOTE: The bases can be numbers, variables, or any expressions, but they must be the same.

EXAMPLE 2   *Using the Product Rule for Exponents*

Use the Product Rule for Exponents, if possible, to simplify each of the following. Leave the result in exponential form.

(a) $3^7 \cdot 3^9$ (b) $x^{18} \cdot x^{13}$ (c) $(2x^2y)(5x^5y^4)$

(d) $(x + 3)^2 \cdot (x + 3)^5$ (e) $(x + 3)^2 \cdot (x - 3)^5$

Solution

In parts (a)–(d), the Product Rule for Exponents applies because the bases are the same.

(a) $3^7 \cdot 3^9 = 3^{7+9} = 3^{16}$

(b) $x^{18} \cdot x^{13} = x^{18+13} = x^{31}$

(c) $(2x^2y)(5x^5y^4) = (2 \cdot 5)(x^2 \cdot x^5)(y \cdot y^4) = 10x^{2+5}y^{1+4} = 10x^7y^5$

(d) $(x + 3)^2 \cdot (x + 3)^5 = (x + 3)^{2+5} = (x + 3)^7$

(e) Because the bases are not the same, the Product Rule for Exponents does not apply. ∎

Power to a Power Rule

We now consider a method for raising an exponential expression to a power.

EXPLORATION 2

Raising an Exponential Expression to a Power

Use your calculator to evaluate each pair of expressions. Compare the results and answer the questions that follow.

(a) $(2^3)^4$ 2^{12}

(b) $[(-3)^4]^2$ $(-3)^8$

(c) $\left[\left(\dfrac{2}{5}\right)^2\right]^3$ $\left(\dfrac{2}{5}\right)^6$

(d) Based on your observations, propose a rule for raising an exponential expression to a power.

(e) Use the rule to determine $(b^3)^2$.

Discovery

(a) $(2^3)^4 = 4096$ $2^{12} = 4096$

(b) $[(-3)^4]^2 = 6561$ $(-3)^8 = 6561$

(c) $\left[\left(\dfrac{2}{5}\right)^2\right]^3 = 0.004096$ $\left(\dfrac{2}{5}\right)^6 = 0.004096$

(d) The values in each pair are the same. In each pair the product of the exponents in the first expression is the exponent in the second expression. Thus, to raise a power to a power, multiply the exponents.

(e) $(b^3)^2 = b^{3 \cdot 2} = b^6$

We can confirm the result in part (e) by using the Product Rule for Exponents.

$$(b^3)^2 = b^3 \cdot b^3 = b^{3+3} = b^6$$

Power to a Power Rule

For any real number b and positive integers m and n,

$$(b^m)^n = b^{mn}.$$

EXAMPLE 3   *Using the Power to a Power Rule*

Use the Power to a Power Rule to remove the parentheses.

(a) $(y^5)^4$

(b) $(a^5)^3(a^4)^2$

Solution

(a) $(y^5)^4 = y^{5 \cdot 4} = y^{20}$

(b) $(a^5)^3(a^4)^2 = a^{5 \cdot 3} \cdot a^{4 \cdot 2} = a^{15} \cdot a^8 = a^{15+8} = a^{23}$ ■

Products and Quotients to a Power

In Exploration 3 we consider exponential expressions in which the base is a product or a quotient.

EXPLORATION 3   *Products and Quotients Raised to a Power*

Use your calculator to evaluate each pair of expressions. Compare the results and answer the questions that follow.

(a) $[2 \cdot 5]^4$ $2^4 \cdot 5^4$

(b) $[(-3) \cdot 4]^3$ $(-3)^3 \cdot 4^3$

(c) $\left(\dfrac{2}{5}\right)^5$ $\dfrac{2^5}{5^5}$

(d) What rules for raising a product and a quotient to a power are suggested by the results in parts (a)–(c)?

(e) Use your proposed rules to determine $(ab)^5$ and $\left(\dfrac{k}{c}\right)^7$, $c \neq 0$.

Discovery

(a) $[2 \cdot 5]^4 = 10,000$ $2^4 \cdot 5^4 = 10,000$

(b) $[(-3) \cdot 4]^3 = -1728$ $(-3)^3 \cdot 4^3 = -1728$

(c) $\left(\dfrac{2}{5}\right)^5 = 0.01024$ $\dfrac{2^5}{5^5} = 0.01024$

(d) The values in each pair are the same.

> A product raised to a power is the product of each factor raised to the power.
>
> A quotient raised to a power is the quotient of the numerator and denominator each raised to the power.

(e) $(ab)^5 = a^5 b^5$ $\left(\dfrac{k}{c}\right)^7 = \dfrac{k^7}{c^7}$

We summarize our findings in Exploration 3 in the following two rules.

Raising Products and Quotients to a Power

For any real numbers a and b and a positive integer n,

$$(ab)^n = a^n b^n \qquad\qquad \text{Product to a Power Rule}$$

$$\left(\frac{a}{b}\right)^n = \frac{a^n}{b^n}, \; b \neq 0 \qquad \text{Quotient to a Power Rule}$$

EXAMPLE 4   *Powers of Products and Quotients*

Simplify by removing the parentheses. Assume that no divisor has a value of 0.

(a) $(gt)^8$ (b) $(-3y^2)^3$

(c) $\left(\dfrac{x}{y}\right)^{11}$ (d) $\left(\dfrac{2a^3}{b^4}\right)^4$

Solution

(a) $(gt)^8 = g^8 t^8$ Product to a Power Rule

(b) $(-3y^2)^3 = (-3)^3 (y^2)^3 = -27y^6$ Product to a Power Rule

(c) $\left(\dfrac{x}{y}\right)^{11} = \dfrac{x^{11}}{y^{11}}$ Quotient to a Power Rule

(d) $\left(\dfrac{2a^3}{b^4}\right)^4 = \dfrac{(2a^3)^4}{(b^4)^4}$ Quotient to a Power Rule

$$= \dfrac{2^4 (a^3)^4}{(b^4)^4} \qquad\qquad \text{Product to a Power Rule}$$

$$= \dfrac{16a^{12}}{b^{16}} \qquad\qquad\quad \text{Power to a Power Rule}$$

Quotient Rule for Exponents

We conclude our list of basic exponent rules by considering quotients of exponential expressions.

EXPLORATION 4   *Dividing Exponential Expressions*

Use your calculator to evaluate and compare each pair of expressions. Then, answer the questions that follow.

(a) $\dfrac{4^7}{4^4}$ 4^3

(b) $\dfrac{(-5)^8}{(-5)^3}$ $(-5)^5$

(c) $\dfrac{(4.37)^3}{4.37}$ $(4.37)^2$

(d) Propose a division rule based on your observations.

(e) Use the rule to determine $\dfrac{b^5}{b^2}$.

Discovery

(a) $\dfrac{4^7}{4^4} = 64$ $4^3 = 64$

(b) $\dfrac{(-5)^8}{(-5)^3} = -3125$ $(-5)^5 = -3125$

(c) $\dfrac{(4.37)^3}{4.37} = 19.0969$ $(4.37)^2 = 19.0969$

(d) The results in each pair are the same. To divide exponential expressions with the same base, subtract the exponents.

(e) $\dfrac{b^5}{b^2} = b^{5-2} = b^3$

We can confirm the result in part (e) algebraically.

$$\frac{b^5}{b^2} = \frac{b^{2+3}}{b^2} = \frac{b^2 \cdot b^3}{b^2} = \frac{b^2}{b^2} \cdot \frac{b^3}{1} = 1 \cdot b^3 = b^3$$

Quotient Rule for Exponents

For any nonzero real number b and natural numbers m and n,
$$\frac{b^m}{b^n} = b^{m-n}.$$

EXAMPLE 5   *Using the Quotient Rule for Exponents*

Use the Quotient Rule for Exponents to perform the indicated operations. Assume that no divisor has a value of 0.

(a) $\dfrac{20^{12}}{20^8}$

(b) $\dfrac{6x^9}{9x^6}$

(c) $\dfrac{(x-2)^9}{(x-2)^5}$

(d) $\dfrac{5a^7b^6}{10a^6b^2}$

(e) $\dfrac{n^3}{n^3}$

(f) $\dfrac{b^4}{b^7}$

Solution

(a) $\dfrac{20^{12}}{20^8} = 20^{12-8} = 20^4 = 160{,}000$

(b) $\dfrac{6x^9}{9x^6} = \dfrac{3}{3} \cdot \dfrac{2}{3} \cdot \dfrac{x^9}{x^6} = 1 \cdot \dfrac{2}{3} \cdot x^{9-6} = \dfrac{2}{3}x^3$

(c) $\dfrac{(x-2)^9}{(x-2)^5} = (x-2)^{9-5} = (x-2)^4$

(d) $\dfrac{5a^7b^6}{10a^6b^2} = \dfrac{5}{5} \cdot \dfrac{1}{2} \cdot a^{7-6}b^{6-2} = \dfrac{1}{2}ab^4$

(e) $\dfrac{n^3}{n^3} = n^{3-3} = n^0$

Note that we applied the Quotient Rule for Exponents properly, but the result n^0 has not yet been defined. The only exponents we have used have been natural numbers, a set that does not include 0. We will consider this situation in the next section.

(f) $\dfrac{b^4}{b^7} = b^{4-7} = b^{-3}$

Again, applying the Quotient Rule for Exponents has resulted in an exponent that is not a natural number. We will interpret this result in the next section.

2.3 Quick Reference

Bases

- An **exponential expression** is an expression that involves exponents. For natural number exponents, the **base** is the number being multiplied, and the **exponent** indicates the number of factors of the base.

Product Rule for Exponents

- For any real number b and for positive integers m and n, $b^m \cdot b^n = b^{m+n}$.

- To multiply exponential expressions with like bases, add the exponents.

Power to a Power Rule	• For any real number b and for positive integers m and n, $(b^m)^n = b^{mn}$. • To raise a power to a power, multiply the exponents.

Products and Quotients to a Power	• Product to a Power Rule
	  1. For any real numbers a and b and for a positive integer n, $(ab)^n = a^n b^n$.
	  2. To raise a product to a power, raise each factor of the product to the power.
	• Quotient to a Power Rule
	  1. For any real numbers a and b, $b \neq 0$, and for a positive integer n,
	  $$\left(\frac{a}{b}\right)^n = \frac{a^n}{b^n}.$$
	  2. To raise a quotient to a power, raise the numerator and denominator to the power.

Quotient Rule for Exponents	• For any nonzero real number b and natural numbers m and n, $\dfrac{b^m}{b^n} = b^{m-n}$. • To divide exponential expressions with like bases, subtract the exponents.

2.3   *Exercises*

 1. Identify the bases and the exponents in the expressions -3^2, $(-3)^2$, -2^3, and $(-2)^3$. Explain how the parentheses affect the order of operations.

 2. Explain why -5^3 and $(-5)^3$ have the same value and why -5^2 and $(-5)^2$ do not have the same value.

In Exercises 3 and 4, simplify each part by removing parentheses. Then, identify the parts that are equivalent.

3. (a) $-(-b)^3$ (b) $-b^3$

    (c) $(-b)^3$ (d) b^3

    (e) $-(b)^3$

4. (a) $-(-b)^2$ (b) $-b^2$

    (c) b^2 (d) $(-b)^2$

 5. Write $2^5 \cdot 2^3$ in expanded form. What is the total number of factors of 2? How does this illustrate the Product Rule for Exponents?

 6. Explain why we can multiply $2x^2$ by $3x$ but we cannot combine the terms in the sum $2x^2 + 3x$.

In Exercises 7–22, use the Product Rule for Exponents and express the result in exponential form.

7. $2^4 \cdot 2^7$ **8.** $(-3)^2 \cdot (-3)^7$

9. $t \cdot t^4$ **10.** $x^5 \cdot x^7$

11. $-5x^4 \cdot x^2$ **12.** $y^3 \cdot 2y^5$

13. $6t \cdot 3t^7 \cdot t^2$ **14.** $-2y^5 \cdot y^2 \cdot y^3$

15. $(6+t)^5(6+t)^4$ **16.** $(x-y)^2(x-y)^3$

17. $y^5(3y^2)$ **18.** $2x^2(3x)$

19. $(-2x^7)(3x)$ **20.** $(-8t^9)(5t^7)$

21. $(2xy^2)(-3x^3y)$ **22.** $(-x^3y^7)(-2x^4y^2)$

In Exercises 23–28, perform the following operations.

(a) If possible, add the given expressions.

(b) Multiply the given expressions.

23. $8y^5, 8y^4$ **24.** $6t^5, -7t^5$

25. $n, 3n, -6n$ **26.** $m^2, 2m^2, -3m^2$

27. $y^3, 5y^3, -4y^3$ **28.** x, x^2, x^5

In Exercises 29 and 30, simplify the given expression by using the

(a) Order of Operations.

(b) Distributive Property.

29. $x^2(3x^3 + 5x^3)$ **30.** $3y(y^2 + 3y^2)$

31. When performing operations with exponential expressions, sometimes we multiply exponents, sometimes we add exponents, and sometimes we do nothing with the exponents. Use the following expressions to explain the differences.

$$x^3 \cdot x^3 \qquad x^3 + x^3 \qquad (x^3)^3$$

32. From the following list, identify the two expressions that are equivalent to $(2x^3)^5$ and explain why the third expression is not equivalent to $(2x^3)^5$.

(i) $2^5(x^3)^5$ (ii) $2x^{15}$ (iii) $32x^{15}$

33. The expression $(3 \cdot 5)^3$ means "multiply 3 and 5; then raise the result to the third power." Describe an alternate way to evaluate $(3 \cdot 5)^3$.

34. The expression $(\frac{5}{2})^4$ means "divide 5 by 2; then raise the result to the fourth power." Describe an alternate way to evaluate $(\frac{5}{2})^4$.

In Exercises 35–52, simplify by removing the parentheses. Assume that no divisor has a value of 0.

35. $(x^4)^5$ **36.** $(y^5)^7$

37. $(x^3)^4$ **38.** $(t^5)^2$

39. $[(x-3)^3]^4$ **40.** $[(2-5x)^6]^2$

41. $(xy)^5$ **42.** $(-2t)^3$

43. $(-x^2y)^3$ **44.** $(a^5b^9)^6$

45. $(-2x^2y^4)^3$ **46.** $(3x^3y^5)^4$

47. $\left(\dfrac{a}{b}\right)^6$ **48.** $\left(-\dfrac{3}{v}\right)^5$

49. $\left(\dfrac{4x^6}{y^5}\right)^2$ **50.** $\left(\dfrac{xy^2}{z}\right)^3$

51. $\left(\dfrac{7x^3}{5y^5}\right)^2$ **52.** $\left(\dfrac{5a^5}{-2b^6}\right)^4$

In Exercises 53–70, use the Quotient Rule for Exponents and express the result in exponential form. Assume all expressions are defined.

53. $\dfrac{2^7}{2^4}$ **54.** $\dfrac{-7^{11}}{7^5}$

55. $\dfrac{x^9}{x^3}$ **56.** $\dfrac{z^5}{z^2}$

57. $\dfrac{x^{12}}{x}$ **58.** $\dfrac{y^8}{y^7}$

59. $\dfrac{(x+1)^5}{(x+1)^3}$ **60.** $\dfrac{(2x-1)^6}{(2x-1)^5}$

61. $\dfrac{-5x^4}{10x^3}$ **62.** $\dfrac{-9y^4}{-3y}$

63. $\dfrac{-3z^5}{-4z^2}$ **64.** $\dfrac{24x^3}{-40x}$

65. $\dfrac{a^4b^5}{a^2b^4}$ **66.** $\dfrac{x^9y^{12}}{x^6y^5}$

67. $\dfrac{a^4b^5}{b^2}$ **68.** $\dfrac{x^{15}y^{10}}{x^{12}y^2}$

69. $\dfrac{6x^{12}y^8}{12x^6y^4}$ **70.** $\dfrac{48a^{17}b^{10}}{6a^4b^2}$

In Exercises 71–74, insert grouping symbols so that the expression has the given value.

71. $7 - 2^2$; 25

72. $9 \div 3^2$; 9

73. $8 \cdot 5^2$; 1600

74. $5 + 3^2 \div 4$; 16

In Exercises 75–80, determine whether the statement is true or false. Assume x and y are both positive.

75. $\dfrac{x^{12}}{x^3} = x^4$

76. $(x^2)^3 = x^8$

77. $(xy)^2 = x^2y^2$

78. $(x^3)^7 = (x^7)^3$

79. $\dfrac{x^{20}}{x^{19}} > 0$

80. $x^6 \cdot x^2 = x^{12}$

In Exercises 81–98, perform the indicated operations. Assume all expressions are defined.

81. $(-5x^3)(4x^5)$

82. $(x^3y^5)(x^2y)(x^7y^9)$

83. $(3s^3t^8)^3$

84. $(-2ab^7c^4)^5$

85. $(-t)^3(t^2)^4$

86. $(x^3)^2 \cdot x^2$

87. $(-2a^4)(a^2)^3$

88. $(-2x^3)^4(3x^5)$

89. $(xy^2)^2(xy)^2$

90. $3x^5 \cdot (2x^3)^3$

91. $\dfrac{-3x^{12}}{12x^4}$

92. $\dfrac{6s^7t^8}{4s^4t^3}$

93. $\dfrac{(-2x^3)^4}{8x^6}$

94. $\dfrac{9y^{14}}{(3y^3)^3}$

95. $\dfrac{(2x^3)^2}{10x^2}$

96. $\dfrac{8y^{13}}{(2y^4)^3}$

97. $\dfrac{(a^5b^7)^2}{(a^2b^4)^3}$

98. $\dfrac{(5x^3y^4)^2}{x^5y^6}$

 99. If $x^2 = 5$, what is x^6? Explain your reasoning.

 100. Evaluate $(5 - 3)^2$ and $(3 - 5)^2$. Explain why the order of subtraction does not matter if the results are squared.

 In Exercises 101–104, write an expression for the area of each figure.

101.

c^3

c^3

102.

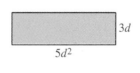

$3d$

$5d^2$

103.

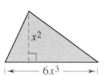

x^2

$6x^3$

104.

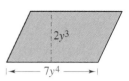

$2y^3$

$7y^4$

In Exercises 105–108, determine the value of n.

105. $(x^n)^3 = x^{12}$

106. $x^n \cdot x^3 = x^{12}$

107. $\dfrac{x^n}{x^4} = x^2$

108. $(x^5)^n = x^{20}$

 109. We know that $(cd)^2 = c^2d^2$ because of the Product to a Power Rule. However, there is no corresponding exponent rule for a sum raised to a power. For example, how can you show that $(c + d)^2$ is not the same as $c^2 + d^2$?

 110. Explain how you can determine the signs of the results in advance when you simplify $(-x^4y^6)^7$ and $(-x^3y^5)^6$.

In Exercises 111–114, show that the statement is true. Assume x represents a nonzero number.

111. $\dfrac{x^5}{x^4} + \dfrac{x^4}{x^3} + \dfrac{x^3}{x^2} + \dfrac{x^2}{x} + x = 5x$

112. $(x^3)^4 + x^7 \cdot x^5 + \dfrac{x^{20}}{x^8} = 3x^{12}$

113. $(2x)^3 - x^3 \cdot x^2 - 8x^3 + x^5 = 0$

114. $(-x)^2 + (-x^2) - x^2 + (x^1)^2 = 0$

Challenge

The Quotient Rule for Exponents cannot be applied to $4^8/8^4$ because the bases are not the same. However, in cases like this, we can sometimes change all of the bases to the same base.

For example: $4^8 = (2^2)^8 = 2^{16}$ and $8^4 = (2^3)^4 = 2^{12}$.

Then $\dfrac{4^8}{8^4} = \dfrac{2^{16}}{2^{12}} = 2^4 = 16$.

In Exercises 115–120, use this technique to perform the indicated operations.

115. $8^4 \cdot 2^3$

116. $9^4 \cdot 3^5$

117. $9^5 \cdot 27^2$

118. $8^3 \cdot 4^5$

119. $\dfrac{8^5}{2^{10}}$

120. $\dfrac{9^5}{3^8}$

In Exercises 121–126, insert $<$, $=$, or $>$ to make the statement true.

121. If $b > 0$, then b^n _____ 0.

122. If n is odd and $b < 0$, then b^n _____ 0.

123. If n is even and $b > 0$, then $-b^n$ _____ 0.

124. If n is even and $b < 0$, then $-b^n$ _____ $(-b)^n$.

125. If n is odd and $b < 0$, then $-b^n$ _____ $(-b)^n$.

126. If $b > 1$ and $m > n$, then b^m _____ b^n.

In Exercises 127–130, use the exponent rules to simplify the expression.

127. $y^{3n} \cdot y^2$

128. $\dfrac{y^{5m+1}}{y^{2m}}$

129. $(a^{3n})^{4n}$

130. $\left(\dfrac{b^{2n}}{a^m}\right)^n$

In Exercises 131–140, simplify.

131. $(-2a^5b^6)^5(-a^2b^5)^4$

132. $(-4x^4y^6)(2x^5y^3)^3(x^6y^5)$

133. $\dfrac{(3x^5y^2)^2(2x^4y^3)}{6x^5y^2}$

134. $\dfrac{(-3xy^5)^2(8x^4y)}{6x^3(xy^7)}$

135. $\dfrac{(x^7y)(2xy^3)^3}{10x^8y^3}$

136. $\dfrac{3(xy^2)^5}{(9xy^5)(x^3y^3)}$

137. $\dfrac{(x^3y^4)^3}{(x^4y^3)^4} \cdot (x^2y)^5$

138. $\dfrac{(x^4y^5)^2}{(x^2y^4)^3} \cdot (xy^3)^4$

139. $\left(\dfrac{-3x^4y^5}{z^3}\right)^3 \cdot \left(\dfrac{5x^2y^3}{z^5}\right)^2$

140. $\left(\dfrac{9x^3y}{z^4}\right)^2 \cdot \left(\dfrac{2x^3y^2}{z^4}\right)^3$

2.4 Zero and Negative Exponents

Zero Exponent ▪ *Negative Exponents* ▪ *Properties of Exponents*

Zero Exponent

In the previous section, we applied the Quotient Rule for Exponents to write $\dfrac{n^3}{n^3} = n^{3-3} = n^0$.

An alternate approach to the problem uses the fact that any nonzero number divided by itself is 1. Therefore, $\dfrac{n^3}{n^3} = 1$.

Because n^0 and 1 are both results of simplifying n^3/n^3, n^0 and 1 must have the same value. This is the motivation for the following definition.

> ### Definition of the Zero Exponent
> For any nonzero real number n, $n^0 = 1$.

NOTE: The base can be any quantity except 0, that is, 0^0 is undefined. In all expressions involving a variable base and zero exponent, we will assume that the base is not zero unless otherwise indicated. ■

All the following quantities have a value of 1.

$$x^0 \qquad 5^0 \qquad (-827)^0 \qquad (x+7)^0 \qquad \pi^0 \qquad \left(\frac{5}{3}\right)^0$$

EXAMPLE 1 *Exponential Expressions with Zero Exponents*

Perform the following operations. Use your calculator to verify parts (a)–(c).

(a) 5^0

(b) $(-5)^0$

(c) -5^0

(d) $(4y)^0$

(e) $4y^0$

Solution

(a) The base is 5: $5^0 = 1$.

(b) The base is -5: $(-5)^0 = 1$.

(c) The quantity -5^0 can be written $-1 \cdot 5^0 = -1 \cdot 1 = -1$.

(d) The base is $(4y)$: $(4y)^0 = 1$.

(e) The base is y: $4y^0 = 4 \cdot y^0 = 4 \cdot 1 = 4$. ■

Negative Exponents

In the previous section, we applied the Quotient Rule for Exponents to write $\dfrac{b^4}{b^7} = b^{4-7} = b^{-3}$.

An alternate approach to the problem uses the Product Rule for Exponents and the fact that any nonzero number divided by itself is 1.

$$\frac{b^4}{b^7} = \frac{1 \cdot b^4}{b^3 \cdot b^4} = \frac{1}{b^3} \cdot \frac{b^4}{b^4} = \frac{1}{b^3} \cdot 1 = \frac{1}{b^3}$$

This suggests that $b^{-3} = \dfrac{1}{b^3}$.

EXPLORATION 1   *Negative Exponents*

Use your calculator to evaluate the three expressions in each part. Then, answer the questions that follow.

(a) $\dfrac{2^3}{2^5}$ 2^{-2} $\dfrac{1}{2^2}$

(b) $\dfrac{(-5)^4}{(-5)^7}$ $(-5)^{-3}$ $\dfrac{1}{(-5)^3}$

(c) Formulate a rule for evaluating a number raised to a negative exponent.

(d) Use the result to write b^{-3} with a positive exponent.

Discovery

(a) Figure 2.3   (b) Figure 2.4

```
2^3/2^5
              .25
2^(-2)
              .25
1/2^2
              .25
```

```
(-5)^4/(-5)^7
             -.008
(-5)^(-3)
             -.008
1/(-5)^3
             -.008
```

(c) The expression in the second column has a negative exponent and is the result of using the Quotient Rule for Exponents.

The expression in the third column is the reciprocal of the expression in the second column but with a positive exponent.

A number raised to a negative exponent is the same as the reciprocal of the number raised to a positive exponent.

(d) $b^{-3} = \dfrac{1}{b^3}$

An additional definition is needed to include negative integers as exponents.

Definition of the Negative Exponent

For any nonzero real number b and any integer n, $b^{-n} = \dfrac{1}{b^n}$.

NOTE: The base cannot be zero. In all expressions involving a variable base and a negative exponent, we will assume that the base is not zero unless otherwise indicated. ■

Because n can be any integer, the definition may be expressed in the following way.

$$b^{-n} = \frac{1}{b^n} \quad \text{or} \quad b^n = \frac{1}{b^{-n}}.$$

EXAMPLE 2 *Exponential Expressions with Negative Exponents*

Evaluate each expression. Use your calculator to verify the results.

(a) $(-4)^{-2}$ (b) -4^{-2}

(c) $-(-5)^{-1}$ (d) $3^{-1} + 2^{-1}$

(e) $(3 + 2)^{-1}$

Solution

(a) $(-4)^{-2} = \dfrac{1}{(-4)^2} = \dfrac{1}{16} = 0.0625$ The sign of the exponent changes, but the sign of the base does not.

(b) $-4^{-2} = -1 \cdot 4^{-2} = -1 \cdot \dfrac{1}{4^2} = -\dfrac{1}{16} = -0.0625$

(c) $-(-5)^{-1} = -1 \cdot (-5)^{-1} = \dfrac{-1}{(-5)^1} = \dfrac{1}{5}$

(d) $3^{-1} + 2^{-1} = \dfrac{1}{3} + \dfrac{1}{2} = \dfrac{2}{6} + \dfrac{3}{6} = \dfrac{5}{6}$

(e) $(3 + 2)^{-1} = 5^{-1} = \dfrac{1}{5}$ ■

EXAMPLE 3 *Writing Exponential Expressions with Positive Exponents*

Write each of the following with positive exponents.

(a) x^{-5} (b) $3k^{-5}$

(c) $\dfrac{1}{b^{-3}}$ (d) $\dfrac{3}{p^{-8}}$

Solution

(a) $x^{-5} = \dfrac{1}{x^5}$

(b) $3k^{-5} = 3 \cdot \dfrac{1}{k^5} = \dfrac{3}{k^5}$

(c) $\dfrac{1}{b^{-3}} = b^3$

(d) $\dfrac{3}{p^{-8}} = 3 \cdot \dfrac{1}{p^{-8}} = 3p^8$ ■

Example 3 suggests that we can move exponential factors in a fraction from the numerator to the denominator or from the denominator to the numerator if we change the sign of the exponent. For example,

$$\frac{7^{-2}}{2^{-5}} = \frac{7^{-2}}{1} \cdot \frac{1}{2^{-5}} = \frac{1}{7^2} \cdot \frac{2^5}{1} = \frac{2^5}{7^2}.$$

In general, for nonzero real numbers a and b and integers m and n,

$$\frac{a^{-m}}{b^{-n}} = \frac{b^n}{a^m}.$$

NOTE: This rule applies only to exponential *factors,* not exponential terms. Writing $a^{-2} + b^{-2}$ as

$$\frac{1}{a^2 + b^2}$$

is incorrect because a^{-2} and b^{-2} are terms, not factors. The correct expression is

$$\frac{1}{a^2} + \frac{1}{b^2}.$$ ■

EXAMPLE 4   *Writing Exponential Expressions with Positive Exponents*

Write each expression with positive exponents.

(a) $\dfrac{s^{-6}}{t^{-9}}$ (b) $\dfrac{c^2 d^{-3}}{5k^{-2}}$ (c) $6x^{-1}y$ (d) $x^{-2}y + xy^{-2}$

Solution

(a) $\dfrac{s^{-6}}{t^{-9}} = \dfrac{t^9}{s^6}$

(b) $\dfrac{c^2 d^{-3}}{5k^{-2}} = \dfrac{c^2 k^2}{5d^3}$ The factors c^2 and 5 have positive exponents and are not moved.

(c) $6x^{-1}y = \dfrac{6x^{-1}y}{1} = \dfrac{6y}{x}$ The factors 6 and y have positive exponents and are not moved.

(d) $x^{-2}y + xy^{-2} = \dfrac{y}{x^2} + \dfrac{x}{y^2}$ Apply the definition of negative exponent to each term.   ■

Properties of Exponents

Assuming the expression is defined, all the properties of exponents discussed in Section 2.3 are valid for an expression with any integer exponents.

EXAMPLE 5   *Products and Quotients with Integer Exponents*

Perform the indicated operations and express the results with positive exponents. Assume that all variables represent nonzero numbers.

(a) $(3x^5)(-2x^{-3})$

(b) $(-3x^{-5}y^5z^{-3})(5x^4yz^4)$

(c) $\dfrac{4x^{-2}}{6x^{-5}}$

(d) $\dfrac{x^4y^{-3}z^0}{x^{-1}y}$

Solution

(a) $(3x^5)(-2x^{-3}) = -6x^{5+(-3)} = -6x^2$

(b) $(-3x^{-5}y^5z^{-3})(5x^4yz^4) = -15x^{-5+4}y^{5+1}z^{-3+4} = -15x^{-1}y^6z^1 = \dfrac{-15y^6z}{x}$

(c) $\dfrac{4x^{-2}}{6x^{-5}} = \dfrac{4x^5}{6x^2} = \dfrac{2}{3} \cdot x^{5-2} = \dfrac{2}{3} \cdot x^3 = \dfrac{2x^3}{3}$

(d) $\dfrac{x^4y^{-3}z^0}{x^{-1}y} = x^{4-(-1)}y^{-3-1}z^0 = x^5y^{-4} \cdot 1 = \dfrac{x^5}{y^4}$

EXAMPLE 6   *Using the Power Rules for Integer Exponents*

Perform the indicated operations and express the results with positive exponents. Assume that all variables represent nonzero numbers.

(a) $(2r^{-3}t^2)^{-3}$

(b) $\left(\dfrac{m^2}{n^{-3}}\right)^{-4}$

(c) $\dfrac{2cd^{-1}}{(2cd)^{-1}}$

(d) $\dfrac{(3x^{-5}y^2)^3(4x^3y^3)^{-2}}{(2x^2y^5)^3(4x^{-5}y^7)^0}$

Solution

(a) $(2r^{-3}t^2)^{-3} = 2^{-3}(r^{-3})^{-3}(t^2)^{-3} = 2^{-3}r^9t^{-6} = \dfrac{r^9}{2^3t^6} = \dfrac{r^9}{8t^6}$

(b) $\left(\dfrac{m^2}{n^{-3}}\right)^{-4} = \dfrac{(m^2)^{-4}}{(n^{-3})^{-4}} = \dfrac{m^{-8}}{n^{12}} = \dfrac{1}{m^8n^{12}}$

(c) $\dfrac{2cd^{-1}}{(2cd)^{-1}} = \dfrac{2c(2cd)^1}{d^1} = \dfrac{4c^2d}{d} = 4c^2$

(d) $\dfrac{(3x^{-5}y^2)^3(4x^3y^3)^{-2}}{(2x^2y^5)^3(4x^{-5}y^7)^0} = \dfrac{(3x^{-5}y^2)^3}{(2x^2y^5)^3(4x^3y^3)^2 \cdot 1}$

$= \dfrac{27x^{-15}y^6}{(8x^6y^{15})(16x^6y^6)}$

$= \dfrac{27x^{-15}y^6}{128x^{12}y^{21}}$

$= \dfrac{27}{128} \cdot x^{-15-12} \cdot y^{6-21}$

$= \dfrac{27}{128}x^{-27}y^{-15}$

$= \dfrac{27}{128x^{27}y^{15}}$

2.4   Quick Reference

Zero Exponent

- A zero exponent arises when we use the Quotient Rule for Exponents to divide $\dfrac{b^n}{b^n} = b^{n-n} = b^0$, $b \neq 0$.

- For any nonzero real number n, $n^0 = 1$.

Negative Exponents

- A negative exponent arises when we use the Quotient Rule for Exponents to divide $\dfrac{b^m}{b^n}$, $b \neq 0$, where $m < n$.

- For any nonzero real number b and any integer n, $b^{-n} = \dfrac{1}{b^n}$ and $b^n = \dfrac{1}{b^{-n}}$.

- For nonzero real numbers a and b and integers m and n, $\dfrac{a^{-m}}{b^{-n}} = \dfrac{b^n}{a^m}$.

Other Properties of Exponents

- Assuming the expression is defined, all the properties of exponents discussed in Section 2.3 are valid for an expression with any integer exponents.

2.4   Exercises

 1. In the definition of n^0, why must n be a nonzero number?

 2. Explain why $2x^0$ and $(2x)^0$ have different values.

In Exercises 3–22, evaluate each expression.

3. 4^0

4. $\dfrac{1}{9^0}$

5. $3t^0$

6. $-7x^0$

7. -6^0

8. $(2y)^0$

9. 5^{-2}

10. 7^{-1}

11. $(-2)^{-3}$

12. $(-4)^{-2}$

13. $\dfrac{1}{15^{-1}}$

14. $\dfrac{1}{-3^{-2}}$

15. $\left(\dfrac{1}{4}\right)^{-1}$

16. $\left(-\dfrac{2}{3}\right)^{-3}$

17. $\dfrac{2^{-2}}{2}$

18. $\dfrac{3^0}{4^{-2}}$

19. $\dfrac{5^0}{6 \cdot 3^{-2}}$

20. $\dfrac{8 \cdot 4^{-2}}{3^{-3}}$

21. $3^{-1} + 2^{-2}$

22. $2^{-1} - 5^{-1}$

 23. In the definition of b^{-n}, why must b be a nonzero number?

24. Explain how to evaluate a number that has a negative exponent.

In Exercises 25–40, write the expressions with positive exponents. Assume all expressions are defined.

25. x^{-7}

26. $\dfrac{1}{y^{-3}}$

27. $4x^{-4}$

28. $-5y^{-1}$

29. $2^{-5}x^3$

30. $4u^{-3}v^2$

31. $-x^{-3}y^{-4}$

32. yz^{-3}

33. x^0y^{-2}

34. $-10x^0y^{-2}$

35. $\dfrac{1}{-5k^{-6}}$

36. $\dfrac{1}{3t^{-5}}$

37. $(2 + y)^{-3}$

38. $\dfrac{3}{(x + 1)^{-5}}$

39. $\dfrac{-3a^{-1}}{2b^{-1}}$

40. $\dfrac{-2a^{-4}}{b^{-3}}$

In Exercises 41–48, insert $<$, $=$, or $>$ to make the statement true.

41. 5^0 _____ -5^0

42. -3^0 _____ $(-3)^0$

43. $(-8)^0$ _____ 8^0

44. $(x + 100)^0$ _____ $(x + 99)^0$

45. 5^{-2} _____ $(-5)^{-2}$

46. -5^{-2} _____ $(-5)^{-2}$

47. $(5 + 6)^{-1}$ _____ $(5 + 7)^{-1}$

48. $\left(\dfrac{2}{3}\right)^{-1}$ _____ $\dfrac{3}{2}$

In Exercises 49 and 50, list the given expressions in order from least to greatest. Assume x represents a nonzero number.

49. $(-10)^0$, 0, $-|8|^0$, $-1 - 4^0$, $4^0 + x^0$

50. $-x^0 + 1$, $(-x)^0 + 1$, $|-6|^0$, $(-2)(-x)^0$

In Exercises 51–56, use the Product Rule for Exponents and express the result in exponential form with positive exponents. Assume all expressions are defined.

51. $x^5 \cdot x^{-5}$

52. $y^{-6} \cdot y^5 \cdot y^0$

53. $(-4x^{-4})(3x)$

54. $(x^{-4})(-3x^{-6})(-x^7)$

55. $7x^{-4}y^5(2x^{-3}y^{-4})$

56. $(x^{-4}y^7)(xy^2)(x^8y^{-10})$

In Exercises 57–74, use the Quotient Rule for Exponents and express the result in exponential form with positive exponents. Assume all expressions are defined.

57. $\dfrac{5^3}{5^6}$

58. $\dfrac{3^0}{3^{-2}}$

59. $\dfrac{z^{-2}}{z^5}$

60. $\dfrac{x^6}{x^{-7}}$

61. $\dfrac{y^{-8}}{y^{-3}}$

62. $\dfrac{-2t^5}{t^{-7}}$

63. $\dfrac{3z^{-8}}{2z^{-5}}$

64. $\dfrac{21t^4}{7t^{10}}$

65. $\dfrac{-(3u)^0}{v^{-5}}$

66. $\dfrac{x^{-3}}{3y^0}$

67. $\dfrac{x^{-3}y^8}{x^2y^{-2}}$

68. $\dfrac{x^{-2}y^{-3}}{x^{-3}y^{-1}}$

69. $\dfrac{4^{-1}x^{-3}}{-4^2x^{-2}}$

70. $\dfrac{3^{-2}y^{-5}}{3^{-1}y^{-7}}$

71. $\dfrac{3^{-2}x^{-5}y^{-7}}{4x^5y^{-9}}$

72. $\dfrac{10x^{-5}y^3}{5x^{-7}y^7}$

73. $\dfrac{x^{21}y^{-22}}{x^{-18}y^{26}}$

74. $\dfrac{8x^{-3}y^0}{24x^2y^{-5}}$

75. The quotient of what expression and x^{-3} is x^{-5}?

76. The quotient of what expression and x^{-2} is 3?

77. The product of what expression and x^{-4} is 1?

78. The product of what expression and x^5 is x?

In Exercises 79–90, simplify and express the result with positive exponents. Assume all expressions are defined.

79. $(x^{-6})^5$

80. $(y^2)^{-4}$

81. $(2a^{-5})^{-4}$

82. $(4t)^{-3}$

83. $(a^{-4}b^5)^{-6}$

84. $(xy^{-3})^2$

85. $(-2x^{-2}y^4)^3$

86. $(2x^6y^{-8})^{-2}$

87. $\left(-\dfrac{2}{k}\right)^{-5}$

88. $\left(\dfrac{x^3}{y^{-4}}\right)^{-3}$

89. $\left(\dfrac{-4x^{-3}}{5^{-1}y^4}\right)^3$

90. $\left(\dfrac{5x^{-3}}{3y^2}\right)^{-2}$

In Exercises 91–94, insert grouping symbols so that the expression has the given value.

91. $4 - 3^{-9}$; 1

92. $5 \cdot 2^{-1}$; 0.1

93. $2^5 + 1^0$; 1

94. $5 \div 3^{-1}$; $\dfrac{3}{5}$

In Exercises 95–98, insert grouping symbols in the expression on the left side of the equation to make the statement true.

95. $2xy^{-1} = \dfrac{2}{xy}$

96. $2xy^{-2} = \dfrac{2}{x^2y^2}$

97. $x^{-1} + y^{-1} = \dfrac{1}{\dfrac{1}{x} + y}$

98. $x^5y^{-3}z^0 = x^5$

In Exercises 99–118, perform the indicated operations and express the result with positive exponents. Assume all expressions are defined.

99. $(3x^{-2})^0(2x^{-3})^{-4}$

100. $(-5x^{-2}y)^{-1}$

101. $-4x^{-2}(x^3)^{-1}$

102. $(-t)^3(t^{-2})^4$

103. $-6x^3(2x^{-3}y^{-2})$

104. $(2x^{-2}y^3)(-3x^4y^{-7})$

105. $(2y^3)^4(-4y^2)^{-2}$

106. $(a^{-5}b^{-6})^5(a^{-2}b^5)^{-4}$

107. $(a^4b^{-3})^{-2}(a^{-3}b^{-2})^4$

108. $(-2s^{-3}t^5)^4(-s^{-1}t^{-15})$

109. $\dfrac{-4x^2y^{-2}}{8y^5}$

110. $\dfrac{6x^{-3}y^4}{-2x^2y^{-3}}$

111. $\left(\dfrac{-2x^{-2}}{y^5}\right)^{-3}$

112. $\left(\dfrac{3x^{-3}}{x^5}\right)^{-3}$

113. $\dfrac{(x^{-2})^0x^{-3}}{x^5}$

114. $\dfrac{(t^{-6})^2}{t^4(t^3)^{-4}}$

115. $\dfrac{(a^2b^{-3})^{-4}}{(a^{-3}b^5)^2}$

116. $\dfrac{9a^2b^{-2}}{(3a^{-2}b)^{-1}}$

117. $\dfrac{(3x^2y^{-3})^{-2}}{27^{-1}x^{-6}y^2}$

118. $\dfrac{8s^4t^{-2}}{(2s^2t^{-3})^{-3}}$

In Exercises 119–122, determine the number or expression that satisfies the given conditions.

119. The square of the number is 5^{-4}.

120. The number raised to the -2 power is $\frac{9}{16}$.

121. The expression raised to the -2 power is $\dfrac{x^6}{9}$.

122. When the expression is squared and then multiplied by $3x^{-2}$, the result is $3x^6$.

123. Show that

$$\dfrac{x^{-4}}{x^{-5}} - \dfrac{x^{-3}}{x^{-4}} + \dfrac{x^{-2}}{x^{-3}} - \dfrac{x^{-1}}{x^{-2}} + \dfrac{x^0}{x^{-1}} - x = 0.$$

124. Which of the following are the same?

$$x, \; x^{-1}, \; -x, \; \frac{1}{x}, \; -\left(\frac{1}{x^{-1}}\right), \; \left(\frac{1}{x^{-1}}\right)^{-1},$$

reciprocal of x, opposite of x

Exploring with Real Data

In Exercises 125–128, refer to the following background information and data.

In the mid-1980s, Americans began switching from whole milk to low fat or skim milk to reduce fats in their diets. The accompanying figure shows the percentages of consumers who chose these two products in their purchases of liquid dairy products. (Source: NFO Research.)

Figure for 125–128

125. We can write 26% (the percentage of consumers who bought whole milk in 1992) as $\frac{26}{100}$ or $\frac{26}{10^2}$. Write this expression with a negative exponent.

126. Write each of the percentages shown for 1986 as expressions with negative exponents.

127. Use the Distributive Property to show that the sum of the three expressions in Exercise 126 is 1.

128. Assuming they do not process their own milk, to what extent, if any, do dairy farmers benefit from the results of research such as this?

Challenge

129. Show that $\dfrac{x^{-m}}{y^{-n}} - \dfrac{y^n}{x^m} = 0.$

130. Show that $\dfrac{(ab)^n}{a^n b^n} = x^0.$

131. Show that $\left(\dfrac{a}{b}\right)^{-n}\left(\dfrac{b}{a}\right)^n = \left(\dfrac{b}{a}\right)^{2n}.$

132. Show that $\dfrac{b^{m+n}}{b^{m-n}} = b^{2n}.$

In Exercises 133–138, determine the values of m and n that make the statement true. Assume all expressions are defined.

133. $\dfrac{a^3 b^n}{a^5 b^2} \cdot \dfrac{a^4 b^3}{a^m b^{-5}} = 1$

134. $\dfrac{c^{-2} d^{-3}}{c^{-9} d^n} \cdot \dfrac{c^m d^5}{c^4 d^{-1}} = 1$

135. $x^6 (x^n)^3 = x^{12}$

136. $(xy^{-3})^2 (x^n y^m) = \dfrac{x}{y}$

137. $\left(\dfrac{x^{-2} y}{x^2 y^{-4}}\right)(x^m y^n) = x^{12}$

138. $\left(\dfrac{x^3 y^{-3}}{x^{-2} y^{-3}}\right)\left(\dfrac{x^m y^{-1}}{x^2 y^n}\right) = x$

In Exercises 139–146, use all necessary exponent rules to simplify the expression. Write your result with positive exponents. Assume all expressions are defined.

139. $\dfrac{(-3ab^{-5})^3 \cdot a^{-2} b^4}{(a^7 b^{-6})^2}$

140. $\dfrac{9(-2ab^{-3})^3 \cdot a^2 b^3}{3^{-1}(a^{-2} b^4)^2}$

141. $\dfrac{(5x^2 y^{-3})^{-2}(6x^{-2} y^4)^2}{(9x^{-5} y^7)^0 (2x^{-1} y^{-2})^{-3}}$

142. $\dfrac{(3x^{-4} y^3)^3 (4x^{-3} y^4)^{-2}}{(2x^{-4} y^{-5})^{-3}}$

143. $\left(\dfrac{2a^{-1}bc^{-2}}{3a^0b^{-3}c}\right)^{-3} \cdot \left(\dfrac{4a^{-5}b^0c^2}{9a^4c^{-4}}\right)^2$

144. $\left(\dfrac{a^2b^{-4}}{a^{-3}b^0c}\right)^5 \cdot \left(\dfrac{c^{-2}a^{-4}b^{-1}}{c^{-5}a^3}\right)^{-6}$

145. $\left(\dfrac{6x^{-2}y^0z^3}{8y^5z^{-3}}\right)\left(\dfrac{4x^2y^{-2}z^3}{5z^5}\right)$

146. $\left(\dfrac{x^2y^{-4}}{x^{-3}y^{-2}z}\right)\left(\dfrac{x^2yz^{-3}}{x^{-3}z^{-1}}\right)$

2.5 Scientific Notation

Powers of 10 ▪ *Applications*

Powers of 10

From arithmetic we know that multiplying a number by 10 has the effect of moving the decimal point one place to the right.

$$2.587 \cdot 10 = 25.87$$
$$2.587 \cdot 10 \cdot 10 = 258.7$$
$$2.587 \cdot 10 \cdot 10 \cdot 10 = 2587$$

Reversing the process, and using an exponent to represent repeated multiplication, we can write 2587 as follows.

$$2587 = 2.587 \cdot 10 \cdot 10 \cdot 10 = 2.587 \cdot 10^3$$

We say that 2587 is in **decimal notation** and that $2.587 \cdot 10^3$ is in **scientific notation.**

A number in scientific notation has the form $n \cdot 10^p$, where n is a number such that $1 \le n < 10$ and the exponent p is an integer. The exponent p indicates the number of places the decimal point must be shifted to return to its original location.

EXAMPLE 1 *Writing Large Numbers in Scientific Notation*

Write the following numbers in scientific notation.

(a) 250,000

(b) 63,420,000

Solution

(a) Place the decimal point after the first nonzero digit (2) to obtain a number between 1 and 10.

$$250,000 \rightarrow 2.50000$$

To return to its original location, the decimal point must be shifted five places to the right.

2.50000   250,000.

5 places

Therefore, $p = 5$, and $250,000 = 2.5 \cdot 10^5$.

(b) Place the decimal point after the first nonzero digit (6) to obtain a number between 1 and 10.

$$63{,}420{,}000 \rightarrow 6.3420000$$

To return to its original location, the decimal point must be shifted seven places to the right.

6.3420000   63,420,000.

7 places

Therefore, $p = 7$, and $63{,}420{,}000 = 6.342 \cdot 10^7$.   ■

We also know from arithmetic that dividing a number by 10 has the effect of moving the decimal point one place to the left.

$$\frac{6.297}{10} = 0.6297$$

$$\frac{6.297}{10 \cdot 10} = 0.06297$$

$$\frac{6.297}{10 \cdot 10 \cdot 10} = 0.006297$$

By reversing the process, we can write the following.

$$0.006297 = \frac{6.297}{10 \cdot 10 \cdot 10}$$

$$= \frac{6.297}{10^3}$$

$$= 6.297 \cdot 10^{-3}$$

This number is in the scientific notation form $n \cdot 10^p$, where $1 \le n < 10$ and $p = -3$. This time, the negative exponent indicates that the decimal point must be shifted three places to the left to return to its original location.

| EXAMPLE 2 | *Writing Small Numbers in Scientific Notation* |

Write each of the following numbers in scientific notation.

(a) 0.007   (b) 0.00000058

Solution

(a) Place the decimal point after the first nonzero digit (7) to obtain a number between 1 and 10.

$$0.007 \rightarrow 7.0$$

To return to its original location, the decimal point must be shifted three places to the left.

0.007   7.0

3 places

Therefore, $p = -3$, and $0.007 = 7.0 \cdot 10^{-3}$.

(b) Place the decimal point after the first nonzero digit (5) to obtain a number between 1 and 10.

$$0.00000058 \rightarrow 5.8$$

To return to its original location, the decimal point must be shifted seven places to the left.

0.00000058   5.8

7 places

Therefore, $p = -7$, and $0.00000058 = 5.8 \cdot 10^{-7}$.

Using the methods illustrated by Examples 1 and 2, we can write any number in scientific notation.

General Scientific Notation

The scientific notation for a number is

$$n \cdot 10^p$$

where $1 \le n < 10$ and p is an integer.

NOTE: For consistency, we use the multiplication dot when writing numbers in scientific notation. However, $\times$ is also commonly used as the multiplication symbol. The result in part (b) of Example 2 can be written 5.8×10^{-7}.

You should become equally skillful in changing a number from decimal notation to scientific notation and from scientific notation to decimal notation.

EXAMPLE 3   *Changing from One Notation to the Other*

In parts (a) and (b), write the given number in decimal notation. In parts (c) and (d), write the given number in scientific notation.

(a) $1.78 \cdot 10^4$ (b) $7.2 \cdot 10^{-6}$

(c) $4,000,000,000,000$ (d) 0.000000089

Solution

(a) $1.78 \cdot 10^4 = 17,800$

(b) $7.2 \cdot 10^{-6} = 0.0000072$

(c) $4,000,000,000,000 = 4.0 \cdot 10^{12}$
(The national debt in 1993 was \$4 trillion!)

(d) $0.000000089 = 8.9 \cdot 10^{-8}$

SCIENTIFIC

Because a calculator can display only a limited number of digits, scientific notation is used to display very large and very small numbers.

EXAMPLE 4 *Scientific Notation on a Calculator*

Use a calculator to perform the following operations. Write the result in scientific notation $n \cdot 10^p$ with n rounded to the nearest tenth.

(a) 500^{20} (b) $0.00005 \div 600,000$ (c) $857,000 \cdot 26,984$

Solution

(a) $500^{20} \approx 9.5 \cdot 10^{53}$ (b) $0.00005 \div 600,000 \approx 8.3 \cdot 10^{-11}$

(c) $857,000 \cdot 26,984 \approx 2.3 \cdot 10^{10}$

Applications

As the name implies, scientific notation is used in the sciences as a compact way of working with very large and very small numbers.

EXAMPLE 5 *Distance in Light Years*

The speed of light is approximately 186,000 miles per second. Use your calculator to compute one light year. (A *light year* is the distance light travels in 1 year.)

Solution

There are 60 seconds in a minute, 60 minutes in an hour, 24 hours in a day, and 365 days in a year.

$$186,000 \, \frac{\text{miles}}{\text{second}} \cdot 60 \, \frac{\text{seconds}}{\text{minute}} \cdot 60 \, \frac{\text{minutes}}{\text{hour}} \cdot 24 \, \frac{\text{hours}}{\text{day}} \cdot 365 \, \frac{\text{days}}{\text{year}} \approx 5.87 \cdot 10^{12} \, \frac{\text{miles}}{\text{year}}$$

Light travels approximately $5.87 \cdot 10^{12}$ miles in a year.

EXAMPLE 6 *The Doomsday Gap*

If the sun were suddenly to burn out, how long would it take for earthlings to find out about it? (Use 93,000,000 miles as the distance from the sun to the earth, and use 186,000 miles per second as the speed of light.)

Solution

$$\text{Time} = \frac{\text{Distance from sun to earth}}{\text{Speed of light}}$$

$$= \frac{9.3 \cdot 10^7 \, \text{miles}}{1.86 \cdot 10^5 \, \dfrac{\text{miles}}{\text{second}}}$$

$$= \frac{9.3}{1.86} \cdot 10^{7-5} \, \text{seconds}$$

$$= 5 \cdot 10^2 \, \text{seconds}$$

$$= 500 \, \text{seconds}$$

It would take 500 seconds or approximately 8.3 minutes before disaster would strike the earth.

2.5 Quick Reference

Powers of 10
- A number in **scientific notation** has the form $n \cdot 10^p$, where n is a number such that $1 \leq n < 10$ and the exponent p is an integer.

- The exponent p indicates the number of places the decimal point must be shifted to return to its original location.

- The following steps are used to write a number in scientific notation.
  1. Place the decimal point after the first nonzero digit. The result is n.
  2. Count the number of places the decimal point is from its original location. This is the absolute value of p.
  3. If the decimal point must be moved to the right to return to its original location, p is positive. If the decimal point must be moved to the left to return to its original location, p is negative.

- Because a calculator can display only a limited number of digits, scientific notation is used to display very large and very small numbers.

2.5 Exercises

 1. A number is in scientific notation when it is written in the form $n \cdot 10^p$. What must be true about the numbers n and p?

 2. From the following list, identify the one number that is in scientific notation. Explain why the other two numbers are not in scientific notation.

    (i) $35 \cdot 10^{-4}$ (ii) $1 \cdot 10^{-9}$ (iii) $10 \cdot 10^3$

In Exercises 3–18, write the given number in scientific notation.

3. 125,000

4. 15,000

5. 2570.4

6. 110.25

7. −537,600,000

8. 1,000,000,000

9. 456,700,000,000

10. −35,795,000

11. 5

12. 2.1

13. −0.4

14. 0.025

15. 0.00000645

16. −0.00000001

17. 0.0000003749

18. 0.000000246

In Exercises 19–30, write the given number in decimal form.

19. $1.34 \cdot 10^6$

20. $7.358 \cdot 10^{-6}$

21. $-4.214 \cdot 10^{-7}$

22. $6.87 \cdot 10^4$

23. 10^{-6}

24. 10^9

25. $3 \cdot 10^4$

26. $-5 \cdot 10^{-2}$

27. $-5.72 \cdot 10^9$

28. $6.3 \cdot 10^{-9}$

29. $8.9 \cdot 10^{-4}$

30. $-4.83 \cdot 10^7$

In Exercises 31–38, use a calculator to perform the indicated operations. Write the result in scientific notation $n \cdot 10^p$ where n is rounded to the nearest tenth.

31. 428^{20}

32. $(-249)^{15}$

33. -376^{-9}

34. 426^{-23}

35. $0.0000006 \div 50,000$

36. $0.0000047637 \div 423,300$

37. $8,765,432 \cdot 654,321$

38. $-2,763,742 \cdot 2,060,000$

In Exercises 39–42, insert $<$, $=$, or $>$ to make the statement true.

39. $2.84 \cdot 10^9$ _____ $2.84 \cdot 10^8$

40. $1.73 \cdot 10^5$ _____ $17.3 \cdot 10^4$

41. $6.58 \cdot 10^{-3}$ _____ $6.58 \cdot 10^{-4}$

42. 10^7 _____ $1.0 \cdot 10^7$

In Exercises 43–46, write the given expression as a single number in decimal form.

43. $7 \cdot 10^3 + 1 \cdot 10^2 + 5 \cdot 10^1 + 3$

44. $5 \cdot 10^3 + 4 \cdot 10^2 + 9 \cdot 10^1 + 1$

45. $2 + 3 \cdot 10^{-1} + 4 \cdot 10^{-2}$

46. $9 + 2 \cdot 10^{-1} + 8 \cdot 10^{-2}$

In Exercises 47–52, estimate the value of each expression. Then, verify your estimate by performing the operations on your calculator.

47. $(2 \cdot 10^2)(3 \cdot 10^1)$

48. $(2.5 \cdot 10^{-5})(1.2 \cdot 10^3)$

49. $\dfrac{4.2 \cdot 10^8}{2.1 \cdot 10^6}$ **50.** $\dfrac{1}{10^{-3}}$

51. $(2 \cdot 10^{-1})^{-1}$ **52.** $(4 \cdot 10^{-2})^{-1}$

In Exercises 53–66, perform the indicated operations. Express the result in scientific notation.

53. $(3.72 \cdot 10^7)(9.8 \cdot 10^3)$

54. $(5.87 \cdot 10^{-4})(6.73 \cdot 10^7)$

55. $(2.35 \cdot 10^{-7})(4.8 \cdot 10^5)(5.26 \cdot 10^{-4})$

56. $(4.3 \cdot 10^{12})(2.65 \cdot 10^{-7})(6.4 \cdot 10^4)$

57. $\dfrac{4.23 \cdot 10^8}{5.2 \cdot 10^{-5}}$ **58.** $\dfrac{8.6 \cdot 10^{-4}}{4.9 \cdot 10^6}$

59. $(5.7 \cdot 10^3)^4$ **60.** $(6.89 \cdot 10^{-5})^3$

61. $(8.31 \cdot 10^5)^{-1}$ **62.** $(4.5 \cdot 10^{-5})^{-1}$

63. $\dfrac{(4.9 \cdot 10^5)(3 \cdot 10^{-7})}{3.79 \cdot 10^8}$

64. $\dfrac{(9.5 \cdot 10^5)(6.52 \cdot 10^{-7})}{7.3 \cdot 10^{12}}$

65. $\dfrac{(5.43 \cdot 10^6)(3.9 \cdot 10^{-4})}{(7 \cdot 10^{-4})(1.35 \cdot 10^{-7})}$

66. $\dfrac{(6.5 \cdot 10^{12})(4.18 \cdot 10^{-15})}{(1.12 \cdot 10^{10})(1.3 \cdot 10^{-4})}$

67. The distance between an electron and a proton in a hydrogen atom is $5.3 \cdot 10^{-11}$ meters. Write this distance in decimal form.

68. The radius of the nucleus of a heavy atom is 0.000000000007 millimeters. Write the radius in scientific notation.

69. If a computer can do one addition problem in $6 \cdot 10^{-6}$ seconds, how many such problems can the computer do in one-half minute?

70. The mass of one hydrogen atom, in grams, is approximately $1.67339 \cdot 10^{-24}$. If this number were written in decimal notation, how many 0's would there be between the decimal point and the first nonzero digit?

71. Property tax rates are expressed in mils, where one mil is one dollar for each thousand dollars of assessed property value. If the millage rate for a county is 43 mils and the total property value in the county is $5.3 \cdot 10^7$ dollars, what will the annual tax revenue be for the year? Express your answer in decimal notation and in scientific notation.

72. One way to convert feet into miles is to multiply the number of feet by $1.894 \cdot 10^{-4}$. To the nearest thousandth of a mile, how long is a 100-yard football field?

SCIENTIFIC   In Exercises 73–78, write the given number in scientific notation. Then, set your calculator for scientific notation mode, enter the given number, and verify your answer.

73. $3456 \cdot 10^6$

74. $423 \cdot 10^9$

75. $0.0001234 \cdot 10^{12}$

76. $0.00258 \cdot 10^{-10}$

77. $0.0000975 \cdot 10^{-7}$

78. $75,394 \cdot 10^{-9}$

In Exercises 79–84, set your calculator for scientific notation mode and determine the displayed result when you perform the following tasks.

79. ENTER 85962

80. STORE 7000 → X

81. $2.56 \div 1250$

82. 2^{36}

83. 2^{-36}

84. $25 \cdot 960$

Challenge

In Exercises 85–88, determine the value of n that makes the statement true.

85. $\dfrac{(4 \cdot 10^7)(8 \cdot 10^{-3})}{(16 \cdot 10^{-2})(2 \cdot 10^n)} = 1$

86. $\dfrac{(3 \cdot 10^4)(8 \cdot 10^n)}{12 \cdot 10^{-2}} = 2$

87. $\dfrac{(15 \cdot 10^{-8})(4 \cdot 10^n)}{(24 \cdot 10^{12})(5 \cdot 10^{-7})} = \dfrac{1}{2}$

88. $\dfrac{(18 \cdot 10^5)(5 \cdot 10^{-3})}{(8 \cdot 10^n)(15 \cdot 10^7)} = \dfrac{3}{4}$

 ENGINEER   In **engineering notation**, numbers are written in the form $n \cdot 10^p$ where $1 \le n < 1000$ and p is an integer that is a multiple of 3.

In Exercises 89–94, set your calculator for engineering notation mode and determine the displayed result when you perform the following tasks.

89. ENTER 1234

90. STORE 50000 → X

91. 3^{-28}

92. $1 \div 1000000$

93. 5^{30}

94. $8745 \cdot 9579$

---

2   Chapter Review Exercises

Throughout the following exercises, assume that all expressions are defined.

Section 2.1

 1. Describe two ways of using your calculator to evaluate $x^2 + x + 1$ for $x = 3$.

2. What do we mean by a *solution* of an equation?

In Exercises 3 and 4, evaluate each expression for the given values of the variables. Use your calculator only to verify your results.

3. $5a - 4b$ for $a = 3$ and $b = -2$

4. $2x^2 - 3y + 4xy$ for $x = -3$ and $y = 2$

5. Use your calculator to evaluate the expression $\sqrt{x^2 + 3y} - 2xy$ for $x = 5$ and $y = -2$. Round your result to the nearest hundredth.

In Exercises 6 and 7, determine which member of the following set is a solution of the given equation.

$$\left\{ -7, -4, -\frac{2}{3}, 3, \frac{5}{4}, 6 \right\}$$

6. $6x + 5 = 1$

7. $15 - 2x = 3$

8. Suppose that you store -3 in your calculator for x. What value would you need to store for y in order for x/y to have a value of 3?

9. (a) Translate the following into an equation: 3 less than twice a number is 7.

   (b) Show whether -2 is a solution of your equation.

10. The formula for converting Fahrenheit temperatures to Celsius is $C = \frac{5}{9}(F - 32)$. Convert 95°F to Celsius.

11. The three numbers given in each part are the lengths of the sides of a triangle. Determine whether the triangle is a right triangle.

    (a) 5, 4, 3

    (b) 12, 15, 18

12. A 15-foot post is supported by a 17-foot guy wire that is attached to the ground 8 feet from the base of the post. Show whether the post is perpendicular to the ground.

Section 2.2

 13. Explain why $3x^2y$ and $3xy^2$ are not like terms.

 14. Explain how to identify the terms of an expression.

In Exercises 15 and 16, identify the terms and coefficients of the expression.

15. $3x - 2y - z$ **16.** $3x^2 - 7(h - 5)$

In Exercises 17 and 18, state whether the given terms are like terms.

17. $7cb, 7cd$ **18.** $4x^2y^3, -2x^2y^3$

In Exercises 19 and 20, simplify the expression by removing the grouping symbols.

19. $-(2a + 3b - 4c)$

20. $-2(3x - 4y) + 3(2h - 5k)$

In Exercises 21 and 22, simplify the expressions by combining like terms.

21. $14x^2 - 3x^3 + 2x^2 - 7x^3 + x^3$

22. $12ab^2 - 10a^2b + 5ab^2 - 10a^2b$

In Exercises 23 and 24, translate the given verbal expressions into algebraic expressions. Let n represent the number. Simplify the expression, if possible.

23. The product of -5 and the sum of 4 and 3 times the number

24. Ten more than the number less 4 times the number

In Exercises 25 and 26, simplify the given expression.

25. $3 - 3(2x - 1) + 4x$

26. $4y - 2[y - 5(6 - y)]$

Section 2.3

 27. Explain why we simplify the product $x^2 \cdot x^3$ by adding the exponents rather than by multiplying them.

 28. To simplify $(x^2y^3)^4$, we use the Product to a Power Rule. Explain why the Power to a Power Rule must also be used.

In Exercises 29 and 30, use the Product Rule for Exponents to simplify. Express the result in exponential form.

29. $-3 \cdot 3^4 \cdot 3^7$ **30.** $k^4 \cdot k^5 \cdot k \cdot k^2$

In Exercises 31 and 32, use the Quotient Rule for Exponents to simplify. Express the result in exponential form.

31. $\dfrac{4^{11}}{4^7}$ **32.** $\dfrac{x^5y^9}{xy^6}$

In Exercises 33–38, simplify.

33. $\left(-\dfrac{2}{5}\right)^3$ **34.** $(c^7)^6$

35. $\left(\dfrac{2h^3}{k^2}\right)^3$ **36.** $(-x^3)^3(2x^7)$

37. $(-7x^5y)(5x^7y^8)$ **38.** $\dfrac{30c^{18}}{6c^6}$

Section 2.4

 39. If $b \neq 0$, then $b^n/b^n = b^{n-n} = b^0$. Explain what prompted us to define $b^0 = 1$ for $b \neq 0$.

 40. If $b^{-n} = a$, where $n > 0$, explain why neither a nor b can be 0.

In Exercises 41–46, evaluate each expression. Use your calculator to verify the results.

41. 1776^0 **42.** -2^{-2}

43. $(-3)^{-3}$ **44.** $5^{-1} + 3^{-1}$

45. $(5 + 3)^{-1}$ **46.** $7^0 \cdot 4^{-2}$

In Exercises 47 and 48, write each expression with positive exponents.

47. $\left(-\dfrac{h}{k}\right)^{-1}$

48. $\dfrac{-4h^{-2}}{3k^{-3}}$

In Exercises 49 and 50, use the Product Rule for Exponents to simplify. Express the result in exponential form with positive exponents.

49. $r^0 \cdot r^{-3} \cdot r \cdot r^{-7}$

50. $(-4x^{-2})(x^{-8})(x^7)$

In Exercises 51 and 52, use the Quotient Rule for Exponents to simplify. Express the result in exponential form with positive exponents.

51. $\dfrac{y^{-7}}{y^{-5}}$

52. $\dfrac{2^{-3}x^{-4}}{2^2x^{-5}}$

In Exercises 53–56, simplify. Express the result with positive exponents.

53. $\left(\dfrac{c^2}{d}\right)^{-7}$

54. $(2x^4y^{-5})^{-3}$

55. $\left(\dfrac{2h^{-5}}{k^3}\right)^{-4}$

56. $\dfrac{y^4(y^3)^{-5}}{y^{-7}}$

In Exercises 57–62, simplify. Express the result with positive exponents.

57. $(-2x^{-4}y^0z)(3x^{-5}y^{-6}z^{-3})(x^6y^{-4}z^9)$

58. $\dfrac{(6x^{-2}y^3)^{-2}}{2^{-3}x^{-3}y^7}$

59. $(3^{-1} - 4^{-1})^{-1}$

60. $\left(\dfrac{7x^{-3}y^4}{z^{-5}}\right)^{-2}\left(\dfrac{2x^{-8}y^2}{z^{-5}}\right)^4$

61. $(a^{-2}b^3)^{-3}(a^{-7}b^5)^4(a^{-5}b)^0$

62. $\dfrac{4(-3c^5d^{-4})^{-3}c^2d^3}{2^{-1}(c^{-3}d^{-4})^2}$

Section 2.5

63. We can write 23,000 as $2.3 \cdot 10^p$. Explain how to determine p.

64. Use your calculator to compute 5^{25}. Explain how to interpret the displayed result.

In Exercises 65 and 66, write the given number in scientific notation.

65. 24,500,000,000

66. 0.00000563

In Exercises 67 and 68, write the given number in decimal notation.

67. $6.83 \cdot 10^{-7}$

68. $8.49 \cdot 10^5$

In Exercises 69 and 70, use your calculator to perform the indicated operations. Write the result in scientific notation $n \cdot 10^p$ where n is rounded to the nearest tenth.

69. $(-439)^{33}$

70. $\dfrac{7.4 \cdot 10^{-9}}{4.3 \cdot 10^7}$

71. The land area of Missouri is 68,945 square miles. The population of Missouri in 1989 was 4,916,766.

(a) Write both of these numbers in scientific notation $n \cdot 10^p$ with n rounded to the nearest tenth.

(b) Use the numbers in part (a) to calculate the average population per square mile.

2   Chapter Test

For all questions in this test, assume all expressions are defined.

1. Identify the elements of $\{-2, 0, 2, 3\}$ that are solutions of the equation $3x - 4 = -10$.

2. Identify the terms and coefficients of the expression $5x^2 - 4x + 2$.

 3. Explain the difference in meaning of the words *term* and *factor*.

In Questions 4–6, simplify the given product. Express the results with positive exponents.

4. $(x^3y^2)(xy^4)$ 5. $t^3 \cdot t^{-2} \cdot t$ 6. $(2a^{-5}b^4)^2$

In Questions 7–9, simplify the given quotient. Express the results with positive exponents.

7. $\left(\dfrac{a^2}{b^{-3}}\right)^{-2}$ 8. $\dfrac{4a^{-3}}{5b^{-2}}$ 9. $\dfrac{c^{-3}d^4}{c^{-5}d^{-2}}$

In Questions 10–12, evaluate the given expression.

10. $\dfrac{(-2)^{-3}}{(-2)^2}$ 11. $3x^0$ 12. $(-2)^{-3}$

In Questions 13 and 14, write the given number in decimal notation.

13. $2.7 \cdot 10^7$ 14. $4.56 \cdot 10^{-5}$

15. The wavelength of sunlight is approximately 0.00000057 meters. Write this number in scientific notation.

In Questions 16–18, evaluate the expression for the given value(s) of the variable(s).

16. $3 - 5x$ for $x = 1.4$

17. $x^2 + 1$ for $x = 2$

18. $\frac{1}{2}bh$ for $b = 5.32$ and $h = 7.6$

In Questions 19–21, simplify the given expression by removing parentheses and combining like terms.

19. $(3x - 2y) + 7y$

20. $(2s - 5t) + 2(s + 2t)$

21. $4 - 2[3x - (x - 1)]$

22. Determine whether the numbers 5, 12, and 13 are the lengths of the sides of a right triangle.

23. Write an algebraic expression for the width of a rectangle if the width is 3 less than half the length. (Let L represent the length.)

24. (a) Translate the following information into an equation.

The sum of a number and 3 times the number is 2 less than twice the number.

(b) Determine whether -1 is a solution of your equation.

25. In 1988 Chicago's O'Hare Airport served 56,700,000 passengers while Atlanta's Hartsfield Airport served $4.59 \cdot 10^7$ passengers. Write the total number of passengers served by the two airports in decimal notation and in scientific notation.

1-2 Cumulative Test

1. Graph the set $\{x \mid -2 < x \le 3\}$ on a number line. Then, describe the set in interval notation.

2. For each of the following, determine whether the statement is true or false.

   (a) All decimal numbers are irrational numbers.

   (b) If $a < -2$ and $-2 < b$, then $a < b$.

   (c) The inequalities $x < 4$ and $4 > x$ have the same meaning.

   (d) The distance between -3 and 5 on the number line can be represented either by $|-3 - 5|$ or by $|5 - (-3)|$.

   (e) If $n < 0$, then $-n$ represents a negative number.

In Questions 3–10, perform the indicated operations, if possible.

3. $-6 + [5 - (-4)]$

4. $\dfrac{3}{7} + \left(-\dfrac{2}{3}\right)$

5. $2 - |-5| + (-7)$

6. $-\dfrac{2}{5} - \dfrac{2}{3}$

7. $\dfrac{1}{2}(-1)(-4)(3)(-2)$

8. $\left(-\dfrac{1}{4}\right)\left(-\dfrac{6}{5}\right)\left(-\dfrac{15}{2}\right)$

9. $-1 \div 0$

10. $-24 \div 6 \div (-2)$

In Questions 11–16, evaluate the given expression, if possible.

11. -5^2

12. $15 \div \sqrt{16 - 7}$

13. $8 - 2 \cdot 3 + 10 \div 5$

14. $-1 - \sqrt{-1}$

15. $\dfrac{2^2 + 4^2}{2 - 2(5 - 6)}$

16. $\dfrac{-2 + \sqrt{(-2)^2 - 4(1)(-8)}}{2(1)}$

In Questions 17–20, use the given property of real numbers to write the expression in an alternate form.

17. Distributive Property: $9x + 12 = $ _____

18. Multiplication Property of -1: $-1x = $ _____

19. Commutative Property of Multiplication: $x(2 + 5) = $ _____

20. Property of Multiplicative Inverses: $(\tfrac{1}{5} \cdot 5)y = $ _____

In Questions 21 and 22, evaluate the expression for the given values(s) of the variable(s).

21. $\dfrac{\sqrt{20 - 2x}}{x^2 + 4}$ for $x = 2$

22. $\dfrac{x^2 y}{(x - y)^2}$ for $x = 5$, $y = -2$

23. Which member of the set $\{-2,\ 0,\ 3,\ 6\}$ is a solution of the equation $\dfrac{x}{2} = -(3 - x)$?

24. In each part, the three given numbers are the lengths of the sides of a triangle. Determine whether the triangle is a right triangle.

    (a) 10, 24, 27   (b) 3, 5, $\sqrt{34}$

In Questions 25 and 26, simplify the given expression.

25. $-2(x - 5) - (4 - x)$

26. $x^2 y - xy^2 - x(xy - y^2)$

In Questions 27–32, use the properties of exponents to simplify the given expression. Write the result with positive exponents. Assume all expressions are defined.

27. $(x - 3)^0$

28. $\dfrac{x^{-3} \cdot x^5}{x^3}$

29. $(2x^2)^{-3}$

30. $\left(\dfrac{x^{-1}}{y}\right)^{-2}$

31. $\dfrac{18x^{-3}}{27z^{-2}}$

32. $\dfrac{(2x^{-2})^2(3x^4)}{(3x^2)^{-2}}$

33. Write $2.7 \cdot 10^{-5}$ in decimal form.

34. At the equator, the circumference of the earth is approximately 25,000 miles. Write this number in scientific notation.

3

The Coordinate Plane and Functions

Since the Endangered Species Act of 1973, the gray wolf population in the continental United States has increased. The bar graph shows the estimated population of gray wolves between 1960 and 1990.

Superimposed on the bar graph is the graph of a **function** that models the data. Does the shape of the graph suggest that the gray wolf population will increase indefinitely? What natural factors might change the shape of the graph in the future? (To learn more about how this model function can be used, see Exercises 75–78 at the end of Section 3.5.)

In Chapter 3 we introduce the coordinate plane, relations, and functions. We also examine the many advantages of a graphing calculator in producing the graphs of functions. The chapter concludes with notation, evaluation and analysis of functions.

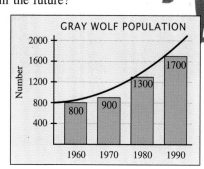

GRAY WOLF POPULATION

(Source: U.S. Fish and Wildlife Service.)

117

3.1   The Coordinate Plane

Rectangular Coordinate System ▪ *The Graphing Calculator* ▪ *Ordered Pairs and Points* ▪ *Distance Between Points* ▪ *Midpoints of Line Segments* ▪ *Applications*

Rectangular Coordinate System

In Section 1.1 we used a number line to visualize the real number system. In this section we use a system of *two* number lines to visualize ordered *pairs* of real numbers.

Two numbers written in parentheses, where the order in which they are written is significant, form an **ordered pair.** The following are some examples.

$$(5, 10) \qquad (3, -6) \qquad (-6, 3) \qquad (-1, 2) \qquad (-2, -4) \qquad (0, 3)$$

Because a different order represents a different pair of numbers, $(3, -6)$ and $(-6, 3)$ are considered different ordered pairs.

Ordered pairs of numbers can be represented on a **rectangular** (or **Cartesian**) **coordinate system,** which is sometimes called the **coordinate plane.** (See Fig. 3.1.) The rectangular coordinate system consists of two number lines that are perpendicular to each other and that intersect at the zero point of each line.

The horizontal number line is called the *x*-axis, and the vertical number line is called the *y*-axis. The point of intersection is the **origin.**

The number lines divide the plane into four regions called **quadrants.** The quadrants are numbered I to IV counterclockwise beginning with the upper right-hand region.

Figure 3.1

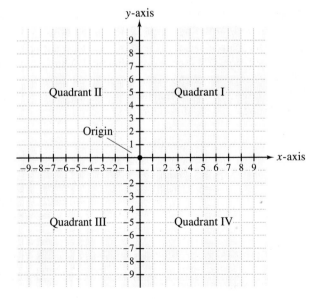

The Graphing Calculator

We can use a graphing calculator to display the rectangular coordinate system or some particular portion of it.

Each time you begin a new graphing problem, you will probably want to erase old graphs from the *graph screen*.

The *default viewing rectangle* for most calculators is oriented so that the origin is centered on the graphing screen. (See Fig. 3.2.) The tick marks along the *x*-axis are evenly distributed on both sides of the origin. The same is true for the tick marks along the *y*-axis.

Figure 3.2

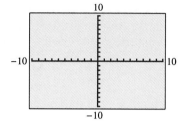

We may need to customize the viewing rectangle to meet the needs of a particular graph. Important features may be more apparent if the graph is drawn over a wider range. You may use a smaller range to see the behavior of the graph near particular points.

EXPLORATION 1   *Changing the Range on a Calculator*

To represent the first quadrant with 20 units along the *x*-axis, 12 units along the *y*-axis, and each tick mark indicating one unit, we use the following *range-and-scale notation.*

$$[\ X\!: 0, 20, 1 \ ] \qquad [\ Y\!: 0, 12, 1 \ ]$$

The X and the Y indicate the *x*- and *y*-axes. The first two numbers in each block are the *range values*. The first numbers are the minimum *x*- and *y*-values, and the second numbers are the maximum *x*- and *y*-values. The third numbers represent the distance between the tick marks on each axis. We call these numbers the *scale values*.

Now enter the given range values and display the viewing rectangle.

Figure 3.3

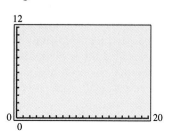

Discovery

Note that the *y*-axis is along the left edge of the screen and the *x*-axis is along the bottom edge of the screen. (See Fig. 3.3.) In this case, the 20 tick marks along the *x*-axis and the 12 tick marks along the *y*-axis represent one unit each.

In addition to adjusting the minimum and maximum values along either axis, we can adjust the scale along either axis so that each tick mark represents something other than one unit.

EXPLORATION 2   *Changing the Scale on a Calculator*

Compare the displays for the following two settings.

(a) [X: −20, 20, 1]   [Y: −10, 20, 1]

(b) [X: −20, 20, 4]   [Y: −10, 20, 2]

Discovery

(a) In Fig. 3.4 the tick marks are so close together they are difficult to distinguish.

Figure 3.4

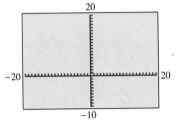

(b) In Fig. 3.5 the tick marks along the *x*-axis represent 4 units, and the tick marks along the *y*-axis represent 2 units. This setting makes the tick marks more visible.

Figure 3.5

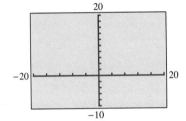

Ordered Pairs and Points

Figure 3.6

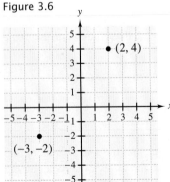

There is a one-to-one correspondence between ordered pairs of real numbers and the points of the Cartesian coordinate system. For this reason, we often use the words *ordered pair* and *point* to represent the ordered pair interchangeably.

We can completely describe each point's position with just two numbers—one for the *x*-axis and one for the *y*-axis. For example, the pair (2, 4) corresponds to the point that is 2 units to the right of the origin and 4 units up. (See Fig. 3.6.)

Figure 3.6 also shows how the pair (−3, −2) is represented by the point that is 3 units to the left of the origin and 2 units down.

Locating the point corresponding to a specific ordered pair is called **plotting,** or **graphing,** the point. The numbers in the ordered pair are called the **coordinates** of the point. The first number in the ordered pair is called the **x-coordinate,** and the second number in the ordered pair is called the **y-coordinate.** Officially, it is a point that is being plotted, but it is common to refer to the process as plotting an ordered pair.

EXPLORATION 3   *Exploring the Coordinate Plane*

 CURSOR

 INTEGER

Your calculator has a *general cursor,* which you can move around on the viewing screen. The position of the cursor is indicated by displayed coordinates.

For the following questions, we use the integer setting so that the cursor will stop only at points whose *x*-coordinates are integers.

(a) Move your cursor around in Quadrant I. How would you describe the coordinates of all of the points? Do the same for Quadrants II, III, and IV.

(b) Move the cursor back and forth along the *x*-axis. What do you observe about the coordinates?

(c) Move the cursor up and down along the *y*-axis. What do you observe about the coordinates?

Discovery

(a) Both coordinates are positive in Quadrant I. In the other three quadrants, the signs of the coordinates have the following patterns.

Quadrant II: $(-, +)$

Quadrant III: $(-, -)$

Quadrant IV: $(+, -)$

(b) On the *x*-axis the *y*-coordinate is always 0.

(c) On the *y*-axis the *x*-coordinate is always 0.

In Exploration 3 we observed that a point may lie in one of the four quadrants or it may be a point of one of the axes, in which case the point does not belong to any quadrant.

We often give letter names to points for convenient reference.

EXAMPLE 1   *Plotting Points*

Plot the points associated with the given ordered pairs. State in which quadrant each point lies.

(a) $A(2, 5)$ (b) $B(-3, -4)$ (c) $C(0, -2)$

(d) $D(-2, 0)$ (e) $E(4, -3)$

Figure 3.7

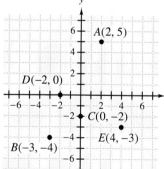

Solution

(a) Point *A* is in Quadrant I.

(b) Point *B* is in Quadrant III.

(c) Because Point *C* lies on the *y*-axis, it does not belong to any quadrant.

(d) Because Point *D* lies on the *x*-axis, it does not belong to any quadrant.

(e) Point *E* is in Quadrant IV.

In Example 1 we were given ordered pairs, and we plotted the points representing those pairs. Given a point, we can estimate the ordered pair represented by the point.

EXAMPLE 2 *Determining Coordinates of Points in a Plane*

Determine the coordinates of each point in Fig. 3.8.

Figure 3.8

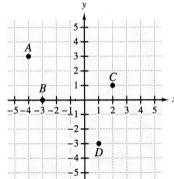

Solution

(a) $A(-4, 3)$ (b) $B(-3, 0)$

(c) $C(2, 1)$ (d) $D(1, -3)$

It is customary to use the word **graph** to mean the plotted points corresponding to a set of ordered pairs.

Distance Between Points

We have seen that if the coordinates of two points of a number line are a and b, then the distance between the points can be found by calculating $|a - b|$ or $|b - a|$.

To find the distance between two points in a rectangular coordinate system, we use the same principle along with the Pythagorean Theorem.

EXPLORATION 4 *Distance Between Points in a Coordinate Plane*

Figure 3.9 shows two points $A(-3, 2)$ and $B(5, 8)$. Suppose we need to know the distance AB between the points.

Figure 3.9

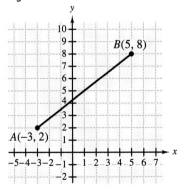

Figure 3.10

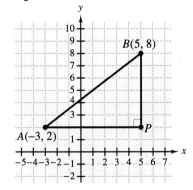

We begin by adding a third point P, which is on a horizontal line containing point A and on a vertical line containing point B. (See Fig. 3.10.)

(a) What are the coordinates of point P?

(b) What is the distance AP? What is the distance BP?

(c) Because $\angle P$ is a right angle, $\triangle APB$ is a right triangle, and we can use the Pythagorean Theorem to calculate the length AB of the hypotenuse. Calculate AB.

Discovery

(a) Point P has the same first coordinate as point B and the same second coordinate as point A. Therefore, point P represents the pair $(5, 2)$.

(b) $AP = |-3 - 5| = |-8| = 8$
$BP = |8 - 2| = |6| = 6$

(c) $(AB)^2 = (AP)^2 + (BP)^2 = 8^2 + 6^2 = 64 + 36 = 100$
$AB = 10$

We can use the techniques in Exploration 4 to obtain a general method for finding the distance d between any two points $A(x_1, y_1)$ and $B(x_2, y_2)$.

The horizontal distance between A and B is $|x_2 - x_1|$, and the vertical distance is $|y_2 - y_1|$. According to the Pythagorean Theorem,

$$d^2 = |x_2 - x_1|^2 + |y_2 - y_1|^2.$$

Therefore,

$$d = \sqrt{|x_2 - x_1|^2 + |y_2 - y_1|^2}.$$

This formula is called the **distance formula** for the distance between two points in the Cartesian coordinate system. Because the differences are being squared, the order in which we subtract the coordinates is not important, and it is not necessary to take the absolute value of each difference.

> ### The Distance Formula
>
> If $A(x_1, y_1)$ and $B(x_2, y_2)$ are points in a Cartesian coordinate system, then the distance d between the points is given by
> $$d = \sqrt{(x_2 - x_1)^2 + (y_2 - y_1)^2}.$$

EXAMPLE 3 *Using the Distance Formula*

Find the distance between the points $E(-2, 1)$ and $F(3, 13)$.

Solution

$$d = \sqrt{(x_2 - x_1)^2 + (y_2 - y_1)^2}$$
$$d = \sqrt{(-2 - 3)^2 + (1 - 13)^2}$$
$$= \sqrt{(-5)^2 + (-12)^2}$$
$$= \sqrt{25 + 144}$$
$$= \sqrt{169}$$
$$= 13$$

We can use a calculator to determine d in one keying sequence. Remember, though, that the differences are quantities and must be enclosed in parentheses. Also, the square root symbol serves as a grouping symbol, so the entire quantity under the square root symbol must also be enclosed in parentheses.

EXAMPLE 4   *Computing the Distance Between Two Points with a Calculator*

Use your calculator to determine the distance between the given points. Round your result to the nearest hundredth.

(a) $R(6.1, -3.5)$ and $S(-2.3, 9.8)$

(b) $M(137, -23)$ and $N(79, 44)$

Solution

The following are typical keying sequences.

(a) $\sqrt{((6.1 - (-2.3))\wedge 2 + (-3.5 - 9.8)\wedge 2)} \approx 15.73$

(b) $\sqrt{((137 - 79)\wedge 2 + (-23 - 44)\wedge 2)} \approx 88.62$ ◼

Midpoints of Line Segments

The **midpoint** of a line segment is the point of the segment that is equidistant (the same distance) from the endpoints of the segment.

> ### The Midpoint of a Line Segment
>
> For the line segment with endpoints $A(x_1, y_1)$ and $B(x_2, y_2)$, it can be shown that the midpoint M has coordinates
>
> $$x_m = \frac{x_1 + x_2}{2} \quad \text{and} \quad y_m = \frac{y_1 + y_2}{2}.$$

In words, the midpoint coordinates are the averages of the endpoint coordinates.

EXAMPLE 5   *Finding the Midpoint of a Line Segment*

(a) Determine the midpoint M of the line segment whose endpoints are $P(4, 5)$ and $Q(-8, 9)$.

(b) Use the distance formula to verify that $MP = MQ$.

Solution

(a) $x_m = \dfrac{4 + (-8)}{2} = -2 \qquad y_m = \dfrac{5 + 9}{2} = 7$

The coordinates of midpoint M are $(-2, 7)$. (See Fig. 3.11.)

Figure 3.11

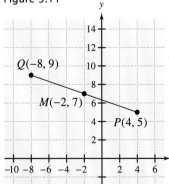

(b) $MP = \sqrt{(-2-4)^2 + (7-5)^2} = \sqrt{36+4} = \sqrt{40}$

$MQ = \sqrt{[-2-(-8)]^2 + (9-7)^2} = \sqrt{36+4} = \sqrt{40}$

Therefore, $MP = MQ$.

Applications

We can apply the distance and midpoint formulas to problems involving geometric figures drawn in a coordinate plane.

In Example 6, we use the following definition: If two line segments have the same midpoint, then the line segments **bisect** each other.

EXAMPLE 6 *The Diagonals of a Rectangle Bisect Each Other*

Figure 3.12 shows a grid map of forest ranger facilities. Observation towers are located at points A, B, C, and D, which are the vertices of a rectangle. The main office is located at point P where the diagonals of the rectangle intersect. Each tick mark represents 1 mile.

Figure 3.12

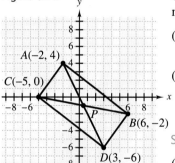

(a) Show that the main office is equidistant from the four observation towers by proving that the diagonals of the rectangle bisect each other.

(b) Find the distance from the main office to any one of the four towers.

Solution

(a) For the diagonal with endpoints A and D, the coordinates of the midpoint are as follows.

$$x_m = \frac{-2+3}{2} = \frac{1}{2} \qquad y_m = \frac{4+(-6)}{2} = -1$$

For the diagonal with endpoints B and C, the coordinates of the midpoint are as follows.

$$x_m = \frac{-5+6}{2} = \frac{1}{2} \qquad y_m = \frac{0+(-2)}{2} = -1$$

Because $(\frac{1}{2}, -1)$ is the midpoint of both diagonals, the diagonals bisect each other. Therefore, the main office is equidistant from all four towers.

(b) The distance from the main office at $P(\frac{1}{2}, -1)$ to the tower at $A(-2, 4)$ is calculated with the distance formula.

$$d = \sqrt{[-2-0.5]^2 + [4-(-1)]^2} \approx 5.6 \text{ miles}$$

The distance from the office to any of the other towers is the same.

3.1   Quick Reference

Rectangular Coordinate System

- An **ordered pair** consists of two numbers written in parentheses with the order in which the numbers are written being significant.

- Ordered pairs of real numbers are represented by points plotted in a **rectangular** (or **Cartesian**) **coordinate system.** This system consists of two perpendicular number lines called **axes** intersecting at the **origin.** The horizontal number line is called the **x-axis;** the vertical number line is called the **y-axis.**

- The axes divide the plane into four regions called **quadrants.** The quadrants are numbered counterclockwise from I to IV beginning with the upper right-hand quadrant.

The Graphing Calculator

- We can use a graphing calculator to display graphs on a coordinate system. The *default viewing rectangle* is the manufacturer's default setting.

- We can view specific portions of the coordinate system by adjusting the range. Also, either of the axes can be scaled to give the tick marks values other than 1.

Ordered Pairs and Points

- The two numbers of an ordered pair are called the **x-** and **y-coordinates.** These coordinates indicate the location of the plotted point that represents the pair.

- The *x*-coordinate of a point indicates its horizontal distance from the origin; the *y*-coordinate of a point indicates its vertical distance from the origin.

- We use the word **graph** to mean all of the plotted points corresponding to a given set of ordered pairs.

- Given an ordered pair, we can plot the point representing the pair. Given a plotted point, we can estimate the ordered pair represented by the point.

Distance Between Points

- If $A(x_1, y_1)$ and $B(x_2, y_2)$ are points in a Cartesian coordinate system, then the distance d between the points is given by
$$d = \sqrt{(x_2 - x_1)^2 + (y_2 - y_1)^2}.$$

- When using your calculator to compute the distance between two points, be sure to enclose both differences and the entire radicand in parentheses.

Midpoints of Line Segments

- For the line segment with endpoints $A(x_1, y_1)$ and $B(x_2, y_2)$, the midpoint M has coordinates
$$x_m = \frac{x_1 + x_2}{2} \quad \text{and} \quad y_m = \frac{y_1 + y_2}{2}.$$

- In words, the midpoint coordinates are the averages of the endpoint coordinates.

3.1 *Exercises*

 1. Explain in your own words how to plot the point associated with the ordered pair $A(-2, 4)$.

 2. Give an example of a point that is not in any quadrant and describe its location.

In Exercises 3–6, plot the points associated with the given ordered pairs.

3. $A(4, -6)$, $B(-6, 4)$, $C(3, 5)$

4. $A(-2, -3)$, $B(-3, -2)$, $C(-5, 6)$

5. $A(0, -3)$, $B(-5, 0)$, $C(-2, -1)$

6. $A(3, 1)$, $B(3, 7)$, $C(4, -5)$

In Exercises 7–12, plot the points associated with the ordered pairs in the given set. From the resulting pattern, what do you think is the next ordered pair?

7. $\{(-4, -3), (-2, -2), (0, -1), (2, 0),$
$(4, 1), \ldots\}$

8. $\{(-5, 6), (-2, 4), (1, 2), (4, 0), (7, -2), \ldots\}$

9. $\{(-3, 5), (-2, 5), (-1, 5), (0, 5), (1, 5), \ldots\}$

10. $\{(4, 2), (4, 1), (4, 0), (4, -1), (4, -2), \ldots\}$

11. $\{(-4, 0), (-2, 3), (0, 0), (2, 3), (4, 0), \ldots\}$

12. $\{(-1, 1), (0, 0), (1, -1), (2, 0), (3, 1), \ldots\}$

In Exercises 13–18, without plotting the given point, name the quadrant in which the point lies.

13. $P\left(-2, -\dfrac{4}{5}\right)$ **14.** $B\left(\dfrac{3}{2}, -9\right)$

15. $P(2.4, 6.5)$ **16.** $B\left(-\sqrt{2}, -\sqrt{3}\right)$

17. $A\left(\sqrt{5}, -7\right)$ **18.** $P\left(-\sqrt{6}, 10\right)$

In Exercises 19–24, refer to the accompanying figure and determine the coordinates of the points.

19. A **20.** B **21.** C
22. D **23.** E **24.** F

Figure for 19–24

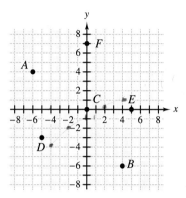

 25. Give an example of an ordered pair for which (x, y) is the same as (y, x). Explain why, in general, (x, y) is *not* the same as (y, x).

26. Determine the number of points in the graph of each of the following sets. If the number cannot be determined, explain why.

$A = \{(3, -2), (-4, -5), (-2, 7), (0, 9)\}$
$B = \{(x, y) \mid x = 3 \text{ and } -3 \leq y \leq 5\}$
$C = \{(x, y) \mid y = x\}$
$D = \{(x, y) \mid x = 3 \text{ and } -3 \leq y \leq 5 \text{ and } y$
    is an integer$\}$

In Exercises 27–32, draw the graph of each set.

27. $A = \{(x, y) \mid x \text{ is an integer}, -3 \leq x \leq 3,$
    $y = 2\}$

28. $B = \{(x, y) \mid y \text{ is an integer}, -2 \leq y \leq 5,$
    $x = -2\}$

29. $C = \{(x, y) \mid y \text{ is an integer}, -4 \leq y \leq 2,$
    $x = -y\}$

30. $D = \{(x, y) \mid x \text{ is an integer}, -3 \leq x \leq 4,$
    $y = -x + 1\}$

31. $C = \{(x, y) \mid x \text{ and } y \text{ are integers}, -2 \leq x \leq 1,$
    $1 \leq y \leq 3\}$

32. $D = \{(x, y) \mid x \text{ and } y \text{ are integers}, 1 \leq x \leq 4,$
    $-3 \leq y \leq 2\}$

In Exercises 33–42, in which *possible* quadrants or on which axes could the point lie?

33. (x, y), $x > 0$

34. $(x, 0)$

35. (x, y), $y < 0$

36. (x, y), $xy < 0$

37. $(-3, y)$

38. (x, y), $xy > 0$

39. (x, y), $x < 0$

40. $(0, y)$

41. (x, y), $y > 0$

42. $(x, 5)$

In Exercises 43–46, what is true about x and y for the given location of point P?

43. $P(x, y)$ lies on the x-axis to the left of the origin.

44. $P(x, y)$ lies on the y-axis below the origin.

45. $P(x, y)$ lies on the y-axis above the origin.

46. $P(x, y)$ lies on the x-axis to the right of the origin.

47. Suppose you know that $P(a, b)$ lies in Quadrant II. In which quadrant does $R(b, a)$ lie? (Note that R has the same coordinates as P, but in reverse order.)

48. Suppose you know that $A(c, d)$ lies in Quadrant III. In which quadrant does $B(d, c)$ lie? (Note that B has the same coordinates as A, but in reverse order.)

49. Suppose you plot the point $P(c, d)$. What is the location of the point $Q(-c, d)$ in reference to the location of P?

50. Suppose you plot the point $P(a, b)$. What is the location of the point $Q(a, -b)$ in reference to the location of P?

51. Suppose you customize a viewing rectangle according to these specifications:
[X: 0, 10, 2] [Y: −4, 12, 4].

(a) What are the smallest x- and y-coordinates that will be displayed?

(b) What are the largest x- and y-coordinates that will be displayed?

(c) How many tick marks will there be along each axis?

52. Suppose you customize a viewing rectangle so that the smallest x- and y-values are −2 and −10,

respectively; the largest x- and y-values are 12 and 8, respectively; and the x- and y-scales are 1 and 2, respectively.

(a) Write the above information in range notation.

(b) How many tick marks will there be along each axis?

In Exercises 53–56, determine an appropriate range and scale for graphing each set. (There is no one correct answer.)

53. $A = \{(300, 430), (250, 500), (400, 325), (125, 150)\}$

54. $B = \{(0.5, 0.25), (-0.7, 0.4), (0.1, -0.9), (-0.2, 0.45)\}$

55. $C = \{(-7, 9), (-5, -14), (-2, 17), (-10, -19), (-1, 18)\}$

56. $D = \{(125, 32), (150, 43), (160, 45), (136, 50)\}$

 57. In your own words, explain how to determine the distance between two points in a coordinate plane.

 58. In your own words, explain how to determine the coordinates of the midpoint of a line segment drawn between two points.

In Exercises 59–68, determine the distance (to the nearest hundredth) between the given points P and Q.

59. $P(5, 2)$, $Q(-12, 2)$

60. $P(7, -4)$, $Q(-17, -4)$

61. $P(3, 8)$, $Q(3, -8)$

62. $P(-5, -14)$, $Q(-5, 13)$

63. $P(-3, 2)$, $Q(5, 8)$

64. $P(-5, -6)$, $Q(4, 3)$

65. $P(-7, -2)$, $Q(-3, -5)$

66. $P(-8, 5)$, $Q(7, -9)$

67. $P(-4, 3)$, $Q(-6, -7)$

68. $P(-5, -7)$, $Q(5, -4)$

In Exercises 69–72, use the distance formula to determine if the three points are collinear. (Points are **collinear** if they are points of a line.) *mean same line*

69. $A(2, 4)$, $B(0, -2)$, $C(-1, -5)$

70. $A(-4, 3)$, $B(2, 0)$, $C(4, -1)$

71. $A(-4, 1)$, $B(-1, 2)$, $C(4, -1)$

$12\sqrt{}$

72. $A(5, -4)$, $B(3, -2)$, $C(-1, 2)$

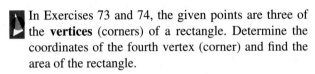 In Exercises 73 and 74, the given points are three of the **vertices** (corners) of a rectangle. Determine the coordinates of the fourth vertex (corner) and find the area of the rectangle.

73. $A(0, 0)$, $B(0, 5)$, $C(8, 0)$

74. $A(-1, 0)$, $B(-1, -6)$, $C(-5, 0)$

In Exercises 75–80, determine the midpoint of the line segment whose endpoints are given.

75. $(4, -3)$, $(4, -1)$ 76. $(-5, 6)$, $(5, -6)$

77. $(3, 7)$, $(-1, 3)$ 78. $(3, -2)$, $(-5, 5)$

79. $\left(2, \dfrac{1}{2}\right)$, $(-2, 3)$

80. $\left(\dfrac{1}{3}, \dfrac{-11}{3}\right)$, $\left(\dfrac{5}{3}, \dfrac{2}{3}\right)$

In Exercises 81 and 82, the given points are the four **vertices** (corners) of a rectangle. Show that the diagonals bisect each other.

81. $A(2, 5)$, $B(8, 1)$, $C(2, -8)$, $D(-4, -4)$

82. $A(2, 4)$, $B(6, -4)$, $C(0, -7)$, $D(-4, 1)$

In Exercises 83 and 84, show that the points A, B, C, and D are vertices of a **parallelogram;** that is, show that both pairs of opposite sides are equal in length.

83. $A(-2, 5)$, $B(4, 3)$, $C(-2, -3)$, $D(-8, -1)$

84. $A(0, 7)$, $B(8, 3)$, $C(4, -1)$, $D(-4, 3)$

In Exercises 85–88, determine if the three given points are vertices of a right triangle.

85. $A(-2, 4)$, $B(0, 9)$, $C(3, 2)$

86. $A(-2, -1)$, $B(4, -9)$, $C(2, 2)$

87. $A(2, 4)$, $B(-2, 1)$, $C(3, -5)$

88. $A(-5, 0)$, $B(2, 6)$, $C(1, -7)$

89. A train carrying hazardous waste derails and leaks poisonous gas. Authorities use a grid map to help them develop evacuation plans. They issue evacuation orders for all people living within a radius of 6 miles of the derailment. On the grid map, the scene of the accident is located at $(3, -2)$, and your house is located at $(-1, 4)$. Do you need to evacuate? (All coordinate units are in miles.)

90. The weather service is tracking Hurricane Jeffrey on a grid map with coordinate units representing 10 miles. At 10:00 A.M., the center of the hurricane is located at $(3, -4)$. Hurricane-force winds extend 70 miles out from the center. It is predicted that the eye of Jeffrey will be located at $(-1, -2)$ in two hours. Your cabin by Lake Helen is located at $(-7, -5)$ on the grid map. Will you feel the force of hurricane winds at noon? What is the predicted rate (in miles per hour) of the movement of the eye of Jeffrey?

91. When a rectangular coordinate system is superimposed on an aerial photograph, a church is at the origin and a school is at $(3, 2)$. Coordinate units are in miles. A cartographer uses the grid photograph to produce a map whose scale is 6 inches = 1 mile. To the nearest tenth of an inch, how far apart are the church and the school on the map?

92. On a grid map, a Navy cruiser is at $P(-20, 5)$ and is steaming toward Subic Bay located at $S(30, -15)$. To remain in the required sea lane, the cruiser must first travel to $Q(1, 1)$ and then, from there, head directly to port. How much farther will the cruiser travel than a helicopter that flies from point P directly to Subic Bay? (All coordinate units are in miles.)

Exploring with Real Data

In Exercises 93–96, refer to the following background information and data.

The accompanying figure shows a bar graph of population trends in the Midwest and South. (Source: U.S. Bureau of the Census.)

Figure for 93–96

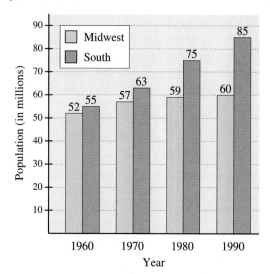

93. Write the four ordered pairs represented in the graph for the Midwest.

94. Write the four ordered pairs represented in the graph for the South.

95. Compare the population growth trends in these two regions of the country.

96. The land that is now Ohio was once part of the Western Reserve. In fact, Ohio is still considered a Midwestern state. Why is this geographically odd but historically logical?

Challenge

With most calculators, you can plot individual points. The steps for doing this are as follows.

 ERASE DATA

    (a) Erase unwanted graphs and set an appropriate range and scale.

    (b) Erase any previously stored ordered pairs.

 ENTER DATA

    (c) Enter your data in the form of ordered pairs.

 PLOT DATA

    (d) Plot the points.

97. Use your calculator to plot the points associated with the following ordered pairs.

    $A(-4, -3)$ $B(-3, 4)$ $C(2, 5)$
    $D(3, 0)$ $E(5, -2)$

98. In Exercise 97, why is point D not visible?

 LINE GRAPH   You may also be able to connect plotted points to create a **line graph.**

99. Use your calculator to produce a line graph of the data in Exercise 97.

 HISTOGRAM   A **histogram** is a special kind of bar graph. You may be able to use your calculator to produce a histogram.

100. Select a suitable range and scale and use your calculator to draw a histogram of the following data.

x	y
1	25
2	112
3	258
4	173
5	81

3.2 Relations and Functions

Relations ▪ *Functions*

Relations

The Parks and Recreation Board suspects that the number of people who go to White Sands Beach on any given day depends on the temperature on that day. The following table summarizes some of the data collected over one summer.

Temperature (°F)	60	70	75	80	85	90
People	27	94	138	221	347	462

The data in this table can also be described with ordered pairs. The first coordinate of each pair is the temperature, and the second coordinate is the number of people at the beach. The set of ordered pairs can be written as follows.

{(60, 27), (70, 94), (75, 138), (80, 221), (85, 347), (90, 462)}

This set is an example of a relation.

Definition of Relation

A **relation** is a set of ordered pairs.

The set can be finite or infinite. Any set whose members are ordered pairs is a relation.

The following are some other examples of relations.

$A = \{(3, 6), (-2, 7), (5, 3), (7, 3), (-5, -2)\}$
$R = \{(3, 6), (3, 7), (5, 3), (7, 3)\}$
$S = \{(1, 4), (2, 5), (3, 6), (4, 7), \ldots\}$
$F = \{(x, y) \mid y \text{ is 3 more than } x\}$

There are two other important definitions involving relations.

Definitions of Domain and Range

The **domain** of a relation is the set of first coordinates of the ordered pairs in the relation.

The **range** of a relation is the set of second coordinates of the ordered pairs in the relation.

EXAMPLE 1 *The Domain and Range of a Relation*

State the domain and range for the relations A, R, S, and F given previously.

Solution

Set A:	Domain:	$\{-5, -2, 3, 5, 7\}$
	Range:	$\{-2, 3, 6, 7\}$
Set R:	Domain:	$\{3, 5, 7\}$
	Range:	$\{3, 6, 7\}$
Set S:	Domain:	$\{1, 2, 3, 4, 5, 6, \ldots\}$
	Range:	$\{4, 5, 6, 7, 8, 9, \ldots\}$
Set F:	Domain:	All real numbers
	Range:	All real numbers

A relation is often described by an algebraic rule or formula rather than by a listing of the members of the set or by a description of them in words. For instance, in Example 1 the relation F can be written as $\{(x, y) \mid y = x + 3\}$.

Plotting the ordered pairs of a relation on a coordinate system is called **graphing the relation.**

EXAMPLE 2 *Graphs of Relations*

Graph the following relations.

(a) $A = \{(3, 2), (6, 2), (1, 5), (-1, 3), (0, -2)\}$

(b) $B = \{(1, -1), (2, -2), (3, -3), (4, -4), \ldots\}$

Solution

(a) Figure 3.13   (b) Figure 3.14

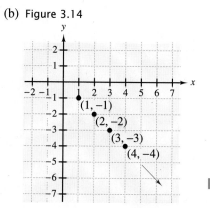

As in Fig. 3.14, arrowheads are placed on a graph to indicate that the graph extends forever in the indicated direction.

A graph of a relation provides a visual suggestion about the domain and range of the relation.

EXAMPLE 3   *Estimating the Domain and Range of a Relation from its Graph*

From each graph, estimate the domain and range of the relation.

(a) Figure 3.15

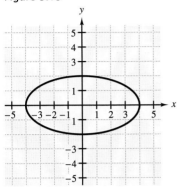

(b) Figure 3.16

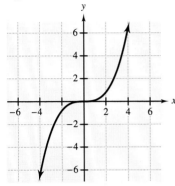

Solution

(a) The graph in Fig. 3.15 includes points whose first coordinates extend from -4 to 4, inclusive. Therefore, the domain is $\{x \mid -4 \le x \le 4\}$. The range is $\{y \mid -2 \le y \le 2\}$.

(b) The graph in Fig. 3.16 appears to be spreading out forever and extending upward and downward forever. Therefore, we judge the domain and the range to be the set of all real numbers.   ■

We will refer to the graph of a relation as a *complete graph* if all of the important features of the graph are shown. Graphs can have unexpected turns and breaks that affect the determination of the domain and range. Moreover, a "hole" in a graph is not always readily apparent in a calculator graph. For these reasons, when we wish to determine domains and ranges, we must regard graphs as suggestive, not conclusive.

Functions

A topic central to the study of mathematics involves a special kind of relation, which we will now define.

> ### Definition of Function
>
> A **function** is a relation in which no two different ordered pairs have the same first coordinate.

Because a function is a relation, we can describe it by listing its ordered pairs. Another way to describe a function is to state a rule that associates with each member of the domain *exactly one* member of the range.

A relation can be portrayed with a **mapping diagram,** and from it, we can determine whether the relation is a function.

Figure 3.17

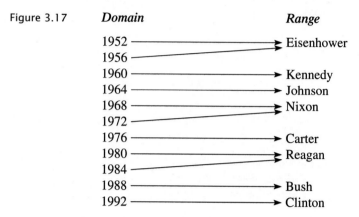

For the mapping diagram in Fig. 3.17, the domain is the set of presidential election years from 1952 to 1992. The range is the set of presidents elected in those years. The arrows show how each member of the domain is associated with *exactly one* member of the range. This relation is a function.

The following is a mapping diagram showing the major league baseball teams that won the World Series during the years 1958–1964.

Figure 3.18

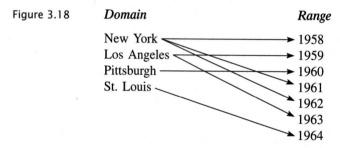

Figure 3.18 shows that two members of the domain (New York and Los Angeles) are associated with more than one member of the range. Therefore, this relation is *not* a function.

EXAMPLE 4   *Recognizing Functions*

Which of the following relations is a function?

(a) $A = \{(-2, 3), (1, 4), (4, 3), (5, -2)\}$

(b) $B = \{(1, 3), (2, 5), (-2, 5), (2, -1)\}$

(c) $C = \{(-1, 1), (-2, 2), (-3, 3), \ldots\}$

Solution

(a) Relation *A* is a function because no two ordered pairs have the same first coordinate. [There are two pairs, (−2, 3) and (4, 3), with the same second coordinate, but that does not violate the definition of a function.]

(b) Relation *B* is not a function because two pairs, (2, 5) and (2, −1), have the same first coordinate.

(c) If we assume that the indicated pattern continues, relation *C* is a function. ▪

In Example 4, we found that the relation

$$B = \{(1, 3), (2, 5), (-2, 5), (2, -1)\}$$

is not a function because two of its members, (2, 5) and (2, −1), have the same first coordinate. Plotted on a coordinate system, the point *P* representing (2, 5) lies directly above the point *Q* representing (2, −1). (See Fig. 3.19.)

Figure 3.19

Figure 3.20

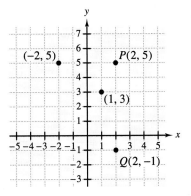
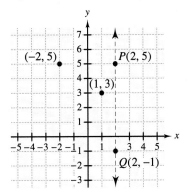

Note that the line containing points *P* and *Q* is a vertical line, as shown in Fig. 3.20. In general, if a vertical line can be drawn so that it intersects the graph of a relation in more than one point, the points of intersection have the same first coordinate. This implies that the relation does not satisfy the definition of a function.

This visual method for determining whether a relation is a function is called the **Vertical Line Test.**

The Vertical Line Test

To test the graph of a relation to determine if the relation is a function, imagine (or draw) a series of vertical lines. If any vertical line can be drawn so that it intersects the graph at more than one point, then the graph does not represent a function. If no such line can be drawn, the graph does represent a function.

EXAMPLE 5   *Using the Vertical Line Test*

Determine whether each graph represents a function.

(a) Figure 3.21

(b) Figure 3.22

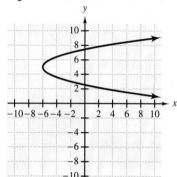

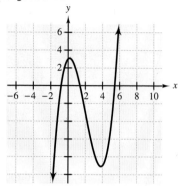

Solution

(a) There are many locations where a vertical line would intersect the graph twice. Figure 3.23 shows one example. The graph does not pass the Vertical Line Test and is not the graph of a function.

(b) Figure 3.24 shows that no vertical line can be drawn to intersect the graph more than once. The graph passes the Vertical Line Test and is the graph of a function.

Figure 3.23

Figure 3.24

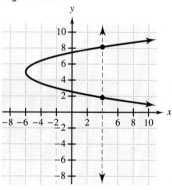

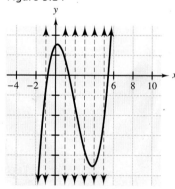

If a relation is expressed in the form of a rule, such as an equation, determining whether the relation is a function requires more careful study.

EXPLORATION 1   *Equations That Describe Functions*

Consider the following two relations.

$$G = \{(x, y) \mid x = y^2\} \qquad H = \{(x, y) \mid y = x^2\}$$

(a) Is it possible to list two pairs in set G for which the first coordinate is associated with more than one second coordinate?

(b) Is it possible to list two pairs in set H for which the first coordinate is associated with more than one second coordinate?

(c) Which of the two relations, if either, violates the definition of function?

(d) If a relation is given in the form of a rule, such as an equation, what question should we ask in order to test whether the relation is a function?

Discovery

(a) For relation G, if $x = 9$, then the equation becomes $y^2 = 9$. This equation is satisfied by two values of y, 3 and -3. Two pairs are $(9, 3)$ and $(9, -3)$.

(b) Because every real number has only one square, each x-value of relation H is associated with a unique (one and only one) y-value.

(c) Because $(9, 3)$ and $(9, -3)$ are both members of the set, relation G is not a function. Relation H is a function.

(d) If a relation is given in the form of a rule, such as an equation, the question to ask is, "Is there an x-value associated with more than one y-value?" If the answer to the question is "Yes," then the relation is *not* a function.

EXAMPLE 6 *Identifying a Function When the Rule of the Relation Is Given*

Determine whether each of the given relations is a function.

(a) $A = \{(x, y) \mid y = |x|\}$

(b) $B = \{(x, y) \mid x = |y|\}$

Solution

(a) Because every real number has exactly one absolute value, each x-value of relation A is associated with a unique (one and only one) y-value. Therefore, relation A is a function.

(b) For relation B, if $x = 5$, then the equation becomes $|y| = 5$. This equation is satisfied by two values of y, 5 and -5. Because $(5, 5)$ and $(5, -5)$ are both members of the set, relation B is not a function. ■

We have seen various ways in which relations can be described.

1. a set of ordered pairs

2. a mapping diagram

3. a graph

4. a rule or equation

Regardless of the way a relation is given, we determine whether the relation is a function in essentially the same way. For a relation to be a function, each member of the domain of the relation must be associated with exactly one member of the range.

3.2   *Quick Reference*

Relations	▪ A **relation** is a set of ordered pairs.
	▪ The **domain** of a relation is the set of first coordinates of the ordered pairs; the **range** is the set of second coordinates of the ordered pairs.

Functions	▪ A **function** is a relation in which no two different ordered pairs have the same first coordinate.
	▪ A **mapping diagram** is a method for associating the members of the domain of a relation with the members of the range and for determining whether the relation is a function.
	▪ The **Vertical Line Test** is a visual way of determining if a graph is the graph of a function. If it is possible to draw a vertical line that intersects the graph in more than one point, the graph does not represent a function. If it is not possible to draw such a line, then the graph does represent a function.
	▪ To determine if a relation expressed as an equation represents a function, determine if any replacement for x will result in more than one y-value. If so, the equation does not represent a function. If each replacement for x results in exactly one y-value, the equation does represent a function.

3.2   *Exercises*

 1. In your own words, explain the difference between a relation and a function.

 2. Explain the meaning of *domain* and *range*.

The following table shows the transition cost (in millions of dollars) for each newly elected president from 1952 to 1992.

President	Transition Cost (in millions)
Eisenhower (1952–1953)	2.1
Kennedy (1960–1961)	6.2
Nixon (1968–1969)	6.1
Carter (1976–1977)	4.7
Reagan (1980–1981)	5.2
Bush (1988–1989)	4.2
Clinton (1992–1993)	5.0

(Source: *Congressional Quarterly's Guide to the Presidency.*)

3. In the following set C, the first coordinate of each pair is the name of a president, and the second coordinate of each pair is the total of private and public funds spent for his transition.

$$C = \{(\underline{\quad}, 2.1), (\text{Nixon}, \underline{\quad}),$$
$$(\underline{\quad}, 6.2), (\underline{\quad}, 5.2),$$
$$(\text{Carter}, \underline{\quad}), (\text{Bush}, \underline{\quad}),$$
$$(\underline{\quad}, 5.0)\}$$

(a) Use the data in the table to fill in the missing coordinates.

(b) Does the relation C represent a function?

The following table shows the most popular pets owned by Americans in 1993.

Pet	Number Owned (in millions)
Cats	57
Dogs	53
Birds	11
Horses	5

(Source: American Veterinary Medical Association.)

4. In the following set P, the first coordinate is a type of pet, and the second coordinate is the number of pets owned (in millions).

$$P = \{(\text{Birds}, \underline{\quad}), (\underline{\quad}, 53),$$
$$(\underline{\quad}, 5), (\text{Cats}, \underline{\quad})\}$$

   (a) Use the data in the table to fill in the missing coordinates.

   (b) Does the relation P represent a function?

In Exercises 5–10, write a set of ordered pairs that describes each situation. Give the domain and range of each relation.

5. The first coordinate is an integer between -3 and 3, inclusive; the second coordinate is the square of the first coordinate.

6. The first coordinate is a positive integer less than 8; the second coordinate is the sum of the first coordinate and -4.

7. The first coordinate is the length of the side of a square, where the length is a positive integer not exceeding 5; the second coordinate is the perimeter of the square.

8. The first coordinate is the radius of a circle, where the radius is a positive integer not exceeding 4; the second coordinate is the area of the circle.

9. The first coordinate is the number of hours worked (10, 20, 30, 40); the second coordinate is the salary at $7 per hour.

10. The first coordinate is the number of toppings on a pizza (up to 5); the second coordinate is the price of a pizza, which is $7 plus $1 per topping.

In Exercises 11–20, sketch the graph of each relation and give the domain and range. Identify any relations that are functions.

11. $\{(1, 2), (3, 4), (5, 6), (7, 8), (9, 10)\}$

12. $\{(-1, 2), (-3, 4), (-5, 6), (-7, 8), (-9, 10)\}$

13. $\{(3, 1), (4, 1), (5, 1)\}$

14. $\{(2, 3), (5, 3), (6, 3)\}$

15. $\{(-2, 4), (-2, 5), (-2, 7)\}$

16. $\{(-1, -3), (-1, -5), (-1, -7)\}$

17. $\{(-2, 2), (-2, -3), (3, 4), (2, -3)\}$

18. $\{(4, -4), (-2, 3), (4, 3), (5, 0)\}$

19. $\{(-3, -4), (-4, -5), (-5, -6), \ldots\}$

20. $\{(-3, 4), (-4, 5), (-5, 6), \ldots\}$

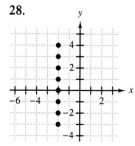 21. What feature of a complete graph of a relation helps in estimating the domain of the relation?

22. What feature of a complete graph of a relation helps in estimating the range of the relation?

23. In your own words, state the Vertical Line Test and explain the reasoning behind it.

24. Are all functions relations? Are all relations functions? Explain.

In Exercises 25–36, estimate the domain and range of the relation whose graph is given. Identify any graphs that represent functions.

25.
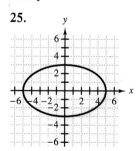

26.
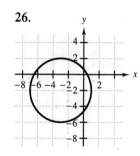

27.
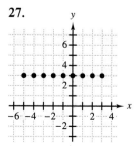

28.
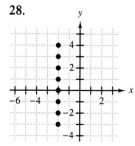

29.

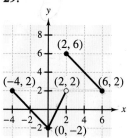

30.

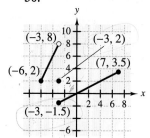

31.

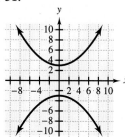

32.

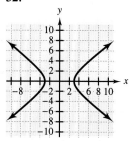

33.

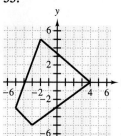

34.

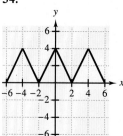

35.

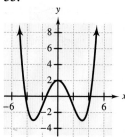

36.

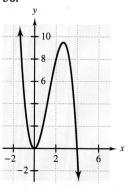

In Exercises 37–40, determine whether the mapping diagram represents a function.

37.

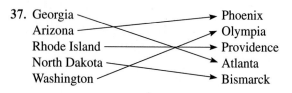

38.

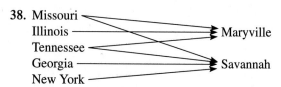

39.

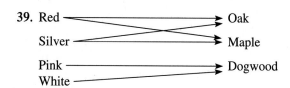

40.

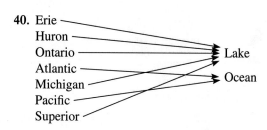

41. Using the following list, complete a mapping diagram to show that the numbers in the left column are members of the sets in the right column. Is the relation a function?

1	{1, 4, 11}
2	{5, 6}
3	{3, 12}
4	{2, 7, 9}
5	

42. Using the following list, complete a mapping diagram to show that the sets in the left column contain the numbers in the right column. Is the relation a function?

{0, 2}	5
{−3, 4}	4
{3, 5, 7}	3
{−6, 0, 2}	2

In Exercises 43–50, determine if the given relation is a function.

43. $\{(x, y) \mid x = y + 3\}$

44. $\{(x, y) \mid y = x - 5\}$

45. $\{(x, y) \mid y = (x - 4)^2\}$

46. $\{(x, y) \mid x = (y + 1)^2\}$

47. $\{(x, y) \mid x = y^3\}$ 48. $\{(x, y) \mid y = x^3\}$

49. $\{(x, y) \mid y = |2x|\}$ 50. $\{(x, y) \mid x = 3|y|\}$

Challenge

In Exercises 51–56, the variables x and y are used to represent human beings. Determine whether the given relation is a function.

51. $A = \{(x, y) \mid y \text{ is an ancestor of } x\}$

52. $B = \{(x, y) \mid x \text{ is the mother of } y\}$

53. $C = \{(x, y) \mid y \text{ is the mother of } x\}$

54. $D = \{(x, y) \mid x \text{ is a brother of } y\}$

55. $E = \{(x, y) \mid y \text{ is an aunt of } x\}$

56. $F = \{(x, y) \mid x \text{ is a grandfather of } y\}$

3.3 Graphing on a Calculator

Tracing a Graph ▪ *Magnifying a Graph*

Tracing a Graph

Consider the function $F = \{(x, y) \mid y = x + 3\}$. Some of the ordered pairs in F are (1, 4), (−2, 1), (0, 3), (−7, −4), and (−3, 0). Of course, there are many other pairs of numbers in F whose coordinates are not integers. A few such pairs are (2.58, 5.58), $(\frac{1}{2}, 3\frac{1}{2})$, and (1000.4, 1003.4).

Because function F is an infinite set, it is not possible to plot every point of the graph. A compromise approach to graphing the function is to plot enough points to establish a pattern, assume that all other points would fit the pattern, and then draw a smooth line or curve to represent the graph.

Another approach to graphing the function is to take advantage of a graphing calculator's ability to plot many points very quickly. It is important to remember, however, that a calculator is no more capable of plotting infinitely many points than we are. Therefore, what we see in the viewing rectangle is only the set of points plotted by the calculator, not all of the points of the graph. Nevertheless, we will refer to the result as a *graph.**

In the following exploration, we will learn how to produce a graph with a calculator and how to read the coordinates of the plotted points.

* Calculator models vary in the resolution with which graphs are displayed, and graphs may have a jagged appearance. For improved readability we use smooth curves in our figures.

EXPLORATION 1   *Producing and Tracing a Calculator Graph*

For the purposes of this exploration, use your calculator's integer setting.

ENTER FUNCTION   Suppose we wish to produce the graph of the function $y = x + 15$ on a calculator. We must begin by entering the function.

GRAPH   Next, produce the graph. (See Fig. 3.25.)

Figure 3.25

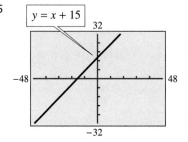

TRACE   The calculator's trace feature allows us to move a *tracing* cursor along the graph and read the coordinates of the points that have been plotted.

(a) Trace the graph and compare the y-coordinates with the corresponding x-coordinates.

(b) Trace to the point where the graph crosses the y-axis. What are the coordinates? What are the coordinates of the point where the graph crosses the x-axis?

Discovery

(a) Figure 3.26 shows the tracing cursor at $(6, 21)$. For this point and for every other point of the graph, the y-coordinate is 15 greater than the corresponding x-coordinate, that is, each point represents a solution of the equation $y = x + 15$.

(b) The graph crosses the y-axis at $(0, 15)$ and the x-axis at $(-15, 0)$. (See Figs. 3.27 and 3.28.)

Figure 3.26

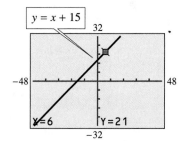

Figure 3.27

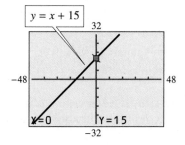

Figure 3.28

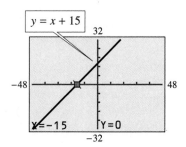

In Exploration 1 we were able to move the cursor to the points where the graph crosses the two axes because the *x*-coordinates of those points are integers. As we will see in Exploration 2, we cannot always trace to the desired point.

EXPLORATION 2   *A Calculator Does Not Plot Every Point*

Use the integer setting to produce the graph of $y = -2x - 11$. (Be sure to erase the old graph.) Now trace to the point where the graph crosses the *x*-axis. What do you observe?

Discovery

We know that the point at which the graph crosses the *x*-axis must have a *y*-coordinate of 0. As we trace near the *x*-axis, the *y*-coordinates jump between -1 and 1. (See Figs. 3.29 and 3.30.)

Figure 3.29

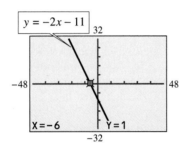

Figure 3.30

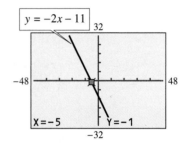

No plotted point has a *y*-coordinate of 0.

We conclude from Exploration 2 that the point where the graph crosses the *x*-axis does not have integer coordinates. We must redefine the viewing rectangle in order to locate the point.

Magnifying a Graph

ZOOM IN

If we are close to a point, but not exactly on the point, we can use a calculator feature that allows us to magnify the graph at that particular point.

EXPLORATION 3   *Zoom and Trace*

Return to the graph of $y = -2x - 11$. Trace as close as possible to the point where the graph crosses the *x*-axis and note the *y*-coordinate. Then zoom in, trace as close as possible to the point, and observe the *y*-coordinate.

Is the *y*-coordinate closer to 0 than it was before? Are you ready to estimate the *x*-coordinate? Repeat the process a few times.

Figure 3.31

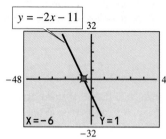

Discovery

In Fig. 3.31, we see an initial display of the graph.

Figures 3.32 and 3.33 show the graphs after the first and second zoom-and-trace cycles. Your graphs may look slightly different.

Figure 3.32

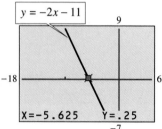

Figure 3.33

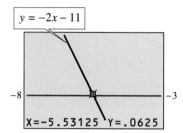

Each time we repeat the zoom-and-trace process, the *y*-coordinate becomes closer and closer to 0. The *x*-coordinate appears to be nearing -5.5.

NOTE: During the zoom-and-trace process, as a coordinate becomes closer to 0, the value may eventually be displayed in scientific notation. ■

We can use zoom and trace with either the default or integer definition of the viewing rectangle. We usually use the integer setting if we need only integer information, and we use the default setting for points whose coordinates are not integers.

EXAMPLE 1   *Graphing the Function $y = 0.1x^2$*

(a) Produce the graph of $H = \{(x, y) \mid y = 0.1x^2\}$ with the integer setting.

(b) What does the graph suggest about the domain and range?

(c) Trace to a point with an *x*-coordinate of -15. What is the corresponding *y*-coordinate?

(d) Trace to a point with a *y*-coordinate of 14.4. What is the corresponding *x*-coordinate?

(e) Trace to a point with a *y*-coordinate of -4. Explain this situation.

(f) Trace to a point whose *x*-coordinate is 7.5. What is the corresponding *y*-coordinate?

Figure 3.34

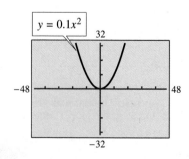

Solution

(a) Observe that the Vertical Line Test confirms that set H is a function. (See Fig. 3.34.)

(b) Because the graph in Fig. 3.34 spreads out indefinitely to the right and left, the domain of H appears to be the set of all real numbers. For the function $y = 0.1x^2$, we can replace *x* with any real number.

Figure 3.35

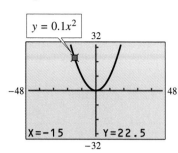

The lowest point of the graph appears to be the origin, but the graph extends upward forever. Therefore, the range of H is $\{y \mid y \geq 0\}$. We confirm this by noting that the square of any real number is nonnegative and, because $y = 0.1x^2$, $y \geq 0$.

(c) The point whose x-coordinate is -15 appears to be $(-15, 22.5)$. (See Fig. 3.35.)

We can verify by substitution.

$$y = 0.1x^2$$
$$22.5 = 0.1(-15)^2$$
$$22.5 = 22.5$$

(d) There are two points with a y-coordinate of 14.4. They are $(-12, 14.4)$ and $(12, 14.4)$. (See Figs. 3.36 and 3.37.)

Figure 3.36

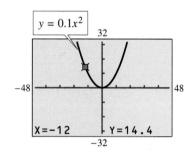

Figure 3.37

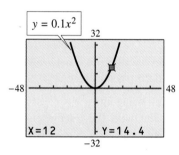

We can verify these points by substitution.

$$y = 0.1x^2 \qquad\qquad y = 0.1x^2$$
$$14.4 = 0.1(-12)^2 \qquad 14.4 = 0.1(12)^2$$
$$14.4 = 14.4 \qquad\qquad 14.4 = 14.4$$

(e) Because the range of H is $\{y \mid y \geq 0\}$, there is no member of H with a y-coordinate of -4. It is not possible to trace to such a point on the graph.

(f) Because 7.5 is not an integer, we cannot trace to the point when the graph is displayed with the integer setting. However, we can zoom and trace to obtain improved estimates of the coordinates. After some zoom-and-trace cycles, we estimate the y-coordinate to be 5.625.

$$y = 0.1x^2$$
$$5.625 = 0.1(7.5)^2$$
$$5.625 = 5.625$$

The graph of a function is a visual representation of the function, and from it, we can learn much about the domain, the range, and the general behavior of the function. However, we cannot count on a calculator to provide perfectly accurate details about specific points of the graph. Estimated coordinates are merely suggestive and need to be confirmed algebraically. You must realize that your calculator will be an effective tool only to the extent that you balance it with a solid knowledge of the definitions, rules, and procedures of algebra.

3.3   *Quick Reference*

Tracing a Graph	▪ With a calculator, we can enter a function on the function screen and produce the graph of the function on the graphing screen.
	▪ By tracing to a point, we can read the estimated coordinates of the point.

Magnifying a Graph	▪ We can magnify a graph by zooming in at any point.
	▪ Typically, we zoom in and trace to investigate the details of a graph near a certain point and to improve the estimates of the coordinates of a point.

3.3   *Exercises*

1. Explain how to enter the function $\{(x, y) \mid y = \frac{3}{4}x - 1\}$ in your calculator.

2. Explain the difference between the general cursor and the tracing cursor.

3. The following three displays are of the same graph but with the tracing cursor in different locations.

Use the graphs to complete the following table.

x	y

(a)

(b)

(c)

Figures for 3

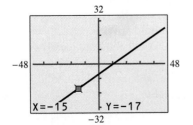

4. The following three displays are of the same graph but with the tracing cursor in different locations.

Figures for 4

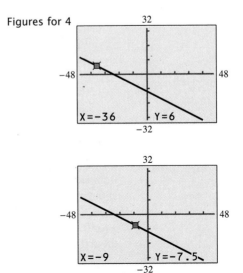

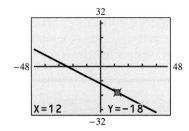

Use the graphs to complete the following table.

	x	y
(a)		
(b)		
(c)		

 5. Describe how you would use your calculator to determine the x-coordinate for $y = 23$ on the graph of $\{(x, y) \mid y = 9x + 5\}$.

 6. Describe how you would use your calculator to determine the y-coordinate for $x = 8$ on the graph of $\{(x, y) \mid y = \frac{3}{4}x - 1\}$.

In Exercises 7–14, use the integer setting to graph the given function. Then use the graph to complete the table.

7. $y = \frac{1}{3}x - 5$

x	y
	-10
	-8
	-5
	0
	4
	8

8. $y = \frac{1}{2}x + 3$

x	y
	-12
	-8
	0
	2
	10
	22

9. $y = 0.1x^2 + 0.2x - 12.3$

x	y
11	
17	
1	
-9	
-19	

10. $y = 10 + 0.2x - 0.1x^2$

x	y
-8	
-18	
0	
10	
20	

11. $y = 15 - |x + 7|$

x	y
-5	
-15	
-30	
15	
20	

12. $y = |x - 8| - 12$

x	y
	17
	27
	34
	-3
	-11

13. $y = 14 - \sqrt{3x + 102}$

x	y
	2
	8
-31	
	14
	-1

14. $y = \sqrt{8x + 256} - 17$

x	y
-32	
	-13
	-5
18	
	7

In Exercises 15–18, graph the given function on your calculator.

(a) Determine the x-coordinate corresponding to the given y-coordinate and satisfying the given condition.

(b) Determine the coordinates of the points where the graph crosses the axes.

Round decimal results to the nearest hundredth.

15. $\{(x, y) \mid y = x^2 + 2x - 3\}$; $y = 2$ and $x < 0$

16. $\{(x, y) \mid y = x^2 - 2x - 8\}$; $y = -3$ and $x > 0$

17. $\{(x, y) \mid y = 8 - 2x - x^2\}$; $y = 3$ and $x > 0$

18. $\{(x, y) \mid y = 3 + 2x - x^2\}$; $y = -4$ and $x < 0$

19. In the accompanying figure, the graph of one of the following equations is given. Identify the equation.

$$y = -x + 5 \qquad y = x - 5 \qquad y = \tfrac{1}{2}x - 2$$

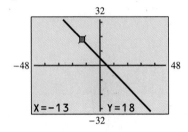

20. In the accompanying figure, the graph of one of the following equations is given. Identify the equation.

$$y = x + 2$$
$$y = \frac{3}{4}x - 4 \qquad\qquad y = -\frac{3}{4}x + 4$$

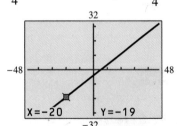

In Exercises 21 and 22, use the integer setting to produce the graph of the given function. Then trace the graph to determine two *x*-coordinates corresponding to the given *y*-value.

21. $y = 9 - 1.2x - 0.1x^2$; $y = -13$

22. $y = 9 - 1.2x - 0.1x^2$; $y = 11$

23. The graph of $y = 0.1x^3 - 0.8x^2 - 2.8x + 14$ has three points whose *y*-coordinate is 6. Use the integer setting to display the graph on your calculator and determine the *x*-coordinates of the three points.

24. The graph of $y = 1.2x^2 + 0.4x - 0.1x^3 - 12$ has three points whose *y*-coordinate is 12. Use the integer setting to display the graph on your calculator and determine the *x*-coordinates of the three points.

 25. Explain how you would decide whether to graph with the integer setting or the default setting.

 26. Explain the difference in the displayed coordinates when you trace the graph of $y = x - 2$ with an integer setting and with a default setting.

27. If you graph $y = x + 3$ with an integer setting, which of the following would not be displayed points of the graph?

   (a) (0, 3)   (b) (3.5, 6.5)

   (c) (−3, 0)   (d) (5, 2)

28. If you graph $y = 4 - 2x$ with an integer setting, which of the following would not be displayed points of the graph?

   (a) (1.5, 1)   (b) (2, 0)

   (c) (8, −2)   (d) (0, 4)

In Exercises 29–32, zoom and trace to estimate the two *x*-coordinates where the graph of the given function crosses the *x*-axis. Round your answers to the nearest hundredth.

29. $y = x^2 + 3x - 5$ 30. $y = x^2 + 4x - 7$

31. $y = 2x + 5 - x^2$ 32. $y = 3x + 7 - x^2$

 33. Explain how to estimate the domain and range of a function by inspection of its graph.

 34. What can be said about the graph of a function whose domain is the set of all real numbers?

In Exercises 35–40, use the displayed graph to estimate the domain and range of the function.

35.

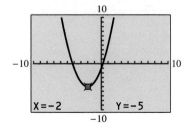

36.

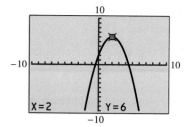

37.

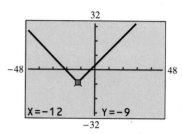

38.

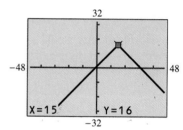

39.

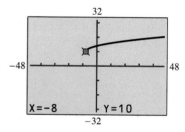

40.

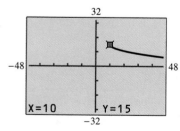

In Exercises 41–44, use the integer setting to graph each function. Then trace to estimate the domain and range.

41. $y = 12 - \sqrt{7x + 28}$

42. $y = 6 + \sqrt{8x - 56}$

43. $y = 12 - |x - 8|$ **44.** $y = 7 + |x + 9|$

45. Suppose your calculator displays 6.1538E-5. When this is converted to a decimal number, how many zeros will there be between the decimal point and the first nonzero digit?

46. Suppose your calculator displays 2.3164E-7. When this is converted to a decimal number, how many zeros will there be between the decimal point and the first nonzero digit?

47. What decimal number is a calculator giving when it displays the following?

(a) 5.6239E-4 (b) 7.1812E-2 (c) 1.1111E-3

48. What decimal number is a calculator giving when it displays the following?

(a) 7.8532E-5 (b) 8.3421E-3 (c) 2.2222E-4

 49. Use your calculator's integer setting to graph $y = \sqrt{4x + 100} + 5$.

(a) What happens if you try to trace to a point whose x-coordinate is -26? Why?

(b) What are the coordinates of the endpoint of the graph?

50. Use the integer setting to graph both of the following functions on the same coordinate system.

$$y = 0.5x + 12 \qquad y = -x - 9$$

(a) What are the coordinates of the point at which the graphs intersect?

(b) Using the word *solution,* interpret the meaning of this ordered pair.

 DOT MODE   A function can be entered in the calculator and graphed for a specified domain.

In Exercises 51–54, use the integer setting and dot mode to graph the function $y = x + 17$ for the specified domain. Then trace to the endpoint(s) of the graph and verify the x-coordinate(s).

51. $x \geq -25$ **52.** $x \leq 6$

53. $-23 \leq x \leq -8$ **54.** $x \leq -15$ or $x \geq -5$

PIECE-WISE   A **piece-wise function** is described by two or more rules, each for a specified domain.

In Exercises 55 and 56, graph the given piece-wise function and trace to the endpoints to verify the *x*-coordinates.

55. $y = \begin{cases} x + 17, & x \le -15 \\ -x + 8, & x > -15 \end{cases}$

56. $y = \begin{cases} x + 17, & x \le -15 \\ -x + 8, & -15 < x \le 20 \end{cases}$

Exploring with Real Data

In Exercises 57–60, refer to the following background information and data.

The accompanying figure shows the number of countries in which McDonald's restaurants were established during the period 1980–1990. (Source: McDonald's Corporation.) Letting *x* represent the number of years since 1980, an equation that models the data in the figure is

$$y = 2.8x + 26.4$$

where *y* is the number of countries.

Figure for 57–60

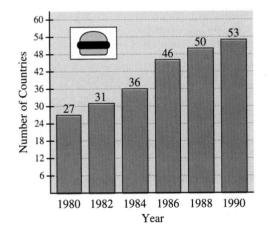

57. Select suitable range and scale settings and graph the equation.

58. By tracing the graph, complete the following table.

x	*y*	*Difference Between Actual and Estimated y*
0		
2		
4		
6		
8		
10		

59. Determine the average of the six entries in the Difference column in Exercise 58. What does the result suggest about the accuracy of the equation as a model of the data?

60. Name one factor that would prevent the trend indicated in the figure from continuing indefinitely.

Challenge

In Exercises 61–66, use the integer setting to graph each function. Then write the set of *x*-values for which

    (a) $y \le 0$.

    (b) $y \ge 0$.

61. $y = x - 3$

62. $y = x + 4$

63. $y = x^2 - 9$

64. $y = 9 - x^2$

65. $y = \sqrt{x}$

66. $y = |x| + 1$

3.4 Functions: Notation and Evaluation

Function Notation ▪ Evaluating Functions ▪ Evaluating Functions with a Calculator

Function Notation

When a relation is described by an algebraic rule, set notation is frequently omitted.

The relation	*can be written*
$g = \{(x, y) \mid x = y^2\}$	$x = y^2$
$f = \{(x, y) \mid y = x^2\}$	$y = x^2$

When the relation is a function, **function notation** is used. The function $f = \{(x, y) \mid y = x^2\}$ is written $f(x) = x^2$.

The notation $f(x)$ is read "*f* of *x*." It indicates that the name of the function is f, that the variable used for the first coordinate is x, and that the second coordinate is determined by the expression x^2. Any letter may be used for the name of the function or for the variables.

NOTE: Note that $f(x)$ does *not* mean "*f* times *x*." The symbol f is used simply to name the function. ■

Compare the rule for the function f with its function notation.

$$y = x^2$$
$$f(x) = x^2$$

From this, we can see that the symbols $f(x)$ and y refer to the same thing, namely, the second coordinate of the ordered pair. Thus, f is the name of the function, whereas $f(x)$ is the second coordinate of the ordered pair whose first coordinate is x.

Evaluating Functions

In the previous section, we used a calculator to trace the graph of a function to determine the y-coordinate corresponding to a particular x-coordinate.

We can use the rule of a function to determine any y-coordinate without reference to a graph. For the function $y = x^2$, to determine the value of y when $x = 4$, we replace x with 4 to obtain $y = (4)^2 = 16$.

The process of replacing x with a number and determining y is called **evaluating** the function.

With function notation, to ask for the value of the function when $x = 4$, we write

"For $f(x) = x^2$, what is $f(4)$?"

The symbol $f(4)$ means the value of the second coordinate when the first coordinate is 4. We can determine the value $f(4)$ by replacing the first coordinate with 4.

$$f(4) = (4)^2 = 16$$

EXAMPLE 1   *Evaluating a Function*

Let $g(x) = x^2 - 7$. Evaluate each of the following.

(a) $g(3)$ (b) $g(-2)$ (c) $g(6.3)$

(d) Verify that the graph of function g contains each of the points in parts (a)–(c).

Solution

(a) $g(3) = 3^2 - 7 = 9 - 7 = 2$

(b) $g(-2) = (-2)^2 - 7 = 4 - 7 = -3$

(c) $g(6.3) = (6.3)^2 - 7 = 39.69 - 7 = 32.69$

(d) The points in parts (a)–(c) are (3, 2), (−2, −3), and (6.3, 32.69). For the first two points, we can use the integer setting to graph $y = x^2 - 7$ and trace to the exact locations of the points. (See Figs. 3.38 and 3.39.)

Figure 3.38

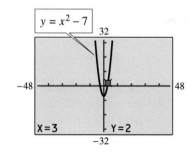

Figure 3.39

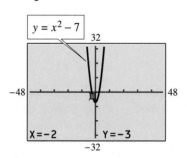

Figure 3.40

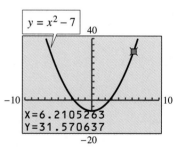

For (6.3, 32.69) we redefine the viewing rectangle and trace as close as possible to the point. (See Fig. 3.40.) Then we perform enough zoom-and-trace cycles to satisfy ourselves that (6.3, 32.69) is a point of the graph. ∎

EXAMPLE 2   *Evaluating Functions*

Evaluate each function for the given value. Use your calculator to verify your results.

(a) $f(x) = -x^2 - x - 1; f(-1)$

(b) $g(x) = 4 - 2(3 - 3x); g(1)$

Solution

(a) $f(x) = -x^2 - x - 1$ Remember that $-x^2 = -1x^2$.
$$f(-1) = -1 \cdot (-1)^2 - 1(-1) - 1$$
$$= -1 \cdot 1 - 1(-1) - 1$$
$$= -1 + 1 - 1$$
$$= -1$$

(b) $g(x) = 4 - 2(3 - 3x)$
$$g(1) = 4 - 2(3 - 3 \cdot 1)$$
$$= 4 - 2(3 - 3)$$
$$= 4 - 2(0)$$
$$= 4$$

Evaluating Functions with a Calculator

In Example 2 we used a calculator to evaluate a function in the same way that we used it to evaluate an algebraic expression. We store the value of the variable and then enter the expression on the home screen.

EVALUATE Y

An alternative method is to enter the function on the function screen and evaluate the corresponding value of *y*. This method is often more efficient.

If we are using the calculator to graph the function, then we have already entered the function on the function screen. If we are evaluating the function for more than one value of the variable, this method eliminates the need to enter the expression repeatedly.

EXAMPLE 3   *Evaluating a Function with a Calculator*

Let $f(x) = \sqrt{x - 5}$. Evaluate each of the following.

(a) $f(30)$ (b) $f(20)$ (c) $f(1)$

(d) Produce the graph of f to explain the result in part (c).

Solution

(a) Enter function f on the function screen as shown in Fig. 3.41. Note the use of parentheses when entering the function. Without them, the calculator will interpret the function as $\sqrt{x} - 5$.

To evaluate $f(30)$, store 30 in x and evaluate y. (See Fig. 3.42.) From the display, we see that $f(30) = 5$.

Figure 3.41   Figure 3.42

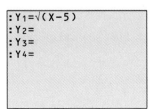

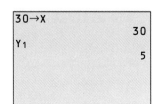

Figure 3.43

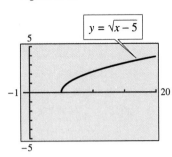

(b) To evaluate $f(20)$, store 20 in x and evaluate y. The result is approximately 3.87.

(c) Your calculator reports an error message when you try to evaluate $f(1)$.

(d) Because function f is already entered, we can immediately produce its graph. (See Fig. 3.43.)

From the graph, we see that the domain of f appears to be $\{x \mid x \geq 5\}$. The calculator displayed an error when we tried to evaluate $f(1)$ because 1 is not in the domain of the function. Note, too, that $f(1) = \sqrt{1 - 5} = \sqrt{-4}$, which is not a real number. ■

EXAMPLE 4   *Evaluating a Function with a Calculator*

Let
$$f(x) = \frac{x^2 + x^{-2}}{5 - 2(x - 4)^3}.$$

Use your calculator to determine $f(4)$.

Solution   Figure 3.44 shows the entry of function f. Note the parentheses around the numerator and denominator. The calculator will not treat the expression as a fraction without them.

Figure 3.44   Figure 3.45

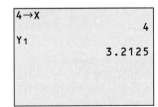

In Fig. 3.45, we store 4 in x and evaluate y. From the display, we see that $f(4) = 3.2125$. ■

The hardest part of using a calculator to evaluate a function is keying in the function. But using the calculator avoids the arithmetic that would be needed to evaluate the function by hand.

As the next two examples illustrate, a function can be evaluated for a variable expression instead of a specific value.

EXAMPLE 5   *Evaluating a Function for a Variable Expression*

Let $g(x) = \sqrt{x^2 - 4}$. Find $g(t)$ and $g(3t)$.

Solution

$g(t) = \sqrt{(t)^2 - 4} = \sqrt{t^2 - 4}$ Replace x with t.

$g(3t) = \sqrt{(3t)^2 - 4} = \sqrt{9t^2 - 4}$ Replace x with $3t$.   ■

EXAMPLE 6   *Operations with Functions*

Let $f(x) = 3x - 4$. Determine each of the following.

(a) $f(5)$ (b) $f(5 + h)$ (c) $f(5 + h) - f(5)$ (d) $\dfrac{f(5 + h) - f(5)}{h}$

Solution

(a) $f(5) = 3(5) - 4 = 15 - 4 = 11$

(b) $f(5 + h) = 3(5 + h) - 4$ Substitute $5 + h$ for x.
$\quad\quad\quad\;\; = 15 + 3h - 4$
$\quad\quad\quad\;\; = 11 + 3h$

(c) $f(5 + h) - f(5) = (11 + 3h) - 11$ Use the results of parts (a) and (b).
$\quad\quad\quad\quad\quad\quad\;\; = 3h$

(d) $\dfrac{f(5 + h) - f(5)}{h} = \dfrac{3h}{h}$ Use the result of part (c).
$\quad\quad\quad\quad\quad\quad\;\; = 3$

3.4   *Quick Reference*

Function Notation

- The **function notation** $f(x) = x + 7$ indicates three things.

  1. The name of the function is f.
  2. The first coordinate variable is x.
  3. The second coordinate is determined by $x + 7$.

Evaluating Functions

- For the function f, the question, "What is the second coordinate when the first coordinate is 3?" can be stated more succinctly as "What is $f(3)$?"

- The process of determining the value of $f(x)$ for a given value of x is called **evaluating** the function.

- If $y = f(x)$, then (x, y) can be represented by a point of the graph of function f.

Evaluating Functions with a Calculator

- We can use a calculator to evaluate a function with methods previously discussed for evaluating algebraic expressions.

- An alternative method is to enter the function, store the value for which the function is to be evaluated, and evaluate y.

- The alternative method is better if the problem also involves graphing the function or evaluating the function for multiple values of the variable.

- When entering functions in a calculator, be sure to include grouping symbols when they are needed. The following are examples of expressions that may need to be enclosed in parentheses:

  1. fractions
  2. numerators and denominators
  3. quantities under radical signs (radicands)
  4. exponential bases with more than one symbol

3.4   *Exercises*

 1. If $f(x) = 2x + 1$, explain how to evaluate $f(3)$.

2. If $g(x) = 1 - 3x$, explain how to evaluate $g(t + 4)$.

In Exercises 3–16, fill in the blank.

3. $f(x) = 5x + 4; f(3) = 5(\underline{\hspace{0.5cm}}) + 4$

4. $h(x) = 7x - 6; h(1) = 7(\underline{\hspace{0.5cm}}) - 6$

5. $R(x) = 4 - x^2; R(2) = 4 - (\underline{\hspace{0.5cm}})^2$

6. $P(x) = 25 - x^2; P(0) = 25 - (\underline{\hspace{0.5cm}})^2$

7. $f(x) = 2x^3 - 4x + 5;$
$f(\underline{\hspace{0.5cm}}) = 2(2)^3 - 4(2) + 5$

8. $g(x) = 4x^3 + 3x^2 - 1;$
$g(\underline{\hspace{0.5cm}}) = 4(1)^3 + 3(1)^2 - 1$

9. $h(x) = \sqrt{3x + 4}; h(-1) = \sqrt{3(\underline{\hspace{0.5cm}}) + 4}$

10. $g(x) = \sqrt{5 - 4x}; g(1) = \sqrt{5 - 4(\underline{\hspace{0.5cm}})}$

11. $a(x) = |x - 5|; a(\underline{\hspace{0.5cm}}) = |1 - 5|$

12. $b(x) = |2x + 9|; b(\underline{\hspace{0.5cm}}) = |2(4.5) + 9|$

13. $g(x) = 2x - 7; g(3 + t) = 2(\underline{\hspace{0.5cm}}) - 7$

14. $f(x) = 9x + 2; f(b - 4) = 9(\underline{\hspace{0.5cm}}) + 2$

15. $h(x) = -4x^2 + 3x - 2;$
$h(\underline{\hspace{0.5cm}}) = -4a^2 + 3a - 2$

16. $P(x) = 7x^2 - 9x + 1; P(\underline{\hspace{0.5cm}}) = 7a^2 - 9a + 1$

In Exercises 17–20, let $f(x) = 7x - 2$ and $g(x) = 2x - 5$. Evaluate each of the following.

17. $g(-4)$ **18.** $f(-3)$

19. $g(3) - f(2)$ **20.** $f(0) + g(1)$

In Exercises 21–24, let $P(x) = x^2 - 3x + 5$ and $Q(x) = x^2 + 4x - 8$. Evaluate each of the following.

21. $P(-1)$

22. $Q(1)$

23. $Q(2) + P(0)$

24. $P(2) - Q(0)$

 25. Given $f(x) = -x^2$, $g(x) = x^2$, and $h(x) = (-x)^2$, determine the values in parts (a) and (b). Then answer the questions in parts (c)–(e).

(a) $f(7)$, $g(7)$, and $h(7)$

(b) $f(-7)$, $g(-7)$, and $h(-7)$

(c) Let A represent the step "multiply by -1"; let B represent the step "square." Using A and B, write the sequence to be followed in evaluating each of the three functions.

(d) Will any of the three functions have the same graph? Explain.

(e) Verify your prediction by graphing the functions.

 26. Given $f(x) = -x^3$, $g(x) = x^3$, and $h(x) = (-x)^3$, determine the values in parts (a) and (b). Then answer the questions in parts (c) and (d).

(a) $f(2)$, $g(2)$, and $h(2)$

(b) $f(-2)$, $g(-2)$, and $h(-2)$

(c) Which of the following is/are suggested by the results? Explain your answer.
  (i) For all x, $f(x) = g(x)$.
  (ii) For all x, $g(x) = h(x)$.
  (iii) For all x, $f(x) = h(x)$.

(d) Verify your prediction by graphing the functions.

In Exercises 27–30, let $g(x) = -x^2 - 3x$ and $h(x) = x^4 - x$. Evaluate each of the following.

27. $g(-1)$ **28.** $h(-1)$

29. $h(-2)$ **30.** $g(-3)$

In Exercises 31–34, let $s(x) = -x^3 - 5x + 3$ and $T(x) = x^3 + 2x^2 - 5$. Evaluate each of the following.

31. $s(1)$ **32.** $T(1)$

33. $T(-2)$ **34.** $s(-2)$

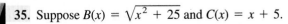 **35.** Suppose $B(x) = \sqrt{x^2 + 25}$ and $C(x) = x + 5$.

(a) Evaluate $B(0)$ and $C(0)$.

(b) Evaluate $B(3)$ and $C(3)$.

(c) Evaluate $B(-6)$ and $C(-6)$.

(d) Would you expect the two functions to have the same graph? Explain.

 36. Suppose $f(x) = \sqrt{x^2}$, $g(x) = x$, and $h(x) = |x|$.

(a) Evaluate $f(3)$, $g(3)$, and $h(3)$.

(b) Evaluate $f(-5)$, $g(-5)$, and $h(-5)$.

(c) Which two of these three functions do you think have the same graph? Why?

In Exercises 37–40, let $h(x) = \sqrt{3x + 4}$ and $g(x) = \sqrt{5 - 4x}$. Evaluate each of the following.

37. $h(-1)$ **38.** $h(4)$

39. $g(1)$ **40.** $g(-5)$

 41. Suppose $f(x) = |x + 7|$ and $g(x) = |x| + 7$.

(a) If we replace x with only positive values, would the graphs of the two functions be the same?

(b) If we replace x with any real number, would the graphs of the two functions be the same?

(c) Test your conclusions in parts (a) and (b) by producing the graphs of f and g on your calculator.

42. Suppose $A(x) = \left(\sqrt{x - 5}\right)^2$ and $B(x) = |x - 5|$.

(a) Evaluate $A(14)$ and $B(14)$.

(b) Evaluate $A(1)$ and $B(1)$.

(c) Would you expect the two functions to have the same graph?

(d) Test your conclusion in part (c) by producing the graphs of A and B.

In Exercises 43–46, let $f(x) = |2x - 1|$ and $g(x) = |3 - 2x|$. Evaluate each of the following.

43. $f(0)$ **44.** $f(-1)$

45. $g(-2)$ **46.** $g(2)$

47. Suppose $h(x) = \sqrt{3x - 6} \div 1 - x$ and $k(x) = \sqrt{3x - 6} \div (1 - x)$.

(a) Evaluate $h(5)$ and $k(5)$.

(b) Evaluate $h(14)$ and $k(14)$.

(c) Explain the role of the parentheses.

 48. Suppose $B(x) = 4x + 8 \div 2 - x$ and $C(x) = (4x + 8) \div (2 - x)$.

(a) Evaluate $B(1)$ and $C(1)$.

(b) Explain the role of the parentheses.

In Exercises 49–56, evaluate the given function as indicated. Verify the result with your calculator.

49. $f(x) = x^2 - (4 - x)$; $f(-2)$

50. $g(z) = -z^2 - (3 - z)$; $g(-2)$

51. $h(c) = 5 - 5(5 - c)$; $h(5)$

52. $f(t) = 3 - 3(8 - t)$; $f(1)$

53. $k(m) = 7(3m - 1) - (8 - m)$; $k(-2)$

54. $f(p) = 6(2p - 3) - (1 - 3p)$; $f(2)$

55. $m(s) = [4 - (s - 1)] + 8s$; $m(3)$

56. $n(t) = -[4 - (t - 2)] + 8t$; $n(3)$

 57. Suppose $G(x) = 3\{x - [x - (x - 1)]\} - 3x$.

(a) Evaluate $G(1)$, $G(-4)$, and $G(2.3)$.

(b) What is your conjecture about $G(x)$ for any value of x?

(c) Show that your conjecture is correct by simplifying the expression.

 58. Suppose $F(x) = 2[(x - 1) - 2] - 2(x - 2)$.

(a) Evaluate $F(3)$, $F(1)$, and $F(-3.7)$.

(b) What is your conjecture about $F(x)$ for any value of x?

(c) Show that your conjecture is correct by simplifying the expression.

 59. Explain how to use your calculator to evaluate a function with only the home screen.

 60. Explain how to use your calculator to evaluate a function with the function screen.

In Exercises 61–64, let $g(x) = 3x - 5$ and $f(x) = 8x - 3$. Evaluate each of the following.

61. $g(-9.3)$ **62.** $g\left(\dfrac{3}{4}\right)$

63. $f\left(\dfrac{5}{8}\right)$

64. $f(50)$

In Exercises 65–68, let $p(x) = -6x^2 - 8x + 3$ and $h(x) = -4x^2 + 3x - 5$. Evaluate each of the following.

65. $h\left(\dfrac{2}{5}\right)$

66. $h(-11.4)$

67. $p(25.7)$

68. $p\left(-\dfrac{1}{7}\right)$

In Exercises 69–72, let $f(x) = 2x^3 - 4x - 15$ and $g(x) = 4x^3 + 3x - 1$. Evaluate each of the following.

69. $f(-2.3)$

70. $f(-1.9)$

71. $g\left(\dfrac{3}{7}\right)$

72. $g\left(-\dfrac{1}{7}\right)$

73. Suppose $f(x) = 2x - 3\left[x - \sqrt{x - 1}\right]$.

(a) Determine $f(1)$ without your calculator.

(b) Insert any parentheses necessary to evaluate f with your calculator.
2 X – 3 X – √ X – 1

(c) Verify the result in part (a) by using your calculator.

74. Suppose $g(x) = (5x + 2)^2 - \sqrt{12 - x} - (x - 1)$.

(a) Determine $g(3)$ without your calculator.

(b) Insert any parentheses necessary to evaluate g with your calculator.
5 X + 2 ^ 2 – √ 12 – X – X – 1

(c) Verify the result in part (a) by using your calculator.

In Exercises 75–78, evaluate the given function as indicated.

75. $h(x) = \sqrt{3x + 4}$

(a) $h(20.8)$

(b) $h(0.9)$

76. $g(x) = \sqrt{5 - 4x}$

(a) $g(-2.4)$

(b) $g(-4.9)$

77. $k(x) = \sqrt{x^2 - 2x - 15}$

(a) $k(11.3)$

(b) $k(-4.5)$

78. $A(x) = \sqrt{2x^2 - x - 1}$

(a) $A(3.4)$

(b) $A(-5.7)$

In Exercises 79 and 80, insert any parentheses necessary to evaluate the function with your calculator. Then use your calculator to evaluate the function as indicated.

79. $K(x) = 3 - 5 \cdot |x - 3|$

(a) 3 – 5 * ABS X – 3

(b) $K(9)$

80. $n(x) = |x + 2| - |x - 2|$

(a) ABS X + 2 – ABS X – 2

(b) $n(5)$

In Exercises 81–84, evaluate each function as indicated.

81. $c(x) = |x^2 + 2x - 7|$

(a) $c(106)$

(b) $c(-89)$

82. $d(x) = |2 - x - x^2|$

(a) $d(57)$

(b) $d(-104)$

83. $f(x) = \dfrac{x + x^{-1}}{2x - x^{-2}}$

(a) $f(-1)$

(b) $f(1)$

(c) $f\left(\dfrac{5}{9}\right)$

(d) $f\left(-\dfrac{4}{9}\right)$

(e) $f(0.7)$

84. $g(x) = \dfrac{x^{-2} + x^{-1}}{3x + x^{-1}}$

(a) $g(-1)$

(b) $g(1)$

(c) $g\left(\dfrac{7}{11}\right)$

(d) $g\left(-\dfrac{5}{11}\right)$

(e) $g(-0.3)$

In Exercises 85–88, let $f(x) = 5x - 7$ and $g(x) = 6x + 2$. Evaluate each of the following.

85. $f(2a)$

86. $g(3z)$

87. $g(3 + z)$

88. $f(a + 2)$

In Exercises 89–92, let $h(x) = x^2 - 5x + 6$ and $g(x) = x^2 + 4x - 8$. Evaluate each of the following.

89. $h(t)$

90. $g(t)$

91. $g(-3t)$

92. $h(2t)$

In Exercises 93–96, let $f(x) = \sqrt{x^2 + 5}$ and $g(x) = \sqrt{6 - x^2}$. Evaluate each of the following.

93. $f(a)$

94. $g(m)$

95. $g(-3b)$

96. $f(-2a)$

In Exercises 97–100, let $f(x) = |x^3 - 2|$ and $h(x) = |x^3 + 4|$. Evaluate each of the following.

97. $f(-s)$

98. $h(-r)$

99. $h(2r)$

100. $f(3s)$

101. Let $f(x) = 2x - 5$. Evaluate f as indicated.

(a) $f(3)$

(b) $f(3 + h)$

(c) $f(3 + h) - f(3)$

(d) $\dfrac{f(3 + h) - f(3)}{h}$

102. Let $f(x) = 5 - 3x$. Evaluate f as indicated.

(a) $f(-2)$

(b) $f(-2 + h)$

(c) $f(-2 + h) - f(-2)$

(d) $\dfrac{f(-2 + h) - f(-2)}{h}$

103. Let $g(x) = 7 + 4x$. Evaluate g as indicated.

(a) $g(-8)$

(b) $g(-8 + h)$

(c) $g(-8 + h) - g(-8)$

(d) $\dfrac{g(-8 + h) - g(-8)}{h}$

104. Let $g(x) = 5x + 7$. Evaluate g as indicated.

(a) $g(4)$

(b) $g(4 + h)$

(c) $g(4 + h) - g(4)$

(d) $\dfrac{g(4 + h - g(4)}{h}$

Exploring with Real Data

In Exercises 105–108, refer to the following background information and data.

The decade of the 1980s saw a great many bank failures. The accompanying figure shows the number of bank failures during the period 1985–1992. (Source: Federal Deposit Insurance Corporation.) If we let x represent the number of years since 1985, a function that approximates the number of bank failures during the period 1985–1992 is

$$F(x) = \begin{cases} \dfrac{100}{3}x + 120 & (0 \le x \le 3). \\ -25x + 295 & (3 \le x \le 7). \end{cases}$$

Figure for 105–108

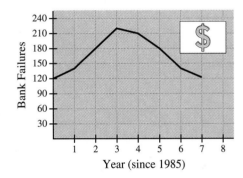

105. Select a suitable range and scale and produce the graph of function F on your calculator.

106. Trace the graph to estimate the year in which the number of bank failures reached its peak (absolute maximum of the function). Estimate the number of bank failures that year.

107. If the number of bank failures were to continue to decline according to function F, in what year would there be no bank failures? Explain whether you think function F is a reliable model for predicting the number of bank failures beyond 1992.

108. In order to "get the government off our backs," the Reagan administration deregulated the banking industry in the early 1980s. The ensuing fraud and corruption cost American taxpayers hundreds of billions of dollars. What measures were taken by the federal government to prevent another occurrence of this scandal?

Challenge

109. Suppose $f(x) = 3/x$.

(a) Evaluate each of the following.

(i) $f(1)$

(ii) $f(0.1)$

(iii) $f(0.01)$

(iv) $f(0.001)$

(v) $f(0.0001)$

(b) As the x-values become closer to zero, what is happening to the values of $f(x)$?

(c) Use a calculator to evaluate $f(0)$. Explain the result.

110. Suppose $g(x) = x^2$.

(a) Evaluate each of the following.

(i) $g(11)$

(ii) $g(111)$

(iii) $g(1111)$

(iv) $g(11,111)$

(b) Based on the pattern in part (a), predict the value of $g(11,111,111)$.

(c) Can you confirm your prediction by evaluating $g(11,111,111)$ with your calculator? Why?

3.5 Analysis of Functions

Goals of Analysis ▪ Maximums, Minimums, and Intercepts
Estimating Domain and Range ▪ Points of Intersection

Goals of Analysis

Roughly speaking, **analyzing a function** refers to learning how a function behaves. That is, to analyze a function is to ask questions like the following.

1. What are the highest and lowest points of the graph?

2. What are the domain and range of the function?

3. Where does the graph cross the x- and y-axes, if at all?

We can often estimate the answers to such questions by exploring graphs and visually observing the behavior of the function.

Maximums, Minimums, and Intercepts

In the following, we consider the flight of a thrown ball. It would be very difficult to make physical measurements of times and distances while the ball is in the air. Instead, we use a function to model the flight of the ball.

EXPLORATION 1

Analyzing the Flight Path of a Thrown Ball

A man is standing at the top of a 20-story building. He throws a ball upward and outward. (See Fig. 3.46.)

Figure 3.46

The flight of the ball is described by the following function where h is the height of the ball (in feet) from the ground after x seconds.

$$h(x) = -16x^2 + 100x + 200$$

Display the graph of the function and use it to estimate the answers to the following questions.

(a) From what height did the man release the ball?

(b) How high does the ball go?

(c) How long does it take for the ball to reach the ground?

(d) After how many seconds is the ball 300 feet from the ground?

Discovery

Because h represents height and x represents time, we restrict both variables to nonnegative values. Figure 3.47 shows the graph of function h.

Figure 3.47

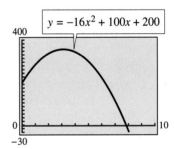

(a) The man releases the ball when $x = 0$. In Fig. 3.48, we trace to the point whose x-coordinate is 0. The corresponding y-coordinate is approximately 200 (feet).

Figure 3.48

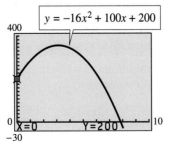

Figure 3.49

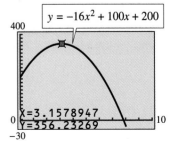

(b) In Fig. 3.49, we trace to the highest point of the graph. The y-coordinate is approximately 356 (feet).

Figure 3.50

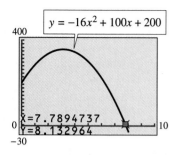

(c) When the ball reaches the ground, the height y is zero. In Fig. 3.50, we trace to the point whose y-coordinate is 0. The corresponding x-coordinate is approximately 7.8 (seconds).

(d) In Figs. 3.51 and 3.52, we trace to the point whose y-coordinate is 300 and estimate the corresponding x-coordinate. We find that there are actually two such points. On the way up, the ball reaches 300 feet in approximately 1.3 seconds. As the ball falls, the height is 300 feet after approximately 5 seconds.

Figure 3.51

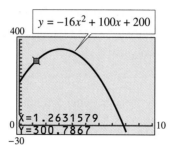

Figure 3.52

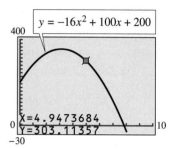

In Exploration 1, the maximum height of the ball corresponds to the y-coordinate of the highest point of the graph. This y-coordinate is called the **absolute maximum** of the function.

Similarly, the y-coordinate of the lowest point of a graph is called the **absolute minimum** of the function.

We call the point where the graph intersects the x-axis the **x-intercept.** At this point, $y = 0$. In Exploration 1 the x-intercept corresponds to the time when the height is 0, that is, the time when the ball reaches the ground.

Similarly, we call the point where the graph crosses the y-axis the **y-intercept.** At this point, $x = 0$. In Exploration 1 the y-intercept corresponds to the height when the time is 0, that is, the height at which the man released the ball.

Note that the domain of the function consists of time values between 0 seconds and approximately 7.8 seconds. The range of the function consists of all height values between 0 feet and approximately 356 feet.

Summary of Definitions

The **absolute maximum** of a function is the y-coordinate of the highest point of the graph of the function.

The **absolute minimum** of a function is the y-coordinate of the lowest point of the graph of the function.

An **x-intercept** of a graph is a point at which the graph intersects the x-axis. At this point, $y = 0$.

A **y-intercept** of a graph is a point at which the graph intersects the y-axis. At this point, $x = 0$.

EXPLORATION 2   *More on Maximums and Minimums*

Using the default viewing rectangle, graph the function

$$f(x) = -3x^4 + 4x^3 + 12x^2 - 10.$$

(a) We have said that a graph is a complete graph if all of the important features of the graph are visible on the viewing screen. Experiment with range and scale adjustments until you obtain a complete graph.

(b) Estimate the absolute maximum and the absolute minimum of the function.

(c) Note the *relatively* high point of the graph in Quadrant III and the *relatively* low point of the graph near the *y*-axis. Estimate the coordinates of these two points.

Discovery

(a) The graph of function *f* appears to extend downward forever on both sides. Figure 3.53 shows the important features of the graph.

Figure 3.53

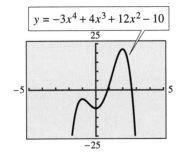

Figure 3.54

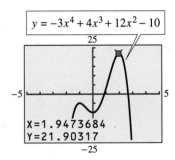

(b) In Fig. 3.54 we trace to the highest point of the graph. The point is approximately (2, 22). Therefore, the absolute maximum is approximately 22. The function has no absolute minimum.

(c) In Fig. 3.55 the coordinates of the relatively high point in Quadrant III are approximately $(-1, -5)$.

Figure 3.55

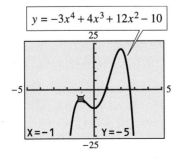

Figure 3.56

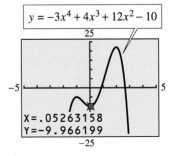

In Fig. 3.56 the point at the bottom of the dip near the *y*-axis has the approximate coordinates $(0, -10)$.

Informally, we say that the relatively high point in Quadrant III is the highest point in its *neighborhood,* and we call the y-coordinate of the point a **local maximum.** Similarly, if a point is the lowest point in the neighborhood of a dip, we call the y-coordinate of the point a **local minimum.**

When speaking of maximums and minimums, we frequently refer to the point itself as a maximum or minimum. Note that the absolute maximum is also a local maximum because it is the highest point in its neighborhood as well as the highest point of the graph.

Estimating Domain and Range

We can use estimates of absolute maximums and minimums to approximate the range of a function. We can also approximate the domain from the graph.

EXPLORATION 3   *Domain and Range*

(a) Use suitable range and scale settings to display a complete graph of $f(x) = \sqrt{x + 3}$.

(b) Estimate the absolute minimum and the absolute maximum. Use these estimates to approximate the range of the function.

(c) Estimate the least and greatest x-coordinates. Use these estimates to approximate the domain of the function.

Discovery

(a) Figure 3.57

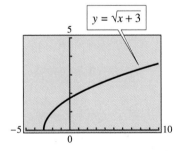

$$y = \sqrt{x + 3}$$

(b) The lowest point of the graph shown in Fig. 3.57 is the x-intercept, where the y-coordinate (absolute minimum) is 0. There is no absolute maximum. The range appears to be $\{y \mid y \geq 0\}$.

(c) The smallest x-coordinate occurs at the x-intercept and is approximately -3. There is no greatest x-coordinate. The domain appears to be $\{x \mid x \geq -3\}$.

Points of Intersection

It is frequently useful to display the graphs of two functions on the same set of axes. In most calculators we can store two functions and produce the graph of either one or both of them.

ACTIVATE

DEACTIVATE

A function whose graph is displayed is called an *active function*. A function that has been entered but whose graph is not displayed is called an *inactive function*.

EXAMPLE 1   *Two Graphs on One Set of Axes*

Let $f(x) = x^3$ and $g(x) = -x + 10$.

(a) Activate f, deactivate g, and produce the graph.

(b) Activate g, deactivate f, and produce the graph.

(c) Activate f and g and produce both graphs.

(d) Estimate the x-coordinate of the point where $f(x) = g(x)$.

Solution

(a) Figure 3.58   (b) Figure 3.59

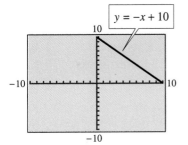

(c) Figure 3.60

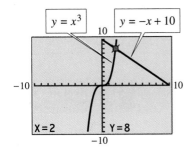

(d) Figure 3.60 shows that the point of intersection of the two graphs is the point at which $f(x) = g(x)$. Trace to the point and estimate the coordinates: $(2, 8)$.

3.5 *Quick Reference*

Goals of Analysis	▪ Analyzing a function involves learning how a function behaves.
	▪ We can use the graph of a function to estimate its high and low points, its domain and range, and the points at which the graph crosses the axes.

Maximums, Minimums, and Intercepts	▪ The **absolute maximum** of a function is the y-coordinate of the highest point of the graph; the **absolute minimum** is the y-coordinate of the lowest point of the graph.
	▪ The point at which the graph of a function crosses the x-axis is called the ***x-intercept;*** the point at which the graph crosses the y-axis is called the **y-intercept.**
	▪ Informally, we call the y-coordinate of the point that is highest in the neighborhood of a peak of a graph a **local maximum.** The y-coordinate of the point that is lowest in the neighborhood of a dip is called a **local minimum.**

Estimating Domain and Range	▪ We can use estimates of absolute maximums and minimums to approximate the range of a function.
	▪ We can use the graph of a function to estimate the domain by determining the smallest and greatest x-coordinates.

Points of Intersection	▪ We can store two or more functions in a calculator and display their graphs simultaneously.
	▪ Having stored two functions, we can activate one so that its graph will be displayed, and we can deactivate the other so that its graph will not be displayed.
	▪ By displaying the graphs of two different functions, we can estimate the coordinates of their **point(s) of intersection.**

3.5 *Exercises*

Many of the exercises in this section require estimates. Round all decimal answers to the nearest hundredth.

 1. Explain how you can estimate or determine the y-intercept of a graph

    (a) with your calculator and

    (b) algebraically.

 2. Explain how to estimate the x-intercept of a graph with your calculator.

 3. Explain why a function has at most one y-intercept.

 4. In parts (a)–(d), sketch an example of a graph having the given number of x-intercepts. Then answer the question in part (e).

    (a) 0   (b) 1

    (c) 2   (d) 3

(e) Use the definition of function to explain why the graph of a function can have more than one *x*-intercept.

In Exercises 5–8, use the integer setting to graph the given function on your calculator. Then use your graph to estimate the answers to the following questions. Verify parts (a) and (b) by substitution.

(a) What is the *x*-value for the given value of $f(x)$?

(b) What are the *x*- and *y*-intercepts?

5. $f(x) = x - 4; f(x) = 6$

6. $f(x) = 4 - x; f(x) = -3$

7. $f(x) = \frac{1}{3}x - 8; f(x) = -5$

8. $f(x) = -\frac{1}{3}x - 7; f(x) = -8$

In Exercises 9–16, use the integer setting to graph the given function. Then use your graph to estimate the values in parts (a)–(c).

9. $h(x) = \frac{2}{5}x - 9$

(a) $h(-15)$ (b) *x*, if $h(x) = -5$

(c) *x*, if $h(x) = 11$

10. $c(x) = -\frac{1}{3}x + 5$

(a) $c(-9)$ (b) *x*, if $c(x) = -5$

(c) *x*, if $c(x) = 16$

11. $g(x) = x^2 - 2x - 8$

(a) $g(7)$ (b) *x*, if $g(x) = 7$

(c) *x*, if $g(x) = -10$

12. $k(x) = -x^2 - x + 56$

(a) $k(-2)$ (b) *x*, if $k(x) = 14$

(c) *x*, if $k(x) = 50$

13. $f(x) = 10 - \sqrt{8x}$

(a) $f(8)$ (b) *x*, if $f(x) = -2$

(c) *x*, if $f(x) = 10$

14. $h(x) = \sqrt{12x} + 3$

(a) $h(12)$ (b) *x*, if $h(x) = 0$

(c) *x*, if $h(x) = 3$

15. $f(x) = |x| + 5$

(a) $f(6)$ (b) *x*, if $f(x) = 19$

(c) *x*, if $f(x) = 4$

16. $p(x) = 15 - |2x|$

(a) $p(17)$ (b) *x*, if $p(x) = 7$

(c) *x*, if $p(x) = 16$

 17. Explain how part (a) in Exercises 9–16 could have been done without reference to the graph of the function.

18. Suppose $g(x) = 7x - 2$.

(a) What is the easiest method for calculating $g(5)$?

(b) What is our present method for estimating the value of *x* for which $g(x) = 5$?

In Exercises 19–24, use the integer setting to graph the given function on your calculator. Trace the graph to estimate the function's absolute maximum, if any, and the function's absolute minimum, if any.

19. $f(x) = x^2 + 10x + 25$

20. $g(x) = -x^2 + 8x - 20$

21. $h(x) = 9 - |x|$ 22. $m(x) = |x| - 9$

23. $p(x) = 6 + \sqrt{10x}$ 24. $r(x) = 6 - \sqrt{10x}$

In Exercises 25–28, use the default setting to graph the given function on your calculator. Then use your graph to estimate each of the following.

(a) The *x*-value for the given value of $f(x)$.

(b) The *x*- and *y*-intercepts.

25. $f(x) = 1.6x - 5.9; f(x) = 3$

26. $g(x) = 7 - 0.8x; g(x) = 2$

27. $h(x) = \frac{1}{2} - 3x; h(x) = -3.6$

28. $q(x) = \frac{1}{2}x - 3; q(x) = -3.6$

In Exercises 29–32, use the default setting to graph the given function. Then use your graph to estimate the values in parts (a)–(c).

29. $f(x) = 0.95x$

  (a) $f(7)$ (b) x, if $f(x) = 3.1$

  (c) x, if $f(x) = -4.5$

30. $g(x) = x + 4 - 0.2x^2$

  (a) $g(-6)$ (b) x, if $g(x) = 2.8$

  (c) x, if $g(x) = 6$

31. $h(x) = x^3 - 5$

  (a) $h(0.95)$ (b) x, if $h(x) = 5.8$

  (c) x, if $h(x) = -5.4$

32. $n(x) = \sqrt{5 - 2x}$

  (a) $n(4)$ (b) x, if $n(x) = -1$

  (c) x, if $n(x) = 2.7$

In Exercises 33–36, use the default setting to graph the given function on your calculator. Trace the graph to estimate the function's absolute maximum, if any, and the function's absolute minimum, if any.

33. $f(x) = 0.7x^2 + 3x$ **34.** $g(x) = 4.5x - x^2$

35. $c(x) = -\sqrt{3x^2}$ **36.** $h(x) = |x^2 - 5|$

37. Explain how it is possible that a local maximum of a function can also be the absolute maximum.

38. Explain how it is possible that a local minimum of a function can also be the absolute minimum.

For Exercises 39–42, graph the given function on your calculator. Then zoom and trace to estimate

  (a) the coordinates of the local minimum of the graph in Quadrant IV.

  (b) the coordinates of the local maximum of the graph in Quadrant II.

39. $f(x) = x^3 - 2x$

40. $h(x) = x^3 - 7x$

41. $g(x) = x^3 - 3x^2 - 5x + 6$

42. $h(x) = x^3 + 2x^2 - 5x - 6$

In Exercises 43 and 44, use the default setting to graph the given function. For each exercise, perform the following.

  (a) Change the range so that the local minimum or maximum point is visible.

  (b) Estimate the coordinates of the point representing the local minimum or maximum.

43. $f(x) = x^3 - x^2 - 9x + 9$

44. $f(x) = x^3 - x^2 - 10x - 8$

45. Suppose you want to know only how *many* x-intercepts the graph of a function has. Describe an easy way to determine this.

46. Explain how you can determine the number of x-intercepts of $f(x) = -(2 + |x|)$. How many are there?

In Exercises 47 and 48, which members of the set $\{1, 3, -2\}$ are members of the domain of the given function.

47. $f(x) = \dfrac{1}{x + 2}$ **48.** $f(x) = \sqrt{2 - x}$

49. Suppose function f has both an absolute minimum and an absolute maximum. Assuming there are no breaks in the graph of f, explain how these numbers can be used to state the range of f.

50. Suppose the range of function g is $\{y \mid y \geq 3\}$. What can you conclude about the absolute minimum and absolute maximum, if any, of function g?

In Exercises 51–54, sketch an example of a graph that has the given characteristics.

51. The graph has an absolute maximum and an absolute minimum.

52. The graph has an absolute maximum but no absolute minimum.

53. The graph has an absolute minimum but no absolute maximum.

54. The graph has no absolute maximum and no absolute minimum.

In Exercises 55 and 56, sketch an example of a graph that meets the given conditions.

55. The graph has three x-intercepts: $(-3, 0)$, $(1, 0)$, and $(5, 0)$.

The graph has local maximums at $(-1, 4)$ and at $(3, 7)$.

The graph has a local minimum at $(1, 0)$.

56. The graph has an absolute minimum at $(-3, -5)$.

The graph has one x-intercept: the origin.

The graph has a local maximum at $(2, 4)$.

The graph has a local minimum at $(5, 3)$.

 57. For function f, $f(1) = 5$ and $f(3) = -2$. Explain how you can conclude that there is at least one x-intercept $(a, 0)$ such that $1 \leq a \leq 3$. (Assume that the graph of function f is a smooth curve with no breaks in it.)

 58. Suppose function g has a local maximum at $(-1, 7)$ and another local maximum at $(1, 5)$. If the y-intercept is $(0, b)$, can we conclude that $5 \leq b \leq 7$? Sketch an example of a graph that supports your answer.

59. At the local hardware store, the cash register is programmed to add a 5% sales tax to the price x of all goods purchased. The total cost C is then computed as follows.

$$C(x) = x + 0.05x = 1.05x$$

Use the integer setting to graph function C on your calculator. Then answer the following questions.

(a) Describe the apparent shape of the graph.

(b) What is $C(26)$?

(c) What is $C(11)$?

(d) What do your results in parts (b) and (c) represent?

(e) Calculate $\dfrac{C(26) - C(11)}{26 - 11}$.

(f) Compare your result in part (e) to $C(x) = 1.05x$. What do you observe?

In Exercises 60 and 61, produce a complete graph of the given function on your calculator. Then answer the following questions.

(a) How many x- and y-intercepts are there?

(b) How many local minimum and maximum points are there?

(c) What are your estimates of the domain and range?

(d) What are the approximate coordinates of the x-intercept that is the farthest to the right?

60. $f(x) = x^4 + 3x^3 - 2x^2 - 5x + 3$

61. $g(x) = 4x^2 - 2x^3 - x^4 + 3x - 5$

 62. Explain why -5 is not in the domain of $y = g(x) = \sqrt{2x + 3}$.

63. Explain why -2 is not in the range of $y = h(x) = |x + 2|$.

In Exercises 64–67, use the integer setting to graph the given function on your calculator. Then answer the following questions.

(a) What are the x- and y-intercepts?

(b) What is the absolute minimum and the absolute maximum?

(c) What are your estimates of the domain and range?

64. $f(x) = |x + 2| + 3$ **65.** $g(x) = 5 - |x - 2|$

66. $h(x) = \sqrt{x - 1}$ **67.** $h(x) = 5 - \sqrt{x + 4}$

 68. Explain why function f has an absolute minimum but function g does not. What is the absolute minimum of function f?

$f(x) = x + 1$ Domain: $\{x \mid x \geq 0\}$

$g(x) = x + 1$ Domain: $\{\text{all real numbers}\}$

69. Explain why function f has an absolute maximum but function g does not. What is the absolute maximum of function f?

$f(x) = x^2$ Domain: $\{x \mid 0 \leq x \leq 1\}$

$g(x) = x^2$ Domain: $\{\text{all real numbers}\}$

In Exercises 70–73, use the integer setting to graph the two given functions on the same coordinate axes. Then name the graphs to which the given points belong.

70. $f(x) = x + 1$; $g(x) = 2x - 3$

    (a) $(3, 4)$ (b) $(0, -3)$ (c) $(4, 5)$

71. $f(x) = \dfrac{1}{2}x - 2$; $g(x) = -x + 1$

    (a) $(-5, 6)$ (b) $(6, 1)$ (c) $(2, -1)$

72. $f(x) = \dfrac{1}{2}x$; $g(x) = |x|$

    (a) $(10, 5)$ (b) $(10, 10)$ (c) $(0, 0)$

73. $f(x) = \dfrac{1}{10}x^2$; $g(x) = 2x - 10$

    (a) $(8, 6)$ (b) $(8, 6.4)$ (c) $(10, 10)$

74. Visual Graphics, Inc., is designing an advertisement that will be placed on a rectangular board. The customer has specified that the length L (in feet) must be twice the width W, and the area and the perimeter of the board must be the same. The designer makes the following calculations.

    Area $= LW = (2W) \cdot W = 2W^2$

    Perimeter $= 2L + 2W = 2(2W) + 2W$
                   $= 4W + 2W = 6W$

Enter these two functions in your calculator and use the default setting to graph the functions on the same axes.

    (a) To which graph does $(2, 8)$ belong? Interpret the meaning of that ordered pair.

    (b) To which graph does $(2, 12)$ belong? Interpret the meaning of that ordered pair.

    (c) What is the point of intersection?

    (d) How wide and how long should the board be?

    (e) What is the area of the board and what is the perimeter?

Exploring with Real Data

In Exercises 75–78, refer to the following background information and data.

Since the Endangered Species Act of 1973, the gray wolf population in the continental United States has increased. The figure shows the estimated populations of gray wolves between 1960 and 1990. (Source: U.S. Fish and Wildlife Service.) If we let x represent the number of years since 1960, a function that approximates the population P of gray wolves during this 30-year period is

$$P(x) = 0.74x^2 + 7.75x + 800.$$

Figure for 75–78

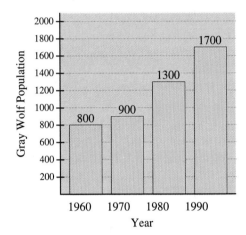

75. Select a suitable range and scale and graph function P on your calculator. During the period 1960–1990, when was the gray wolf population at its lowest (the absolute minimum of the function)? Estimate the population that year.

76. During this period, when was the gray wolf population at its highest (the absolute maximum of the function)? Estimate the population that year.

77. If the population continues to grow according to the function P, what is the projected population in the year 2000?

78. Does the shape of the graph suggest that the gray wolf population will increase indefinitely? What natural factors might change the shape of the graph in the future?

Challenge

79. (a) Draw a graph of the following piece-wise function.

$$f(x) = \begin{cases} x + 3, & x < 0 \\ 5, & x = 0 \\ -x, & x > 0 \end{cases}$$

(b) What is the absolute maximum of the function?

(c) In words, how would you describe your estimate of the range of the function?

80. Consider the **constant function** $g(x) = 5$.

(a) What are the domain and range of g?

(b) We say that a function f is an **increasing function** if $f(x_2) \geq f(x_1)$ when $x_2 > x_1$ for all x in the domain of f. Explain why function g is an increasing function.

<table>
<tr><td>**3**</td><td>## Chapter Review Exercises</td></tr>
</table>

Section 3.1

In Exercises 1–6, determine the coordinates of the points shown in the accompanying figure.

Figure for 1–6

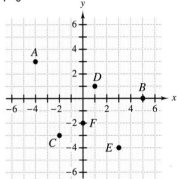

1. A **2.** B **3.** C **4.** D **5.** E **6.** F

In Exercises 7–11, state the quadrant in which the given point lies.

7. $P(-\pi, \pi)$ **8.** $Q(5, -6)$

9. $R(a, b)$ if $a < 0$ and $b > 0$

10. $S(3.25, 4.56)$

11. $T(a, b)$ if $a + b = 3$ and $a > 3$

12. On which axis is $P(c, d)$ if $c = 0$?

13. In which quadrant is $P(a, b)$ if $Q(b, a)$ is known to be in Quadrant IV?

14. Draw the graph of $\left\{ \left(\frac{1}{2}, -2\right), (-3, 5), (0, -6), (7, 0), \left(-\frac{5}{3}, -1\right) \right\}$.

In Exercises 15 and 16, determine each of the following.

(a) The distance (to the nearest hundredth) between the points A and B.

(b) The midpoint of the line segment whose endpoints are A and B.

15. $A(-5, 4)$ and $B(-8, -5)$

16. $A(7, -1)$ and $B(13, -9)$

17. Find the perimeter of the parallelogram $ABCD$ if the vertices are $A(-6, 1)$, $B(-3, 5)$, $C(9, 0)$, and $D(6, -4)$.

18. Determine if $A(-3, -2)$, $B(-1, 8)$, and $C(3, 4)$ are vertices of a right triangle.

Section 3.2

In Exercises 19–22, state the domain and range for each relation. Identify any relation that is a function.

19. $\{(0, 1), (3, 5), (-4, 7), (5, -6)\}$

20. $\{(x, y) \mid x = |y + 2|\}$

21.

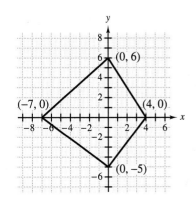

22.

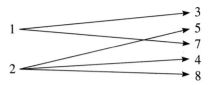

Section 3.3

23. The graph of $y = 0.2x^2 + 2x - 15$ contains the point (10, 25). Use the integer setting to display the graph on your calculator. Then determine another x-coordinate for which $y = 25$.

24. Which of the following equations is an equation of the line displayed in the accompanying figure?

$$y = 3x + 11 \qquad\qquad y = \frac{1}{3}x - 11$$

$$y = \frac{1}{4}x + 11$$

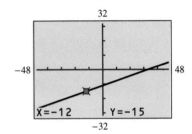

25. Produce the graph of $y = 1.7x - 3.9$. Zoom and trace to estimate the x-coordinate (to the nearest hundredth) of the point whose y-coordinate is 11.4.

Section 3.4

In Exercises 26–33, evaluate the given function as indicated.

26. $f(x) = -x^2 - 5x + 2$; $f(-2)$

27. $g(a) = a^3 - 6a$; $g(-1)$

28. $h(c) = \sqrt{c^2 + 4c}$; $h(3)$

29. $A(x) = |2x - 3|$; $A(-5)$

30. $B(r) = \dfrac{r^{-2} + r^{-3}}{5r - r^{-1}}$; $B(2)$

31. $Q(a) = 5a - 3(5a - 3) - a$; $Q(-1)$

32. $P(c) = c^3 - 3 \cdot |2c^{-1} - 5|$; $P(2)$

33. $R(d) = \sqrt{5d + 4} - d^2$; $R(-0.23)$

34. Determine $f(2 + h)$ if $f(x) = 7 - 3x$.

35. Let

$$f(x) = \frac{\sqrt{x^2 - 3} - 4}{|x^{-1} - 2|}.$$

Show how you would insert parentheses to enter f in your calculator.

$\sqrt{}$ X ∧ 2 − 3 − 4 ÷ ABS X ∧ −1 − 2

Section 3.5

In Exercises 36–39, use the integer setting to produce the graph of the given function. Use the graph to answer the questions in each part.

36. $f(x) = 20 - \sqrt{7x + 4}$

  (a) Estimate the domain and range.

  (b) What is $f(11)$?

  (c) For what value of x is $f(x) = 15$?

37. $g(x) = -0.2x^2 + 3.6x + 8$

  (a) What are the x- and y-intercepts?

  (b) What is $g(3)$?

  (c) For what value of x is $g(x) = 25$?

38. $h(x) = x + 3$;   Domain: $\{x \mid 0 \le x \le 7\}$

  (a) Determine the absolute minimum and absolute maximum of h.

  (b) What is $h(2)$?

  (c) For what value of x is $h(x) = 7$?

39. $k(x) = 0.1x^3 - 7x$

  (a) Estimate the local maximum and local minimum.

  (b) Estimate the domain and range.

  (c) What is the y-intercept?

40. Use the integer setting to produce the graphs of $f(x) = 0.75x - 1$ and $g(x) = 4.25 - x$ on the same coordinate axes. Determine the coordinates of their point(s) of intersection.

3   Chapter Test

For Questions 1–4, determine the coordinates of the points in the accompanying figure.

Figure for 1–4

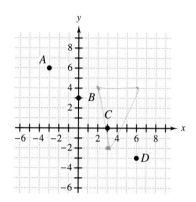

1. *A*
2. *B*
3. *C*
4. *D*

For Questions 5 and 6, use the given conditions to determine in which quadrant(s) or on what axis the point $P(a, b)$ could be located.

5. $a > 0$ 6. $Q(b, a)$ is in Quadrant IV

Use the points $A(-2, -1)$, $B(3, -2)$, and $C(2, 4)$ to answer Questions 7 and 8.

7. Determine the midpoint of the line segment $\overline{BC}$.

8. If $\overline{AB}$ is a side of a triangle, how long is that side?

For Questions 9 and 10, suppose a weather radar system can detect a tornado within 75 miles of the radar location. The radar is located 30 miles east and 20 miles north of your home and a tornado is located 40 miles west and 10 miles south of your home. (See figure.)

Figure for 9–10

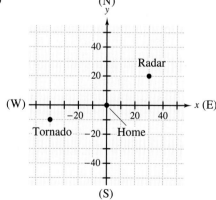

9. How far is the tornado from the radar site? Is it within range of the radar?

10. Show whether your home is on the line connecting the tornado and the radar location.

In Questions 11 and 12, determine whether the given graph represents a function and explain your answer.

11.

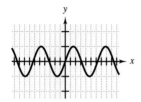

12.

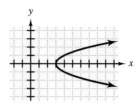

In Questions 13 and 14, determine the domain and range for the given relation.

13.

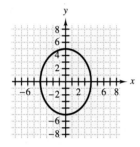

14. $\{(1, 3), (2, 7), (3, -1), (5, 3)\}$

In Questions 15–17, evaluate the function as indicated.

15. $g(x) = \dfrac{-x^{-2} + x^{-1}}{|x - 3|}$; $g(2)$

16. $f(x) = \sqrt{2x - 1}$; $f(5)$

17. $h(x) = 3x + 2$; $h(1 + t) - h(1)$

Use the integer setting to graph the function $f(x) = x^2 - 9x - 10$. Then use the graph to answer Questions 18 and 19.

18. Estimate the range of f.

19. Determine the x-intercept(s).

Use the integer setting to graph the function $g(x) = \sqrt{4 - x}$. Then use the graph to answer Questions 20 and 21.

20. For what value of x is $g(x) = 3$?

21. Estimate the domain of g.

Use the integer setting to graph the functions $f(x) = |x - 2|$ and $g(x) = -\frac{1}{2}x + 4$ on the same coordinate axes. Then use the graphs to answer Questions 22 and 23.

22. Determine the point(s) of intersection of the graphs.

23. What is the y-intercept of f?

24. In the accompanying figure, what name is given to the following points?

(a) *A*   (b) *B*   (c) *C*   (d) *D*   (e) *E*   (f) *F*

Figure for 24

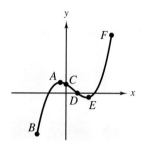

For Questions 25 and 26, use the following information.

The profit *P* per day at a theater is approximated by the function

$$P(a) = 500a - 50a^2 - 450$$

where *a* is the price of admission. Use the graph of function *P* to estimate the answers to Questions 25 and 26.

25. To the nearest whole number, what is the absolute maximum of function *P*? Interpret your answer.

26. On the graph of *P*, what is the first coordinate (to the nearest hundredth) of the point whose second coordinate is 500? Interpret your answer.

4 Linear Equations and Inequalities

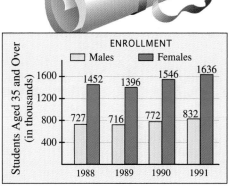

During the period 1988–1991, college and university students in the age group 35 and older represented a substantial portion of the total enrollment. However, as we can see from the bar graph, women in this age group outnumbered men by about 2 to 1.

Linear functions, whose graphs are straight lines, can be used to model such data. Imagine lines drawn as close as possible to the tops of the two sets of bars. How might the difference between female and male enrollments in this age group be explained? (More detailed data can be found in Exercises 95–98 at the end of Section 4.7.)

In Chapter 4 we introduce linear equations, methods for solving them, and applications. We then consider solution techniques and applications for linear and compound inequalities as well as absolute value equations and inequalities.

ENROLLMENT

Students Aged 35 and Over (in thousands)

☐ Males   ■ Females

	1988	1989	1990	1991
Males	727	716	772	832
Females	1452	1396	1546	1636

1600
1200
800
400

(Source: U.S. Bureau of Census.)

4.1 Introduction to Linear Equations

Equations ▪ *Linear Equations in One Variable* ▪ *The Graphing Method* ▪ *Inconsistent Equations and Identities*

Equations

Consider the function $f(x) = 2x - 5$. In Chapter 3 we considered two types of problems involving such functions.

(a) What is $f(3)$? (b) For what value of x is $f(x) = 15$?

It is important to understand the difference between the two problems. In (a) we are given an x-value, and we are to find the corresponding y-value. We know how to *estimate* the y-value by producing the graph of f, tracing to the point whose x-coordinate is 3, and reading the corresponding y-coordinate. However, we also know how to determine the *exact* value of y algebraically by evaluating $f(3)$.

In (b) we are given a y-value, and we are to find the corresponding x-value. Again, we know how to *estimate* the x-value graphically, but we have not considered algebraic methods for determining the *exact* value of x.

Before we consider such methods, some definitions are needed.

Definition of Equation

An **equation** is a statement that two algebraic expressions have the same value.

For the equation $2x - 5 = 15$, the expression $2x - 5$ is called the *left side* of the equation. Similarly, the expression 15 is called the *right side* of the equation.

Determining the exact value of x for which $2x - 5 = 15$ is called **solving the equation.**

Linear Equations in One Variable

There are many different types of equations. In this chapter we will focus on just one type called a **linear equation in one variable.**

Definition of a Linear Equation in One Variable

A **linear equation in one variable** is an equation that can be written in the form $Ax + B = 0$ where A and B are real numbers and $A \neq 0$.

The following are examples of linear equations.

$$3x + 5 = 0 \qquad 5(x + 3) = 4 - x$$

The first equation is in the form $Ax + B = 0$. It can be shown that the other equation can also be written in that form.

The following are examples of equations that are *not* linear equations.

(a) $\dfrac{3}{x} = 9$ The variable cannot appear in a denominator.

(b) $x + 2 = \sqrt{2x} - 1$ The variable cannot appear under a radical sign.

(c) $x^2 - 25 = 0$ The variable cannot have an exponent other than 1.

When we solve an equation, we are finding the **solution(s)** of the equation.

Definitions of Solution and Solution Set

A **solution** of an equation is a replacement for the variable that makes the equation true. A **solution set** is the set of all solutions of an equation.

As we saw in Chapter 2, we verify that a number is a solution of an equation by replacing every occurrence of the variable with that number. We then perform the necessary arithmetic to see if the two sides of the equation have the same value. If they do, then the number is indeed a solution.

The Graphing Method

Verifying a solution involves testing whether a replacement for the variable makes the equation true. *Determining* what the solution is requires other methods. One approach is to estimate the solution of an equation by using a graph.

 EXPLORATION 1   *Estimating a Solution by Graphing*

Consider the equation $x + 15 = 30 - 2x$. We can give the left and right sides of the equation the names y_1 and y_2.

$$y_1 = x + 15 \qquad y_2 = 30 - 2x$$

Now we can consider the equation $x + 15 = 30 - 2x$ in a new way:

$$y_1 = y_2.$$

(a) Use the integer setting to produce the graph of each side of the equation.

(b) Use the graph to estimate the solution of the equation.

Discovery

(a) Figure 4.1

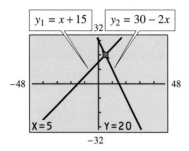

(b) The graph of y_1 in Fig. 4.1 consists of all points whose coordinates satisfy the function $y_1 = x + 15$. The graph of y_2 consists of all points whose coordinates satisfy the function $y_2 = 30 - 2x$.

Trace to the point where the two graphs intersect. The point is $(5, 20)$. The coordinates of this point satisfy *both* functions. In other words, when $x = 5$, $y_1 = y_2$.

The solution can be verified algebraically. We can use a calculator to store 5 for x and then either evaluate the two sides of the equation (see Fig. 4.2) or test the truth of the equation (see Fig. 4.3).

Figure 4.2   Figure 4.3

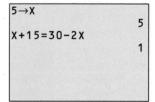

In Exploration 1, the second coordinate of the point of intersection is 20. This *y*-coordinate is the value of both the left side and the right side of the original equation when x is replaced with the solution. In fact, that is why the *x*-coordinate *is* the solution; it makes the left and right sides of the equation equal.

Here is a summary of the graphing approach to solving an equation.

Equation Solving: The Graphing Method

1. Graph y_1 = left side of the original equation.

2. Graph y_2 = right side of the original equation.

3. Trace to the point of intersection.

4. The *x*-coordinate of that point is the solution of the equation.

5. The *y*-coordinate is the value of both the left side and the right side of the original equation when x is replaced with the solution.

EXAMPLE 1   *Solving an Equation by the Graphing Method*

Use the graphing method to solve the equation $\frac{1}{2}x - 19 = -2x + 6$.

Solution

We use the integer setting to produce the graphs of $y_1 = \frac{1}{2}x - 19$ and $y_2 = -2x + 6$. (See Fig. 4.4.)

Figure 4.4

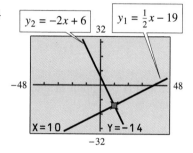

Trace to the point of intersection. The *x*-coordinate of the point of intersection is 10.

Verify that 10 is a solution of the equation.

$$\frac{1}{2}x - 19 = -2x + 6$$

$$\frac{1}{2}(10) - 19 = -2(10) + 6 \qquad \text{Replace } x \text{ with 10.}$$

$$5 - 19 = -20 + 6 \qquad \text{Simplify.}$$

$$-14 = -14 \qquad \text{True}$$

This verifies that 10 is a solution. ■

Inconsistent Equations and Identities

A linear equation in one variable always has exactly *one* solution. As we see in the next exploration, there are other equations that have no solution or that have infinitely *many* solutions.

EXPLORATION 2   *Inconsistent Equations and Identities*

(a) Use the graphing method to solve the equation $x + 3 = 1 + x$. What does the graph suggest about the solution set?

(b) Use the graphing method to solve the equation $x + 3 = 3 + x$. What does the graph suggest about the solution set?

Discovery

(a) Figure 4.5 shows the graphs of $y_1 = x + 3$ and $y_2 = 1 + x$. The two graphs *appear* to be parallel lines.

The solution of the equation is the x-coordinate of the point where the graphs intersect. But if the graphs really do not intersect, then there is no value of x that will make the equation true. This equation appears to have no solution.

(b) Figure 4.6 shows the graphs of $y_1 = x + 3$ and $y_2 = 3 + x$.

There appears to be only one line, but the calculator is actually displaying both graphs. One graph lies directly on top of the other. We say that the graphs *coincide*.

The goal is to find the point of intersection of the two graphs, but in this case, *every* point along the line is a point of intersection. No matter what x-value is chosen, it is a solution.

Because any replacement for x is a solution, the solution set is the set of all real numbers.

Figure 4.5

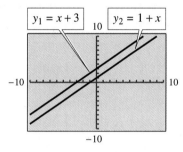

Figure 4.6

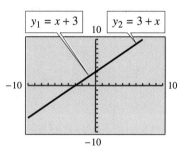

An equation is called an **identity** if every real number for which both sides of the equation are defined is a solution of the equation. The graphs of the left and right sides of the equation coincide.

A **conditional equation** is an equation with at least one solution, but the equation is not an identity. A linear equation in one variable is an example of a conditional equation.

An equation that has no solution is called an **inconsistent equation.** The graphs of the left and right sides of the equation are parallel lines. The solution set for an inconsistent equation is the empty set.

While the graphing method is visually helpful in understanding the concept of solving an equation, algebraic techniques are often more useful when an exact solution is required. We will learn about such techniques in the next section.

4.1 Quick Reference

Equations
- An **equation** is a statement that two algebraic expressions have the same value.

Linear Equations in One Variable
- A **linear equation in one variable** is an equation that can be written in the form $Ax + B = 0$, where A and B are real numbers and $A \neq 0$.
- A **solution** of an equation is a replacement for the variable that makes the equation true.
- The **solution set** of an equation is the set of all solutions of the equation.
- We verify a solution by replacing every occurrence of the variable by the proposed solution. If the replacement makes the equation true, it is a solution.

The Graphing Method
- We estimate the solution of a linear equation by graphing the left and right sides of the equation on the same coordinate axes. The x-coordinate of the point of intersection, if any, is the estimated solution.
- We verify the estimated solution by testing it in the equation to see if the replacement makes the equation true.

Inconsistent Equations and Identities
- An equation is called an **identity** if every real number for which both sides of the equation are defined is a solution of the equation. The graphs of the left and right sides of the equation coincide.
- A **conditional equation** is an equation with at least one solution, but the equation is not an identity.
- An equation that has no solution is called an **inconsistent equation.** The graphs of the left and right sides of the equation are parallel lines. The solution set for an inconsistent equation is the empty set.

4.1 Exercises

 1. In your own words, state the definition of an equation.

 2. State the definition of a linear equation in one variable. Give two examples of equations that are not linear equations in one variable.

In Exercises 3–12, determine whether each of the following is an equation.

3. $x - 3 + 2x = 5$

4. $10 = x$

5. $4(x - 5) - 7(2x + 3)$

6. $\dfrac{9}{5} \cdot 5 - 6 + 7x$

7. $0 = -7$

8. $3 + 4 - 5 = 6$

9. $3x - 5 + 7x + 12$

10. $5t - 6 = 7$

11. $x^2 = 5x + 6$

12. $t^2 - 5t + 6$

In Exercises 13–20, state whether each equation is a linear equation in one variable.

13. $\frac{4}{7}x + 3 = 0$

14. $3x - 2 = 7$

15. $x^3 - 2x = 4$

16. $2x^2 + 5x = 6$

17. $2\pi^2 + 5x = 6$

18. $3\sqrt{2} - 7x = 0$

19. $\sqrt{x} = 5$

20. $6x = \sqrt{5}$

21. Explain the difference between a solution of an equation and a solution set.

22. Explain how to verify that 3 is a solution of the equation $5x - 7 = 14 - 2x$

  (a) without using your calculator and

  (b) with your calculator.

In Exercises 23–32, verify that the given number is a solution of the given equation.

23. $3x - 5 = 7x + 7$; -3

24. $15 - 2x = 3x$; 3

25. $4(x - 2) + 5 = 5(3 - x) + 3(x + 4)$; 5

26. $29 + 4(3x - 5) = 2(6x + 7) - 5$; $\frac{2}{3}$

27. $-2[2x - 3(1 - x)] = -5(x + 3) + 11$; 2

28. $\frac{2}{3}(x - 4) + \frac{3}{4}(x + 5) = 2(x - 1)$; $\frac{37}{7}$

29. $\frac{3}{5}y = -2$; $-\frac{10}{3}$

30. $\frac{x}{2} - \frac{2x}{5} + \frac{1}{2} = \frac{7}{10}x - \frac{2}{5}$; $\frac{3}{2}$

31. $6x - (5x - 8) = 1$; -7

32. $-0.2t = 5.4$; -27

In Exercises 33–40, determine which of the given numbers is a solution of the given equation.

33. $3x - 4 = 11$; $3, 5, 7$

34. $4x + 7 = 17 - x$; $-2, 1, 2$

35. $\frac{x}{2} + \frac{1}{3} = \frac{x}{3} - \frac{1}{2}$; $-1, -4, -5$

36. $\frac{t}{4} - 2 = 3 - \frac{3}{4}t$; $2, 5, 8$

37. $3(x - 2) + 4 = 3x - 2$; $-3, 0, 2$

38. $4 - 2(3x + 2) = -4x - 2x$; $-5, -1, 4$

39. $5(t + 2) = 3(t - 4)$; $-11, -1, 0$

40. $2y - 5 = 5(y - 1) - y$; $-2, 0, 2$

41. Describe the graph and solution set of an inconsistent equation. Give an example of an equation that is an inconsistent equation.

42. Describe the graph and solution set of an identity. Give an example of an equation that is an identity.

In Exercises 43–48, each figure shows the graphs of the left and right sides of an equation. From the figure estimate the solution of the equation.

43.

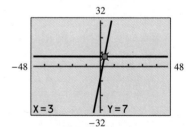

44.

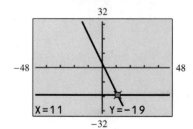

45.

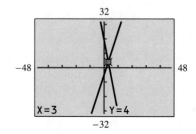

46.

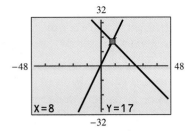

47.

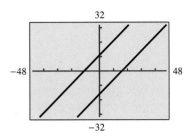

48.

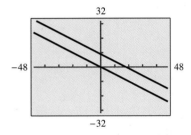

In Exercises 49–68, use the graphing method to solve each equation. Then verify the solution.

49. $4x - 3 = 5$

50. $4x + 13 = -3$

51. $2 - 3x = -4$

52. $5x = 3x - 4$

53. $2x + 3 = 3 + 2x$

54. $x - 2 = -(2 - x)$

55. $4 - x = 3x + 12$

56. $2x - 3 = 7 - 3x$

57. $2(x + 3) = 6 + 5x$

58. $2(x - 5) = 8$

59. $2 - 3x = -3(x + 4)$

60. $2x + 5 - x = 7 + x$

61. $58 - (4 - 3x) = 0$

62. $3(x - 2) - 5 = 10$

63. $1 - 2(x - 1) = -(2x - 3)$

64. $x - (2x - 5 - x) = 5$

65. $\dfrac{1}{2}x + 2 = -\dfrac{1}{6}x + 6$

66. $1 - \dfrac{t}{3} = 5$

67. $\dfrac{t - 3}{3} = \dfrac{t}{6}$

68. $x - 5 = \dfrac{9 - x}{3}$

 69. When we solve a linear equation with the graphing method, we graph the left and right sides of the equation and then determine the coordinates (a, b) of the point of intersection.

(a) What is the significance of the number a?

(b) What is the significance of the number b?

70. To solve $x = 5$ with the graphing method, we would graph $y_1 = x$ and $y_2 = 5$. Even without producing the graphs, explain how you know the coordinates of the point of intersection.

In Exercises 71–78, use the integer setting on your calculator to solve each equation with the graphing method. Then verify the solution. In each case, write the coordinates of the point of intersection of the graphs of the left and right sides of the equation.

71. $2x + 3 = 3x - 8$

72. $-2x - 2 = 12 - x$

73. $\dfrac{3}{2}(x + 7) = 3$

74. $-(x - 3) = 2(10 - x)$

75. $3.75 - 0.5x = -0.25x$

76. $4x - 7 = 3x$

77. $\dfrac{5}{6}x + 3 = -7$

78. $\dfrac{1}{3}x + \dfrac{5}{3} = \dfrac{x}{3} - 1$

 79. Explain how the graphing method would reveal that $-(x + 1) = -x + 1$ is an inconsistent equation.

 80. Explain how the graphing method would reveal that $-(3 - x) + 5 = x + 2$ is an identity.

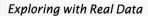

Exploring with Real Data

In Exercises 81–84, refer to the following background information and data.

The accompanying line graph shows the number of cellular phone customers (in thousands) served by Southwestern Bell for selected years in the period 1984–1990. (Source: Southwestern Bell Corporation.) The line segment connecting the two data points for 1986 and 1988 has the equation $y = 101.5x - 568$, where y is the number of phone customers (in thousands) and x is the number of years since 1980.

Figure for 81–84

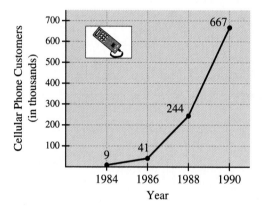

81. Is $y = 101.5x - 568$ a linear equation in one variable? Why?

82. Replace y with 41. Is the resulting equation a linear equation in one variable? Why?

83. If you replace y with 244, how can you tell from the line graph what the corresponding x-value is? Confirm your answer by using the graphing method to solve the equation $244 = 101.5x - 568$.

84. State three factors that you think might explain the trend shown in this figure.

Challenge

In Exercises 85–88, use the graphing method to estimate the solution of the given equation. When necessary, round your estimates to the nearest tenth.

85. $3x + 40 = 3.5x - 20$

86. $2(x + 15) = x - 40$

87. $\sqrt{3}x - 12 = \pi x$ 88. $\dfrac{x}{\sqrt{2}} = 8 - x$

89. Give two ways to restrict the domain so that $2|x| + 3 = 7$ is a linear equation in one variable.

90. If B and C are constants, is the equation $0 \cdot x + B = C$ a linear equation in one variable? Why? Under what condition would the equation be an identity?

91. Show that a linear equation in one variable cannot be an inconsistent equation.

4.2 Solving Linear Equations

Properties of Equations ▪ *Applying the Properties* ▪ *Simplifying Both Sides* ▪ *Clearing Fractions* ▪ *Special Cases* ▪ *The Algebraic Solving Routine*

Properties of Equations

To determine the approximate solution of an equation, we can use the graphing method discussed in Section 4.1. To obtain exact solutions, we must use algebraic techniques. This section begins with the basic properties of all equations. These properties are then used to obtain exact solutions of linear equations.

> ### *Definition of Equivalent Equations*
>
> **Equivalent equations** are equations that have exactly the same solution sets.

For example, the equations $2x + 1 = 9$, $2x = 8$, and $x = 4$ are equivalent equations because the solution set of each is $\{4\}$.

The following properties of equations may be applied to produce equivalent equations.

> ### *Properties of Equations*
>
> Suppose A, B, and C represent algebraic expressions.
>
> 1. If $A = B$, then $A + C = B + C$.   Addition Property of Equations
>   Adding (or subtracting) the same quantity to both sides of an equation produces an equivalent equation.
>
> 2. If $A = B$ and $C \neq 0$, then $AC = BC$.   Multiplication Property of Equations
>   Multiplying (or dividing) both sides of an equation by the same nonzero quantity produces an equivalent equation.
>
> 3. If $A = B$, then $B = A$.   Symmetric Property of Equations
>   The two sides of an equation can be swapped to produce an equivalent equation.
>
> 4. If $A = B$, then A may be replaced with B in any equation. Any expression in an equation may be replaced with an equivalent expression.
>   Substitution Property of Equations

Applying the Properties

Solving an equation involves using the previously stated properties to write an equivalent equation whose solution set is easy to determine. Usually the goal is to find an equivalent equation in which the variable stands alone (is isolated) on one side of the equation so that the solution is obvious. To accomplish this, the Addition Property of Equations is used to isolate the variable term and the Multiplication Property of Equations is used to isolate the variable itself.

Our first example requires only the addition and multiplication properties of equations. However, a full summary of the algebraic method can be found at the end of this section.

EXAMPLE 1   *Solving with the Addition and Multiplication Properties of Equations*
Solve the equation $7x + 8 = -22 - 3x$ and verify the solution by substitution.

Solution

$$7x + 8 = -22 - 3x$$
$$7x + 8 + 3x = -22 - 3x + 3x \qquad \text{Add } 3x \text{ to both sides of the equation.}$$
$$10x + 8 = -22 \qquad \text{Combine like terms.}$$
$$10x + 8 - 8 = -22 - 8 \qquad \text{Subtract 8 from both sides of the equation.}$$
$$10x = -30$$
$$\frac{10x}{10} = \frac{-30}{10} \qquad \text{Divide both sides of the equation by 10.}$$
$$x = -3$$

To verify by substitution, replace x with -3 in the original equation.

$$7x + 8 = -22 - 3x$$
$$7(-3) + 8 = -22 - 3(-3) \qquad \text{Replace } x \text{ with } -3.$$
$$-21 + 8 = -22 + 9$$
$$-13 = -13 \qquad \text{True} \qquad ■$$

Simplifying Both Sides

The next example illustrates how the expressions on the two sides of an equation may need to be simplified. Recall that the Distributive Property allows us to remove parentheses and combine like terms.

EXAMPLE 2   *Simplifying Both Sides of an Equation*
Solve the following equations.

(a) $4 - 2(x - 5) = 3x - 4(x + 2)$

(b) $1.09 - 0.3t = 0.4t - 1.7 + 0.2t$

Solution

(a) $4 - 2(x - 5) = 3x - 4(x + 2)$

$\qquad 4 - 2x + 10 = 3x - 4x - 8 \qquad$ Remove parentheses.

$\qquad\qquad 14 - 2x = -x - 8 \qquad$ Combine like terms.

$\qquad 14 - 2x + x = -x - 8 + x \qquad$ Add x to both sides of the equation.

$\qquad\qquad 14 - x = -8 \qquad$ Combine like terms.

$\qquad 14 - x - 14 = -8 - 14 \qquad$ Subtract 14 from both sides of the equation.

$\qquad\qquad -1x = -22$

$$\dfrac{-1x}{-1} = \dfrac{-22}{-1} \qquad \text{Divide both sides of the equation by } -1.$$

$\qquad\qquad x = 22$

(b) $\qquad 1.09 - 0.3t = 0.4t - 1.7 + 0.2t$

$\qquad 1.09 - 0.3t = 0.6t - 1.7 \qquad$ Combine like terms.

$1.09 - 0.3t - 0.6t = 0.6t - 0.6t - 1.7 \qquad$ Subtract $0.6t$ from both sides.

$\qquad 1.09 - 0.9t = -1.7 \qquad$ Combine like terms.

$1.09 - 1.09 - 0.9t = -1.7 - 1.09 \qquad$ Subtract 1.09 from both sides.

$\qquad\qquad -0.9t = -2.79$

$$\dfrac{-0.9t}{-0.9} = \dfrac{-2.79}{-0.9} \qquad \text{Divide both sides by } -0.9.$$

$\qquad\qquad t = 3.1 \qquad\qquad\qquad\qquad\qquad \blacksquare$

In part (b) of Example 2, we could have eliminated (or *cleared*) the decimals from the original equation by multiplying both sides by 100.

$$1.09 - 0.3t = 0.4t - 1.7 + 0.2t$$
$$100(1.09 - 0.3t) = 100(0.4t - 1.7 + 0.2t)$$
$$109 - 30t = 40t - 170 + 20t$$

This makes the resulting equivalent equation somewhat easier to manage, although the use of a calculator to do the routine arithmetic makes this step unnecessary.

Clearing Fractions

When equations involve fractions, it is often helpful to clear the fractions by using the Distributive Property to multiply both sides of the equation by the least common multiple (LCM).

NOTE: Remember to multiply each term by the LCM—even terms that are not fractions. $\qquad\blacksquare$

EXAMPLE 3   *Clearing Fractions from an Equation*

Solve the following equations.

(a) $\dfrac{3}{5}x - \dfrac{2}{3} = \dfrac{9}{10} + \dfrac{1}{15}x$

(b) $\dfrac{1}{2} - \dfrac{1}{3}(x - 2) = 2(x - 1) + \dfrac{2}{3}$

Solution

(a) $\dfrac{3}{5}x - \dfrac{2}{3} = \dfrac{9}{10} + \dfrac{1}{15}x$

Because the LCM of 5, 3, 10, and 15 is 30, multiply each term of both sides by 30.

$$30 \cdot \frac{3}{5}x - 30 \cdot \frac{2}{3} = 30 \cdot \frac{9}{10} + 30 \cdot \frac{1}{15}x$$

$$\left(30 \cdot \frac{3}{5}\right)x - 30 \cdot \frac{2}{3} = 30 \cdot \frac{9}{10} + \left(30 \cdot \frac{1}{15}\right)x \qquad \text{Associative Property of Multiplication}$$

$$18x - 20 = 27 + 2x$$

$$18x - 20 - 2x = 27 + 2x - 2x \qquad \text{Subtract } 2x \text{ from both sides.}$$

$$16x - 20 = 27 \qquad \text{Combine like terms.}$$

$$16x - 20 + 20 = 27 + 20 \qquad \text{Add 20 to both sides.}$$

$$16x = 47$$

$$\frac{16x}{16} = \frac{47}{16} \qquad \text{Divide both sides by 16.}$$

$$x = \frac{47}{16}$$

The solution can be verified with a calculator. (See Fig. 4.7.) Note the essential use of the parentheses.

Figure 4.7

```
47/16→X
              2.9375
(3/5)X-2/3
         1.095833333
9/10+(1/15)X
         1.095833333
```

Replacing x with $\frac{47}{16}$ (or 2.9375) makes both the left and right sides of the equation equal to 1.095833333. The solution is verified.

(b)

$$\frac{1}{2} - \frac{1}{3}(x - 2) = 2(x - 1) + \frac{2}{3} \qquad \text{The LCM is 6.}$$

$$6 \cdot \frac{1}{2} - 6 \cdot \frac{1}{3}(x - 2) = 6 \cdot 2(x - 1) + 6 \cdot \frac{2}{3} \qquad \text{Multiply both sides by 6.}$$

$$6 \cdot \frac{1}{2} - \left(6 \cdot \frac{1}{3}\right)(x - 2) = (6 \cdot 2)(x - 1) + 6 \cdot \frac{2}{3} \qquad \begin{array}{l}\text{Associative Property of}\\ \text{Multiplication}\end{array}$$

$$3 - 2(x - 2) = 12(x - 1) + 4$$

$$3 - 2x + 4 = 12x - 12 + 4 \qquad \text{Distributive Property}$$

$$7 - 2x = 12x - 8$$

$$7 - 2x - 12x = 12x - 8 - 12x \qquad \text{Subtract 12x from both sides.}$$

$$7 - 14x = -8 \qquad \text{Combine like terms.}$$

$$7 - 14x - 7 = -8 - 7 \qquad \text{Subtract 7 from both sides.}$$

$$-14x = -15$$

$$\frac{-14x}{-14} = \frac{-15}{-14} \qquad \text{Divide both sides by } -14.$$

$$x = \frac{15}{14}$$

Figure 4.8

```
15/14→X
          1.071428571
1/2-(1/3)(X-2)
           .8095238095
2(X-1)+2/3
           .8095238095
```

We verify this solution with a calculator in the usual way. (See Fig. 4.8.) ■

Special Cases

In Exploration 2 of the previous section, we used the graphing method to solve $x + 3 = 1 + x$ and found that the graphs of the two sides appeared to be parallel lines. We can verify that an equation is inconsistent with the algebraic method.

EXAMPLE 4   *Verifying an Inconsistent Equation Algebraically*

Solve $x + 3 = 1 + x$ algebraically.

Solution

$$x + 3 = 1 + x$$

$$x - x + 3 = 1 + x - x \qquad \text{Subtract x from both sides.}$$

$$0 + 3 = 1 + 0 \qquad \text{Combine like terms.}$$

$$3 = 1 \qquad \text{False}$$

The equation $x + 3 = 1 + x$ is equivalent to the false equation $3 = 1$. This means that $x + 3 = 1 + x$ is also false, no matter what value is used to replace x. Therefore, the equation has no solution.

This also confirms that the graphs of the two sides of the equation are indeed parallel lines because there is no point of intersection. ■

We also attempted to solve $x + 3 = 3 + x$ in Exploration 2 of the previous section. We found that the graphs of the two sides of the equation appeared to coincide. We can use the algebraic method to verify an identity.

| EXAMPLE 5 | *Verifying an Identity Algebraically* |

Solve $x + 3 = 3 + x$ algebraically.

Solution

$$x + 3 = 3 + x$$
$$x - x + 3 = 3 + x - x \qquad \text{Subtract } x \text{ from both sides.}$$
$$0 + 3 = 3 + 0 \qquad \text{Simplify.}$$
$$3 = 3 \qquad \text{True}$$

The equation $x + 3 = 3 + x$ is equivalent to the true equation $3 = 3$. This means that $x + 3 = 3 + x$ is also true, no matter what value is used to replace x. Therefore, the solution set of the equation is the set of all real numbers.

This also confirms that the two graphs are indeed coinciding lines because all their points are points of intersection. ■

The Algebraic Solving Routine

The following is a general summary of the steps for solving a linear equation in one variable algebraically.

Algebraic Routine for Solving a Linear Equation in One Variable

1. If necessary, clear fractions by multiplying every term of both sides of the equation by the LCM of the denominators of all the fractions.

2. Simplify both sides of the equation by

   (a) removing grouping symbols and/or

   (b) combining like terms.

3. If the variable is on both sides, use the Addition Property of Equations to eliminate the variable term from one side or the other.

4. Use the Addition Property of Equations to isolate the variable term.

5. Use the Multiplication Property of Equations to isolate the variable itself.

6. Verify the solution.

The algebraic method of solving an equation has the advantage of producing an exact solution.

Nevertheless, graphic interpretations are extremely valuable in the development of a conceptual knowledge of equations and their solutions. Before computers and graphing calculators, the task of producing graphs was time-consuming and laborious. Now the graphing calculator can be an excellent tool for promoting a visual understanding of the relationship between equations and graphs.

4.2 *Quick Reference*

Properties of Equations

- **Equivalent equations** are equations that have exactly the same solution sets.
- An equivalent equation results when
  1. the same quantity is added to or subtracted from both sides of an equation (Addition Property of Equations).
  2. both sides of an equation are multiplied or divided by the same nonzero quantity (Multiplication Property of Equations).
  3. the two sides of an equation are swapped (Symmetric Property of Equations).
  4. an expression in an equation is replaced with an equivalent expression (Substitution Property of Equations).

Applying the Properties

- The Addition Property of Equations can be used to eliminate a variable term from one side of an equation and to isolate a variable term.
- The Multiplication Property of Equations can be used to isolate a variable by eliminating the coefficient of the isolated variable term.

Simplifying Both Sides

- Before applying the properties of equations, it is usually best to simplify both sides of the equation by removing parentheses and combining like terms.

Clearing Fractions

- Remove (clear) fractions from an equation by multiplying every term of both sides of the equation by the LCM of the denominators.

Special Cases

- The algebraic routine for solving an equation will reveal special cases.
  1. If a resulting equivalent equation is a false statement, the original equation is **inconsistent,** and the solution set is empty.
  2. If a resulting equivalent equation is a true statement, the original equation is an **identity,** and the solution set is the set of all real numbers.

Algebraic Solving Routine

- The summary at the end of this section lists the suggested steps for solving a linear equation in one variable algebraically.

4.2   *Exercises*

1. What are equivalent equations?

2. What is an important advantage of the algebraic method over the graphing method for solving equations?

In Exercises 3–6, state the property of equations that justifies the given statement.

3. If $x + y = 7$ and $y = 3x$, then $x + 3x = 7$.

4. If $5x + 2 = 4x$, then $x + 2 = 0$.

5. If $-1 = 8 - 5x$, then $8 - 5x = -1$.

6. If $\frac{1}{5}x = 4$, then $x = 20$.

In Exercises 7–22, simplify when necessary and use the Addition Property of Equations to solve the equations.

7. $x + 3 = 7$ **8.** $r + 4 = -10$

9. $z - 3 = -11$ **10.** $a - 5 = 8$

11. $9x = 8x$ **12.** $-6y = 8 - 7y$

13. $7 - 2s = 10 - 3s$ **14.** $14 - 3s = 19 - 4s$

15. $8 + 3z = 2z - 9$ **16.** $4k + 4 = 4 + 3k$

17. $4x - 3 + 3x = 7 + 6x - 4$

18. $7x - 7 - 4x - 2x + 12 = 8 - 3$

19. $3 - 2(x - 5) = 4 - 3(x + 1)$

20. $5(4 - x) + 3(2x - 1) = 0$

21. $6(x + 2) = 12 + 5(x + 7)$

22. $7x + 2 - 4(x - 3) = 2(x + 7)$

23. The following shows some first steps in solving the given equation.

$$2x + 8 = 5$$
$$2x + 8 - 5 = 5 - 5$$
$$2x + 3 = 0$$

Has the Addition Property of Equations been used correctly? Is this the best way to start solving the equation? If not, describe a better first step.

24. Consider the following two equations.

$$x + 2 = 3$$
$$-2x = 6$$

When we solve these equations, explain why we *add* -2 to both sides of the first equation and *divide* both sides by -2 in the second equation.

In Exercises 25–36, simplify when necessary and use the Multiplication Property of Equations to solve the equations.

25. $7x = 0$ **26.** $3x = 12$

27. $-x = 9$ **28.** $-5m = 30$

29. $-3x = \dfrac{2}{5}$ **30.** $\dfrac{2}{5}x = \dfrac{9}{10}$

31. $\dfrac{y}{3} = 12$ **32.** $4 = \dfrac{-x}{5}$

33. $x + 3x + x = 30$

34. $9x - 4x + 3x = -72$

35. $4 - 2[2x - (x - 1)] = 6x - 3(2x - 4)$

36. $3(x + 1) + 4(x - 1) = (1 - 2x) + 2(x + 10)$

37. (a) Explain why $\dfrac{3x}{5}$ is equivalent to $\dfrac{3}{5}x$.

      (b) Explain why $\dfrac{3}{5x}$ is *not* equivalent to $\dfrac{3}{5}x$.

38. The following are possible first steps in solving the equation $\frac{2}{3}x = 10$. Explain why each approach is correct.

   (i) Multiply both sides by $\frac{3}{2}$.

   (ii) Divide both sides by $\frac{2}{3}$.

   (iii) Multiply both sides by 3 and then divide both sides by 2.

In Exercises 39–62, solve each equation algebraically. Verify the solutions.

39. $5x + 8 = 23$

40. $4n - 7 = -35$

41. $6x - 8 = 2x$

42. $4 = 3d - 17$

43. $7x - 5 = 8x + 7$

44. $7 + 8x = 4x - 13$

45. $9k - 7 = 3k + 5$

46. $6y + 17 = 2y - 3$

47. $3(3 - 2x) = 33 - 2x$

48. $2(3y + 8) = -2(2 - y)$

49. $2y - 7 + 3y = 5 - 4y + 7$

50. $9(x - 1) = 3(x + 1)$

51. $7w - 5 = 11w - 5 - 4w$

52. $9w + 7 = 16w + 7(1 - w)$

53. $9x + 5 = 5x + 3(x - 1)$

54. $11x + 2(4 - 3x) = 3(3 - x) + 15$

55. $16 + 7(6 - x) = 15 - 4(x + 2)$

56. $7 - 4(t - 3) = 6t - 3(t + 3)$

57. $2(3a + 2) = 2(a + 1) + 4a$

58. $8t - 7 - 3t = 5t + 2$

59. $-2(x + 5) = 5(1 - x) + 3(7 - x)$

60. $8(x - 7) = 3[1 + (x + 4)]$

61. $2y - 3(2y - 3) = 2y - 5(3y - 4)$

62. $7(3x - 5) - x = 2(3x - 2)$

 63. Consider solving the equation $\frac{1}{3} - 2x = \frac{2}{5}$. To clear fractions, we multiply both sides by 15. Why is it necessary to multiply $2x$ by 15?

 64. To solve $\frac{1}{3}(x + 5) = \frac{1}{2}$, we can multiply both sides by 6 to clear the fractions. Is the result on the left side $2(x + 5)$ or $2(6x + 30)$? Why?

In Exercises 65–74, solve the given equation and use your calculator to verify the solution.

65. $2x - \dfrac{3}{4} = -\dfrac{5}{6}$

66. $\dfrac{2}{15}y + \dfrac{3}{5} = 2 - \dfrac{2}{3}y$

67. $\dfrac{y}{2} + \dfrac{y}{4} = \dfrac{7}{8} - \dfrac{y}{8}$

68. $\dfrac{5 + 3x}{2} + 7 = 2x$

69. $\dfrac{5}{4}(x + 2) = \dfrac{x}{2}$

70. $\dfrac{1}{4}(x - 4) = \dfrac{1}{3}(x + 6)$

71. $x + \dfrac{3x - 1}{9} = 4 + \dfrac{3x + 1}{3}$

72. $\dfrac{2x}{5} - \dfrac{2x - 1}{2} = \dfrac{x}{5}$

73. $\dfrac{1}{4} + \dfrac{1}{6}(4a + 5) = 2(a - 3) + \dfrac{5}{12}$

74. $\dfrac{5}{6}(1 - t) - \dfrac{t}{2} = -\dfrac{1}{3}(1 - 3t)$

In Exercises 75–80, solve the given equation and use your calculator to verify the solution.

75. $2.6 = 0.4z + 1$

76. $1 - 0.3c = -0.5$

77. $x + 0.4x = 74$

78. $0.3(x + 20) = 0.45(20)$

79. $0.08x + 0.15(x + 200) = 7$

80. $0.6(2x + 1) - 0.4(x - 2) = 1$

 81. What is the difference between a conditional equation and an inconsistent equation?

 82. What is the difference between a conditional equation and an identity?

In Exercises 83–88, determine whether the given equations are equivalent equations.

83. $x + 4 = 7$ and $3x - 5 = 4$

84. $\dfrac{1}{3}x = 2$ and $x + 3 = 9$

85. $x + 3 = 4 + x$ and $2x - 3 = 4 + 2x$

86. $x + 3 = 3 + x$ and $2x - 5 = -(5 - 2x)$

87. $\dfrac{1}{4}x = -2$ and $8 - x = 0$

88. $3x = 0$ and $x - 3 = 0$

In Exercises 89–96, state whether the given equation is an identity, an inconsistent equation, or a conditional equation. Determine the solution set.

89. $x - 6 = 6 - x$ **90.** $2x - 7 = x + 4$

91. $x + 6 = 6 + x$ **92.** $x + 3 = 5 + x$

93. $2x - 5 = 7 + 3x - (x - 3)$

94. $x + 3 = 5x - [2x - (3 - 2x)]$

95. $\dfrac{1}{5}x - 2 = \dfrac{1}{3}x - 4$

96. $-\dfrac{1}{3}x + 8 = 0$

 97. What is the solution set of $2x = 8$? Now multiply both sides of $2x = 8$ by 0. What is the solution set of the resulting equation? Explain why the Multiplication Property of Equations excludes multiplying both sides by 0.

 98. What is the solution set of $3x = 2x$? Now divide both sides of $3x = 2x$ by x. What is the solution set of the resulting equation? Why are the equations not equivalent?

In Exercises 99–102, determine a value of c so that the equations are equivalent.

99. $5 - 3x = c + 2x$ and $2x - 3 = 5$

100. $7 - 2x = 4x - c$ and $3x - 8 = 7x$

101. $2x + c = 3x + 2$ and $4x + 7 = 6x - 5$

102. $3x - c = 5x - 3$ and $6 - 5x = 3x + 14$

In Exercises 103–106, determine a value of k so that the equations are inconsistent.

103. $3x - 2 = 4 + kx$ **104.** $5 - kx = 7x + 2$

105. $2x - 4 = x - 1 + kx$

106. $kx - 3x = 2x + 1$

In Exercises 107–110, determine a value of c so that the equations are identities.

107. $2x + 3 = c + 2x$ **108.** $x - 4 = -(c - x)$

109. $cx + 2 = 2 + 5x$

110. $4 - 2x = 3x - (cx - 4)$

In Exercises 111–114, determine a value of m so that the solution set of the equation is $\{7\}$.

111. $3x - 5 = m$ **112.** $8 - 2x = m + 4$

113. $3x + m - 2 = 4 - x$

114. $mx - 7 = 13$

In Exercises 115–136, solve the equations.

115. $2x - 5 = 5 - 2x$

116. $19 - 5z = 6z + 41$

117. $17 - 4z = 3z + 52$ **118.** $3y + 2 = 4 + 3y$

119. $7 + 12t + 6 = 2t + 3t - 15$

120. $4x - 14x = 16 - 10 - 11x$

121. $3(x + 1) - 2 = 2(x + 1) + x - 1$

122. $x + 3(2 - x) = 6 - 2x$

123. $5 + 3[t - (1 - 4t)] = 7$

124. $13 - 3(x + 2) = 5x - 9 + 2(x - 7)$

125. $3(2x + 1) = 8x - 2(x - 2)$

126. $3(3 - 5x) - (1 + 7x)$
$$= 3(3x + 2) - 5(2x - 1)$$

127. $5 - 3[5 - 3(4x + 3)]$
$$= 5x - 3[5x - 3(5x + 3)]$$

128. $3 + 2[3 - (7t - 4)] = 2t - [3 - (3t + 5)]$

129. $\dfrac{4}{9}t - \dfrac{1}{6} = \dfrac{1}{3}t + 1$

130. $\dfrac{t}{5} + \dfrac{7}{15} = 2t + \dfrac{2}{3}$

131. $\dfrac{3}{4}(x - 3) - \dfrac{1}{2}(3x - 5) = 2(3 - x)$

132. $\dfrac{3t - 8}{2} = -8 - \dfrac{t}{3}$

133. $0.12 - 0.05y = 0.04y - 0.15$

134. $1 - 0.3x = 0.6x - 1.7$

135. $x(x + 6) = 5(x + 1) + x^2$

136. $x^2 - 4 = x(x + 1)$

Exploring with Real Data

In Exercises 137–140, refer to the following background information and data.

The accompanying bar graph shows the number of registered shareholders (in thousands) of Exxon stock at the end of each year between 1986 and 1990. (Source: Exxon Corporation.) The data in this figure can be modeled by the equation $y = 778 - 26x$, where y is the number of registered shareholders (in thousands) of stock at the end of year x, and x is the number of years since 1985.

Figure for 137–140

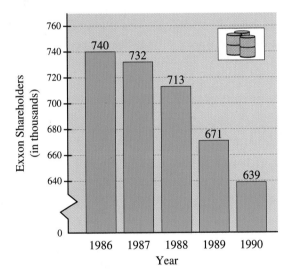

137. For which year is the model most accurate?

138. If the model were to remain valid indefinitely, in what year would there be no shareholders?

139. How can you tell from the model that y will continue to decrease as x increases?

140. What dramatic event involving Exxon Corporation might help to explain the downward trend in the number of shareholders?

Challenge

In Exercises 141–146, solve for x.

141. $axb = c$

142. $\dfrac{x}{a} = b$

143. $b - ax = c$

144. $a(x - b) = c$

145. $ax + b = cx + d$

146. $\dfrac{x + a}{3} = b$

In Exercises 147–150, solve the equation.

147. $(5 \cdot 10^{-3})x - (3.5 \cdot 10^{10}) = 0$

148. $x - 0.\overline{3} = 0.5$

149. $\dfrac{x}{2} + \dfrac{x + 1}{3} + \dfrac{x + 2}{4} + \dfrac{x + 3}{5} = 1$

150. $x - 2\{(x - 1) - 3[(x - 2) - 4(x - 3)]\}$
$= 19$

In Exercises 151–154, determine whether the given equations are equivalent.

151. $|x| = 5$ and $x = 5$

152. $\sqrt{x} = 5$ and $x = 25$

153. $h^2 = 25$ and $h = 5$

154. $\dfrac{3}{x} = 0$ and $x = 0$

4.3 Formulas

Solving Formulas ▪ *Applications*

Solving Formulas

A **formula** is an equation that states how to calculate the value of one variable in terms of one or more other variables.

Formulas occur in every walk of life—science, business, geometry, statistics, medicine, and so on. These formulas are usually written in the form in which they are most often used. But it is not uncommon to convert a formula into one that can be used to compute one of the other variables in the formula. This process is called **solving the formula** for the desired variable.

The formula for the perimeter P of a rectangle is $P = 2L + 2W$, where L represents the length of the rectangle and W represents the width. For given values of L and W, it is easy to calculate P. On the other hand, if the values of P and W are given, the task of calculating L involves solving an equation. If we must perform this task repeatedly, it is easier if we have a formula for L in terms of P and W.

EXAMPLE 1 *Solving the Perimeter Formula*

(a) Solve the formula $P = 2W + 2L$ for L.

(b) Use a calculator and the formula from part (a) to find the length of a rectangle whose perimeter is 110.4 inches and whose width is 20.7 inches.

Solution

(a) Use the Symmetric Property of Equations to swap the sides of the formula. (This is merely a step of convenience so that the variable L will appear on the left side.)

$$2W + 2L = P$$

Because the task is to solve for L, we treat L as the variable and W and P as constants.

$$2W + 2L = P$$
$$2W - 2W + 2L = P - 2W \qquad \text{Subtract } 2W \text{ from both sides.}$$
$$2L = P - 2W$$
$$\frac{2L}{2} = \frac{P - 2W}{2} \qquad \text{Divide both sides by 2.}$$
$$L = \frac{P - 2W}{2}$$

In the last step we used the Multiplication Property of Equations to divide both *sides* by 2. We might choose to divide *each term* by 2. Therefore, an alternate way to write the formula is

$$L = \frac{P}{2} - \frac{2W}{2} = \frac{P}{2} - W.$$

 ALPHA

(b) We store the given values for P and W and use the derived formula to evaluate L. (See Fig. 4.9.)

In Fig. 4.10, we verify that the alternate formula gives the same value for L.

Figure 4.9

```
110.4→P
            110.4
20.7→W
            20.7
(P-2W)/2
            34.5
```

Figure 4.10

```
110.4→P
            110.4
20.7→W
            20.7
P/2-W
            34.5
```

It is important to understand that a formula is an equation and that solving a formula involves the same rules and procedures as solving any other equation.

In Example 1 solving the formula for L, in general, involves exactly the same steps as solving the equation for L using specific values of P and W. By solving the formula for L, we have to go through the steps only once, and the new formula provides an arithmetic means of calculating L.

EXAMPLE 2 *A Formula for the Height of a Trapezoid*

A **trapezoid** is a four-sided figure with one pair of sides parallel. (See Fig. 4.11.) The parallel sides are called **bases** with lengths b_1 and b_2. The **height** h of a trapezoid is the distance between the parallel sides.

Figure 4.11

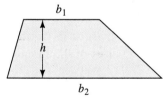

The formula for calculating the area of a trapezoid is

$$A = \frac{1}{2}(b_1 + b_2)h.$$

Solve this formula for h.

Solution   In the formula $\frac{1}{2}(b_1 + b_2)h = A$, treat h as the variable and regard all other variables as constants.

$$2 \cdot \frac{1}{2}(b_1 + b_2)h = 2 \cdot A \qquad \text{Multiply both sides by 2 to clear the fraction.}$$

$$(b_1 + b_2)h = 2A \qquad \text{The coefficient of } h \text{ is } (b_1 + b_2).$$

$$\frac{(b_1 + b_2)h}{(b_1 + b_2)} = \frac{2A}{(b_1 + b_2)} \qquad \text{Divide both sides by } (b_1 + b_2).$$

$$h = \frac{2A}{b_1 + b_2}$$

The formula for converting Fahrenheit temperatures F to Celsius temperatures C is $C = \frac{5}{9}(F - 32)$.

EXAMPLE 3   *Converting Celsius Temperatures to Fahrenheit Temperatures*

(a) Solve the formula $C = \frac{5}{9}(F - 32)$ for F.

(b) Use a calculator and the formula for F to convert 0°C and 100°C to Fahrenheit temperatures.

Figure 4.12

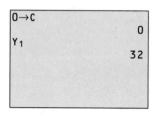

Figure 4.13

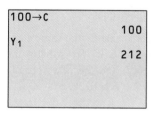

Solution

(a) Swap sides to place F on the left.

$$\frac{5}{9}(F - 32) = C$$

$$\frac{9}{5} \cdot \frac{5}{9}(F - 32) = \frac{9}{5} \cdot C \qquad \text{Multiply both sides by } \frac{9}{5} \text{ to clear the fraction.}$$

$$F - 32 = \frac{9}{5}C$$

$$F = \frac{9}{5}C + 32 \qquad \text{Add 32 to both sides.}$$

(b) To avoid having to enter the formula each time, we enter it as Y_1. Then, we can evaluate the formula for both of the given values of C. (See Figs. 4.12 and 4.13.)

Note that 32°F, the freezing temperature of water, corresponds to 0°C and that 212°F, the boiling temperature of water, corresponds to 100°C. ∎

Applications

When we use a calculator to graph a function, the function must be in the form $y = $ (expression). In other words, the equation must be solved for y.

EXAMPLE 4   *Solving an Equation for y*

Solve the equation $3x - y = 4$ for y.

Solution

We treat x as if it were a constant and isolate the variable y.

$$3x - y = 4$$

$$3x - y - 3x = -3x + 4 \qquad \text{Subtract } 3x \text{ from both sides.}$$

$$-1y = -3x + 4$$

$$\frac{-1y}{-1} = \frac{-3x}{-1} + \frac{4}{-1} \qquad \text{Divide both sides by } -1.$$

$$y = 3x - 4$$

Figure 4.14

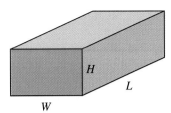

If a formula is to be used only once, there is no advantage to solving the formula for one of its variables. Solving a formula for a particular variable is most useful when we need to calculate values for that variable repeatedly. Nevertheless, we will continue to illustrate the process of solving formulas.

The volume V of a rectangular solid is given by the formula $V = LWH$, where L is the length, W is the width, and H is the height. (See Fig. 4.14.) This formula is used in the next example.

When we use a formula, it is important that the units be compatible. For example, a cubic yard refers to a cube that is one yard long, one yard wide, and one yard high. (See Fig. 4.15.)

Figure 4.15

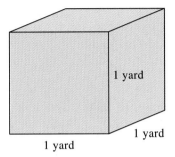

Because 1 yard is the same as 3 feet,

1 cubic yard $= 3 \cdot 3 \cdot 3 = 27$ cubic feet.

EXAMPLE 5   *A Concrete Slab in the Form of a Rectangular Solid*

In order to install an underground electric cable underneath an existing parking lot, a strip of concrete 200 feet long and 4 feet wide had to be removed. (See Fig. 4.16.) After the installation, the contractor ordered 8 cubic yards of concrete to repair the parking lot.

Figure 4.16

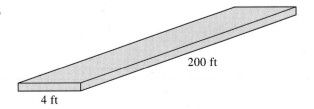

200 ft

4 ft

(a) Given the formula $V = LWH$, for which variable does the formula need to be solved in order to determine the thickness of the new slab of concrete? Solve the formula for that variable.

(b) Use the new formula to determine the thickness of the slab of concrete.

Solution

(a) The thickness of the concrete is represented by H, so we solve the formula $V = LWH$ for H.

$$LWH = V \qquad \text{The coefficient of } H \text{ is } LW.$$

$$\frac{LWH}{LW} = \frac{V}{LW} \qquad \text{Divide both sides by } LW.$$

$$H = \frac{V}{LW}$$

(b) Because the length and width are expressed in feet, the volume of 8 cubic yards of concrete ordered must be expressed in cubic feet.

$$8 \text{ cubic yards} \cdot 27 \frac{\text{cubic feet}}{\text{cubic yard}} = 216 \text{ cubic feet}$$

$$H = \frac{216}{(200)\,(4)} = 0.27 \text{ feet}$$

The concrete slab will be about 0.27 feet or

$$0.27 \text{ feet} \cdot 12 \frac{\text{inches}}{\text{foot}} \approx 3 \text{ inches thick.}$$

It is impossible to be a good equation solver and a poor formula solver. The procedures are identical in both cases.

No matter what your current or future line of work may be, you may very well find that being able to solve a formula has the most practical application of all the mathematics you learn.

4.3 *Quick Reference*

Solving Formulas

- A **formula** is an equation that states how to calculate the value of one variable in terms of one or more other variables.

- Formulas are written in the form in which they are most commonly used. However, we sometimes need to know the value of another variable in the formula. If we need to calculate that variable repeatedly, it is usually best to **solve the formula** for that variable.

- When solving a formula for a particular variable, treat all other variables as constants. The methods for solving a formula are identical to the algebraic methods for solving any other equation.

4.3   *Exercises*

 1. In your own words, explain what is meant by solving a formula for a given variable.

 2. If we solve the formula $I = Prt$ for r, which letter do we regard as the variable? How do we treat the other letters?

In Exercises 3–32, solve the formula for the given variable.

3. $F = ma$ for m

4. $C = \pi d$ for d

5. $I = Prt$ for r

6. $V = LWH$ for W

7. $v = \dfrac{s}{t}$ for t

8. $r = \dfrac{d}{t}$ for t

9. $A = \dfrac{1}{2}bh$ for b

10. $V = \dfrac{1}{3}Bh$ for h

11. $A = \dfrac{1}{2}h(a + b)$ for a

12. $A = \dfrac{1}{2}h(a + b)$ for h

13. $A = \pi r^2$ for π

14. $x^2 = 4py$ for y

15. $E = mc^2$ for m

16. $V = \dfrac{4}{3}\pi r^3$ for π

17. $a = \dfrac{v - w}{t}$ for v

18. $z = \dfrac{x - u}{s}$ for u

19. $A = P + Prt$ for r

20. $IR + Ir = E$ for r

21. $P = \dfrac{nRT}{V}$ for R

22. $P = \dfrac{nRT}{V}$ for V

23. $R_T = \dfrac{R_1 + R_2}{2}$ for R_1

24. $A = \dfrac{r^2\theta}{2}$ for θ

25. $ax + by + c = 0$ for y

26. $y = mx + b$ for m

27. $s = \dfrac{1}{2}gt^2 + vt$ for v

28. $s = \dfrac{1}{2}gt^2 + vt$ for g

29. $A = a + (n - 1)d$ for n

30. $A = a + (n - 1)d$ for d

31. $S = \dfrac{n}{2}[2a + (n - 1)d]$ for a

32. $S = \dfrac{n}{2}[2a + (n - 1)d]$ for d

 33. The following shows two different ways of solving the equation $2x + 3y = 6$ for y. Explain why both methods are correct.

$$2x + 3y = 6 \qquad\qquad 2x + 3y = 6$$
$$3y = -2x + 6 \qquad\qquad 3y = -2x + 6$$
$$y = -\dfrac{2}{3}x + 2 \qquad\qquad y = \dfrac{-2x + 6}{3}$$

 34. In terms of graphing, what might be our motivation for solving the equation $2x + 3y = 6$ for y?

In Exercises 35–52, solve the given equation for y.

35. $2x + 5y = 10$

36. $3x - y = 9$

37. $x - y + 5 = 0$

38. $x - 5y + 10 = 0$

39. $2x + 3y - 9 = 0$

40. $5x - 6y + 12 = 0$

41. $3x = 4y + 11$

42. $5x = 7y - 15$

43. $3x - y = 5x + 3y$

44. $3(2x - y) = 4x - y - 2$

45. $0.4x - 0.3y + 12 = 0$

46. $10 = 0.4y - 0.5x$

47. $y - 4 = \dfrac{1}{2}(x + 6)$

48. $y + 7 = -\dfrac{3}{4}(x - 12)$

49. $\dfrac{4}{5}x - \dfrac{2}{3}y + 8 = 0$

50. $\dfrac{y}{8} - \dfrac{x}{4} = 0$

51. $\dfrac{y - 2}{3} = \dfrac{x + 3}{4}$

52. $\dfrac{3 - x}{2} + 5 = \dfrac{y + 1}{3}$

In Exercises 53–54, round all decimal results to the nearest hundredth.

53. The surface area A of a right circular cylinder is given by the formula $A = 2\pi r^2 + 2\pi rh$ where r is the radius of the circular base and h is the height of the cylinder.

(a) What is the surface area of a right circular cylinder if the radius is 2.7 inches and the height is 5.83 inches?

(b) Solve the formula for h.

(c) What is the height of a right circular cylinder if its surface area is 2858.85 square centimeters and its radius is 13 centimeters?

54. The volume V of a right circular cone is given by the formula $V = \frac{1}{3}\pi r^2 h$ where r is the radius of the circular base and h is the height of the cone.

(a) What is the volume of a right circular cone if the radius of the base is 2.7 meters and the height is 11.3 meters?

(b) Solve the formula for h.

(c) What is the height of a right circular cone if the volume is 1256.64 cubic feet and the radius is 10 feet?

In Exercises 55 and 56, solve the formula for the appropriate variable. Use your calculator for the repeated evaluations.

55. A satellite travels in a circular orbit of the earth. The circumference of the earth is approximately 25,000 miles. (The formula for the circumference of a circle with radius r is $C = 2\pi r$.) How high above the earth is the satellite if the length of one orbit is

(a) 25,800 miles?   (b) 26,000 miles?

(c) 26,500 miles?

56. Simple interest I on an investment P is calculated by the formula $I = Prt$, where r is the interest rate and t is the time in years. The following investments are available. Which has the best interest rate?

	Amount Invested	Time	Interest
(i)	$10,000	6 months	$375.00
(ii)	$ 5,000	5 months	$170.83
(iii)	$ 5,000	3 months	$100.00
(iv)	$ 7,500	8 months	$350.00
(v)	$ 8,000	1 year	$640.00

57. An amount of $2560 is invested at a simple interest rate. The investment yields $179.20 in interest for the year. What is the interest rate?

58. A runner jogged 4 miles in ten laps around a circular track. What is the greatest distance straight across the track?

59. A person needs to buy a tablecloth for a circular table that has a diameter of 5 feet. The tablecloth is to hang over the edge of the table by 6 inches. A store advertises circular tablecloths with an area of 28.3 square feet. Is the tablecloth the proper size?

60. The perimeter of a rectangular swimming pool is 180 feet. If the length is twice the width, how far is it across the diagonal of the pool? (Hint: Use the Pythagorean Theorem.)

61. Suppose you want to put fringe around a circular tablecloth whose diameter is 6 feet. If you already have 6 yards of fringe, do you need to buy more? Explain.

62. Suppose you have two fish tanks. One tank is 5 feet by 2 feet, and the depth of the water is 18 inches. Will the water be deeper if you pour it into the other tank, which is 6 feet by 1.5 feet? Explain.

63. The area of a soup can label is 30.19 square inches. The height of the label is $3\frac{11}{16}$ inches.

(a) What is the radius of the can?

(b) What is the area of the top of the can?

64. A regulation baseball has a circumference between 9 and $9\frac{1}{4}$ inches. Between what two values is the radius of the baseball?

65. The front of a concrete dam is in the shape of a trapezoid and has dimensions as shown in the accompanying figure. If the surface area of the front of the dam is 482,850 square feet, what is the height of the dam?

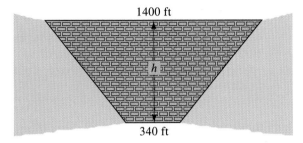

1400 ft

h

340 ft

66. A garden in the shape of a trapezoid with the dimensions shown in the accompanying figure has an area of 1386 square feet. What is the length of the side perpendicular to the bases?

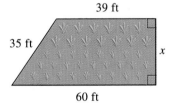

39 ft

35 ft

x

60 ft

67. If a couple decides to paddle their canoe directly across (through the center of) a circular pond whose circumference is 200 meters, how far will they travel?

68. The **Rule of 72** is often used to estimate the time or rate required for money to double with interest compounded continuously. The rule states that the product of the time and rate required for money to double is 72. The formula is $tr = 72$, where t represents the time in years and r represents the annual percentage rate. For example, at 9% interest it would take about 8 years to double an investment.

(a) Solve the formula $tr = 72$ for r and approximate what interest rate is required for an investment to double in six years.

(b) Solve the formula $tr = 72$ for t and approximate how long it will take for money invested at 7% compounded continuously to double in value.

69. Determine a formula for the surface area A of the rectangular box shown in the accompanying figure. (The surface area is the sum of areas of the six sides.)

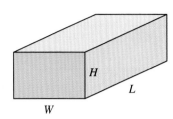

H

L

W

70. Which of the following formulas are equivalent to the formula $A = \frac{1}{2}bh$?

(i) $b = \dfrac{2A}{h}$

(ii) $b = 2A \cdot \dfrac{1}{h}$

(iii) $b = \dfrac{2}{h} \cdot A$

(iv) $b = \dfrac{A}{\frac{1}{2}h}$

(v) $b = \dfrac{A}{2h}$

71. (a) Figure (a) shows a metal band placed tightly around a baseball of radius 1.44 inches. The band is then removed and lengthened by 1 foot. Figure (b) shows the longer band placed around the baseball. How far will the band be from the surface of the baseball? (Hint: Compare the radius of the lengthened band with the radius of the baseball.)

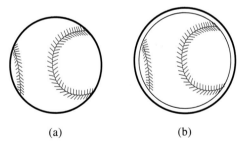

(a)   (b)

(b) Similarly, suppose a metal band is placed around the earth whose radius is 4000 miles. If the metal band is removed, lengthened by 1 foot, and replaced, how far will the band be from the surface of the earth?

(c) If exactly the same procedure is followed with a metal band around the Astrodome, can you answer the same question even without knowing the radius of the Astrodome? Why?

72. A student drove 20 miles from home to college to take her math test. Her average speed was 40 mph. When she arrived at the college, she realized she had forgotten her calculator. Traveling at an average speed of 50 mph, she returned home to get her calculator. Finally, driving at an average speed of 60 mph, she drove back to the college to take the test. To the nearest tenth, what was her average speed for the entire journey?

Exploring with Real Data

In Exercises 73–76, refer to the following background information and data.

From a modest beginning, Wal-Mart Stores has risen to become one of the largest retailing chains in America. The accompanying bar graph shows the number of stores at the end of selected years in the period 1983–1991. (Source: Wal-Mart Stores, Inc.) The figure also shows the graph of $y = 128x + 382$, where y represents the number of stores at the end of year x, and x is the number of years since 1982.

Figure for 73–76

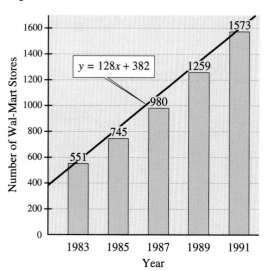
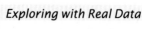

73. According to the equation, how many stores were added from the end of 1985 to the end of 1987? How close is your result to the actual data shown in the figure?

74. Based on just a visual inspection of the figure, how accurate do you judge the equation to be?

75. According to the equation, during what year was the first Wal-Mart store established? Do you think the equation is a valid model prior to 1983?

76. Consider the impact chain stores such as Wal-Mart have had on large urban department stores. How would the trend in the number of such urban stores compare to the trend indicated in the figure?

Challenge

In Exercises 77–80, solve the given equation for x.

77. $a(x + c) = b(x - c)$

78. $3b - (b - x) = b(x - 2)$

79. $c = \dfrac{x - a}{x + a}$

80. $\dfrac{2x - 1}{x + 3} = y$

In Exercises 81–84, solve the given equation for the indicated variable.

81. $S = \dfrac{a}{1 - r}$ for r **82.** $F = \dfrac{mv^2}{r}$ for r

83. $\dfrac{P_1V_1}{T_1} = \dfrac{P_2V_2}{T_2}$ for V_2

84. $\dfrac{P_1V_1}{T_1} = \dfrac{P_2V_2}{T_2}$ for T_1

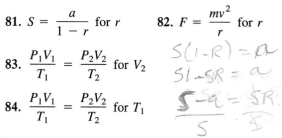

In Exercises 85–88, solve the given formula for the indicated variable.

85. $E = IR + Ir$ for I

86. $A = 2LW + 2LH + 2WH$ for L

87. $A = P + Prt$ for P

88. $A = \frac{1}{2}b_1h + \frac{1}{2}b_2h$ for h

4.4 Applications

A General Approach to Problem Solving ▪ *Consecutive Integers* ▪ *Piece Lengths* ▪ *Angles of a Triangle* ▪ *Rectangles* ▪ *Things of Value*

A General Approach to Problem Solving

From a practical standpoint, it makes little sense to learn mathematics and not apply it to anything. This section and the next provide the opportunity to bring together several skills and to use them in solving various kinds of application problems.

Many of the problems involve common, everyday situations. Others, while not as realistic, are important because they help you obtain a conceptual understanding of how mathematics can be used to describe a problem and lead to the development of problem-solving strategies.

Although we will illustrate various types of application problems, real-life problems rarely fall into a specific category. It is extremely important to concentrate on the general approach and to develop skills and strategies that can be applied to any problem.

The following is a summary of a general approach to problem solving. We will elaborate on these strategies as we discuss examples.

A General Approach to Problem Solving

1. On your first reading of the problem, make sure you understand what the problem is about and what question(s) you need to answer.

2. Assign a variable to the unknown quantity. Be specific about what the variable represents.

3. If there are other unknown quantities, represent them in terms of the same variable assigned in step 2.

4. Write an equation that describes, in symbols, the information given in the problem.

5. Solve the equation.

6. Use the solution to answer the question(s) asked in the problem.

7. Check the answer(s) by confirming that the conditions stated in the problem are met.

Steps 2 and 3 of this general approach can sometimes be facilitated with a picture, table, or other visual aid.

The following examples illustrate how this general approach can be used in a variety of applications. Naturally, there is no way to include every possible type of problem that could ever be encountered. But these examples will provide a good start toward developing methods that will generally work well for any problem.

Consecutive Integers

Two or more integers that appear immediately after one another on the number line are called **consecutive integers.** Problems can also involve **consecutive even integers** or **consecutive odd integers.** Here are examples.

Consecutive integers:	2, 3, 4, 5, 6, 7
Consecutive even integers:	$-4, -2, 0, 2, 4, 6$
Consecutive odd integers:	$-5, -3, -1, 1, 3, 5$

If we let n represent the first integer of a sequence, the rest of the representation looks like this.

Consecutive integers:	$n, n + 1, n + 2, n + 3, \ldots$
Consecutive even integers:	$n, n + 2, n + 4, n + 6, \ldots$
Consecutive odd integers:	$n, n + 2, n + 4, n + 6, \ldots$

Note that consecutive even integers and consecutive odd integers are represented the same way, but the variable n represents a different kind of number in the two cases.

EXAMPLE 1   *Consecutive Odd Integers*

The lockers on the right side of the hall are numbered consecutively with even integers; the lockers on the left side of the hall are numbered consecutively with odd integers. Two people have adjoining lockers on the left side of the hall. They note that their locker numbers have a sum of 60. What are their locker numbers?

Solution

Because the lockers are on the left side of the hall, the numbers are consecutive odd integers. The goal is to determine the two locker numbers. Let

x = first odd integer and

$x + 2$ = second odd integer.

The sum of the locker numbers can be represented by $x + (x + 2)$. Because the sum is given as 60, the equation is

$$x + (x + 2) = 60.$$

Solve the equation.

$x + (x + 2) = 60$	
$x + x + 2 = 60$	Remove parentheses.
$2x + 2 = 60$	Combine like terms.
$2x + 2 - 2 = 60 - 2$	Subtract 2 from both sides.
$2x = 58$	
$\dfrac{2x}{2} = \dfrac{58}{2}$	Divide both sides by 2.
$x = 29$	

Because x (the first locker number) is 29, $x + 2$ (the second locker number) is 31. The locker numbers are 29 and 31. Is the sum of the locker numbers 60? Yes, $29 + 31 = 60$. ◼

At the end of Example 1, we verified the answers by confirming that they meet the conditions of the problem. In future examples we will leave this verification to you.

NOTE: It is *not* helpful to substitute answers into the equation, because the equation you have written might be in error. Instead, verify an answer by confirming that it meets the conditions described in the problem. ◼

Piece Lengths

In this type of problem, material of some known length is to be cut into two or more pieces. We are given the relative sizes of the pieces, and the task is to determine the actual length of each piece.

EXAMPLE 2   *Lengths of Pieces of Rope*

A rope is 72 feet long and is to be cut into three pieces. The first piece must be 8 feet shorter than the second piece, and the third piece must be twice as long as the second piece. How long should the first piece be?

Solution

The rope 72 feet long is being cut into three pieces. (See Fig. 4.17.) Although the goal is to determine how long the first piece should be, we will determine all three lengths so that the results can be verified.

Figure 4.17

Note that the first and third pieces are both described in comparison to the second piece, so we assign the variable to the second piece. We indicate this assignment on the picture. (See Fig. 4.18.) Your picture does not have to be to scale. Its purpose is just to give some visual aid.

Figure 4.18
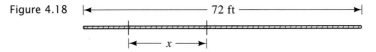

Because the first piece is 8 feet shorter than the second piece, its length can be represented by $x - 8$. (See Fig. 4.19.) The third piece is twice as long as the second piece, so its length can be represented by $2x$.

Figure 4.19
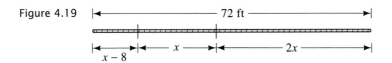

Figure 4.19 also shows the overall length of 72, which is the sum of the three piece lengths. The equation is

$$(x - 8) + x + 2x = 72.$$

Solve the equation.

$$(x - 8) + x + 2x = 72$$
$$x - 8 + x + 2x = 72 \qquad \text{Remove parentheses.}$$
$$4x - 8 = 72 \qquad \text{Combine like terms.}$$
$$4x - 8 + 8 = 72 + 8 \qquad \text{Add 8 to both sides.}$$
$$4x = 80$$
$$\frac{4x}{4} = \frac{80}{4} \qquad \text{Divide both sides by 4.}$$
$$x = 20$$

Because x represents the length of the *second* piece, the length of the second piece is 20 feet. The first piece is $x - 8$, which is $20 - 8$ or 12 feet. The third piece is $2x$, which is $2 \cdot 20$ or 40 feet.

But the question asks for the length of the first piece, so the answer is 12 feet. ■

Angles of a Triangle

Problems of this kind usually describe the relative measures of the three angles of a triangle. The task is to determine the measure of each angle. The basis for the equation in such problems is the theorem from geometry that states that the sum of the measures of the angles of a triangle is 180°.

EXAMPLE 3   *The Measures of the Angles of a Triangle*

A surveyor is mapping a plot of land in the form of a triangle. The second angle of the triangle is 21° greater than the first angle. The third angle is 11° more than twice the first angle. What are the measures of the three angles?

Solution

The goal is to determine the measures of all three angles of the triangle.

Note that the second and third angles are both described in comparison to the first angle. Therefore, assign the variable to the measure of the first angle.

Figure 4.20

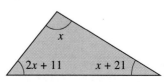

A picture is useful in indicating this assignment as well as the representations of the other two angles. (See Fig. 4.20.) The triangle does not have to be to scale, and it does not matter which angle is considered the first angle.

Because the sum of the measures of the angles is 180°, we write the equation by adding the represented sizes of the angles and letting the sum equal 180.

$$x + (x + 21) + (2x + 11) = 180$$

Now solve the equation.

$$x + x + 21 + 2x + 11 = 180 \qquad \text{Remove parentheses.}$$
$$4x + 32 = 180 \qquad \text{Combine like terms.}$$
$$4x + 32 - 32 = 180 - 32 \qquad \text{Subtract 32 from both sides.}$$
$$4x = 148$$
$$\frac{4x}{4} = \frac{148}{4} \qquad \text{Divide both sides by 4.}$$
$$x = 37$$

The measure of the first angle is $37°$. The second angle is $x + 21 = 37 + 21$ or $58°$. The third angle is $2x + 11 = 2 \cdot 37 + 11$ or $85°$. ■

Rectangles

The basic quantities involving a rectangle are the length, width, perimeter, and area. Rectangle problems typically describe the relationship between the length and width and provide either the perimeter or the area.

EXAMPLE 4   *Carpet for a Rectangular Room*

A carpet layer has been to a home to provide an estimate of the cost of carpeting a rectangular dining room. When he returns to his office, he finds that he has misplaced the dimensions. He remembers that the perimeter is 52 feet and that the width is 4 feet less than the length. If the carpet costs \$12.95 per square yard, how can he still provide the estimate, and what is it?

Solution   The total cost of the carpeting is \$12.95 times the number of square yards needed. The area of a rectangle is given by $A = LW$. Therefore, the primary goal is to determine the dimensions that he misplaced.

Figure 4.21

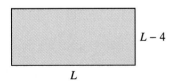

$L - 4$

L

Because the width is described in comparison to the length, it is easier to assign the variable L to the length. Since the width is 4 feet less than the length, the width can be represented by $L - 4$. (See Fig. 4.21.)

The perimeter of a rectangle is given by the formula $P = 2L + 2W$. Because we know the perimeter is 52, the equation is

$$2L + 2(L - 4) = 52.$$

Solve the equation.

$$2L + 2(L - 4) = 52$$
$$2L + 2L - 8 = 52 \qquad \text{Distributive Property}$$
$$4L - 8 = 52 \qquad \text{Combine like terms.}$$
$$4L - 8 + 8 = 52 + 8 \qquad \text{Add 8 to both sides.}$$
$$4L = 60$$
$$\frac{4L}{4} = \frac{60}{4} \qquad \text{Divide both sides by 4.}$$
$$L = 15$$

The carpet layer now knows that the length of the dining room, represented by L, is 15 feet. The width, represented by $L - 4$, is $15 - 4$ or 11 feet.

Knowing the dimensions of the room, the carpet layer can now produce the estimate. The area of the room is found by the formula $A = LW$.

$$A = 15 \cdot 11 = 165 \text{ square feet}$$

One square yard is the same as 9 square feet. Therefore, the area is $165 \div 9 \approx 18.3$ square yards. At \$12.95 per square yard, the cost is $12.95 \cdot 18.3 \approx \$237.$ ■

Things of Value

There are many applications involving things of value. The objects may be coins, tickets to a play, or pounds of bananas.

The problems usually involve more than one of an item. Typically, the per-unit value is given, as is the total value of all the items. The total value for a collection of items is determined by multiplying the number of items times their per-unit value. To organize the information in the problem, a table is particularly useful.

EXAMPLE 5   *Two Kinds of Tickets with Different Values*

At the local ball game, children's tickets cost \$2.00 each and adult tickets cost \$4.75 each. There were six fewer adults than children at the game. If the total cost of the tickets was \$93.00, how many adults went to the game?

Solution   We can summarize the information in a table. The table also shows how the variable is assigned and how other unknown quantities are represented.

	Number of Tickets	*Per-Ticket Cost*	*Cost of Tickets*
Children	x	2.00	$2.00x$
Adults	$x - 6$	4.75	$4.75(x - 6)$
TOTAL			93.00

The cost of the children's tickets plus the cost of the adults' tickets equals the total cost of the tickets.

$$2.00x + 4.75(x - 6) = 93.00$$

Solve the equation.

$$2.00x + 4.75(x - 6) = 93.00$$
$$2.00x + 4.75x - 28.50 = 93.00 \qquad \text{Distributive Property}$$
$$6.75x - 28.50 = 93.00 \qquad \text{Combine like terms.}$$
$$6.75x - 28.50 + 28.50 = 93.00 + 28.50 \qquad \text{Add 28.50 to both sides.}$$
$$6.75 = 121.50$$
$$\frac{6.75x}{6.75} = \frac{121.50}{6.75} \qquad \text{Divide both sides by 6.75.}$$
$$x = 18$$

Remember that x represents the number of *children* who went to the game. The question asks for the number of *adults,* which we represented with $x - 6$. The number of adults who went to the game was $18 - 6 = 12$. ■

Note that the monetary units used in Example 5 were dollars. The units could have been cents to avoid the decimals, but the arithmetic involved in solving the equation is virtually the same either way. When solving applications involving monetary units, just be sure your units are the same in all expressions.

An application problem can have explicit or implied conditions that cannot be met. The following are some examples of applications with solutions that must be disqualified.

Application Problem	*Disqualified Solution*
Consecutive integers	Not an integer
Consecutive even integers	Odd integer
Consecutive odd integers	Even integer
Dimensions of a figure	Not a positive number
Number of things of value	Not a whole number
Time, rate, or distance	Not a positive number

Remember that a solution of an equation is not necessarily the answer to the question. Make sure your answer meets all of the conditions of the problem.

4.4 *Quick Reference*

A General Approach to Problem Solving

- Although the application examples in this section are presented in categories, it is important to understand and use a general approach that can be applied to any problem.

  1. On your first reading of the problem, make sure you understand what the problem is about and what question(s) you need to answer.

  2. Assign a variable to the unknown quantity. Be specific about what the variable represents.

  3. If there are other unknown quantities, represent them in terms of the same variable assigned in step 2.

  4. Write an equation that describes, in symbols, the information given in the problem.

  5. Solve the equation.

  6. Use the solution to answer the question(s) asked in the problem.

  7. Check the answer(s) by confirming that the conditions stated in the problem are met.

- Pictures, tables, and other visual aids are often useful for labeling unknown quantities and organizing information.

4.4   *Exercises*

Numbers

1. Suppose the total of two numbers is *T*. If one of the numbers is *n*, how can you represent the other number in terms of *n*? Show that the sum of these representations is *T*.

2. Let *x* represent a number, and write a word problem leading to the equation $2x - 1 = 19$.

3. The sum of a number and 6 more than the number is 32. What is the number?

4. The sum of a number and twice the number is 57. What is the number?

5. One number is 12 more than another number. If one-half of the smaller number equals one-third of the larger number, what is the larger number?

6. Increasing a number by 3 and multiplying the result by 4 is the same as decreasing the number by 2 and multiplying the result by 9. What is the number?

7. In the election of 1952, Eisenhower received 353 more electoral votes than Stevenson. If 531 electoral votes were cast, how many electoral votes did Eisenhower receive?

8. In the election of 1948, Truman received 114 more electoral votes than Dewey. If 492 electoral votes were cast, how many electoral votes did Truman receive?

9. There are 39 students in a political science class. The instructor notices that there are half as many men as women. How many women are in the class?

10. Sixty people attended a college luncheon. The number of women was one-third the number of men. How many men attended?

11. A brick balances with half of a brick and 20 ounces. How much does the brick weigh?

12. An uncle wills $3000 to his two nieces and one nephew. The older niece is to receive twice what the younger niece receives, and the nephew is to receive an amount equal to the sum of the nieces' amounts. How much will each person receive?

13. In a poker game two people began with equal sums of money. When the loser lost $50 more than a quarter of his money, the winner had twice as much money as the loser. How much money did each person have at the end of the game?

14. An appliance company had four more washers than dryers in its inventory. On Friday half the dryers and one-third of the washers were shipped out. If a total of eight appliances were shipped, how many were dryers?

15. A retail store ordered some sofas and chairs from a manufacturer. The number of chairs ordered was four more than the number of sofas. The store received a partial shipment consisting of half of the sofas and one-third of the chairs that were ordered. If the total number of pieces was the same as the number of sofas originally ordered, how many sofas did the store receive?

16. Some guests had arrived at a party by 8:00. At 8:30, two more arrived. By 9:00 the 21 guests in attendance were three times the number who were there at 8:30. How many had arrived by 8:00?

17. By 10:00 A.M., the Dow-Jones average had dropped from the opening average. By noon the Dow-Jones average slipped another 8 points. At the close of trading, the 50-point drop was five times the amount by which the average had dropped at noon. By how much had the average dropped at 10:00 A.M.?

18. The government's farm price support program pays a farmer not to plant any crops this year, so he must decide how not to plant his 100 acres. He decides not to plant corn on three times as many acres as he will not plant strawberries. The number of acres on which he will not harvest hay is equal to the total number of acres on which he will plant neither corn nor strawberries. On how many acres will the farmer not plant corn?

Consecutive Integers

19. Explain why *x* and $x + 2$ can be used to represent either two consecutive *even* numbers or two consecutive *odd* integers.

20. For three consecutive integers, the sum of the smallest and largest integers equals twice the middle integer. What are the integers? Explain the result.

21. The sum of two consecutive integers is 95. What are the two integers?

22. The sum of three consecutive integers is 3. What are the three integers?

23. The sum of two consecutive odd integers is 0. What are the two integers?

24. The sum of two consecutive even integers is 174. What are the two integers?

25. Two people are reading the same book, but one is two pages ahead of the other. If the two people add their page numbers, the sum is 216. On what page is the faster reader?

26. At the sports club, two people have adjacent lockers, and the sum of their locker numbers is 434. Show how you know that the lockers are not numbered consecutively?

27. Three sisters were born 2 years apart. If the sum of their current ages is 108, how old is the elder sister?

28. A precinct worker issues voting passes to three voters consecutively. The serial numbers on the passes total 237. What was the serial number of the second voter's pass?

 29. For any three consecutive integers, show that the average of the smallest and largest integers is the middle integer.

 30. Show that the absolute value of the difference between any two consecutive integers is always 1.

 31. The sum of three consecutive even integers is 19. What are the integers? Could you have anticipated your answer without solving an equation?

 32. The sum of three consecutive odd integers is 19. What are the integers? Can the conditions of the problem be met? Explain.

33. After three college basketball games, a forward, center, and guard have a total of 143 points. The forward has scored the most points, and the guard has scored the fewest. The forward's and center's points are consecutive integers. The guard's and center's points are consecutive even integers. How many points has the forward scored?

34. A bookmark is placed in a book between two pages whose sum is 193. On what page did the reader stop reading? (There are two possible answers.)

Piece Lengths

 35. Suppose an item of length L is cut into three pieces whose lengths are a, b, and c. Write an equation involving the expression $a + b + c$.

 36. Suppose an item of length L is cut into three pieces whose lengths are consecutive integers. If the length of the shortest piece is x, write an expression for L.

37. A scout leader is on a camping trip with his troop. He wants to cut a 40-meter rope into three lengths. The second length must be 2 meters longer than the first length. The third length must be 6 meters longer than twice the first length. What are the three lengths of rope?

38. A 45-foot rope is to be cut into three pieces. The second piece must be twice as long as the first piece, and the third piece must be 9 feet longer than three times the length of the second piece. How long should each of the three pieces be?

39. A fabric store worker has a 106-yard bolt of cloth that is to be cut into three lengths. The second piece is to be 4 yards longer than the first. The third piece is to be three times as long as the second piece. What is the length of each piece?

40. A 54-foot log is to be cut into three smaller logs whose lengths are consecutive even integers. How long should each piece be?

Angles of a Triangle

 41. Suppose the measures of the three angles of a triangle are d, e, and f. Write an equation involving the expression $d + e + f$.

 42. If the measures of three angles are 52°, 84°, and 43°, explain how you know that the angles are *not* angles of a triangle.

 43. One angle of a triangle is 20° greater than the first angle. The third angle is twice as large as the first angle. What are the measures of the three angles?

44. The second angle of a triangle is 7° more than three times the first angle. The third angle is 12° greater than the second angle. What are the measures of the three angles?

45. A **right triangle** is a triangle with one angle measuring 90°. If the second angle of a triangle is twice the first angle, and the third angle is one and a half times greater than the second angle, determine whether the triangle is a right triangle.

46. An **isosceles triangle** is a triangle with at least two angles having the same measure. (See figure.) If the first angle of a triangle is 30° smaller than the third angle, and the third angle is 20° less than twice the second angle, determine whether the triangle is an isosceles triangle.

Figure for 46   Figure for 47

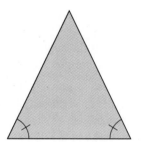 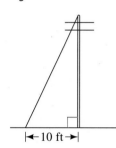

47. A supporting guy wire is attached to the top of a utility pole and is fastened at the ground 10 feet from the base of the pole. (See figure.) If the angle between the guy wire and the ground is eight times the angle between the guy wire and the pole, what is the measure of the angle between the guy wire and the ground?

48. An Atlanta Braves pennant is a triangle with two of the angles having the same measure. The third angle is one-fifth the sum of the two equal angles. What is the measure of the smallest angle?

Rectangles

49. What is the difference between the perimeter of a rectangle and the area of the rectangle?

50. Explain why we should check the answer to an application problem by verifying that the answer meets the stated conditions of the problem rather than by substituting into the original equation.

51. The width of a rectangle is 8 feet less than the length. If the perimeter of the rectangle is 64 feet, what are the dimensions of the rectangle?

52. The width of a rectangle is 16 inches less than the length. If the perimeter of the rectangle is 88 inches, what are the dimensions of the rectangle?

53. The length of a rectangle is 4 yards greater than the width. If the perimeter of the rectangle is 60 yards, what are the dimensions of the rectangle?

54. The length of a rectangle is 7 meters greater than the width. If the perimeter of the rectangle is 134 meters, what are the dimensions of the rectangle?

55. The height of a book is 4 inches greater than its width. If the perimeter of the book is 48 inches, can the book be placed upright on a bookcase shelf 12 inches high?

56. An investor owns a rectangular plot of land. The length of the land is 240 feet more than twice the width, and the perimeter is exactly 1 mile. How wide is the land?

57. A rectangular garden has a perimeter of 96 feet. The width is 12 feet less than the length. Each row of vegetables, planted the long way, is 3 feet wide. How many rows can be planted?

58. The width of one poster is 7 inches more than half the length. The width of a second poster is 9 inches less than the length of the first poster, and the length of the second poster is 10 inches greater than its width. If the two posters have the same perimeter, which poster is wider?

59. A rectangle is 5 inches longer than it is wide. If its width is increased by 8 inches and its length is doubled, the new rectangle has a perimeter of 60 inches. How long is the original rectangle?

60. The width of a sheet of paper is 2.5 inches less than its length. A photocopy is made that is 90% of the original size. The reduced copy has a perimeter of 35.1 inches. How wide was the original paper?

61. Curtains 40 inches long are to be hung on a rectangular window. The height of the window is one and a half times its width, and the perimeter of the window is 150 inches. Will the curtains cover the full height of the window?

62. A computer is 2 inches taller than it is wide. The perimeter of the front of the computer is 60 inches. Will the computer fit under a shelf that is 18 inches above the top of the desk?

Things of Value

63. If *n* represents a number of quarters, write an expression for the value of the coins in cents and another expression for the value of the coins in dollars.

64. If 10*x* represents the value of a number of coins in cents, what does *x* represent? What kind of coins are they?

65. A collection of nickels and dimes contains nine more nickels than dimes. If the total value of the collection is $2.25, how many nickels are there?

66. A collection of dimes and quarters has a value of $5.20. If there are four more quarters than dimes, how many quarters are there in the collection?

67. A collection of dimes and quarters has a value of $5.15. If there are six more dimes than quarters, how many dimes are there in the collection?

68. A collection of nickels and dimes is valued at $2.50. If there are eight more nickels than dimes, how many nickels are there in the collection?

69. A large family went to a school play and bought seven more tickets for children than for adults. A child's ticket cost $1.25 and an adult's ticket cost $2.75. The family spent a total of $28.75 for the tickets. How many children went to the play?

70. The total receipts for a college basketball game were $674, with a student's ticket selling for $1.25 and an adult's ticket selling for $2.75. There were 136 more students than adults in attendance. How many people attended the game?

71. Hamburgers sell for $2 each and hot dogs sell for $1.25 each. On a certain day, the number of hamburgers sold was twice the number of hot dogs sold. Determine the number of hot dogs sold if the total income was $304.50.

72. Mama's Pizzeria sells individual pizzas for $8 each and an order of bread sticks for $1 each. During a 2-week period, the number of pizzas sold was four times the number of bread stick orders sold. How many pizzas were sold if the total revenue was $2970?

Miscellaneous Applications

73. A person invested one-fourth of an inheritance in bank stock, one-fifth in bonds, and one-half in a mutual fund. If the sum of her investments was $38,000, how much did she inherit?

74. There was a fish whose head weighed 12 pounds. Its tail weighed as much as the head plus half the weight of the body. The body weighed 26 pounds more than the head and tail together. What was the weight of the fish?

75. An **obtuse triangle** is a triangle with one angle having a measure that is greater than 90°. (See figure.) If one angle of a triangle is four times the second angle, and the third angle is 2° less than twice the second angle, determine whether the triangle is an obtuse triangle.

Figure for 75   Figure for 76

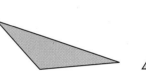 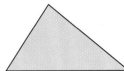

76. An **acute triangle** is a triangle with all three angles having measures that are less than 90°. (See figure.) If the second angle of a triangle is six-sevenths as large as the third angle, and the first angle is five-sixths as large as the second angle, determine whether the triangle is an acute triangle.

77. The sum of a number and three times the number is 68. What is the number?

78. One number is 21 less than another number. If one-half of the smaller number equals one-fifth of the larger number, what is the smaller number?

79. A collection of nickels and dimes contains four more dimes than nickels. If the total value of the collection is $2.50, how many dimes are there?

80. A collection of nickels and dimes has a value of $2.50. If there are seven more dimes than nickels, how many dimes are there in the collection?

81. In the election of 1884, Blaine (from Maine) received 37 fewer electoral votes than Cleveland. If 401 electoral votes were cast, how many electoral votes did each receive?

82. In the election of 1880, Hancock received 59 fewer electoral votes than Garfield. If 369 electoral votes were cast, how many electoral votes did each receive?

83. The sum of two consecutive odd integers is 224. What are the two integers?

84. The sum of two consecutive even integers is 250. What are the two integers?

85. A 105-yard bolt of cloth is to be cut into three lengths. The second piece is to be 5 yards shorter than the first piece, and the third piece is to be twice the length of the second piece. What are the lengths of the three pieces?

86. A 32-foot pipe is cut into three pieces. The first two lengths are consecutive integers, and the third piece is 6 feet longer than the second. How long are the three pieces?

87. The sum of a number and 7 less than the number is 165. What is the number?

88. Increasing a number by 5 and multiplying the result by 7 is the same as multiplying the number by -5 and subtracting 1 from the result. What is the number?

89. Two angles are **supplementary angles** if the sum of their measures is $180°$. What is the measure of the smaller of two supplementary angles if the measure of the larger is $10.8°$ more than twice the measure of the smaller?

90. Two angles are **complementary angles** if the sum of their measures is $90°$. What is the measure of the larger of two complementary angles if the measure of the smaller angle is $15°$ less than half the measure of the larger angle?

91. A commercial developer reads the zoning laws on how he must allot his 50 acres of land. The number of acres of lawn and landscaping must equal the number of acres used for buildings. The number of acres used for parking must be two-fifths the number of acres used for buildings. The number of acres for roads and utilities must be one-tenth the number of acres for lawn and landscaping. How many acres of parking lot must the developer provide?

92. There are 266 men, women, and children at the company picnic. There are four times as many men as children and twice as many women as children. How many of each came to the picnic?

93. The sum of three consecutive odd integers is 369. What are the three integers?

94. The sum of three consecutive even integers is 180. What are the three integers?

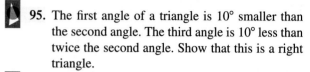

95. The first angle of a triangle is $10°$ smaller than the second angle. The third angle is $10°$ less than twice the second angle. Show that this is a right triangle.

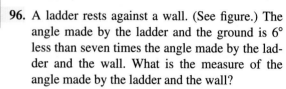

96. A ladder rests against a wall. (See figure.) The angle made by the ladder and the ground is $6°$ less than seven times the angle made by the ladder and the wall. What is the measure of the angle made by the ladder and the wall?

Figure for 96   Figure for 97

   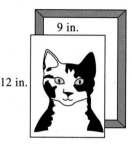

9 in.

12 in.

97. A 9-inch by 12-inch picture is to be mounted in a frame. (See figure.) The width of the frame is 5 inches less than the length, and the perimeter of the frame is 42 inches. Will the picture fit in the frame?

98. The width of a rectangle is 2 feet more than half the length. If the perimeter of the rectangle is 88 feet, what is the width?

99. The ages of three sisters are consecutive odd integers. Their brother is 13 years older than the middle sister. If the sum of their four ages is 97, how old is the oldest sister?

100. A member of a three-person bowling team tells his wife that he bowled over 200 last night, but he will not give the actual score. His wife asks another member of the team about it, but she learns only that the team's three scores were consecutive odd integers. His wife then checks with the third member of the team, who says that the team score was 597. Could her husband have bowled over 200? Can she know for sure? Explain.

101. A child has saved $3.55 in nickels and dimes. If there are ten fewer nickels than dimes, how many nickels does the child have?

102. Soft drinks sell for 75 cents each and milk shakes sell for $1.25 each. On a certain day, the number of soft drinks sold was three times the number of milk shakes sold. How many soft drinks were sold if the total income was $280?

Exploring with Real Data

In Exercises 103 and 104, refer to the following background information and data.

A flea can jump 30,000 times in a row for 72 continuous hours without resting. Each horizontal leap is 12 inches and has an acceleration that is about 50 times that of a space shuttle. (Source: *Atlanta Journal and Constitution.*)

103. According to this data, what is the elapsed time between jumps?

104. How many miles can a flea travel in 72 hours?

Challenge

105. George P. Burdell enjoys entertaining friends with number games. In one game he asks a friend to choose any number. Next he tells his friend to subtract one from the number, multiply the result by three, and then subtract twice the number increased by one. Finally, he tells his friend to add five to the result. George asks what the resulting number is and tells his friend that it is the chosen number. Write an equation to describe George's trick. Explain why George's trick works for any number.

106. George P. Burdell enjoys entertaining friends with number games. In one game he asks a friend to choose any number and add eight to it; multiply the result by three and then subtract three from the result; multiply by two and then divide by six; and finally, subtract the original number. George knows that the result is 7. Write an equation to describe George's trick. Explain why George's trick works for any number.

4.5 More Applications

Percentages ▪ *Fixed and Variable Costs* ▪ *Simple Interest* ▪
Distance, Rate, and Time ▪ *Mixtures* ▪ *Liquid Solutions*

For most of the application problems in the previous section, equations were based on information provided by the problem itself.

This section deals with problems for which certain common knowledge is needed. In a sense, the use of this common knowledge makes the applications in this section more practical.

Percentages

Ad valorem taxes are taxes based on the value of property. The valuation of the property, for tax purposes, is usually some percentage of the market or list value of the property. The ad valorem tax is then some percentage of the valuation.

EXAMPLE 1   *Ad Valorem Tax on an Automobile*

Suppose the valuation of an automobile for tax purposes is 40% of the list value of the car, and the ad valorem tax rate is 2%. If the tax due on an automobile is $27, what is the list value of the car?

Solution

Let x = the list value of the car.

$$\text{Valuation} = 0.40x \qquad\qquad \text{Valuation = 40\% of list value}$$

$$\text{Tax due} = (0.02)(0.40x) = 0.008x \qquad \text{Tax due = 2\% of the valuation}$$

Because the tax due is $27, we must solve the following equation.

$$0.008x = 27.00$$

$$\frac{0.008x}{0.008} = \frac{27.00}{0.008}$$

$$x = 3375$$

The list value of the car is $3375.00.   ■

When an item is placed on sale, the amount by which the price is reduced is called the **discount,** which is expressed as a percentage of the **original price.**

Retail store owners buy their goods at **wholesale prices.** Then, to cover overhead and the profit they want to make, they add an amount called **markup,** which is expressed as a percentage of the wholesale price. The result is the **retail price.**

EXAMPLE 2   *Wholesale Price Plus Markup Equals Retail Price*

All the goods at a hobby store are marked up 40%. If the retail price of one item is $32.13, what was the wholesale cost to the store owner?

Solution

Let w = the wholesale cost.

$$\text{Markup} = 0.40w \qquad \text{Markup} = 40\% \text{ of wholesale cost}$$

Because the retail price is $32.13, we must solve the following equation.

$$
\begin{aligned}
w + 0.40w &= 32.13 \qquad \text{Wholesale Cost + Markup = Retail Price}\\
1.00w + 0.40w &= 32.13\\
1.40w &= 32.13\\
\frac{1.40w}{1.40} &= \frac{32.13}{1.40}\\
w &= 22.95
\end{aligned}
$$

The wholesale cost was $22.95.

Fixed and Variable Costs

The cost of doing business can be separated into two components. One is the **fixed cost,** which is a cost that is incurred even if no products are sold or no service is rendered. The other component is the **variable cost,** which is a cost that depends on the number of products produced or sold or on the amount of service rendered.

EXAMPLE 3   *The Cost of an Auto Rental*

An employee submits an expense account for a 1-day round trip to a town 40 miles away. The expense account includes a car rental for $82.00. The fixed daily rate is $28.00 and the mileage rate is $0.20 per mile. Is there reason to question the expense of the car rental?

Solution   Let M = the number of miles driven. The fixed cost is $28.00. The variable cost is $0.20 times the number of miles driven, or $0.20M$. The total cost is $82.00, so we solve the following equation.

$$
\begin{aligned}
28.00 + 0.20M &= 82.00 \qquad \text{Fixed Cost + Variable Cost = Total Cost}\\
28.00 + 0.20M - 28.00 &= 82.00 - 28.00\\
0.20M &= 54\\
\frac{0.20M}{0.20} &= \frac{54}{0.20}\\
M &= 270
\end{aligned}
$$

The employee has been charged for 270 miles when the trip should have been only 80 miles. Either the employee or the car rental agency has made a mistake.

Simple Interest

When money is deposited in a bank account, the bank pays for the use of the money. The amount deposited is called the **principal** (*P*). The amount paid by the bank is called the **interest** (*I*). Interest is expressed as a percentage (*r*) of the principal.

When the bank calculates interest just once a year, it is called **simple interest.** To calculate the interest that accumulates over a period of *t* years, we use the formula *I = Prt*.

EXAMPLE 4 *Loan Plus Interest Equals Loan Payoff*

A used car buyer borrows money for 1 year at a simple interest rate of 12%. If the payoff of the loan is $5040, what was the amount of the original loan?

Solution

The amount of the loan is the principal *P*. The interest is 12% of *P*, or 0.12*P*. At the end of the year, the lender will receive the original amount of the loan plus interest, or *P* + 0.12*P*.

Because the loan payoff is $5040, the equation is as follows.

$$P + 0.12P = 5040 \qquad \text{Loan Amount + Interest = Loan Payoff}$$
$$1.00P + 0.12P = 5040$$
$$1.12P = 5040$$
$$\frac{1.12P}{1.12} = \frac{5040}{1.12}$$
$$P = 4500$$

The original loan was $4500. ■

In a *dual-investment* problem, a given amount of money is divided into two different investments, each yielding a given interest percentage. With the total amount of interest earned given, the problem is to determine the individual investment amounts.

A table can be very useful in arranging the information for such problems.

EXAMPLE 5 *Investments in Two Bond Funds*

A total of $10,000 was invested in two bond mutual funds, a junk bond fund and a government bond fund. The junk bond fund is risky and yields 11% interest. The safer government bond fund yields only 5%. The year's total income from the two investments was $740. How much was invested in each fund?

Solution

Let *x* = the amount of money invested in the junk bond fund. The remaining money, 10,000 − *x*, is the amount invested in the government bond fund.

	Amount Invested	*Interest Rate*	*Interest Earned*
Junk Bonds	x	0.11	0.11x
Government Bonds	10,000 − x	0.05	0.05(10,000 − x)
TOTALS	10,000		740

The total interest earned is the sum of the interest earned from each fund.

$$0.11x + 0.05(10{,}000 - x) = 740$$
$$0.11x + 500 - 0.05x = 740$$
$$0.06x + 500 = 740$$
$$0.06x + 500 - 500 = 740 - 500$$
$$0.06x = 240$$
$$\frac{0.06x}{0.06} = \frac{240}{0.06}$$
$$x = 4000$$

The amount invested in the junk bond fund was $4000, and 10,000 − 4000 or $6000 was invested in the government bond fund.   ■

Distance, Rate, and Time

Distance d is related to the speed or rate r and time t by the formula $d = rt$. Again, a table is often useful for organizing the information.

EXAMPLE 6   *Distance, Rate, and Time*

Two friends drove from Deadwood, South Dakota, to Cody, Wyoming, a distance of 370 miles. The average driving speed of one driver was 16 mph slower than that of the other driver. The faster driver drove for 2.5 hours and the other driver half an hour longer. What was the average speed of each driver?

Solution

Let r = the average speed for the faster driver.

	Rate	*Time*	*Distance*
Fast Driver	r	2.5	2.5r
Slow Driver	$r - 16$	3	3(r − 16)
TOTAL			370

The total distance driven by the two drivers was $2.5r + 3(r - 16)$, and that total was 370.

$$2.5r + 3(r - 16) = 370$$
$$2.5r + 3r - 48 = 370$$
$$5.5r - 48 = 370$$
$$5.5r = 418$$
$$\frac{5.5r}{5.5.} = \frac{418}{5.5}$$
$$r = 76$$

The faster driver averaged 76 mph, and the slower driver averaged 60 mph. ■

Mixtures

Mixture problems usually involve the mixing or blending of two items, each with its own unit price or cost. The resulting mixture has a unit value that is *between* the unit prices of the individual items.

NOTE: The unit price of a mixture is *not* the sum of the unit prices of the items that are mixed. ■

EXAMPLE 7 *A Mixture of Nursery Stock*

The owner of a small nursery specializes in iris and day lily starter plants. He has been selling irises at a price of $3.00 per dozen and day lilies at a price of $8.00 per dozen.

A landscaper wants to buy 15 dozen of a mixture of irises and day lilies, and she is willing to pay $5.00 per dozen for the mixture. How many dozen of each type should the nursery owner include in the mixture?

Solution

Let x = the number of dozen of irises to include. The number of dozen of day lilies is the total of 15 dozen minus the number of dozen of irises, or $15 - x$.

The table summarizes the information in the problem.

	Number of Dozen	*Price per Dozen*	*Total Cost*
Iris	x	3.00	$3.00x$
Day Lily	$15 - x$	8.00	$8.00(15 - x)$
TOTALS	15	5.00	75.00

The total cost of the mixture is the sum of the costs for each type.

$$3x + 8(15 - x) = 75$$

This time we use the graphing method to estimate the solution of the equation. We use the integer setting to produce the graph of $y_1 = 3x + 8(15 - x)$. Then we trace to the point whose y-coordinate is 75. (See Fig. 4.22.)

Figure 4.22

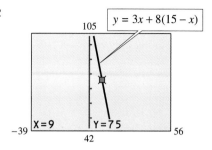

$$y = 3x + 8(15 - x)$$

Because $x = 9$ when $y = 75$, we conclude that the mixture should consist of 9 dozen irises and $15 - 9 = 6$ dozen day lilies. This result can be confirmed by solving the equation $3x + 8(15 - x) = 75$.

By tracing to other points along the graph, we can estimate the cost of *any* iris and day lily combination. ◼

Liquid Solutions

Figure 4.23

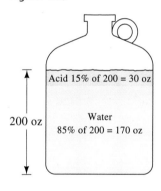

When two liquids are mixed together, the result is a **liquid solution.**

Figure 4.23 shows a bottle containing 200 ounces of a 15% acid solution, meaning that 15% of the volume is pure acid. We say that the **concentration** of acid in the solution is 15%.

Because 15% of the total volume is acid, the number of ounces of acid in the bottle is $(0.15)(200) = 30$ ounces. The remaining 85% of the solution is water, which means that $(0.85)(200) = 170$ ounces of water are in the bottle. Of course, the total is $30 + 170 = 200$ ounces.

Liquid solution problems often involve the mixing of two solutions of different concentrations. The result is a solution whose concentration is between the concentrations of the original two solutions.

NOTE: When two solutions are mixed, the concentration of the result is *not* the sum of the concentrations of the two solutions. ◼

In Fig. 4.24, 50 liters of a 30% acid solution are mixed with 100 liters of a 75% solution. The result is a 150-liter solution.

Figure 4.24

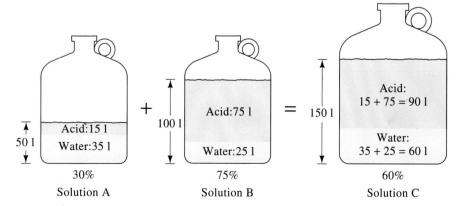

Because solution A has $(0.30)(50) = 15$ liters of acid and solution B has $(0.75)(100) = 75$ liters of acid, the resulting solution C will have $15 + 75 = 90$ liters of acid. The resulting concentration is $90 \div 150 = 0.60$ or 60%.

EXAMPLE 8 *Mixing Two Acid Solutions*

An embossing shop etches metal with acid solutions. The owner is discussing a certain job with his foreman, and they decide to use 30 ounces of a 40% acid solution.

Upon checking the acid solution supplies, the foreman has found some 25% acid solution and some 50% acid solution, but there is no 40% acid solution. The owner explains to his foreman how the two existing solutions can be mixed to obtain the 40% solution. How much of each solution is needed?

Solution

Let $x =$ the number of ounces of the 25% solution. The remaining $30 - x$ ounces will be 50% solution.

The table summarizes the information in the problem.

	Ounces of Solution	Concentration of Acid	Ounces of Acid
25% Solution	x	0.25	$0.25x$
50% Solution	$30 - x$	0.50	$0.50(30 - x)$
Mixed Solution	30	0.40	12

The total amount of acid in the mixture is the sum of the amounts of acid in each individual solution.

$$0.25x + 0.5(30 - x) = 12$$
$$0.25x + 15 - 0.5x = 12$$
$$-0.25x + 15 = 12$$
$$-0.25x + 15 - 15 = 12 - 15$$
$$-0.25x = -3$$
$$\frac{-0.25x}{-0.25} = \frac{-3}{-0.25}$$
$$x = 12$$

By mixing 12 ounces of the 25% solution and $30 - 12$ or 18 ounces of the 50% solution, the foreman obtains the required 30 ounces of 40% solution. ∎

Although the sample problems in these sections have been organized into typical problem categories, they are offered only as illustrations of the general problem-solving strategy presented in the previous section.

The examples are not meant to suggest that every problem can be wedged into some known category and solved according to a fixed routine. It is more important to *think* about the problem rather than to seek out quick, ready-made prescriptions.

4.5 Exercises

Percentages

1. How do we translate the word *of* in mathematics? Translate the phrase *15% of 30* into a mathematical expression.

2. Suppose there is a 5% sales tax in your area. In which equation below does x represent the cost of an item *before* the sales tax? In which equation does x represent the cost *including* the sales tax?

   (i) $(1.05)(32.00) = x$ (ii) $1.05x = 32.00$

3. A boy delivers newspapers for The Daily News, Inc., and he keeps 12% of all the money he collects. One week he kept $28. How much money did he turn in to The Daily News, Inc.?

4. Washington County collects a 6% sales tax. If an automobile costs $15,498 before taxes, what is the total cash paid for the car? (Ignore all other fees associated with the purchase.)

5. On Monday an investor bought some stock. By Wednesday the stock had increased in value by 4%. However, on Friday the stock dropped 4% from Wednesday's value. The investor immediately sold the stock for $349.44. What did the investor pay for the stock? Did she make money, lose money, or break even?

6. A student correctly answered 15 test questions of equal value. If his percentage grade was 83% (rounded off), how many questions were on the test?

7. Each month a worker's salary is reduced by 7% for social security, 20% for federal taxes, 6% for her pension, and 7% for state taxes. If the worker's monthly take-home pay is $1200, what is her monthly salary?

8. A person's weekly wages are reduced by 36% for taxes and other deductions. He decides to spend his entire week's take-home pay on some stereo equipment costing $220 plus a 4% sales tax. What is this person's weekly pay?

9. For a 20% markup, in which of the following equations does x represent a wholesale price? In which does x represent a retail price?

   (i) $1.20x = 50.00$ (ii) $1.20(50.00) = x$

10. If a price tag is marked $12.00, which of the following equations represents a markup? Which represents a discount?

    (i) $0.85c = 12.00$ (ii) $1.15c = 12.00$

11. A $60 blouse is on sale for $48. What is the discount percentage?

12. A salesperson receives a base salary of $360 a week and a commission of 4% on all sales. What was the amount of his sales during the first week of December if his weekly check was for $500?

13. A teapot is on sale for $17.50. If the teapot is marked down 30%, what was the original price?

14. A coffee maker is on sale for $30.60. If the price of the coffee maker is marked down 15%, what was the original price?

15. A sign in a store window reads, "Everything marked down 20%." A customer found a suit on sale for $239.20. What was the price of the suit before the markdown?

16. Cardigans Galore and The Sweater Shop both claim that their goods are priced exactly the same. When The Sweater Shop held a "20% Off" sale, a shopper found a sweater on sale for $68.00. Cardigans Galore sells the same sweater for $80.00. Was the original price at The Sweater Shop the same as the current price at Cardigans Galore?

Fixed and Variable Costs

17. Classify each of the following costs of running a business as a fixed cost or a variable cost.

    (a) expenditures for materials

    (b) annual insurance

    (c) electric bill

    (d) wages of hourly employees

    (e) annual salary of the manager

18. If the cost C of producing x items is given by the function $C(x) = ax + b$, which term represents the fixed cost? Which term represents the variable cost?

19. Ace Construction Company rents a bulldozer for $100 per day plus $50 for each hour that the bulldozer is in use. If the bulldozer rental bill on April 18 was $480, how many hours was the bulldozer used?

20. Sport Car Rentals rents a car for $35 per day and 12 cents per mile. After 2 days, a bill (before tax) was $127.84. How many miles were driven?

21. EMH Corporation has leased a photocopier. Because the company must pay $300 per month plus 3 cents per copy made, it makes a policy that no more than 3600 copies can be made per month. At the end of the first month, the accounting department receives a photocopying bill for $414. By how many copies has the company exceeded its limit?

22. Last year, Evans Company produced furnace vents with a variable cost of $5.00. This year, the fixed cost has risen by 11%. If the cost of producing 100 furnace vents is now $508.88, what was the fixed cost last year?

Simple Interest

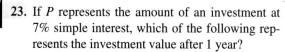

23. If P represents the amount of an investment at 7% simple interest, which of the following represents the investment value after 1 year?

    (i) $P + 0.07$ (ii) $0.07P$

    (iii) $P + 0.07P$ (iv) $0.07 + 0.07P$

24. If L is the amount of a one-year loan at 12% simple interest, and if A is the amount that is owed at the end of the year, which of the following equations is correct?

    (i) $L = A + 0.12A$ (ii) $A = 0.12L + L$

    (iii) $A - L = 0.12$ (iv) $L = A + 0.12L$

25. If money is invested at 5% simple interest for 1 year, what initial investment is required to obtain a year-end value of $600?

26. If an investment pays 5.5% simple interest, determine the initial investment required to obtain a year-end value of $1000.

27. How much did a person invest at 7% simple interest if she has $1177 at the end of 1 year?

28. Suppose a relative borrows money from you for 1 year at a simple interest rate of 9%. If the payoff of the loan is $1362.50, what was the amount of the original loan?

29. The owner of a sporting goods store had $15,000 cash that he could apply toward the cost of expanding his store. He needed to borrow the rest of the cost at 8% simple interest for 1 year. If his total cash outlay at the end of the year was $36,600, what was the amount of his loan?

30. A homeowner took out an 8% simple interest equity loan on his home. At the end of the first year of the loan, he paid the interest due plus $3800 toward the principal. At the end of the second year, before making any payment, he still owed $33,000. What was the amount of the equity loan?

31. If m dollars are borrowed at a simple interest rate r for 1 year, which of the following is an expression for the interest owed?

    (i) $\dfrac{mr}{100}$ (ii) $m + \dfrac{r}{100}$

    (iii) $\dfrac{100m}{r}$ (iv) $m + \dfrac{mr}{100}$

32. A big spender has two credit cards, one that carries an 18% finance charge and another that carries a 20% finance charge. This person uses only one of the cards to avoid paying 38% in finance charges. What is your opinion of this reasoning?

33. A person invests $6500 in two investments, one at 9% simple interest and the other at 6% simple interest. If the total income for the year is $465, how much was invested at each rate?

34. An investor puts $26,000 into two funds, one at 10% simple interest and the other at 6% simple interest. If the total income for the year is $2000, how much was invested at each rate?

35. To purchase a car costing $10,000, the buyer borrowed part of the money from the bank at 9% simple interest and the rest from her mother-in-law at 12% simple interest. If her total interest for the year was $1080, how much did she borrow from the bank?

36. A person invested $5000, part at a bank paying 8% simple interest and part at a bank paying 8% compounded continuously to yield the equivalent of 8.3% simple interest. At the end of the year the combined income was $408.10. How much was invested at each rate?

Distance, Rate, and Time

37. To use the formula $d = rt$ to solve a problem in which the rate is given as 45 mph, in what units must d and t be expressed?

38. Suppose you use the formula $d = rt$ to solve a problem in which the rate is given as 45 mph and the distance is given as 210 miles. Which of the following is the proper calculation of time?

(i) $210 \cdot 45$ (ii) $\dfrac{45}{210}$

(iii) $45 + 210$ (iv) $\dfrac{210}{45}$

39. A boy left home at 8:30 A.M. to return his friend's bike. He rode at 15 mph and arrived at his friend's house at 9:10 A.M. After chatting awhile, he began his walk home at 9:30 A.M. How fast did he walk if he arrived home at noon?

40. Two travelers left a restaurant in Salina, Kansas, and traveled in opposite directions on Interstate 70. If one driver averaged 65 mph and the other averaged 60 mph, how long was it before they were 400 miles apart?

41. A student left her college campus by bus to travel home, a distance of 472 miles. At the same time, her father left home in his car to meet the bus. Her father met the bus in 4 hours and his average speed was 6 mph faster than that of the bus. Determine the speeds of the bus and the car.

42. Two cars left the Ranch Motel at the same time and traveled in opposite directions around a circular scenic loop. The first car averaged 40 mph and the other car averaged 35 mph. Determine the distance around the loop if the first car returned to the Ranch Motel 30 minutes ahead of the second car.

43. One car left Valdosta averaging 55 mph and another car left Atlanta averaging 70 mph. They traveled toward each other on Interstate 75. Atlanta is north of Valdosta by 300 miles. When did they meet if they both left at noon? How far south of Atlanta did they meet?

44. A mother jogs at a rate of 4.5 mph and her daughter jogs at a rate of 5 mph. The mother begins her jog, and 15 minutes later her daughter begins her jog following the same route. How long will it take the daughter to overtake her mother, and how far will they have jogged?

Mixtures

45. If the unit value of an item is v and if n represents the number of items purchased, which expression represents the total cost of the items?

(i) $\dfrac{v}{n}$ (ii) $v + n$

(iii) nv (iv) $\dfrac{n}{v}$

46. If the unit value of one item is $5.00 and the unit value of another item is $9.00, explain why a mixture of the two items must have a unit value that is between $5.00 and $9.00.

47. The owner of The Nuttery wishes to mix 20 pounds of pecans selling for $1.80 per pound with some cashews selling for $2.40 per pound to make a mixture that will sell for $2 per pound. How many pounds of cashews should be used?

48. A seed company has fescue grass seed worth 55 cents per pound and some bluegrass seed worth 95 cents per pound. How many pounds of each should be mixed to produce 100 pounds of seed mix that will sell for 70 cents per pound?

49. The Nature Center sells bird seed for 75 cents per pound and sunflower seed for $1.10 per pound. To the nearest pound, how many pounds of sunflower seed should be used to create a 50-pound mixture that sells for $0.90 per pound?

50. The Indian Tea Company makes a tea blend of cinnamon tea worth $2.50 per kilogram and black tea worth $2.00 per kilogram. How many kilograms of each should be used to produce a 20-kilogram blend whose total worth is $46.00?

51. A special grade of lime worth 90 cents per pound is to be mixed with 5 pounds of standard lime worth 50 cents per pound. If the mixture is worth 65 cents per pound, how many pounds of the special grade of lime should be used?

52. Two parts of pea gravel worth $10.00 per ton are mixed with one part of sand worth $8.00 per ton to form a base for road paving. If the bill for a project is $728.00, what is the approximate price per ton for the mixture?

Liquid Solutions

53. Explain how to determine the concentration of a liquid mixed with water. If the concentration of a substance is known, how do you determine the amount of substance in a solution?

54. What is the concentration of acid in n liters of pure water? What is the concentration of acid in n liters of pure acid? If c_1 and c_2 are the concentrations of acid in two different solutions, describe the concentration of acid when the two solutions are mixed together.

55. Cagle's Dairy mixed two grades of milk containing 3% and 4.5% butterfat, respectively, to obtain 150 gallons of milk that contained 4% butterfat. How many gallons of each were used in the mixture?

56. Two blocks of alloy, one containing 50% silver and the other containing 25% silver, are melted together to obtain 160 pounds of a new alloy that is 35% silver. Determine the weight of each original block.

57. A pharmacist has two decongestants, one containing 0.8% pseudoephedrine HCl, the other containing 1.4% of the same substance. A prescription calls for a decongestant containing 1.1% pseudoephedrine HCl. How much of each should be used to produce 4 ounces of a 1.1% solution?

58. A biologist wishes to have 15 gallons of an 80% formaldehyde solution. In her inventory, she has pure formaldehyde and some 50% formaldehyde solution. How many gallons of each should she mix to obtain the desired solution?

59. A worker has 64 ounces of a trisodium phosphate (TSP) to clean a wooden deck. However, the 20% concentration is too strong, so he must add water to dilute the solution to a 10% concentration. How many ounces of water should he add?

60. An 8-ounce solution of 12% boric acid is too weak and the concentration needs to be increased to 20%. How many ounces of pure boric acid should be added?

Miscellaneous Applications

61. How much was invested at 6% simple interest if the value of the account is $1563.50 at the end of 1 year?

62. A person borrows money for 1 year at a simple interest rate of 11%. If the payoff of the loan is $1470.75, what was the amount of the original loan?

63. A college student bought his textbooks at the bookstore for $119. He spent $4 more on his biology book than on his English book. His math book cost $16 more than his English book. How much did he spend on each book?

64. The cash register of Soft Hardware contains $6.65 in just nickels and dimes. If there is a total of 85 coins, how many nickels are in the drawer?

65. Party Ideas has two grades of confetti. How many pounds of plain confetti worth $4.50 per pound should be mixed with 6 pounds of sparkling confetti worth $6 per pound in order to obtain a mixture worth $5.40 per pound?

66. Candy Corner has two kinds of candy creams and wishes to make a mixture of 100 pounds to sell at $3 per pound. If the two types are priced at $2.50 per pound and $3.75 per pound, how many pounds of each must be mixed to produce the desired amount?

67. Two angles are supplementary. The smaller is 20° less than one-fourth of the larger. What is the measure of each angle?

68. Two angles are complementary. One angle is 30° less than twice the other angle. What is the measure of each angle?

69. A chef has a 12% vinegar solution and a 25% vinegar solution. How much of each should he use to make 2 cups of a 15% vinegar solution?

70. A chemist added a solution that was 90% acid to an 18-gallon solution that was 60% acid. She obtained a solution that was 66% acid. How many gallons of 90% acid did she add?

71. A college student leaves her dorm at 10:00 A.M. and walks to the campus bookstore 2 miles away. She arrives at 10:30 A.M., buys her books, and leaves at 11:15 A.M. She then returns to the dorm, but she passes it and keeps walking an additional mile to the sorority house. What time does she arrive?

72. The small digital computers used on road bikes measure time and distance, but the clock is active only when the bike is actually moving. The computer also calculates and displays the average speed for the entire trip. A cyclist rides 15 miles from 2:00 to 3:00 and stops for a sundae. At 3:30, she leaves for home and arrives at 4:20. To the nearest tenth, what average speed will the computer display?

73. To obtain an $80,000 mortgage on their home, a couple had to pay 2.5 discount points. A **discount point** is 1% of the mortgage amount. If the points were added to the mortgage amount, how much was the total loan?

74. To be able to buy a home, the purchaser had to borrow 90% of the purchase price. If her mortgage loan was $108,000, what was the purchase price of the home?

Challenge

75. You have decided to sell your business. Your buyer has agreed to purchase your inventory at what it cost you to buy (wholesale price). The buyer asks, "What was your markup on all of your goods?" You reply that you marked up everything 25%. The buyer says, "Just give me a 25% discount on everything, and we'll be even." Will you agree to this plan? Why or why not?

76. Big Ed's Car Rental rents cars for $25 a day plus $0.40 per mile driven. Pop's Car Rental rents cars for $33 a day plus $0.30 per mile driven.

(a) Which agency provides the lowest cost for a 100-mile trip?

(b) Let f represent the total cost of renting a car from Big Ed's, and let g represent the total cost of renting a car from Pop's. Let m represent the number of miles driven. Write f and g as functions of m.

(c) Use your calculator to display both graphs. What is the significance of the point of intersection?

(d) Set the two functions equal to each other, and solve the resulting equation for m.

(e) What are the values of the left and right sides of the equation in part (d) for the solution? What do these values represent?

(f) From parts (d) and (e), what are the coordinates of the point of intersection of the graphs in part (c)?

(g) How would you use the information obtained in parts (d) and (e) to make decisions about which car rental agency to use for *any* trip?

4.6 Linear Inequalities

Notation ▪ *Linear Inequalities in One Variable* ▪ *Solving Inequalities by Graphing* ▪ *Properties of Inequalities* ▪ *Solving Inequalities Algebraically* ▪ *Special Cases*

Notation

In Section 1.1 we used **inequalities** to describe the order of real numbers on a number line. As is true for equations, an inequality may be true or false.

$-2 > -5$ True

$x < 9$ False if, for example, $x = 15$.

$z \geq 0$ True if, for example $z = 0$.

The last inequality is true if z is replaced with 0 or any number greater than 0. We say that these replacements **satisfy** the inequality.

There are several ways to describe numbers that satisfy an inequality. Here is a summary of the methods we have used.

Set-Builder Notation	*Graph*	*Interval Notation*
$\{x \mid a \leq x \leq b\}$		$[a, b]$
$\{x \mid a < x < b\}$		(a, b)
$\{x \mid a \leq x < b\}$		$[a, b)$
$\{x \mid a < x \leq b\}$		$(a, b]$
$\{x \mid x \leq b\}$		$(-\infty, b]$
$\{x \mid x < b\}$		$(-\infty, b)$
$\{x \mid x \geq a\}$		$[a, \infty)$
$\{x \mid x > a\}$		(a, ∞)
$\{\text{Real numbers}\}$		$(-\infty, \infty)$

Recall that the **infinity symbol** ∞ does *not* represent a number. It is used to indicate that the interval extends to the right (∞) or left ($-\infty$) without end.

Linear Inequalities in One Variable

There are many different types of inequalities. In this section we discuss one type called a **linear inequality in one variable.**

Definition of Linear Inequality in One Variable

A **linear inequality in one variable** is an inequality that can be written in the form $Ax + B < 0$, where A and B are real numbers and $A \neq 0$.

This definition, and all others in this section, is also valid for the symbols $>$, $\leq$, and $\geq$.

The following are examples of linear inequalities.

$$3x + 6 \geq 0 \qquad t > -3 \qquad 3(x - 5) \leq 6 - 3x$$

The first example is written in the form $Ax + B \geq 0$. It can be shown that the other two can also be written in the defined form. An inequality written with the symbol $<$ or $>$ is called a **strict inequality.**

The following are examples of inequalities that are *not* linear inequalities.

$$x^2 - 5 < 0 \qquad \text{No exponent can exceed 1.}$$

$$\frac{3}{x} + 2x > 5 \qquad \text{No variable can be in a denominator.}$$

$$\sqrt{x} - 9 > 0 \qquad \text{There can be no variable in a radicand.}$$

The definitions of **solution** and **solution set** for inequalities are similar to those for equations.

Definition of Solution and Solution Set of an Inequality

A **solution** of an inequality is any replacement for the variable that makes the inequality true. The **solution set** of an inequality is the set of all solutions of the inequality.

A solution of an inequality can be verified by substitution or with the same calculator techniques used for verifying solutions of equations.

Solving Inequalities by Graphing

We can use a calculator to estimate solutions of inequalities. The techniques are very similar to those used in estimating solutions of equations.

EXPLORATION 1   *Estimating Solutions by Graphing*

Consider the inequality $15 - 2x < -11$.

If we let $y_1 = 15 - 2x$ and $y_2 = -11$, then we can write the inequality in a new way: $y_1 < y_2$.

(a) Use the integer setting and produce the graphs of y_1 and y_2 on the same coordinate axes.

(b) What is the point of intersection of the graphs? What does this point represent?

(c) Where is the graph of y_1 below the graph of y_2?

(d) For what values of x is $y_1 < y_2$?

(e) Write the solution set of the inequality with interval notation.

Discovery

(a) Figure 4.25 shows the graphs of y_1 and y_2.

(b) From Fig. 4.25 we estimate the point of intersection to be $(13, -11)$. This is the point for which $y_1 = y_2$. [We can verify that $(13, -11)$ is the point of intersection by substituting 13 for x in the equation $15 - 2x = -11$.]

(c) The graph of y_1 is below the graph of y_2 to the right of the point of intersection $(13, -11)$.

(d) All points to the right of $(13, -11)$ have x-coordinates that are greater than 13.

(e) The solution set of the inequality is $(13, \infty)$.

Figure 4.25

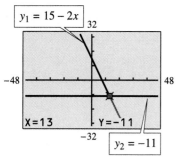

In Exploration 1, the inequality is of the form $y_1 < y_2$. Therefore, we looked for the interval of x-values for which the graph of y_1 was *below* the graph of y_2. For inequalities of the form $y_1 > y_2$, we look for the interval of x-values for which the graph of y_1 is *above* the graph of y_2.

For inequalities involving the symbols $\le$ and $\ge$, the x-coordinate of the point of intersection is included in the solution set.

As always, graphing methods provide only estimates of solution sets. For accuracy, we must use algebraic methods. Before we discuss such methods, we consider some properties of inequalities.

Properties of Inequalities

As was true for equations, inequalities are **equivalent** if they have the same solution set. The Addition and Multiplication Properties of Inequalities may be applied to produce equivalent inequalities.

■ *Addition Property of Inequalities*

If $A < B$, then $A + C < B + C$.

Adding (or subtracting) the same quantity to (from) both sides of an inequality produces an equivalent inequality.

Before stating the Multiplication Property of Inequalities, let's experiment.

EXPLORATION 2

Multiplying Both Sides of an Inequality

Consider the following true inequalities.

$$3 < 7 \qquad 4 > -3 \qquad -5 < -2$$

(a) Multiply both sides of each inequality by 3 and insert the correct inequality symbol in each result.

(b) Multiply both sides of each inequality by -3 and insert the correct inequality symbol in each result.

(c) What conclusions can you draw?

Discovery

(a) $\quad 3 \ < \ 7 \qquad\qquad 4 \ > \ -3 \qquad\qquad -5 \ < \ -2$

$\quad 3 \cdot 3 \ \underline{\ ?\ } \ 3 \cdot 7 \qquad 3 \cdot 4 \ \underline{\ ?\ } \ 3 \cdot (-3) \qquad 3 \cdot (-5) \ \underline{\ ?\ } \ 3 \cdot (-2)$

$\qquad 9 \ < \ 21 \qquad\qquad 12 \ > \ -9 \qquad\qquad -15 \ < \ -6$

(b) $\quad 3 \ < \ 7 \qquad\qquad 4 \ > \ -3 \qquad\qquad -5 \ < \ -2$

$-3 \cdot 3 \ \underline{\ ?\ } \ -3 \cdot 7 \qquad -3 \cdot 4 \ \underline{\ ?\ } \ -3 \cdot (-3) \qquad -3 \cdot (-5) \ \underline{\ ?\ } \ -3 \cdot (-2)$

$\quad -9 \ > \ -21 \qquad\qquad -12 \ < \ 9 \qquad\qquad 15 \ > \ 6$

(c) In part (a) multiplying both sides by a positive number did not affect the direction of the inequality symbol. In part (b) multiplying both sides by a negative number reversed the direction of the inequality symbol.

An experiment involving division would result in the same conclusions as those in Exploration 2. Here is the formal statement of the property.

■ *Multiplication Property of Inequalities*

If $A < B$ and $C > 0$, then $AC < BC$.

Multiplying or dividing both sides of an inequality by the same *positive* number produces an equivalent inequality.

If $A < B$ and $C < 0$, then $AC > BC$.

Multiplying or dividing both sides of an inequality by the same *negative* number produces an equivalent inequality if the inequality symbol is reversed.

Note that multiplying both sides of an inequality by zero does *not* result in an equivalent inequality, and, of course, division by zero is undefined.

Solving Inequalities Algebraically

We can use the Addition and Multiplication Properties of Inequalities to solve inequalities algebraically. The solving routine is nearly identical to the solving routine for equations. The one exception is the need to reverse the inequality symbol when we multiply or divide both sides of an inequality by a negative number.

EXAMPLE 1 *Solving an Inequality Algebraically*

Use algebraic methods to solve the inequality $15 - 2x < -11$.

Solution

$$15 - 2x < -11$$
$$15 - 2x - 15 < -11 - 15 \qquad \text{Addition Property of Inequalities}$$
$$-2x < -26$$
$$\frac{-2x}{-2} > \frac{-26}{-2} \qquad \text{Multiplication Property of Inequalities—dividing both sides by } -2 \text{ reverses the inequality symbol.}$$
$$x > 13$$

The solution set is $(13, \infty)$, which confirms the result in Exploration 1. (See Fig. 4.26.)

Figure 4.26

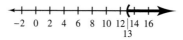

As with equations, it may be necessary to simplify both sides of an inequality by removing grouping symbols and combining like terms.

EXAMPLE 2 *Simplifying Both Sides of an Inequality*

Estimate the solution of $2 + 3(x - 4) > x - 10$ graphically. Then, solve the inequality.

Solution

Figure 4.27

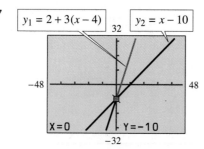

From Fig. 4.27, we see that the graph of y_1 is above the graph of y_2 for all x-values greater than 0. Thus our estimate of the solution set is $(0, \infty)$.

To solve algebraically, begin by simplifying the left side.

$$2 + 3(x - 4) > x - 10$$
$$2 + 3x - 12 > x - 10 \qquad \text{Distributive Property}$$
$$3x - 10 > x - 10$$
$$3x - 10 - x > x - 10 - x \qquad \text{Addition Property of Inequalities}$$
$$2x - 10 > -10 \qquad \text{Combine like terms.}$$
$$2x - 10 + 10 > -10 + 10 \qquad \text{Addition Property of Inequalities}$$
$$2x > 0$$
$$\frac{2x}{2} > \frac{0}{2} \qquad \text{Multiplication Property of Inequalities}$$
$$x > 0$$

Note that dividing both sides by 2 did not affect the inequality symbol. The solution set is $(0, \infty)$. (See Fig. 4.28.)

Figure 4.28
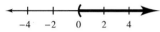

When solving equations, it is often useful to clear fractions. The same is true when solving inequalities.

EXAMPLE 3   *Clearing Fractions from an Inequality*

Solve $\frac{1}{2}x + \frac{2}{3} < \frac{5}{6}x + 1$.

Solution

The LCM of all of the denominators is 6. To clear the fractions, multiply every term of both sides by 6.

$$\frac{6}{1} \cdot \frac{1}{2}x + \frac{6}{1} \cdot \frac{2}{3} < \frac{6}{1} \cdot \frac{5}{6}x + 6 \cdot 1 \qquad \text{Multiplication Property of Inequalities}$$
$$3x + 4 < 5x + 6$$
$$3x + 4 - 5x < 5x + 6 - 5x \qquad \text{Addition Property of Inequalities}$$
$$-2x + 4 < 6 \qquad \text{Combine like terms.}$$
$$-2x + 4 - 4 < 6 - 4 \qquad \text{Addition Property of Inequalities}$$
$$-2x < 2$$
$$\frac{-2x}{-2} > \frac{2}{-2} \qquad \text{Multiplication Property of Inequalities—dividing by } -2 \text{ reverses the inequality symbol.}$$
$$x > -1$$

The solution set is $(-1, \infty)$. (See Fig. 4.29.)

Figure 4.29
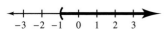

Special Cases

When the graphs of the left and right sides of an equation do not intersect, the equation has no solution. In the following exploration, we consider the same kind of special case for inequalities.

EXPLORATION 3   *A Special Case*

(a) Consider $x + 12 < x - 9$. Produce the graphs of $y_1 = x + 12$ and $y_2 = x - 9$. For what x-values is the graph of y_1 *below* the graph of y_2?

(b) Consider $x + 12 > x - 9$. The graphs are the same as in part (a). For what x-values is the graph of y_1 *above* the graph of y_2?

(c) What is your conjecture about the solutions of an inequality if the graphs of the left and right sides are parallel?

Discovery

(a) Figure 4.30 shows the graphs of y_1 and y_2. Note that the graph of y_1 is *never* below the graph of y_2. Therefore, there are no x-values that satisfy the inequality. The solution set is empty.

(b) From Figure 4.30, we see that the graph of y_1 is *always* above the graph of y_2. Every x-value satisfies the inequality. The solution set is all real numbers.

(c) If the graphs of the left and right sides of an inequality are parallel, then the solution set is either the empty set or the set of all real numbers.

The results in Exploration 3 can also be obtained with algebraic methods.

Figure 4.30

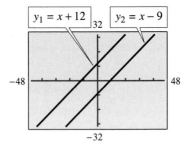

$y_1 = x + 12$ $y_2 = x - 9$

EXAMPLE 4   *Detecting Special Cases Algebraically*

Solve the following inequalities.

(a) $x + 12 < x - 9$ (b) $x + 12 > x - 9$

Solution

(a) $x + 12 < x - 9$

   $x - x + 12 < x - x - 9$ Addition Property of Inequalities

   $12 < -9$ False inequality

The resulting false inequality is equivalent to the original inequality, so we conclude that there is no replacement that makes the original inequality true. The solution set is empty.

(b) $x + 12 > x - 9$

   $x - x + 12 > x - x - 9$ Addition Property of Inequalities

   $12 > -9$ True inequality

The resulting true inequality is equivalent to the original inequality, so we conclude that every replacement makes the original inequality true. The solution set is all real numbers.

4.6 *Quick Reference*

Notation	■ Set-builder notation, number line graphs, and interval notation can all be used to represent numbers that satisfy an inequality.
	■ In interval notation, the infinity symbol is used to indicate that an interval extends to the right (∞) or to the left ($-\infty$) without end.

Linear Inequalities in One Variable	■ A **linear inequality in one variable** is an inequality that can be written in the form $Ax + B < 0$, where A and B are real numbers and $A \neq 0$. The definitions involving $>$, $\geq$, and $\leq$ are similar.
	■ A **solution** of an inequality is any replacement for the variable that makes the inequality true. The **solution set** of an inequality is the set of all solutions.

Solving Inequalities by Graphing	■ To estimate the solution set of an inequality graphically, let y_1 represent the left side of the inequality and y_2 represent the right side. Produce the graphs of y_1 and y_2 on the same coordinate axes.

1. For $y_1 < y_2$, estimate the interval of x-values for which the graph of y_1 is below the graph of y_2.

2. For $y_1 > y_2$, estimate the interval of x-values for which the graph of y_1 is above the graph of y_2.

3. For inequalities involving $\leq$ and $\geq$, the x-coordinate of the point of intersection is included in the solution set.

Properties of Inequalities	■ Inequalities are **equivalent** if they have the same solution set.
	■ Addition Property of Inequalities: If $A < B$, then $A + C < B + C$. Adding (or subtracting) the same quantity to (from) both sides of an inequality produces an equivalent inequality.
	■ Multiplication Property of Inequalities:

1. If $A < B$ and $C > 0$, then $AC < BC$. Multiplying or dividing both sides of an inequality by the same *positive* number produces an equivalent inequality.

2. If $A < B$ and $C < 0$, then $AC > BC$. Multiplying or dividing both sides of an inequality by the same *negative* number produces an equivalent inequality if the inequality symbol is reversed.

Solving Inequalities Algebraically	■ The routine for solving inequalities is nearly identical to that for solving equations. The exception is the need to reverse the inequality symbol when we multiply or divide both sides by a negative number.
	■ As with equations, simplifying both sides and clearing fractions are steps in the solving process.

Special Cases   ▪ If the graphs of the left and right sides of an inequality do not intersect, the solution set is either the empty set or the set of all real numbers.

▪ These special cases can be detected algebraically when we obtain an equivalent inequality that is false (in which case the solution set is empty) or true (in which case the solution set is the set of all real numbers).

4.6   *Exercises*

 1. Explain the difference between [*a*, *b*] and (*a*, *b*).

 2. For {*x* | *x* ≥ 2}, is it correct to write [2, ∞] or [2, ∞)?

In Exercises 3–8, represent the solution set with interval notation, set notation, and a number line graph.

3. $x < 2$ **4.** $x > -5$

5. $x \geq 3$ **6.** $x \leq -4$

7. $-3 < x \leq 2$ **8.** $-2 \leq x < 4$

 9. Compare the graphing method for solving equations with the graphing method for solving inequalities.

 10. When you use the graphing method to solve an inequality, under what condition does the point of intersection represent a solution?

In Exercises 11–16, the given calculator display shows the graphs of the left and right sides of the given inequality. Write the solution set of the inequality with interval notation.

11. $y_1 \leq y_2$

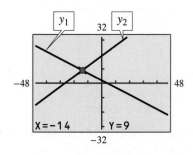

12. $y_1 > y_2$

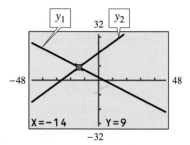

13. $y_1 < y_2$

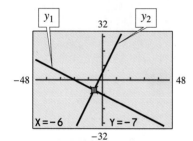

14. $y_1 \geq y_2$

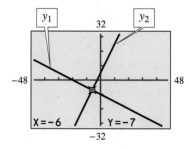

15. $y_1 < y_2$

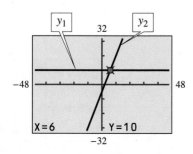

16. $y_1 \geq y_2$

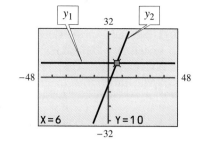

In Exercises 17–26, use the graphing method to estimate the solution set of the given inequality. Write your estimate in interval notation.

17. $3x - 16 < 8$ **18.** $10 - 2x < 5x - 11$

19. $3x + 9 \geq -2x - 16$

20. $-x \geq 2x - 12$ **21.** $3x - 8 \leq 3(2 + x)$

22. $6(1 + x) > 2(3x - 5)$

23. $20 - 3x \geq 2x$ **24.** $2x - 7 \geq 11 - x$

25. $2(x - 5) > 2x + 1$ **26.** $2 - 3x \geq 15 - 3x$

In Exercises 27–34, solve the inequality algebraically. Write the solution set with interval notation.

27. $x + 3 < 8$ **28.** $x - 7 > -2$

29. $-7 + x \leq -7$ **30.** $x + 6 \geq -9$

31. $5x + 12 < 5 + 4x$ **32.** $4x \geq 3x + 2$

33. $3 - 6x \leq 2 - 7x$ **34.** $7 + 4x \geq 3x + 7$

 35. Explain how the Multiplication Property of Inequalities applies to writing $-x < 3$ as $x > -3$.

36. If $-x < -z$, what is true about x? Explain.

In Exercises 37–44, assume that $a < b$ and insert the correct inequality symbol.

37. $2a$ _____ $2b$ **38.** $a - 6$ _____ $b - 6$

39. b _____ a **40.** $-\dfrac{1}{2}a$ _____ $-\dfrac{1}{2}b$

41. $-a$ _____ $-b$ **42.** $\dfrac{a}{4}$ _____ $\dfrac{b}{4}$

43. $-3 + a$ _____ $-3 + b$

44. $a \div (-13)$ _____ $b \div (-13)$

In Exercises 45–50, assume that $a < b$ and determine the values of c for which the inequality is true.

45. $ac > bc$ **46.** $ac < bc$

47. $a + c < b + c$ **48.** $a - c < b - c$

49. $\dfrac{a}{c} < \dfrac{b}{c}$ **50.** $\dfrac{a}{c} > \dfrac{b}{c}$

51. Compare the algebraic method for solving an inequality with the algebraic method for solving an equation.

52. Explain the difference between the solution sets of $x < a$ and $x \leq a$.

In Exercises 53–60, solve the inequality algebraically. Write the solution set with interval notation.

53. $3x \leq 12$ **54.** $5x > 35$

55. $-x < 5$ **56.** $-4t \geq -32$

57. $0.2y \geq 1.6$ **58.** $0.3n < -2.7$

59. $-\dfrac{2}{5}t > -6$ **60.** $6x < \dfrac{2}{3}$

61. Describe the graphs of the left and right sides of an inequality whose solution set is

(a) the empty set.
(b) the set of all real numbers.

62. In each part, the result of solving an inequality is given. What is the solution set? Why?

(a) $-2 < 3$ (b) $3 < -5$ (c) $4 > 4$
(d) $-7 \leq -7$ (e) $8 \geq -8$

In Exercises 63–96, solve the inequality algebraically. Write the solution set with interval notation and graph the solution set on a number line.

63. $3x - 4 < 14$ **64.** $7x + 1 \geq 15$

65. $7 - 3t \leq -8$ **66.** $5 - y > 0$

67. $5x + 3 < 7 + 5x$ **68.** $3 - 2x \geq -2x + 1$

69. $4x - 4 \leq 7x - 13$ **70.** $2x + 7 < 7 + 5x$

71. $8 - x > x - 8$ **72.** $5 - t < t + 5$

73. $3 - x \geq 4x + 1$

74. $6 + 4x < 11 - 2x$

75. $3x - x < 2x$ **76.** $1 + 2x \leq 2x + 1$

77. $-(x - 2) \leq 2(x + 1)$

78. $3(x + 4) > 2x - 3$ **79.** $5 - 4x \geq 4(2 - x)$

80. $2 - 5x < -5(1 + x)$

81. $x - 4(x - 4) - 4x > x - 4$

82. $3(x - 1) - 4(2x + 3) > 0$

83. $5 + 4(x - 3) \geq 2x + 3(x - 1) + 4$

84. $5 - 5(5 - x) \leq 5(x - 1) + 2x$

85. $\dfrac{2}{3}x - 5 < \dfrac{5}{6}$ **86.** $\dfrac{5}{4}x - \dfrac{1}{2} \geq x - \dfrac{5}{4}$

87. $\dfrac{2x + 1}{-3} < 6$ **88.** $\dfrac{5x - 1}{-5} > \dfrac{3}{10}$

89. $x - \dfrac{1}{3} + \dfrac{5}{6}x \leq \dfrac{1}{2} - 3x$

90. $\dfrac{5}{12}x + \dfrac{1}{3} - x \leq \dfrac{7}{12} - \dfrac{x}{2}$

91. $\dfrac{3}{4}(x + 3) + 2x < 1$

92. $\dfrac{1}{4}(x - 2) < \dfrac{1}{2}x - \dfrac{2}{3}$

93. $0.2x + 1.7 \geq 2.94 - 5.3x$

94. $0.75x + 0.25 \leq x - 1.5$

95. $2.8(1.3x - 0.9) \leq 1.2 - 9.97x$

96. $0.5(0.6x + 3.1) \geq x - 1.73$

 97. The lengths of the sides of a triangle are represented by a, b, and c. What must be true about any two sides with respect to the third side? Write three true inequalities involving a, b, and c.

Challenge

In Exercises 98–101, solve the inequality for x.

98. $5x + a \leq b$ **99.** $cx < 10$ where $c < 0$

100. $a^2 x \geq 8$, $a \neq 0$ **101.** $b - x > a$

 102. If $a < b$, then is it true that $a^2 < b^2$?

 103. If $a < b$, then is it true that $\dfrac{1}{a} < \dfrac{1}{b}$?

104. Suppose you graph the left and right sides of a linear inequality and you find that the lines coincide. Explain your answers to the following.

(a) If you correctly conclude that the solution set is empty, what is the inequality symbol? (There are two possibilities.)

(b) If you correctly conclude that the solution set is the set of real numbers, what is the inequality symbol? (There are two possibilities.)

4.7 Compound Inequalities

Intersection and Union ▪ Solution Sets ▪ Double Inequalities ▪
Algebraic Methods

Intersection and Union

The **intersection** of two sets A and B is the set of all elements that belong to both A and B. Using set notation, we write the intersection of sets A and B in the following way.

$$A \cap B = \{x \mid x \in A \quad \text{and} \quad x \in B\}$$

The word *and* means that the element x must belong to *both* sets A and B.

If $A = \{x \mid x \leq 5\}$ and $B = \{x \mid x \geq 1\}$, then

$$A \cap B = \{x \mid x \geq 1 \quad \text{and} \quad x \leq 5\}.$$

This set may also be written $A \cap B = \{x \mid 1 \leq x \leq 5\}$. (See Fig. 4.31.)

Figure 4.31

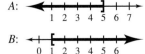

The **union** of two sets A and B is the set of all elements of A together with all elements of B. Using set notation, we write the union of sets A and B in the following way.

$$A \cup B = \{x \mid x \in A \quad or \quad x \in B\}$$

The word *or* means that the element x belongs to *either A or B or both*.

Figure 4.32

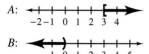

If $A = \{x \mid x \geq 3\}$ and $B = \{x \mid x < 0\}$, then

$$A \cup B = \{x \mid x < 0 \quad or \quad x \geq 3\}.$$

(See Fig. 4.32.)

Solution Sets

A **compound inequality** is two inequalities connected with the words *and* or *or*.

1. The solution set for a compound inequality connected with *and* is the *intersection* of the solution sets for each individual inequality.

2. The solution set for a compound inequality connected with *or* is the *union* of the solution sets for each individual inequality.

EXAMPLE 1 *Intersection of Solution Sets*

Describe the solution set of the compound inequality

$$x < 4 \quad and \quad x > -3.$$

Solution The solution set is the intersection of the individual sets. The graph of the solution set is as shown in Fig. 4.33. In interval notation the solution set is $(-3, 4)$.

Figure 4.33

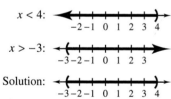

EXAMPLE 2 *Union of Solution Sets*

Describe the solution set of the compound inequality

$$x > 1 \quad or \quad x < -2.$$

Solution The solution set is the union of the individual sets. The graph of the solution set is as shown in Fig. 4.34. In interval notation the solution set is $(-\infty, -2) \cup (1, \infty)$.

Figure 4.34

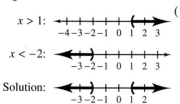

Double Inequalities

The value of an expression may be *between* two values. We state this relationship by using a **double inequality.**

A double inequality is a special kind of compound inequality in which the connective *and* is implied. For example, all numbers between 2 and 8 can be described by the double inequality $2 < x < 8$. This means that x is greater than 2 *and* x is less than 8. All numbers between 2 and 8 *inclusive* can be described by the double inequality $2 \leq x \leq 8$.

Another example of a double inequality is $-26 \leq 3x - 8 \leq 16$. This means the quantity $3x - 8$ lies between the numbers -26 and 16, inclusive. We can state this double inequality by using two single inequalities: $-26 \leq 3x - 8$ *and* $3x - 8 \leq 16$.

The solutions of $-26 \leq 3x - 8 \leq 16$ are those replacements for x that make the double inequality true. Graphing a double inequality can provide a visual idea of what the solutions are.

 E X P L O R A T I O N 1   *Graphing Double Inequalities*

Consider the double inequality $-26 \leq 3x - 8 \leq 16$. Let $y_1 = -26$, $y_2 = 3x - 8$, and $y_3 = 16$. Use the integer setting to graph all three functions on the same coordinate axes.

(a) Determine the point of intersection of the graphs of y_1 and y_2.

(b) Determine the point of intersection of the graphs of y_3 and y_2.

(c) Trace along that portion of the graph of y_2 that is between the graphs of y_1 and y_3. Describe the x-coordinates of the points.

(d) What is the solution set of the double inequality?

Discovery

(a) The point of intersection of y_1 and y_2 is $(-6, -26)$. (See Fig. 4.35.)

(b) The point of intersection of y_3 and y_2 is $(8, 16)$. (See Fig. 4.36.)

Figure 4.35   Figure 4.36

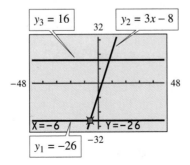

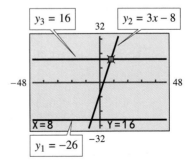

(c) The graph of y_2 lies between the graphs of y_1 and y_3 for all values of x between -6 and 8, inclusive.

(d) The solution set is $[-6, 8]$.

When solving a double inequality algebraically, our goal is to isolate the variable in the middle.

EXAMPLE 3 *Solving a Double Inequality Algebraically*

Solve $-26 \le 3x - 8 \le 16$ algebraically.

Solution

$$-26 \le 3x - 8 \le 16$$
$$-26 + 8 \le 3x - 8 + 8 \le 16 + 8 \qquad \text{Addition Property of Inequalities}$$
$$-18 \le 3x \le 24$$
$$\frac{-18}{3} \le \frac{3x}{3} \le \frac{24}{3} \qquad \text{Multiplication Property of Inequalities}$$
$$-6 \le x \le 8$$

Figure 4.37

As in Exploration 1, the solution set is all numbers between -6 and 8, inclusive. The solution set is $[-6, 8]$. (See Fig. 4.37.)   ■

Algebraic Methods

To solve a compound inequality, we begin by solving each inequality individually. Then we join the solution sets with the appropriate connective, *and* or *or*.

A compound inequality whose connective is *and* is called a **conjunction.**

EXAMPLE 4 *Solving a Conjunction*

Solve the conjunction $3 - x < 5$ and $2x - 3 \le 7$. That is, find all values of x for which both inequalities are true.

Solution

Solve each inequality individually.

$$3 - x < 5 \qquad \text{and} \qquad 2x - 3 \le 7$$
$$3 - x - 3 < 5 - 3 \quad \text{and} \quad 2x - 3 + 3 \le 7 + 3$$
$$-x < 2 \qquad \text{and} \qquad 2x \le 10$$
$$x > -2 \qquad \text{and} \qquad x \le 5$$

Figure 4.38

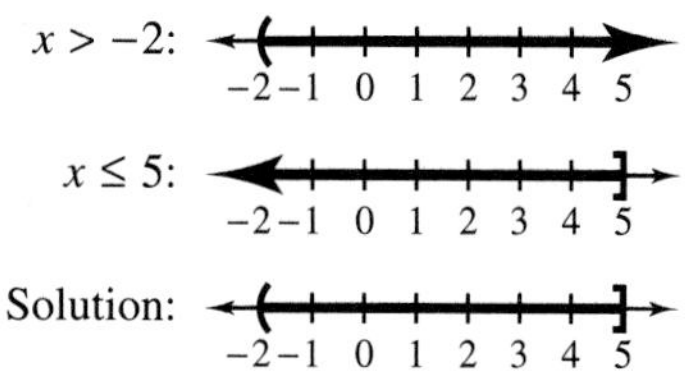

$x > -2$:

$x \le 5$:

Solution:

The connective is *and,* so the solution set for the compound inequality is the intersection of the two sets. The solution set is $(-2, 5]$. (See Fig. 4.38.)   ■

EXAMPLE 5   *Solving a Conjunction*

Solve the conjunction $x - 4 \geq -3$ and $2x - 6 > 0$.

Solution

Solve each inequality individually.

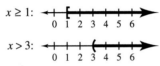

$$x - 4 \geq -3 \qquad \text{and} \qquad 2x - 6 > 0$$
$$x - 4 + 4 \geq -3 + 4 \quad \text{and} \quad 2x - 6 + 6 > 0 + 6$$
$$x \geq 1 \qquad \text{and} \qquad 2x > 6$$
$$x \geq 1 \qquad \text{and} \qquad x > 3$$

From the graphs we can visually determine the intersection by noting the points common to both graphs. The solution set is $(3, \infty)$. (See Fig. 4.39.)

Figure 4.39 $x \geq 1$:

      0 1 2 3 4 5 6

$x > 3$:

      0 1 2 3 4 5 6

Solution:

      0 1 2 3 4 5 6

EXAMPLE 6   *Solving a Conjunction*

Solve the conjunction $x + 2 \leq 0$ and $5 - x \leq 3$.

Solution

Solve each inequality individually.

$$x + 2 \leq 0 \qquad \text{and} \qquad 5 - x \leq 3$$
$$x + 2 - 2 \leq 0 - 2 \quad \text{and} \quad 5 - x - 5 \leq 3 - 5$$
$$x \leq -2 \qquad \text{and} \qquad -x \leq -2$$
$$x \leq -2 \qquad \text{and} \qquad x \geq 2$$

Figure 4.40 $x \leq -2$:

      -2 -1 0 1 2 3 4

$x \geq 2$:

      -3 -2 -1 0 1 2 3

Because there are no points common to both graphs in Fig. 4.40, there are no solutions of the compound inequality. The solution set is the empty set.

A compound inequality whose connective is *or* is called a **disjunction.** To solve a disjunction we solve each individual inequality and then form the **union** of the individual solution sets.

EXAMPLE 7   *Solving a Disjunction*

Solve the disjunction $3x - 2 < 1$ or $x - 3 \geq 1$. That is, find all values of x for which at least one of the inequalities is true.

Solution

Solve each inequality individually.

$$
\begin{array}{ccc}
3x - 2 < 1 & \text{or} & x - 3 \geq 1 \\
3x - 2 + 2 < 1 + 2 & \text{or} & x - 3 + 3 \geq 1 + 3 \\
3x < 3 & \text{or} & x \geq 4 \\
x < 1 & \text{or} & x \geq 4
\end{array}
$$

Figure 4.41 $x < 1$: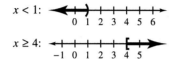

$x \geq 4$:

The connective is *or*, so the solution set for the compound inequality is the union of the two sets. From the graphs in Fig. 4.41, we see that the points that belong to at least one of the sets are as shown in Fig. 4.42. The solution set is $(-\infty, 1) \cup [4, \infty)$.

Figure 4.42   Solution: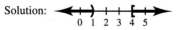

EXAMPLE 8   *Solving a Disjunction*

Solve the disjunction $x + 4 \geq 4$ or $x - 5 \leq 0$.

Solution

Solve each inequality individually.

$$
\begin{array}{ccc}
x + 4 \geq 4 & \text{or} & x - 5 \leq 0 \\
x + 4 - 4 \geq 4 - 4 & \text{or} & x - 5 + 5 \leq 0 + 5 \\
x \geq 0 & \text{or} & x \leq 5
\end{array}
$$

Figure 4.43 $x \geq 0$: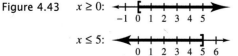

$x \leq 5$:

The connective is *or*, so the solution set for the compound inequality is the union of the two sets. We can see from the graphs in Fig. 4.43 that every real number belongs to at least one of the two sets. The solution set for the compound inequality is the set of all real numbers.

4.7   *Quick Reference*

Intersection and Union	▪ If *A* and *B* are sets, then the **intersection** of the two sets $A \cap B$ is the set of all numbers belonging to both *A* and *B*: $A \cap B = \{x \mid x \in A \text{ and } x \in B\}$.
	▪ If *A* and *B* are sets, then the **union** of the two sets, $A \cup B$, is the set of all numbers belonging to either *A* or *B* (or both): $A \cup B = \{x \mid x \in A \text{ or } x \in B\}$.
Solution Sets	▪ A **compound inequality** consists of two inequalities connected with the words *and* or *or*.
	▪ If the connective is *and*, the solution set is the *intersection* of the individual solution sets.
	▪ If the connective is *or*, the solution set is the *union* of the individual solution sets.
Double Inequalities	▪ A **double inequality** is a statement that the value of an expression is between two values. It is a special kind of compound inequality with the implied connective *and*.
	▪ To solve a double inequality graphically, graph the three components of the double inequality on the same coordinate axes. Then, estimate the values of *x* for which the graph of the middle expression is between the graphs of the outer expressions.
	▪ To solve a double inequality algebraically, apply the properties of inequalities to all three components at once.
Algebraic Methods	▪ A **conjunction** is a compound inequality whose connective is *and*. To solve a conjunction, solve each individual inequality and then form the *intersection* of the individual solution sets.
	▪ A **disjunction** is a compound inequality whose connective is *or*. To solve a disjunction, solve each individual inequality and then form the *union* of the individual solution sets.

4.7   *Exercises*

 1. Explain the use of the word *and* as it is used in compound inequalities.

 2. Explain the use of the word *or* as it is used in compound inequalities.

In Exercises 3–10, determine whether the given numbers are solutions of the compound inequality.

3. $5x < 12$ and $3x > 3$; 2, 1

4. $-3x < 15$ or $2x > -5$; -8, 4

5. $2x + 7 < -3$ or $3x + 5 > 11$; 5, -7

6. $4 - 3x < 2$ and $3 + 2x > -1$; 3, -6

7. $0.5x > 4.5$ or $0.3x > 1.8$; 7, 11

8. $3 - 0.2x > 7$ and $1.2x > 14$; 8, 12

9. $0.5 \le 2x - 3 < 5$; 3.75, 5

10. $-3 < x + 2 \le 5$; -5, 3, 0

11. What is the difference between a conjunction and a disjunction?

12. For each of the given pairs of compound inequalities, explain the difference between their solution sets.

(a) $x < -7$ or $x \ge 1$ $x < -7$ and $x \ge 1$

(b) $x < 1$ or $x \ge -7$ $x < 1$ and $x \ge -7$

In Exercises 13–18, graph the solution set of each conjunction.

13. $x \ge -5$ and $x < 3$

14. $x \ge -6$ and $x \le 2$

15. $x < 5$ and $x \le 1$

16. $x \ge 0$ and $x > -3$

17. $x \le 4$ and $x > 6$

18. $x > -1$ and $x \le -7$

In Exercises 19–24, graph the solution set of each disjunction.

19. $x < -5$ or $x > 4$

20. $x \le -7$ or $x \ge 8$

21. $x \ge -3$ or $x > 4$

22. $x \le 2$ or $x < 7$

23. $x \le 4$ or $x > 0$

24. $x \ge -3$ or $x < 5$

In Exercises 25 and 26, explain how to interpret the given graphics display to determine the solution set of each of the following. Write the solution set in interval notation.

(a) $y_1 \le y_2 \le y_3$ (b) $y_2 \le y_1$ or $y_2 \ge y_3$

25.

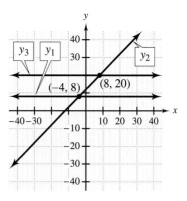

26.

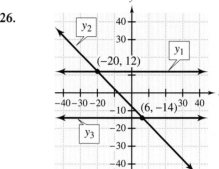

In Exercise 27–30, solve the given inequalities with the graphing method and write the solution sets with interval notation.

27. $-12 \le 3x - 2 \le 8$ **28.** $0 < 3x - 1 < 8$

29. $-2 < \dfrac{x - 1}{-3} < 2$

30. $-4 \le \dfrac{3 - x}{2} < 4$

31. If $-3 < -x < 2$, what is true about x? Why?

32. Explain whether each of the following pairs of compound inequalities are equivalent or not equivalent and why.

(a) $x > -3$ and $x < 7$ $-3 < x < 7$

(b) $3 < x$ or $x < -2$ $3 < x < -2$

(c) $1 \le x$ and $x \ge 5$ $1 \le x \ge 5$

In Exercises 33–46, solve the inequalities algebraically. Graph the solution set on a number line.

33. $-4 < x + 3 < 7$ **34.** $-3 < x - 2 < 5$

35. $2 \le -3t \le 18$

36. $-14 \le -2y < -6$

37. $9 \ge 2x + 1 \ge -2$

38. $7 > 3x - 2 \ge -5$

39. $-5 \le -2x - 7 < 2$ **40.** $-3 < 5 - x < 4$

41. $-\dfrac{7}{4} \le \dfrac{3}{4}x - 1 < 2$ **42.** $3 > \dfrac{t}{2} + 1 \ge 0$

43. $-\dfrac{17}{6} < \dfrac{2}{3}x - \dfrac{1}{6} \le \dfrac{11}{3}$

44. $-2 < \dfrac{1}{2}x - \dfrac{1}{4} < \dfrac{7}{4}$

45. $6.5 \ge 0.75x + 0.5 > -2.5$

46. $-3.25 < 0.75 - 0.8x < 4.75$

47. Describe how the solution set of a disjunction can be all real numbers but that the solution set of the corresponding conjunction is the empty set.

48. Explain why the solution set of a disjunction of linear inequalities can never be the empty set.

In Exercises 49–54, solve each conjunction algebraically and write the solution set with interval notation. Graph the solution set on a number line.

49. $-x \le 5$ and $-x \ge -7$

50. $-x \ge -12$ and $-x \le 10$

51. $x + 1 \ge 5$ and $x - 2 < 12$

52. $x - 2 > 3$ and $x - 5 < 4$

53. $2x - 5 \le 7$ and $x + 4 > -7$

54. $2x + 3 \ge -1$ and $3x + 6 < 21$

In Exercises 55–60, solve each disjunction algebraically and write the solution set with interval notation. Graph the solution set on a number line.

55. $-x \ge 4$ or $-x \le -3$

56. $-x \ge 8$ or $-x \le -2$

57. $x + 4 < -3$ or $x + 4 > 3$

58. $x - 5 < -4$ or $x - 5 > 4$

59. $-4x - 5 > 0$ or $3x - 2 \ge 1$

60. $-3x - 6 > 3$ or $4x + 3 \ge 5$

In Exercises 61–76, solve each compound inequality and write the solution set with interval notation.

61. $2x - 5 \le -17$ and $3x > -27$

62. $-8x \le 24$ and $-2x \ge -8$

63. $x - 2 > 3$ or $x + 5 < 4$

64. $4 - x \ge 5$ or $2x - 3 > 5$

65. $7 - x \ge 10$ or $3x - 5 > 7$

66. $x + 3 > 5$ or $x + 4 < 0$

67. $-4x - 5 \ge 0$ and $3x - 2 \ge 1$

68. $2x - 3 \ge 5$ and $3x + 4 < -2$

69. $\dfrac{1}{2}x - 2 \ge 1$ or $\dfrac{2}{3}x - 1 \ge \dfrac{5}{3}$

70. $-\dfrac{1}{3}x + 1 \ge \dfrac{5}{3}$ or $\dfrac{4}{9}x - 1 < \dfrac{1}{3}$

71. $\dfrac{2}{5}x + \dfrac{1}{2} \ge \dfrac{1}{2}$ and $\dfrac{1}{4}x - \dfrac{1}{3} \le \dfrac{1}{6}$

72. $\dfrac{1}{3}x - 1 > -\dfrac{5}{3}$ and $\dfrac{1}{4}x + \dfrac{1}{2} < \dfrac{1}{4}$

73. $0.8 - 0.2x < -0.4$ or $0.75x - 0.25 < -0.25$

74. $-0.2x + 0.6 < 0$ or $4.6x - 11.5 \ge 6.9$

75. $1.8x + 3.6 < 5.4$ and $2.35x + 7.05 > -4.7$

76. $-19.2 < 4.8x - 8$ and $8.1x - 13.5 \le -10.8$

77. For any real number c, describe the solution set of each compound inequality.

(a) $x \le c$ and $x \ge c$ (b) $x > c$ and $x < c$

78. For any real number c, describe the solution set of each compound inequality.

(a) $x \le c$ or $x \ge c$ (b) $x > c$ or $x < c$

In Exercises 79–94, solve the compound inequality algebraically and write the solution set with interval notation.

79. $-2x > x - 6$ and $2x - 4 \ge 1 + x$

80. $2 - x > 2 + 2x$ and $-2x > -3x$

81. $5x - 7 \ge 8$ and $2 - 3x < -4$

82. $-4x > x - 10$ and $x + 7 \leq 4$

83. $5 - 2x < -9$ or $6x - 2 > 5x$

84. $-5x \leq 10$ or $-3x \geq -4x - 2$

85. $x - 3 \leq 1$ and $x - 4 \geq 0$

86. $4x \leq 0$ and $x - 5 \geq -5$

87. $7 - x < 0$ or $x + 2 > 0$

88. $2x \geq 0$ or $2x \leq 12 + 5x$

89. $-2x + 6 \geq -4$ and $2x - 1 > 9$

90. $3x - 5 < -5$ and $2(x - 4) > -8$

91. $-x \leq 20 - 3x$ and $4 - x \leq -6$

92. $3(2x + 1) \geq 5x$ and $2x - 5 \leq -11$

93. $5 - x \leq 5$ or $3(2x + 1) < 3 + 5x$

94. $1 - 2x < 7$ or $2x + 8 < 7 + x$

Exploring with Real Data

In Exercises 95–98, refer to the following background information and data.

The following table shows the 1988–1991 enrollment (in thousands) in college by age and gender.

Student	1988	1989	1990	1991
Male	5950	5950	6192	6439
18–24	3770	3717	3922	3954
25–34	1395	1443	1412	1605
35 and older	727	716	772	832
Female	7166	7231	7427	7618
18–24	4021	4085	4042	4218
25–34	1568	1637	1749	1680
35 and older	1452	1396	1546	1636

(Source: U.S. Bureau of Census.)

For the year 1991, let $M = \{$male students$\}$, $F = \{$female students$\}$, $A = \{$students between the ages of 25 and 34$\}$, and $B = \{$students 35 and older$\}$.

95. How many elements are in the set $M \cap A$? Interpret your answer.

96. How many elements are in the set $F \cup (M \cap B)$? Interpret your answer.

97. Describe the sets $F \cup M$ and $F \cap M$.

98. In 1991 there were approximately twice as many female students in the 35 and older age group than male students. State two social factors that might explain this large difference.

Challenge

In Exercises 99–102, solve for x.

99. $1 < x - c < 6$

100. $b < \dfrac{x + 1}{a} < c, a < 0$

101. $-2 \leq kx < 5, k < 0$

102. $-1 \leq a(x + 2) \leq 7, a > 0$

103. Given $x + k < 5$ and $k - x < 1$, for what values of k is the solution set empty?

104. Given $x + k > -2$ or $k - x > -1$, for what values of k is the solution set all real numbers?

In Exercises 105–110, use interval notation to write the solution set of the compound inequality.

105. $2 \leq x \leq 7$ and $x > 6$

106. $-3 < x \leq 4$ or $x < 0$

107. $-3 \leq x < 10$ and $x < -3$

108. $4 < x \leq 8$ and $x \geq 8$

109. $0 < x \leq 5$ or $1 < x < 7$

110. $-1 < x \leq 2$ or $5 \leq x < 8$

111. Write a complete explanation of how to solve the following compound inequality graphically.

$$0.5x - 15 < x - 5 < -x + 21$$

Include these items in your explanation.

(a) What functions are entered in the calculator and graphed?

(b) What is the significance of the points of intersection?

(c) What portion of the total graph reveals the solutions? What is the solution set?

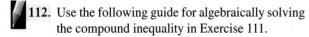

112. Use the following guide for algebraically solving the compound inequality in Exercise 111.

(a) Write the double inequality in the form of a conjunction.

$$0.5x - 15 < x - 5 \text{ and } x - 5 < -x + 21$$

(b) Solve each inequality.

(c) Form the intersection of the two solution sets in part (b).

(d) Compare your result with the solution set obtained in Exercise 111.

In Exercises 113–116, solve the compound inequality algebraically and write the solution set with interval notation.

113. $3x + 2 \leq x + 8 \leq 2x + 15$

114. $2x - 1 < x + 1 < 7 - x$

115. $x < x + 1 < x + 2$

116. $x + 1 < x < x + 2$

4.8 Absolute Value: Equations and Inequalities

Graphs and Number Lines ▪ *Algebraic Interpretation* ▪ *Special Cases* ▪ *Algebraic Methods*

Graphs and Number Lines

Equations and inequalities sometimes involve absolute values. Graphic interpretations will be very helpful in developing some generalizations about the solution sets.

EXPLORATION 1   *Graphing Methods*

Use the integer setting to produce the graphs of $y_1 = |x|$ and $y_2 = 15$ on the same coordinate axes.

(a) Trace the graph and estimate the solutions of the equation $|x| = 15$.

(b) We can write the inequality $|x| < 15$ as $y_1 < y_2$. Trace the graph and determine the x-coordinates of the points where the graph of y_1 is below the graph of y_2.

(c) We can write the inequality $|x| > 15$ as $y_1 > y_2$. Trace the graph and determine the x-coordinates of the points where the graph of y_1 is above the graph of y_2.

Discovery

(a) In Fig. 4.44 the graphs of y_1 and y_2 intersect at $(-15, 15)$ and at $(15, 15)$. The solution set is $\{-15, 15\}$. Note that these two numbers are 15 units from the origin on the number line. (See Fig. 4.45.)

Figure 4.44

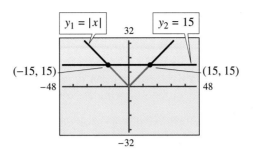

Figure 4.45

(b) In Fig. 4.44 the graph of y_1 is below the graph of y_2 for all x-coordinates between -15 and 15. Therefore, the solution set is $\{x \mid -15 < x < 15\}$. Note that these numbers are all less than 15 units from the origin on the number line. (See Fig. 4.46.)

(c) In Fig. 4.44 the graph of y_1 is above the graph of y_2 for all x-coordinates less than -15 and all x-coordinates greater than 15. Therefore, the solution set is $\{x \mid x < -15 \text{ or } x > 15\}$. Note that these numbers are all more than 15 units from the origin on the number line. (See Fig. 4.47.)

Figure 4.46

Figure 4.47

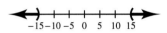

The equation $|x| = 15$ in Exploration 1 is an example of a **linear absolute value equation in one variable.**

Definition of Linear Absolute Value Equation in One Variable

A **linear absolute value equation in one variable** is an equation that can be written in the form $|Ax + B| = C$ where A, B, and C are real numbers and $A \neq 0$.

The inequalities $|x| < 15$ and $|x| > 15$ in Exploration 1 are examples of a **linear absolute value inequality in one variable.**

Definition of Linear Absolute Value Inequality in One Variable

A **linear absolute value inequality in one variable** can be written in the form $|Ax + B| < C$ where A, B, and C are real numbers and $A \neq 0$. A similar definition may be stated for $>$, $\geq$, and $\leq$.

Algebraic Interpretation

In Exploration 1 we found a graphing method for solving an absolute value equation or inequality. We also observed that the solutions can be interpreted in terms of their distance from the origin on the number line.

The methods in Exploration 1 can be generalized. For any expression A and any positive number C, the following are true.

1. If $|A| = C$, then A represents a number that is C units from the origin.

2. If $|A| > C$, then A represents a number that is more than C units from the origin.

3. If $|A| < C$, then A represents a number that is less than C units from the origin.

Later in this section, we will discuss algebraic methods for solving absolute value equations and inequalities. But these methods are best understood with the visual assistance that graphs and number lines can provide.

EXPLORATION 2   *Combining Graphs and Algebraic Methods*

Use the integer setting to produce the graphs of $y_1 = |3 - 2x|$ and $y_2 = 17$ on the same coordinate axes. Use the graphs to estimate the solution set for each of the following.

     (a) $|3 - 2x| = 17$ (b) $|3 - 2x| > 17$ (c) $|3 - 2x| < 17$

Discovery

Figure 4.48

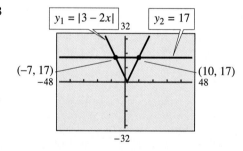

(a) In Fig. 4.48, there are two points of intersection, and the solutions appear to be -7 and 10.

Because $|3 - 2x| = 17$, the expression $3 - 2x$ represents a number that is 17 units from the origin. (See Fig. 4.49.)

Figure 4.49

$$\begin{array}{c} 3 - 2x \qquad\qquad 3 - 2x \\ \xleftarrow{\hspace{1cm}} \blacklozenge \underset{-17}{} \;\;+\;\; \blacklozenge \underset{17}{} \xrightarrow{\hspace{1cm}} \\ -17 \quad\;\; 0 \quad\;\; 17 \end{array}$$

This suggests the following equations. Solve each one.

$$
\begin{aligned}
3 - 2x &= 17 \quad \text{or} \quad 3 - 2x = -17 \\
-2x &= 14 \quad \text{or} \qquad\;\; -2x = -20 \\
x &= -7 \quad \text{or} \qquad\quad\;\; x = 10
\end{aligned}
$$

The solution set is $\{-7, 10\}$.

(b) In Fig. 4.48, the graph of y_1 is above the graph of y_2 for x-values less than -7 and greater than 10.

Because $|3 - 2x| > 17$, the expression $3 - 2x$ represents a number that is more than 17 units from the origin. (See Fig. 4.50.)

Figure 4.50

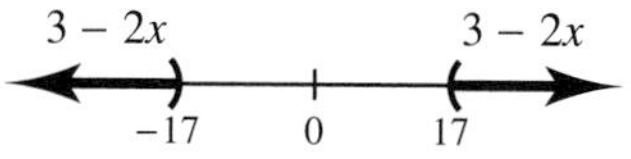

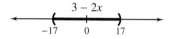

This suggests the following inequalities. Solve each one.

$$3 - 2x < -17 \quad \text{or} \quad 3 - 2x > 17$$
$$-2x < -20 \quad \text{or} \quad\quad -2x > 14$$
$$x > 10 \quad \text{or} \quad\quad\quad x < -7$$

The solution set is $(-\infty, -7) \cup (10, \infty)$.

(c) In Fig. 4.48, the graph of y_1 is below the graph of y_2 for x-values between -7 and 10.

Because $|3 - 2x| < 17$, the expression $3 - 2x$ represents a number that is less than 17 units from the origin. (See Fig. 4.51.)

Figure 4.51

This suggests the following inequalities. Solve each one.

$$3 - 2x > -17 \quad \text{and} \quad 3 - 2x < 17$$
$$-2x > -20 \quad \text{and} \quad\quad -2x < 14$$
$$x < 10 \quad \text{and} \quad\quad\quad x > -7$$

The solution set is $(-7, 10)$.

The following summarizes the algebraic principles illustrated in Exploration 2.

> **Compound Equations or Inequalities Arising from Linear Absolute Value Equations or Inequalities**
>
> For a linear absolute value equation or inequality with $C > 0$:
>
> 1. $|Ax + B| = C$ implies $Ax + B = C$ or $Ax + B = -C$
>
> 2. $|Ax + B| < C$ implies $Ax + B > -C$ and $Ax + B < C$ (which can be written $-C < Ax + B < C$).
>
> 3. $|Ax + B| > C$ implies $Ax + B < -C$ or $Ax + B > C$.

Special Cases

Because the absolute value of a number is never negative, no portion of the graph of $|Ax + B|$ can be below the x-axis. This limitation can present some interesting special situations when we solve absolute value equations and inequalities.

EXPLORATION 3

Special Cases

Use the graphing method to solve the following equations and inequalities. In each case note the value of the constant on the right side.

(a) $|x - 6| = -12$ (b) $|x - 6| > -12$

(c) $|2x + 14| = 0$ (d) $|2x + 14| < 0$

Discovery

Figure 4.52

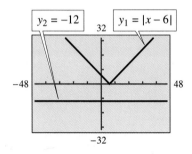

(a) In Fig. 4.52 we produce the graphs of $y_1 = |x - 6|$ and $y_2 = -12$ on the same coordinate axes.

There is no point of intersection, so the equation has no solution.

     Generalization: An equation $|Ax + B| = C$, where $C < 0$, has no solution.

(b) In Fig. 4.52 every point of the graph of y_1 is above the graph of y_2, so every real number is a solution.

     Generalization: An inequality $|Ax + B| > C$, where $C < 0$, is satisfied by all real numbers.

(c) In Fig. 4.53 we produce the graphs of $y_1 = |2x + 14|$ and $y_2 = 0$. Note that the graph of $y_2 = 0$ is the *x*-axis.

Figure 4.53

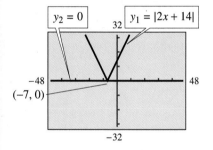

There is one point of intersection, $(-7, 0)$, so the equation has one solution, -7.

     Generalization: An equation $|Ax + B| = 0$ has one solution.

(d) In Fig. 4.53 no point of the graph of y_1 is below the graph of y_2, so there is no solution.

     Generalization: An inequality $|Ax + B| < 0$ has no solution.

Algebraic Methods

The graphing method is useful for estimating the number of solutions or for estimating the solutions themselves. However, to determine exact solutions, we use algebraic methods.

We can use algebraic methods to solve the compound equations or inequalities arising from absolute value equations or inequalities. The possibilities were summarized earlier.

In addition, generalizations such as those formed in Exploration 3 might be needed to determine solutions algebraically.

EXAMPLE 1   *Solving Absolute Value Equations and Inequalities Algebraically*

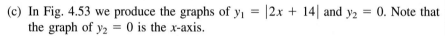

Solve each of the following algebraically.

(a) $|2x + 5| - 4 = -2$ (b) $|4 - x| = |3 - 2(5 - x)|$

(c) $|1 - 3x| - 6 \geq 1$ (d) $9 - |x - 2| > 1$

Solution

(a) We write the equation in the form $|Ax + B| = C$.

$$|2x + 5| - 4 = -2$$
$$|2x + 5| - 4 + 4 = -2 + 4 \qquad \text{Add 4 to both sides.}$$
$$|2x + 5| = 2$$

Now write $|2x + 5| = 2$ as two linear equations and solve each one.

$$2x + 5 = 2 \qquad \text{or} \quad 2x + 5 = -2$$
$$2x = -3 \quad \text{or} \qquad 2x = -7$$
$$x = -\frac{3}{2} \quad \text{or} \qquad x = -\frac{7}{2}$$

The solution set is $\{-\frac{3}{2}, -\frac{7}{2}\}$.

(b) We begin by simplifying the expression on the right.

$$|4 - x| = |3 - 2(5 - x)|$$
$$|4 - x| = |3 - 10 + 2x| \qquad \text{Distributive Property}$$
$$|4 - x| = |-7 + 2x|$$

From the definition of absolute value (see Section 1.1), $|A| = A$ or $|A| = -A$. Therefore, for an equation of the form $|A| = |B|$, there are four possibilities:

$$A = B \quad \text{or} \quad -A = -B \quad \text{or} \quad A = -B \quad \text{or} \quad -A = B.$$

Because the first two equations are equivalent and the last two equations are equivalent, we reduce the possibilities to two: $A = B$ or $A = -B$. This means that $|4 - x| = |-7 + 2x|$ can be written as a disjunction.

$$4 - x = -7 + 2x \quad \text{or} \quad 4 - x = -(-7 + 2x)$$
$$-3x = -11 \qquad \text{or} \quad 4 - x = 7 - 2x$$
$$x = \frac{11}{3} \qquad \text{or} \qquad x = 3$$

The solution set is $\{\frac{11}{3}, 3\}$.

(c) We write $|1 - 3x| - 6 \geq 1$ in the form $|Ax + B| \geq C$.

$$|1 - 3x| - 6 \geq 1$$
$$|1 - 3x| - 6 + 6 \geq 1 + 6 \qquad \text{Add 6 to both sides.}$$
$$|1 - 3x| \geq 7$$

The resulting inequality can be written as a disjunction.

$$1 - 3x \leq -7 \quad \text{or} \quad 1 - 3x \geq 7$$
$$-3x \leq -8 \quad \text{or} \qquad -3x \geq 6$$
$$x \geq \frac{8}{3} \quad \text{or} \qquad x \leq -2$$

The solution set is $(-\infty, -2] \cup [\frac{8}{3}, \infty)$. (See Fig. 4.54.)

Figure 4.54

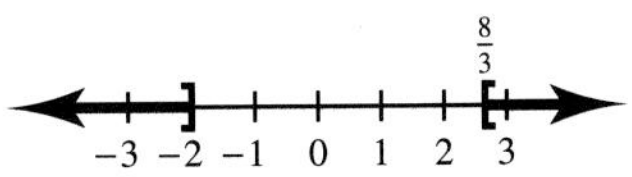

(d) We begin by isolating the absolute value expression.

$$9 - |x - 2| > 1$$

$9 - |x - 2| - 9 > 1 - 9$ Subtract 9 from both sides.

$-|x - 2| > -8$ Multiply both sides by -1 and reverse the inequality symbol.

$|x - 2| < 8$

Write $|x - 2| < 8$ as two linear inequalities and solve each one.

$x - 2 > -8$ and $x - 2 < 8$

$x > -6$ and $x < 10$

The solution set is $(-6, 10)$. (See Fig. 4.55.)

Figure 4.55

$-8\ -6\ -4\ -2\ \ 0\ \ 2\ \ 4\ \ 6\ \ 8\ \ 10\ 12$

4.8   *Quick Reference*

Graphs and Number Lines

- A **linear absolute value equation in one variable** is an equation that can be written in the form $|Ax + B| = C$ where A, B, and C are real numbers and $A \neq 0$.

- A **linear absolute value inequality in one variable** can be written in the form $|Ax + B| < C$ where A, B, and C are real numbers and $A \neq 0$. A similar definition may be stated for $>$, $\geq$, and $\leq$.

- To estimate the solution set of an absolute value equation or inequality graphically, graph the left and right sides of the equation or inequality. Determine the point(s) of intersection.

  1. If the symbol is $=$, $\leq$, or $\geq$, the x-coordinate of each point of intersection is a solution.

  2. If the symbol is $<$, determine the x-values of all points for which the graph of the left side is *below* the graph of the right side.

  3. If the symbol is $>$, determine the x-values of all points for which the graph of the left side is *above* the graph of the right side.

- We can interpret the absolute value of a number as its distance from zero on the number line.

Algebraic Interpretation

- If $C > 0$, $|Ax + B| = C$ is equivalent to the disjunction

  $Ax + B = C$ or $Ax + B = -C$.

- If $C > 0$, $|Ax + B| < C$ is equivalent to the conjunction

  $Ax + B < C$ and $Ax + B > -C$.

- If $C > 0$, $|Ax + B| > C$ is equivalent to the disjunction

  $Ax + B > C$ or $Ax + B < -C$.

Special Cases
- If $C = 0$, then
  1. $|Ax + B| = C$ is equivalent to $Ax + B = 0$.
  2. $|Ax + B| < C$ has no solution.
  3. $|Ax + B| > C$ is true for every real number except for the solution of $Ax + B = 0$.
- If $C < 0$, then
  1. $|Ax + B| = C$ has no solution.
  2. $|Ax + B| < C$ has no solution.
  3. $|Ax + B| > C$ is true for every real number.

Algebraic Methods
- To solve a compound absolute value equation or inequality algebraically:
  1. Isolate the absolute value expression.
  2. Translate the absolute value equation or inequality into an equivalent conjunction or disjunction.
  3. Solve each component of the compound equation or inequality and form the union or intersection of the solution sets.

4.8   *Exercises*

  1. Interpret $|x - 2| = 5$ in terms of distance on a number line.

2. How does the solution set for $|x + 3| = 5$ differ from the solution set for $x + 3 = 5$?

In Exercises 3–10, solve the given equation by graphing.

3. $|x| = 12$ **4.** $|x - 8| = 9$

5. $|2x - 10| + 7 = 7$ **6.** $|6 - x| - 9 = 5$

7. $-|x + 10| = -15$ **8.** $|x + 7| = -8$

9. $10 - |x + 8| = 17$ **10.** $15 - |x - 6| = 6$

In Exercises 11 and 12, use the graphics display to determine the solution set of the equation $y_1 = y_2$.

11.

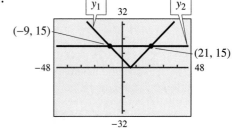

12.

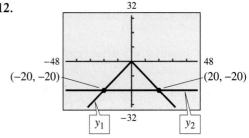

In Exercises 13–18, solve the equation algebraically.

13. $|2x - 5| = 10$ **14.** $|3 - 4x| = 12$

15. $|5x - 2| = 0$ **16.** $|5 - x| = 0$

17. $|x + 4| - 3 = 7$ **18.** $6 + |3 - x| = 7$

In Exercises 19 and 20, use the graphics display to determine the solution set of the following inequalities.

     (a) $y_1 < y_2$ (b) $y_1 > y_2$

19.

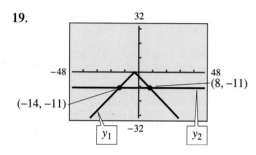

20.

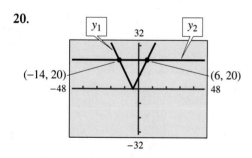

In Exercises 21–28, solve the inequality by graphing. Write the solution set with interval notation.

21. $|x| < 12$ **22.** $|x - 8| \geq 9$

23. $15 - |x - 6| > 6$ **24.** $-|x + 10| \leq -15$

25. $|2x - 10| + 7 \geq 7$ **26.** $10 - |x + 8| \leq 17$

27. $|x + 7| < -8$ **28.** $10 - |x + 8| \geq 17$

 29. What is the difference between the solution sets of $|x| \geq 5$ and $|x| \leq 5$?

 30. Explain how the graphing method shows that $|x| < -5$ has no solutions.

In Exercises 31–38, solve the inequality algebraically. Write the solution set with interval notation and graph the solution set on a number line.

31. $|1 - 2t| < 9$ **32.** $|3y + 2| < 14$

33. $|7 + x| \leq 0$ **34.** $|x - 1| \leq 0$

35. $|4 - x| - 5 \leq 2$ **36.** $|2x - 1| - 3 \leq 10$

37. $|x + 5| < -3$ **38.** $|4 + 3x| < -5$

In Exercises 39–46, solve the inequality algebraically. Write the solution set with interval notation and graph the solution set on a number line.

39. $|3x + 2| \geq 7$ **40.** $|5 - 2x| \geq 11$

41. $|x - 5| > 0$ **42.** $|3 + x| > 0$

43. $|3x - 2| \geq -5$ **44.** $|5 - 7x| \geq -1$

45. $|4t + 3| + 2 > 10$ **46.** $|2 - 3t| - 9 > 8$

In Exercises 47–50, x represents any real number. Write an absolute value inequality in x to describe each situation.

47. The distance from 5 to x is less than 7.

48. The distance from -5 to x is less than or equal to 4.

49. The distance from -7 to x is greater than or equal to 10.

50. The distance from 7 to x is greater than 12.

In Exercises 51–54, write an absolute value inequality in x to describe each graph.

51.

$-5\ -4\ -3\ -2\ -1\ \ 0\ \ 1\ \ 2\ \ 3\ \ 4\ \ 5\ \ 6$

52.

$-5\ -4\ -3\ -2\ -1\ \ 0\ \ 1\ \ 2\ \ 3\ \ 4\ \ 5\ \ 6$

53.

$-4\ -3\ -2\ -1\ \ 0\ \ 1\ \ 2\ \ 3\ \ 4\ \ 5$

54.

$-4\ -3\ -2\ -1\ \ 0\ \ 1\ \ 2\ \ 3\ \ 4\ \ 5$

In Exercises 55–78, solve each equation algebraically.

55. $|x - 4| = 2$ **56.** $|x + 3| = 6$

57. $|2k - 3| = 8$　　　**58.** $|4 - 3m| = 15$

59. $|0.23x - 0.45| = 0.56$

60. $|3.67x - 2.39| = 2.58$

61. $|y + 1| - 3 = 7$

62. $|2x + 3| - 5 = 12$

63. $|t| + 8 = 5$　　　**64.** $|-t| - 7 = -12$

65. $5 - 3|x - 5| = -13$

66. $2 - |x + 1| = -7$

67. $|t + 4| - 3 = -23$

68. $|m - 23| + 9 = 4$

69. $|3x - 4| - 4 = -4$　　**70.** $|5 - 2x| + 5 = 5$

71. $5 + 3|2x + 1| = 20$

72. $7 - 2|1 - 3x| = -13$

73. $0.24 + |0.46k + 0.48| = -0.73$

74. $1.37 - |9.37r + 8.25| = 3.38$

75. $\left| \dfrac{5 - 3x}{4} \right| = 3$

76. $\left| \dfrac{4x - 5}{3} \right| = 5$

77. $\left| x - \dfrac{2}{3} \right| = \dfrac{3}{4}$

78. $\left| \dfrac{1}{2}x + 2 \right| = \dfrac{2}{3}$

In Exercises 79–82, solve the given equation.

79. $|x - 4| = x + 2$　　　**80.** $|t - 3| = t + 3$

81. $|x - 1| = 7 - (x - 1)$

82. $|t + 5| = -t$

In Exercises 83–88, solve the given equation.

83. $|x - 1| = |x - 19|$　　**84.** $|x + 3| = |x - 3|$

85. $|2x - 5| = |x|$　　　**86.** $|t| = |3 - 4t|$

87. $|x - 2| = |2 - x|$

88. $|1 - 2x| = |2x - 1|$

 89. What is the first step to solve $-3|x - 2| \geq 12$?

 90. Explain why each given pair of inequalities is equivalent or not equivalent.

　(a) $|x| > 5$　　　　$x > 5$

　(b) $|x - 1| \leq 3$　　　$-3 \leq x - 1 \leq 3$

In Exercises 91–120, solve each inequality algebraically. Write the solution set with interval notation.

91. $|k + 3| < 4$　　　**92.** $|y + 5| \leq 9$

93. $|6 - x| \geq 8$　　　**94.** $|3 - k| \leq 5$

95. $-5|2x + 3| < -35$

96. $-2|7 - x| > -16$

97. $5 - |4 - 3x| > 2$

98. $12 - |6 - 5p| < 9$

99. $|4x + 3| + 9 \geq 4$　　**100.** $|3x + 2| \leq -3$

101. $|3y - 5| + 4 \geq 10$　　**102.** $|2x - 7| - 3 \leq 8$

103. $|3y - 5| + 6 \leq 6$

104. $|2k + 7| - 7 \geq 8 - 15$

105. $|5 - 3x| + 7 > 9$　　**106.** $|4 - 3x| - 5 < 2$

107. $5 - |x - 2| \geq 7$

108. $24 - |m + 3| \leq 27$

109. $|0.26x + 0.65| + 0.32 \leq 2.69$

110. $|3.56t - 0.37| - 5.6 \geq 7.97$

111. $5 - 2|2x - 1| > -1$

112. $4 - 3|x + 5| < -8$

113. $\left| \dfrac{2x + 1}{4} \right| > 1$

114. $\left| \dfrac{3x - 1}{2} \right| > 3$

115. $\left| \dfrac{4x - 3}{2} \right| + 5 \leq 7$

116. $\left| \dfrac{3t - 1}{4} \right| - 6 > 5$

117. $\left| \dfrac{1}{3}x - 1 \right| + \dfrac{5}{3} \geq 1$

118. $\left| \dfrac{3}{4}x - \dfrac{1}{2} \right| - 5 < 1$

119. $8|x - 2| \geq 3|x - 2| + 10$

120. $5|x| < 2|x| + 12$

In Exercises 121–124, use the graphing method to estimate the solution set of the given inequality.

121. $|x| \geq x + 3$ **122.** $|x| \geq x$

123. $|x - 3| < 1 - 2x$ **124.** $2x + 7 < |x + 2|$

In Exercises 125–128, use the graphing method to estimate the solution set of the given inequality.

125. $|x + 6| < |x - 4|$ **126.** $|x + 5| > |x - 5|$

127. $|t| \geq |15 - 2t|$

128. $|2x - 11| \leq |x + 6|$

Challenge

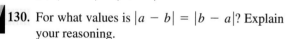 **129.** For what values are the following true? Explain your reasoning.

(a) $|x - 2| = x - 2$

(b) $|x - 2| = -(x - 2)$

(c) $|x - 2| = |2 - x|$

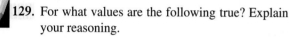 **130.** For what values is $|a - b| = |b - a|$? Explain your reasoning.

In Exercises 131–134, solve for x.

131. $|2x + 1| = k, k > 0$

132. $|x + b| = 4$

133. $c|x| \geq b, b > 0, c < 0$

134. $|x| + b = -4, b < -4$

In Exercises 135 and 136, determine the values of x for which the given equation is true.

135. $x + |x| = 2x$ **136.** $x + |x| = 0$

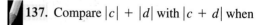 **137.** Compare $|c| + |d|$ with $|c + d|$ when

(a) both c and d are positive.

(b) both c and d are negative.

(c) c and d have opposite signs.

(d) Which of the following do you think is true for all numbers c and d?

(i) $|c| + |d| \leq |c + d|$

(ii) $|c| + |d| \geq |c + d|$

(iii) $|c| + |d| = |c + d|$

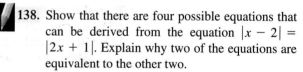 **138.** Show that there are four possible equations that can be derived from the equation $|x - 2| = |2x + 1|$. Explain why two of the equations are equivalent to the other two.

4.9 Applications

Linear Inequalities ▪ *Compound Inequalities* ▪ *Absolute Value Inequalities*

We use inequalities to represent the conditions of an applied problem when words such as "at least," "no more than," and "at most" are used in the problem.

The general approach for solving such applications is identical to the strategy for solving applications involving equations. Assign a variable to the unknown quantity and represent other unknown quantities in terms of that variable. Translate the conditions of the problem into an inequality and solve.

In this section we consider examples of applications involving the types of inequalities we have discussed in this chapter.

Linear Inequalities

Example 1 illustrates an application that can be solved with a single linear inequality.

EXAMPLE 1 *Minimum Bowling Average*

To qualify for the semifinals in a bowling tournament, a bowler must have at least a 160 average for five games. Suppose the scores for your first four games were 172, 150, 148, and 162. What is the lowest score you can bowl in your fifth game and qualify for the semifinals?

Solution Your final average will be determined by adding the five scores and dividing the sum by 5. Letting x represent the fifth score, your average will be computed as follows.

$$\frac{172 + 150 + 148 + 162 + x}{5}$$

Because your average must be *at least* 160, this quantity must be *greater than or equal to* 160.

$$\frac{172 + 150 + 148 + 162 + x}{5} \geq 160$$

$$\frac{632 + x}{5} \geq 160$$

$$\frac{5}{1} \cdot \frac{632 + x}{5} \geq 5 \cdot 160 \qquad \text{Multiply both sides by 5.}$$

$$632 + x \geq 800$$

$$632 + x - 632 \geq 800 - 632 \qquad \text{Subtract 632 from both sides.}$$

$$x \geq 168$$

You must bowl *at least* 168 for your fifth game in order to qualify.

Compound Inequalities

In everyday situations, there is sometimes more information given in a problem than is needed. Problem solvers must know how to sort out what is relevant.

In Example 2 we consider an application involving a compound inequality.

EXAMPLE 2   *Fencing a Pasture*

Farm Resources is a firm that specializes in installing fencing. A customer has called the sales representative about installing a 4-foot-high fence around a rectangular field where she keeps her horse.

She tells the representative that the width of the field must be 60% of the length and the perimeter must be at least 800 feet. She says that she cannot spend more than $6000.

The customer wants to know the minimum and maximum length that the field can be. If the fencing costs $5.00 per linear foot, what is the sales representative's reply?

Solution

The representative begins by dividing the customer's $6000 budget by $5 to determine the maximum amount of fencing that she can buy.

$$\frac{6000}{5} = 1200$$

She can buy at most 1200 feet of fencing, and she has said that the minimum amount is 800 feet. Therefore, the perimeter of the field must be between 800 feet and 1200 feet, inclusive.

$$800 \leq \text{perimeter} \leq 1200$$

The sales representative lets L represent the length of the field. The width is 60% of L or 0.60L. (See Fig. 4.56.)

Figure 4.56

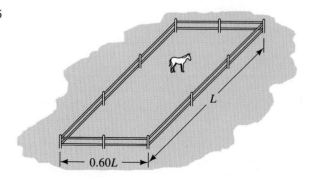

The perimeter of the rectangular field is given by $2L + 2W$ or, in this case, $2L + 2(0.60L)$.

$$800 \le 2L + 2(0.60L) \le 1200$$
$$800 \le 2L + 1.2L \le 1200$$
$$800 \le 3.2L \le 1200$$
$$\frac{800}{3.2} \le \frac{3.2L}{3.2} \le \frac{1200}{3.2}$$
$$250 \le L \le 375$$

Thus the length of the field will be at least 250 feet and at most 375 feet. ■

In Example 2 the fact that the fence was to be 4 feet high had nothing to do with the problem, except, perhaps, as it affected the unit cost of the fencing.

Absolute Value Inequalities

Absolute value inequalities typically arise in applied situations in which the difference between two quantities is less than or greater than a fixed amount. The following example illustrates how absolute value plays a role in matters of **tolerance** in measurements.

EXAMPLE 3   *Temperature Tolerances*

A nuclear reactor shuts down automatically if the temperature in the reactor differs from 1200°F by more than 30°. For what temperatures will the reactor continue to operate?

Solution

Let T = reactor temperature. The difference between T and 1200 cannot be greater than 30. In other words, the difference between T and 1200 must be less than or equal to 30 in order for the reactor to operate.

$$|T - 1200| \le 30$$
$$T - 1200 \ge -30 \quad \text{and} \quad T - 1200 \le 30$$
$$T \ge 1170 \quad \text{and} \qquad\qquad T \le 1230$$

In interval notation, the solution set is [1170, 1230]. The reactor will operate as long as the temperature is between 1170° and 1230°, inclusive. ■

4.9   *Exercises*

In Exercises 1–10, translate each of the following into an algebraic expression or an inequality. In the case of an inequality, write the solution set with interval notation.

1. A number x is less than 5.

2. Five less than a number x

3. The temperature t is at least 70°.

4. Nine degrees more than the temperature t

5. Three points higher than a test score s

6. The test score s is no more than 92.

7. Two hundred dollars less than the profit p

8. The profit p exceeds $2000.

9. A number x is between -5 and 7.

10. A number x exceeds -5 but is no more than 7.

$$-5 > x \leq 7$$

Linear Inequalities

11. Your exam scores so far are 85, 78, and 80. You have one exam left to take. If the passing grade for the course is 65, do you even need to show up for the last test?

12. Your grades on tests 1, 2, and 4 are 82, 76, and 90. Unfortunately, you cut the third test and received a 0. If you have one test left to take, and if the passing grade for the course is 70, can you still pass the course?

13. Cross-country runners are awarded these quality points according to how the runners finish in a race.

Finish Position	Points
1	10
2	7
3	5
4	3
5	1

To win a trophy at the end of ten races, a runner must have an average of at least 7 points.

Here are your finish positions for your first nine races.

1	4	1	2	2	1	2	2	3

What is the minimum position in which you can finish the tenth race in order to win a trophy?

14. The length of a rectangle is 6 inches more than the width. The perimeter of the rectangle can be no more than 48 inches. What is the maximum width?

15. The width of a rectangle is 10 feet less than the length. The perimeter is at least 80 feet. What is the minimum length?

16. A car rental agency rents cars for $26.20 per day plus $0.22 per mile driven. If your travel budget is $200, what is the maximum number of miles you can drive during a 1-day rental?

17. You want to make a 1-day round trip to a city that is 200 miles away. You can spend at most $90 for a car rental. If the 30–30 Car Rental Agency rents cars for $30 per day plus $0.30 per mile driven, can you make the trip?

18. A pipe is at least 21 feet long, and you want to cut it into three pieces. The second piece is to be twice as long as the first piece, and the third piece is to be 1 foot longer than the second piece. What is the minimum length of the first piece?

19. You know that a rope is no more than 100 feet long. You need to cut the rope into three pieces. The second piece is to be three times as long as the first piece, and the third piece must be 18 feet long. What is the maximum length of the second piece?

20. "How much of the book have you read?" you ask your friend. She replies, "I'll give you a clue. If you add the number of the page I'm on plus the number of the next page plus the page after that, the sum will be at least 108." What is the lowest page number of the book your friend could be reading now?

21. Your instructor offers a one-million-dollar prize to anyone who can solve this puzzle. There are three consecutive even integers, and at least two of them are positive. The sum of the first and third integers is no more than the second integer. Show why your instructor will never have to pay.

22. At the big game your friends pool their money and collect $18 for the concession stand. Your mission is to buy six soft drinks and as many hot dogs as you have money left to buy. If soft drinks are $1.00 and hot dogs are $1.75, what is the maximum number of hot dogs you can buy?

23. You are given a gift certificate worth $50 at the local pizza parlor. You can use the certificate to buy pizzas for $4.75 and drinks for $1.25. You and seven friends decide to use your certificate. If all eight of you want a full pizza each, what is the maximum number of drinks you will be able to buy?

24. A rich, eccentric aunt leaves some money in her will for Calvin, John, and Carrie. Calvin is to receive the least, Carrie is to receive the most. The aunt requires the three relatives to solve a puzzle before they can receive their inheritance. The will reads as follows.

> Calvin and John are given consecutive even integers. Carrie's number is 5 more than Calvin's number. If one-seventh of Calvin's number, one-third of John's number, and one-sixth of Carrie's number are added together, the sum exceeds 5. Determine the smallest numbers that satisfy these conditions and multiply them by $1000. The result is your inheritance.

How much will Calvin inherit?

25. The ages of a brother and sister are consecutive integers. Their cousin is 6 years younger than the brother. If one-fourth of the brother's age, one-third of the sister's age, and one-half of the cousin's age are added together, the sum is no more than the brother's age. What is the brother's maximum age?

26. Suppose you have a gift certificate worth $20 for one long-distance phone call. If the charge is $1.10 for the first minute and $0.42 for each additional minute, what is the longest that you can talk?

Compound Inequalities

27. A bank kept track of the number of people who used the automated teller machine during the past month. The fewest number was 28 and the greatest number was 51. Explain why this information can be translated into either a double inequality or a conjunction.

28. Which of the following is the correct interpretation of $3 < x > 8$? Explain your answer.

    (i) x is between 3 and 8.

    (ii) x is greater than 3 and x is greater than 8.

29. The enrollment at a college is between 2000 and 2100 students. The number of women is 300 less than twice the number of men. At least how many men are enrolled? What is the maximum number of men enrolled?

30. A doll manufacturer makes happy dolls and grumpy dolls. Each day, the number of grumpy dolls made is two-thirds the number of happy dolls made. If the total daily production is between 75 and 100, what is the least number of happy dolls made? What is the maximum number of grumpy dolls made?

31. A senior, a junior, and a freshman work on the student newspaper. Each month, the junior works two-thirds the number of hours that the senior works, and the freshman works one-fifth the number of hours that the senior does. If their combined hours for the month are at least 56, what is the minimum number of hours that the junior works?

32. A father can read a page in three-fourths the time it takes his son. If their combined time is no more than 1.5 minutes, what is the maximum time it takes the son?

33. A freezer is most efficient if the Fahrenheit temperature is between 0° and 10°, inclusive. What range of *Celsius* temperatures is most efficient?

34. On interstate highways (not in urban areas) the minimum speed is 40 mph and the maximum speed is 65 mph. What is the legal range of speeds in *kilometers per hour*. (1 mile ≈ 1.61 kilometers.)

35. A rectangular plot, whose length is to be 3 feet more than twice the width, needs to be fenced.

The fencing available is at least 600 feet and at most 900 feet in length. What is the range of possible lengths?

36. An automobile repair shop charges a flat rate of $50 plus $32 per hour for the mechanic's time. If a customer receives an estimate of between $130 and $170 for fixing his car, what is the estimate of the mechanic's time?

37. When we say that the length L of a rectangle is no more than 12, we write $L \leq 12$. However, there is another implied inequality that also applies to L. Explain why the conjunction $0 \leq L \leq 12$ is a more thorough translation.

38. If x is a real number such that $0.99 < x < 1$, would you describe the number of solutions as none, one, or infinitely many? Why?

39. The distance from a lawyer's home to the shopping center is twice the distance from her home to her office. The office and the shopping center are 4 miles apart. The lawyer leaves her home, goes to her office and to the shopping center, and returns home. If the total distance traveled was between 11.8 and 12.4 miles, how far is it from the lawyer's home to her office?

40. The cost C of producing x items is given by $C(x) = 0.10x + 2$. The revenue R is given by $R(x) = 0.50x$. If the expected profit is between $300 and $320, how many items must be produced?

41. A drugstore sells toothbrushes for $1.30 and toothpaste for $3.00. For an average month the store sells twice as much toothpaste as toothbrushes, and combined sales for these items are between $206.00 and $257.50. How many toothbrushes are sold in an average month?

42. A builder constructs rectangular swimming pools. For every design the width is one-half of the length. If the perimeter of the pool can range between 120 feet and 192 feet, what pool widths does the builder offer?

43. In each of a player's ten basketball games, the number of free throws he made was two-thirds the number of two-point field goals he made. The player never scored less than 8 points and never scored more than 16 points. What was the minimum and maximum number of two-point field goals the player made?

44. In a small community the number of registered Republicans is 1.2 times the number of registered Democrats. It is known that 60% of the Democrats and 40% of the Republicans will vote in the special election. If the expected voter turnout is between 216 and 243, how many registered Democrats are there?

45. A packet of marigold seeds has twice as many seeds as a packet of delphinium seeds. The germination rate for marigolds is 90%, and the germination rate for delphiniums is 50%. A gardener plants all the seeds and expects to have a total of between 92 and 115 plants. How many marigold seeds were planted?

46. A light-bulb manufacturer finds that the number of defective 60-watt bulbs averages 1.4 times the number of defective 100-watt bulbs. On a given day the total number of defective bulbs is between 36 and 48. How many 100-watt bulbs are defective?

Absolute Value Inequalities

47. If x_1 and x_2 are two numbers on a number line, the distance between them is $|x_1 - x_2|$. Why is it necessary to take the absolute value of the difference?

48. Solving the inequality $|-x| < 3$, a student states that $|-x| = x$ and therefore $x < 3$. Explain the error in the student's reasoning.

In Exercises 49–52, describe the given conditions with an absolute value inequality and write the corresponding double inequality.

49. A carpenter cuts a board to a length of 12 feet with no more than a 2% error. (Let L = the length of the cut board.)

50. A scale is never off its measurement by more than 3 pounds. The scale indicates that a person weighs 135 pounds. (Let w = the person's weight.)

51. A pipe 3 inches in diameter must have a tolerance of one-sixteenth of an inch so that the pipe can fit into a specific hole in the wall. (Let d = the diameter of the pipe.)

52. A freezer is most efficient when the temperature is 5°F plus or minus 6°. (Let t = temperature of the freezer.)

53. On a psychological test the actual score *s* differs from the expected score of 50 by no more than 6 points. What is the range of actual scores?

54. For submarine duty a sailor's height *h* can differ from 6 feet by no more than 2 inches. What range of heights is required?

55. A medical clinic considers a patient's body temperature to be normal if it deviates from 98.6°F by no more than 0.5°. What is the normal temperature range?

56. A housing inspector approves residential wiring if the incoming voltage deviates from 220 volts by less than 5 volts. What range of voltages would pass inspection?

Exploring with Real Data

In Exercises 57–60, refer to the following background information and data.

The accompanying figure compares the working capital and long-term debt for the Walt Disney Company during the period 1987–1991. (Source: Walt Disney Company.) With *y* representing millions of dollars and *x* representing a year, where $87 \leq x \leq 91$, the following two functions can be used to model the data in the figure.

Working capital: $y_1 = 149x - 12{,}139$
Long-term debt: $y_2 = 423x - 36{,}537$

Figure for 57–60

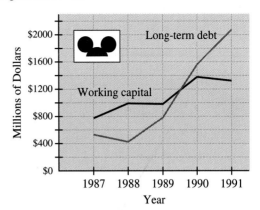

57. Equate the two functions and solve for *x* to determine the year when working capital and long-term debt were equal.

58. Use y_1 and y_2 to write an inequality that states that long-term debt exceeds working capital. Estimate the solution of this inequality by inspecting the graphs in the figure.

59. Use your calculator to produce the graphs of y_1 and y_2. Trace the graphs to estimate the dollar amount when working capital and long-term debt were equal.

60. Why do you think a company might be attractive to investors even if its long-term debt exceeds its working capital?

4   Chapter Review Exercises

Section 4.1

In Exercises 1 and 2, state whether each equation is a linear equation.

1. $3x^2 - 4x = 7$

2. $\frac{2}{3}x - 5 = 9$

In Exercises 3 and 4, verify that the given number is a solution of the equation.

3. $3x - 7 = 2(x - 5); -3$

4. $\frac{2}{7}(x - 5) + \frac{1}{3}(x + 7) = x + 2; -\frac{23}{8}$

5. Explain how to use your graphing calculator to solve $2x - 5 = x + 3$ and to check the solution.

6. If the graphs of the left and right sides of an equation are parallel, what can you conclude?

In Exercises 7–10, solve the equations using the graphing method.

7. $4 - x = -2$

8. $4(2x + 1) = 8x - 1$

9. $2(x + 3) = 2(x + 1) + 4$

10. $\frac{3}{2}x = \frac{5}{2} + x$

Section 4.2

In Exercises 11–18, solve each equation.

11. $\frac{-y}{3} = 10$

12. $4(5x - 6) - 6(3x - 2) = 4$

13. $0.2n + 0.3 = 0.3n - 5.2$

14. $\frac{3}{2}p + \frac{1}{6} = \frac{5}{3}$

15. $7x - 5 = 12x - 5 - 4x$

16. $\frac{x - 2}{6} = \frac{x}{3}$

17. $4(2s + 1) = s + 3(2s - 1)$

18. $\frac{2}{3}(k + 2) + \frac{1}{4}(k - 4) = k - \frac{1}{6}$

In Exercises 19–22, state whether the given equation is an identity, an inconsistent equation, or a conditional equation. State the solution set.

19. $3x - 2(1 - 4x) = 6x + 5$

20. $3c - 4(1 + c) = 5 - c$

21. $2 - (2y - 3) = 5 - 2y$

22. $0.4q - 1.4 + 0.6q = 1 - 0.8q + 1.4$

In Exercises 23 and 24, determine k so that the equation meets the given condition.

23. $3 - 2x = kx - 2 + 3x$;
the equation is inconsistent.

24. $-2(k - x) = 2x + 6$;
the equation is an identity.

Section 4.3

In Exercises 25–30, solve each formula for the given variable.

25. $V = LWH$ for H

26. $R_T = \frac{R_1 + R_2}{2}$ for R_1

$2R+ = R_1 + R_2$
$2R+ - R_2 = R_1$

27. $\frac{ac}{x} + \frac{cd}{y} = 6$ for c

28. $A = \frac{1}{2}h(a + b)$ for b

29. $E = IR$ for I

30. $h = vt - 16t^2$ for v

In Exercises 31 and 32, solve each equation for y.

31. $3x - 4y - 25 = 0$

32. $\frac{4}{5}x = \frac{2}{3}y - \frac{2}{5}$

33. Solve the formula $A = P + Prt$ for r. Then use the formula to determine the simple interest rate (r) required for an investment of \$2400 (P) to be worth \$2556 (A) at the end of one year (t).

34. To obtain a building permit for a small office building, the developer must have at least 4500 square feet of paved parking area. (See figure.) If the trapezoidal area in front of the building is paved for parking, will the building permit requirement be met? What is the actual paved area?

Figure for 34

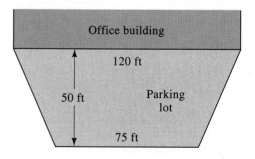

Figure for 35

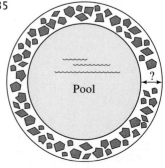

35. A circular goldfish pool is surrounded by a sidewalk. (See figure.) The circumference of the outside of the sidewalk is 43.98 feet, and the circumference of the inside of the sidewalk is 31.42 feet. What is the width of the sidewalk?

Section 4.4

36. One number is 3 less than another number. If one-fifth of the larger number is added to one-half of the smaller number, the result is 9. What is the smaller number?

37. A 100-foot rope is to be cut into three pieces whose lengths are consecutive integers. One foot of rope will be left over. How long should each piece of rope be?

38. A sailboat jib is not quite a right triangle with angle A being 10° less than one-half of angle B. Angle C is 15° more than angle B. What are the measures of the three angles? (See figure.)

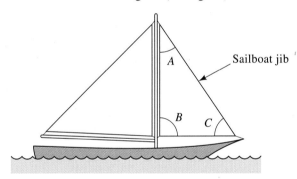

Sailboat jib

A

B *C*

39. A coin collection jar is full of just nickels and dimes. If there are 355 coins valued at $24.25, how many nickels are in the jar?

40. Determine two consecutive odd integers so that the difference between twice the first and 20 more than the second is 31.

Section 4.5

41. Menswear Unlimited advertised, "All suits are 40% off." If a suit is on sale for $93, what was the original price?

42. How much money was invested at 6% simple interest if the value of the investment was $1653.60 at the end of 1 year?

43. Stewart Construction Company rents a large crane for $275 per day and provides an operator for $20 per hour. The crane rental bill for July 29 was $455. How long did the operator work?

44. A 10% salt solution is to be mixed with a 20% salt solution to obtain 15 gallons of a 12% salt solution. How many gallons of each should be used?

45. Two trucks leave Kansas City at midnight and travel in opposite directions. (See figure.) The truck traveling east is averaging 6 mph less than the truck going west. At 3:30 A.M. the trucks are 469 miles apart. What is the average rate of each truck?

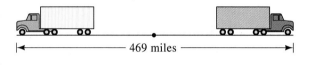

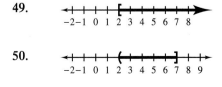

←————— 469 miles —————→

46. Two investments yield 7.5% and 9.3%, respectively. If a total of $8000 is invested and the income for the year is $695.40, how much is invested at each rate?

47. Two angles are complementary. The difference between one-third of the larger angle and the smaller angle is 10. What is the measure of each angle?

48. At a chemical plant, a mixing vat receives water from a large pipe and an alkaline solution from a smaller pipe. The larger pipe delivers liquid 1.5 times as fast as the small pipe. The small pipe is opened at noon and the larger pipe is opened at 1:45 P.M. At 5:30 P.M., the tank contains 133.5 gallons of liquid. At what rate does each pipe deliver liquid?

Section 4.6

In Exercises 49–51, for each graph write the set with interval notation.

49.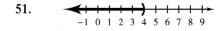

    −2 −1 0 1 2 3 4 5 6 7 8

50.
    −2 −1 0 1 2 3 4 5 6 7 8 9

51.
    −1 0 1 2 3 4 5 6 7 8 9

In Exercises 52–54, illustrate each set on the number line.

52. $(-\infty, 7]$ **53.** $[0, 3)$ **54.** $(-2, \infty)$

In Exercises 55 and 56, translate each verbal description into an inequality.

55. A number x is a positive real number.

56. A number n is a negative number greater than -7.

In Exercises 57–60, use the graphing method to solve each inequality and write the solution set with interval notation.

57. $2x - 1 > 7$ **58.** $-2(x - 2) < 1 - 2x$

59. $5x + 7 - 3x \geq 7 + 2x$

60. $-8 < \dfrac{1}{2}x - 2 \leq 7$

In Exercises 61–72, solve each inequality and write the solution set with interval notation. Draw a number line graph of the solution set.

61. $3x - 4 < 14$ **62.** $7 - 3t \leq -8$

63. $4x - 4 \leq 7x - 13$ **64.** $5x - 3 \geq 6x + 5$

65. $-17 < 4x - 7 \leq -8$

66. $6 + 3(2x - 5) \geq 2x + 3(x - 2) - 7$

67. $-\dfrac{2}{3}(x - 4) \leq \dfrac{2x - 3}{-4}$

68. $-3 < \dfrac{2x - 1}{-1} < 5$

69. $7 > x + 3 > -4$

70. $\dfrac{1}{6}t + \dfrac{2}{3} < \dfrac{1}{3}t - \dfrac{1}{2}$

71. $-8 \leq -4x < 20$ **72.** $\dfrac{t}{-3} \geq -9$

Section 4.7

In Exercises 73–84, solve each compound inequality and write the solution set with interval notation.

73. $-t \geq -2$ or $\dfrac{t}{4} > -1$

74. $-5 \leq t - 7 < 5$

75. $\dfrac{x + 2}{4} \leq 1$ and $\dfrac{x + 2}{4} \geq -1$

76. $2x - 3 \geq 5$ and $x > 0$

77. $x + 1 > 0$ or $3x - 4 < 0$

78. $t + 3 \leq 5$ and $4t < -4$

79. $-5x + 2 \geq 12$ or $3x + 4 \geq 25$

80. $x - 6 > 4$ or $x + 5 \leq 20$

81. $3x + 4 < x$ or $x + 2 > 8 - x$

82. $4x - 3 \leq 3x$ and $5 - 2x \leq 1$

83. $3x - 4 > 5x - 2$ and $3x - 2 < 2x + 3$

84. $x + \dfrac{2}{3} < -\dfrac{1}{3}$ and $\dfrac{3}{2}x + 1 > \dfrac{1}{2}$

Section 4.8

In Exercises 85–90, solve each equation.

85. $|-y| = 8$ **86.** $|3 + 2x| = 18$

87. $|2x - 3| - 7 = 5$ **88.** $|3x + 5| = -2$

89. $2 + |x - 2| = 3 - |x - 2|$

90. $|-x| + 5 = 1$

In Exercises 91 and 92, solve each equation by graphing.

91. $|x - 3| = \dfrac{1}{2}x - 1$

92. $|x + 2| = 4 - |2x - 3|$

In Exercises 93–98, solve each absolute value inequality. Write the solution set with interval notation.

93. $|3x + 5| < 14$ **94.** $|4 - x| > 5$

95. $7 - |c - 5| \leq 4$ **96.** $8 - |5 - d| \leq 12$

97. $|-t| \geq 7$ **98.** $|t| \leq -7$

99. Write a double inequality that is equivalent to $|x - 3| \leq 5$.

100. Write an absolute value inequality whose solution set is $(-\infty, -4) \cup (8, \infty)$.

In Exercises 101–104, solve the inequality. Write the solution set with interval notation.

101. $|x + 3| > 2x + 1$ **102.** $|x + 2| \le |3 - x|$

103. $|3 + x| < x + 4$ **104.** $|x + 2| < 3x$

Section 4.9

105. The width of a rectangular lot is 7 meters less than the length. The perimeter of the lot can be no more than 254 meters. What is the maximum width?

106. How much money must be invested at 8% simple interest to realize a yearly income between $150 and $200?

107. Two brothers were born exactly 3 years apart. If one boy is 9 years old, how old is his brother? Write an absolute value equation and solve it.

108. A shipping company requires that the total of the length, width, and height of a box not exceed 153 inches. If the height is 75% of the length and the length is twice the width, what are the possible values for the dimensions?

109. Frugal Car Rental charges $20 per day plus 10 cents per mile. Big C Car Rental charges $26 per day plus 9 cents per mile. For what range of miles is Frugal Car Rental the best choice for a 1-day rental?

110. Equal amounts of money are invested in two funds that pay 7% and 8.5% simple interest. The first investment must earn at least $119 per year and interest from the other investment must not exceed $238 per year. How much should be invested in each fund?

111. A wrestler can remain in his weight class as long as his present weight does not vary by more than 5 pounds. If his current weight is 136 pounds, what is the range of weights in his weight class?

112. If 3 is subtracted from twice a positive integer, the result is negative. Show that there is only one solution.

113. If a number is reduced by 1, the absolute value of the result is less than the original number. Show that the number cannot be negative.

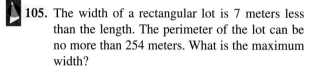

4   Chapter Test

1. State the property that justifies each step.

$$2x - 3 = 7$$
$$2x - 3 + 3 = 7 + 3 \quad \text{(a)} \underline{\quad\quad}$$
$$2x = 10 \quad \text{(b)} \underline{\quad\quad}$$
$$\frac{2x}{2} = \frac{10}{2} \quad \text{(c)} \underline{\quad\quad}$$
$$x = 5 \quad \text{(d)} \underline{\quad\quad}$$

In Questions 2–4, state whether the given equation is a conditional equation, an identity, or an inconsistent equation. Write the solution set.

2. $10 - 3x = 6x - 17$ **3.** $3x - 4 = 5x - 4 - 2x$

4. $3 - 2(x - 7) = 3(x + 1) - 5x$

5. Solve the formula $A = P + Prt$ for t.

6. Solve the equation $2x + 3y = 12$ for y.

7. What was the original investment at a simple interest rate of 8% if the value of the investment at the end of a year is $2484?

8. A person has 200 meters of fencing to use to enclose a rectangular garden. If the width is three-fifths of the length, what are the dimensions of the garden?

9. A radiator with a 6-gallon capacity is filled with a 50% antifreeze solution. How much must be drained so that adding pure antifreeze gives a 60% antifreeze solution?

10. Two cars leave Wall, South Dakota, at the same time. The first car travels west while the other travels east at 12 mph less than the first car. After 1.5 hours they are 144 miles apart. What is the speed of each car?

11. Use interval notation to describe the set whose graph is given.

In Questions 12–15, solve the given inequality and draw a number line graph of the solution set.

12. $3x + 1 > 13$

13. $4 - 2(x - 3) < 1 - 5x$

14. $\dfrac{2x + 13}{-2} \geq 1$

15. $x + \dfrac{2}{3}(x - 3) < \dfrac{1}{2}$

In Questions 16–19, solve the compound inequality and draw a number line graph of the solution set.

16. $-3 \leq 2x + 1 \leq 5$

17. $3 - x < 5$ and $2x - 1 < 5$

18. $3 - 2x \leq 9$ or $x + 1 \geq 0$

19. $2x \geq 0$ or $x - 3 \leq 1$

20. Solve $|x + 2| - 5 = -3$.

In Questions 21–23, solve the given absolute value inequality and draw a number line graph of the solution set.

21. $3 - |x - 1| < 1$

22. $|2x - 1| \leq 5$

23. $|x - 3| < -2$

24. The charge for cellular phone service is $25 per month plus 50 cents per minute that the phone is used. If the phone bill is between $31 and $46 per month, what is the most and least number of minutes the phone was used?

25. The average snowfall for the first week of January is 8.7 inches. If the snowfall totals for the first 6 days are 1.2, 0, 4.25, 1.5, 0, and 0.4 inches, how much snow could fall on January 7 without the total exceeding the average?

3-4   Cumulative Test

1. If $a < 0$ and $b > 0$, name the quadrant or axis that contains the given point.

(a) $P(a, b)$

(b) $Q(b, 0)$

(c) $R(0, -a)$

(d) $S(b, a)$

2. From the following list, identify the relations that are functions.

(i) $\{(0, 1), (1, 2), (2, 3), \ldots\}$

(ii)

(iii) $\{(x, y) \mid x = 3\}$

(iv) $\{(x, y) \mid y = 3\}$

3. For $P(-6, 3)$ and $Q(8, -7)$, determine the following.

(a) PQ (b) The midpoint of $\overline{PQ}$

 4. Explain how to use the Vertical Line Test to determine whether the graph of a relation represents a function.

5. Use the integer setting to produce the graph of $y = 0.5x^2 + 8x - 7$. What are the x-coordinates of the two points whose y-coordinates are both 11?

6. Evaluate each function as indicated. Use your calculator in part (b) but not in part (a).

(a) $g(x) = 4(3x - 1) - (2 - x)$; $g(-1)$

(b) $f(x) = \sqrt{2x + 7} - x^2$; $f(3)$

7. If $f(x) = \dfrac{1}{x - 1}$, explain the result of evaluating $f(1)$ with your calculator.

8. Produce the graph of $y = x^3 - 5x$ on your calculator. Estimate each of the following to the nearest tenth.

(a) All intercepts   (b) Local maximum and minimum

 9. To solve $5x - 3 = 2x + 9$, we graph $y_1 = 5x - 3$ and $y_2 = 2x + 9$ and note that the point of intersection of the two graphs is $(4, 17)$. Explain the significance of the 4 and 17.

10. Solve each equation algebraically.

    (a) $2(3x - 1) - 2x = -(x - 4)$ (b) $\frac{1}{3}(x + 1) - \frac{1}{2} = \frac{3}{5}(2x - 3)$

11. Solve the formula $A = \frac{1}{2}\,bh$ for b.

12. After three tests of equal weight, a student has an 81 average. If the first and last test scores were 80 and 91, what was the second test score?

13. The second angle of a triangle is 2° larger than the first angle. If the third angle is 2° more than one-fourth the second angle, what are the measures of the angles?

14. A car and a bicycle are 39 miles apart. The car driver and the cyclist leave at 1:00 P.M. and head directly toward each other. If the driver averages 40 mph and the cyclist averages 12 mph, at what time will they meet?

15. For three consecutive integers, the sum of the first two integers is 10 more than the largest integer. What are the integers?

16. A person invests $8000 in two funds, one at 5% simple interest and the other at 6.5% simple interest. If the total income for the year is $475, how much was invested at each rate?

17. Solve each inequality algebraically and graph the solution set on a number line.

    (a) $x - 2(x - 2) < 5x + 8$ (b) $\dfrac{3 - x}{-4} \geq x$

18. Solve each compound inequality. Write the solution set with interval notation and graph the solution set on a number line.

    (a) $-5x + 1 > 6$ or $2x - 1 \geq 5$ (b) $4 - x < 1$ and $\frac{1}{3}x < 2$

19. Solve each of the following.

    (a) $|x - 3| < 8$ (b) $2|x + 1| - 3 = 5$

    (c) $-2|3 + x| \leq -4$

20. The length of a rectangle is 3 more than twice the width. If the perimeter of the rectangle does not exceed 42 inches, what is the maximum length of the rectangle?

5

Properties of Lines

In 1969 only 4% of all state legislators were women. By 1993 the figure had risen to 20%. In the figure a **linear equation in two variables** has been used to model the data from 1969 to 1993.

One property of such a graph is its **slope,** which in this case gives us a measure of the rate at which the percentage of female state legislators increased during the given period. Assuming the model remains valid in future years, by what year would half of all state legislators be women? (More about this real-data application can be found in Exercises 111–114 at the end of Section 5.2.)

In Chapter 5 we consider linear equations in two variables and the properties of their graphs. In particular, we study a line's slope and its many applications. We conclude with a discussion of linear inequalities and their graphs.

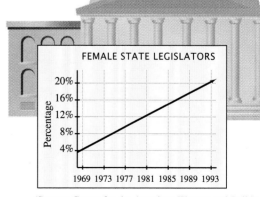

FEMALE STATE LEGISLATORS

(Source: Center for the American Woman and Politics.)

5.1 Linear Equations in Two Variables

The Standard Form ▪ *Solutions and Graphs* ▪ *Intercepts* ▪
Special Cases

The Standard Form

Linear equations in one variable typically arise when a problem involves a single unknown quantity or when all unknown quantities can be easily represented with just one variable.

When problems involve two unknown quantities, it is often more convenient to use two different variables and to write equations in those variables. One such type of an equation is a **linear equation in two variables.**

Definition of Linear Equation in Two Variables

A **linear equation in two variables** is an equation that can be written in the *standard form* $Ax + By = C$, where A, B, and C are real numbers and A and B are not both zero.

Each of the following equations is a linear equation in two variables.

Equation	*Standard Form*
$3x + 2y = 12$	$3x + 2y = 12$
$y = 2x + 3$	$-2x + y = 3$
$y = 3$	$0x + y = 3$
$x = -2$	$x + 0y = -2$

An equation may have two variables but not be a linear equation. The following are some examples of equations that *cannot* be written in the standard form $Ax + By = C$.

$x^2 + 3y = 5$ The exponents on both variables must be 1.

$\dfrac{2}{x} + y = -2$ No variable can appear in the denominator of a fraction.

$xy = 1$ The variables cannot be part of the same term.

Solutions and Graphs

To solve a linear equation in two variables, we must find a replacement for each variable so that the resulting equation is true. We say that such replacements *satisfy* the equation.

Definition of a Solution of a Linear Equation in Two Variables

A **solution** of a linear equation in two variables is a pair of numbers (x, y) that satisfies the equation.

EXPLORATION 1   *Solutions and Graphs*

Consider the equation $2x + y = 8$.

(a) Complete the following table of solutions. To determine the y-values, solve the equation for y and use your calculator to evaluate y.

x		3	-1	-4	0	9
y	0					

(b) What is your conjecture about the number of solutions of the equation?

(c) From the table in part (a), write the solutions as ordered pairs (x, y).

(d) Plot the points that represent the ordered pairs in part (c). What is your conjecture about the complete graph of the equation?

(e) Use the integer setting to produce the graph of the function on your calculator. Then trace the graph to verify the points that you plotted in part (d).

Discovery

(a) For the first entry, replace y with 0 and solve for x.

$$2x + y = 8$$
$$2x + 0 = 8$$
$$2x = 8$$
$$x = 4$$

Solving for y, we obtain $y = -2x + 8$. Now we can use a calculator to determine the y-value for each given x-value.

x	4	3	-1	-4	0	9
y	0	2	10	16	8	-10

(b) Because either variable may be replaced with any real number, the equation has infinitely many solutions.

(c) The solutions can be written as ordered pairs.

$(4, 0)$ $(3, 2)$ $(-1, 10)$ $(-4, 16)$ $(0, 8)$ $(9, -10)$

(d) Figure 5.1 shows the points associated with the ordered pairs in part (c). Because the equation has infinitely many solutions, the complete graph consists of infinitely many points that form a straight line.

(e) Figure 5.2 is a calculator display of the complete graph of $y = -2x + 8$. The tracing cursor is on the point $(9, -10)$. By tracing, we can see that the other points in part (c) are also points of the graph.

Figure 5.1

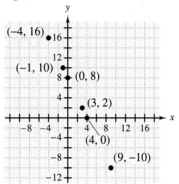

Figure 5.2

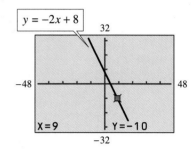

Exploration 1 suggests the following generalizations about a linear equation in two variables.

1. The graph is a straight line.

2. There is a **one-to-one correspondence** between the solutions of the equation and the points of the graph. In other words,

   (a) Every solution of the equation is represented by a point of the graph.

   (b) Every point of the graph represents a solution of the equation.

Intercepts

The graph of a linear equation in two variables may intersect the y-axis and/or the x-axis. In Chapter 3, we defined these points as **intercepts.** Because of the importance of these points, we repeat the definitions.

> ### Definitions of Intercepts
>
> A point where a graph intersects the y-axis is called a **y-intercept.** A point where a graph intersects the x-axis is called an **x-intercept.**

Although we define y-intercept and x-intercept as *points,* the words are often used to refer to the coordinates of the points.

EXAMPLE 1 *Estimating Intercepts with a Calculator*

Use the integer setting to produce the graph of $y = 2x - 10$. Then, trace the graph to estimate the intercepts.

Solution

The x-intercept is $(5, 0)$; the y-intercept is $(0, -10)$. [See Fig. 5.3(a) and (b).]

Figure 5.3

(a)

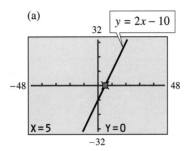

(b)

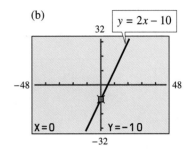

While we can use graphing methods to provide reasonable estimates of the x- and y-intercepts, we can determine the exact coordinates algebraically.

y-intercept: The y-intercept is the point whose x-coordinate is 0. Replace x with 0 and calculate the corresponding value of y.

x-intercept: The x-intercept is the point whose y-coordinate is 0. Replace y with 0 and calculate the corresponding value of x.

EXAMPLE 2 *Determining Intercepts Algebraically*

Use algebraic methods to determine the intercepts for $y = 2x - 3$.

Solution

To determine the x-intercept, replace y with 0 and solve for x.

$$y = 2x - 3$$
$$0 = 2x - 3$$
$$3 = 2x$$
$$x = \frac{3}{2}$$

The x-intercept is $(\frac{3}{2}, 0)$.

To determine the y-intercept, replace x with 0 and solve for y.

$$y = 2x - 3$$
$$y = 2(0) - 3$$
$$y = -3$$

The y-intercept is $(0, -3)$.

EXPLORATION 2   *Determining the y-Intercept from the Equation*

(a) Use the integer setting to graph each of the following on the same coordinate axes.

$$y_1 = x + 12 \qquad y_2 = x + 5 \qquad y_3 = x \qquad y_4 = x - 8$$

(b) Compare the y-intercept of each graph with the corresponding equation. What conclusion is suggested?

Discovery

(a)

Figure 5.4

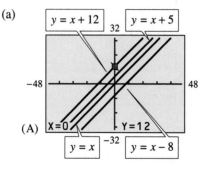

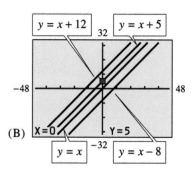

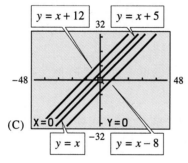

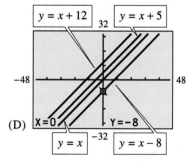

(b) **Equation   y-Intercept**

$y = x + 12$	$(0, 12)$	See Fig. 5.4(A).
$y = x + 5$	$(0, 5)$	See Fig. 5.4(B).
$y = x + 0$	$(0, 0)$	See Fig. 5.4(C).
$y = x - 8$	$(0, -8)$	See Fig. 5.4(D).

In each case, the y-coordinate of the y-intercept is the same as the constant term of the equation.

The conclusion in Exploration 2 can be supported as follows. When we solve a linear equation in two variables for y, the resulting equation has the form $y = mx + b$, where m and b are real numbers. We determine the y-intercept by replacing x with 0.

$$y = mx + b$$
$$y = m(0) + b$$
$$y = 0 + b$$
$$y = b$$

The y-intercept is $(0, b)$.

In summary, when a linear equation in two variables is written in the form $y = mx + b$, the y-intercept $(0, b)$ can be determined from the constant term of the equation. (The use of the letter m will be discussed in the next section.)

EXAMPLE 3   *Determining the y-Intercept from the Equation*

Write each of the following equations in the form $y = mx + b$. Then determine the y-intercept in each case.

(a) $y = x - 4$ (b) $5x + 7y = 9$

Solution

(a) The equation is already solved for y. The constant term is -4. Therefore, the y-intercept is $(0, -4)$.

(b) Solve the equation for y.

$$5x + 7y = 9$$
$$5x + 7y - 5x = -5x + 9 \qquad \text{Subtract } 5x \text{ from both sides.}$$
$$7y = -5x + 9 \qquad \text{Combine like terms.}$$
$$\frac{7y}{7} = \frac{-5x}{7} + \frac{9}{7} \qquad \text{Divide each term on both sides by 7.}$$
$$y = -\frac{5}{7}x + \frac{9}{7}$$

The constant term is $\frac{9}{7}$. Therefore, the y-intercept is $(0, \frac{9}{7})$. ■

Determining the y-intercept is easy when the equation is written in the form $y = mx + b$. However, the y-intercept can be determined from any equation form by replacing x with 0 and computing the corresponding y-value.

Consider again the equation in part (b) of Example 3.

$$5x + 7y = 9$$
$$5(0) + 7y = 9 \qquad \text{Replace } x \text{ with 0.}$$
$$7y = 9$$
$$y = \frac{9}{7}$$

The y-intercept is $(0, \frac{9}{7})$.

Special Cases

At the beginning of this section, we stated that $y = 3$ and $x = -2$ are both examples of linear equations in two variables.

Equation	*Standard Form*
$y = 3$	$0 \cdot x + 1 \cdot y = 3$
$x = -2$	$1 \cdot x + 0 \cdot y = -2$

In both equations, one of the variables is missing. For this reason, we consider the equations to be special cases.

 **EXPLORATION 3**

Special Case: y = Constant

(a) Use the integer setting to produce the graphs of the following equations.

$$y = 4 \qquad y = 10 \qquad y = -17$$

Describe the similarity in the graphs.

(b) Move the tracing cursor back and forth along the graphs. What do you observe about the y-coordinates?

(c) What are the intercepts?

Discovery

Figure 5.5

(a) The lines appear to be horizontal. (See Fig. 5.5.)

(b) The points of each line all have the same y-coordinate.

(c) There is no x-intercept. The y-intercepts are, in order, (0, 4), (0, 10), and (0, −17).

The following characteristics apply to equations of the form $y = $ constant.

1. All solutions are ordered pairs in which the second number is the constant.

2. The graph is a horizontal line. The constant indicates how far above or below the origin to draw the line.

3. With the exception of $y = 0$ (for which every point of the graph is an x-intercept), equations of this form have no x-intercept. The y-intercept is (0, b), where b is the constant.

Another special case of a linear equation in two variables is an equation of the form $x = $ constant.

 **EXPLORATION 4**

Special Case: x = Constant

Consider the equation $x = -2$.

(a) Complete the following table of solutions.

x						
y	−3	−1	0	2	5	9

(b) Plot the points that represent the ordered pairs in part (a). What is your conjecture about the complete graph? Now draw the graph.

(c) Does the graph represent a function? Can you use your calculator to produce the graph?

Discovery

(a) The equation $x = -2$ indicates that every ordered pair must have an x-coordinate of −2.

x	−2	−2	−2	−2	−2	−2
y	−3	−1	0	2	5	9

(b) The complete graph is a vertical line. (See Fig. 5.6.)

Figure 5.6

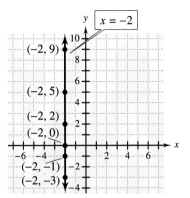

(c) Because a vertical line does not pass the Vertical Line Test, the graph does not represent a function. Because the equation $x = -2$ cannot be solved for y, it cannot be entered on the function screen of your calculator, and the graph cannot be produced in this way.

The following characteristics apply to equations of the form $x = $ constant.

1. All solutions are ordered pairs in which the first number is the constant.

2. The graph is a vertical line. The constant indicates how far to the right or left of the origin to draw the line.

3. With the exception of $x = 0$ (for which every point of the graph is a y-intercept), equations of this form have no y-intercept. The x-intercept is $(a, 0)$, where a is the constant.

It will be very useful to keep these two special cases in mind. As is true for all linear equations in two variables, the graph will be a straight line. Knowing that the line is horizontal when $y = $ constant and vertical when $x = $ constant simplifies sketching the graph.

Here is a summary of what is known so far about the two special cases.

Summary of Special Cases of Linear Equations

Equation	Graph	Intercepts	Function?
$x = a$	vertical line	x-intercept: $(a, 0)$ y-intercept: if $a \neq 0$, none.	no
$y = b$	horizontal line	x-intercept: if $b \neq 0$, none. y-intercept: $(0, b)$	yes

5.1   *Quick Reference*

The Standard Form
- A **linear equation in two variables** is an equation that can be written in the *standard form* $Ax + By = C$, where A, B, and C are real numbers and A and B are not both zero.

Solutions and Graphs
- A **solution** of a linear equation in two variables is a pair of numbers (x, y) that satisfies the equation.

- Solutions can be determined by replacing one variable with some value and solving the equation for the other variable.

- There exists a one-to-one correspondence between the solutions of an equation and the points of its graph.

- The graph of a linear equation in two variables is a straight line.

- To produce the graph of a linear equation on a calculator, solve the equation for y and enter the resulting function.

Intercepts
- A point where a graph intersects the y-axis is called a **y-intercept.** A point where a graph intersects the x-axis is called an **x-intercept.**

- We can estimate the x- and y-intercepts with a calculator by locating the points where the graph intersects the axes.

- We determine the exact x-intercept algebraically by replacing y with 0 and solving for x. Similarly, we determine the exact y-intercept algebraically by replacing x with 0 and solving for y.

- If a linear equation is solved for y, the form of the equation is $y = mx + b$. From this, we can see that the y-intercept is $(0, b)$.

Special Cases
- The graph of the equation $y = b$ is a horizontal line. The constant b indicates how far the line is above or below the origin. Excluding the graph of $y = 0$, the graph has the y-intercept $(0, b)$ and no x-intercepts.

- The graph of the equation $x = a$ is a vertical line. The constant a indicates how far the line is to the right or left of the origin. Excluding the graph of $x = 0$, the graph has the x-intercept $(a, 0)$ and no y-intercepts.

5.1   *Exercises*

1. Why is $Ax + By = C$ called a linear equation?

2. Explain why $x = -5$ can be a linear equation in two variables even though there is no y-term.

In Exercises 3–8, state whether the equation is a linear equation in two variables.

3. $\dfrac{x}{2} + \dfrac{y}{3} - \dfrac{7}{8} = 0$ 4. $y = \dfrac{2x}{3} + 5$

5. $y = x^2 - 3$

6. $4y = \dfrac{3}{x} = \dfrac{4}{5}$

7. $-6 = 2xy$

8. $4y + 5 = -7$

In Exercises 9–14, determine whether the given ordered pair is a solution of the given equation.

9. $(4, 5)$; $y = 2x - 3$

10. $(-2, 2)$; $y = 3x + 4$

11. $\left(2, \dfrac{3}{2}\right)$; $3x - 4y = 12$

12. $(4, 5)$; $\dfrac{1}{4}x + \dfrac{2}{5}y = 3$

13. $(-2, -1)$; $x - 5 = 3$

14. $(-2, -3)$; $7 + y = 4$

In Exercises 15–20, determine k so that the given ordered pair is a solution of the equation.

15. $2x - ky = 7$; $(1, -1)$

16. $y + 3x = k$; $(2, -5)$

17. $y = -\dfrac{2}{3}x + k$; $(12, -3)$

18. $kx + y = 0$; $(-2, 6)$

19. $y = -2x + k$; $(k, k + 1)$

20. $kx - y = 9$; $(3, k - 5)$

In Exercises 21–26, find the values of a, b, and c so that the ordered pairs are solutions of the given equation.

21. $y = -2x + 5$; $(-3, a)$, $(4, b)$, $(c, -7)$

22. $y = 4x - 3$; $(a, 5)$, $(b\ -3)$, $(3, c)$

23. $x - 5 = -2$; $(a, 5)$, $(b, 9)$, $(c, -6)$

24. $5 - 2y = -1$; $(-1, a)$, $(4, b)$, $(-12, c)$

25. $x + 3y - 4 = 0$; $(a, 3)$, $(7, b)$, $(c, -4)$

26. $5y - 3x = -10$; $(a, -2)$, $(-5, b)$, $(10, c)$

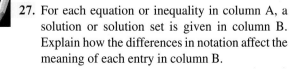 **27.** For each equation or inequality in column A, a solution or solution set is given in column B. Explain how the differences in notation affect the meaning of each entry in column B.

Column A	Column B
$\lvert x \rvert = 2$	$\{-2, 2\}$
$\lvert x \rvert < 2$	$(-2, 2)$
$y = x + 4$	$(-2, 2)$

 28. Explain the difference between the solution set of $x = 1$, a linear equation in one variable, and $x = 1$, a linear equation in two variables.

In Exercises 29–34, complete the table.

29. $y = 4 - x$

x	10	-6	
y			-3

30. $x = 2y + 3$

x	5	-13	
y			-6

31. $y - 2 = 1$

x	-3	0	4
y			

32. $x - 2 = -2$

x			
y	-4	0	2

33. $f(x) = 3x + 1$

x	0	3	-5
$f(x)$			
(x, y)			

34. $g(x) = -2x - 5$

x	-2	-7	3
$g(x)$			
(x, y)			

 35. How are the points of a graph of an equation related to the solutions of the equation?

36. What is the minimum number of points that need to be plotted to graph a line?

37. (a) Which of the following equations is graphed in the accompanying figure?

$$x - 2y - 5 = 0$$
$$2x - 3y - 6 = 0$$
$$5x - 6y - 18 = 0$$

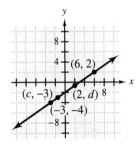

(b) What are the values of c and d in the figure?

38. (a) Which of the following equations is graphed in the accompanying figure?

$$3x + 4y + 9 = 0$$
$$5x + 8y + 11 = 0$$
$$7x + 8y + 13 = 0$$

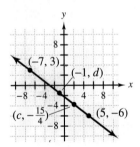

(b) What are the values of c and d in the figure?

In Exercises 39–54, determine three solutions of the given equation and sketch the graph. If possible, produce the graph of the equation on your calculator. Compare your sketch with the calculator's graph.

39. $y = 3x - 4$

40. $y = 4x + 3$

41. $y = \dfrac{2}{3}x - 2$

42. $y = -\dfrac{3}{4}x + 5$

43. $3x - 5y = 15$

44. $2x - 3y = 6$

45. $y - 2 = 3$

46. $y + 5 = 7$

47. $x - 9 = -11$

48. $x + 7 = 9$

49. $y = x$

50. $y = -x$

51. $2x - 3 = 11$

52. $3y - 7 = 8$

53. $3x = 21 - 7y$

54. $5x + 2y - 10 = 0$

 55. Explain how to determine a line's x-intercept and y-intercept graphically.

 56. Describe the algebraic method for determining intercepts.

In Exercises 57–62, use the integer setting to produce the graph of the given equation. Then, trace the graph to determine the intercepts.

57. $y = 12 - 3x$

58. $y = -12 + x$

59. $x = 3y + 30$

60. $x = 15 - y$

61. $3x - 4y = 36$

62. $2x + 3y = 30$

63. (a) Which of the following equations is graphed in the accompanying figure?

$$2x + 3y = 6 \qquad 2x - 3y = 6$$
$$2x + 3y = -6$$

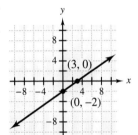

(b) Produce the graph of this equation on your calculator and compare it to the one in the figure.

64. (a) Which of the following equations is graphed in the accompanying figure?

$$3x + 2y = -6 \qquad 3x - 2y = 6$$
$$3x - 2y = -6$$

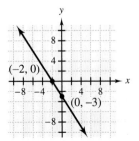

(b) Produce the graph of this equation on your calculator and compare it to the one in the figure.

In Exercises 65 and 66, select from the given equations the one whose graph has the given y-intercept. (There could be more than one answer.)

65. The y-intercept is $(0, 3)$.

$$5x + 3y - 9 = 0 \qquad 6x + 2y - 6 = 0$$
$$3x + 4y + 12 = 0$$

66. The y-intercept is $(0, -2)$.

$$5x + 3y - 6 = 0 \qquad 5x + 3y + 6 = 0$$
$$2x + 4y + 8 = 0$$

In Exercises 67–72, write each equation in the form $y = mx + b$. Then determine the y-intercept.

67. $2x + 5y = 14$

68. $4x - 3y = 15$

69. $y - 2 = -4(x + 3)$

70. $y + 4 = 17$

71. $\dfrac{1}{3}x + \dfrac{3}{4}y - 9 = 0$

72. $\dfrac{3}{4}x + \dfrac{3}{5}y - 8 = 0$

In Exercises 73–80, determine the intercepts algebraically.

73. $y = -2x$

74. $y = x$

75. $3x - 4y = 12$

76. $2x - 3y = 12$

77. $\dfrac{x}{3} + \dfrac{y}{4} = 1$

78. $\dfrac{x}{5} - \dfrac{y}{3} = 1$

79. $y + 2 = 17$

80. $2x + 9 = 7$

 81. Draw a line whose x-intercept and y-intercept are the same point. If the equation of the line is $y = mx + b$, what is the value of b?

 82. What is the point of intersection of all of the graphs of the following equations?

$$y = \dfrac{1}{3}x + 8 \qquad y = -\dfrac{1}{2}x + 8$$
$$y = 2x + 8 \qquad y = -x + 8$$

In Exercises 83 and 84, produce the graph of the given equation. Then determine the equations of the lines described in parts (a) and (b). To check your equations, produce their graphs on the same axes.

83. $y = x$

(a) A line shifted upward 4 units from the given line

(b) A line shifted downward 3 units from the given line

84. $y = -1$

(a) A line shifted upward 5 units from the given line

(b) A line shifted downward 4 units from the given line

In Exercises 85–88, determine a, b, and c so that the given equation has the specified intercepts.

85. $ax + 4y = c$; $(0, 2)$, $(-4, 0)$

86. $x - 2by = c$; $(0, 5)$, $(-1, 0)$

87. $ax + by = 15$; $(0, -5)$, $(3, 0)$

88. $\dfrac{x}{a} + \dfrac{y}{b} = 1$; $(0, 1)$, $(1, 0)$

 89. Describe the graphs of the equations $x = $ constant and $y = $ constant.

 90. Describe the form of an equation whose graph is a

(a) vertical line.

(b) horizontal line.

In Exercises 91–98, determine whether the graph of the given equation is horizontal, vertical, or neither.

91. $3x = 12$

92. $x - 5 = 6$

93. $x - y = 0$

94. $y = x$

95. $-2y = -8$

96. $y + 3 = -2$

97. $x = 2y$

98. $y = 0x$

In Exercises 99–104, write two ordered pairs whose coordinates satisfy the given conditions. Then, translate the conditions into an equation. Produce the graph of the equation and trace it to verify that your two ordered pairs are represented by points of the graph.

99. One number y is the opposite of another number x.

100. The numbers x and y are equal.

101. One number y is half of another number x.

102. One number y is 3 less than twice another number x.

103. The sum of two numbers x and y is 5.

104. The difference of two numbers x and y is 2.

 PIECE-WISE

In Exercises 105–108, graph using the specified domain.

105. $y = 2x + 1, x \geq 0$

106. $y = 1 - x, x \leq 0$

107. $y = x - 5, 10 \leq x \leq 15$

108. $y = 3x, -2 \leq x \leq 3$

In Exercises 109–112, translate the given information into an equation with an appropriate domain. Graph the equation and trace the graph to answer the question(s).

109. The cost of a pizza is $8.00 plus $1.50 per topping, with a maximum of five toppings. What are all of the possible costs of a pizza?

110. A taxi fare is $4.00 plus 15 cents per tenth of a mile. What are all of the possible fares for a trip of between 2 and 2.5 miles, inclusive?

111. A used car salesperson makes $135 per week plus $75 per car sold. How many cars will the salesperson have sold if the week's income is between $360 and $510?

112. When the daily high temperature is 75°F, the number of people using the local jogging path is 120 per day. For each 1° temperature increase over 75°F (up to 100°F), the number of joggers decreases by 4. For what temperature range will the number of joggers be between 88 and 100, inclusive?

Exploring with Real Data

In Exercises 113–116, refer to the following background information and data.

The bar graph shows the number of 1-pound loaves of bread consumed each year per person from 1988 to 1992. (Source: Annual Survey of Manufacturers.)

The data can be modeled by the linear equation

$$L = 0.625Y - 1193.8, \quad 1988 \leq Y \leq 1992,$$

where L is the number of loaves of bread and Y is the year.

Figure for 113–116

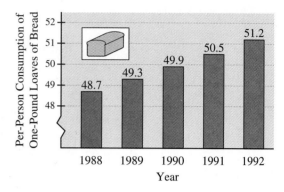

113. Assuming the model is valid for the future, predict the number of loaves of bread that will be consumed per person in 1999.

114. According to the model, what is the average annual increase in per-person consumption of bread during the indicated period?

115. What is the significance of the x-intercept of the graph of the linear equation? What year is represented by this point?

116. If the model were to remain accurate indefinitely, in what year would the per-person consumption of bread be 1 pound per day?

Challenge

117. Interpret the roles of g and h in the equation

$$\frac{x}{g} + \frac{y}{h} = 1.$$

(Hint: Consider where the graph crosses the axes.)

In Exercises 118–120, determine and plot the intercepts of the graph of the given equation. Be sure to take into account the conditions on a and b. Then sketch the graph.

118. $ax + by = ab; a > 0, b < 0$

119. $2x + by = 4b; b < 0$

120. $ax + 2y = -2a; a > 0$

5.2 Slope of a Line

The Slope Formula ▪ *The Slope–Intercept Form* ▪ *Graphing with Slope*

The Slope Formula

In the previous section, we discovered the significance of the constant term b when a linear equation is written in the form $y = mx + b$. The y-intercept of the graph of the equation is $(0, b)$.

In the following exploration, we investigate the coefficient m to learn how to interpret the number graphically.

EXPLORATION 1 · *Effect of the Coefficient of x*

(a) Display the graphs of the following equations on the same coordinate axes.

$$y_1 = \frac{1}{2}x + 3 \qquad y_2 = 1x + 3$$

$$y_3 = 2x + 3 \qquad y_4 = 3x + 3$$

(b) Display the graphs of the following equations on the same coordinate axes.

$$y_1 = -\frac{1}{2}x + 3 \qquad y_2 = -1x + 3$$

$$y_3 = -2x + 3 \qquad y_4 = -3x + 3$$

(c) What is the y-intercept of each graph?

(d) How do the coefficients of x affect the graphs?

Discovery

(a) Figure 5.7   (b) Figure 5.8

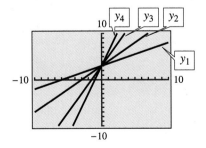 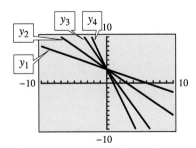

(c) The y-intercept is $(0, 3)$ for each graph.

(d) In Fig. 5.7, the coefficients of x are positive and the lines rise from left to right. As the coefficients increase, the lines are steeper.

In Fig. 5.8, the coefficients of x are negative and the lines fall from left to right. As the coefficients increase in absolute value, the lines are steeper.

In Exploration 1 each equation is in the form $y = mx + b$, and we see that the coefficient of x seems to influence the steepness of the line.

Steepness is a relative word, and it is too vague to describe the orientation of a line specifically. What we need is a numerical measure of steepness. That measurement is called the **slope** of the line.

Consider the two lines in Figs. 5.9 and 5.10.

Figure 5.9

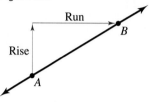

Figure 5.10

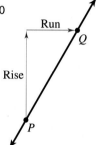

One way to travel from point A to point B in Fig. 5.9 is to start at A, travel vertically (**rise**) upward to the point directly across from B, and then travel horizontally (**run**) to the right to reach B. We can travel from point P to point Q in Fig. 5.10 in a similar way.

We can see the difference in the steepness of the two lines by comparing the rise to the run in each case. For the line that is rising more gradually (Fig. 5.9), the rise is relatively small compared to the run. For the steeper line (Fig. 5.10), the rise is relatively large compared to the run.

This suggests a way of describing the steepness, or slope, of a line. We compare the rise to the run with a ratio.

In Fig. 5.11, $P(x_1, y_1)$ and $Q(x_2, y_2)$ are two general points of a line. Point $R(x_1, y_2)$ is on a vertical line with P and a horizontal line with Q.

Figure 5.11

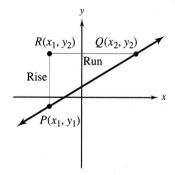

The rise PR is the difference between the y-coordinates.

$$\text{Rise} = y_2 - y_1$$

The run QR is the difference between the x-coordinates.

$$\text{Run} = x_2 - x_1$$

With these formulas for rise and run, we can now define the slope of a line.

Definition of Slope

The **slope** m of a line containing $P(x_1, y_1)$ and $Q(x_2, y_2)$ is defined as follows.

$$m = \frac{\text{Rise}}{\text{Run}} = \frac{y_2 - y_1}{x_2 - x_1}, \qquad x_2 \neq x_1$$

Note that the conventional symbol for slope is the letter *m*.

There are several important observations to be made from this definition. First, we can use the formula to calculate the slope of any line except a vertical line. The slope of a vertical line is undefined.

Next, the order in which we subtract the coordinates does not matter as long as we are consistent. We can choose to compute $y_2 - y_1$ for the rise, but then we must compute $x_2 - x_1$ for the run. Or we can choose to compute $y_1 - y_2$ for the rise, but then we must compute $x_1 - x_2$ for the run.

Finally, the definition refers to any two points of the line. No matter which two points we use, the calculated slope will be the same.

EXAMPLE 1 *Determining Slope Given Two Points of a Line*

Sketch the line determined by the given points and determine the slope of the line.

(a) $A(2, -1)$ and $B(4, 5)$ (b) $C(-2, 5)$ and $D(2, -3)$

Solution

Figure 5.12   Figure 5.13

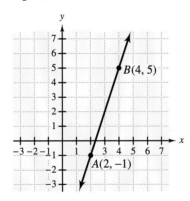 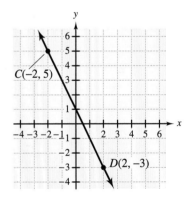

(a) In Fig. 5.12 the line rises from left to right, which indicates that the slope is positive.

$$m = \frac{y_2 - y_1}{x_2 - x_1} = \frac{5 - (-1)}{4 - 2} = \frac{6}{2} = 3$$

(b) In Fig. 5.13 the line falls from left to right, which indicates that the slope is negative.

$$m = \frac{-3 - 5}{2 - (-2)} = \frac{-8}{4} = -2$$

EXAMPLE 2   *Special Cases*

Sketch the line determined by the given points and determine the slope of the line.

(a) $A(2, 5)$ and $B(-3, 5)$

(b) $E(1, 3)$ and $F(1, 7)$

Solution

Figure 5.14

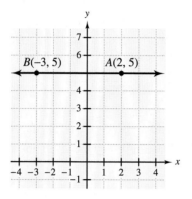

Figure 5.15

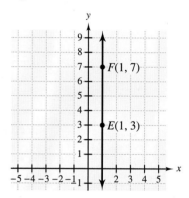

(a) In Fig. 5.14 the line is horizontal because points A and B have the same y-coordinate.

$$m = \frac{y_2 - y_1}{x_2 - x_1} = \frac{5 - 5}{-3 - 2} = \frac{0}{-3 - 2} = \frac{0}{-5} = 0$$

(b) In Fig. 5.15, the line is vertical because points E and F have the same x-coordinate. The slope formula does not apply because the ratio is

$$\frac{y_2 - y_1}{x_2 - x_1} = \frac{7 - 3}{1 - 1} = \frac{4}{0},$$

which is undefined. For this reason, the slope of any vertical line is undefined.

NOTE: For these special cases, it is important to know the difference between *zero* and *undefined*. A horizontal line has a defined slope, and it is 0. A vertical line does not have a defined slope. *Zero* **does not mean the same thing as** *undefined.*

One of the special cases, that of a vertical line, is handled in the definition of slope. The definition states that the slope of a vertical line is undefined.

In Example 2 we found that the slope of a horizontal line is 0. Because each point of a horizontal line has the same y-coordinate, $y_2 - y_1 = 0$ for any two points. Therefore,

$$m = \frac{y_2 - y_1}{x_2 - x_1} = \frac{0}{x_2 - x_1} = 0.$$

We can now add this new knowledge about slopes to our summary for special cases.

> **Summary of Special Cases**
>
> | Equation | Graph | Intercepts | Slope |
> |----------|-------|------------|-------|
> | $x = a$ | vertical | x-intercept: $(a, 0)$
y-intercept: If $a \neq 0$, none. | undefined |
> | $y = b$ | horizontal | y-intercept: $(0, b)$
x-intercept: If $b \neq 0$, none. | 0 |

The Slope–Intercept Form

When a linear equation is written in the form $y = mx + b$, we have seen that the coefficient of the x-term affects the slope of the line. In the next exploration, we see why we use m to represent the coefficient of x.

 EXPLORATION 2   *The Coefficient of x and Slope*

(a) For each equation, obtain two solutions and use them to determine the slope of the line.

$$y = \frac{1}{2}x + 1 \qquad y = 2x + 1$$

$$y = -1x + 1 \qquad y = -3x + 1$$

(b) For each equation compare the coefficient of x with the slope of the corresponding graph.

Discovery

(a) Note that the y-intercept of each graph is $(0, 1)$, and so $(0, 1)$ is one solution of each equation.

Equation	Second Solution	Slope
$y = \frac{1}{2}x + 1$	$(2, 2)$	$m = \dfrac{2 - 1}{2 - 0} = \dfrac{1}{2}$
$y = 2x + 1$	$(1, 3)$	$m = \dfrac{3 - 1}{1 - 0} = \dfrac{2}{1} = 2$
$y = -1x + 1$	$(1, 0)$	$m = \dfrac{0 - 1}{1 - 0} = \dfrac{-1}{1} = -1$
$y = -3x + 1$	$(-1, 4)$	$m = \dfrac{4 - 1}{-1 - 0} = \dfrac{3}{-1} = -3$

(b) These results suggest the following generalization: If an equation is written in the form $y = mx + b$, the coefficient of x is the same as the slope m.

The proof of the generalization in Exploration 2 is left as an exercise.

The $y = mx + b$ form of a linear equation is known as the **slope–intercept form** for reasons that are now apparent.

> ### Definition of Slope–Intercept Form
>
> The **slope–intercept form** of a linear equation is the form $y = mx + b$.
>
> 1. The constant term b is the y-coordinate of the y-intercept.
>
> 2. The coefficient m of the x-term is the slope of the graph.

Graphing with Slope

If we know the slope and one point of a line, we can sketch the line. To do this, we need to recall the geometric interpretation of slope. Slope is defined as the ratio of rise to run between any two points of the line.

EXAMPLE 3   *Using Slope to Draw a Line*

Draw the line that contains the point $P(-3, 4)$ and whose slope is -2.

Solution   Plot the point $P(-3, 4)$. We can write the slope as $\frac{-2}{1}$, which means that the rise is -2 and the run is 1.

Starting at $P(-3, 4)$, move *down* 2 units and to the *right* 1 unit to locate a second point Q of the line. (See Fig. 5.16.) Points P and Q determine the line.

Figure 5.16

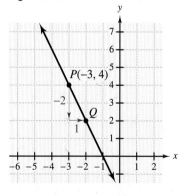

Figure 5.17

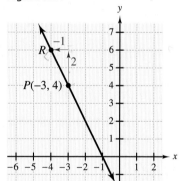

Note that the slope can also be written $\frac{2}{-1}$, which means that the rise is 2 and the run is -1. Starting at $P(-3, 4)$, move *up* 2 units and to the *left* 1 unit to locate a second point R of the line. (See Fig. 5.17.) The line determined by points P and R is the same as the line determined by points P and Q.   ∎

The triangle formed by the rise, the run, and the line is sometimes called the **slope triangle**.

NOTE: When using a point and the slope to graph a line, always start the slope triangle at the plotted point, not at the origin. ■

In Example 3, we were given the slope of the line and a point of the line. If we are given just the equation of the line, we can write the equation in slope–intercept form, from which it is easy to determine the slope and y-intercept. Using this information, we can then draw the graph as we did in Example 3.

The following summarizes the approach to graphing a line with information about its slope and y-intercept.

> ### Using a Line's Slope and y-Intercept to Graph the Line
>
> 1. Write the equation in the form $y = mx + b$.
>
> 2. The y-intercept is $(0, b)$. Plot that point.
>
> 3. The slope is m. Starting at the plotted point, move up or down as indicated by the rise. Then move right or left as indicated by the run. The destination is a second point of the line. Plot it.
>
> 4. Draw the line through the two plotted points.

EXAMPLE 4   *Using the Slope and y-Intercept to Draw a Line*

Sketch the graph of $3x - 2y = 8$.

Solution   First, we write the equation in slope–intercept form.

$$3x - 2y = 8$$
$$-2y = -3x + 8$$
$$y = \frac{3}{2}x - 4$$

Plot the y-intercept, $P(0, -4)$.

The slope is $\frac{3}{2}$, which means the rise is 3 and the run is 2. From the y-intercept, draw the slope triangle by moving up 3 units and right 2 units. The destination is a second point Q of the line. (See Fig. 5.18.) The points P and Q determine the line. ■

Figure 5.18

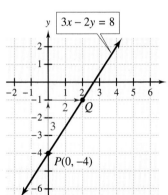

To graph an equation with its slope and y-intercept, we need to write the equation in the form $y = mx + b$. But if we need to solve for y, why not then just enter $y = mx + b$ in a calculator and produce the graph electronically? Because this is the quickest of all methods, it is usually what we will want to do.

There will be times, however, when we will have information about slopes and points, but we will not know the equation. Because we will still want to be able to produce the graph, learning pencil and paper methods is important. Although using a graphing calculator has many advantages, you should try to avoid becoming so dependent on a calculator that you fail to understand the underlying mathematical concepts.

5.2   *Quick Reference*

The Slope Formula

- Given points A and B of a line, the steepness, or **slope,** of the line is measured by the ratio of the vertical distance between A and B (**rise**) and the horizontal distance between A and B (**run**).

$$\text{Slope} = \frac{\text{Rise}}{\text{Run}}$$

- The slope of a line does not depend on the points of the line used to calculate it or on the points designated as the starting and ending points.

- For two points (x_1, y_1) and (x_2, y_2) of a line, $x_1 \neq x_2$, the slope m is defined as follows.

$$m = \frac{y_2 - y_1}{x_2 - x_1} = \frac{y_1 - y_2}{x_1 - x_2}$$

- The slope of any horizontal line is 0; the slope of any vertical line is undefined.

- A line with a positive slope rises from left to right; a line with a negative slope falls from left to right.

The Slope–Intercept Form

- The **slope–intercept form** of a linear equation is the form $y = mx + b$.

- The constant term b is the y-coordinate of the y-intercept of the line. The coefficient m of the x-term is the slope of the line.

Graphing with Slope

- If we know the slope of a line and one of its points, we can use the following procedure to draw the line.

  1. Plot the given point.

  2. Interpreting slope as $\frac{\text{rise}}{\text{run}}$, and starting at the plotted point, draw a slope triangle. The destination point is another point of the line.

  3. Draw the line through the starting and ending points.

- If the equation is given, write it in the form $y = mx + b$ to determine the starting point (y-intercept) and slope.

5.2   *Exercises*

 1. What is the geometric interpretation of the slope of a line?

 2. Describe a line that has a

   (a) positive slope.

   (b) negative slope.

In Exercises 3–10, determine from the given graph whether the slope of the line is positive, negative, zero, or undefined.

3.

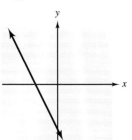

4.

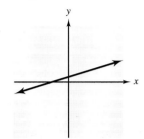

5.

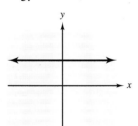

6.

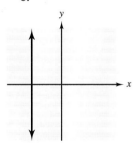

7.

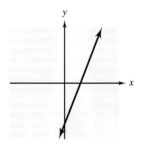

8.

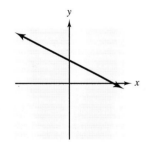

9.

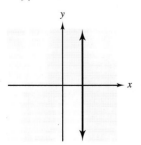

10.

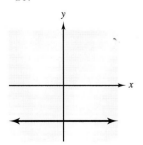

In Exercises 11–18, refer to the graph to determine the slope of the line.

11.

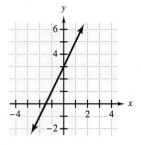

12.

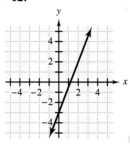

13.

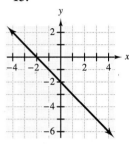

14.

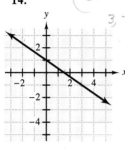

15.

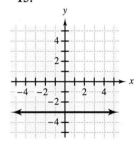

16.

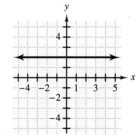

17.

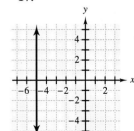

18.

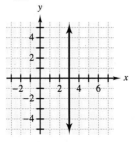

 19. What is the algebraic definition of the slope of a line?

20. What is the difference between a zero slope and an undefined slope?

In Exercises 21–34, determine the slope of the line that contains the given points.

21. $(8, -2)$ and $(6, -12)$

22. $(-3, -4)$ and $(-6, -10)$

23. $(3, 4)$ and $(5, 11)$

24. $(-2, 4)$ and $(5, 8)$

25. $(5, -6)$ and $(5, 7)$

26. $(2, -4)$ and $(4, -4)$

27. $(2.3, -5.9)$ and $(-1.7, 5.4)$

28. $(0, 6.94)$ and $(2.45, -3.5)$

29. $(-2, 5)$ and $(8, -3)$

30. $(-1, -6)$ and $(-7, -4)$

31. $(-3, 7)$ and $(6, 7)$

32. $(-3, 7)$ and $(-3, 4)$

33. $\left(\dfrac{1}{2}, \dfrac{1}{3}\right)$ and $\left(\dfrac{2}{3}, \dfrac{3}{4}\right)$

34. $\left(\dfrac{1}{3}, \dfrac{2}{3}\right)$ and $\left(\dfrac{3}{4}, \dfrac{1}{2}\right)$

 35. If the slope of a line is -3, two of the following procedures are correct ways to draw a slope triangle from a given point of the line. Identify the two correct procedures and explain why the third procedure is incorrect.

   (i) Move down 3 units and right 1 unit.

   (ii) Move down 3 units and left 1 unit.

   (iii) Move up 3 units and left 1 unit.

 36. State the slope of the x-axis and y-axis and explain your answers.

In Exercises 37–42, determine the unknown coordinate so that the line containing the points has the given slope.

37. $(-3, y)$, $(2, -2)$; $m = -1$

38. $(0, 8)$, $(2, y)$; $m = -\dfrac{5}{4}$

39. $(2, -1)$, $(x, 0)$; slope is undefined

40. $(x, -3)$, $(-5, 7)$; slope is undefined

41. $(-3, 4)$, $(2, y)$; $m = 0$

42. $(0, y)$, $(6, 5)$; $m = 0$

 43. A line contains the points $P(3, -2)$ and $Q(-11, 6)$. Which of the following is a correct way to calculate the slope of the line? Why?

   (i) $m = \dfrac{6 - (-2)}{3 - (-11)} = \dfrac{8}{14} = \dfrac{4}{7}$

   (ii) $m = \dfrac{-2 - 6}{3 - (-11)} = \dfrac{-8}{14} = -\dfrac{4}{7}$

 44. The definition of the slope of a line containing the points $P(x_1, y_1)$ and $Q(x_2, y_2)$ was given as

$$m = \dfrac{y_2 - y_1}{x_2 - x_1}.$$

  (a) Explain why

$$\dfrac{y_1 - y_2}{x_1 - x_2}$$

    is an equivalent expression for slope.

  (b) Explain why neither

$$\dfrac{y_2 - y_1}{x_1 - x_2} \quad \text{nor} \quad \dfrac{y_1 - y_2}{x_2 - x_1}$$

    is an equivalent expression for slope.

In Exercises 45–50, determine two solutions of the given equation and use the points to determine the slope of the line.

45. $3x + 5y = 15$ **46.** $x - 2y - 6 = 0$

47. $3 + y = 12$ **48.** $7 + 3x = 22$

49. $y = \dfrac{3}{4}x - 3$ **50.** $y = \dfrac{2}{3}x + 2$

 51. The slope of a line containing the points (x_1, y_1) and (x_2, y_2) is $\dfrac{4}{5}$. Does this mean that $y_2 - y_1 = 4$ and $x_2 - x_1 = 5$? Explain.

 52. If a linear equation is in the form $Ax + By = C$, describe the steps we would take to determine the slope of the graph with the slope formula.

In Exercises 53–66, write the equation in slope–intercept form, if possible. Then determine the slope of the line.

53. $y + 2x = 17$

54. $y - 3x = 10$

55. $2x - y = 0$

56. $y + 3x = 0$

57. $y + 2 = 17$

58. $y - 3 = 5$

59. $x + 3y - 6 = 0$

60. $3x + 4y = 10$

61. $4x - 3 = 13$

62. $2x + 9 = 7$

63. $3y = 5x - 7$

64. $7x = 5y + 8$

65. $\dfrac{3}{4}x - \dfrac{2}{3}y = 12$

66. $\dfrac{1}{4}x + \dfrac{2}{5}y = 10$

67. If a linear equation is in the form $Ax + By = C$, $B \neq 0$, show that the slope of the graph is $-(A/B)$.

68. If a linear equation is in the form $Ax + By = C$, $B \neq 0$, show that the y-intercept of the graph is C/B.

69. A line has a slope of -2. Which of the following could be an equation of the line?

(i) $6x + 3y - 7 = 0$

(ii) $4x - 2y + 3 = 0$

(iii) $x + \dfrac{1}{2}y - 1 = 0$

70. A line has a slope of $\frac{1}{3}$. Which of the following could be an equation of the line?

(i) $\dfrac{1}{9}x = \dfrac{1}{3}y + 3$

(ii) $9x - 3y = 4$

(iii) $12x + 2 - 4y = 0$

71. What is the rise of a line if the line's slope is zero?

72. What is the run of a line if the line's slope is undefined?

In Exercises 73–84, draw a line that contains the given point and has the given slope.

73. $(0, 3)$; $m = \dfrac{3}{4}$

74. $(0, -2)$; $m = -\dfrac{3}{4}$

75. $(0, -1)$; $m = -4$

76. $(0, 3)$; $m = 2$

77. $(-5, 0)$; m is undefined

78. $(0, 2)$; $m = 0$

79. $(-2, -3)$; $m = \dfrac{2}{3}$

80. $(-3, -5)$; $m = -\dfrac{3}{5}$

81. $(3, 7)$; $m = 0$

82. $(-4, -3)$; m is undefined

83. $(2, 6)$; $m = -5$

84. $(-5, 2)$; $m = 4$

In Exercises 85–96, use what you know about the slope and y-intercept to sketch the graph of the given equation. When possible, produce the graph on your calculator and compare it to your sketch.

85. $y = \dfrac{3}{4}x + 5$

86. $y = -\dfrac{2}{3}x - 2$

87. $y = -3x + 4$

88. $y = 2x + 3$

89. $y = 2$

90. $y = -2$

91. $y = x$

92. $y = -x$

93. $y = -\dfrac{3}{4}x$

94. $y = \dfrac{2}{5}x$

95. $x = 3$

96. $x = -4$

 97. An equation is in the form $y = 3x + b$. What is the effect on the graph if the value of b increases?

 98. An equation is in the form $y = mx + b$, where $m > 0$. What is the effect on the graph if the value of m increases?

In Exercises 99–108, if possible, write the given equation in the form $y = mx + b$. Use what you know about the slope and y-intercept to sketch the graph.

99. $2x + 5y = 10$

100. $4x - 5y = 20$

101. $x - 3 = 4$

102. $x + 6 = 2$

103. $2x - y - 6 = 2$

104. $x + 3y - 6 = 0$

105. $y + 3 = 10$

106. $y - 2 = 4$

107. $y - 2x = 0$

108. $y + 3x = 0$

109. A student is trying to produce the graph of $y = -2x + 3$ on a calculator. The accompanying figure shows the result. What error has the student probably made?

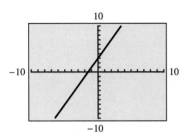

110. A student is trying to produce the graph of $y = -2x + 3$ on a calculator. The accompanying figure shows the result. What error has the student probably made?

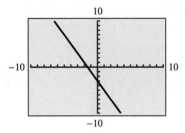

Figure for 111–114

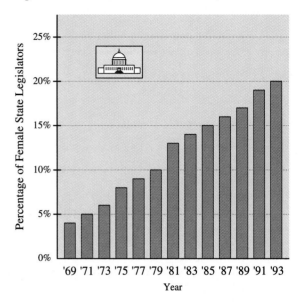

113. If a corresponding linear equation were written to model the percentage of male legislators during the same period, what can you predict about the coefficient of X? Why?

114. According to the model, one-half of all state legislators will be female by the year 2037. How valid do you think the model is for years beyond those reported here?

Exploring with Real Data

In Exercises 111–114, refer to the following background information and data.

The bar graph shows the percentage of female state legislators during the period 1969 through 1993. (Source: Center for the American Woman and Politics.) The data can be modeled by the linear equation $P = -1314.94 + 0.67X$, $X \geq 1969$, where P is the percentage of female legislators and X is the year.

111. Assuming the model is valid for the future, predict the percentage of female state legislators in the year 2001.

112. In the model the coefficient of X is positive. What does this tell you about the graph of the equation? Is this consistent with the data shown in the bar graph?

Challenge

115. If the slope of a line is 50, the line may appear to be vertical in the default viewing rectangle. Which range setting, X or Y, do you need to adjust so that the line does not appear to be vertical and so that the intercepts are more clearly displayed? Why?

In Exercises 116–121, adjust the range and scale so that both intercepts of the graph are clearly displayed and the line does not appear vertical or horizontal.

116. $y = 100x - 7$ **117.** $y = 90x + 200$

118. $y = x - 200$ **119.** $y = 70 + 0.1x$

120. $y = 0.02x + 0.05$ **121.** $y = 35 - 0.02x$

122. Suppose $P(x_1, y_1)$ and $Q(x_2, y_2)$ are two points of the graph of $y = ax + b$.

(a) Write an equation expressing the fact that (x_1, y_1) is a solution of $y = ax + b$.

(b) Write an equation expressing the fact that (x_2, y_2) is a solution of $y = ax + b$.

(c) Subtract the equation in part (b) from the equation in part (a) and solve for a.

(d) Explain why a is the slope of the graph of $y = ax + b$.

5.3 Applications of Slope

Geometric Applications ▪ *Rate of Change* ▪ *Direct Variation*

Geometric Applications

As we have seen, if linear equations in the form $y = mx + b$ have different x-terms, but the constant terms are the same, then the graphs of the equations have the same y-intercept.

Similarly, if such equations have different constant terms, but the x-terms are the same, then the graphs have different y-intercepts, but they have the same slope.

EXPLORATION 1   *Different Lines with the Same Slope*

Use your calculator to display the graphs of the following equations on the same coordinate axes.

$$y_1 = 2x - 9 \qquad y_2 = 2x - 1$$
$$y_3 = 2x + 3 \qquad y_4 = 2x + 8$$

(a) How are the graphs different?

(b) How are the graphs similar?

(c) What is your conjecture about graphs such as these?

Discovery

(a) Because the constant terms of the four equations are all different, the graphs have different y-intercepts.

$$(0, -9) \qquad (0, -1) \qquad (0, 3) \qquad (0, 8)$$

(b) The coefficients of the x-terms are all 2. Therefore, the slope of each line is 2.

(c) In geometry, if two lines in the same plane do not intersect, then the lines are **parallel.** From the graphs in Fig. 5.19, it appears that the four lines are parallel. We conjecture that distinct lines with the same slope are parallel.

Figure 5.19

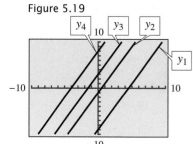

> ### Parallel Lines and Slopes
>
> Two distinct, nonvertical lines are **parallel** if and only if they have the same slope. Also, any two distinct vertical lines are parallel.

The word *distinct* means that the lines must be different lines. If two lines have the same slope as well as the same y-intercept, the lines *coincide*. We do not want to think of such lines as being parallel because they intersect everywhere.

Also, because the slopes of vertical lines are undefined, we cannot say that their slopes are equal. We simply define vertical lines as parallel.

EXAMPLE 1   *Determining Whether Lines Are Parallel*

(a) Produce the graphs of the following equations on your calculator.

   (i) $2x - 3y = 6$

   (ii) $y = \dfrac{2}{3}x + 11$

   (iii) $3y = 6 - 2x$

   (iv) $-y = 2 - \dfrac{2}{3}x$

(b) From the calculator display, predict which equations have graphs that are parallel lines.

(c) Confirm your predictions in part (b) by noting the slopes and y-intercepts of the graphs of the equations.

Solution

(a) Figure 5.20

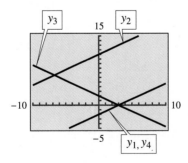

(b) From Fig. 5.20, it appears that the graphs of equations (i) and (ii) are parallel.

(c)

Equation	Slope–Intercept Form	Slope	y-Intercept
(i) $2x - 3y = 6$	$y = \dfrac{2}{3}x - 2$	$\dfrac{2}{3}$	$(0, -2)$
(ii) $y = \dfrac{2}{3}x + 11$	$y = \dfrac{2}{3}x + 11$	$\dfrac{2}{3}$	$(0, 11)$
(iii) $3y = 6 - 2x$	$y = -\dfrac{2}{3}x + 2$	$-\dfrac{2}{3}$	$(0, 2)$
(iv) $-y = 2 - \dfrac{2}{3}x$	$y = \dfrac{2}{3}x - 2$	$\dfrac{2}{3}$	$(0, -2)$

Only equations (i) and (ii) have the same slope and different y-intercepts. Therefore, these are the only equations whose graphs are parallel lines. ■

Two lines are **perpendicular** if they intersect to form a right angle.

Depending on the configuration of your calculator's viewing rectangle, perpendicular lines may not appear perpendicular because of the different spacing of the tick marks along the axes. It is as if the figure were drawn on a piece of elastic and then the elastic were stretched sideways. In such a case, the figure would be distorted.

 SQUARE

If your calculator model requires it, you can correct this distortion by making the tick mark spacing the same along both axes.

 **EXPLORATION 2**   *Perpendicular Lines*

Use a square viewing rectangle, if necessary, and produce the graphs of the following pairs of equations on the calculator.

(i) $y_1 = \dfrac{1}{2}x + 3$ (ii) $y_1 = 3x - 5$

     $y_2 = -2x - 4$ $y_2 = -\dfrac{1}{3}x + 8$

For each pair describe the apparent relationship between the lines. Do you notice any relationship between the slopes of the lines in each pair?

Discovery

Figure 5.21 shows the graphs of the pair of equations in (i), and Fig. 5.22 shows the graphs of the pair of equations in (ii). The lines appear to be perpendicular.

Figure 5.21   Figure 5.22

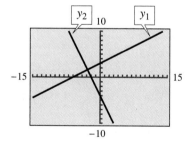   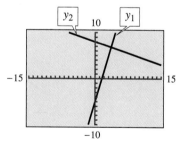

In pair (i) the slopes of the lines are $\frac{1}{2}$ and -2. Observe that the product of the slopes is -1.

$$\frac{1}{2} \cdot (-2) = -1$$

In pair (ii) the slopes are 3 and $-\frac{1}{3}$. Again, the product of the slopes is -1.

$$3 \cdot \left(-\frac{1}{3}\right) = -1$$

Exploration 2 suggests that the product of the slopes of two nonvertical perpendicular lines is -1. It can be shown that this is true in general. We state the result without proof.

> ### Perpendicular Lines and Slopes
>
> If neither of two lines is vertical, then the lines are **perpendicular** if and only if the product of their slopes is -1. Also, a vertical line and a horizontal line are perpendicular.

We cannot refer to the product of the slopes of two lines if one of the lines is vertical because the slope of a vertical line is undefined. We simply define a vertical line and a horizontal line to be perpendicular.

EXAMPLE 2   *Determining Whether Lines Are Perpendicular*

Use a square viewing rectangle (if necessary) to produce the graphs of the equations $7x - 12y = -36$ and $11x + 6y = -24$ on your calculator. Do the lines appear to be perpendicular? Verify your conclusion.

Solution

Write the equations in the slope–intercept form and produce their graphs.

$$y = \frac{7}{12}x + 3$$

$$y = -\frac{11}{6}x - 4$$

$$-12y = \frac{-7x}{-12} - \frac{36}{-12}$$

Figure 5.23

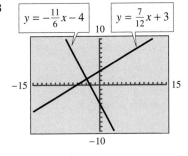

$$y = -\frac{11}{6}x - 4 \qquad y = \frac{7}{12}x + 3$$

The lines in Fig. 5.23 appear to be perpendicular, but a picture, while it is suggestive, does not prove anything. The only way to know for sure is to check the slopes. The slopes are $\frac{7}{12}$ and $-\frac{11}{6}$.

$$\frac{7}{12} \cdot \left(-\frac{11}{6}\right) = -\frac{77}{72} \approx -1.07$$

Because the product of the slopes is not -1, the lines are not perpendicular. ∎

EXAMPLE 3 *Slopes of Perpendicular Lines*

Suppose the equation of line L is $y = \frac{2}{3}x - 5$. If line N is perpendicular to line L, what is the slope of line N?

Solution   From the equation of line L, the slope of line L is $\frac{2}{3}$. Let m represent the slope of line N.

Because lines L and N are perpendicular, the product of their slopes is -1.

$$\frac{2}{3} \cdot m = -1$$

$$\frac{3}{2} \cdot \frac{2}{3} \cdot m = \frac{3}{2} \cdot (-1)$$ Multiply both sides by $\frac{3}{2}$ to solve for m.

$$m = -\frac{3}{2}$$

The slope of line N is $-\frac{3}{2}$.

EXAMPLE 4 *Determining Whether a Triangle Is a Right Triangle*

The points $A(2, 5)$, $B(3, 7)$, and $C(6, 3)$ are the vertices of a triangle. (See Fig. 5.24.) Is the triangle a right triangle?

Figure 5.24

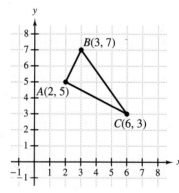

Solution   Previously, we determined whether a triangle was a right triangle by applying the Pythagorean Theorem. Now we can determine if any two sides are perpendicular by examining their slopes.

From Fig. 5.24 it appears line segments $\overline{AB}$ and $\overline{AC}$ are perpendicular. We use the slope formula to find the slopes of these line segments.

Points	*Slope*
$A(2, 5)$ and $B(3, 7)$	$m = \dfrac{7 - 5}{3 - 2} = \dfrac{2}{1} = 2$
$A(2, 5)$ and $C(6, 3)$	$m = \dfrac{3 - 5}{6 - 2} = \dfrac{-2}{4} = -\dfrac{1}{2}$

Because the product of the slopes is $2 \cdot (-\frac{1}{2}) = -1$, the line segments are perpendicular. Therefore, the triangle is a right triangle.

Rate of Change

An important concept in mathematics involves the rate at which a function changes. Specifically, for the function $y = mx + b$, the **rate of change** of y with respect to x is the amount by which y changes for a unit change in x.

EXPLORATION 3   *Rate of Change*

Use the integer setting to produce the graph of $y = 2x - 8$ on your calculator.

(a) Slowly move the tracing cursor up the line. By how much does the y-coordinate change for each unit change in x? Does your answer depend on the points that you trace?

(b) What is the ratio of the change in y to the change in x?

(c) What is the slope of the line?

(d) What is your conjecture about the rate of change of y with respect to x for an equation $y = mx + b$?

Discovery

(a) For any points of the line, as x increases by 1 unit, y increases by 2 units.

(b) The ratio of the change in y to the change in x is $\frac{2}{1}$ or 2.

(c) For $y = 2x - 8$, the slope of the line is 2.

(d) The rate of change of y with respect to x is the same as the slope of the line.

Exploration 3 suggests that the slope of a line represents the rate of change of y with respect to x.

Slope as a Rate of Change

For a linear equation in two variables, the rate of change of y with respect to x is the slope of the graph of the equation. This rate of change, like the slope, is constant.

For many functions, the rate of change is not constant. However, linear functions do have a constant rate of change, and it is the slope of the line.

EXAMPLE 5   *Hourly Pay as Rate of Change*

A salesperson earns a base salary of $300 per week plus a 5% commission on all sales.

(a) Write an equation for the salesperson's weekly income I as a function of weekly sales s.

(b) What is the rate of change in the weekly income with respect to the weekly sales?

Solution

(a) $I = 0.05s + 300$. The variables are I and s instead of y and x, but the equation is in slope–intercept form.

(b) A graph of the equation is a straight line with a y-intercept of 300 and a slope of 0.05. The slope indicates that the rate of change of I with respect to s is 0.05. Note that the rate of change is simply the commission.   ■

Direct Variation

For the slope–intercept form $y = mx + b$, a special case arises when the y-intercept is the origin, that is, when $b = 0$. The form of the equation then reduces to $y = mx$, and $(0, 0)$ is a solution for all values of m.

In this case, we say that y **varies directly** with x.

> ### Definition of Direct Variation
>
> The value of y **varies directly** with the value of x if there is a constant k such that $y = kx$.

The constant k is called the **constant of variation.**

Sometimes we say y is **proportional** to x, and if $x \neq 0$, we write $\dfrac{y}{x} = k$.

In this form, the constant k is called the **constant of proportionality.**

Consider the following direct variation. Suppose a car travels at 50 mph. The distance traveled is given by $d = 50t$, where d is the distance in miles and t is the time in hours. As the time increases, the distance increases at a rate of 50 mph. We say that the distance varies directly with the time or that the distance is directly proportional to the time. The constant of variation is the speed, 50 mph.

There are many such examples of one variable varying directly with another variable. In each case, the slope m of the graph is the same as the constant k of variation.

EXAMPLE 6 *Proportional Distances*

The actual distance between two towns is directly proportional to the distance between them on a map. On a certain map, the distance between Tortilla Flat, Arizona, and Tombstone, Arizona, is $6\frac{3}{4}$ inches. Suppose that on this map, $\frac{1}{3}$ of an inch represents 7 miles. What is the actual distance between these towns?

Solution　Let d represent the actual distance in miles and let x represent the map distance in inches. Because d is directly proportional to x, the relation is $d = kx$.

A map distance of $\frac{1}{3}$ of an inch corresponds to an actual distance of 7 miles. Replace x with $\frac{1}{3}$ and d with 7 to determine k.

$$d = kx$$
$$7 = k \cdot \frac{1}{3}$$
$$21 = k$$

The constant of variation is 21 and so $d = 21x$. Since the map distance between the two towns is 6.75,

$$d = 21(6.75) = 141.75.$$

The distance from Tortilla Flat to Tombstone is 141.75 miles.

5.3   *Quick Reference*

Geometric
Applications

- Two distinct, nonvertical lines are **parallel** if and only if they have the same slope. Any two distinct vertical lines are parallel.

- If neither of two lines is vertical, then the lines are **perpendicular** if and only if the product of their slopes is -1. A vertical line and a horizontal line are perpendicular.

Rate of
Change

- For a linear function $y = mx + b$, the **rate of change** of y with respect to x is constant, and it is the slope of the graph of the function.

Direct
Variation

- We say that y **varies directly** with x if there is a constant k such that $y = kx$. The linear function $y = kx$ is called a **direct variation** where k is the **constant of variation.**

- If $x \neq 0$, a direct variation can be written

$$\frac{y}{x} = k,$$

where k is called the **constant of proportionality.**

- For direct variations of the form $y = kx$, $(0, 0)$ is always a solution. Therefore, only one other solution is needed to determine k.

5.3   *Exercises*

In all exercises in this section, L_1 refers to line 1 and L_2 refers to line 2. The symbol m_1 is used to represent the slope of line 1, and m_2 is used to represent the slope of line 2.

 1. What is the geometric definition of perpendicular lines?

 2. What is the geometric definition of parallel lines?

In Exercises 3–8, the slopes of L_1 and L_2 are given. Determine if L_1 and L_2 are parallel, perpendicular, or neither.

3. $m_1 = 2$; $m_2 = -2$

4. $m_1 = \dfrac{1}{4}$; $m_2 = 4$

5. $m_1 = \dfrac{1}{2}$; $m_2 = 0.5$

6. $m_1 = -4$; $m_2 = 0.25$

7. $m_1 = \dfrac{3}{5}$; $m_2 = -\dfrac{5}{3}$

8. $m_1 = \dfrac{7}{9}$; $m_2 = \dfrac{9}{7}$

In Exercises 9 and 10, for the given slope, determine m_2 so that

(a) L_1 is parallel to L_2.

(b) L_1 is perpendicular to L_2.

9. $m_1 = -3$ **10.** $m_1 = \dfrac{5}{8}$

In Exercises 11–16, two points of L_1 and two points of L_2 are given. Determine if the lines are parallel, perpendicular, or neither.

11. L_1: (3, 5) and (−2, 1)
    L_2: (−6, 9) and (−1, 13)

12. L_1: (−6, 4) and (−2, −8)
    L_2: (108, 120) and (100, 144)

13. L_1: (0, 0) and (8, 8)
    L_2: (−1, 4) and (3, −7)

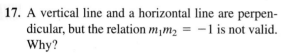

14. L_1: (4, 8) and (1, −2)
    L_2: (−3, 5) and (2, 2)

15. L_1: (−4, 2) and (−4, 7)
    L_2: (0, 5) and (3, 5)

16. L_1: (−2, 3) and (1, 5)
    L_2: (0, 4) and (2, 1)

17. A vertical line and a horizontal line are perpendicular, but the relation $m_1 m_2 = -1$ is not valid. Why?

18. What can you conclude about L_1 and L_2 if $m_1 = -(1/m_2)$, where $m_2 \neq 0$? Why?

In Exercises 19–24, the equations of L_1 and L_2 are given. Determine if the lines are parallel, perpendicular, or neither.

19. L_1: $3y = 5 - 2x$ L_2: $y = -\dfrac{2}{3}x + 9$

20. L_1: $y = \dfrac{2}{3}x + 1$ L_2: $3x + 2y = 0$

21. L_1: $x = -4$ L_2: $y = -4$

22. L_1: $y - 3 = 12$ L_2: $5 - y = 14$

23. L_1: $y = 5x - 3$ L_2: $y = -5x + 3$

24. L_1: $\dfrac{1}{2}x - \dfrac{1}{3}y = 0$ L_2: $\dfrac{5}{6}x - \dfrac{5}{9}y = 1$

In Exercises 25–30, the equation of one line is given and two points of another line are given. Are the lines parallel, perpendicular, or neither?

25. L_1: $x - y = 2$ L_2: (5, 9) and (6, 8)

26. L_1: (−3, 1) and (4, 2) L_2: $y = \dfrac{1}{7}x - 2$

27. L_1: (−2, −5) and (0, 4) L_2: $y = \dfrac{9}{2}x + 4$

28. L_1: $-2x + 5y = 10$ L_2: (−6, 6) and (−4, 1)

29. L_1: $y = 5$ L_2: (2, 3) and (−6, 0)

30. L_1: (2, −1) and (−7, −1) L_2: $x = -1$

31. If L_1 is parallel to the y-axis, then $m_1 = $ _____ .

32. If L_2 is perpendicular to the x-axis, then $m_2 = $ _____ .

33. If L_1 is perpendicular to the y-axis, then $m_1 = $ _____ .

34. If L_2 is parallel to the x-axis, then $m_2 = $ _____ .

In Exercises 35–38, determine the slope of the line and sketch the graph.

35. The line contains (−1, −4) and is parallel to the line containing (−2, 3) and (4, −1).

36. The line contains (−3, 1) and is perpendicular to the line $y = 2x - 3$.

37. The x-intercept of the line is (3, 0), and the line is parallel to the y-axis.

38. The y-intercept of the line is (0, −6), and the line is parallel to the x-axis.

In Exercises 39 and 40, information is given about two lines L_1 and L_2. Determine the value of k so that

  (a) L_1 and L_2 are parallel.

  (b) L_1 and L_2 are perpendicular.

39. L_1 contains the points (3, −4) and (−1, 2). L_2 contains the points (5, 0) and (−2, k).

40. The equation of L_1 is $3x + 12y = 15$. The equation of L_2 is $y = kx + 7$.

In Exercises 41 and 42, show that the triangles in the figures are right triangles.

41.
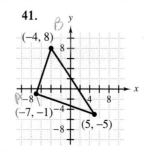

42.
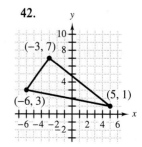

In Exercises 43 and 44, determine if the three given points are collinear. (Points are **collinear** if they are points of the same straight line.)

43. $P(-5, 8)$ $Q(-1, 3)$ $R(7, -7)$

44. $A(-7, -2)$ $B(0, 0)$ $C(21, 5)$

 45. Describe the slope of a line in terms of rate of change.

46. For a linear function, why is the rate of change constant?

In Exercises 47–50, determine the rate of change of y with respect to x.

47. $y = \dfrac{1}{2}x + 3$ 48. $y = -\dfrac{2}{3}x + 5$

49. $y = -4x + 9$ 50. $y = 3x - 2$

In Exercises 51–54, determine the rate of change of y with respect to x for the line containing the given points.

51. $(-1, -2)$ and $(3, 4)$ 52. $(-2, 3)$ and $(3, 1)$

53. $(2, -2)$ and $(-1, 5)$ 54. $(-3, 1)$ and $(2, 5)$

In Exercises 55–62, the rate of change of a line $y = mx + b$ is given. For the given increase/decrease of one variable, determine the corresponding increase/decrease of the other variable.

55. $m = 3$; x increased by 2

56. $m = -2$; y increased by 6

57. $m = -\dfrac{5}{4}$; y increased by 5

58. $m = \dfrac{1}{3}$; x increased by 9

59. $m = 5$; x decreased by 2

60. $m = -4$; y decreased by 12

61. $m = -\dfrac{3}{2}$; y decreased by 6

62. $m = \dfrac{3}{5}$; x decreased by 20

63. Ace Auto Rental rents a car for $29 a day plus 10 cents per mile. For a 1-day rental, let m represent the number of miles driven and let c represent the total cost.

   (a) Write an equation for total rental cost.

   (b) What is the rate of change in the total cost with respect to the number of miles driven?

64. A telemarketer earns $100 per week plus a 30% commission on all sales. Let I represent the weekly income and let s represent the weekly sales.

   (a) Write an equation for the total weekly income.

   (b) What is the rate of change in income with respect to sales?

In Exercises 65 and 66, assume that the described relationship is linear.

65. In 1950 the cost of a soft drink was 5 cents. In 1990 the cost of the same soft drink was 55 cents. During this period, what was the annual rate of change in the cost of the drink?

66. On a certain January day in Minnesota, the temperature at 6:00 A.M. was $-15°F$. By noon, the temperature was up to $12°F$. What was the hourly rate of temperature change?

 67. Explain the meaning of the statement, "*A* varies directly with *B*."

 68. Suppose y varies directly with x. If $x = a$ when $y = b$ and $x = c$ when $y = d$, explain why $a/b = c/d$, where a, b, c, and d are nonzero numbers.

In Exercises 69–74, determine whether the given equation represents a direct variation.

69. $y = 4x$ 70. $4 = xy$

71. $\dfrac{y}{4} = x$ 72. $y = 2x + 3$

73. $y = 5x - 3$ 74. $y = 3$

75. Suppose y varies directly with x. If x increases, then y _____ , and if y decreases, then x _____ .

76. Suppose y varies directly with x and k is the constant of proportionality. Then, $y =$ _____ and $k =$ _____ .

77. The weight of a load of bricks varies directly with the number of bricks in the load. If a load of 500 bricks weighs 1175 pounds, what will a load of 1200 bricks weigh?

78. The weight that can be lifted by an automobile jack varies directly with the force exerted downward on the jack handle. If a force of 9 pounds will lift 954 pounds, what weight will be lifted by a force of 15 pounds?

79. If the height of a triangle remains constant, then the area of the triangle varies directly with the length of the base. If a triangular sail with a fixed height has an area of 165 square feet when the base is 12 feet, what would the area be if the base were increased to 18 feet?

80. In physics, Hooke's law states the following: The distance d a hanging spring is stretched varies directly with the weight of an attached object. If a weight of 4.5 pounds stretches a spring a distance of 3 inches, how far will a weight of 9.5 pounds stretch this spring?

81. The amount paid for typing varies directly with the number of pages typed. If $100 is paid for typing 16 pages, how much will be paid to type 25 pages?

82. The discount on a blouse varies directly with the marked (selling) price. If the discount is $11.84 on a blouse marked $37, what is the discount on a blouse marked $50?

Exploring with Real Data

In Exercises 83–86, assume that the described relationship is linear.

83. The per-share dividend for Hewlett-Packard increased from 23 cents per share in 1987 to 90 cents per share in 1993. If the rate of increase had been constant, what would have been the yearly rate of change in per-share dividends? (Source: Hewlett-Packard.)

84. Diebold, Inc., is a leading manufacturer of automatic teller machines. The per-share earnings changed from $1.59 in 1990 to $2.35 in 1993. Assuming a constant rate of change, what was the yearly rate of change in per-share earnings? (Source: Diebold, Inc.)

85. The per-share earnings of Wang Labs changed from a profit of $0.56 in 1988 to a loss of $1.46 in 1990. Assuming a constant rate of change, what was the yearly rate of change in per-share earnings? (Source: Wang Labs.)

86. Hubbell, Inc., makes electrical equipment. The sales in 1988 were 614.2 million dollars. In 1992 the sales were 786.1 million dollars. Assuming a constant rate of change, what was the rate of change in annual sales? (Source: Hubbell, Inc.)

Challenge

In Exercises 87–92, the equation of a line is given. Determine k so that the given condition is met.

87. $kx + 3y = 10$; the slope of the line is -2.

88. $2x - ky = 2$; the line is parallel to $y = \frac{1}{2}x - 5$.

89. $3kx - 3y = 4$; the line is perpendicular to $x + y = 0$.

90. $5x + 2ky = 7$; the slope of the line is undefined.

91. $kx - 5y = 20$; the line is parallel to the x-axis.

92. $kx + ky = 4$; the slope of the line is -1.

 93. Suppose A, B, C, and D are consecutive vertices of a quadrilateral. Explain how you would show that the quadrilateral is a parallelogram. (A **parallelogram** is a quadrilateral whose opposite sides are parallel.)

 94. Suppose A, B, C, and D are consecutive vertices of a parallelogram. Explain how you would show that the parallelogram is a rectangle. (A **rectangle** is a parallelogram with one right angle.)

95. A **rhombus** is a parallelogram whose sides are all the same length. The vertices of a rhombus are $A(0, 0)$, $B(3, 4)$, $C(8, 4)$, and $D(5, 0)$. Show that the diagonals of the rhombus are perpendicular.

96. If L_1, L_2, and L_3 are three lines in the same plane such that L_1 and L_2 are both perpendicular to L_3, show that L_1 and L_2 are parallel.

5.4   Equations of Lines

Slope–Intercept Form   ▪   *Point–Slope Form*   ▪   *Special Cases*   ▪
Analyzing Conditions   ▪   *Linear Regression*

Slope–Intercept Form

Until now, we began with a linear equation and used it to draw its graph or to determine certain characteristics of the graph such as its slope and intercepts. In this section, we reverse the process. Now we will begin with information about the graph of a linear equation and determine an equation of the line.

If we *know* the equation of a line, it is useful to write the equation in the slope–intercept form $y = mx + b$, because of the information it reveals. The coefficient of x is the slope m of the line, and the constant term b gives us the y-intercept $(0, b)$.

If we wish to *determine* an equation of a line, we need a model (or formula) for writing the equation.

Figure 5.25 shows a line that contains some point $P(x, y)$ and whose y-intercept is $Q(0, b)$.

Figure 5.25

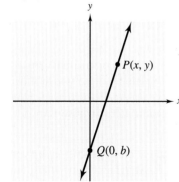

The slope of the line can be found by using the slope formula.

$$m = \frac{y - b}{x - 0} = \frac{y - b}{x}$$

The equation

$$\frac{y - b}{x} = m$$

is in the form of a direct variation, and so, as we saw in Section 5.3, it is equivalent to

$$y - b = mx.$$

Adding b to both sides, we obtain the model we are seeking.

$$y = mx + b$$

Thus, the model for *writing* an equation is the same as the slope–intercept form of a *known* equation.

Because $y = mx + b$ can be used as a model for writing an equation, the best information we can be given about a line is the slope and y-intercept, as Example 1 illustrates.

| EXAMPLE 1 | *Writing an Equation of a Line Given the Slope and the y-Intercept* |

Write an equation of a line whose slope is 2 and whose y-intercept is $(0, -3)$.

Solution

We are given that $m = 2$ and $b = -3$.

$y = mx + b$ Slope–intercept form of the required equation

$y = 2x - 3$ Replace *m* with 2 and *b* with −3.

The equation of the line is $y = 2x - 3$. ■

Suppose we are given the slope of a line along with some point other than the *y*-intercept. To determine an equation of the line, we use the key concept that relates equations and graphs.

> Every point of a line has coordinates that are solutions of the equation of the line.

EXAMPLE 2   *Writing an Equation of a Line Given the Slope and One Point*

Write an equation of the line that contains the point $P(1, 3)$ and whose slope is 2.

Solution

Because $m = 2$, we can start the equation as follows.

$y = mx + b$ Slope–intercept form of a linear equation

$y = 2x + b$ Replace *m* with 2.

Because $P(1, 3)$ is a point of the line, the coordinates satisfy the equation.

$3 = 2(1) + b$ Replace *x* with 1 and *y* with 3.

$3 = 2 + b$

$1 = b$

Now we know that $b = 1$, and so the equation can be completed.

$y = 2x + 1$ Replace *b* with 1. ■

Point–Slope Form

As we have seen, the form $y = mx + b$ can be used to write an equation of a line even if the given point is not the *y*-intercept.

An alternate model can be used in such cases. Figure 5.26 shows a line with a *fixed* point $Q(x_1, y_1)$ and point $P(x, y)$, which represents *any* other point of the line.

The slope of the line is given by

$$m = \frac{y - y_1}{x - x_1}.$$

Because this equation is in the form of a direct variation, it is equivalent to

$$y - y_1 = m(x - x_1).$$

This result is known as the **point–slope form** of a linear equation.

Figure 5.26

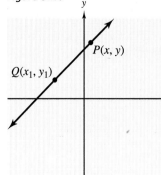

> ### The Point–Slope Form of a Linear Equation
>
> The **point–slope form** of a linear equation is
>
> $$y - y_1 = m(x - x_1)$$
>
> where m is the slope of the line and (x_1, y_1) is a fixed point of the line.

The point–slope form is particularly useful for writing an equation of a line when the slope and one point of the line are given.

EXAMPLE 3   *Writing an Equation of a Line Given the Slope and One Point*

Use the point–slope form to write an equation of a line that contains $P(-5, -3)$ and whose slope is $-\frac{2}{5}$.

Solution

$$y - y_1 = m(x - x_1)$$ Point–slope form of a linear equation

$$y - (-3) = -\frac{2}{5}[x - (-5)]$$ Replace m with $-\frac{2}{5}$, x_1 with -5, and y_1 with -3.

$$y + 3 = -\frac{2}{5}(x + 5)$$

$$y + 3 = -\frac{2}{5}x - 2$$ Distributive Property

$$y = -\frac{2}{5}x - 5$$ Subtract 3 from both sides.   ■

The slope–intercept form and the point–slope form both require that we know the slope of a line in order to write its equation. If we are given just two points of a line, we can calculate the slope and write an equation of the line with either of the forms.

EXAMPLE 4   *Writing an Equation of a Line Given Two Points*

Write an equation of the line containing the points $P(4, 2)$ and $Q(-3, -5)$.

Solution

We use the slope formula to compute the slope of the line.

$$m = \frac{y_2 - y_1}{x_2 - x_1}$$

$$m = \frac{-5 - 2}{-3 - 4} = \frac{-7}{-7} = 1$$

We use the point–slope form to begin writing the equation.

$$y - y_1 = m(x - x_1)$$

$$y - y_1 = 1(x - x_1)$$ Replace m with 1.

Because points P and Q both represent solutions, we can use either one as the fixed point. We use point $P(4, 2)$.

$$y - 2 = 1(x - 4) \qquad \text{Replace } x_1 \text{ with 4 and } y_1 \text{ with 2.}$$
$$y - 2 = x - 4$$
$$y = x - 2 \qquad \text{Add 2 to both sides.}$$

Special Cases

As we have seen, the special cases of linear equations have graphs with the following characteristics.

Equation	*Graph Characteristics*
y = constant	horizontal line, slope is 0
x = constant	vertical line, slope is undefined

Until now, we have started with the special equation y = constant or x = constant and have determined the characteristics of the graph. Now we start with information about the line and determine its equation.

EXAMPLE 5   *Special Cases*

(a) Write an equation of a line whose slope is 0 and that contains the point $R(2, 4)$.

(b) Write an equation of the line containing the points $C(4, 2)$ and $D(4, 7)$.

Solution

(a) Because the slope is 0, the line is horizontal. (See Fig. 5.27.) Therefore, the equation is of the form y = constant, where the constant is the y-coordinate of every point of the line.

Because $R(2, 4)$ is a point of the line and its y-coordinate is 4, every point of the line has a y-coordinate of 4. The equation is $y = 4$.

(b) Because the x-coordinates of points C and D are the same, the line is vertical. (See Fig. 5.28.) Therefore, the equation is of the form x = constant. Points C and D and all other points of the line have an x-coordinate of 4. The equation is $x = 4$.

Figure 5.27

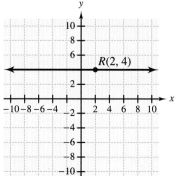

Figure 5.28

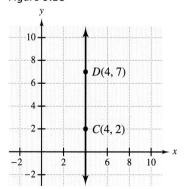

Analyzing Conditions

Sometimes a line is described in reference to one or more other lines. If enough information is furnished about the given line(s), we can deduce the features of the required line and write its equation.

EXAMPLE 6   *Describing a Line Relative to Other Lines*

Figure 5.29

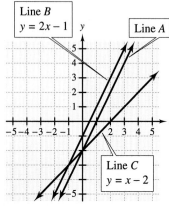

The equation of line B is $y = 2x - 1$ and the equation of line C is $y = x - 2$. Line A is parallel to line B and has the same y-intercept as line C. (See Fig. 5.29.) Write an equation of line A.

Solution   The equation of line B is $y = 2x - 1$, which means that the slope of line B is 2. Because line A is parallel to line B, the slope of line A is also 2. Thus we can begin writing the equation of line A.

$$y = mx + b$$
$$y = 2x + b \qquad \text{Replace } m \text{ with 2.}$$

The equation of line C is $y = x - 2$, which means that the y-intercept of line C is $(0, -2)$. Because line A has the same y-intercept as line C, the y-intercept of line A is also $(0, -2)$. Now the equation of line A can be completed.

$$y = 2x - 2 \qquad \text{Replace } b \text{ with } -2.$$

■

EXAMPLE 7   *Horizontal and Vertical Lines*

The equation of line L is $y = -2$. Line M is perpendicular to line L and contains the point $(3, 4)$. (See Fig. 5.30.) Write an equation of line M.

Figure 5.30

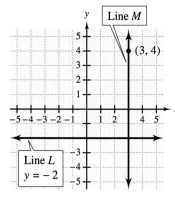

Solution   The equation of line L is of the form $y = $ constant. Therefore, line L is a horizontal line. Because line M is perpendicular to line L, line M must be a vertical line. Therefore, its equation is of the form $x = $ constant, where the constant is the x-coordinate of every point of the line.

Because $(3, 4)$ is a point of the line and its x-coordinate is 3, every point of the line has an x-coordinate of 3. The equation of line M is $x = 3$.

■

Sometimes we need to write a linear equation to model all or part of the conditions in an applied problem.

EXAMPLE 8   *Modeling with a Linear Equation*

At age 8, a boy competed in the high jump and managed a personal best of 3 feet. By age 12, the boy was jumping 5 feet. Because he was also interested in mathematics, he decided that his progress in the high jump could be modeled by a linear equation. He let x represent his age and y represent the height of his jump.

(a) What linear equation did the boy write?

(b) If he uses this equation to predict how high he can jump at age 20, what will his result be?

Solution

(a) If we write ordered pairs in the form (age, height) or (x, y), two solutions of the equation are $(8, 3)$ and $(12, 5)$.

The slope is $m = \frac{5-3}{12-8} = \frac{2}{4} = \frac{1}{2}$. Thus we can begin to write the equation.

$$y = \frac{1}{2}x + b$$

Use either ordered pair as replacements for x and y.

$$3 = \frac{1}{2} \cdot 8 + b \qquad \text{Replace } x \text{ with 8 and } y \text{ with 3.}$$

$$3 = 4 + b$$

$$b = -1$$

Now we can write the complete equation.

$$y = \frac{1}{2}x - 1$$

(b) The boy wants to know the height y he will jump when his age x is 20.

$$y = \frac{1}{2} \cdot 20 - 1 = 10 - 1 = 9 \text{ feet}$$

Using the linear equation as a model, he predicts that he will be able to jump 9 feet at age 20. ▪

NOTE: Try using the model in Example 8 to predict the height the boy will be able to jump at age 60. Because of the nature of a given problem, a mathematical model may be reasonably accurate only for a limited domain. ■

Linear Regression

Regression analysis is a process of fitting a graph as closely as possible to two or more given points. If there are only two points, the best **regression model** is a linear equation.

 ERASE DATA

ENTER DATA

LINREG

By entering the coordinates of the two points in your calculator, you can have the calculator determine the equation of the line. The calculator does everything we have done: It calculates the slope and y-intercept of the line, and it forms the equation.

EXAMPLE 9   *Using Linear Regression to Write an Equation of a Line*

Two points of a line are $P(75, 45)$ and $Q(120, 54)$. Use the linear regression capability of your calculator to determine an equation of the line containing P and Q. Verify your result algebraically.

Solution   Clear the calculator's memory of old data and enter the coordinates of P and Q as new data. Now execute the linear regression procedure. The calculator will display either the linear equation or the information (slope and y-intercept) needed to write the equation $y = 0.2x + 30$.

Using points P and Q, we can calculate the slope algebraically: $m = \frac{54 - 45}{120 - 75} = \frac{9}{45} = 0.2$.

$$y = mx + b$$
$$y = 0.2x + b \qquad \text{Replace } m \text{ with } 0.2.$$
$$45 = 0.2(75) + b \qquad \text{Using point } P, \text{ replace } x \text{ with } 75 \text{ and } y \text{ with } 45.$$
$$45 = 15 + b$$
$$30 = b$$

Now we can complete the equation: $y = 0.2x + 30$, the same result obtained from the linear regression procedure.

5.4   *Quick Reference*

Slope-Intercept Form

- The **slope–intercept form** $y = mx + b$ can be used as a model for writing an equation of a line.

- If the slope m and y-intercept b of a line are given, we can write an equation of the line by substituting directly into the slope–intercept form $y = mx + b$.

- We can also use the slope–intercept form if the slope and a point (x_1, y_1) of the line are given. Replace m with the given slope and replace x and y with x_1 and y_1, respectively. Then, we can solve for b.

Point-Slope Form

- An equivalent model for writing a linear equation is the **point–slope form** $y - y_1 = m(x - x_1)$. The point–slope form is particularly useful when you are trying to write an equation of a line for which the given point is not the y-intercept.

- If two points of a line are given, we can write an equation of the line by using the slope formula to calculate the slope. Then, using the point–slope form, we can replace m with the slope and x_1 and y_1 with the coordinates of either of the given points.

Special Cases

- The equation of a horizontal line can be found with either the slope–intercept form or the point–slope form, but it is much easier to use the model $y = $ constant. The constant is the y-coordinate of any point of the line.

- The equation of a vertical line cannot be found with the slope–intercept or point–slope models because the slope is undefined. Instead, use the model $x =$ constant, where the constant is the x-coordinate of any point of the line.

Analyzing Conditions

- If a nonhorizontal or nonvertical line is described in reference to one or more other lines, we use the given information to determine the slope of the required line and one of its points. Then, we can write the equation of the line with the slope–intercept or point–slope model.

- If the given conditions indicate that a line is horizontal or vertical, then we can write the equation of the line with the simpler models $y =$ constant or $x =$ constant.

Linear Regression

- If two points of a line are given, we can use the **linear regression** capability of a calculator to determine the equation of the line.

5.4   *Exercises*

 1. What information is needed to write an equation of a line with the slope–intercept form?

 2. Why can the slope–intercept form be used to write an equation of a horizontal line but not an equation of a vertical line?

In Exercises 3–6, write the equation of the line.

3. $m = -4, b = -3$ **4.** $m = 3, b = 4$

5. $m = \dfrac{1}{2}, b = 3$ **6.** $m = -\dfrac{2}{3}, b = -2$

In Exercises 7–12, write an equation of the line having the given slope and containing the given y-intercept.

7. $m = -5, (0, 1)$ **8.** $m = 6, (0, -6)$

9. $m = -\dfrac{3}{4}, (0, -5)$ **10.** $m = \dfrac{5}{8}, (0, 8)$

11. $m = 7, \left(0, -\dfrac{2}{3}\right)$ **12.** $m = -4, \left(0, \dfrac{1}{2}\right)$

In Exercises 13–18, write an equation for the line whose graph is shown in the figure.

13.
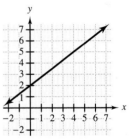

14.
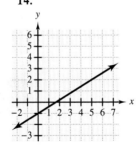

15.
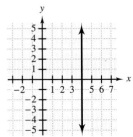

16.
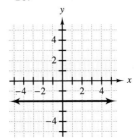

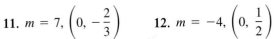

17.

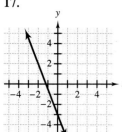

18.

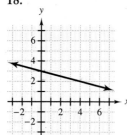

 19. Why can the point–slope form be used to write an equation of a horizontal line but not an equation of a vertical line?

 20. What information is needed to write an equation of a line with the point–slope form?

In Exercises 21–30, write an equation of the line with the given slope and containing the given point. Express the equation in the form $y = mx + b$ if possible.

21. $m = 2$, $(1, 3)$ **22.** $m = -3$, $(2, 5)$

23. $m = \dfrac{2}{5}$, $(10, -4)$ **24.** $m = -\dfrac{1}{3}$, $(-6, 3)$

25. $m = -\dfrac{2}{3}$, $(-2, -5)$ **26.** $m = \dfrac{4}{5}$, $(-1, -2)$

27. $m = 0$, $(2, -5)$ **28.** $m = 0$, $(-3, -4)$

29. m is undefined, $(-3, 4)$

30. m is undefined, $(-4, 5)$

In Exercises 31–36, write an equation of the line with the given slope and containing the given point. Express the equation in the form $Ax + By = C$.

31. $m = 1$, $(7, 4)$ **32.** $m = 2$, $\left(2, \dfrac{1}{2}\right)$

33. $m = \dfrac{5}{3}$, $(2, 0)$ **34.** $m = -2$, $\left(-1, \dfrac{2}{3}\right)$

35. $m = -\dfrac{1}{4}$, $\left(\dfrac{1}{2}, -1\right)$

36. $m = -\dfrac{5}{6}$, $(-3, 3)$

 37. If we are given two points of a line but not the slope of the line, how can we use the point–slope or slope–intercept model to write an equation of the line?

 38. If a line contains $A(2, -3)$ and $B(2, 5)$, why is the point–slope form not an appropriate model for writing an equation of the line?

In Exercises 39–52, write an equation of the line that contains the given pair of points.

39. $(3, 4)$ and $(1, 8)$ **40.** $(4, 7)$ and $(6, 11)$

41. $(-2, 7)$ and $(6, -9)$ **42.** $(5, -9)$ and $(-1, 9)$

43. $(2, 3)$ and $(-3, -2)$ **44.** $(12, 1)$ and $(4, -1)$

45. $(2, -3)$ and $(-4, 1)$ **46.** $(3, -2)$ and $(-5, 4)$

47. $(-4, 2)$ and $(-4, 5)$ **48.** $(2, -3)$ and $(2, 4)$

49. $(-3, -4)$ and $(1, -2)$ **50.** $(5, -4)$ and $(2, -2)$

51. $(-3, 5)$ and $(4, 5)$ **52.** $(2, -3)$ and $(5, -3)$

 53. Describe the equation of a line that is perpendicular to a horizontal line.

 54. Describe the equation of a line that is parallel to the y-axis.

In Exercises 55–66, write an equation of the described line.

55. The line is horizontal and contains the point $(3, -4)$.

56. The line is horizontal and contains the point $(-5, 6)$.

57. The line contains the point $(-2, 5)$ and has the same slope as the line whose equation is $y = 3$.

58. The line contains the point $(5, -3)$ and has the same slope as the line whose equation is $y = -5$.

59. The line contains the point $(-2, -3)$ and has the same slope as the line whose equation is $2x + 3y = 1$.

60. The line contains the point $(3, 5)$ and has the same slope as the line whose equation is $3x - 4y = 12$.

61. The line is vertical and contains the point $(-2, 5)$.

62. The line is vertical and contains the point $(7, -4)$.

63. The line contains the point $(1, 5)$ and has the same y-intercept as the line whose equation is $2x + 3y = 6$.

64. The line contains the point $(2, 3)$ and has the same y-intercept as the line whose equation is $3x - 2y = 8$.

65. The line contains the point $(-2, -3)$ and has an undefined slope.

66. The line contains the point $(3, 5)$ and has an undefined slope.

In Exercises 67–70, use the given information to write an equation of L_1.

67. The equation of L_2 is $y = 4x - 7$. L_1 is parallel to L_2, and its y-intercept is $(0, 2)$.

68. The equation of L_2 is $2x - 3y = 1$. L_1 is perpendicular to L_2, and its y-intercept is $(0, -4)$.

69. The equation of L_2 is $y = 4$. L_1 is perpendicular to L_2 and contains the point $(-2, 7)$.

70. The equation of L_2 is $x = -1$. L_1 is parallel to L_2 and contains the point $(3, 5)$.

71. If L_1 and L_2 are perpendicular and the slope of L_1 is m_1, write a formula for m_2.

72. If the slope of L_1 is m_1, what additional information is needed in order to write an equation of L_1?

In Exercises 73 and 74, L_1 is perpendicular to L_2. Refer to the figures and write an equation for L_1 and L_2.

73.

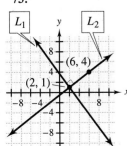

74.

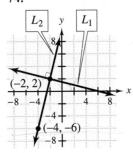

75. Consider the equations of L_1 and L_2.

L_1: $y = 3x - 7$ $\qquad$ L_2: $x + y = 3$

Write an equation of L_3 where L_3 is parallel to L_2 and L_3 has the same y-intercept as L_1.

76. Consider the equations of L_1 and L_2.

L_1: $y = 3x - 7$ $\qquad$ L_2: $x + y = 3$

Write an equation of L_3 where L_3 is perpendicular to L_1 and L_3 has the same y-intercept as L_2.

In Exercises 77 and 78, the equations of L_1 and L_2 are given. Write an equation of L_3 and an equation of L_4 according to the given conditions.

77. L_1: $3x + 4y = 12$ $\qquad$ L_2: $x - y = 3$

    (a) L_3 is perpendicular to L_2 and has the same y-intercept as L_1.

    (b) L_4 is parallel to L_1 and has the same y-intercept as L_2.

78. L_1: $x - 2y = 6$ $\qquad$ L_2: $4 - y = x$

    (a) L_3 is parallel to L_1 and has the same y-intercept as L_2.

    (b) L_4 is perpendicular to L_1 and has the same x-intercept as L_2.

In Exercises 79–86, determine the value of k for which the given condition is met.

79. The line $kx + 3y = 10$ is parallel to the line $y = 3 - 2x$.

80. The line $3y + 2kx - 8 = 0$ is parallel to the line $y = -2 + x$.

81. The line $2x - ky = 2$ is perpendicular to the line

$$y = \frac{3 - 4x}{2}.$$

82. The line $x = 5 - 3ky$ is perpendicular to the line

$$y = \frac{15 - 2x}{-3}.$$

83. The line $x + 2y = 6k$ has the same y-intercept as

    $$y = \frac{1 - x}{-2}.$$

84. The line $3x - 5y - k = 0$ has the same y-intercept as $y = 2 - 4x$.

85. The line $x - y = 2k$ has the same x-intercept as $y = 4x - 8$.

86. The line $kx - 5y - 20 = 0$ has the same x-intercept as $y = 3x + 12$.

87. In 1994 a person bought a computer for $1500. She estimated the useful life of this computer to be 7 years with a remaining value of $100 at the end of this period. She used the straight line (linear) method of depreciation to estimate the loss in value of this equipment over time.

    (a) Write the equation for the value v of the computer t years from 1994. (Let $t = 0$ for year 1994.)

    (b) Use the equation in part (a) to determine the value of the computer in 1998.

88. The population of a small town grew from 35 in 1980 to 62 in 1990. Assuming the growth is linear, write an equation to describe the population for any year. What is the rate of change in the population?

89. The net loans at a certain bank increased from $92.1 million in 1990 to $105.7 million in 1992. Using a linear equation, predict the net loans in 1999. What is the rate of change of the net loans?

90. A library needs one children's librarian for each 1000 children attending story hour during a calendar quarter. The number of children attending story hour increased from 1153 during the first quarter of 1991 to 1732 during the same quarter of 1992. Using a linear equation, determine when the library can fully support three children's librarians.

Exploring with Real Data

91. Archer Daniels Midland reported that the shareholder equity per common share in 1987 was $6.82 and in 1993 was $14.27. Assuming the growth in shareholder equity was linear, determine an equation to predict the shareholder equity in any year. Using the equation, what is the predicted shareholder equity per common share in 2000? (Source: Archer Daniels Midland Annual Report 1993.)

92. Coca-Cola Enterprises reported the net income per share as $0.63 in 1987 and $-$1.45 in 1992. Write a linear equation to describe the income per share in any year. In 1989 the net income per share was 41 cents. How close is the linear model to the actual value? (Source: Coca-Cola Enterprises Annual Report 1992.)

93. The deposits at Etowah Bank in Georgia grew from $130 million in 1988 to $195 million in 1992. Assume a linear relation and write an equation for deposits in any year. Use the equation to predict the deposits in the bank in 1999. (Source: Etowah Bank.)

94. During the first quarter of 1991, the circulation at the Gilmer Library was 11,824 volumes. During the same quarter of 1992, the circulation was 13,215 volumes. The librarian must project circulation for the next 3 years to justify a request for increased funding. Using a linear equation, predict the circulation for the first quarter of each of the next 3 years. (Source: Sequoyah Regional Library Statistical Report.)

In Exercises 95–98, use the linear regression capability of your calculator to determine an equation of the line containing P and Q.

95. $P(75, 30)$ and $Q(-245, -130)$

96. $P(692, -536)$ and $Q(-108, 64)$

97. $P(-173, -212)$ and $Q(-573, 288)$

98. $P(250, 140)$ and $Q(450, -20)$

In Exercises 99 and 100, use the linear regression capability of your calculator to determine an equation of the line containing P and Q and of the line containing P and R. Then state whether P, Q, and R are collinear and explain why.

99. $P(-3, 1)$, $Q(6, 4)$, $R(-9, -1)$

100. $P(-1.9, -4.9)$, $Q(1.1, 4.1)$, $R(2.9, 9.8)$

Challenge

101. If (x_1, y_1) and (x_2, y_2) are points of a line, show that an equation of the line can be written as

$$y - y_1 = \frac{y_2 - y_1}{x_2 - x_1}(x - x_1).$$

102. Show that the equation of the line with intercepts $(a, 0)$ and $(0, b)$ can be written as

$$\frac{x}{a} + \frac{y}{b} = 1, \ a \neq 0 \text{ and } b \neq 0.$$

In Exercises 103–106, use the results of Exercise 102 to determine the intercepts directly from the given equation.

103. $\dfrac{x}{3} + \dfrac{y}{2} = 1$

104. $\dfrac{x}{-4} - \dfrac{y}{2} = 1$

105. $\dfrac{2x}{5} - \dfrac{3y}{4} = 1$

106. $4x + 3y = 12$

107. Write an equation of the line that contains the point (a, b) and is perpendicular to the line $ax + by = c$.

108. Write an equation of the line that contains the point (b, a) and is parallel to the line whose equation is $ax + by = c$.

5.5   Graphs of Linear Inequalities

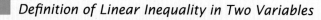

Linear Inequalities   ▪   Graphs of Solution Sets   ▪   Graphs of Compound Inequalities

Linear Inequalities

The definition of a linear inequality in two variables is very similar to the definition of a linear equation in two variables.

> ### Definition of Linear Inequality in Two Variables
>
> A **linear inequality in two variables** is an inequality that can be written in the form $Ax + By < C$, where A, B, and C represent any real numbers, and A and B are not both zero. Similar definitions can be stated with $>$, $\geq$, and $\leq$.

The following are some examples of linear inequalities along with their standard forms.

Inequality	Standard Form
$2x - 3y < 1$	$2x - 3y < 1$
$y \leq 4x - 1$	$-4x + y \leq -1$
$x > 3$	$1x + 0y > 3$
$y + 3 \geq 5$	$0x + 1y \geq 2$

> ### Definitions of a Solution and a Solution Set
>
> A **solution** of a linear inequality in two variables is a pair of numbers (x, y) that makes the inequality true. The **solution set** is the set of all solutions of the inequality.

As is true with equations, it is a simple matter to determine whether a particular pair of numbers is a solution of a linear inequality.

EXAMPLE 1   *Verifying Solutions of an Inequality*

Given the linear inequality $y \leq 4x - 3$, determine whether the following are solutions.

(a) $(0, -3)$ (b) $(1, 2)$

Solution

(a) $y \leq 4x - 3$

$-3 \leq 4(0) - 3$ Replace x with 0 and y with -3.

$-3 \leq -3$ True

The pair $(0, -3)$ is a solution.

(b) $y \leq 4x - 3$

$2 \leq 4(1) - 3$ Replace x with 1 and y with 2.

$2 \leq 1$ False

The pair $(1, 2)$ is not a solution.

Graphs of Solution Sets

The **graph** of the solution set of an inequality is a picture of all of the solutions of the inequality.

EXPLORATION 1   *Graphing a Solution Set*

Consider the inequality $y \leq x - 9$. Use the integer setting to produce the graph of the companion equation $y = x - 9$.

Figure 5.31

X=8   Y=15

(a) Do the points of the line represent solutions of the inequality $y \leq x - 9$?

(b) Move the general cursor to various points *above* the line. (See Fig. 5.31.) Are these points solutions of the inequality?

(c) Now move the general cursor to various points *below* the line. (See Fig. 5.32.) Are these points solutions of the inequality?

(d) What is your conjecture about the set of points that represents the solutions of the inequality?

Figure 5.32

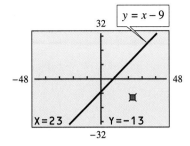

Discovery

(a) Because the inequality is satisfied by any ordered pair for which $y = x - 9$, the points of the line represent solutions.

(b) No matter what point you select above the line, the coordinates do not satisfy the inequality.

(c) Every point you select below the line has coordinates that satisfy the inequality.

(d) The solution set of the inequality is represented by all points of the line and below the line.

To depict the solutions described in Exploration 1, we shade the region below the line.

We call the line shown in Fig. 5.33 the **boundary line.** It divides the entire coordinate plane into two regions called **half-planes.** As we shall see, a boundary line may or may not be part of a half-plane.

Figure 5.33

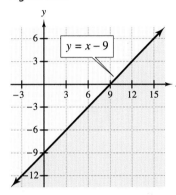

We can select any point on either side of the boundary line and test the coordinates of that point to see whether they satisfy the inequality. For this reason, we call the selected point a **test point.**

If the coordinates of a test point satisfy the inequality, then we shade the half-plane in which the test point lies. Otherwise, we shade the other half-plane.

The boundary line is a solid line if the points of the line represent solutions. If the points of the line do not represent solutions, then the boundary line is a dashed line.

The following is a summary of the method for sketching the graph of a linear inequality.

Sketching the Graph of a Linear Inequality

1. Either with a calculator or by hand, produce the graph of the companion equation. The boundary line is solid for $\leq$ and $\geq$ inequalities and dashed for $<$ and $>$ inequalities.

2. Select any test point in either half-plane (not on the boundary line). By substitution, determine whether the point represents a solution. If the test point represents a solution, shade the half-plane in which the test point lies. If the test point does not represent a solution, shade the other half-plane.

When a linear inequality is in the form $y < mx + b$, all points below the boundary line represent solutions. Similarly, when a linear inequality is in the form $y > mx + b$, all points above the boundary line represent solutions. This suggests an alternative graphing technique.

1. Solve the inequality for y.

2. Draw the boundary line as in step 1 of the procedure for sketching the graph of a linear inequality.

3. (a) For inequalities of the form $y < mx + b$ or $y \leq mx + b$, shade the region below the line.

   (b) For inequalities of the form $y > mx + b$ or $y \geq mx + b$, shade the region above the line.

For the special cases $x <$ constant, $x \leq$ constant, $x >$ constant, and $x \geq$ constant, the method is as follows.

1. Draw the vertical boundary line.

2. For inequalities of the form $x <$ constant or $x \leq$ constant, shade to the left of the line.

3. For inequalities of the form $x >$ constant or $x \geq$ constant, shade to the right of the line.

EXAMPLE 2   *Graphing Linear Inequalities*

Sketch the graph of each inequality.

(a) $y > 2x$ (b) $2x - 3y \geq 12$

(c) $y - 5 \leq 2$ (d) $x < -2$

Solution

(a) First, draw the graph of $y = 2x$ as a dashed line. Because the inequality symbol is $>$, we shade above the line. (See Fig. 5.34.)

Figure 5.34   Figure 5.35

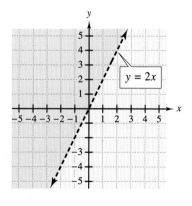

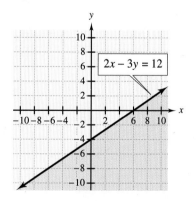

(b) Solve the inequality for y.

$$2x - 3y \geq 12$$

$$-3y \geq -2x + 12 \qquad \text{Subtract } 2x \text{ from both sides.}$$

$$\frac{-3y}{-3} \leq \frac{-2x}{-3} + \frac{12}{-3} \qquad \begin{array}{l}\text{Divide both sides by } -3 \text{ and reverse the inequality} \\ \text{symbol.}\end{array}$$

$$y \leq \frac{2}{3}x - 4$$

Draw the solid line $y = \frac{2}{3}x - 4$. Because the inequality symbol is $\leq$, the solutions are below the line. (See Fig. 5.35.)

(c) Solve the inequality for y.

$$y - 5 \leq 2$$
$$y \leq 7$$

The boundary line is the solid horizontal line $y = 7$. Because the inequality symbol is $\leq$, we shade below the line. (See Fig. 5.36.)

Figure 5.36

Figure 5.37

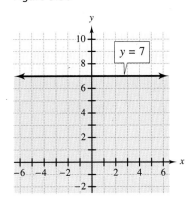

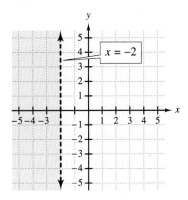

(d) Draw the dashed vertical line $x = -2$. Observe that the inequality symbol is $<$, so we shade to the left of the line. (See Fig. 5.37.) ∎

We can use a graphing calculator to produce the graph of the solution set of a linear inequality.

EXAMPLE 3 *Graphing a Linear Inequality with a Calculator*

Use your calculator to graph the solution set of $3x + 2y \leq -10$.

Solution We must first solve the inequality for y.

$$3x + 2y \leq -10$$
$$2y \leq -3x - 10$$
$$y \leq -\frac{3}{2}x - 5$$

SHADE

Figure 5.38

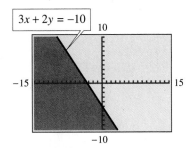

Graphs of Compound Inequalities

In Section 4.7 we solved compound inequalities in one variable. Linear inequalities in two variables can also be connected with *and* or *or* to form compound inequalities.

To produce the graph of a compound inequality, we graph each inequality on the same coordinate system. For inequalities connected with *and* (a conjunction), we shade the intersection of the solution sets, and for inequalities connected with *or* (a disjunction), we shade the union.

EXAMPLE 4   *Graphing a Conjunction*

Graph the compound inequality $y \geq 2$ and $y \leq 2x + 5$.

Solution   The graph of $y \geq 2$ is the region on and above the line $y = 2$. The graph of $y \leq 2x + 5$ is the region on and below the line $y = 2x + 5$. (See Fig. 5.39.)

Figure 5.39

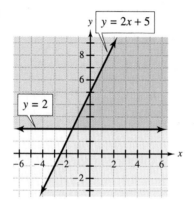

Because the inequalities are connected with *and*, the graph of the compound inequality is the intersection of the two graphs. (See Fig. 5.40.)

Figure 5.40

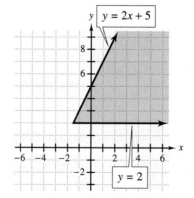

EXAMPLE 5   *Graphing a Disjunction*

Graph the compound inequality $y \geq x - 3$ or $2x + y < 1$.

Solution   The graph of $y \geq x - 3$ is the region on and above the line $y = x - 3$. The graph of $2x + y < 1$ is the region below the line $y = -2x + 1$. (See Fig. 5.41.)

Figure 5.41   Figure 5.42

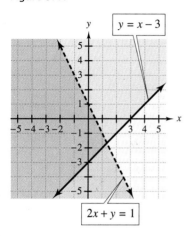   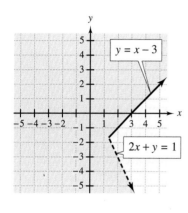

Because the inequalities are connected with *or*, the graph of the compound inequality is the union of the two graphs. (See Fig. 5.42.)   ■

To graph inequalities involving absolute value, we must recall the following for any expression A and any positive number c:

1. If $|A| < c$, then $-c < A < c$.

2. If $|A| > c$, then $A < -c$ or $A > c$.

Similar rules hold for the symbols $\leq$ and $\geq$.

EXAMPLE 6   *Graphing an Absolute Value Inequality*

Graph the absolute value inequality $|x + y| > 2$.

Figure 5.43

Solution   The inequality means $x + y > 2$ or $x + y < -2$. We solve both inequalities for y to obtain $y > -x + 2$ or $y < -x - 2$.

The graph of $y > -x + 2$ is the region above the line $y = -x + 2$. The graph of $y < -x - 2$ is the region below the line $y = -x - 2$.

Because the inequalities are connected with *or*, the graph of the absolute inequality is the union of the two graphs. (See Fig. 5.43.)   ■

EXAMPLE 7   *Graphing an Absolute Value Inequality*

Graph the absolute value inequality $|y + x| \le 4$.

Solution   The absolute value inequality means that $y + x \le 4$ and $y + x \ge -4$. Solving each inequality for y, we obtain $y \le -x + 4$ and $y \ge -x - 4$.

The graph of $y \le -x + 4$ is the region below the line $y = -x + 4$. The graph of $y \ge -x - 4$ is the region above the line $y = -x - 4$. (See Fig. 5.44.)

Figure 5.44   Figure 5.45

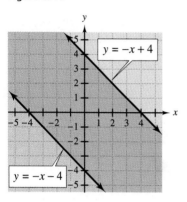   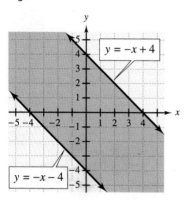

Because the inequalities are connected with *and*, the graph of the absolute value inequality is the intersection of the two graphs. (See Fig. 5.45.) ■

5.5   *Quick Reference*

Linear Inequalities

- A **linear inequality in two variables** is an inequality that can be written in the form $Ax + By < C$, where A, B, and C represent any real numbers, and A and B are not both zero. A similar definition can be stated with $>$, $\ge$, and $\le$.

- A **solution** of a linear inequality in two variables is a pair of numbers (x, y) that makes the inequality true. The **solution set** is the set of all solutions of the inequality.

Graphs of Solution Sets

- The **graph** of a linear inequality in two variables is a picture of its solutions. It consists of a boundary line (solid if it represents solutions, dashed if it does not) and a shaded region (a **half-plane**) on one side of the boundary line.

- The recommended method for graphing a linear inequality is as follows.

  1. Solve the inequality for y and draw the solid (for $\ge$ and $\le$) or dashed (for $>$ and $<$) boundary line that represents the corresponding equation.

  2. Determine the region to shade by one of these methods.

     (a) If the coordinates of a test point satisfy the inequality, shade the half-plane in which the test point lies; otherwise, shade the other half-plane.

(b) For $<$ or $\le$ inequalities, shade below the boundary line; for $>$ or $\ge$ inequalities, shade above the boundary line.

Graphs of Compound Inequalities

- Two or more linear inequalities in two variables can be connected by *and* (conjunction) or *or* (disjunction).

- To sketch a graph of a compound inequality, draw the graph of each component of the inequality on the same coordinate system. For a conjunction, shade the intersection of the two graphs; for a disjunction, shade the union of the two graphs.

- To sketch a graph of an absolute value inequality, translate the inequality into a conjunction or disjunction, and graph accordingly.

5.5 Exercises

 1. How do the solution sets of $x + 2y < 6$ and of $x + 2y \le 6$ differ?

2. Does the inequality $y > x + 2$ define a function? Explain.

In Exercises 3–8, determine whether each ordered pair is a solution of the given inequality.

3. $y < 3x - 5$; (1, 0), (0, −6), (2, 1)

4. $y \ge 1 - 2x$; (5, −9), (−3, 8), (2, −7)

5. $2x - 7y \ge 12$; (12, 1), (6, 0), (−4, −2)

6. $4 - 3x - 5y < 0$; (4, 3), (−5, −4), (7, 3.4)

7. $y \le 4$; (1, 10), (0, 4), (4, 5)

8. $x > -6$; (−5, −7), (−5.99, 3), (−6.01, 4)

 9. For the inequality $x + 3 > 7$, we might write the solution set as $(4, \infty)$, or we might graph the solution set as a shaded half-plane. What assumptions would we be making in each case?

10. When graphing a linear inequality, when do we use a solid line and when do we use a dashed line for the boundary line?

In Exercises 11–16, the figure shows the boundary line for the graph of the given inequality. Complete the graph by shading the correct half-plane.

11. $y \le 4x - 3$

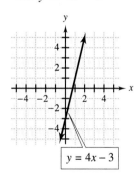

12. $y \ge -2x$

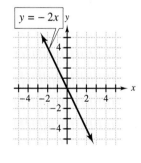

13. $2x + y > 6$

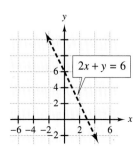

14. $3y - 2x < 9$

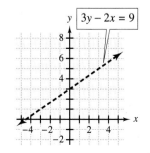

15. $y < -3$

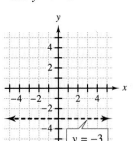

16. $2y - x \geq 0$

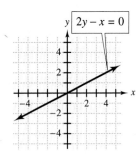

In Exercises 39–44, write an inequality for the given graph.

39.

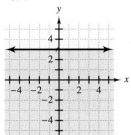

40.

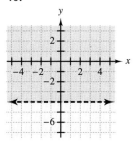

 17. How do we determine which half-plane to shade by

  (a) testing a point?

  (b) observing the inequality symbol?

 18. To sketch the graph of $x - 2y < 8$, you must draw the boundary line and then decide which half-plane to shade. A classmate tells you to shade below the line because the inequality symbol is $<$. Is your classmate correct? Why, or why not?

41.

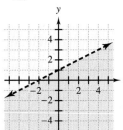

42.

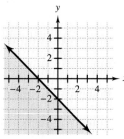

43.

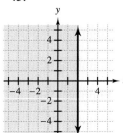

44.

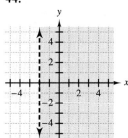

In Exercises 19–38, graph the inequality.

19. $y < 7 - 3x$

20. $y \geq \dfrac{1}{2}x - 5$

21. $-y \leq -2x$

22. $y \geq x$

23. $3x + 2y \leq 6$

24. $4x + 5y \geq 20$

25. $-2y \geq 4x - 7$

26. $2x \geq 3 - y$

27. $-6 \leq x - 2y$

28. $x - 3y \leq 0$

29. $2x - 5y \geq 10$

30. $x - 6 < -y$

31. $y - 3 \leq -2(x - 3)$

32. $y + 2 - 4(x - 1) > 0$

33. $\dfrac{x}{5} + \dfrac{y}{2} < 1$

34. $\dfrac{x}{3} - \dfrac{y}{2} \geq 1$

35. $y \geq 20$

36. $y - 3 \leq 2$

37. $x < 5$

38. $2x + 6 \geq 0$

In Exercises 45–50, determine an ordered pair of numbers that satisfies the given condition. Then, translate the given information into a linear inequality, produce its graph, and test whether your pair is a solution.

45. One number x is at least 1 more than another number y.

46. One number y is at most twice another number x.

47. One number y is not less than 4 less than 3 times another number x.

48. One number x is no greater than half of another number y.

49. The sum of a number x and twice another number y exceeds 10.

50. The difference of a number x and two-thirds of another number y exceeds 7.

In Exercises 51 and 52, when you produce the graph, be aware that the variables represent nonnegative numbers.

51. The main hold of a supply ship has an area of 6000 square feet. Pallets of two sizes are used to store materials in the hold. One pallet is 4 feet by 4 feet, and the other is 3 feet by 5 feet. Write a linear inequality that describes the number of pallets that can be used and produce its graph.

52. A hospital food service can serve at most 1000 meals per day. Patients on a normal diet receive three meals each day, and patients on special [...]lay. Write a linear [...]umber of patients [...] its graph.

[...]nction.

$[...] > -3$ and $x > 2$

[...]nction.

$[...] x \geq 2$ or $y > -3$

[...]graph of the given

1

67. $3x + y < 5$ or $x - y < 2$

68. $3x + y < 5$ and $x - y < 2$

69. $2x - y - 5 < 0$ and $4x - 2y > 0$

70. $\dfrac{y}{2} - \dfrac{x}{3} \leq 1$ or $\dfrac{x}{3} + \dfrac{y}{2} \geq 1$

71. $5y + 3x < 30$ and $5y + 3x < -20$

72. $2x - 3y < 9$ or $2x - 3y < -12$

73. $x \geq y$ and $x \leq y$

74. $x - 3 > y$ or $x - 3 < y$

75. $x \geq 0$, $y \geq 0$, and $y \leq 10 - x$

76. $x \geq 0$, $y \geq 0$, and $y \geq 12 - 3x$

In Exercises 77–82, write a compound inequality that describes the graph.

77.

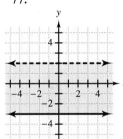

78.

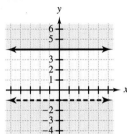

79.

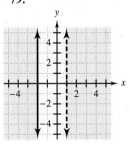

80.

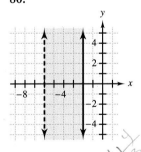

81.

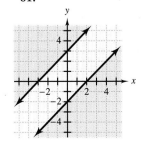

82.

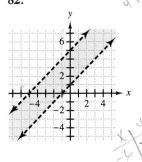

In Exercises 83 and 84, when you produce the graph, be aware that the variables represent nonnegative numbers.

83. A county animal shelter houses only cats and dogs. The maximum number of animals that can be kept is 100, but the county will close the shelter if the population falls below 20. Write a compound inequality that describes these conditions and produce its graph.

84. Two nonnegative numbers have a sum of at least 50 or a difference of at most 10. Write a compound inequality that describes these conditions and produce its graph. Use the graph to show that not all nonnegative pairs of numbers satisfy the given conditions.

In Exercises 85–96, sketch the graph of the given absolute value inequality.

85. $|y| \leq 3$

86. $|x| \geq 4$

87. $|x - 2y| > 4$

88. $|2x + y| < 5$

89. $|y - x| \geq 3$

90. $|y + x| \leq 6$

91. $|x| \leq -2$

92. $|y| \geq -1$

93. $|x - y| \leq 0$

94. $|2x + 3y| \leq 0$

95. $|x + y| > 0$

96. $|y - 2x| > 0$

In Exercises 97–100, graph the given compound inequality.

97. $|x| > 3$ and $|y| \geq 2$ **98.** $|x| < 5$ or $|y| < 3$

99. $|x| > 3$ and $|y| < -2$

100. $|y| < 4$ or $|x| < 0$

101. Two basketball teams are considered evenly matched if the difference in their scores does not exceed 5 points. Write an absolute value inequality to describe this condition and produce its graph.

102. A community organization has determined that social harmony among its members is greatest when the difference between the number of male and female members is less than 10. Write an absolute value inequality to describe this condition and produce its graph.

Challenge

In Exercises 103–106, graph the given compound inequality.

103. $|x + y| > -2$ or $|x| < -5$

104. $|y + 2x| < -1$ and $|y| > -1$

105. $y > |x + 5|$

106. $x \geq |2x + y|$

In Exercises 107–114, draw an example of a graph of the given inequality.

107. $y < b, b > 0$ **108.** $x \geq b, b < 0$

109. $y \geq mx, m < 0$ **110.** $y \leq mx, m > 0$

111. $y < mx + b, m > 0, b < 0$

112. $y > mx + b, m < 0, b > 0$

113. $ax + by \leq ab, a > 0, b < 0$

114. $2x + 3y \geq 12b, b < 0$

 TEST

You may be able to use your calculator to determine if a given ordered pair is a solution of a linear inequality. The following is a typical routine for testing $(0, -3)$ in the inequality $y \leq 4x - 3$.

1. Enter the function Y_1: $y \leq 4x - 3$ on the function screen.

2. Store 0 in X and -3 in Y.

3. Evaluate Y_1. The calculator returns a 1 to indicate that the inequality is true for $(0, -3)$. The calculator returns a 0 if the given pair is not a solution of the inequality.

In Exercises 115–118, use this calculator routine to test the given ordered pairs in the inequality.

115. $y > -3x + 1$; $(-3, -2)$, $(1, 7)$

116. $y < x + 4$; $(1, -2)$, $(-5, 3)$

117. $y \leq 2x - 3$; $(1, 0)$, $(2, -2)$

118. $y \geq -x - 2$; $(-1, 4)$, $(-4, -2)$

5   Chapter Review Exercises

Section 5.1

In Exercises 1–4, state whether the equation is a linear equation in two variables.

1. $3x = \dfrac{2}{y} + \dfrac{3}{2}$
           2. $3x - 8 = 15$

3. $x = \dfrac{y}{3} - \dfrac{4}{5}$
           4. $5 = \dfrac{-3xy}{4}$

In Exercises 5 and 6, verify that the given ordered pair is a solution of the given equation.

5. $(5, -11);\ y = -3x + 4$

6. $(-14, -6);\ 2x - 7y = 14$

In Exercises 7 and 8, find a, b, and c so that the ordered pairs are solutions of the given equation.

7. $y = 7 - x;\ (-2, a),\ (b, 14),\ (c, -5)$

8. $4y - 3x = 7;\ (1, a),\ (b, 4),\ (-9, c)$

In Exercises 9 and 10, complete the table of solutions of the given equation.

9. $2y - 3x = 9$

x	3	7	-7
y			

10. $2x - 4 = 3y$

x	2	11	-4
y			

In Exercises 11–14, determine the x- and y-intercepts.

11. $7x - 3y = 21$
           12. $y = -2x + 3$

13. $y - 2 = 18$
           14. $2x - 5 = 7$

In Exercises 15–18, write each equation in the form $y = mx + b$. Then determine the y-intercept.

15. $15 - 3x + 6y = 0$
      16. $4x - 5y = 19$

17. $2x - y = 8$
           18. $\dfrac{1}{2}x - \dfrac{2}{5}y = 6$

In Exercises 19–22, determine three solutions to the equation and sketch the graph.

19. $5x + 3y = 15$
           20. $\dfrac{1}{4}x - \dfrac{2}{5}y = 10$

21. $y + 3 = 9$
           22. $2x + 1 = 7$

23. Which of the following equations does the graph represent?

   (i) $3x + 4y = 12$
     (ii) $3x + 4y = -12$

   (iii) $3x - 4y = -12$

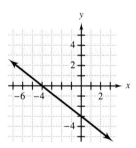

Section 5.2

24. In the accompanying figure name the line whose slope is

   (a) negative.
          (b) positive.

   (c) undefined.
        (d) zero.

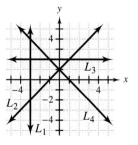

In Exercises 25–28, determine the slope of the line that contains the given points.

25. $(-5, -8)$ and $(2, -3)$

26. $(3, -4)$ and $(3, 5)$

27. $(2, -4)$ and $(-7, 6)$

28. $(-4, 5)$ and $(4, 5)$

In Exercises 29 and 30, write each equation in the form $y = mx + b$. Then, determine the slope.

29. $3x - 4y = 15$ 30. $x + 2y = 4$

In Exercises 31 and 32, draw a line through the given point with the given slope.

31. $(0, -4); m = -\dfrac{3}{5}$ 32. $(2, -3); m = 0.5$

In Exercises 33–36, use what you know about the slope and y-intercept to draw a sketch of the graph of the given equation.

33. $y = -x + 3$ 34. $3y + 2x = 3$

35. $y = 3$ 36. $x - 2 = 0$

Section 5.3

In Exercises 37–42, L_1 refers to line 1 with slope m_1 and L_2 refers to line 2 with slope m_2.

In Exercises 37–40, the slopes of L_1 and L_2 are given. Determine if L_1 and L_2 are parallel, perpendicular, or neither.

37. $m_1 = \dfrac{1}{4}; m_2 = 0.25$

38. $m_1 = 5; m_2 = -5$

39. $m_1 = \dfrac{1}{3}; m_2 = -3$

40. $m_1 = 0; m_2$ is undefined

In Exercises 41 and 42, the equations of L_1 and L_2 are given. Determine if the lines are parallel, perpendicular, or neither.

41. $L_1: y = 0.5x - 3$ $L_2: y = -2x + 4$

42. $L_1: \dfrac{2}{3}x + \dfrac{3}{5}y = 1$ $L_2: 15 - 10x = 9y$

43. Determine whether a triangle with vertices $A(-3, 6)$, $B(10, 0)$, and $C(2, -4)$ is a right triangle.

44. Determine whether the points $P(-4, 3)$, $Q(2, 1)$, and $R(5, -1)$ are collinear.

In Exercises 45 and 46, determine the rate of change of y with respect to x.

45. $y = -3x + 5$ 46. $y = \dfrac{2}{3}x + 4$

In Exercises 47 and 48, the constant rate of change of a line is given along with an increase or decrease in one variable. Determine the increase or decrease in the other variable.

47. $m = -5; x$ increased by 3

48. $m = \dfrac{5}{3}; y$ increased by 10

In Exercises 49 and 50, assume a constant rate of change.

49. The price of a ticket to a college football game increased from $5.00 in 1980 to $20.00 in 1992. What was the annual rate of change of the price of the ticket?

50. If the temperature fell from 20°F at 5:00 P.M. to −6°F at 6:00 A.M., what was the hourly rate of change in temperature?

In Exercises 51–54, determine whether the given equation represents a direct variation.

51. $y = 4x - 7$ 52. $y = \dfrac{x}{-3}$

53. $y = 7x$ 54. $xy = 4$

55. The number n of toothpicks produced by a machine varies directly with the number of hours h the machine is operating. If it produces 15,000 toothpicks in 6 hours, how many toothpicks are made in 45 hours by this one machine?

56. The per-share dividend that a company declares varies directly with the earnings per share. If a company pays a $1.50 per-share dividend when earnings are $4.00 per share, what would the earnings per share be for a dividend of $2.00 per share?

Section 5.4

In Exercises 57 and 58, write an equation of the line having the given slope and containing the given point.

57. $m = -3; (-2, 5)$ **58.** $m = \dfrac{4}{3}; (1, 2)$

In Exercises 59–62, write an equation of the line containing the given pair of points.

59. $(2, -4)$ and $(-5, 3)$

60. $(2, -5)$ and $(2, 4)$

61. $(-3, -6)$ and $(1, -1)$

62. $(-2, -5)$ and $(4, -5)$

In Exercises 63–68, write an equation of the line with the following conditions.

63. The line is horizontal and contains the point $(-3, 6)$.

64. The line is vertical and contains the point $(6, -3)$.

65. The line contains the point $(1, 5)$ and is parallel to the line whose equation is $2x + 3y = 6$.

66. The line contains the point $(2, -3)$ and is perpendicular to the line $2x - 3y = 7$.

67. The line is graphed in the accompanying figure.

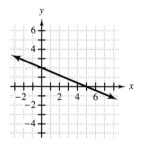

68. The line is perpendicular to the line whose equation is $y = 3x + 4$ and has the same y-intercept as the line whose equation is $y = 3x - 5$.

69. A store owner decides to expand the facilities when annual sales reaches $800,000. Sales were $150,000 in 1992 and $400,000 in 1994. Assuming the increase in sales is linear, predict the year in which the store owner will expand.

70. Suppose a public school system will lose its accreditation if its student–teacher ratio rises above 30:1. In 1993 the system had 1000 students and 40 teachers. During the following year, there were 1100 students and 45 teachers. Assuming a linear change in the student–teacher ratio, determine whether the system is headed for a loss of accreditation.

Section 5.5

In Exercises 71–74, determine whether each of the given ordered pairs is a solution of the given inequality.

71. $4x - y \geq -2$; $(3, 14), (-2, 0), (-4, -20)$

72. $2x - 3y < 0$; $(0, 0), (5, 4), (-4, -4)$

73. $y \leq -3$; $(1, 10), (0, -4), (4, 5)$

74. $x > 5$; $(-5, -7), (4.99, 5), (5.01, 8)$

In Exercises 75–78, sketch the graphs of the following inequalities.

75. $3x - 4y \leq 12$

76. $x \leq 2y$

77. $y + 4 \leq 7$

78. $8 - 2x \geq 0$

In Exercises 79–82, sketch the graph of the compound inequality.

79. $5y - 2x > 7$ or $y + 3x < -2$

80. $y \leq 2x - 3$ and $y \geq 4 - x$

81. $y \leq 4$ and $x + 2 \leq 0$

82. $x \leq -3$ or $y < 2x - 1$

In Exercises 83–86, sketch the graph of the absolute value inequality.

83. $|y| \geq 2$

84. $|4x - 3y| \geq 5$

85. $|x - 2| - 2 < 5$

86. $|3x + 2y| \leq -1$

87. At a minor league stadium, grandstand seats cost $6 and bleacher seats cost $4. The total receipts from the ticket sales for a certain game were less than $11,400. Write a linear inequality that describes the number of each kind of ticket sold.

88. An **isosceles triangle** is a triangle with two sides of equal length. The perimeter of a certain isosceles triangle is at most 42 inches. Write a linear inequality that describes the lengths of the sides of the triangle.

89. A community athletics council sponsors coed soccer teams. The number of boys and girls enrolled in the league cannot exceed 400, and there must be at least 100 girls. Write a compound inequality that describes the number of youngsters who can participate in the program.

90. In a survey a certain percentage of those polled were in favor of a local bond issue. The survey is known to be accurate to within 3 percentage points. Write an absolute value inequality that describes the relationship between the percentage of the sample who were in favor and the actual percentage of all voters who were in favor.

5 Chapter Test

1. Complete the ordered pairs so that they correspond to points on the graph of $2x - y = -2$.

   (a) (_____ , 0)   (b) (0, _____)

   (c) (−2, _____)   (d) (_____ , 3)

In Questions 2–5, sketch the graph of each equation. Give the x- and y-intercepts and the slope of each.

2. $y = x$ 3. $y = -x - 2$

4. $2x - 3y = 9$ 5. $x - 2 = 3$

6. Refer to the graph and fill in the blanks with the symbols $<$ or $>$:

   m _____ 0 and b _____ 0.

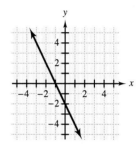

In Questions 7–12, determine an equation of the line that satisfies the given information.

7. The line contains the points (2, −3) and (−1, −5).

8. The line contains the point (−1, 3) and has a slope of 2.

9. The line is vertical and contains the point (2, 5).

10. The line contains the point (−1, 3) and is parallel to the line whose equation is $y = -4$.

11. The line contains the point (1, 4) and is perpendicular to the line whose equation is $y = -\frac{1}{2}x + 3$.

12. The graph of the line is shown in the figure.

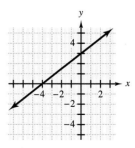

In Questions 13 and 14, graph the given inequalities.

13. $2x - y \leq 5$ 14. $y - 3 > 0$

In Questions 15 and 16, graph the given compound inequalities.

15. $y \geq 2x + 1$ and $x + y \geq 3$

16. $x - 2y < 2$ or $2y - x > 6$

In Questions 17 and 18, graph the given absolute value inequalities.

17. $|x + y| \leq 3$ **18.** $|x + 2y| > 5$

19. Show that the points $(1, -2)$, $(-2, 1)$, and $(5, 2)$ are the vertices of a right triangle.

20. Draw a line through $(-1, 3)$ with a slope of -2.

In Questions 21 and 22, determine if the lines are perpendicular, parallel, or neither.

21. $2x + 3y = 3$ and $2y - 3x = -4$

22. $y = 3x + 1$ and $x - 3y = -1$

23. The velocity of an object dropped from rest varies directly with the time of the fall. After 0.5 seconds, the speed is 16 feet per second. Write the relation between the velocity and the time. What is the velocity after 2 seconds?

24. The earnings for a company dropped from \$1.35 per share in 1990 to \$0.45 in 1992. Assuming that the relationship is linear, what was the rate of change in per-share earnings per year?

25. At 6:00 A.M., the temperature in Death Valley was 88°F. At 10:00 A.M., the temperature was 98°F. Assuming the rise in temperature is linear, write an equation to describe the rise in temperature. Use the equation to predict the temperature at 3:00 P.M.

6 Systems of Linear Equations

The manufacture and sales of toys is a multibillion-dollar industry in America. The bar graph shows the retail sales (in billions of dollars) of toys from 1988 to 1992.

The figure shows a curve drawn across the tops of the bars. This graph appears to be a good model for the retail sales of toys during this period. But what is the function whose graph has been drawn? To determine the function, we must solve a **system of equations.** (To learn more about writing this function, see Exercises 55–58 at the end of Section 6.5.)

In this chapter we introduce systems of linear equations in two and three variables. We discuss both graphic and algebraic methods for solving systems. These methods include Cramer's Rule, for which we need to learn about matrices and determinants. After using our new skills to solve application problems, we conclude with the topic of systems of linear inequalities.

RETAIL SALES OF TOYS

Sales (in billions)

$20, 16, 12, 8, 4

12.97 13.10 13.13 15.15 17.00

1988 1989 1990 1991 1992

(Source: Toy Manufacturers of America.)

6.1 Graphing and Substitution Methods

Systems of Equations ▪ *Solving by Graphing* ▪ *Substitution Method* ▪
Special Cases ▪ *Applications*

Systems of Equations

We write equations to describe the conditions and relationships given in applied problems. To this point, applications have led to a single equation in one or two variables. However, applications often lead to more than one equation.

 **EXPLORATION 1**   *Two Equations Involving Numbers*

(a) Write an equation to describe two numbers whose sum is 6. Produce the graph that represents all the solutions of the equation.

(b) Write an equation to describe two numbers whose difference is 22. Produce the graph that represents all the solutions of this second equation.

(c) Produce the graphs in parts (a) and (b) on the same coordinate system. Trace to the point of intersection of the two lines. What is the significance of this point?

Discovery

(a) The equation is $x + y = 6$. We write $y = -x + 6$ and produce the graph. (See Fig. 6.1.) The equation has infinitely many solutions.

Figure 6.1

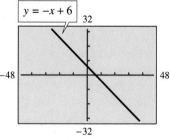

Figure 6.2

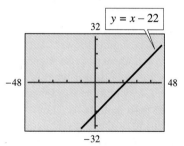

Figure 6.3

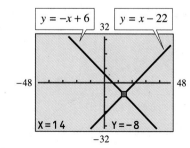

(b) The equation is $x - y = 22$. We write $y = x - 22$ and produce the graph. (See Fig. 6.2.) This equation also has infinitely many solutions.

(c) Figure 6.3 shows both graphs on the same coordinate system. The point of intersection is $(14, -8)$.

By substitution, we can easily verify that the point of intersection has coordinates that satisfy the conjunction $x + y = 6$ *and* $x - y = 22$.

Considered simultaneously, the pair of equations in Exploration 1 is an example of a **system of equations.** A system of equations is a *conjunction,* but we usually write the system, for example,

$$x + y = 6$$
$$x - y = 22$$

without explicitly writing the word *and.* The ordered pair $(14, -8)$ is a **solution** of this system because it satisfies *both* equations.

Definition of a System of Equations

Two or more equations considered simultaneously form a **system of equations.**

In certain applications there can be many equations and many variables. In this section we consider only systems of two linear equations in two variables. The following are some examples.

$$x + y = 6 \qquad y = 2x - 1 \qquad 2x - y = 9$$
$$x - y = 8 \qquad y = 3 \qquad 3x - 4y = -24$$

Definition of Solution

A **solution** of a system of two linear equations in two variables is an ordered pair (a, b) that satisfies both equations of the system.

EXAMPLE 1 *Verifying Solutions of a System of Equations*

For the following system of equations, determine whether the given pair of numbers is a solution.

$$x - y = 2$$
$$x - 2y = -2$$

(a) $(5, 3)$ (b) $(6, 4)$

Solution

(a) In each equation replace x with 5 and y with 3.

$x - y = 2$	$x - 2y = -2$
$5 - 3 = 2$	$5 - 2(3) = -2$
$2 = 2$ True	$5 - 6 = -2$
	$-1 = -2$ False

Because $(5, 3)$ satisfies the first equation but not the second equation, the pair is not a solution of the system of equations.

(b) In each equation replace x with 6 and y with 4.

$$x - y = 2 \qquad\qquad x - 2y = -2$$
$$6 - 4 = 2 \qquad\qquad 6 - 2(4) = -2$$
$$2 = 2 \quad \text{True} \qquad 6 - 8 = -2$$
$$-2 = -2 \quad \text{True}$$

Because (6, 4) satisfies both equations, the pair is a solution of the system of equations. ■

Solving by Graphing

As always, verifying solutions is easy because it involves only some substitution and arithmetic. Determining a solution requires other methods.

EXAMPLE 2   *Solving a System of Equations by Graphing*

Using a calculator, solve the following system of equations by graphing.

$$2x - y = 9$$
$$3x - 4y = -24$$

Solution   Solve each equation for y.

$$y = 2x - 9 \qquad \text{First equation}$$
$$y = \frac{3}{4}x + 6 \qquad \text{Second equation}$$

Enter the functions, produce both graphs, and trace to the point of intersection. (See Fig. 6.4.)

Figure 6.4

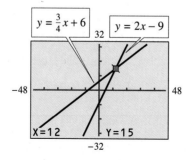

Figure 6.5

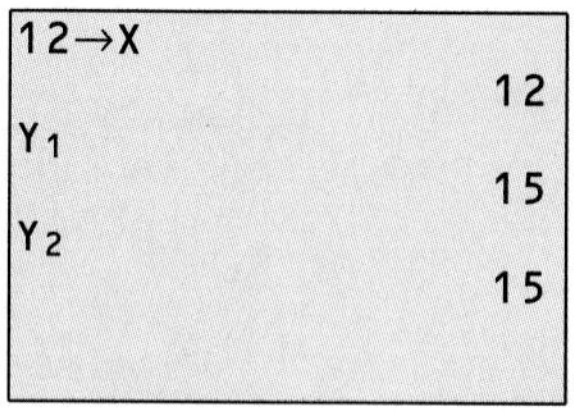

In Fig. 6.5, we verify that (12, 15) is the solution by showing that when x is 12, y is 15 in both equations of the system. ■

Substitution Method

When we use the graphing method to solve a system of equations, the point of intersection of the two graphs corresponds to the solution of the system. At that point, one value of x gives the same value for y in each equation. The substitution method is an algebraic method that uses this same idea.

The substitution method is a particularly useful method when it is easy to solve one of the equations for a certain variable. The following is a summary of the substitution method for solving systems of equations.

The Substitution Method for Solving a System of Two Equations

1. Solve one equation for one of the variables. Either variable may be chosen.

2. Substitute the expression for the solved variable into the same variable in the other equation. This results in an equation in one variable.

3. Solve the equation.

4. Substitute the solution into the equation that has been solved for the other variable. Compute the corresponding value of the other variable.

5. Check the solution of the system by substituting the solution into both equations.

EXAMPLE 3   *Solving a System with the Substitution Method*

Solve the following system with the substitution method.

$$2x - y = 9$$
$$3x - 4y = -24$$

Solution   We choose to solve the first equation for y.

$$2x - y = 9$$

$\qquad -y = -2x + 9$ Subtract $2x$ from both sides.

$\qquad\quad y = 2x - 9$ Multiply both sides by -1.

Substitute $2x - 9$ for y in the second equation.

$$3x - 4y = -24$$

$3x - 4(2x - 9) = -24$ Replace y with $2x - 9$.

$3x - 8x + 36 = -24$ Distributive Property

$\qquad -5x + 36 = -24$ Combine like terms.

$\qquad\qquad -5x = -60$ Subtract 36 from both sides.

$\qquad\qquad\quad x = 12$ Divide both sides by -5.

Now determine the *y*-value of the solution by substituting 12 for *x* in the equation that is already solved for *y*.

$$y = 2x - 9$$
$$y = 2(12) - 9 \qquad \text{Replace } x \text{ with 12.}$$
$$y = 24 - 9$$
$$y = 15$$

The solution is (12, 15). This system was solved graphically in the previous example. Recall that the coordinates of the point of intersection were (12, 15).   ■

Special Cases

When we use the graphing method to solve a system of two linear equations in two variables, one possible result is two lines that intersect at exactly one point. The pair of coordinates of that point of intersection is the solution of the system. However, two lines may not intersect at exactly one point. The next two explorations illustrate such situations.

EXPLORATION 2

Lines That Do Not Intersect

(a) Produce the graphs of the following equations.

$$x - 2y = 5$$
$$2x - 4y = 4$$

(b) What can you predict about the solution of the system of equations?

(c) How does the substitution method confirm your prediction in part (b)?

Discovery

(a) Solving the two equations for *y* results in the following.

$$y = \frac{1}{2}x - \frac{5}{2} \qquad \text{First equation}$$

$$y = \frac{1}{2}x - 1 \qquad \text{Second equation}$$

Figure 6.6 shows the graphs of the two equations.

(b) Since the slope of each line is $\frac{1}{2}$, the lines are parallel. Because there is no point of intersection, the system has no solution.

(c) To solve the system with the substitution method, we choose to solve the first equation for *x*.

$$x - 2y = 5 \qquad \text{First equation}$$
$$x = 2y + 5 \qquad \text{Add } 2y \text{ to both sides.}$$

Figure 6.6

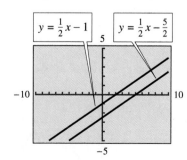

$$y = \frac{1}{2}x - 1 \qquad y = \frac{1}{2}x - \frac{5}{2}$$

Substitute $2y + 5$ for x in the second equation and solve for y.

$2x - 4y = 4$	Second equation
$2(2y + 5) - 4y = 4$	Replace x with $2y + 5$.
$4y + 10 - 4y = 4$	Distributive Property
$10 = 4$	False

Because the resulting equation is false, it has no solution. Therefore, the system of equations has no solution.

Systems of equations whose graphs do not intersect have no solution. Such systems are called **inconsistent.** If a system of equations has at least one solution, the system is called **consistent.** The system in Exploration 2 is an inconsistent system. The system in Example 3 is consistent.

EXPLORATION 3

Lines That Coincide

(a) Produce the graphs of the following equations.

$$2x + y = 3$$
$$4x + 2y = 6$$

(b) What can you predict about the solution of the system of equations?

(c) How does the substitution method confirm your prediction in part (b)?

Discovery

(a) Solve each equation for y.

$y = -2x + 3$	First equation
$y = -2x + 3$	Second equation

Figure 6.7

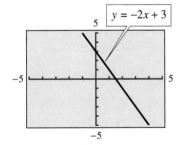

$y = -2x + 3$

Figure 6.7 shows the graphs of the two equations.

(b) The two equations are exactly the same. Therefore, the lines coincide. A solution of a system of equations is a point of intersection of the graphs. But in this case, all of the points of the lines are points of intersection. There are infinitely many solutions, and they are described by the equation $y = -2x + 3$.

(c) In part (a) we solved both equations for y. This makes it easy to replace y in the first equation with the expression for y in the second equation.

$y = -2x + 3$	First equation
$-2x + 3 = -2x + 3$	Replace y with $-2x + 3$.
$-2x + 2x + 3 = -2x + 2x + 3$	Add $2x$ to both sides.
$3 = 3$	True

The resulting equation is an identity that has infinitely many solutions. Thus, the solution set of this system contains the infinitely many solutions of the equation $y = -2x + 3$.

Systems of equations whose graphs coincide have infinitely many solutions. Such equations are called **dependent.** If the lines do not coincide, the equations are called **independent.**

If an algebraic method of solving a system results in an identity, as in Exploration 3, then the equations are dependent and the system has infinitely many solutions.

Figure 6.8 summarizes the three possible outcomes when we solve a system of two linear equations in two variables.

Figure 6.8

(a)

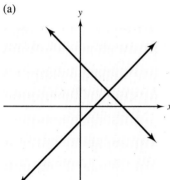

Solution is unique.
Equations are independent.
System is consistent.

Substitution method:

Equation in one variable
has a unique solution.

(b)

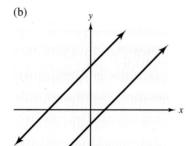

No solution.
Equations are independent.
System is inconsistent.

Equation in one variable
has no solution.

(c)

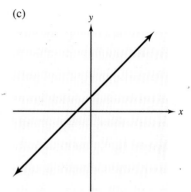

Infinitely many solutions.
Equations are dependent.
System is consistent.

Equation in one variable
is an identity.

Applications

Many applied problems can be modeled and solved with an equation having a single variable. However, when the problem involves two unknown quantities, it is often more convenient to use two variables.

> **EXAMPLE 4** *Using a System of Equations to Solve an Application Problem*

Two angles are supplementary. One angle is 20° less than the other. What are the measures of the two angles?

Solution

Let x = measure of one angle and
 y = measure of the other angle.

The following system of equations describes the given conditions.

$$x + y = 180 \qquad \text{The angles are supplementary.}$$
$$y = x - 20 \qquad \text{One angle is 20° less than the other angle.}$$

Replace y in the first equation with $x - 20$.

$$x + (x - 20) = 180$$
$$2x - 20 = 180 \qquad \text{Remove parentheses and combine like terms.}$$
$$2x = 200 \qquad \text{Add 20 to both sides.}$$
$$x = 100 \qquad \text{Divide both sides by 2.}$$

Now we use the second equation of the system to determine y.

$$y = x - 20$$
$$y = 100 - 20 = 80 \qquad \text{Replace } x \text{ with 100.}$$

The measures of the angles are $100°$ and $80°$.

6.1 *Quick Reference*

Systems of Equations

- Two or more equations considered simultaneously form a **system of equations.**

- A **solution** of a system of two linear equations in two variables is an ordered pair (a, b) that satisfies both equations of the system.

Solving by Graphing

- To estimate the solution of a system of two linear equations in two variables, use the following procedure.

  1. Solve each equation for y.

  2. Produce the graphs of the resulting functions.

  3. Trace to the point of intersection, if any, and estimate the coordinates. This ordered pair is the estimated solution of the system.

Substitution Method

- The substitution method is an algebraic method for solving a system of equations.

- To use the substitution method to determine the solution of a system of two linear equations in two variables, use the following procedure.

  1. Solve one equation for one of the variables. Either variable may be chosen.

  2. Substitute the expression for the solved variable into the same variable in the other equation. This results in an equation in one variable.

  3. Solve the equation.

  4. Substitute the solution into the equation that has been solved for the other variable. Compute the corresponding value of the other variable.

  5. Check the solution of the system by substituting the solution into both equations.

Special Cases ▪ Two special cases may arise when solving a system of two linear equations in two variables.

1. The system of equations is **inconsistent.** The graphs of the equations are parallel, and the system has no solution.

2. The equations of the system are **dependent.** The graphs of the equations coincide, and the solution set of the system contains the infinitely many solutions of either equation.

6.1   Exercises

---

1. Explain how to determine whether an ordered pair is a solution of a system of equations in two variables.

2. The pair (1, 2) does not satisfy the first equation of the following system.

$$2x + 3y = 7$$
$$x - 2y = 1$$

Why can you conclude, even without testing the pair in the second equation, that (1, 2) is not a solution of the system?

In Exercises 3–6, determine whether each ordered pair is a solution of the given system of linear equations.

3. (0, 1), (2, −3), (4, 0)

$$2x + y = 1$$
$$3x - 2y = 12$$

4. $\left(4, -\dfrac{5}{3}\right)$, (−0.5, 0), $\left(\dfrac{1}{2}, \dfrac{2}{3}\right)$

$$2x + 3y = 3$$
$$8x = 12y - 4$$

5. (1, 1), $\left(-2, \dfrac{8}{7}\right)$, (2, 0)

$$2x + 7y = 4$$
$$14y + 4x = 8$$

6. (4, −2), (−2, 1), (−6, 13)

$$3x = 2(4 - y)$$
$$2y + 1 = -3(x - 3)$$

In Exercises 7–10, determine the value of *a, b,* or *c* so that the given ordered pair is a solution of the system of equations.

7. $ax - 2y = 5$ (3, −1)
   $x + by = 1$

8. $ax + y = 5$ (1, 2)
   $x + by = 7$

9. $ax + 3y = 9$ (3, −1)
   $2x - y = c$

10. $3x - y = c$ (8, 2)
    $x + by = 2$

11. Explain why two linear equations in two variables cannot have exactly two solutions.

12. Describe the graph of a system that has

   (a) no solution.

   (b) a unique solution.

   (c) infinitely many solutions.

In Exercises 13–16, from the given graph, what is your conjecture about the number of solutions of the system of equations?

13.

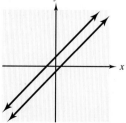

14.

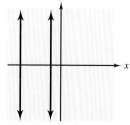

15.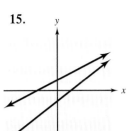

16.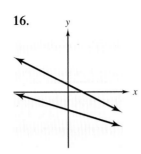

In Exercises 17–22, determine the number of solutions of the system without graphing or solving.

17. $y = 3x - 4$
$y = 3x + 5$

18. $y = -2x - 3$
$y + 2x = 4$

19. $y + 2x = 6$
$y - 2x = 8$

20. $y = \frac{2}{3}x + 5$

$y = -\frac{3}{2}x - 4$

21. $y + x = 5$
$y = -x + 5$

22. $y + 3x = 4$
$y = -3x + 4$

23. (a) How do the graphs of the following systems of equations differ?

   (i) The equations are independent.

   (ii) The equations are dependent.

   (b) Is it possible for a system of two independent equations to have no solution? Explain.

24. (a) How do the graphs of the following systems of equations differ?

   (i) The system is consistent.

   (ii) The system is inconsistent.

   (b) Is it possible for a consistent system to have infinitely many solutions? Explain.

In Exercises 25–28, estimate the solution of the system of equations.

25.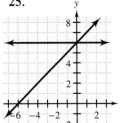

26.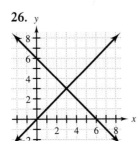

27.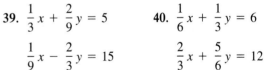

28.

In Exercises 29–40, use the graphing method to estimate the solution of the given system of equations. Then use the substitution method to determine the solution algebraically.

29. $y = 2x - 2$
$y = -3x + 13$

30. $y = 4x - 13$
$3x + 2y = 29$

31. $y = 0.6x - 10$
$5y - 3x = 4$

32. $3x + y = 12$
$2y + 10 = -6x$

33. $2x + y = 8$
$y + 2 = 0$

34. $y = 3x - 2$
$y = 4$

35. $x - y + 6 = 0$
$2x - y - 5 = 0$

36. $3x + 2y + 8 = 0$
$x + 2y - 16 = 0$

37. $y = -2(4 - x)$
$y + 8 = 2x$

38. $2y - 3x = 10$
$-1.5x = 5 - y$

39. $\frac{1}{3}x + \frac{2}{9}y = 5$

$\frac{1}{9}x - \frac{2}{3}y = 15$

40. $\frac{1}{6}x + \frac{1}{3}y = 6$

$\frac{2}{3}x + \frac{5}{6}y = 12$

In Exercises 41–44, sketch the graphs of the equations of the given system and estimate the solution.

41. $x + 3 = 0$
$2x + 3y = 6$

42. $x - 2 = 0$
$4x + 5y = 23$

43. $x = -4$
$y = 7$

44. $x = 3$
$y = -5$

45. Suppose you plan to solve the following system of equations with the substitution method.

$$3x - 4y = 16$$
$$2x + y = 7$$

   (a) Would it be correct to begin by solving one of the equations for x?

   (b) How would you begin the solution? Why?

46. When solving a system of equations with the substitution method, suppose you obtain the equation $-3 = -3$. What does this result tell you about the system and its solution?

In Exercises 47–58, use the substitution method to solve the system of equations. Identify inconsistent systems and dependent equations.

47. $x - y - 5 = 0$
$2x - y + 17 = 0$

48. $2x + 3y + 15 = 0$
$x + 2y + 18 = 0$

49. $y + 3x - 4 = 0$
$12 = 9x + 3y$

50. $2x = y - 5$
$3y = 6x + 15$

51. $x = 3y + 9$
$3y - x = 6$

52. $x - 2y = 5$
$3x = 6y + 8$

53. $y = 2x - 5$
$4x - 2y - 10 = 0$

54. $x - 2y = 5$
$2x = 4y + 10$

55. $y = 2x - 7$
$3x + 2y = 0$

56. $x = 2y - 5$
$2x - 3y = -6$

57. $6x = 5y$
$12x + 7 = 10y$

58. $2x + 3y + 2 = 0$
$0 = 6x + 9y$

 59. When solving a system of equations with the substitution method, suppose you obtain the equation $13 = 7$. What does this result tell you about the system and its solution?

 60. Use the integer setting to produce the graph of the following system of equations.

$y = \quad x + 10$
$y = 0.9x - 10$

What does the appearance of the graph suggest about the system of equations? Look closely at the two equations and explain how the graph is deceiving.

In Exercises 61–64, use your calculator to produce the graph of the given system of equations. From the graph, predict the number of solutions of the system. Then, use the substitution method to solve the system.

61. $y = -2x + 5$
$21x + 10y = 50$

62. $1.5x + y = 9$
$y = -1.52x + 9$

63. $x = y$
$11x - 10y = 80$

64. $y = \dfrac{7}{20}x$
$3y - x = -36$

In Exercises 65–68, determine the value of k so that the equations are dependent.

65. $y = \dfrac{2}{3}x + 1$
$kx + 3y = 3$

66. $3x + ky = 15$
$y = 5 - x$

67. $y = 3x + k$
$y + (2 - k)x = 5$

68. $kx - 4y = -2$
$4y - 3x = k - 1$

In Exercises 69–72, determine the value of k so that the system is inconsistent.

69. $kx + y = 2k$
$y = 2x + 5$

70. $2y - x = 6$
$x + ky = 0$

71. $kx + 3y = 12$
$y = \dfrac{4}{3}x + (k + 1)$

72. $x = ky - 6$
$3x + y = 9k$

In Exercises 73–86, translate the given information into a system of equations. Use the graphing method to solve the system and then answer the question.

73. The sum of two numbers is 16 and their difference is 4. What are the two numbers?

74. The difference of two numbers is one-third of the smaller number. The sum of the numbers is 6 less than twice the larger number. What are the two numbers?

75. One number exceeds a second number by 8. Three-fourths of the larger number is 3 more than the smaller number. What are the two numbers?

76. A number is 6 less than a larger number. Two-thirds of the larger number plus one-fourth of the smaller number is 15. What are the two numbers?

 77. Two angles are complementary. The difference between twice the larger angle and 5 times the smaller angle is 19°. What is the measure of each angle?

 78. Two angles are complementary. The smaller angle is one-third of 10° less than the larger angle. What is the measure of each angle?

$y = x + 9$
$y = Large$
$y = small$
$3x + 3 = 8$
$y + 4.$

SECTION 6.1   *Graphing and Substitution Methods*   **355**

79. Two angles are supplementary. The larger angle is 24° less than twice the smaller angle. What is the measure of each angle?

80. Two angles are supplementary. Two-thirds of the larger angle is 30° more than the smaller angle. What is the measure of each angle?

81. The perimeter of a rectangle is 74 feet. The length is 7 feet more than twice the width. What are the dimensions of the rectangle?

82. The perimeter of a rectangle is 44 meters. The width is 4 meters less than the length. What are the dimensions of the rectangle?

83. The perimeter of a rectangle is 80 inches. The width is one-third of the length. What are the dimensions of the rectangle?

84. The perimeter of a rectangle is 32 meters. The length is 4 meters less than 3 times the width. What are the dimensions of the rectangle?

85. Tees Unlimited sold 35 more T-shirts than sweatshirts last Monday. T-shirts sell for $11 each, while sweatshirts sell for $20 each. In all, $1718 worth of shirts were sold that day. How many of each type of shirt were sold?

86. A farmer received 90 cents per bushel more for his wheat than he did for his corn. If he sold 6000 bushels of wheat and 4000 bushels of corn for a total of $30,400, what was the price per bushel of each?

Exploring with Real Data

In Exercises 87–90, refer to the following background information and data.

From 1985 to 1992 health care costs incurred by the Ford Motor Company nearly doubled. During the same period expenditures on steel declined.

	Expenditures (in billions)	
	1985	1992
Health Care	0.77	1.40
Steel	1.30	1.20

(Source: Ford Motor Company.)

The number of millions of dollars y spent during a given year x ($85 \leq x \leq 92$) can be described by the following equations.

Health Care Costs:	$y = -6880 + 90x$
Steel Expenditures:	$y = 2514 - 14x$

87. Treat these two equations as a system of equations and estimate the solution of the system graphically.

88. Solve the system algebraically.

89. Interpret the solution of the system of equations.

90. Opponents of health care reform fear that higher taxes will be levied to pay for the spiraling increase in medical costs. How will we pay health care costs incurred by the auto industry even if there are no higher taxes?

Challenge

In Exercises 91 and 92, use the substitution method to solve the given system in terms of a, b, c, and m.

91. $ax + by = c$
    $y = mx$

92. $y = mx + b$
    $x = \dfrac{y}{m} - \dfrac{c}{m}$ $(b \neq c)$

In Exercises 93 and 94, determine the value of c so that the given equations are independent.

93. $y = c - 3x$
    $3x + y = -2$

94. $4y - x = 5$
    $x = 4y + c$

In Exercises 95 and 96, determine the value of a so that the given system is consistent.

95. $5y = x$
    $y = ax - 2$

96. $y = ax + 4$
    $y = 3x + (a + 1)$

In Exercises 97–100, determine conditions on the slope and y-intercept so that the system has the specified number of solutions.

97. $y = m_1x + b_1$
    $y = m_2x + b_2$
    No solution

98. $y = m_1x + b_1$
    $y = m_2x + b_2$
    Infinitely many solutions

99. $y = 3x + b_1$
    $y = -3x + b_2$
    One solution

100. $y = m_1x$
     $y = m_2x$
     No solution

101. Show whether there is a value of k such that the given system has exactly one solution.

$$x - y = k$$
$$3x - 3y = 2$$

102. Is it possible to determine a and c so that the given system of equations has

(a) no solution?

(b) infinitely many solutions?

$$2ax + ay = 1$$
$$y - x = c$$

6.2 The Addition Method

Addition Method ▪ *Special Cases* ▪ *Applications*

Addition Method

The substitution method is not always the most convenient method for algebraically solving a system of equations. An alternative algebraic method is the **addition method,** sometimes called the **elimination method.**

Consider the system of equations

$$A = B$$
$$C = D$$

where A, B, C, and D each represent some expression. Because C and D have the same value, the Addition Property of Equations allows us to add C to the left side of $A = B$ and to add D to the right side of $A = B$. The resulting equation is $A + C = B + D$. We will refer to this process as *adding the equations.*

The idea behind the addition method is to eliminate one of the variables by adding the two equations, with the result being a linear equation in only one variable.

EXAMPLE 1 *Solving a System of Equations with the Addition Method*

Use the addition method to solve the following system of equations.

$$x + y = 6$$
$$x - y = 2$$

Solution

$$
\begin{array}{ll}
x + y = 6 & \text{Add the two equations.} \\
\underline{x - y = 2} & \\
2x \quad\quad = 8 & \text{The variable } y \text{ is eliminated.}
\end{array}
$$

The resulting equation has only one variable. When we solve this equation we obtain $x = 4$. Replace x with 4 in either of the original equations and solve for y.

$$x + y = 6 \qquad \text{First equation}$$
$$4 + y = 6 \qquad \text{Replace } x \text{ with 4.}$$
$$y = 2$$

The solution of the system is (4, 2).

Adding two equations results in the elimination of a variable only when the variable appears in both equations and has coefficients that are opposites. If this is not the case, the Multiplication Property of Equations makes it possible to change the coefficient(s) of one of the variables in one or both of the equations so that the coefficients are opposites.

EXAMPLE 2   *Using the Multiplication Property of Equations with the Addition Method*

Solve the following system of equations.

$$2x - y = 9$$
$$3x - 4y = -24$$

Solution   If we choose to eliminate y, its coefficients must be opposites. Therefore, we multiply both sides of the first equation by -4 and then add the equations.

$$-4(2x - y) = -4(9) \quad \rightarrow \quad -8x + 4y = -36 \qquad \text{Multiply first equation by } -4.$$
$$3x - 4y = -24 \quad \rightarrow \quad \underline{\quad 3x - 4y = -24 \quad} \qquad \text{Second equation}$$
$$-5x = -60 \qquad \text{Add the equations.}$$
$$x = 12 \qquad \text{Solve for } x.$$

Replace x with 12 in either equation to determine y.

$$2x - y = 9 \qquad \text{First equation}$$
$$2(12) - y = 9 \qquad \text{Replace } x \text{ with 12.}$$
$$24 - y = 9$$
$$-y = -15$$
$$y = 15$$

The solution is (12, 15).

Deciding which variable to eliminate is arbitrary. Usually, we choose the variable whose elimination requires the least amount of work.

Writing each equation in the same form is a good idea when using the addition method because then the like terms are lined up in columns. It is common to write both equations in the standard form $Ax + By = C$.

The Multiplication Property of Equations requires us to multiply both sides of an equation by the same number. However, we can multiply both sides of one equation by one number and both sides of the other equation by a *different* number. It is not necessary to multiply both equations by the same number.

| EXAMPLE 3 | *Using the Multiplication Property of Equations with the Addition Method* |

Solve the following system.

$$3x = 33 - 5y$$
$$3y = 4x - 15$$

Solution First, we write the equations in standard form so that the like terms are aligned.

$$3x + 5y = 33$$
$$-4x + 3y = -15$$

To eliminate x, multiply the first equation by 4 and the second equation by 3. Then, add the equations.

$$4(3x + 5y) = 4(33) \quad \rightarrow \quad 12x + 20y = 132$$
$$3(-4x + 3y) = 3(-15) \quad \rightarrow \quad \underline{-12x + 9y = -45}$$
$$29y = 87$$
$$y = 3$$

Replace y with 3 in either equation.

$$3x = 33 - 5y \qquad \text{First equation}$$
$$3x = 33 - 5(3) \qquad \text{Replace } y \text{ with 3.}$$
$$3x = 18$$
$$x = 6$$

The solution is (6, 3).

When the equations involve fractions, we can multiply by the LCM to clear them.

| EXAMPLE 4 | *Clearing Fractions* |

Solve the following system.

$$\frac{3}{2}x - \frac{3}{4}y = \frac{15}{4}$$

$$\frac{4}{3}x - \frac{5}{3}y = 3$$

Solution Clear the fractions by multiplying each term by the LCM. For the first equation, the LCM is 4; for the second equation, the LCM is 3.

$$4 \cdot \frac{3}{2}x - 4 \cdot \frac{3}{4}y = 4 \cdot \frac{15}{4} \quad \rightarrow \quad 6x - 3y = 15$$

$$3 \cdot \frac{4}{3}x - 3 \cdot \frac{5}{3}y = 3 \cdot 3 \quad \rightarrow \quad 4x - 5y = 9$$

Now use the Multiplication Property of Equations again, this time to adjust the coefficients. To eliminate x, multiply the first equation by 2 and the second equation by -3. Then, add the equations.

$$2(6x - 3y) = 2(15) \quad \rightarrow \quad 12x - 6y = 30$$
$$-3(4x - 5y) = -3(9) \quad \rightarrow \quad \underline{-12x + 15y = -27}$$
$$9y = 3$$
$$y = \frac{3}{9} = \frac{1}{3}$$

Now substitute $\frac{1}{3}$ for y in either equation.

$$\frac{3}{2}x - \frac{3}{4}y = \frac{15}{4} \qquad \text{First equation}$$

$$\frac{3}{2}x - \left(\frac{3}{4}\right)\left(\frac{1}{3}\right) = \frac{15}{4} \qquad \text{Replace } y \text{ with } \tfrac{1}{3}.$$

$$\frac{3}{2}x - \frac{1}{4} = \frac{15}{4}$$

$$\frac{3}{2}x = 4$$

$$x = \frac{8}{3}$$

The solution is $(\frac{8}{3}, \frac{1}{3})$. ■

There is an alternative to substituting the known value of a variable in order to determine the value of the other variable. We can go back to the system in which the fractions have been cleared, eliminate the known variable, and solve for the unknown variable.

In Example 4 we could have returned to the system

$$6x - 3y = 15$$
$$4x - 5y = 9$$

to eliminate y and solve for x. Multiply the first equation by 5 and the second equation by -3.

$$30x - 15y = 75$$
$$\underline{-12x + 15y = -27}$$
$$18x = 48$$
$$x = \frac{48}{18} = \frac{8}{3}$$

Special Cases

In Section 6.1 we saw how the substitution method detected the two special cases: lines that are parallel and lines that coincide. When both variables were eliminated, the result was either a false equation or an identity.

The addition method also detects these special cases, as our next exploration illustrates.

EXPLORATION 1

Special Cases

(a) Use the addition method to solve the following systems of equations.

System (1)	*System (2)*
$x - 2y = 5$	$2x + y = 3$
$2x - 4y = 4$	$4x + 2y = 6$

(b) How would you interpret the result in each case?

(c) How would writing the equations in the slope–intercept form confirm your interpretations in part (b)?

Discovery

(a) *System (1)*

$$-2(x - 2y) = -2(5) \quad \rightarrow \quad -2x + 4y = -10 \qquad \text{Multiply first equation by } -2.$$
$$\underline{2x - 4y = \qquad 4 \quad \rightarrow \quad 2x - 4y = \qquad 4} \qquad \text{Second equation}$$
$$0 = -6 \qquad \text{False}$$

System (2)

$$-2(2x + y) = -2(3) \quad \rightarrow \quad -4x - 2y = -6 \qquad \text{Multiply first equation by } -2.$$
$$\underline{4x + 2y = \qquad 6 \quad \rightarrow \quad 4x + 2y = \qquad 6} \qquad \text{Second equation}$$
$$0 = 0 \qquad \text{True}$$

(b) For System (1) the equation $0 = -6$ is false. Therefore, the system is inconsistent and has no solution. For System (2) the equation $0 = 0$ is true. Therefore, the equations are dependent and the system has infinitely many solutions.

(c) In slope–intercept form, the equations of the two systems are as follows.

System (1)	*System (2)*
$y = \dfrac{1}{2}x - \dfrac{5}{2}$	$y = -2x + 3$
$y = \dfrac{1}{2}x - 1$	$y = -2x + 3$

In System (1) the graphs have the same slope and different y-intercepts. The lines are parallel, and the system has no solution.

In System (2) the graphs have the same slope and the same y-intercept. The graphs coincide, and every point of the line represents a solution.

Exploration 1 shows that special cases can be identified algebraically with the addition method in exactly the same way that they are identified with the substitution method.

Applications

When solving applied problems involving two variables, the initial work is always the same. We assign variables and write a system of equations. The only decision that remains concerns our choice of method for solving the system.

EXAMPLE 5   *A Dual Investment Application*

Part of a $5000 account was invested in a fund yielding 7% and the remainder was invested in a fund yielding 9%. If the income for the year is $380, how much was invested at each rate?

Solution

Let x = the amount invested at 7% and

     y = the amount invested at 9%.

The system of equations is as follows.

$$x + y = 5000 \qquad \text{The total investment is \$5000.}$$
$$0.07x + 0.09y = 380 \qquad \text{The total income is \$380.}$$

Multiply the first equation by -0.07 and add the equations.

$$-0.07x - 0.07y = -350$$
$$\underline{0.07x + 0.09y = 380}$$
$$0.02y = 30$$
$$y = 1500$$

Replace y in the first equation with 1500 and solve for x.

$$x + y = 5000$$
$$x + 1500 = 5000$$
$$x = 3500$$

The investments were $3500 at 7% and $1500 at 9%.   ∎

We now have three methods for solving a system of equations. The graphing method has the advantage of being visually helpful, and it is a quick method for obtaining an estimate of the solution.

The substitution method and the addition method are both algebraic methods. Choose the one that is most convenient for the system. Either method will work for all linear systems.

The ideal solving process is a combination of graphing and algebraic methods. This will help you to remain aware of the conceptual nature of the problem and to avoid falling into a mere symbol manipulation routine.

6.2   *Quick Reference*

Addition Method

- The addition method is another algebraic method for solving a system of two equations. The following is a summary of the method.

  1. Write both equations in the form $Ax + By = C$.

  2. If necessary, multiply one or both of the equations by appropriate numbers so that the coefficients of one of the variables are opposites.

  3. Add the equations to eliminate a variable.

  4. Solve the resulting equation.

  5. Substitute that value in either of the original equations and solve for the other variable.

- It is usually best to clear fractions before using the addition method.

Special Cases

- After you add the equations to eliminate one variable, the resulting equation may be false, or it may be an identity.

  1. If false, then the system is inconsistent and has no solution.

  2. If an identity, then the two equations of the system are dependent, and the system has infinitely many solutions.

6.2   *Exercises*

 1. Suppose you wish to solve the following system.

$$3x - \ y = 8$$
$$x + 4y = 7$$

    (a) Describe the operation you would perform to eliminate x.

    (b) Describe the operation you would perform to eliminate y.

 2. Suppose you wish to use the addition method to solve the following system. Describe the easiest way to eliminate one of the variables.

$$2x + \ y = \ 3$$
$$3x + 4y = -2$$

In Exercises 3–10, use the addition method to solve each system.

3. $x + y = 10$
    $x - y = \ 2$

4. $2x - y = \ \ 7$
    $2x + y = 13$

5. $\ \ x + 2y = 4$
    $-x - \ y = 3$

6. $3x - 3y = \ 14$
    $3x + 3y = -2$

7. $3x + 2y = 4$
    $4y = 3x + 26$

8. $x - 3y = 9$
    $5y = x - 17$

9. $4x = 3y + 6$
    $12 = 5x + 3y$

10. $2x = y + 9$
    $3y = 2x - 19$

 11. Suppose you solve a system of equations and the resulting equation is $0 = 0$. Does this mean that the solution of the system is $(0, 0)$? Explain.

 12. Suppose you solve a system of equations in x and y, and you eliminate y to obtain $x = 4$. Is 4 the solution of the system? Explain.

In Exercises 13–20, use the addition method to solve each system.

13. $2x + 3y + 1 = 0$
$5x + 3y = 29$

14. $5x - 2y = 0$
$4y = 3x + 14$

15. $3x - 2y = 21$
$5x + 4y = 13$

16. $2x = 3y - 14$
$4x + 5y = 16$

17. $2x + 3y = 3$
$4x = 6y - 2$

18. $3x + 4y + 1 = 0$
$6x = 8y - 14$

19. $6x = 5y + 10$
$3x + 2y = 23$

20. $4x + 3y = 2$
$5x + 6y = 7$

 21. Suppose the solution of a system of two linear equations is (3, 2). Try to visualize the general appearance of the graph of the system. Now explain why there are many other systems whose solutions are also (3, 2).

 22. Explain why the addition method cannot be used to solve the following system of equations. What is the solution of the system?

$x = -3$
$y = 7$

In Exercises 23–30, use the addition method to solve each system.

23. $6x + 11y = 17$
$4x - 5y = -1$

24. $5x - 3y = 2$
$3x - 2y = -1$

25. $-6x + 5y = 10$
$5x + 4y = 8$

26. $-3x + 7y = 10$
$5x - 2y = -6$

27. $2x - 7y + 11 = 0$
$7x + 4y + 10 = 0$

28. $3x + 2y + 7 = 0$
$2x - 3y = 4$

29. $5y - 2x = 5$
$5x + 2y = 2$

30. $4x = 3y + 8$
$3x = 4y + 6$

 31. When you use the addition method, how can you tell when the equations of a system are dependent?

 32. When you use the addition method, how can you tell when a system is inconsistent?

In Exercises 33–44, use the addition method to solve the system.

33. $2x + 3y + 18 = 0$
$5y = 6x - 2$

34. $y = 3x - 5$
$2x + 3y = 6$

35. $3x - 2y = 8$
$6x = 4y + 17$

36. $4x + 12y = 32$
$7 - x = 3y$

37. $3x - y = 7$
$2y = 6x - 14$

38. $4x + 12y = 36$
$9 - x = 3y$

39. $2 + 7y = 5x$
$15x = 21y - 6$

40. $5 = x - 4y$
$12y - 3x = -8$

41. $6y + 7x = 16$
$3x = 2y + 16$

42. $5y = 2(3x - 2)$
$8 + 3y = -2x$

43. $2y = 3x - 5$
$-12x = -20 - 8y$

44. $8x + 9y = -3$
$6 + 18y = -16x$

In Exercises 45–56, use the addition method to solve each system.

45. $4(x - 2) - 5(y - 3) = 14$
$3(x + 4) - 2(y + 5) = 13$

46. $3(x - 5) + (y + 7) = 7$
$2(x + 3) - 5(y + 1) = 11$

47. $\dfrac{1}{2}x + \dfrac{3}{10}y = \dfrac{1}{2}$
$-\dfrac{5}{3}x - y = \dfrac{4}{3}$

48. $-0.4x + 1.2y = 1$
$-0.6x + 1.8y = -1$

49. $x - 1.5y = 0.25$
$0.9y - 0.6x = -0.15$

50. $\dfrac{3}{2}x - y = \dfrac{3}{4}$
$2x - \dfrac{4}{3}y = 1$

51. $0.1x - 0.25y = 1.05$
$0.625x + 0.4y = 0.675$

52. $0.4x - 0.3y = -0.1$
$0.5x + 0.2y = 1.6$

53. $\dfrac{x + 2}{9} + \dfrac{y + 2}{6} = 1$
$\dfrac{x - 1}{4} - \dfrac{y + 1}{3} = -1$

54. $\dfrac{x - 2}{12} - \dfrac{y - 2}{4} = 1$

$\dfrac{x + 3}{2} + \dfrac{y - 4}{4} = 1$

55. $\dfrac{x - y}{6} + \dfrac{x + y}{3} = 1$

$y = 6 - 3x$

56. $\dfrac{x + y}{2} = \dfrac{1}{3} + \dfrac{x - y}{2}$

$\dfrac{x + 2}{2} - \dfrac{y + 4}{2} = 4$

In Exercises 57–60, determine values of a, b, and c so that the system has the given solution.

57. $x + by = -2a$

$ax - 4y = \quad b$

$(-2, -2)$

58. $5x + 2ay = \quad c$

$ax - 3y = -6c$

$(2, -2)$

59. $ax - by = -17$

$bx + ay = -1$

$(-2, 5)$

60. $ax - \qquad y = c + 6$

$x - (3a + 4)y = c$

$(7, 1)$

In Exercises 61–64, determine k so that the system of equations will be inconsistent.

61. $x + 2y = 5$

$2x - ky = 7$

62. $3x - 2y = 10$

$kx + 6y = 5$

63. $y = -2x + 3$

$kx - 3y = 4$

64. $x = 2y - 5$

$4x + ky = 2$

In Exercises 65–68, determine a, b, or c so that the equations will be dependent.

65. $x + 2y = \quad c$

$3x + 6y = 12$

66. $2x + by = \quad 5$

$-6x - 12y = -15$

67. $ax + by = 3$

$x - y = 1$

68. $2x + 3y = -4$

$ax - 6y = \quad c$

69. If 8 hamburgers and 4 milk shakes cost \$20 and 3 hamburgers and 2 milk shakes cost \$8.10, then what are the prices of a hamburger and a milk shake?

70. If 5 root beers and 7 orders of French fries cost \$7.25 and 4 root beers and 5 orders of French fries cost \$5.50, then what are the prices of one root beer and one order of French fries?

71. A parking meter contains \$3.85 in nickels and dimes. If there is a total of 50 coins, how many dimes are in the meter?

72. A coin jar contains 34 coins. If the coins are all quarters and dimes and their total value is \$6.55, how many quarters are in the jar?

73. There are 64 coins in a cash register and their total value is \$4.25. If the coins are all nickels and dimes, how many of each coin are there?

74. A vending machine attendant fills a coin dispenser with \$6.85 in nickels and quarters. If there is a total of 61 coins in the dispenser, how many of each coin are there?

75. An accountant invested \$6000, part at 7% simple interest and the remainder at 9% simple interest. How much did she invest at each rate if her total income at the end of 1 year was \$455?

76. To assist with the down payment on a home, a broker loaned a total of \$7000 to a home buyer. Part of the loan was made at 6% simple interest and the remainder at 8.5% simple interest. How much did the buyer borrow at each rate if the total interest at the end of 1 year was \$477?

In Exercises 77–84, solve the given system and obtain a solution in terms of a, b, and c.

77. $x + y = b$
    $x - y = c$

78. $2x + y = 3$
    $x + y = c$

79. $ax + y = 3$
    $x - y = c$

80. $x + by = c$
    $x - by = c$

81. $2ax + \ y = 3$
    $-ax + 3y = 2$

82. $ax + by = c$
    $bx + ay = c$

83. $ax + by = a$
    $x - \ \ y = 1$

84. $2x + y = 4 + c$
    $x - 2 = y$

Exploring with Real Data

In Exercises 85–88, refer to the following background information and data.

Census figures during the 1980s showed some shifts in certain age-group populations. The following table indicates the trends for two such groups.

Age Group	Population (in millions)	
	1980	1986
15–24	42,743	39,261
Over 65	25,704	29,173

(Source: U.S. Bureau of Census.)

Assuming linear relationships, the age-group population y (in millions) for a given year x ($80 \le x \le 86$) is given by the following equations.

Age Group 15–24: $y = \ \ \ 89{,}170 - 580x$
Age Group Over 65: $y = -20{,}550 + 578x$

85. Using a suitable range and scale, produce the graphs of the two equations on the same coordinate axes.

86. Assuming the equations are valid beyond 1986, use the graph to estimate the year in which the populations of the two age groups will be the same.

87. Verify your result in Exercise 86 by solving the system of equations with the addition method.

88. If the indicated trends were to continue, what implication might there be on our nation's Social Security system?

Challenge

In Exercises 89–92, you are given a system of *nonlinear* equations in x and y. To solve a system that is in the given form, we can begin by letting

$$u = \frac{1}{x} \text{ and } v = \frac{1}{y}.$$

Example: Solve the following system.

$$\frac{2}{x} + \frac{3}{y} = \ \ 5$$
$$\frac{5}{x} - \frac{1}{y} = -2$$

If we let

$$u = \frac{1}{x} \text{ and } v = \frac{1}{y},$$

we obtain the following system.

$$2u + 3v = \ \ 5$$
$$5u - \ \ v = -2$$

Now use the addition method to solve this system for u and v. Those values can then be used to determine x and y.

89. $\dfrac{3}{x} + \dfrac{2}{y} = 2$
    $\dfrac{1}{x} - \dfrac{6}{y} = -1$

90. $\dfrac{3}{x} + \dfrac{2}{y} = 0$
    $\dfrac{9}{x} - \dfrac{4}{y} = 5$

91. $\dfrac{5}{x} + \dfrac{10}{y} = 3$
    $\dfrac{1}{x} + \dfrac{3}{y} = 1$

92. $\dfrac{1}{x} + \dfrac{1}{y} = -1$
    $\dfrac{5}{x} - \dfrac{2}{y} = -1$

93. Using the technique described in Exercises 89–92, solve the following system. Explain the result.

$$\frac{3}{x} + \frac{5}{y} = 10$$
$$\frac{2}{x} + \frac{1}{y} = \ \ 2$$

In Exercises 94–97, determine a, b, and c so that the system will be consistent.

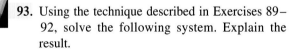

94. $x - 2y = 3$
    $ax - 2y = 4$

95. $-3x + 2y = 5$
    $6x + by = 7$

96. $3x - 2y = c$
$-6x + 4y = -10$

97. $4x - 8y = 20$
$x - 2y = c$

98. $4x - 7y = -1$
$by - 8x = 2$

99. $ax + 3y = 2$
$2x + 6y = 4$

100. $x - 2y = c$
$6y - 3x = 3$

101. $x + y = 5$
$2x + 2y = c$

In Exercises 98–101, determine a, b, and c so that the equations will be independent.

6.3 Systems of Equations in Three Variables

Linear Equations in Three Variables ▪ *Graphical Interpretation* ▪
Algebraic Methods ▪ *Special Cases* ▪ *Applications*

Linear Equations in Three Variables

Although we have concentrated on linear equations in two variables, linear equations can be defined for more than two variables.

Definition of a Linear Equation in Three Variables

A **linear equation in three variables** is any equation that can be written in the *standard form* $Ax + By + Cz = D$, where A, B, C, and D are real numbers and A, B, and C are not all zero.

Similar definitions may be stated for linear equations in more than three variables.

The following are some examples of linear equations in three variables.

Equation	*Standard Form*
$3x - y + 2z = 5$	$3x + (-1)y + 2z = 5$
$y = 2x + 1$	$-2x + 1y + 0z = 1$
$x = 1$	$1x + 0y + 0z = 1$

A solution of a linear equation in *two* variables is an ordered *pair* (x, y) that satisfies the equation. We can extend this definition for linear equations in three variables.

Definition of Solution

A **solution** of a linear equation in three variables is an **ordered triple** (x, y, z) of numbers that satisfies the equation.

EXAMPLE 1   *Verifying a Solution*

Verify that (5, 3, 6) is a solution of the equation $x + y - z = 2$.

Solution

$$x + y - z = 2$$
$$5 + 3 - 6 = 2 \qquad \text{Replace } x \text{ with 5, } y \text{ with 3, and } z \text{ with 6.}$$
$$2 = 2 \qquad \text{True}$$

The ordered triple (5, 3, 6) is a solution.

The following is an example of a system of three linear equations in three variables.

$$3x - y + 2z = 0$$
$$2x + 3y + 8z = 8$$
$$-x + y + 6z = 0$$

A solution of such a system is an ordered triple (x, y, z) that satisfies each equation of the system.

In previous sections we developed graphic and algebraic methods for solving systems of two linear equations in two variables. We can extend those methods to solve systems of three linear equations in three variables.

Graphical Interpretation

We know that the graph of a linear equation in *two* variables is a *line* drawn in a *two*-dimensional (rectangular) coordinate system.

The graph of a linear equation in *three* variables is a *plane* drawn in a *three*-dimensional coordinate system, also called a **coordinate space.** (See Fig. 6.9.)

Note that a coordinate space includes three axes that are all perpendicular to each other. We must visualize the x-axis as coming out from the page toward the reader.

Figure 6.10 shows the point $P(4, 2, 3)$ drawn in a coordinate space.

Figure 6.9

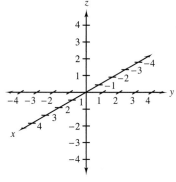

Figure 6.10

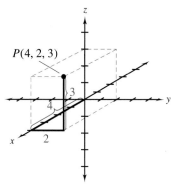

Figure 6.11 shows one of many planes that contains point P. Planes, like lines, extend without end. However, a plane in a coordinate space is conventionally drawn as a bounded figure so that its location can be visualized.

For a system of three linear equations in three variables, the graph of each equation of the system is a plane. The solution set of such a system is represented by the intersection of the graphs (planes) of the three equations. Any point in common to all three planes represents a solution of the system. There are three possible outcomes.

First, the solution could be unique. This means that the three planes intersect at one point P. (See Fig. 6.12.)

Second, there could be no solution. This can occur when at least two of the planes are parallel (see Fig. 6.13) or when one plane is parallel to the line L of intersection of the other two planes (see Fig. 6.14).

Third, there could be infinitely many solutions. This can occur when the planes coincide (see Fig. 6.15) or when they intersect in a common line L (see Fig. 6.16).

Figure 6.11

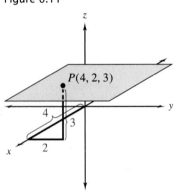

Figure 6.12

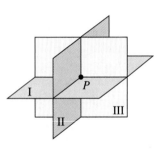

Figure 6.13

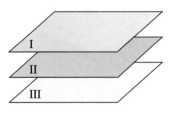

Figure 6.14

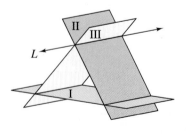

Figure 6.15

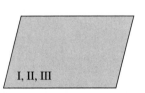

Figure 6.16

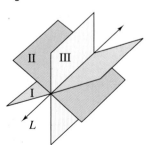

Because of the difficulties inherent in drawing planes in coordinate space, we will not use graphing methods for estimating solutions of systems of three or more linear equations. Our focus will be on algebraic methods. Nevertheless, your ability to visualize the various ways in which planes can be oriented in space will provide you with a geometric interpretation of the system and its solution.

Algebraic Methods

The substitution method we used for solving systems of two linear equations in two variables can also be used to solve systems of three linear equations in three variables.

Although the substitution method can be used for any system, the addition method is more generally applicable. We will limit our discussion to that one algebraic method.

EXAMPLE 2   *Solving a System of Three Equations with the Addition Method*

Use the addition method to solve the following system.

$$3x - y + 2z = 0 \quad (1)$$
$$2x + 3y + 8z = 8 \quad (2)$$
$$x + y + 6z = 0 \quad (3)$$

Solution   The plan is to eliminate a variable from one pair of equations. Then, we eliminate the same variable from another pair of equations. In the following, we eliminate y from equations (1) and (3) and from equations (2) and (3).

$$3x - y + 2z = 0 \quad (1) \qquad\qquad 2x + 3y + 8z = 8 \quad (2)$$
$$x + y + 6z = 0 \quad (3) \qquad\qquad x + y + 6z = 0 \quad (3)$$

In the first pair, eliminate y by adding the equations. In the second pair, multiply both sides of equation (3) by -3 and then add the equations.

$$
\begin{array}{rr}
3x - y + 2z = 0 & 2x + 3y + 8z = 8 \\
\underline{x + y + 6z = 0} & \underline{-3x - 3y - 18z = 0} \\
4x \quad\quad + 8z = 0 & -x \quad\quad\quad - 10z = 8
\end{array}
$$

The result is two equations in two variables.

$$4x + 8z = 0 \quad (4)$$
$$-x - 10z = 8 \quad (5)$$

Now we apply the addition method again to solve this system. Multiply equation (5) by 4 and add the equations.

$$
\begin{array}{r}
4x + 8z = 0 \\
\underline{-4x - 40z = 32} \\
-32z = 32 \\
z = -1
\end{array}
$$

Now substitute -1 for z in equation (4) to determine x.

$$4x + 8z = 0 \qquad \text{Equation (4)}$$
$$4x + 8(-1) = 0 \qquad \text{Replace } z \text{ with } -1.$$
$$4x - 8 = 0$$
$$4x = 8$$
$$x = 2$$

Finally, to determine the value of y, substitute 2 for x and -1 for z in any of the three original equations.

$$x + y + 6z = 0 \qquad \text{Equation (3)}$$
$$2 + y + 6(-1) = 0 \qquad \text{Replace } x \text{ with 2 and } z \text{ with } -1.$$
$$2 + y - 6 = 0$$
$$y - 4 = 0$$
$$y = 4$$

The solution is the ordered triple $(2, 4, -1)$. ∎

Our work is simplified when one or more variables are missing from the equations of the system. In such cases a single application of the addition method will usually produce a value for one of the variables.

EXAMPLE 3 *Systems with Missing Variables*

Solve the following system of equations.

$$3x + 4y = 4 \qquad (1)$$
$$-3x + 2z = -3 \qquad (2)$$
$$6x - 8z = 0 \qquad (3)$$

Solution Note that the y-variable is missing from equations (2) and (3). Therefore, we can choose to eliminate x in those equations and immediately obtain a value for z. Multiply equation (2) by 2 and add.

$$2(-3x + 2z) = 2(-3) \qquad \rightarrow \qquad -6x + 4z = -6$$
$$6x - 8z = 0 \qquad \rightarrow \qquad \underline{6x - 8z = 0}$$
$$-4z = -6$$
$$z = \frac{3}{2}$$

Replace z with $\frac{3}{2}$ in equation (2) to determine x.

$$-3x + 2z = -3 \qquad \text{Equation (2)}$$
$$-3x + 2\left(\frac{3}{2}\right) = -3 \qquad \text{Replace } z \text{ with } \frac{3}{2}.$$
$$-3x + 3 = -3$$
$$-3x = -6$$
$$x = 2$$

Finally, replace x with 2 in equation (1) to determine y.

$$3x + 4y = 4 \qquad \text{Equation (1)}$$
$$3(2) + 4y = 4 \qquad \text{Replace } x \text{ with 2.}$$
$$6 + 4y = 4$$
$$4y = -2$$
$$y = -\frac{1}{2}$$

The solution is the ordered triple $(2, -\frac{1}{2}, \frac{3}{2})$. ∎

Special Cases

The addition method will detect the special cases in which the system has no solution or has infinitely many solutions.

EXAMPLE 4   *An Inconsistent System of Three Linear Equations*

Solve the following system.

$$2x - y - 5z = 3 \quad (1)$$
$$5x - y + 14z = -11 \quad (2)$$
$$7x - 2y + 9z = -5 \quad (3)$$

Solution   We eliminate y in two pairs of equations.

$$2x - y - 5z = 3 \quad (1) \qquad\qquad 5x - y + 14z = -11 \quad (2)$$
$$7x - 2y + 9z = -5 \quad (3) \qquad\qquad 7x - 2y + 9z = -5 \quad (3)$$

Multiply both sides of equations (1) and (2) by -2 and then add each pair of equations.

$$
\begin{aligned}
-4x + 2y + 10z &= -6 \\
\underline{7x - 2y + 9z} &= \underline{-5} \\
3x \qquad + 19z &= -11
\end{aligned}
\qquad\qquad
\begin{aligned}
-10x + 2y - 28z &= 22 \\
\underline{7x - 2y + 9z} &= \underline{-5} \\
-3x \qquad - 19z &= 17
\end{aligned}
$$

When we add the two resulting equations, both variables are eliminated.

$$
\begin{aligned}
3x + 19z &= -11 \\
\underline{-3x - 19z} &= \underline{17} \\
0 &= 6 \qquad \text{False}
\end{aligned}
$$

The resulting false statement indicates that there is no solution of the system of equations. The solution set is the empty set, and the system is inconsistent.   ∎

Determining the orientation of the three planes in Example 4 is a matter that we must leave for more advanced courses. As a point of interest, the planes representing equations (2) and (3) intersect in a line. The plane representing equation (1) is parallel to that line. This situation was illustrated in Fig. 6.14.

EXAMPLE 5   *A System of Three Dependent Equations*

Solve the following system of equations.

$$6x - 4y + 2z = 12 \quad (1)$$
$$18x - 12y + 6z = 36 \quad (2)$$
$$-3x + 2y - z = -6 \quad (3)$$

Solution   Notice that multiplying equation (1) by 3 results in equation (2). Multiplying equation (3) by -6 also results in equation (2). In other words, the equations are all equivalent.

Graphically, it can be shown that the three planes coincide, as was illustrated in Fig. 6.15. Therefore, the solution set contains all ordered triples that satisfy any one of the equations. We choose equation (3). The solution set is

$$\{(x, y, z) \mid -3x + 2y - z = -6\}. \qquad \blacksquare$$

Applications

An applied problem involving three unknown quantities may be solved most conveniently with a system of three linear equations in three variables.

EXAMPLE 6 *Three Different Exhibit Fees*

Each fall the Riverfest Craft Fair rents three types of space to exhibitors. The first year, there were 40 spaces for food vendors, 50 spaces to sell crafts, and 20 spaces to demonstrate crafts. The total receipts were $3700.

The next year, there were 30 food vendors, 60 spaces for selling crafts, and 15 demonstration booths, and the total receipts were $3900. The third year, with 25 food vendors, 80 spaces for selling crafts, and 30 demonstration booths, the total receipts were $4925.

Assuming that the fees did not change over these 3 years, what was the fee for each type of space?

Solution   Let $x =$ the fee for a food booth,

$$y = \text{the fee for a craft booth, and}$$
$$z = \text{the fee for a demonstration booth.}$$

The equations are as follows.

$$40x + 50y + 20z = 3700 \qquad (1)$$
$$30x + 60y + 15z = 3900 \qquad (2)$$
$$25x + 80y + 30z = 4925 \qquad (3)$$

Because z has the smallest coefficients, we choose to eliminate z. We select equations (1) and (3) for one pair, and equations (2) and (3) for the other pair.

$$40x + 50y + 20z = 3700 \quad (1) \qquad\qquad 30x + 60y + 15z = 3900 \quad (2)$$
$$25x + 80y + 30z = 4925 \quad (3) \qquad\qquad 25x + 80y + 30z = 4925 \quad (3)$$

In the first pair, multiply equation (1) by -3 and equation (2) by 2. In the second pair multiply equation (1) by -2.

$$
\begin{array}{ll}
-120x - 150y - 60z = -11100 & \qquad -60x - 120y - 30z = -7800 \\
\underline{50x + 160y + 60z = 9850} & \qquad \underline{25x + 80y + 30z = 4925} \\
-70x + 10y \; = -1250 & \qquad -35x - 40y \; = -2875
\end{array}
$$

Now we have a system of two equations in two variables.

$$-70x + 10y = -1250$$
$$-35x - 40y = -2875$$

To eliminate y, multiply the first equation by 4.

$$\begin{array}{rcl} -280x + 40y & = & -5000 \\ \underline{-35x - 40y} & = & \underline{-2875} \\ -315x & = & -7875 \\ x & = & 25 \end{array}$$

Verify that the corresponding values for y and z are $y = 50$ and $z = 10$. The food vendors paid $25, the craft vendors paid $50, and the demonstrators paid $10. ■

6.3 *Quick Reference*

Linear Equations in Three Variables

- A **linear equation in three variables** is any equation that can be written in the standard form $Ax + By + Cz = D$, where A, B, C, and D are real numbers and A, B, and C are not all zero.

- A **solution** of a linear equation in three variables is an **ordered triple** (x, y, z) of numbers that satisfies the equation.

- A **solution** of a system of linear equations in three variables is an ordered triple (x, y, z) that satisfies each equation of the system.

Graphical Interpretation

- The graph of a linear equation in three variables is a plane drawn in a three-dimensional **coordinate space.**

- For a system of three linear equations in three variables, any point that is common to all three planes represents a solution.

- When solving a system of three linear equations in three variables, there are three possible outcomes.

  1. *Unique solution.* The three planes intersect at exactly one point.

  2. *No solution.* At least two of the planes are parallel or one plane is parallel to the line of intersection of the other two planes.

  3. *Infinitely many solutions.* The three planes coincide or the three planes intersect in a common line.

Algebraic Methods

- The addition method is the algebraic method discussed in this section. To use this method, we do the following.

  1. Write all equations in the form $Ax + By + Cz = D$.

  2. Add two pairs of equations to eliminate the same variable in each pair. It may be necessary to multiply one or both of the equations in each pair so that the variable will be eliminated when you add the equations.

  3. Solve the resulting system of two equations in two variables.

  4. Substitute to determine the values of the other two variables.

Special Cases   ▪ When we use the addition method, we can detect special cases after the first elimination of a variable.

1. If the resulting system of two linear equations in two variables is inconsistent, then the system has no solution.

2. If the two linear equations of the resulting system are dependent, then the system has infinitely many solutions.

6.3   *Exercises*

 1. The equation $0 = 0$ can be written $0x + 0y + 0z = 0$. Does this mean that $0 = 0$ is a linear equation in three variables? Explain.

 2. Written beside each equation of the following system is a solution of that equation.

$$x + y + z = 1 \qquad (0, 0, 1)$$
$$x - y + z = 1 \qquad (1, 0, 0)$$
$$x + y - z = 1 \qquad (1, 1, 1)$$

Is $\{(0, 0, 1), (1, 0, 0), (1, 1, 1)\}$ the solution set of the system? Explain.

In Exercises 3–6, determine whether each of the given ordered triples is a solution of the system of equations.

3. $\quad 5x + 7y - 2z = -1$
$\qquad x - 2y + z = 8$
$\qquad 3x - y + 3z = 14$
$\qquad (3, -2, 1), (2, -1, 2)$

4. $\quad 3x - 2y - z = 5$
$\qquad x - y + 2z = -5$
$\qquad 2x + y - z = 18$
$\qquad (0, -1, -3), (5, 6, -2)$

5. $\quad 3x + 4y - z = -12$
$\qquad -x + 2y + 3z = 8$
$\qquad 2x + 6y + z = -8$
$\qquad (-3, 1, 7), (0, -2, 4)$

6. $\quad -3x + 4y + z = -13$
$\qquad 2x - 2y + 3z = -2$
$\qquad 5x + 6y - 2z = 29$
$\qquad \left(4, \dfrac{1}{2}, -3\right), (2, -1, -3)$

In Exercises 7–10, determine the value of a, b, c, or d so that the ordered triple is a solution to the system.

7. $\quad ax - 2y + z = 7$
$\qquad x - y + cz = -2$
$\qquad 2x + by - z = 0$
$\qquad (1, -1, 2)$

8. $\quad x + y + cz = 0$
$\qquad 3x + by + z = 2$
$\qquad ax - y - z = 6$
$\qquad (2, -2, -4)$

9. $\quad 2x + y + z = d$
$\qquad x + by - z = 7$
$\qquad -3x + 2y + cz = 4$
$\qquad (0, 2, -1)$

10. $\quad ax - y + z = 0$
$\qquad x - 3y + z = d$
$\qquad x + y + cz = 8$
$\qquad (3, 2, -1)$

 11. Compare the graph of a linear equation in two variables with the graph of a linear equation in three variables.

 12. If a system of three linear equations in three variables has a unique solution, describe the graph of the system.

 13. Describe one way in which the graph of a system of three linear equations in three variables would indicate that the system has

(a) no solution.

(b) infinitely many solutions.

14. Suppose you are solving the following system.

$$2x + y - 3z = 12$$
$$x - y + z = 0$$
$$3x + 2y - z = 7$$

It is easy to eliminate y in the first two equations and to eliminate z in the last two equations. If you do this, are you making progress toward solving the system? Explain.

In Exercises 15–18, use the addition method to solve the system of equations.

15.
$$x + y + z = 0$$
$$3x + y = 0$$
$$ y - 2z = 7$$

16.
$$2x + 4y + 5z = -3$$
$$ y + 4z = 1$$
$$3x - y = 9$$

17.
$$x + 2y - 3z = 5$$
$$x + 2z = 15$$
$$ 2y - z = 6$$

18.
$$x - y + z = 0$$
$$ 2y + 3z = 1$$
$$x + 2z = 0$$

In Exercises 19–30, use the addition method to solve the system of equations.

19.
$$2x - y + z = 9$$
$$3x + 2y - z = 4$$
$$4x + 3y + 2z = 8$$

20.
$$2x + 3y - 3z = 1$$
$$x - y + 2z = 7$$
$$3x + 2y + z = 6$$

21.
$$x + 2y - 3z = 1$$
$$x + y + 2z = -1$$
$$3x + 3y - z = 4$$

22.
$$x + 3y - 5z = -8$$
$$-2x + y + 3z = 9$$
$$3x - 2y + z = 2$$

23.
$$4x + 5y - 2z = 23$$
$$-6x + 2y + 7z = -14$$
$$8x + 3y + 3z = 11$$

24.
$$-x + y + z = 0$$
$$x - 2y + 3z = 7$$
$$-2x + y - 2z = 9$$

25.
$$7x - 2y + 3z = 19$$
$$x + 8y - 6z = 9$$
$$2x + 4y + 9z = 5$$

26.
$$x + y - 4z = -3$$
$$2x - 3y - 2z = 5$$
$$x + y + z = 2$$

27.
$$2x + 3y + 2z = 2$$
$$-3x - 6y - 4z = -3$$
$$-\frac{1}{6}x + y + z = 0$$

28.
$$3x + 4y - 2z = -4$$
$$-5x + 3y + 4z = 6$$
$$-2x - 2y + 7z = -3$$

29.
$$4x + 3y - 3z = 5$$
$$2x - 6y + 9z = -7$$
$$6x + 6y - 3z = 7$$

30.
$$2x + 4y + 3z = 1$$
$$2x - 8y - 9z = 6$$
$$4x + 12y + 9z = 0$$

31. How do we determine algebraically if a system of equations has

  (a) no solution?

  (b) infinitely many solutions?

32. It is possible to solve the following system of equations without the addition method. Describe the method.

$$x + y - z = 8$$
$$2y + z = 1$$
$$z = 3$$

In Exercises 33–40, determine the solution of the given system. Identify inconsistent systems or dependent equations and describe the solution set for these special cases.

33.
$$x + 2y - z = 5$$
$$x - 2y + z = 2$$
$$2x + 4y - 2z = 7$$

34. $3x - 3y - 4z = -7$
$7x - 6y + 6z = 4$
$4x - 3y + 10z = -5$

35. $x + 2y - z = 3$
$-4x - 8y + 4z = -12$
$3x + 6y - 3z = 9$

36. $2x - y + 4z = 6$
$6x - 3y + 12z = 18$
$-4x + 2y - 8z = -12$

37. $x + y = 1$
$2x + 3y - z = -1$
$3x - y - 4z = 7$

38. $x - 2z = 8$
$2x + y - z = 5$
$3x - 2y - 4z = 6$

39. $x - y + z = 2$
$2x + y - 2z = -2$
$3x - 2y + z = 2$

40. $x + y - z = 2$
$2x - 3y + z = 5$
$3x + 2y - 4z = 3$

In Exercises 41–52, describe the solution set of each system of equations.

41. $x + y + z = 0$
$2x + 3y + 2 = 0$
$y - 4z = 0$

42. $3x + y + z = 2$
$x - 2z = 7$
$y + z + 1 = 0$

43. $x + y = 1$
$y - 2z = 2$
$x - 3z = 14$

44. $x + y = 1$
$y - 2z = 5$
$x - 3z = 1$

45. $x + z = y$
$\dfrac{1}{2}y + \dfrac{3}{4}z = \dfrac{1}{4}$
$\dfrac{1}{2}x + z + \dfrac{1}{2} = 0$

46. $x + y = 3 + z$
$\dfrac{1}{4}y - \dfrac{3}{4}z = 2$
$\dfrac{1}{2}x - z = \dfrac{3}{2}$

47. $3x + 2z = y + 3$
$2(x + y) = 2 - z$
$4 + 3y = x + z$

48. $6x + 3z = 1 + 3y$
$4 + y = 2x + z$
$4x - 2y = 7 - 2z$

49. $4x + 6z = 12 + 3y$
$6 + 1.5y = 2x + 3z$
$0.5x - 0.375y + 0.75z = 1.5$

50. $5x - 2y + 15z = 20$
$x - 0.4y + 3z = 4$
$2.5x - y + 7.5z = 10$

51. $\dfrac{2}{3}x + \dfrac{5}{2}y - z = -21$
$x - \dfrac{1}{2}y + \dfrac{1}{8}z = \dfrac{1}{2}$
$y + 2z = 8 + 2x$

52. $\dfrac{1}{4}x + \dfrac{1}{3}y - \dfrac{1}{6}z = \dfrac{1}{3}$
$x + y + \dfrac{1}{2}z = 1$
$\dfrac{1}{3}x - \dfrac{1}{2}y + \dfrac{1}{12}z = \dfrac{1}{4}$

In Exercises 53–60, translate the given information into a system of equations, solve the system, and answer the question.

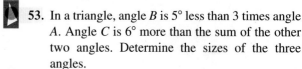
53. In a triangle, angle B is 5° less than 3 times angle A. Angle C is 6° more than the sum of the other two angles. Determine the sizes of the three angles.

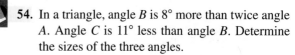
54. In a triangle, angle B is 8° more than twice angle A. Angle C is 11° less than angle B. Determine the sizes of the three angles.

55. A 100-yard rope is cut into three pieces. The second piece is 12 feet longer than the sum of the lengths of the other two pieces. The third piece is 6 yards less than the length of the first piece. How long is each piece?

56. A 14-foot board is cut into three pieces. The second piece is 2 feet less than twice the length of the first piece. The third piece is 36 inches less than the sum of the lengths of the other two pieces. How long is each piece?

57. A collection of 155 coins consists of nickels, dimes, and quarters. The total value of the coins is $24.95. If the number of dimes is two more than twice the number of nickels, how many of each coin is in the collection?

58. A cash register contains a total of 90 coins consisting of pennies, nickels, dimes, and quarters. There are only three pennies and the total value of the coins is $9.28. If there are three times as many dimes as quarters, how many of each coin are in the cash register?

59. A store has a sale on T-shirts, sweatshirts, and tank tops. A purchase of 3 T-shirts, 2 sweatshirts, and 5 tank tops costs $77. It costs $60 for 5 T-shirts and 1 sweatshirt or for 4 T-shirts and 6 tank tops. What is the sale price for each shirt?

60. A total of $10,000 is to be invested in three funds. The following table shows three possible combinations of investments along with the total return for each combination.

Growth Fund	Income Fund	Money Market	Total Return
$2000	$5000	$3000	$460
5000	3000	2000	510
7000	2000	1000	550

What is the percentage rate of return of each fund?

Challenge

In Exercises 61–64, extend the substitution method to solve the given system of three equations in three variables.

61.
$$2x + y \quad\ = -4$$
$$x \quad\ + z = \ 2$$
$$2y - z = -1$$

62.
$$x - y \quad\ = \ 0$$
$$3x \quad\ + z = 11$$
$$y - z = \ 1$$

63.
$$x + y + z = -2$$
$$x - 3y = 8$$
$$z - 2y = 8$$

64.
$$2x - 3y = 0$$
$$2z + y = -2$$
$$2x - 2y + 3z = -4$$

In Exercises 65 and 66, extend the addition method to solve a system of four linear equations in four variables.

65.
$$x + y + z - w = -1$$
$$2x - y + 2z + w = \ 1$$
$$x - y - z + 2w = \ 6$$
$$x + 2y + 3z - 2w = -4$$

66.
$$x + y + z = \ 2$$
$$y + z + w = \ 1$$
$$x + z + w = -1$$
$$x + y + w = -2$$

In Exercises 67 and 68, make an appropriate substitution to solve the systems of equations. (See Section 6.2, Exercises 89–92.)

67.
$$\frac{3}{x} + \frac{2}{y} - \frac{1}{z} = -1$$
$$\frac{1}{x} + \frac{1}{y} + \frac{1}{z} = \ 2$$
$$\frac{1}{x} - \frac{1}{y} + \frac{2}{z} = \ 6$$

68.
$$\frac{2}{x} + \frac{3}{y} = 2$$
$$\frac{3}{y} - \frac{1}{z} = 0$$
$$\frac{4}{x} - \frac{2}{z} = 0$$

69. Solve the given system for x, y, and z. Express your solutions in terms of a, b, and c.

$$x + y \quad\;\; = a$$
$$\quad\;\; y + z = b$$
$$x \quad\;\; + z = c$$

70. For what value(s) of a is the solution of the system unique?

$$ax + 2y + \;\; z = 0$$
$$2x - \;\; y + 2z = 0$$
$$x + \;\; y + 3z = 0$$

6.4 Matrices

Matrices ▪ *Matrices and Systems of Equations* ▪ *Special Cases* ▪ *Automatic Solving with a Calculator*

Matrices

A **matrix** is an array of numbers. As is true with ordered pairs, the position of each of the numbers is meaningful. The following are examples of matrices.

$$A = \begin{bmatrix} 3 & -5 & 0 \\ 2 & 4 & 1 \\ 2 & 1 & 6 \end{bmatrix} \qquad B = \begin{bmatrix} 2 & 4 & -5 & -1 \\ 0 & 2 & 7 & -2 \end{bmatrix}$$

The matrices are given letter names for easy reference. The numbers in the array are called the **elements** of the matrix. The **rows** of the matrix are read horizontally, and the **columns** are read vertically. In matrix B the elements in the first row are 2, 4, -5, and -1. The elements in the first column are 2 and 0.

We refer to the rows and columns of a matrix with the symbols R and C. For example, R_2 refers to row 2 and C_1 refers to column 1.

The **dimensions** of a matrix are the number of rows and the number of columns. Matrix A has 3 rows and 3 columns, and we refer to it as a 3×3 (read "3 by 3") matrix. Matrix B has 2 rows and 4 columns, and we refer to it as a 2×4 (read "2 by 4") matrix. In general, an $m \times n$ matrix has m rows and n columns.

If the number of rows and number of columns is the same, the matrix is a **square matrix.** The 3×3 matrix A is an example of a square matrix.

Matrices and Systems of Equations

We have discussed two algebraic methods for solving systems of equations. Another solving method involves the use of matrices.

Recall that we solve systems by using the properties of equations to produce equivalent equations. However, when we apply those properties, only the coefficients and constants change. Therefore, we can use matrices of these numbers along with techniques analogous to algebraic operations to solve the same systems.

Consider the following system of equations.

$$2x + 3y = -5$$
$$5x - 2y = 16$$

We define the **coefficient matrix** to be a matrix whose elements are the coefficients of the variables in the equations. For this system the coefficient matrix is

$$A = \begin{bmatrix} 2 & 3 \\ 5 & -2 \end{bmatrix}.$$

The **constant matrix** is a matrix whose elements are the constants on the right side of the system of equations. For the given system the constant matrix is

$$B = \begin{bmatrix} -5 \\ 16 \end{bmatrix}.$$

The **augmented matrix** is the combination of the coefficient matrix and the constant matrix. For the given system the augmented matrix is

$$M = \left[\begin{array}{cc|c} 2 & 3 & -5 \\ 5 & -2 & 16 \end{array} \right].$$

To solve the original system of equations, our strategy is to try to transform the system's augmented matrix into a matrix of the form

$$\left[\begin{array}{cc|c} 1 & 0 & n_1 \\ 0 & 1 & n_2 \end{array} \right]$$

where n_1 and n_2 are numbers. If we can do this, then, when we translate this matrix back into a system of equations, we obtain

$$1x + 0y = n_1$$
$$0x + 1y = n_2.$$

With the system written in this way, it is then clear that the solutions are $x = n_1$ and $y = n_2$.

In order to transform a matrix in this way, we use **row operations.** Row operations correspond to the algebraic operations we perform on the equations of a system. Just as the properties of equations allow us to produce equivalent systems, row operations can be used to produce **row-equivalent** matrices.

The following is a summary of permissible row operations. Included in this summary are examples of the notation we use to indicate each operation.

The Elementary Row Operations on Matrices

For any two rows R_1 and R_2,

1. The rows may be interchanged: $R_2 \leftrightarrow R_1$.

2. A row may be multiplied by a nonzero number n and the original row may be replaced by the result: $n \cdot R_1 \rightarrow R_1$.

3. A row may be replaced by the sum of itself and another row:
   $R_1 + R_2 \rightarrow R_1$.

4. A row may be replaced by the sum of itself and a multiple of another row: $R_1 + n \cdot R_2 \rightarrow R_1$.

EXAMPLE 1 *Using Matrices to Solve a System of Two Equations in Two Variables*

Solve the following system of equations.

$$2x + y = 1$$
$$x - y = 5$$

Solution

$$\begin{bmatrix} 2 & 1 & | & 1 \\ 1 & -1 & | & 5 \end{bmatrix}$$

Augmented matrix

$$2x + 1y = 1$$
$$1x - 1y = 5$$

Row Operation	**Row-Equivalent Matrix**		**System of Equations**
$R_2 \leftrightarrow R_1$	$\begin{bmatrix} 1 & -1 & \| & 5 \\ 2 & 1 & \| & 1 \end{bmatrix}$	Interchange rows to obtain a 1 in row 1, column 1.	$1x - 1y = 5$ $\quad$ $2x + 1y = 1$
$-2 \cdot R_1 + R_2 \rightarrow$	$\begin{bmatrix} 1 & -1 & \| & 5 \\ 0 & 3 & \| & -9 \end{bmatrix}$	Multiply row 1 by -2 and add the result to row 2. Now we have a 0 in row 2, column 1.	$1x - 1y = 5$ $\quad$ $0x + 3y = -9$
$R_2 \div 3 \rightarrow$	$\begin{bmatrix} 1 & -1 & \| & 5 \\ 0 & 1 & \| & -3 \end{bmatrix}$	Divide row 2 by 3 to obtain a 1 in row 2, column 2.	$1x - 1y = 5$ $\quad$ $0x + 1y = -3$
$R_1 + R_2 \rightarrow$	$\begin{bmatrix} 1 & 0 & \| & 2 \\ 0 & 1 & \| & -3 \end{bmatrix}$	Replace row 1 with the sum of rows 1 and 2. Now we have a 0 in row 1, column 2.	$1x + 0y = 2$ $\quad$ $0x + 1y = -3$

The system of equations represented by this matrix is equivalent to the original system. The solution of the system is the ordered pair $(2, -3)$. ∎

The matrix method for solving a system of three equations in three variables involves the same strategy as was illustrated in Example 1, but, because the augmented matrix is bigger (3×4), more row operations are usually needed to achieve the result.

The goal remains the same. The coefficient matrix must have 1's down its diagonal and 0's elsewhere. Then, the solution (ordered triple) is determined from the numbers in the right-hand column.

EXAMPLE 2 *Using Matrices to Solve a System of Three Equations in Three Variables*

Solve the following system of equations.

$$x + y + 2z = -1$$
$$3x + y - 6z = 7$$
$$-x + 2y + 2z = 0$$

Solution

$$\begin{bmatrix} 1 & 1 & 2 & | & -1 \\ 3 & 1 & -6 & | & 7 \\ -1 & 2 & 2 & | & 0 \end{bmatrix}$$ Augmented matrix

$-3 \cdot R_1 + R_2 \rightarrow$ $$\begin{bmatrix} 1 & 1 & 2 & | & -1 \\ 0 & -2 & -12 & | & 10 \\ -1 & 2 & 2 & | & 0 \end{bmatrix}$$ Multiply row 1 by -3 and add the result to row 2.

$R_1 + R_3 \rightarrow$ $$\begin{bmatrix} 1 & 1 & 2 & | & -1 \\ 0 & -2 & -12 & | & 10 \\ 0 & 3 & 4 & | & -1 \end{bmatrix}$$ Replace row 3 with the sum of rows 1 and 3.

$R_2 \div (-2) \rightarrow$ $$\begin{bmatrix} 1 & 1 & 2 & | & -1 \\ 0 & 1 & 6 & | & -5 \\ 0 & 3 & 4 & | & -1 \end{bmatrix}$$ Divide row 2 by -2.

$-1 \cdot R_2 + R_1 \rightarrow$ $$\begin{bmatrix} 1 & 0 & -4 & | & 4 \\ 0 & 1 & 6 & | & -5 \\ 0 & 3 & 4 & | & -1 \end{bmatrix}$$ Multiply row 2 by -1 and add the result to row 1.

$-3 \cdot R_2 + R_3 \rightarrow$ $$\begin{bmatrix} 1 & 0 & -4 & | & 4 \\ 0 & 1 & 6 & | & -5 \\ 0 & 0 & -14 & | & 14 \end{bmatrix}$$ Multiply row 2 by -3 and add the result to row 3.

$R_3 \div (-14) \rightarrow$ $$\begin{bmatrix} 1 & 0 & -4 & | & 4 \\ 0 & 1 & 6 & | & -5 \\ 0 & 0 & 1 & | & -1 \end{bmatrix}$$ Divide row 3 by -14.

$4 \cdot R_3 + R_1 \rightarrow$ $$\begin{bmatrix} 1 & 0 & 0 & | & 0 \\ 0 & 1 & 6 & | & -5 \\ 0 & 0 & 1 & | & -1 \end{bmatrix}$$ Multiply row 3 by 4 and add the result to row 1.

$-6 \cdot R_3 + R_2 \rightarrow$ $$\begin{bmatrix} 1 & 0 & 0 & | & 0 \\ 0 & 1 & 0 & | & 1 \\ 0 & 0 & 1 & | & -1 \end{bmatrix}$$ Multiply row 3 by -6 and add the result to row 2.

The solution is $(0, 1, -1)$.

Special Cases

Matrix solution methods reveal inconsistent systems and dependent equations in the same way as algebraic methods.

EXAMPLE 3   *An Inconsistent System*

Solve the following system of equations.

$$x - 2y = 5$$
$$2x - 4y = 4 \ ,$$

Solution

$$\begin{bmatrix} 1 & -2 & \bigg| & 5 \\ 2 & -4 & \bigg| & 4 \end{bmatrix}$$ Augmented matrix

$$-2R_1 + R_2 \rightarrow \begin{bmatrix} 1 & -2 & \bigg| & 5 \\ 0 & 0 & \bigg| & -6 \end{bmatrix}$$ Multiply row 1 by -2 and add the result to row 2.

The second row corresponds to the equation $0x + 0y = -6$ or $0 = -6$. This false equation indicates that the system has no solution. ∎

EXAMPLE 4 *Dependent Equations*

Solve the following system of equations.

$$2x + y = 3$$
$$4x + 2y = 6$$

Solution

$$\begin{bmatrix} 2 & 1 & \bigg| & 3 \\ 4 & 2 & \bigg| & 6 \end{bmatrix}$$ Augmented matrix

$$-2R_1 + R_2 \rightarrow \begin{bmatrix} 2 & 1 & \bigg| & 3 \\ 0 & 0 & \bigg| & 0 \end{bmatrix}$$ Multiply row 1 by -2 and add the result to row 2.

The second row corresponds to the equation $0x + 0y = 0$ or $0 = 0$, which is true. The system has infinitely many solutions. ∎

Automatic Solving with a Calculator

These same matrix methods can be used to solve larger systems of equations. Of course, larger systems require more row operations.

Using matrix operations that are beyond the scope of this book, a calculator can solve a system automatically.

EXAMPLE 5 *Automatic Solving with a Calculator*

 SOLVE

Use your calculator to solve the system of equations in Example 2.

$$x + y + 2z = -1$$
$$3x + y - 6z = 7$$
$$-x + 2y + 2z = 0$$

Figure 6.17

```
[C]
[ 0 ]
[ 1 ]
[ -1]
```

Solution   Let A = coefficient matrix and B = constant matrix.

$$A = \begin{bmatrix} 1 & 1 & 2 \\ 3 & 1 & -6 \\ -1 & 2 & 2 \end{bmatrix} \qquad B = \begin{bmatrix} -1 \\ 7 \\ 0 \end{bmatrix}$$

Figure 6.17 shows a typical screen display of the result. The solution of the system is $(0, 1, -1)$. ∎

6.4 *Quick Reference*

Matrices

- A **matrix** is an array of numbers called **elements.**

- The **dimensions** of a matrix are the number of rows and the number of columns. (An $m \times n$ matrix has m rows and n columns.) If the dimensions are the same, the matrix is a **square matrix.**

Matrices and Systems of Equations

- For a system of equations written in standard form,

  1. the **coefficient matrix** is the matrix of coefficients of the variables,

  2. the **constant matrix** is the matrix of constants on the right sides of the equations, and

  3. the **augmented matrix** is the combination of the coefficient and constant matrices.

- For any matrix the permissible row operations are as follows.

  1. Rows may be interchanged.

  2. A row may be multiplied by a number.

  3. A row may be replaced by the sum of itself and another row.

  4. A row may be replaced by the sum of itself and a multiple of another row.

- When solving systems with matrices, the goal is to obtain 1's along the diagonal of the coefficient matrix and 0's elsewhere. The solution is then determined from the numbers in the last column.

Special Cases

- When solving a system of equations with matrix methods, a special case exists if at any time a row of the coefficient matrix is all 0's.

  1. If the associated element of the constant matrix is not 0, the system is inconsistent and has no solution.

  2. If the associated element of the constant matrix is 0, at least two equations of the system are dependent, and the system has infinitely many solutions.

Automatic Solving with a Calculator

- Most sophisticated calculators can solve a system of equations automatically.

- After the solving procedure has been executed, the calculator typically displays a solution matrix.

6.4   Exercises

 1. What is a 2 × 4 matrix?

2. For the matrix $\begin{bmatrix} 0 & 5 \\ 1 & 9 \end{bmatrix}$,

are the elements of the second column 1 and 9 or 5 and 9? Why?

In Exercises 3–8, state the dimensions of the matrix.

3. $\begin{bmatrix} 2 & -3 & -5 & 4 \\ 1 & 0 & -6 & 7 \end{bmatrix}$ **4.** $\begin{bmatrix} 2 & -1 \\ -1 & 4 \\ 5 & -7 \end{bmatrix}$

5. $\begin{bmatrix} 2 & 4 \\ -5 & 0 \end{bmatrix}$ **6.** $\begin{bmatrix} 1 & 2 & -3 \\ 0 & -4 & 5 \\ 6 & 0 & -7 \end{bmatrix}$

7. $\begin{bmatrix} 2 & -3 & 4 & -5 \\ 0 & 1 & -7 & 6 \\ 5 & 8 & 0 & -9 \end{bmatrix}$ **8.** $\begin{bmatrix} 1 \\ 2 \\ -3 \\ 0 \end{bmatrix}$

 9. In your own words, describe each of the following.

(a) A coefficient matrix

(b) A constant matrix

(c) An augmented matrix

 10. For a system of n linear equations in n variables, explain why the coefficient matrix must be a square matrix.

In Exercises 11–14, write the augmented matrix for the given system of linear equations.

11. $3x + 2y = 6$
$x - 4y = 9$

12. $3x - 2y = 8$
$6x = 4y + 17$

13. $x + 2y - 3z = 5$
$x + 2z = 15$
$2y - z = 6$

14. $x + y - z = 0$
$2y + 3z = 1$
$x + 2z = 0$

In Exercises 15–18, write a system of linear equations associated with the augmented matrix.

15. $\begin{bmatrix} 2 & 1 & | & 1 \\ 3 & -2 & | & 12 \end{bmatrix}$ **16.** $\begin{bmatrix} 2 & 1 & | & 8 \\ 0 & 1 & | & 2 \end{bmatrix}$

17. $\begin{bmatrix} 1 & 1 & -1 & | & 2 \\ 2 & -3 & 1 & | & 5 \\ 3 & 2 & -4 & | & 3 \end{bmatrix}$

18. $\begin{bmatrix} 1 & 1 & -1 & | & 3 \\ 0 & 1 & -3 & | & 8 \\ 1 & 0 & -2 & | & 3 \end{bmatrix}$

 19. When using matrices to solve a system of linear equations, what is your goal with respect to the coefficient matrix?

 20. For a system of n linear equations in n variables, what are the dimensions of the constant matrix? Why?

In Exercises 21–24, state the row operation that has been performed in each step of the matrix solution of the given system of linear equations. Then, state the solution.

21. $3x + 6y = 12$
$2x - 3y = 1$

Matrix	Row Operation
$\begin{bmatrix} 3 & 6 & \vert & 12 \\ 2 & -3 & \vert & 1 \end{bmatrix}$	Augmented matrix
$\begin{bmatrix} 1 & 2 & \vert & 4 \\ 2 & -3 & \vert & 1 \end{bmatrix}$	(a) _____
$\begin{bmatrix} 1 & 2 & \vert & 4 \\ 0 & -7 & \vert & -7 \end{bmatrix}$	(b) _____
$\begin{bmatrix} 1 & 2 & \vert & 4 \\ 0 & 1 & \vert & 1 \end{bmatrix}$	(c) _____
$\begin{bmatrix} 1 & 0 & \vert & 2 \\ 0 & 1 & \vert & 1 \end{bmatrix}$	(d) _____

(e) The solution is _____.

22. $2x - 6y = -18$
$3x + 4y = \ -1$

Matrix	*Row Operation*
$\begin{bmatrix} 2 & -6 & \vert & -18 \\ 3 & 4 & \vert & -1 \end{bmatrix}$	Augmented matrix
$\begin{bmatrix} 1 & -3 & \vert & -9 \\ 3 & 4 & \vert & -1 \end{bmatrix}$	(a) _____
$\begin{bmatrix} 1 & -3 & \vert & -9 \\ 0 & 13 & \vert & 26 \end{bmatrix}$	(b) _____
$\begin{bmatrix} 1 & -3 & \vert & -9 \\ 0 & 1 & \vert & 2 \end{bmatrix}$	(c) _____
$\begin{bmatrix} 1 & 0 & \vert & -3 \\ 0 & 1 & \vert & 2 \end{bmatrix}$	(d) _____

(e) The solution is _____ .

23. $x + y - \ z = \ \ 4$
$2x - y + \ z = -1$
$x + y - 2z = \ \ 5$

Matrix	*Row Operation*
$\begin{bmatrix} 1 & 1 & 1 & \vert & 4 \\ 2 & -1 & 1 & \vert & -1 \\ 1 & 1 & -2 & \vert & 5 \end{bmatrix}$	Augmented matrix
$\begin{bmatrix} 1 & 1 & -1 & \vert & 4 \\ 0 & -3 & 3 & \vert & -9 \\ 0 & 0 & -1 & \vert & 1 \end{bmatrix}$	(a) _____ (b) _____
$\begin{bmatrix} 1 & 1 & -1 & \vert & 4 \\ 0 & 1 & -1 & \vert & 3 \\ 0 & 0 & 1 & \vert & -1 \end{bmatrix}$	(c) _____ (d) _____
$\begin{bmatrix} 1 & 1 & 0 & \vert & 3 \\ 0 & 1 & 0 & \vert & 2 \\ 0 & 0 & 1 & \vert & -1 \end{bmatrix}$	(e) _____ (f) _____
$\begin{bmatrix} 1 & 0 & 0 & \vert & 1 \\ 0 & 1 & 0 & \vert & 2 \\ 0 & 0 & 1 & \vert & -1 \end{bmatrix}$	(g) _____

(h) The solution is _____ .

24. $x + \ y - \ z = 4$
$2y - 5z = 2$
$7z = 0$

Matrix	*Row Operation*
$\begin{bmatrix} 1 & 1 & -1 & \vert & 4 \\ 0 & 2 & -5 & \vert & 2 \\ 0 & 0 & -7 & \vert & 0 \end{bmatrix}$	Augmented matrix $-\!1 \cdot R_3 \rightarrow R_3$
$\begin{bmatrix} 1 & 1 & -1 & \vert & 4 \\ 0 & 2 & -5 & \vert & 2 \\ 0 & 0 & 1 & \vert & 0 \end{bmatrix}$	(a) _____
$\begin{bmatrix} 1 & 1 & 0 & \vert & 4 \\ 0 & 2 & 0 & \vert & 2 \\ 0 & 0 & 1 & \vert & 0 \end{bmatrix}$	(b) _____ (c) _____
$\begin{bmatrix} 1 & 1 & 0 & \vert & 4 \\ 0 & 1 & 0 & \vert & 1 \\ 0 & 0 & 1 & \vert & 0 \end{bmatrix}$	(d) _____
$\begin{bmatrix} 1 & 0 & 0 & \vert & 3 \\ 0 & 1 & 0 & \vert & 1 \\ 0 & 0 & 1 & \vert & 0 \end{bmatrix}$	(e) _____

(f) The solution is _____ .

In Exercises 25–28, fill in the blanks to indicate the results of performing row operations. State the solution.

25. $\begin{bmatrix} 4 & 8 & \vert & 52 \\ 3 & -1 & \vert & -17 \end{bmatrix}$

$\begin{bmatrix} 1 & \text{(a)} & \vert & 13 \\ 3 & -1 & \vert & -17 \end{bmatrix}$

$\begin{bmatrix} 1 & \text{(b)} & \vert & 13 \\ 0 & -7 & \vert & \text{(c)} \end{bmatrix}$

$\begin{bmatrix} 1 & \text{(d)} & \vert & 13 \\ 0 & \text{(e)} & \vert & 8 \end{bmatrix}$

$\begin{bmatrix} 1 & 0 & \vert & \text{(f)} \\ 0 & \text{(g)} & \vert & 8 \end{bmatrix}$

(h) Solution: _____

26. $\begin{bmatrix} 6 & 2 & \vert & -2 \\ 1 & -1 & \vert & 1 \end{bmatrix}$

$\begin{bmatrix} \dfrac{(a)}{6} & -1 & \vert & \dfrac{(b)}{-2} \\ & 2 & \vert & \end{bmatrix}$

$\begin{bmatrix} \dfrac{(c)}{0} & -1 & \vert & \dfrac{(d)}{-8} \\ & (e) & \vert & \end{bmatrix}$

$\begin{bmatrix} \dfrac{(f)}{0} & -1 & \vert & \dfrac{(g)}{(h)} \\ & 1 & \vert & \end{bmatrix}$

$\begin{bmatrix} (i) & 0 & \vert & (j) \\ 0 & 1 & \vert & (k) \end{bmatrix}$

(l) Solution: _____

27. $\begin{bmatrix} 2 & -3 & 1 & \vert & 0 \\ 1 & 1 & -1 & \vert & 3 \\ 3 & -2 & -2 & \vert & 7 \end{bmatrix}$

$\begin{bmatrix} 1 & 1 & -1 & \vert & (a) \\ 2 & (b) & 1 & \vert & 0 \\ 3 & -2 & -2 & \vert & 7 \end{bmatrix}$

$\begin{bmatrix} 1 & 1 & -1 & \vert & 3 \\ 0 & -5 & (c) & \vert & -6 \\ 0 & -5 & 1 & \vert & (d) \end{bmatrix}$

$\begin{bmatrix} 1 & (e) & -1 & \vert & 3 \\ 0 & -5 & 3 & \vert & -6 \\ 0 & 0 & (f) & \vert & 4 \end{bmatrix}$

$\begin{bmatrix} 1 & 1 & -1 & \vert & 3 \\ 0 & -5 & 3 & \vert & (g) \\ 0 & 0 & 1 & \vert & (h) \end{bmatrix}$

$\begin{bmatrix} 1 & 1 & 0 & \vert & (i) \\ 0 & -5 & (j) & \vert & 0 \\ 0 & 0 & 1 & \vert & -2 \end{bmatrix}$

$\begin{bmatrix} 1 & (k) & 0 & \vert & 1 \\ 0 & 1 & 0 & \vert & (l) \\ 0 & 0 & 1 & \vert & -2 \end{bmatrix}$

$\begin{bmatrix} 1 & 0 & (m) & \vert & 1 \\ (n) & 1 & 0 & \vert & 0 \\ 0 & 0 & 1 & \vert & -2 \end{bmatrix}$

(o) Solution: _____

28. $\begin{bmatrix} 3 & 6 & -3 & \vert & 9 \\ 1 & -2 & 1 & \vert & 3 \\ 2 & -1 & -1 & \vert & 0 \end{bmatrix}$

$\begin{bmatrix} \underline{(a)} & 2 & 1 & \vert & 3 \\ \dfrac{(b)}{2} & -2 & 1 & \vert & 3 \\ 2 & -1 & -1 & \vert & 0 \end{bmatrix}$

$\begin{bmatrix} 1 & 2 & -1 & \vert & 3 \\ 0 & (c) & 2 & \vert & 0 \\ 0 & (d) & 1 & \vert & -6 \end{bmatrix}$

$\begin{bmatrix} 1 & -3 & (e) & \vert & -3 \\ 0 & 6 & 0 & \vert & (f) \\ 0 & -5 & 1 & \vert & -6 \end{bmatrix}$

$\begin{bmatrix} 1 & -3 & 0 & \vert & -3 \\ 0 & (g) & 0 & \vert & 2 \\ 0 & -5 & 1 & \vert & (h) \end{bmatrix}$

$\begin{bmatrix} 1 & 0 & 0 & \vert & (i) \\ 0 & 1 & 0 & \vert & 2 \\ 0 & 0 & 1 & \vert & (j) \end{bmatrix}$

(k) Solution: _____

 29. For the system

$$2x = 3y + 2$$
$$y + 1 = x,$$

what must be done before the coefficient matrix and constant matrix can be written?

 30. Each of the following shows the last row of an augmented matrix after row operations are performed. In each part, describe the solution set of the associated system of equations.

(a) 0 0 | 6

(b) 0 0 | 0

(c) 0 1 | 0

In Exercises 31–42, use matrix methods to solve the given system of equations. Identify inconsistent systems and dependent equations.

31. $2x - 3y = 18$
    $x + 2y = -5$

32. $3x + 4y = 0$
    $x + 2y = 2$

33. $4x + 8y = 0$
$x - 2y = 1$

34. $y - x = 1$
$2x + y = 0$

35. $2x - 4y = 7$
$2y - x = 4$

36. $2x - 3y = 7$
$4x = 6y + 7$

37. $3x - 2y = 6$
$6y - 9x = -18$

38. $4x - 5y = 7$
$10y = 8x - 14$

39. $x + z = 1$
$y - z = 5$
$x - y = 2$

40. $x - y = 8$
$x + z = 7$
$y + 2z = 1$

41. $x + y - z = -4$
$-x - 2y + z = 7$
$2x + 2y + z = -2$

42. $-x + 2y - z = -7$
$3x - 7y - 2z = 23$
$-2x + 2y + 3z = -10$

51. $x - 2y - 2z = -2$
$2x + 3y + z = -1$
$3x + y + z = 1$

52. $x + y + z = 0$
$2x - 2y + 3z = 6$
$5x - y + 2z = 0$

53. $x + y - z = 0$
$2x - 3y + 4z = 4$
$3x - 2y + z = 6$

54. $3x - 2y + 5z = -16$
$2x + 5y + 4z = 0$
$4x - 6y - 7z = 13$

In Exercises 43–46, write the coefficient matrix and the constant matrix that are associated with the given system of equations.

43. $2x - 5y = -5$
$2y + 5x = 2$

44. $x - 3y = 9$
$5y = x - 17$

45. $z + 3x = 4y + 2$
$1 - y = x + 3z$
$z = 2x + 3y$

46. $z = x + y$
$y - 2 = 2x + 3z$
$z - 1 = 3x + 2y$

In Exercises 47–54, use your calculator to solve the given system of equations.

47. $3x + 4y = -5$
$x + 3 = 0$

48. $y - 2 = 4$
$2x - y = -2$

49. $3x + y + z = 5$
$2x - y + z = 6$
$y = 2z + 2$

50. $x + 3y + z = 2$
$2y + z = 1$
$x - 2y - 3z = 4$

Exploring with Real Data

In Exercises 55–58, refer to the following background information and data.

The following table shows the number (in millions) of delivered pieces of mail, by class, for the years 1980, 1985, and 1990.

Class of Mail	1980	1985	1990
First Class	60,332	72,517	89,917
Second Class	10,221	10,380	10,680
Third Class	30,381	52,170	63,725
Fourth Class	633	576	663

(Source: U.S. Postal Service.)

55. Write the data in this table as a matrix. What are the dimensions of the matrix?

56. What does the entry in the third row, second column, represent?

57. In what row or column of the matrix would you look to determine the trend in the number of second class mailings?

58. Some argue that postal rates have climbed steadily because of the amount of "junk mail" being delivered. Which class of mail experienced the greatest percentage increase from 1980 to 1990?

Challenge

59. (a) Suppose you use your calculator to solve the following system.

$$z + x - 2y = 11$$
$$3z - 2x + y = -22$$
$$-z + 4x - y = 24$$

You enter the matrices

$$\begin{bmatrix} 1 & 1 & -2 \\ 3 & -2 & 1 \\ -1 & 4 & -1 \end{bmatrix} \qquad \begin{bmatrix} 11 \\ -22 \\ 24 \end{bmatrix}$$

and the calculator result is

$$\begin{bmatrix} -3 \\ 4 \\ -5 \end{bmatrix}.$$

What is the solution?

(b) Now write each equation of the system in the form $Ax + By + Cz = D$ and solve the resulting system. Is the solution the same as the one you obtained in part (a)?

In Exercises 60–63, use a calculator to solve the system.

60.
$$2x + y - z + w = -7$$
$$x - 2y + z - 2w = 16$$
$$-x + 2y + 3z + 4w = -15$$
$$3x - y - 2z - w = 3$$

61.
$$x + y + z + w = -2$$
$$5x - 2y + 3z + w = 1$$
$$-x + 2y - z - 2w = -9$$
$$3x + 4y + 2z + 3w = -10$$

62.
$$x - z = 6$$
$$y + w + 2 = 0$$
$$x - y + w + 6 = 0$$
$$y - x = 1$$

63.
$$x + y = 0$$
$$y + 2z + 3 = 0$$
$$2z + 3w - 4 = 0$$
$$3w + 4x = 10$$

64. Use matrix methods to solve the following system.

$$x + y = 1$$
$$x + cy = 5$$

For what value(s) of c is the solution unique? For what value(s) of c is there no solution?

65. Use matrix methods to solve the following system.

$$x + 3y = 2$$
$$ax + y = 1$$

For what value(s) of a is the solution unique? For what value(s) of a is there no solution?

66. Use matrix methods to determine p, q, and r so that the given system of equations has

(a) a unique solution.

(b) no solution.

(c) infinitely many solutions.

$$x - y = p$$
$$y - z = q$$
$$z - x = r$$

6.5 Applications

2 × 2 Systems   ■   *3 × 3 Systems*   ■   *Curve Fitting*

2 × 2 Systems

In this section we use our ability to solve systems of equations for solving a variety of application problems. While our discussion focuses on specific examples, our goal remains to develop a general understanding of problem-solving strategies. We begin with examples of applications that lead to systems of two equations in two variables.

Mixture problems typically involve two or more items, each with a different cost. The items are combined to create a mixture whose cost is between the costs of the individual items.

EXAMPLE 1   *A Mixture of Cashews and Peanuts*

Cashews costing $6.00 per pound and peanuts costing $2.00 per pound are mixed together. The mixture weighs 10 pounds and costs $3.00 per pound. How many pounds of cashews and how many pounds of peanuts are used in the mixture?

Solution   Let c = the number of pounds of cashews and

p = the number of pounds of peanuts.

We can use a table to organize the information.

	Weight in Pounds	×	*Cost per Pound*	=	*Total Cost*
Cashews	c		6		$6c$
Peanuts	p		2		$2p$
Mixture	10		3		30

From this table we write the following system of equations.

$$c + p = 10 \qquad \text{The total weight of the mixture is 10 pounds.}$$
$$6c + 2p = 30 \qquad \text{The total cost of the mixture is \$30.00.}$$

$$
\begin{aligned}
-2c - 2p &= -20 \\
6c + 2p &= 30
\end{aligned}
\qquad
\begin{array}{l}\text{Multiply the first equation by } -2 \\ \text{and add the equations.}\end{array}
$$

$$4c = 10$$
$$c = \frac{10}{4} = 2.5$$

$$c + p = 10 \qquad \text{First equation}$$
$$2.5 + p = 10 \qquad \text{Replace } c \text{ with 2.5.}$$
$$p = 7.5$$

The mixture consists of 2.5 pounds of cashews and 7.5 pounds of peanuts.   ■

Dual investment problems involve different amounts of money in two or more investments, each paying a different simple interest rate.

EXAMPLE 2   *Dual Investments with Simple Interest*

A bank customer invested in a 2-year certificate paying 8%. This year, she invested in a 1-year certificate, but it will pay only 3.8%. The total of her two investments is $10,000. If the certificates pay yearly simple interest, and the total interest earned at the end of next year will be $1234, how much did this customer invest in each certificate?

Solution   Let $x =$ the amount invested in the 8% certificate and
$y =$ the amount invested in the 3.8% certificate.

We use a table to organize the given information.

	Amount Invested	×	Interest Rate	=	Interest Earned
CD Paying 8%	x		0.08		$0.08x$
CD Paying 3.8%	y		0.038		$0.038y$
TOTALS	10,000				1234

From the table, we can write a system of equations.

$x + y = 10,000$ The total amount invested is $10,000.

$0.08x + 0.08x + 0.038y = 1234$ The total interest earned will be $1234.

Note that the last column of the table shows the interest earned on the 8% CD for 1 year only. Therefore, in the second equation, we need to include that term twice, once for each of the 2 years the customer has held the certificate.

The two simplified equations form the following system of equations.

$$x + \quad\quad y = 10,000$$
$$0.16x + \quad 0.038y = \quad 1234$$

$$
\begin{aligned}
-0.16x - 0.160y &= -1600 \\
0.16x + 0.038y &= \quad 1234
\end{aligned}
$$
Multiply the first equation by -0.16 and add the equations.

$$-0.122y = -366$$

$$y = \frac{-366}{-0.122} = 3000$$

$x + y = 10,000$ First equation

$x + 3000 = 10,000$ Replace y with 3000.

$x = 7000$

The bank customer invested $7000 in the 8% certificate and $3000 in the 3.8% certificate.   ∎

Recall that a liquid solution is some substance, such as acid, mixed with water. The *concentration* of the substance is the amount of the substance compared to the total amount of solution. Refer to Section 4.5 to review the details concerning this type of problem.

EXAMPLE 3   *Creating a Liquid Solution with a Specified Concentration*

A nurse has a 20% alcohol solution and a 45% alcohol solution. How many liters of each should be used to make 5 liters of a 30% alcohol solution?

Solution   Let x = the number of liters of the 20% (weak) solution and
y = the number of liters of the 45% (strong) solution.

	Liters of Solution	×	Concentration of Alcohol	=	Liters of Alcohol
Weak	x		0.20		$0.20x$
Strong	y		0.45		$0.45y$
Mix	5		0.30		1.5

From the table we write the following system of equations.

$$x + \quad y = 5 \qquad \text{The total amount of solution is 5 liters.}$$
$$0.20x + 0.45y = 1.5 \qquad \text{The total amount of alcohol in the solution is 1.5 liters.}$$

This time, we use the substitution method to solve the system.

$$x = 5 - y \qquad \text{Solve the first equation for } x.$$
$$0.20(5 - y) + 0.45y = 1.5 \qquad \text{Replace } x \text{ in the second equation with } 5 - y.$$
$$1 - 0.20y + 0.45y = 1.5 \qquad \text{Distributive Property}$$
$$1 + 0.25y = 1.5 \qquad \text{Combine like terms.}$$
$$0.25y = 0.5$$
$$y = \frac{0.5}{0.25} = 2$$

$$x = 5 - y \qquad \text{Replace } y \text{ with 2 in the equation solved for } x.$$
$$x = 5 - 2$$
$$x = 3$$

The nurse should use 3 liters of the 20% solution and 2 liters of the 45% solution.

■

EXAMPLE 4   *Determining Boat Speed and Rate of Current*

A boat travels 72 miles upstream in 6 hours and returns in 4 hours. What is the speed of the boat in still water and what is the rate of the current?

Solution   Let b = speed of the boat in still water and
c = rate of the current.

As it travels upstream, the boat is slowed down by the current. The net rate is the boat's rate minus the rate of the current and is given by the expression $b - c$.

The boat is helped by the current as it returns downstream. The net rate is the boat's rate plus the rate of the current and is given by the expression $b + c$.

	Rate (mph)	×	Time (hours)	=	Distance (miles)
Upstream	$b - c$		6		72
Downstream	$b + c$		4		72

Because Rate × Time = Distance, we multiply across the rows to obtain the following system of equations.

$$6(b - c) = 72$$
$$4(b + c) = 72$$

$b - c = 12$	Divide both sides of the first equation by 6.
$b + c = 18$	Divide both sides of the second equation by 4.

We can estimate the solution of the system by solving the two equations for c and producing their graphs.

$$c = b - 12$$
$$c = 18 - b$$

Figure 6.18

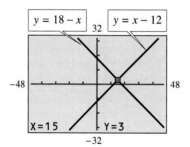

Because we solved the equations for c, points of the lines have coordinates of the form (b, c). From Fig. 6.18 we estimate the solution to be (15, 3). This solution can easily be verified by substitution.

The speed of the boat in still water is 15 mph, and the rate of the current is 3 mph. ∎

When we model the conditions of a problem with a system of equations, we may learn that the conditions cannot be met.

EXAMPLE 5 *Inventory of New and Used Vehicles*

An auto dealer sells new and used cars and trucks. He is planning his inventory for next month. His plan is to double the total number of cars and have a total of 190 cars and trucks. He also wants 25% of his cars and 25% of his trucks to be used, and the total of used cars and trucks is to be 42. How many cars and trucks will he have in his inventory?

Solution   Let x = the current total number of new and used cars and
y = the current total number of new and used trucks.

If the dealer doubles the number of cars, his total number of vehicles can be described by

$$2x + y = 190.$$

The number of used cars and trucks can be described by

$$0.25(2x) + 0.25y = 42.$$

We solve each equation for y.

$$2x + y = 190 \qquad\qquad 0.25(2x) + 0.25y = 42$$
$$y = -2x + 190 \qquad\qquad 0.50x + 0.25y = 42$$
$$0.25y = -0.50x + 42$$
$$y = -2x + 168$$

From the equations in the form $y = mx + b$, we can see that the slopes of their graphs are the same and the y-intercepts are different. Therefore, the graphs will be parallel. This system of equations has no solution, and the auto dealer cannot carry out his plan. ■

3 × 3 Systems

Example 6 shows how an application involving three unknowns can be modeled with a system of three linear equations in three variables.

EXAMPLE 6 *A Furniture Production Schedule*

The Wood Shop makes wooden porch furniture. Rockers require 5 hours of sawing time, 3 hours of assembly time, and 3 hours of finishing time. Chairs require 4 hours of sawing time, 2.5 hours of assembly time, and 2 hours of finishing time. Tables require 3 hours of sawing time, 2 hours of assembly time, and 2 hours of finishing time. The shop employs four full-time people to saw; two full-time and one half-time for assembly; and two full-time and one part-time (10 hours per week) for finishing. Assuming full time is 40 hours per week, how many rockers, chairs, and tables can they produce per week?

Solution Let $r =$ the number of rockers produced per week,
$c =$ the number of chairs produced per week, and
$t =$ the number of tables produced per week.

Four full-time workers provide $4(40) = 160$ available hours for sawing. Two full-time workers and one half-time worker provide $2.5(40) = 100$ available hours for assembly. Two full-time workers and one part-time (one-fourth of full time) worker provide $2.25(40) = 90$ hours for finishing.

	Rocker	*Chair*	*Table*	*Available Hours*
		Hours Required		
Sawing	5	4	3	160
Assembly	3	2.5	2	100
Finishing	3	2	2	90

The system of equations is as follows.

$$5r + 4c + 3t = 160$$
$$3r + 2.5c + 2t = 100$$
$$3r + 2c + 2t = 90$$

To use the automatic solving procedure on a calculator, we need the following matrices. The coefficient matrix is

$$A = \begin{bmatrix} 5 & 4 & 3 \\ 3 & 2.5 & 2 \\ 3 & 2 & 2 \end{bmatrix}.$$

The constant matrix is

$$B = \begin{bmatrix} 160 \\ 100 \\ 90 \end{bmatrix}.$$

 SOLVE

The reported solution is (10, 20, 10). The shop can produce 10 rockers, 20 chairs, and 10 tables each week.

Curve Fitting

In Chapter 5 we saw that two given points uniquely determine a straight line, and we learned how to find an equation of the line.

Similarly, three given points uniquely determine the graph of a function whose form is $y = ax^2 + bx + c$. We can use the three given points to create a system of three equations. Then we can solve the system for a, b, and c and thereby learn what function the graph represents.

This process of determining a function whose graph contains given points is known as **curve fitting.**

EXAMPLE 7 *Curve Fitting*

Determine the values of a, b, and c so that the graph of $y = ax^2 + bx + c$ contains the points $(-2, 4)$, $(1, -5)$, and $(4, 4)$. Use your calculator to produce the graph of the equation and trace to verify that the points are on the graph.

Solution   Substitute each ordered pair into $y = ax^2 + bx + c$.

$(-2, 4)$: $4 = a(-2)^2 + b(-2) + c$
$(1, -5)$: $-5 = a(1)^2 + b(1) + c$
$(4, 4)$: $4 = a(4)^2 + b(4) + c$

After simplifying each equation, the system of equations is as follows.

$$4a - 2b + c = 4$$
$$a + b + c = -5$$
$$16a + 4b + c = 4$$

Figure 6.19

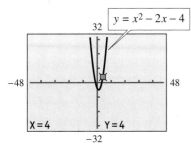

To solve with your calculator, enter the coefficient matrix and the constant matrix.

$$A = \begin{bmatrix} 4 & -2 & 1 \\ 1 & 1 & 1 \\ 16 & 4 & 1 \end{bmatrix} \qquad B = \begin{bmatrix} 4 \\ -5 \\ 4 \end{bmatrix}$$

The solution matrix shows that $a = 1$, $b = -2$, and $c = -4$.

The function is $f(x) = x^2 - 2x - 4$. The graph and the point $(4, 4)$ are shown in Fig. 6.19. The other two given points can also be traced.

6.5 Exercises

Mixtures

1. A candy store offers a mixture of candies for $3.25 per pound. A 20-pound mixture is produced by mixing assorted creams worth $2.95 per pound with chocolate-covered almonds worth $3.70 per pound. How many pounds of assorted creams are in the 20-pound mixture?

2. A 45-pound nut mixture has been made to sell for $5.40 per pound. The mixture was produced by mixing cashews worth $6.00 per pound with pecans worth $5.00 per pound. How many pounds of each nut were used to make the mixture?

3. A total of 8 pounds of fruit costs $6.12. The apples cost $0.45 per pound and the grapes cost $1.29 per pound. How many pounds of apples are there?

4. A landscaper specializes in azaleas and rhododendrons. A commercial customer has a budget of $2250.00 and wants the landscaper to plant a mixture of 100 azaleas and rhododendrons around a new office building. Including planting costs, the landscaper charges $40.00 per rhododendron and $15.00 per azalea. How many of each shrub can the landscaper plant?

5. During a bake sale, an organization sold all of its cakes and cookies and three-fourths of its pies. The cakes were priced at $3.50 each, the pies at $4.00 each, and the cookies at $1.40 per dozen.

A sum of $361.00 was earned at the sale. How many baked items were donated if there were six more pies than cakes and if 128 items were sold? (A dozen cookies represents one baked item.)

6. A caterer made 30 pounds of a party mix to sell for $4.05 a pound. Peanuts selling for $2.95 a pound, cashews selling for $6.25 a pound, and coated chocolates selling for $3.50 a pound were all used in the party mix. If two more pounds of peanuts than cashews were used, how many pounds of each item were in the party mix?

Tickets

7. A play was attended by 456 people. Patron's tickets cost $2 and all other tickets cost $3. If the total box office receipts were $1131, how many of each kind of ticket was sold?

8. The college basketball game was attended by 1243 people. Student tickets cost $2 and all other tickets cost $8. If the total receipts were $5018, how many students and how many nonstudents attended the game?

9. The attendance at the Atlanta Braves and Los Angeles Dodgers game was 42,245. All seats were either box seats costing $12.00 or reserved seats costing $8.00. The total receipts for the game were $395,004. How many box seats were sold?

10. Tickets for a charity ball were $10 apiece or $15 per couple. If 802 people attended the dance and a total of $6530 was collected from ticket sales, how many couples attended?

11. The attendance at the opera was 5500. Orchestra seats were $65 each, loge seats were $50 each, and balcony seats were $30 each. The total receipts for the performance were $246,750. If there were 100 more loge seats than orchestra seats sold, how many of each type of seat were sold?

12. At the state fair the total receipts for one evening were $17,000 for the 9000 people who attended. The entrance tickets were $3.00 for each adult, $1.50 for each student, and $1.00 for each child under 10 years old. If there were twice as many students as children under 10, how many adults attended the state fair that evening?

Investments

13. A lottery winner invested $11,000, part at 6.5% simple interest and the remainder at 9% simple interest. How much was invested at each rate if the total interest income at the end of 1 year was $830.85?

14. A financial adviser invested $10,000, part at 8.5% simple interest and the remainder at 6.9% simple interest. How much was invested at each rate if the total interest income at the end of 1 year was $745.84?

15. A used car buyer borrowed $7000 from two sources. One loan was to be repaid at 6% simple interest and the other loan at 7% simple interest. If the borrower paid a total of $452 in interest for the year, how much was borrowed at each rate?

16. Newlyweds purchased kitchen appliances for $8500. They borrowed the money, part at 9.5% and the remainder at 8% simple interest. If the couple owed a total of $741.50 in interest at the end of the year, how much did they borrow at each rate?

17. On behalf of a client, a stockbroker invested a total of $16,500 in a stock fund, a bond fund, and a mutual fund with yields of 6%, 7.5%, and 8% simple interest, respectively. At the end of 1 year, the total income from all three funds was

$1191.50. If the broker invested $1200 less in the mutual fund than twice the amount that was invested in the bond fund, how much was invested in each fund?

18. The following shows three orders placed at a fast food restaurant.

Order	Cost
2 hamburgers, 3 fries, 2 shakes	$ 9.75
4 hamburgers, 2 shakes	$10.80
5 fries, 3 shakes	$ 9.25

What is the price of each item?

Liquid Solutions

19. A brick mason has a 34% solution of muriatic acid and a 55% solution of muriatic acid. How much of each solution should be mixed together to produce 70 liters of a 40% acid solution?

20. Solution *A* is 34% insecticide while solution *B* is 60% insecticide. How much of each solution should be mixed together to produce 26 liters that is 49% insecticide?

21. How much 3.5% chlorine solution should be added to a 2% chlorine solution to obtain 600 gallons of a 2.5% chlorine solution?

22. A nurse needs 15 liters of a 40% alcohol solution. A solution of 20% alcohol and another solution of 70% alcohol are available. How much of each solution should be mixed?

23. An infirmary has solutions of 10%, 20%, and 50% alcohol content. A nurse needs 18 liters of a 30% alcohol solution. How much of each solution is needed if twice as much of the 50% alcohol solution as the 20% alcohol solution is to be used?

24. A chemist is analyzing three acid solutions labeled *A*, *B*, and *C*. Each solution was formed by mixing together two of three other acid solutions labeled 1, 2, and 3. The following table summarizes the amounts of solutions 1, 2, or 3 that were used to make solutions *A*, *B*, and *C*. All amounts are measured in cubic centimeters. The numbers in parentheses indicate the acid concentrations in solutions *A*, *B*, and *C*.

	Amount of Solution 1	Amount of Solution 2	Amount of Solution 3
A (28%)	2	8	0
B (31%)	0	12	3
C (46%)	9	0	6

What are the acid concentrations in solutions 1, 2, and 3?

Distance, Rate, and Time

25. A man paddles his canoe 21 miles upstream in 7 hours. Then, he paddles downstream to his starting point in 3 hours. What is the speed of the canoe in still water and what is the speed of the current?

26. A motor boat travels 137.5 miles upstream in 12.5 hours. The next day, it travels downstream to its starting point in 5.5 hours. What is the speed of the boat in still water and what is the speed of the current?

27. An airplane travels 3360 miles with the wind in 7 hours and against the wind in 8 hours. What is the speed of the airplane in calm air and what is the speed of the wind?

28. An airplane travels 2860 miles with the wind in 5.5 hours and against the wind in 6.5 hours. What is the speed of the airplane in calm air and what is the speed of the wind?

29. A vacationer paddled his canoe downstream in 2 hours and paddled upstream back to his starting point in 6 hours. At noon the next day, he decided to drift downstream. At 5:00 P.M. he fell asleep. He woke up at 7:00 P.M. to discover that he had drifted 2 miles past his turn-around point of the previous day. Determine the rate of the current and the rate of the canoe in still water.

30. An athlete is training for the triathlon and has discovered that her rates seem to be staying the same. On Monday she jogged for 30 minutes and rode the bike for 30 minutes and covered 21 miles. On Tuesday she rode the bike for one-half hour and swam for one hour and covered 18.4 miles. On Wednesday she jogged for 15 minutes and biked for 45 minutes and covered 25.5 miles. What was her rate for each sport?

Curve Fitting

31. Determine the values of a and b so that the graph of $y = ax + b$ contains the points $(3, -2)$ and $(-1, 6)$.

32. Determine the values of a and b so that the graph of $y = ax + b$ contains the points $(-4, 5)$ and $(-6, -1)$.

33. Determine the values of a, b, and c so that the graph of $y = ax^2 + bx + c$ contains the points $(-3, 22)$, $(4, 15)$, and $(1, -6)$. Write the equation.

34. Determine the values of a, b, and c so that the graph of $y = ax^2 + bx + c$ contains the points $(-2, -11)$, $(3, -6)$, and $(1, -2)$. Write the equation.

35. Determine the values of a, b, and c so that the graph of $y = ax^2 + bx + c$ contains the points $(-1, 2)$, $(2, -7)$, and $(3, -26)$. Write the equation.

36. Determine the values of a, b, and c so that the graph of $y = ax^2 + bx + c$ contains the points $(-1, 2)$, $(3, 10)$, and $(-2, 15)$. Write the equation.

Miscellaneous

37. When a father and his son play golf, the father agrees to pay twice as much as his son pays. The total cost of one round of golf is $51. How much does each pay for one round of golf?

38. A recipe calls for three times as much flour as sugar. Six cups of the mixture are used. How many cups of flour are to be used?

39. A cash register drawer contains $505 in five- and ten-dollar bills. If there is a total of 70 bills, how many bills of each denomination are in the drawer?

40. A girl is 5 years older than her brother. In 4 years, the brother will be three-fourths his sister's age then. Determine their present ages.

41. Your building-supply store stocks two brands of electric drills. The retail price of the heavy-duty drill is $46.50, and the price of the homeowner's drill is $18.95. You have hired a part-time employee to help with your annual inventory. You neglect to tell the employee that you need to know how many of each kind of drill you have in your inventory. Therefore, one line of your inventory report is this.

Item	*Quantity*	*Retail Value*
Drills	31	$918.05

From this report, determine how many of each kind of drill you have.

42. A city school system wants to ensure a balance of minority students and teachers in the schools. At Hamilton Middle School, there is a combined total of 435 students and faculty of whom 167 are minorities. If 40% of the students and 20% of the faculty are minorities, how many students attend the school?

43. A paint store has a total of 75 one-gallon cans of paint with high-gloss enamel selling at $19.95 per gallon and flat latex selling at $10.95 per gallon. If the total value of the one-gallon cans is $1109.25, how many of each type are in stock?

44. The owner of a precious metals shop wishes to produce 69.8 ounces of an alloy that is 25% gold. Some 20% gold alloy and some 28% gold alloy are available. How many ounces of each alloy should be melted and mixed to produce the desired alloy?

45. A florist sells orchid corsages for $23 each and carnation corsages for $14 each. If the total receipts for 95 corsages were $1663, how many of each type of corsage were sold?

46. At a concession stand, students sell candy bars for 75 cents each and popcorn for $1.25 a box. If the total receipts for 365 items were $352.75, how many of each item did they sell?

47. At a local election, 536 of the 700 voters voted in favor of a bond referendum. If 88% of the Democrats and 72% of the Republicans supported the referendum, how many Republicans voted against it?

48. A computer programmer has produced two programs, one in the BASIC language and one in the Pascal language. Together, the programs total 950 lines of code of which a total of 15 lines contain errors. If 1% of the BASIC program lines contain errors, and 2% of the Pascal program lines contain errors, how many lines of code were written in Pascal?

49. A hiking club of 40 members climbed a mountain on three different trails. Part of the group climbed an 8-mile trail, another part climbed a rougher 6-mile trail, and the third group climbed the steepest trail of 5.5 miles. The second group had two more people than the third group. If the members of the club walked a total of 280 person-miles, how many hikers climbed each trail? (Note: If a group of 25 people walks 10 miles, the group has walked a total of $10 \cdot 25$ or 250 person-miles.)

50. The church secretary, the organist, and a volunteer addressed envelopes for a bulk mailing. The organist can address 120 envelopes per hour, the volunteer 116 per hour, and the secretary 140 per hour. They worked a total of 6 person-hours and addressed a total of 740 envelopes. The volunteer worked 1 hour longer than the secretary. How long did each work?

51. At a certain college, A's are worth 4 points, B's are worth 3 points, and C's are worth 2 points. A student's *quality points* are determined by multiplying the number of credit hours for each course times the point value of the grade earned in the course. One semester, a student earned 52 quality points for 17 credit hours of course work. The number of credit hours in which the student received B's was twice the number of credit hours in which the student received C's. If none of the grades was below C, what was the number of credit hours the student earned for each grade?

52. The sum of four digits on a license plate is 18. The first and last digits are the same, and the sum of the first and the second digits is 10. The second digit is twice the third. What is the number on the license plate?

53. A bookshelf 4 feet in length exactly holds three sets of books totaling 35 volumes. The books in the first set are 1 inch thick, those in the second set are 1.5 inches thick, and those in the third set are 2 inches thick. If the number of books in the first set is one more than in the second set, how many volumes are in each set?

54. The perimeter of a triangle is 40 inches. The length of the shortest side is 4 inches more than the positive difference between the lengths of the other two sides. The length of the longest side is 4 inches less than the sum of the lengths of the other two sides. Which of the three sides can you determine from this information? What can you say about the other two sides?

Exploring with Real Data

In Exercises 55–58, refer to the following background information and data.

The following table shows the number of billions of dollars of retail toy sales (excluding video games) for the period 1988–1992.

	1988	1989	1990	1991	1992
Retail Sales (in billions)	12.97	13.10	13.13	15.15	17.00

(Source: Toy Manufacturers of America.)

55. Let t represent the number of years since 1985 and let S represent retail toy sales (in billions). Write the ordered pairs (t, S) that represent the data in the table for the years 1988, 1990, and 1992.

56. Assume that each of the ordered pairs in Exercise 55 satisfies a function of the form $S(t) = at^2 + bt + c$. Use the three ordered pairs to write a system of three equations.

57. Use your calculator to solve the system in Exercise 56 for a, b, and c. By substituting these values in $S(t) = at^2 + bt + c$, write the specific function S.

58. Write the ordered pairs (t, S) that represent the data in the table for 1989 and 1991. How well does function S model the data for these two years?

Challenge

59. A two-cycle engine requires an oil and gas mixture that is 2.5% oil. A man mixed 6 gallons of gas and oil, but then discovered he had mistakenly used a 2% mixture. How many *ounces* of oil does he need to add to the mixture to correct the problem? (There are 128 ounces in a gallon.)

60. On a pan balance, two baseballs and three softballs balance with one softball and five baseballs. In another experiment, two softballs and one baseball balance with four baseballs. How many softballs are needed to balance with three baseballs?

61. A track inspector and a crew foreman were standing in a tunnel whose length is 0.6 miles. They departed in opposite directions to leave the tunnel before the next train came through. Both men exited their ends of the tunnel in 3 minutes, and the inspector reached the end of the tunnel just as the train arrived. The train passed the foreman, who had walked 0.12 miles beyond the other end of the tunnel. If the train was traveling at 30 mph, how fast did each man walk?

62. A woman is applying to be a salesperson for Ace Auto Company. She must decide if she wants to have a base salary of $20,000 per year with a 12% commission on annual sales or to have a base salary of $30,000 per year with an 8% commission on annual sales. If x represents the annual sales and y_1 and y_2 represent the first and second annual earnings options, write a system of equations and use your calculator to solve the system graphically. Recommended range and scale:

[X : 100000, 400000, 100000] and
[Y : 20000, 60000, 10000].

At what level of annual sales will the annual earnings be the same? For what range of sales should the salesperson choose the first option? For what range of annual sales should she choose the second option?

6.6 Systems of Linear Inequalities

Systems of Inequalities ▪ *Applications*

Systems of Inequalities

When a problem involving inequalities requires two or more conditions to be satisfied, a system of inequalities is needed.

Two or more inequalities considered simultaneously is a **system of inequalities.** In this section we consider systems of linear inequalities in two variables. For brevity, we refer to such systems simply as *systems of inequalities.*

The inequalities of a system are implicitly joined by *and.* Therefore, a system of inequalities is an instance of a compound linear inequality, a concept that was introduced in Section 5.5.

> **Definition of Solution and Solution Set**
>
> A **solution** of a system of inequalities is a pair of numbers that satisfies each inequality. The **solution set** of a system is the set of all solutions.

EXAMPLE 1 *Verifying Solutions of a System of Inequalities*

Determine whether each ordered pair is a solution of the given system of inequalities.

$$x - 2y < 5$$
$$y \geq 3 - x$$

(a) (6, 11) (b) (−5, 3)

Solution

(a)

Inequality	Replace x with 6 and y with 11	Result	
$x - 2y < 5$	$6 - 2(11) < 5$	$-16 < 5$	True
$y \geq 3 - x$	$11 \geq 3 - 6$	$11 \geq -3$	True

Because the ordered pair (6, 11) satisfies both inequalities, it is a solution of the system.

(b)

Inequality	Replace x with −5 and y with 3	Result	
$x - 2y < 5$	$-5 - 2(3) < 5$	$-11 < 5$	True
$y \geq 3 - x$	$3 \geq 3 - (-5)$	$3 \geq 8$	False

Although (−5, 3) satisfies the first inequality, it does not satisfy the second inequality. Therefore, (−5, 3) is not a solution of the system. ▪

A graph of a system of inequalities is a picture of all the solutions of the system.

Graphing the Solution Set of a System of Inequalities

1. Graph each inequality.

2. Determine the *intersection* of the solution sets.

EXAMPLE 2 *The Graph of a System of Inequalities*

Graph the following system of inequalities.

$$3x - y \leq 9$$
$$y + 6 > -x$$

Solution To produce the graphs, we solve each inequality for y.

$$y \geq 3x - 9$$
$$y > -x - 6$$

The graphs of these inequalities are the half-planes shown in Figs. 6.20 and 6.21.

Figure 6.20

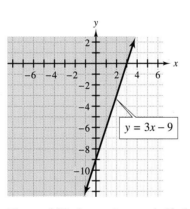

Figure 6.21

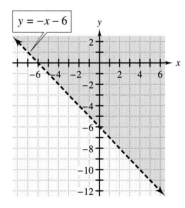

Figure 6.22 shows the two half-planes in the same coordinate system. Their intersection is shown in Fig. 6.23.

Figure 6.22

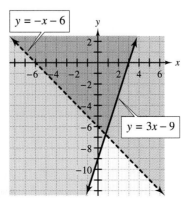

Figure 6.23

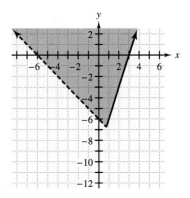

EXAMPLE 3 *The Graph of a System of Inequalities*

Figure 6.24

Graph the following system of inequalities.

$$y \leq x - 7$$
$$x \leq y + 3$$

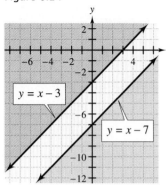

$y = x - 3$

$y = x - 7$

Solution   To produce the graphs, we solve each inequality for y.

$$y \leq x - 7$$
$$y \geq x - 3$$

The graph of each inequality is shown in Fig. 6.24. Because the intersection is empty, the system of inequalities has no solution.

EXAMPLE 4 *A Graph of a System of Three Inequalities*

Graph the following system of inequalities.

Figure 6.25

$$x - y < 3$$
$$2x + y > 1$$
$$x - 4 \leq 0$$

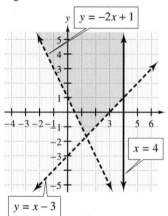

$y = -2x + 1$

$x = 4$

$y = x - 3$

Solution   To produce the graphs, we solve the first two inequalities for y and the third inequality for x.

$x - y < 3$	$2x + y > 1$	$x - 4 \leq 0$
$y > x - 3$	$y > -2x + 1$	$x \leq 4$

The graph of each inequality is a half-plane. The solution set of the system of inequalities is the intersection of the three half-planes. (See Fig. 6.25.)

Applications

Application problems may require two or more inequalities to describe the conditions of the problem.

EXAMPLE 5 *Feasible Production of Recycled Products*

A recycling company collects type 1 and type 2 plastics for the manufacture of recycling bins and litter boxes. A recycling bin requires 5 units of type 1 plastic and 1 unit of type 2 plastic. A litter box requires 3 units of type 1 plastic and 2 units of type 2 plastic. If there are 75 units of type 1 plastic available and 36 units of type 2 plastic available, describe the number of recycling bins and litter boxes that can be manufactured.

Solution

	Units Required	
	Type 1	*Type 2*
Recycling Bins	5	1
Litter Boxes	3	2
Available	75	36

Let x = the number of recycling bins manufactured and

y = the number of litter boxes manufactured.

The amount of plastic available gives two inequalities.

$$5x + 3y \leq 75 \quad \text{or} \quad y \leq -\frac{5}{3}x + 25$$

$$x + 2y \leq 36 \quad \text{or} \quad y \leq -\frac{1}{2}x + 18$$

Because it is impossible to make a negative number of items, each variable must be nonnegative. Thus the complete system of inequalities is as follows.

$$y \leq -\frac{5}{3}x + 25$$

$$y \leq -\frac{1}{2}x + 18$$

$$x \geq 0$$
$$y \geq 0$$

The shaded region in Fig. 6.26 represents the solution set of the system.

Figure 6.26

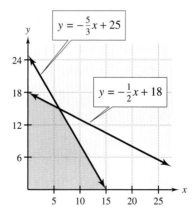

Note that every point in the shaded area, as well as every point of the line segments that bound the shaded area, represents a solution of the system. However, because of the nature of the problem, we can consider only those points whose coordinates are integers.

6.6 *Quick Reference*

Systems of Inequalities

- A **system of inequalities** consists of two or more inequalities considered simultaneously. In this section we consider systems of linear inequalities in two variables.

- A **solution** of a system of inequalities is a pair of numbers that satisfies each inequality of the system.

- To graph the solution set of a system of inequalities,

  1. graph each inequality and

  2. determine the intersection of the solution sets.

6.6 *Exercises*

1. Use the word *half-plane* to describe the graph of the solution set of a system of linear inequalities.

2. Explain why a system of inequalities is a conjunction.

In Exercises 3–14, graph the solution set of the given system of linear inequalities.

3. $x \geq 3$
  $y \geq 2$

4. $y > 1$
  $x < -2$

5. $16 \geq 3x - 4y$
  $y < 6$

6. $5y + 3x > -20$
  $x \geq -2$

7. $3y - 4x < 15$
  $y > 0$

8. $x \geq 0$
  $x + 3y \leq 12$

9. $y \geq \dfrac{1}{2}x$
  $y \leq 3x$

10. $y < 2x - 5$
  $y > -\dfrac{2}{3}x - 5$

11. $6 \leq 2x + 3y$
  $2y - 5x \geq 14$

12. $x - y > 3$
  $2x + y \leq 3$

13. $y - 2x > 7$
  $y + 3 < 2x$

14. $2x + 3y \leq -15$
  $y \geq -\dfrac{2}{3}x - 1$

In Exercises 15–18, write the system of linear inequalities whose solution set is represented by the shaded region of the graph.

15.

16.

17.

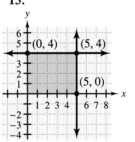

18.

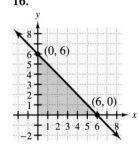

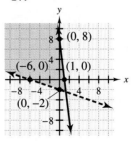

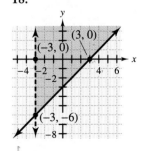

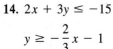

19. Describe the graph of the solution set of the following system of linear inequalities.

$y \leq 5$
$y \geq -5$
$x \leq 5$
$x \geq -5$

20. Suppose you want to determine if (a, b) is a solution of a system of three inequalities, and you find that the pair does not satisfy the first inequality. Why is it not necessary to test the pair in the other two inequalities?

In Exercises 21–26, graph the system of linear inequalities.

21. $y + 3x \leq 3$
$y + x + 1 \geq 0$
$y < x + 7$

22. $y \leq x + 5$
$y - 3x - 2 > 0$
$y + x + 2 \geq 0$

23. $x + 5 \geq 0$
$y - x > 1$
$2x + y < 6$

24. $x + y \leq 7$
$2y > x$
$y < 5x$

25. $y - 2x < 1$
$y + x + 2 \geq 0$
$y + 7 \geq 0$

26. $y \geq 0$
$x + 2y \leq 8$
$3x - 2y < 4$

27. For an application involving a number x of men and a number y of women, suppose the given conditions can be translated into the following two linear inequalities. State two additional implied inequalities.

$2x + y \leq 30$
$x + y \geq 10$

28. For the application in Exercise 27, every point in the intersection of the graphs of the inequalities represents a solution of the system. However, not all of these points can be used to solve the problem. Why?

In Exercises 29–40, graph the system of linear inequalities.

29. $2x + 3y \leq 18$
$x \geq 0$
$y \geq 0$

30. $3x + y \leq 10$
$x \geq 0$
$y \geq 0$

31. $y \geq x$
$y \leq 8$
$x \geq 0$
$y \geq 0$

32. $x \leq 6$
$y \leq 4$
$x \geq 0$
$y \geq 0$

33. $-2x + y \leq 7$
$x \leq 5$
$x \geq 0$
$y \geq 0$

34. $2x - y \leq 5$
$y \leq 7$
$x \geq 0$
$y \geq 0$

35. $-x + y \leq 1$
$3x + 4y \geq 32$
$x \geq 0$
$y \geq 0$

36. $2x - y \leq 7$
$x + y \geq 2$
$x \geq 0$
$y \geq 0$

37. $3x + y \geq 5$
$x + y \geq 3$
$y \geq \dfrac{1}{2}x$
$x \geq 0$
$y \geq 0$

38. $4x + 3y \geq 18$
$x - y \leq 8$
$2x + 3y \geq 12$
$x \geq 0$
$y \geq 0$

39. $-x + y \geq 3$
$x + y \leq -2$
$x \geq 0$
$y \geq 0$

40. $x - y \geq 5$
$x + y \leq 0$
$x \geq 0$
$y \geq 0$

41. A bank invests at most $10 million in either short-term notes or in mortgages. If the amount in mortgages cannot exceed $7 million, describe the distribution of funds with a system of inequalities and graph the system. If $2 million is invested in short-term notes, what is the most that can be invested in mortgages? If $6 million is invested in short-term notes, what is the most that can be invested in mortgages?

42. A farmer allocates at most 2000 acres of land for either corn or soybeans. At most 40% of the total available acres can be used for soybeans. The number of acres for corn cannot exceed the number of acres of soybeans by more than 700 acres. Write and graph the system of inequalities that describes these conditions. If 1100 acres is in corn, what is the greatest and fewest number of acres that can be planted in soybeans?

43. A restaurant prepares tuna fruit salads and chicken Caesar salads for luncheon parties. The number of people who request tuna fruit salad never exceeds two-thirds of the number who order chicken Caesar salads. The smallest group that the restaurant books is 50. Write a system of inequalities that describes the number of salads to be made and graph the system. If 60 chicken Caesar salads are prepared, what is the possible number of tuna fruit salads prepared?

44. A coffee company produces at least 100 pounds of Irish cream coffee and Hawaiian macadamia coffee. The number of pounds of Irish cream cannot exceed twice the number of pounds of Hawaiian macadamia. Describe and graph the number of pounds of each that can be produced. What is the fewest number of pounds of Hawaiian macadamia that can be produced?

45. A builder specializes in the construction of small barns and tool buildings. The following time-table shows the number of hours needed to frame and roof the buildings and the maximum number of hours available for each kind of construction.

	Hours to Frame	Hours to Roof
Barn	15	7
Tool Shed	5	6
Maximum Hours Available	90	84

(a) Write a system of linear inequalities and draw the graph of the system. Shade the region that represents the possible number of barns and tool buildings that can be built under these conditions.

(b) Can the builder construct 4 barns and 5 tool buildings?

(c) Can the builder construct 5 barns and 4 tool buildings?

46. A businesswoman runs a full-service catering business. She prepares food ahead of time and, on the day of the event, she sets up the tables and provides decorations. Her time requirements and constraints are summarized in the following table.

	Food Preparation Hours	Set-Up Hours
Banquet	13	8
Reception	3	5
Maximum Hours Available	39	40

(a) Write a system of linear inequalities and draw the graph of the system. Shade the region showing the possible number of banquets and receptions that can be catered.

(b) Can she cater 3 banquets and 2 receptions?

(c) Can she cater 2 banquets and 3 receptions?

Challenge

47. Refer to Exercise 41. If mortgages yield 7% and short-term notes yield 5%, how should the funds be invested to maximize the return to the bank?

48. Refer to Exercise 42. Suppose the profit on soybeans is 70 cents per bushel and the profit on corn is 40 cents per bushel. If the farmer can produce 35 bushels of soybeans per acre and 150 bushels of corn per acre, what land allocation would produce a maximum profit?

6   Chapter Review Exercises

Section 6.1

 1. Suppose you draw the graphs of $y_1 = ax + b$ and $y_2 = cx + d$ on the same coordinate axes. If the graphs intersect only at point Q, what is the significance of that point?

 2. Suppose you use the substitution method to solve the following system.

$$2x + 3y = 3$$
$$y = 7x + 1$$

What is the first equation in one variable that you would write? Why?

In Exercises 3 and 4, determine whether the given ordered pair is a solution of the given system of equations.

3. $(2, 1);$ $3x - 5y = 1$
           $y = 2x - 3$

4. $(-4, 5);$ $3x + y = -7$
            $2x - 3y = 12$

In Exercises 5–12, use the graphing method to solve the given systems of equations.

5. $y = 2x - 5$ **6.** $x - 2y = 5$
    $y = -x + 4$ $y = -3x + 1$

7. $x - 3y = 7$
    $3x = 9y + 2$

8. $2x - 5y = 10$ **9.** $4x - 3y + 18 = 0$
    $10y - 4x + 20 = 0$ $x + 2y - 1 = 0$

10. $2x + y + 2 = 0$
     $3x + 2y - 1 = 0$

11. $5x - 2y = 1$ **12.** $2x - 3y = 12$
     $3x - 4y = 9$ $3x - y = 11$

In Exercises 13–22, state whether the graphs of the given systems intersect at exactly one point, are parallel, or are coincident.

13. $y = -3x + 5$
     $y + 3x = 7$

14. $0.25x + 0.25y = -0.5$
     $-2x = 4 + 2y$

15. $y = -x + 4$
     $y = x - 3$

16. $7x - 6y = 5$
     $7x - y = 5(y + 1)$

17. $2x - 3y = 6$ **18.** $y = 5$
     $6y = 4x - 12$ $x - 2 = 4$

19. $-4a - 5b = -\dfrac{5}{8}$ **20.** $3x = y + 2$
                           $3y = 9x + 21$
     $5b + 4a = \dfrac{1}{8}$

21. $y = 5 + 1.8x$ **22.** $\dfrac{1}{2}y = -5$
     $14 = y$ $0.6x + y = -7$

In Exercises 23–32, use the substitution method to solve the given system of equations.

23. $y = x + 1$
     $2x + 3y = 6$

24. $y = -2x + 3$
     $3x - 4y + 11 = 0$

25. $x = -2y + 3$ **26.** $y = x + 3$
     $2y + x = 4$ $2x - 2y + 6 = 0$

27. $x - 2y = 6$ **28.** $y - x = 8$
     $2x + y = 4$ $5y + 5x = 8$

29. $3x + y = 2$ **30.** $2x + y = 0$
     $4x + 8y = 1$ $5x + y = -2$

31. $\dfrac{5}{3}x + \dfrac{3}{2}y = 1$
     $15x + 18y = 6$

32. $\frac{3}{5}x - \frac{5}{3}y = \frac{2}{15}$

$\frac{9}{4}x + \frac{15}{8}y + \frac{9}{8} = 0$

Section 6.2

 33. Suppose that as you solve a system of equations with the addition method, you obtain $x = \frac{17}{31}$. Explain how you can determine the y-value without using substitution.

 34. Suppose that as you solve a system of equations with the addition method, you obtain an equation with no variable. Explain how to interpret this result.

In Exercises 35–42, use the addition method to solve the given system of equations.

35. $3x - 4y = 22$
$2x + 5y = -16$

36. $x - 2y + 4 = 0$
$3x + y + 5 = 0$

37. $y - 3x = 4$
$6x - 2y + 8 = 0$

38. $y = 2x + 3$
$4x - 2y - 6 = 0$

39. $\frac{3}{4}x + \frac{1}{3}y - 4 = 0$

$\frac{1}{2}x = \frac{2}{3}y + 8$

40. $\frac{3}{4}x + \frac{1}{3}y - 3 = 0$

$9x = 36 - 4y$

41. $\frac{1}{5}x - \frac{2}{3}y = 2$

$3x = 10y + 19$

42. $\frac{2}{5}x + \frac{1}{2}y = 2$

$\frac{3}{10}x + \frac{1}{4}y = 0$

In Exercises 43 and 44, determine c so that the given system of equations will be inconsistent.

43. $x + 3y = 6$
$4x - cy = 8$

44. $y - 4x = c$
$y - (c + 1)x = -7$

In Exercises 45 and 46, determine c so that the equations of the given system will be dependent.

45. $x - 2y = 5$
$cx + 6y + 15 = 0$

46. $\frac{1}{2}x + y - 3 = 0$
$cx + 4y - 12 = 0$

Section 6.3

 47. What geometric figure represents the solutions of a linear equation in three variables? What do we call the coordinate system in which such a figure is drawn?

 48. Suppose you want to solve the following system of equations.

$2x + y + 3z = 7$
$3x - y + z = 1$
$x + 4y - z = 2$

You can easily eliminate y from the first two equations and z from the last two equations. What is your opinion of this initial approach?

In Exercises 49 and 50, determine whether the given ordered pair is a solution of the given system of equations.

49. $x + y + z = 2$
$2x - 3y + 3z = 2$
$x - 2y + z = 1$
$(-2, 1, 3)$

50. $3x - z = 3$
$y + 2x - 2 = 0$
$x + y + z = 5$
$(2, 0, 3)$

In Exercises 51–62, use the addition method to solve the given system of equations.

51. $x + 2y + z = 5$
$2x + y - 2z = -3$
$4x + 3y - z = 0$

52. $y = x + 2$
$x + 2y + 4z = 9$
$x + y - z = 9$

53. $5x - 2y + 7z = -19$
$x + 3y - 2z = 3$
$4x - y - z = 16$

54. $x + 2y + 3z = -10$
$3x + 2y + z = -2$
$x + 3y + 2z = -8$

55. $2y = x + z - 3$
$7x + 5z = 3y + 1$
$2x + 2z + 5 = 4y$

56. $2x - 5y + 7z = -3$
$x + 2y - 4z = 0$
$2x - 5y + 3z = -7$

57. $2x - 2y + 10z = 12$
$x - 2y + 3z = -1$
$4x - 3y + z = -32$

58. $3x + 4z = 2y - 6$
$2(y - x) = 4z + 1$
$2(z - 2) = y - x$

59. $x + y - z = 8$
$y + 2z + 8 = 0$
$x - 2y + 3 = 0$

60. $x - y + z = 9$
$2x + y - z = 6$
$5x + 2y - 2z = -4$

61. $x + \frac{3}{2}y - \frac{5}{4}z = 3$

$\frac{1}{6}x + \frac{1}{2}y - \frac{1}{3}z = 1$

$\frac{1}{2}x + y - \frac{6}{5}z = 2$

62. $3(x + 2z) = 8 + y$
$x - 3 = 2(2y + z)$
$z + 7 = x + y$

Section 6.4

63. What is the difference between a coefficient matrix and an augmented matrix?

64. Describe the row operation used to transform
$\begin{bmatrix} 2 & 5 \\ -1 & 3 \end{bmatrix}$ into $\begin{bmatrix} 0 & 11 \\ -1 & 3 \end{bmatrix}$.

In Exercises 65 and 66, write the augmented matrix for the given system of equations.

65. $4x - 3y = 12$
$x - 2y + 3 = 0$

66. $y - 2z - 7 = -11$
$2z - 2x = 5$
$2y - 6x = 1$

In Exercises 67 and 68, write a system of linear equations associated with the given augmented matrix.

67. $\begin{bmatrix} 1 & 2 & -3 & | & 4 \\ 3 & 1 & -2 & | & 5 \\ 1 & 1 & 0 & | & 0 \end{bmatrix}$ **68.** $\begin{bmatrix} 5 & -2 & | & -2 \\ 6 & 0 & | & 7 \end{bmatrix}$

In Exercises 69–76, use the matrix method to solve the given system of equations.

69. $2x - y = 5$
$x - 2y = 4$

70. $y + 9 = 1.2x$
$15 - x = 0$

71. $x - y - 5 = 0$
$2x - y + 17 = 0$

72. $2x + 3y + 15 = 0$
$x + 2y + 18 = 0$

73. $4x - 3y - 6z = 6$
$x - 3y + 6z = -1$
$x - 6y + 12z = -1$

74. $x - y - 2z = -2$
$x + y + 6z = 6$
$y - z = -1$

75. $3x - 4y + 2z = 1$
$-3x + 3y - z = 0$
$-6x + 9y - 3z = 3$

76. $2x - z = 3$
$2y + z = 5$
$2x + y = 6$

In Exercises 77–82, use your calculator to solve the given system of equations.

77. $x + 10 = 0$
$y + 0.3x = 6$

78. $4x - 3y = -12$
$5x + 2y = 1$

79. $x + y + z = 0$
$y - 2z = 3$
$x + z = 1$

80. $1.6x + 1.2y - z = 3.2$
$x - 3y - 0.75z = -13$
$0.5x + y + z = 3.5$

81. $2x + y - 3z = 6$
$4x + 4y - 5z = 3$
$-2x + y + 2z = -4$

82. $x - y = 1$
$x + 3z = 6$
$y - 2z = 0$

Section 6.5

In Exercises 83–88, assign variables to the unknown quantities and write a system of equations to describe the situation. Use any method to solve the system.

83. A cash register drawer contains $705 in five- and twenty-dollar bills. If there is a total of 60 bills, how many bills of each denomination are in the drawer?

84. The difference between one number and twice a smaller number is 20. The smaller number plus 10 is one-half the larger number. What are the numbers?

85. A hardware store stocks two types of hammers. The store has a total of 42 hammers, with sledge hammers selling at $22.95 each and claw hammers selling at $10.95 each. If the total value of the hammers is $639.90, how many of each type are in stock?

86. A forestry ranger had to hike from the park office up into the mountains to reach the newly planted seedlings that she had to inspect. She hiked at an average rate of 3.5 mph to reach this area. The ranger then averaged 2.5 mph through the inspection area. On her return to the park office, she again averaged 3.5 mph. The total length of her hike was 17 miles, which she covered in 6 hours. How long was she in the planted area? How far from the park office is the inspection site?

87. The sum of the three digits on a license plate is 12. The sum of the first two digits is 7 and the sum of the last two digits is 8. What is the three-digit number on the license plate?

88. Determine the values of a, b, and c so that the graph of $y = ax^2 + bx + c$ contains the points (4, 8), (2, 2.5), and (−8, 35). Then write the specific equation.

Section 6.6

In Exercises 89–100, graph the given system of inequalities.

89. $y < 4$
$2x - 5y < 10$

90. $x \geq -3$
$2x + 3y \leq 5$

91. $y < 3x + 1$
$2y + 3x \leq 10$

92. $y + x < -5$
$y - x \geq -5$

93. $y \leq 3x$
$3y \geq x$
$x + y \leq 9$

94. $4y - 3x < 16$
$y > -5$
$6y < 5x + 36$

95. $y \geq 2$
$y \geq -3$
$x \geq -5$

96. $y - x > 3$
$y < x - 5$
$3x + 7y < 14$

97. $x + y \leq 10$
$x + y \geq 2$
$x \geq 0$
$y \geq 0$

98. $x + 4y \leq 24$
$2y - 3x \geq -4$
$x \geq 0$
$y \geq 0$

99. $2x - 3y \leq 6$
$y \leq 7$
$x \leq 9$
$x \geq 0$
$y \geq 0$

100. $3y - 2x \leq 15$
$x \leq 12$
$y - x \leq 3$
$x \geq 0$
$y \geq 0$

101. For the following application problem, write a system of inequalities that describes the given conditions and draw a graph of the solution set.

A certain pickup truck can carry no more than 1000 pounds and no more than 20 bags of mortar mix and/or lime. A bag of mortar mix weighs 80 pounds and a bag of lime weighs 40 pounds. How many bags of each type can the truck carry?

6   Chapter Test

1. Use the graphing method to estimate the solution of the following system of equations. Show your verification of the solution by substitution.

$$y = -2x + 3$$
$$x - y = 3$$

In Questions 2 and 3, use the substitution method to solve the system of equations.

2. $x + 2y = 10$
$3x + 4y = 8$

3. $2x + y = 7$
$x - y = 8$

In Questions 4 and 5, use the addition method to solve the system of equations.

4. $x - 2y = 3$
$3x = 6y + 9$

5. $x - \dfrac{3}{2}y = 13$

$\dfrac{3}{2}x - y = 17$

6. Use the addition method to solve the system of equations.

$$x - 2y + 3z = 7$$
$$2x + y + z = 4$$
$$-3x + 2y - 2z = -10$$

In Questions 7 and 8, use matrix methods to solve the system of equations.

7. $3x - 6y = 24$
$5x + 4y = 12$

8. $x + y - z = 6$
$3x - 2y + z = -5$
$x + 3y - 2z = 14$

9. Use your calculator to solve the system of equations.

$$x + 2y - z = -3$$
$$2x - 4y + z = -7$$
$$-2x + 2y - 3z = 4$$

In Questions 10–12, classify the given system of equations as a consistent system with a unique solution, an inconsistent system, or a system of dependent equations.

10. $2x + 3y = 5$
$4x + 6y = 10$

11. $2x + y = 2$
$8x + 4y = 7$

12. $x + y = 6$
$x - y = 4$

13. Describe the graph of a system of two linear equations in two variables for each of the following.

   (a) The equations are dependent.

   (b) The system has a unique solution.

   (c) The system is inconsistent.

In Questions 14 and 15, graph the solution set of the system of inequalities.

14. $y < -2x$

   $y > \dfrac{2}{3}x$

15. $2x + y \leq 12$

   $y \leq 8$

   $x \geq 0$

   $y \geq 0$

In Questions 16–19, define variables, translate the problem into a system of equations, solve the system of equations, and state the answer to the question.

16. With a tail wind an airplane flies 600 miles in 3 hours, and against a head wind the plane flies the same distance in 4 hours. What is the wind speed and the airspeed of the plane?

17. Two angles are complementary. The measure of one angle is 5° less than twice the measure of the other. What is the measure of each angle?

18. A chemistry lab has two containers of hydrochloric acid. One is a 15% solution and the other a 40% solution. How many liters of each is required to make 100 liters of a 30% solution?

19. For a seafood party the caterer knows that the number of pounds of shrimp must be half the combined weight of the crabs and lobster. The number of pounds of crabs should be 5 less than the number of pounds of lobster. The cost of shrimp is $5 per pound, crab is $7 per pound, and lobster is $9 per pound. How many pounds of each can the caterer purchase for $960.50?

In Question 20, write a system of inequalities that describes the given conditions and draw a graph of the solution set.

20. On a vacant lot the total number of pines and dogwoods does not exceed 20. There are at least twice as many pines as dogwoods. What are the possible numbers of pines and dogwoods?

5-6   Cumulative Test

1. Find the values of a and b so that the ordered pairs are solutions of the given equation.

    $2x + 3y = 7$ $(3, a)$ $(b, 1)$

2. Determine the intercepts of the graph of $5x + y = -10$.

3. For each of the given equations, state whether the slope of the graph is positive, negative, zero, or undefined.

    (a) $x + 2y = 3$ (b) $x + 2 = 3$ (c) $0 \cdot x + 2y = 3$ (d) $x = 2y + 3$

4. Write $x - 2y + 4 = 0$ in slope–intercept form. Then, use what you know about the slope and y-intercept to draw a graph of the equation.

5. Consider the equation $y = \frac{2}{3}x - 1$. Describe the rate of change of y with respect to x.

6. Line L_1 contains point $P(-4, 3)$. Line L_2 is perpendicular to L_1. If L_1 and L_2 intersect at point $Q(3, -2)$, what is the slope of L_2?

7. Write an equation of a line that contains points $A(-2, -5)$ and $B(6, 3)$. Express the equation in the form $Ax + By = C$.

8. The equation of line L_1 is $2x + 3y = 6$. Line L_2 is parallel to L_1 and contains the point $P(1, -4)$. Write the equation of L_2 in slope–intercept form.

9. The graph of $x - y < 7$ is a shaded half-plane.
    (a) Describe the boundary line.
    (b) Describe two methods for determining which half-plane to shade.

10. Draw the graph of $|x - y| \le 5$.

11. Each of the following is a graph of a system of two linear equations in two variables. From the graph, describe the solution set of the system.

(a)
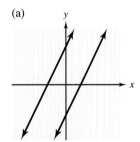

(b)
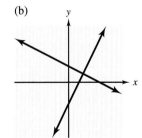

(c)
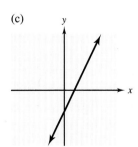

12. Use the substitution method to solve the following system of equations.

    $x + 7y = 40$
    $3x - 2y = -18$

13. Use the addition method to solve each of the following systems. Indicate if a system is inconsistent or if the equations are dependent.

    (a) $2x - 3y = -21$

    $\quad\;\; y = \dfrac{2}{3}x + 7$

    (b) $4x - y = -9$

    $\quad\;\; 2y - x = 11$

    (c) $y = 5x + 1$

    $\quad\;\; x - \dfrac{y}{5} = 0.2$

14. A college bookstore placed a textbook order for Mathematics 101 and English 122. The mathematics textbooks cost $35 each and the English textbooks cost $28 each. If a total of 420 books were ordered and the total cost was $13,160, how many of each book were ordered?

15. Use the addition method to solve the following system of equations.

$$2x + 3y + 5z = 18$$
$$3x - 2y + 4z = 13$$
$$4x - 3y + 3z = 8$$

16. Two angles are supplementary. A third angle is half as large as the first angle and 30° larger than the second angle. Find the measures of the three angles.

17. In each part, fill in the blank to indicate the result of performing a row operation.

(a) $\begin{bmatrix} -1 & 6 & | & 3 \\ 1 & 5 & | & -2 \end{bmatrix} \rightarrow \begin{bmatrix} 0 & \underline{(a)} & | & 1 \\ 1 & 5 & | & -2 \end{bmatrix}$

(b) $\begin{bmatrix} 8 & -16 & | & 24 \\ -3 & 1 & | & 0 \end{bmatrix} \rightarrow \begin{bmatrix} 1 & -2 & | & \underline{(b)} \\ -3 & 1 & | & 0 \end{bmatrix}$

(c) $\begin{bmatrix} 1 & 0 & | & 0 \\ -3 & 2 & | & -1 \end{bmatrix} \rightarrow \begin{bmatrix} 1 & 0 & | & 0 \\ 0 & \underline{(c)} & | & -1 \end{bmatrix}$

18. Use matrix methods to solve the following system of equations. Then, verify the solution with your calculator's automatic solve procedure.

$$x + y + z = 2$$
$$2x - y + z = 9$$
$$-x + 2y + 2z = -5$$

19. A cyclist rides 32 miles against the wind in 2 hours. The return trip takes 40 minutes less time. What is the wind speed?

20. A cruise ship has a total of 500 living accommodations on decks A, B, and C. For a 4-day cruise, A-deck spaces cost $1200, B-deck spaces cost $900, and C-deck spaces cost $500. The combined number of A- and B-deck spaces equals the number of C-deck spaces. If the total receipts from a sold-out cruise are $380,000, how many B-deck spaces are there?

21. Graph the solution set of the following system of inequalities.

$$x + y \leq 9$$
$$2x - y \leq 0$$
$$y \geq 2$$

22. The perimeter of a rectangle cannot exceed 40 inches. The width of the rectangle can be no longer than half the length. Write a system of *all* inequalities that describe these conditions.

7 Polynomials

As the two bar graphs show, attendance at men's and women's college basketball games has increased steadily since 1986.

The data in these graphs can be modeled by functions described by **polynomials.** We can, for example, add the two polynomials to determine a model function for the total attendance at college basketball games. (In Exercises 111–114 at the end of Section 7.1, you will find that the attendance at women's games is increasing at a faster rate than at men's games.)

In Chapter 7 we learn how to perform the four basic operations with polynomials. We also learn numerous techniques for factoring a polynomial, and we apply this skill to equation solving. We conclude with division, including synthetic division, and two important theorems about polynomials.

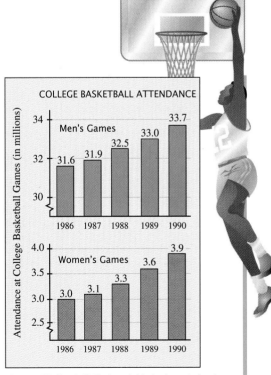

COLLEGE BASKETBALL ATTENDANCE

Attendance at College Basketball Games (in millions)

Men's Games

| 1986 | 1987 | 1988 | 1989 | 1990 |
| 31.6 | 31.9 | 32.5 | 33.0 | 33.7 |

Women's Games

| 1986 | 1987 | 1988 | 1989 | 1990 |
| 3.0 | 3.1 | 3.3 | 3.6 | 3.9 |

(Source: National Collegiate Athletic Association.)

415

7.1 Addition and Subtraction

Polynomials • Evaluating Polynomials • Vocabulary • Addition of Polynomials • Subtraction of Polynomials • Polynomial Functions

Polynomials

Recall that **terms** are parts of an expression that are separated by plus signs. In this section we consider one kind of term called a **monomial,** and we consider expressions called **polynomials** whose terms are monomials.

>   *Definition of Monomial and Polynomial*
>
> A **monomial** is a number or a product of a number and one or more variables with nonnegative integer exponents. A **polynomial** is a finite sum of monomials.

An example of a polynomial is $4x^5 - 5x^3 + 7x + 2$. We can write this expression as a *sum* of terms: $4x^5 + (-5x^3) + 7x + 2$. The terms are $4x^5$, $-5x^3$, $7x$, and 2 (or $2x^0$).

The **numerical coefficient** (usually just called the **coefficient**) of a term is the numerical factor of the term. In our example, the coefficients are 4, −5, 7, and 2. We also say that 2 is the **constant term.**

The following are some examples of algebraic expressions that are not polynomials.

$5x^2 + 3x^{-5}$ Exponents must be nonnegative integers.

$\dfrac{5}{x} - x + 7$ The term $5/x = 5x^{-1}$, and exponents must be nonnegative.

We can use function notation to give a polynomial a name and to indicate the variable used in the polynomial. For example, to write a polynomial in t, we give the polynomial a name such as p and use the function notation $p(t)$. (We can also use function notation for polynomials in more than one variable, but that topic is beyond the scope of this book.)

Evaluating Polynomials

Function notation is especially useful when we evaluate a polynomial. We can write "the value of the polynomial p when x has a value of 3" as $p(3)$.

EXAMPLE 1   *Evaluating a Polynomial*

For $p(t) = -t^3 + 3t - 6$, calculate $p(-2)$.

Solution

$$p(t) = -t^3 + 3t - 6$$
$$p(-2) = -1(-2)^3 + 3(-2) - 6$$
$$= -1(-8) - 6 - 6$$
$$= 8 - 6 - 6$$
$$= -4$$

Figure 7.1

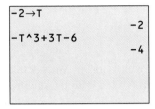

You can also use your calculator to evaluate the polynomial. Figure 7.1 shows a typical screen display. ■

Vocabulary

A polynomial consists of one or more terms (monomials). When a polynomial has only one term, we usually refer to the expression simply as a monomial. A polynomial with two nonzero terms is called a **binomial,** and a polynomial with three nonzero terms is called a **trinomial.**

Polynomial	Number of Terms	Name
$t^2 - 6 + 3t$	3	trinomial
$16 - x^2$	2	binomial
$5x^4$	1	monomial
-15	1	monomial

The **degree of a term** in one variable is the value of the exponent on the variable. (The constant term 0 does not have a degree.) The **degree of a polynomial** in one variable is the value of the exponent of the nonzero term with the largest degree.

Polynomial	Term	Degree of the Term	Degree of the Polynomial
$4x^5 - 5x^3 + 7x + 2$	$4x^5$	5	5
	$-5x^3$	3	
	$7x$	1	
	$2 \ (2x^0)$	0	
$5x^4$	$5x^4$	4	4
$-15 \ (\text{or} \ -15x^0)$	-15	0	0

The conventional way to write a polynomial is to write the terms in order of descending (or decreasing) degrees. In other words, we begin with the term with the largest exponent. Then, we proceed to the term with the next largest exponent until all terms are included. We say that the polynomial is written in **descending order.**

In some instances, it is convenient to write the polynomial in **ascending** (or increasing) **order.** In such cases, we begin with the term with the lowest exponent. Then the terms that follow have increasing degrees.

	Descending	*Ascending*
Polynomial	*Order*	*Order*
$4x^5 - 5x^3 + 7x + 2$	$4x^5 - 5x^3 + 7x + 2$	$2 + 7x - 5x^3 + 4x^5$
$-2x - x^5 + 7x^3$	$-x^5 + 7x^3 - 2x$	$-2x + 7x^3 - x^5$

In the first polynomial in this example, there is no term with x^4 or x^2. We say that these are *missing terms.* The polynomial could have been written with the missing terms: $4x^5 + 0x^4 - 5x^3 + 0x^2 + 7x + 2$.

A polynomial can have more than one variable in a term. For example, the expression $5x^2y^4 + x^5y^2 + 3xy - 6x + 3$ is a polynomial in two variables.

The degree of a term involving more than one variable is the sum of the exponents on the variables in that term. The degree of a polynomial is the degree of the nonzero term with the highest degree.

EXAMPLE 2   *Determining the Degree of a Polynomial*

Determine the degree of the polynomial $3x^2y^4 - 4xy^3 + x^3y^5$.

Solution

Term	*Degree of Term*	
$3x^2y^4$	6	Add the exponents: $2 + 4 = 6$.
$-4xy^3$	4	Note that $-4xy^3 = -4x^1y^3$. Add the exponents: $1 + 3 = 4$.
x^3y^5	8	Add the exponents: $3 + 5 = 8$.

The degree of the term with the highest degree is 8. Therefore, the degree of the polynomial is 8. ■

EXAMPLE 3   *Evaluating a Polynomial in Two Variables*

Evaluate the polynomial $3x^2y^4 - 4xy^3 + x^3y^5$ for $x = 2$ and $y = -1$.

Solution

$$\begin{aligned}
3x^2y^4 - 4xy^3 + x^3y^5 &= 3(2)^2(-1)^4 - 4(2)(-1)^3 + (2)^3(-1)^5 \\
&= 3(4)(1) - 4(2)(-1) + 8(-1) \\
&= 12 + 8 - 8 \\
&= 12
\end{aligned}$$

Figure 7.2

```
2→X
                      2
-1→Y
                     -1
3X^2Y^4-4XY^3+X^
3Y^5
                     12
```

As before, we can use a calculator to evaluate the polynomial. Figure 7.2 shows a typical display. ■

Addition of Polynomials

Terms of a polynomial are **like terms** if they have the same variables with the same exponents. To add polynomials, we simply combine the like terms.

EXAMPLE 4   *Adding Polynomials*

Add the polynomials $5x^3 - 4x + 2$ and $-x^3 + 5x^2 - 2x + 5$.

Solution

$$(5x^3 - 4x + 2) + (-x^3 + 5x^2 - 2x + 5)$$
$$= 5x^3 - 4x + 2 - x^3 + 5x^2 - 2x + 5 \qquad \text{Remove parentheses.}$$
$$= 4x^3 + 5x^2 - 6x + 7 \qquad \text{Combine like terms.}$$

Sometimes we write a polynomial addition problem as column addition. This approach has the advantage that like terms are aligned in columns. When you use the column approach, it is helpful to write any missing terms with a coefficient of 0.

$$
\begin{array}{ll}
\quad 5x^3 + 0x^2 - 4x + 2 & \text{First polynomial} \\
\underline{-1x^3 + 5x^2 - 2x + 5} & \text{Second polynomial} \\
\quad 4x^3 + 5x^2 - 6x + 7 & \text{Add the polynomials.}
\end{array}
$$

Subtraction of Polynomials

The definition of subtraction of real numbers applies to the subtraction of polynomials. If P and Q represent polynomials, then $P - Q = P + (-Q)$.

EXAMPLE 5   *Subtracting Polynomials*

Subtract the polynomial $2x^3 - x + 4$ from $-2x^3 + x^2 - 4x + 1$.

Solution

$$(-2x^3 + x^2 - 4x + 1) - (2x^3 - x + 4)$$
$$= -2x^3 + x^2 - 4x + 1 - 2x^3 + x - 4 \qquad \text{Remove parentheses.}$$
$$= -4x^3 + x^2 - 3x - 3 \qquad \text{Combine like terms.}$$

Polynomial subtraction problems can also be done with columns. Again, it is helpful to write any missing terms with the coefficient of 0.

$$
\begin{array}{ll}
-2x^3 + 1x^2 - 4x + 1 & \\
\underline{2x^3 + 0x^2 - 1x + 4} & \text{Change signs.}
\end{array}
\qquad
\begin{array}{l}
-2x^3 + 1x^2 - 4x + 1 \\
\underline{-2x^3 - 0x^2 + 1x - 4} \\
-4x^3 + 1x^2 - 3x - 3
\end{array}
$$

$$\text{Add the polynomials.}$$

Examples 4 and 5 show that we add and subtract polynomials simply by removing parentheses and combining like terms. We can simplify any polynomial expression in the same way.

EXAMPLE 6   *Simplifying Polynomial Expressions*

(a) $2(x^3 - 2x^2 - 6x - 1) - 3(-2x^3 + x^2 - 4x - 2)$

$\qquad = 2x^3 - 4x^2 - 12x - 2 + 6x^3 - 3x^2 + 12x + 6$ Distributive Property

$\qquad = 8x^3 - 7x^2 + 4$ Combine like terms.

(b) $(2x^3y^4 - x^2y^3 + 3xy - 7) + (x^3y^4 + 3x^2y^3 + 5x - 2)$

$\qquad = 2x^3y^4 - x^2y^3 + 3xy - 7 + x^3y^4 + 3x^2y^3 + 5x - 2$ Remove parentheses.

$\qquad = 3x^3y^4 + 2x^2y^3 + 3xy - 9 + 5x$ Combine like terms.

(c) $(x^2y^3 - 2xy^2 - 5xy - 7) - (x^2y^3 - xy^2 - 5xy + 3)$

$\qquad = x^2y^3 - 2xy^2 - 5xy - 7 - x^2y^3 + xy^2 + 5xy - 3$ Remove parentheses.

$\qquad = -xy^2 - 10$ Combine like terms.

Polynomial Functions

When two polynomial *expressions* are added or subtracted, the result is another polynomial expression. If $p(x)$ and $q(x)$ are two polynomial *functions,* then $p(x) + q(x)$, the sum of the two functions, is also a function. The functional notation used for $p(x) + q(x)$ is $(p + q)(x)$. Similarly, we write $p(x) - q(x) = (p - q)(x)$.

EXAMPLE 7   *Evaluating Sums of Polynomial Functions*

Given $p(x) = x^2 - 5$ and $q(x) = 3x + 2$, write $(p + q)(x)$. Using your calculator to evaluate the expressions, verify that

$\qquad (p + q)(3) = p(3) + q(3).$

Solution

$\qquad (p + q)(x) = p(x) + q(x)$

$\qquad\qquad = (x^2 - 5) + (3x + 2)$

$\qquad\qquad = x^2 + 3x - 3$

To evaluate the expressions with a calculator, we enter $Y_1 = p(x) = x^2 - 5$, $Y_2 = q(x) = 3x + 2$, and $Y_3 = (p + q)(x) = x^2 + 3x - 3$. Then, we let $x = 3$ and show that $Y_1 + Y_2 = Y_3$. Figure 7.3 shows a typical display.

Figure 7.3

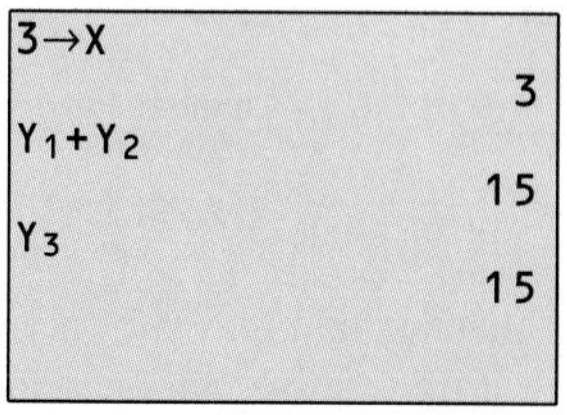

```
3→X
                    3
Y1+Y2
                   15
Y3
                   15
```

7.1 Quick Reference

Polynomials	▪ A **monomial** is a number or the product of a number and one or more variables raised to nonnegative integer powers. A **polynomial** is a finite sum of monomials. ▪ The **numerical coefficient** of a term is the numerical factor of the term. A **constant term** of a polynomial is a term that is just a number.
Evaluating Polynomials	▪ The value of a polynomial $P(x)$ when x has a value of c is expressed as $P(c)$. ▪ Using a calculator is an efficient way of evaluating a polynomial.
Vocabulary	▪ Polynomials with one, two, and three terms are called **monomials, binomials,** and **trinomials,** respectively. ▪ The **degree of a term** in one or more variables is the sum of the exponents on the variables. The **degree of a polynomial** is the highest degree of any of its nonzero terms. ▪ A polynomial is in **descending order** when the terms are written from highest to lowest degree. It is in **ascending order** when the terms are written from lowest to highest degree.
Addition of Polynomials	▪ **Like terms** are terms that are identical except for their coefficients. We combine like terms by using the Distributive Property to combine the coefficients. ▪ To add polynomials, we remove parentheses and combine the like terms. Column addition has the advantage that like terms are aligned in columns.
Subtraction of Polynomials	▪ To subtract polynomials, remove parentheses by changing the sign of each term of the polynomial that is being subtracted. Then, combine like terms. ▪ Column subtraction has the advantage that like terms are aligned in columns. However, remember to change the sign of each term of the polynomial being subtracted. Then, add the like terms. ▪ All polynomial expressions are simplified by removing parentheses and combining like terms.
Polynomial Functions	▪ When two polynomial functions $p(x)$ and $q(x)$ are added or subtracted, the result is another polynomial function, which can be written as $(p + q)(x)$ for addition or $(p - q)(x)$ for subtraction.

7.1 Exercises

1. Explain why each of the following is not a polynomial.

(a) $\dfrac{5}{x + 2}$

(b) $\dfrac{1}{3} x^{-4}$

2. Two of the following statements are true. Explain why the other two statements are not true.

(i) The term $5x^2$ is both a monomial and a polynomial.

(ii) The coefficients of $-2x^3 + 5x - 8$ are -2 and 5.

(iii) An example of a third-degree term is 2^3.

(iv) The expression $5 + 4x + 3x^2 + 2x^3$ is in ascending order.

In Exercises 3–6, state whether the expression is a polynomial.

3. $5x^2 + 4x - 6$

4. $\dfrac{5}{x^2} + 4x - 6$

5. $\sqrt{7} - 4x$

6. $4x^3 - 8x + 5x^2$

In Exercises 7–12, give the coefficient and degree of each term.

7. 5

8. $-y$

9. $-4x^3$

10. $2\pi z^2$

11. xy

12. $4ab^2$

13. What is the difference between the degree of a term and the degree of a polynomial with more than one term?

14. The expression $x^2 + 5$ can be written in the form $x^2 + 0x + 5$. Does this mean that $x^2 + 5$ is a trinomial? Why?

In Exercises 15–18, determine the degree of the polynomial.

15. $2x + 3y$

16. $4xy - 3y$

17. $2x^4 + xy^2 - x^2y^3$

18. $7a^4 - 3a^2b$

In Exercises 19–24, simplify each polynomial and determine its degree. Classify the simplified polynomial as a monomial, binomial, trinomial, or none of these.

19. $7x + 3x^5 - 9$

20. $5x^6$

21. $5x^3 + 7x - 2x^2 - 3$

22. $2x^2 + 3x - 1 - 2x^2$

23. $x^2y^3 + xy^2$

24. $3xy^2 + 2^3 - 3xy^2$

In Exercises 25–30, determine the degree of the polynomial. Then, write the polynomial in descending order and in ascending order.

25. $6 - 3y^2 + 6y$

26. $7 - 2y^3$

27. $x^3 - 5x^6 + 4x^9 - 7x^8$

28. $3x^4 + 5x^7 - 6x^2 + 4x - 7x^{11}$

29. $4n^8 - 3n^4 + n^{12} - 2n^6$

30. $8p^5 - p^9 + 22p^4 - 5^7$

31. Suppose that you need to evaluate the function $f(x) = x^2 - 5x + 3$ for ten different values of x. Explain the advantage of using a calculator.

32. Suppose that you need to evaluate $xy - 3$ for $x = 2$ and $y = -1$. Is this an appropriate calculator problem? Why?

In Exercises 33–38, evaluate the function as indicated.

33. $P(x) = 3 - x$
$P(-2), P(3)$

34. $P(x) = -2x + 3$
$P(0), P(-1)$

35. $Q(x) = -x^2 + 3x - 1$
$Q(0), Q(-2)$

36. $P(x) = 3x^3 - 2x^2 + 6x + 4$
$P(-3), P(1)$

37. $R(x) = x^3 - 4x^5 + 5x^2 + 2$
$R(2.1), R(-1.7)$

38. $R(x) = x^4 + 5x^3 - 2x^2 - 5x$
$R(-3.1), R(1.2)$

In Exercises 39–44, evaluate the polynomial for the given values of the variables.

39. $2x^2 + 3xy$
$x = -1, y = 2$

40. $x^2y - 3y^2$
$x = 2, y = -1$

41. $ab^2 - 2ab$
$a = 1, b = -3$

42. $ab^2 + a^2b$
$a = -2, b = 2$

43. $x^2y + 3xy - 2y^2$
$x = -2, y = 3$

44. $x^3y^2 - 4x^2y + 5y^3$
$x = 1, y = -1$

In Exercises 45–50, add the polynomials.

45. $(3x + 6) + (4x - 5)$

46. $(3y - 6) + (2y + 4)$

47. $(-a^2 + 2a - 1) + (a^2 + 2a + 1)$

48. $(x^2 - 5x) + (3x^2 - 4x - 2)$

49. $(2xy^2 - y^3) + (y^3 + xy^2)$

50. $(m^2n + mn) + (mn + 1)$

In Exercises 51–56, subtract the polynomials.

51. $(-5x - 3) - (2x - 9)$

52. $(5x - 6) - (-4x + 8)$

53. $(2x^2 - 4x + 7) - (-2x^2 - 3x - 7)$

54. $(x^2 + 6x + 3) - (4x - x^2)$

55. $(-3t + 7) - (2t^2 + 6)$

56. $(m^2 + 3) - (4 - 3m)$

In Exercises 57–64, perform the indicated operations.

57. $(x^2 + 5x) + (4 - 5x^2)$

58. $(3 - x^2) + (x^2 - 3x)$

59. $(4mn + n^2) - (m^2 - 3mn)$

60. $(x^2y + 3) - (2 - x^2y)$

61. $(x^2y + 2xy + 3y^2) + (-x^2y - 4xy + 2y^2)$

62. $(ab^2 + a^2b - 3ab) + (3ab^2 - 2ab - a^2b)$

63. $(9a^2b + 2ab - 5a^2) - (2ab - 3a^2 + 4a^2b)$

64. $(4y^3 - 3xy^2 + 4) - (2xy^2 - 3y^3 - 2)$

In Exercises 65–72, add or subtract as indicated.

65. Subtract
$$\begin{array}{r} 4x^3 - 10x \\ 6x^3 - \ \ 2x \\ \hline \end{array}$$

66. Subtract
$$\begin{array}{r} 3y - 6 \\ -5y + 2 \\ \hline \end{array}$$

67. Add
$$\begin{array}{r} x^2 + 3x - 7 \\ -5x^2 + 2x - 1 \\ \hline \end{array}$$

68. Add
$$\begin{array}{r} 4x^3 - 5x^2 \ \ \ \ \ \ + 6 \\ 2x^2 - 5x + 7 \\ \hline \end{array}$$

69. Subtract
$$\begin{array}{r} 4x^2 - x - 6 \\ 3x^2 - x + 6 \\ \hline \end{array}$$

70. Subtract
$$\begin{array}{r} 2x^4 + 3x^2 \ \ \ \ \ \ - 6 \\ 3x^4 \ \ \ \ \ \ - 2x + 5 \\ \hline \end{array}$$

71. Add
$$\begin{array}{r} x^2 + 3x - 2 \\ -2x^2 - \ \ x \\ 3x^2 \ \ \ \ \ \ + 4 \\ \hline \end{array}$$

72. Add
$$\begin{array}{r} y^3 + 2y^2 \ \ \ \ \ \ - 5 \\ -4y^3 \ \ \ \ \ \ - 3y + 2 \\ 6y^2 + 2y - 5 \\ \hline \end{array}$$

 73. For the problem "Subtract $x - 2$ from $3x + 4$," which of the following is correct? Why?

(i) $(x - 2) - (3x + 4)$

(ii) $(3x + 4) - (x - 2)$

(iii) $3x + 4 - x - 2$

 74. To solve the equation $\frac{1}{2}x - \frac{1}{3} = \frac{3}{4}$, we clear fractions by multiplying by the LCD. Can we do the same for the polynomial $\frac{1}{2}x - \frac{1}{3} - \frac{3}{4}$? Why?

75. Find the sum of $3x^2 - 6x$ and $4x + 7$.

76. Find the sum of $y^3 - 2y + 4$ and $y^2 + 2y + 1$.

77. Subtract $t + 3$ from $t - 3$.

78. Subtract $4 - x$ from $x - 4$.

79. Subtract $7x^2 - 2x + 9$ from $-5x - 2x^2$.

80. Subtract $2x^3 + 3x - 1$ from $x^3 + 3x - 1$.

81. What must be added to $x - 7$ to obtain 0?

82. What must be added to $3x + 5$ to obtain $x + 9$?

83. What must be subtracted from $x - 3$ to obtain 0?

84. What must be subtracted from $5x - 9$ to obtain $3x + 2$?

In Exercises 85–88, perform the indicated operations.

85. $\left(y^2 - \frac{1}{5}y - \frac{2}{3}\right) - \left(\frac{1}{2}y^2 - \frac{1}{5}y + \frac{1}{3}\right)$

86. $\left(\frac{1}{4}x^2 + \frac{x}{2} + 2\right) + \left(\frac{1}{2}x^2 + \frac{1}{6}x - \frac{3}{2}\right)$

87. $(2.3x^2 - 1.45x + 3.7) - (6.42 - 1.2x^2)$

88. $(5.78a - 4.6a^2) + (0.65a^2 + 4.5)$

In Exercises 89–102, simplify.

89. $(2x - 3) - (3x^2 + 4x - 5) - (3x - 4x^2)$

90. $(-x^2 - x - 1) + (2x^2 + 3x + 4) - (2x - 3x^2)$

91. $(4a^3 - 7a^2 + 4a) - (7a^2 + 5a^3 + 5) - (-a^2 - 7a + 6)$

92. $(6x^2 - 7x^3 + 6x) - (-x^3 + 5x^2 - 3) - (4x^4 + 3x^2 - 6x)$

93. $-[3x - (2x - 1)]$

94. $-(2y + 4) - (2y - 4)$

95. $(x - 3) + [2x - (3x - 1)]$

96. $(2 - 3x) - [(3x + 1) - (1 - 2x)]$

97. $[3 - (y - 1)] - (2y + 1)$

98. $[(a^2 - 2) + (a + 2)] - (a^2 - 3a)$

99. $(3x^2 - 2xy + y^2) - (4xy + 3x^2 - y^2) + (4xy^2 - 2x^2)$

100. $(4a^2 + 5ab - 7b^2) + (5ab - a^2 + 3b^2) - (2a^2 - 7ab + 3a^2b)$

101. $(2x^{2a} - 3x^a + 1) + (-3x^{2a} + x^a - 6)$

102. $(-3y^{2n} + y^n + 8) - (y^{2n} + y^n - 8)$

In Exercises 103–106, let $P(x) = 4x + 3$, $Q(x) = x^2 - 5$, and $R(x) = 2x^2 - 7x - 8$. Perform the indicated operations.

103. $P(x) + Q(x)$

104. $Q(x) - R(x)$

105. $R(x) - [P(x) - Q(x)]$

106. $[R(x) - P(x)] - Q(x)$

107. The attendance for a game at a minor league baseball stadium is approximated by $A(t) = -2t^2 + 280t - 8000$, where t is the noonday temperature (Fahrenheit). For the following noonday temperatures, what would be the expected attendance?

   (a) 80°F   (b) 50°F   (c) 98°F

   (d) 40°F   (e) 100°F

   (f) Using the results in parts (d) and (e), and taking into account what the function represents, what is the domain of $A(t)$?

108. The number of people swimming at a certain Florida beach is approximated by $N(t) = -t^2 + 140t - 4500$, where t is the Fahrenheit temperature of the water. For the following water temperatures, what would be the expected number of people swimming?

   (a) 60°F   (b) 75°F   (c) 86°F

   (d) 50°F   (e) 90°F

   (f) Using the results in parts (d) and (e), and taking into account what the function represents, what is the domain of $N(t)$?

109. The following functions describe costs and revenue (in dollars) from the production and sale of x units of light bulbs.

Cost C of Manufacturing	$C(x) = 20x + 12$
Cost M of Marketing	$M(x) = x^2 - 3x + 10$
Revenue R from Sales	$R(x) = 2x^2 + 8x$

   (a) What is the cost of manufacturing 100 units of bulbs?

   (b) What is the cost of marketing 100 units of bulbs?

(c) Write a function $T(x) = C(x) + M(x)$ that represents the total cost of manufacturing and marketing x units of bulbs.

(d) What is the total cost of manufacturing and marketing 100 units of bulbs?

(e) Show that $C(100) + M(100) = T(100)$.

(f) What is the revenue from selling 100 units of bulbs?

(g) Assuming profit is revenue minus total cost, write a function $P(x) = R(x) - T(x)$ that represents the profit from selling x units of bulbs.

(h) What is the profit from selling 100 units of bulbs?

(i) Show that $P(100) = R(100) - T(100)$.

110. The following functions describe costs and revenue (in dollars) from the production and sale of x units of radios.

Cost C of Manufacturing	$C(x) = 25x + 700$
Cost M of Marketing	$M(x) = x^2 - 2x + 15$
Revenue R from Sales	$R(x) = 4x^2 + 3x$

(a) What is the cost of manufacturing 70 units of radios?

(b) What is the cost of marketing 70 units of radios?

(c) Write a function $T(x) = C(x) + M(x)$ that represents the total cost of manufacturing and marketing x units of radios.

(d) What is the total cost of manufacturing and marketing 70 units of radios?

(e) Show that $C(70) + M(70) = T(70)$.

(f) What is the revenue from selling 70 units of radios?

(g) Assuming profit is revenue minus total cost, write a function $P(x) = R(x) - T(x)$ that represents the profit from selling x units of radios.

(h) What is the profit from selling 70 units of radios?

(i) Show that $P(70) = R(70) - T(70)$.

Exploring with Real Data

In Exercises 111–114, refer to the following background information and data.

Attendance at men's and women's college basketball games has increased steadily since 1986. Figure (a) shows attendance at men's games, and Fig. (b) shows attendance at women's games. Both figures are for the period 1986–1990, with attendance reported in millions. (Source: National Collegiate Athletic Assoc.)

Figures for 111–114

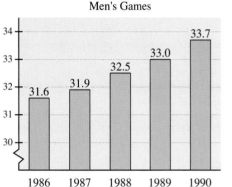
(a)

(b)

The attendance $A_M(t)$ for men's games and the attendance $A_W(t)$ for women's games can be approximated by the following functions, where t is the number of years since 1985.

$$A_M(t) = 31 + 0.5t$$
$$A_W(t) = 0.04t^2 - 0.01t + 3$$

111. Write the function $(A_M + A_W)(t)$ representing the total attendance at men's and women's college basketball games.

112. Evaluate $A_M(2) + A_W(2)$.

113. Evaluate $(A_M + A_W)(2)$. How does your result compare to your result in Exercise 112?

114. Calculate

$$\frac{A_M(5) - A_M(1)}{A_M(1)} \cdot 100 \quad \text{and}$$

$$\frac{A_W(5) - A_W(1)}{A_W(1)} \cdot 100.$$

What is the significance of these two results and what conclusion can you draw from them?

Challenge

In Exercises 115–118, determine a, b, and c so that the sum or difference on the left is equal to the polynomial on the right.

115. $(ax^2 + bx - 3) + (x^2 + 2x + c) = 3x + 1$

116. $(by + c) - (ay^3 - 2y + 5) = y^3 - 1$

117. $(x^2 + bx - c) + ax^2 = x^2 + 6x + 9$

118. $(2x^2 + bxy + 3y^2) - (ax^2 - xy - cy^2)$
$= x^2 + xy - 4y^2$

 119. If $(ax^2 + 5) + (bx^2 - 7) = 4x^2 - 2$, explain why it is not possible to determine unique values for a and b.

 120. If the degree of $p(x)$ is $n > 1$, what is the degree of $p(3)$?

In Exercises 121–124, evaluate the polynomial for the indicated values.

121. $P(x) = -x^2 + 5x - 3$
$P(-a)$, $P(2a)$

122. $H(x) = 3x + 5$
$H(-3a)$, $H(a + b)$

123. $xy^2 - 2xy$
$x = 2a$, $y = b$

124. $x^2y + xy^2$
$x = a$, $y = -a$

125. Suppose that $P(x) = x^3 - 5x^2 + 6x + 10$ and $P(0) = a + b + c$, where a, b, and c are constants. If c is the sum of a and b, and if b is one less than twice a, what are a, b, and c?

7.2 Multiplication and Special Products

Multiplying by a Monomial ▪ Multiplying Polynomials ▪ Multiplying Binomials ▪ Special Products ▪ Simplifying Expressions

Multiplying by a Monomial

In Chapter 2 we learned the multiplication rule for exponential quantities having the same base.

$$b^m \cdot b^n = b^{m+n}$$

In words, if the bases are the same, then we add the exponents. To multiply polynomials, we use this rule, often repeatedly.

The following is an example of a product of two monomials.

$$(4x^4)(5x^2)$$

To multiply these monomials, we use the associative and commutative properties with the multiplication rule for exponents.

$$(4x^4)(5x^2) = (4 \cdot 5)(x^4 \cdot x^2) = 20(x^{4+2}) = 20x^6$$

To multiply a monomial times a polynomial we use the Distributive Property.

EXAMPLE 1 *The Product of a Monomial and a Polynomial*

Multiply $3x^2(5x^3 - x^2 - 4x + 2)$.

Solution   We use the Distributive Property to multiply the monomial times each term of the polynomial.

$$3x^2(5x^3 - x^2 - 4x + 2)$$
$$= 3x^2(5x^3) - 3x^2(x^2) - 3x^2(4x) + 3x^2(2)$$
$$= 15x^5 - 3x^4 - 12x^3 + 6x^2$$ ■

Multiplying Polynomials

Multiplying a polynomial times another polynomial requires repeated use of the Distributive Property.

EXAMPLE 2 *The Product of Two Polynomials*

Multiply $(2x - 3)(x^2 + 3x - 5)$.

Solution   Use the Distributive Property. First, multiply each term of the second polynomial by $2x$. Next, multiply each term of the second polynomial by -3.

$$(2x - 3)(x^2 + 3x - 5)$$
$$= 2x(x^2 + 3x - 5) - 3(x^2 + 3x - 5)$$
$$= 2x(x^2) + 2x(3x) - 2x(5) - 3(x^2) - 3(3x) + 3(5)$$
$$= 2x^3 + 6x^2 - 10x - 3x^2 - 9x + 15$$
$$= 2x^3 + 3x^2 - 19x + 15$$ ■

Another way to approach the problem in Example 2 is to write it like a multiplication problem in arithmetic. Write one polynomial under the other.

$$
\begin{array}{r}
x^2 + 3x - 5 \\
2x - 3 \\
\hline
-3x^2 - 9x + 15 \\
2x^3 + 6x^2 - 10x \\
\hline
2x^3 + 3x^2 - 19x + 15
\end{array}
$$

$-3(x^2 + 3x - 5)$

$2x(x^2 + 3x - 5)$

Add the results.

EXAMPLE 3   *The Product of Two Trinomials*

Multiply $(x^2 + 3x - 1)(2x^3 - x + 1)$.

Solution

Use the Distributive Property to remove the parentheses. Then, combine the like terms.

$$(x^2 + 3x - 1)(2x^3 - x + 1)$$
$$= x^2(2x^3 - x + 1) + 3x(2x^3 - x + 1) - 1(2x^3 - x + 1)$$
$$= 2x^5 - x^3 + x^2 + 6x^4 - 3x^2 + 3x - 2x^3 + x - 1$$
$$= 2x^5 + 6x^4 - 3x^3 - 2x^2 + 4x - 1$$

With practice, you will soon find that you do not need to write the first step. Also, you may find the vertical method to be especially appealing when the polynomials are large. ■

EXAMPLE 4   *The Product of Two Binomials*

Multiply $(2x - 1)(3x + 5)$.

Solution

Again, we use the Distributive Property.

$$(2x - 1)(3x + 5) = 2x(3x + 5) - 1(3x + 5)$$
$$= 6x^2 + 10x - 3x - 5$$
$$= 6x^2 + 7x - 5$$ ■

Multiplying Binomials

To multiply binomials efficiently, shortcuts or mnemonic devices are often used.

Look carefully at the result in Example 4.

$$(2x - 1)(3x + 5) = 6x^2 + 10x - 3x - 5$$

1. The product of the *first terms* ($2x$ and $3x$) of the binomials is the first term of the result ($2x \cdot 3x = 6x^2$).

2. The products of the *outer terms* ($2x$ and 5) and the *inner terms* (-1 and $3x$) of the binomials are the next two terms of the result [$2x \cdot 5 + (-1)(3x) = 10x - 3x$].

3. The product of the *last terms* (-1 and 5) of the binomials is the last term of the result [$(-1)(5) = -5$].

Using the letters in the word FOIL might help you remember which terms are multiplied and allow you to quickly compute the product.

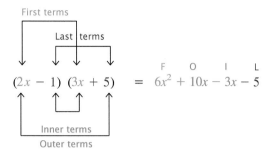

$$(2x - 1)\ (3x + 5) \quad = \quad 6x^2 + 10x - 3x - 5$$

When it is possible to do so, you should be able to combine like terms mentally.

EXAMPLE 5   *Using the FOIL Pattern to Multiply Binomials*

(a) $(x - 3)(4x - 1) = 4x^2 - x - 12x + 3$
$$= 4x^2 - 13x + 3$$

(b) $(x + 4)^2 = (x + 4)(x + 4)$
$$= x^2 + 4x + 4x + 16$$
$$= x^2 + 8x + 16$$

Note that the product $(x + 4)^2$ is the product of the same two binomials.

(c) $(3x + 5)(3x - 5) = 9x^2 - 15x + 15x - 25$
$$= 9x^2 - 25$$

Note that the product $(3x + 5)(3x - 5)$ is the product of the sum of two terms and the difference of the same two terms. ■

Special Products

Parts (b) and (c) of Example 5 illustrate two special products of binomials.

When you let A and B represent terms (monomials), the special product of the sum and difference of those two terms can be written as $(A + B)(A - B)$. We can use the FOIL method to find the product.

$$(A + B)(A - B) = A^2 - AB + AB - B^2$$
$$= A^2 - B^2$$

In words, the product of the sum and difference of the same two terms is the difference of their squares.

> **Product of the Sum and Difference of the Same Two Terms**
>
> $$(A + B)(A - B) = A^2 - B^2$$

EXAMPLE 6 *The Product of a Sum and Difference of the Same Two Terms*

Multiply.

(a) $(3x + 5)(3x - 5)$ (b) $(5 + x^2)(5 - x^2)$

Solution

(a) $(A\ + B)(A\ - B) = A^2\ - B^2$
$$\downarrow\quad\downarrow\ \downarrow\quad\downarrow\quad\ \downarrow\quad\ \downarrow$$
$$(3x + 5)\,(3x - 5) = (3x)^2 - 5^2 = 9x^2 - 25$$

(b) $(5 + x^2)(5 - x^2) = 5^2 - (x^2)^2 = 25 - x^4$

The square of a binomial is another special product. Letting A and B represent terms and using the FOIL method, we find that the square of a binomial is a trinomial of a specific form.

$$(A + B)^2 = (A + B)(A + B) = A^2 + AB + AB + B^2 = A^2 + 2AB + B^2$$
$$(A - B)^2 = (A - B)(A - B) = A^2 - AB - AB + B^2 = A^2 - 2AB + B^2$$

The first term of the result is the square of the first term of the binomial. The last term of the result is the square of the last term of the binomial. The middle term of the result is double the product of the two terms of the binomial.

> **The Square of a Binomial**
>
> $$(A + B)^2 = A^2 + 2AB + B^2$$
> $$(A - B)^2 = A^2 - 2AB + B^2$$

EXAMPLE 7 *The Square of a Binomial*

Determine the following products.

(a) $(x + 4)^2$ (b) $(2x^3 + 5)^2$ (c) $(3 - 2y)^2$

Solution

(a) $(A + B)^2 = A^2 + 2 \cdot A \cdot B + B^2$
$$\downarrow\quad\downarrow\quad\ \downarrow\quad\downarrow\ \downarrow\quad\downarrow\quad\ \downarrow$$
$$(x + 4)^2 = x^2 + 2 \cdot (x) \cdot (4) + 4^2 = x^2 + 8x + 16$$

(b) $(2x^3 + 5)^2 = (2x^3)^2 + 2(2x^3)(5) + 5^2 = 4x^6 + 20x^3 + 25$

(c) $(3 - 2y)^2 = 9 - 12y + 4y^2$

Simplifying Expressions

Sometimes, we must multiply polynomials as part of a larger simplification problem. In the following examples, we use the Distributive Property, we use the FOIL pattern, and we recognize special products in simplifying the expressions.

EXAMPLE 8   *Simplifying Expressions*

Simplify the following expressions.

(a) $(x + 3)(2x + 1) - (x + 3)(x - 3)$

(b) $[(x - 2) + y][(x - 2) - y]$

Solution

(a) The first product can be found with the FOIL method. The second product is the sum of two terms times the difference of the same two terms, and the result is the difference of their squares.

$$(x + 3)(2x + 1) - (x + 3)(x - 3)$$
$$= (2x^2 + 7x + 3) - (x^2 - 9)$$
$$= 2x^2 + 7x + 3 - x^2 + 9$$
$$= x^2 + 7x + 12$$

In the first step, the product $(x + 3)(x - 3)$ is the *quantity* $(x^2 - 9)$. The quantity must remain in parentheses until the subtraction is performed.

(b) The first factor is the *sum* of $x - 2$ and y and the second factor is the *difference* of $x - 2$ and y. Therefore, the product is the difference of the squares of $x - 2$ and y.

$$[(x - 2) + y][(x - 2) - y] = (x - 2)^2 - y^2$$
$$= x^2 - 4x + 4 - y^2$$

7.2   *Quick Reference*

Multiplying by a Monomial

- To multiply a monomial by a monomial, multiply the coefficients and the bases. If the bases are the same, multiply them by adding the exponents.

- To multiply a monomial times a polynomial, use the Distributive Property to multiply each term of the polynomial by the monomial.

Multiplying Polynomials

- To multiply two polynomials, multiply each term of one polynomial by each term of the other polynomial and then combine like terms.

- For larger polynomials, multiplying in columns helps to align like terms for easier addition.

| *Multiplying Binomials* | • The letters in the word FOIL (**F**irst–**O**utside–**I**nside–**L**ast) give us a pattern for remembering how to multiply two binomials quickly. |

| *Special Products* | • The product of the sum and difference of two terms has the form $(A + B)(A - B)$, where A and B are terms. |

$$(A + B)(A - B) = A^2 - B^2.$$

• The square of a binomial has the form $(A + B)^2$ or $(A - B)^2$. The products have the following forms.

$$(A + B)^2 = A^2 + 2AB + B^2$$
$$(A - B)^2 = A^2 - 2AB + B^2$$

| *Simplifying Expressions* | • The Distributive Property, the FOIL pattern, and the special product forms are useful in simplifying algebraic expressions that involve products of polynomials. |

7.2   Exercises

 1. Which of the following statements is false? Why?

(i) The simplified product of a nonzero monomial and a binomial is a binomial.

(ii) The simplified product of a binomial and a binomial is a trinomial.

 2. Give an example of two binomials whose simplified product has the following number of terms.

(a) 4   (b) 3   (c) 2

In Exercises 3–12, multiply the monomial times the polynomial.

3. $3x^2(2x - 3)$ **4.** $2x^3(3x + 4)$

5. $-4x^3(2x^2 + 5)$ **6.** $a^4(a^2 - 4a + 3)$

7. $-xy^2(x^2 - xy - 2)$

8. $-6x^2y(x^2 - 2xy + 7xy^2)$

9. $x^n(x + 5)$ **10.** $x^n(x - 4)$

11. $x^2(x^{2n} - 4x^n - 3)$

12. $2x^n(x^2 - 3x + 4)$

In Exercises 13–34, multiply the binomials.

13. $(x - 6)(x + 4)$ **14.** $(x + 7)(x - 3)$

15. $(y - 7)(y - 2)$ **16.** $(x - 6)(x - 5)$

17. $(y + 5)(3 - y)$ **18.** $(8 - x)(x + 2)$

19. $(5x - 6)(4x + 3)$ **20.** $(2x + 3)(4x - 5)$

21. $(9x + 2)(x + 3)$ **22.** $(4y - 7)(7 - 4y)$

23. $(4 - 5y)(2y + 1)$ **24.** $(1 - 4x)(2 + 7x)$

25. $(x - 9y)(x - 4y)$ **26.** $(x - 5y)(x - 3y)$

27. $(a + 5b)(a - 2b)$ **28.** $(m + 4n)(m - n)$

29. $(6x - 5y)(2x - 3y)$ **30.** $(3x + 2y)(4x + y)$

31. $(2x^2 - 3)(3x^2 - 2)$ **32.** $(4x^2 - 5)(3x^2 - 7)$

33. $\left(3x - \dfrac{3}{2}\right)\left(x + \dfrac{1}{2}\right)$

34. $\left(\dfrac{1}{4}x + 1\right)\left(\dfrac{3}{4}x + 1\right)$

In Exercises 35–44, multiply the polynomials.

35. $(x + 2)(x^2 - 4x + 5)$

36. $(x - 3)(x^2 + 5x - 3)$

37. $(2x - 3)(3x^2 + x - 1)$

38. $(3x + 2)(2x^2 - 2x + 1)$

39. $(x + y)(x^2 - xy + y^2)$

40. $(x - 2y)(x^2 + 2xy + 4y^2)$

41. $(2x - 1)(2x^3 - x^2 + 3x - 4)$

42. $(3x + 2)(3x^3 - 2x^2 + 5x - 2)$

43. $(x^2 - 2x + 1)(x^2 + 2x + 1)$

44. $(x^2 + 4x + 4)(x^2 - 4x + 4)$

In Exercises 45–48, multiply the polynomials.

45. $x^2 + 2x + 4$
$\underline{\qquad x - 2}$

46. $x^2 - 3x + 9$
$\underline{\qquad x + 3}$

47. $2x^2 - 4x + 5$
$\underline{\qquad x^2 - 2x + 3}$

48. $2x^2 + 3x - 2$
$\underline{\qquad x^2 - 2x + 1}$

 **49.** Can the FOIL method be used to multiply $(x + 3)(x - 3)$? What is the advantage of recognizing this as a special product instead?

 50. Can the FOIL method be applied to multiply $x - 3(x - 3)$? Why?

In Exercises 51–60, square the binomial.

51. $(x + 3)^2$

52. $(x - 5)^2$

53. $(2 - y)^2$

54. $(4 + x)^2$

55. $(7x - 3)^2$

56. $(9 + 2a)^2$

57. $(3a - b)^2$

58. $(m - 6n)^2$

59. $(2x + 5y)^2$

60. $(3x - 4y)^2$

In Exercises 61–70, multiply the binomials.

61. $(x - 4)(x + 4)$

62. $(x + 3)(x - 3)$

63. $(7 - y)(7 + y)$

64. $(3 + n)(3 - n)$

65. $(2x - 3)(2x + 3)$

66. $(6 + 5a)(6 - 5a)$

67. $(4x + 5y)(4x - 5y)$

68. $(2x - 7y)(2x + 7y)$

69. $(9n - 4m)(9n + 4m)$

70. $(a - 2b)(a + 2b)$

 71. For $x^2(x + 1)(x - 3)$, two approaches are

(i) multiply the two binomials and then multiply the result by x^2.

(ii) multiply $x + 1$ by x^2 and then multiply the results times $x - 3$.

Explain why both approaches are correct. Which approach is easier?

 72. Explain why each of the following can be written as a special product.

(a) $(a - b)(b + a)$ (b) $(b - a)(a - b)$

In Exercises 73–98, multiply and simplify.

73. $\left(\dfrac{2}{3}x + 3\right)\left(\dfrac{2}{3}x - 3\right)$

74. $\left(2x + \dfrac{1}{3}\right)\left(2x - \dfrac{1}{3}\right)$

75. $(a^3 - b^3)(a^3 + b^3)$

76. $(a^2 + b^2)(a^2 - b^2)$

77. $(4 - pq^2)(4 + p^2q)$

78. $(a^2b^2 - 1)(a^2b^2 - 3)$

79. $\left(\dfrac{1}{2}x + 1\right)^2$ **80.** $\left(\dfrac{2}{3}x - 3\right)^2$

81. $(x^2y^2 + 2)^2$ **82.** $(x^3 + 4yz^2)^2$

83. $2x^3(2x - 5)(3x + 1)$

84. $-3y(y^2 + 1)(y^2 - 1)$

85. $5x - 6(4x + 3)$ **86.** $2x + 3(4x - 5)$

87. $(b - 3)(2b + 5)(b - 2)$

88. $(a - 5)(3a + 7)(a + 2)$

89. $(2x + 5)(2x - 5) - (2x + 5)^2$

90. $(a + 2b)^2 - (a + b)(a - b)$

91. $(x + 7)(x - 7) - (x + 3)(x - 3)$

92. $(m - 3)(3 - m) - (m + 3n)(m - 3n)$

93. $(x + 6)(x - 4) - (2x - 3)(x + 2)$

94. $(3x + 5)(2x + 4) - (x - 6)(4x - 5)$

95. $(x + 3)^2 - (x - 3)^2$

96. $(x - 2)^2 + (5 - 2x)(1 + x)$

97. $4x^2 - x[5x - 3(x - 2) + 2x]$

98. $2 - 3x[(x - 1)(x + 3) - x^2]$

99. Multiply the square of the quantity $4x - 3$ by $2x^3$.

100. Multiply the square of the quantity $2 + 3y$ by $y + 1$.

101. Subtract the square of the quantity $5 - 2y$ from the product of the sum and difference of $3y$ and 2.

102. Subtract the product of $3 - 5x$ and $x + 4$ from $7x^2 + 5x + 6$.

In Exercises 103–106, determine a and b so that the product is equal to the polynomial.

103. $(ax + 3)(2x + b) = 8x^2 - 2x - 6$

104. $(ax + b)(3x + 1) = 6x^2 - 7x + b$

105. $(ax - 2)(3x + b) = 9x^2 - 4$

106. $(ax - b)^2 = 49 + 14x + x^2$

107. The length of a rectangle is $(2x - 3)$ feet and the width is $(x + 3)$ feet. Write a polynomial expression representing the area of the rectangle in terms of x.

108. Write a polynomial expression that represents the product of three consecutive integers.

109. A gardener uses 300 feet of fencing to enclose three sides of a rectangular garden. With w representing the width of the garden, write a polynomial expression that represents the area of the garden.

110. At noon, a car left a gas station and traveled north at an average of r miles per hour. The car reached its destination t hours later. At 1:00 P.M., a van left the same gas station and traveled south. Its average speed was 5 mph greater than the car's average speed. Write a polynomial expression (in two variables) that represents the total distance between the car and the van at the time that the car reached its destination.

111. The length of a side of a square is $(3x - 5)$ feet. Write a polynomial expression representing the area of the square in terms of x.

112. The lengths of the two legs of a right triangle are $(3x + 2)$ and $(x - 5)$. Use the Pythagorean Theorem to derive a polynomial expression for the square of the hypotenuse.

In Exercises 113 and 114, write a polynomial expression in x that represents the area of the given figure.

113.   **114.**

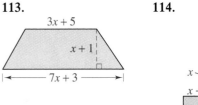

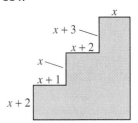

In Exercises 115 and 116, write a polynomial expression in x that represents the volume of the given figure.

115.   **116.**

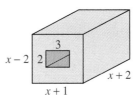

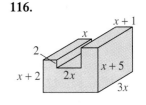

Exploring with Real Data

In Exercises 117–120, refer to the following background information and data.

The following table shows the number of college textbooks sold during the period 1985–1990. Also shown are the average prices for college textbooks during the same period.

	1985	1987	1988	1989	1990
Number Sold (in millions)	110	119	130	136	137
Average Book Price (in dollars)	14.31	15.15	15.37	15.76	16.93

(Source: Book Industry Study Group, Inc.)

The number $B(t)$ of textbooks sold can be approximated by $B(t) = 103.96 + 5.91t$, where t is the number of years since 1984 (for 1985, $t = 1$).

The average book price $P(t)$ for textbooks can be approximated by $P(t) = 0.16t^2 - 0.61t + 14.75$, where t is the number of years since 1984.

117. Use the data in the table to calculate the total expenditure for college textbooks in 1990.

118. The total expenditure $E(t)$ for college textbooks for any year t is the product of the number sold and the average price. Write a function for $E(t) = B(t) \cdot P(t)$.

119. Use your function $E(t)$ to approximate the total expenditures for college textbooks in 1990.

120. Use the expression

$$\frac{[E(6) - \text{Actual Total Expenditures in 1990}]}{\text{Actual Total Expenditures in 1990}}$$

to show that the model is in error by only about 1%.

Challenge

In Exercises 121 and 128, multiply.

121. $(x^n + 2)(x^n + 4)$

122. $(x^a + y^b)(2x^a - y^b)$

123. $(x^n - 2)^2$ **124.** $(2x^n + y^n)^2$

125. $(x^n + 4)(x^n - 4)$

126. $(2x^n - 3y^n)(2x^n + 3y^n)$

127. $x^{n-3}(x^{n+6} - x^3)$ **128.** $(x + 1)(x^{2n} - x^n + 1)$

 129. In the accompanying figure, the total area of the largest rectangle is the sum of the areas of the four inner rectangles. Use this fact to show the following.

$$(a + b)^2 = a^2 + 2ab + b^2$$

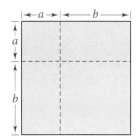

 130. In the accompanying figure, the total area of the largest rectangle is the sum of the areas of the three inner rectangles. Use this fact to show the following.

$$(a - b)(a + b) = a^2 - b^2$$

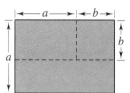

 131. Explain why the following two products are the same. Which is easier to multiply?

(i) $(x - a)[(x + a)(x + a)]$

(ii) $[(x - a)(x + a)](x + a)$

 132. Explain how to use a special product to find the product of the trinomials

$$(x + 3y - 4)(x + 3y + 4).$$

In Exercises 133 and 134, decide the best way to group the factors and then perform the multiplication.

133. $(2x + 1)(4x^2 + 1)(2x - 1)$

134. $(1 + 3a)(9a^2 - 1)(3a - 1)$

In Exercises 135–138, decide how to rearrange terms and to group within the factors to obtain a special product. Then perform the multiplication.

135. $(x + y + 2)(2 + x + y)$

136. $(x + y + 2)(-2 + x - y)$

137. $(x^2 + 5x + 3)(x^2 + 5x - 3)$

138. $(x^2 - 5x + 3)(x^2 + 5x - 3)$

In Exercises 139–142, multiply.

139. $[2y - (x - 3)]^2$

140. $[(2a + 3) - b]^2$

141. $[(x - 3) + 5y][(x - 3) - 5y]$

142. $[2 + (a + 3b)][2 - (a + 3b)]$

7.3 Common Factors and Grouping

Greatest Common Factor ▪ *Factoring by Grouping*

Greatest Common Factor

To factor a number, we write the number as a product of numbers. Sometimes there is more than one way to factor a number.

$$24 = 1 \cdot 24 \qquad 24 = 2 \cdot 12 \qquad 24 = 3 \cdot 8 \qquad 24 = 4 \cdot 6$$

The numbers on the right are numbers that can be divided into 24 with a remainder of 0. These numbers are called **factors** of 24.

To factor a number *completely* means to write the number as the product of prime numbers. Recall that a **prime number** is a natural number greater than 1 that has exactly two different natural number factors, 1 and itself. To factor 24 completely, we write

$$24 = 3 \cdot 2 \cdot 2 \cdot 2 = 3(2)^3.$$

The prime factors of 24 are 3 and 2. We use exponents for repeated factors.

The **greatest common factor (GCF)** of two numbers a and b is the largest integer that is a factor of both a and b.

EXAMPLE 1   *Finding the Greatest Common Factor*
Find the GCF for 36, 48, and 120.

Solution

$$36 = 3 \cdot 3 \cdot 2 \cdot 2 = 3^2 2^2 \qquad \text{Factor each number completely.}$$
$$48 = 3 \cdot 2 \cdot 2 \cdot 2 \cdot 2 = 3(2)^4$$
$$120 = 3 \cdot 2 \cdot 2 \cdot 2 \cdot 5 = 3(2)^3 5$$

Observe that 3 and 2 are the only factors that are common to all three numbers. One factor of 3 is common to all three numbers, and two factors of 2 are common to all three numbers.

The GCF is $(3)^1(2)^2 = 12$. ∎

To factor a polynomial means to write the polynomial as a product of polynomials. A polynomial is **prime over the integers** (or simply **prime**) if it cannot be factored into polynomials with integer coefficients.

When we factor a polynomial, we begin by determining whether the terms of the polynomial have a GCF other than 1. The GCF consists of every factor that is common to all of the terms.

> ### Finding the GCF of a Polynomial
>
> 1. Factor each term completely.
>
> 2. Determine the factors that are common to all the terms. For a repeated factor, use the smallest exponent that appears on it in the terms.
>
> 3. The GCF is the product of these factors.

EXAMPLE 2 *Determining the GCF in Monomials*

Determine the GCF for the monomials $75x^3y^4$, $45x^2y^5$, and $225x^4y^7$.

Solution

$$75x^3y^4 = 3 \cdot 5^2 \cdot x^3 \cdot y^4 \qquad \text{Factor the monomials completely.}$$
$$45x^2y^5 = 3^2 \cdot 5 \cdot x^2 \cdot y^5$$
$$225x^4y^7 = 3^2 \cdot 5^2 \cdot x^4 \cdot y^7$$

The factors common to all three monomials are 3, 5, x, and y. For each of these factors, use its smallest exponent.

The GCF $= 3 \cdot 5 \cdot x^2 \cdot y^4 = 15x^2y^4$. ∎

In Section 7.2 we used the Distributive Property to multiply a monomial times a polynomial. For example,

$$12x^4(2x^3 - 5) = (12x^4)(2x^3) - (12x^4)(5) = 24x^7 - 60x^4.$$

When we multiply, we are using the following form of the Distributive Property.

$$a(b + c) = ab + ac$$

To factor a polynomial, we also use the Distributive Property, but in reverse order. For example,

$$\begin{array}{ccc} a\,b + a\,c & = & a\,(b + c) \\ \downarrow \quad \downarrow & & \downarrow \\ 5\,x + 5\,y & = & 5\,(x + y). \end{array}$$

To use the Distributive Property to factor a polynomial, we determine the GCF for the terms in the polynomial. Then, we write the polynomial as the product of the GCF and another polynomial.

A polynomial is factored completely if each factor other than a monomial factor is prime. Note that $12x - 20 = 2(6x - 10)$, but the factor $6x - 10$ is not prime because it still has a common factor of 2. Because 4 is the GCF in $12x - 20$, we factor completely by writing $12x - 20 = 4(3x - 5)$.

EXAMPLE 3 *Factoring Out the Greatest Common Monomial Factor*

Factor completely.

(a) $10x^3 - 5x^2$ (b) $-4x^3 - 8y^2$ (c) $-12x^2y^3 + 10xy^3$

Solution

(a) $10x^3 - 5x^2 = (5x^2)(2x) - (5x^2)(1)$ Determine the GCF of each term.

$= 5x^2(2x - 1)$ Distributive Property

(b) $-4x^3 - 8y^2 = (-4)x^3 + (-4)(2y^2)$ We regard the common numerical factor as *negative.*

$= -4(x^3 + 2y^2)$ Distributive Property

(c) $-12x^2y^3 + 10xy^3 = (-2xy^3)(6x) - (-2xy^3)(5)$ GCF $= -2xy^3$

$= -2xy^3(6x - 5)$ Distributive Property ◼

NOTE: After we factor the GCF from an expression, we can (and should) multiply the indicated product to verify that we obtain the original expression. ◼

EXAMPLE 4   *Factoring Out the Greatest Common Monomial Factor*

Factor the trinomial $54x^3y^5 + 36x^4y^2 - 90x^6y^3$.

Solution

$54x^3y^5 = 3^3 2^1 x^3 y^5$ Factor the monomials.

$36x^4y^2 = 3^2 2^2 x^4 y^2$

$90x^6y^3 = 3^2 2^1 5^1 x^6 y^3$

GCF $= 3^2 2^1 x^3 y^2 = 18x^3y^2$ Use the smallest exponent on each common factor.

In factored form, $54x^3y^5 + 36x^4y^2 - 90x^6y^3 = 18x^3y^2(3y^3 + 2x - 5x^3y)$. ◼

EXAMPLE 5   *Factoring Out the Greatest Common Binomial Factor*

Factor each of the following.

(a) $3x(x + 4) - y(x + 4)$

(b) $5(2x - 1) - 2y(1 - 2x) + z(2x - 1)$

(c) $x^2(x + 5) + 3x^4(x + 5) - 6x^3(x + 5)$

(d) $6x^3(y - 1)^2 + 9x(y - 1)^5$

Solution

(a) $3x(x + 4) - y(x + 4)$ The binomial $(x + 4)$ is common to both terms.

$= (x + 4)(3x - y)$

(b) $5(2x - 1) - 2y(1 - 2x) + z(2x - 1)$ Use the rule $a - b = -(b - a)$ to rewrite $1 - 2x$.

$= 5(2x - 1) + 2y(2x - 1) + z(2x - 1)$

$= (2x - 1)(5 + 2y + z)$ Now $(2x - 1)$ is common to all three terms.

(c) $x^2(x + 5) + 3x^4(x + 5) - 6x^3(x + 5)$ Note that x^2 and $(x + 5)$ are common to all three terms.

$= x^2(x + 5)(1 + 3x^2 - 6x)$

(d) $6x^3(y - 1)^2 + 9x(y - 1)^5$ Note that 3, x, and $(y - 1)^2$ are common to both terms.

$= 3x(y - 1)^2[2x^2 + 3(y - 1)^3]$ ◼

Factoring by Grouping

In some cases we must be creative to find the GCF. At first glance, the polynomial $x^3 - 3x^2 + 2x - 6$ does not appear to have a common factor. A factor of x appears in three of the terms but not in the last term.

However, suppose the first two terms and last two terms are grouped together.

$$x^3 - 3x^2 + 2x - 6 = (x^3 - 3x^2) + (2x - 6)$$

The first group has a GCF of x^2; the second group has a GCF of 2.

$$x^3 - 3x^2 + 2x - 6 = x^2(x - 3) + 2(x - 3)$$

The polynomial is still not factored because it is a sum of two terms, not a product. Observe that the two terms have a common binomial factor $(x - 3)$.

$$x^3 - 3x^2 + 2x - 6 = x^2(x - 3) + 2(x - 3)$$
$$= (x - 3)(x^2 + 2)$$

In factored form, $x^3 - 3x^2 + 2x - 6 = (x - 3)(x^2 + 2)$.

This method of factoring is called the **grouping method.**

> ### *Factoring by the Grouping Method*
>
> 1. Group the terms so that each group has a common factor.
>
> 2. Factor out the GCF in each group.
>
> 3. If there is a common factor in the resulting terms, factor it out. If there is no common factor, a different grouping may produce the desired result.

EXAMPLE 6 *Factoring by the Grouping Method*

Factor completely.

(a) $ax + bx + ay + by$

(b) $6x - 3y - 2ax + ay$

(c) $x^3 + 2x^2 - 5x - 10$

Solution

(a) $ax + bx + ay + by$

$\qquad = (ax + bx) + (ay + by)$ Group the expression in pairs.

$\qquad = x(a + b) + y(a + b)$ Factor out the GCF in each pair.

$\qquad = (a + b)(x + y)$ In the resulting two terms, the GCF is $(a + b)$.

(b) When the third term is negative, care must be taken when we group the terms. Just as we change signs when we *remove* parentheses from $-(m - n) = -m + n$, we must change signs when we *insert* parentheses in $-m + n = -(m - n)$.

$$6x - 3y - 2ax + ay$$

$= (6x - 3y) - (2ax - ay)$ Group the expression in pairs.

$= 3(2x - y) - a(2x - y)$ Factor out the GCF in each pair.

$= (2x - y)(3 - a)$ In the resulting two terms, the GCF is $(2x - y)$.

(c) $x^3 + 2x^2 - 5x - 10$

$= (x^3 + 2x^2) - (5x + 10)$ Change signs in the second group.

$= x^2(x + 2) - 5(x + 2)$ Factor out the GCF in each pair.

$= (x + 2)(x^2 - 5)$ In the resulting two terms, the GCF is $(x + 2)$. ■

The grouping method for factoring is a method worth trying, but it does not guarantee success. Keep in mind that a polynomial may be a prime polynomial having no factors other than 1 and itself.

7.3   Quick Reference

Greatest Common Factor

- We say that m is a **factor** of n if m can be divided into n with a remainder of 0.

- A **prime number** is a natural number greater than 1 that has exactly two different natural number factors, 1 and itself.

- The **greatest common factor (GCF)** of two numbers a and b is the largest number that is a factor of both a and b.

- The first step in factoring a polynomial is to determine the GCF of the terms of the polynomial.

  1. Factor each term completely.

  2. Determine the factors that appear in all terms. For a repeated factor, use the smallest exponent that appears on it in the terms.

  3. The GCF is the product of these factors.

- If the GCF is not 1, use the Distributive Property to write the polynomial as the product of the GCF and another polynomial.

- A polynomial is **prime** if it cannot be factored into polynomials with integer coefficients.

Factoring by Grouping

- To factor polynomials with more than three terms, try using the grouping method.

  1. Group the terms so that each group has a common factor.

  2. Factor out the GCF in each group.

  3. If the resulting terms have a common factor, factor it out. If there is no common factor, a different grouping may produce the desired result.

- If there is no common factor after grouping, the polynomial cannot be factored by grouping.

7.3   *Exercises*

 1. In the expression $2x + 3y$, what are $2x$ and $3y$ called? In the expression $(2x)(3y)$, what are $2x$ and $3y$ called?

2. Is the GCF of an expression always a monomial? Give an example that supports your answer.

In Exercises 3–10, determine the GCF.

3. $24, 60x$

4. $21x, 35$

5. $12x, 16y, 24z$

6. $25a, 30b, 15c$

7. $3x^5, 4x^7, 5x^3$

8. $2y^{10}, 5y^4, 6y^5$

9. m^5n^3, mn^6

10. p^5q, pq^5

In Exercises 11–16, complete the factorization.

11. $-6xy + 3y = -3y(\underline{\hspace{1cm}})$

12. $-8x^2 + 6x = -2x(\underline{\hspace{1cm}})$

13. $5x^2 + 10x + 15 = \underline{\hspace{1cm}}(x^2 + 2x + 3)$

14. $30 - 15a + 9a^2 = \underline{\hspace{1cm}}(10 - 5a + 3a^2)$

15. $-xy^2 + 2x^2y = \underline{\hspace{1cm}}(\underline{\hspace{1cm}} - 2x)$

16. $-2m^3n^5 + 4m^2n^2 = \underline{\hspace{1cm}}(\underline{\hspace{1cm}} - 2)$

In Exercises 17–36, determine the GCF and factor the given expression completely.

17. $3x + 12$

18. $2x + 6$

19. $5x + 10y - 30$

20. $6 - 24a - 18b$

21. $12x^2 - 4x$

22. $6x^2 + 2x$

23. $7x^2 - 10y$

24. $5a^3 - 8b^2$

25. $x^3 - x^2$

26. $c^4 + 3c^3$

27. $3x^3 - 9x^2 + 12x$

28. $15b^5 + 5b^7 - 10b^4$

29. $x^5y^4 - x^4y^3 + xy^2$

30. $x^7y^8 + xy^5 - x^5y$

31. $3x(a + by) - 2y(a + by)$

32. $2x(x - 3) - 5y(x - 3)$

33. $4x^2(2x + 3) - 7y^2(2x + 3)$

34. $5x^5(x - 2) - 9y^3(x - 2)$

35. $x(x + 5) - 4(x + 5)$

36. $y(y - 3) + 2(y - 3)$

 37. Determine whether each of the following is a correct factorization of $4x - 12$. If it is incorrect, explain why.
(a) $4(x - 3)$ (b) $-4(-x + 3)$
(c) $4(-3 + x)$ (d) $-4(3 - x)$

 38. Explain how the Multiplication Property of -1 can be used to factor $x - 2y$ so that one of the factors is $2y - x$.

In Exercises 39–44, factor out the GCF and the opposite of the GCF.

39. $x - 3$

40. $5 - 2x$

41. $6 + 3x - x^2$

42. $-a^2 + 5$

43. $-4y^3 - 2y^2$

44. $3x^2 + 9x^4$

In Exercises 45–50, factor the given expression.

45. $y(y - 3) + 2(3 - y)$

46. $x(x - 2) - 3(2 - x)$

47. $5(3 - z) - z(z - 3)$

48. $4(6 - a) + a(a - 6)$

49. $(x + 3)(x - 4) - (x + 2)(4 - x)$

50. $(3 - a)(2a - 1) + (a - 2)(a - 3)$

 51. If $(2x - 5)(x + 3)$ is the result of factoring an expression by grouping, what were the four terms in the original expression?

 52. Which of the following is the correct way to group $x^3 + x^2 - x - 1$ in pairs? Explain.
(i) $(x^3 + x^2) - (x - 1)$
(ii) $(x^3 + x^2) - (x + 1)$

In Exercises 53–72, use the grouping method to factor the given expression completely.

53. $ab + bx + ay + xy$

54. $ax + 3a + 2x + 6$

55. $cd + 3d - 4c - 12$

56. $ac + 5a - 2c - 10$

57. $ab - ay - bx - xy$

58. $ab - by + ax + xy$

59. $ax^2 + 3x^2 + ay + 3y$

60. $ax^2 + 2a - x^2y - 2y$

61. $3ax + 6bx - 4ay - 8by$

62. $6xy + 4bx + 9ay + 6ab$

63. $x^3 + 2x^2 + 3x + 6$

64. $x^3 + 4x^2 - 3x - 12$

65. $3y^3 + 9y^2 + 12y + 36$

66. $4y^3 - 12y^2 + 20y - 60$

67. $a^2b + a^2y + b + y$

68. $ab^2 - abx - b + x$

69. $x^3 - 15 + 3x - 5x^2$

70. $2x^3 + 12 - 3x - 8x^2$

71. $6x^3 - 12 + 8x - 9x^2$

72. $6x^3 + 6 + 9x + 4x^2$

 73. Suppose you use the grouping method to factor as follows.

$$x^7 + x^6 + x^3 + x^2 = (x^7 + x^6) + (x^3 + x^2)$$
$$= x^4(x^3 + x^2) + (x^3 + x^2)$$
$$= (x^3 + x^2)(x^4 + 1)$$

Looking at the first factor, you realize that you have not factored completely. Must you start over, or can you write the correct factorization in just one more step? Explain.

74. It is not possible to use the grouping method to factor $2x^3 + 6x^2 + 8x + 4$. Is it correct to conclude that the polynomial is prime?

In Exercises 75–78, factor completely.

75. $18x^3 + 6x^2 - 45x - 15$

76. $2a^3 + 2a^2b + 6ab + 6b^2$

77. $3xy^3 - 6xy^2 + 7xy - 14x$

78. $r^2t^2 + r^2t + rt^2 + rt$

In Exercises 79–82, complete the factorization of each expression.

79. $\dfrac{2}{3}y - \dfrac{5}{6} = \dfrac{1}{6}(\underline{})$

80. $1 - \dfrac{3}{4}x = \dfrac{1}{4}(\underline{})$

81. $\dfrac{1}{2}x^2 + \dfrac{1}{3}x + 1 = \dfrac{1}{6}(\underline{})$

82. $x^2 + \dfrac{1}{4}x - \dfrac{1}{2} = \dfrac{1}{4}(\underline{})$

In Exercises 83–86, factor the given expression as in Exercises 79–82.

83. $\dfrac{1}{3}x + 5$

84. $\dfrac{1}{4}a - \dfrac{3}{2}b$

85. $\dfrac{1}{6}x^2 + \dfrac{1}{4}x + 3$

86. $x^2 - \dfrac{1}{2}x + \dfrac{1}{5}$

 87. For the monomials x^5y^3, x^7y^2, and x^4y^6, explain how you can determine the GCF by inspecting exponents.

 88. Which of the following is the correct way to factor $(a + b)^{10} + (a + b)^6$? Explain.

(i) $(a + b)^6(a + b)^4$

(ii) $(a + b)^6[(a + b)^4 + 1]$

In Exercises 89–100, factor out the GCF in the expression.

89. $d^{20} - d^{15}$

90. $y^{12} + y^{20}$

91. $88x^{48} + 11x^{51}$

92. $100a^{50} + 10a^{25}$

93. $24a^5b^3 - 6ab^4 + 12a^3b^2$

94. $16c^6d^4 + 8c^2d^3 - 12c^2d$

95. $48x^3y^4z^3 - 60x^2y^6z^8 + 72x^4y^5z^7$

96. $50a^4b^6c^3 + 70a^4b^2c^5 - 20a^7b^5c^7$

97. $4a^2b^3(x - 4) + 12a^3b^2(x - 4)$

98. $15c^4d^5(x + 3) + 5c^5d^4(x + 3)$

99. $15x^4(b - 2)^3 + 5x^3(b - 2)^3$

100. $24x^2y^4(a - 4)^4 - 8x^2y^2(a - 4)^4$

In Exercises 101–104, write the given formula in completely factored form.

101. If P dollars are invested at a simple interest rate r for one year, the account value A is given by $A = P + Pr$.

102. If an object is propelled upward at an initial velocity of 560 feet per second, the height h of the object after t seconds is given by $h = 560t - 16t^2$.

103. The total surface area A of a right circular cylinder is given by $A = 2\pi rh + 2\pi r^2$, where h is the height of the cylinder and r is its radius.

104. The area A of a trapezoid is given by $A = \frac{1}{2}b_1h + \frac{1}{2}b_2h$, where h is the altitude of the trapezoid and b_1 and b_2 are the lengths of the bases.

Exploring with Real Data

In Exercises 105–108, refer to the following background information and data.

The following table shows the number of Americans who traveled to Europe during the period 1984–1990. The table also shows their average expenditures.

	1984	1986	1988	1990
Number of Travelers (in thousands)	5623	5887	7438	8043
Average Expenditures (in dollars)	1408	1560	1521	1791

(Source: U.S. Travel and Tourism Administration.)

The number $N(t)$ of travelers can be approximated by the function $N(t) = 21.31t^2 + 312.68t + 5511.35$, where t is the number of years since 1984.

The average expenditure $A(t)$ can be approximated by the function $A(t) = 1380.65 + 61.44t$, where t is the number of years since 1984.

105. Total expenditures $T(t)$ during this period is the product of the number of travelers $N(t)$ and their average expenditure $A(t)$. Write a function for $T(t)$ in factored form.

106. If you were to write an expression for $T(t)$ in polynomial form, what would be the degree of the polynomial?

107. Explain how you know that the graph of $T(t)$ for the given period would rise from left to right.

108. Based on the way $T(t)$ was formed in Exercise 91, state two circumstances that would cause the value of $T(t)$ to decrease in the future.

Challenge

109. Explain how to verify that $3x^{-4} + 2x^{-6} = x^{-6}(3x^2 + 2)$.

110. Although the expression $2x^{-3} + 5x^{-4}$ is not a polynomial, can we still determine the GCF by using the smaller exponent on x? Is the GCF x^{-4} in this case? If so, what is the other factor?

In Exercises 111–114, complete the factorization of each expression.

111. $2x^{-3} - 5x^{-4} = x^{-4}(\underline{\hspace{1cm}})$

112. $x^{-5} + x^5 = x^{-5}(\underline{\hspace{1cm}})$

113. $6x^{-6}y^{-3} - 2x^{-4}y^{-4} = 2x^{-6}y^{-4}(\underline{\hspace{1cm}})$

114. $(3 - x)^{-2} + (3 - x)^{-1} = (3 - x)^{-2}(\underline{\hspace{1cm}})$

In Exercises 115–118, factor completely as in Exercises 97–100.

115. $2x^3 - 10x^{-4}$ **116.** $y^{-2} + 3y^{-4}$

117. $5x^{-4}y^7 - 15x^3y^{-2}$

118. $(2x - 1)^{-3} - (2x - 1)^{-4}$

In Exercises 119–124, factor out the GCF. Assume that n is a positive integer.

119. $a^{2n} + 2a^n$ **120.** $y^{5n+6} - y^6$

121. $8c^{2n+3} - 4c^{n+2} + 12c^n$

122. $x^{4n} - x^{3n} - 5x^{2n}$

123. $(a + 3)^{2n} - (a + 3)^n$

124. $(x + 5)^n + (x + 5)^{n-1}$

125. Use the grouping method to factor the following expression.

$$xy(x + 1) + x(x + 1) + y^2(x + 1) + y(x + 1)$$

126. Use the grouping method to factor the following expression.

$$a^2x + 3a^2 + bx + 3b - 5x - 15$$

127. If n is an integer, show that when $4n^2 + 2n$ is factored completely, one factor is an even number and the other factor is an odd number.

7.4   Special Factoring

Difference of Two Squares   ▪   Perfect Square Trinomials   ▪
Sum and Difference of Two Cubes

Difference of Two Squares

We can reverse the special product $(A + B)(A - B) = A^2 - B^2$ and write

$$A^2 - B^2 = (A + B)(A - B),$$

where A and B represent terms.

We recognize a **difference of two squares** as two perfect square terms separated by a minus sign. If the binomial we want to factor matches the pattern $A^2 - B^2$, then the factors are $A + B$ and $A - B$.

EXAMPLE 1   *Factoring a Difference of Two Squares*

Factor completely.

(a) $81 - 4y^2$ (b) $25x^2 - 49y^2$ (c) $(3x - 1)^2 - 25$

Solution   The binomial in each part is in the form $A^2 - B^2$.

(a) $81 - 4y^2 = 9^2 - (2y)^2 = (9 + 2y)(9 - 2y)$

(b) $25x^2 - 49y^2 = (5x)^2 - (7y)^2 = (5x + 7y)(5x - 7y)$

(c) $(3x - 1)^2 - 25 = (3x - 1)^2 - 5^2$

$\qquad\qquad\qquad\quad = [(3x - 1) + 5][(3x - 1) - 5]$

$\qquad\qquad\qquad\quad = (3x - 1 + 5)(3x - 1 - 5)$

$\qquad\qquad\qquad\quad = (3x + 4)(3x - 6)$ The factor $(3x - 6)$ has a common factor of 3.

$\qquad\qquad\qquad\quad = 3(3x + 4)(x - 2)$ ◼

EXAMPLE 2   *Factoring Out a GCF with the Other Factor Being a Difference of Two Squares*

Factor completely.

(a) $5x^4 - 80$ (b) $8x^3 - 2x$

Solution

(a) $5x^4 - 80 = 5(x^4 - 16)$ Factor out the GCF 5.

$\qquad\qquad\quad = 5[(x^2)^2 - 4^2]$ The other factor is a difference of two squares.

$\qquad\qquad\quad = 5(x^2 + 4)(x^2 - 4)$ $A^2 - B^2 = (A + B)(A - B)$

$\qquad\qquad\quad = 5(x^2 + 4)(x^2 - 2^2)$ The last factor is a difference of two squares.

$\qquad\qquad\quad = 5(x^2 + 4)(x + 2)(x - 2)$

Note that the factor $(x^2 + 4)$ cannot be factored further.

(b) $8x^3 - 2x = 2x(4x^2 - 1) = 2x(2x + 1)(2x - 1)$ ■

Perfect Square Trinomials

Previously, we have seen the results of squaring binomials.

$$(A + B)^2 = A^2 + 2AB + B^2$$
$$(A - B)^2 = A^2 - 2AB + B^2$$

The trinomials on the right are called **perfect square trinomials** because they are trinomials that are the squares of binomials. By reversing these special products, we can establish two factoring patterns.

$$A^2 + 2AB + B^2 = (A + B)^2$$
$$A^2 - 2AB + B^2 = (A - B)^2$$

EXAMPLE 3   *Factoring Perfect Square Trinomials*

Factor completely.

(a) $x^2 - 12x + 36$ (b) $25x^2 + 70xy + 49y^2$ (c) $x^6 + 10x^3y^2 + 25y^4$

Solution

$$A^2 \;-\; 2\;A\;B\;+\;B^2 \;=\;(A - B)^2$$

(a) $x^2 - 12x + 36 = (x)^2 - 2(x)(6) + (6)^2 = (x - 6)^2$

(b) $25x^2 + 70xy + 49y^2 = (5x)^2 + 2(5x)(7y) + (7y)^2 = (5x + 7y)^2$

(c) $x^6 + 10x^3y^2 + 25y^4 = (x^3)^2 + 2(x^3)(5y^2) + (5y^2)^2 = (x^3 + 5y^2)^2$ ■

Sum and Difference of Two Cubes

Sometimes a factoring pattern can be discovered by experimenting with multiplication.

EXPLORATION 1   *Multiplications That Lead to Factoring Patterns*

(a) Perform the following multiplications.

   (i) $(A + B)(A^2 - AB + B^2)$

   (ii) $(A - B)(A^2 + AB + B^2)$

(b) In words, how would you describe the results in each part.

(c) Write two factoring patterns that are revealed by these results.

Discovery

(a) $(A + B)(A^2 - AB + B^2) = A^3 - A^2B + AB^2 + A^2B - AB^2 + B^3$
$$= A^3 + B^3$$
$(A - B)(A^2 + AB + B^2) = A^3 + A^2B + AB^2 - A^2B - AB^2 - B^3$
$$= A^3 - B^3$$

(b) The first result is a **sum of two cubes,** and the second result is a **difference of two cubes.**

(c) By reversing the order of the results, we have the following:
$$A^3 + B^3 = (A + B)(A^2 - AB + B^2)$$
$$A^3 - B^3 = (A - B)(A^2 + AB + B^2)$$

These patterns give us a way of factoring a sum of two cubes or a difference of two cubes.

EXAMPLE 4   *Factoring Sums and Differences of Two Cubes*

Factor completely.

(a) $y^3 + 64$ (b) $54x^3 + 2y^3$ (c) $t^6 - 125$

Solution

$$A^3 + B^3 = (A + B)(A^2 - A \cdot B + B^2)$$

(a) $y^3 + 64 = (y)^3 + (4)^3 = (y + 4)(y^2 - y \cdot 4 + 4^2)$
$$= (y + 4)(y^2 - 4y + 16)$$

(b) $54x^3 + 2y^3$
$$= 2(27x^3 + y^3) \qquad\qquad \text{Note the GCF 2.}$$
$$= 2[(3x)^3 + y^3] \qquad\qquad \text{The other factor is a sum of two cubes.}$$
$$= 2(3x + y)[(3x)^2 - (3x)(y) + y^2] \qquad A^3 + B^3 = (A + B)(A^2 - AB + B^2)$$
$$= 2(3x + y)(9x^2 - 3xy + y^2)$$

(c) $t^6 - 125 = (t^2)^3 - 5^3 \qquad\qquad \text{The expression is a difference of two cubes.}$
$$= (t^2 - 5)[(t^2)^2 + t^2 \cdot 5 + 5^2] \qquad A^3 - B^3 = (A - B)(A^2 + AB + B^2)$$
$$= (t^2 - 5)(t^4 + 5t^2 + 25)$$

7.4   Quick Reference

Difference of Two Squares

- An expression consisting of two perfect square terms separated by a minus sign is called a **difference of two squares.**

- The special factoring pattern for a difference of two squares is
$$A^2 - B^2 = (A + B)(A - B).$$

Perfect Square Trinomials

- A **perfect square trinomial** is a trinomial of the form

$$A^2 + 2AB + B^2 \quad \text{or} \quad A^2 - 2AB + B^2.$$

- The special factoring patterns for perfect square trinomials are as follows.

$$A^2 + 2AB + B^2 = (A + B)^2$$
$$A^2 - 2AB + B^2 = (A - B)^2$$

Sum and Difference of Two Cubes

- A **sum of two cubes** is a binomial of the form $A^3 + B^3$; a **difference of two cubes** is a binomial of the form $A^3 - B^3$.

- The special factoring patterns for the sum of two cubes and the difference of two cubes are as follows.

$$A^3 + B^3 = (A + B)(A^2 - AB + B^2)$$
$$A^3 - B^3 = (A - B)(A^2 + AB + B^2)$$

7.4 *Exercises*

1. In words, how do we recognize a difference of two squares?

2. Why can we factor $9x^2 + 36$, but not $9x^2 + 16$?

In Exercises 3–14, factor completely.

3. $x^2 - 9$

4. $9x^2 - 16$

5. $1 - 4x^2$

6. $25 - a^2$

7. $16x^2 - 25y^2$

8. $49x^2 - 64y^2$

9. $\frac{1}{4}x^2 - 1$

10. $\frac{1}{9}y^2 - \frac{1}{16}$

11. $a^6 - b^{16}$

12. $x^4 - y^2$

13. $16x^{2n} - 25$

14. $81 - 49y^{2n}$

15. In words, how do we recognize a perfect square trinomial?

16. For $x^2 + 6xy + 9y^2$, how do we check the suspected factorization $(x + 3y)^2$?

In Exercises 17–28, factor completely.

17. $a^2 - 4a + 4$

18. $x^2 + 8x + 16$

19. $4x^2 + 12x + 9$

20. $9x^2 - 12x + 4$

21. $x^2 - 14xy + 49y^2$

22. $x^2 + 12xy + 36y^2$

23. $x^2 + x + \frac{1}{4}$

24. $y^2 - \frac{4}{3}y + \frac{4}{9}$

25. $x^4 - 10x^2 + 25$

26. $y^4 + 6y^2 + 9$

27. $x^{2n} - 14x^n + 49$

28. $36 + 12y^{2n} + y^{4n}$

29. In words, how do we recognize a difference of two cubes?

30. Show how to factor $b^3 - a^3$ so that one of the factors is $a - b$.

In Exercises 31–42, factor completely.

31. $x^3 + 1$

32. $8 - a^3$

33. $27y^3 - 8$

34. $64y^3 + 27$

35. $125 - 8x^3$

36. $64 - 27z^3$

37. $27x^3 + y^3$

38. $8a^3 - b^3c^3$

39. $x^6 + 1$

40. $b^6 - 125$

41. $x^{3a} - 27$

42. $x^{3a} + 8$

In Exercises 43–72, factor completely.

43. $x^2 + 2x + 1$

44. $x^2 - 10x + 25$

45. $0.25x^2 - 0.49$

46. $1.44a^2 - 0.0016$

47. $x^3 - 1$

48. $27 + b^3$

49. $c^2 + 100 - 20c$

50. $d^2 + 81 + 18d$

51. $2x^3 + 54$

52. $3x^3 - 24$

53. $81 + y^2$

54. $32a^2 + 50$

55. $36 - 84y + 49y^2$

56. $25 + 60p + 36p^2$

57. $4x^2 - 64$

58. $16 - 4y^2$

59. $25x^2 - 40xy + 16y^2$

60. $16x^2 + 40xy + 25y^2$

61. $x^4 - 16$

62. $81a^4 - b^4$

63. $8 + 125x^6$

64. $125 - 27x^6$

65. $0.04a^2 - 0.04a + 0.01$

66. $2.25c^2 - 24.6c + 67.24$

67. $x^{2n} - y^{4m}$

68. $x^{2n} - y^{2n}$

69. $y^8 - 6y^4 + 9$

70. $x^8 + 4x^4 + 4$

71. $216x^3 - y^3$

72. $a^3b^3 + 8$

In Exercises 73–102, factor completely.

73. $x^6 - 16x^3 + 64$

74. $x^6 + 64 + 16x^3$

75. $x^{12} - 4y^{20}$

76. $49 - 81x^4$

77. $x^6 + 4 - 4x^3$

78. $y^6 + 9 + 6y^3$

79. $98 - 2y^6$

80. $125x^2 - 5x^8$

81. $125x^5y^3 - x^2$

82. $27x^9y^7 - 8y^4$

83. $49x^7 - 70x^5 + 25x^3$

84. $9x^6 - 42x^4 + 49x^2$

85. $x^4 + 36y^4$

86. $16a^4 + 25b^6$

87. $64 + 27x^3y^9$

88. $1 - 27x^6y^3$

89. $x^4 - 8x^2 + 16$

90. $x^4 - 18x^2 + 81$

91. $16x^4 - 25y^2$

92. $c^8 - 1$

93. $a^{21} + b^9$

94. $c^{12} - d^{15}$

95. $64x^4 + 80x^2y + 25y^2$

96. $49x^6 + 126x^3y^2 + 81y^4$

97. $16a^4 - 81d^4$

98. $x^4 - c^4$

99. $2x^4 + 16x$

100. $3y^5 - 81y^2$

101. $125x^6 - 27y^{12}$

102. $27x^9 + 8y^{15}$

In Exercises 103–108, factor completely.

103. $4a(x^2 - y^2) + 8b(x^2 - y^2)$

104. $10x(a^2 - b^2) + 5y(a^2 - b^2)$

105. $x^2(a + b) - y^2(a + b)$

106. $4a^2(x - y) - b^2(x - y)$

107. $(a^2 - 1)(m + 1) - (a^2 - 1)(m + 3)$

108. $(b^2 - 4)(n + 2) + (b^2 - 4)(3 - n)$

There are two ways to factor

$$(x + 3)^2 + 8(x + 3) + 16.$$

One is to simplify the expression and then factor the result. A more efficient method is to recognize that the expression matches the pattern $A^2 + 2AB + B^2$. Thus,

$$(x + 3)^2 + 8(x + 3) + 16$$
$$(\ \ A\ \ )^2 + 2 \cdot \quad A \quad \cdot B + B^2$$
$$\downarrow \qquad \downarrow \quad \downarrow \quad \downarrow \quad \downarrow$$
$$= (x + 3)^2 + 2 \cdot (x + 3) \cdot 4 + 4^2$$
$$[(\ \ A\ \ ) + B]^2$$
$$\downarrow \qquad \downarrow$$
$$= [(x + 3) + 4]^2$$
$$= (x + 7)^2$$

In Exercises 109–114, use this method to factor the expression completely.

109. $(2x + 1)^2 - 10(2x + 1) + 25$

110. $(3 - x)^2 + 14(3 - x) + 49$

111. $36(x - 1)^2 + 36(x - 1) + 9$

112. $81(2 + x)^2 - 18(2 + x) + 1$

113. $(x^2 - 2)^2 - 4(x^2 - 2) + 4$

114. $(x^2 - 9)^2 + 16(x^2 - 9) + 64$

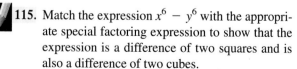 **115.** Match the expression $x^6 - y^6$ with the appropriate special factoring expression to show that the expression is a difference of two squares and is also a difference of two cubes.

116. The expression $x^6 + y^6$ is both a sum of two squares and a sum of two cubes. Which way should you regard the expression if you want to factor it?

In Exercises 117–120, factor completely. (Hint: In each case, begin by factoring the expression as a difference of two squares.)

117. $x^6 - 64$

118. $c^6 - 1$

119. $a^6 - b^6$

120. $x^6 - 64y^6$

In Exercises 121–126, determine a, b, or c so that the trinomial is a perfect square.

121. $9x^2 + bx + 1$

122. $x^2 + bx + 100$

123. $ax^2 - 10x + 25$

124. $ax^2 + 12x + 36$

125. $9x^2 - 24x + c$

126. $4x^2 - 20x + c$

In some cases, a trinomial that is not a perfect square trinomial can be written as a sum of two squares.

$$x^2 + 12x + 40 = (x^2 + 12x + 36) + 4$$
$$= (x + 6)^2 + 2^2$$

In Exercises 127–130, fill in the blanks to convert the given trinomial into a sum of two squares.

127. $x^2 + 4x + 5 = (\quad) + 1 = (\quad)^2 + (\quad)^2$

128. $x^2 - 6x + 13 = (\quad) + 4 = (\quad)^2 + (\quad)^2$

129. $x^2 - 2x + 10 = (\quad) + (\quad) = (\quad)^2 + (\quad)^2$

130. $x^2 + 8x + 20 = (\quad) + (\quad) = (\quad)^2 + (\quad)^2$

131. A circular swimming pool of radius r is surrounded by a circular deck of radius R. (See figure.) Write an expression for the area of the deck. Then write the expression in factored form.

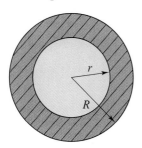

132. The product of three numbers is given by the expression $x^3 + 10x^2 + 25x$. By factoring, show that two of the numbers are the same and that the third number is 5 less than the other two numbers.

133. In physics, kinetic energy is the energy an object has by virtue of its motion. The formula for kinetic energy K is $K = \frac{1}{2}mv^2$, where m is the mass of the object and v is the velocity of the object. Suppose two objects of equal mass have velocities v_1 and v_2, respectively. Write an expression for the difference in their kinetic energies. Then write the expression in factored form.

134. A pharmaceutical company determines that a drug costing x dollars to develop and market will yield x^4 dollars in revenue. Write an expression for the net earnings (revenue minus cost) of the drug. Then write the expression in factored form.

Exploring with Real Data

In Exercises 135–138, refer to the following background information and data.

The following table shows the sales (in millions of dollars) of tennis shoes and tennis equipment for the period 1988–1991.

	Sales (in millions)			
	1988	1989	1990	1991
Tennis Shoes	353	508	582	599
Tennis Equipment	264	315	281	294

(Source: National Sporting Goods Association.)

The sales $E(t)$ of tennis equipment during this period can be modeled by $E(t) = 10.5t + 255$, where t is the number of years since 1987.

The *total* sales $T(t)$ of tennis shoes and equipment during this period can be modeled by the function $T(t) = -32.5t^2 + 245t + 396$, where t is the number of years since 1987.

135. Write a modeling function for the sales $S(t)$ of tennis shoes.

136. Explain how you know at a glance that the expression for $S(t)$ is not a perfect square trinomial.

137. If $T(5)$ is greater than $T(4)$, can we conclude that there was an increase in the sales of tennis equipment? Why?

138. It would be reasonable to assume that sales of tennis shoes and equipment would rise or fall together. However, we can see from the table that sales of tennis shoes during this period consistently increased while sales of tennis equipment peaked in 1989. How can this be explained?

Challenge

In Exercises 139–144, completely factor the expression as a difference of two squares.

139. $(a + b)^2 - 4$ **140.** $16 - (x + y)^2$

141. $16 - (x - y)^2$

142. $(2x + y)^2 - (a + b)^2$

143. $(3 + y)^2 - (y - 3)^2$ **144.** $25 - (x - 4)^2$

In Exercises 145–148, completely factor the expression as a sum of two cubes or as a difference of two cubes.

145. $(3 + 2x)^3 + 27y^3$

146. $(3 + 2x)^3 + (2 - 3y)^3$

147. $64y^3 - (4 + y)^3$

148. $(5 - 2x)^3 - (3 + 4y)^3$

In Exercises 149–156, factor completely. (Hint: Start with the grouping method.)

149. $x^5 - x^3 + 8x^2 - 8$

150. $4x^5 - x^3 - 32x^2 + 8$

151. $x^5 - 32 + 8x^2 - 4x^3$

152. $x^5 + 1 - x^3 - x^2$

153. $ax^3 - ay^3 + bx^3 - by^3$

154. $x^2y^3 - x^2z^3 - y^3 + z^3$

155. $y^2 - 9x^2 + 12x - 4$

156. $x^2 - y^2 - 12y - 36$

In Exercises 157–164, factor completely. Assume that a, b, m, and n are positive integers.

157. $x^{3n} - 4x^n$ **158.** $x^{2n+3} - x$

159. $50x^{6a-4} - 2y^{10b+6}$

160. $48x^{4n+6} - 75y^{2m-8}$

161. $5x^{6n+8} - 20x^4$ **162.** $28y^{5n+4} - 7y^n$

163. $x^{3a} + y^{6a}$ **164.** $x^{9a} - y^{3b}$

7.5 Factoring Trinomials

Leading Coefficient a = 1 ▪ *Leading Coefficient a ≠ 1* ▪
Trial-and-Error Method ▪ *General Strategy*

Leading Coefficient $a = 1$

A **quadratic trinomial** is a trinomial of the form $ax^2 + bx + c$. We call the number a the **leading coefficient.**

When the leading coefficient a is 1, the trinomial has the form $x^2 + bx + c$. From our experience with multiplication, we may recall that $x^2 + bx + c$ was often the result of multiplying two binomials of the form $(x + m)(x + n)$, where m and n are constants. Thus, it seems reasonable to try to factor the trinomial with that pattern in mind.

$$x^2 + bx + c = (x + m)(x + n)$$

We can factor the trinomial in this way if we can find numbers m and n such that $mn = c$ and $m + n = b$.

EXAMPLE 1 *Factoring a Quadratic Trinomial $x^2 + bx + c$*
Factor $x^2 + 8x + 12$.

Solution

Because $c = 12$, we find pairs of numbers whose product is 12.

1 and 12	2 and 6	3 and 4
−1 and −12	−2 and −6	−3 and −4

Because $b = 8$, we choose the pair 2 and 6 whose sum is 8.

$$x^2 + 8x + 12 = (x + 2)(x + 6)$$

We can (and should) verify that the factorization is correct by confirming that $(x + 2)(x + 6) = x^2 + 8x + 12$. ▪

EXAMPLE 2 *Factoring a Quadratic Trinomial $x^2 + bx + c$*
Factor $x^2 - x - 20$.

Solution

Because $c = -20$, we find pairs of numbers whose product is -20.

−1 and 20	−2 and 10	−4 and 5
1 and −20	2 and −10	4 and −5

452 CHAPTER 7 *Polynomials*

Because $b = -1$, we choose the pair 4 and -5 whose sum is -1.

$$x^2 - x - 20 = (x - 5)(x + 4)$$

Verify by multiplying $(x - 5)(x + 4)$ to obtain $x^2 - x - 20$. ∎

EXAMPLE 3 *Factoring Quadratic Trinomials*

Factor completely.

(a) $x^2 - 5x + 6$ (b) $3x^2 + 18x - 21$

(c) $t^2 + 11st + 18s^2$ (d) $y^4 - 5y^2 - 14$

Solution

(a) $x^2 - 5x + 6 = (x - 2)(x - 3)$

(b) $3x^2 + 18x - 21 = 3(x^2 + 6x - 7)$ The GCF is 3.
$$= 3(x + 7)(x - 1)$$

(c) $t^2 + 11st + 18s^2 = (t + 2s)(t + 9s)$

(d) Note that $y^4 - 5y^2 - 14$ is a fourth-degree trinomial, not a quadratic trinomial. In this case, however, the method still applies.

$$y^4 - 5y^2 - 14 = (y^2 - 7)(y^2 + 2)$$

Both factors have a degree greater than 1, so they should be inspected to see if they can be factored further. In this case, they cannot. ∎

Leading Coefficient $a \neq 1$

When the leading coefficient of a quadratic trinomial is a number other than 1, an approach to factoring the trinomial is to extend the method used in Examples 1–3.

1. Find all pairs of numbers whose product is ac.

2. Choose the pair whose sum is b.

3. Write the middle term bx as the sum of two terms with the pair of numbers as the coefficients.

4. Factor the expression with the grouping method.

EXAMPLE 4 *Factoring $ax^2 + bx + c$ When $a \neq 1$*

Factor $6x^2 - x - 2$.

Solution Because $a = 6$ and $c = -2$, the product $ac = -12$. The following are all pairs of numbers whose product is -12.

-1 and 12	-2 and 6	-3 and 4
1 and -12	2 and -6	3 and -4

Because $b = -1$, we choose the pair 3 and -4 whose sum is -1.

$$\begin{aligned} 6x^2 - x - 2 &= 6x^2 + 3x - 4x - 2 &&\text{Write } -x \text{ as } 3x - 4x.\\ &= (6x^2 + 3x) - (4x + 2) &&\text{Group in pairs.}\\ &= 3x(2x + 1) - 2(2x + 1) &&\text{Factor out the GCF from each pair.}\\ &= (2x + 1)(3x - 2) &&\text{Now the GCF is } (2x + 1). \end{aligned}$$

To verify that $6x^2 - x - 2 = (2x + 1)(3x - 2)$, we multiply $(2x + 1)(3x - 2)$ and obtain $6x^2 - x - 2$. ■

EXAMPLE 5 *Factoring $ax^2 + bx + c$ When $a \neq 1$*

Factor $10x^2 - 21x + 9$.

Solution   Because $a = 10$ and $c = 9$, we seek all pairs of numbers whose product is 90. However, ac is positive, which means the signs of the pair of numbers must be the same. In fact, because b is negative, both signs are negative. Therefore, we need to list only the factors that are negative.

$$\begin{array}{ccc} -1 \;\;\text{and}\;\; -90 & -2 \;\;\text{and}\;\; -45 & -3 \;\;\text{and}\;\; -30 \\ -6 \;\;\text{and}\;\; -15 & -9 \;\;\text{and}\;\; -10 & \end{array}$$

Of these, only -6 and -15 have a sum of $b = -21$.

$$\begin{aligned} 10x^2 - 21x + 9 &= 10x^2 - 15x - 6x + 9 &&\text{Write } -21x \text{ as } -15x - 6x.\\ &= (10x^2 - 15x) - (6x - 9) &&\text{Group in pairs.}\\ &= 5x(2x - 3) - 3(2x - 3) &&\text{Factor out the GCFs.}\\ &= (2x - 3)(5x - 3) &&\text{Now the GCF is } (2x - 3). \end{aligned}$$

To verify, $(2x - 3)(5x - 3) = 10x^2 - 21x + 9$. ■

Trial-and-Error Method

An alternate method for factoring a trinomial is the trial-and-error method.

Suppose we want to factor $6x^2 + 7x - 3$. To write the polynomial as the product of binomials, we recall that the product of the first terms must be $6x^2$, and the product of the last terms must be -3.

Here are the possibilities.

$$6x^2\text{: } 6x \cdot 1x \quad \text{or} \quad 2x \cdot 3x \qquad -3\text{: } 1(-3) \quad \text{or} \quad -1 \cdot 3$$

Using all the combinations of these possibilities, we can list the possible factors of $6x^2 + 7x - 3$. Then, we check the middle term resulting from multiplying the factors.

Possible Factors	*Middle Term*
$(6x + 1)(x - 3)$	$-17x$
$(6x - 3)(x + 1)$	common factor
$(6x - 1)(x + 3)$	$17x$
$(6x + 3)(x - 1)$	common factor
$(2x + 1)(3x - 3)$	common factor
$(2x - 3)(3x + 1)$	$-7x$
$(2x - 1)(3x + 3)$	common factor
$(2x + 3)(3x - 1)$	$7x$

Note that we can eliminate four of the eight possible factorizations because one of the binomial factors has a common factor. Because $6x^2 + 7x - 3$ does not have a common factor, none of the factors of $6x^2 + 7x - 3$ has a common factor.

The last possible factorization in the list produces the correct middle term $7x$. Therefore,

$$6x^2 + 7x - 3 = (2x + 3)(3x - 1).$$

As always, factoring out the GCF, if any, is the first step in factoring. Once you have done so, it is not necessary to list or try any combination in which a binomial factor has a common factor.

EXAMPLE 6   *Factoring Quadratic Trinomials by Trial and Error*

Factor completely.

(a) $4x^2 + 11x + 6$ (b) $12x^2 + 7x - 12$

(c) $30x^2 - 65x + 10$ (d) $-2t^2 + 3t + 1$

(e) $3y^2 + yz - 10z^2$

Solution

(a) $4x^2 + 11x + 6$

$a = 4: 4 \cdot 1$ or $2 \cdot 2$ $c = 6: 6 \cdot 1$ or $3 \cdot 2$

The possible factorizations have the following forms.

$$(4x + \underline{\quad})(x + \underline{\quad})   \text{ or }   (2x + \underline{\quad})(2x + \underline{\quad})$$

The only possible factorizations that do not have a common factor are as follows.

Possible Factors	*Middle Term*
$(4x + 1)(x + 6)$	$25x$
$(4x + 3)(x + 2)$	$11x$

Because $b = 11$, the correct factorization is $(4x + 3)(x + 2)$.

(b) $12x^2 + 7x - 12$

$a = \quad 12: \quad 12 \cdot 1 \quad$ or $\quad 6 \cdot 2 \quad$ or $\quad 4 \cdot 3$

$c = -12: -12 \cdot 1 \quad$ or $\quad -6 \cdot 2 \quad$ or $\quad -4 \cdot 3 \quad$ or

$\quad\quad\quad\quad\quad -1 \cdot 12 \quad$ or $\quad -2 \cdot 6 \quad$ or $\quad -3 \cdot 4$

Eliminating combinations that have a common factor, we have the following possible factorizations.

Possible Factors	Middle Term
$(12x + 1)(x - 12)$	$-143x$
$(12x - 1)(x + 12)$	$143x$
$(4x + 3)(3x - 4)$	$-7x$
$(4x - 3)(3x + 4)$	$7x$

Because $b = 7$, the correct factorization is $(4x - 3)(3x + 4)$.

(c) First, we factor out the GCF 5.

$$30x^2 - 65x + 10 = 5(6x^2 - 13x + 2)$$

The other factor is a quadratic trinomial, which may be factorable. Because the sign of the constant term is $+$ and the sign of the middle term is $-$, the binomial factors must be of the form

$$(\underline{\quad\quad} - 2)(\underline{\quad\quad} - 1).$$

The only possibilities are as follows.

Possible Factors	Middle Term
$(x - 2)(6x - 1)$	$-13x$
$(3x - 2)(2x - 1)$	$-7x$

Because $b = -13$, the correct factorization of the trinomial $6x^2 - 13x + 2$ is $(x - 2)(6x - 1)$. Thus, the complete factorization of $30x^2 - 65x + 10$ is $5(x - 2)(6x - 1)$.

(d) $-2t^2 + 3t + 1$

The only possibilities are as follows.

Possible Factors	Middle Term
$(-2t + 1)(t + 1)$	$-1t$
$(2t + 1)(-t + 1)$	$1t$

Because neither possibility gives us the correct middle term $3t$, we conclude that the trinomial is prime.

(e) $3y^2 + yz - 10z^2$

$a = 3: 3 \cdot 1 \qquad c = -10: -10 \cdot 1 \quad \text{or} \quad -5 \cdot 2 \quad \text{or}$
$$-1 \cdot 10 \quad \text{or} \quad -2 \cdot 5$$

The possible factorizations are as follows.

Possible Factors	Middle Term	Possible Factors	Middle Term
$(3y - 10z)(y + 1z)$	$-7yz$	$(3y + 10z)(y - 1z)$	$7yz$
$(3y - 5z)(y + 2z)$	$1yz$	$(3y + 5z)(y - 2z)$	$-1yz$
$(3y - 1z)(y + 10z)$	$29yz$	$(3y + 1z)(y - 10z)$	$-29yz$
$(3y - 2z)(y + 5z)$	$13yz$	$(3y + 2z)(y - 5z)$	$-13yz$

Because $b = 1$, the correct factorization is $(3y - 5z)(y + 2z)$. ■

When selecting a method for factoring quadratic trinomials, we should consider the number of possible factorizations that are likely to arise.

The trial-and-error method is useful for a trinomial such as $2x^2 + 5x + 3$ because 2 and 3 are prime numbers and the number of possible binomial factors is very limited.

When many factorizations are possible, using the method of finding the factors of ac whose sum is b, and then using the grouping method, is the more systematic approach. Bear in mind, though, that not every trinomial is factorable.

With practice and experience, your instincts will become more reliable, and you will be able to arrive at correct factorizations quickly.

General Strategy

As we solve problems in algebra, we sometimes must factor a polynomial and need to decide which of the methods of factoring we have studied is appropriate. It is helpful to have a general strategy to guide us in making the decision.

A polynomial is factored when it is written as the product of polynomials. A polynomial is factored completely if each factor other than a monomial factor is prime. Sometimes more than one of the following methods is needed to factor the polynomial completely.

1. When factoring a polynomial, always look for a greatest common factor first. Even if other methods of factoring are required, factoring out the GCF will make other factoring methods easier to use.

2. When factoring a polynomial with just two terms, determine whether it matches one of the special product patterns.

3. When factoring a polynomial having three terms, check to see if it is a perfect square trinomial and, if so, use the special product for the square of a binomial.

4. When factoring a trinomial that is not a perfect square trinomial, use the grouping method or the trial-and-error method.

5. When factoring a polynomial having more than three terms, try the grouping method.

7.5 *Quick Reference*

Leading Coefficient a = 1

- A trinomial of the form $ax^2 + bx + c$ is called a **quadratic trinomial.** The number a is called the **leading coefficient.**

- To factor a quadratic trinomial $x^2 + bx + c$ (the leading coefficient is 1), one approach is as follows.

  1. List the pairs of numbers whose product is the constant term c.

  2. Choose the pair whose sum is b, the coefficient of the middle term.

  3. These numbers are the constant terms of the binomial factors.

Leading Coefficient a ≠ 1

- To factor a quadratic trinomial $ax^2 + bx + c$ when the leading coefficient a is not 1, one approach is as follows.

  1. Find the product of the leading coefficient a and the constant term c.

  2. List the pairs of numbers whose product is ac.

  3. Choose the pair whose sum is b, the coefficient of the middle term.

  4. Use these numbers to write the middle term as a sum of two terms.

  5. Factor by grouping.

Trial-and-Error Method

- To factor a quadratic trinomial $ax^2 + bx + c$ by the trial-and-error method, follow these steps.

  1. List all the pairs of numbers whose product is the leading coefficient a.

  2. List all the pairs of numbers whose product is the constant term c.

  3. Use these two sets of numbers to form possible binomial factors of the trinomial. (Eliminate all possibilities that would produce a common factor in the terms of either of the binomial factors.)

  4. Multiply the remaining possibilities to determine the correct factorization.

General Strategy

- The first step in factoring is to factor out the greatest common factor.

- If the polynomial has just two terms, determine if it matches the pattern of one of the special products.

  1. Difference of two squares
  $$A^2 - B^2 = (A - B)(A + B)$$

  2. Difference of two cubes
  $$A^3 - B^3 = (A - B)(A^2 + AB + B^2)$$

  3. Sum of two cubes
  $$A^3 + B^3 = (A + B)(A^2 - AB + B^2)$$

- If the polynomial has three terms, determine if it is a perfect square trinomial.

  1. $A^2 + 2AB + B^2 = (A + B)^2$
  2. $A^2 - 2AB + B^2 = (A - B)^2$

- If the trinomial to be factored is not a perfect square trinomial, use the grouping method or the trial-and-error method.

- If the polynomial has more than three terms, try the grouping method. It might be necessary to rearrange terms or to group in ways other than in pairs.

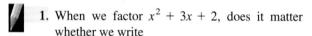

7.5 Exercises

1. When we factor $x^2 + 3x + 2$, does it matter whether we write

   $(x + 1)(x + 2)$ or $(x + 2)(x + 1)$?

   Why?

2. After you factor an expression, how can you check the accuracy of your work?

In Exercises 3–12, factor the quadratic trinomial completely.

3. $x^2 + 3x + 2$
4. $x^2 - 4x + 3$
5. $x^2 + 18 - 9x$
6. $x^2 - 18 + 3x$
7. $14 + 9x + x^2$
8. $21 + 10x + x^2$
9. $x^2 + 2x + 2$
10. $x^2 + 3x + 5$
11. $x^2 - 6xy - 16y^2$
12. $x^2 - 4xy - 45y^2$

In Exercises 13–22, factor the trinomial completely.

13. $2x^2 + 5x + 2$
14. $2x^2 + 7x + 6$
15. $12 + 4x - x^2$
16. $4 + 3x - x^2$
17. $3x^2 - 8x - 3$
18. $3x^2 - 4x - 15$
19. $4x^2 - 17x - 15$
20. $12 + 7x - 10x^2$
21. $12x^2 + 19x + 21$
22. $12x^2 + 13x + 14$

In Exercises 23–58, factor the trinomial completely.

23. $c^2 + c - 2$
24. $x^2 - 2x - 3$
25. $6x^2 + 13x - 15$
26. $15x^2 + 8x - 7$
27. $x^2 + 5x - 24$
28. $3x + x^2 - 40$
29. $24 + 2x - x^2$
30. $15 + 2x - x^2$
31. $b^2 - 2b - 15$
32. $c^2 - 4c - 12$

33. $3x^2 + 5x + 2$
34. $3x^2 - 7x - 6$
35. $15 + 6x^2 - 19x$
36. $6 + x - 12x^2$
37. $x^2 + 25 + 10x$
38. $16 + y^2 + 8y$
39. $2 + x - 15x^2$
40. $2 - 11x + 15x^2$
41. $18 - 11x - 10x^2$
42. $3x^2 + 10x - 8$
43. $24 + a^2 - 10a$
44. $b^2 - 2b - 24$
45. $9x^2 - 9x - 10$
46. $6x^2 + x - 15$
47. $x^2 - 16x + 63$
48. $x^2 - 2x - 63$
49. $9x^2 - 9x - 18$
50. $9x^2 - 9x - 54$
51. $12x^2 + 16x - 3$
52. $13x + 35x^2 - 12$
53. $8x^2 - 2xy - 21y^2$
54. $8x^2 - 2xy - 15y^2$
55. $2x^2 - 12 + 2x$
56. $3x^2 - 12 + 9x$
57. $24x + 36 + 4x^2$
58. $48x + 40 + 8x^2$

 59. Explain why it is easier to factor $17x^2 + 392x + 23$ than $24x^2 + 11x - 18$.

 60. Which of the following is a correct factorization of $6 - 5x + x^2$? Explain.

   (i) $(2 - x)(3 - x)$ (ii) $(x - 2)(x - 3)$

In Exercises 61–66, factor completely.

61. $8x^3y - 24x^2y^2 - 80xy^3$
62. $6x^6 - 60x^5 + 96x^4$
63. $45x + 12x^2 - 12x^3$
64. $10y^6 + 55y^5 + 75y^4$
65. $28x^2y - 30xy - 18y$
66. $15y^3 - 84y^2 - 36y$

In Exercises 67–78, factor the expression completely.

67. $a^2b^2 - 3ab - 10$ **68.** $c^2d^2 - 4cd - 21$

69. $12x^2y + 36xy + x^3y$

70. $28x^2 + 98x + 2x^3$

71. $x^2y^2 - 2xy - 48$

72. $36 - 9ab - a^2b^2$

73. $18x^2 - 12xy + 2y^2$

74. $8x^2y^2 + 18xy - 5$

75. $6m^2 + 5mn - n^2$

76. $16m^2n^4 + 12mn^5 + 2n^6$

77. $16x^3y^2 - 12x^2y^3 + 2xy^4$

78. $30ab^4 + 24a^2b^3 - 6a^3b^2$

In Exercises 79–106, factor completely.

79. $5x^2 - 7x - 6$ **80.** $5x^2 - 31x + 6$

81. $25 - 81y^2$ **82.** $9x^2 - 81$

83. $2x^3 + x^2 - 6x - 3$

84. $3x^3 + 9x^2 + x + 3$

85. $3x(x - 2) + 5y(x - 2)$

86. $y^2(y - 3) - 5(y - 3)$

87. $125x^3 + 1$ **88.** $64 - 27a^3$

89. $6x + x^2 + 8$ **90.** $x^2 - x - 12$

91. $x^2 + 9$ **92.** $x^2 - 5x + 36$

93. $4x^2 + 20x + 25$ **94.** $45a^2 - 30a + 5$

95. $ax - 3a - bx + 3b$

96. $xy + 3y - 5x - 15$

97. $25y^3 - 25$ **98.** $a^3b^3 + 8$

99. $y^4 - 16x^4$ **100.** $5x^2 - 20$

101. $a^2b^7 + 3a^3b^5 - a^2b^4$

102. $3x^4 + 15x^3 - 3x^2$

103. $x + 21x^2 - 2$ **104.** $3 - 22y + 7y^2$

105. $16 - 8a + a^2$ **106.** $x^2 + 6xy + 9y^2$

In Exercises 107–134, factor completely.

107. $50 - 18x^4$ **108.** $3x^4 + 14x^2 + 8$

109. $2y + 10 - 3xy - 15x$

110. $2x^3 - 3x^2 - 18x + 27$

111. $24x^4 - 24x^2 + 20x^3$

112. $8a^2b - 27ab + 9b$

113. $c^{20} - d^8$ **114.** $81 - t^4$

115. $x^2 + xy + y^2$ **116.** $x^2 + 4x + 16$

117. $a^2x^3 - a^2y^3 - 4x^3 + 4y^3$

118. $18a^3 - 8a + 45a^2 - 20$

119. $a^5b^4 - a^7b^8 + a^6b^3$

120. $(a + b)(c + d) - 2(a + b)(c - d)$

121. $m^2 - 10m + 16$ **122.** $8 + 7y - y^2$

123. $4x^4 - 64y^4$ **124.** $8x^2 - 98y^2$

125. $(a + b)^3 - 8$ **126.** $(c + d)^2 - 16$

127. $20x^2 - 60x + 45$

128. $x^2y^2 + 3x^2y - 88x^2$

129. $30a^6 - 104a^5 + 90a^4$

130. $12a^2 - 22a - 42$

131. $c^5 - 4c^3 - c^2 + 4$ **132.** $16x^5 - 54x^2$

133. $x^4 + 2x^2y^2 - 99y^4$ **134.** $a^2 - 13a + 40$

In Exercises 135–140, factor the expression completely.

135. $x^3(x - 1) + 11x^2(x - 1) - 42x(x - 1)$

136. $x^3(2x - 1) + 4x^2(2x - 1) - 21x(2x - 1)$

137. $x^4(x^2 - 25) - 10x^2(x^2 - 25) + 9(x^2 - 25)$

138. $3x^4(y^2 - 4) + 7x^2(y^2 - 4) - 6(y^2 - 4)$

139. $2x^2y^2(xy + 1) + 3xy(xy + 1) - 9(xy + 1)$

140. $2x^2(2 - y) + xy(2 - y) - 10y^2(2 - y)$

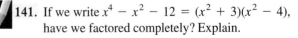

141. If we write $x^4 - x^2 - 12 = (x^2 + 3)(x^2 - 4)$, have we factored completely? Explain.

142. How can you tell at a glance that $x^4 + 7x + 10$ cannot be factored with our current methods for factoring trinomials?

In Exercises 143–154, factor completely.

143. $x^4 + 28x^2 - 60$ **144.** $x^4 - 17x^2 - 60$

145. $x^6 + 17x^3 + 60$

146. $x^6 - 16x^3 + 60$

147. $x^4 - 13x^2 + 36$

148. $x^4 - 5x^2 + 4$

149. $5x^8 + 3x^4 - 14$

150. $9x^8 + 77x^4 - 36$

151. $4x^4 - 37x^2 + 9$

152. $4x^4 + 11x^2 - 3$

153. $9x^4 - 13x^2 + 4$

154. $9x^4 - 31x^2 - 20$

In Exercises 155–160, factor completely.

155. $x^{2n} + 11x^n + 24$

156. $y^{2n} + 11y^n + 30$

157. $y^{4n} - 20y^{2n} + 99$

158. $x^{4n} - 17x^{2n} + 72$

159. $5x^{2n} + 2x^n - 3$

160. $7x^{2n} + 16x^n - 15$

A method for factoring $12(y - 3)^2 - 29(y - 3) + 15$ is to make a temporary substitution for $y - 3$. Let $u = y - 3$.

$12(y - 3)^2 - 29(y - 3) + 15$

$= 12u^2 - 29u + 15$ Replace $y - 3$ with u.

$= (3u - 5)(4u - 3)$ Factor the trinomial.

$= [3(y - 3) - 5][4(y - 3) - 3]$ Replace u with $y - 3$.

$= [3y - 9 - 5][4y - 12 - 3]$ Distributive Property

$= (3y - 14)(4y - 15)$

In Exercises 161–166, use this method to factor completely.

161. $(x - 1)^2 + 11(x - 1) + 28$

162. $(x + 2)^2 + 11(x + 2) + 24$

163. $(2x + 1)^2 - 15(2x + 1) + 56$

164. $12(2 - 3y)^2 - 29(2 - 3y) + 14$

165. $2x^2(x + 2)^2 + x(x + 2) - 6$

166. $2x^2(2 + y)^2 - x(2 + y) - 15$

167. The product of three numbers is given by the expression $x^3 + 3x^2 + 2x$, where x is an integer and each of the numbers is a polynomial with integer coefficients. Factor the expression to show that the numbers are consecutive integers.

 168. If the area of a rectangular office is given by $x^2 + 10x + 21$ square feet, where the dimensions are binomials with integer coefficients, factor the expression to show that the length of the office is 4 feet greater than the width.

169. It is rare that real-data applications can be modeled with polynomials whose coefficients are integers. For example, an application might be modeled by $1.01x^2 + 4.99x - 24.05$, a polynomial that cannot be factored with our present methods. Round off the coefficients of this expression and approximate a factorization.

 170. The area of a triangle is $\frac{1}{2}x^2 + 7x + 20$, where the base and height are polynomials with integer coefficients. Factor the expression to determine the difference between the base and the height of the triangle.

Exploring with Real Data

In Exercises 171–174, refer to the following background information and data.

The following table shows the annual average food costs (in dollars) for couples 20–50 years old. The data are reported for the period 1986–1990.

Year	Average Annual Food Costs
1986	3140.80
1987	3302.00
1988	3468.40
1989	3660.80
1990	3884.40

(Source: U.S. Department of Agriculture.)

A function that models these data is $f(t) = 184.60t + 2937.48$, where $f(t)$ is the average annual food costs and t is the number of years since 1985.

171. Use your calculator to factor 52 out of the expression $184.60t + 2937.48$, and write the function in the form $f(t) = 52(\underline{})$.

172. What does the other factor represent?

173. Use both functions to calculate the average annual food costs for the year 1989. Do you obtain the same value in each case? By how much does this value differ from the actual reported figure for 1989?

174. Would you expect the average annual food costs to increase or decrease after age 50? Give two reasons for your answer.

Challenge

In Exercises 175–182, determine all the integers k such that the trinomial can be factored.

175. $x^2 + kx + 10$ **176.** $x^2 + kx + 12$

177. $x^2 + kx - 12$ **178.** $x^2 + kx - 10$

179. $3x^2 + kx + 5$ **180.** $2x^2 + kx + 7$

181. $7x^2 + kx - 2$ **182.** $5x^2 + kx - 3$

In Exercises 183 and 184, determine a positive number k so that the polynomial can be factored.

183. $x^2 + 3x + k$ **184.** $x^2 + 2x + k$

185. Factor $3x^{4n} + 5x^{3n} + 2x^{2n}$.

186. Factor $6x^2 - x^4 + 27$ so that one factor is $x - 3$.

187. If x is an odd integer, show that the expression $3x^2 + 12x + 20$ is the sum of the squares of three consecutive odd integers. (Hint: Write $3x^2 = x^2 + x^2 + x^2$, write $12x = 4x + 8x$, and write $20 = 4 + 16$.)

188. Given that $x + 1$ and $x - 1$ are factors, completely factor $x^6 - 14x^4 + 49x^2 - 36$.

7.6 Solving Equations by Factoring

Quadratic Equations ▪ *Solving by Factoring* ▪ *Applications*

Quadratic Equations

In Chapter 4 we developed a routine for solving linear equations in one variable. Equations involving polynomials of degree two or higher require different solution techniques.

An equation involving a polynomial of degree 2 is called a **second-degree** or **quadratic equation.**

Definition of a Quadratic Equation

A **quadratic equation** can be written in the *standard form*

$$ax^2 + bx + c = 0$$

where a, b, and c are real numbers and $a \neq$ zero.

When a quadratic equation is written in standard form, the term ax^2 is called the **quadratic term,** the term bx is called the **linear term,** and the term c is called the **constant term.**

EXPLORATION 1   *Solving a Quadratic Equation by Graphing*

Use the integer setting to produce the graph of $p(x) = \frac{1}{4}x^2 + x - 15$. Then, trace the graph to determine the value(s) of x for which the following are true.

(a) $p(x) = 20$ (b) $p(x) = 0$

(c) $p(x) = -16$ (d) $p(x) = -22$

(e) What is your conjecture about the number of solutions a quadratic equation can have?

Discovery

(a) Figure 7.4

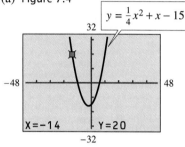 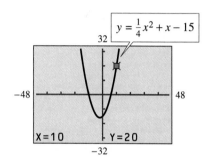

From Fig. 7.4, we see that $p(x) = 20$ has two solutions. We estimate them to be -14 and 10. We can evaluate $p(-14)$ and $p(10)$ to verify that each is equal to 20.

(b) Figure 7.5

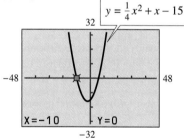 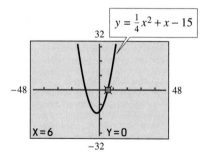

Fig. 7.5 shows that $p(x) = 0$ also has two solutions. They appear to be -10 and 6. By evaluaing $p(-10)$ and $p(6)$, we can verify that each is equal to 0.

(c) Figure 7.6

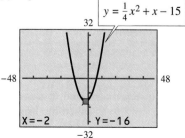

From Fig. 7.6, -2 appears to be the only value of x for which $p(x) = -16$. Verify that $p(-2) = -16$.

(d) There is no value of x for which $p(x) = -22$.

(e) From these results, we conclude that a quadratic equation can have zero, one, or two real number solutions.

Solving by Factoring

Even though we can use a graph to estimate solutions of an equation, we want to develop an algebraic method to determine the solutions exactly, just as we did for linear equations.

A simple but important property of real numbers is central to the method we will develop. The Zero Factor Property states that if the product of two numbers is zero, then one of the numbers must be zero.

> **Zero Factor Property**
>
> If $ab = 0$, then $a = 0$ or $b = 0$.

NOTE: The expression on one side of the equation must be in the form of a *product* or else the Zero Factor Property does not apply. Also, to use the Zero Factor Property, the *product must be zero.* ▪

In general, when a polynomial $p(x)$ can be factored, we can use the Zero Factor Property to solve the equation $p(x) = 0$.

EXAMPLE 1   *Solving Quadratic Equations by Factoring*

Solve each equation.

(a) $2x^2 - 9x - 5 = 0$ (b) $9x^2 + 4 = 12x$ (c) $x^2 + 4x - 3 = 4$

Solution

(a) In Fig. 7.7, the graph of $y = 2x^2 - 9x - 5$ has two x-intercepts, that is, points for which $y = 0$. This indicates that the equation has two solutions.

$$2x^2 - 9x - 5 = 0 \qquad \text{The equation is in standard form.}$$
$$(2x + 1)(x - 5) = 0 \qquad \text{Factor the expression on the left side.}$$
$$2x + 1 = 0 \quad \text{or} \quad x - 5 = 0 \qquad \text{Zero Factor Property}$$
$$x = -\frac{1}{2} \quad \text{or} \qquad x = 5 \qquad \text{Solve each case for } x.$$

These exact solutions are consistent with the solutions approximated by the graph.

Figure 7.7

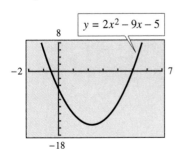

$y = 2x^2 - 9x - 5$

Figure 7.8

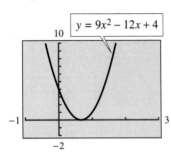

$y = 9x^2 - 12x + 4$

(b)
$$9x^2 + 4 = 12x$$
$$9x^2 - 12x + 4 = 0 \qquad \text{Write the equation in standard form.}$$

The graph of $y = 9x^2 - 12x + 4$ is shown in Fig. 7.8. There appears to be only one x-intercept, which suggests that the equation has only one solution. To determine the solution algebraically, we solve by factoring.

$$9x^2 - 12x + 4 = 0$$
$$(3x - 2)(3x - 2) = 0 \qquad \text{Factor the left side.}$$
$$3x - 2 = 0 \qquad \text{Zero Factor Property}$$
$$x = \frac{2}{3} \qquad \text{Solve for } x.$$

Because the two factors are the same, we obtain just one distinct solution (or root). We say that $\frac{2}{3}$ is a **double root.**

Figure 7.9

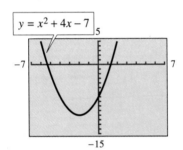

$y = x^2 + 4x - 7$

(c)
$$x^2 + 4x - 3 = 4$$
$$x^2 + 4x - 7 = 0 \qquad \text{Write the equation in standard form.}$$

Because we cannot factor $x^2 + 4x - 7$, we are unable to use our current algebraic method for solving the equation. Note, however, that this does not necessarily mean that the equation has no solution. In fact, the graph of $p(x) = x^2 + 4x - 7$ clearly shows that there are two solutions, one between -6 and -5 and another between 1 and 2. (See Fig. 7.9.)

■

Part (c) of Example 1 illustrates an important point. Even if we are unable to use the factoring method to solve a quadratic equation $p(x) = 0$, we can use the graph of $y = p(x)$ to determine how many solutions, if any, the equation has.

EXAMPLE 2   *Solving Quadratic Equations Algebraically*

Solve each equation.

(a) $12x^2 = 3x$

(b) $x(6x + 1) = 12$

Solution

(a)
$$12x^2 = 3x$$
$$12x^2 - 3x = 0 \qquad \text{Write the equation in standard form.}$$
$$3x(4x - 1) = 0 \qquad \text{Factor the left side.}$$

Note that the product on the left side has three factors. They are 3, x, and $4x - 1$. The Zero Factor Property states that one of the factors must be 0. However, the factor 3 is definitely not 0. Therefore, either x or $4x - 1$ is 0.

$$x = 0 \quad \text{or} \quad 4x - 1 = 0 \qquad \text{Zero Factor Property}$$
$$x = 0 \quad \text{or} \qquad x = \frac{1}{4} \qquad \text{Solve each case for } x.$$

The two solutions are 0 and $\frac{1}{4}$.

(b) $$x(6x + 1) = 12$$

$$6x^2 + x = 12 \qquad \text{Distributive Property}$$

$$6x^2 + x - 12 = 0 \qquad \text{Write the equation in standard form.}$$

$$(2x + 3)(3x - 4) = 0 \qquad \text{Factor the left side.}$$

$$2x + 3 = 0 \quad \text{or} \quad 3x - 4 = 0 \qquad \text{Zero Factor Property}$$

$$x = -\frac{3}{2} \quad \text{or} \quad x = \frac{4}{3} \qquad \text{Solve each case for } x.$$

The two solutions are $-\frac{3}{2}$ and $\frac{4}{3}$. ∎

While the factoring method is a classic technique for solving quadratic equations, it can also be used for other types of equations.

EXAMPLE 3 *Solving Other Types of Equations by Factoring*

Solve the following equations.

(a) $4x^3 + 12x^2 - x - 3 = 0$ (b) $|x^2 - 13| = 12$

Solution

Figure 7.10

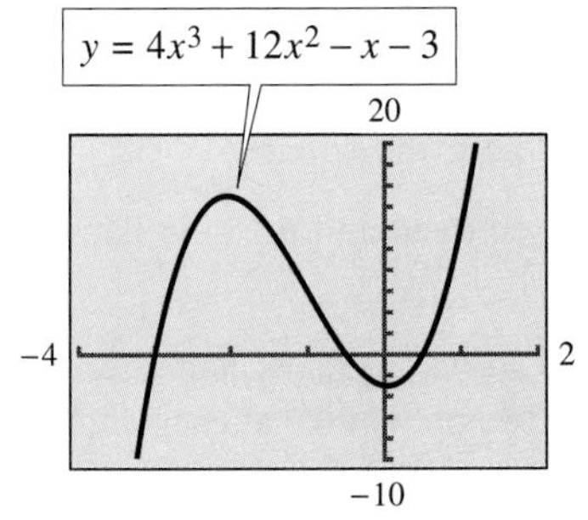

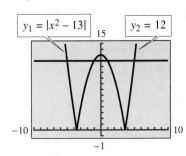

$y = 4x^3 + 12x^2 - x - 3$

(a) From the graph of $y = 4x^3 + 12x^2 - x - 3$ in Fig. 7.10, we anticipate that the equation has three solutions. Note that this is a third-degree equation, not a quadratic equation. However, if we can factor the left side, the factoring method can still be used to solve the equation.

$$4x^3 + 12x^2 - x - 3 = 0$$

$$(4x^3 + 12x^2) - (x + 3) = 0 \qquad \text{Group the left side in pairs.}$$

$$4x^2(x + 3) - (x + 3) = 0 \qquad \text{Factor out the GCF in the first pair.}$$

$$(x + 3)(4x^2 - 1) = 0 \qquad \text{Factor out the GCF } x + 3.$$

$$(x + 3)(2x + 1)(2x - 1) = 0 \qquad \text{Factor the difference of two squares.}$$

$$x + 3 = 0 \quad \text{or} \quad 2x + 1 = 0 \quad \text{or} \quad 2x - 1 = 0 \qquad \text{Zero Factor Property}$$

$$x = -3 \quad \text{or} \quad x = -\frac{1}{2} \quad \text{or} \quad x = \frac{1}{2} \qquad \text{Solve each case for } x.$$

The solutions are -3, $-\frac{1}{2}$, and $\frac{1}{2}$.

Figure 7.11

$y_1 = |x^2 - 13|$ $y_2 = 12$

(b) To use the graphing method, we let $y_1 = |x^2 - 13|$ and $y_2 = 12$. In Fig. 7.11, we see that the two graphs intersect in four points. Thus, we expect the equation to have four solutions.

The equation $|x^2 - 13| = 12$ states that $x^2 - 13$ represents a number whose absolute value is 12. Therefore, $x^2 - 13$ represents either 12 or -12.

$$x^2 - 13 = 12 \quad \text{or} \quad x^2 - 13 = -12$$

Each possibility is a quadratic equation that we will try to solve by the factoring method.

$$x^2 - 13 = 12 \qquad\qquad\qquad x^2 - 13 = -12$$
$$x^2 - 25 = 0 \qquad\qquad\qquad x^2 - 1 = 0$$
$$(x + 5)(x - 5) = 0 \qquad\qquad\qquad (x + 1)(x - 1) = 0$$
$$x + 5 = 0 \quad\text{or}\quad x - 5 = 0 \qquad\qquad x + 1 = 0 \quad\text{or}\quad x - 1 = 0$$
$$x = -5 \quad\text{or}\quad x = 5 \qquad\qquad\qquad x = -1 \quad\text{or}\quad x = 1$$

The four solutions are 5, −5, 1, and −1. ■

Applications

Many applications can be modeled by a second-degree (or higher) equation. At this point, we can solve such applied problems only if we can write the equation in the form $p(x) = 0$ and factor $p(x)$. (Later, we will learn other techniques for solving polynomial equations.)

Sometimes a solution of an equation that models an application must be disqualified because of the nature of the problem. The next example illustrates a case like this.

EXAMPLE 4   *Constructing an Open Box*

We can construct an open box from a square piece of metal by cutting a 2-inch square from each corner and folding the sides up. If the volume of the box must be 72 cubic inches, how large should the square piece of metal be?

Figure 7.12

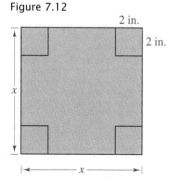

Figure 7.13

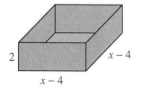

Solution   Figure 7.12 shows the square piece of metal with the square corners to be cut out. We let x represent the length of the side of the metal square.

Figure 7.13 shows the box after the corners have been removed and the side flaps have been folded up.

The volume of the box is given by the formula $V = LWH$, where L, W, and H are the length, width, and height, respectively. Because 2 inches are cut off at each end, the width and length of the box are both $x - 4$. The height of the box is 2. Therefore,

$$V = (x - 4)(x - 4)(2) \quad\text{or}\quad V = 2(x - 4)^2.$$

Because the volume must be 72, we have the following equation.

$2(x - 4)^2 = 72$	
$(x - 4)^2 = 36$	Divide both sides by 2.
$x^2 - 8x + 16 = 36$	Square the binomial.
$x^2 - 8x - 20 = 0$	Write the equation in standard form.
$(x - 10)(x + 2) = 0$	Factor the left side.
$x - 10 = 0 \quad\text{or}\quad x + 2 = 0$	Zero Factor Property
$x = 10 \quad\text{or}\quad x = -2$	Solve each case for x.

Although −2 is a solution of the equation, we know the length of the metal square cannot be a negative number. Therefore, the length is 10 inches. ■

7.6   Quick Reference

Quadratic Equations

- A **quadratic equation** can be written in the *standard form*

  $$ax^2 + bx + c = 0,$$

  where a, b, and c are real numbers and $a \neq 0$.

- The term ax^2 is the **quadratic term,** the term bx is the **linear term,** and the term c is the **constant term.**

Solving by Factoring

- The **Zero Factor Property** is as follows.

  If $ab = 0$, then $a = 0$ or $b = 0$.

  In words, if a product of two or more factors is equal to zero, then one of the factors must equal zero.

- For the Zero Factor Property to apply,

  1. one side of the equation must be a product, and

  2. the other side of the equation must be 0.

- To solve a quadratic equation by factoring:

  1. Write the equation in standard form.

  2. Factor the polynomial completely.

  3. Set each factor equal to zero.

  4. Solve each resulting equation.

  5. Check the solution(s).

- When solving a quadratic equation, if two factors are the same, each results in the same solution. The solution is called a **double root.**

7.6   Exercises

 1. The equation $x^2 + 3x - 1 = 0$ has two solutions, but they cannot be determined by factoring. Describe a method that can be used to estimate the solutions.

 2. Describe the graphs of second-degree equations that have 0, 1, and 2 solutions.

3. Let $p(x) = x^2 - 4x - 12$. Use graphing to estimate the solutions of the following equations.

   (a) $p(x) = 0$ (b) $p(x) = -18$

   (c) $p(x) = -16$ (d) $p(x) = 20$

4. Let $p(x) = x^2 - 2x - 8$. Use graphing to estimate the solutions of the following equations.

   (a) $p(x) = -9$ (b) $p(x) = -16$

   (c) $p(x) = 0$ (d) $p(x) = 16$

 5. For the equation $3(x - 1)(x + 2) = 0$, the Zero Factor Property states that $3 = 0$, $x - 1 = 0$, or $x + 2 = 0$. Which of these equations leads to a solution of the original equation? Explain.

 6. Describe the graph and the algebraic circumstances in which a second-degree equation has a double root.

In Exercises 7–14, use the Zero Factor Property to solve the equation.

7. $(x - 4)(x + 3) = 0$

8. $3(2x + 5)(x + 1) = 0$

9. $5x(6 - x) = 0$ **10.** $x(1 - 2x) = 0$

11. $(x - 5)^2 = 0$ **12.** $4(x + 4)^2 = 0$

13. $(x + 5)(2x - 6)(3x + 3) = 0$

14. $x(3 + 2x)(1 - x) = 0$

15. How do we distinguish between a first-degree equation and a second-degree equation?

16. Verify that solving $x^2 = 7x$ by factoring results in two solutions. Then explain why the following *incorrect* method results in only one solution.

$$x^2 = 7x$$
$$\frac{x^2}{x} = \frac{7x}{x}$$
$$x = 7$$

In Exercises 17–24, solve algebraically.

17. $x^2 + 7x = 0$ **18.** $2c^2 - 5c = 0$

19. $2g^2 - 18 = 0$ **20.** $h^2 - 49 = 0$

21. $x^2 - 5x + 6 = 0$

22. $-3b^2 - 7b + 6 = 0$

23. $\frac{2}{5}x^2 - \frac{3}{5}x - 1 = 0$

24. $x^2 - \frac{7}{6}x - \frac{1}{2} = 0$

25. Explain why the Zero Factor Property does not apply to the equation $(x + 2)(x - 3) = -6$.

26. Given $(x + 2)(x - 3) = 7$, explain what is wrong with this conclusion: $x + 2 = 7$ or $x - 3 = 7$.

In Exercises 27–42, solve algebraically.

27. $a^2 = 5a$ **28.** $x^2 = 25$

29. $(x - 2)^2 = 25$

30. $(a + 3)^2 - 64 = 0$

31. $x^2 + 5x = 6$ **32.** $y^2 - 2y = 8$

33. $x^2 + 35 = 12x$ **34.** $x^2 + 12 = 7x$

35. $12 = 11d - 2d^2$ **36.** $15 = 26g - 7g^2$

37. $-8 = -9y^2 - 14y$ **38.** $-x - 6 = -12x^2$

39. $x(x + 1) = 30$ **40.** $x(x - 1) = 42$

41. $a(a - 32) + 60 = 0$

42. $c(c + 23) + 60 = 0$

In Exercises 43–54, solve algebraically.

43. $x(4 + x) = 5$ **44.** $x(3 - x) = 2$

45. $3x(2 + x) = 24$ **46.** $x(2 - 3x) = \frac{1}{3}$

47. $\frac{2}{5}x(x - 2) = x - 2$

48. $3(x + 1)^2 = 9 - x$

49. $(x - 3)(x + 2) = 6$

50. $(x - 7)(x - 1) = -8$

51. $(3x - 2)(x + 2) = (x - 2)(x + 1) + 10$

52. $3(x - 3)(x + 1) = (2x - 1)(x - 1) + 8$

53. $(4x - 3)^2 + (4x - 3) = 6$

54. $5(x + 2)^2 - 17(x + 2) - 12 = 0$

55. Produce the graph of $p(x) = x^3 + 1$. The graph suggests only one value of c for which $p(c) = 0$. Determine that number c by solving $x^3 + 1 = 0$ by factoring. Can the quadratic factor be factored more? Produce the graph of the quadratic factor to support your conclusion.

56. Produce the graph of $p(x) = 2x^4 - 3x^2 + 2$. Explain how the graph tells you that $p(x)$ cannot be factored.

In Exercises 57–72, solve algebraically.

57. $(x - 5)(x^2 + 3x - 18) = 0$

58. $(2x - 5)(8x^2 + 14x - 15) = 0$

59. $3x^3 - 27x = 0$ **60.** $8x^2 - 50x^4 = 0$

61. $6x^3 + 2x^2 = 4x$ **62.** $6y^2 = 3y^3 + 9y^4$

63. $x^3 - 5x^2 - x + 5 = 0$

64. $x^3 + 3x^2 - x - 3 = 0$

65. $2x^3 - 3x^2 - 8x + 12 = 0$

66. $3x^3 + 4x^2 - 27x - 36 = 0$

67. $x^4 + 36 = 13x^2$

68. $x^4 - 26x^2 + 25 = 0$

69. $6(y - 2)^3 - (y - 2)^2 - (y - 2) = 0$

70. $y^2(4y^2 - 9) - 16y(4y^2 - 9) + 48(4y^2 - 9) = 0$

71. $3x^2(3 - x) + 3x(3 - x) + 18(x - 3) = 0$

72. $10(3 - 2x)x^2 - 27(3 - 2x)x - 5(2x - 3) = 0$

In Exercises 73–76, one solution of the equation is given. Determine a, b, or c and another solution of the equation.

73. $2x^2 + bx + 2 = 0$; $x = -2$

74. $(2a + 1)x^2 + x + 10 = 0$; $x = 2$

75. $2x^2 + x + (3 - 4c) = 0$; $x = 3$

76. $-x^2 + x + c = 0$; $x = -1$

In Exercises 77–82, write an equation with the given solutions.

77. 3, -4

78. -2, 5

79. 4 (double root)

80. -2 (double root)

81. 0, $\dfrac{1}{2}$

82. $\dfrac{2}{3}$, -2

In Exercises 83–88, solve for x.

83. $x^2 - cx - 2c^2 = 0$

84. $x^2 + xy - 2y^2 = 0$

85. $3y^2 - 4xy = 4x^2$

86. $cx + 2c^2 = 6x^2$

87. $x^2 + 2ax = 3a^2$

88. $4a^2 = 10x^2 + 3ax$

In Exercises 89–94, solve graphically.

89. $|x^2 - 4x| = 3$

90. $|2x^2 - 5x| = 2$

91. $|2x^2 - 3x| = 2$

92. $|x^2 - x| = 6$

93. $|6x^2 - 5x - 2| = 2$

94. $|2x^2 - 9x - 2| = 3$

In Exercises 95–100, solve algebraically.

95. $|x^2 + 5x| = 6$

96. $|3x^2 - 5x| = 2$

97. $|x^2 + 10x + 15| = 6$

98. $|x^2 + 7x + 8| = 2$

99. $|x^2 - 3x - 1| = 3$

100. $|x^2 - 5x + 5| = 1$

In Exercises 101–116, write an equation describing the conditions of the problem, solve the equation, and then answer the question.

101. The sum of a number and its square is 72. Determine the number.

102. The sum of twice a number and its square is 63. Determine the number.

103. The sum of the squares of three consecutive integers is 50. Determine the numbers.

104. The sum of the squares of three consecutive integers is 77. Determine the numbers.

105. The sum of the cube of a number and 6 times the number is equal to 5 times the square of the number. Determine the number.

106. The sum of the cube of a number and 12 times the number is equal to 8 times the square of the number. Determine the number.

107. Two numbers have a sum of 3 and a product of -10. Determine the numbers.

108. The sum of two numbers is 12 and the sum of their squares is 74. Determine the numbers.

109. A college sophomore is 3 years older than his sister. The product of their ages is 304. How old is the sister?

110. Two cousins were born 4 years apart. If the product of their ages is 357, how old is the younger cousin?

111. The length of a rectangle is 5 meters more than the width. The area of the rectangle is 594 square meters. What are the dimensions of the rectangle?

112. The width of a rectangular living room is 7 feet less than the length. The area of the room is 294 square feet. What are the dimensions of the rectangle?

113. A path of uniform width surrounds a rectangular garden whose width is 14 feet and whose length is 20 feet. If the area of the path is 152 square feet, how wide is the path?

114. A farmer wants to use 500 feet of fencing to enclose a rectangular area of 15,000 square feet. To meet these conditions, what should be the dimensions of the fenced area?

115. One leg of a right triangle is 7 inches more than the other leg. The area of the triangle is 60 square inches. What are the lengths of the three sides?

116. One leg of a right triangle is 5 feet more than the other leg. The area of the triangle is 150 square feet. What are the lengths of the three sides?

Exploring with Real Data

In Exercises 117–120, refer to the following background information and data.

The following table shows the total retail value (in millions of dollars) of phonograph records manufactured during selected years in the period 1975–1990.

	1975	1980	1985	1990
Manufacturers' Value (in millions)	1697	2560	1562	181

(Source: Recording Industry Association of America.)

The data in the table can be modeled by the function $V(t) = -27.4t^2 + 583.1t - 534.9$, where $V(t)$ represents the manufacturers' value (in millions of dollars) and t is the number of years since 1970.

117. Use a suitable range and scale setting to produce the graph of the model function $V(t)$.

118. What equation would you write to predict the year in which the manufacturers' value will be zero? Describe the relative merits of solving the equation by factoring and by graphing.

119. From the shape of the graph of $V(t)$ and given the nature of the data, would you say that the model would be more reasonably accurate for 1970 or for 1990?

120. In your opinion, what is the explanation for the dramatic trend shown by the table and by the graph of the function?

Challenge

121. Write a second-degree equation whose roots are a and b. Write the equation in standard form.

122. Write a third-degree equation whose roots are a, b, and c. Write the equation in standard form.

123. Solve

$$(x + y + 1)(1 - x) + (x + y - 1)(x - 1) = 0.$$

In Exercises 124 and 125, solve the equation. (Hint: Recall the substitution technique: $u = 1/x$.)

124. $3\left(\dfrac{1}{x}\right)^2 + 14\left(\dfrac{1}{x}\right) - 24 = 0$

125. $\dfrac{2}{x^2} - \dfrac{1}{x} - 1 = 0$

126. A rectangular swimming pool is surrounded by a concrete deck. The diagram shows the relative dimensions of the pool and deck.

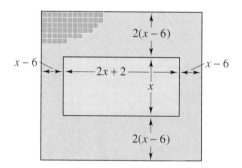

The combined area of the pool and deck is 2116 square feet. The owner plans to cover the concrete deck with tile that costs $80 per hundred square feet. Determine the total cost of the job.

7.7 Division

Monomial Divisor ▪ *Long Division*

Long division of numbers is a familiar operation from arithmetic. The following simple problem is a quick review of the procedure and of the related vocabulary.

$$
\begin{array}{r}
5 \quad \text{Quotient} \\
\text{Divisor} \quad 41 \ \overline{)\ 211} \quad \text{Dividend} \\
\underline{205} \\
6 \quad \text{Remainder}
\end{array}
$$

To check the result, we use the relation

(Quotient)(Divisor) + Remainder = Dividend.

Check: 5(41) + 6 = 211.

Division of polynomials involves many of the same principles as division in arithmetic.

Monomial Divisor

To divide a polynomial by a monomial, we use the property for dividing exponential forms. Recall that we simplify the coefficients as fractions and subtract exponents when the bases are the same. For example,

$$
\frac{15x^5}{9x^3} = \frac{15}{9} \cdot \frac{x^5}{x^3} = \frac{5}{3} \cdot \frac{x^{5-3}}{1} = \frac{5}{3} \cdot \frac{x^2}{1} = \frac{5x^2}{3}
$$

and

$$
\frac{4y^3}{20y^5} = \frac{4}{20} \cdot \frac{y^3}{y^5} = \frac{1}{5} \cdot \frac{y^{3-5}}{1} = \frac{1}{5} \cdot \frac{y^{-2}}{1} = \frac{1}{5y^2} .
$$

These examples are special cases of division of polynomials. Each is a monomial divided by a monomial.

To divide a polynomial by a monomial, we use the rule for adding fractions with a common denominator.

$$
\frac{a}{c} + \frac{b}{c} = \frac{a + b}{c}
$$

For division, we use this property in reverse order.

$$
\frac{a + b}{c} = \frac{a}{c} + \frac{b}{c}
$$

EXAMPLE 1 | *Dividing a Polynomial by a Monomial*

Find the following quotients.

(a) $\dfrac{12x^3 - 8x^2 + 6x}{2x}$

(b) $\dfrac{15x^6 + 21x^4 - 2x^2 + 3x}{3x^2}$

Solution

(a) $\dfrac{12x^3 - 8x^2 + 6x}{2x} = \dfrac{12x^3}{2x} - \dfrac{8x^2}{2x} + \dfrac{6x}{2x}$ Write the quotient as a sum of quotients.

$\qquad\qquad\qquad\quad = 6x^2 - 4x + 3$ Simplify each quotient.

The quotient is $6x^2 - 4x + 3$ and the remainder is zero.

Check: $(6x^2 - 4x + 3)(2x) + 0 = 12x^3 - 8x^2 + 6x$.

(b) $\dfrac{15x^6 + 21x^4 - 2x^2 + 3x}{3x^2} = \dfrac{15x^6}{3x^2} + \dfrac{21x^4}{3x^2} - \dfrac{2x^2}{3x^2} + \dfrac{3x}{3x^2}$

$\qquad\qquad\qquad\qquad\qquad\quad = 5x^4 + 7x^2 - \dfrac{2}{3} + \dfrac{1}{x}$ ■

Long Division

To divide a polynomial by a polynomial with more than one term, we use a procedure similar to long division in arithmetic.

EXAMPLE 2 | *Dividing a Polynomial by a Binomial*

Divide $(2x^3 - 3x^2 - x + 2)$ by $(x - 2)$.

Solution

$x - 2 \;\overline{\big)\; 2x^3 - 3x^2 - x + 2}$ Write the problem in long division format.

$\; \overset{\textstyle 2x^2}{\overline{}}$
$x - 2 \;\overline{\big)\; 2x^3 - 3x^2 - x + 2}$ Divide the first term of the dividend $2x^3$ by the first term of the divisor x.

$\; \overset{\textstyle 2x^2}{\overline{}}$
$x - 2 \;\overline{\big)\; 2x^3 - 3x^2 - x + 2}$ Multiply the result $2x^2$ times the divisor $x - 2$.
$\; 2x^3 - 4x^2$ Write the result under the dividend.

$$
\begin{array}{r}
2x^2 \\
x - 2 \enclose{longdiv}{2x^3 - 3x^2 - x + 2} \\
\underline{2x^3 - 4x^2} \\
x^2 - x
\end{array}
$$

Subtract (by changing signs and adding) and then bring down the next term of the dividend $-x$.

$$
\begin{array}{r}
2x^2 + x \\
x - 2 \enclose{longdiv}{2x^3 - 3x^2 - x + 2} \\
\underline{2x^3 - 4x^2} \\
x^2 - x
\end{array}
$$

Divide the first term of the result x^2 by the first term of the divisor x.

$$
\begin{array}{r}
2x^2 + x \\
x - 2 \enclose{longdiv}{2x^3 - 3x^2 - x + 2} \\
\underline{2x^3 - 4x^2} \\
x^2 - x \\
x^2 - 2x
\end{array}
$$

Multiply the result x times the divisor $x - 2$.

Write the result under the dividend.

$$
\begin{array}{r}
2x^2 + x \\
x - 2 \enclose{longdiv}{2x^3 - 3x^2 - x + 2} \\
\underline{2x^3 - 4x^2} \\
x^2 - x \\
x^2 - 2x \\
\underline{} \\
x + 2
\end{array}
$$

Subtract (by changing signs and adding) and then bring down the next term of the dividend 2.

$$
\begin{array}{r}
2x^2 + x + 1 \\
x - 2 \enclose{longdiv}{2x^3 - 3x^2 - x + 2} \\
\underline{2x^3 - 4x^2} \\
x^2 - x \\
x^2 - 2x \\
\underline{} \\
x + 2 \\
x - 2 \\
\underline{} \\
4
\end{array}
$$

Divide the first term of the result x by the first term of the divisor x. Then, multiply the result 1 by the divisor $x - 2$.

Then subtract.

When $2x^3 - 3x^2 - x + 2$ is divided by $x - 2$, the quotient is $2x^3 + x + 1$ and the remainder is 4. As in arithmetic, we check by multiplying $(2x^2 + x + 1)$ by $(x - 2)$ and adding the remainder 4.

$$
\begin{array}{cc}
\text{Quotient} & \text{Divisor} \\
\downarrow & \downarrow
\end{array}
$$

Check: $\quad (2x^2 + x + 1)(x - 2) + 4 \quad \leftarrow$ Remainder

$\qquad = 2x^3 + x^2 + x - 4x^2 - 2x - 2 + 4$

$\qquad = 2x^3 - 3x^2 - x + 2 \quad \leftarrow$ Dividend

NOTE: When using long division to divide polynomials, we continue the division operation until the remainder is 0 or the degree of the remainder is less than the degree of the divisor.

EXAMPLE 3   *Missing Terms in the Dividend*

Divide $x^3 + 27$ by $x + 3$.

Solution As we write the problem as a long division problem, it is helpful to include the missing terms in the dividend with zero coefficients.

$$
\begin{array}{r}
x^2 - 3x + 9 \\
x + 3 \overline{\smash{\big)}\ x^3 + 0x^2 + 0x + 27} \\
\underline{x^3 + 3x^2} \\
-3x^2 + 0x \\
\underline{-3x^2 - 9x} \\
9x + 27 \\
\underline{9x + 27} \\
0
\end{array}
$$

Include the missing x^2 and x terms.

$x^2(x + 3) = x^3 + 3x^2$

Subtract and bring down $0x$.

$-3x(x + 3) = -3x^2 - 9x$

Subtract and bring down 27.

$9(x + 3) = 9x + 27$

Subtract. The remainder is 0.

The quotient is $x^2 - 3x + 9$ and the remainder is 0.

$$\frac{x^3 + 27}{x + 3} = x^2 - 3x + 9$$

To check, multiply $(x^2 - 3x + 9)$ by $(x + 3)$. The result is $x^3 + 27$. ∎

EXAMPLE 4   *Missing Terms in the Dividend and the Divisor*

Divide $x^3 + 2x - 1$ by $2x^2 + 4$.

Solution Write the problem in the long division format with zero for the coefficients of the missing terms in the dividend and the divisor.

$$2x^2 + 0x + 4 \ \overline{\smash{\big)}\ x^3 + 0x^2 + 2x - 1}$$

$$
\begin{array}{r}
\tfrac{1}{2}x \\
2x^2 + 0x + 4 \overline{\smash{\big)}\ x^3 + 0x^2 + 2x - 1} \\
\underline{x^3 + 0x^2 + 2x} \\
- 1
\end{array}
$$

Divide x^3 by $2x^2$.

$\tfrac{1}{2}x(2x^2 + 0x + 4) = x^3 + 0x^2 + 2x$

Subtract and bring down -1.

We stop dividing because the degree of the remainder is less than the degree of the divisor. The quotient is $\tfrac{1}{2}x$; the remainder is -1. ∎

7.7 *Quick Reference*

Monomial Divisor

- To divide a monomial by a monomial, use the exponent property for dividing exponential forms.

- To divide a polynomial by a monomial, divide each term of the polynomial by the monomial.

Long Division
- To divide a polynomial by a polynomial, we can use a method similar to long division in arithmetic.
- To keep like terms in columns when dividing polynomials, use coefficients of 0 for missing terms.
- When dividing polynomials, continue the division operation until the remainder is 0 or the degree of the remainder is less than the degree of the divisor.
- To check the answer, multiply the quotient times the divisor and add the remainder. The result should be the dividend.

7.7 *Exercises*

1. Identify the dividend, divisor, quotient, and remainder in the following long division problem.

$$x \enclose{longdiv}{\begin{array}{c} 1 \\ \overline{x + 7} \\ \underline{x} \\ 7 \end{array}}$$

2. Suppose that you divide $x^2 + 5x + 9$ by $x + 1$ and obtain the quotient $x + 4$ and a remainder of 5. Explain how you can check your work. Then, perform the check to determine if this result is correct.

In Exercises 3–14, determine the quotient.

3. $\dfrac{21x^5}{14x^3}$

4. $\dfrac{5x^6}{20x^3}$

5. $\dfrac{10x^4 + 20x^3}{5x^2}$

6. $\dfrac{4x^3 - 2x^2}{2x^2}$

7. $\dfrac{4x^5 + 6x^3 + x^2}{2x^2}$

8. $\dfrac{6x^6 - 3x^5 + 2x^3}{3x^3}$

9. $\dfrac{16a^4 + 12a^3 - 6a^2 + 8a}{2a}$

10. $\dfrac{18a^4 - 12a^3 - 15a^2 + 9a}{3a}$

11. $\dfrac{20x^5 - 12x^4 + 6x^2 - 8x}{4x^2}$

12. $\dfrac{20x^5 + 10x^4 - 3x^2 + 15x}{5x^2}$

13. $\dfrac{15c^5d^7 + 6c^4d^5 - 9cd^3}{3c^2d^2}$

14. $\dfrac{18c^4d^7 - 10c^2d^5 + 8cd^2}{2cd^3}$

In Exercises 15–18, determine the quotient. (Hint: In Exercises 17 and 18, begin by simplifying the numerator.)

15. $\dfrac{6x^3(3x + 1) + 4x^2(2x - 3)}{2x^2}$

16. $\dfrac{9x^2(2x - 1) - 3x(x - 2)}{3x}$

17. $\dfrac{(x + 4)^2 - (x - 4)^2}{2x}$

18. $\dfrac{(3 - x)^2 - (3 + x)^2}{3x}$

In Exercises 19–22, determine the quotient.

19. $\dfrac{16x^{4n} - 64x^{2n}}{4x^{2n}}$

20. $\dfrac{6x^{6n} + 12x^{5n} - 9x^{4n}}{3x^{2n}}$

21. $\dfrac{x^{2n} - x^n}{x^n}$

22. $\dfrac{x^{6n} - x^{4n}}{x^{3n}}$

23. When we perform long division, when do we stop dividing?

 24. If a polynomial of degree $n > 2$ is divided by a polynomial of degree 2, what are the possible degrees of the remainder? Explain.

In Exercises 25–34, use long division to determine the quotient and remainder.

25. $\dfrac{2x + 1}{x - 3}$

26. $\dfrac{4x + 3}{x + 2}$

27. $\dfrac{x^2 + 9x + 20}{x + 4}$

28. $\dfrac{x^2 - 8x + 16}{x - 4}$

29. $(x^2 + 4x + 3) \div (x + 2)$

30. $(x^2 - x - 6) \div (x + 3)$

31. $(3x^2 + 5x + 2) \div (x - 3)$

32. $(2x^2 - 4x - 1) \div (x + 2)$

33. $\dfrac{3x^2 + 2}{x - 1}$

34. $\dfrac{x^2 - 2}{x + 1}$

In Exercises 35–40, use long division to determine the quotient and remainder.

35. $(4a^3 - a^2 + 5a + 4) \div (4a + 3)$

36. $(3x^4 + 2x^3 - 4x^2 + 7x - 5) \div (3x - 1)$

37. $(6a^3 + 5a^2 + 3) \div (3a - 2)$

38. $(4x^4 + 5x^2 - 7x) \div (2x + 1)$

39. $\dfrac{12x^3 + 13x^2 - 1}{4x - 1}$

40. $\dfrac{10x^3 + x^2 + 12x + 9}{5x + 3}$

 41. What potential difficulty could arise in the following long division problem?

$$x + 1 \enclose{longdiv}{} \quad \begin{array}{r} x^2 \hphantom{} \\ \overline{x^3 + 2x + 5} \\ \underline{x^3 + x^2} \end{array}$$

What could have been done to prevent this difficulty?

 42. Consider the following long division problem.

$$x + 1 \enclose{longdiv}{} \quad \begin{array}{r} x^2 + 1 \hphantom{} \\ \overline{x^3 + x^2 + x} \\ \underline{x^3 + x^2} \\ x \\ \underline{x + 1} \end{array}$$

The next step is to subtract, but there is no constant in the dividend from which we can subtract 1. Explain how to proceed.

In Exercises 43–48, use long division to determine the quotient and remainder.

43. $\dfrac{x^2 + 3}{x^2 - 1}$

44. $\dfrac{x^2 - 4}{x^2 + 1}$

45. $\dfrac{x^2 + 3x - 4}{x^2 + 2x}$

46. $\dfrac{3x^3 + 2x - 1}{x^2 + 2}$

47. $(a^4 + 7a - a^2 - 3) \div (a^2 - 2a + 3)$

48. $(2c^4 + 3 - 8c) \div (c^2 - 3c + 1)$

In Exercises 49–52, use long division to determine the quotient and remainder.

49. $\dfrac{x^2 + 4xy + 3y^2}{x - 2y}$

50. $\dfrac{4x^2 - 2xy - y^2}{2x + y}$

51. $\dfrac{4x^3 - 3xy^2 - 2y^3}{x + y}$

52. $\dfrac{x^3 - x^2y - y^3}{x - 2y}$

In Exercises 53–56, perform the indicated operations.

53. $(2x + 1)(x - 5) \div (x + 2)$

54. $(x^2 + 7x + 12) \div (x + 3) \div (x - 2)$

55. $(3x + 1)^2 \div (x - 1)$

56. $(x^2 + 2)^2 \div (x^2 + 1)$

In Exercises 57–60, determine the missing quantity.

	Quotient	Divisor	Remainder	Dividend
57.	$x + 3$	$2x - 1$	2	____
58.	$x^2 + 1$	$x + 1$	-3	____
59.	$x - 3$	____	-1	$x^2 - x - 7$
60.	$x - 2$	____	1	$x^3 - 2x^2 + x - 1$

61. When we say, "Find the quotient of $x + 2$ and $x^2 + 5x - 3$," do we divide $x + 2$ by $x^2 + 5x - 3$ or do we divide $x^2 + 5x - 3$ by $x + 2$?

62. Letting $P(x) = x^2 + 2x - 15$, $D(x) = x - 3$, and $Q(x) = x + 5$, show that

$$\frac{P(x)}{D(x)} = Q(x).$$

Now try to verify the following.

(a) $\dfrac{P(2)}{D(2)} = Q(2)$ (b) $\dfrac{P(3)}{D(3)} = Q(3)$

In which part were you unsuccessful? Why?

63. Find the quotient of $x^3 + 2x - 3$ and the product of x and $x - 3$.

64. Divide the product of $x^2 + 2$ and $x - 1$ by $x + 2$.

65. Subtract $4x - 1$ from the quotient of $6x^2 + 11x - 7$ and $2x - 1$.

66. If the sum of $x^2 + 3x - 2$ and $-2x^2 + x - 1$ is divided by $x - 1$, what is the quotient?

67. What polynomial divided by $3x^2$ results in $3x^3 + 4x^2 + x - 4$?

68. When $x^2 - 2x - 3$ is divided by $P(x)$, the quotient is $x + 2$ and the remainder is 5. What is $P(x)$?

69. The width of a rectangle is $x - 3$. The area of the rectangle is $x^2 + 5x - 24$. What expression represents the length of the rectangle? By how much does the length exceed the width?

70. A box of width w and length $w + 8$ has a volume of $w^3 + 10w^2 + 16w$. What is the height of the box? By how much does the height of the box exceed the width?

71. The length in feet of a roll of fencing is represented by $x^3 - 65x - 6$. The fencing is to be stretched across $x + 9$ posts, which are evenly spaced.

(a) Write an expression that represents the distance between the posts. (Hint: The number of spans between the posts is 1 less than the number of posts.)

(b) How much fencing will be left over?

72. The total cost of producing n items is $n^2 + 52n$ dollars. If 50 additional items are produced, the total cost is increased by \$100.

(a) What is the unit cost of the original n items?

(b) Does producing 50 additional items increase or decrease the unit cost? By how much?

Exploring with Real Data

In Exercises 73–76, refer to the following background information and data.

The following table shows an increase in the number of cable television systems in operation from 1983 through 1991. Also shown in the table is the number (in thousands) of cable TV subscribers for the same period.

Year	Cable TV Systems	Subscribers (in thousands)
1983	5,600	25,000
1985	6,600	32,000
1987	7,900	41,100
1989	9,050	47,500
1991	10,704	51,000

(Source: *Television and Cable Factbook*, Warren Publishing Co., Washington, D.C.)

The number $S(t)$ of cable TV systems can be modeled by the function $S(t) = 612.0t + 3677.2$, where t is the number of years since 1980.

The number $N(t)$ of cable TV subscribers can be modeled by $N(t) = -143.6t^2 + 5333.6t + 10{,}491.7$, where t is the number of years since 1980.

73. Calculate $N(t) \div S(t)$. Use your calculator to perform the necessary arithmetic, and round all results to the nearest tenth. Let $Q(t)$ represent the quotient and ignore the remainder.

74. Interpret the meaning of $Q(t)$.

75. Note that $Q(t)$ is in slope–intercept form. By inspecting the slope, judge whether $Q(t)$ is increasing rapidly, decreasing rapidly, or remaining relatively constant.

76. Of the three quantities $S(t)$, $N(t)$, and $Q(t)$, which do you expect will increase significantly in the future? Why?

Challenge

In Exercises 77–80, determine k so that the quotient has the specified remainder.

77. $(x^2 + 4x + k) \div (x - 2)$, $R = 4$

78. $(3x^2 + 7x + k) \div (3x + 1)$, $R = 0$

79. $(4x^2 + kx - 30) \div (x + 3)$, $R = 0$

80. $(x^2 + kx + 3) \div (x + 1)$, $R = 3$

In Exercises 81 and 82, divide.

81. $(x^{4n} + 3x^{3n} - x^{2n} + 2x^n + 1) \div (x^n + 3)$

82. $(x^{2n} + 2x^n - 4) \div (x^n + 1)$

83. For $(x^2 + bx + b^2) \div (x - b)$, determine the positive value of b for which the remainder is 12.

84. Suppose that when $ax^2 + 3x + c$ is divided by $x + 1$, the remainder is 0. If a is twice c, what are the values of a and c?

7.8 Synthetic Division

A method called **synthetic division** is a shortcut for dividing a polynomial by a binomial of the form $x - c$.

Consider the following long division problem on the left and compare it to the numerical arrangement on the right.

$$
\begin{array}{r}
2x^2 + 3x + 8 \\
x - 3 \overline{\smash{\big)}\, 2x^3 - 3x^2 - 1x + 2} \\
\underline{2x^3 - 6x^2} \\
3x^2 - x \\
\underline{3x^2 - 9x} \\
8x + 2 \\
\underline{8x - 24} \\
26
\end{array}
$$

$$
\begin{array}{c|rrrr}
③ & 2 & -3 & -1 & 2 \\
  &   & 6 & 9 & 24 \\
\hline
  & 2 & 3 & 8 & \boxed{26}
\end{array}
$$

We can make the following observations about the numbers on the right.

1. The circled number 3 is the same as the 3 in the divisor $x - 3$. We refer to this number as the *divider*.

2. The numbers 2, -3, -1, and 2 along the first row are the coefficients of the dividend $2x^3 - 3x^2 - x + 2$.

3. The numbers 2, 3, and 8 along the bottom row are the coefficients of the quotient $2x^2 + 3x + 8$.

4. The boxed number 26 is the remainder.

Although we still need to examine this numerical arrangement more closely, it is already apparent that the quotient and remainder have been obtained more easily than with the long division method on the left.

Two questions remain: How were the numbers in the bottom row determined? How were the numbers in the middle row determined?

The answer to the first question is easy. Each number in the bottom row is the sum of the two numbers above it.

To answer the second question about the numbers in the middle row, we observe that the first number in the bottom row is simply the number directly above it.

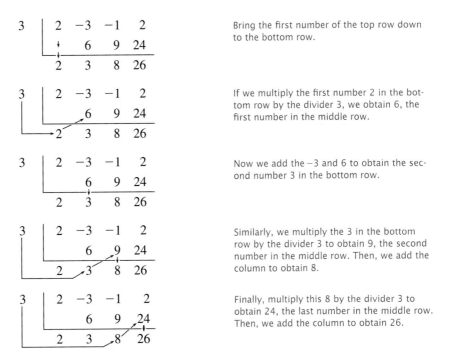

Bring the first number of the top row down to the bottom row.

If we multiply the first number 2 in the bottom row by the divider 3, we obtain 6, the first number in the middle row.

Now we add the −3 and 6 to obtain the second number 3 in the bottom row.

Similarly, we multiply the 3 in the bottom row by the divider 3 to obtain 9, the second number in the middle row. Then, we add the column to obtain 8.

Finally, multiply this 8 by the divider 3 to obtain 24, the last number in the middle row. Then, we add the column to obtain 26.

Once we complete this numerical routine, we can read across the bottom row to determine the coefficients of the quotient. The last number of the bottom row is the remainder.

Quotient: $2x^2 + 3x + 8$ Remainder: 26

Note that the degree of the quotient is 1 less than the degree of the dividend. Because the degree of the dividend is 3, the degree of the quotient is 2. That is how we knew to begin the quotient with $2x^2$.

Synthetic division is a valid shortcut to long division provided certain requirements are met.

1. The divisor must be of the form $x - c$. For a divisor such as $x - 7$, the divider is 7. For a divisor such as $x + 5$, we write $x - (-5)$, and the divider is −5.

2. The dividend must be written in descending order.

3. We write 0 as the coefficient of any missing term.

The advantages of synthetic division are the ease and speed with which it can be performed. The method eliminates much of the repetition involved in long division. For example, it is unnecessary to keep writing the variable factors of the terms.

EXAMPLE 1   *Dividing with Synthetic Division*

Use synthetic division to divide $x^4 - 3x^3 + 2x - 4$ by $x - 3$.

Solution

$$
\begin{array}{r|rrrrr}
3 & 1 & -3 & 0 & 2 & -4 \\
  &   & 3 & 0 & 0 & 6 \\
\hline
  & 1 & 0 & 0 & 2 & 2 \\
\end{array}
$$

Include a 0 as the coefficient of the missing x^2 term.

The remainder is 2.

$$1x^3 + \ 0x^2 + 0x + 2 \qquad \text{Quotient}$$

The degree of the dividend is 4, so the degree of the quotient is 3. The quotient is $x^3 + 2$ and the remainder is 2.

To check, multiply the quotient by the divisor and add the remainder. The result is the dividend.

$$(x^3 + 2)(x - 3) + 2 = x^4 - 3x^3 + 2x - 4.$$
■

EXAMPLE 2   *Dividing with Synthetic Division*

Show that when we divide $x^3 - 6x^2 - x + 30$ by $x + 2$, the remainder is 0.

Solution   The divisor must be written as $x - (-2)$ with $c = -2$.

$$
\begin{array}{r|rrrr}
-2 & 1 & -6 & -1 & 30 \\
   &   & -2 & 16 & -30 \\
\hline
   & 1 & -8 & 15 & 0 \\
\end{array}
$$

The remainder is 0.

$$1x^2 - \ 8x + 15 \qquad \text{Quotient}$$

Because the degree of the dividend is 3, the degree of the quotient is 2. The quotient is $x^2 - 8x + 15$. The remainder is 0.

To check, we multiply the quotient by the divisor to obtain the dividend.

$$(x + 2)(x^2 - 8x + 15) = x^3 - 6x^2 - x + 30$$
■

7.8 *Quick Reference*

Synthetic Division

- To divide a polynomial (written in descending order) by $x - c$, where c is a constant called the *divider,* follow these steps.

  1. In the first row, write the coefficients of the dividend with zero as the coefficient of any missing term.

  2. Write the divider c to the left of the first row in step 1.

  3. Bring the first coefficient in the top row down to the bottom row.

  4. Multiply the divider c times the new entry in the bottom row and place the result in the middle row of the next column.

5. Add the column and write the result underneath in the bottom row.

6. Repeat steps 4 and 5 until the bottom row is completely filled.

7. The bottom row then gives the coefficients of the quotient. The last number in the bottom row is the remainder. The degree of the quotient will be 1 less than the degree of the dividend.

7.8   *Exercises*

1. To use synthetic division to divide a polynomial by $x + 3$, show how to write the divisor in order to determine the divider and state what the divider is.

2. Suppose that you want to use synthetic division to divide

$$\frac{x - x^5}{x - 2}.$$

How should you write the row of dividend coefficients?

In Exercises 3–20, use synthetic division to determine the quotient and remainder.

3. $\dfrac{3x^2 - 10x - 8}{x - 4}$

4. $\dfrac{x^2 - 2x - 15}{x + 3}$

5. $\dfrac{x^2 - 2x - 9}{x - 6}$

6. $\dfrac{3x^2 + 14x + 5}{x + 3}$

7. $(6x^2 + x + 1) \div \left(x - \dfrac{1}{2}\right)$

8. $(2x^2 - 3x - 10) \div \left(x + \dfrac{5}{2}\right)$

9. $(3x^2 - 5x - 2) \div \left(x + \dfrac{1}{3}\right)$

10. $(3x^2 - 17x + 10) \div \left(x - \dfrac{2}{3}\right)$

11. $(2x^3 + 5x^2 - 3x + 2) \div (x + 2)$

12. $(3x^4 + x^3 - 5x^2 + 2x - 3) \div (x + 1)$

13. $(2x^4 - 3x^3 + x^2 - 3x + 4) \div (x - 2)$

14. $(3x^3 - 4x^2 - 2x - 5) \div (x - 3)$

15. $(x^2 + x^4 - 14) \div (x + 2)$

16. $(x^2 - 18x + 2x^3 + 42) \div (x + 4)$

17. $\dfrac{2x^2 - 6 + x^3 - 3x}{x - 2}$

18. $\dfrac{4x + x^3 - 13}{x - 3}$

19. $\dfrac{x + 5 - 2x^3 + x^5}{x + 2}$

20. $\dfrac{2x^4 - 25x + x^5 - 5}{x + 3}$

In Exercises 21–26, divide synthetically. Then, write the equation Dividend = (Quotient)(Divisor) + Remainder.

21. $(x^2 - 3 + x + 3x^6) \div (x + 1)$

22. $(4x^5 + x + 2x^6 + 2) \div (x + 2)$

23. $(x^5 + 32) \div (x + 2)$

24. $(x^5 + 243) \div (x + 3)$

25. $(x^4 + 16) \div (x + 2)$

26. $(x^4 + 81) \div (x - 3)$

In Exercises 27 and 28, by inspecting the given synthetic division problem, determine a, b, c, d, and e.

27.

$$\begin{array}{r|rrrrr} 2 & a & b & c & d & e \\ & & 2 & 0 & 0 & 2 \\ \hline & 1 & 0 & 0 & 1 & 0 \end{array}$$

28.
$$
\begin{array}{r|rrrrr}
-3 & a & b & c & d & e \\
  &   & -6 & 15 & 3 & -12 \\
\hline
  & 2 & -5 & -1 & 4 & -6
\end{array}
$$

In Exercises 29 and 30, the given synthetic division problem represents the division of a third-degree polynomial by a divisor of the form $x - c$. State the dividend and the divisor.

29.
$$
\begin{array}{r|rrrr}
? & 1 & 3 & -2 & -4 \\
  &   & -1 & -2 & 4 \\
\hline
  & 1 & 2 & -4 & 0
\end{array}
$$

30.
$$
\begin{array}{r|rrrr}
? & 1 & -4 & 3 & -11 \\
  &   & 4 & 0 & 12 \\
\hline
  & 1 & 0 & 3 & 1
\end{array}
$$

31. What is the quotient of $2x - 4 - 4x^5 + x^7$ and $x - 2$?

32. Divide $x - 3$ into $x - 12 - 17x^2 + x^5 - 3x^3$.

33. Divide $x^4 + x^2 + 2$ by $x + 1$.

34. What are the quotient and remainder when $x - 2$ is divided into $x^4 - 2x^2 - 5x + 4$?

 35. Can you use synthetic division to divide the polynomial $x^4 + x - 2$ by $x^2 - 1$? Why?

 36. For $\dfrac{3}{x - 2}$,

what are the quotient and remainder? Write the expression in the form

Dividend = (Quotient)(Divisor) + Remainder.

37. For $P(x) = x^3 - 2x^2 + x - 5$, determine $P(2)$. Then, divide $P(x)$ by $x - 2$ and compare the remainder to $P(2)$.

38. For $P(x) = x^3 - 3x^2 + 4x + 2$, determine $P(3)$. Then, divide $P(x)$ by $x - 3$ and compare the remainder to $P(3)$.

39. For $P(x) = 2x^3 - x^2 + 5x + 3$, determine $P(-2)$. Then, divide $P(x)$ by $x + 2$ and compare the remainder to $P(-2)$.

40. For $P(x) = x^4 + 3x^2 - 5$, determine $P(-3)$. Then, divide $P(x)$ by $x + 3$ and compare the remainder to $P(-3)$.

Challenge

In Exercises 41–44, determine the value of k so that the remainder will be zero.

41. $(x^3 - 4x^2 - x + k) \div (x - 3)$

42. $(x^3 + 3x^2 + k) \div (x + 2)$

43. $(x^3 + kx^2 - 2k + 3) \div (x - 1)$

44. $(4x^3 + kx^2 + 7) \div (x + 1)$

7.9 The Remainder and Factor Theorems

The Remainder Theorem ▪ *The Factor Theorem* ▪ *A New Factoring Method*

The Remainder Theorem

In Sections 7.7 and 7.8 we discussed methods for dividing a polynomial by a binomial in the form $x - c$. In the following exploration we use synthetic division to divide a polynomial $P(x)$ by $x - c$. As we will see, there is a relationship between the remainder and the value of $P(c)$.

EXPLORATION 1 *Dividing P(x) by x − c and Comparing Remainders with P(c)*

For each of the following, a polynomial $P(x)$ is to be divided by a divisor $x - c$. Determine the remainder. Then, calculate $P(c)$ and compare the result to the remainder.

(a) $\dfrac{2x^3 - 3x^2 - x + 2}{x - 2}$

(b) $\dfrac{x^3 + 27}{x + 3}$

(c) $\dfrac{x^4 - 3x^3 + 2x - 4}{x - 3}$

(d) $\dfrac{x^3 - 6x^2 - x + 30}{x + 2}$

Discovery The following is a summary of our results. Entries in the remainder column can be verified by synthetic division. Verify the values in the last column with your calculator.

Dividend $P(x)$	Divisor $x - c$	c	Remainder	$P(c)$
$2x^3 - 3x^2 - x + 2$	$x - 2$	2	4	4
$x^3 + 27$	$x - (-3)$	-3	0	0
$x^4 - 3x^3 + 2x - 4$	$x - 3$	3	2	2
$x^3 - 6x^2 - x + 30$	$x - (-2)$	-2	0	0

In each part, we observe that dividing $P(x)$ by $x - c$ results in a remainder that is equal to $P(c)$.

Exploration 1 suggests the following theorem.

The Remainder Theorem

If a polynomial $P(x)$ is divided by $x - c$, where c is a constant, then the remainder is $P(c)$.

We can justify the Remainder Theorem with the following reasoning. Suppose that when $P(x)$ is divided by $x - c$, the quotient is $Q(x)$, and the remainder is R. We know that the relationship among these quantities can be expressed in the following way.

$$P(x) = Q(x) \cdot (x - c) + R$$

When we evaluate $P(x)$ for $x = c$, we obtain the following.

$$P(c) = Q(c) \cdot (c - c) + R$$
$$= 0 + R$$
$$= R$$

Therefore, the value of $P(c)$ is the same as the remainder when $P(x)$ is divided by $x - c$.

NOTE: The Remainder Theorem applies only if the divisor is of the form $x - c$. Divisors of the form $x + c$ must first be changed to $x - (-c)$. ■

The Remainder Theorem gives us a second way to evaluate a polynomial for $x = c$ and to determine the remainder when $P(x)$ is divided by $x - c$. We can evaluate $P(c)$ directly, or we can use synthetic division.

EXAMPLE 1   *Using the Remainder Theorem to Determine a Remainder*

What is the remainder when $p(x) = x^4 - 2x^3 + x^2 - 3x - 2$ is divided by $x + 1$?

Solution   Because $x + 1 = x - (-1)$, the Remainder Theorem states that the remainder is $p(-1)$.

$$p(x) = x^4 - 2x^3 + x^2 - 3x - 2$$
$$p(-1) = (-1)^4 - 2(-1)^3 + (-1)^2 - 3(-1) - 2$$
$$= 1 + 2 + 1 + 3 - 2$$
$$= 5$$

If you prefer, you can enter

$$y_1 = x^4 - 2x^3 + x^2 - 3x - 2$$

in your calculator and compute $p(-1)$. Figure 7.14 shows the results.

We can verify these results with synthetic division.

$$
\begin{array}{r|rrrrr}
-1 & 1 & -2 & 1 & -3 & -2 \\
   &   & -1 & 3 & -4 & 7 \\
\hline
   & 1 & -3 & 4 & -7 & 5 \quad \text{Remainder} \\
   & \downarrow & \downarrow & \downarrow & \downarrow \\
   & 1x^3 - & 3x^2 + & 4x - & 7 & \quad \text{Quotient}
\end{array}
$$

Figure 7.14

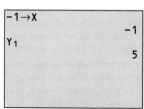

When you divide $P(x)$ by $x - c$, determining the remainder by evaluating $P(c)$ is a good use of a calculator. However, if you also want to know the quotient, then synthetic division is the best approach.

The Factor Theorem

When dividing a polynomial $P(x)$ by $x - c$, we obtain additional information if the remainder is 0.

> ### The Factor Theorem
>
> If a polynomial $P(x)$ is divided by $x - c$, where c is a constant, and if the remainder is 0, then $x - c$ is a factor of $P(x)$.

To see why the Factor Theorem is true, refer again to $P(x) = Q(x) \cdot (x - c) + R$. If the remainder is 0, then we can write $P(x) = Q(x) \cdot (x - c)$, which shows that $x - c$ is a factor of $P(x)$.

Moreover, the Remainder Theorem states that $R = P(c)$. Therefore, if $R = 0$, then $P(c) = 0$. Here, then, is another way of stating the Factor Theorem.

> ### The Factor Theorem
>
> For the polynomial $P(x)$ and constant c, if $P(c) = 0$, then $x - c$ is a factor of $P(x)$.

This alternate wording of the Factor Theorem implies that if we can find a value of c so that $P(c) = 0$, then $x - c$ is a factor of $P(x)$. Conversely, if $x - c$ is a factor of $P(x)$, then $R = P(c) = 0$.

 **EXPLORATION 2**

Using a Graph of $y = P(x)$ to Determine Factors of $P(x)$

Consider $p(x) = x^3 + x^2 - 17x + 15$.

(a) Use your calculator to compute $p(c)$ for $c = -5$, $c = 1$, and $c = 3$.

(b) According to the Factor Theorem, what do your results in part (a) imply?

(c) Produce the graph of $y = p(x)$ and determine the points for which $p(x) = 0$. What are these points called?

(d) Compare the x-coordinates of the points in part (c) with the values of c in part (a).

(e) What is your conjecture about how the graph of a polynomial can be used to factor the polynomial?

Discovery

(a) $p(-5) = p(3) = p(1) = 0$

(b) Because -5, 3, and 1 are values of c such that $p(c) = 0$, the Factor Theorem states that $x + 5$, $x - 3$, and $x - 1$ are all factors of $p(x)$.

Figure 7.15

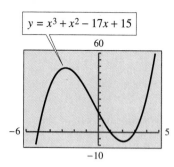

$y = x^3 + x^2 - 17x + 15$

(c) From the graph of $y = p(x)$, which is shown in Fig. 7.15, the points for which $p(x) = 0$ are $(-5, 0)$, $(3, 0)$, and $(1, 0)$. These are the x-intercepts of the graph.

(d) The x-coordinates of the x-intercepts are the same as the values used to evaluate $p(x)$ in part (a).

(e) To estimate the factors of a polynomial $p(x)$, we can produce the graph of $y = p(x)$ and locate the x-intercepts. Each x-intercept $(c, 0)$ is associated with a factor of $p(x)$, $x - c$.

Another implication of the Factor Theorem involves equations. Consider an equation whose left side is a polynomial $P(x)$ and whose right side is 0: $P(x) = 0$.

If $x - c$ is a factor of $P(x)$, then $P(c) = 0$. In other words, c is a **solution** (or **root**) of the equation.

In summary, the following statements are all equivalent. Each one implies all of the others.

> ## Equivalent Statements Implied by the Factor Theorem
>
> 1. For a polynomial $P(x)$ and constant c, $x - c$ is a factor of $P(x)$.
>
> 2. $P(c) = 0$.
>
> 3. The remainder $R = 0$ when $P(x)$ is divided by $x - c$.
>
> 4. The number c is a solution (or root) of $P(x) = 0$.
>
> 5. An x-intercept of the graph of $y = P(x)$ is $(c, 0)$.

A New Factoring Method

Exploration 2 gives us another method for factoring a polynomial $P(x)$. Producing the graph of $y = P(x)$ and locating the x-intercepts is a way to estimate values of c such that $P(c) = 0$. Once we have verified that $P(c) = 0$, the Factor Theorem states that $x - c$ is a factor of $P(x)$.

The factoring methods discussed earlier in this chapter are generally the best methods to use for first- and second-degree polynomials. For third- and higher-degree polynomials, the use of a graph in conjunction with the Factor Theorem is a good approach.

EXAMPLE 2   *Using a Graph and the Factor Theorem to Factor a Polynomial*
Factor $p(x) = 2x^3 - 5x^2 - 4x + 3$ completely.

Solution

We produce the graph of $p(x) = 2x^3 - 5x^2 - 4x + 3$.

Figure 7.16

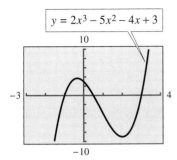

$$y = 2x^3 - 5x^2 - 4x + 3$$

From the graph in Fig. 7.16, two of the x-intercepts appear to be $(-1, 0)$ and $(3, 0)$. We can easily verify that $p(-1) = 0$ and $p(3) = 0$. Therefore, according to the Factor Theorem, two factors of $p(x)$ are $x + 1$ and $x - 3$.

We could also use the graph to estimate the third x-intercept, but in the following, we use an alternate approach.

Synthetically divide $p(x)$ by the factor $x - 3$.

$$
\begin{array}{r|rrrr}
3 & 2 & -5 & -4 & 3 \\
  &   & 6 & 3 & -3 \\
\hline
  & 2 & 1 & -1 & 0
\end{array}
$$

The quotient is $2x^2 + x - 1$ and the remainder is 0 [as we knew it should be because $x - 3$ is a factor of $p(x)$]. Therefore, we can write $p(x)$ as follows.

$$
\begin{aligned}
p(x) &= (2x^2 + x - 1)(x - 3) \\
&= (2x - 1)(x + 1)(x - 3)
\end{aligned}
$$

Note the factor $x + 1$, which we had previously anticipated. Also, $p(x) = 0$ if $2x - 1 = 0$ or $x = \frac{1}{2}$. This indicates that the x-intercept that we did not attempt to estimate is $(\frac{1}{2}, 0)$. ∎

7.9   Quick Reference

The Remainder Theorem

- The Remainder Theorem states that if a polynomial $P(x)$ is divided by $x - c$, where c is a constant, then the remainder is $P(c)$. The divisor must be in the form $x - c$.

- The Remainder Theorem is useful if only the remainder from a division problem is sought. If the quotient is also needed, then synthetic division is the best method to use.

The Factor Theorem

- The Factor Theorem states that if a polynomial $P(x)$ is divided by $x - c$, where c is a constant, and if the remainder is 0 [that is, if $P(c) = 0$], then $x - c$ is a factor of $P(x)$.

- The following are all equivalent statements.

  1. For a polynomial $P(x)$ and constant c, $x - c$ is a factor of $P(x)$.

  2. $P(c) = 0$.

  3. The remainder $R = 0$ when $P(x)$ is divided by $x - c$.

  4. The number c is a solution (or root) of $P(x) = 0$.

  5. An x-intercept of the graph of $y = P(x)$ is $(c, 0)$.

A New Factoring Method	▪ The following procedure can be used to factor a polynomial $P(x)$.

1. Produce the graph of $y = P(x)$ and estimate the x-intercepts.

2. Each x-intercept has the form $(c, 0)$. Verify that $P(c) = 0$. Then the Factor Theorem states that $x - c$ is a factor of $P(x)$.

▪ A variation of this procedure is as follows.

1. Estimate and verify one x-intercept $(c, 0)$ in order to determine one factor $x - c$.

2. Divide $P(x)$ synthetically by $x - c$ to determine the quotient $Q(x)$.

3. Write $P(x) = Q(x) \cdot (x - c)$ and then factor $Q(x)$.

4. Repeat this process until $P(x)$ is factored completely.

7.9   *Exercises*

1. If $P(x) = x^4 + 3x^3 + 4x - 1$ is divided by $x - c$, what is the remainder if $P(c) = 5$? Why?

2. If $x - c$ is a factor of $P(x) = x^3 + 6x^2 - 5$, explain why $c^3 + 6c^2 - 5 = 0$.

In Exercises 3–12, use synthetic division to determine the remainder when $P(x)$ is divided by $x - a$, where a is given. Determine $P(a)$ to verify the result.

3. $P(x) = 3x^3 - 4x^2 + 5x - 6$; $a = 1$

4. $P(x) = 2x^3 - 3x^2 - 5x + 7$; $a = -1$

5. $P(x) = 4x^3 - 5x + 7$; $a = -2$

6. $P(x) = 5x^3 - 3x^2 + x$; $a = 2$

7. $P(x) = x^4 - 3x^2 + 5$; $a = 2$

8. $P(x) = 2x^4 - 3x^3 + 2x$; $a = 3$

9. $P(x) = x^5 + 3x^3 - 7$; $a = -3$

10. $P(x) = x^6 - 3x$; $a = 2$

11. $P(x) = -8x^4 - x^3 - x$; $a = \dfrac{1}{2}$

12. $P(x) = 24x^6 - 3x^2$; $a = -\dfrac{1}{2}$

13. If $P(x) = 2x^4 + x^3 - x^2 + 5x - 6 = 0$ and $P(a) = 0$, what do we call a? Why?

14. If $P(x) = 2x^4 + x^3 - x^2 + 5x - 6 = 0$ and $P(c) = 0$, what do we call $x - c$? Why?

In Exercises 15–18, produce the graph of the polynomial and verify that the ordered pair $(a, 0)$ is an x-intercept. Then, verify algebraically that a is a solution of the equation.

15. $x^4 + 3x^3 - 2x^2 + 5x - 7 = 0$; $a = 1$

16. $x^5 - 3x^4 + 2x^3 + 2x - 4 = 0$; $a = 2$

17. $x^4 - 16x^2 - 3x - 12 = 0$; $a = -4$

18. $3x^3 - 7x^2 + 7x - 10 = 0$; $a = 2$

In Exercises 19–26, use synthetic division and the Factor Theorem to show that the given binomial is a factor of the given polynomial.

19. $x + 2$; $2x^3 + 7x^2 + 14x + 16$

20. $x - 3$; $x^3 - x^2 - x - 15$

21. $x - 1$; $2x^3 - 7x^2 + 11x - 6$

22. $x + 3$; $x^3 + x^2 - 4x + 6$

23. $x + 1$; $x^3 - x^2 + 3x + 5$

24. $x + 4$; $x^3 + 3x^2 - 2x + 8$

25. $x - \dfrac{3}{2}$; $x^3 - \dfrac{3}{2}x^2 - 16x + 24$

26. $x + \dfrac{3}{4}$; $16x^3 + 12x^2 + x + \dfrac{3}{4}$

In Exercises 27–34, use synthetic division and the Factor Theorem to show that the given binomial $D(x)$ is a factor of the given polynomial $P(x)$. Write the polynomial in the form $P(x) = Q(x)D(x)$, where $Q(x)$ is the quotient.

27. $x + 3$; $x^3 - 5x + 12$

28. $x + 2$; $x^3 - 2x^2 - 7x + 2$

29. $x + 2$; $2x^3 + 7x^2 + 12x + 12$

30. $x - 2$; $3x^3 - 5x^2 + 3x - 10$

31. $x - \dfrac{1}{2}$; $2x^3 - 5x^2 - 4x + 3$

32. $x + \dfrac{2}{3}$; $3x^3 - 4x^2 - x + 2$

33. $x - 4$; $x^4 - 4x^3 - x^2 + 6x - 8$

34. $x + 3$; $x^5 + 3x^4 - x^3 - 3x^2 - 3x - 9$

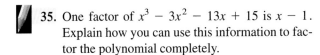 **35.** One factor of $x^3 - 3x^2 - 13x + 15$ is $x - 1$. Explain how you can use this information to factor the polynomial completely.

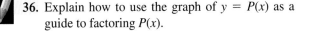 **36.** Explain how to use the graph of $y = P(x)$ as a guide to factoring $P(x)$.

In Exercises 37–48, use synthetic division, the given factor, and the Factor Theorem to factor the polynomial completely.

37. $x + 3$; $x^3 + 8x^2 + 21x + 18$

38. $x + 2$; $x^3 - 8x^2 + 5x + 50$

39. $x + 5$; $x^3 - 21x + 20$

40. $x + 4$; $x^3 - 13x + 12$

41. $x - 1$; $2x^3 + x^2 - 5x + 2$

42. $x - 2$; $3x^3 - 7x^2 - 2x + 8$

43. $x - 3$; $4x^3 - 7x^2 - 21x + 18$

44. $x + 2$; $6x^3 + x^2 - 19x + 6$

45. $x + 2$; $8x^3 + 6x^2 - 17x + 6$

46. $x - 2$; $10x^3 - 21x^2 - x + 6$

47. $x - \dfrac{2}{3}$; $3x^3 - 8x^2 - 5x + 6$

48. $x - \dfrac{5}{6}$; $6x^3 - 11x^2 - 31x + 30$

In Exercises 49–62, factor the polynomial completely.

49. $x^3 - 4x^2 + x + 6$

50. $x^3 + 5x^2 + 2x - 8$

51. $x^3 - 4x^2 + 5x - 6$

52. $x^3 - x + 6$

53. $2x^3 - 5x^2 - x + 6$

54. $2x^3 + x^2 - 13x + 6$

55. $6x^3 - 7x^2 - 7x + 6$

56. $6x^3 + 7x^2 - 7x - 6$

57. $3x^3 + 8x^2 - 5x - 6$

58. $3x^3 + 7x^2 - 2x - 8$

59. $x^4 + 3x^3 - 7x^2 - 27x - 18$

60. $x^4 + 4x^3 - x^2 - 16x - 12$

61. $x^5 - x^4 - 3x^3 + 3x^2 - 4x + 4$

62. $x^5 - 2x^4 + x^3 - 2x^2 - 2x + 4$

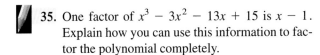 **63.** Suppose that the graph of $y = P(x)$ has three x-intercepts: $(a, 0)$, $(b, 0)$, and $(c, 0)$. What are three factors of $P(x)$? Can you be certain that $P(x)$ is of degree 3? Why?

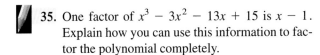 **64.** Suppose $P(x)$ is of degree 2 and two factors of $P(x)$ are $x + 2$ and $x - 3$. Can you be certain that $P(x) = (x + 2)(x - 3) = x^2 - x - 6$? Why?

In Exercises 65–68, write a polynomial with the given degree and x-intercepts.

65. degree 2; $(3, 0)$ and $(-2, 0)$

66. degree 2; $(2, 0)$ and $(-1, 0)$

67. degree 3; $(-4, 0)$, $(0, 0)$, and $(3, 0)$

68. degree 3; $(-2, 0)$, $(1, 0)$, and $(4, 0)$

69. In a division problem, the dividend is x^2, the quotient is $x + 3$, and the remainder is 9. What is the divisor?

70. If the quotient is $x^2 + 2x + 5$, the divisor is $x - 2$, and $P(2) = 5$, what is the dividend $P(x)$?

71. Suppose that $x - c$ is a factor of $P(x)$. What is a solution of the equation $P(x) = 0$? Why?

72. Using graphing and the Factor Theorem, solve the following equation.

$$x^6 + x^5 - 4x^4 - 12x^3 - 8x^2 + 32x + 32 = 0$$

Observe that at one of the x-intercepts, the graph does not cross the x-axis. How many factors correspond to this x-intercept?

In Exercises 73–76, one root of the equation is given. Determine the other solutions.

73. $x^4 + x^3 - 11x^2 - 9x + 18 = 0$; $x = 1$

74. $x^4 + x^3 - 7x^2 - x + 6 = 0$; $x = 2$

75. $x^5 + x^4 - 10x^3 - 10x^2 + 9x + 9 = 0$; $x = -1$ (double root)

76. $x^5 + 2x^4 - 13x^3 - 26x^2 + 36x + 72 = 0$; $x = -2$ (double root)

77. Explain how to determine the remainder when $x^{100} - x^{37} - 1$ is divided by $x + 1$ without using synthetic division. Then, determine the remainder.

78. What is the remainder when $P(x) = x^{201} - x^{55} - 3$ is divided by $x + 1$? Is the remainder different when $P(x)$ is divided by $x - 1$? Why?

79. The volume of a rectangular box is given by the polynomial $x^3 + 12x^2 + 35x + 24$, and the dimensions of the box are polynomials with integer coefficients. If the height of the box is $x + 8$, show that the length of the box is 2 greater than the width.

80. The product of four integers is given by the expression $x^4 + 5x^3 - x^2 - 5x$. Each integer is represented by a polynomial with integer coefficients. If one of the numbers is $x + 5$, show that the other three numbers are consecutive integers.

81. Suppose that a card is drawn from a standard deck (52 possible outcomes), a coin is flipped (2 possible outcomes), and a die is rolled (6 possible outcomes). One possible outcome is 6♣ and "tails" and 4. The Counting Principle states that the total number of outcomes is the product of the number of outcomes for each task. In other words, $52 \cdot 2 \cdot 6 = 624$.

Suppose that an experiment involving four tasks has a total of $x^4 - 6x^3 - 4x^2 + 54x - 45$ outcomes, where the number of outcomes of each task is represented by a polynomial with integer coefficients. Factor the polynomial to determine the number of outcomes of each task.

82. Suppose $P(x) = p_1(x) \cdot p_2(x)$, where p_1 and p_2 are polynomials. If $x + 2$ is a factor of both p_1 and p_2, and if $P(x) = x^5 + 2x^4 - 5x^3 - 10x^2 + 4x + 8$, what is the complete factorization of $P(x)$?

Exploring with Real Data

In Exercises 83–86, refer to the following background information and data.

Following 1985, the circulation of morning daily newspapers fell on hard times. The bar graph shows the circulation (in thousands) of morning daily newspapers in cities with populations between one-half million and one million people.

Figure for 83–86

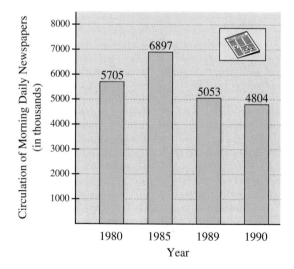

The circulation $C(t)$ (in thousands) can be modeled by the function $C(t) = -65.3t^2 + 541.8t + 5725.8$, where t is the number of years since 1980.

83. Treat -65.3 as a common factor and write the function $C(t)$ in the form $C(t) = -65.3(\underline{\hspace{1cm}})$.

84. Use a suitable range and scale setting to produce the graph of the function $C(t) = -65.3t^2 + 541.8t + 5725.8$ on your calculator. Then, estimate the x-intercepts to the nearest integer. Use this information to write an approximate factorization of the polynomial in parentheses in Exercise 83.

85. For what values of t is $C(t) = 0$? To what years do these values of t correspond? If the function were to remain valid in the future, what would be the significance of the positive value of t?

86. What are two plausible explanations for the downturn in the circulation of morning daily newspapers in large cities?

Challenge

In Exercises 87–92, determine k so that the polynomial is divisible by the binomial.

87. $x^4 - 3x^3 + kx + 3$; $x - 3$

88. $x^3 + kx^2 + 2x - 4$; $x + 2$

89. $k^2x^3 - 7kx + 6$; $x - 1$

90. $2k^2x^4 - 5kx^3 + 3x$; $x + 1$

91. $x^4 - 16$; $x + k$

92. $x^3 - 4x^2 + 3x - 12$; $x + k$

93. For what integers n is $x + 1$ a factor of $x^n + 1$?

94. For what integers n is $x - c$ a factor of $x^n - c^n$?

95. Determine whether there is a number c such that $x + c$ is a factor of $x^4 + x^2 + 1$.

---

7   Chapter Review Exercises

Section 7.1

In Exercises 1–4, answer each of the following.

  (a) State whether the given polynomial is a monomial, binomial, trinomial, or none of these.

  (b) State the degree of the polynomial.

  (c) Write the polynomial in descending order.

  (d) State the coefficient of the x^2 term.

1. $3x - 2x^2$ 2. x^4

3. $x^2 + 9x^5 - 3$

4. $x^6 - 7x^2 + 5x^5 - 4x$

In Exercises 5 and 6, determine the degree of the polynomial.

5. $4x^2 + y^2 + x^2y^4$ 6. $a^3b - b^3$

In Exercises 7 and 8, evaluate the given function as indicated.

7. $P(x) = x^5 - x^3 + 2x^2$; $P(1)$

8. $P(x) = -x^2 - 3x - 5$; $P(-2)$

In Exercises 9 and 10, evaluate the given polynomial as indicated.

9. $x^5y^3 - 4x^3y^4 + x^2$ at $x = 1$ and $y = 2$

10. $a^3 - a^2b + ab^2 - b^3$ at $a = 1$ and $b = -1$

In Exercises 11–14, perform the indicated operations.

11. $(5x^2 - 3x + 4) + (-5x - 3x^2 + 9)$

12. $(3x - 2) - (x^2 - 3x + 2) + (x^2 - 4x - 5)$

13. Find the sum of $3x^2 - 6x$ and $4x + 7$.

14. Subtract $-2x^2 + 3x - 7$ from $5x^2 - x$.

Section 7.2

In Exercises 15–28, multiply.

15. $4x^5(-2x^3)$

16. $(3cd^5)(-c^2d^7)(-3c^5d^3)$

17. $2x^3(3x^2 - 4)$ 18. $-b^2(b^2 + 5b - 3)$

19. $2a^2b(5 - b^2 + 3b)$

20. $-4xy^2(x^3 + 4xy^2 - 7x^2)$

21. $(x - 4)(x + 3)$ 22. $(2x - 5)(3x + 7)$

23. $(2x + 3)(8 - 3x)$ 24. $(x^n - 4)(3x^n + 5)$

25. $(x + 2)(2x^2 - 3x + 1)$

26. $(a^2 - 5)(a^3 + 2a^2 + 1)$

27. $(x + 3)(2x - 5) - (3x + 1)(x + 4)$

28. $2x - 5[2x - 5(x + 1) - 5x]$

29. Write a polynomial expression for the area of a rectangle with a length of $4x - 5$ and a width of $2x - 3$.

30. If three numbers are represented by $x + y$, $y - 3$, and $4 - x$, write a polynomial expression for the product of the numbers.

 31. In words, describe how to recognize a difference of two squares.

32. For what special product is the following template used for multiplying?

$$(\quad)^2 + 2(\quad)(\quad) + (\quad)^2$$

In Exercises 33–42, perform the indicated operations.

33. $(3x - 2)(3x + 2)$ 34. $(1 - 3y)(1 + 3y)$

35. $(3y - 2)^2$ 36. $(ab + 4)^2$

37. $(3 - nm^2)^2$ 38. $(x^h + y^g)^2$

39. $(2y + 1)^3$ 40. $(x - 5)^3$

41. $(2x - 1)(x - 3) - (x - 3)^2$

42. $[(x + 2y) + 5][(x + 2y) - 5]$

43. Write a polynomial expression for the product of three consecutive integers where x represents the second integer.

 44. The width and height of a rectangular carton are the same. The length is 1 foot longer than the height. This carton is then placed inside a second carton in the shape of a cube with the same length as the first carton. (See figure.) Write a polynomial expression for the space that remains inside the second carton.

Figure for 44

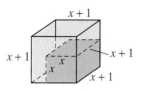

Section 7.3

 45. Suppose that you want to factor $-x^3y^2 - 2x^2y^3$ so that one of the factors is a binomial with a positive leading coefficient. What would you select for the GCF?

 46. Explain why $2x + 4y + 3x + 9y$ cannot be factored with the grouping method.

In Exercises 47–54, factor completely.

47. $x^{25} - x^{35}$ 48. $6x^3 - 36x^2$

49. $x^6y^3 - x^7y^4 - x^3y^5$

50. $3x(x - a) - 4y(x - a)$

51. $x^3 + 3x^2 - 2x - 6$

52. $ac - af + bf - bc$

53. $2x^{n+4} - 5x^{n+2} + 6x^n$

54. $(3x - 2)^n + 2(3x - 2)^{n-1}$

55. An odd number can be represented by $2n - 1$, where n is an integer. Using this representation, show that the difference between the square of an odd number and the odd number itself can be written $2(2n - 1)(n - 1)$.

56. The total cost of z items is given by $xy^2z - 4z$. What binomial represents the unit cost of the items?

Section 7.4

 57. How can you quickly tell that $9x^2 - 30x + 24$ is not a perfect square trinomial?

 58. Suppose that you factor an expression and obtain the following: $(x + 1)(x^2 - x + 1)$. What expression did you factor?

In Exercises 59–66, factor completely.

59. $49x^2 - 16y^2$ 60. $a^{20} - b^{10}$

61. $16 - a^4$

62. $8a^3 - 1$

63. $27b^6 + 8a^3$

64. $4x^2 + 12xy + 9y^2$

65. $c^2 + 64 + 16c$

66. $c^4 + 6c^2 + 9$

67. The base and height of a parallelogram are represented by completely factored binomials. If the area of the parallelogram is given by $a^2 - 16$ square feet, show that the base and height differ by 8 feet.

68. The total revenue from selling n items is given by $4x^2 + 9y^2$. If the total cost of the n items is $12xy$, write a completely factored expression for the profit.

Section 7.5

In Exercises 69–72, factor completely.

69. $x^2 - 5x - 6$

70. $x^2 + 5x + 4$

71. $2x^2 + 12x - 14$

72. $x^4 + 7x^2 - 60$

73. What is a disadvantage of using the trial-and-error method to factor $6x^2 + 25x + 24$?

74. One way to factor $x(2x - 1) + 3(2x - 1)$ is to simplify the expression to obtain $2x^2 + 5x - 3$. Then, using the trial-and-error method, we can write $2x^2 + 5x - 3 = (2x - 1)(x + 3)$. Is this approach correct? What is a more efficient approach?

In Exercises 75–82, factor completely.

75. $3x^2 + 14x + 15$

76. $8x^2 - 3x - 5$

77. $12x^2 + 8x - 15$

78. $12x^2 - 12x - 9$

79. $2x - 8 + x^2$

80. $125x^2 - 80y^2$

81. $125x^3 - 64$

82. $3x^2 - 3x - 60$

83. While factoring $4x^2 + 16xy + 16y^2$, suppose you overlook the GCF of 4 and write $(2x + 4y)^2$. Can you now factor out 2 and write $2(x + 2y)^2$? Why?

84. Last year, a billboard designer created a sign L feet long and W feet wide. The area of the sign was A square feet. This year, the designer changed the dimensions of the sign, and the new area is given by $A - 5W + 2L - 10$ square feet. If the dimensions of the new sign can be represented by binomials, what changes did the designer make to the sign?

In Exercises 85–92, factor completely.

85. $3x^2 + 4x - 15$

86. $2x^3 + 5x^2 - 18x - 45$

87. $x^9 + 27y^6$

88. $y^4 - x^4$

89. $x^3y^4 - x^4y^3$

90. $50x^3 - 25x^4$

91. $6x^3 + 25x^2 + 21x$

92. $12(2x - 3)^2 + 15(3 - 2x) - 18$

Section 7.6

 93. In which of the following equations is the value of a or b definitely known?

$$a + b = 0 \qquad ab = 0$$

94. The left side of $5(x - 2)(x + 4) = 0$ has three factors, but the equation has only two solutions. Why?

In Exercises 95–104, solve the equation algebraically.

95. $x^2 = 8x$

96. $x^2 - 81 = 0$

97. $x^2 + x = 42$

98. $4x^2 - 23x + 15 = 0$

99. $x(x + 2) = 24$

100. $(x + 3)(x - 2) = 6$

101. $x^3 + 3x^2 - 4x = 12$

102. $(x - 1)^2 - 2(x - 1) = 15$

103. $|3x^2 + 7x - 2| = 4$

104. $|5x^2 + 2x + 2| = 5$

105. The sum of twice a number and its square is 63. Determine the number.

106. Standing on the roof of an eight-story building, a person throws a ball upward with an initial velocity of 64 feet per second. If the release point is 80 feet above the ground, then the height h of the ball after t seconds is determined by $h(t) = -16t^2 + 64t + 80$. Use this equation to find the time it takes for the ball to reach the ground.

Section 7.7

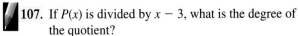 **107.** If $P(x)$ is divided by $x - 3$, what is the degree of the quotient?

108. Explain how to check the result of a long division problem.

In Exercises 109 and 110, determine the quotient.

109. $\dfrac{28x^5y^4}{7x^3y}$

110. $\dfrac{6a^5b^4 - 4a^3b^3 + 8a^2b}{2a^2b^3}$

In Exercises 111–114, use long division to find the quotient and remainder.

111. $(9x^3 - 18x^2 - 7x + 5) \div (3x - 2)$

112. $(12x^4 - 4x^3 + 2x - 17) \div (2x^2 + 3)$

113. $(x^3 - 3x^2y + 3xy^2 - y^3) \div (x - 3y)$

114. $(x^4 - 1) \div (x - 1)^2$

115. If $P(x)$ is divided by $x + 1$, the quotient is $2x^2 - x + 1$ and the remainder is 3. Write a polynomial expression for $P(x)$.

116. A grandmother left $x^4 + x^3 - 2x^2 + 3x + 5$ dollars in her will to be divided equally among her x grandchildren. However, she miscounted and there were actually $x + 1$ grandchildren, all entitled to an equal share. Write a polynomial expression that represents the monetary bequest to each grandchild.

Section 7.8

 117. If you use synthetic division to divide $P(x)$ by $x + 7$, what is the divider?

 118. Interpret the bottom row of numbers in the following synthetic division problem.

$$
\begin{array}{r|rrrr}
-1 & 2 & 1 & 0 & 4 \\
   &   & -2 & 1 & -1 \\
\hline
   & 2 & -1 & 1 & 3
\end{array}
$$

In Exercises 119–122, use synthetic division to find the quotient and remainder.

119. $(2x^3 - 3x^2 + 5) \div (x - 1)$

120. $(3x - 4x^3 + x^5 + 2) \div (x + 2)$

121. $(x^5 - 32) \div (x - 2)$

122. $(2x^3 + 3x^2 - 5) \div (x - 0.5)$

123. For the following synthetic division problem, determine a, b, c, and d.

$$
\begin{array}{r|rrrr}
3 & a & b & c & d \\
  &   & 3 & 3 & 6 \\
\hline
  & 1 & 1 & 2 & 13
\end{array}
$$

124. For the following synthetic division problem, what is the divider and what is the divisor?

$$
\begin{array}{r|rrrr}
-5 & a & b & c & d \\
   &   & e & f & g \\
\hline
   & a & h & j & k
\end{array}
$$

Section 7.9

 125. Suppose that when $P(x)$ is divided by $x - 2$, the remainder is 3. Explain how you know the value of $P(2)$.

126. Suppose that when $P(x)$ is divided by $x + 3$, the remainder is 0. What does this imply about $x + 3$?

127. Use synthetic division and the Remainder Theorem to determine the value of $P(1)$, where $P(x) = x^5 - 2x^4 + 3x^3 - 4x^2 + 5$.

128. Use synthetic division and the Factor Theorem to show that $x + 2$ is a factor of $2x^3 + x^2 - 4x + 4$.

129. Suppose you divide $P(x)$ by $x - 5$, by $x - 2$, and by $x + 4$, and each time the remainder is 0. What can you conclude about the graph of $y = P(x)$? Why?

130. Suppose $P(x) = x^3 - 4x + 1$ and c is a number such that $c^3 - 4c + 1 = 0$. What does this imply about $x - c$? Why?

In Exercises 131 and 132, use any method to factor the given polynomial completely.

131. $x^3 - 4x^2 + x + 6$

132. $2x^3 - 5x^2 - 14x + 8$

7   Chapter Test

1. Evaluate $-x^2y^3 + 2xy^2$ for $x = -1$ and $y = 2$.

2. If $p(x) = 3x^4 - 4x^3 - 5x^2 + 6x - 7$ and $q(x) = 5x^3 - 8x + 2$, show two ways to evaluate each of the following.

   (a) $(p + q)(1)$

   (b) $(p - q)(-1)$

In Questions 3–8, determine the product.

3. $-2x^3(x^2 - 5x + 3)$ 4. $(2t + s)(3t + 4s)$

5. $(x - 3)(x^2 + 3x - 2)$ 6. $(x^2 + 3y)^2$

7. $(5x - 7z)(5x + 7z)$ 8. $(-3a^2b^5)(2a^3b)$

9. Suppose $p(x) = x^2 + 1$ and $q(x) = x^3 - 1$. Write the polynomial $P(x)$, where $P(x) = p(x) \cdot q(x)$. Produce the graph of $y = P(x)$ and determine the quadrant(s), if any, in which the graph is not located.

10. If $P(x)$ is divided by $D(x) = x + 3$, the quotient $Q(x)$ is $2x^2 - x + 3$, and the remainder R is -10. Determine $P(x)$. Then, write an equation (with no fractions) that relates $P(x)$, $Q(x)$, $D(x)$, and R.

In Questions 11–14, divide by any method to determine the quotient and remainder.

11. $\dfrac{3x^3 - 2x^2 + 5x - 7}{x + 2}$ 12. $\dfrac{15x^6y^4 - 21x^5y^8}{-3xy^2}$

13. $\dfrac{x^3 + 3x^2 - 3x + 1}{x^2 - 1}$ 14. $\dfrac{x^5 - 1}{x + 2}$

15. For the following synthetic division calculation, what is the significance of the number k?

$$\begin{array}{r|rrrr} -2 & a & b & c & d \\ & & e & f & g \\ \hline & a & h & j & k \end{array}$$

16. Show that when the sum of the squares of three consecutive even integers is divided by the second integer, the remainder is 8.

In Questions 17–22, factor completely.

17. $18r^3s^5 - 12r^5s^4$ 18. $xy + 3y - x - 3$

19. $27 - 12x^2$ 20. $4y^2 - 20xy + 25x^2$

21. $1 - 8x^3$ 22. $6x^2 + 11x + 4$

23. Use a graph and the Factor Theorem to factor the polynomial.

   $x^3 - 3x^2 - 10x + 24$

24. The figure shows the graph of $y = p(x)$. Estimate three factors of $p(x)$.

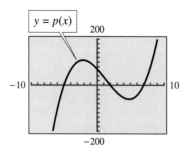

In Questions 25–28, solve.

25. $3x^2 - 4x = 0$

26. $5x^2 + 13x = 6$

27. $2x^3 - x^2 - 8x + 4 = 0$

28. $|2x^2 - 3x - 2| = 3$

29. Verify that $x - 5$ is a factor of $x^3 - 2x^2 - 13x - 10$.

30. A rectangular garden is 3 feet longer than it is wide. If the garden were widened by 3 feet and lengthened by 5 feet, its area would be 10 square feet more than twice the area of the original garden. Find the dimensions of the original garden.

31. A city lot is in the shape of a trapezoid with two right angles. The figure shows the relative lengths of the four sides of the lot. If the area of the lot is 5500 square feet, what are the lengths of the parallel sides? (The area A of a trapezoid is $A = \frac{1}{2}h(b_1 + b_2)$, where h is the height and b_1 and b_2 are the lengths of the parallel sides.)

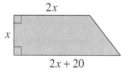

8 Rational Expressions

In professional sports, revenues come from sources other than just ticket sales. The two line graphs compare the trends in average major league baseball players' salaries and total season attendance at games.

If salaries were paid only from ticket sales, what would be the price of a ticket? Using polynomials to model the data in the graphs, we can approximate the answer with a **rational expression.** If you were a player's agent, would you argue that average salaries have not kept pace with attendance figures? (For more on these questions, see Exercises 115–118 at the end of Section 8.1.)

In this chapter we learn how to perform the basic operations with rational expressions, and we develop methods for solving equations and application problems that involve rational expressions.

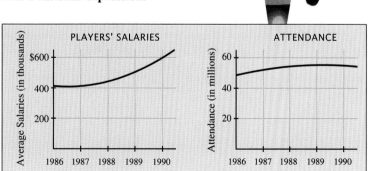

(Source: Major League Baseball Players Association.)

497

8.1   Introduction to Rational Expressions

Basic Definitions ▪ *Simplifying Rational Expressions* ▪ *Equivalent Rational Expressions*

Basic Definitions

Recall that a **rational number** is a number which can be written as p/q, where p and q are integers and $q \neq 0$.

In algebra **rational expressions** play the same role as rational numbers or fractions in arithmetic. In fact, we sometimes refer to rational expressions as algebraic fractions. As we will see, the rules for performing operations with rational expressions are the same as the rules for rational numbers.

> ### Definitions of Rational Expression and Rational Function
>
> A **rational expression** can be written as
>
> $$\frac{P(x)}{Q(x)},$$
>
> where P and Q are polynomials and $Q(x) \neq 0$.
>
> A **rational function** is a function of the form $y = R(x)$, where $R(x)$ is a rational expression.

The following are some examples of rational expressions.

$$\frac{3 + x}{x - 5} \qquad \frac{x^2 - 2x - 3}{3x^3 - 6x^2} \qquad x^3$$

Note that x^3 satisfies the definition of a rational expression because it can be written

$$\frac{x^3}{1}.$$

Recall that the **domain** of a function is the set of all permissible replacements for the variable. In the following exploration, we will consider ways of determining the domain of a rational function.

EXPLORATION 1   *The Domain of a Rational Function*

Consider the rational function

$$r(x) = \frac{3 + x}{x - 5}.$$

(a) Evaluate $r(6)$, $r(0)$, and $r(5)$.

(b) Are 6, 0, and 5 all in the domain of function r? Why?

(c) Propose a method for determining the domain of a rational function.

Discovery

(a) $r(6) = \dfrac{3 + 6}{6 - 5} = 9$; $r(0) = \dfrac{3 + 0}{0 - 5} = -\dfrac{3}{5}$; $r(5) = \dfrac{3 + 5}{5 - 5} = \dfrac{8}{0}$ (undefined)

(b) Replacing x with 5 makes the denominator of the rational expression 0. Therefore, 5 is not in the domain of r.

(c) The domain of a rational function is the set of all real numbers except those for which the denominator is 0.

The results of Exploration 1 suggest a method for determining the domain of a rational function

$$R(x) = \dfrac{P(x)}{Q(x)}.$$

Find all values of x for which $Q(x) = 0$ and exclude these values from the domain. These excluded values are often called **restricted values.**

EXAMPLE 1

Determining the Domain of a Rational Function

Determine the domain of each rational function.

(a) $f(x) = \dfrac{2x}{x + 3}$ (b) $g(x) = \dfrac{x^2 + 5}{2x^2 + 3x}$ (c) $h(x) = \dfrac{x}{x^2 + 1}$

Solution

(a) $f(x) = \dfrac{2x}{x + 3}$

$x + 3 = 0$ Set the denominator equal to 0.

$x = -3$ Solve for x and exclude the result from the domain.

The domain is $\{\, x \mid x \neq -3 \,\}$.

(b) $g(x) = \dfrac{x^2 + 5}{2x^2 + 3x}$

$2x^2 + 3x = 0$ Set the denominator equal to 0.

$x(2x + 3) = 0$ Solve with the factoring method.

$x = 0$ or $2x + 3 = 0$ Zero Factor Property

$x = 0$ or $2x = -3$ Solve each case for x and exclude the results from the domain.

$x = 0$ or $x = -\dfrac{3}{2}$

The domain is $\{x \mid x \neq 0 \text{ and } x \neq -\frac{3}{2}\}$.

(c) $h(x) = \dfrac{x}{x^2 + 1}$

$\qquad x^2 + 1 = 0$ Set the denominator equal to 0.

$\qquad\quad x^2 = -1$ This equation has no real number solution.

There are no values of x to exclude from the domain. Thus, the domain of h is the set of all real numbers.   ∎

Simplifying Rational Expressions

In arithmetic we can sometimes reduce a fraction to lowest terms. Sometimes a rational expression can also be reduced. The following basic property for rational numbers is also used for simplifying rational expressions.

> ### Reduction Property for Fractions
>
> If a, b, and c are real numbers, where $b \neq 0$ and $c \neq 0$, then
>
> $$\frac{ac}{bc} = \frac{a}{b}.$$

If we let a, b, and c represent algebraic expressions (with the same restrictions), then the Reduction Property is also valid for rational expressions.

To apply the Reduction Property, the numerator and denominator must be products, that is, they must be in *factored form*. Then, if the numerator and denominator have a common factor, the property allows us to divide both the numerator and the denominator by that factor. We will refer to this process as "dividing out the common factor."

NOTE: A rational expression is in **lowest terms,** or is **simplified**, if there is no factor other than 1 or -1 that is common to the numerator and denominator.   ∎

EXAMPLE 2   *Simplifying a Rational Expression*

Simplify each of the following rational expressions.

(a) $\dfrac{5x^4 + 10x^3}{15x^2}$ (b) $\dfrac{x^2 - 9}{x^2 - 8x + 15}$

Solution

(a) $\dfrac{5x^4 + 10x^3}{15x^2} = \dfrac{5x^3\,(x + 2)}{15x^2}$ Factor.

$\qquad\qquad\quad = \dfrac{x(x + 2)}{3}$ Divide out the common factors.

(b) $\dfrac{x^2 - 9}{x^2 - 8x + 15} = \dfrac{(x + 3)(x - 3)}{(x - 5)(x - 3)}$ Factor.

$\qquad\qquad\qquad\quad = \dfrac{x + 3}{x - 5}$ Divide out the common factor.   ∎

If we wish to list restricted values, the best time to do that is when the rational expression is in factored form. If we look only at the simplified result in part (b) of Example 2, we might mistakenly identify 5 as the only restricted value. From the factored form of the expression, we can see that 5 and 3 are restricted values.

When simplifying rational expressions, we sometimes need to use the fact that $b - a = -1(a - b)$.

EXAMPLE 3 *Using $a - b = -(b - a)$ to Simplify a Rational Expression*

(a) $\dfrac{y - 4}{4 - y} = \dfrac{1(y - 4)}{-1(y - 4)}$ Factor.

$\qquad\quad = \dfrac{1}{-1}$ Divide out the common factor.

$\qquad\quad = -1$

(b) $\dfrac{t^2 - 25}{5 - t} = \dfrac{(t + 5)(t - 5)}{-1(t - 5)}$ Factor.

$\qquad\quad = \dfrac{t + 5}{-1}$ Divide out the common factor.

$\qquad\quad = -(t + 5)$ or $-t - 5$ ■

Because the main step in simplifying a rational expression is the factoring step, it is necessary to remember the many techniques we have developed for factoring.

EXAMPLE 4 *Simplifying Rational Expressions*

(a) $\dfrac{4x^4 + 6x^3 - 10x^2}{6x^4 - 6x^2} = \dfrac{2x^2(2x^2 + 3x - 5)}{6x^2(x^2 - 1)}$ Factor out the GCF.

$\qquad\quad = \dfrac{2x^2(2x + 5)(x - 1)}{6x^2(x + 1)(x - 1)}$ Factor the polynomials.

$\qquad\quad = \dfrac{2x + 5}{3(x + 1)}$ Divide out the common factors.

(b) $\dfrac{x^3 - 1}{x^2 - 1} = \dfrac{(x - 1)(x^2 + x + 1)}{(x + 1)(x - 1)}$ Difference of two cubes and difference of two squares

$\qquad\quad = \dfrac{x^2 + x + 1}{x + 1}$ Divide out the common factors.

(c) $\dfrac{16 - 8x + x^2}{(x + y)(x - 4) - (x - 4)} = \dfrac{x^2 - 8x + 16}{(x - 4)[(x + y) - 1]}$ GCF is $(x - 4)$.

$\qquad\quad = \dfrac{(x - 4)(x - 4)}{(x - 4)(x + y - 1)}$ Perfect square trinomial

$\qquad\quad = \dfrac{x - 4}{x + y - 1}$ Divide out the common factor.   ■

Equivalent Rational Expressions

We use the rule $\dfrac{ac}{bc} = \dfrac{a}{b}$ to simplify a fraction. In reverse, the rule is $\dfrac{a}{b} = \dfrac{ac}{bc}$.

This means that we can multiply the numerator and denominator by the same number c ($c \neq 0$). The appearance of the fraction is thereby changed, but its value is not; that is, the fractions are **equivalent.** This process is sometimes called *renaming* the fraction.

Being able to write an equivalent rational expression with a specified denominator will be essential to our later discussion about addition and subtraction of rational expressions.

 **EXPLORATION 2**   *Renaming a Fraction*

(a) Suppose that we wish to rename $\frac{2}{5}$ so that its denominator is 15. By what factor must we multiply the original denominator 5 to obtain the new denominator 15?

(b) What other multiplication must we perform?

(c) What is the renamed fraction?

Discovery

(a) We wish to write $\dfrac{2}{5} = \dfrac{?}{15}$.

$$\dfrac{-}{3} \cdot \dfrac{2}{5} = \dfrac{?}{15} \qquad \text{Multiply the original denominator by 3 to obtain 15.}$$

(b) $\dfrac{3}{3} \cdot \dfrac{2}{5} = \dfrac{?}{15} \qquad \text{We must also multiply the numerator by 3.}$

(c) The renamed fraction is $\frac{6}{15}$.

For a given rational expression we can apply a method similar to that used in Exploration 2 to write an equivalent rational expression with a specified denominator.

EXAMPLE 5   *Writing Equivalent Rational Expressions*

Supply the missing numerator.

(a) $\dfrac{2}{3x} = \dfrac{?}{12x^2y^4}$

(b) $\dfrac{a + 2}{a^2 - 2a - 3} = \dfrac{?}{(a + 1)^2(a - 3)}$

Solution

(a) $\dfrac{2}{3x} = \dfrac{?}{12x^2y^4} = \dfrac{?}{3x(4xy^4)}$ Factor the desired denominator.

Because $4xy^4$ is a factor of the desired denominator, but not of the original denominator, we multiply the numerator and denominator of the original expression by $4xy^4$.

$$\dfrac{2(4xy^4)}{3x(4xy^4)} = \dfrac{8xy^4}{12x^2y^4}$$

The missing numerator is $8xy^4$.

(b) $\dfrac{a+2}{a^2-2a-3} = \dfrac{a+2}{(a+1)(a-3)} = \dfrac{?}{(a+1)^2(a-3)}$ Factor the original denominator.

There is one factor of $(a+1)$ in the desired denominator that is not in the original denominator. Therefore, multiply the numerator and denominator of the given expression by $(a+1)$.

$$\dfrac{(a+1)(a+2)}{(a+1)(a+1)(a-3)} = \dfrac{a^2+3a+2}{(a+1)^2(a-3)}$$

The missing numerator is $a^2 + 3a + 2$.

8.1 *Quick Reference*

Basic Definitions

- A **rational expression** can be written as

$$\dfrac{P(x)}{Q(x)}$$

where P and Q are polynomials and $Q(x) \neq 0$.

- A **rational function** is a function of the form $y = R(x)$ where $R(x)$ is a rational expression.

- The **domain** of a rational function is the set of all real numbers that are permissible replacements for the variable.

- To determine the domain of a rational function, set the denominator equal to zero and solve the resulting equation. The solutions are replacements for the variable that are *not* permissible (**restricted values**) and, therefore, must be *excluded* from the domain.

Simplifying Rational Expressions

- If a, b, and c represent algebraic expressions, where $b \neq 0$ and $c \neq 0$, then

$$\dfrac{ac}{bc} = \dfrac{a}{b}.$$

- A rational expression is **simplified** if there is no factor other than 1 or -1 that is common to the numerator and denominator.

- Follow these steps to simplify a rational expression.

  1. Factor the numerator and the denominator completely.

  2. Divide out any factor that is common to both the numerator and the denominator.

Equivalent Rational Expressions

- Given a rational expression, you can write an equivalent rational expression with a specified denominator by following these steps.

  1. Completely factor the denominator of the given expression and of the desired expression.

  2. Identify those factors in the desired denominator that do not appear in the original denominator.

  3. Multiply both the numerator and denominator of the original expression by those factors.

8.1 Exercises

 1. In general, what numbers are excluded from the domain of a rational function?

 2. What is the first step in determining the domain of

$$f(x) = \frac{x - 6}{x^2 - 2x - 15} \, ?$$

In Exercises 3 and 4, determine the domain of the function. Then produce the graph of the function on your calculator. What feature of the graph indicates the domain?

3. $y = \dfrac{10}{x - 5}$

4. $y = \dfrac{x^2 - 4}{x - 2}$

In Exercises 5–8, evaluate the rational expression, if possible, for each of the given values of the variable.

5. $\dfrac{x}{x - 3}$; $-3, 0, 3$

6. $\dfrac{x - 2}{2x + 4}$; $-2, 0, 2$

7. $\dfrac{t - 3}{4 - t}$; $-4, 3, 4$

8. $\dfrac{t - 6}{6 - 2t}$; $-3, 3, 6$

In Exercises 9–20, determine the domain of the given rational function.

9. $y = \dfrac{2x + 3}{x - 4}$

10. $y = \dfrac{x + 3}{x + 3}$

11. $y = \dfrac{x + 9}{3}$

12. $y = \dfrac{x - 2}{2}$

13. $y = \dfrac{x - 4}{(x + 2)^2}$

14. $y = \dfrac{3x - 1}{x^2}$

15. $y = \dfrac{2x - 6}{x^2 - 9}$

16. $y = \dfrac{x + 1}{2x^2 + 6x}$

17. $y = \dfrac{3}{x^2 + 2x - 8}$

18. $y = \dfrac{4x - 3}{5x^2 - 9x - 2}$

19. $y = \dfrac{2x - 1}{x^2 + 4}$

20. $y = \dfrac{1 - 2x}{1 + 2x^2}$

In Exercises 21–26, fill in the blank so that the given number(s) is/are not in the domain of the function.

21. $y = \dfrac{5}{x \underline{\quad} 2}; -2$

22. $y = \dfrac{x}{x \underline{\quad} 3}; 3$

23. $y = \dfrac{4x}{2x \underline{\quad}}; \dfrac{3}{2}$

24. $y = \dfrac{x + 1}{3x \underline{\quad}}; -\dfrac{1}{3}$

25. $y = \dfrac{x}{x^2 \underline{\quad}}; 5, -5$

26. $y = \dfrac{2x}{x^2 \underline{\quad}}; 3, -3$

In the application problems in Exercises 27 and 28, you will need your calculator to answer the questions.

27. Suppose that you are in charge of cleaning up an oil spill from a tanker and that one of your jobs is to estimate the total cost. The model for estimating the total cost C (in millions of dollars) is

$$C(x) = \frac{100}{100 - x}, \ 0 \le x < 100,$$

where x is the percentage of the oil that is cleaned up.

(a) What is the cost of cleaning up 50% of the oil spill?

(b) According to the given model, can you remove 100% of the oil? Why?

(c) Use a suitable range and scale to produce the graph of $C(x)$. Trace to a point where the cost begins to rise rapidly. At that point, what percentage of the oil would be cleaned up and at what cost?

28. Use the x^{-1} key on your calculator to produce the graph of $y = x^{-1}$.

(a) What does the graph suggest about the domain of $y = x^{-1}$?

(b) Use the word *reciprocal* to interpret your answer in part (a).

In Exercises 29 and 30, identify the *terms* in the numerator and denominator.

29. $\dfrac{2x + 3}{x + 3}$

30. $\dfrac{5x - 2}{x - 2}$

In Exercises 31 and 32, identify the *factors* in the numerator and denominator.

31. $\dfrac{2(x + 3)}{7(x + 3)}$

32. $\dfrac{5(x - 2)}{3(x - 2)}$

33. Explain why the first of the following expressions can be reduced but the second cannot.

(i) $\dfrac{2(x + 3)}{7(x + 3)}$ (ii) $\dfrac{2x + 3}{7x + 3}$

34. If $x \ne 3$, then we know that $\dfrac{3(x - 3)}{x - 3} = 3$

is true. Is $\dfrac{3x - 3}{x - 3} = 3$ also true? Why?

In Exercises 35–48, simplify the rational expression.

35. $\dfrac{15x}{120x^2}$

36. $\dfrac{8a^8}{24a^3}$

37. $\dfrac{20c^2 d^5}{24cd^3}$

38. $\dfrac{15a^2 b^7}{-12a^3 b^2}$

39. $\dfrac{-x(2x)^2}{4(-x)^3}$

40. $\dfrac{-3x(2x^2)}{6x(-x^2)}$

41. $\dfrac{6x^3 - 3x^2}{9x^2}$

42. $\dfrac{4a^5 + 8a^7}{12a^4}$

43. $\dfrac{2(3x + 1)}{2 + 6x}$

44. $\dfrac{-3(2x - 1)}{3 - 6x}$

45. $\dfrac{x^2 - 81}{x + 9}$

46. $\dfrac{x + 8}{x^2 - 64}$

47. $\dfrac{a^2 + 5a}{4a - 20}$

48. $\dfrac{3b - 9}{b^2 + 3b}$

In Exercises 49–54, determine whether the correct method has been used to simplify the rational expression.

49. $\dfrac{x + 3}{3} = x$

50. $\dfrac{x + 3}{3} = x + 1$

51. $\dfrac{x-3}{x+3} = -1$

52. $\dfrac{3+x}{x+3} = 1 + 1 = 2$

53. $\dfrac{x-3}{3-x} = -1$

54. $\dfrac{3x+1}{3x+2} = \dfrac{1}{2}$

In Exercises 55–66, simplify the rational expression.

55. $\dfrac{2-x}{2+9x-5x^2}$

56. $\dfrac{x+3}{12+x-x^2}$

57. $\dfrac{x^2-16}{x^2-x-12}$

58. $\dfrac{10x^2-x-2}{6x^2+x-2}$

59. $\dfrac{a^3-a}{6a-6}$

60. $\dfrac{5a+10}{a^3-4a}$

61. $\dfrac{x-3}{2x^2-5x-3}$

62. $\dfrac{2+x}{6-x-2x^2}$

63. $\dfrac{x^2-14x+45}{x^2-17x+72}$

64. $\dfrac{4a^2-13a-12}{a^2-a-12}$

65. $\dfrac{2x^2+6x-1}{2x(x+3)-1}$

66. $\dfrac{a+3}{a(a+6)+9}$

In Exercises 67–72, simplify the rational expression.

67. $\dfrac{ab-3a}{b^2-9}$

68. $\dfrac{5ab+25b}{a^2-25}$

69. $\dfrac{x+3y}{x^2+2xy-3y^2}$

70. $\dfrac{a+4b}{a^2+3ab-4b^2}$

71. $\dfrac{c^2-5cd+6d^2}{2c^2-5cd-3d^2}$

72. $\dfrac{2x^2+13xy+20y^2}{2x^2+17xy+30y^2}$

In Exercises 73–78, determine the unknown numerator or denominator.

73. $\dfrac{2}{a-b} = \dfrac{?}{b-a}$

74. $\dfrac{2x}{1-3x} = \dfrac{?}{3x-1}$

75. $\dfrac{5}{x-3} = \dfrac{?}{3-x}$

76. $\dfrac{2x-3}{3x-2} = \dfrac{?}{2-3x}$

77. $\dfrac{x}{x-3} = -\dfrac{x}{?}$

78. $\dfrac{x+2}{2-x} = -\dfrac{x+2}{?}$

In Exercises 79–82, determine whether the given expressions have a value of 1, −1, or neither.

79. (a) $\dfrac{x+3}{x-3}$ (b) $\dfrac{3+x}{x+3}$ (c) $\dfrac{x-3}{3-x}$

80. (a) $\dfrac{x-y}{-x+y}$ (b) $\dfrac{x+y}{-y+x}$ (c) $-\dfrac{y-x}{x-y}$

81. (a) $\dfrac{-2x+5}{-(5-2x)}$ (b) $\dfrac{2x-5}{-5+2x}$ (c) $\dfrac{2x-5}{2x+5}$

82. (a) $\dfrac{x+y}{y+x}$ (b) $\dfrac{x-y}{y-x}$ (c) $\dfrac{x-y}{x+y}$

In Exercises 83–92, simplify the rational expression.

83. $\dfrac{5x-2}{2-5x}$

84. $\dfrac{x-y}{y-x}$

85. $\dfrac{5x-3}{-15x+9}$

86. $\dfrac{-3x+6}{4x-8}$

87. $\dfrac{a^2-49}{7-a}$

88. $\dfrac{x^2-2x-8}{4-x}$

89. $\dfrac{z-5w}{25w^2-z^2}$

90. $\dfrac{x^2-4}{(2-x)^2}$

91. $\dfrac{-2x^2-3x+20}{2x^2+x-15}$

92. $\dfrac{16-x^2}{2x^2-9x+4}$

In Exercises 93–98, simplify the rational expression.

93. $\dfrac{x^3+3x^2-9x-27}{x^2+6x+9}$

94. $\dfrac{8-27y^3}{2-3y+2x-3xy}$

95. $\dfrac{x^3+8}{x^2-4}$

96. $\dfrac{1+x+x^2}{1-x^3}$

97. $\dfrac{18x^5-39x^4+18x^3}{8x^4-18x^2}$

98. $\dfrac{16 - x^4}{x^3 + 2x^2 + 4x + 8}$

In Exercises 99–110, supply the missing numerator.

99. $\dfrac{3}{2xy^2} = \dfrac{?}{8x^3y^3}$ 100. $\dfrac{-3}{5ab^2} = \dfrac{?}{15a^2b^5}$

101. $\dfrac{5}{x - 5} = \dfrac{?}{5x - 25}$

102. $\dfrac{-5}{x - 6} = \dfrac{?}{7x - 42}$

103. $x + 5 = \dfrac{?}{x + 5}$ 104. $3x = \dfrac{?}{2x^2}$

105. $\dfrac{5}{4x + 12} = \dfrac{?}{4x^2 - 36}$

106. $\dfrac{9}{5x + 5} = \dfrac{?}{5x^2 - 5}$

107. $\dfrac{-3}{x + 1} = \dfrac{?}{x^2 + 5x + 4}$

108. $\dfrac{5}{x + 1} = \dfrac{?}{x^2 - 3x - 4}$

109. $\dfrac{9}{x^2 - 5x - 6} = \dfrac{?}{(x - 6)(x - 5)(x + 1)}$

110. $\dfrac{8}{x^2 - 7x + 6} = \dfrac{?}{(x - 6)(x^2 + 3x - 4)}$

In Exercises 111–114, supply the denominator.

111. $\dfrac{5t}{6} = \dfrac{-10t^2}{?}$ 112. $\dfrac{4}{5x^2} = \dfrac{12x^3y}{?}$

113. $\dfrac{2x - 1}{1 - x} = \dfrac{1 - x - 2x^2}{?}$

114. $\dfrac{x + 2}{2 - 3x} = \dfrac{3x^2 + 4x - 4}{?}$

Exploring with Real Data

In Exercises 115–118, refer to the following background information and data.

In professional sports, revenue comes from television rights, promotions, and other marketing ventures as well as from ticket sales. If the dramatic rise in salaries for professional athletes were paid just from ticket sales, we might wonder at the cost of a ticket.

The following table shows the total attendance (in thousands) at major league baseball games for the regular seasons from 1986 through 1990. Also shown are the average players' salaries (in thousands of dollars) for each season.

Year	Total Attendance (thousands)	Average Salary (thousands)
1986	47,506	$413
1987	52,011	412
1988	52,999	439
1989	55,173	497
1990	54,824	598

(Source: Major League Baseball Players Association.)

The attendance $A(t)$ and the average salary $S(t)$, where t is the number of years since 1985, are approximated by the following functions.

$$A(t) = -643t^2 + 5441t + 43,702$$
$$S(t) = 15.75t^2 - 48.25t + 445.5$$

There were 26 major league teams, each with a roster of 25 players.

115. If the players' salaries were paid only from ticket sales, write a rational expression that would describe how much of the cost of a ticket would be for salaries.

116. Use the rational expression derived in Exercise 115 to determine what the ticket prices would have been in 1986 and in 1990 if all revenue from ticket sales had been used to pay the players' salaries.

117. Calculate the percentage increase in total attendance and in average salaries for the period 1986–1990.

118. If you were a player's agent, would you argue that increases in average salaries have not kept pace with increases in attendance? Why?

Challenge

In Exercises 119–122, simplify the given rational expression.

119. $\dfrac{2x^{-2} - 3x^{-1}}{3x^2 - 2x}$ 120. $\dfrac{3x^{-2} + 5x^{-4}}{3x + 5x^{-1}}$

121. $\dfrac{x^2 + 3x}{3x^{-4} + x^{-3}}$

122. $\dfrac{2x^5 - x^2}{2x - x^{-2}}$

In Exercises 123–126, determine the domain of the rational function where n is a positive integer.

123. $\dfrac{3}{x^n - 1}$

124. $\dfrac{4}{x^n + 1}$

125. $\dfrac{x}{x^{2n} + 1}$

126. $\dfrac{x + 3}{x^{2n} - 1}$

In Exercises 127–130, simplify the given rational expression.

127. $\dfrac{(3x - 1)^{2n}}{(3x - 1)^n}$

128. $\dfrac{5^{n+1} (x^2 + 1)^n}{5^n (x^2 + 1)^{n-1}}$

129. $\dfrac{x^{2n} - 9}{x^n + 3}$

130. $\dfrac{x^{2n} + 3x^n - 4}{x^{n+1} - x}$

131. Evaluate the expression without doing any significant arithmetic. (Hint: Let $x = 5250$.)

$$\frac{5251^2 - 5248^2}{5250(5254) - 5251^2}$$

8.2 Multiplication and Division

Multiplication ▪ *Division* ▪ *Combined Operations*

Multiplication

To multiply fractions, recall that we multiply the numerators, multiply the denominators, and then reduce the answer.

$$\frac{10}{21} \cdot \frac{15}{8} = \frac{150}{168} = \frac{25 \cdot 6}{28 \cdot 6} = \frac{25}{28}$$

An easier method is to reduce first and then multiply the numerators and the denominators.

$$\frac{10}{21} \cdot \frac{15}{8} = \frac{5 \cdot 2}{7 \cdot 3} \cdot \frac{5 \cdot 3}{4 \cdot 2} = \frac{25}{28}$$

In arithmetic, reducing before multiplying is a convenience. For rational expressions it is nearly a necessity.

The definition of multiplication of rational expressions is similar to the definition of multiplication of arithmetic fractions.

Definition of Multiplication of Rational Expressions

If a, b, c, and d represent algebraic expressions, where b and d are not 0, then

$$\frac{a}{b} \cdot \frac{c}{d} = \frac{ac}{bd}.$$

We are guided by our experience with arithmetic in stating the routine for multiplying rational expressions.

> ## To Multiply Rational Expressions
>
> 1. Factor each numerator and each denominator completely.
>
> 2. Divide out any factors that are common to both a numerator and a denominator.
>
> 3. Multiply the numerators and multiply the denominators.

As in arithmetic, we can divide out common factors from the numerator of one fraction and the denominator of another fraction.

EXAMPLE 1   *Multiplying Rational Expressions*

Determine the product of the rational expressions.

(a) $\dfrac{8x^2 - 4x}{2x^2 + 5x - 3} \cdot \dfrac{x^2 - 9}{4x^2}$

(b) $(x^2 - 4xy + 4y^2) \cdot \dfrac{6}{2x^2 - 8y^2}$

(c) $\dfrac{x^2 - y^2}{y - x} \cdot \dfrac{1}{x^3 + y^3}$

Solution

(a) $\dfrac{8x^2 - 4x}{2x^2 + 5x - 3} \cdot \dfrac{x^2 - 9}{4x^2}$

$= \dfrac{4x(2x - 1)}{(2x - 1)(x + 3)} \cdot \dfrac{(x + 3)(x - 3)}{4 \cdot x \cdot x}$ Factor numerator and denominator.

$= \dfrac{x - 3}{x}$ Divide out common factors.

(b) $(x^2 - 4xy + 4y^2) \cdot \dfrac{6}{2x^2 - 8y^2}$

$= \dfrac{(x - 2y)(x - 2y)}{1} \cdot \dfrac{2 \cdot 3}{2(x + 2y)(x - 2y)}$ Factor.

$= \dfrac{3(x - 2y)}{x + 2y}$ Divide out common factors.

(c) $\dfrac{x^2 - y^2}{y - x} \cdot \dfrac{1}{x^3 + y^3}$

$= \dfrac{(x + y)(x - y)}{-1(x - y)} \cdot \dfrac{1}{(x + y)(x^2 - xy + y^2)}$ Factor.

$= \dfrac{1}{-(x^2 - xy + y^2)}$ or $-\dfrac{1}{x^2 - xy + y^2}$ Divide out common factors.

Division

Two rational expressions are **reciprocals** of each other if their product is 1. The following are some examples of rational expressions and their reciprocals.

Rational Expression	*Reciprocal*	
$\dfrac{x}{2x + 3}$	$\dfrac{2x + 3}{x}$	
x^2	$\dfrac{1}{x^2}$	
$\dfrac{1}{x - 1}$	$x - 1$	
$\dfrac{0}{3}$	None	Division by zero is undefined.

In Chapter 1 division was defined as multiplication by the reciprocal of the divisor.

$$10 \div \frac{2}{3} = \frac{10}{1} \cdot \frac{3}{2} = \frac{5 \cdot 2}{1} \cdot \frac{3}{2} = \frac{15}{1} = 15$$

Division of rational expressions is similar to division of fractions in arithmetic.

Definition of Division of Rational Expressions

If a, b, c, and d represent algebraic expressions, where b, c, and d are not 0, then

$$\frac{a}{b} \div \frac{c}{d} = \frac{a}{b} \cdot \frac{d}{c}.$$

EXAMPLE 2 *Division of Rational Expressions*

(a) $\dfrac{3x^2 - 12}{x^2} \div \dfrac{3x + 6}{x} = \dfrac{3x^2 - 12}{x^2} \cdot \dfrac{x}{3x + 6}$ Definition of division

$\qquad\qquad\qquad\qquad = \dfrac{3(x + 2)(x - 2)}{x \cdot x} \cdot \dfrac{x}{3(x + 2)}$ Factor.

$\qquad\qquad\qquad\qquad = \dfrac{x - 2}{x}$ Divide out common factors.

(b) $\dfrac{6x^2 + 3x}{3x - 4} \div (6x^2 - 5x - 4)$

$\qquad = \dfrac{6x^2 + 3x}{3x - 4} \cdot \dfrac{1}{6x^2 - 5x - 4}$ Definition of division

$\qquad = \dfrac{3x(2x + 1)}{3x - 4} \cdot \dfrac{1}{(3x - 4)(2x + 1)}$ Factor.

$\qquad = \dfrac{3x}{(3x - 4)^2}$ Divide out common factors.

(c) $\dfrac{x^3 - x^2 + 2x - 2}{2x^2 - 5x + 3} \div \dfrac{x^4 - 4}{2x^3 + 3x^2 - 4x - 6}$

$= \dfrac{x^3 - x^2 + 2x - 2}{2x^2 - 5x + 3} \cdot \dfrac{2x^3 + 3x^2 - 4x - 6}{x^4 - 4}$ Definition of division

$= \dfrac{x^2(x - 1) + 2(x - 1)}{(2x - 3)(x - 1)} \cdot \dfrac{x^2(2x + 3) - 2(2x + 3)}{(x^2 + 2)(x^2 - 2)}$ Factor.

$= \dfrac{(x - 1)(x^2 + 2)}{(2x - 3)(x - 1)} \cdot \dfrac{(2x + 3)(x^2 - 2)}{(x^2 + 2)(x^2 - 2)}$

$= \dfrac{2x + 3}{2x - 3}$ Divide out common factors.

Combined Operations

Multiplication and division operations can appear in the same expression.

EXAMPLE 3 *Combined Operations*

$\dfrac{y - 3}{y^2 + y - 2} \div (y + 2) \cdot \dfrac{y^2 + 2y - 3}{9 - y^2}$

$= \dfrac{y - 3}{y^2 + y - 2} \cdot \dfrac{1}{y + 2} \cdot \dfrac{y^2 + 2y - 3}{9 - y^2}$ Definition of division

$= \dfrac{-1(3 - y)}{(y + 2)(y - 1)} \cdot \dfrac{1}{y + 2} \cdot \dfrac{(y + 3)(y - 1)}{(3 + y)(3 - y)}$ Factor.

$= \dfrac{-1}{(y + 2)^2}$ Divide out common factors.

A division problem can be written as a fraction whose numerator or denominator (or both) is a rational expression.

EXAMPLE 4 *Division Written as a Fraction*

$\dfrac{\dfrac{x + 5}{5}}{\dfrac{x^2 - 25}{10}} = \dfrac{x + 5}{5} \div \dfrac{x^2 - 25}{10}$ The fraction indicates division.

$= \dfrac{x + 5}{5} \cdot \dfrac{10}{x^2 - 25}$ Definition of division

$= \dfrac{x + 5}{5} \cdot \dfrac{5 \cdot 2}{(x + 5)(x - 5)}$ Factor.

$= \dfrac{2}{x - 5}$ Divide out common factors.

8.2   *Quick Reference*

Multiplication

- If a, b, c, and d represent algebraic expressions, where b and d are not 0, then

$$\frac{a}{b} \cdot \frac{c}{d} = \frac{ac}{bd}.$$

- Follow these steps to multiply rational expressions.

  1. Factor each numerator and denominator completely.

  2. Divide out any factors that are common to both a numerator and a denominator.

  3. Multiply the numerators and multiply the denominators.

Division

- Two rational expressions are **reciprocals** of each other if their product is 1.

- If a, b, c, and d represent algebraic expressions, where b, c, and d are not 0, then

$$\frac{a}{b} \div \frac{c}{d} = \frac{a}{b} \cdot \frac{d}{c}.$$

- Follow these steps to divide rational expressions.

  1. Determine the reciprocal of the divisor.

  2. Change the operation from division to multiplication by the reciprocal.

  3. Follow the steps for multiplication.

Combined Operations

- In expressions involving both multiplication and division of rational expressions, follow the Order of Operations.

- Division of rational expressions can be indicated with a fractional expression.

$$\frac{\dfrac{a}{b}}{\dfrac{c}{d}} = \frac{a}{b} \div \frac{c}{d}$$

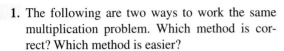

8.2 Exercises

1. The following are two ways to work the same multiplication problem. Which method is correct? Which method is easier?

(i) $\dfrac{x + 3}{x - 2} \cdot \dfrac{x - 2}{x + 1} = \dfrac{x^2 + x - 6}{x^2 - x - 2}$

$= \dfrac{(x + 3)(x - 2)}{(x + 1)(x - 2)}$

$= \dfrac{x + 3}{x + 1}$

(ii) $\dfrac{x + 3}{x - 2} \cdot \dfrac{x - 2}{x + 1} = \dfrac{x + 3}{x + 1}$

2. Consider the following multiplication problem.

$\dfrac{x + 1}{x^2 - 2x - 15} \cdot \dfrac{x^2 - 5x}{x^2 - 1}$

$= \dfrac{(x + 1)(x^2 - 5x)}{(x^2 - 2x - 15)(x^2 - 1)}$

$= \dfrac{x^3 - 4x^2 - 5x}{x^4 - 2x^3 - 16x^2 + 2x + 15}$

The next step is to simplify the result, but the numerator and denominator will be difficult to factor. How could the problem have been done more easily?

In Exercises 3–14, multiply and simplify.

3. $\dfrac{5a^2 b}{b^3 c^4} \cdot \dfrac{b^2 c^4}{25a^3}$

4. $\dfrac{4x^3 y^2}{z^3} \cdot \dfrac{y^3 z^4}{2x^5}$

5. $(a + 4) \cdot \dfrac{a - 4}{3a + 12}$

6. $\dfrac{2x + 1}{x - 5} \cdot (15 - 3x)$

7. $\dfrac{5x - 7}{12 - 3x} \cdot \dfrac{3x - 12}{7 - 5x}$

8. $\dfrac{5p - 25}{2p^2} \cdot \dfrac{6p}{p - 5}$

9. $\dfrac{4c^2}{5c + 30} \cdot \dfrac{c^2 - 36}{16c^3}$

10. $\dfrac{2v}{v + 7} \cdot \dfrac{v^2 + 11v + 28}{8v^2}$

11. $\dfrac{7w^2 - 14w}{w^2 + 3w - 10} \cdot \dfrac{w + 5}{21w}$

12. $\dfrac{z^2 - 11z + 18}{6z^3} \cdot \dfrac{z^2 + 2z}{z^2 - 4}$

13. $\dfrac{y^2 - 1}{y^2 + 14y + 48} \cdot \dfrac{y^2 + 5y - 6}{(y - 1)^2}$

14. $\dfrac{z^2 + 5z - 14}{z^2 - 8z + 12} \cdot \dfrac{z^2 + z - 42}{z^2 + 9z + 14}$

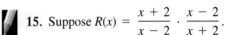 15. Suppose $R(x) = \dfrac{x + 2}{x - 2} \cdot \dfrac{x - 2}{x + 2}.$

Because $\dfrac{x + 2}{x - 2} \cdot \dfrac{x - 2}{x + 2} = 1,$

would you expect the graph of $y = R(x)$ to be the horizontal line $y = 1$? Produce the graph of $y = R(x)$ and trace to the points whose x-coordinates are 2 and -2. In each case, what is the corresponding y-coordinate? Why?

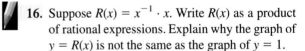 16. Suppose $R(x) = x^{-1} \cdot x$. Write $R(x)$ as a product of rational expressions. Explain why the graph of $y = R(x)$ is not the same as the graph of $y = 1$.

In Exercises 17–20, fill in the blank to make the resulting equation true.

17. $\dfrac{x - 3}{3x - 21} \cdot \dfrac{7 - x}{?} = \dfrac{1}{3}$

18. $\dfrac{?}{x + 1} \cdot \dfrac{x + 1}{x - 2} = x + 2$

19. $\dfrac{3x^2 + 8x - 3}{2x + 1} \cdot \dfrac{?}{3x^2 - 4x + 1} = x + 3$

20. $\dfrac{x^2 - 16}{?} \cdot \dfrac{x^2 + 3x - 10}{x^2 - 6x + 8} = \dfrac{x + 4}{x + 2}$

21. What is the difference between the *opposite* of $x + 3$ and the *reciprocal* of $x + 3$?

22. For $R(x) = \dfrac{5}{x - 1} \div \dfrac{x - 2}{3}$,

why is 2 excluded from the domain of R?

In Exercises 23–34, divide and simplify.

23. $\dfrac{24x^3y^7}{16x^4y^3} \div \dfrac{9y^6x}{x^5}$

24. $\dfrac{25x^5y^4}{15y} \div \dfrac{x^3y^3}{3x}$

25. $\dfrac{2x + 6}{5x^2} \div \dfrac{x + 3}{15x}$

26. $\dfrac{(x + 1)^2}{8x + 8} \div \dfrac{7x^2 + 7x}{8x^2}$

27. $\dfrac{5x - x^2}{x + 2} \div \dfrac{x^2 - 6x + 5}{x^2 - 3x + 2}$

28. $\dfrac{x^2 - 64}{x^2 - 9} \div \dfrac{6x^2 + 48x}{2x - 6}$

29. $\dfrac{x^2 + 3x - 4}{8x} \div \dfrac{x^2 + 10x + 24}{x + 6}$

30. $\dfrac{3x - 12}{x^2 - 9x + 20} \div \dfrac{9x + 54}{x^2 + x - 30}$

31. $\dfrac{x^2 - 3x - 10}{x^2 - 25} \div \dfrac{x^2 - 4}{x^2 - x - 30}$

32. $\dfrac{x^2 - 7x - 8}{x^2 - 2x - 48} \div \dfrac{6x^2 + 6x}{x^2 + 7x + 6}$

33. $\dfrac{x^2 - 16}{2x^2 - 8x} \div (3x^2 + 10x - 8)$

34. $\dfrac{x^2 + 12x + 36}{x^2 + 3x - 18} \div (x^2 + 9x + 18)$

In Exercises 35–38, fill in the blank to make the resulting equation true.

35. $\dfrac{?}{5} \div \dfrac{3x - 4}{10} = 4x - 2$

36. $\dfrac{x^2 + 7x}{x - x^2} \div \dfrac{?}{7x - 7} = -1$

37. $\dfrac{2x^2 + 4x}{x^2 + 2x + 4} \div \dfrac{x^2 - 4}{?} = 2x$

38. $\dfrac{x^3 - 5x^2}{?} \div \dfrac{x^2 - 4x - 5}{x^2 + 2x} = \dfrac{x + 2}{x^2 + x}$

39. If A, B, and C are polynomials, write a rational expression for the quotient of A/B and C.

40. If A, B, and C are polynomials, write a rational expression for the quotient of A and B/C.

In Exercises 41–58, perform the indicated operation and simplify.

41. $\dfrac{32x^5}{3y^4} \cdot \dfrac{5y^3}{-8x^2}$

42. $\dfrac{6x^3}{7y^4} \div \dfrac{36x^7}{49y^5}$

43. $-2x \cdot \dfrac{x + 3}{4x - x^2}$

44. $\dfrac{x - 3x^2}{3 - x} \div -3x^2$

45. $\dfrac{x^2 - 4}{x^2 - 1} \div \dfrac{12x^2 - 24x}{4x + 4}$

46. $\dfrac{b^2 - 16}{b^2 - 9} \cdot \dfrac{4b + 12}{8b^2 + 32b}$

47. $\dfrac{4x - 8}{x^2 - 5x + 6} \div \dfrac{8x + 48}{x^2 + 3x - 18}$

48. $\dfrac{b^2 - 7b + 10}{2b - 10} \cdot \dfrac{4b^2 + 8b}{b^2 - 4}$

49. $\dfrac{a^2 - 81}{a^2 - 5a - 36} \cdot \dfrac{a^2 + 10a + 24}{a^2 - 3a - 54}$

50. $\dfrac{x^2 - 5x - 24}{x^2 - 3x - 18} \div \dfrac{x^2 - 64}{3x - 18}$

51. $\dfrac{x^2 - 3x - 4}{x^2 - 10x + 24} \div \dfrac{x^2 + 5x + 4}{x^2 - 5x - 6}$

52. $\dfrac{t^2 + t - 20}{t^2 + 2t - 24} \cdot \dfrac{t^2 + 11t + 30}{t^2 + 9t + 20}$

53. $\dfrac{3x^3 + 6x^2}{2x^2 + x - 6} \div (6x^2 - 15x)$

54. $(w^2 - w - 12) \cdot \dfrac{9w^2}{3w^4 - 27w^2}$

55. $\dfrac{xy + 3x - 2y - 6}{x^2 - 6x + 8} \cdot \dfrac{xy - 4y + 2x - 8}{y^2 + 5y + 6}$

56. $\dfrac{ab - ad + bc - cd}{b^2 + bd - 2d^2} \div \dfrac{a^2 - ac - 2c^2}{ab + 2ad - bc - 2cd}$

57. $\dfrac{x^3 - 27}{x^2 - 9} \div \dfrac{x^2 + 3x + 9}{x^2 + 8x + 15}$

58. $\dfrac{x^3 + y^3}{x^2 - y^2} \cdot \dfrac{x^2 - 3xy + 2y^2}{x - 2y}$

59. In baseball, a pitcher's earned run average (ERA) is determined by the formula ERA $= R/(I/9)$, where R is the total number of earned runs surrendered and I is the total number of innings pitched. Write an alternate formula for ERA that does not contain a fraction in the numerator or denominator.

60. One number is 3 less than another number. If the product of their reciprocals is given by a rational expression whose denominator is $x^3 - x^2 - 6x$, what is the numerator of the expression?

In Exercises 61–68, perform the indicated operations.

61. Divide $\dfrac{x - 3}{8x}$ into $\dfrac{3x - 9}{2x^2}$.

62. Divide $\dfrac{6x^2 - 24x}{4x^2}$ into $\dfrac{(x - 4)^2}{4x - 16}$.

63. Divide $\dfrac{x^2 - 25}{x^2 - 3x - 10}$ by $\dfrac{x^2 + x - 30}{x^2 + 8x + 12}$.

64. Divide $\dfrac{x^2 - 9}{x^2 - 11x + 24}$ by $\dfrac{x^2 + 3x - 18}{x^2 - 2x - 48}$.

65. Multiply $\dfrac{3m + 6}{5m^2}$ by $\dfrac{25m}{m + 2}$.

66. Multiply $\dfrac{6x^2}{6x - 18}$ by $\dfrac{x^2 - 9}{5x^3}$.

67. Determine the product of
$$\dfrac{2p + 16}{p^2 + 3p - 40} \text{ and } \dfrac{p^2 - 11p + 30}{6p - 36}.$$

68. Determine the product of
$$\dfrac{x^2 - 9}{x^2 + 2x - 24} \text{ and } \dfrac{x^2 + 9x + 18}{(x + 3)^2}.$$

In Exercises 69–78, perform the indicated operations.

69. $\dfrac{6x - 18}{3x - 6} \cdot \dfrac{5 + 5x}{25x + 25} \div \dfrac{12 - 4x}{9x - 18}$

70. $\dfrac{x^2 - 2x}{4y} \cdot \dfrac{28xy^2}{3x - 6} \div \dfrac{2x^2 + 2x}{21x^2y^2}$

71. $\dfrac{2x + 6}{x - 5} \div \left[\dfrac{x}{4x + 12} \cdot \dfrac{2(x + 3)^2}{x^2 - 5x} \right]$

72. $(2x^2 + 3x - 5) \div \left(\dfrac{6x + 15}{x - 2} \div \dfrac{3}{x - 2} \right)$

73. $\dfrac{x^3 - 1}{x + 1} \div \dfrac{x^2 + 2x + 1}{x^2 + x + 1} \div \dfrac{x + 1}{3x^2}$

74. $\dfrac{x^3 - 8}{x^2 - 4} \div \dfrac{x^2 + 2x + 4}{x^3 - 3x^2} \cdot \dfrac{x^2 + 6x + 8}{x^2 + x - 12}$

75. $\dfrac{2x^2 + x - 6}{x^3 - 3x^2} \div \dfrac{xy + 3x + 2y + 6}{15x^3 - 10x^2} \cdot$
$$\dfrac{xy + 3x - 3y - 9}{6x^2 - 13x + 6}$$

76. $\dfrac{x^3 + 2x^2 - x - 2}{2x + 4} \div \dfrac{2x^2 - 4x - 6}{x - 3} \cdot$
$$\dfrac{x^3 + 4x^2}{x^2 + 3x - 4}$$

77. $\dfrac{x^2 + x - 2}{x^2 - 1} \div (x + 2) \cdot \dfrac{x^2 - 9}{x^2 + 3x}$

78. $\dfrac{x^2 + x - 6}{x^2 - 16} \cdot \dfrac{x^2 - 4x}{x^2 - 4} \div (x^2 + 3x)$

 79. Explain how to determine the domain of a rational expression of the form
$$\dfrac{P/Q}{R/S},$$
where P, Q, R, and S are polynomials in x.

80. Suppose that
$$R(x) = \dfrac{\dfrac{x^2}{x + 1}}{\dfrac{x - 3}{2 - x}}.$$
Explain why $R(3)$ is undefined.

In Exercises 81–86, divide and simplify.

81. $\dfrac{\dfrac{4x^2}{6x + 54}}{\dfrac{32x^3}{x^2 - 81}}$

82. $\dfrac{\dfrac{6x}{x - 5}}{\dfrac{12x^2}{x^2 + 3x - 40}}$

83. $\dfrac{\dfrac{8x}{x+5}}{\dfrac{16x^2}{x^2+x-20}}$

84. $\dfrac{\dfrac{7x-35}{4x^3}}{\dfrac{x^2-25}{4x^2+20x}}$

85. $\dfrac{\dfrac{5x}{x^2-4}}{\dfrac{35x^2}{x^2-x-6}}$

86. $\dfrac{\dfrac{2x}{x^2-4}}{\dfrac{8x^2}{x^2-7x-18}}$

Exploring with Real Data

In Exercises 87–90, refer to the following background information and data.

The following table shows the number T (in thousands) of foreign travelers to the United States from 1986 to 1991. The table also shows the total expenditures E (in millions of dollars) by these travelers during the same period.

Year	Number of Travelers (thousands)	Total Expenditures (millions)
1986	26,158	$20,273
1987	29,489	23,366
1988	34,238	28,935
1989	36,605	35,173
1990	39,089	40,579
1991	42,114	45,551

(Source: U.S. Bureau of Economic Analysis.)

The number of travelers T can be approximated by the function $T(t) = 23{,}520.8 + 3169.9t$, where t is the number of years since 1985.

The total expenditures E can be approximated by the function $E(t) = 13{,}682.9 + 5286.5t$, where t is again the number of years since 1985.

The average amount that each traveler spent while in the United States can be modeled by the rational expression

$$\frac{E(t) \cdot 1000}{T(t)}.$$

87. Use the actual data from the table to determine the average expenditure in the year 1991.

88. Use the model for average expenditures to estimate the average expenditure for the year 1991.

89. Calculate the difference between the actual average expenditure in Exercise 87 and the estimated average expenditure in Exercise 88.

90. If the model is reasonably accurate, the difference found in Exercise 89 should be relatively small. However, multiply this difference by the actual total number of travelers in 1991. What is the resulting discrepancy in total expenditures?

Challenge

In Exercises 91–94, answer the question by determining the unknown quantity.

91. What rational expression multiplied by

$\dfrac{x+3}{x-5}$ is $\dfrac{x^2+x-6}{x^2-2x-15}$?

92. The product of what rational expression and $x^2 + x - 1$ is 1?

93. What rational expression divided by

$\dfrac{2x^2+x-15}{2x^2-3x-5}$ is $\dfrac{2x-3}{x+3}$?

94. What rational expression divided into

$\dfrac{x+4}{x^2-1}$ is $\dfrac{1}{x^2-5x+4}$?

In Exercises 95–98, perform the indicated operations.

95. $\dfrac{x^2(2x-1) - 10x(1-2x) + 21(2x-1)}{2x^2+13x-7} \div \dfrac{x}{x+3}$

96. $\dfrac{x^2(x+4) + x(x+4) - 6(x+4)}{2x^2+9x+4} \cdot \dfrac{3}{6+2x}$

97. $\dfrac{2x^{2n}+6x^n+4}{x^{2m}-4x^m+3} \cdot \dfrac{x^{2m}-x^m-6}{4x^{2n}-4x^n-8}$

98. $\dfrac{x^{2n}-4x^n-12}{x^{2m}-4x^m-5} \div \dfrac{x^{2n}-3x^n-18}{x^{2m}-7x^m+10}$

For Exercises 99 and 100, produce the graph of $y = 8/x$ and observe the behavior of the function in Quadrant I.

99. For x-values closer and closer to 0, what is true about the corresponding y-values? Describe how you can select x-values to make the y-values as large as you want.

100. For larger and larger x-values, what is true about the corresponding y-values? Describe how you can select x-values to make the y-values as close to 0 as you want.

8.3 Addition and Subtraction

Like Denominators ▪ *Least Common Multiple* ▪ *Unlike Denominators* ▪ *Combined Operations* ▪ *Expressions with Negative Exponents*

Like Denominators

Unlike multiplication and division, addition and subtraction of fractions requires that the fractions have a common denominator. Recall that to add fractions with the same denominator, we simply add the numerators, retain the common denominator, and reduce the result to lowest terms.

$$\frac{2}{9} + \frac{4}{9} = \frac{6}{9} = \frac{2 \cdot 3}{3 \cdot 3} = \frac{2}{3}$$

The procedure for subtracting fractions with like denominators is similar. We use the same principle for addition and subtraction of rational expressions with the same denominator.

> ### Definition of Addition and Subtraction of Rational Expressions
>
> If a, b, and c are algebraic expressions, where $c \neq 0$, then
>
> $$\frac{a}{c} + \frac{b}{c} = \frac{a + b}{c} \quad \text{and} \quad \frac{a}{c} - \frac{b}{c} = \frac{a - b}{c}.$$

The following are the steps for adding or subtracting rational expressions with like denominators.

> ### To Add or Subtract Rational Expressions with Like Denominators
>
> 1. Add (or subtract) the numerators.
>
> 2. Retain the common denominator.
>
> 3. Simplify the result.

When we add or subtract rational expressions, each numerator must be treated as if it were enclosed in parentheses. For a subtraction problem, in particular, we will insert parentheses around each numerator. Then, when we subtract the numerators, we will remove the parentheses to simplify.

EXAMPLE 1   *Combining Rational Expressions with Like Denominators*

Perform the indicated operations.

(a) $\dfrac{3}{r^2 s} + \dfrac{2}{r^2 s}$ (b) $\dfrac{3x + 1}{x + 3} - \dfrac{3x + 2}{x + 3}$

(c) $\dfrac{x - 2}{x - 1} - \dfrac{4x - 5}{x - 1}$ (d) $\dfrac{6x}{x^2 - x - 12} - \dfrac{4x + 7}{x^2 - x - 12} + \dfrac{x - 5}{x^2 - x - 12}$

Solution

(a) $\dfrac{3}{r^2 s} + \dfrac{2}{r^2 s} = \dfrac{5}{r^2 s}$

(b) $\dfrac{3x + 1}{x + 3} - \dfrac{3x + 2}{x + 3}$

$\quad = \dfrac{(3x + 1) - (3x + 2)}{x + 3}$ Insert parentheses.

$\quad = \dfrac{3x + 1 - 3x - 2}{x + 3}$ Subtract by removing parentheses.

$\quad = \dfrac{-1}{x + 3} \quad \text{or} \quad -\dfrac{1}{x + 3}$ Combine like terms.

(c) $\dfrac{x - 2}{x - 1} - \dfrac{4x - 5}{x - 1}$

$\quad = \dfrac{(x - 2) - (4x - 5)}{x - 1}$ Insert parentheses.

$\quad = \dfrac{x - 2 - 4x + 5}{x - 1}$ Subtract by removing parentheses.

$\quad = \dfrac{-3x + 3}{x - 1}$ Combine like terms.

$\quad = \dfrac{-3(x - 1)}{x - 1}$ Factor out the GCF −3.

$\quad = -3$ Divide out the common factor.

(d) $\dfrac{6x}{x^2 - x - 12} - \dfrac{4x + 7}{x^2 - x - 12} + \dfrac{x - 5}{x^2 - x - 12}$

$= \dfrac{6x - (4x + 7) + (x - 5)}{x^2 - x - 12}$ ⠀⠀⠀Insert parentheses.

$= \dfrac{6x - 4x - 7 + x - 5}{x^2 - x - 12}$ ⠀⠀⠀Simplify by removing parentheses.

$= \dfrac{3x - 12}{x^2 - x - 12}$ ⠀⠀⠀Combine like terms.

$= \dfrac{3(x - 4)}{(x - 4)(x + 3)}$ ⠀⠀⠀Factor.

$= \dfrac{3}{x + 3}$ ⠀⠀⠀Divide out the common factor. ■

Least Common Multiple

In arithmetic, if we wish to add or subtract fractions with unlike denominators, we must first rewrite each fraction with a common denominator.

 **EXPLORATION 1**⠀⠀*Determining a Least Common Multiple*

Suppose that we wish to add $\frac{1}{72} + \frac{1}{48}$. We must first find the least common denominator (LCD). This number is called the **least common multiple (LCM)** of 72 and 48.

(a) Completely factor 72 and 48 and write the results in exponential form.

(b) For each base obtained in part (a), write the factor with the largest exponent that appears in either product.

(c) To obtain the LCM, write the product of the factors in part (b). What is the LCM?

Discovery

(a) $72 = 2^3 \cdot 3^2$; $48 = 2^4 \cdot 3$

(b) For the base 2, the largest exponent is 4: 2^4.
For the base 3, the largest exponent is 2: 3^2.

(c) LCM $= 2^4 \cdot 3^2 = 16 \cdot 9 = 144$

For polynomial denominators, we determine the LCM in a similar way. The following describes the routine.

> ## To Determine the LCM of Two or More Polynomials
>
> 1. Factor each polynomial completely. Write the result in exponential form.
>
> 2. List every base that appears in at least one polynomial. For each base, use the largest exponent that appears in any of the polynomials.
>
> 3. The LCM is the product of the factors in the list.

EXAMPLE 2   *Determining the LCM of Polynomials*

Determine the LCM of the given polynomials.

(a) $3xy^4$, $4x^3y$, $6x^5y^2$ (b) y, $y - 2$

(c) $3x^2 - 3x$, $6x - 6$ (d) $2x^2 - x - 3$, $4x^2 - 4x - 3$

(e) $t^2 + 4t + 4$, $t^2 - t - 6$

Solution

(a) $3xy^4 = 3xy^4$ Factor completely.

$4x^3y = 2^2x^3y$ The bases that appear in at least one expression are 2, 3, x, and y.

$6x^5y^2 = 2 \cdot 3x^5y^2$

$\text{LCM} = 2^2 \cdot 3x^5y^4 = 12x^5y^4$ For each base, use the largest exponent that appears in any expression.

(b) $\text{LCM} = y(y - 2)$

(c) $3x^2 - 3x = 3x(x - 1)$ Factor completely.

$6x - 6 = 6(x - 1) = 2 \cdot 3(x - 1)$ The bases that appear in at least one expression are 2, 3, x, and $x - 1$.

$\text{LCM} = 2 \cdot 3x(x - 1) = 6x(x - 1)$

(d) $2x^2 - x - 3 = (2x - 3)(x + 1)$ Factor completely.

$4x^2 - 4x - 3 = (2x + 1)(2x - 3)$

$\text{LCM} = (2x - 3)(x + 1)(2x + 1)$

(e) $t^2 + 4t + 4 = (t + 2)^2$ Factor completely.

$t^2 - t - 6 = (t - 3)(t + 2)$ The bases that appear in at least one expression are $t + 2$ and $t - 3$.

$\text{LCM} = (t + 2)^2(t - 3)$ For the base $t + 2$, the largest exponent that appears in either expression is 2. ∎

Unlike Denominators

If the rational expressions to be added or subtracted have unlike denominators, then the first task is to write equivalent rational expressions with the same denominator.

The routine begins with determining the LCD, that is, the LCM of the denominators. Then we rename the fractions so that each one has the LCD as its denominator. Finally, we perform the addition or subtraction.

EXAMPLE 3 *Combining Rational Expressions with Unlike Denominators*

Perform the indicated operations.

(a) $\dfrac{2}{r^2 s} - \dfrac{3}{rs^2}$

(b) $\dfrac{x-1}{6} + \dfrac{2x+1}{9}$

(c) $\dfrac{1}{x-3} - \dfrac{2}{3-x}$

(d) $x + \dfrac{3}{x+2}$

Solution

(a) $\dfrac{2}{r^2 s} - \dfrac{3}{rs^2}$ The LCD is $r^2 s^2$.

$= \dfrac{2 \cdot s}{r^2 s \cdot s} - \dfrac{3 \cdot r}{rs^2 \cdot r}$ Write equivalent fractions.

$= \dfrac{2s}{r^2 s^2} - \dfrac{3r}{r^2 s^2}$

$= \dfrac{2s - 3r}{r^2 s^2}$ Combine the numerators.

(b) $\dfrac{x-1}{6} + \dfrac{2x+1}{9}$ The LCD is 18.

$= \dfrac{(x-1) \cdot 3}{6 \cdot 3} + \dfrac{(2x+1) \cdot 2}{9 \cdot 2}$ Write equivalent fractions.

$= \dfrac{3x-3}{18} + \dfrac{4x+2}{18}$

$= \dfrac{3x - 3 + 4x + 2}{18}$ Combine the numerators.

$= \dfrac{7x-1}{18}$ Combine like terms.

(c) $\dfrac{1}{x-3} - \dfrac{2}{3-x}$

$= \dfrac{1}{x-3} - \dfrac{(-1) \cdot 2}{(-1) \cdot (3-x)}$ $-1(a-b) = b-a$

$= \dfrac{1}{x-3} - \dfrac{-2}{x-3}$ Now the denominators are the same.

$= \dfrac{1-(-2)}{x-3}$ Combine the numerators.

$= \dfrac{3}{x-3}$

(d) $x + \dfrac{3}{x+2} = \dfrac{x}{1} + \dfrac{3}{x+2}$ The LCD is $x + 2$.

$\qquad = \dfrac{x(x+2)}{1(x+2)} + \dfrac{3}{x+2}$ Write equivalent fractions.

$\qquad = \dfrac{x^2 + 2x}{x+2} + \dfrac{3}{x+2}$

$\qquad = \dfrac{x^2 + 2x + 3}{x+2}$ Combine the numerators.

In Example 3 it was not necessary to factor any of the denominators in order to determine the LCD. In this next example, both of the problems require factoring.

EXAMPLE 4   *Factoring to Determine the LCD*

(a) $\dfrac{3x-2}{3x^2 + 3x} - \dfrac{x}{x^2 + 2x + 1}$ Factor the denominators.

$\qquad = \dfrac{3x-2}{3x(x+1)} - \dfrac{x}{(x+1)^2}$ The LCD is $3x(x+1)^2$.

$\qquad = \dfrac{(3x-2)(x+1)}{3x(x+1)(x+1)} - \dfrac{x \cdot 3x}{3x(x+1)^2}$ Write equivalent fractions.

$\qquad = \dfrac{3x^2 + x - 2}{3x(x+1)^2} - \dfrac{3x^2}{3x(x+1)^2}$ Simplify the numerators.

$\qquad = \dfrac{3x^2 + x - 2 - 3x^2}{3x(x+1)^2}$ Combine the numerators.

$\qquad = \dfrac{x-2}{3x(x+1)^2}$ Combine like terms.

(b) $\dfrac{x+4}{x^2 + x - 6} - \dfrac{2x+1}{x^2 - 6x + 8}$ Factor the denominators.

$\qquad = \dfrac{x+4}{(x+3)(x-2)} - \dfrac{2x+1}{(x-4)(x-2)}$ LCD is $(x+3)(x-2)(x-4)$.

$\qquad = \dfrac{(x+4)(x-4)}{(x+3)(x-2)(x-4)} - \dfrac{(2x+1)(x+3)}{(x-4)(x-2)(x+3)}$ Write equivalent fractions.

$\qquad = \dfrac{x^2 - 16}{(x-4)(x-2)(x+3)} - \dfrac{2x^2 + 7x + 3}{(x-4)(x-2)(x+3)}$ Simplify the numerators.

$\qquad = \dfrac{(x^2 - 16) - (2x^2 + 7x + 3)}{(x-4)(x-2)(x+3)}$ Combine the numerators.

$\qquad = \dfrac{x^2 - 16 - 2x^2 - 7x - 3}{(x-4)(x-2)(x+3)}$ Remove parentheses.

$\qquad = \dfrac{-x^2 - 7x - 19}{(x-4)(x-2)(x+3)}$ Combine like terms.

Combined Operations

We may find it necessary to perform both addition and subtraction in the same problem.

EXAMPLE 5 *Combined Operations with Rational Expressions*

$$\frac{-2x}{x+3} + \frac{3}{3-x} - \frac{8x-12}{x^2-9}$$

$$= \frac{-2x}{x+3} + \frac{-1(3)}{-1(3-x)} - \frac{8x-12}{(x+3)(x-3)} \qquad -1(a-b) = b-a$$

$$= \frac{-2x}{x+3} + \frac{-3}{x-3} - \frac{8x-12}{(x+3)(x-3)} \qquad \text{LCD is } (x+3)(x-3).$$

$$= \frac{-2x(x-3)}{(x+3)(x-3)} + \frac{-3(x+3)}{(x+3)(x-3)} - \frac{8x-12}{(x+3)(x-3)} \qquad \begin{array}{l}\text{Write}\\\text{equivalent}\\\text{fractions.}\end{array}$$

$$= \frac{-2x^2+6x}{(x+3)(x-3)} + \frac{-3x-9}{(x+3)(x-3)} - \frac{8x-12}{(x+3)(x-3)} \qquad \begin{array}{l}\text{Simplify the}\\\text{numerators.}\end{array}$$

$$= \frac{(-2x^2+6x) + (-3x-9) - (8x-12)}{(x+3)(x-3)} \qquad \text{Combine the numerators.}$$

$$= \frac{-2x^2+6x-3x-9-8x+12}{(x+3)(x-3)} \qquad \text{Remove parentheses.}$$

$$= \frac{-2x^2-5x+3}{(x+3)(x-3)} \qquad \text{Combine like terms.}$$

$$= \frac{-1(2x^2+5x-3)}{(x+3)(x-3)} \qquad \text{Factor the numerator.}$$

$$= \frac{-1(2x-1)(x+3)}{(x+3)(x-3)}$$

$$= \frac{-(2x-1)}{x-3} \quad \text{or} \quad -\frac{2x-1}{x-3} \qquad \begin{array}{l}\text{Divide out the common}\\\text{factor.}\end{array} \qquad ■$$

Expressions with Negative Exponents

Expressions with negative exponents sometimes lead to addition or subtraction of rational expressions.

EXAMPLE 6   *Expressions with Negative Exponents*

Add $x^{-1} + y^{-1}$. Express the result with positive exponents.

Solution

$$x^{-1} + y^{-1} = \frac{1}{x} + \frac{1}{y} \qquad \text{Definition of negative exponent}$$

$$= \frac{y}{xy} + \frac{x}{xy} \qquad \text{The LCD is } xy.$$

$$= \frac{x + y}{xy} \qquad \text{Combine the numerators.}$$

NOTE: In the expression $x^{-1} + y^{-1}$, the definition of negative exponent applies to each term:

$$x^{-1} + y^{-1} = \frac{1}{x} + \frac{1}{y}.$$

Compare this with

$$(x + y)^{-1} = \frac{1}{x + y}.$$

8.3 Quick Reference

Like Denominators

■ If a, b, and c are algebraic expressions, where $c \neq 0$, then

$$\frac{a}{c} + \frac{b}{c} = \frac{a + b}{c} \quad \text{and} \quad \frac{a}{c} - \frac{b}{c} = \frac{a - b}{c}.$$

■ To add or subtract rational expressions with like denominators, follow these steps.

1. Add or subtract the numerators.

2. Retain the common denominator.

3. Simplify the result.

■ When subtracting rational expressions, be sure to treat each numerator as a quantity in parentheses.

Least Common Multiple

■ To determine the **least common multiple (LCM)** of two or more polynomials, follow these steps.

1. Factor each polynomial completely. Write the result in exponential form.

2. List every base that appears in at least one polynomial. For each base, use the largest exponent that appears in any of the polynomials.

3. The LCM is the product of the factors in the list.

Unlike Denominators

- Follow these steps when adding or subtracting rational expressions with unlike denominators.

    1. Determine the least common denominator (LCD), which is the LCM of the denominators.

    2. Rewrite each rational expression as an equivalent expression with the LCD as its denominator.

    3. Combine and simplify the numerators.

    4. Simplify the result.

Combined Operations

- The procedure for adding and subtracting rational expressions can be extended to addition and subtraction involving three or more rational expressions.

Expressions with Negative Exponents

- Using the definition of a negative exponent, we can write expressions involving negative exponents as rational expressions.

8.3 Exercises

 1. Suppose that two rational expressions have the same denominator. When we add the expressions, how do we determine the numerator and denominator of the sum?

 2. So far, the following work is correct.

$$\frac{x^2}{x-3} - \frac{15}{x-3} + \frac{2x}{x-3} =$$

$$\frac{x^2 - 15 + 2x}{x-3}$$

Explain why the problem is not yet done.

In Exercises 3–14, perform the indicated operations.

3. $\dfrac{5}{a} + \dfrac{3}{a}$

4. $\dfrac{5}{t} - \dfrac{1}{t}$

5. $\dfrac{5x-4}{x} - \dfrac{8x+5}{x}$

6. $\dfrac{8a-2b}{b} + \dfrac{4a-5b}{b}$

7. $\dfrac{4x-45}{x-9} + \dfrac{2x-9}{x-9}$

8. $\dfrac{3x-63}{x-6} + \dfrac{8x-3}{x-6}$

9. $\dfrac{3c+51d}{c-8d} - \dfrac{9c+3d}{c-8d}$

10. $\dfrac{3x-24y}{x+7y} - \dfrac{6x-3y}{x+7y}$

11. $\dfrac{2x+5}{x^2-2x-8} + \dfrac{x+1}{x^2-2x-8}$

12. $\dfrac{3t+4}{t^2+10t+25} - \dfrac{2t-1}{t^2+10t+25}$

13. $\dfrac{x^3+6x^2-20x}{x(x-2)} - \dfrac{x^2-6x}{x(x-2)}$

14. $\dfrac{x(x-1)}{(2x+3)(x+1)} - \dfrac{2(2x+3)}{(2x+3)(x+1)}$

In Exercises 15–20, perform the indicated operations.

15. $\dfrac{x}{x^2 - x - 6} - \dfrac{x + 1}{x^2 - x - 6} + \dfrac{x - 2}{x^2 - x - 6}$

16. $\dfrac{x + 5}{2x^2 + 5x + 3} + \dfrac{3x + 3}{2x^2 + 5x + 3} -$

$$\dfrac{2x + 5}{2x^2 + 5x + 3}$$

17. $\dfrac{6}{2x + 3} - \left[\dfrac{x}{2x + 3} - \dfrac{3 - x}{2x + 3} \right]$

18. $\dfrac{x}{4 - x} - \left[\dfrac{x + 4}{4 - x} + \dfrac{x}{4 - x} \right]$

19. $\left[\dfrac{3x + 1}{x - 2} - \dfrac{2x - 2}{x - 2} \right] \div (x + 3)$

20. $\left[\dfrac{3x + 1}{5 - x} + \dfrac{1 - x}{5 - x} \right] \cdot \dfrac{x - 5}{x + 1}$

In Exercises 21–26, fill in the blank to make the resulting equation true.

21. $\dfrac{2x - 5}{x + 4} + \dfrac{?}{x + 4} = \dfrac{3x + 1}{x + 4}$

22. $\dfrac{x - 4}{2x - 3} - \dfrac{?}{2x - 3} = \dfrac{2}{2x - 3}$

23. $\dfrac{?}{t^2 + 3} - \dfrac{3t - t^2}{t^2 + 3} = \dfrac{t^2 - 3}{t^2 + 3}$

24. $\dfrac{x^2 - 5x + 2}{x^2 + x - 1} + \dfrac{?}{x^2 + x - 1}$

$$= \dfrac{3x^2 + x - 4}{x^2 + x - 1}$$

25. $\dfrac{?}{(x + 2)(x - 5)} - \dfrac{x - x^2}{(x + 2)(x - 5)}$

$$= \dfrac{3x^2 - 2x}{(x + 2)(x - 5)}$$

26. $\dfrac{?}{a - 2b} + \dfrac{a^2 + b}{a - 2b} = \dfrac{a^2 + ab}{a - 2b}$

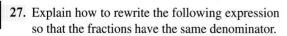

27. Explain how to rewrite the following expression so that the fractions have the same denominator.

$$\dfrac{15}{x} + \dfrac{7}{-x}$$

28. Explain how to rewrite the following expression so that the fractions have the same denominator.

$$\dfrac{x}{2x - 1} - \dfrac{1}{1 - 2x}$$

In Exercises 29–32, fill in the blanks so that the resulting equation is true.

29. $\dfrac{a}{3} - \dfrac{5}{-3} = \dfrac{a}{3} + \dfrac{?}{3} = \dfrac{?}{3}$

30. $\dfrac{7}{c} + \dfrac{2}{-c} = \dfrac{7}{c} + \dfrac{?}{c} = \dfrac{?}{c}$

31. $\dfrac{3x + 9}{x - 2} - \dfrac{8x}{2 - x} = \dfrac{3x + 9}{x - 2} + \dfrac{?}{x - 2}$

$$= \dfrac{?}{x - 2}$$

32. $\dfrac{2x - 1}{x - 2} + \dfrac{7x}{2 - x} = \dfrac{2x - 1}{x - 2} + \dfrac{?}{x - 2}$

$$= \dfrac{?}{x - 2}$$

In Exercises 33–38, perform the indicated operations.

33. $\dfrac{2x + 9}{7x - 8} - \dfrac{7x - 6}{8 - 7x}$

34. $\dfrac{5x + 3}{9x - 2} + \dfrac{6x - 6}{2 - 9x}$

35. $\dfrac{x^2}{x - 4} + \dfrac{16}{4 - x}$

36. $\dfrac{x^2}{x - 2y} + \dfrac{4y^2}{2y - x}$

37. $\dfrac{a^2}{a - b} - \dfrac{b^2}{b - a}$

38. $\dfrac{x - 4}{x^2 - 25} - \dfrac{x - 6}{25 - x^2}$

In Exercises 39–48, determine the LCM of the given expressions.

39. $5x^2y,\ 10xy^2,\ 15x^3y$ 40. $3x^5y,\ 6xy^4,\ 9x^3y^7$

41. $6x,\ 3x - 12$ 42. $2t,\ 4t + 8$

43. $t + 5,\ -3t - 15,\ 10t$ 44. $x - 7,\ 4x - 28,\ 6x$

45. $3x^2 + x - 2$, $4x^2 + 5x + 1$

46. $3x^2 + 4x - 15$, $x^2 + 5x + 6$

47. $x^2 + 3x + 2$, $x^2 - 4$, $3x + 6$

48. $y^2 + 2y - 3$, $y^2 + y - 2$, $3y - 3y^2$

49. Suppose the LCD of two fractions is $x(x - 4)$. If one of the fractions is

$$\frac{x}{x - 4},$$

explain how to rewrite the fraction so that its denominator is the LCD.

50. Suppose the GCF of two denominators is 1. Explain how to determine the LCD for the two fractions.

In Exercises 51–64, perform the indicated operations.

51. $\dfrac{5}{a^3b} + \dfrac{7}{ab^2}$

52. $\dfrac{3}{5c^3d^2} - \dfrac{4}{15c^2d^4}$

53. $x - \dfrac{3}{x}$

54. $3 + \dfrac{5x}{x - 2}$

55. $\dfrac{3}{x} + \dfrac{8}{x - 4}$

56. $\dfrac{3}{t + 3} - \dfrac{2}{t}$

57. $\dfrac{1}{x - 5} - \dfrac{1}{x + 7}$

58. $\dfrac{3}{x - 8} + \dfrac{x}{x - 3}$

59. $\dfrac{2}{t + 1} + t + 1$

60. $1 - 2t - \dfrac{2t}{t + 3}$

61. $\dfrac{3}{6(t - 5)} - \dfrac{2 - t}{9(t - 5)}$

62. $\dfrac{2x + 1}{3(2x - 1)} + \dfrac{2 - 3x}{2(1 - 2x)}$

63. $\dfrac{5t + 2}{t(t + 1)^2} - \dfrac{3}{t(t + 1)}$

64. $\dfrac{3}{y(y - 2)} + \dfrac{2y - 1}{y^2(y - 2)}$

In Exercises 65–68, determine the sum or difference as indicated.

65. Find the sum of

$$\frac{9}{x^2 + 7x + 10} \quad \text{and} \quad \frac{9}{x^2 - 4}.$$

66. Subtract

$$\frac{1}{x^2 - 5x - 6} \quad \text{from} \quad \frac{7}{x^2 - 4x - 12}.$$

67. Find the difference of

$$\frac{5}{x + 9} \quad \text{and} \quad \frac{x}{(x + 9)^2}.$$

68. Add

$$\frac{9}{x - 3} \quad \text{and} \quad \frac{x}{x^2 - 6x + 9}.$$

In Exercises 69–82, perform the indicated operations.

69. $\dfrac{2}{3t^2} + \dfrac{3}{2t^2 + t - 1}$

70. $\dfrac{3}{2x} - \dfrac{x - 3}{2x^2 - x - 6}$

71. $\dfrac{2x^2 - 7}{10x^2 - 3x - 4} - \dfrac{x - 3}{5x - 4}$

72. $\dfrac{9}{x - 3} + \dfrac{x}{x^2 - 6x + 9}$

73. $\dfrac{x + 9}{4x - 36} - \dfrac{x - 9}{x^2 - 18x + 81}$

74. $\dfrac{x - 3}{2x + 6} - \dfrac{x + 3}{x^2 + 6x + 9}$

75. $\dfrac{7}{x^2 + 15x + 56} + \dfrac{1}{x^2 - 64}$

76. $\dfrac{5}{x^2 - 4x - 12} + \dfrac{9}{x^2 - 15x + 54}$

77. $\dfrac{x}{x + 6} - \dfrac{72}{x^2 - 36}$

78. $\dfrac{x}{x + 3} - \dfrac{18}{x^2 - 9}$

79. $\dfrac{x}{x^2 + 4x + 3} - \dfrac{2}{x^2 - 2x - 3}$

80. $\dfrac{x}{x^2 + 5x + 4} + \dfrac{3}{x^2 + 11x + 10}$

81. $\dfrac{x}{x^2 + 8x + 15} + \dfrac{15}{x^2 + 4x - 5}$

82. $\dfrac{x}{x^2 + 4x + 3} - \dfrac{3}{x^2 - 4x - 5}$

83. Write each of the following as a single rational expression with no negative exponents. Are the two expressions equivalent?

$$(x + y)^{-1} \qquad x^{-1} + y^{-1}$$

84. Write each of the following as a single rational expression with no negative exponents. Are the two expressions equivalent?

$$(xy)^{-1} \qquad x^{-1}y^{-1}$$

In Exercises 85–90, simplify and express the result as a single rational expression with no negative exponents.

85. $x^{-1} - y^{-1}$

86. $3x^{-1} + 4y^{-1}$

87. $(x - y)^{-1}$

88. $(x + y)^{-1}$

89. $x(x + 3)^{-1} + 3(x - 2)^{-1}$

90. $4(x - 3)^{-1} + x(x - 3)^{-2}$

91. Suppose that one *cycle* of a sound wave can be modeled by the function

$$y = \frac{8x}{x^2 + 1},$$

where $-3 \le x \le 3$.

(a) Use the given domain of x and the calculator setting [Y: $-5, 5, 1$] to produce the graph of one cycle of the sound wave.

(b) Roughly speaking, the *amplitude* of a sound wave is the vertical distance between the highest and lowest points of the wave. Trace to these points on the graph and estimate the amplitude of this sound wave.

92. A quality control manager has measured the average number of errors that employees make while assembling electronic circuit boards. She has modeled the information with $E(x) = 5/x$, where $E(x)$ is the average number of errors made after x hours of training ($x \ge 1$).

(a) Use the domain of x and the range setting [Y: $0, 5, 1$] to produce the graph of $E(x)$. Does the graph suggest that increased hours of training reduces errors? Can the average number of errors be reduced to zero if enough training is provided?

(b) Using certain incentives along with training, the manager believes she can reduce the average error rate by

$$\frac{5}{x + 1}, \quad \text{that is,} \quad E(x) = \frac{5}{x} - \frac{5}{x + 1}.$$

Write this new function $E(x)$ as a single rational expression and produce its graph along with the original graph. Does the graph suggest that the average number of errors would be reduced with the incentive plan?

Exploring with Real Data

In Exercises 93–96, refer to the following background information and data.

The following table shows the number (in thousands) of Americans using food stamps during the period 1980–1990. Also shown are the costs (in millions of dollars) of the food stamp program during the same period.

	1980	1985	1990
Number of People Using Food Stamps (thousands)	22,028	19,198	20,498
Total Cost (millions)	8,721	10,744	14,205

(Source: U.S. Department of Agriculture.)

With $P(t)$ representing the number of people receiving food stamps and $C(t)$ representing the total cost, the following functions model the data in the table.

$$P(t) = 82.6t^2 - 1805t + 28{,}988$$
$$C(t) = 28.8t^2 - 27t + 8136$$

For both functions, t is the number of years since 1975.

93. For 1990, what was the actual number of people using food stamps and what was the actual cost in dollars?

94. Using $P(t)$ and $C(t)$, write a function $A(t)$ for the average cost per person during the given period. [Remember that $P(t)$ is in thousands and $C(t)$ is in millions.]

95. Evaluate $A(5)$, $A(10)$, and $A(15)$, that is, determine the average cost per person for 1980, 1985, and 1990.

96. Suppose that in the future the budgeted amount for food stamps remains at the 1990 level, but the average cost per person decreases. Would this imply that the number of food stamp users has increased or decreased? Why?

Challenge

In Exercises 97 and 98, perform the indicated operations.

97. $\left(\dfrac{1}{x + h} - \dfrac{1}{x} \right) \div h$

98. $\left[\dfrac{1}{(x + h)^2} - \dfrac{1}{x^2} \right] \cdot \dfrac{1}{h}$

In Exercises 99 and 100, perform the indicated operations.

99. $\dfrac{x^n}{x^n + 3} - \dfrac{9}{x^{2n} + 3x^n}$

100. $\dfrac{x^{2n} - 19}{x^{2n} + 4x^n - 5} - \dfrac{x^n - 4}{x^n - 1}$

In Exercises 101 and 102, perform the indicated operations.

101. $\dfrac{\dfrac{1}{x} - \dfrac{2x}{x^2 + 1}}{\dfrac{1}{x}}$

102. $\dfrac{\dfrac{x}{x + 1} + \dfrac{1}{x}}{\dfrac{1}{x} - 1}$

8.4 Complex Fractions

Introduction ▪ *Simplifying Complex Fractions* ▪ *Expressions with Negative Exponents*

Introduction

In Example 4 of Section 8.2, we simplified the following fraction.

$$\dfrac{\dfrac{x + 5}{5}}{\dfrac{x^2 - 25}{10}}$$

This is an algebraic expression whose numerator and denominator are rational expressions. We call the expression a **complex fraction.**

Definition of Complex Fraction

A **complex fraction** is a fraction with a rational expression in the numerator, in the denominator, or in both.

The following are some other examples of complex fractions.

$$\dfrac{\dfrac{a+b}{b}}{\dfrac{a-b}{2}} \qquad \dfrac{\dfrac{3}{y}+2}{\dfrac{5}{y}-3} \qquad \dfrac{\dfrac{1}{p-1}+p}{\dfrac{1}{p-1}-p} \qquad \dfrac{\dfrac{3}{x+3}}{\dfrac{1}{x+3}-\dfrac{1}{x}}$$

Note that the numerator or denominator (or both) can have more than one fraction in it.

In the following discussion, we will learn how to simplify complex fractions.

Simplifying Complex Fractions

A complex fraction can be simplified in two ways. One method begins with combining the terms in both the numerator and the denominator. Then, we divide by multiplying the numerator by the reciprocal of the denominator. This method is consistent with the normal Order of Operations.

EXAMPLE 1 *Using the Order of Operations to Simplify a Complex Fraction*

Simplify

$$\dfrac{\dfrac{3}{y}+2}{\dfrac{5}{y}-3}.$$

Solution

$$\dfrac{\dfrac{3}{y}+2}{\dfrac{5}{y}-3} = \dfrac{\dfrac{3}{y}+\dfrac{2}{1}}{\dfrac{5}{y}-\dfrac{3}{1}} \qquad \text{The LCD of the numerator is } y.$$

$$\text{The LCD of the denominator is } y.$$

$$= \dfrac{\dfrac{3}{y}+\dfrac{2y}{y}}{\dfrac{5}{y}-\dfrac{3y}{y}} \qquad \text{Write equivalent fractions.}$$

$$= \dfrac{\dfrac{3+2y}{y}}{\dfrac{5-3y}{y}} \qquad \text{Combine the fractions.}$$

$$= \dfrac{3+2y}{y}\cdot\dfrac{y}{5-3y} \qquad \text{Multiply by the reciprocal of the denominator.}$$

$$= \dfrac{3+2y}{5-3y}$$

■

Another way to simplify a complex fraction is to multiply the numerator and denominator by the LCD of *all the fractions*. Since multiplying the numerator and denominator of the complex fraction by the same number is equivalent to multiplying by 1, the resulting expression will be equivalent to the original complex fraction.

To compare the two methods, we use this approach on the same complex fraction that we simplified in Example 1.

EXAMPLE 2 *Using the LCD to Simplify a Complex Fraction*

Simplify

$$\frac{\dfrac{3}{y} + 2}{\dfrac{5}{y} - 3}.$$

Solution   Multiply the numerator and denominator of the complex fraction by y, which is the LCD of all of the fractions.

$$\frac{\dfrac{3}{y} + 2}{\dfrac{5}{y} - 3} = \frac{y \cdot \left(\dfrac{3}{y} + 2\right)}{y \cdot \left(\dfrac{5}{y} - 3\right)}$$

$$= \frac{y \cdot \dfrac{3}{y} + 2 \cdot y}{y \cdot \dfrac{5}{y} - 3 \cdot y} \qquad \text{Distributive Property}$$

$$= \frac{3 + 2y}{5 - 3y} \qquad \text{Simplify the numerator and denominator.}$$

Note that the result is the same as in Example 1.

Generally, multiplying both the numerator and denominator by the LCD is the easier method. It is the method we will use for the remaining examples.

EXAMPLE 3 *Simplifying a Complex Fraction*

Simplify

$$\frac{\dfrac{1}{p - 1} + p}{\dfrac{1}{p - 1} - p}.$$

Solution

Multiply the numerator and denominator of the complex fraction by $p - 1$, which is the LCD of all of the fractions.

$$\frac{\dfrac{1}{p-1} + p}{\dfrac{1}{p-1} - p} = \frac{(p-1) \cdot \left(\dfrac{1}{p-1} + p\right)}{(p-1) \cdot \left(\dfrac{1}{p-1} - p\right)}$$

$$= \frac{(p-1) \cdot \dfrac{1}{p-1} + p(p-1)}{(p-1) \cdot \dfrac{1}{p-1} - p(p-1)} \qquad \text{Distributive Property}$$

$$= \frac{1 + p^2 - p}{1 - p^2 + p} \qquad \qquad \text{Simplify the numerator and denominator.}$$

$$\text{or } \frac{1 - p + p^2}{1 + p - p^2}$$

∎

EXAMPLE 4   *Simplifying a Complex Fraction*

Simplify

$$\frac{\dfrac{3}{x+3}}{\dfrac{1}{x+3} - \dfrac{1}{x}}.$$

Solution   The LCD of all of the fractions is $x(x + 3)$. Multiply the numerator and denominator of the complex fraction by the LCD.

$$\frac{\dfrac{3}{x+3}}{\dfrac{1}{x+3} - \dfrac{1}{x}} = \frac{x(x+3) \cdot \dfrac{3}{x+3}}{x(x+3) \cdot \left(\dfrac{1}{x+3} - \dfrac{1}{x}\right)}$$

$$= \frac{x(x+3) \cdot \dfrac{3}{x+3}}{x(x+3) \cdot \dfrac{1}{x+3} - x(x+3) \cdot \dfrac{1}{x}} \qquad \text{Distributive Property}$$

$$= \frac{3x}{x - (x+3)} \qquad \qquad \text{Simplify.}$$

$$= \frac{3x}{x - x - 3} \qquad \qquad \text{Remove parentheses.}$$

$$= \frac{3x}{-3} \qquad \qquad \text{Combine like terms.}$$

$$= -x \qquad \qquad \text{Divide out the common factor 3.}$$

∎

NOTE: The complex fraction in Example 4 has two restricted values, 0 and -3. The very simple expression $-x$ is equivalent to the original complex fraction only if x is not replaced with 0 or -3. ∎

Expressions with Negative Exponents

Recall the definition of a negative exponent.

$$a^{-n} = \frac{1}{a^n} \quad \text{for} \quad a \neq 0$$

When an expression with negative exponents is rewritten with positive exponents, the result is sometimes a complex fraction.

EXAMPLE 5 *Simplifying Expressions with Negative Exponents*

Rewrite

$$\frac{x^{-1} + y^{-1}}{x^{-1}y^{-1}}$$

with only positive exponents. Then, simplify the resulting expression.

Solution

$$\frac{x^{-1} + y^{-1}}{x^{-1}y^{-1}} = \frac{\dfrac{1}{x} + \dfrac{1}{y}}{\dfrac{1}{xy}} \qquad \text{Definition of negative exponent}$$

$$= \frac{xy \cdot \left(\dfrac{1}{x} + \dfrac{1}{y}\right)}{xy \cdot \dfrac{1}{xy}} \qquad \text{Multiply the numerator and denominator by the LCD } xy.$$

$$= \frac{xy \cdot \dfrac{1}{x} + xy \cdot \dfrac{1}{y}}{xy \cdot \dfrac{1}{xy}} \qquad \text{Distributive Property}$$

$$= \frac{y + x}{1}$$

$$= x + y$$

EXAMPLE 6 *Simplifying Expressions with Negative Exponents*

$$\frac{2r^{-2} + c^{-1}}{r^{-1} + 3c^{-2}} = \frac{\dfrac{2}{r^2} + \dfrac{1}{c}}{\dfrac{1}{r} + \dfrac{3}{c^2}}$$ Definition of negative exponent

$$= \frac{r^2c^2 \cdot \left(\dfrac{2}{r^2} + \dfrac{1}{c}\right)}{r^2c^2 \cdot \left(\dfrac{1}{r} + \dfrac{3}{c^2}\right)}$$ The LCD is r^2c^2.

$$= \frac{r^2c^2 \cdot \dfrac{2}{r^2} + r^2c^2 \cdot \dfrac{1}{c}}{r^2c^2 \cdot \dfrac{1}{r} + r^2c^2 \cdot \dfrac{3}{c^2}}$$ Distributive Property

$$= \frac{2c^2 + r^2c}{rc^2 + 3r^2}$$ Simplify.

8.4 *Quick Reference*

Introduction	▪ A **complex fraction** is a fraction with a rational expression in the numerator, the denominator, or both.

Simplifying Complex Fractions	▪ To simplify a complex fraction, perform the indicated operations in the numerator and denominator. Then, divide the numerator by the denominator (multiply by the reciprocal of the denominator).
	▪ An alternate method for simplifying a complex fraction is to follow these steps.
	1. Determine the LCD for all fractions in the complex fraction.
	2. Multiply the numerator and denominator of the complex fraction by the LCD.
	3. Simplify the result if possible.

Expressions with Negative Exponents	▪ When expressions with negative exponents are rewritten with positive exponents, the result is sometimes a complex fraction.
	▪ Use the rule
	$$a^{-n} = \frac{1}{a^n}$$
	to rewrite the expression with positive exponents. Then, simplify the result.

8.4 *Exercises*

 1. If A, B, C, and D represent polynomials, explain why each of the following is a complex fraction.

$$\dfrac{\dfrac{A}{B}}{C} \qquad \dfrac{\dfrac{A}{B}}{C} \qquad \dfrac{\dfrac{A}{B}}{\dfrac{C}{D}}$$

2. Do the following expressions mean the same thing? Why? Evaluate each expression to support your answer.

$$\dfrac{\dfrac{2}{3}}{4} \quad \text{and} \quad \dfrac{2}{\dfrac{3}{4}}$$

 3. How can the complex fraction

$$\dfrac{\dfrac{w}{w+1}}{\dfrac{w^2}{w-1}}$$

be written with the $\div$ symbol?

4. How can

$$\dfrac{a+2}{a-1} \div \dfrac{a-3}{a+5}$$

be written as a complex fraction?

In Exercises 5–12, simplify the complex fraction.

5. $\dfrac{\dfrac{3}{x}}{\dfrac{9}{x^3}}$

6. $\dfrac{-\dfrac{42}{y^5}}{\dfrac{35}{y^2}}$

7. $\dfrac{\dfrac{2x}{x+3}}{\dfrac{6x^2}{x+3}}$

8. $\dfrac{\dfrac{c+d}{4}}{\dfrac{c-d}{3}}$

9. $\dfrac{\dfrac{5}{x+5}}{\dfrac{10}{x^2-25}}$

10. $\dfrac{-\dfrac{2}{x^2-1}}{\dfrac{2}{1-x}}$

11. $\dfrac{\dfrac{2x^2+3x-2}{x^2-x-6}}{\dfrac{1-x-2x^2}{x^2+3x+2}}$

12. $\dfrac{\dfrac{3x^2-13x-30}{6x^2+13x+5}}{\dfrac{x^2-4x-12}{2x^2+x}}$

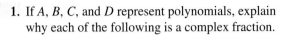 **13.** One method for simplifying a complex fraction is to use the Order of Operations. Describe this method.

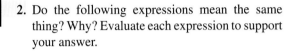 **14.** We can write the expression $(\tfrac{3}{4} + \tfrac{2}{3}) \div \tfrac{5}{8}$ as the complex fraction

$$\dfrac{\frac{3}{4} + \frac{2}{3}}{\frac{5}{8}}.$$

Explain why there is no need for parentheses in the complex fraction.

In Exercises 15–26, simplify the complex fraction.

15. $\dfrac{\dfrac{2}{3} + \dfrac{4}{t}}{\dfrac{1}{t} + \dfrac{1}{2}}$

16. $\dfrac{\dfrac{3}{x} - \dfrac{5}{9}}{\dfrac{4}{3} - \dfrac{4}{x}}$

17. $\dfrac{\dfrac{2}{y^2} + \dfrac{1}{y}}{\dfrac{2}{y} + 1}$

18. $\dfrac{4 - \dfrac{1}{x^2}}{\dfrac{2}{x} - \dfrac{1}{x^2}}$

19. $\dfrac{4 - \dfrac{1}{x^2}}{2 + \dfrac{1}{x}}$

20. $\dfrac{1 + \dfrac{3}{x}}{x - \dfrac{9}{x}}$

21. $\dfrac{\dfrac{1}{c} - \dfrac{1}{cd}}{\dfrac{1}{cd} - \dfrac{1}{c}}$

22. $\dfrac{\dfrac{1}{x} + \dfrac{1}{y}}{\dfrac{1}{x} + \dfrac{1}{xy}}$

23. $\dfrac{\dfrac{5}{x^4} - \dfrac{2}{x^3}}{\dfrac{5}{x^3} - \dfrac{2}{x^2}}$

24. $\dfrac{4 - \dfrac{5}{x}}{\dfrac{4}{x} - \dfrac{5}{x^2}}$

25. $\dfrac{x+y}{\dfrac{1}{x} + \dfrac{1}{y}}$

26. $\dfrac{\dfrac{3}{a^2} - \dfrac{2}{b}}{a + 2b}$

In Exercises 27–38, simplify the complex fraction.

27. $$\frac{\dfrac{2}{x+3} - \dfrac{3}{x-3}}{2 + \dfrac{1}{x^2-9}}$$

28. $$\frac{\dfrac{4}{4-x^2} - 1}{\dfrac{1}{x+2} + \dfrac{1}{x-2}}$$

29. $$\frac{\dfrac{3}{2x+1} - (x+1)}{(x-3) - \dfrac{2}{2x+1}}$$

30. $$\frac{(2x-1) + \dfrac{1}{x-2}}{x + \dfrac{1}{x-2}}$$

31. $$\frac{\dfrac{1}{1-y} - \dfrac{1}{1+y}}{\dfrac{1+y}{1-y} - \dfrac{1-y}{1+y}}$$

32. $$\frac{\dfrac{x+4}{x-4} + \dfrac{x-4}{x+4}}{\dfrac{1}{x-4} + \dfrac{1}{x+4}}$$

33. $$\frac{3 + \dfrac{1}{x} - \dfrac{14}{x^2}}{6 + \dfrac{11}{x} - \dfrac{7}{x^2}}$$

34. $$\frac{6 - \dfrac{1}{x} - \dfrac{35}{x^2}}{2 - \dfrac{11}{x} + \dfrac{15}{x^2}}$$

35. $$\frac{5 + \dfrac{2x}{y} + \dfrac{2y}{x}}{\dfrac{2x}{y} - \dfrac{2y}{x} - 3}$$

36. $$\frac{\dfrac{12x}{y^2} + \dfrac{13}{xy} - \dfrac{35}{x^3}}{\dfrac{10}{x^3} - \dfrac{3}{xy} - \dfrac{4x}{y^2}}$$

37. $$\frac{x + \dfrac{8}{x^2}}{1 - \dfrac{2}{x} + \dfrac{4}{x^2}}$$

38. $$\frac{\dfrac{8}{y^3} - 27}{\dfrac{4}{y^3} - \dfrac{12}{y^2} + \dfrac{9}{y}}$$

39. A second method for simplifying a complex fraction involves the LCD of all the fractions in the numerator and denominator. Describe this method.

40. When using the method described in Exercise 39, why is the value of the complex fraction not changed?

In Exercises 41–48, simplify the complex fraction.

41. $$\frac{\dfrac{4y^2}{2y+x} + (3x-y)}{\dfrac{5x^2}{2y+x} - (3y+2x)}$$

42. $$\frac{x+1+\dfrac{2}{2x-3}}{x+2+\dfrac{3}{2x-3}}$$

43. $$\frac{a+b}{1 + \dfrac{2b}{a-b}}$$

44. $$\frac{1 + \dfrac{a+6}{a}}{\dfrac{a}{9} - \dfrac{1}{a}}$$

45. $$\frac{x+3+\dfrac{4}{x-2}}{x^2+3x+2}$$

46. $$\frac{1 + \dfrac{c}{a+c}}{1 + \dfrac{3c}{a-c}}$$

47. $$\frac{\dfrac{1}{a^3} + \dfrac{1}{b^3}}{b^2 - ab + a^2}$$

48. $$\frac{\dfrac{1}{27} - \dfrac{1}{a^3}}{a^2 + 3a + 9}$$

49. For

$$\frac{x}{x+3} + \frac{5}{x-3},$$

why is it incorrect to multiply by the LCD as we do when we simplify a complex fraction?

50. Simplify the expression

$$\frac{x^{-3} - x^{-2}}{2x^{-3} + x^{-2}}$$

with the methods described in parts (a) and (b). Then answer part (c).

(a) Multiply the numerator and denominator by x^3.

(b) Write the expression with positive exponents and then simplify the result.

(c) Which of the two methods did you find easier?

In Exercises 51–58, rewrite the expression with positive exponents and simplify.

51. $\dfrac{a^{-1}b^{-1}}{a^{-1}+b^{-1}}$

52. $\dfrac{a^{-1}-b^{-1}}{a^{-2}-b^{-2}}$

53. $\dfrac{x^{-1}+x}{x^{-2}+x^2}$

54. $\dfrac{a^{-1}+b}{a+b^{-1}}$

55. $\dfrac{1-x^{-2}}{3-x^{-1}-4x^{-2}}$

56. $\dfrac{1-4x^{-2}}{2+5x^{-1}+2x^{-2}}$

57. $\dfrac{xy^{-3} + x^{-2}y}{x - y}$ **58.** $\dfrac{-2a}{a^{-2} - a^{-1}}$

59. Write and simplify a complex fraction whose numerator is

$$3 - \frac{1}{x}$$

and whose denominator is $9 - \dfrac{1}{x^2}$.

60. Let $f(x) = \dfrac{2}{x}$.

Write and simplify a complex fraction whose numerator is $f(2 + h) - f(2)$ and whose denominator is h.

61. Write and simplify a complex fraction that is the reciprocal of

$$x + \frac{5}{x + 1}.$$

62. Write and simplify a complex fraction that is the average of

$$\frac{1}{a} \quad \text{and} \quad \frac{1}{a - 1}.$$

63. Consider any three consecutive integers.

(a) Write a complex fraction whose numerator is the ratio of the first integer to the second integer and whose denominator is the ratio of the second integer to the third integer.

(b) Simplify the complex fraction and show that the denominator of the result is always one more than the numerator.

64. For the fall quarter, x male students and y female students were accepted at a certain college. Of these students, m male students and n female students decided not to attend.

(a) Write a fraction that represents the ratio of accepted male students who actually enrolled. Then, do the same for the ratio of accepted female students who actually enrolled.

(b) Write a fraction that represents the ratio of all accepted students who actually enrolled. Is this fraction the sum of the fractions in part (a)?

(c) Write and simplify a complex fraction that is the ratio (male to female) of the two fractions in part (a). Interpret the meaning of this complex fraction.

Exploring with Real Data

In Exercises 65–68, refer to the following background information and data.

The bar graph shows the percentages of American households that participated in vegetable gardening and indoor houseplant activities from 1985 to 1990. (Source: National Gardening Association.)

Figure for 65–68

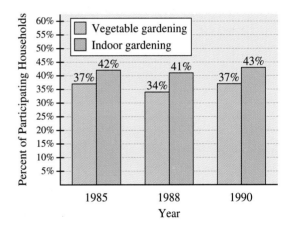

With t representing the number of years since 1980, the data in the graph can be modeled by the following functions.

$$V(t) = 0.5t^2 - 7.5t + 62$$
$$H(t) = 5t^2 - 89t + 433$$

For these functions, $V(t)$ is the percentage of households that participated in vegetable gardening, and $H(t)$ is the percentage of households that participated in indoor houseplant activities.

65. Let $R(t) = \dfrac{H(t)}{V(t)}$.

Which function, $V(t)$ or $H(t)$, determines whether the rational expression has any restricted values?

66. Use suitable range and scale settings to produce the graph of the function that you identified in Exercise 65. Does the rational expression appear to have any restricted values?

67. Use the actual 1990 data in the graph to calculate the percentage amount by which the percentage of indoor houseplant participants exceeded the percentage of vegetable gardeners. Now calculate $R(10)$ and compare the results.

68. Give two feasible explanations for the fact that a greater percentage of Americans were involved with houseplants than with vegetable gardening.

Challenge

In Exercises 69–72, use the given function to write

$$\frac{f(x + h) - f(x)}{h}.$$

Then, simplify the expression.

69. $f(x) = \dfrac{1}{2x - 1}$

70. $f(x) = \dfrac{x}{x + 2}$

71. $f(x) = \dfrac{x + 1}{x - 2}$

72. $f(x) = \dfrac{2 - 3x}{x}$

In Exercises 73 and 74, simplify the complex fraction.

73. $\dfrac{\dfrac{a + 1}{3^a}}{\dfrac{a}{3^{a+1}}}$

74. $\dfrac{\dfrac{x^2 - 1}{2x^{n+1}}}{\dfrac{x + 1}{x^n}}$

In Exercises 75–80, simplify the complex fraction.

75. $\dfrac{\dfrac{1}{x + h} - \dfrac{1}{x}}{h}$

76. $\dfrac{\dfrac{1}{x + 3} - \dfrac{1}{3}}{x}$

77. $2x - \dfrac{5 + \dfrac{3}{x}}{7 - \dfrac{2}{x}}$

78. $3x - \dfrac{4 + \dfrac{2}{x}}{\dfrac{3}{x} - 5}$

79. $1 + \dfrac{1}{1 + \dfrac{1}{1 + \frac{1}{2}}}$

80. $\dfrac{1 - \dfrac{1}{1 - \frac{1}{3}}}{1 + \dfrac{1}{1 + \frac{1}{3}}}$

8.5 Equations with Rational Expressions

The Algebraic Method ▪ *Solving Formulas*

The Algebraic Method

The following are some examples of equations containing rational expressions.

$$\frac{2}{x} = x + 1 \qquad \frac{4}{x + 1} - \frac{5}{x} = \frac{20}{x^2 + x} \qquad \frac{1}{x} - \frac{8}{2x - x^2} = \frac{x + 2}{x - 2}$$

When solving such equations, our goal is to determine the permissible replacements for the variable that make the equation true.

In Chapter 4 we solved linear equations containing one or more fractions, but the denominators did not contain variables. The technique was to multiply both sides of the equation by the LCD. This eliminated the fractions and made the rest of the solving process easier.

Our first example is a review of that method.

EXAMPLE 1 *Using the Technique of Clearing Fractions to Solve an Equation*

Solve the equation

$$\frac{t}{8} + \frac{t + 1}{2} = 3.$$

Solution

$$\frac{t}{8} + \frac{t + 1}{2} = 3$$

$$8 \cdot \frac{t}{8} + 8 \cdot \frac{t + 1}{2} = 8 \cdot 3 \qquad \text{Multiply both sides by the LCD 8.}$$

$$t + 4(t + 1) = 24 \qquad \text{Simplify each term.}$$

$$t + 4t + 4 = 24 \qquad \text{Distributive Property}$$

$$5t + 4 = 24 \qquad \text{Combine like terms.}$$

$$5t = 20 \qquad \text{Solve for } t.$$

$$t = 4$$

We will proceed in a similar way with equations that contain rational expressions with variables in the denominator. The Multiplication Property of Equations allows us to multiply both sides of an equation by the same *nonzero* quantity. We can use this property to eliminate the fractions just as we did in Example 1.

However, there is a potential problem. A particular replacement for the variable may be a restricted value, that is, it may cause the denominator of a rational expression to be zero. Because restricted values are not permissible replacements for the variable, they cannot be solutions. We must take this into account as we solve equations containing rational expressions.

| EXAMPLE 2 | *Solving an Equation That Contains a Rational Expression* |

Solve the equation

$$\frac{15}{x} + 5 = 2.$$

Figure 8.1

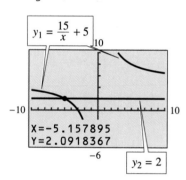

Solution   As always, a graph can help us to estimate the number of solutions of the equation and to approximate the solutions. Figure 8.1 shows the graphs of

$$y_1 = \frac{15}{x} + 5 \quad \text{and} \quad y_2 = 2.$$

From the graph we estimate that there is one solution, and it appears to be approximately -5.

Now we solve the equation algebraically.

$$\frac{15}{x} + 5 = 2, \quad x \neq 0 \qquad \text{Note that 0 is a restricted value.}$$

$$x \cdot \frac{15}{x} + 5 \cdot x = 2 \cdot x \qquad \text{Multiply both sides by the LCD } x.$$

$$15 + 5x = 2x \qquad \text{Simplify each term.}$$

$$15 = -3x \qquad \text{Solve for } x.$$

$$x = -5$$

Because -5 is not a restricted value, the solution of the equation is -5, as we predicted from the graph. ∎

| EXAMPLE 3 | *Solving an Equation That Contains a Rational Expression* |

Figure 8.2

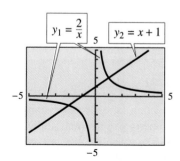

Solve the equation

$$\frac{2}{x} = x + 1.$$

Solution

Figure 8.2 shows the graphs of $y_1 = 2/x$ and $y_2 = x + 1$.

The graphs appear to intersect at two points, $(-2, -1)$ and $(1, 2)$. Thus, we estimate that the equation has two solutions, -2 and 1.

The following is an algebraic solution.

$$\frac{2}{x} = x + 1, \quad x \neq 0$$ Note that 0 is a restricted value.

$$x \cdot \frac{2}{x} = x \cdot (x + 1)$$ Multiply both sides by the LCD x.

$$2 = x^2 + x$$ The result is a quadratic equation.

$$0 = x^2 + x - 2$$ Write the equation in standard form.

$$0 = (x + 2)(x - 1)$$ Factor the trinomial.

$$x + 2 = 0 \quad \text{or} \quad x - 1 = 0$$ Zero Factor Property

$$x = -2 \quad \text{or} \quad x = 1$$ Solve each case for x.

Because neither number is a restricted value, the solutions are -2 and 1, which confirms our estimates from the graph. ◼

To clear fractions, it may be necessary to factor the denominators in order to determine the LCD.

EXAMPLE 4   *Factoring Denominators to Determine the LCD*

Solve the equation

$$\frac{4}{x + 1} - \frac{5}{x} = \frac{20}{x^2 + x}.$$

Solution

$$\frac{4}{x + 1} - \frac{5}{x} = \frac{20}{x^2 + x}$$ Begin by factoring the denominators.

$$\frac{4}{x + 1} - \frac{5}{x} = \frac{20}{x(x + 1)}, \quad x \neq 0, -1$$ Note that 0 and -1 are restricted values.

Now we see that the LCD is $x(x + 1)$. Multiply both sides by the LCD.

$$x(x + 1) \cdot \frac{4}{x + 1} - x(x + 1) \cdot \frac{5}{x} = x(x + 1) \cdot \frac{20}{x(x + 1)}$$

$$4x - 5(x + 1) = 20$$ Simplify each term.

$$4x - 5x - 5 = 20$$ Distributive Property

$$-x - 5 = 20$$ Combine like terms.

$$-x = 25$$ Solve for x.

$$x = -25$$

Because it is not a restricted value, -25 is the solution. You can check the solution by hand or with your calculator. ◼

EXAMPLE 5 *Solving an Equation with Rational Expressions*

Solve the equation

$$\frac{2}{x + 3} - \frac{1}{x} = \frac{-6}{x^2 + 3x}.$$

Solution

$$\frac{2}{x + 3} - \frac{1}{x} = \frac{-6}{x^2 + 3x} \qquad \text{Begin by factoring the denominators.}$$

$$\frac{2}{x + 3} - \frac{1}{x} = \frac{-6}{x(x + 3)}, \quad x \neq 0, -3 \qquad \text{Note that 0 and } -3 \text{ are restricted values.}$$

The LCD is $x(x + 3)$. Multiply both sides of the equation by the LCD to clear the fractions.

$$x(x + 3) \cdot \frac{2}{x + 3} - x(x + 3) \cdot \frac{1}{x} = x(x + 3) \cdot \frac{-6}{x(x + 3)}$$

$$2x - (x + 3) = -6 \qquad \text{Simplify each term.}$$

$$2x - x - 3 = -6 \qquad \text{Distributive Property}$$

$$x - 3 = -6 \qquad \text{Combine like terms.}$$

$$x = -3 \qquad \text{Solve for } x.$$

Our solution process leads us to -3 as the apparent solution, but -3 *is a restricted value*. Therefore, -3 cannot be a solution and the solution set is empty. ∎

NOTE: In Example 5 the solution of the equation appeared to be -3 until we realized that -3 was a restricted value and had to be discarded. We call such values **extraneous solutions.** ∎

EXAMPLE 6 *An Equation with an Extraneous Solution*

Solve the equation

$$\frac{1}{x} - \frac{8}{2x - x^2} = \frac{x + 2}{x - 2}.$$

Solution

$$\frac{1}{x} - \frac{(-1)8}{(-1)(2x - x^2)} = \frac{x + 2}{x - 2} \qquad -1(a - b) = b - a$$

$$\frac{1}{x} - \frac{-8}{x^2 - 2x} = \frac{x + 2}{x - 2} \qquad \text{Now the denominators are in descending order.}$$

$$\frac{1}{x} - \frac{-8}{x(x - 2)} = \frac{x + 2}{x - 2}, \quad x \neq 0, 2 \qquad \text{Factor the denominators.}$$

Note that the LCD is $x(x - 2)$ and the restricted values are 0 and 2. Multiply both sides of the equation by the LCD.

$$\frac{1x(x-2)}{x} - \frac{-8x(x-2)}{x(x-2)} = \frac{x(x-2)(x+2)}{x-2}$$

$(x-2) - (-8) = x(x+2)$ Simplify each term.

$x - 2 + 8 = x^2 + 2x$ Distributive Property

$x + 6 = x^2 + 2x$ The result is a quadratic equation.

$0 = x^2 + x - 6$ Write the equation in standard form.

$0 = (x+3)(x-2)$ Factor the trinomial.

$x + 3 = 0$ or $x - 2 = 0$ Zero Factor Property

$x = -3$ or $x = 2$ Solve each case for x.

The apparent solutions are -3 and 2. However, because 2 is a restricted value, it is an extraneous solution. Therefore, the only solution is -3. ∎

Solving Formulas

When solving a formula for a specified variable, we regard the specified variable as the only variable and treat all other variables as if they were constants.

Sometimes, formulas contain rational expressions. The method for solving such formulas for specified variables is the same as the method for solving equations containing rational expressions.

EXAMPLE 7   *Solving a Formula with a Rational Expression*

In electronics, a *resistor* is any device that offers resistance to the flow of electric current. When a light bulb is on, it acts as a resistor. Suppose two light bulbs are placed in a parallel circuit. (See Fig. 8.3.) Let r_1 and r_2 represent their resistances.

Figure 8.3

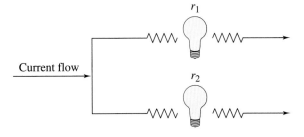

The formula for the total resistance R in this parallel circuit is

$$\frac{1}{R} = \frac{1}{r_1} + \frac{1}{r_2}.$$

Solve this formula for r_1.

Solution

$$\frac{1}{R} = \frac{1}{r_1} + \frac{1}{r_2}$$ The LCD is Rr_1r_2.

$$Rr_1r_2 \cdot \frac{1}{R} = Rr_1r_2 \cdot \frac{1}{r_1} + Rr_1r_2 \cdot \frac{1}{r_2}$$ Multiply both sides by the LCD.

$$r_1r_2 = Rr_2 + Rr_1$$ Simplify each term.

$$r_1r_2 - Rr_1 = Rr_2$$ Place the terms containing r_1 on the same side.

$$(r_2 - R)r_1 = Rr_2$$ Factor out the common factor r_1.

$$r_1 = \frac{Rr_2}{r_2 - R}$$ Divide both sides by $r_2 - R$.

8.5   *Quick Reference*

The Algebraic Method

- By using graphs, we can estimate the number of solutions of an equation with rational expressions, and we can approximate the solutions.

- If a replacement for a variable in a rational expression causes the denominator of the expression to be zero, the replacement is a **restricted value.**

- To solve an equation containing rational expressions algebraically, follow these steps.

    1. Factor each denominator completely and note the restricted values, if any.

    2. Determine the LCD for the fractions.

    3. Clear the fractions by multiplying each term of both sides of the equation by the LCD.

    4. Simplify each side of the equation.

    5. Use previously discussed techniques to solve the resulting equation.

    6. Verify the solution in the original equation.

- If the value obtained in step 5 is a restricted value, it cannot be a solution, and it is called an **extraneous solution.**

Solving Formulas

- To solve a *formula* containing rational expressions, treat the specified variable as the only variable and all other variables as constants. Solve the formula for the specified variable with the same steps as outlined for solving an *equation* containing rational expressions.

8.5 Exercises

 1. For the equation

$$\frac{1}{x + 3} - \frac{1}{x} = 5,$$

why are -3 and 0 not permissible replacements for x? What do we call -3 and 0 in this case?

 2. Even before we start to solve the equation

$$3 + \frac{2}{1 - x} = \frac{1}{2x + 1},$$

how do we know that $-\frac{1}{2}$ and 1 are not solutions?

In Exercises 3–6, state the restricted value(s) for the given equation.

3. $\dfrac{x}{x + 3} - \dfrac{2}{x - 4} = \dfrac{1}{x}$

4. $\dfrac{3}{a^2} + \dfrac{a}{a - 2} = 4$

5. $\dfrac{y}{y^2 - 9} + \dfrac{7}{y} = \dfrac{3y + 2}{y + 3}$

6. $\dfrac{2x + 1}{x^2 + 2x - 15} - \dfrac{x - 3}{x^2 - 25} = \dfrac{x + 5}{2}$

 7. To solve $\dfrac{x^2 - 4}{x - 2} = 12,$

we clear fractions and obtain $x^2 - 12x + 20 = 0$, which has the solutions 2 and 10. If we graph

$$y_1 = \frac{x^2 - 4}{x - 2} \quad \text{and} \quad y_2 = 12,$$

will the graphs intersect at two points? Why?

 8. The equation $\dfrac{x}{x + 1} = \dfrac{2}{x}$

has no x^2 terms in it. Does this mean that when we clear the fractions, the result will not be a quadratic equation?

In Exercises 9–14, solve by graphing. Verify the solutions.

9. $\dfrac{3}{x} = x + 2$

10. $\dfrac{5}{x + 4} = x$

11. $\dfrac{x + 3}{x + 2} = x + 3$

12. $\dfrac{x - 3}{x + 5} = x$

13. $\dfrac{12}{x} = x^2 + 3x - 4$

14. $\dfrac{16}{x} = 4 + 4x - x^2$

In Exercises 15–26, solve the equation.

15. $\dfrac{1}{8} = \dfrac{1}{2t} + \dfrac{2}{t}$

16. $\dfrac{5}{4} + \dfrac{4}{3} = \dfrac{7}{x}$

17. $\dfrac{5}{x - 2} = 3 - \dfrac{1}{x - 2}$

18. $4 + \dfrac{6}{5 - x} = \dfrac{2x}{5 - x}$

19. $\dfrac{7}{12} + \dfrac{1}{2y - 10} = \dfrac{5}{3y - 15}$

20. $\dfrac{4}{3x + 2} = 1 - \dfrac{2}{9x + 6}$

21. $\dfrac{5}{x - 2} = \dfrac{1}{x + 2} + \dfrac{4}{x^2 - 4}$

22. $\dfrac{2}{t - 5} + \dfrac{3}{t + 5} = \dfrac{10}{t^2 - 25}$

23. $\dfrac{t}{t + 4} - \dfrac{2}{t - 3} = 1$

24. $\dfrac{1}{x} + \dfrac{1}{x - 1} = \dfrac{2}{x + 2}$

25. $\dfrac{3}{x + 3} = \dfrac{2}{x - 4} - \dfrac{10}{x^2 - x - 12}$

26. $\dfrac{3}{x + 1} - \dfrac{4}{2x - 1} = \dfrac{5}{2x^2 + x - 1}$

27. Suppose that before solving

$$\frac{x}{3} + \frac{3}{x} = 0$$

algebraically, you decide to produce the graph of

$$y = \frac{x}{3} + \frac{3}{x}.$$

Explain how this would allow you to write the solution set immediately.

28. In Exercise 27, if you use the algebraic method and clear the fractions, how would the resulting equation confirm what the graph suggests?

In Exercises 29–36, solve the equation.

29. $2x - \dfrac{15}{x} = 7$

30. $12 - \dfrac{72}{x^2} = \dfrac{5}{x}$

31. $\dfrac{x}{4} - \dfrac{5}{x} = \dfrac{1}{4}$

32. $\dfrac{x}{2} - \dfrac{2}{x} = 0$

33. $\dfrac{t}{t - 2} + 1 = \dfrac{8}{t - 1}$

34. $\dfrac{8}{t} + \dfrac{3}{t - 1} = 3$

35. $\dfrac{3}{x + 2} + 1 = \dfrac{6}{4 - x^2}$

36. $\dfrac{x + 2}{x - 12} = \dfrac{1}{x} + \dfrac{14}{x^2 - 12x}$

37. Neither of the following equations has a solution, but the reason there is no solution is different in each case. Explain why each equation has no solution.

    (i) $\dfrac{x - 4}{x - 9} = \dfrac{5}{x - 9}$ (ii) $\dfrac{4}{x + 1} = \dfrac{4}{x - 1}$

38. The solution of $\dfrac{x}{x + 1} - \dfrac{x - 1}{x + 1} = \dfrac{1}{x + 1}$ is as follows.

$$\frac{(x + 1)}{1} \cdot \frac{x}{(x + 1)} - \frac{(x + 1)}{1} \cdot \frac{(x - 1)}{(x + 1)}$$

$$= \frac{(x + 1)}{1} \cdot \frac{1}{(x + 1)}$$

$$x - (x - 1) = 1$$
$$x - x + 1 = 1$$
$$1 = 1$$

Because the last equation is true, should we conclude that all real numbers are solutions? Why?

In Exercises 39–62, solve the equation.

39. $\dfrac{1}{x + 1} + \dfrac{2}{3x + 3} = \dfrac{1}{3}$

40. $\dfrac{4}{x - 2} + \dfrac{3}{x} = \dfrac{15}{x^2 - 2x}$

41. $\dfrac{x + 2}{x + 3} = \dfrac{x - 1}{x + 1}$

42. $\dfrac{-2}{3x + 2} = \dfrac{3}{1 - 2x}$

43. $\dfrac{5}{x + 5} = 4 - \dfrac{x}{x + 5}$

44. $\dfrac{4}{5x} = \dfrac{1}{3x}$

45. $\dfrac{10}{(2x - 1)^2} = 4 + \dfrac{3}{2x - 1}$

46. $\dfrac{1}{x + 3} + \dfrac{21}{(x + 3)^2} = 2$

47. $\dfrac{2}{x^2 + x - 2} + \dfrac{3x}{x^2 + 5x + 6}$

$$= \frac{5x}{x^2 + 2x - 3}$$

48. $\dfrac{4}{2x^2 - 7x - 15} - \dfrac{2}{2x^2 + 13x + 15}$

$$= \frac{-2}{x^2 - 25}$$

49. $2x + 15 = \dfrac{8}{x}$

50. $x = \dfrac{4 - 11x}{3x}$

51. $\dfrac{x}{x - 3} + \dfrac{2}{x} = \dfrac{3}{x - 3}$

52. $\dfrac{x + 8}{x + 2} + \dfrac{12}{x^2 + 2x} = \dfrac{2}{x}$

53. $\dfrac{8r}{r - 4} - \dfrac{7}{r^2 - 16} = 8$

54. $\dfrac{8}{q - 3} + \dfrac{9}{q^2 - 11q + 24} = \dfrac{2}{q - 8}$

55. $\dfrac{5}{x - 7} + \dfrac{2}{x + 5} = \dfrac{1}{x^2 - 2x - 35}$

56. $\dfrac{3}{x + 3} = \dfrac{4}{x + 1} - \dfrac{5}{x^2 + 4x + 3}$

57. $\dfrac{x + 4}{6x^2 + 5x - 6} + \dfrac{x}{2x + 3} = \dfrac{x}{3x - 2}$

58. $\dfrac{2 - 6x}{x^2 - x - 6} = \dfrac{x - 1}{x + 2} - \dfrac{x + 1}{3 - x}$

59. $\dfrac{2}{x + 1} - \dfrac{1}{x - 3} = \dfrac{-8}{x^2 - 2x - 3}$

60. $\dfrac{6}{z - 5} + \dfrac{6}{z^2 - 11z + 30} = \dfrac{2}{z - 6}$

61. $\dfrac{x}{x + 2} + \dfrac{2}{x + 3} + \dfrac{2}{x^2 + 5x + 6} = 0$

62. $\dfrac{1}{9x^2 + 3x - 2} = \dfrac{x}{3x - 1} - \dfrac{1}{9x + 6}$

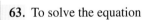 **63.** To solve the equation

$$\dfrac{x}{x - 1} + \dfrac{x + 3}{x + 2} = 3,$$

and to perform the addition

$$\dfrac{x}{x - 1} + \dfrac{x + 3}{x + 2} + 3,$$

we begin in the same way: Determine the LCD. However, what we do with the LCD is completely different for the two problems. Without actually doing the problems, explain what to do with the LCD in each case.

 64. In the following problems, identify the one for which an LCD is not needed and explain why.

(i) $\dfrac{2}{x} - \dfrac{x^2}{7 - x}$

(ii) $\dfrac{2}{x} = \dfrac{x^2}{7 - x}$

(iii) $\dfrac{2}{x} \div \dfrac{x^2}{7 - x}$

(iv) $\dfrac{2}{x} + \dfrac{x^2}{7 - x}$

In Exercises 65–76, solve the equation or simplify the expression.

65. $\dfrac{3}{x - 8} + \dfrac{x}{x - 3}$

66. $\dfrac{3}{x + 4} + \dfrac{6}{x} + \dfrac{12}{x^2 + 4x}$

67. $4 + \dfrac{19}{2t - 3} = \dfrac{t + 2}{3 - 2t}$

68. $\dfrac{2x + 1}{x - 2} = \dfrac{2x - 3}{x - 3}$

69. $\dfrac{y^2 - 49}{8y + 56} \cdot \dfrac{y^2 - 7y}{(y - 7)^2}$

70. $\dfrac{t^2 - 4}{t^2 - 25} \cdot \dfrac{3t - 15}{6t^2 - 12t}$

71. $1 - 2t - \dfrac{2t}{t + 3}$

72. $\dfrac{3}{6(t - 5)} - \dfrac{2 - t}{9(t - 5)}$

73. $\dfrac{x + 5}{x - 1} = \dfrac{8}{x - 3} - \dfrac{7}{x^2 - 4x + 3}$

74. $3\left[3 + \dfrac{1}{x + 1} \right] = \dfrac{-4}{x^2 + x}$

75. $\dfrac{x^2 + x - 6}{x^2 + 9x + 18} \div \dfrac{x^2 - 4}{4x + 24}$

76. $\dfrac{3x^3 + 6x^2}{2x^2 + x - 6} \div (6x^2 - 15x)$

77. The current I flowing in an electrical circuit is given by $I = E/R$, where E is the voltage and R is the resistance. Solve this formula for R.

78. The electrostatic force F between two charged bodies is given by

$$F = \frac{q_1 q_2}{kr^2},$$

where q_1 and q_2 are the electric charges on the two bodies, r is the distance between the two bodies, and k is a constant. Solve this formula for q_1.

79. In chemistry, Charles's law is given by

$$\frac{P_1 V_1}{T_1} = \frac{P_2 V_2}{T_2},$$

where P is pressure, T is absolute temperature, and V is volume. The subscript 1 might be used for initial conditions, and the subscript 2 might be used for final conditions. Solve this formula for T_1.

80. When a metal bar is heated, its length increases according to the formula $L_t = L_0(kt + 1)$, where L_0 is the length of the bar at $0°C$, L_t is the length of the bar at $t°C$, and k is the coefficient of linear expansion. Solve this formula for k.

81. An invested amount P can be calculated with the formula

$$P = \frac{I}{rt},$$

where I is interest earned, r is the simple interest rate, and t is time. Solve this formula for r.

82. In the accompanying figure, triangle ABC is a right triangle with altitude $\overline{AP}$. In geometry, it can be shown that $a/h = h/b$. Solve this equation for b.

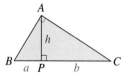

In Exercises 83–88, solve the formula for the specified variable.

83. $F = \dfrac{m_1 m_2}{\mu r^2}$; m_2

84. $p = \dfrac{2T}{r}$; r

85. $S = \dfrac{a(1 - r^n)}{1 - r}$; a

86. $\dfrac{1}{f} = (k - 1)\left(\dfrac{1}{p} + \dfrac{1}{q} \right)$; f

87. $\dfrac{x}{a} + \dfrac{y}{b} = 1$; a

88. $\dfrac{x}{a} + \dfrac{y}{b} + \dfrac{z}{c} = 1$; c

In Exercises 89–92, solve for y.

89. $\dfrac{y - 4}{x + 3} = -2$

90. $\dfrac{y + 2}{x - 5} = \dfrac{2}{3}$

91. $\dfrac{y}{x + 4} = -\dfrac{4}{5}$

92. $\dfrac{y - 1}{x} = -\dfrac{3}{4}$

Exploring with Real Data

In Exercises 93–96, refer to the following background information and data.

The bar graph shows the amounts (in billions of dollars) spent on children's books during the period 1987–1992. (Source: Book Industry Study Group, Inc.)

Figure for 93–96

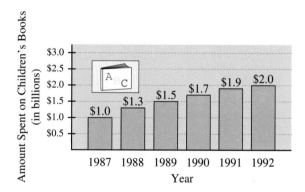

A rational function that models the data in the figure is

$$A(t) = \frac{1}{0.0186t^2 + 0.2304t + 1.21176},$$

where $A(t)$ is the amount spent (in billions of dollars) and t is the number of years since 1986.

93. Complete the following table. Round $A(t)$ values to the nearest tenth.

t	Year	A(t)	Actual Data from Figure
1			
2			
3			
4			
5			
6			

94. Compare the $A(t)$ values in your table with the actual data. How well does the function model the actual data?

95. According to the model function, how much will be spent on children's books in the year 2000? Beyond 1992, do you think the graph of $A(t)$ will continue to reflect the trend indicated in the figure?

96. Do the data in the figure necessarily prove that the number of children's books being sold is increasing? Why?

Challenge

In Exercises 97–100, solve the equation.

97. $\dfrac{\dfrac{x+3}{x-2}}{\dfrac{2}{x-2}} = 5$

98. $\dfrac{1+\dfrac{6}{x}}{1+\dfrac{1}{x}} = 2x$

99. $\dfrac{1}{18} + \dfrac{1}{9x} = \dfrac{1}{2x^2} + \dfrac{1}{x^3}$

100. $1 + \dfrac{2}{x} = \dfrac{1}{x^2} + \dfrac{2}{x^3}$

In Exercises 101 and 102, solve for y.

101. $\dfrac{y - y_1}{x - x_1} = m$

102. $\dfrac{y - b}{x} = m$

In Exercises 103–106, solve for x.

103. $\dfrac{a}{3x} - \dfrac{b}{x} = \dfrac{1}{2}$

104. $\dfrac{a}{b} = \dfrac{x}{b + x}$

105. $\dfrac{3x - 5c}{2x} = \dfrac{x}{5c}$

106. $\dfrac{4c - x}{3x} = \dfrac{x}{2c}$

In Exercises 107 and 108, use substitution to solve the given equation.

107. $\left(\dfrac{1}{x+1}\right)^2 + 3\left(\dfrac{1}{x+1}\right) - 4 = 0$

108. $\left(\dfrac{x-1}{x+6}\right)^2 + 4\left(\dfrac{x-1}{x+6}\right) - 12 = 0$

8.6 Applications

Work Rate ▪ Inverse Variation ▪ Ratio and Proportion ▪
Numbers ▪ Distance, Rate, Time ▪ Percentages

In Chapter 4 we developed a general approach to application problem-solving. This approach is appropriate regardless of the type of equation used to model the problem. In this section we present a variety of examples of applications that lead to equations with rational expressions.

Work Rate

Work rate problems nearly always involve two or more people or machines performing tasks at different rates.

If a machine can perform a task in 5 hours, then it completes $\frac{1}{5}$ of the task each hour. This is the machine's *work rate*. The combined work rate of two or more machines is the sum of their individual work rates.

EXAMPLE 1   *Work Rate*

At a juice-processing plant, one pipe can fill a storage tank with cranberry juice in 2 hours. Another pipe can fill the same tank with apple juice in 3 hours. If both pipes are open, how long does it take to fill the tank with cran-apple juice?

Solution   Let x = the number of hours to fill the tank with both pipes open.

	Hours to Fill Tank	*Work Rate*
Cranberry Juice	2	$\frac{1}{2}$
Apple Juice	3	$\frac{1}{3}$
Combined	x	$\frac{1}{x}$

The combined work rate is the sum of the individual work rates.

$$\frac{1}{2} + \frac{1}{3} = \frac{1}{x}$$

$$6x \cdot \frac{1}{2} + 6x \cdot \frac{1}{3} = 6x \cdot \frac{1}{x} \qquad \text{Multiply by the LCD } 6x.$$

$$3x + 2x = 6 \qquad \text{Simplify each term.}$$

$$5x = 6 \qquad \text{Combine like terms.}$$

$$x = \frac{6}{5} \quad \text{or} \quad 1.2$$

It takes 1.2 hours to fill the tank with cran-apple juice. ■

Inverse Variation

In Chapter 5 we studied direct variation. Another type of variation is **inverse variation.**

> *Definition of Inverse Variation*
>
> A quantity y **varies inversely** with x if there is a constant k so that
>
> $$y = \frac{k}{x}.$$
>
> We sometimes say that y is **inversely proportional** to x. The constant k is called the **constant of variation.**

EXAMPLE 2 *Inverse Variation*

If the temperature is constant, the volume of a gas varies inversely with the pressure. If the pressure of 2 liters of a gas is 12 newtons per square centimeter, what is the pressure when the volume is 1.6 liters?

Solution

Let P = the pressure and V = the volume. Then, the inverse variation is given by

$$V = \frac{k}{P}.$$

Substitute the initial conditions of $V = 2$ and $P = 12$ to determine the value of k.

$$2 = \frac{k}{12}$$

$$k = 24$$

Therefore,

$$V = \frac{24}{P}.$$

When $V = 1.6$, the inverse variation is

$$1.6 = \frac{24}{P}.$$

$1.6P = 24$ Multiply both sides by the LCD P.

$P = 15$ Solve for P.

The pressure is 15 newtons per square centimeter. ■

Ratio and Proportion

A **ratio** is a comparison of the values of two quantities. The comparison of x to y, for example, can be expressed with the fraction

$$\frac{x}{y}.$$

Another notation sometimes used is $x{:}y$.

A **proportion** is an equation stating that two ratios are equal. If the ratio of x to y is equal to the ratio of a to b, we can write

$$\frac{x}{y} = \frac{a}{b},$$

or we can use the alternate notation $x{:}y :: a{:}b$.

In our discussion of ratios and proportions, we will use the fraction notation.

In plane geometry, **similar triangles** are triangles whose corresponding angles have the same measure. In the next example, we use the theorem from plane geometry that states that the corresponding sides of similar triangles are proportional.

EXAMPLE 3 *Similar Triangles*

In Fig. 8.4, the corresponding angles of the two triangles have the same measures, that is, $m\angle A = m\angle D$, $m\angle B = m\angle F$, and $m\angle C = m\angle E$. Determine the lengths of the sides $\overline{AB}$ and $\overline{DE}$.

Figure 8.4

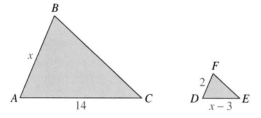

Solution   Because the measures of the corresponding angles are equal, the two triangles are similar triangles, and the corresponding sides are proportional:

$$\frac{AB}{DF} = \frac{BC}{FE} = \frac{AC}{DE}.$$

Using the proportion

$$\frac{AB}{DF} = \frac{AC}{DE}$$

and the relative dimensions shown in Fig. 8.4, we have the following.

$$\frac{x}{2} = \frac{14}{x - 3}$$

$$2(x - 3) \cdot \frac{x}{2} = 2(x - 3) \cdot \frac{14}{x - 3} \qquad \text{Multiply by the LCD } 2(x - 3).$$

$$x(x - 3) = 28 \qquad \text{Simplify each term.}$$

$$x^2 - 3x = 28 \qquad \text{Distributive Property}$$

$$x^2 - 3x - 28 = 0 \qquad \text{Write the quadratic equation in standard form.}$$

$$(x - 7)(x + 4) = 0 \qquad \text{Factor the trinomial.}$$

$$x - 7 = 0 \quad \text{or} \quad x + 4 = 0 \qquad \text{Zero Factor Property}$$

$$x = 7 \quad \text{or} \qquad x = -4 \qquad \text{Solve each case for } x.$$

Because the length of a side is not negative, the only solution is $x = 7$. In the first triangle, *AB* is 7, and in the second triangle, *DE* is 4. ■

NOTE: In Example 3 the solution -4 is not an extraneous solution; it is a permissible replacement for x. We disqualify -4 simply because it is not a meaningful solution for the application. ■

EXAMPLE 4 *Win-to-Loss Ratio*

At the All-Star break, the ratio of wins to losses for the Pittsburgh Pirates was 3 to 2. If the same ratio holds for the entire 162-game season, how many games will the Pirates win?

Solution Let $x =$ the number of games that will be won. Then $162 - x =$ the number of games that will be lost.

Because the ratio of wins to losses is 3 to 2, the equation is as follows.

$$\frac{3}{2} = \frac{x}{162 - x}$$

$$2(162 - x) \cdot \frac{3}{2} = 2(162 - x) \cdot \frac{x}{162 - x} \qquad \text{Multiply by the LCD } 2(162 - x).$$

$$3(162 - x) = 2x \qquad \text{Simplify each term.}$$

$$486 - 3x = 2x \qquad \text{Distributive Property}$$

$$486 = 5x \qquad \text{Add } 3x \text{ to both sides.}$$

$$97.2 = x \qquad \text{Solve for } x.$$

If the 3 to 2 win–loss ratio holds, the Pirates' season record will be about 97 wins and 65 losses, probably good enough to win the division championship. ■

Numbers

Equations with rational expressions may arise when you work application problems involving numbers and their reciprocals.

EXAMPLE 5   *Numbers and Their Reciprocals*

The difference between a number and six times its reciprocal is -1. What is the number?

Solution

Let $x =$ the number. Then its reciprocal is $1/x$ and the following equation describes the conditions.

$$x - \frac{6}{x} = -1$$

$$x \cdot x - x \cdot \frac{6}{x} = -1 \cdot x \qquad \text{Multiply by the LCD } x.$$

$$x^2 - 6 = -x \qquad \text{Simplify each term.}$$

$$x^2 + x - 6 = 0 \qquad \text{Write the quadratic equation in standard form.}$$

$$(x + 3)(x - 2) = 0 \qquad \text{Factor the trinomial.}$$

$$x + 3 = 0 \quad \text{or} \quad x - 2 = 0 \qquad \text{Zero Factor Property}$$

$$x = -3 \quad \text{or} \qquad x = 2 \qquad \text{Solve each case.}$$

There are two numbers that satisfy the conditions, -3 and 2.

Distance, Rate, Time

In our previous work with distance, rate, and time, we used the formula $d = rt$. If the focus of an applied problem is on time, then we need a way to represent time in terms of distance and rate. We can do this by solving the formula $d = rt$ for t.

$$rt = d$$

$$t = \frac{d}{r}$$

EXAMPLE 6   *Distance, Rate, and Time*

From Front Royal, Virginia, a couple drove 210 miles on the scenic route to Roanoke, Virginia. From Roanoke they traveled Interstate 81 for 120 miles to Wytheville, Virginia. The average speed on the interstate was 25 mph faster than along the scenic route. The entire trip took 8 hours. How long did they drive on the scenic route and how long on the interstate? What was their average speed on each?

Solution

Let $s =$ the average speed on the scenic route. Then $s + 25$ is the average speed on the interstate.

	d	r	$t = \dfrac{d}{r}$
Scenic	210	s	$\dfrac{210}{s}$
Interstate	120	$s + 25$	$\dfrac{120}{s + 25}$

Because the total time is 8 hours, we have the following equation.

$$\frac{210}{s} + \frac{120}{s + 25} = 8$$

$$s(s + 25) \cdot \frac{210}{s} + s(s + 25) \cdot \frac{120}{s + 25} = 8 \cdot s(s + 25) \qquad \text{Multiply by the LCD } s(s + 25).$$

$$210(s + 25) + 120s = 8s(s + 25) \qquad \text{Simplify each term.}$$

$$210s + 5250 + 120s = 8s^2 + 200s \qquad \text{Distributive Property}$$

$$330s + 5250 = 8s^2 + 200s \qquad \text{Combine like terms.}$$

$$8s^2 - 130s - 5250 = 0 \qquad \text{Write the quadratic equation in standard form.}$$

$$2(4s^2 - 65s - 2625) = 0 \qquad \text{GCF is 2.}$$

$$2(4s + 75)(s - 35) = 0 \qquad \text{Factor the trinomial.}$$

$$4s + 75 = 0 \quad \text{or} \quad s - 35 = 0 \qquad \text{Zero Factor Property}$$

$$4s = -75 \quad \text{or} \quad s - 35 = 0 \qquad \text{Solve each case for } s.$$

$$s = -18.75 \quad \text{or} \quad s = 35$$

Because speed is not negative, the only solution is $s = 35$. The average speed on the scenic route was 35 mph and the average speed on the interstate was 60 mph.

The driving time on the scenic route was 210/35 or 6 hours and the driving time on the interstate was 120/60 or 2 hours. ■

EXAMPLE 7   *Distance, Rate, and Time*

A canoeist can paddle 2 miles up the Olentangy River in the same time she can paddle 6 miles downstream. The speed of the current is 2 mph. What is her speed in still water?

Solution   Let r = rate of canoe in still water.

	d	r	$t = \dfrac{d}{r}$
Upstream	2	$r - 2$	$\dfrac{2}{r - 2}$
Downstream	6	$r + 2$	$\dfrac{6}{r + 2}$

Because the time is the same for the upstream and downstream trips, the equation is as follows.

$$\frac{2}{r-2} = \frac{6}{r+2}$$

$$(r+2)(r-2) \cdot \frac{2}{r-2} = (r+2)(r-2) \cdot \frac{6}{r+2} \qquad \text{Multiply by the LCD } (r+2)(r-2).$$

$$2(r+2) = 6(r-2) \qquad \text{Simplify each term.}$$

$$2r+4 = 6r-12 \qquad \text{Distributive Property}$$

$$-4r = -16 \qquad \text{Solve for } r.$$

$$r = 4$$

The canoeist paddles at a rate of 4 mph in still water.

Percentages

Because percentages are actually ratios (multiplied by 100), applications involving percentages sometimes lead to equations with rational expressions.

EXAMPLE 8 *Basketball Free Throw Percentage*

A basketball player has made 32 free throws in 45 attempts for a current success rate of 71%. How many consecutive free throws must the player make to raise the average to 75%?

Solution Let x = required number of consecutive successful free throws. The success rate is determined by the expression

$$\frac{\text{Free Throws Made}}{\text{Free Throws Attempted}}.$$

The numerator of this formula is 32 plus the number of free throws the player needs to make: $32 + x$.

The denominator of the formula is 45 plus the number of free throws the player will attempt: $45 + x$.

Because the goal is a 75% success rate, the equation is as follows.

$$\frac{32+x}{45+x} = 0.75$$

$$(45+x) \cdot \frac{32+x}{45+x} = 0.75 \cdot (45+x) \qquad \text{Multiply by the LCD } 45 + x.$$

$$32+x = 0.75(45+x) \qquad \text{Simplify each term.}$$

$$32+1.00x = 33.75+0.75x \qquad \text{Distributive Property}$$

$$0.25x = 1.75 \qquad \text{Solve for } x.$$

$$x = 7$$

Seven consecutive successful free throws will raise the player's average to 75%.

8.6 *Exercises*

Work Rate

1. If it takes 5 hours to complete a certain job, how much of the job is completed in 1 hour? in 3 hours? in 5 hours? In general, if it takes h hours to complete a task, what portion of the task is completed in 1 hour?

2. If two machines can complete a certain job in 12 and 15 minutes, respectively, write an expression that represents the portion of the job completed in 1 minute with the two machines working together.

3. Working together, two people can wash their car in 10 minutes. One person, working alone, can wash the car in a half hour. How long does it take the other person to do the job working alone?

4. Using a 21-inch mower, a person can mow the front lawn in 45 minutes. It takes only a half hour with a 36-inch mower. If two people work together, one with the smaller mower and the other with the larger mower, how long will it take to mow the lawn?

5. A professional painter can paint the outside trim on a certain house as fast as two apprentices working together. One apprentice can do the painting alone in 15 hours; the other apprentice can finish the job in 10 hours. How long would it take the painter to paint the trim without help?

6. Working together, two front-end loaders can fill a railroad car with gravel in 20 minutes. Working alone, the smaller machine would take 9 minutes longer than the larger machine would take working alone. How long would it take each of the loaders working alone?

Inverse Variation

7. Which of the following problem situations can be modeled with an inverse variation?

   (i) As the demand for a certain item increases, the cost of the item also increases.

   (ii) As the temperature in Michigan decreases, the number of tourists in Florida increases.

8. If r is inversely proportional to m, which of the following is true? What do we call k in either case?

   (i) $r = \dfrac{k}{m}$ (ii) $r = km$

9. If y varies inversely as x, and if $y = 9$ when $x = 5$, what is y when $x = 3$?

10. If a varies inversely as c, and if $a = 12$ when $c = 2$, what is a when $c = 10$?

11. If w varies inversely as z^2, and if $z = 2$ when $w = 24$, what is w when $z = 4$?

12. If y varies inversely as x^2, and if $x = 2$ when $y = 150$, what is y when $x = 10$?

13. Suppose that it takes 9 hours for four workers to roof a house. If the time required to do the job varies inversely as the number of people working on it, how long does it take for five workers to finish a roof of the same size?

14. Suppose that it takes three 8-hour days for 5 workers to landscape a yard completely. If the time required to do the job varies inversely as the number of people working on it, how long would it take for 16 workers to landscape the same yard?

15. Suppose the price of oil varies inversely with the supply. If an OPEC nation can sell oil for $26.00 per barrel when daily production is 3 million barrels, what will the price of oil be if the daily production is increased to 4 million barrels?

16. The average number of words that can be printed on standard paper varies inversely with the font size. If 320 words can be printed with a 10-point font, how many words can be printed with a 12-point font?

Ratio and Proportion

17. An arithmetic rule states that if $\dfrac{a}{b} = \dfrac{c}{d}$,

    then the **cross-products** ad and bc are equal. Explain why this is so.

18. For the two similar triangles, six students wrote equations that they believed to be true. Which are correct? Why are the others incorrect?

(i) $\dfrac{x}{y} = \dfrac{c}{d}$ (ii) $\dfrac{x}{c} = \dfrac{y}{d}$ (iii) $\dfrac{d}{y} = \dfrac{c}{x}$

(iv) $dx = cy$ (v) $\dfrac{y}{c} = \dfrac{d}{x}$ (vi) $cx = dy$

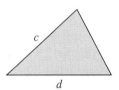

19. On a map, $\frac{1}{2}$ inch represents 10 miles. If the distance from Minneapolis to Duluth is 7.7 inches on the map, how many miles separate the two cities?

20. The two triangles *ABC* and *DEF* are similar triangles. Determine the lengths of the sides $\overline{AC}$ and $\overline{DE}$.

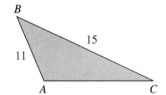

21. How many minutes will it take a train traveling at 50 mph to travel three-fifths of a mile?

22. What is the average speed of an automobile that travels 40 miles in 1 hour and 20 minutes?

Numbers

23. Consider the following two problems.

  (i) Find the sum of $\frac{1}{3}$ and the reciprocal of a number.

  (ii) If the sum of $\frac{1}{3}$ and the reciprocal of a number is 2, what is the number?

For both problems, the LCD is $3x$. Compare the two different ways in which the LCD is used to solve the problems.

24. Consider two positive numbers represented by x and $x + 3$. Suppose the difference between the reciprocals of these two numbers is positive. Is this fact expressed with

$$\frac{1}{x} - \frac{1}{x+3} \quad \text{or} \quad \frac{1}{x+3} - \frac{1}{x}?$$

Why?

25. Twice a number plus three times its reciprocal is 7. Find the number.

26. The difference between three times a number and eight times its reciprocal is 10. Find the number.

27. One number is 4 less than the other. The sum of the reciprocal of the larger number and three times the reciprocal of the smaller number is $\frac{3}{5}$. What are the numbers?

28. One number is three more than another number. The sum of the reciprocal of the smaller number and two times the reciprocal of the larger number is $\frac{9}{20}$. What are the numbers?

Distance, Rate, Time

29. The formula $d = rt$ can be rewritten as $r = d/t$. If we treat d as a constant, explain why r varies inversely with t.

30. If a person travels 200 miles at an average speed of r mph, and then travels another 100 miles at an average speed that is 5 mph faster, write an expression that represents the total time for the trip. What additional information would be needed in order to determine the average speeds?

31. A small plane travels 90 mph faster than a train. The plane travels 525 miles in the same time it takes the train to travel 210 miles. Determine the rate of each.

32. A private plane takes 4 hours less time than a car to make a trip of 495 miles. If the plane's rate is $\frac{9}{5}$ of the car's rate, what is the rate of each?

33. Two cyclists competed in a 5-mile bicycle race. The more experienced cyclist gave the novice cyclist a head start of 0.6 miles and still won the race by 2 minutes. If the experienced cyclist's average speed was 3 mph faster than the novice's average speed, what was the novice's average speed?

34. A pitcher throws a fastball to a catcher standing 66 feet away. At the same instant, the shortstop throws a ball to the first baseman standing 124 feet away. The first baseman catches the throw from the shortstop a half second after the catcher catches the pitch. If the pitcher can pitch 8 feet per second faster than the shortstop can throw, how long does a batter have to react to the pitcher's fastball?

Percentages

35. Samples of light bulbs from each production lot are tested. A production lot passes inspection if at least 99% of the sampled bulbs work properly. If a production lot consists of 2000 bulbs, and if the sample accurately reflects their working condition, explain how to predict the number of bulbs from each lot that will not work.

36. A baseball player's batting average is calculated by the formula

$$\text{Batting Average} = \frac{\text{Number of Hits}}{\text{At Bats}}.$$

However, batting averages, such as 0.315, are usually reported as if they were integers, such as 315. A batter who averages 300 is considered an excellent hitter. Explain how to determine the percentage of at bats at which such a batter is unsuccessful.

37. A softball player has a batting average of 0.252 with 112 hits out of 444 times at bat. If the batter goes on a hitting streak, how many hits in a row would raise the batting average to 0.267?

38. A baseball player has a batting average of 0.295 with 105 hits out of 356 times at bat. The batter then goes into a hitless streak and the batting average drops to 0.269. How many consecutive at bats without a hit did the batter have?

39. In 1980, 360 male students attended a certain college. By 1990, the number of male students had increased by 300 and the number of female students had increased by 700. If the percentage of male students was the same in 1990 as it was in 1980, what was the total enrollment in 1980?

40. A group of people suffering from a particular ailment agreed to participate in an experimental testing of a new medication. Some of the people received the medication while the others received a placebo (a preparation containing no medicine and given for its psychological effect). Of the group receiving the medication, 35

claimed it helped. Ten people in the group that received the placebo also said they felt much better. For both groups combined, 50% reported no beneficial effect. How many people were tested?

Miscellaneous

41. Consider two data items *A* and *B*. If an increase in *A* causes a corresponding decrease in *B*, or if a decrease in *A* causes a corresponding increase in *B*, we say that *A* and *B* are inversely related. Which of the following would you suspect are inversely related?

 (a) Humidity (*A*) and drying time of varnish (*B*)

 (b) Number of amateur skiers (*A*) and the price of ski equipment (*B*)

 (c) Illiteracy rate (*A*) and library circulation (*B*)

 (d) Number of auto accidents (*A*) and insurance premiums (*B*)

 (e) Unemployment (*A*) and home ownership (*B*)

42. A soft-drink bottling company uses a vat that can be filled with carbonated water in 1 hour. The vat can be filled with syrup in 9 hours. The vat of soft-drink mixture can be emptied to the bottling apparatus in 3 hours. If the operator opens all three valves and then takes a 90-minute lunch break, describe the operator's state of mind when he returns from lunch and show why.

43. In the accompanying figure, a yardstick is placed perpendicular to the ground, and it casts a shadow 15 inches long. If the nearby oak tree casts a 25-foot shadow, how tall is the oak tree?

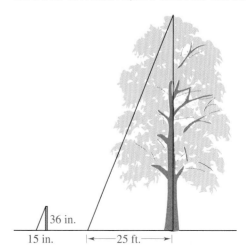

36 in.

15 in. |—— 25 ft. ——|

44. It takes one person twice as long to shovel snow from the driveway as it takes another using a snow blower. If the two of them together can clear the driveway in 8 minutes, how long does it take the person shoveling alone?

45. Two self-employed house cleaners can wash the windows of a particular client in a half hour. Working alone, one cleaner could wash all the windows in 11 minutes less time than it would take the other, but the faster cleaner refuses to do windows. Working alone, how long will it take the other cleaner to wash the windows?

46. The number of hairs in an animal's coat varies inversely with the average winter temperature. One year, the average winter temperature was 10° lower than the previous year, and an animal was found to have 1.5 times as many hairs as it had the previous year. What was the average winter temperature during the previous year?

47. A motorist took the interstate from Cincinnati to Columbus, a distance of 80 miles. On the return trip the motorist took some back roads, which reduced the average speed by 15 mph. If the total driving time for the round trip was $3\frac{1}{9}$ hours, what was the average speed on the interstate?

48. A sprinter ran the 100-yard dash for the college track team. It took 1 second longer to run the final 50 yards than it took to run the first 50 yards. If the sprinter's average speed for the total distance was $9\frac{1}{11}$ yards per second, what was the time for the final 50 yards?

49. Two people inherited a total of $76,000. The will specified that the inheritance was to be divided in a 3:2 ratio. How much did each receive?

50. The reciprocal of the product of two consecutive even integers is $\frac{1}{48}$. Find the two integers.

51. The reciprocal of the product of two consecutive odd integers is $\frac{1}{35}$. Find the two integers.

52. When the wind speed is 40 mph, a plane can fly 200 miles against the wind in the same time it can fly 300 miles with the wind. What is the speed of the plane?

53. A 1200-bushel grain bin can be filled in 3 hours and emptied in 2 hours. If grain is added to a full bin at the same time that the bin is being emptied, how long will it be before there is no grain in the bin?

54. An elevated water tank serves both commercial and residential customers. The tank can be filled in 14 hours. Commercial usage alone can empty the tank in 7 hours and residential usage alone can empty the tank in 9 hours. If the intake valve is open at the same time both outtake valves are open, how long will it take to exhaust the water supply if the tank is initially full?

55. A boat can travel 15 miles upstream and 15 miles back downstream in a total of 4 hours. If the speed of the current is 5 mph, what is the speed of the boat in still water?

56. Two cats can eat a 7-pound bag of cat food in 24 days. One cat can eat a 3.5 pound bag of food in 3 weeks. How long would it take the other cat to eat a 3.5 pound bag?

57. A clothing chain has noticed that the number of down parkas sold by any one of their stores is inversely proportional to the average January temperature in the city where the store is located. If the store in Houston, Texas, where the average January temperature is 50°F, sells 500 parkas, how many parkas are sold in Great Falls, Montana, where the average January temperature is 21°F? What would the average January temperature be in a city that sold 4000 parkas?

58. One number is 3 more than the other. The sum of their reciprocals is $\frac{1}{2}$. What are the numbers?

59. Twice a number plus 1 is divided by 1 less than half the number. The result is 5. What is the number?

Exploring with Real Data

60. During the year 1992 a dividend of $79.20 was paid on 60 shares of AT&T stock. At this rate, how many additional shares were needed to earn a dividend of $176.88? (Source: AT&T.)

61. During the year 1992, a dividend of $165.00 was paid on 75 shares of General Electric stock. At this rate, how many additional shares were needed to earn a dividend of $347.60? (Source: General Electric Company.)

Challenge

62. In Woodville Notch, all registered voters are either Democrats, Republicans, or Independents. The number of Democrats and Republicans are equal. There are 11 Independents of which 9 are women. In a recent election, five-sixths of the Democrats voted, seven-ninths of the Republicans voted, and all Independents voted. Half of the Democrats are women, and two-thirds of the Republicans are women. All the women voted, which means that 75% of the voters were women. How many registered voters are in Woodville Notch?

63. Pecans worth $30.00 are mixed with cashews worth $45.00. There are 3 pounds more of cashews than of pecans. If the unit price of the cashews is 50 cents greater than the unit price of pecans, how many pounds of each kind of nut were used?

64. A builder bought three sizes of lag bolts to build a gazebo. The builder needed 4 pounds more of medium bolts than of small bolts and 3 pounds fewer of large bolts than of small bolts. The sales invoice showed the following.

Lag Bolts, Small	$3.20
Lag Bolts, Medium	8.40
Lag Bolts, Large	5.50

If the sum of the unit prices for the small and medium bolts is equal to the unit price for the large bolts, how many pounds of each size did the builder buy?

65. The two legs of a right triangle are represented by $\dfrac{3}{x}$ and $\dfrac{12}{x+1}$. If the hypotenuse is represented by $\dfrac{5}{x}$, determine the lengths of the sides of the triangle.

66. For the right triangle ABC in the accompanying figure, a geometry theorem states that $(AQ)^2 = BQ \cdot QC$. Use the given representations of lengths to determine AQ.

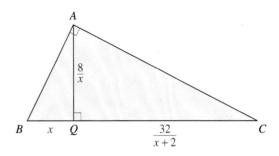

67. In the figure the square on the left and the rectangle on the right have dimensions as shown. If the area of the rectangle is 3 square inches greater than the area of the square, find the dimensions of each figure.

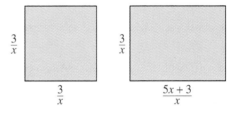

8   Chapter Review Exercises

Section 8.1

1. For $f(x) = \dfrac{x - 7}{x^2 - 2x - 15}$,

find two numbers a_1 and a_2 for which $f(a_1)$ and $f(a_2)$ are undefined. What do we call a_1 and a_2 in this case?

2. Consider the function $g(x) = \dfrac{1}{x - 5}$.

Use your calculator to evaluate each of the following.

   (a) $g(6)$ (b) $g(5.5)$ (c) $g(5.01)$

   (d) $g(5.0001)$ (e) $g(5)$

Explain the result in part (e).

In Exercises 3 and 4, determine the domain of the rational expression.

3. $\dfrac{5x - 2}{2x - 5}$ **4.** $\dfrac{3x - 4}{3x^2 + 11x - 20}$

5. Explain why 3 can be divided out of the numerator and denominator of

$$\dfrac{3(x + 1)}{3(x + 5)}, \quad \text{but not of} \quad \dfrac{3x + 1}{3x + 5}.$$

In Exercises 6–9, simplify the rational expression.

6. $\dfrac{2x - 3}{4x^2 - 9}$ **7.** $\dfrac{x^2 + x - 12}{2x^2 - 9x + 9}$

8. $\dfrac{9 - x^2}{x^2 + x - 12}$

9. $\dfrac{ab + 4a - 3b - 12}{ab + 2a - 3b - 6}$

10. General Appliance Company has x refrigerators in its inventory. After receiving a shipment of five refrigerators, the total value of the inventory (in thousands of dollars) is $x^2 - 2x - 35$. Write a simplified expression for the average unit value of the refrigerators in the inventory.

In Exercises 11 and 12, supply the missing quantity.

11. $\dfrac{5}{3x + 12} = \dfrac{?}{3x^2 - 48}$

12. $\dfrac{3}{x^2 + 2x - 8} = \dfrac{?}{(x - 2)(x^2 - x - 20)}$

Section 8.2

In Exercises 13 and 14, multiply and simplify.

13. $\dfrac{8x - 40}{4x^3} \cdot \dfrac{4x^2 + 20x}{x^2 - 25}$

14. $\dfrac{x^2 - x - 56}{x^2 - 49} \cdot \dfrac{8x + 64}{x^2 - 64}$

15. What is the product of

$$\dfrac{x^2 + 3x + 2}{x^2 + 8x + 12} \quad \text{and} \quad \dfrac{x^2 + 7x + 6}{x^2 - x - 2} ?$$

16. Multiply $\dfrac{2x^3}{4x^4 - 36x^2}$ by $x^2 - 2x - 15$.

17. Consider two nonzero numbers x and y.

   (a) What is the reciprocal of their product?

   (b) What is the product of their reciprocals?

   (c) Are the results found in parts (a) and (b) equivalent?

In Exercises 18 and 19, divide and simplify.

18. $\dfrac{3x}{x + 9} \div \dfrac{15x^2}{x^2 + 7x - 18}$

19. $\dfrac{x^2 - 5x + 6}{x^2 + 3x - 18} \div \dfrac{x^2 - 4}{6x + 36}$

20. Divide

$$\dfrac{3x - 3}{8x^3} \quad \text{by} \quad \dfrac{x^2 - 1}{8x^2 + 8x}.$$

21. What is the quotient of

$$\dfrac{x^2 + 3x - 4}{x^2 - 2x - 24} \quad \text{and} \quad \dfrac{x^2 - 5x + 4}{x^2 - 7x + 6} ?$$

22. List all the restricted values for the problem in Exercise 21.

23. The length and width of a rectangle are, respectively,

$$\frac{x^2 + 5x + 6}{x^2 - 3x - 18} \quad \text{and} \quad \frac{2x - 12}{x^2 - 4}.$$

Write a simplified expression for the area of the rectangle.

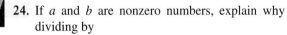 **24.** If a and b are nonzero numbers, explain why dividing by

$$\frac{a^{-1}}{b^{-1}}$$

is the same as multiplying by $\dfrac{a}{b}$.

Section 8.3

In Exercises 25 and 26, determine the LCD.

25. $\dfrac{1}{x^2 - 49}, \ \dfrac{1}{9x - 63}$

26. $\dfrac{1}{x^2 + 3x - 2}, \ \dfrac{1}{x^2 - 4}, \ \dfrac{1}{x^2 - x - 2}$

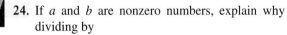 **27.** To add $\frac{2}{3}$ and $\frac{5}{3}$, we use the Distributive Property as follows.

$$\frac{2}{3} + \frac{5}{3} = \frac{1}{3} \cdot 2 + \frac{1}{3} \cdot 5 = \frac{1}{3}(2 + 5)$$

$$= \frac{1}{3} \cdot 7 = \frac{7}{3}$$

With this as background, explain why it is not possible to add fractions that have unlike denominators.

In Exercises 28–31, perform the indicated operations.

28. $\dfrac{5x + 3}{x - 1} - \dfrac{9x - 1}{x - 1}$

29. $\dfrac{2x - 1}{5x - 4} + \dfrac{6x - 6}{4 - 5x}$

30. $\dfrac{3}{10x} + \dfrac{8}{5x^2}$

31. $\dfrac{7}{x^2 + x - 12} + \dfrac{3}{x^2 - 16}$

32. Determine the unknown numerator.

$$\frac{x}{5x + 10} = \frac{?}{5x^2 - 25x - 70}$$

33. Write $2x^{-1} + 3(x + 2)^{-1}$ as a single rational expression with no negative exponents.

34. What is the sum of

$$\frac{x}{x^2 + 8x + 15} \quad \text{and} \quad \frac{15}{x^2 + 4x - 5}?$$

35. Subtract

$$\frac{1}{x^2 - x - 30} \quad \text{from} \quad \frac{7}{x^2 + 2x - 48}.$$

36. Consider any two consecutive odd integers. Find the sum of their reciprocals and show that the numerator of the sum is always an even number.

Section 8.4

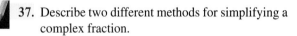 **37.** Describe two different methods for simplifying a complex fraction.

In Exercises 38 and 39, simplify the complex fraction.

38. $\dfrac{9 - \dfrac{1}{x^2}}{3 + \dfrac{1}{x}}$

39. $\dfrac{\dfrac{a}{9} - \dfrac{1}{a}}{\dfrac{1}{3} + \dfrac{a + 4}{a}}$

40. The area of a rectangle is $1 + \dfrac{2}{x} - \dfrac{3}{x^2}$.

If the length of the rectangle is $1 - \dfrac{9}{x^2}$,

write a simplified expression for the width.

41. For nonzero numbers m and n, show that

$$\frac{\dfrac{m^{-1}}{n^{-1}}}{\dfrac{n}{m}} = 1.$$

42. Rewrite $\dfrac{1 - 9x^{-2}}{2 + 3x^{-1} - 9x^{-2}}$

with positive exponents and simplify.

In Exercises 43 and 44, simplify the complex fraction.

43. $\dfrac{\dfrac{x+4}{x-4} + \dfrac{x-4}{x+4}}{\dfrac{1}{x+4} + \dfrac{1}{x-4}}$

44. $\dfrac{\dfrac{12x}{y^2} + \dfrac{13}{xy} - \dfrac{35}{x^3}}{\dfrac{10}{x^3} - \dfrac{3}{xy} - \dfrac{4x}{y^2}}$

Section 8.5

 45. Suppose that you make no errors in solving an equation containing rational expressions. How is it possible that the apparent solution does not check? What do we call the apparent solution in this case?

 46. Suppose that you are solving an equation containing rational expressions. After you clear the fractions and simplify, you obtain an equation of the form $a = b$, where a and b are constants. How do you interpret this result?

In Exercises 47–54, solve.

47. $10x - \dfrac{2}{x} = 1$

48. $\dfrac{x}{5} + \dfrac{5}{4x} = \dfrac{5}{4}$

49. $\dfrac{x+3}{x-5} = \dfrac{8}{x-5}$

50. $\dfrac{5}{x+3} + \dfrac{8}{x-3} = \dfrac{5}{x^2-9}$

51. $\dfrac{7}{x+1} - \dfrac{2}{x-1} = \dfrac{1}{x^2-1}$

52. $\dfrac{9}{3-x} + \dfrac{9}{x+3} = \dfrac{-54}{x^2-9}$

53. $\dfrac{2x}{x-2} - \dfrac{16}{x^2-4} = 2$

54. $\dfrac{6}{x-1} + \dfrac{7}{x^2-5x+4} = \dfrac{2}{x-4}$

55. Use a calculator graph to estimate the number of solutions of the equation
$$x^2 + 1 = \dfrac{3}{x-2}.$$

56. Solve $\dfrac{1}{x} + \dfrac{1}{y} = \dfrac{1}{3}$ for x.

Section 8.6

57. Working alone, it takes twice as long for a helper to do a welding job as it would take for an experienced welder. If it takes 8 hours for the two to complete the job working together, how long does it take the helper working alone?

58. Seven members of the Future Farmers of America painted a small barn in 6 hours. If two additional members had helped, how long would it have taken to paint the barn?

59. A chemical tank truck can be filled through an intake pipe in 2.5 hours. With the outtake pipe open, the tank can be emptied in 3 hours. If the tank is initially empty, how long does it take to fill the tank with both pipes open?

60. A baseball player has a batting average of 0.244 with 42 hits out of 172 times at bat. If he goes on a hitting streak, how many hits in a row would raise his batting average to 0.278?

61. The total travel time for a family to reach a campsite is 2 hours. They drive 81 miles and hike 2 miles. If they drive 50 mph faster than they hike, what is their average driving speed?

62. During the first half of the season, a basketball player missed 7 free throw attempts. During the second half, she made twice as many free throws as she made during the first half, but she had 16 more attempts than she had during the first half. If her free throw percentage at the end of the season was 75%, how many free throws did she make for the season?

8 Chapter Test

1. For $\dfrac{x - 3}{x^2 + 5x - 24}$, explain why 3 is a restricted value.

2. Determine the domain of $\dfrac{2x - 1}{x^2 + 3x}$.

For Questions 3–5, simplify.

3. $\dfrac{6x^2 + 12x}{3x^3 - 6x^2}$

4. $\dfrac{(x + 5)^2}{x^2 - 4x - 45}$

5. $\dfrac{x^2 + xy - 2y^2}{y^2 - x^2}$

For Questions 6 and 7, perform the indicated multiplication or division.

6. $\dfrac{t^2 + 3t - 4}{t^2 + 2t - 3} \cdot \dfrac{t^2 - t - 6}{t^2 + t - 12}$

7. $\dfrac{y^2 - 25}{2y + 10} \div (y^2 - 10y + 25)$

8. Determine the missing numerator.

$$\frac{3x}{x^2 - x - 2} = \frac{?}{(x - 2)^2(x + 1)}$$

9. Explain why

$$\frac{(m + n)^{-1}}{r^{-1}} - \frac{r - 1}{m + n}$$

is an easier problem than it appears to be at first and show that the result does not involve r.

For Questions 10–12, perform the indicated addition or subtraction.

10. $\dfrac{7x - 1}{3x + 4} - \dfrac{4x - 5}{3x + 4}$

11. $\dfrac{3}{t^2} - \dfrac{5}{t}$

12. $\dfrac{x - 7}{2x^2 + 9x - 5} + \dfrac{4 - x}{4x^2 + 23x + 15}$

13. Which of the following is a complex fraction? Show that two of the expressions are equivalent.

(i) $\dfrac{\dfrac{m}{n}}{p}$ (ii) $\dfrac{m}{\dfrac{n}{p}}$ (iii) $\dfrac{\dfrac{m}{n}}{\dfrac{1}{p}}$

For Questions 14 and 15, simplify the complex fraction.

14. $\dfrac{\dfrac{x + 3y}{y}}{\dfrac{x^2 - 9y^2}{6y}}$

15. $\dfrac{x - 3 + \dfrac{x - 3}{x^2}}{x + 2 + \dfrac{1}{x} + \dfrac{2}{x^2}}$

16. Write

$$\frac{a^{-2}}{a^{-2} - b^{-2}}$$

as a single rational expression with positive exponents.

For Questions 17–20, solve the equation.

17. $\dfrac{3}{a + 3} = \dfrac{2}{a - 2}$

18. $1 - \dfrac{3}{x} = \dfrac{10}{x^2}$

19. $\dfrac{x}{x - 2} + \dfrac{2}{3} = \dfrac{2}{x - 2}$

20. $\dfrac{8t}{t^2 - 16} = \dfrac{3}{t + 4} + \dfrac{5}{4 - t}$

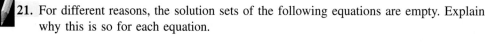

21. For different reasons, the solution sets of the following equations are empty. Explain why this is so for each equation.

(a) $\dfrac{x - 2}{x - 1} = \dfrac{1}{1 - x}$

(b) $\dfrac{x - 2}{x - 1} = \dfrac{x + 3}{x - 1}$

22. Solve the formula $\dfrac{1}{A} + \dfrac{1}{B} = 2$ for B.

In Questions 23–25, assign a variable, write an equation to describe the conditions of the problem, solve the equation, and answer the question.

23. In a target-shooting competition, scores are found to vary inversely with the wind speed. If the success rate is 60% when the wind speed is 20 mph, what is the success rate when the wind speed is 30 mph?

24. In the spring a cat sheds so fast that the carpet is covered with hair in 4 hours. If the cat is out of the house, it takes 1.5 hours to clean the carpet. How long does it take to clean the carpet with the cat in the house?

25. The time it takes a person to paddle a kayak 2 miles downstream is the same as the time it takes to paddle a half mile upstream. If the rate of the current is 3 mph, what is this person's paddling rate in still water?

7-8   Cumulative Test

1. Perform the indicated operations.

$(3x^2 - 6x + 8) - (x^3 - 2x^2 - 7) + (2x^3 - 4x^2 + 7x - 16)$

2. Multiply and simplify.

$(x - 3)^2 - (x + 3)(x - 3)$

3. Factor completely.

(a) $x^4y^3 - x^3y^4$ (b) $ax + 3ay - 7bx - 21by$

4. Factor completely.

(a) $(2x + 1)^2 - 16$ (b) $2x^2 - 12x + 18$ (c) $y^3 + 8$

5. Factor completely.

(a) $z^2 + 2z - 63$ (b) $a^2 - 4ab - 45b^2$

(c) $12 + 9x - 3x^2$

6. Explain how the Zero Factor Property can be used to solve $x^2 - 3x = 0$.

7. Solve.

(a) $(x - 4)(x - 3) = 12$ (b) $a^4 + 25 = 26a^2$

8. If the square of a number is decreased by 24, the result is 5 times the number. What is the number?

9. Find the quotient and remainder for the following.

$(4c^4 + 5c^2 - 7c) \div (2c - 1)$

10. Divide synthetically. Then use the results to write the equation Dividend = (Quotient)(Divisor) + Remainder.

$(x^5 - 3x^3 - 17x^2 + x - 13) \div (x - 3)$

11. Explain how the Factor Theorem can be used to determine whether $x - c$ (where c is a constant) is a factor of a polynomial $P(x)$.

12. If $P(x) = 2x^3 - 5x^2 + x - 1$, use the Remainder Theorem to evaluate $P(2)$.

13. Simplify.

$$\frac{x^2 - x - 12}{4x^2 - 13x - 12}$$

14. Multiply.

$$\frac{a^2 - 16}{a^2 - 3a - 4} \cdot \frac{a^2 + 3a + 2}{a^2 - 2a - 8}$$

15. Add.

$$\frac{3}{y^2 + 11y + 10} + \frac{y}{y^2 + 5y + 4}$$

16. Simplify.

$$\frac{\dfrac{1}{x+3}}{\dfrac{1}{x} - \dfrac{1}{x+3}}$$

17. Solve.

$$\frac{2}{x} - \frac{x+8}{x+2} = \frac{12}{x^2 + 2x}$$

18. A landscape worker can lay sod in a yard in 6 hours. It would take a helper 8 hours to do the same job. How long would it take to sod the yard if both people work together?

19. A driver of a car averaged 5 mph faster during the final 120 miles of a 340-mile trip. If the entire trip took 6 hours, at what average speed did the driver travel during the first 220 miles?

9 Radical Expressions

Suppose that you are a staff member with the National Park Service. Faced with budgetary cuts, you assemble data on the number of visitors to national monuments (see graph) and on the cost of administering these parks.

The data in the graph can be modeled by an expression with a **rational exponent.** This model, along with cost data, can be used to show the cost per visitor to national monuments and to project budgetary requirements for future years. (See Exercises 85–88 at the end of Section 9.3 for more on this application.)

In Chapter 9 we discuss radical and rational exponential expressions. We learn about the properties of such expressions and about how to simplify and perform operations with them. Next, we develop methods for solving equations with radicals and exponents. We conclude by extending the number system to the set of complex numbers.

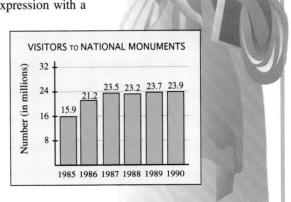

VISITORS TO NATIONAL MONUMENTS

(Source: U.S. National Park Service.)

9.1 Radicals

Higher Order Roots ▪ *Evaluating $\sqrt[n]{b^n}$*

Higher Order Roots

In Chapter 1 we defined a to be a **square root** of b if $a^2 = b$. We saw, for example, that 5 and -5 are both square roots of 25 because $5^2 = 25$ and $(-5)^2 = 25$. We also defined the **principal square root** of b to be the *positive* number a such that $a^2 = b$. The notation $\sqrt{b}$ represents the principal square root of b, where $b \geq 0$.

The concept of square root can be generalized to higher order roots. For example, 3 is the **third root** or **cube root** of 27 because $3^3 = 27$; 2 and -2 are **fourth roots** of 16 because $2^4 = 16$ and $(-2)^4 = 16$.

> ### Definition of nth Root
>
> For any real numbers a and b and any positive integer $n > 1$, if $a^n = b$, then a is an **nth root** of b.

In the following exploration, we consider the number of nth roots, if any, that a number can have.

EXPLORATION 1   *nth Roots*

(a) What is the fourth root of 81? the sixth root of 64? If n is even and $b > 0$, what is your conjecture about the number of nth roots of b?

(b) What is the fourth root of -16? the eighth root of -1? If n is even and $b < 0$, what is your conjecture about the number of nth roots of b?

(c) What is the third root of 27? of -27? What is the fifth root of 32? of -32? If n is odd, what is your conjecture about the number of nth roots of b?

(d) What are the seventh and twelfth roots of 0? What is the nth root of 0 for any $n > 1$?

Discovery

(a) Because $3^4 = 81$ and $(-3)^4 = 81$, 3 and -3 are both fourth roots of 81. Because $2^6 = 64$ and $(-2)^6 = 64$, 2 and -2 are both sixth roots of 64. If n is even and $b > 0$, b has two nth roots.

(b) There is no real number a such that $a^4 = -16$. Nor is there a real number a such that $a^8 = -1$. If n is even and $b < 0$, b has no nth roots.

(c) Because $3^3 = 27$, 3 is the third root of 27. By similar reasoning, -3 is the third root of -27, 2 is the fifth root of 32, and -2 is the fifth root of -32. If n is odd, then any real number b has one nth root.

(d) The seventh and twelfth roots of 0 are 0 because $0^7 = 0$ and $0^{12} = 0$. In fact, for *any* $n > 1$, the nth root of 0 is 0 because $0^n = 0$.

The following is a summary of the results in Exploration 1.

Number of Real nth Roots of a Number

Assume b is a real number and let n be a positive integer greater than 1.

1. If $b = 0$, then b has *one* nth root: 0.

2. If n is even, and

    (a) if $b > 0$, then b has *two* nth roots;

    (b) if $b < 0$, then b has *no* real number nth roots.

3. If n is odd, then any real number b has *one* nth root.

When b has two nth roots (that is, when n is even and $b > 0$), one root is positive and the other root is negative. We will refer to the *positive* nth root as the **principal nth root.**

In Chapter 1 we used the symbol $\sqrt{b}$ for the principal square root of b. A similar symbol is used for nth roots of b.

Definition of $\sqrt[n]{b}$

For any real number b and any positive integer $n > 1$, if b has one nth root, then $\sqrt[n]{b}$ is the nth root of b. If b has two nth roots, then $\sqrt[n]{b}$ is the principal nth root of b.

The symbol $\sqrt{}$ is called the **radical sign.** The number n is called the **index of the radical,** and b is called the **radicand.** For a square root, the index is usually omitted. It is understood that $\sqrt{b}$ is the same as $\sqrt[2]{b}$.

For an even index, the radicand cannot be negative. For an odd index, the radicand can represent any real number.

For even integers $n > 1$, $\sqrt[n]{b}$ is defined for $b \geq 0$. For odd integers $n > 1$, $\sqrt[n]{b}$ is defined for all real numbers.

EXAMPLE 1 *Evaluating nth Roots*

Evaluate.

(a) $\sqrt[3]{-64}$ (b) $\sqrt{-64}$ (c) $-\sqrt{64}$

(d) $\sqrt[4]{81}$ (e) $\sqrt[6]{-1}$ (f) $\sqrt[9]{-1}$

Solution

		n	*b*	*Justification*
(a)	$\sqrt[3]{-64} = -4$	3	-64	$(-4)^3 = -64$
(b)	$\sqrt{-64}$ is not a real number	2	-64	*n* even, $b < 0$
(c)	$-\sqrt{64} = -8$	2	64	$-1 \cdot \sqrt{64} = -1 \cdot 8 = -8$
(d)	$\sqrt[4]{81} = 3$	4	81	$3^4 = 81$
(e)	$\sqrt[6]{-1}$ is not a real number	6	-1	*n* even, $b < 0$
(f)	$\sqrt[9]{-1} = -1$	9	-1	$(-1)^9 = -1$

Evaluating $\sqrt[n]{b^n}$

For *even* integers $n > 1$, $\sqrt[n]{b^n}$ is the *principal* *n*th root of b^n. Because $(b)^n = b^n$ and $(-b)^n = b^n$ when *n* is even, $\sqrt[n]{b^n}$ is either *b* or $-b$, depending on which one is positive. To ensure a positive result, we write $|b|$.

For *odd* integers $n > 1$, $\sqrt[n]{b^n}$ has *only one* *n*th root and the absolute value signs are not necessary.

> ### Evaluating $\sqrt[n]{b^n}$
>
> Let *b* be any real number.
>
> For even integers $n > 1$, $\sqrt[n]{b^n} = |b|$.
>
> For odd integers $n > 1$, $\sqrt[n]{b^n} = b$.

EXAMPLE 2 *Evaluating $\sqrt[n]{b^n}$ with Numerical Radicands*

(a) $\sqrt{(-5)^2} = |-5| = 5$ Because *n* is even, absolute value symbols are needed.

(b) $\sqrt[5]{(-4)^5} = -4$ Because *n* is odd, absolute value symbols are not used.

(c) $\sqrt[4]{3^4} = |3| = 3$ Because *n* is even, absolute value symbols are needed.

When the radicand is a variable expression and the index is even, we must use absolute value symbols around the result if there is a possibility that the result could be negative.

EXAMPLE 3 *Evaluating $\sqrt[n]{b^n}$ with Variable Radicands*

(a) $\sqrt{t^6} = \sqrt{(t^3)^2} = |t^3|$ Because t^3 could be negative, absolute value symbols are needed.

(b) $\sqrt{(x-3)^2} = |x-3|$ Because $x - 3$ could be negative, absolute value symbols are needed.

(c) $\sqrt{a^4} = \sqrt{(a^2)^2} = |a^2| = a^2$ Because a^2 cannot be negative, absolute value symbols are not needed.

(d) $\sqrt[3]{t^3} = t$ Because the index is odd, absolute value symbols are not used.

9.1   *Quick Reference*

Higher Order Roots
- For any real numbers a and b and any positive integer $n > 1$, if $a^n = b$, then a is an ***n*th root** of b.

- Assume b is a real number and let n be a positive integer greater than 1.
  1. If $b = 0$, then b has *one* nth root: 0.
  2. If n is even, and
     (a) if $b > 0$, then b has *two* nth roots;
     (b) if $b < 0$, then b has *no* real number nth roots.
  3. If n is odd, then any real number b has *one* nth root.

- For any real number b and any positive integer $n > 1$, if b has one nth root, then $\sqrt[n]{b}$ is the nth root of b. If b has two nth roots, then $\sqrt[n]{b}$ is the **principal *n*th root** of b.

- The symbol $\sqrt{}$ is called the **radical sign.** The number n is called the **index of the radical,** and b is called the **radicand.**

- For an even index, the radicand cannot be negative. For an odd index, the radicand can represent any real number.

Evaluating $\sqrt[n]{b^n}$
- If b is any real number, then for even integers $n > 1$, $\sqrt[n]{b^n} = |b|$. For odd integers $n > 1$, $\sqrt[n]{b^n} = b$.

- When the radicand is a variable expression and the index is even, we must use absolute value symbols around the result if there is a possibility that the result could be negative.

9.1   *Exercises*

 1. Why do the following two questions have different answers?
   (a) What are the square roots of 9?
   (b) What is $\sqrt{9}$?

 2. In the expression $\sqrt[n]{b}$, what do we call n and b?

In Exercises 3–14, determine the specified root(s), if any, of the given number.

3. Square root of 36

4. Square root of $\frac{16}{25}$

5. Square root of -4

6. Square root of -25

7. Third root of 27

8. Third root of -64

9. Fourth root of 625

10. Fourth root of 16

11. Fifth root of -243

12. Fifth root of -32

13. Sixth root of -64

14. Sixth root of 64

In Exercises 15–36, evaluate.

15. $\sqrt{16}$

16. $\sqrt{64}$

17. $\sqrt[3]{-64}$

18. $\sqrt[5]{32}$

19. $\sqrt{\dfrac{4}{9}}$

20. $-\sqrt{\dfrac{25}{144}}$

21. $\sqrt[4]{-16}$ **22.** $\sqrt[6]{-64}$

23. $-\sqrt{16}$ **24.** $-\sqrt{36}$

25. $4\sqrt{25}$ **26.** $-2\sqrt{9}$

27. $\sqrt{(-4)^2}$ **28.** $\sqrt{2^6}$

29. $\sqrt[3]{64}$ **30.** $\sqrt[4]{16}$

31. $-\sqrt[5]{-32}$ **32.** $-\sqrt[3]{-125}$

33. $\sqrt[3]{3^{-9}}$ **34.** $\sqrt[3]{-5^6}$

35. $\sqrt[5]{-2^{10}}$ **36.** $\sqrt[4]{(-3)^4}$

 37. The following statements are correct.

$$\sqrt{x^6} = |x^3| \qquad \sqrt{x^4} = x^2$$

Why are absolute value symbols required for one result, but not the other?

38. Describe the conditions under which $\left(\sqrt[n]{x}\right)^n = \sqrt[n]{x^n} = x$.

In Exercises 39–52, evaluate the radical, if possible. Classify the number as rational, irrational, or not a real number. If the number is irrational, give its value to the nearest hundredth.

39. $\sqrt[6]{64}$ **40.** $\sqrt[3]{-27}$

41. $-\sqrt{10}$ **42.** $\sqrt{8}$

43. $\sqrt{0.0625}$ **44.** $\sqrt{0.36}$

45. $\dfrac{\sqrt{18^2}}{6}$ **46.** $\dfrac{\sqrt{20^2}}{5}$

47. $\sqrt[3]{12}$ **48.** $\sqrt[3]{-15}$

49. $-\sqrt[4]{-16}$ **50.** $-\sqrt[3]{-8}$

51. $\sqrt{(-5)^2}$ **52.** $\sqrt{-4^{-2}}$

In Exercises 53–56, evaluate each expression.

53. $\sqrt{4^2 - 4(3)(1)}$ **54.** $\sqrt{9^2 - 4(2)(9)}$

55. $\sqrt{9^2 - 4(10)(-9)}$ **56.** $\sqrt{19^2 - 4(6)(-7)}$

In Exercises 57–64, translate the given phrase into a radical expression. Assume all variables represent positive numbers.

57. Twice the square root of x.

58. Five less than the cube root of y.

59. The difference of the fifth root of a and 9.

60. The square root of c plus the square root of b.

61. The square root of c plus b.

62. The square root of the quantity $c + b$.

63. The fourth root of the product of x and y.

64. The quotient of the square root of x and the cube root of x.

In Exercises 65–80, determine the indicated root. Assume all variables represent positive numbers.

65. $\sqrt{x^{10}}$ **66.** $\sqrt{x^{16}}$ **67.** $4\sqrt{y^6}$

68. $-3\sqrt{z^8}$ **69.** $\sqrt[3]{x^{21}}$ **70.** $\sqrt[3]{y^{12}}$

71. $-\sqrt[3]{y^{15}}$ **72.** $4\sqrt[3]{a^6}$ **73.** $\sqrt[5]{x^{35}}$

74. $\sqrt[7]{x^{35}}$ **75.** $-\sqrt[4]{x^8}$ **76.** $5\sqrt[6]{y^{18}}$

77. $\sqrt{(3x)^6}$ **78.** $\sqrt{(5ac)^4}$

79. $\sqrt{(x + 3)^{10}}$ **80.** $\sqrt{(x - 2)^{16}}$

In Exercises 81–102, determine the indicated root. Assume all variables represent any real number.

81. $\sqrt{x^6}$ **82.** $\sqrt{t^{10}}$

83. $-\sqrt{x^8}$ **84.** $3\sqrt{y^{12}}$

85. $5\sqrt{y^{14}}$ **86.** $-\sqrt{a^{18}}$

87. $\sqrt[5]{x^{15}}$ **88.** $\sqrt[7]{y^{21}}$

89. $\sqrt[4]{a^4}$ **90.** $\sqrt[6]{y^6}$

91. $\sqrt{(x + 2)^2}$ **92.** $\sqrt{(3 - x)^2}$

93. $\sqrt{(x^2 - 9)^2}$ **94.** $\sqrt{(1 - a^2)^2}$

95. $-2\sqrt{(x - 1)^2}$ **96.** $-\sqrt{(1 + 2x)^2}$

97. $\sqrt{(3x^3 - 5x)^2}$　　　　**98.** $\sqrt{(a^4 + a)^2}$

99. $\sqrt[3]{(1 - x)^3}$　　　　**100.** $\sqrt[5]{(-x + 4)^5}$

101. $\sqrt[8]{(x - 7)^8}$　　　　**102.** $\sqrt[4]{(1 - 3x)^4}$

In Exercises 103–108, insert =, <, or > to describe the correct relationship between the two given quantities. Use your calculator to evaluate the quantities.

103. $\dfrac{2}{\sqrt{3}}$ ____ $\dfrac{3}{\sqrt{2}}$

104. $\dfrac{2\sqrt{5}}{3}$ ____ $\dfrac{10}{3\sqrt{5}}$

105. $\sqrt{7 - 5}$ ____ $\sqrt{7} - \sqrt{5}$

106. $\sqrt{10 + 3}$ ____ $\sqrt{10} + \sqrt{3}$

107. $\sqrt{25 - 9}$ ____ $\sqrt{25} - \sqrt{9}$

108. $\sqrt{9} + \sqrt{16}$ ____ $\sqrt{25}$

In Exercises 109–116, supply the missing number(s) to make each statement true.

109. $\sqrt{x^{(\ )}} = x^2$　　　**110.** $\sqrt[3]{y^{(\ )}} = y^4$

111. $\sqrt[(\ )]{y^{24}} = |y^3|$　　　**112.** $\sqrt[(\ )]{y^4} = |y|$

113. $(\ )\sqrt[4]{16} = -6$　　　**114.** $(\ )\sqrt[3]{-8} = 2$

115. $(\ )\sqrt[5]{x^{(\ )}} = 4x^2$　　　**116.** $(\ )\sqrt[6]{y^{(\ )}} = -y^2$

117. For a rectangle whose length is L and whose width is W, the length d of a diagonal can be found with the formula $d = \sqrt{L^2 + W^2}$. What is the length of a diagonal of a television screen whose width is 21 inches and whose height is 16 inches?

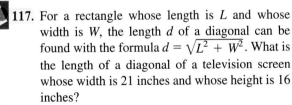

118. If the volume V and the height h of a cylindrical metal drum are known, then the radius r of the drum can be found with the formula

$$r = \sqrt{\dfrac{V}{\pi h}}.$$

Determine the radius of a metal drum whose height is 20 inches and whose volume is 2200 cubic inches.

Exploring with Real Data

In Exercises 119–122, refer to the following background information and data.

The graph in the accompanying figure shows the sales (in millions of dollars) of golf equipment during the period 1985–1990. (Source: National Sporting Goods Association.)

Figure for 119–122

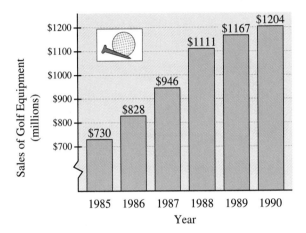

The equipment sales E can be modeled with the function

$$E(t) = 314.6 + 363.1\sqrt{t},$$

where t is the number of years since 1984.

119. For which two years is the model most accurate?

120. For which year is the model least accurate?

121. What *percentage* increase over 1990 does the model project for sales of golf equipment in 1991?

122. Based on the given data, increases in which of the following items might be reasonably expected to have occurred during this same period of time?

(a) Greens fees (for using golf course)

(b) Disposable income in all classes of society

(c) The percentage of female golfers

(d) The cost of manufacturing golf equipment

Challenge

In Exercises 123–126, determine the indicated root.

123. $\sqrt{t^{2n}}$

124. $\sqrt[3]{t^{3n}}$

125. $\sqrt[n]{t^{2n}}$

126. $\sqrt[n]{t^{n^2}}$

127. We define $\sqrt[n]{b}$ for $n > 1$.

(a) Why do we exclude $n = 1$? What meaning would $\sqrt[1]{b}$ have?

(b) Why do we exclude $n = 0$? What meaning would $\sqrt[0]{b}$ have? (Hint: Consider the two cases $b = 1$ and $b \neq 1$.)

9.2   Rational Exponents

Definition of $b^{1/n}$ ▪ *Definition of $b^{m/n}$* ▪ *Negative Rational Exponents*

Definition of $b^{1/n}$

Initially, we used an exponent as a convenient way to write repeated multiplication: $5 \cdot 5 \cdot 5 \cdot 5 = 5^4$. However, to perform operations with exponents, we had to define a zero exponent and a negative exponent.

This may lead us to wonder if a rational exponent could have any meaning.

EXPLORATION 1   *Rational Exponents*

In parts (a)–(d) use your calculator to perform the operation in column A and compare the result to the value of the radical in column B. Then answer the questions that follow.

Column A	Column B
(a) $9^{1/2}$	$\sqrt{9}$
(b) $64^{1/3}$	$\sqrt[3]{64}$
(c) $81^{1/4}$	$\sqrt[4]{81}$
(d) $32^{1/5}$	$\sqrt[5]{32}$

(e) What do you observe about the denominator of the exponent and the index of the radical?

(f) Propose a meaning for $b^{1/n}$.

Discovery

(a) $9^{1/2} = 3$; $\sqrt{9} = 3$

(b) $64^{1/3} = 4$; $\sqrt[3]{64} = 4$

(c) $81^{1/4} = 3$; $\sqrt[4]{81} = 3$

(d) $32^{1/5} = 2$; $\sqrt[5]{32} = 2$

In each part, the exponential expression and the radical expression have the same value.

(e) The denominator of the exponent is the same as the index of the radical.

(f) These results suggest that $b^{1/n} = \sqrt[n]{b}$.

Our conjecture in Exploration 1 is consistent with the Power to a Power Rule for Exponents: $(a^m)^n = a^{mn}$, where m and n are integers. If we extend this rule to rational exponents, replacing a^m with $b^{1/n}$, we have $(b^{1/n})^n = b^{1/n \cdot n} = b^1 = b$. By our definition of nth root, this suggests that $b^{1/n}$ must be an nth root of b.

> ### Definition of $b^{1/n}$
>
> For any positive integer $n > 1$ and for any real number b for which $\sqrt[n]{b}$ is defined, $b^{1/n}$ is defined as $\sqrt[n]{b}$.

NOTE: This definition implies that b cannot be negative when n is even. ■

EXAMPLE 1 *Evaluating Expressions with Rational Exponents*

Justification

(a) $25^{1/2} = \sqrt{25} = 5$ $5^2 = 25$

(b) $(-25)^{1/2}$ is not a real number n is even and $b < 0$

(c) $-25^{1/2} = -\sqrt{25} = -5$ $-1 \cdot \sqrt{25} = -1 \cdot 5 = -5$

(d) $\left(\dfrac{4}{9}\right)^{1/2} = \sqrt{\dfrac{4}{9}} = \dfrac{2}{3}$ $\left(\dfrac{2}{3}\right)^2 = \dfrac{4}{9}$

(e) $8^{1/3} = \sqrt[3]{8} = 2$ $2^3 = 8$

(f) $(-8)^{1/3} = \sqrt[3]{-8} = -2$ $(-2)^3 = -8$

(g) $256^{1/4} = \sqrt[4]{256} = 4$ $4^4 = 256$

(h) $32^{1/5} = \sqrt[5]{32} = 2$ $2^5 = 32$ ■

 CUBE ROOT

Most calculators have keys only for square roots and cube roots. However, by writing a radical as an expression with a rational exponent, we can use a calculator to find higher order roots.

EXAMPLE 2   *Finding Higher Order Roots with a Calculator*

Use your calculator to evaluate each quantity. Round your results to two decimal places.

(a) $\sqrt[5]{16{,}807}$ (b) $\sqrt[4]{10}$ (c) $\sqrt[7]{-17}$

Solution

In each part, we write the radical in exponential form.

(a) $\sqrt[5]{16{,}807} = 16{,}807^{1/5}$

(b) $\sqrt[4]{10} = 10^{1/4}$

(c) $\sqrt[7]{-17} = (-17)^{1/7}$

Figure 9.1 is a typical screen display showing the results for all three parts.

Note that the exponents are enclosed in parentheses. Also, if the base is negative, as in part (c), then the base must be enclosed in parentheses. ■

Figure 9.1

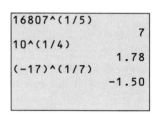

Definition of $b^{m/n}$

So far, the rational exponents we have used have had 1 as the numerator. However, extending the Power to a Power Rule for Exponents to rational exponents motivates a logical definition of *any* rational exponent.

Suppose m and n are positive integers with no common factor except 1. Then for all real numbers b for which $b^{1/n}$ is defined,

$$b^{m/n} = (b^{1/n})^m \qquad \text{Power to a Power Rule for Exponents}$$
$$= \left(\sqrt[n]{b}\right)^m. \qquad \text{Definition of } b^{1/n}$$

> ### Definition of $b^{m/n}$
> If m and n are positive integers with no common factor except 1, then $b^{m/n} = \left(\sqrt[n]{b}\right)^m$ for all real numbers b for which $b^{1/n}$ is defined.

NOTE: Because it is also true that $b^{m/n} = (b^m)^{1/n} = \sqrt[n]{b^m}$ for all real numbers b for which $b^{1/n}$ is defined, an alternate definition of $b^{m/n}$ is $\sqrt[n]{b^m}$. ■

We can use this definition to write expressions with rational exponents as radicals. To do so, note the roles of the numbers m and n in the exponent of $b^{m/n}$. The number n indicates the index of the radical, and the number m indicates the power to which the radical is to be raised.

The procedure for evaluating $b^{m/n}$ can be summarized as follows.

1. Determine the nth root of b.

2. Raise the result to the m power.

EXAMPLE 3 *Evaluating Expressions with Rational Exponents*

(a) $8^{2/3} = \left(\sqrt[3]{8}\right)^2 = (2)^2 = 4$

(b) $(-32)^{4/5} = \left(\sqrt[5]{-32}\right)^4 = (-2)^4 = 16$ With parentheses, the base is −32.

(c) $-32^{4/5} = -\left(\sqrt[5]{32}\right)^4 = -1 \cdot (2)^4 = -16$ Without parentheses, the base is 32.

(d) $36^{3/2} = \left(\sqrt{36}\right)^3 = (6)^3 = 216$ ◼

Even if we cannot evaluate an expression, we can use the definition of $b^{m/n}$ to write the expression in the form of a radical.

EXAMPLE 4 *Writing Exponential Expressions as Radicals*

Write the following exponential expressions as radicals.

(a) $x^{3/5}$ (b) $(5x)^{3/4}$ (c) $5x^{3/4}$

Solution

(a) $x^{3/5} = \left(\sqrt[5]{x}\right)^3$ or $\sqrt[5]{x^3}$

(b) $(5x)^{3/4} = \sqrt[4]{(5x)^3} = \sqrt[4]{125x^3}$ The base of the exponent is 5x.

(c) $5x^{3/4} = 5\sqrt[4]{x^3}$ The base of the exponent is x.   ◼

Exponential expressions with numerical bases can be evaluated without a calculator as long as the roots and powers are familiar. In Example 5 we do need a calculator.

EXAMPLE 5 *Using a Calculator to Evaluate Exponential Expressions*

Evaluate the following expressions to two decimal places.

Figure 9.2

(a) $6^{2/3}$ (b) $(-6)^{2/3}$ (c) $6^{3/2}$ (d) $(-6)^{3/2}$

```
6^(2/3)
              3.30
(-6)^(1/3)^2
              3.30
6^(3/2)
             14.70
```

Solution

Figure 9.2 is a typical screen display for parts (a)–(c).

When a base is negative, as in part (b), some calculators require the rational exponent to be separated, as in $(-6)^{2/3} = \left[(-6)^{1/3}\right]^2$.

The expression in part (d) is not a real number because $(-6)^{1/2}$ is not defined. ◼

Negative Rational Exponents

For integer exponents, we defined

$$a^{-n} = \frac{1}{a^n}$$

provided $a \neq 0$. We can extend the definition to negative rational exponents.

> ### Definition of a Negative Rational Exponent
>
> If m and n are positive integers with no common factor other than 1, and if $b \neq 0$, then
>
> $$b^{-m/n} = \frac{1}{b^{m/n}}$$
>
> for all real numbers b for which $b^{1/n}$ is defined.

EXAMPLE 6   *Evaluating Exponential Expressions with Negative Exponents*

Evaluate each of the following. Use a calculator for parts (d) and (e) and for verifying parts (a)–(c).

(a) $16^{-3/4}$ (b) $(-125)^{-2/3}$ (c) $\left(\dfrac{16}{25}\right)^{-1/2}$

(d) $150^{-3/4}$ (e) $(-7)^{-3/5}$

Solution

Figure 9.3

```
150^(-3/4)
              .023
(-7)^(1/5)^(-3)
             -.311
```

(a) $16^{-3/4} = \dfrac{1}{16^{3/4}} = \dfrac{1}{\left(\sqrt[4]{16}\right)^3} = \dfrac{1}{2^3} = \dfrac{1}{8}$

(b) $(-125)^{-2/3} = \dfrac{1}{(-125)^{2/3}} = \dfrac{1}{\left(\sqrt[3]{-125}\right)^2} = \dfrac{1}{(-5)^2} = \dfrac{1}{25}$

(c) $\left(\dfrac{16}{25}\right)^{-1/2} = \dfrac{1}{\left(\dfrac{16}{25}\right)^{1/2}} = \dfrac{1}{\sqrt{\dfrac{16}{25}}} = \dfrac{1}{\dfrac{4}{5}} = \dfrac{5}{4}$

(d)–(e) Figure 9.3 is a typical screen display for parts (d) and (e). The results are rounded to three decimal places.   ■

9.2   *Quick Reference*

Definition of $b^{1/n}$

- For any positive integer $n > 1$ and for any real number b for which $\sqrt[n]{b}$ is defined, $b^{1/n}$ is defined as $\sqrt[n]{b}$.

- If n is an odd integer, then b can be any real number. If n is an even integer, then b must be nonnegative.

- By writing a radical as an expression with a rational exponent, we can use a calculator to find higher order roots.

Definition of $b^{m/n}$	▪ If m and n are positive integers with no common factor except 1, then $b^{m/n} = \left(\sqrt[n]{b}\right)^m$ for all numbers b for which $b^{1/n}$ is defined.

▪ The number n indicates the index of the radical, and the number m indicates the power to which the radical is to be raised. To evaluate $b^{m/n}$, follow these steps.

1. Determine the nth root of b.

2. Raise the result to the m power.

Negative Rational Exponents	▪ If m and n are positive integers with no common factor other than 1, and if $b \neq 0$, then

$$b^{-m/n} = \frac{1}{b^{m/n}}$$

for all real numbers b for which $b^{1/n}$ is defined.

9.2 Exercises

1. One of the following is false. Identify the false statement and explain why it is not true.

  (i) The square roots of 16 are 4 and -4.

  (ii) $\sqrt{16} = 4$

  (iii) $16^{1/2} = 4$ or -4

  (iv) $64^{1/3} = 4$

2. Only one of the following expressions represents a real number. Identify and evaluate this expression. Then, explain why the other expressions do not represent real numbers.

  (i) $(-4)^{1/2}$ (ii) $(-4)^{-1/2}$

  (iii) $4^{-1/2}$ (iv) $\left(-\dfrac{1}{4}\right)^{-1/2}$

In Exercises 3–20, evaluate the given expression.

3. $49^{1/2}$ **4.** $(-32)^{1/5}$

5. $(-64)^{1/3}$ **6.** $81^{1/4}$

7. $\left(\dfrac{9}{16}\right)^{1/2}$ **8.** $\left(\dfrac{1}{49}\right)^{1/2}$

9. $(-16)^{1/4}$ **10.** $(-36)^{1/2}$

11. $\left(\dfrac{8}{27}\right)^{1/3}$ **12.** $\left(\dfrac{1}{32}\right)^{1/5}$

13. $(-27)^{1/3}$ **14.** $16^{1/4}$

15. $-32^{1/5}$ **16.** $-64^{1/3}$

17. $-(-625)^{1/4}$ **18.** $(-64)^{1/2}$

19. $-625^{1/4}$ **20.** $-36^{1/2}$

In Exercises 21–26, use a calculator to evaluate the given expression. Round results to two decimal places.

21. $15^{1/4}$ **22.** $23^{1/3}$

23. $(-12)^{1/2}$ **24.** $(-7)^{1/4}$

25. $-4^{1/4}$ **26.** $-20^{1/2}$

27. When you convert $9^{2/5}$ to a radical, how do you interpret the numerator and denominator of the exponent?

28. Which of the following is the correct way to write $\left(\sqrt[3]{2}\right)^5$ in exponential form? Use the words *index, base,* and *power* to explain how you know.

  (i) $2^{5/3}$ (ii) $5^{2/3}$ (iii) $3^{5/2}$

In Exercises 29–40, write the given exponential expression as a radical. Assume all variables represent positive numbers.

29. $10^{1/4}$ **30.** $15^{1/5}$

31. $3x^{2/3}$ **32.** $-4y^{5/6}$

33. $(3x)^{2/3}$ **34.** $(2ab)^{3/4}$

35. $(x+2)^{1/2}$ **36.** $(2x-1)^{1/3}$

37. $(2x+3)^{2/3}$ **38.** $(t+2)^{5/4}$

39. $3x^{1/2}+(2y)^{1/2}$ **40.** $(-2a)^{1/3}-2b^{1/3}$

In Exercises 41–54, write the given radical expression in exponential form. Assume all variables represent positive numbers.

41. $\sqrt[5]{20}$ **42.** $\sqrt[4]{12}$

43. $\sqrt[3]{y^2}$ **44.** $\sqrt[5]{z^3}$

45. $\sqrt[4]{3x^3}$ **46.** $\sqrt[5]{7x^2y}$

47. $\sqrt{x^2+4}$ **48.** $\sqrt{x+3}$

49. $\sqrt{t}-\sqrt{5}$ **50.** $\sqrt{2y}+1$

51. $\sqrt[3]{(2x-1)^2}$ **52.** $\sqrt[4]{(x+1)^3}$

53. $3\sqrt{x}+\sqrt{3y}$ **54.** $\sqrt{5y}-7\sqrt{2z}$

In Exercises 55–64, evaluate the given radical expression. Round your results to two decimal places.

55. $\sqrt[5]{-90}$ **56.** $\sqrt[4]{20}$

57. $-\sqrt[6]{740}$ **58.** $4\sqrt[7]{4020}$

59. $\sqrt[8]{230^5}$ **60.** $\sqrt[7]{49^3}$

61. $\sqrt[4]{10}-27\sqrt[5]{3.95}$ **62.** $\sqrt[5]{5}-9\sqrt[6]{1.3}$

63. $\sqrt[7]{10}\cdot\sqrt[5]{11.2}$ **64.** $\sqrt[4]{200}\cdot\sqrt[5]{90}$

In Exercises 65–90, evaluate the given expression without using a calculator.

65. $9^{-1/2}$ **66.** $(-27)^{-1/3}$

67. $\left(\dfrac{1}{9}\right)^{-1/2}$ **68.** $\left(\dfrac{1}{27}\right)^{-1/3}$

69. $27^{2/3}$ **70.** $36^{3/2}$

71. $(-27)^{-2/3}$ **72.** $-16^{-3/2}$

73. $32^{3/5}$ **74.** $25^{3/2}$

75. $(-16)^{-1/2}$ **76.** $(-4)^{3/2}$

77. $-16^{3/4}$ **78.** $-64^{5/6}$

79. $8^{-2/3}$ **80.** $8^{-4/3}$

81. $\left(\dfrac{8}{27}\right)^{-2/3}$ **82.** $\left(\dfrac{32}{243}\right)^{-4/5}$

83. $\left(\dfrac{1}{32}\right)^{3/5}$ **84.** $\left(\dfrac{16}{625}\right)^{3/4}$

85. $-(-8)^{-4/3}$ **86.** $(-8)^{5/3}$

87. $(-36)^{5/2}$ **88.** $-64^{5/6}$

89. $-4^{-5/2}$ **90.** $(-8)^{-2/3}$

In Exercises 91–100, evaluate the given expression.

91. $16^{1/2}+49^{1/2}$ **92.** $16^{3/2}-16^{3/4}$

93. $4^{-1/2}+9^{-1/2}$ **94.** $8^{-1/3}-16^{-1/2}$

95. $27^{1/3}+4^{-1/2}$ **96.** $81^{-3/4}+81^{-3/2}$

97. $(-8)^{2/3}(16^{1/4})$ **98.** $(9^{-3/2})(-27^{1/3})$

99. $(-7^3)^{1/3}(7^{-2})^{-1/2}$ **100.** $(10^6)^{1/3}(125)^{-2/3}$

Recall that every rational number can be expressed as a terminating decimal or as a nonterminating, repeating decimal.

In Exercises 101–104, evaluate the given expression without using your calculator.

101. $16^{0.25}$ **102.** $(-27)^{0.\overline{3}}$

103. $9^{-0.5}$ **104.** $49^{1.5}$

In Exercises 105–110, use a calculator to evaluate the given expression. Round results to two decimal places.

105. $16^{2/3}$ **106.** $16^{-2/3}$

107. $\left(\dfrac{5}{9}\right)^{5/9}$ **108.** $10^{4/5}$

109. $(-5)^{-2/5}$ **110.** $(-9)^{2/3}$

In Exercises 111–120, supply the missing number to make each statement true.

111. $(\ \ )^{1/2}=3$ **112.** $(\ \ )^{1/3}=-2$

113. $(\quad)^{-1/2} = \dfrac{1}{4}$

114. $(\quad)^{-1/4} = \dfrac{1}{2}$

115. $(\quad)^{5/3} = -32$

116. $(\quad)^{2/3} = 9$

117. $\left(\dfrac{4}{9}\right)^{(\ )} = \dfrac{3}{2}$

118. $(64)^{(\ )} = \dfrac{1}{4}$

119. $(-27)^{(\ )} = 9$

120. $(-8)^{(\ )} = \dfrac{1}{16}$

121. Use [X : 0, 2, 1] and [Y : 0, 2, 1] to produce the graphs of $y_1 = x^{1/2}$, $y_2 = x^{1/4}$, and $y_3 = x^{1/8}$.

   (a) Where do the three graphs intersect? Why would *all* graphs of $y = x^{1/n}$, $n > 1$, intersect at these points?

   (b) For what values of x is $x^{1/2} < x^{1/4} < x^{1/8}$?

   (c) For what values of x is $x^{1/2} > x^{1/4} > x^{1/8}$?

122. Show that $16^{(1/2)/(2/3)} = 8$.

123. A researcher has developed a mathematical model for predicting a final exam grade: $G = 60T^{2/9}$. In this model G is the predicted grade, and T is the number of hours spent studying for the exam. Use [X : 1, 10, 1] and [Y : 0, 100, 10] to produce the graph of the model function.

   (a) If the passing grade is 70, how many hours of study are required to pass the exam?

   (b) How many additional hours of study would be needed to earn a 90 on the exam?

   (c) What grade is predicted if a student does not study at all? According to the model, can a student earn a 100? How realistic is the model at these extreme points of the graph?

124. Suppose that the Department of Labor uses a mathematical model to relate annual federal funding F (in billions of dollars) for jobs creation to the unemployment rate R: $F = 10R^{-1/3}$. (If the current unemployment rate is 6%, $R = 6$.) Use [X : 1, 100, 10] and [Y : 0, 10, 1] to produce the graph of the model function.

   (a) Approximately what total of annual funding would be required to maintain an unemployment rate of 5%?

   (b) If $6.8 billion were spent annually for jobs creation, what would be the expected unemployment rate?

   (c) Why is the model not valid for zero funding?

Exploring with Real Data

In Exercises 125–128, refer to the following background information and data.

The following formula is used to calculate windchill temperatures.

$$Y = 91.4 - (91.4 - T)[0.478 + 0.301(x^{1/2} - 0.02x)]$$

In this formula, T is the actual air temperature in Fahrenheit degrees, x is the wind speed in miles per hour, and Y is the windchill temperature. (Source: U.S. Meteorological Service.)

125. Suppose you want to know the wind speed x when the temperature T is 10°F and the windchill temperature Y is -20°F. Enter the function in your calculator. (Note that the formula has two variables T and x.) Now store 10 in T and produce the graph. Trace the graph to the point where $Y = -20$ and read the corresponding x-value.

126. Suppose that the temperature is 20°F. What would the wind speed have to be for the windchill temperature to be 50° colder?

127. Suppose that the temperature is 12°F. For what wind speed would there be no windchill?

128. Suppose that the temperature T is 15°F. Trace the graph to determine the wind speed x_0 for which $Y = T$. If the wind speed is *less* than x_0, compare Y and T. Does this mean that at these wind speeds we would feel *warmer* than it actually is? What restrictions would you place on x to make the formula valid?

Challenge

In Exercises 129–132, evaluate.

129. $\sqrt{16^{-1/2}}$

130. $\left(\sqrt{81}\right)^{-1/2}$

131. $\left(\sqrt[3]{64}\right)^{-1/2}$

132. $\sqrt[3]{729^{-1/2}}$

133. Use the integer setting on your calculator to produce the graph of $y = (-2)^{1/x}$. Explain why the graph is displayed as a set of distinguishable points rather than as a smooth curve.

134. Store 2 in N and produce the graph of $y = (x^N)^{1/N}$. Now store 3 in N and produce the graph again. Explain the difference in the graphs.

9.3 Properties of Rational Exponents

The properties of exponents that were stated for integer exponents are also valid for rational exponents.

Properties of Rational Exponents

Let m and n be rational numbers. For nonzero real numbers a and b for which the expressions are defined, the following are properties of exponents.

$$a^m a^n = a^{m+n}$$ Product Rule

$$\frac{a^m}{a^n} = a^{m-n}$$ Quotient Rule

$$a^{-n} = \frac{1}{a^n}$$ Definition of Negative Exponent

$$(a^m)^n = a^{mn}$$ Power to a Power Rule

$$(ab)^n = a^n b^n$$ Power of a Product Rule

$$\left(\frac{a}{b}\right)^n = \frac{a^n}{b^n}$$ Power of a Quotient Rule

$$\left(\frac{a}{b}\right)^{-n} = \left(\frac{b}{a}\right)^n$$ Negative Power of a Quotient Rule

---

EXAMPLE 1 *Evaluating Exponential Expressions*

Use the properties of exponents to evaluate each of the following.

(a) $(5^6)^{1/2}$

(b) $2^{1/2} \cdot 8^{1/2}$

(c) $\left[\dfrac{3^6}{5^3}\right]^{-2/3}$

(d) $16^{1/3} \cdot 16^{5/12}$

Solution

(a) $(5^6)^{1/2} = 5^{6 \cdot (1/2)} = 5^3 = 125$ Power to a Power Rule

(b) $2^{1/2} \cdot 8^{1/2} = [2 \cdot 8]^{1/2} = 16^{1/2} = 4$ Power of a Product Rule

(c) $\left[\dfrac{3^6}{5^3}\right]^{-2/3} = \dfrac{3^{6 \cdot (-2/3)}}{5^{3 \cdot (-2/3)}} = \dfrac{3^{-4}}{5^{-2}} = \dfrac{5^2}{3^4} = \dfrac{25}{81}$ Power of a Quotient Rule

An alternate method is to use the rule

$$\left(\frac{a}{b}\right)^{-n} = \left(\frac{b}{a}\right)^{n}.$$

$$\left[\frac{3^6}{5^3}\right]^{-2/3} = \left[\frac{5^3}{3^6}\right]^{2/3} = \frac{5^{3\cdot(2/3)}}{3^{6\cdot(2/3)}} = \frac{5^2}{3^4} = \frac{25}{81}$$

(d) $16^{1/3} \cdot 16^{5/12} = 16^{1/3+5/12} = 16^{(4/12)+(5/12)}$

$$= 16^{9/12} = 16^{3/4} = \left(\sqrt[4]{16}\right)^3 = 2^3 = 8 \qquad \text{Product Rule} \qquad \blacksquare$$

Recall that $\sqrt[n]{b^n} = |b|$ if n is even and $\sqrt[n]{b^n} = b$ if n is odd. But $\sqrt[n]{b^n} = (b^n)^{1/n}$. Thus, we can state the following generalization.

> ### Simplifying Expressions of the Form $(b^n)^{1/n}$
>
> For an even positive integer n, $(b^n)^{1/n} = |b|$.
>
> For an odd positive integer n, $n > 1$, $(b^n)^{1/n} = b$.

EXAMPLE 2   *Simplifying Expressions of the Form $(b^n)^{1/n}$*

(a) $\left[(-3)^2\right]^{1/2} = |-3| = 3$ Because n is even, absolute value symbols are needed.

(b) $\left[(-3)^5\right]^{1/5} = -3$ Because n is odd, absolute value symbols are not used.

(c) $(y^3)^{1/3} = y$ Because n is odd, absolute value symbols are not used.

(d) $(z^8)^{1/8} = |z|$ Because n is even, absolute value symbols are needed. $\blacksquare$

EXAMPLE 3   *Simplifying Exponential Expressions*

Use the properties of exponents to simplify each of the following. Write all results with positive exponents. Assume that all variables represent positive numbers.

(a) $\dfrac{x^{1/3}}{x^{4/3}}$ (b) $y^{1/3} \cdot y^{1/6}$ (c) $[a^3 b^{-1}]^{-1/2}$

(d) $\left[\dfrac{x^{1/2}}{y^3}\right]^{-1/3}$ (e) $a^{1/3}(a^{5/3} - a^{-2/3})$

Solution

(a) $\dfrac{x^{1/3}}{x^{4/3}} = x^{1/3-4/3} = x^{-3/3} = x^{-1} = \dfrac{1}{x}$ Quotient Rule

(b) $y^{1/3} \cdot y^{1/6} = y^{1/3+1/6} = y^{2/6+1/6} = y^{3/6} = y^{1/2}$ Product Rule

(c) $[a^3 b^{-1}]^{-1/2} = a^{3\cdot(-1/2)} b^{-1\cdot(-1/2)} = a^{-3/2} b^{1/2} = \dfrac{b^{1/2}}{a^{3/2}}$ Power of a Product Rule

(d) $\left[\dfrac{x^{1/2}}{y^3}\right]^{-1/3} = \left[\dfrac{y^3}{x^{1/2}}\right]^{1/3}$ $\left(\dfrac{a}{b}\right)^{-n} = \left(\dfrac{b}{a}\right)^{n}$

$\qquad\qquad = \dfrac{y^{3\cdot(1/3)}}{x^{(1/2)\cdot(1/3)}}$ Power of a Quotient Rule

$\qquad\qquad = \dfrac{y}{x^{1/6}}$

(e) $a^{1/3}(a^{5/3} - a^{-2/3}) = a^{1/3} \cdot a^{5/3} - a^{1/3} \cdot a^{-2/3}$ Distributive Property

$\qquad\qquad = a^{1/3+5/3} - a^{1/3-2/3}$ Product Rule

$\qquad\qquad = a^{6/3} - a^{-1/3}$

$\qquad\qquad = a^2 - \dfrac{1}{a^{1/3}}$ Definition of negative exponent ■

Use the summary of exponent properties at the beginning of this section for the Quick Reference.

9.3   *Exercises*

1. If a and b are positive real numbers, only one of the following equations is true.

(i) $(a^2b^2)^{1/2} = ab$ (ii) $(a^2 + b^2)^{1/2} = a + b$

Identify the equation that is true and state a rational exponent rule that supports your claim. For the equation that is not true, select values for a and b for which the left and right sides are not equal.

2. Explain why it is easier to simplify $\left[(x^{1/3})^{1/5}\right]^{1/2}$ than it is to simplify $x^{1/3} \cdot x^{1/5} \cdot x^{1/2}$.

In Exercises 3–28, use the properties of exponents to evaluate the given expression.

3. $(6^3)^{2/3}$ **4.** $(3^{-6})^{1/3}$

5. $(2^{2/7})^{14}$ **6.** $(3^{2/5})^{10}$

7. $\left(-\dfrac{1}{25}\right)^{-3/2}$ **8.** $\left(-\dfrac{1}{16}\right)^{3/4}$

9. $\left(-\dfrac{27}{125}\right)^{-1/3}$ **10.** $\left(-\dfrac{1}{32}\right)^{-3/5}$

11. $12^{1/2} \cdot 3^{1/2}$ **12.** $2^{2/3} \cdot 4^{2/3}$

13. $\left(\dfrac{4}{3}\right)^{1/3} \cdot \left(\dfrac{2}{9}\right)^{1/3}$

14. $\left(\dfrac{8}{27}\right)^{-1/4} \cdot \left(\dfrac{2}{3}\right)^{-1/4}$

15. $\left(\dfrac{7^3}{2^6}\right)^{-2/3}$ **16.** $\left(\dfrac{5^6}{3^9}\right)^{1/3}$

17. $65^{1/3} \cdot 65^{2/3}$ **18.** $7^{1/4} \cdot 7^{-1/4}$

19. $49^{7/10} \cdot 49^{-1/5}$ **20.** $16^{-1} \cdot 16^{1/2}$

21. $8^{2/9} \cdot 8^{1/9}$ **22.** $16^{5/8} \cdot 16^{-3/8}$

23. $6^{0.6} \cdot 6^{0.4}$ **24.** $5^{0.22} \cdot 5^{1.78}$

25. $\dfrac{27^{1/3}}{27^{2/3}}$ **26.** $\dfrac{16^{-3/8}}{16^{-1/8}}$

27. $8^{1/6} \div 8^{1/2}$ **28.** $25^{1/4} \div 25^{3/4}$

29. Which of the following is true? Explain your choice.

(i) $x^{2/5} \cdot x^{5/2} = x^1 = x$

(ii) $x^{2/5} \cdot x^{5/2} = x^{29/10}$

30. Describe the conditions under which each of the following is true.

(a) $(x^n)^{1/n} = x$ (b) $(x^n)^{1/n} = |x|$

In Exercises 31–46, use the properties of exponents to simplify the given expression. Write the result with only positive exponents. Assume all variables represent positive numbers.

31. $x^{2/3} \cdot x^{1/3}$

32. $a^{-2/5} \cdot a^{4/5}$

33. $t \cdot t^{5/7}$

34. $b^{-1/3} \cdot b^{1/4} \cdot b^{5/6}$

35. $(a^{-1/2}b)(a^{3/4}b^{1/2})$

36. $(x^{1/3}y^{-1/3})(x^{2/3}y^{-2/3})$

37. $\dfrac{z^{1/2}}{z^{1/3}}$

38. $\dfrac{m^{1/2}}{m^2}$

39. $\dfrac{a^{1/3}b}{b^{-1/3}}$

40. $\dfrac{x^{1/2}y^{-2}}{xy^{1/2}}$

41. $(a^{-2})^{1/2}$

42. $(y^{-3/4})^{-4}$

43. $(y^{2/3})^{3/2}$

44. $(z^{3/4})^{-4/3}$

45. $(a^6b^4)^{3/2}$

46. $\left(\dfrac{a^6}{b^8}\right)^{-1/2}$

In Exercises 47–50, supply the missing number.

47. $8^{(\ )} = 2$

48. $x^{1/2} \cdot x^{(\ )} = x^{3/2}, x \geq 0$

49. $(x^{2/3})^{(\ )} = x, x \geq 0$

50. $\left(\dfrac{x}{3}\right)^{(\ )} = \dfrac{9}{x^2}, x \neq 0$

In Exercises 51–58, simplify the expression. All variables represent any real number.

51. $\left[(-2)^6\right]^{1/2}$

52. $\left[(-3)^3\right]^{1/3}$

53. $(a^{15})^{1/5}$

54. $(x^4)^{1/4}$

55. $(a^4b^2)^{1/2}$

56. $(x^3y^6)^{1/3}$

57. $\left[(x+y)^2\right]^{1/2}$

58. $\left[(2x-1)^2\right]^{1/2}$

In Exercises 59–76, use the properties of exponents to simplify the given expression. Write the result with only positive exponents. Assume all variables represent positive numbers.

59. $(a^{-4}b^{1/3})^{1/2}$

60. $(x^2y^{-1})^{-1/2}$

61. $(a^6b^{-9})^{2/3}$

62. $(25a^4b^6)^{1/2}$

63. $(-3^{2/3}x^{1/2}y^{-1/3})^3$

64. $(x^{3/4}y^{-3/8}z^{-1/4})^8$

65. $\left(\dfrac{a^{4/5}}{b^{8/9}}\right)^{-15/4}$

66. $\left(\dfrac{x^6}{y^{-3/2}}\right)^{-2/3}$

67. $\left(\dfrac{x^{-1/3}}{y^{-1/2}}\right)^{-6}$

68. $\left(\dfrac{a^{-4}}{b^{-2}}\right)^{-5/2}$

69. $\left(\dfrac{16x^{-12}y^8}{z^4}\right)^{1/4}$

70. $\left(\dfrac{27x^{-12}y^9}{z^{-6}}\right)^{1/3}$

71. $(a^{-2/3}b^3)^6(a^{5/7}b^{-2})^7$

72. $(-27x^9)^{1/3}(2x^{-1/2})^2$

73. $\dfrac{(a^{5/6}b^{-2/5}c^{7/60})^{30}}{a^{21}b^{-5}}$

74. $\dfrac{(x^{3/4}y^{-5/6}z^{-5/24})^{12}}{x^7z^{1/2}}$

75. $\dfrac{(9x^4y^{-2})^{-1/2}}{(x^6y^{-3})^{1/3}}$

76. $\left(\dfrac{x^{-2}y^{2/3}}{z^{-1}}\right)\left(\dfrac{8x^3y}{z^6}\right)^{1/3}$

In Exercises 77–84, find the product.

77. $a^{3/4}(a^{5/4} + a^{-3/4})$

78. $y^{-2/3}(y^{5/3} + y^{2/3})$

79. $x^{-1/4}(x^{9/4} - x^{2/3})$

80. $b^{7/5}(b^{-2/5} - b^{4/15})$

81. $(x^{1/2} + y^{1/2})(x^{1/2} - y^{1/2})$

82. $(3^{1/2} + 2^{1/2})(3^{1/2} - 2^{1/2})$

83. $(x^{1/2} - 3^{1/2})^2$

84. $(x^{1/2} + y^{1/2})^2$

Exploring with Real Data

In Exercises 85–88, refer to the following background information and data.

Suppose that an influential member of Congress is arguing against the budget for the National Park Service. In particular, she claims that the cost per visitor to national monuments more than tripled during 1985–1990. As a staff member for the National Park Service, you have gathered the following data on the number of visitors to national monuments.

	1985	1986	1987	1988	1989	1990
Visitors (millions)	15.9	21.2	23.5	23.2	23.7	23.9

(Source: U.S. National Park Service.)

A function that models these data is $V(t) = 17.05t^{0.22}$, where V is the number (in millions) of visitors during a year t and t is the number of years since 1984.

Suppose that the Office of Management and Budget has also supplied you with a function that models the total expenditures E (in millions) for the operation and maintenance of national monuments: $E(t) = 50.5t^{0.33}$, where t is defined as before.

85. Use the two given functions to write a simplified function for the average expenditure A per visitor for any year t.

86. Make a table showing the average expenditure per visitor for the years 1985–1990 by evaluating the function given in Exercise 85 for $t = 1$, $2, \dots, 6$.

87. Does this set of data support the claim of the member of Congress? Why or why not?

88. If 24 million visitors to national monuments had been projected for 1991, what should the appropriation have been to keep the average expenditure per visitor the same as it was in 1990?

Challenge

In Exercises 89–94, simplify the expression. Assume all variables represent positive numbers.

89. $\dfrac{x^{4/n}}{x^{3/n}}$

90. $\dfrac{y^{-2/n}}{y^{-1/n}}$

91. $x^{-n/3} \cdot x^{n/2}$

92. $y^{1/n} \cdot y^{2/n}$

93. $\left(\dfrac{x^{2n}}{y^{4n}}\right)^{1/2}$

94. $\left(\dfrac{a^n}{b^{2n}}\right)^{1/n}$

In Exercises 95–98, simplify the expression. Assume all variables represent positive numbers.

95. $(a^{3n}b^{5n})^{1/n}$

96. $(x^n y^{4n})^{2/n}$

97. $\dfrac{(x^{2/3})^{3n}}{x^{n-1}}$

98. $\dfrac{(x^{n+4})^{1/2}}{(x^{1/2})^n}$

Sometimes expressions with rational exponents have a common factor that can be factored out of the expression. Remember to look for the smallest exponent.

Example: $5x^{-2/3} - 8x^{1/3} = x^{-2/3}(5 - 8x)$

In Exercises 99–102, use this technique to factor the expression.

99. $3x^{-1/2} + 5x^{3/2}$

100. $a^{-1/2} + a^{-5/2}$

101. $5a^{1/2} - 10a^{-1/2}$

102. $6x^{1/3} - 3x^{4/3}$

Sometimes an expression in the form of a trinomial with rational exponents can be factored.

Example:
$$x^{2/3} - 5x^{1/3} + 4 = \left[x^{1/3}\right]^2 - 5\left[x^{1/3}\right] + 4$$
Let $u = x^{1/3}$.
$$\begin{aligned}\text{Then } \left[x^{1/3}\right]^2 - 5\left[x^{1/3}\right] + 4 &= u^2 - 5u + 4\\ &= (u - 4)(u - 1)\\ &= (x^{1/3} - 4)(x^{1/3} - 1)\end{aligned}$$

In Exercises 103–106, use this technique to factor the expression.

103. $t^{2/5} + 4t^{1/5} - 5$

104. $3x^{4/3} + 2x^{2/3} - 5$

105. $2x^{1/3} - 7x^{1/6} - 15$

106. $6y^{2/3} - 5y^{1/3} + 1$

9.4 Properties of Radicals

Domain of a Radical Function ▪ *Product and Quotient Rules* ▪
Other Properties

Domain of a Radical Function

A **radical expression** is an algebraic expression that contains a radical. A **radical function** is a function that is defined by a radical expression. The domain of a radical function is the set of values of the variable for which the expression is defined.

EXAMPLE 1 *Determining the Domain of a Radical Function*

Use a graph to estimate the domain of each function. Then determine the domain algebraically.

(a) $y = \sqrt{1 - 2x}$ (b) $y = \sqrt[3]{x + 3}$ (c) $y = \sqrt[4]{x^2 + 1}$

Solution

(a) In Fig. 9.4, the graph of $y = \sqrt{1 - 2x}$ appears to have an *x*-intercept $\left(\frac{1}{2}, 0\right)$, and the rest of the graph is to the left of that point. We estimate the domain to be $\{x \mid x \le \frac{1}{2}\}$.

Figure 9.4

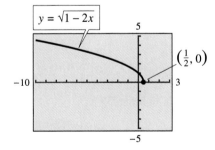

The expression $\sqrt{1 - 2x}$ is defined for all numbers for which $1 - 2x \ge 0$. To determine the domain, we solve the following inequality.

$$1 - 2x \ge 0$$
$$-2x \ge -1$$
$$x \le \frac{1}{2}$$

The domain of the function is $\{x \mid x \le \frac{1}{2}\}$.

(b) In Fig. 9.5, the graph of $y = \sqrt[3]{x + 3}$ appears to extend forever to the left and right. This suggests that the domain is the set of all real numbers.

Because the index is odd, $\sqrt[3]{x + 3}$ is defined for all real numbers. Therefore, the domain of the function is the set of all real numbers.

(c) In Fig. 9.6, the graph of $y = \sqrt[4]{x^2 + 1}$ apparently extends to the left and right forever. Thus, the domain appears to be the set of all real numbers.

Figure 9.5   Figure 9.6

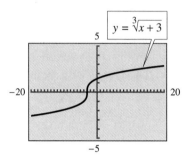

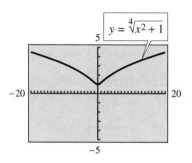

Although the index is even, $x^2 + 1$ will be positive for any replacement for x. Therefore, the domain of the function is the set of all real numbers. ■

Product and Quotient Rules

In the following exploration, we investigate the rules for multiplying and dividing radicals.

 **EXPLORATION 1**   *Product and Quotient Rules for Radicals*

Use your calculator to evaluate the product or quotient in column A and the corresponding radical in column B.

Column A	*Column B*
$\sqrt{6} \cdot \sqrt{2}$	$\sqrt{12}$
$\dfrac{\sqrt{24}}{\sqrt{8}}$	$\sqrt{3}$
$\sqrt[3]{15} \cdot \sqrt[3]{3}$	$\sqrt[3]{45}$
$\dfrac{\sqrt[3]{30}}{\sqrt[3]{6}}$	$\sqrt[3]{5}$

From these experiments, what is your conjecture about the rules for multiplying and dividing radicals?

Discovery

In each case, the expression in column A has the same value as the corresponding radical in column B. For multiplication, it appears that we multiply the radicands. For division, it appears that we divide the radicands.

We can use the Product and Quotient Rules for Exponents to support our conjectures in Exploration 1.

$$\sqrt[n]{a} \cdot \sqrt[n]{b} = a^{1/n} \cdot b^{1/n} = (ab)^{1/n} = \sqrt[n]{ab}$$

$$\frac{\sqrt[n]{a}}{\sqrt[n]{b}} = \frac{a^{1/n}}{b^{1/n}} = \left(\frac{a}{b}\right)^{1/n} = \sqrt[n]{\frac{a}{b}}$$

Product and Quotient Rules for Radicals

For any integer $n > 1$, and for all real numbers a and b for which the operations are defined,

$$\sqrt[n]{a} \cdot \sqrt[n]{b} = \sqrt[n]{ab} \qquad \text{Product Rule for Radicals}$$

and

$$\frac{\sqrt[n]{a}}{\sqrt[n]{b}} = \sqrt[n]{\frac{a}{b}}. \qquad \text{Quotient Rule for Radicals}$$

NOTE: The Product and Quotient Rules for Radicals apply only if the indices are the same. ■

EXAMPLE 2 *Using the Product Rule for Radicals*

Use the Product Rule for Radicals to multiply the radical expressions. Assume that all variables represent positive numbers.

(a) $\sqrt{2a} \cdot \sqrt{7b}$ (b) $\sqrt[4]{\dfrac{1}{x}} \cdot \sqrt[4]{\dfrac{2}{y}}$ (c) $\sqrt[3]{4y} \cdot \sqrt[3]{3y}$ (d) $\sqrt{2} \cdot \sqrt[3]{3}$

Solution

(a) $\sqrt{2a} \cdot \sqrt{7b} = \sqrt{14ab}$

(b) $\sqrt[4]{\dfrac{1}{x}} \cdot \sqrt[4]{\dfrac{2}{y}} = \sqrt[4]{\dfrac{2}{xy}}$

(c) $\sqrt[3]{4y} \cdot \sqrt[3]{3y} = \sqrt[3]{12y^2}$

(d) The Product Rule for Radicals does not apply to $\sqrt{2} \cdot \sqrt[3]{3}$ because the indices are not the same. ■

EXAMPLE 3   *Using the Quotient Rule for Radicals*

Use the Quotient Rule for Radicals to divide the radical expressions. Assume that all variables represent positive numbers.

(a) $\dfrac{\sqrt[5]{64}}{\sqrt[5]{2}}$ (b) $\dfrac{\sqrt{x^5}}{\sqrt{x^3}}$ (c) $\dfrac{\sqrt{15a}}{\sqrt{3a}}$ (d) $\dfrac{\sqrt{8}}{\sqrt[3]{2}}$

Solution

(a) $\dfrac{\sqrt[5]{64}}{\sqrt[5]{2}} = \sqrt[5]{\dfrac{64}{2}} = \sqrt[5]{32} = 2$

(b) $\dfrac{\sqrt{x^5}}{\sqrt{x^3}} = \sqrt{\dfrac{x^5}{x^3}} = \sqrt{x^2} = x$

(c) $\dfrac{\sqrt{15a}}{\sqrt{3a}} = \sqrt{\dfrac{15a}{3a}} = \sqrt{5}$

(d) The Quotient Rule for Radicals does not apply to $\dfrac{\sqrt{8}}{\sqrt[3]{2}}$ because the indices are not the same. ∎

In some cases, the Product and Quotient Rules for Radicals can be used to determine the nth root of an expression. For this purpose, we use the rules as follows.

$\sqrt[n]{ab} = \sqrt[n]{a}\,\sqrt[n]{b}$ Product Rule for Radicals

$\sqrt[n]{\dfrac{a}{b}} = \dfrac{\sqrt[n]{a}}{\sqrt[n]{b}}$ Quotient Rule for Radicals

EXAMPLE 4   *Using the Product and Quotient Rules for Radicals to Find the nth Root*

(a) $\sqrt{16x^{16}} = \sqrt{16}\,\sqrt{x^{16}}$ Product Rule for Radicals

$= 4x^8$ Because $(x^8)^2 = x^{16}$

(b) $\sqrt[3]{\dfrac{8}{x^3y^6}} = \dfrac{\sqrt[3]{8}}{\sqrt[3]{x^3y^6}}$ Quotient Rule for Radicals

$= \dfrac{\sqrt[3]{8}}{\sqrt[3]{x^3}\,\sqrt[3]{y^6}}$ Product Rule for Radicals

$= \dfrac{2}{xy^2}$ Because $(y^2)^3 = y^6$

(c) $\sqrt{x^2 + 12x + 36} = \sqrt{(x+6)^2}$ Factor the radicand.

$= |x + 6|$ Absolute value because n is even ∎

NOTE: If a radicand has more than one term, we must factor the radicand before we can simplify. *We cannot take the square root of each term.* In part (c) of Example 4, we factored the trinomial and then used the rule $\sqrt[n]{x^n} = |x|$, where n is even. ∎

Other Properties

If the index of a radical and the exponent of the radicand have a common factor, the expression can be rewritten with a smaller index. We call the process **reducing the index.**

As we will see in Example 5, the technique begins with writing the radical in exponential form.

EXAMPLE 5   *Reducing the Index*

Write the given expression with a smaller index. Assume that all variables represent positive numbers.

(a) $\sqrt[12]{9^6}$ (b) $\sqrt[15]{t^{10}}$ (c) $\sqrt[6]{8y^3}$

Solution

(a) $\sqrt[12]{9^6} = 9^{6/12} = 9^{1/2} = \sqrt{9} = 3$

(b) $\sqrt[15]{t^{10}} = t^{10/15} = t^{2/3} = \sqrt[3]{t^2}$

(c) $\sqrt[6]{8y^3} = \sqrt[6]{(2y)^3} = (2y)^{3/6} = (2y)^{1/2} = \sqrt{2y}$ ∎

The properties of exponents can also be used to write a term with more than one radical as a term involving only one radical.

EXAMPLE 6   *Reducing the Number of Radicals in a Term*

Write each expression with only one radical symbol.

(a) $\sqrt[3]{\sqrt{2}}$ (b) $\sqrt[3]{x} \cdot \sqrt[4]{x}$ $(x \geq 0)$

Solution

(a) $\sqrt[3]{\sqrt{2}} = \left[2^{1/2}\right]^{1/3} = 2^{1/6} = \sqrt[6]{2}$

(b) $\sqrt[3]{x} \cdot \sqrt[4]{x} = x^{1/3} \cdot x^{1/4} = x^{4/12} \cdot x^{3/12} = x^{7/12} = \sqrt[12]{x^7}$ ∎

9.4   *Quick Reference*

Domain of a
Radical
Function

- A **radical expression** is an algebraic expression that contains a radical. A **radical function** is a function that is defined by a radical expression.

- The domain of a radical function is the set of values of the variable for which the expression is defined.

- If the index of a radical is odd, the domain of the radical function is the set of all real numbers for which the radicand is defined.

- To determine the domain of a radical function for which the index is even and the radicand is R, solve the inequality $R \geq 0$. The solution set of the inequality is the domain of the function.

Product and
Quotient
Rules

- Product Rule for Radicals: For any integer $n > 1$ and all real numbers a and b for which the operations are defined,
$$\sqrt[n]{a} \cdot \sqrt[n]{b} = \sqrt[n]{ab}.$$

- Quotient Rule for Radicals: For any integer $n > 1$ and all real numbers a and b for which the operations are defined,
$$\frac{\sqrt[n]{a}}{\sqrt[n]{b}} = \sqrt[n]{\frac{a}{b}}.$$

Other
Properties

- If the index of a radical and the exponent of the radicand have a common factor, the expression can be rewritten with a smaller index.

- Some operations involving radicals with different indices can be performed by writing the radicals in exponential form, using the exponent rules to perform the operations, and then converting the exponential expression back to a radical.

9.4   *Exercises*

1. Why is the domain of $f(x) = \sqrt[3]{x}$ the set of all real numbers, but the domain of $g(x) = \sqrt{x}$ is $\{x \mid x \geq 0\}$?

2. Consider the radical function $y = \sqrt{R}$, where R represents the radicand. Explain how to determine the domain of the function.

In Exercises 3–16, determine the domain of the given radical expression.

3. $\sqrt{x - 2}$

4. $\sqrt{x + 7}$

5. $\sqrt{2x - 3}$

6. $\sqrt{2x}$

7. $\sqrt{2 - x}$

8. $\sqrt{1 - 2x}$

9. $\sqrt{1 + 3x^2}$

10. $\sqrt{x^2 + 5}$

11. $\sqrt[3]{3 - 2x}$

12. $\sqrt[5]{5x + 7}$

13. $\sqrt[4]{x + 3}$

14. $\sqrt[6]{5 - x}$

15. $\sqrt[6]{1 + x^2}$

16. $\sqrt[4]{2x^2 + 3}$

 17. If x is any real number, why is it necessary to use absolute value notation when we simplify $\sqrt{x^6}$, but not when we simplify $\sqrt[3]{x^9}$?

 18. If $f(x) = \sqrt[n]{x^n} = x$, what is the domain of f?

In Exercises 19–30, determine the indicated root. Assume all variables represent positive numbers.

19. $\sqrt{25y^2}$

20. $\sqrt{16x^6y^{10}}$

21. $\sqrt{49w^8t^4}$

22. $\sqrt{36x^8y^{12}}$

23. $\sqrt{\dfrac{81x^4}{y^6}}$

24. $\sqrt{\dfrac{25x^6}{9y^2}}$

25. $\sqrt[3]{8t^6}$

26. $\sqrt[4]{16x^8y^{16}}$

27. $\sqrt[4]{81x^{12}y^{20}}$

28. $\sqrt[3]{27x^{27}y^{12}}$

29. $\sqrt[3]{\dfrac{a^9}{125b^3}}$

30. $\sqrt[3]{\dfrac{8x^{12}}{27y^6}}$

In Exercises 31–38, supply the missing numbers to make each statement true. Assume all variables represent positive numbers.

31. $\sqrt{(\ )x^{(\ )}y^8} = 3x^3y^4$

32. $\sqrt{(\ )x^4y^{(\ )}} = 5x^2y$

33. $(\ )\sqrt{4t^{(\ )}} = -2t^5$

34. $(\ )\sqrt{x^{(\ )}} = 3x^2$

35. $\sqrt[3]{(\ )x^{(\ )}} = 4x$

36. $\sqrt[4]{(\ )y^{(\ )}} = 2y^3$

37. $\sqrt[(\ )]{(\ )x^8} = 7x^4$

38. $\sqrt[(\ )]{(\ )y^{12}} = -5y^4$

 39. Explain why the following is an incorrect application of the Product Rule for Radicals.
$$\sqrt{-3}\,\sqrt{-12} = \sqrt{36} = 6$$

 40. Explain whether the Product Rule for Radicals applies to the problem $\sqrt[3]{2}\,\sqrt{2}$.

In Exercises 41–56, use the Product Rule for Radicals to multiply the radicals. Assume all variables represent positive numbers.

41. $\sqrt{2}\,\sqrt{3}$

42. $\sqrt{3x}\,\sqrt{5y}$

43. $\left(2\sqrt{5}\right)^2$

44. $\left(-4\sqrt{2}\right)\left(3\sqrt{2}\right)$

45. $\sqrt[3]{x-2}\,\sqrt[3]{x+3}$

46. $\sqrt{x}\,\sqrt{x+2}$

47. $\sqrt{27t^5}\,\sqrt{3t^3}$

48. $\left(\sqrt{7x^3}\right)^2$

49. $\left(\sqrt{2x+1}\right)^2$

50. $\sqrt{x+1}\,\sqrt{x+1}$

51. $\sqrt[4]{2x}\,\sqrt[4]{4y^2}$

52. $\sqrt[5]{3xy^2}\,\sqrt[5]{9x^2y}$

53. $\sqrt{5}\,\sqrt[3]{x}$

54. $\sqrt[3]{2}\,\sqrt{3}$

55. $\sqrt{\dfrac{1}{xy}}\,\sqrt{\dfrac{x}{y}}$

56. $\sqrt{\dfrac{48x}{y}}\,\sqrt{\dfrac{y}{3x^5}}$

 57. Explain whether the Quotient Rule for Radicals applies to the problem
$$\dfrac{\sqrt[3]{16}}{\sqrt{2}}.$$

58. If it is known that $\sqrt[4]{x^9}$ represents a nonzero real number, why is it not necessary to use absolute value when simplifying the expression
$$\dfrac{\sqrt[4]{x^9}}{\sqrt[4]{x^5}}?$$

In Exercises 59–72, use the Quotient Rule for Radicals to divide the radicals. Assume all variables represent positive numbers.

59. $\dfrac{\sqrt{48}}{\sqrt{3}}$

60. $\dfrac{\sqrt{10}}{\sqrt{360}}$

61. $\dfrac{\sqrt{t^3}}{\sqrt{t^9}}$

62. $\dfrac{\sqrt{x^7}}{\sqrt{x}}$

63. $\dfrac{\sqrt[3]{54}}{\sqrt[3]{2}}$

64. $\dfrac{\sqrt[4]{3}}{\sqrt[4]{48}}$

65. $\dfrac{\sqrt{10}}{\sqrt[3]{2}}$

66. $\dfrac{\sqrt[3]{12}}{\sqrt{6}}$

67. $\dfrac{\sqrt[3]{x^{11}}}{\sqrt[3]{x^8}}$

68. $\dfrac{\sqrt[4]{x^{15}}}{\sqrt[4]{x^{11}}}$

69. $\dfrac{\sqrt{3a^3}}{\sqrt{75a^5}}$

70. $\dfrac{\sqrt{16xy}}{\sqrt{9x^3y^5}}$

71. $\dfrac{\sqrt{49a^7b^4}}{\sqrt{4a}}$

72. $\dfrac{\sqrt{ab^3}}{\sqrt{a^7b^5}}$

In Exercises 73–80, reduce the index. Assume all variables represent positive numbers.

73. $\sqrt[6]{x^3}$

74. $\sqrt[4]{x^2}$

75. $\sqrt[9]{8x^3}$

76. $\sqrt[8]{81x^4}$

77. $\sqrt[10]{x^4y^6}$

78. $\sqrt[8]{x^2y^4}$

79. $\left(\sqrt[4]{a}\right)^2$

80. $\left(\sqrt[6]{a}\right)^2$

 81. What is the first step in simplifying the radical $\sqrt{x^2 + 10x + 25}$? Why is this step necessary? Can this same step be used to simplify the radical $\sqrt{x^2 + 10x + 21}$? Explain.

82. When simplifying the following two radicals, only one of the results must be enclosed in absolute value symbols. Explain why this is so.

(i) $\sqrt{x^4 + 2x^2 + 1}$ (ii) $\sqrt{x^2 + 2x + 1}$

In Exercises 83–98, determine the indicated root. Assume all variables represent any real number.

83. $-\sqrt{4x^8}$

84. $-3\sqrt{9y^{12}}$

85. $\sqrt{25x^{10}y^8}$

86. $\sqrt{x^6y^{12}z^{16}}$

87. $\sqrt[6]{x^6y^{18}z^{12}}$

88. $\sqrt[4]{16x^{12}y^4}$

89. $\sqrt{x^2(y-3)^4}$

90. $\sqrt{y^4(2x+7)^2}$

91. $\sqrt{4y^2 - 4y + 1}$

92. $\sqrt{9t^2 + 12t + 4}$

93. $\sqrt{x^6 + 2x^5 + x^4}$

94. $\sqrt{9y^4 - 30y^5 + 25y^6}$

95. $\sqrt{9 \cdot 10^{-8}}$

96. $\sqrt[3]{8 \cdot 10^6}$

97. $\sqrt{3.6 \cdot 10^3}$

98. $\sqrt{4.9 \cdot 10^{-7}}$

In Exercises 99–106, write the given expression with just one radical symbol. Assume all variables represent positive numbers.

99. $\sqrt[3]{x^2}\,\sqrt[4]{x}$

100. $\sqrt[3]{x}\,\sqrt[5]{x}$

101. $\dfrac{\sqrt[4]{x}}{\sqrt[3]{x^2}}$

102. $\dfrac{x}{\sqrt[3]{x}}$

103. $\sqrt[4]{\sqrt[3]{7}}$

104. $\sqrt[4]{\sqrt[5]{x^3}}$

105. $\sqrt{2}\,\sqrt[3]{5}$

106. $\sqrt[3]{7}\,\sqrt{3}$

107. (a) Evaluate $\sqrt[4]{6}$ by entering $6\wedge(1 \div 4)$ in your calculator.

(b) Now enter $\sqrt{\sqrt{6}}$ and compare your result to part (a). What single radical expression are you evaluating when you enter $\sqrt{\sqrt{6}}$?

(c) Use rational exponents to show why the results are the same.

108. (a) Use your calculator to evaluate each of the given expressions. Round results to three decimal places.

$$\sqrt{2} \quad \sqrt{\sqrt{2}} \quad \sqrt{\sqrt{\sqrt{2}}} \quad \sqrt{\sqrt{\sqrt{\sqrt{2}}}}$$

$$\sqrt{\sqrt{\sqrt{\sqrt{\sqrt{2}}}}}$$

(b) As we continue to take square roots, what number does the value of the expression seem to be approaching?

(c) Now write each expression in exponential form. What number are the exponents approaching? Explain how this fact supports your conjecture in part (b).

109. Consider two positive numbers such that one number is 6 greater than the other number. If the product of the numbers is increased by 9, what is the square root of the result? Write your answer in simplified form.

110. Consider three nonzero numbers x, y, and z. How many of these numbers must be positive for the following to be defined in the real number system?

(a) $\sqrt{x}\,\sqrt{y}\,\sqrt{z}$ (b) $\sqrt{xyz}$

Exploring with Real Data

In Exercises 111–114, refer to the following background information and data.

In recent years, the percentage of unionized workers has steadily declined. (See figure.) (Source: Bureau of Labor Statistics.)

Figure for 111–114

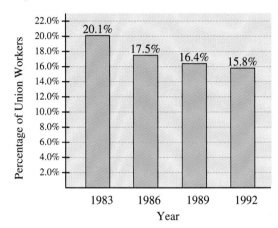

The data in the figure can be modeled by the function

$$M(t) = 20.13t^{-0.10},$$

where M is the percentage of the workforce belonging to unions and t is the number of years since 1982.

111. The figure does not show the union percentage for 1990. Use the model function to approximate this information.

112. Use a suitable range and scale setting to produce the graph of the model function. Trace the graph to estimate the year when the percentage of union membership was 18%.

113. Assuming the model is valid beyond 1992, what is the predicted percentage of union membership in 1999?

114. Suppose the federal government passed a bill banning companies from permanently replacing striking employees. How might this action invalidate the model in succeeding years?

Challenge

In Exercises 115 and 116, use a graph to estimate the domain of the given expression.

115. $\sqrt{x^2 + x - 12}$ **116.** $\sqrt{-x^2 - 6x - 5}$

In Exercises 117 and 118, write the given expression with just one radical symbol. Assume that x is positive.

117. $\sqrt{\sqrt{\sqrt{x}}}$ **118.** $\sqrt[4]{\sqrt[3]{\sqrt{x}}}$

In Exercises 119–124, evaluate the given expression, if possible.

119. $\sqrt[3]{-\sqrt[4]{1}}$ **120.** $\sqrt[4]{-\sqrt[3]{1}}$

121. $\sqrt[3]{\sqrt{64}}$ **122.** $\sqrt[3]{-\sqrt{64}}$

123. $\sqrt{\sqrt[3]{64}}$ **124.** $\sqrt{-\sqrt[3]{64}}$

In Exercises 125–128, state if the equation is always true, sometimes true, or never true. If sometimes, state the conditions under which the equation is true.

125. $\sqrt{n}\,\sqrt{n} = n$ **126.** $\sqrt[3]{n}\,\sqrt[3]{n}\,\sqrt[3]{n} = n$

127. $\sqrt{-n}\,\sqrt{n} = -n \quad (n \neq 0)$

128. $\sqrt[3]{\sqrt{n}} = \sqrt{\sqrt[3]{n}}$

129. Describe the values of x for which $\sqrt{(x-2)^2} = x - 2$ and for which $\sqrt{(x-2)^2} = 2 - x$.

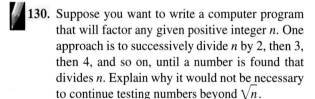

130. Suppose you want to write a computer program that will factor any given positive integer n. One approach is to successively divide n by 2, then 3, then 4, and so on, until a number is found that divides n. Explain why it would not be necessary to continue testing numbers beyond $\sqrt{n}$.

9.5 Simplifying Radicals

Simplifying with the Product Rule

In the previous section, we found that we could determine the indicated root for expressions such as the following.

$$\sqrt{36x^8} = 6x^4$$

$$\sqrt[3]{8y^9} = 2y^3$$

For the first radical, the radicand $36x^8$ is a *perfect square* because $36x^8 = 6^2 \cdot (x^4)^2$, and the exponent on each factor is divisible by 2. Similarly, for the second radical, the radicand $8y^9$ is a *perfect cube* because $8y^9 = 2^3 \cdot (y^3)^3$ and the exponent on each factor is divisible by 3.

In general, if the exponent on each factor of an expression is divisible by n, we say that the expression is a **perfect *n*th power.**

Of course, radicands are not always perfect nth powers. For example, we are not able to use the methods of Section 9.4 for $\sqrt{50}$ or for $\sqrt[3]{x^5}$ because 50 is not a perfect square and x^5 is not a perfect cube. However, as we will see, such radicals can be *simplified.*

To be considered simplified, a radical must be written in a form that meets certain conditions. As we proceed through this section, we will discuss what these conditions are. The following is the first condition.

Simplifying a Radical: Condition 1

If the index of a radical is n, then in order for the radical to be in simplified form, the radicand must contain no factor that is a perfect nth power.

We can use the Product Rule for Radicals to satisfy this first condition. If the radicand is an integer, we write it as a product, one of whose factors is the largest perfect nth power factor of the integer.

EXAMPLE 1 *Using the Product Rule to Simplify a Radical*

Simplify each radical. Use a calculator to verify that your result has the same value as the original radical.

(a) $\sqrt{75}$ (b) $\sqrt{32}$ (c) $\sqrt[3]{24}$ (d) $\sqrt[4]{48}$ (e) $\sqrt[3]{-54}$

Solution

(a) $\sqrt{75} = \sqrt{25 \cdot 3}$ The largest perfect square factor of 75 is 25.

$\quad\quad = \sqrt{25} \cdot \sqrt{3}$ Product Rule for Radicals

$\quad\quad = 5\sqrt{3}$

(b) $\sqrt{32} = \sqrt{16 \cdot 2}$ The largest perfect square factor of 32 is 16.

$\quad\quad = \sqrt{16} \cdot \sqrt{2}$ Product Rule for Radicals

$\quad\quad = 4\sqrt{2}$

(c) $\sqrt[3]{24} = \sqrt[3]{8 \cdot 3}$ The largest perfect cube factor of 24 is 8.

$\quad\quad = \sqrt[3]{8} \cdot \sqrt[3]{3}$ Product Rule for Radicals

$\quad\quad = 2\sqrt[3]{3}$

(d) $\sqrt[4]{48} = \sqrt[4]{16 \cdot 3}$ The largest perfect fourth power factor of 48 is 16.

$\quad\quad = \sqrt[4]{16} \cdot \sqrt[4]{3}$ Product Rule for Radicals

$\quad\quad = 2\sqrt[4]{3}$

(e) $\sqrt[3]{-54} = \sqrt[3]{-27 \cdot 2}$ We choose the negative perfect cube factor -27.

$\quad\quad = \sqrt[3]{-27} \cdot \sqrt[3]{2}$ Product Rule for Radicals

$\quad\quad = -3\sqrt[3]{2}$ ■

A similar procedure is used when the radicand contains a variable raised to a power.

EXAMPLE 2 *Using the Product Rule to Simplify a Radical*

Simplify. Assume all variables represent positive numbers.

(a) $\sqrt{y^7}$ (b) $\sqrt[3]{a^4 b^8}$

Solution

(a) $\sqrt{y^7} = \sqrt{y^6 \cdot y}$ The largest perfect square factor of y^7 is y^6.

$\quad\quad = \sqrt{y^6} \cdot \sqrt{y}$ Product Rule for Radicals

$\quad\quad = y^3 \sqrt{y}$

(b) $\sqrt[3]{a^4 b^8} = \sqrt[3]{a^3 \cdot a \cdot b^6 \cdot b^2}$ The largest perfect cube factors are a^3 and b^6.

$\quad\quad = \sqrt[3]{a^3 b^6} \cdot \sqrt[3]{ab^2}$ Product Rule for Radicals

$\quad\quad = ab^2 \sqrt[3]{ab^2}$ ■

To save a step in the simplifying process, we can just write the given radical as a product of two radicals with one of the radicands containing all of the perfect nth power factors.

EXAMPLE 3 *Using the Product Rule to Simplify a Radical*

In the following, assume all variables represent positive numbers.

(a) $\sqrt{49a^7} = \sqrt{49a^6} \cdot \sqrt{a} = 7a^3\sqrt{a}$

(b) $\sqrt{300r^{12}s^9} = \sqrt{100r^{12}s^8} \cdot \sqrt{3s} = 10r^6s^4\sqrt{3s}$

(c) $\sqrt[3]{24x^9y^{11}} = \sqrt[3]{8x^9y^9} \cdot \sqrt[3]{3y^2} = 2x^3y^3\sqrt[3]{3y^2}$

(d) $\sqrt[5]{64p^{12}q^{23}} = \sqrt[5]{32p^{10}q^{20}} \cdot \sqrt[5]{2p^2q^3} = 2p^2q^4\sqrt[5]{2p^2q^3}$

(e) $\sqrt{2x^2 + 12x + 18} = \sqrt{2(x^2 + 6x + 9)}$ Factor out the GCF 2.

$\qquad\qquad\qquad\quad = \sqrt{2(x + 3)^2}$ Factor the trinomial.

$\qquad\qquad\qquad\quad = \sqrt{(x + 3)^2} \cdot \sqrt{2}$ Product Rule for Radicals

$\qquad\qquad\qquad\quad = (x + 3)\sqrt{2}$ Note that parentheses are needed. ■

Sometimes we need to use the Product Rule for Radicals twice in the same problem. First, we use it to multiply the radicals. Then, we use it again to simplify the result.

NOTE: Although it is often possible to do some simplifying before multiplying, it is generally less confusing to find the product first and then simplify. ■

EXAMPLE 4 *Multiplying and Simplifying Radicals*

Multiply as indicated and simplify the result. Assume all variables represent positive numbers.

(a) $\sqrt{12a^4b^4} \cdot \sqrt{3a^2b^6}$ 　　　　　　　　(b) $\sqrt{6x^3y^7} \cdot \sqrt{8x^4y^3}$

Solution

(a) $\sqrt{12a^4b^4} \cdot \sqrt{3a^2b^6} = \sqrt{36a^6b^{10}}$ Product Rule for Radicals

$\qquad\qquad\qquad\qquad\quad = 6a^3b^5$ The radicand is a perfect square.

(b) $\sqrt{6x^3y^7} \cdot \sqrt{8x^4y^3} = \sqrt{48x^7y^{10}}$ Product Rule for Radicals

$\qquad\qquad\qquad\qquad = \sqrt{16x^6y^{10}} \cdot \sqrt{3x}$ Product Rule for Radicals

$\qquad\qquad\qquad\qquad = 4x^3y^5\sqrt{3x}$ ■

Index Reduction and Rational Exponents

In Section 9.4 we saw how we can sometimes use rational exponents to reduce the index of a radical. The second condition for a simplified radical is that the index must be as small as possible.

> **Simplifying a Radical: Condition 2**
>
> In order for a radical to be in simplified form, the index must be reduced to its smallest value.

Example 5 is a review of the procedure.

EXAMPLE 5   *Reducing the Index of a Radical*

Simplify $\sqrt[8]{16x^4y^6}$. Assume that x and y are positive.

Solution

$$\sqrt[8]{16x^4y^6} = (2^4x^4y^6)^{1/8} \qquad \text{Definition of } b^{1/n}$$

$$= 2^{2/4}x^{2/4}y^{3/4} \qquad \text{Power to a Power Rule for Exponents}$$

$$= (2^2x^2y^3)^{1/4} \qquad \text{Note that the denominator of each exponent is 4.}$$

$$= \sqrt[4]{4x^2y^3} \qquad \text{Definition of } b^{1/n}$$

Simplifying with the Quotient Rule

The following is the third condition for a simplified radical.

> ### Simplifying a Radical: Condition 3
>
> In order for a radical to be simplified, the radicand must contain no fractions.

The easiest fractional radicand to simplify is one in which the denominator is a perfect power. Then the Quotient Rule for Radicals can be applied directly to simplify the radical.

$$\sqrt[n]{\frac{a}{b}} = \frac{\sqrt[n]{a}}{\sqrt[n]{b}}$$

EXAMPLE 6   *Simplifying a Radical with a Fractional Radicand*

Use the Quotient Rule for Radicals to simplify the radical.

(a) $\sqrt{\dfrac{7}{9}}$ (b) $\sqrt[3]{\dfrac{9z}{125}}$ (c) $\sqrt[4]{\dfrac{a^5}{16x^8}}$ $(a > 0, x \neq 0)$

Solution

(a) $\sqrt{\dfrac{7}{9}} = \dfrac{\sqrt{7}}{\sqrt{9}} = \dfrac{\sqrt{7}}{3}$

(b) $\sqrt[3]{\dfrac{9z}{125}} = \dfrac{\sqrt[3]{9z}}{\sqrt[3]{125}} = \dfrac{\sqrt[3]{9z}}{5}$

(c) $\sqrt[4]{\dfrac{a^5}{16x^8}} = \dfrac{\sqrt[4]{a^5}}{\sqrt[4]{16x^8}} = \dfrac{\sqrt[4]{a^4}\cdot\sqrt[4]{a}}{2x^2} = \dfrac{a\sqrt[4]{a}}{2x^2}$ $(a > 0, x \neq 0)$

If the radicand is a fraction, it might be possible to perform a simple division to eliminate the fraction. If so, then we can use previously discussed procedures to complete the simplification.

EXAMPLE 7 *Simplifying a Radical with a Fractional Radicand*

Simplify $\sqrt{\dfrac{45x^9y}{x^3}}$, where $x > 0$ and $y > 0$.

Solution

$$\sqrt{\frac{45x^9y}{x^3}} = \sqrt{45x^6y} \qquad \text{Perform the indicated division in the radicand.}$$

$$= \sqrt{9x^6}\,\sqrt{5y} \qquad \text{Product Rule for Radicals}$$

$$= 3x^3\sqrt{5y}$$

Radicals in the Denominator

The following is the fourth condition for a simplified radical.

> **Simplifying a Radical: Condition 4**
>
> In order for a radical expression to be in simplified form, there must be no radicals in the denominator of a fraction.

In some cases, we can satisfy this condition simply by applying the Quotient Rule for Radicals and then using previously discussed techniques to simplify the resulting radical.

EXAMPLE 8 *Simplifying Expressions with Radicals in the Denominator*

Simplify each radical expression. Assume all variables represent positive numbers.

(a) $\dfrac{\sqrt{6x^3}}{\sqrt{2x^2}}$ (b) $\dfrac{\sqrt{a^{11}b^8c^5}}{\sqrt{a^2bc}}$ (c) $\dfrac{\sqrt[3]{48x^9y^{12}}}{\sqrt[3]{3xy^2}}$

Solution

(a) $\dfrac{\sqrt{6x^3}}{\sqrt{2x^2}} = \sqrt{\dfrac{6x^3}{2x^2}} \qquad \text{Quotient Rule for Radicals}$

$= \sqrt{3x} \qquad \text{Divide the numerator by the denominator.}$

(b) $\dfrac{\sqrt{a^{11}b^8c^5}}{\sqrt{a^2bc}} = \sqrt{\dfrac{a^{11}b^8c^5}{a^2bc}}$ Quotient Rule for Radicals

$= \sqrt{a^9b^7c^4}$ Divide the numerator by the denominator.

$= \sqrt{a^8b^6c^4}\sqrt{ab}$ Product Rule for Radicals

$= a^4b^3c^2\sqrt{ab}$

(c) $\dfrac{\sqrt[3]{48x^9y^{12}}}{\sqrt[3]{3xy^2}} = \sqrt[3]{\dfrac{48x^9y^{12}}{3xy^2}}$ Quotient Rule for Radicals

$= \sqrt[3]{16x^8y^{10}}$ Divide the numerator by the denominator.

$= \sqrt[3]{8x^6y^9}\sqrt[3]{2x^2y}$ Product Rule for Radicals

$= 2x^2y^3\sqrt[3]{2x^2y}$ ■

After applying the Quotient Rule for Radicals, we may find that the denominator does not divide evenly into the numerator. However, we may be able to simplify the radicand.

EXAMPLE 9 *Simplifying Expressions with Radicals in the Denominator*

Simplify the radical expression. Assume all variables represent positive numbers.

(a) $\dfrac{\sqrt{45xy}}{\sqrt{x^3}}$ (b) $\dfrac{\sqrt[3]{72x^2y^7}}{\sqrt[3]{x^{11}}}$ (c) $\dfrac{\sqrt{9}}{\sqrt{5}}$

Solution

(a) $\dfrac{\sqrt{45xy}}{\sqrt{x^3}} = \sqrt{\dfrac{45xy}{x^3}}$ Quotient Rule for Radicals

$= \sqrt{\dfrac{45y}{x^2}}$ Simplify the radicand.

$= \dfrac{\sqrt{9}\,\sqrt{5y}}{\sqrt{x^2}}$ Product Rule for Radicals

$= \dfrac{3\sqrt{5y}}{x}$

(b) $\dfrac{\sqrt[3]{72x^2y^7}}{\sqrt[3]{x^{11}}} = \sqrt[3]{\dfrac{72x^2y^7}{x^{11}}}$ Quotient Rule for Radicals

$= \sqrt[3]{\dfrac{72y^7}{x^9}}$ Simplify the radicand.

$= \dfrac{\sqrt[3]{8y^6}\cdot\sqrt[3]{9y}}{\sqrt[3]{x^9}}$ Product Rule for Radicals

$= \dfrac{2y^2\sqrt[3]{9y}}{x^3}$

(c) The expression $\dfrac{\sqrt{9}}{\sqrt{5}}$

presents some difficulty. We cannot divide 9 by 5 evenly, and 5 is not a perfect square. We cannot simplify this expression with our present methods. ■

In Section 9.7 we will develop techniques for simplifying radical expressions such as the one in part (c) of Example 9.

9.5 Quick Reference

Simplifying with the Product Rule	▪ If the exponent of each factor of an expression is divisible by n, then the expression is a **perfect nth power.**
	▪ Condition 1 for a simplified radical: If the index of a radical is n, the radicand must contain no factor that is a perfect nth power.
	▪ To satisfy Condition 1, we can use the Product Rule for Radicals to write the radical as a product of radicals with one of the radicands containing all of the perfect nth power factors.
Index Reduction and Rational Exponents	▪ Condition 2 for a simplified radical: The index must be reduced to its smallest value.
	▪ To satisfy Condition 2, we use the previously discussed method of writing the radical with fractional exponents and reducing the index.
Simplifying with the Quotient Rule	▪ Condition 3 for a simplified radical: The radicand must contain no fractions.
	▪ If the denominator of a fractional radicand is a perfect nth power, then the Quotient Rule for Radicals can be applied directly to simplify the radical.
	▪ It may be possible to simplify the radicand. This may eliminate the fraction, or it may result in a denominator that is a perfect nth power.
Radicals in the Denominator	▪ Condition 4 for a simplified radical: There must be no radicals in the denominator of a fraction.
	▪ If dividing the numerator by the denominator eliminates the fraction, or if the denominator is a perfect power, then the resulting radical can be simplified. Otherwise, we cannot simplify the radical with our present methods.

9.5 Exercises

 1. Explain why $\sqrt{32}$ can be simplified but $\sqrt{30}$ cannot.

 2. To simplify $\sqrt{16x^{16}}$, we treat the coefficient and the exponent differently even though they are the same number. Explain what to do with each.

In Exercises 3–14, simplify the radical. Use your calculator to verify that the result has the same value as the original radical.

3. $\sqrt{12}$ **4.** $\sqrt{20}$

5. $\sqrt{50}$ **6.** $\sqrt{18}$

7. $\sqrt[3]{54}$ **8.** $\sqrt[3]{192}$

9. $\sqrt[3]{-48}$ **10.** $\sqrt[3]{-54}$

11. $\sqrt[4]{32}$ **12.** $\sqrt[4]{162}$

13. $\sqrt[5]{256}$ **14.** $\sqrt[5]{128}$

In Exercises 15–20, simplify. Assume all variables represent positive numbers.

15. $\sqrt{a^5 b^3}$ **16.** $\sqrt{x^9 y^4 x^7}$

17. $\sqrt[3]{a^9 b^8}$ **18.** $\sqrt[3]{x^{17} y^{15}}$

19. $\sqrt[4]{x^{16} y^9}$ **20.** $\sqrt[4]{a^5 b^7}$

In Exercises 21–34, simplify. Assume all variables represent positive numbers.

21. $\sqrt{60t}$ **22.** $\sqrt{20x^{10}}$

23. $\sqrt{25x^{25}}$ **24.** $\sqrt{18a^9}$

25. $\sqrt{500a^7 b^{14}}$ **26.** $\sqrt{100a^3 b}$

27. $-\sqrt{49x^6 y^7}$ **28.** $-\sqrt{8r^5 t^9}$

29. $\sqrt{28w^6 t^9}$ **30.** $\sqrt{75x^5 y^7 z^9}$

31. $\sqrt[3]{81x^9 y^{11}}$ **32.** $\sqrt[3]{54x^8}$

33. $\sqrt[5]{-32x^7 y^9}$ **34.** $\sqrt[4]{32x^6 y^8}$

In Exercises 35–42, supply the missing numbers to make the statement true. Assume all variables represent positive numbers.

35. $\sqrt{(\ )a^{(\ )} b^{(\ )}} = 4a^2 b \sqrt{b}$

36. $(\ )\sqrt{4x^{(\ )} y^{(\ )}} = -6x^3 y^2 \sqrt{xy}$

37. $\sqrt{3x^{(\ )} y^3}\ \sqrt{(\ )x^2 y^{(\ )}} = 3x^2 y^3$

38. $\sqrt{6a^{(\ )} b^3}\ \sqrt{(\ )ab^{(\ )}} = 3a^2 b^4 \sqrt{2a}$

39. $\dfrac{\sqrt{(\ )x^{(\ )} y^8}}{\sqrt{3x^2 y^{(\ )}}} = 2x^3 \sqrt{3xy}$

40. $\dfrac{\sqrt{90a^{(\ )} b^9}}{\sqrt{(\ )ab^{(\ )}}} = 3ab \sqrt{5b}$

41. $\sqrt[3]{(\ )x^{(\ )} y^{(\ )}} = 2xy \sqrt[3]{2x}$

42. $\sqrt[3]{(\ )x^{(\ )} y^{(\ )}} = 5x^2 y \sqrt[3]{2y^2}$

 43. It would be incorrect to simplify $\sqrt{4x^2 + 16}$ by taking the square root of each term. What step must first be taken before the radical can be simplified?

 44. The following are two correct simplifications. Explain why it is necessary in one case, but not in the other, to use absolute value in the result.

$$\sqrt{y^4 + 4y^2 + 4} = y^2 + 2$$
$$\sqrt{y^2 + 4y + 4} = |y + 2|$$

In Exercises 45–50, simplify. Assume all variables represent positive numbers.

45. $\sqrt{12x + 8y}$ **46.** $\sqrt{ab^2 + b^3}$

47. $\sqrt{4x^2 + 16}$ **48.** $\sqrt{x^4 + x^2}$

49. $\sqrt{9x^7y^6 + 9x^6y^7}$

50. $\sqrt{50a^3 + 75a^4b}$

In Exercises 51–58, simplify. Assume all variables represent any real number.

51. $\sqrt{9y^2 + 9y^4}$

52. $\sqrt{t^6 + t^4}$

53. $\sqrt{r^4t^2 + r^6t^4}$

54. $\sqrt{a^4b^6 + a^2b^8}$

55. $\sqrt{y^2 - 10y + 25}$

56. $\sqrt{2t^2 - 28t + 98}$

57. $\sqrt{9n^6 + 6n^5 + n^4}$

58. $\sqrt{t^4 + 12t^3 + 36t^2}$

59. If P and Q are monomials, describe two circumstances in which it would be possible to simplify

$$\sqrt{\frac{P}{Q}}$$ with methods discussed in this section.

60. Explain why none of our current methods can be used to simplify

$$\sqrt{\frac{7}{x}}, \, x > 0.$$

In Exercises 61–66, simplify the radical. Assume all variables represent positive numbers.

61. $\sqrt{\dfrac{28}{9}}$

62. $\sqrt{\dfrac{125a}{144}}$

63. $\sqrt{\dfrac{60x^{11}y^3}{3y}}$

64. $\sqrt{\dfrac{245x^7y^9z^{12}}{5x^3y^2z^5}}$

65. $\sqrt[3]{\dfrac{-16a}{1000}}$

66. $\sqrt[4]{\dfrac{243w}{16}}$

In Exercises 67–80, multiply and simplify. Assume all variables represent positive numbers.

67. $\left(-2\sqrt{5}\right)\left(2\sqrt{5}\right)$

68. $\left(4\sqrt{6x}\right)\left(-2\sqrt{2x}\right)$

69. $\sqrt[3]{3}\ \sqrt[3]{54}$

70. $\sqrt[3]{2}\ \sqrt[3]{20}$

71. $\sqrt{7x}\ \sqrt{14x}$

72. $\sqrt{3y^2}\ \sqrt{15y}$

73. $\sqrt{3x^3}\ \sqrt{2x^6}$

74. $\left(\sqrt{3x^2y}\right)^2$

75. $\sqrt{6xy^3}\ \sqrt{12x^3y^2}$

76. $\sqrt{3c^{11}d^{13}}\ \sqrt{15c^{12}d^{14}}$

77. $\sqrt[3]{25x^2y^2}\ \sqrt[3]{5x^3y^4}$

78. $\left(\sqrt[3]{9x^2y^4}\right)^2$

79. $\sqrt[4]{54x^3y}\ \sqrt[4]{3x^2y^4}$

80. $\sqrt[4]{27x^2y^5}\ \sqrt[4]{3x^3y^3}$

81. Explain why $\sqrt[6]{x^4}$ can be simplified but $\sqrt[6]{x^5}$ cannot.

82. To simplify $\sqrt[6]{z^{10}}$, we can proceed as follows.
$$\sqrt[6]{z^{10}} = z^{10/6} = z^{5/3} = \sqrt[3]{z^5}$$

Is the radical now simplified? Why?

In Exercises 83–92, simplify. Assume all variables represent positive numbers.

83. $\sqrt[15]{x^{20}}$

84. $\sqrt[6]{x^9}$

85. $\sqrt[8]{16x^{12}y^{20}}$

86. $\sqrt[6]{27x^{15}y^{12}}$

87. $\sqrt[12]{27x^6y^9}$

88. $\sqrt[15]{32x^{10}y^5}$

89. $\sqrt[21]{64x^{18}y^{15}}$

90. $\sqrt[12]{16x^4y^8}$

91. $\sqrt[6]{8x^{15}y^9}$

92. $\sqrt[4]{9x^6y^{10}}$

In Exercises 93–100, use the Quotient Rule for Radicals to simplify. Assume all variables represent positive numbers.

93. $\sqrt{\dfrac{5}{49}}$

94. $\dfrac{\sqrt{30}}{\sqrt{5}}$

95. $\dfrac{\sqrt{a^5}}{\sqrt{b^{10}}}$

96. $\sqrt{\dfrac{x^{100}}{100}}$

97. $\dfrac{\sqrt{18x^9d^{13}}}{\sqrt{2x^4d^{10}}}$

98. $\sqrt{\dfrac{75x^3y^5}{x^7y^{13}}}$

99. $\sqrt[3]{\dfrac{16x^2}{2x^{14}}}$

100. $\dfrac{\sqrt[3]{54x^{13}}}{\sqrt[3]{x^2}}$

In Exercises 101–106, determine the value of a. Assume all variables represent positive numbers.

101. $4\sqrt{x} = \sqrt{ax}$

102. $10\sqrt{x^2y} = \sqrt{ax^2y}$

103. $\dfrac{6}{\sqrt{x}} = \sqrt{\dfrac{a}{x}}$

104. $\dfrac{2}{\sqrt{y}} = \sqrt{\dfrac{a}{y}}$

105. $\sqrt{x^3} = \sqrt[a]{x^6}$

106. $\sqrt[3]{x^5} = \sqrt[a]{x^{10}}$

 107. If a, b, and c are the lengths of the sides of a triangle, then Hero's Formula is a formula for the area of the triangle.

$$A = \dfrac{}{\frac{1}{4}\sqrt{(a + b + c)(a + b - c)(a - b + c)(-a + b + c)}}$$

If the lengths of the sides of a triangle are represented by $3x$, $4x$, and $5x$, where $x > 0$, use Hero's Formula to show that the area of the triangle is $6x^2$.

108. Use your calculator to show that $\dfrac{7}{\sqrt{7}} = \sqrt{7}$.

From this, you may conclude that dividing a number by its square root always results in the square root of the number. Use rational exponents to prove that your conclusion is true for any positive number.

Challenge

In Exercises 109–116, simplify the expression. Assume all variables represent positive numbers.

109. $\sqrt{t^{2n}}$

110. $\sqrt[3]{t^{3n}}$

111. $\sqrt[n]{t^{2n}}$

112. $\sqrt[n]{t^{n^2}}$

113. $\sqrt{x^{2a}y^{4b}}$

114. $\sqrt[3]{x^{6a}y^{3b}}$

115. $\sqrt[3]{\dfrac{x^{3m}}{8^{3n}}}$

116. $\sqrt{\dfrac{y^{16n}}{9^{6m}}}$

 117. What is the least positive value of n for which it is possible to simplify $\sqrt{\sqrt{(a + b)^n}}$ so that there is no radical in the result?

118. Use your calculator to evaluate each of the following expressions.

$$\sqrt{2} \qquad \sqrt{2 + \sqrt{2}} \qquad \sqrt{2 + \sqrt{2 + \sqrt{2}}}$$

$$\sqrt{2 + \sqrt{2 + \sqrt{2 + \sqrt{2}}}}$$

$$\sqrt{2 + \sqrt{2 + \sqrt{2 + \sqrt{2 + \sqrt{2}}}}}$$

What number do the values of these expressions seem to be approaching? Do you think that this pattern holds true for any positive integer n? If so, repeat the experiment for $n = 3$.

9.6   Operations with Radicals

Addition and Subtraction   ▪   Multiplication   ▪   Conjugates

Addition and Subtraction

In a manner similar to adding like terms, we use the Distributive Property to add or subtract radicals. For example, recall that we add $3x + 5x$ as follows.

$$3x + 5x = (3 + 5)x = 8x$$

To add $3\sqrt{7} + 5\sqrt{7}$, we use the Distributive Property in the same way.

$$3\sqrt{7} + 5\sqrt{7} = (3 + 5)\sqrt{7} = 8\sqrt{7}$$

To add or subtract radical expressions, the index and the radicands must be the same. We say that radicals with the same radicand and the same index are **like radicals.** Thus only like radicals can be combined.

Sometimes one or both of two unlike radicals can be simplified so that they become like radicals and can then be combined.

EXAMPLE 1   *Adding and Subtracting Radicals*
Perform the indicated operation, if possible.

(a) $2\sqrt{3} + 5\sqrt{3}$ (b) $\sqrt{45} - \sqrt{80}$ (c) $3\sqrt{75} - 2\sqrt{48}$

(d) $\sqrt[3]{32} + \sqrt[3]{4}$ (e) $4\sqrt{2} + 3\sqrt[3]{2}$ (f) $\sqrt{3} + \sqrt{2}$

Solution

(a) $2\sqrt{3} + 5\sqrt{3} = (2 + 5)\sqrt{3} = 7\sqrt{3}$ Like radicals. Use the Distributive Property.

(b) $\sqrt{45} - \sqrt{80} = \sqrt{9}\,\sqrt{5} - \sqrt{16}\,\sqrt{5}$ Product Rule for Radicals
$\phantom{(b)\ \sqrt{45} - \sqrt{80}} = 3\sqrt{5} - 4\sqrt{5}$
$\phantom{(b)\ \sqrt{45} - \sqrt{80}} = (3 - 4)\sqrt{5}$ Distributive Property
$\phantom{(b)\ \sqrt{45} - \sqrt{80}} = -\sqrt{5}$

(c) $3\sqrt{75} - 2\sqrt{48} = 3\sqrt{25}\,\sqrt{3} - 2\sqrt{16}\,\sqrt{3}$ Product Rule for Radicals
$\phantom{(c)\ 3\sqrt{75} - 2\sqrt{48}} = 3 \cdot 5\sqrt{3} - 2 \cdot 4\sqrt{3}$
$\phantom{(c)\ 3\sqrt{75} - 2\sqrt{48}} = 15\sqrt{3} - 8\sqrt{3}$
$\phantom{(c)\ 3\sqrt{75} - 2\sqrt{48}} = (15 - 8)\sqrt{3}$ Distributive Property
$\phantom{(c)\ 3\sqrt{75} - 2\sqrt{48}} = 7\sqrt{3}$

(d) $\sqrt[3]{32} + \sqrt[3]{4} = \sqrt[3]{8}\,\sqrt[3]{4} + \sqrt[3]{4}$ Product Rule for Radicals

$\qquad\qquad\quad = 2\sqrt[3]{4} + 1\sqrt[3]{4}$

$\qquad\qquad\quad = (2 + 1)\sqrt[3]{4}$ Distributive Property

$\qquad\qquad\quad = 3\sqrt[3]{4}$

(e) Because the indices are different, the radicals are not like radicals, and so the terms cannot be combined.

(f) Because the radicands are different and the radicals cannot be simplified, $\sqrt{3}$ and $\sqrt{2}$ cannot be combined.   ■

In the future, we will not specifically include the Distributive Property step, but it is important to remember that it is this property that permits the combining of terms.

EXAMPLE 2   *Adding and Subtracting Radical Expressions*

Perform the indicated operation. Assume all variables represent positive numbers.

(a) $\sqrt{48x} - \sqrt{27x}$ (b) $4\sqrt{a^3} + 2a\sqrt{9a}$ (c) $3\sqrt[3]{16t^7} + 4t\sqrt[3]{2t^4}$

Solution

(a) $\sqrt{48x} - \sqrt{27x} = \sqrt{16}\cdot\sqrt{3x} - \sqrt{9}\cdot\sqrt{3x}$ Product Rule for Radicals

$\qquad\qquad\qquad = 4\sqrt{3x} - 3\sqrt{3x}$

$\qquad\qquad\qquad = \sqrt{3x}$ Combine the terms.

(b) $4\sqrt{a^3} + 2a\sqrt{9a} = 4\sqrt{a^2}\cdot\sqrt{a} + 2a\sqrt{9}\cdot\sqrt{a}$ Product Rule for Radicals

$\qquad\qquad\qquad = 4a\sqrt{a} + 2a\cdot 3\sqrt{a}$

$\qquad\qquad\qquad = 4a\sqrt{a} + 6a\sqrt{a}$

$\qquad\qquad\qquad = 10a\sqrt{a}$ Combine the terms.

(c) $3\sqrt[3]{16t^7} + 4t\sqrt[3]{2t^4} = 3\sqrt[3]{8t^6}\cdot\sqrt[3]{2t} + 4t\sqrt[3]{t^3}\cdot\sqrt[3]{2t}$ Product Rule for Radicals

$\qquad\qquad\qquad = 3\cdot 2t^2\sqrt[3]{2t} + 4t\cdot t\sqrt[3]{2t}$

$\qquad\qquad\qquad = 6t^2\sqrt[3]{2t} + 4t^2\sqrt[3]{2t}$

$\qquad\qquad\qquad = 10t^2\sqrt[3]{2t}$ Combine the terms. ■

NOTE: In parts (b) and (c) of Example 2, we were able to combine the terms because the two terms differ only in their coefficients. That is, not only are the radicals identical, but also the *variable factors are identical*. Even though the radicals are the same in the expression $3t^3\sqrt{4t} + 3t\sqrt{4t}$, we cannot combine the terms because the variable factors are not the same.   ■

The following example illustrates the method for combining radicals whose radicands are fractions or rational expressions.

EXAMPLE 3 *Adding and Subtracting Radical Expressions*

Perform the indicated operation.

(a) $\sqrt{\dfrac{8}{9}} - \sqrt{\dfrac{18}{49}}$
(b) $\sqrt{\dfrac{81}{a^6}} + \sqrt{\dfrac{49}{a^{10}}}$ $(a > 0)$

Solution

(a) $\sqrt{\dfrac{8}{9}} - \sqrt{\dfrac{18}{49}} = \dfrac{\sqrt{8}}{\sqrt{9}} - \dfrac{\sqrt{18}}{\sqrt{49}}$
      Quotient Rule for Radicals

$= \dfrac{\sqrt{8}}{3} - \dfrac{\sqrt{18}}{7}$

$= \dfrac{\sqrt{4} \cdot \sqrt{2}}{3} - \dfrac{\sqrt{9} \cdot \sqrt{2}}{7}$
      Product Rule for Radicals

$= \dfrac{2\sqrt{2}}{3} - \dfrac{3\sqrt{2}}{7}$

$= \dfrac{14\sqrt{2}}{21} - \dfrac{9\sqrt{2}}{21}$
      Rewrite the fractions with 21 as the LCD.

$= \dfrac{5\sqrt{2}}{21}$
      Subtract the numerators.

(b) $\sqrt{\dfrac{81}{a^6}} + \sqrt{\dfrac{49}{a^{10}}} = \dfrac{9}{a^3} + \dfrac{7}{a^5}$ $(a > 0)$
      Both radicands are perfect squares.

$= \dfrac{9a^2}{a^5} + \dfrac{7}{a^5}$
      Rewrite the fractions with a^5 as the LCD.

$= \dfrac{9a^2 + 7}{a^5}$
      Add the numerators. ∎

Multiplication

In Section 9.4, we used the Product Rule for Radicals to multiply radicals with the same index. We can use this rule in conjunction with the Distributive Property to multiply radical expressions that have more than one term.

EXAMPLE 4 *Multiplying Radical Expressions*

(a) $3\sqrt{5}\left(2\sqrt{5} - \sqrt{2}\right) = \left(3\sqrt{5}\right)\left(2\sqrt{5}\right) - \left(3\sqrt{5}\right)\left(\sqrt{2}\right)$ Distributive Property

$= 6 \cdot 5 - 3\sqrt{10}$ Product Rule for Radicals

$= 30 - 3\sqrt{10}$

(b) $\sqrt{6}\left(\sqrt{8} - \sqrt{2}\right) = \sqrt{6}\,\sqrt{8} - \sqrt{6}\,\sqrt{2}$ Distributive Property

$= \sqrt{48} - \sqrt{12}$ Product Rule for Radicals

$= \sqrt{16}\,\sqrt{3} - \sqrt{4}\,\sqrt{3}$ Product Rule for Radicals

$= 4\sqrt{3} - 2\sqrt{3}$ Combine the terms.

$= 2\sqrt{3}$

(c) $\left(\sqrt{2} - 3\right)\left(\sqrt{3} + \sqrt{2}\right) = \sqrt{2}\,\sqrt{3} + \sqrt{2}\,\sqrt{2} - 3\sqrt{3} - 3\sqrt{2}$ FOIL

$\qquad\qquad\qquad\qquad\qquad = \sqrt{6} + 2 - 3\sqrt{3} - 3\sqrt{2}$ Product Rule for Radicals

None of the four terms are like terms, so the result cannot be simplified further.

(d) $\left(\sqrt{3} + \sqrt{6}\right)^2 = \left(\sqrt{3}\right)^2 + 2\cdot\sqrt{3}\,\sqrt{6} + \left(\sqrt{6}\right)^2$ $(A + B)^2 = A^2 + 2AB + B^2$

$\qquad\qquad\qquad = 3 + 2\sqrt{18} + 6$ Product Rule for Radicals

$\qquad\qquad\qquad = 9 + 2\sqrt{18}$

$\qquad\qquad\qquad = 9 + 2\sqrt{9}\,\sqrt{2}$ Product Rule for Radicals

$\qquad\qquad\qquad = 9 + 2\cdot 3\sqrt{2}$

$\qquad\qquad\qquad = 9 + 6\sqrt{2}$ ∎

Conjugates

Recall that the product of a sum of two terms and the difference of the same two terms is the difference of their squares.

$$(A + B)(A - B) = A^2 - B^2$$

The expressions $A + B$ and $A - B$ are called **conjugates** of each other. In the next exploration, we consider the effect of multiplying conjugates that contain square roots.

EXPLORATION 1   *The Product of Square Root Conjugates*

In each of the following, write the conjugate of the given expression and determine the product of the conjugates. Assume all variables are positive numbers.

(a) $4 - \sqrt{5}$ (b) $-2 - 3\sqrt{2}$ (c) $-x + 2\sqrt{y}$

(d) What do you observe about the products?

Discovery

In each case, the product is found according to the pattern

$$(A + B)(A - B) = A^2 - B^2.$$

Number	*Conjugate*	*Product*
(a) $4 - \sqrt{5}$	$4 + \sqrt{5}$	$4^2 - \left[\sqrt{5}\right]^2$
		$= 16 - 5 = 11$
(b) $-2 - 3\sqrt{2}$	$-2 + 3\sqrt{2}$	$(-2)^2 - \left[3\sqrt{2}\right]^2$
		$= 4 - 9\cdot 2$
		$= 4 - 18 = -14$
(c) $-x + 2\sqrt{y}$	$-x - 2\sqrt{y}$	$(-x)^2 - \left[2\sqrt{y}\right]^2$
		$= x^2 - 4y$

(d) There are no square roots in any of the products.

It is easy to see why the results in Exploration 1 are true. If x and y represent nonnegative radicands, then

$$\left(\sqrt{x} + \sqrt{y}\right)\left(\sqrt{x} - \sqrt{y}\right) = \left(\sqrt{x}\right)^2 - \left(\sqrt{y}\right)^2 = x - y.$$

NOTE: The conclusion from Exploration 1 holds true only for square root conjugates. ∎

In the next section, we will see how this fact can be used to remove radicals from the denominators of fractions.

9.6   Quick Reference

Addition and Subtraction

- **Like radicals** have the same radicand and the same index. Only like radicals can be combined by addition or subtraction.

- Simplify all radicals before attempting to add or subtract them.

- Use the Distributive Property to add or subtract like radicals.

Multiplication

- To multiply radical expressions that have more than one term, use the Distributive Property, the FOIL method, and other familiar multiplication techniques that we used with polynomials.

- The special product patterns for polynomials also apply to multiplication of radicals.

$$(A + B)^2 = A^2 + 2AB + B^2$$
$$(A - B)^2 = A^2 - 2AB + B^2$$
$$(A + B)(A - B) = A^2 - B^2$$

Conjugates

- Expressions of the form $A + B$ and $A - B$ are said to be **conjugates** of one another.

- If at least one term of a pair of conjugate expressions is a square root, then the product of the conjugates does not contain a square root.

9.6   Exercises

1. Explain how the Distributive Property is used to add or subtract radical expressions. Demonstrate with $3\sqrt{5} + 4\sqrt{5}$.

2. Consider the following two uses of the Distributive Property.

$$3\sqrt{5} + 7\sqrt{5} = (3 + 7)\sqrt{5} = 10\sqrt{5}$$
$$3\sqrt{5} + x\sqrt{5} = (3 + x)\sqrt{5}$$

Explain why parentheses are needed in the second result, but not in the first.

In Exercises 3–12, perform the indicated operation. Assume that all variables represent positive numbers.

3. $5\sqrt{13} + \sqrt{13}$

4. $8\sqrt{11} - \sqrt{11}$

5. $\sqrt{7} - 2\sqrt{7}$

6. $\dfrac{\sqrt{3}}{3} + \sqrt{3}$

7. $6\sqrt{x} + 10\sqrt{x}$

8. $5\sqrt{ax} - 6\sqrt{ax}$

9. $9\sqrt[3]{y} - 12\sqrt[3]{y}$

10. $3\sqrt[3]{x^2} + 7\sqrt[3]{x^2}$

11. $\sqrt{2x} + 4\sqrt{3} - 5\sqrt{2x} - 3\sqrt{3}$

12. $ab\sqrt{a} + 2\sqrt{b} - 2ab\sqrt{a} + \sqrt{b}$

13. For each of the following pairs, explain why the radicals are not like radicals.

(a) $2\sqrt{5}, 5\sqrt{2}$

(b) $\sqrt{7}, \sqrt[3]{7}$

(c) $\sqrt[3]{4}, \sqrt[4]{3}$

(d) $\sqrt{2x}, \sqrt{3x}$

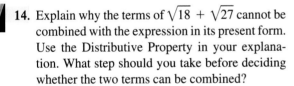

14. Explain why the terms of $\sqrt{18} + \sqrt{27}$ cannot be combined with the expression in its present form. Use the Distributive Property in your explanation. What step should you take before deciding whether the two terms can be combined?

In Exercises 15–26, perform the indicated operation. Use your calculator to verify the result.

15. $\sqrt{27} - \sqrt{12}$

16. $\sqrt{45} + \sqrt{80}$

17. $\sqrt[3]{250} + \sqrt[3]{54}$

18. $\sqrt[3]{24} - \sqrt[3]{192}$

19. $\sqrt[4]{162} - \sqrt[4]{32}$

20. $\sqrt[4]{48} + \sqrt[4]{243}$

21. $5\sqrt{32} + 7\sqrt{72}$

22. $10\sqrt{20} - 3\sqrt{45}$

23. $\dfrac{4}{5}\sqrt{75} - \dfrac{3}{2}\sqrt{12}$

24. $\dfrac{1}{3}\sqrt{45} + \dfrac{1}{4}\sqrt{80}$

25. $4\sqrt{12} - 2\sqrt{27} + 5\sqrt{8}$

26. $3\sqrt{50} - 2\sqrt{20} + 6\sqrt{18}$

In Exercises 27–42, perform the indicated operation. Assume that all variables represent positive numbers.

27. $\sqrt{72x} - \sqrt{50x}$

28. $\sqrt{80c} + \sqrt{20c}$

29. $x\sqrt{48} + 3\sqrt{27x^2}$

30. $3\sqrt{48x^2y} - 2x\sqrt{75y}$

31. $4a\sqrt{20a^2b} + a^2\sqrt{45b}$

32. $5\sqrt{a^5} + 7a\sqrt{4a^3}$

33. $\sqrt{\dfrac{25}{x^2}} + \sqrt{\dfrac{4}{x^4}}$

34. $\sqrt{\dfrac{x^3}{9}} + x\sqrt{\dfrac{9x}{16}}$

35. $x\sqrt{2x} - 3\sqrt{8x^2} + \sqrt{50x^3}$

36. $\sqrt{9x^4} + x\sqrt{12x} - \sqrt{27x^3}$

37. $\sqrt{9xy^3} + 4\sqrt{x^3y} - 5y\sqrt{4xy}$

38. $3\sqrt{x^2} + \sqrt{x^9} - x^3\sqrt{x^3}$

39. $2b\sqrt[3]{54b^5} - 3\sqrt[3]{16b^8}$

40. $\sqrt[4]{32x^3} + \sqrt[4]{2x^3}$

41. $\sqrt[3]{-54x^3} - \sqrt[3]{-16x^3}$

42. $x\sqrt[4]{xy^4} - y\sqrt[4]{x^5}$

43. Explain how the Commutative and Associative Properties of Multiplication are used to determine the product $\left(3\sqrt{2}\right)\left(5\sqrt{7}\right)$.

44. One way to simplify $\sqrt{5}\left(2\sqrt{3} + 7\sqrt{3}\right)$ is as follows.

$$\sqrt{5}\left(2\sqrt{3} + 7\sqrt{3}\right) = \sqrt{5} \cdot 2\sqrt{3} + \sqrt{5} \cdot 7\sqrt{3}$$
$$= 2\sqrt{15} + 7\sqrt{15}$$
$$= 9\sqrt{15}$$

Describe a shorter way to simplify it.

In Exercises 45–56, find the product and simplify your result. Assume all variables represent positive numbers.

45. $\sqrt{15}\ \sqrt{35}$

46. $\sqrt{14}\ \sqrt{21}$

47. $\left(\sqrt{19}\right)^2$

48. $\left(5\sqrt{5}\right)^2$

49. $\left(-4\sqrt{3}\right)\left(3\sqrt{12}\right)$

50. $\left(3\sqrt{5}\right)\left(7\sqrt{20}\right)$

51. $\sqrt{18x}\ \sqrt{8x}$

52. $\sqrt{20t^2}\ \sqrt{8t}$

53. $\left(3\sqrt{5x}\right)\left(2\sqrt{10x}\right)$

54. $\left(10\sqrt{8t}\right)^2$

55. $\sqrt[4]{9x^3}\ \sqrt[4]{9x^2}$

56. $\sqrt[3]{18a^2}\ \sqrt[3]{6a^2}$

In Exercises 57–66, multiply and simplify the result. Assume all variables represent positive numbers.

57. $4\sqrt{3}(2\sqrt{3} - \sqrt{7})$ **58.** $2\sqrt{2}(3\sqrt{8} + \sqrt{5})$

59. $5\sqrt{7}(4\sqrt{3} + 7)$ **60.** $6\sqrt{11}(5\sqrt{5} - 8)$

61. $\sqrt{5}(\sqrt{35} - \sqrt{15})$ **62.** $\sqrt{7x}(\sqrt{28} + \sqrt{63x})$

63. $\sqrt{3}(\sqrt{y} - \sqrt{3y})$ **64.** $\sqrt{5}(\sqrt{10y} - 2y)$

65. $\sqrt{15}\left(\dfrac{\sqrt{5}}{\sqrt{3}} + \dfrac{\sqrt{3}}{\sqrt{5}}\right)$

66. $\sqrt{xy}\left(\dfrac{\sqrt{y}}{\sqrt{x}} - \dfrac{\sqrt{x}}{\sqrt{y}}\right)$

In Exercises 67–74, supply the missing number to make each statement true.

67. $(\ \ )(3 - \sqrt{2}) = 3\sqrt{7} - \sqrt{14}$

68. $(\ \ )(5 + \sqrt{6}) = 5\sqrt{6} + 6$

69. $(\ \ )(3 + \sqrt{2t}) = 3\sqrt{6t} + 2t\sqrt{3}$

70. $(\ \ )(2 - \sqrt{3y}) = 2y\sqrt{3} - 4\sqrt{y}$

71. $4(\ \ ) = 4x + 8\sqrt{3}$

72. $-3(\ \ ) = -6 + 3t\sqrt{7}$

73. $\sqrt{2}(\ \ ) = 4 - \sqrt{6}$

74. $\sqrt{3}(\ \ ) = 3\sqrt{5} - 5\sqrt{3}$

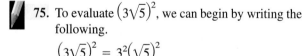

75. To evaluate $(3\sqrt{5})^2$, we can begin by writing the following.

$$(3\sqrt{5})^2 = 3^2(\sqrt{5})^2$$

What exponent rule are we using? Can the same rule be used to evaluate $(3 + \sqrt{5})^2$? Explain.

76. According to the Product Rule for Radicals, the following is true: $\sqrt{x + 2}\,\sqrt{x - 2} = \sqrt{x^2 - 4}$. Produce the graphs of

$$y_1 = \sqrt{x + 2}\,\sqrt{x - 2} \quad \text{and} \quad y_2 = \sqrt{x^2 - 4}.$$

Explain why they are different.

In Exercises 77–86, multiply and simplify the result. Assume all variables represent positive numbers.

77. $(\sqrt{3} - 2)(\sqrt{2} + \sqrt{3})$

78. $(3\sqrt{5} - 2\sqrt{3})(\sqrt{3} - 5)$

79. $(\sqrt{7x} - 3)(\sqrt{3} + \sqrt{7x})$

80. $(\sqrt{3t} - 2)(\sqrt{3} + 4t)$

81. $(2\sqrt{10} + \sqrt{5})(2\sqrt{2} - 5)$

82. $(\sqrt{x} - 5)(2\sqrt{x} + 3)$

83. $(\sqrt{2} - \sqrt{6})^2$ **84.** $(\sqrt{3} + \sqrt{15})^2$

85. $(\sqrt{5} + \sqrt{10})^2$ **86.** $(\sqrt{10} - \sqrt{2})^2$

In Exercises 87–90, supply the missing quantity to make the statement true. Assume all variables represent positive numbers.

87. $(\sqrt{3} + 3)(\ \ ) = \sqrt{15} - 2\sqrt{3} + 3\sqrt{5} - 6$

88. $(\sqrt{3} + 1)(\ \ ) = \sqrt{6} - \sqrt{3} + \sqrt{2} - 1$

89. $(\sqrt{x} + 3\sqrt{y})(\ \ ) = x + 2\sqrt{xy} - 3y$

90. $(2\sqrt{x} - \sqrt{3})(\ \ ) = 2x + \sqrt{3x} - 3$

91. Explain the difference between the conjugate of a binomial and the opposite of a binomial. To illustrate, use $3 - \sqrt{5}$.

92. Which of the following is not a rational number? Explain your choice.

(i) The square of $2\sqrt{5}$.

(ii) The square of $2 + \sqrt{5}$.

(iii) The product of $2 + \sqrt{5}$ and its conjugate.

In Exercises 93–98, determine the product of the expression and its conjugate.

93. $2\sqrt{2} - 3$ **94.** $3\sqrt{2} + 2$

95. $4\sqrt{3} - 3\sqrt{5}$ **96.** $3\sqrt{2} + 4\sqrt{6}$

97. $10 - 5t\sqrt{2}$ **98.** $7w + 3\sqrt{3}$

In Exercises 99–104, determine the missing binomial. Assume all variables represent positive numbers.

99. $\left(\sqrt{x} + \sqrt{2}\right)(\quad) = x - 2$

100. $\left(\sqrt{y} - 3\right)(\quad) = y - 9$

101. $\left(\sqrt{x} - \sqrt{y}\right)(\quad) = x - y$

102. $\left(\sqrt{2a} + \sqrt{b}\right)(\quad) = 2a - b$

103. $\left(\sqrt{3} - 2\right)(\quad) = -1$

104. $\left(5 + \sqrt{3}\right)(\quad) = 22$

105. The width of a rectangle is the square root of a positive number x. The length is $\sqrt{y}$ more than the width. Write simplified expressions for the perimeter and the area of the rectangle.

106. Determine two consecutive integers whose square roots are also consecutive integers.

107. A 50-foot pole is supported by two guy wires, one attached to the top of the pole and the other attached below the top of the pole. The two guy wires, one 10 feet shorter than the other, are attached at the same point on the ground. (See figure.)

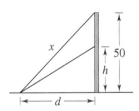

(a) Use the Pythagorean Theorem to write an expression in x for the distance d from the base of the pole to the point at which the wires are attached to the ground.

(b) Write an expression in x for the height h at which the shorter wire is attached to the pole.

108. A rectangular piece of paper is folded so that the top edge is along the side edge. (See figure.) Show that the length d of the diagonal is a multiple of $\sqrt{2}$ and that its value depends only on the width of the paper, not the length.

Figure for 108

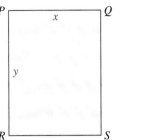 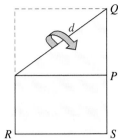

In Exercises 109–112, determine the product.

109. $\dfrac{\sqrt{3} + 2}{\sqrt{3} - 2} \cdot \dfrac{\sqrt{3} + 2}{\sqrt{3} + 2}$

110. $\dfrac{\sqrt{5} - \sqrt{3}}{\sqrt{3} + \sqrt{2}} \cdot \dfrac{\sqrt{3} - \sqrt{2}}{\sqrt{3} - \sqrt{2}}$

111. $\dfrac{\sqrt{7}}{\sqrt{3} + \sqrt{7}} \cdot \dfrac{\sqrt{3} - \sqrt{7}}{\sqrt{3} - \sqrt{7}}$

112. $\dfrac{5}{\sqrt{3} - 1} \cdot \dfrac{\sqrt{3} + 1}{\sqrt{3} + 1}$

In Exercises 113–116, show that the given number and its conjugate are both solutions of the equation.

113. $x^2 - 4x + 1 = 0;\ 2 + \sqrt{3}$

114. $x^2 - 2x - 4 = 0;\ 1 - \sqrt{5}$

115. $4x^2 - 12x + 7 = 0;\ \dfrac{3 - \sqrt{2}}{2}$

116. $9x^2 - 12x - 41 = 0;\ \dfrac{2 + 3\sqrt{5}}{3}$

Challenge

In Exercises 117 and 118, evaluate the expression.

117. $\sqrt[3]{3\sqrt[3]{3}}$

118. $\sqrt{3\sqrt{3\sqrt{3}}}$

In Exercises 119–122, perform the indicated operations. Assume all variables represent positive numbers.

119. $\sqrt{x^{4n+1}} + x^n \sqrt{x^{2n+1}}$

120. $x^{n-1} \sqrt{x^3} + x^{n-3} \sqrt{x^7}$

121. $\sqrt{x^{n-1}}\left(\sqrt{x^{n+3}} + \sqrt{x^{3n+1}}\right)$

122. $\left(\sqrt{x^n} - \sqrt{y^n}\right)\left(\sqrt{x^n} + \sqrt{y^n}\right)$

In Exercises 123–126, perform the indicated operation. Assume that all variables represent positive numbers.

123. $\sqrt[4]{x^2} + \sqrt{x}$

124. $\sqrt{12} + \sqrt[4]{9}$

125. $\sqrt[6]{8} - 2\sqrt[4]{4}$

126. $\sqrt[6]{x^2} - \sqrt[9]{x^3}$

In Exercises 127 and 128, simplify the complex fraction. Assume all variables represent positive numbers.

127. $\dfrac{\dfrac{1}{\sqrt{2}} + \dfrac{2}{\sqrt{10}}}{\dfrac{3}{\sqrt{10}}}$

128. $\dfrac{\dfrac{5\sqrt{x}}{\sqrt{y}} - \dfrac{5\sqrt{y}}{\sqrt{x}}}{\dfrac{\sqrt{y}}{\sqrt{x}} - \dfrac{\sqrt{x}}{\sqrt{y}}}$

9.7 Rationalizing Denominators

Monomial Denominators ▪ *Binomial Denominators*

Monomial Denominators

In Section 9.5 we stated that a simplified radical expression has no radicals in the denominator. Therefore, the expression

$$\frac{3}{\sqrt{5}}$$

is not considered simplified because the radical $\sqrt{5}$ appears in the denominator.

To simplify the expression, we remove the radical from the denominator by **rationalizing the denominator.** In the case of

$$\frac{3}{\sqrt{5}},$$

if we multiply the numerator and denominator by $\sqrt{5}$, the resulting denominator will be a rational number and the expression will be simplified.

$$\frac{3}{\sqrt{5}} = \frac{3 \cdot \sqrt{5}}{\sqrt{5} \cdot \sqrt{5}} = \frac{3\sqrt{5}}{\sqrt{25}} = \frac{3\sqrt{5}}{5}$$

NOTE: We can save a step by recalling the fact that $\sqrt{n}\,\sqrt{n} = n$ for nonnegative numbers n. Also, remember that the $\sqrt{5}$ in the numerator and the 5 in the denominator are different numbers, and they cannot be divided out. ∎

EXAMPLE 1   *Rationalizing a Monomial Denominator*

Rationalize the denominator.

(a) $\dfrac{\sqrt{7}}{\sqrt{3}}$

(b) $\dfrac{4\sqrt{3}}{\sqrt{2}}$

(c) $\dfrac{12}{\sqrt{27}}$

Solution

(a) Multiply the numerator and denominator by $\sqrt{3}$.

$$\frac{\sqrt{7}}{\sqrt{3}} = \frac{\sqrt{7}\cdot\sqrt{3}}{\sqrt{3}\cdot\sqrt{3}} = \frac{\sqrt{21}}{3}$$

(b) Multiply the numerator and denominator by $\sqrt{2}$.

$$\frac{4\sqrt{3}}{\sqrt{2}} = \frac{4\sqrt{3}\cdot\sqrt{2}}{\sqrt{2}\cdot\sqrt{2}} = \frac{4\sqrt{6}}{2} = 2\sqrt{6}$$

(c) We can simplify the expression before rationalizing the denominator.

$$\frac{12}{\sqrt{27}} = \frac{12}{\sqrt{9}\,\sqrt{3}} \qquad \text{Product Rule for Radicals}$$

$$= \frac{12}{3\sqrt{3}}$$

$$= \frac{4}{\sqrt{3}} \qquad \text{Divide out the common factor 3.}$$

$$= \frac{4\cdot\sqrt{3}}{\sqrt{3}\cdot\sqrt{3}} \qquad \text{Multiply the numerator and denominator by } \sqrt{3}.$$

$$= \frac{4\sqrt{3}}{3}$$

EXAMPLE 2   *Rationalizing a Monomial Denominator*

Simplify each expression. Assume all variables represent positive numbers.

(a) $\sqrt{\dfrac{45}{8}}$

(b) $\dfrac{3\sqrt{2a}}{\sqrt{x^3}}$

(c) $\dfrac{x}{\sqrt{x^2+4}}$

Solution

In this example, instead of simplifying radicals first, we multiply the numerator and denominator by a radical selected so that the resulting radicand in the denominator is a perfect square.

(a) $\sqrt{\dfrac{45}{8}} = \dfrac{\sqrt{45}\cdot\sqrt{2}}{\sqrt{8}\cdot\sqrt{2}} = \dfrac{\sqrt{90}}{\sqrt{16}} = \dfrac{\sqrt{9}\,\sqrt{10}}{4} = \dfrac{3\sqrt{10}}{4}$

(b) $\dfrac{3\sqrt{2a}}{\sqrt{x^3}} = \dfrac{3\sqrt{2a}\cdot\sqrt{x}}{\sqrt{x^3}\cdot\sqrt{x}} = \dfrac{3\sqrt{2ax}}{\sqrt{x^4}} = \dfrac{3\sqrt{2ax}}{x^2}$

(c) $\dfrac{x}{\sqrt{x^2+4}} = \dfrac{x \cdot \sqrt{x^2+4}}{\sqrt{x^2+4} \cdot \sqrt{x^2+4}} = \dfrac{x\sqrt{x^2+4}}{x^2+4}$ ■

If the index of the radical in a denominator is 2, we multiply by a number so that the result is a perfect square. If the index is 3, we multiply by a number so that the result is a perfect cube. A similar strategy is used for higher indices.

EXAMPLE 3 *Rationalizing Denominators with Higher Indices*

Rationalize the denominator. Assume all variables represent positive numbers.

(a) $\dfrac{10}{\sqrt[3]{4}}$ (b) $\dfrac{\sqrt[3]{2}}{\sqrt[3]{r^4}}$ (c) $\dfrac{-2}{\sqrt[4]{y^7}}$

Solution

(a) Multiplying by $\sqrt[3]{4}$ would result in $\sqrt[3]{16}$ in the denominator, but 16 is not a perfect cube. Instead, we multiply by $\sqrt[3]{2}$ to obtain the perfect cube radicand 8.

$$\dfrac{10}{\sqrt[3]{4}} = \dfrac{10 \cdot \sqrt[3]{2}}{\sqrt[3]{4} \cdot \sqrt[3]{2}} = \dfrac{10\sqrt[3]{2}}{\sqrt[3]{8}} = \dfrac{10\sqrt[3]{2}}{2} = 5\sqrt[3]{2}$$

(b) Multiply by $\sqrt[3]{r^2}$ to obtain a perfect cube in the denominator.

$$\dfrac{\sqrt[3]{2}}{\sqrt[3]{r^4}} = \dfrac{\sqrt[3]{2} \cdot \sqrt[3]{r^2}}{\sqrt[3]{r^4} \cdot \sqrt[3]{r^2}} = \dfrac{\sqrt[3]{2r^2}}{\sqrt[3]{r^6}} = \dfrac{\sqrt[3]{2r^2}}{r^2}$$

(c) Multiply by $\sqrt[4]{y}$ to obtain a perfect fourth power in the denominator.

$$\dfrac{-2}{\sqrt[4]{y^7}} = \dfrac{-2 \cdot \sqrt[4]{y}}{\sqrt[4]{y^7} \cdot \sqrt[4]{y}} = \dfrac{-2\sqrt[4]{y}}{\sqrt[4]{y^8}} = \dfrac{-2\sqrt[4]{y}}{y^2}$$ ■

Binomial Denominators

To prepare for rationalizing binomial denominators, we consider how we can use the Distributive Property to simplify fractions involving radicals.

EXAMPLE 4 *Simplifying Fractions with the Distributive Property*

Simplify.

(a) $\dfrac{12 + 6\sqrt{5}}{15}$ (b) $\dfrac{4a - \sqrt{48a^7}}{8a} \quad (a > 0)$

Solution

(a) $\dfrac{12 + 6\sqrt{5}}{15} = \dfrac{3(4 + 2\sqrt{5})}{3 \cdot 5}$ Factor the numerator and denominator.

$= \dfrac{4 + 2\sqrt{5}}{5}$ Divide out the common factor 3.

(b) $\dfrac{4a - \sqrt{48a^7}}{8a} = \dfrac{4a - \sqrt{16a^6}\,\sqrt{3a}}{8a}$ Product Rule for Radicals

$\qquad\qquad = \dfrac{4a - 4a^3\sqrt{3a}}{8a}$

$\qquad\qquad = \dfrac{4a\left(1 - a^2\sqrt{3a}\right)}{4a \cdot 2}$ Factor the numerator and denominator.

$\qquad\qquad = \dfrac{1 - a^2\sqrt{3a}}{2}$ Divide out the common factor $4a$.   ∎

In Section 9.6 we observed that the product of square root conjugates is an expression with no radical terms. Therefore, when the denominator of a rational expression contains a binomial with a square root term, we can rationalize the denominator by multiplying both the numerator and denominator by the conjugate of the denominator.

EXAMPLE 5   *Rationalizing a Binomial Denominator*

Rationalize the denominator.

(a) $\dfrac{10}{\sqrt{6} - 2}$ (b) $\dfrac{\sqrt{3} - \sqrt{2}}{\sqrt{5} + \sqrt{3}}$ (c) $\dfrac{\sqrt{x}}{\sqrt{x} + \sqrt{2y}}$ $(x > 0,\ y > 0)$

Solution

(a) $\dfrac{10}{\sqrt{6} - 2} = \dfrac{10 \cdot \left(\sqrt{6} + 2\right)}{\left(\sqrt{6} - 2\right) \cdot \left(\sqrt{6} + 2\right)}$ Multiply the numerator and denominator by the conjugate of $\sqrt{6} - 2$.

$\qquad\qquad = \dfrac{10\left(\sqrt{6} + 2\right)}{\left(\sqrt{6}\right)^2 - 2^2}$ $(A + B)(A - B) = A^2 - B^2$

$\qquad\qquad = \dfrac{10\left(\sqrt{6} + 2\right)}{6 - 4}$

$\qquad\qquad = \dfrac{5 \cdot 2 \cdot \left(\sqrt{6} + 2\right)}{2}$ Factor the numerator.

$\qquad\qquad = 5\left(\sqrt{6} + 2\right)$ Divide out the common factor 2.

(b) $\dfrac{\sqrt{3} - \sqrt{2}}{\sqrt{5} + \sqrt{3}} = \dfrac{\left(\sqrt{3} - \sqrt{2}\right)\left(\sqrt{5} - \sqrt{3}\right)}{\left(\sqrt{5} + \sqrt{3}\right)\left(\sqrt{5} - \sqrt{3}\right)}$ Multiply the numerator and denominator by the conjugate of $\sqrt{5} + \sqrt{3}$.

$\qquad\qquad = \dfrac{\sqrt{15} - \sqrt{9} - \sqrt{10} + \sqrt{6}}{\left(\sqrt{5}\right)^2 - \left(\sqrt{3}\right)^2}$ FOIL

$\qquad\qquad = \dfrac{\sqrt{15} - 3 - \sqrt{10} + \sqrt{6}}{5 - 3}$

$\qquad\qquad = \dfrac{\sqrt{15} - 3 - \sqrt{10} + \sqrt{6}}{2}$

(c) $\dfrac{\sqrt{x}}{\sqrt{x} + \sqrt{2y}} = \dfrac{\sqrt{x}\left(\sqrt{x} - \sqrt{2y}\right)}{\left(\sqrt{x} + \sqrt{2y}\right)\left(\sqrt{x} - \sqrt{2y}\right)}$

Multiply the numerator and denominator by the conjugate of $\sqrt{x} + \sqrt{2y}$.

$= \dfrac{\sqrt{x^2} - \sqrt{2xy}}{\left(\sqrt{x}\right)^2 - \left(\sqrt{2y}\right)^2}$

$(A + B)(A - B) = A^2 - B^2$

$= \dfrac{x - \sqrt{2xy}}{x - 2y}$

We cannot simplify further because the numerator and denominator cannot be factored.

9.7   *Quick Reference*

Monomial Denominators	▪ To rationalize a monomial radical denominator, multiply both the numerator and denominator by a radical selected so that the resulting radicand in the denominator is a perfect power.
Binomial Denominators	▪ The product of a pair of square root conjugates is an expression with no square root terms. ▪ To rationalize a binomial denominator containing a square root term, multiply both the numerator and denominator by the conjugate of the denominator.

9.7   *Exercises*

1. The following shows the beginning of two approaches to simplifying a radical expression.

$$\sqrt{\dfrac{2}{3}} = \dfrac{\sqrt{2}}{\sqrt{3}} = \dfrac{\sqrt{2} \cdot \sqrt{3}}{\sqrt{3} \cdot \sqrt{3}} = \cdots$$

$$\sqrt{\dfrac{2}{3}} = \sqrt{\dfrac{2 \cdot 3}{3 \cdot 3}} = \cdots$$

Do both methods lead to the same correct result? If so, which method do you prefer?

2. When simplifying $\dfrac{5}{\sqrt[3]{2}}$,

explain why it is necessary to multiply the numerator and denominator by $\sqrt[3]{4}$ rather than $\sqrt[3]{2}$.

In Exercises 3–14, rationalize the denominator. Use your calculator to verify that the result and the original expression have the same value.

3. $\dfrac{5}{\sqrt{5}}$

4. $\dfrac{6}{\sqrt{3}}$

5. $\dfrac{-\sqrt{5}}{\sqrt{3}}$

6. $\dfrac{1}{3\sqrt{2}}$

7. $\dfrac{15}{\sqrt{10}}$

8. $\dfrac{\sqrt{6}}{-3\sqrt{12}}$

9. $\dfrac{3\sqrt{7} + 2\sqrt{3}}{2\sqrt{3}}$

10. $\dfrac{4 - \sqrt{10}}{4\sqrt{2}}$

11. $\dfrac{7}{\sqrt[3]{3}}$

12. $\dfrac{9}{\sqrt[3]{25}}$

13. $\dfrac{9}{\sqrt[4]{8}}$

14. $\dfrac{2}{\sqrt[5]{8}}$

 15. To simplify $\dfrac{5}{2\sqrt{3}}$,

should you multiply the numerator and denominator by $2\sqrt{3}$ or by just $\sqrt{3}$? Will you obtain the correct result either way? If so, which method do you prefer?

 16. To simplify $\dfrac{1}{\sqrt[n]{x^m}}$,

we must multiply the numerator and denominator by a radical of the form $\sqrt[n]{x^k}$. For the method to succeed, what must be the relationship among m, k, and n?

In Exercises 17–34, simplify. Assume all variables represent positive numbers.

17. $\dfrac{6}{\sqrt{3x}}$

18. $\dfrac{3}{5\sqrt{t}}$

19. $\dfrac{x^3}{\sqrt{x^3}}$

20. $\sqrt{\dfrac{5}{y^5}}$

21. $\sqrt{\dfrac{a^3}{8}}$

22. $\sqrt{\dfrac{x^4}{20}}$

23. $\dfrac{2}{\sqrt{x+2}}$

24. $\dfrac{4}{\sqrt{4+y^2}}$

25. $\dfrac{\sqrt{x}-3}{\sqrt{x}}$

26. $\dfrac{5+\sqrt{2t}}{2\sqrt{t}}$

27. $18x^2y^3 \div \sqrt{3x^2y^5}$

28. $\sqrt{8a^6b^4} \div \sqrt{32a^4b^5}$

29. $\sqrt{\dfrac{x^5y^2}{20y^3}}$

30. $\sqrt{\dfrac{4xy}{16x^3y^2}}$

31. $\dfrac{2b}{\sqrt[3]{b^2}}$

32. $\dfrac{x}{\sqrt[4]{xy^3}}$

33. $\dfrac{5x}{\sqrt[5]{x^2}}$

34. $\sqrt[4]{\dfrac{4x^5}{64xy^3}}$

In Exercises 35–40, supply the missing number to make the statement true. Assume all variables represent positive numbers.

35. $\dfrac{\sqrt{3}}{\sqrt{5}} = \dfrac{\sqrt{15}}{(\ )}$

36. $\sqrt{\dfrac{3}{8}} = \dfrac{\sqrt{6}}{(\ )}$

37. $\sqrt{\dfrac{2}{7}} = \dfrac{(\ )}{7}$

38. $\dfrac{\sqrt{63}}{\sqrt{5}} = \dfrac{(\ )}{5}$

39. $\dfrac{3}{(\ )} = \sqrt{3}$

40. $\dfrac{5}{(\ )} = \dfrac{\sqrt{5x}}{x}$

 41. Explain why $\dfrac{1}{\sqrt{5}} + \dfrac{\sqrt{2}}{\sqrt{10}}$

cannot be performed with the fractions in their given form. Must the denominators of the two fractions be rationalized before the fractions can be added? Why?

 42. The following shows the start of two different approaches to performing the given addition.

$$\dfrac{1}{\sqrt{2}} + \dfrac{1}{\sqrt{3}} = \dfrac{1}{\sqrt{2}} \cdot \dfrac{\sqrt{2}}{\sqrt{2}} + \dfrac{1}{\sqrt{3}} \cdot \dfrac{\sqrt{3}}{\sqrt{3}} = \cdots$$

$$\dfrac{1}{\sqrt{2}} + \dfrac{1}{\sqrt{3}} = \dfrac{1}{\sqrt{2}} \cdot \dfrac{\sqrt{3}}{\sqrt{3}} + \dfrac{1}{\sqrt{3}} \cdot \dfrac{\sqrt{2}}{\sqrt{2}} = \cdots$$

Complete the two problems and verify that the results are the same. Then comment on the approach that you believe to be more efficient.

In Exercises 43–50, simplify. Assume all variables represent positive numbers.

43. $\dfrac{1}{\sqrt{5}} + \sqrt{45}$

44. $\dfrac{\sqrt{2}}{\sqrt{3}} + \dfrac{1}{\sqrt{6}}$

45. $1 - \dfrac{3}{\sqrt{2}}$

46. $\dfrac{1}{\sqrt{2}} - 2\sqrt{8}$

47. $2\sqrt{x+1} - \dfrac{2x}{\sqrt{x+1}}$

48. $\dfrac{x}{\sqrt{x+3}} - \sqrt{x+3}$

49. $\dfrac{2}{\sqrt{3}+2} - \dfrac{1}{\sqrt{3}-2}$

50. $\dfrac{\sqrt{x}}{\sqrt{x}+5} + \dfrac{2\sqrt{x}}{\sqrt{x}-5}$

In Exercises 51–54, simplify each expression.

51. $\dfrac{15-9\sqrt{3}}{21}$

52. $\dfrac{-4+6\sqrt{5}}{10}$

53. $\dfrac{-12+\sqrt{18}}{6}$

54. $\dfrac{\sqrt{75}+10\sqrt{2}}{15}$

In Exercises 55–58, simplify each expression. Assume all variables represent positive numbers.

55. $\dfrac{\sqrt{45x^5}+15x}{9x}$

56. $\dfrac{\sqrt{72y^9}-18y}{24y}$

57. $\dfrac{\sqrt{x}-3\sqrt{x^3}}{2\sqrt{x}}$

58. $\dfrac{4x\sqrt{x}+\sqrt{4x^3}}{3\sqrt{x^5}}$

In Exercises 59–70, rationalize the denominator. Assume all variables represent positive numbers.

59. $\dfrac{3}{3-\sqrt{6}}$

60. $\dfrac{12}{\sqrt{5}-\sqrt{2}}$

61. $\dfrac{\sqrt{5}-\sqrt{3}}{\sqrt{5}+\sqrt{3}}$

62. $\dfrac{\sqrt{5}-\sqrt{2}}{\sqrt{7}+\sqrt{3}}$

63. $\dfrac{\sqrt{y}+5}{\sqrt{y}-4}$

64. $\dfrac{5z}{\sqrt{z}+z}$

65. $\sqrt{x} \div \left(\sqrt{x}+2\right)$

66. $\sqrt{3} \div \left(\sqrt{3}-\sqrt{x}\right)$

67. $\dfrac{1}{\sqrt{x+1}-2}$

68. $\dfrac{2}{3+\sqrt{x+2}}$

69. $\dfrac{\sqrt{x}-\sqrt{y}}{\sqrt{x}+\sqrt{y}}$

70. $\dfrac{\sqrt{xy}}{\sqrt{x}+\sqrt{y}}$

In Exercises 71–76, fill in the blank. Assume all variables represent positive numbers.

71. $\dfrac{\sqrt{x}}{\sqrt{x}-5} = \dfrac{x+5\sqrt{x}}{(\quad)}$

72. $\dfrac{\sqrt{x}}{\sqrt{x}+2} = \dfrac{x-2\sqrt{x}}{(\quad)}$

73. $\dfrac{(\quad)}{3\sqrt{2}-5} = 3\sqrt{2}+5$

74. $\dfrac{(\quad)}{\sqrt{10}+\sqrt{5}} = \dfrac{2\sqrt{5}-\sqrt{10}}{5}$

75. $\dfrac{10}{(\quad)} = 2\sqrt{7}+2\sqrt{2}$

76. $\dfrac{3}{(\quad)} = \dfrac{3\sqrt{x}-21}{x-49}$

A rationalized *numerator* is not one of the conditions for a simplified radical. However, in more advanced studies, there are occasions when we must rationalize the numerator. The technique is the same as for rationalizing denominators.

In Exercises 77–82, rationalize the *numerator*. Assume all variables represent positive numbers.

77. $\dfrac{\sqrt{5}+\sqrt{7}}{4}$

78. $\dfrac{\sqrt{11}-\sqrt{3}}{8}$

79. $\dfrac{\sqrt{x}+3}{\sqrt{x}-3}$

80. $\dfrac{5-\sqrt{a}}{5+\sqrt{a}}$

81. $\dfrac{\sqrt{x+2}-\sqrt{x}}{2}$

82. $\dfrac{\sqrt{x+h}-\sqrt{x}}{h}$

 83. Explain why it is not possible to simplify the expression

$$\dfrac{\sqrt{x}-\sqrt{x}}{\sqrt{x}+\sqrt{x}}$$

by multiplying the numerator and denominator by the conjugate of $\sqrt{x}+\sqrt{x}$. Describe an easy method for simplifying the expression.

 84. If x and y are positive numbers, explain how $\left(1/\sqrt{x}\right)/\left(\sqrt{x}/\sqrt{y}\right)$ can be simplified without rationalizing any denominators.

 85. If the volume of a right circular cone (see figure) is V and the height is h, then a formula for the radius is $r = \sqrt{\dfrac{3V}{\pi h}}$.

Figure for 85

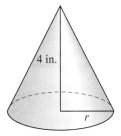

(a) If the volume of the right circular cone in the figure is 120 cubic inches, use this formula to determine the radius.

(b) Rewrite the formula with a simplified radical. Then determine r again and verify that the results are the same.

 86. The volume of a basketball (sphere) is approximately 7421 cubic centimeters. The radius r of a sphere is given by the formula

$$r = \sqrt[3]{\frac{3V}{4\pi}},$$

where V is the volume.

(a) Use this formula to determine the radius of a basketball.

(b) Rewrite the formula with a simplified radical. Determine r again and verify that the results are the same.

 87. The height h of a trapezoid is given by the formula

$$h = \frac{2A}{b_1 + b_2},$$

where A is the area of the trapezoid and b_1 and b_2 are the lengths of the parallel sides. Write a simplified formula for the height of the trapezoid in the accompanying figure.

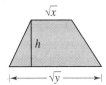

88. Consider two positive consecutive integers. Show that the reciprocal of the sum of their square roots is equal to the difference of their square roots. Also, explain how you know that both expressions are positive.

Exploring with Real Data

In Exercises 89–92, refer to the following background information and data.

The following table shows that the percentage of drivers who wore seat belts increased rapidly during 1988–1991.

	1988	1989	1990	1991
Percentage of Drivers Who Wore Seat Belts	45%	47%	49%	59%

(Source: National Highway Traffic Safety Administration.)

The percentage P of seat belt wearers can be modeled by the function

$$P(t) = \frac{t^{2.5} + 43t^{0.5}}{\sqrt{t}},$$

where t is the number of years since 1987.

89. Rewrite the model function with a rationalized denominator.

90. Explain why there is a difference in the graphs of the model function and of the revised function in Exercise 89.

91. Trace the graph of the model function to determine the value of t for which $P(t) = 100$. Interpret this information. Is it realistic?

92. According to the National Highway Traffic Safety Administration, what statistic is expected to decrease as $P(t)$ increases?

Challenge

 93. Write the special factoring pattern for the difference of two cubes. Then explain how this pattern can be used to rationalize the denominator of

$$\frac{1}{\sqrt[3]{x} - \sqrt[3]{y}}.$$

In Exercises 94 and 95, rationalize the denominator.

94. $\dfrac{3}{\sqrt[3]{x} + \sqrt[3]{2}}$

95. $\dfrac{5}{\sqrt[3]{x} - \sqrt[3]{3}}$

96. Rationalize the denominator of

$$\frac{\sqrt{2} - \sqrt{3}}{\sqrt{2} + \sqrt{3} + \sqrt{5}}.$$

In Exercises 97 and 98, simplify the complex fraction and express the result with a rationalized denominator. Assume all variables represent positive numbers.

97. $\dfrac{\dfrac{1}{\sqrt{3}} + \dfrac{\sqrt{3}}{\sqrt{x}}}{\dfrac{\sqrt{x}}{\sqrt{3}} + \dfrac{3}{\sqrt{3x}}}$

98. $\dfrac{\dfrac{3}{\sqrt{y}} + \dfrac{3}{\sqrt{x}}}{\dfrac{1}{\sqrt{y}} + \dfrac{1}{\sqrt{x}}}$

In Exercises 99–102, rationalize the denominator. Assume that x represents a positive number.

99. $\dfrac{1}{\sqrt{\sqrt{x}}}$

100. $\dfrac{1}{\sqrt{\sqrt[3]{x}}}$

101. $\dfrac{3}{\sqrt[n]{x^{n-3}}}$

102. $\dfrac{x^2}{\sqrt[n]{x^{2n-1}}}$

103. The distance D from a fixed point (x_1, y_1) to a line whose equation is $y = mx + b$ is given by the following formula.

$$D = \frac{|y_1 - mx_1 - b|}{\sqrt{1 + m^2}}$$

Use this formula to find the distance from $P(5, 6)$ to the line whose equation is $y = 2x + 1$. Express the result in simplified radical form.

9.8   Equations with Radicals and Exponents

Equations with Integer Exponents ▪ *Radical Equations* ▪ *Equations with Two Radicals* ▪ *Equations with Rational Exponents* ▪ *Applications*

Equations with Integer Exponents

So far, we have solved equations by writing equivalent equations with the Addition and Multiplication Properties for Equations. For equations with radicals and exponents, we need to expand the operations that produce equivalent equations.

There are two properties that we will use to solve equations with exponents. The first is the Odd Root Property.

Odd Root Property

For real numbers A and B and a positive odd integer n greater than 1, if $A^n = B$, then $A = \sqrt[n]{B}$.

EXAMPLE 1   *Solving Equations with the Odd Root Property*

Solve each equation.

(a) $y^5 = -32$

(b) $(x - 3)^7 = 1$

Solution

(a) $y^5 = -32$

$\quad y = \sqrt[5]{-32}$ Odd Root Property

$\quad y = -2$

(b) $(x - 3)^7 = 1$

$\quad x - 3 = \sqrt[7]{1}$ Odd Root Property

$\quad x - 3 = 1$

$\quad\quad x = 4$

The other property we use for solving equations with exponents is the Even Root Property.

> ### Even Root Property
>
> For real numbers A and B and a positive even integer n, if $A^n = B$, then
>
> 1. for $B > 0$, $A = \pm \sqrt[n]{B}$;
> 2. for $B = 0$, $A = 0$;
> 3. for $B < 0$, there is no real number solution.

The $\pm$ symbol indicates two values, one positive and the other negative.

EXAMPLE 2   *Solving Equations with the Even Root Property*

Solve each equation.

(a) $x^4 = 81$ (b) $(x + 5)^6 = 0$

(c) $(x - 1)^2 - 4 = -1$ (d) $2(t + 1)^2 + 3 = -3$

Solution

(a) $x^4 = 81$

$\quad x = \pm\sqrt[4]{81}$ Even Root Property with $B > 0$

$\quad x = \pm 3$

(b) $(x + 5)^6 = 0$

$\quad x + 5 = 0$ Even Root Property with $B = 0$

$\quad\quad x = -5$

(c) $(x - 1)^2 - 4 = -1$

$\quad (x - 1)^2 = 3$ Add 4 to both sides.

$\quad x - 1 = \pm\sqrt{3}$ Even Root Property with $B > 0$

$\quad\quad x = 1 \pm \sqrt{3}$ Add 1 to both sides.

$\quad\quad x \approx 2.73$ or $x \approx -0.73$

(d) $2(t + 1)^2 + 3 = -3$

$\qquad 2(t + 1)^2 = -6$ Subtract 3 from both sides.

$\qquad\quad (t + 1)^2 = -3$ Divide both sides by 2.

Because $B < 0$, the Even Root Property states that the equation has no real number solution.

Radical Equations

An equation that contains a radical with a variable in the radicand is called a **radical equation.** The following are two examples.

$$\sqrt{1 - 2x} = 3 \qquad \sqrt{6t + 13} = 2t + 1$$

Solving radical equations requires a new technique, that of raising both sides to a power. This procedure is used to eliminate the radical, although it does not always lead to an equivalent equation.

In general, if both sides of an equation are raised to the same power, all solutions of the original equation will be solutions of the new equation. However, the new equation may have solutions that are *not* solutions of the original equation. As in Section 8.5, we call such numbers *extraneous solutions.*

NOTE: Because extraneous solutions can occur when we solve a radical equation, we must check all solutions in the original equation. ■

As with any equation, we can graph the left and right sides of a radical equation to estimate the solution. In the next two examples, we use both graphing and algebraic methods.

EXAMPLE 3   *Solving an Equation with One Radical*

Solve the equation $\sqrt{1 - 2x} = 3$.

Figure 9.7

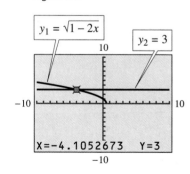

Solution   Figure 9.7 shows the graphs of the left and right sides of the equation. The solution appears to be about -4.

To solve the equation algebraically, we make use of the fact that $\left(\sqrt{b}\right)^2 = b$, $b \geq 0$.

$$\sqrt{1 - 2x} = 3$$

$$\left(\sqrt{1 - 2x}\right)^2 = 3^2 \qquad \text{Square both sides of the equation.}$$

$$1 - 2x = 9 \qquad \text{The radical is eliminated.}$$

$$-2x = 8$$

$$x = -4$$

The solution agrees with the graph. To verify the solution, we can store -4 in X and evaluate Y_1, or we can substitute as follows.

When $x = -4$, $\sqrt{1 - 2x} = \sqrt{1 - 2(-4)} = \sqrt{9} = 3$. ■

EXAMPLE 4   *Solving an Equation with One Radical*

Solve the equation $\sqrt{6x + 13} = 2x + 1$.

Figure 9.8

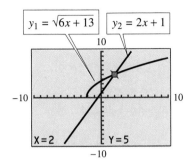

$y_1 = \sqrt{6x + 13}$ $y_2 = 2x + 1$

Solution   Figure 9.8 shows the graphs of the left and right sides of the equation. The solution appears to be about 2.

Now we solve the equation algebraically.

$$\sqrt{6x + 13} = 2x + 1$$

$$\left(\sqrt{6x + 13}\right)^2 = (2x + 1)^2 \qquad \text{Square both sides.}$$

$$6x + 13 = 4x^2 + 4x + 1 \qquad \text{The radical is eliminated.}$$

$$0 = 4x^2 - 2x - 12 \qquad \text{Write the resulting quadratic equation}$$

$$0 = 2(2x^2 - x - 6) \qquad \text{in standard form and factor.}$$

$$0 = 2(2x + 3)(x - 2)$$

$$2x + 3 = 0 \quad \text{or} \quad x - 2 = 0 \qquad \text{Zero Factor Property}$$

$$2x = -3 \quad \text{or} \quad x = 2 \qquad \text{Solve each case for } x.$$

$$x = -\frac{3}{2} \quad \text{or} \quad x = 2$$

From the graph, we expected only *one* solution. Solving algebraically produced *two* values of x.

If we evaluate $y_1 = \sqrt{6x + 13}$ and $y_2 = 2x + 1$ for $x = -\frac{3}{2}$, the calculator display reveals that $y_1 = 2$ and $y_2 = -2$ and so, $y_1 \neq y_2$. Thus, $-\frac{3}{2}$ is an extraneous solution, that is, $-\frac{3}{2}$ is not a solution of the equation.

In the same way, evaluate y_1 and y_2 for $x = 2$ and observe that $y_1 = y_2$. Thus, 2 is the only solution of the equation.   ■

Example 4 shows that a graph can be very helpful in detecting extraneous solutions. Although we will not always show a graph in the examples that follow, combining graphing and algebraic methods continues to be an excellent approach.

Sometimes we must isolate the radical before raising both sides of the equation to a power.

EXAMPLE 5   *Solving an Equation with a Third Root Radical*

Solve the equation $\sqrt[3]{x + 1} - 3 = 1$.

Solution

$$\sqrt[3]{x + 1} - 3 = 1$$

$$\sqrt[3]{x + 1} = 4 \qquad \text{Add 3 to both sides to isolate the radical.}$$

$$\left(\sqrt[3]{x + 1}\right)^3 = 4^3 \qquad \text{Raise both sides to the third power.}$$

$$x + 1 = 64 \qquad \text{The radical is eliminated.}$$

$$x = 63$$

You can use your calculator to check that 63 is the solution.   ■

Equations with Two Radicals

To solve equations with two radicals, it may be necessary to square both sides more than once.

EXAMPLE 6 *Solving an Equation with Two Radicals*

Solve the equation $\sqrt{2a} = 2 + \sqrt{a-2}$.

Solution   We need to isolate one of the radicals. It is usually better to isolate the one with the greater number of terms in the radicand.

$$\sqrt{2a} = 2 + \sqrt{a-2}$$

$$\sqrt{2a} - 2 = \sqrt{a-2} \qquad \text{Isolate the radical on the right.}$$

$$\left(\sqrt{2a} - 2\right)^2 = \left(\sqrt{a-2}\right)^2 \qquad \text{Square both sides.}$$

$$2a - 4\sqrt{2a} + 4 = a - 2 \qquad \text{The radical on the right is eliminated.}$$

$$-4\sqrt{2a} = -a - 6 \qquad \text{Subtract } 2a \text{ and } 4 \text{ to isolate the radical term.}$$

$$\left(-4\sqrt{2a}\right)^2 = (-a-6)^2 \qquad \text{Square both sides.}$$

$$16(2a) = a^2 + 12a + 36 \qquad \text{The radical is eliminated.}$$

$$32a = a^2 + 12a + 36$$

$$0 = a^2 - 20a + 36 \qquad \text{Write the quadratic equation in standard form.}$$

$$0 = (a-18)(a-2) \qquad \text{Factor.}$$

$$a - 18 = 0 \quad \text{or} \quad a - 2 = 0 \qquad \text{Zero Factor Property}$$

$$a = 18 \quad \text{or} \qquad a = 2$$

Use your calculator to check that 18 and 2 are both solutions.   ■

EXAMPLE 7 *Solving an Equation with Two Radicals*

For the equation $1 = \sqrt{t-3} - \sqrt{2t-4}$, estimate the solution graphically. Then solve algebraically.

Solution   Figure 9.9 shows the graph of each side of the equation.

Figure 9.9

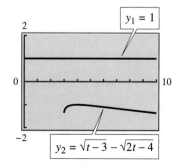

$y_1 = 1$

$y_2 = \sqrt{t-3} - \sqrt{2t-4}$

The graph of the left side is a horizontal line *above* the x-axis. The graph of the right side appears to lie entirely *below* the x-axis. From the graph we conclude that there is no solution.

$$1 = \sqrt{t - 3} - \sqrt{2t - 4}$$

$$1 + \sqrt{2t - 4} = \sqrt{t - 3} \qquad \text{Isolate one radical.}$$

$$\left(1 + \sqrt{2t - 4}\right)^2 = \left(\sqrt{t - 3}\right)^2 \qquad \text{Square both sides.}$$

$$1 + 2\sqrt{2t - 4} + 2t - 4 = t - 3$$

$$2\sqrt{2t - 4} + 2t - 3 = t - 3$$

$$2\sqrt{2t - 4} = -t \qquad \text{Isolate the radical term.}$$

$$\left(2\sqrt{2t - 4}\right)^2 = (-t)^2 \qquad \text{Square both sides.}$$

$$4(2t - 4) = t^2$$

$$8t - 16 = t^2$$

$$0 = t^2 - 8t + 16 \qquad \text{Quadratic equation in standard form}$$

$$0 = (t - 4)^2 \qquad \text{Factor.}$$

$$0 = t - 4 \qquad \text{Even Root Property}$$

$$4 = t$$

Use your calculator to check the value 4. When $t = 4$, the left side of the equation has a value of 1, but the right side has a value of -1. Thus 4 is an extraneous solution. This equation has no solution. ∎

Equations with Rational Exponents

To solve equations of the form $A^{m/n} = B$ for A, raise both sides to the nth power to obtain an integer exponent on A.

EXAMPLE 8 *Solving an Equation with a Rational Exponent*

Solve the equation $(x - 2)^{2/3} = 4$.

Solution

Figure 9.10

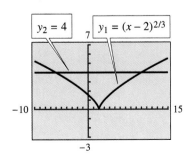

The graph in Fig. 9.10 suggests that we can anticipate two solutions.

$$(x - 2)^{2/3} = 4$$

$$\left[(x - 2)^{2/3}\right]^3 = 4^3 \qquad \text{Raise both sides to the third power.}$$

$$(x - 2)^2 = 64 \qquad \text{Now the exponent is an integer.}$$

$$x - 2 = \pm 8 \qquad \text{Even Root Property}$$

$$x = 2 + 8 = 10 \quad \text{or} \quad x = 2 - 8 = -6$$

Use a calculator to check the solutions. ∎

Applications

It is important to understand how to solve equations with radicals and rational exponents because applied problems can sometimes be modeled by such equations.

EXAMPLE 9   *Auto Speeds Measured by Skid Marks*

The length L (in feet) of the skid marks left by an automobile that has braked to a sudden stop can be used to estimate the speed S (in miles per hour) of the car. The formula we use is $S = 2\sqrt{5L}$.

(a) What was the speed of a car that left a skid mark 90 feet long?

(b) How long will the skid mark be if the car is traveling 50 mph?

Solution

(a) For $L = 90$, $S = 2\sqrt{5L} = 2\sqrt{5(90)} = 2\sqrt{450} \approx 42.43$. The car's speed is approximately 42.43 mph.

(b) For $S = 50$ we must solve the equation $50 = 2\sqrt{5L}$.

$$50^2 = \left(2\sqrt{5L}\right)^2 \qquad \text{Square both sides.}$$
$$2500 = 4(5L) \qquad \text{The radical is eliminated.}$$
$$2500 = 20L \qquad \text{Solve for } L.$$
$$125 = L$$

We can verify that 125 is a solution, and so the skid mark will be 125 feet long. ∎

9.8   *Quick Reference*

Equations with Integer Exponents

- We use two properties to solve equations with integer exponents.

  1. Odd Root Property: For real numbers A and B and a positive odd integer $n > 1$, if $A^n = B$, then $A = \sqrt[n]{B}$.

  2. Even Root Property: For real numbers A and B and a positive even integer n, if $A^n = B$, then

     (a) for $B > 0$, $A = \pm\sqrt[n]{B}$;

     (b) for $B = 0$, $A = 0$;

     (c) for $B < 0$, there is no real number solution.

Radical Equations

- An equation that contains a radical with a variable in a radicand is called a **radical equation.**

- The general technique for solving a radical equation is as follows.

  1. Isolate a radical that contains the variable in the radicand.

  2. Raise both sides of the equation to a power equal to the index of the radical.

3. Simplify by combining like terms.

4. If the equation still contains a term with a variable in a radicand, repeat steps 1–3.

5. After all radicals have been eliminated, solve for the variable and check the solution.

- Because this solving routine does not always produce equivalent equations, an *extraneous solution* might be introduced. Checking solutions is essential.

- Graphing the two sides of a radical equation can help you anticipate the number of solutions and to estimate the solutions.

Equations with Two Radicals

- To solve a radical equation having two radicals, the solving routine may have to be performed twice.

- Remember to check for extraneous solutions.

Equations with Rational Exponents

- To solve equations of the form $A^{m/n} = B$ for A, raise both sides to the nth power to obtain an integer exponent.

Applications

- When solving applications that can be modeled by equations containing radicals or rational exponents, be sure to check for extraneous solutions. Also, remember that a solution of an equation may not be extraneous, but, because of the nature of the problem, it may not qualify as an answer.

9.8 Exercises

1. If all required conditions are met, the Odd Root Property gives us a method for converting an equation of the form $A^n = B$ into a radical equation. Explain how this allows us to solve for A.

2. If A and B are real numbers and n is a positive even integer, the Even Root Property gives three possible results when we solve $A^n = B$ for A. List these possibilities along with the corresponding conditions on B.

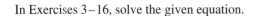

In Exercises 3–16, solve the given equation.

3. $x^2 = 6$

4. $\frac{1}{3}x^2 = 15$

5. $t^2 + 9 = 0$

6. $3x^2 - 27 = 0$

7. $(2x - 5)^2 = 16$

8. $25(x + 2)^2 = 16$

9. $(3x - 1)^2 = 3$

10. $(x + 2)^2 = 18$

11. $x^4 = 16$

12. $x^3 = 64$

13. $(x - 5)^5 = 0$

14. $(x - 3)^3 = -125$

15. $(2x - 3)^5 = -32$

16. $(2x + 3)^4 = -2$

17. Solve the following equations.

(a) $x^3 = 1$ (b) $x^4 = 1$ (c) $x^7 = 1$

(d) $x^8 = 1$ (e) $x^{13} = 1$ (f) $x^{22} = 1$

(g) State a generalization about the solution of $x^n = 1$ where n is a positive integer.

18. Solve the following equations.

(a) $x^3 = -1$ (b) $x^4 = -1$ (c) $x^7 = -1$

(d) $x^8 = -1$ (e) $x^{13} = -1$ (f) $x^{22} = -1$

(g) State a generalization about the solution of $x^n = -1$ where n is a positive integer.

19. If we square both sides of the radical equation $\sqrt{x} = -5$, we obtain $x = 25$. Are the two equations equivalent? Why? Explain why it is essential to check the values obtained when solving a radical equation.

 20. Suppose you are solving the radical equation $\sqrt{x} + 1 = 5$ and you begin by squaring both sides. What is the resulting equation? Have you met your goal of eliminating the radical from the equation? If not, what should your first step have been?

In Exercises 21–28, solve the radical equation.

21. $\sqrt{2x} = 6$

22. $5\sqrt{3x} = 30$

23. $\dfrac{1}{2}\sqrt{y} = 4$

24. $\sqrt{y} + 5 = 2$

25. $\sqrt[3]{z} = 2$

26. $\sqrt[5]{x} = 3$

27. $\dfrac{1}{2}\sqrt[6]{x} = 1$

28. $\sqrt[3]{z} + 5 = 3$

 29. In which of the following equations can we eliminate the radical by squaring both sides once?

　(i) $\sqrt{2x} = x$　　　(ii) $\sqrt{2x} + 1 = x$

　(iii) $\sqrt{2x} = \sqrt{x} + 1$

Describe the resulting equations in (i) and (ii).

 30. Solve the following equations.

　(a) $\sqrt[3]{x} = -1$　(b) $\sqrt[6]{x} = -1$　(c) $\sqrt[11]{x} = -1$

　(d) Write a generalization for the solution of $\sqrt[n]{x} = -1$ where n is a positive integer.

In Exercises 31–60, solve the equation.

31. $\sqrt{5x + 1} = 4$

32. $\sqrt{2x + 1} = \dfrac{1}{2}$

33. $2(3x)^{1/2} = (5x + 7)^{1/2}$

34. $(x^2 - x - 5)^{1/2} - (x^2 - 3x + 7)^{1/2} = 0$

35. $\sqrt[3]{3x - 1} = 2$

36. $\sqrt[3]{4x + 1} + 3 = 0$

37. $\sqrt{1 - 5x} + 6 = 3$

38. $\sqrt{x^2} = 3$

39. $\sqrt[4]{x + 1} = 2$

40. $\sqrt[4]{2x + 1} = \sqrt[4]{x + 6}$

41. $\sqrt{x^2 - 3x} = 2$

42. $\sqrt{3x^2 + 7x - 5} = 1$

43. $\sqrt{x}\,\sqrt{x - 5} = 6$

44. $x = 4 + \sqrt{32 - 2x}$

45. $x\sqrt{2} = \sqrt{5x - 2}$

46. $x\sqrt{3} = \sqrt{9x + 30}$

47. $\sqrt{4x + 13} = 2x - 1$

48. $\sqrt{2x - 3} = 5 - 2x$

49. $\sqrt[3]{x^2 + x + 2} = 2$

50. $\sqrt[3]{x^2 + 6x + 9} = 1$

51. $\sqrt[4]{x + 1} + \sqrt[4]{2x - 3} = 0$

52. $\sqrt[5]{x^2 - 7x + 9} = -1$

53. $\sqrt{x(x - 2)} - x = -10$

54. $\sqrt{x}\,\sqrt{x - 3} = x - 12$

55. $x - 9 = \sqrt{x^2 - x - 4}$

56. $x = \sqrt{45 - 11x} + 5$

57. $\sqrt{\dfrac{t}{3}} = \sqrt{\dfrac{t + 4}{2 + t}}$

58. $\sqrt{t + \dfrac{1}{t}} = \dfrac{\sqrt{17}}{2}$

59. $\dfrac{\sqrt{2}}{\sqrt{x - 1}} = \sqrt{\dfrac{3}{x}}$

60. $\dfrac{\sqrt{2y - 1}}{\sqrt{3y + 4}} = \dfrac{\sqrt{3}}{2}$

 61. For which one of the following equations will it be necessary to square both sides twice in order to solve it?

　(i) $\sqrt{x + 1} = \sqrt{x} + 1$　　(ii) $x + 1 = \sqrt{x} + 1$

Explain how the other equation can be solved by squaring both sides just once.

62. Without solving, explain why the following equation has no solution: $\sqrt{x+1} + 1 = 0$.

In Exercises 63–78, solve the radical equation.

63. $\sqrt{x-3} + 1 = \sqrt{x}$

64. $\sqrt{x+5} = 1 + \sqrt{x}$

65. $\sqrt{4x-3} - \sqrt{3x-5} = 1$

66. $\sqrt{5x+1} - \sqrt{3x-5} = 2$

67. $\sqrt{5x-9} + 3 = \sqrt{x}$

68. $\sqrt{x-8} + \sqrt{x} = 2$

69. $\sqrt{7x+4} - \sqrt{x+1} = 3$

70. $\sqrt{7-2x} - 1 = \sqrt{2}\sqrt{x-1}$

71. $\sqrt{x+3} = 1 + \sqrt{x-2}$

72. $\sqrt{x^2+3x-1} - 1 = \sqrt{x^2+x-2}$

73. $\sqrt{5x-6} = 1 + \sqrt{3x-5}$

74. $\sqrt{2x-1} - 1 = \sqrt{x-1}$

75. $\sqrt{4x+1} - \sqrt{3x-2} = 1$

76. $\sqrt{2x+3} - \sqrt{x-2} = 2$

77. $\sqrt{7x-3} - \sqrt{2x+1} = 2$

78. $\sqrt{5x} - \sqrt{x+4} = 2$

79. Suppose $x^{a/b} = k$. What is the first step in solving the equation for x? What is the purpose of this step?

80. Suppose $x^{-1/2} = 0$. Why would it *not* be permissible to raise both sides of the equation to the -2 power in order to eliminate the rational exponent?

In Exercises 81–88, solve the equations involving rational exponents.

81. $x^{2/3} = 9$

82. $y^{3/4} = -8$

83. $x^{-3/4} = 27$

84. $y^{-2/3} = 25$

85. $(x-3)^{-2/3} = 4$

86. $(x+5)^{-3/5} = 8$

87. $(3x-2)^{3/4} = 8$

88. $(2x-3)^{2/3} = 16$

In Exercises 89–92, formulas for the area and volume of geometric figures and solids are given. In these formulas A represents area, V represents volume, r represents the radius, and h represents the height. Solve each formula for r.

89. $A = \pi r^2$ Area of a circle

90. $V = \pi r^2 h$ Volume of a right circular cylinder

91. $V = \dfrac{4}{3}\pi r^3$ Volume of a sphere

92. $V = \dfrac{1}{3}\pi r^2 h$ Volume of a right circular cone

In Exercises 93–96, use the formulas for r found in Exercises 89–92 to solve the following applications. (Recall that the diameter of an object is twice its radius.)

93. A circular rug was advertised as having an area of 95 square feet. Will the rug fit in a square room 11 feet on each side? Show how you know.

94. The volume of a 3.75-inch-high soup can is 18.41 cubic inches. What is the diameter of the soup can?

95. The volume of a tennis ball is 6.54 cubic inches. What is the diameter of the tennis ball?

96. The volume of the ice cream cone in the accompanying figure is 7.95 cubic inches. What is the diameter of the circular top of the cone?

6 in.

97. The square root of the product of a positive number and its square is twice the number. What is the number?

98. The fourth root of the cube root of the square root of a number is 2. What is the number?

99. Clocks with pendulums can be adjusted to run faster or slower by changing the position of the weight at the end of the pendulum. The time t (in seconds) that it takes a pendulum to swing

through one cycle is given by the formula $t = 2\pi\sqrt{\frac{L}{980}}$, where L is the length (in centimeters) of the pendulum. To the nearest tenth of a centimeter, what is the length of a pendulum that takes 1 second to swing through one cycle?

100. Consider two positive consecutive integers. If the cube root of the smaller integer is multiplied by the square root of the larger integer, the result is the square root of one-half the product of the two integers. What is the smaller integer?

101. Determine the value of x in the accompanying figure.

102. Determine the value of x in the accompanying figure.

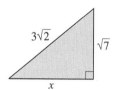

103. A 26-foot ladder rests against a wall with the bottom of the ladder 7 feet from the base of the wall. How far up the wall does the top of the ladder rest? (See figure.)

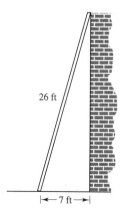

104. A guy wire 44 feet long reaches from the top of a telephone pole to a point on the ground 14 feet from the base of the pole. How tall is the telephone pole? (See figure.)

Figure for 104

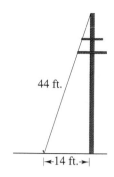

44 ft.

|←14 ft.→|

105. A 30-foot utility pole is supported by a guy wire that is attached to the ground 10 feet from the base of the pole. If the length of the guy wire is represented by $\sqrt[4]{x-1}$, determine x and then determine the length of the guy wire.

106. A 26-foot ladder rests against a wall with the bottom of the ladder 7 feet from the base of the wall. If $x^{2/5}$ represents the distance from the top of the ladder to the base of the wall, determine x and then determine how far up the wall the top of the ladder rests.

107. A surveyor needs to know the distance BC across the pond in the accompanying figure. She places a stake at point C so that the line segments $\overline{AC}$ and $\overline{BC}$ are at right angles. The trees are 150 feet apart, and the distance from the stake to the tree at point A is 125 feet. What is the distance from the stake to the tree at point B?

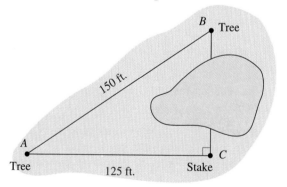

108. The perimeter of the rectangle shown in the figure is equal to 10 feet. Find the dimensions of the rectangle.

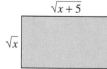

$\sqrt{x+5}$

$\sqrt{x}$

109. Explain how you can tell that $\sqrt{x - 3} = \sqrt{2 - x}$ has no solution without solving it. (Hint: Consider the domains of the radical expressions.)

 110. If x is a nonzero number, then the solution sets of two of the following equations are empty and the solution set of the other two equations is the set of all nonzero real numbers. Explain how each solution set can be determined without solving the equation.

(a) $-\sqrt{x^2} = \sqrt{(-x)^2}$ (b) $-\sqrt[3]{x} = \sqrt[3]{-x}$

(c) $\sqrt{x^2} = \sqrt{(-x)^2}$ (d) $-\sqrt{x^2} = \sqrt{-x^2}$

Exploring with Real Data

In Exercises 111–114, refer to the following background information and data.

The average price of movie tickets increased steadily from 1988 to 1992. The following table shows the average prices for the last 3 years of this period.

	1990	*1991*	*1992*
Average Price of a Movie Ticket	$4.75	$4.89	$5.05

(Source: Motion Picture Association of America.)

The average price $P(t)$ of a movie ticket for the given 3 years can be modeled by $P(t) = \sqrt{t} + 20$, where t is the number of years since 1987.

111. What is the value of t when $P(t) = 5.50$? What is the interpretation of the solution?

112. What average ticket price is predicted by the model for the year 2000?

113. Produce the graph of the model function. Does the graph suggest a continued steady increase in ticket prices, a leveling off of prices, or an eventual decrease in prices?

114. State two factors that could dampen the rapid increases in movie ticket prices.

Challenge

 115. If we square both sides of each of the following equations, the solution set appears to be the set of all real numbers. However, closer examination shows that the solution sets are all different. Solve the equations and compare solution sets.

(a) $\sqrt{x^2} = x$

(b) $\sqrt{x^2} = -x$

(c) $\sqrt{x^4} = x^2$

In Exercises 116–125, solve the equation.

116. $\sqrt[6]{3 + t - t^2} = \sqrt[3]{2t + 3}$

117. $\sqrt{\sqrt{t^2 + 3t + 12} + 3} = \sqrt{t + 5}$

118. $\sqrt{x + 2} - \sqrt{x - 3} = \sqrt{4x - 1}$

119. $\sqrt{x + 3} + \sqrt{x - 3} = \sqrt{4x - 6}$

120. $\sqrt[4]{x + 3} = \sqrt{x + 1}$

121. $\sqrt[4]{8x + 1} = \sqrt{x + 2}$

122. $\sqrt{3 - x^2} + x^2 = 3$

123. $\sqrt{\sqrt{x - 3} + \sqrt{x + 5}} = 3$

124. $\dfrac{\sqrt{x + 1} + \sqrt{x - 1}}{\sqrt{x + 1} - \sqrt{x - 1}} = \dfrac{4x - 1}{2}$

125. $\sqrt{2x^2 - 1} - 1 = \sqrt{x^2 - 1}$

9.9 Complex Numbers

Imaginary Numbers ▪ *Complex Number System* ▪ *Addition and*
Subtraction ▪ *Multiplication* ▪ *Division*

We have defined a to be a square root of b if $a^2 = b$. This implies that b is non-negative because, for any real number a, $a^2 \geq 0$.

This means the equation $x^2 = -1$ has no solution in the set of real numbers. In this section we will define a new kind of number that will allow us to solve equations such as $x^2 = -1$.

Imaginary Numbers

To allow us to define the square root of a negative number, we expand the number system by defining an *imaginary number i*.

> ### *Definition of the Imaginary Number i*
>
> The symbol i represents an **imaginary number** with the properties $i = \sqrt{-1}$ and $i^2 = -1$.

Using this new number i, we can define the square root of any real number and obtain imaginary number solutions to all equations of the form $x^2 = -n$ where n is a positive real number.

> ### *Definition of $\sqrt{-n}$*
>
> For any positive real number n, $\sqrt{-n} = i\sqrt{n}$.

NOTE: The word *imaginary* is used to distinguish these new numbers from real numbers. It should not be inferred from their name that imaginary numbers do not really exist. Like all numbers in mathematics, imaginary numbers exist because we *define* them to exist. ∎

EXAMPLE 1   *Using the Definition of Imaginary Numbers*

Use the definition of $\sqrt{-n}$ to write each square root.

(a) $\sqrt{-49}$ (b) $\sqrt{-7}$ (c) $\sqrt{-\dfrac{4}{9}}$

Solution

(a) $\sqrt{-49} = i\sqrt{49} = 7i$

(b) $\sqrt{-7} = i\sqrt{7}$

(c) $\sqrt{-\dfrac{4}{9}} = i\sqrt{\dfrac{4}{9}} = \dfrac{2}{3}i$ ■

NOTE: In part (b) of Example 1, to avoid confusing $\sqrt{7}i$ with $\sqrt{7i}$, we usually write $i\sqrt{7}$.   ■

It is important to remember that imaginary numbers are not real numbers, and the familiar properties of real numbers do not apply. However, there is a commutative property, an associative property, and a distributive property for this new set of numbers.

NOTE: In particular, the Product Rule for Radicals does *not* apply to imaginary numbers: $\sqrt{-12}\,\sqrt{-3} \neq \sqrt{36}$. As we perform operations with imaginary numbers, we must begin with the definition.   ■

EXAMPLE 2   *Products of Imaginary Numbers*

Determine the products.

(a) $\sqrt{-12}\,\sqrt{-3}$ (b) $\sqrt{-2}\left(\sqrt{-8} - \sqrt{-6}\right)$

Solution

(a) $\begin{aligned}
\sqrt{-12}\,\sqrt{-3} &= i\sqrt{12}\cdot i\sqrt{3} &&\text{Definition of } \sqrt{-n}\\
&= i^2\cdot\sqrt{36} &&\text{Product Rule for Radicals}\\
&= -1\cdot 6 &&i^2 = -1\\
&= -6
\end{aligned}$

(b) $\begin{aligned}
\sqrt{-2}\left(\sqrt{-8} - \sqrt{-6}\right) &= i\sqrt{2}\left(i\sqrt{8} - i\sqrt{6}\right) &&\text{Definition of } \sqrt{-n}\\
&= i^2\sqrt{16} - i^2\sqrt{12} &&\text{Distributive Property}\\
&= -1\cdot 4 - (-1)\sqrt{4}\,\sqrt{3} &&\text{Simplify the radicals.}\\
&= -4 + 2\sqrt{3}
\end{aligned}$ ■

We have defined $i = \sqrt{-1}$ and $i^2 = -1$. To facilitate computations, we need to investigate higher powers of i.

 **EXPLORATION 1** *Powers of i*

(a) Use the definition of i^2 to evaluate $i^3, i^4, \ldots, i^8$.

(b) What pattern is evident in part (a)?

(c) Propose a method for evaluating i to any power.

Discovery

(a)
$$i = i$$
$$i^2 = -1$$
$$i^3 = i^2 \cdot i = -1 \cdot i = -i$$
$$i^4 = (i^2)^2 = (-1)^2 = 1$$
$$i^5 = (i^2)^2 \cdot i = (-1)^2 \cdot i = 1 \cdot i = i$$
$$i^6 = (i^2)^3 = (-1)^3 = -1$$
$$i^7 = (i^2)^3 \cdot i = (-1)^3 \cdot i = -1 \cdot i = -i$$
$$i^8 = (i^2)^4 = (-1)^4 = 1$$

(b) The powers of i cycle through the numbers i, -1, $-i$, and 1.

(c) All powers of i can be written $(i^2)^n$ or $(i^2)^n \cdot i$, where n is an integer.

EXAMPLE 3 *Evaluating Powers of i*

Evaluate.

(a) i^{11} (b) i^{201} (c) i^{48}

Solution

(a) $i^{11} = (i^2)^5 \cdot i = (-1)^5 \cdot i = -1 \cdot i = -i$

(b) $i^{201} = (i^2)^{100} \cdot i = (-1)^{100} \cdot i = 1 \cdot i = i$

(c) $i^{48} = (i^2)^{24} = (-1)^{24} = 1$

Complex Number System

The expanded number system, which includes i, is called the **complex number system.**

The word *complex* means consisting of more than one part. In the definition of a complex number, we refer to numbers consisting of a *real part* and an *imaginary part*.

> **Definition of a Complex Number**
>
> If a and b are real numbers, then $a + bi$ is the standard form of a **complex number.** If $b \neq 0$, then $a + bi$ is an **imaginary number.** If $b = 0$, then $a + bi$ is simply the real number a. For $a + bi$, a is called the **real part** and b is called the **imaginary part.**

Figure 9.11 shows the structure of the complex number system.

Figure 9.11

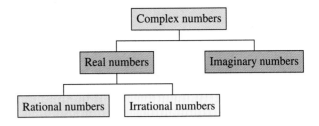

The set of real numbers and the set of imaginary numbers are subsets of the set of complex numbers. Also, a number is either real or imaginary, but not both.

Figure 9.12

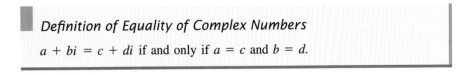

Complex Numbers

In Fig. 9.12 we see that 3, $\sqrt{7}$, $4 + 0i$, $-\sqrt{5}$, and $\frac{7}{8}$ are real numbers. The numbers i, $0 - 3i$, $3 - 7i$, $\sqrt{-5}$, and $\frac{1}{2}i$ are imaginary numbers. Every number in the figure is a complex number.

Two complex numbers are defined to be equal if their real parts are equal and their imaginary parts are equal.

> ### Definition of Equality of Complex Numbers
> $a + bi = c + di$ if and only if $a = c$ and $b = d$.

When working with complex numbers, it is helpful to treat them as binomials. To add, subtract, and multiply complex numbers, we use the same procedures we have used for binomials.

Addition and Subtraction

To add (or subtract) complex numbers, we add (or subtract) the real and imaginary parts.

> ### Definitions of Addition and Subtraction
>
> The sum and difference of two complex numbers $a + bi$ and $c + di$ are defined as follows.
>
> $$(a + bi) + (c + di) = (a + c) + (b + d)i$$
> $$(a + bi) - (c + di) = (a - c) + (b - d)i$$

EXAMPLE 4 *Adding and Subtracting Complex Numbers*

Find the sum or difference.

(a) $(3 - 2i) + (-4 + 7i)$ 　　　　　　　　(b) $8 - (5 - 2i)$

(c) $(1 + i) - (1 - i)$

Solution

(a) $(3 - 2i) + (-4 + 7i) = (3 - 4) + (-2 + 7)i = -1 + 5i$

(b) $8 - (5 - 2i) = 8 - 5 + 2i = 3 + 2i$

(c) $(1 + i) - (1 - i) = 1 + i - 1 + i = 2i$　　　　　■

Multiplication

To find the product of two complex numbers, use the Distributive Property or the FOIL method just as you did for binomials. Remember that $i \cdot i = i^2 = -1$.

> ### Definition of Multiplication
>
> The product of two complex numbers $a + bi$ and $c + di$ is defined as follows.
>
> $$(a + bi)(c + di) = (ac - bd) + (bc + ad)i$$

EXAMPLE 5 *Multiplying Complex Numbers*

Determine the product.

(a) $5i(4 - 3i)$ 　　　　(b) $(3 + 2i)(-1 + 4i)$ 　　　　(c) $(2 + i)^2$

Solution

(a) $5i(4 - 3i) = 20i - 15i^2$ 　　　Distributive Property

$\qquad\qquad\quad = 20i - 15(-1)$

$\qquad\qquad\quad = 15 + 20i$

(b) $(3 + 2i)(-1 + 4i) = -3 + 12i - 2i + 8i^2$ FOIL

$$= -3 + 12i - 2i + 8(-1)$$

$$= -3 + 12i - 2i - 8$$

$$= -11 + 10i$$

(c) $(2 + i)^2 = 2^2 + 4i + i^2$ Square of a binomial

$$= 4 + 4i - 1$$

$$= 3 + 4i$$ ■

Division

To prepare for division, we first consider pairs of complex numbers of the form $a + bi$ and $a - bi$.

In Section 9.6 we defined radical expressions of the form $a + b$ and $a - b$ to be conjugates. Similarly, the complex numbers $a + bi$ and $a - bi$ are called **complex conjugates.**

EXAMPLE 6   *Products of Complex Conjugates*

Write the conjugate of each complex number. Then find the product of the complex number and its conjugate.

(a) $1 - 4i$ (b) $7 + 2i$ (c) $5i$

Solution

(a) The conjugate of $1 - 4i$ is $1 + 4i$.

$$(1 - 4i)(1 + 4i) = 1^2 - (4i)^2$$ $(A - B)(A + B) = A^2 - B^2$

$$= 1 - (-16)$$

$$= 17$$

(b) The conjugate of $7 + 2i$ is $7 - 2i$.

$$(7 + 2i)(7 - 2i) = 7^2 - (2i)^2$$ $(A + B)(A - B) = A^2 - B^2$

$$= 49 - (-4)$$

$$= 53$$

(c) The number $5i$ is the same as $0 + 5i$, and the conjugate is $0 - 5i$, or just $-5i$.

$$(5i)(-5i) = -25i^2 = -25(-1) = 25$$ ■

Notice that the product of complex conjugates is always a *real* number. In fact, for any two complex conjugates $a + bi$ and $a - bi$,

$$(a + bi)(a - bi) = a^2 - (bi)^2$$

$$= a^2 - b^2 i^2$$

$$= a^2 - (-1)b^2$$

$$= a^2 + b^2$$

We summarize this rule for the product of complex conjugates.

> ### Product of Complex Conjugates
> For any two complex conjugates $a + bi$ and $a - bi$,
> $$(a + bi)(a - bi) = a^2 + b^2.$$

Now we can proceed to the methods for dividing complex numbers.

> ### Definition of Division of a Complex Number by a Real Number
> If c is a nonzero real number, then
> $$\frac{a + bi}{c} = \frac{a}{c} + \frac{bi}{c}.$$

To find the quotient

$$\frac{a + bi}{c + di},$$

where $c + di \neq 0$, multiply the numerator and denominator by the conjugate of the denominator.

EXAMPLE 7   *Dividing Complex Numbers*

Determine the quotients.

(a) $\dfrac{4 + 6i}{2}$ (b) $\dfrac{2 - i}{1 + 2i}$ (c) $\dfrac{5 + 2i}{2i}$

Solution

(a) $\dfrac{4 + 6i}{2} = \dfrac{4}{2} + \dfrac{6i}{2} = 2 + 3i$

(b) Multiply the numerator and denominator by the conjugate of the denominator. The product in the denominator is the *sum* of the squares of the coefficients.

$$\begin{aligned}
\frac{2 - i}{1 + 2i} &= \frac{(2 - i)(1 - 2i)}{(1 + 2i)(1 - 2i)} \\
&= \frac{2 - 4i - i - 2}{1 + 4} \qquad \text{FOIL} \\
&= -\frac{5i}{5} \\
&= -i
\end{aligned}$$

(c) Multiply the numerator and denominator by the conjugate of $2i$, which is $-2i$.

$$\frac{5 + 2i}{2i} = \frac{(5 + 2i)(-2i)}{(2i)(-2i)} = \frac{-10i + 4}{4} = 1 - \frac{5}{2}i$$

9.9 *Quick Reference*

Imaginary Numbers
- The symbol i represents an **imaginary number** with the properties $i = \sqrt{-1}$ and $i^2 = -1$.

- For any positive real number n, $\sqrt{-n} = i\sqrt{n}$.

- To perform operations with radicals having negative radicands, use the definition of i to rewrite the radicals in the form of imaginary numbers.

- Powers of i can be evaluated by writing the expression in the form $(i^2)^n$ or $(i^2)^n \cdot i$, where n is an integer.

Complex Number System
- If a and b are real numbers, then $a + bi$ is the standard form of a **complex number.** If $b \neq 0$, then $a + bi$ is an **imaginary number.** If $b = 0$, then $a + bi$ is the real number a.

- For $a + bi$, a is called the **real part** and b is called the **imaginary part.**

- The imaginary numbers and the real numbers are subsets of the set of complex numbers.

- The complex numbers $a + bi$ and $c + di$ are defined to be equal if and only if $a = c$ and $b = d$.

Addition and Subtraction
- The sum of two complex numbers is defined as
$$(a + bi) + (c + di) = (a + c) + (b + d)i.$$

- The difference of two complex numbers is defined as
$$(a + bi) - (c + di) = (a - c) + (b - d)i.$$

Multiplication
- The product of two complex numbers can be found by treating the numbers as binomials and using the FOIL method. The definition is
$$(a + bi)(c + di) = (ac - bd) + (bc + ad)i.$$

Division
- The complex numbers $a + bi$ and $a - bi$ are called **complex conjugates.**

- In general, $(a + bi)(a - bi) = a^2 + b^2$.

- If c is a nonzero real number, then
$$\frac{a + bi}{c} = \frac{a}{c} + \frac{bi}{c}.$$

- To find the quotient of $a + bi$ and $c + di$, multiply the numerator and denominator by the conjugate of the denominator.

9.9 *Exercises*

 1. In order to solve $x^2 = -1$, it is necessary to define a new type of number. What is the number and what two properties does it possess according to its definition?

 2. To perform $\sqrt{-3}\sqrt{-5}$, we begin by writing $\sqrt{-3}\sqrt{-5} = i\sqrt{3} \cdot i\sqrt{5} = (i \cdot i) \cdot (\sqrt{3} \cdot \sqrt{5})$. What two properties are used and for what purpose?

In Exercises 3–10, write the radical expression in the complex number form $a + bi$.

3. $\sqrt{-9}$

4. $\sqrt{-5}$

5. $\sqrt{-18}$

6. $\sqrt{-\dfrac{16}{25}}$

7. $\sqrt{25} + \sqrt{-36}$

8. $\sqrt{80} + \sqrt{-27}$

9. $\dfrac{-4 - \sqrt{-12}}{2}$

10. $\dfrac{-3 + \sqrt{-27}}{-3}$

 11. State the conditions under which two complex numbers are equal.

 12. Explain how to add two complex numbers.

In Exercises 13–16, determine a and b.

13. $(a + 1) + 2bi = 3 - 2i$

14. $(2a + 1) + 3bi = 7 + 6i$

15. $2a + bi = \sqrt{-25} - 2$

16. $4a + 2bi = 3\sqrt{-4}$

17. Which of the following statements is true?

(a) The set of real numbers is a subset of the set of imaginary numbers.

(b) If a and b are nonzero real numbers, then $a + bi$ is an imaginary number.

(c) The number 0 is both a real number and an imaginary number.

18. Classify the following as real numbers or imaginary numbers.

(a) $\sqrt{7}$

(b) $i\sqrt{7}$

(c) $\sqrt{-7}$

(d) $\sqrt{7} + \sqrt{-7}$

(e) $-\sqrt{7} + \sqrt{7}$

(f) $\sqrt{-7} - \sqrt{-7}$

In Exercises 19–30, perform the indicated operation.

19. $(5 + 6i) + (-4 - 5i)$

20. $(-8 + 4i) - (9 - 4i)$

21. $(7 + 3i) - (-8 - 7i)$

22. $(-5 + 2i) + (5 - 3i)$

23. $7i - 2(-3 - 7i)$ **24.** $8 - 3(5 + 2i)$

25. $(7 - 2i) - (3 - 4i) + 2(5 - i)$

26. $(3 - 4i) + 3(2 - 6i) - (3 + 2i)$

27. $\sqrt{-36} + \sqrt{-49}$ **28.** $\sqrt{-32} - \sqrt{-72}$

29. $3 - i - \sqrt{4} - \sqrt{-9}$

30. $\left(4 + \sqrt{-50}\right) - 4\left(3 - \sqrt{-72}\right)$

In Exercises 31–34, determine a and b.

31. $(a + 3i) + (2 + bi) = i$

32. $(a - 2i) - (7 + bi) = -4 - 8i$

33. $\left(5 - \sqrt{-4}\right) - (a + bi) = 6$

34. $(a + bi) + \left(\sqrt{4} - \sqrt{-16}\right) = -1 - 6i$

 35. Explain why the Product Rule for Radicals does *not* apply to such products as $\sqrt{-3}\sqrt{-3}$.

 36. Describe a method for simplifying i^{34}.

In Exercises 37–50, determine the product.

37. $(2i)(5i)$

38. $(-6i)(2i)$

39. $7i(3 - 4i)$

40. $-2i(-3 - 5i)$

41. $(2 - 3i)(3 - 4i)$

42. $(1 + i)(-3 + 5i)$

43. $(3 + i)^2$

44. $(2 - i)(-4 + 3i)$

45. $\sqrt{-8}\sqrt{-2}$

46. $\sqrt{-10}\sqrt{2}$

47. $\sqrt{-3}(\sqrt{-6} + \sqrt{-27})$

48. $\sqrt{-2}(\sqrt{-32} - \sqrt{2})$

49. $(-\sqrt{-1} + 3)(\sqrt{-9} + \sqrt{16})$

50. $(\sqrt{-25} - \sqrt{4})(\sqrt{-4} + \sqrt{25})$

In Exercises 51–54, determine the value of a and b.

51. $2i(a + bi) = 10$

52. $i(a + bi) = -4$

53. $(6 + 5i)(a + bi) = 38 - 9i$

54. $(a + bi)(3 - 5i) = 21 - i$

In Exercises 55–62, evaluate the given power of i.

55. i^{15}

56. i^{45}

57. i^{52}

58. i^{22}

59. i^{30}

60. i^{39}

61. i^{13}

62. i^{24}

63. Suppose a and b are nonzero real numbers. What is the difference in the results of the following operations?

    (a) $(a + b)(a - b)$ (b) $(a + bi)(a - bi)$

64. To find the quotient $\dfrac{a + bi}{c + di}$,

where a, b, c, and d are nonzero real numbers, what step must we take first?

In Exercises 65–70, write the complex conjugate of the given complex number. Then, find the product of the complex number and its conjugate.

65. $7i$

66. $-4i$

67. $2 + 3i$

68. $3 - 4i$

69. $\sqrt{5} - 4i$

70. $7 - i\sqrt{2}$

In Exercises 71 and 72, determine a and b.

71. $(a + bi)(-5 - i) = 26$

72. $(a + bi)(5 - 4i) = 41$

In Exercises 73–84, determine the quotient.

73. $\dfrac{-3}{2i}$

74. $\dfrac{4}{5i}$

75. $\dfrac{9 + 6i}{3i}$

76. $\dfrac{-4 + 10i}{-2i}$

77. $\dfrac{3 - i}{1 + 2i}$

78. $\dfrac{4 + 5i}{3 - 2i}$

79. $\dfrac{3 - 4i}{-7 - 5i}$

80. $\dfrac{5 - 2i}{3 + 5i}$

81. $\dfrac{3}{2 + \sqrt{-3}}$

82. $\dfrac{-4}{1 - \sqrt{-2}}$

83. $\dfrac{\sqrt{-5} - \sqrt{3}}{\sqrt{-3} - \sqrt{9}}$

84. $\dfrac{\sqrt{6} + \sqrt{-3}}{3 - \sqrt{-2}}$

In Exercises 85–88, determine a and b.

85. $\dfrac{a + bi}{i} = 2 + 5i$

86. $\dfrac{a + bi}{3i} = -2 - \dfrac{5}{3}i$

87. $\dfrac{a + bi}{3 + 2i} = \dfrac{8}{13} + \dfrac{12}{13}i$

88. $\dfrac{a + bi}{2 - i} = \dfrac{3}{5} + \dfrac{4}{5}i$

In Exercises 89–110, perform the indicated operations for each expression.

89. $(3 - 4i) + (5 + i)$

90. $(-2 + 3i) - (2 - 7i)$

91. $(4i)(-7i)$

92. $(-3i)(-i)$

93. $\dfrac{1 + 2i}{3i}$

94. $\dfrac{i - 4}{-2i}$

95. $3(2 + i) + 3i(1 - i)$

96. $5 - (4 - 5i)$

97. $(-6 + 5i)^2$

98. $(-2 - i)(3 + 2i)$

99. i^7

100. i^{17}

101. $\dfrac{3 + 4i}{3 - 4i}$

102. $\dfrac{2i}{5 - i}$

103. $7\sqrt{-25} - 4\sqrt{-49}$

104. $5\sqrt{-28} + 2\sqrt{-63}$

105. $\left(1 + \sqrt{-3}\right)\left(-1 - \sqrt{-12}\right)$

106. $\sqrt{-2}\left(\sqrt{18} - \sqrt{-8}\right)$

107. $\dfrac{\sqrt{-75}}{\sqrt{-3}}$

108. $\dfrac{\sqrt{6}\,\sqrt{-10}}{\sqrt{-2}}$

109. $\dfrac{1}{1 - \sqrt{-3}}$

110. $\dfrac{\sqrt{25} - \sqrt{-9}}{\sqrt{-16} - \sqrt{4}}$

In Exercises 111–114, verify that the given number and its conjugate are both solutions of the given equation.

111. $x^2 - 4x + 13 = 0$; $2 + 3i$

112. $x^2 + 9 = 0$; $-3i$

113. $x^2 - 4x + 5 = 0$; $2 - i$

114. $4x^2 + 3 = 0$; $\dfrac{\sqrt{3}}{2}i$

In Exercises 115–118, determine the product.

115. $(x + 6i)(x - 6i)$

116. $(x - 2i)(x + 2i)$

117. $(x - 5i)(x + 5i)$

118. $(x + 3i)(x - 3i)$

The rule for the product of complex conjugates states that

$$(a + bi)(a - bi) = a^2 + b^2.$$

We can use this rule to factor expressions such as $x^2 + 4$ *over the complex numbers.*

Example: $x^2 + 4 = (x + 2i)(x - 2i)$

In Exercises 119–122, use this rule to factor the given expression over the complex numbers.

119. $x^2 + 9$

120. $x^2 + 25$

121. $4x^2 + 1$

122. $49x^2 + 16$

Challenge

123. For a nonzero complex number $a + bi$ and a positive integer n, $(a + bi)^{-n}$ is defined as

$$\frac{1}{(a + bi)^n}.$$

Use this definition to evaluate the following.

(a) i^{-2}

(b) i^{-3}

(c) $(2 + 3i)^{-1}$

(d) $(1 - 2i)^{-2}$

124. Verify that $\dfrac{\sqrt{2}}{2}(1 + i)$

is a square root of i. What is your conjecture about another square root of i? Show that your conjecture is correct.

125. In the complex number system, a nonzero number has two square roots. For example, the square roots of 4 are 2 and -2; the square roots of -4 are $2i$ and $-2i$.

One cube root of -8 is -2. What is your conjecture about the total number of cube roots of a number in the complex number system? To support your conjecture, verify the following.

(a) $\left(1 + i\sqrt{3}\right)^3 = -8$

(b) $\left(1 - i\sqrt{3}\right)^3 = -8$

In Exercises 126–129, perform the indicated operations and express the result in the form $a + bi$.

126. $\dfrac{2i + 1}{i} - \dfrac{1 - 3i}{4 - i}$

127. $\dfrac{2 + 3i}{2 - 3i} + \dfrac{i}{1 - i}$

128. $\left(\dfrac{5i}{2i - 1} - \dfrac{10}{3i - 1}\right)(2i + 3)$

129. $2i\left(\dfrac{i + 1}{2i - 1} - \dfrac{3i}{i + 2}\right)$

9 Chapter Review Exercises

Section 9.1

1. If $x < 0$, then $\sqrt{x^2} = |x|$ and $\sqrt[3]{x^3} = x$. Explain why absolute value symbols are used in one result but not the other.

2. Because $3^2 = 9$ and $(-3)^2 = 9$, we say that 9 has two square roots, 3 and -3. Why, then, do we write $\sqrt{9} = 3$?

In Exercises 3–6, evaluate, if possible.

3. $-\sqrt{64}$

4. $-\sqrt[3]{-64}$

5. $-\sqrt[3]{64}$

6. $\sqrt[4]{-64}$

In Exercises 7–10, simplify the given expression.

7. $\sqrt[4]{x^8}$

8. $\sqrt{(3x)^4}$

9. $\sqrt{(x-4)^2}$, $x \geq 4$

10. $7\sqrt[3]{x^9}$

11. Translate the given phrase into a radical expression.

(a) The square root of 10 minus a.

(b) The square root of the quantity 10 minus a.

12. The side of a cube is given by $s = \sqrt[3]{V}$, where V is the volume of the cube. What is the side of a cube whose volume is 4.096 cubic inches?

Section 9.2

13. Explain why $(-4)^{1/2}$ is not a real number.

14. When converting $b^{m/n}$ into a radical, what are the interpretations of m and n?

In Exercises 15–20, evaluate the given expression, if possible.

15. $16^{3/2}$

16. $\left(-\dfrac{8}{27}\right)^{1/3}$

17. $(-27)^{-2/3}$

18. $18^{1/2} \cdot 2^{1/2}$

19. $-9^{-1/2}$

20. $(-9)^{-1/2}$

21. Write $4x^{6/7}$ and $(4x)^{6/7}$ as radical expressions.

22. Write $\sqrt[3]{\sqrt[5]{x}}$ as an exponential expression.

23. Supply the missing number to make the statement true: $(\quad)^{3/4} = 8$.

24. Show how to evaluate $25^{1.5}$ without a calculator.

Section 9.3

25. Explain why the result of simplifying $(x^4)^{1/4}$ depends on whether x is nonnegative or is any real number.

26. Which of the following is a correct way to begin simplifying the given expression? Explain.

(i) $\left(\dfrac{4}{9}\right)^{-1/2} = \left(\dfrac{9}{4}\right)^{1/2} = \cdots$

(ii) $\left(\dfrac{4}{9}\right)^{-1/2} = \dfrac{4^{-1/2}}{9^{-1/2}} = \cdots$

In Exercises 27–32, simplify the expression and write the result with positive exponents. Assume all variables represent positive numbers.

27. $\dfrac{f^{5/7}}{f^{3/7}}$

28. $(b^{-4/3}c^2)^9$

29. $(a^{-2})^{-1/4}$

30. $\left(\dfrac{27x^6z^{-9}}{y^{-3}}\right)^{1/3}$

31. $(2x^{-3/2})^{-1}$

32. $\dfrac{(3x^{-1})^{-2}}{x^{1/2}}$

33. Supply the missing numbers to make the statement true.

$$(x^{(\ \ )}y^{12})^{1/4} = x^2y^{(\ \ )}$$

34. Simplify $x^{2/3}(x^{-5/3} - x^{7/3})$, $x \neq 0$. Write the result with positive exponents.

35. The length d of a diagonal of a rectangle is given by $d = (L^2 + W^2)^{1/2}$, where L and W are the length and width of the rectangle, respectively. To the nearest hundredth of a foot, how long is the diagonal of a rectangle whose width is 7 feet and whose length is 5 yards?

36. If a number raised to the two-thirds power is 9, what is the number raised to the one-third power?

Section 9.4

37. Explain why the domain of $y_1 = \sqrt{x - 1}$ is $\{x \mid x \geq 1\}$, but the domain of $y_2 = \sqrt[3]{x - 1}$ is the set of all real numbers.

38. Explain why the Product Rule for Radicals cannot be used to multiply $\sqrt[3]{5} \sqrt{5}$. Show a method that can be used to perform the operation.

In Exercises 39 and 40, determine the domain of the expression.

39. $\sqrt{5 - x}$ **40.** $\sqrt[3]{7x + 4}$

In Exercises 41–44, perform the indicated operation. Assume that all variables represent positive numbers.

41. $\sqrt{6} \sqrt{7x}$ **42.** $\sqrt{a} \sqrt{a + 1}$

43. $\dfrac{\sqrt[3]{x^5}}{\sqrt[3]{x^2}}$ **44.** $\dfrac{\sqrt{9x^3 y^4}}{\sqrt{4xy^2}}$

In Exercises 45 and 46, determine the indicated root.

45. $\sqrt[3]{27a^6 b^9}$ **46.** $\dfrac{\sqrt{4z^8}}{\sqrt{25w^6}}$

47. Use the Quotient Rule for Radicals to show that the square root of the reciprocal of a positive number is the reciprocal of the square root of the number.

48. Show that $\sqrt{x^2 - 6x + 9}$ represents the distance between x and 3 on a number line.

Section 9.5

49. State the four conditions under which a radical is simplified.

50. Explain why $\sqrt[3]{24}$ can be simplified but $\sqrt[3]{25}$ cannot.

In Exercises 51–56, simplify, if possible. Assume all variables represent positive numbers.

51. $\sqrt{x^5}$ **52.** $\sqrt{\dfrac{5a^7}{9b^4}}$

53. $\sqrt{c^{17} d^{20}}$ **54.** $\sqrt{\dfrac{3x^{15}}{192x^5}}$

55. $\sqrt[10]{x^8 y^6}$ **56.** $\sqrt{x^2 + 9}$

57. One of the following expressions *cannot* be simplified with just one operation. Identify this expression and then simplify the other two expressions.

(a) $\dfrac{\sqrt{15}}{\sqrt{3}}$ (b) $\dfrac{\sqrt{16}}{\sqrt{3}}$ (c) $\dfrac{\sqrt{15}}{\sqrt{9}}$

58. The area of a rectangle is $10\sqrt{x^2 + 3x}$. If the width of the rectangle is $2\sqrt{x}$, find a simplified expression for the length.

Section 9.6

59. Explain why neither of the following operations can be performed.

(a) $3\sqrt{x} + 3\sqrt{y}$ (b) $2\sqrt{x} + 2\sqrt[3]{x}$

60. Describe the result of multiplying a pair of square root conjugates.

In Exercises 61–64, perform the indicated operations. Assume all variables represent positive numbers.

61. $3\sqrt{45x} + 2\sqrt{20x}$

62. $\sqrt[3]{54} - \sqrt[3]{16}$

63. $\left(3\sqrt{5x}\right)^2$

64. $\left(\sqrt{3} + \sqrt{5}\right)^2$

65. Find the product of $\sqrt{x} + \sqrt{2}$ and its conjugate.

66. Supply the missing number to make the statement true.

$(\ \ )\left(\sqrt{3} + \sqrt{5x}\right) = \sqrt{6} + \sqrt{10x}$

67. Determine which of the following is a rational number by evaluating each expression.

(a) $\left(3\sqrt{5}\right)^2$ (b) $\left(3 + \sqrt{5}\right)\left(3 - \sqrt{5}\right)$

(c) $\left(3 + \sqrt{5}\right)^2$

68. The lengths of the three sides of a triangle are $\sqrt{20}$, $\sqrt{125}$, and $\sqrt{180}$. Express the perimeter of the triangle as a single simplified radical.

Section 9.7

 69. Explain why we cannot rationalize the denominator of

$$\frac{1}{\sqrt[3]{x}}$$

by multiplying the numerator and denominator by $\sqrt[3]{x}$.

 70. Explain why we cannot rationalize the denominator of

$$\frac{1}{3 + \sqrt{2}}$$

by multiplying the numerator and denominator by $\sqrt{2}$. Describe the correct method.

In Exercises 71–76, rationalize the denominator. Assume all variables represent positive numbers.

71. $\dfrac{6}{\sqrt{6}}$ **72.** $\sqrt{\dfrac{13}{3}}$

73. $\dfrac{3x}{\sqrt{x^5}}$ **74.** $\dfrac{5}{5 - \sqrt{3}}$

75. $\dfrac{\sqrt{x} + 7}{\sqrt{3} + 3}$ **76.** $\dfrac{1}{\sqrt[4]{8}}$

In Exercises 77 and 78, simplify.

77. $\dfrac{1}{\sqrt{3}} + \dfrac{5\sqrt{3}}{3}$

78. $\dfrac{1}{\sqrt{5} + a} - \dfrac{1}{\sqrt{5} - a}$

79. The length of a rectangle is one more than the square root of a number x. If the area of the rectangle is $x + \sqrt{x}$, find a simplified expression for the width of the rectangle.

80. For $x > 0$, show that $\dfrac{\sqrt[3]{x^2}}{\sqrt{x}}$ and $\dfrac{\sqrt{x}}{\sqrt[3]{x}}$ both have the same simplified form.

Section 9.8

 81. What is an extraneous solution?

 82. Describe the first step in solving an equation of the form $A^{m/n} = B$.

In Exercises 83–86, solve.

83. $(2x - 3)^3 = 27$

84. $\sqrt{6x + 1} = 2x - 3$

85. $\sqrt{5x - 4} - \sqrt{x + 3} = -1$

86. $(x + 2)^{2/3} = 9$

87. An old totem pole is supported by a 20-foot wire extending from the top of the pole to a point 13 feet from the base of the pole. How tall is the totem pole?

88. The square root of the sum of a number and 4 is 2 more than the square root of the difference of the number and 4. What is the number?

89. One number is 5 more than another number. If the product of their square roots is 6, what are the numbers?

90. A Texas ranch is in the shape of a square 60 miles on each side. If the ranch house is at the exact center of the square, how far is it from the house to any corner of the property?

Section 9.9

 91. In the real number system, $\sqrt{-4}\,\sqrt{-9}$ cannot be performed. Why? What is the result in the complex number system?

 92. What is the difference, if any, between the results of $(a + b)(a - b)$ and $(a + bi)(a - bi)$?

In Exercises 93–100, perform the indicated operations for each expression.

93. $(3 - 5i) - (5 - 7i)$

94. $\sqrt{-9} + (4 - 5i)$

95. $(6 - 7i)(2 + 3i)$

96. $(5 + i)^2$

97. $\dfrac{8 - 4i}{2}$

98. $\left(4 + \sqrt{-7}\right)\left(4 - \sqrt{-7}\right)$

99. i^{57}

100. $\dfrac{3 + 4i}{1 - 2i}$

101. If $i(a + bi) = 6$, determine a and b.

102. Which of the following statements is false?

(a) The number $a + bi$ is an imaginary number if b is not 0.

(b) The number $a + bi$ is a real number if b is 0.

(c) The number $a + bi$ is a complex number if b is 0.

(d) The number $a + bi$ is a real number, an imaginary number, and a complex number if neither a nor b is 0.

9   Chapter Test

1. Determine the domain of $\sqrt{3-x}$.

2. Simplify $(-8x^{-6}y^{12})^{1/3}$.

 3. Explain the difference between $-25^{1/2}$ and $(-25)^{1/2}$.

4. Supply the missing numbers to make the statement true.

$$\sqrt{\frac{(\ \ )}{49}} = \frac{2\sqrt{5}}{(\ \ )}$$

5. Write $\sqrt[3]{\sqrt[4]{x^7}}$ as a single radical.

6. Write $\sqrt[3]{x} \cdot \sqrt{x}$, $x > 0$ as a single radical expression.

7. Add $\dfrac{\sqrt{12}}{4} + \dfrac{\sqrt{27}}{6}$.

8. Multiply $\left(3\sqrt{5} - 4\right)\left(2\sqrt{5} + 1\right)$.

 9. For the expression $\dfrac{\sqrt{b}}{\sqrt{a}}$,

   describe two conditions under which the denominator can be rationalized without multiplying the numerator and denominator by $\sqrt{a}$.

10. Rationalize the denominator of $\dfrac{3\sqrt{2} - 1}{\sqrt{2} + 3}$.

11. For the triangle in the accompanying figure, show that the length of the hypotenuse is $x + 2$, $x \geq 0$.

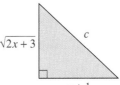

12. Simplify $\sqrt{98x^{11}y^{16}}$.

13. Evaluate $\left(\dfrac{16}{25}\right)^{-3/2}$.

14. Solve $(3t - 1)^{2/3} - 4 = 0$.

 15. Explain why it is essential to check the numbers obtained when solving a radical equation.

16. Solve $3\sqrt{x} + 4 = 13$.

17. An **isosceles triangle** is a triangle with two sides of the same length. If the lengths of these two sides are $\sqrt{1 - 3x}$ and $\sqrt{3x + 1}$, how long are the two sides?

18. Subtract $(6 - 4i) - (7 - 10i)$.

19. What is the square of the complex number $3 + 2i$?

20. Divide $\dfrac{1 - 2i}{3 - i}$.

21. In what form must the numbers $\sqrt{-7}$ and $\sqrt{-3}$ be written before they can be multiplied?

22. Which *two* of the following statements are true?

    (i) Every imaginary number is a complex number, but not every complex number is an imaginary number.

    (ii) The product of complex conjugates is an imaginary number.

    (iii) The radicals $\sqrt{4x^2}$, $\sqrt{x^2 + 4}$, and $\sqrt{x^2 + 4x + 4}$ can all be simplified.

    (iv) The expressions $3 + \sqrt{5}$ and $3 - \sqrt{5}$ are irrational numbers, but their product is a rational number.

23. The volume V of a cube is given by $V = s^3$, where s is the length of every side of the cube. Use a rational exponent to write a formula for s. Then use your calculator to determine the side of a cube whose volume is 200 cubic inches. (Round to two decimal places.)

24. Simplify $\sqrt{3} + \sqrt{-3} + \sqrt{-3}\left(\sqrt{3} + \sqrt{-3}\right)$.

25. In each part, complete the sentence by stating the conditions under which the statement is true.

    (a) If $A^n = B$ and n is an even positive integer, then the equation has no real number solution if _____.

    (b) For a positive integer n, $\sqrt[n]{a^n} = a$ if _____ or if _____.

10 Quadratic Equations

Medical advances and attention to good health practices have brought about a dramatic decrease in deaths due to heart disease since 1985.

By using a **quadratic function** to model the data in the bar graph, we can graphically estimate the years when the number of deaths due to heart disease was above or below a specified level. And we can determine these years exactly by solving **quadratic inequalities.** (To learn more about these activities, see Exercises 91–94 at the end of Section 10.6.)

In Chapter 10 we add the Quadratic Formula and other methods to our techniques for solving quadratic equations, equations that are quadratic in form, and applications. We also learn how to solve quadratic inequalities, higher degree inequalities, and inequalities with rational expressions. The topic of quadratic functions and properties of their graphs concludes the chapter.

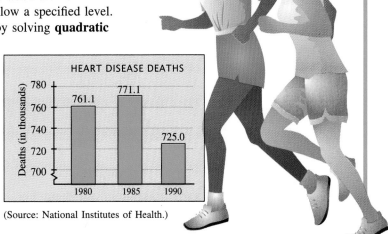

HEART DISEASE DEATHS

(Source: National Institutes of Health.)

10.1 Special Methods

Real Number Solutions   ▪   *Complex Number Solutions*

Real Number Solutions

In Section 7.6 we had our first experience with a class of equations known as **quadratic equations.** The following is a restatement of the definition of a quadratic equation.

> ### Definition of a Quadratic Equation
>
> A **quadratic equation** is an equation that can be written in the *standard form* $ax^2 + bx + c = 0$, where a, b, and c are real numbers and $a \neq 0$.

Associated with a quadratic equation is a **quadratic function.**

> ### Definition of a Quadratic Function
>
> A **quadratic function** is a function of the form
>
> $$f(x) = ax^2 + bx + c$$
>
> where a, b, and c are real numbers and $a \neq 0$.

NOTE: Observe the difference between a quadratic *equation,* which can be written in the form $ax^2 + bx + c = 0$, and a quadratic *function,* which has the form $f(x) = ax^2 + bx + c$. The quadratic equation is the special case $f(x) = 0$.   ◼

The following example provides a review of three previously discussed methods for solving a quadratic equation.

EXAMPLE 1   *Solving Quadratic Equations Graphically and Algebraically*

Solve each quadratic equation graphically and algebraically.

(a) $x^2 - 2x - 8 = 0$ (b) $9x^2 - 12x + 4 = 0$

(c) $5 - x^2 = 0$ (d) $x^2 + 4 = 0$

Solution

(a) To solve the quadratic *equation* $x^2 - 2x - 8 = 0$ graphically, we enter the associated quadratic *function* $y = x^2 - 2x - 8$ and produce the graph.

The solutions of the equation are represented by points of the graph whose y-coordinates are 0, that is, by the x-intercepts.

Figure 10.1

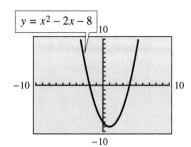

From the graph in Fig. 10.1, we anticipate that the equation has two solutions. By tracing the graph, we estimate that the x-intercepts are $(4, 0)$ and $(-2, 0)$. Thus, the approximate solutions are 4 and -2.

To solve algebraically, we use the factoring method.

$$x^2 - 2x - 8 = 0$$
$$(x - 4)(x + 2) = 0 \qquad \text{Factor the trinomial.}$$
$$x - 4 = 0 \quad \text{or} \quad x + 2 = 0 \qquad \text{Zero Factor Property}$$
$$x = 4 \quad \text{or} \qquad x = -2 \qquad \text{Solve each case for } x.$$

This confirms the solutions suggested by the graphing method. The solutions should be checked by substitution.

Figure 10.2

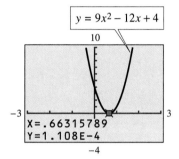

(b) Enter the function $y = 9x^2 - 12x + 4$, produce the graph, and trace to the x-intercepts.

The graph in Fig. 10.2 appears to have only one x-intercept. This means that the equation has only one solution, called a **double root,** and it is approximately 0.66.

We solve algebraically by factoring.

$$9x^2 - 12x + 4 = 0$$
$$(3x - 2)(3x - 2) = 0 \qquad \text{Factor the trinomial.}$$
$$3x - 2 = 0 \qquad \text{Zero Factor Property}$$
$$x = \frac{2}{3}$$

Because $\frac{2}{3} = 0.\overline{6}$, the exact solution and the estimated solution are very close. To verify, check the *exact* solution.

Figure 10.3

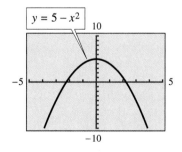

(c) Enter $y = 5 - x^2$, produce the graph, and trace to the x-intercepts.

The graph in Fig. 10.3 indicates that the equation has two solutions. They are approximately 2.24 and -2.24.

To solve algebraically, we can use the Even Root Property: If $A^2 = B$, where A and B are real numbers and $B > 0$, then $A = \sqrt{B}$ or $A = -\sqrt{B}$.

$$5 - x^2 = 0$$
$$x^2 = 5 \qquad \text{Write in the form } A^2 = B.$$
$$x = \sqrt{5} \quad \text{or} \quad x = -\sqrt{5} \qquad \text{Even Root Property}$$

These exact solutions are approximated by the values obtained from the graphing method.

Figure 10.4

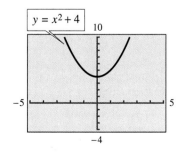

(d) Produce the graph of $y = x^2 + 4$ and observe that there are no x-intercepts. This indicates that the equation has no real number solutions.

Recall that if $B < 0$, then the Even Root Property states that an equation of the form $A^2 = B$ has no real number solutions. Thus, as the graph in Fig. 10.4 suggests, the equation $x^2 = -4$ has no real number solutions. ■

Because a graph provides a picture of all of the real number solutions of $y = ax^2 + bx + c$, we can use it to anticipate the number of solutions a quadratic equation will have, and we can obtain estimates of the solutions.

For complete accuracy, we use an algebraic method for solving the equation. As always, the best approach is a combination of the two methods.

Complex Number Solutions

In Example 1 both the graphing method and the Even Root Property showed that the equation $x^2 = -4$ has no real number solution. However, if we allow A and B to represent *complex numbers* (see Section 9.9), then we can state the following **Square Root Property.**

Square Root Property

For *complex numbers* A and B, if $A^2 = B$, then $A = \sqrt{B}$ or $A = -\sqrt{B}$.

NOTE: Because A and B represent complex numbers in the Square Root Property, the number of solutions does not depend on the sign of B, as is the case in the Even Root Property. ■

Our next example illustrates how we can use the Square Root Property to solve quadratic equations.

EXAMPLE 2 *Using the Square Root Property to Solve Quadratic Equations*
Solve each equation using the Square Root Property.

(a) $(2x + 1)^2 = 5$ (b) $y^2 + 9 = 0$

(c) $(t - 3)^2 = -5$ (d) $7 + 25(2x + 3)^2 = 0$

Solution

(a) $(2x + 1)^2 = 5$

$\qquad 2x + 1 = \pm\sqrt{5}$ Square Root Property with $B > 0$

$\qquad\qquad 2x = -1 \pm \sqrt{5}$ Solve for x.

$$x = \frac{-1 \pm \sqrt{5}}{2}$$

Recall that the $\pm$ symbol indicates two solutions—one evaluated with the plus sign, the other evaluated with the minus sign. When using your calculator to evaluate the two solutions, be sure to enclose the numerator in parentheses. Rounded to two decimal places, the solutions are 0.62 and -1.62.

(b) $y^2 + 9 = 0$

$\qquad y^2 = -9$ — Write in the form $A^2 = B$.

$\qquad y = \pm\sqrt{-9} = \pm i\sqrt{9} = \pm 3i$ — Square Root Property with $B < 0$

(c) $(t - 3)^2 = -5$

$\qquad t - 3 = \pm\sqrt{-5}$ — Square Root Property with $B < 0$

$\qquad t - 3 = \pm i\sqrt{5}$

$\qquad t = 3 \pm i\sqrt{5} \approx 3 \pm 2.24i$ — Add 3 to both sides to solve for t.

(d) $7 + 25(2x + 3)^2 = 0$ — Isolate the squared quantity.

$\qquad 25(2x + 3)^2 = -7$

$\qquad (2x + 3)^2 = -\dfrac{7}{25}$

$\qquad 2x + 3 = \pm\sqrt{-\dfrac{7}{25}}$ — Square Root Property with $B < 0$

$\qquad 2x + 3 = \pm\dfrac{i\sqrt{7}}{5}$ — Subtract 3 and divide by 2 to solve for x.

$\qquad 2x = -3 \pm \dfrac{i\sqrt{7}}{5}$

$\qquad x = -\dfrac{3}{2} \pm \dfrac{\sqrt{7}}{10}i \approx -1.5 \pm 0.26i$

10.1 *Quick Reference*

Real Number Solutions

- A **quadratic equation** is an equation that can be written in the *standard form* $ax^2 + bx + c = 0$, where a, b, and c are real numbers and $a \neq 0$.

- A **quadratic function** is a function of the form $f(x) = ax^2 + bx + c$ where a, b, and c are real numbers and $a \neq 0$.

- Three previously discussed methods for solving a quadratic equation are as follows.

  1. Graphing Method

     (a) The graphing method can be used to learn how many solutions a quadratic equation has and to estimate what the solutions are.

     (b) To use the graphing method, produce the graph of the associated function $f(x) = ax^2 + bx + c$. The x-intercepts of the graph of the quadratic function represent the solutions of the quadratic equation.

  2. Factoring Method

     (a) Write the quadratic equation in standard form and factor the polynomial.

     (b) Apply the Zero Factor Property and set each factor equal to zero.

     (c) Solve the resulting equations.

3. Even Root Property

   (a) For real numbers A and B, where $B \geq 0$, if $A^2 = B$, then $A = \sqrt{B}$ or $A = -\sqrt{B}$.

   (b) If the quadratic equation can be written in the form $A^2 = B$, apply the Even Root Property to solve.

Complex Number Solutions

- If we allow A and B to represent complex numbers, then the **Square Root Property** states that if $A^2 = B$, then $A = \sqrt{B}$ or $A = -\sqrt{B}$.

- Because A and B represent complex numbers in the Square Root Property, the number of solutions does not depend on the sign of B, as is the case in the Even Root Property. Therefore, the property is especially useful in determining complex number solutions of a quadratic equation.

10.1   *Exercises*

 1. When using the graphing method to estimate the solutions of a quadratic equation in standard form, what feature of the graph is relevant? Why?

 2. Describe how to use the graphing method to solve the quadratic equation $x^2 - 2x = 15$ without writing the equation in standard form.

In Exercises 3–14, use the graphing method to solve.

  3. $2x^2 + 5x - 3 = 0$ **4.** $3x^2 + 10x = 8$

  5. $3 - 8x = 3x^2$ **6.** $6 + x - x^2 = 0$

  7. $x^2 - 6x + 9 = 0$ **8.** $x^2 + 8x + 16 = 0$

  9. $10x = 25 + x^2$ **10.** $14x = 49 + x^2$

 11. $x^2 + 3x + 5 = 0$ **12.** $x^2 - 4x + 6 = 0$

 13. $10x - 27 - x^2 = 0$ **14.** $8x - 18 - x^2 = 0$

 15. Explain how to use the Zero Factor Property to solve an equation algebraically.

 16. Explain why it is necessary for a quadratic equation to be in standard form in order to solve it with the Zero Factor Property.

In Exercises 17–26, use the factoring method to solve.

 17. $x^2 + 7x + 6 = 0$ **18.** $4x^2 + 3 = 8x$

 19. $x^2 = x$ **20.** $25 = 10x - x^2$

 21. $\dfrac{3x}{4} = \dfrac{1}{2} + \dfrac{2}{x}$ **22.** $x = \dfrac{28}{x + 3}$

 23. $x(x - 3) = x - 3$

 24. $(3x - 1)^2 - 25x^2 = 0$

 25. $8x^2 + x - 1 = 3x^2 - 2x + 1$

 26. $2x(x + 3) = -5(3 + x)$

 27. The equations $6 - x^2 = 0$ and $6 + x^2 = 0$ are very similar, but their solution sets are quite different. Use the Even Root Property to explain why one equation has two solutions and the other equation has no real number solution.

 28. Suppose that $A^2 = B$, where $B < 0$. Describe the difference between the Even Root Property and the Square Root Property as each property applies to the solution(s) of the equation.

In Exercises 29–36, solve by using the Square Root Property.

 29. $(5x - 4)^2 = 9$ **30.** $(3x - 2)^2 = 15$

 31. $16(x + 5)^2 - 5 = 0$

 32. $9(x + 3)^2 - 4 = 0$

33. $\left(x - \dfrac{3}{2}\right)^2 = \dfrac{9}{4}$ **34.** $(4 - x)^2 = \dfrac{36}{25}$

35. $x(x - 6) = 3(3 - 2x)$

36. $x(x + 12) = 12(3 + x)$

In Exercises 37–54, determine the complex solutions of the given equation.

37. $t^2 + 16 = 0$ **38.** $25 + t^2 = 0$

39. $24 + y^2 = 0$ **40.** $y^2 + 18 = 0$

41. $(x - 3)^2 = -4$ **42.** $(x + 4)^2 = -9$

43. $(t - 1)^2 + 169 = 144$

44. $(2t + 1)^2 + 25 = 16$

45. $(2x + 5)^2 + 6 = 0$

46. $(3x + 7)^2 + 5 = 0$

47. $8(x + 1)^2 + 21 = 3$

48. $25(x + 2)^2 + 11 = 0$

49. $10 + \dfrac{4}{y^2} = 1$ **50.** $40 + \dfrac{25}{y^2} = 4$

51. $4 = \dfrac{9}{(1 - x)^2}$ **52.** $\dfrac{16}{(2 - x)^2} = 49$

53. $x(x + 4) = -4(1 - x)$

54. $x(8 - x) = 8(2 + x)$

 55. The accompanying figure shows the graph of $y = x^2 + 3x + 4$. According to the graph, how many real number solutions are there for the quadratic equation $x^2 + 3x + 4 = 0$? Why? Does this mean that the equation has no solution? Explain.

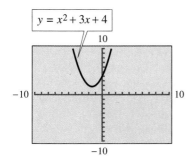

 56. When solving a quadratic equation, how can we recognize when to use the factoring method and when to use the Square Root Property?

In Exercises 57–78, solve.

57. $3x = x^2$ **58.** $12x^2 = 21 + 4x$

59. $t^2 + 45 = 0$ **60.** $x^2 + 20 = 0$

61. $36x^2 + 12x + 1 = 0$

62. $9 - 6t + t^2 = 0$

63. $y^2 = -1.44$ **64.** $x^2 = 1.21$

65. $x(x - 5) = 5(2 - x)$

66. $x^2 + 6x = 6(3 + x)$

67. $x - 2 = \dfrac{35}{x}$ **68.** $\dfrac{7}{x} + 4 = \dfrac{15}{x^2}$

69. $(4x + 5)^2 + 49 = 0$

70. $(3x - 4)^2 + 36 = 0$

71. $\left(\dfrac{3x + 2}{4}\right)^2 = 12$

72. $\left(\dfrac{x - 1}{2}\right)^2 = 6$

73. $t(t + 3) = 3(t - 2)$

74. $5(x^2 + 15) = 2x^2$

75. $x(x - 4) = x - 4$

76. $6x^2 + 9x + 3 = 1 - 4x^2$

77. $(4x - 1)(x - 1) = 7x - 8$

78. $(4x + 1)(5x + 1) = x(-5x - 1)$

Given two numbers, we can write a quadratic equation for which the numbers are solutions. For example, if 2 and −5 are solutions, then

$$x = 2 \quad \text{or} \quad x = -5$$
$$x - 2 = 0 \quad \text{or} \quad x + 5 = 0$$
$$(x - 2)(x + 5) = 0$$
$$x^2 + 3x - 10 = 0$$

Note that the equation is not unique. Multiplying both sides by any nonzero number produces an equivalent equation.

In Exercises 79–86, write a quadratic equation with integer coefficients that has the given pair of numbers as solutions.

79. $3, 5$ **80.** $-1, -2$

81. $3, 3$ **82.** $0, 0$

83. $3, \dfrac{4}{3}$

84. $\dfrac{2}{3}, \dfrac{3}{4}$

85. $-4i, 4i$

86. $i\sqrt{3}, -i\sqrt{3}$

In Exercises 87–96, write an equation that describes the conditions of the problem. Then, solve the equation and answer the question.

87. The product of two consecutive negative integers is 72. What are the two integers?

88. A book lies open so that the product of the facing page numbers is 1806. To what possible pages is the book open?

89. The product of two consecutive odd integers is 24. Try to determine the two integers and interpret the result.

90. Two less than the square of a person's age is 5 less than the person's age. Using the graphing method, determine the person's age.

91. A physical education major is doing his student teaching at an elementary school. One day, he lines up his fifth-grade students and asks them to count off by even numbers beginning with 2. Standing side by side, two students have numbers whose product is 360. What possible numbers could they have?

92. If an object is projected vertically upward, its height h after t seconds is given by $h = 96t - 16t^2$. After how many seconds will the object return to the ground?

93. A sentry walks back and forth from his guard house to the end of a building. His distance d from the guard house at any given time t (in minutes) is given by $d = 18t^2 - 90t$. How long does it take the sentry to complete one round trip?

94. The length of a rectangle is 1 inch less than twice the width. If the area is 45 square inches, what are the dimensions of the rectangle?

95. The height of a triangle is one-third the length of the base. If the area of the triangle is 24 square feet, what is the height?

96. Two trucks left a truck stop together. One traveled north and the other traveled east. When the trucks were 50 miles apart, the eastbound truck had traveled 10 miles farther than the northbound truck. How far had each truck traveled?

Exploring with Real Data

In Exercises 97–100, refer to the following background information and data.

During the period 1970–1985, health-conscious Americans began to reduce their use of sugar. The following table shows the annual per-person consumption of sugar during those years.

	1970	1975	1980	1985
Annual Per-Person Consumption of Sugar (in pounds)	101.8	89.2	83.6	62.7

(Source: Economic Research Service, USDA.)

One function that models the annual per-person sugar consumption S is $S(t) = -0.16t^2 - 0.22t + 101.8$, where t is the number of years since 1970.

97. Produce a graph of the function S and estimate the coordinates of the highest point of the graph.

98. According to the model, when did per-person sugar consumption appear to reach its peak?

99. Use the graph to estimate sugar consumption in 1990 and to estimate the year when sugar consumption would fall to zero. (The actual reported figure for 1990 was 64.2 pounds.) What is your opinion of the function as a model for years that are not in the 1970–1985 period?

100. Assuming that Americans' craving for sweetness did not actually diminish during the reported years, how might the drop in sugar consumption be explained?

Challenge

In Exercises 101–108, solve for x. Assume all variables (other than x) represent positive numbers.

101. $16c^2 = 25x^2$

102. $16x^2 = -25y$

103. $(x + 2)^2 = -y$

104. $y = (x + 3)^2$

105. $(x + a)^2 = 9$

106. $(x + c)^2 = -16$

107. $(ax + 1)^2 + 25 = 9$

108. $(x - b)^2 - 20 = 16$

In Exercises 109–112, use the Zero Factor Property to solve.

109. $(x - 1 + i)(x - 1 - i) = 0$

110. $(x - 2 - 3i)(x - 2 + 3i) = 0$

111. $\left(x - 2 - \sqrt{7}\right)\left(x - 2 + \sqrt{7}\right) = 0$

112. $\left(x - 1 + \sqrt{2}\right)\left(x - 1 - \sqrt{2}\right) = 0$

In Exercises 113–116, use the Square Root Property to solve.

113. $(2x + 5)^2 = (x - 1)^2$

114. $(3 - x)^2 = (1 - 3x)^2$

115. $(2x - 7)^2 = (7 - 2x)^2$

116. $(2x + 1)^2 = (3 + 2x)^2$

10.2 Completing the Square

Square of a Binomial ▪ *Solving Quadratic Equations* ▪ *Complex Solutions*

Square of a Binomial

In the previous section we saw how easily the Square Root Property can be used to solve equations of the form $A^2 = B$. In fact, the method works so well that we might consider trying to write a quadratic equation in the form $A^2 = B$, even if the equation is not originally in that form.

The process of converting a quadratic equation into the form $A^2 = B$ is called **completing the square.** The method is based on the familiar special products for the square of a binomial.

$$(A + B)^2 = A^2 + 2AB + B^2$$
$$(A - B)^2 = A^2 - 2AB + B^2$$

To solve an equation by completing the square, we rewrite the equation so that the polynomial is in the form $A^2 + 2AB + B^2$ or $A^2 - 2AB + B^2$. As we shall see, to write the equation in the desired form, we need to be able to determine the last term of a perfect square trinomial.

EXPLORATION 1   *Completing the Square*

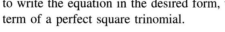

(a) Suppose the first portion of a perfect square trinomial is $x^2 + 8x$. Compare this expression to the model for a perfect square trinomial.

$$x^2 + 8x + \underline{\ ?\ }$$
$$x^2 + 2Bx + B^2$$

For what value of B will the coefficients of the x-terms be equal?

(b) For this value of B, what is B^2? Now write the rest of the trinomial expression with the constant term filled in.

(c) Propose a method for determining the constant term of a perfect square trinomial without referring specifically to the model.

Discovery

(a) For the coefficients of the x terms to be equal, $2B = 8$ and so $B = 4$.

(b) If $B = 4$, then $B^2 = 16$. The perfect square trinomial is $x^2 + 8x + 16$.

(c) The constant term of a perfect square trinomial can be found by taking half the coefficient of the x term and squaring it.

Exploration 1 leads to the following generalization.

Completing a Perfect Square Trinomial

If the first part of a perfect square trinomial is $x^2 + bx$, then the entire perfect square trinomial is

$$x^2 + bx + \left(\frac{b}{2}\right)^2.$$

In words, the constant term is the square of half the coefficient of x.

EXAMPLE 1   *Completing a Perfect Square Trinomial*

Write a perfect square trinomial whose first two terms are given.

(a) $x^2 + 12x$ (b) $x^2 - 5x$

Solution

(a) Half the coefficient of x is 6 and $6^2 = 36$. Thus, the constant term is 36 and the trinomial is $x^2 + 12x + 36$.

(b) Half the coefficient of x is $-\frac{5}{2}$ and $\left(-\frac{5}{2}\right)^2 = \frac{25}{4}$. The constant term is $\frac{25}{4}$ and the trinomial is $x^2 - 5x + \frac{25}{4}$. ∎

Solving Quadratic Equations

To use the method of completing the square to solve an equation, we begin by rewriting the equation with the constant on the right side. Then, on the left side, we complete the square and use the Square Root Property to solve the equation.

NOTE: We can use this method of completing the square to solve any equation in the form $ax^2 + bx + c = 0$. However, if the leading coefficient a is not 1, then it will be necessary to divide both sides of the equation by a before we begin the process of completing the square. ∎

EXAMPLE 2 *Solving Quadratic Equations by the Method of Completing the Square*

Solve each equation by completing the square.

(a) $x^2 + 5x + 2 = 0$

(b) $x(2x + 3) = 6$

(c) $x^2 + 9x + 8 = 0$

Solution

(a) $x^2 + 5x + 2 = 0$

$x^2 + 5x = -2$ | Move the constant term to the right side.

$x^2 + 5x + \dfrac{25}{4} = -2 + \dfrac{25}{4}$ | Add $\left(\dfrac{5}{2}\right)^2 = \dfrac{25}{4}$ to both sides to complete the square.

$x^2 + 5x + \dfrac{25}{4} = -\dfrac{8}{4} + \dfrac{25}{4}$ | The LCD on the right side is 4.

$x^2 + 5x + \dfrac{25}{4} = \dfrac{17}{4}$ | Add the fractions on the right.

$\left(x + \dfrac{5}{2}\right)^2 = \dfrac{17}{4}$ | Factor and write in exponential form.

$x + \dfrac{5}{2} = \pm\dfrac{\sqrt{17}}{2}$ | Square Root Property

$x = -\dfrac{5}{2} \pm \dfrac{\sqrt{17}}{2}$ | Subtract $\dfrac{5}{2}$ from both sides to solve for x.

$x \approx -0.44$ or $x \approx -4.56$

(b) $x(2x + 3) = 6$

$2x^2 + 3x = 6$ | Remove parentheses.

$x^2 + \dfrac{3}{2}x = 3$ | Divide both sides by 2.

$x^2 + \dfrac{3}{2}x + \dfrac{9}{16} = \dfrac{3}{1} + \dfrac{9}{16}$ | Add $\left(\dfrac{1}{2} \cdot \dfrac{3}{2}\right)^2 = \left(\dfrac{3}{4}\right)^2 = \dfrac{9}{16}$ to both sides to complete the square.

$\left(x + \dfrac{3}{4}\right)^2 = \dfrac{57}{16}$ | Add the fractions and factor.

$x + \dfrac{3}{4} = \pm\dfrac{\sqrt{57}}{4}$ | Square Root Property

$x = -\dfrac{3}{4} \pm \dfrac{\sqrt{57}}{4}$ | Subtract $\dfrac{3}{4}$ from both sides to solve for x.

$x \approx 1.14$ or $x \approx -2.64$

(c) $x^2 + 9x + 8 = 0$

$x^2 + 9x \quad\quad = -8$ ⟶ Move the constant term to the right side.

$x^2 + 9x + \dfrac{81}{4} = \dfrac{-8}{1} + \dfrac{81}{4}$ ⟶ Add $\left(\dfrac{9}{2}\right)^2 = \dfrac{81}{4}$ to both sides to complete the square.

$\left(x + \dfrac{9}{2}\right)^2 = \dfrac{49}{4}$ ⟶ Add the fractions and factor.

$x + \dfrac{9}{2} = \pm\sqrt{\dfrac{49}{4}}$ ⟶ Square Root Property

$x + \dfrac{9}{2} = \pm\dfrac{7}{2}$

$x = -\dfrac{9}{2} \pm \dfrac{7}{2}$ ⟶ Subtract $\dfrac{9}{2}$ from both sides to solve for x.

$x = -1 \quad \text{or} \quad x = -8$

NOTE: The equation in part (c) of Example 2 could have been solved by factoring. We used the method of completing the square to illustrate that the method is applicable to any quadratic equation.

Complex Solutions

Recall that the graphing method for solving a quadratic equation will reveal only the *real number* solutions of the equation. If we apply the Square Root Property to solving the equation $A^2 = B$ for $B < 0$, the solutions will be *imaginary numbers*. Although such solutions cannot be found with the graphing method, they can be determined with the method of completing the square.

EXAMPLE 3   *Quadratic Equations with Complex Solutions*

Determine the solutions of each equation.

(a) $x^2 + 4x + 5 = 0$

(b) $3x^2 - 2x + 1 = 0$

Solution

(a) Figure 10.5 shows the graph of $f(x) = x^2 + 4x + 5$. Because the graph has no x-intercepts, we know that the equation has no real number solutions.

Now we solve algebraically by completing the square.

$x^2 + 4x + 5 = 0$

$x^2 + 4x = -5$ ⟶ Move the constant term to the right side.

$x^2 + 4x + 4 = -5 + 4$ ⟶ Add $\left(\dfrac{1}{2} \cdot 4\right)^2 = 4$ to both sides.

$x^2 + 4x + 4 = -1$

$(x + 2)^2 = -1$ ⟶ Factor and write in exponential form.

$x + 2 = \pm\sqrt{-1}$ ⟶ Square Root Property

$x + 2 = \pm i$

$x = -2 \pm i$ ⟶ Subtract 2 from both sides to solve for x.

Figure 10.5

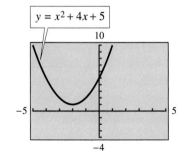
$y = x^2 + 4x + 5$

(b) $3x^2 - 2x + 1 = 0$

$$3x^2 - 2x = -1 \qquad \text{Move the constant term to the right side.}$$

$$x^2 - \frac{2}{3}x = -\frac{1}{3} \qquad \text{Divide both sides by 3.}$$

$$x^2 - \frac{2}{3}x + \frac{1}{9} = -\frac{1}{3} + \frac{1}{9} \qquad \text{Add } \left(\frac{1}{2} \cdot \frac{2}{3}\right)^2 = \frac{1}{9} \text{ to both sides.}$$

$$x^2 - \frac{2}{3}x + \frac{1}{9} = -\frac{3}{9} + \frac{1}{9} \qquad \text{The LCD on the right side is 9.}$$

$$\left(x - \frac{1}{3}\right)^2 = -\frac{2}{9} \qquad \text{Add the fractions and factor.}$$

$$x - \frac{1}{3} = \pm\sqrt{-\frac{2}{9}} \qquad \text{Square Root Property}$$

$$x - \frac{1}{3} = \pm\frac{i\sqrt{2}}{3}$$

$$x = \frac{1}{3} \pm \frac{\sqrt{2}}{3}i \approx 0.33 \pm 0.47i$$

10.2 *Quick Reference*

Square of a Binomial

- If the first part of a quadratic trinomial is $x^2 + bx$, then the entire perfect square trinomial is

$$x^2 + bx + \left(\frac{b}{2}\right)^2.$$

- In words, to determine the constant needed to form a perfect square trinomial (with 1 as the leading coefficient), take half the coefficient of x and square the result. This process is called **completing the square.**

Solving Quadratic Equations

- To solve a quadratic equation $ax^2 + bx + c = 0$ by completing the square, follow these steps.

   1. Rewrite the equation with the constant term on the right side.

   2. If $a \neq 1$, divide both sides of the equation by a.

   3. Take half of the coefficient of x, square it, and add the result to both sides of the equation.

   4. Simplify both sides of the equation.

   5. Factor the polynomial on the left side and write the result in exponential form.

   6. Use the Square Root Property to write equivalent first-degree equations and solve each one.

Complex Solutions

- The method of completing the square could result in an equation of the form $A^2 = B$ where $B < 0$. In this case the solutions will be imaginary numbers.

10.2 Exercises

1. To solve a quadratic equation $ax^2 + bx + c = 0$ by the method of completing the square, we rewrite the equation in the form $A^2 = B$. Describe the form of the expression A.

2. For the expression $x^2 + bx + k$, describe in words how to determine k in order to complete the square.

In Exercises 3–8, determine k so that the trinomial is a perfect square.

3. $x^2 + 6x + k$ **4.** $x^2 - 8x + k$

5. $x^2 + 7x + k$ **6.** $x^2 - 9x + k$

7. $x^2 + \dfrac{4}{3}x + k$ **8.** $x^2 - \dfrac{5}{2}x + k$

9. Two students are solving $x^2 - 8x = -12$. One student rewrites the equation in standard form and solves with the factoring method. The other student adds 16 to both sides to complete the square and then solves with the Square Root Property. Which method is correct? Why?

10. Two students are solving $x^2 + x = -1$. One student rewrites the equation in standard form and tries to solve with the factoring method. The other student tries to solve with the method of completing the square. Which student will be successful? Why?

In Exercises 11–18, solve by factoring and then solve again by completing the square.

11. $x^2 + 4x + 3 = 0$

12. $x^2 + 6x + 8 = 0$

13. $t^2 + 2t = 15$ **14.** $t^2 = 6t + 55$

15. $x^2 - x - 12 = 0$

16. $x^2 + 5x - 6 = 0$

17. $6 = x^2 + x$ **18.** $x^2 = 5x + 14$

In Exercises 19–24, solve by completing the square.

19. $x^2 - 6x + 4 = 0$ **20.** $x^2 + 2x = 5$

21. $x^2 + 2x = 24$ **22.** $x^2 - 4x = 5$

23. $x^2 = 4x + 4$ **24.** $x^2 = 4x + 8$

25. If you want to solve the equation $3x^2 - 5x = 7$ with the method of completing the square, what is the first step you must take?

26. Although we regard the factoring method and the method of completing the square as two different methods for solving a quadratic equation, the method of completing the square does include factoring. At what step of the procedure do we need to factor?

In Exercises 27–30, solve by factoring and then solve again by completing the square.

27. $3y^2 + 4y + 1 = 0$ **28.** $5y^2 + 6y = 8$

29. $15 = 2x^2 + 7x$ **30.** $6x^2 + 3 = 11x$

In Exercises 31–38, solve by completing the square.

31. $4x^2 = 8x + 21$ **32.** $4x^2 = 27 + 12x$

33. $4x^2 + 25 = 20x$ **34.** $16x^2 + 9 = 24x$

35. $x(2x - 3) = 1$ **36.** $4x^2 = 3(x + 1)$

37. $2x^2 + 7x - 1 = 0$

38. $4x^2 + 5x - 3 = 0$

39. Determine k so that $kx^2 + 6x + 1$ is a perfect square.

40. Determine k so that $4x^2 + kx + 25$ is a perfect square.

In Exercises 41–48, determine the imaginary number solutions by completing the square.

41. $x^2 - 4x + 8 = 0$ **42.** $x^2 + 6x + 10 = 0$

43. $x^2 + 6 = 4x$ **44.** $x^2 + 7 = 2x$

45. $x^2 + 3x + 5 = 0$ **46.** $x^2 + 5x + 8 = 0$

47. $2x^2 + x + 3 = 0$

48. $3x^2 + 2x + 7 = 0$

49. Show that the solutions of $x^2 + bx = k$ are imaginary numbers if

$$k < -\left(\frac{b}{2}\right)^2.$$

50. For what value of k does $x^2 + bx = k$ have only one solution?

In Exercises 51–74, solve the equation.

51. $x^2 + 16 = 10x$

52. $2x^2 + x - 6 = 0$

53. $36x^2 - 49 = 0$

54. $\frac{1}{7}y^2 = 49$

55. $x(x - 3) = 10$

56. $b(b + 5) = 36$

57. $4(x + 3)^2 - 7 = 0$

58. $(3x - 2)^2 = 16$

59. $x(x + 1) + 9 = 0$

60. $2a(a + 2) = 3$

61. $(x - 3)(x - 1) = 5x - 6$

62. $(2x - 1)(8x - 1) = -1 - 2x$

63. $x^2 + \frac{3}{2}x = \frac{3}{4}$

64. $x^2 + \frac{4}{3}x = \frac{5}{9}$

65. $x^2 = x - 9$

66. $x^2 = x + 4$

67. $4x^2 + 7x = 3$

68. $3x^2 - x + 8 = 0$

69. $3x^2 + x = \frac{2}{3}$

70. $0.5x^2 - 2x = 1.75$

71. $(x - 2)^2 = -3(x - 2)$

72. $(x + 1)^2 = 3(x + 2)$

73. $4x^2 + 9 = 2x$

74. $2x^2 + 9 = 5x$

In Exercises 75–80, one solution is given. Determine a, b, or c and then determine another solution.

75. $ax^2 + x - 5 = 0$; 1

76. $ax^2 + 9x - 5 = 0$; $\frac{1}{3}$

77. $6x^2 + bx - 3 = 0$; $\frac{1}{3}$

78. $5x^2 + bx - 3 = 0$; 1

79. $x^2 + 2x + c = 0$; 6

80. $2x^2 + 7x + c = 0$; $-\frac{1}{2}$

In Exercises 81–88, use the method of completing the square to solve the application.

81. The sum of the reciprocal of an integer and the square of the reciprocal is $\frac{6}{25}$. What is the integer?

82. The product of the reciprocal of a number and 20 is 8 less than the number. What two numbers satisfy this condition?

83. The square of a positive number is 3 more than 5 times the number. What is the number? (Round to two decimal places.)

84. Seven times a negative number is 1 less than the square of the number. What is the number? (Round to two decimal places.)

85. A platform is in the shape of a right triangle. If the lengths of the legs of the triangle differ by 8 feet, and the length of the longest side of the platform is 50 feet, how long is the shortest side of the platform? (Round to the nearest foot.)

86. Taxes on commercial property have increased by $1.00 per frontage foot along Hayes Street. A store is on a lot in the shape of a right triangle with one leg along Hayes Street. The frontage along Minnow Road is 100 feet. (See figure.)

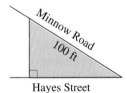

Hayes Street

If the Hayes Street frontage is 30 feet more than twice the length of the other leg of the lot, what is the amount of the tax increase? (Round to the nearest dollar.)

87. A woman uses a rototiller to till her garden. Her neighbor has a larger tiller and could do the job in 30 minutes less time. How long does it take the woman to till the garden if she and her neighbor, working together, can complete the job in 1 hour? (Round to the nearest minute.)

88. A bricklayer can complete a job in 2 hours less time than it takes his apprentice. Working together, the two of them can complete the job in 5 hours. How long would it take the apprentice working alone? (Round to the nearest hour.)

Exploring with Real Data

In Exercises 89–92, refer to the following background information and data.

In the decade of the 1980s, there was a sharp increase in the number of convenience stores. The figure shows the data for selected years between 1980 and 1990. (Source: Economic Research Service, USDA.)

Figure for 89–92

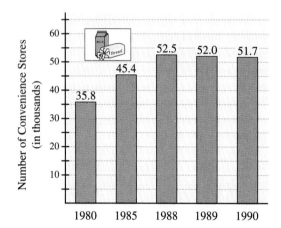

Using only the data from 1985–1990, we can write the quadratic function $N(t) = -0.55t^2 + 9.56t + 11.43$ as a model, where $N(t)$ is the number of stores (in thousands) and t is the number of years since 1980.

89. Produce a graph of the quadratic function and trace to estimate the x-intercepts.

90. Use your calculator to solve the quadratic equation $-0.55t^2 + 9.56t + 11.43 = 0$ by the method of completing the square. Round your results to the nearest hundredth and compare the results to the estimates in Exercise 89.

91. What years are represented by the x-intercepts? What is the interpretation of these points? Do you think the model is valid at these points?

92. The graph in the figure indicates some decline in the number of convenience stores starting in 1989. Name one social condition that might have contributed to this.

Challenge

In Exercises 93–98, use the method of completing the square to solve for x.

93. $x^2 + 10x + c = 0$

94. $x^2 + 4bx + 8 = 0$

95. $x^2 - 2cx + 4c^2 = 0$

96. $x^2 + bx + b^2 = 0$

97. $ax^2 + 2x + a = 0, a \neq 0$

98. $ax^2 + 6x + 9 = 0, a \neq 0$

In Exercises 99–102, solve by completing the square.

99. $x^2 + 2\sqrt{5}x - 4 = 0$

100. $x^2 - 6\sqrt{3}x + 23 = 0$

101. $x^2 + 4\sqrt{2}x - 10 = 0$

102. $x^2 + 2\sqrt{3}x - 9 = 0$

In Exercises 103–106, solve for x. Determine the values of k for which (a) the solutions will be real numbers and (b) the solutions will be imaginary numbers.

103. $x^2 + kx + 1 = 0$

104. $x^2 + 2x + k = 0$

105. $x^2 + kx + k^2 = 0$

106. $x^2 + 4kx - 4 = 0$

In Exercises 107–110, use the method of completing the square to solve the equation.

107. $x^2 + 4ix - 8 = 0$

108. $x^2 - 8ix - 25 = 0$

109. $x^2 + 6ix + 7 = 0$ **110.** $x^2 - 2ix + 3 = 0$

111. The method of completing the square can be illustrated geometrically. Consider the rectangle in the accompanying figure.

(a) Find the total area of the figure by adding the areas of the pieces. Simplify the result.

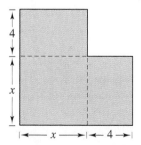

(b) What area needs to be added to "complete the square"? That is, what is the area of the missing corner in the figure?

(c) Find the area of the complete square by adding the areas found in parts (a) and (b).

(d) Use the fact that the lengths of the sides of the complete square are $x + 4$ to calculate the area of the complete square. Do not simplify.

(e) Compare the results in parts (c) and (d).

10.3 The Quadratic Formula

Derivation and Use ▪ *Complex Solutions* ▪ *The Discriminant*

Derivation and Use

Each time we solve a quadratic equation with the method of completing the square, we follow the same steps. By performing the steps just once for the general equation $ax^2 + bx + c = 0$, we can derive a formula for finding the solutions of any quadratic equation.

We begin with the quadratic equation written in standard form, and we complete the square.

$$ax^2 + bx + c = 0$$

$$ax^2 + bx = -c \qquad \text{Move the constant term to the right side.}$$

$$x^2 + \frac{b}{a}x = -\frac{c}{a} \qquad \text{Leading coefficient must be 1. Divide both sides by } a.$$

$$x^2 + \frac{b}{a}x + \frac{b^2}{4a^2} = -\frac{c}{a} + \frac{b^2}{4a^2} \qquad \begin{array}{l}\text{To complete the square, add} \\ \left(\frac{1}{2} \cdot \frac{b}{a}\right)^2 = \frac{b^2}{4a^2} \text{ to both sides.}\end{array}$$

$$x^2 + \frac{b}{a}x + \frac{b^2}{4a^2} = \frac{-4ac}{4a^2} + \frac{b^2}{4a^2} \qquad \text{The LCD on the right side is } 4a^2.$$

$$x^2 + \frac{b}{a}x + \frac{b^2}{4a^2} = \frac{b^2 - 4ac}{4a^2} \qquad \text{Combine the fractions on the right side.}$$

$$\left(x + \frac{b}{2a}\right)^2 = \frac{b^2 - 4ac}{4a^2} \qquad \text{Factor and write in exponential form.}$$

$$x + \frac{b}{2a} = \frac{\pm\sqrt{b^2 - 4ac}}{2a} \qquad \text{Square Root Property}$$

$$x = \frac{-b}{2a} \pm \frac{\sqrt{b^2 - 4ac}}{2a} \qquad \text{Subtract } \frac{b}{2a} \text{ to solve for } x.$$

$$x = \frac{-b \pm \sqrt{b^2 - 4ac}}{2a}$$

Combine the fractions.

This formula is called the **Quadratic Formula.**

> ### The Quadratic Formula
>
> For real numbers a, b, and c, with $a \neq 0$, the solutions of the quadratic equation $ax^2 + bx + c = 0$ are given by the **Quadratic Formula**
>
> $$x = \frac{-b \pm \sqrt{b^2 - 4ac}}{2a}.$$

NOTE: A quadratic equation must be written in standard form before the Quadratic Formula can be used to solve it. ∎

EXAMPLE 1   *Using the Quadratic Formula: Two Rational Number Solutions*
Use the Quadratic Formula to solve $3x^2 - 13x = 30$.

Solution

$$3x^2 - 13x = 30$$

$$3x^2 - 13x - 30 = 0 \qquad \text{Write the equation in standard form.}$$

Substitute $a = 3$, $b = -13$, and $c = -30$ into the Quadratic Formula.

$$x = \frac{-b \pm \sqrt{b^2 - 4ac}}{2a}$$

$$x = \frac{(-13) \pm \sqrt{(-13)^2 - 4(3)(-30)}}{2(3)} \qquad \text{Note that the opposite of } b \text{ is 13.}$$

$$x = \frac{13 \pm \sqrt{169 + 360}}{6} \qquad \text{Simplify the radicand first.}$$

$$x = \frac{13 \pm \sqrt{529}}{6} \qquad \text{Note that 529 is a perfect square.}$$

$$x = \frac{13 \pm 23}{6} \qquad \begin{array}{l}\text{Use the } + \text{ sign for one solution}\\ \text{and the } - \text{ sign for the other}\\ \text{solution.}\end{array}$$

$$x = 6 \quad \text{or} \quad x = -\frac{5}{3}$$

∎

The equation in Example 1 has two real number solutions, and the equation could have been solved by factoring. Note that after substituting into the Quadratic Formula and simplifying, the radicand is a perfect square.

EXAMPLE 2   *Using the Quadratic Formula: Double Root*
Use the Quadratic Formula to solve $16t^2 + 9 = 24t$.

Solution

$$16t^2 + 9 = 24t$$

$$16t^2 - 24t + 9 = 0 \qquad \text{Write the equation in standard form.}$$

Substitute $a = 16$, $b = -24$, and $c = 9$ into the Quadratic Formula.

$$t = \frac{24 \pm \sqrt{(-24)^2 - 4(16)(9)}}{2(16)}$$

$$t = \frac{24 \pm \sqrt{576 - 576}}{32} \qquad \text{Simplify the radicand first.}$$

$$t = \frac{24 \pm 0}{32} = \frac{24}{32} = \frac{3}{4} \qquad \text{Note that the radicand is 0.}$$

The equation in Example 2 has only one distinct solution. The equation could have been solved by factoring, which would have revealed $\frac{3}{4}$ to be a double root. Because the radicand was 0, the $\pm$ symbol instructed us to add 0 for one root and subtract 0 for the other root. Of course, the result is the same in both cases.

EXAMPLE 3 *Using the Quadratic Formula: Two Irrational Number Solutions*
Use the Quadratic Formula to solve $3x^2 - 10 = 0$.

Solution

$$3x^2 - 10 = 0 \qquad \text{Note that } b = 0.$$

Substitute $a = 3$, $b = 0$, and $c = -10$ into the Quadratic Formula.

$$x = \frac{0 \pm \sqrt{0^2 - 4(3)(-10)}}{2(3)}$$

$$x = \frac{0 \pm \sqrt{0 + 120}}{6}$$

$$x = \pm \frac{\sqrt{120}}{6} \qquad \text{Note that 120 is not a perfect square.}$$

$$x \approx 1.83 \quad \text{or} \quad x \approx -1.83$$

The equation in Example 3 has two irrational number solutions and could have been solved with the Square Root Property. The fact that the radicand is not a perfect square indicates that the factoring method could not have been used.

EXAMPLE 4 *Using the Quadratic Formula: Two Irrational Number Solutions*
Use the Quadratic Formula to solve $4x = x^2 + 1$.

Solution

$$4x = x^2 + 1$$

$$-x^2 + 4x - 1 = 0 \qquad \text{Write the equation in standard form.}$$

Substitute $a = -1$, $b = 4$, and $c = -1$ into the Quadratic Formula.

$$x = \frac{-4 \pm \sqrt{4^2 - 4(-1)(-1)}}{2(-1)}$$

$$x = \frac{-4 \pm \sqrt{16 - 4}}{-2}$$

$$x = \frac{-4 \pm \sqrt{12}}{-2}$$

$$x \approx 0.27 \quad \text{or} \quad x \approx 3.73$$

The equation in Example 4 has two irrational number solutions. The factoring method could not have been used. The equation could have been solved by completing the square, but the Quadratic Formula was derived from that method.

Complex Solutions

In each of the previous examples, after substituting a, b, and c into the Quadratic Formula, the radicand was never negative. Therefore, the solutions were always real numbers.

For other equations, it is possible for the radicand in the Quadratic Formula to be negative. To simplify the radicals with negative radicands, recall that $\sqrt{-n} = i\sqrt{n}$ for positive n. In such cases the solutions are imaginary numbers.

The Quadratic Formula can be used to determine all complex number solutions of any quadratic equation.

EXAMPLE 5   *Using the Quadratic Formula: Complex Number Solutions*

Use the Quadratic Formula to solve each equation.

(a) $3x^2 + 2x + 1 = 0$ (b) $3v^2 + 10 = 0$

Solution

(a) $3x^2 + 2x + 1 = 0$

Substitute $a = 3$, $b = 2$, and $c = 1$ into the Quadratic Formula.

$$x = \frac{-2 \pm \sqrt{2^2 - 4(3)(1)}}{2(3)}$$

$$x = \frac{-2 \pm \sqrt{4 - 12}}{6}$$

$$x = \frac{-2 \pm \sqrt{-8}}{6} \qquad \text{The radicand is negative.}$$

$$x = \frac{-2 \pm i\sqrt{8}}{6}$$

$$x = \frac{-2}{6} \pm \frac{\sqrt{8}}{6}i \approx -0.33 \pm 0.47i$$

(b) $3v^2 + 10 = 0$

Substitute $a = 3$, $b = 0$, and $c = 10$ into the Quadratic Formula.

$$v = \frac{0 \pm \sqrt{0^2 - 4(3)(10)}}{2(3)}$$

$$v = \frac{0 \pm \sqrt{0 - 120}}{6}$$

$$v = \frac{0 \pm \sqrt{-120}}{6} \qquad \text{The radicand is negative.}$$

$$v = \pm \frac{i\sqrt{120}}{6} \approx \pm 1.83i$$

In both equations in Example 5, the radicands are negative, and the solutions are complex conjugates.

The Discriminant

In the Quadratic Formula the radicand is $b^2 - 4ac$. This expression is called the **discriminant.** The discriminant provides information about the number and type of solutions of a quadratic equation with rational coefficients.

Information Obtained from the Discriminant $b^2 - 4ac$

Description of Discriminant	Description of Solutions	
	Number of Solutions	*Type of Solutions*
Positive perfect square (Example 1)	2	Rational
Positive and not a perfect square (Examples 3 and 4)	2	Irrational
Zero (Example 2)	1	Rational
Negative (Example 5)	2	Complex conjugates

The discriminant can be calculated quickly and can be used to guide us in deciding which solution method to use. Moreover, application problems, for which imaginary number solutions are not often acceptable, can sometimes be solved immediately just by determining the discriminant.

EXAMPLE 6   *Using the Discriminant to Characterize Solutions of Quadratic Equations*

Use the discriminant to determine the number and type of the solutions of each equation. Do not solve.

(a) $3x^2 + 1 = 2x$ (b) $x(5x + 1) = 12$

(c) $2 + 3x + x^2 = 0$ (d) $25x^2 - 20x + 4 = 0$

Solution

(a) $3x^2 + 1 = 2x$

$3x^2 - 2x + 1 = 0$ Write the equation in standard form.

$b^2 - 4ac = (-2)^2 - 4(3)(1) = 4 - 12 = -8$

Because the discriminant is negative, the solutions of the equation are complex conjugates.

(b) $x(5x + 1) = 12$

$5x^2 + x - 12 = 0$ Simplify and write the equation in standard form.

$b^2 - 4ac = 1^2 - 4(5)(-12) = 1 + 240 = 241$

The discriminant is positive, but it is not a perfect square. Thus, the equation has two irrational solutions.

(c) $2 + 3x + x^2 = 0$

$x^2 + 3x + 2 = 0$

$b^2 - 4ac = 3^2 - 4(1)(2) = 9 - 8 = 1$

Because the discriminant is positive and a perfect square, the equation has two rational solutions.

(d) $25x^2 - 20x + 4 = 0$

$b^2 - 4ac = (-20)^2 - 4(25)(4) = 400 - 400 = 0$

The discriminant is zero. Thus, the equation has one rational solution (double root). ■

We now have five methods available for solving a quadratic equation: the graphing method, the factoring method, the Square Root Property, the method of completing the square, and the Quadratic Formula.

Graphing is an excellent method for estimating solutions and for determining the number of solutions. When exact solutions are required, an algebraic method is necessary, but producing the graph is still a useful visual aid.

When solving an equation algebraically, choose the method that is easiest for that particular equation. Calculating the discriminant can assist in selecting a solution method. The following are some other criteria.

If	*Use*
$b = 0$	Square Root Property
$c = 0$	Factoring method
$b, c \neq 0$	Factoring method if it is easy; otherwise use the Quadratic Formula

The Quadratic Formula can be used to solve any quadratic equation, but the other methods are usually more efficient when they can be used. Having a calculator to perform the arithmetic, however, makes the Quadratic Formula an appealing choice.

10.3 *Quick Reference*

Derivation and Use

- The **Quadratic Formula** for solving a quadratic equation is derived by applying the method of completing the square to the general quadratic equation $ax^2 + bx + c = 0$. The formula is

$$x = \frac{-b \pm \sqrt{b^2 - 4ac}}{2a}.$$

- To use the Quadratic Formula to solve a quadratic equation,

  1. write the equation in standard form,

  2. substitute the values of a, b, and c into the formula, and

  3. evaluate the resulting numerical expression.

Complex Solutions

- When the Quadratic Formula is used to solve a quadratic equation, it is possible for the radicand to be negative. In such cases, the solutions are imaginary numbers.

The Discriminant

- In the Quadratic Formula, the expression $b^2 - 4ac$ is called the **discriminant.**

- If a, b, and c are rational numbers, then the value of the discriminant determines the number and type of solutions as follows.

  1. If the discriminant is positive and

     (a) a perfect square, there are two rational solutions;

     (b) not a perfect square, there are two irrational solutions.

  2. If the discriminant is zero, there is one rational solution (double root).

  3. If the discriminant is negative, the solutions are complex conjugates.

10.3 *Exercises*

 1. To use the Quadratic Formula to solve a quadratic equation, we must know the numbers a, b, and c. What must be true about the equation before these numbers can be determined?

 2. For a quadratic equation $ax^2 + bx + c = 0$, the definition requires that a is not zero. If this were not the case, what difficulty might we encounter with the Quadratic Formula?

In Exercises 3–6, write the equation in standard form and then state the values of a, b, and c.

3. $x(3x - 2) = 7x + 6$

4. $(x - 2)^2 = -4x$

5. $(x + 1)^2 = 3x - 1$

6. $1 = 4x(2x - 5)$

7. Explain why the Quadratic Formula replaces the method of completing the square for solving a quadratic equation.

8. What is your opinion about the best method for solving the equation $x^2 - 2x - 1088 = 0$? Can the factoring method be used? If so, what are the factors?

In Exercises 9–16, solve with the factoring method and then solve again with the Quadratic Formula.

9. $x^2 - x - 2 = 0$

10. $x^2 - 3x + 2 = 0$

11. $x^2 + 16 = 8x$

12. $x^2 + 1 = 2x$

13. $2x^2 + 7x = 4$

14. $2x^2 = 7x - 3$

15. $x^2 = 5x$

16. $4x^2 - 3x = 3x$

In Exercises 17–24, use the Quadratic Formula to solve.

17. $4x^2 = 3 + 9x$

18. $4x^2 + 7x = 3$

19. $x^2 - 2x + 9 = 0$

20. $x^2 + 4x + 10 = 0$

21. $x(x + 2) = 5$

22. $x^2 = 2(2 - x)$

23. $x + 2x^2 + 4 = 0$

24. $9 + 3x^2 + 5x = 0$

25. One way to use your calculator with the Quadratic Formula is to store the values of a, b, and c, and then enter the expressions

$$\frac{-b + \sqrt{b^2 - 4ac}}{2a} \quad \text{and} \quad \frac{-b - \sqrt{b^2 - 4ac}}{2a}.$$

Write the sequence of key strokes for entering the expressions correctly. What expressions will the calculator evaluate if you forget to enclose the numerators in parentheses?

26. When you have a number of quadratic equations to solve with the Quadratic Formula, a better method than that described in Exercise 25 is to enter the two expressions as Y_1 and Y_2 on the function screen. Then, after storing the values of a, b, and c, simply evaluate Y_1 and Y_2. Use this method for the following equations and interpret the results.

(a) $2x^2 + 5x - 1 = 0$

(b) $9x^2 - 6x + 1 = 0$

(c) $4x^2 + x + 3 = 0$

In Exercises 27–52, solve by any method.

27. $6x - 9x^2 = 0$

28. $5 - 13x - 6x^2 = 0$

29. $9 - 49x^2 = 0$

30. $3 - 4x^2 = 0$

31. $49 = 14x - x^2$

32. $9x^2 - 6x = -1$

33. $(5x - 4)^2 = 10$

34. $(4x + 5)^2 = 49$

35. $0.25x^2 - 0.5x - 2 = 0$

36. $0.5x^2 + x = 0.875$

37. $x(x + 4) = 5$

38. $3 = x(x - 6)$

39. $5x^2 + 2x + 7 = 0$

40. $41 + 8x - 4x^2 = 0$

41. $x^2 - 3x + 8 = 0$

42. $x^2 - 5x - 5 = 0$

43. $6x - x^2 - 4 = 0$

44. $x^2 = 4 - 2x$

45. $x(5 + x) + 9 = 0$

46. $5 = 2x(1 - 3x)$

47. $\dfrac{1}{2}x^2 - \dfrac{2}{3}x + \dfrac{4}{3} = 0$

48. $\dfrac{1}{6}x^2 + \dfrac{3}{2} = \dfrac{1}{2}x$

49. $x^2 = \dfrac{5x}{2} + \dfrac{1}{2}$

50. $\dfrac{2}{3}x^2 + 2x = \dfrac{5}{9}$

51. $(x - 3)(2x + 1) = x(x - 4)$

52. $(2x + 1)^2 = 2(3x + 1)$

53. What is the discriminant? In general, what can we learn from the value of the discriminant if a, b, and c are rational numbers?

54. Describe the number and type of solutions of a quadratic equation with rational coefficients when the discriminant is positive, zero, or negative. For the case of a positive discriminant, what type of solution is obtained when the discriminant is a perfect square and when it is not?

In Exercises 55–62, use the discriminant to determine the number and the nature of the solutions of the given equation.

55. $4x^2 + 12x = 7$

56. $x^2 + 12 = 8x$

57. $2x^2 + 5x + 7 = 0$

58. $4x^2 + x + 9 = 0$

59. $36x^2 + 1 = 12x$

60. $4x(x + 1) + 1 = 0$

61. $3x^2 + 4x - 8 = 0$

62. $1 + 6x - 9x^2 = 0$

In Exercises 63–66, determine the values of k so that the equation has two unequal real roots.

63. $x^2 + 8x + k = 0$

64. $3x^2 + 2x + k = 0$

65. $kx^2 - 4x + 1 = 0$

66. $kx^2 - 3x + 2 = 0$

In Exercises 67–70, determine the values of k so that the equation has two imaginary solutions.

67. $x^2 + 6x + k = 0$

68. $5x^2 + 4x + k = 0$

69. $kx^2 - 3x + 1 = 0$

70. $kx^2 - 5x + 4 = 0$

In Exercises 71–74, determine the values of k so that the equation has a double root.

71. $x^2 + kx + 9 = 0$

72. $x^2 + kx + 3 = 0$

73. $x^2 + kx + 2k = 0$

74. $kx^2 + (2k + 4)x + 9 = 0$

75. (a) Explain why $x^2 + ax - 5 = 0$ always has real solutions, regardless of the value of the real number a.

    (b) For any positive real number k, does the equation $x^2 + ax - k = 0$ always have real solutions, regardless of the value of the real number a? Explain.

76. Explain how to use the discriminant to determine whether the factoring method can be used to solve the given equation.

    (a) $2x^2 + 5x = 7$

    (b) $2x^2 + 3x = 8$

In Exercises 77–80, use the Quadratic Formula to factor the polynomial.

77. $x^2 + 4x - 480$

78. $x^2 + 32x + 240$

79. $x^2 + 7x - 450$

80. $x^2 + 4x - 672$

81. Use the Quadratic Formula to show that the sum of the roots of $ax^2 + bx + c = 0$ is $-b/a$. Test the conclusion by solving the equation $2x^2 - 7x - 15 = 0$ and showing that the sum of the two roots is $-b/a$.

82. Use the conclusion in Exercise 81 to explain why 5 and -4 cannot possibly be solutions of the equation $2x^2 - 3x - 20 = 0$ even though the numbers are factors of -20.

83. Use the Quadratic Formula to show that the product of the roots of $ax^2 + bx + c = 0$ is c/a.

84. Use the conclusion in Exercise 83 to explain why 5 and -4 cannot possibly be solutions of the equation $2x^2 - 3x - 20 = 0$ even though the numbers are factors of -20.

In Exercises 85–88, determine the sum and the product of the roots of the given equation without solving the equation.

85. $3x^2 - 2x - 1 = 0$

86. $2x^2 - 2x + 1 = 0$

87. $x^2 + x + 1 = 0$

88. $x^2 - 4x - 1 = 0$

In Exercises 89–94, use a quadratic equation to solve the given application.

89. Four times a number is 5 less than the square of the number. What are the two possible numbers?

90. The square of the reciprocal of a positive number is 3 more than the reciprocal of the number. What is the number?

91. A softball field should be laid out so that the distance between bases is 60 feet. Suppose the infield of a certain softball field is perfectly square and the distance between first and third base is 84.9 feet. To the nearest foot, is this a regulation softball field?

92. An auto dealer wants to string a wire from the top of a pole to the ground and hang colorful pennants from it. The distance from the base of the pole to the point where the wire is attached to the ground is 3 times the height of the pole. If the wire is 63 feet long, how tall is the pole?

93. Four consecutive integers are such that the product of the first and fourth integers is twice the second integer. If the first integer is negative, what is the third integer?

 94. A rectangle is drawn inside a circle so that the diagonals of the rectangle intersect at the center of the circle. (See figure.) The rectangle is 4 feet longer than it is wide. If the radius of the circle is 7 feet, how long is the rectangle?

 95. Explain why $x^2 + kx + (k - 1) = 0$ has two real solutions for any value of k except 2.

Figure for 94

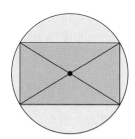

 96. Explain why $k^2x^2 - 4kx + 4 = 0$ has a double root for any nonzero value of k.

Exploring with Real Data

In Exercises 97–100, refer to the following background information and data.

The accompanying figure shows that per capita consumption of beef declined during the period 1980–1990, while per capita consumption of chicken rose. (Source: U.S. Department of Agriculture.)

Figure for 97–100

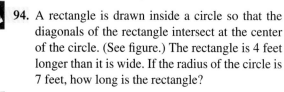

Letting B and C represent per capita consumption of beef and chicken, respectively, the data in the figure can be modeled by the following functions.

Beef: $B(t) = -0.262t^2 + 1.81t + 72.1$

Chicken: $C(t) = 1.49t + 33.23$

In both functions, t is the number of years since 1980.

97. Produce the graph of $y = B(t)$. According to this graph, during what 2 years of this 10-year period was beef consumption per capita at its highest point?

98. Use the Quadratic Formula to determine the positive value of t when $B(t) = C(t)$. Interpret your result.

99. According to the model for per capita beef consumption, what will happen after the year 2000? What does this conclusion indicate about the validity of the model after 1990?

100. What is one factor that could reverse the trends in per capita beef and chicken consumption in the future?

Challenge

In Exercises 101–104, solve for x.

101. $x^2 + 2bx + 3 = 0$

102. $x^2 + ax - 2a = 0$

103. $x^2 - yx - 2 = 0$

104. $yx^2 - 3x + 2y = 0$

In Exercises 105–108, solve for y.

105. $x = y^2 + 2y - 8$

106. $x = 2y - 5 - y^2$

107. $x = y^2 - 3y + 5$

108. $x = 4 + 3y - y^2$

In Exercises 109–112, write a quadratic equation with the given solutions.

109. $\dfrac{2 \pm \sqrt{3}}{4}$

110. $-5 \pm 2\sqrt{5}$

111. $-2 \pm 3i$

112. $\dfrac{1 \pm i\sqrt{2}}{3}$

In Exercises 113–116, use the Quadratic Formula to solve each equation.

113. $x^2 + \sqrt{3}x - 1 = 0$

114. $\sqrt{2}x^2 + \sqrt{5}x - \sqrt{8} = 0$

115. $x^2 - 2\sqrt{2}x + 3 = 0$

116. $x^2 - 2\sqrt{3}x + 7 = 0$

It can be shown that the Quadratic Formula holds for quadratic equations with imaginary coefficients. In Exercises 117–120, solve the following equations and express the solutions in the complex number form $a + bi$.

117. $x^2 - 2ix - 3 = 0$

118. $x^2 + 2ix - 1 = 0$

119. $x^2 + ix + 6 = 0$

120. $ix^2 + 3x - 2i = 0$

10.4 Equations in Quadratic Form

Fractional Equations ▪ *Radical Equations* ▪ *Quadratic Forms* ▪
Complex Solutions

Fractional Equations

In Section 8.5 we discussed methods for solving equations involving rational expressions. The technique began with our clearing any fractions by multiplying both sides of the equation by the LCD. This step sometimes leads to a quadratic equation.

We are now in a position to solve any quadratic equation that might result from clearing fractions.

EXAMPLE 1   *Solving Equations with Rational Expressions*

Solve the equation

$$\frac{2}{t} - \frac{t}{t-2} = 5.$$

Solution

Figure 10.6 shows the graphs of

$$y_1 = \frac{2}{t} - \frac{t}{t-2} \quad \text{and} \quad y_2 = 5.$$

Figure 10.6

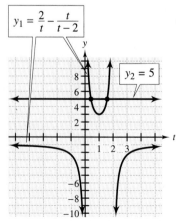

The two points of intersection indicate that the equation has two solutions.

$$\frac{2}{t} - \frac{t}{t-2} = 5 \qquad \text{Note that 0 and 2 are restricted values.}$$

$$\frac{t(t-2)}{1} \cdot \frac{2}{t} - \frac{t(t-2)}{1} \cdot \frac{t}{t-2} = 5t(t-2) \qquad \text{Multiply by the LCD } t(t-2).$$

$$2(t-2) - t^2 = 5t^2 - 10t$$

$$2t - 4 - t^2 = 5t^2 - 10t \qquad \text{Simplify.}$$

$$-6t^2 + 12t - 4 = 0 \qquad \text{Write the equation in standard form.}$$

Now use the Quadratic Formula to solve for *t*.

$$t = \frac{-12 \pm \sqrt{12^2 - 4(-6)(-4)}}{2(-6)}$$

$$t = \frac{-12 \pm \sqrt{144 - 96}}{-12}$$

$$t = \frac{-12 \pm \sqrt{48}}{-12}$$

$$t \approx 1.58 \quad \text{or} \quad t \approx 0.42 \qquad \text{Neither solution is restricted.}$$

Note that these solutions are consistent with those suggested by the graph in Figure 10.6. ■

Radical Equations

In Section 9.8 we developed methods for solving equations involving radicals. An early step in the routine was to raise both sides of the equation to a power equal to the index of the radical. This step can produce a quadratic equation that we can solve with our present methods. Recall, however, that raising both sides of an equation to a power can produce extraneous solutions. We must check all solutions.

EXAMPLE 2   *Solving Equations with Radical Expressions*

Solve the equation $\sqrt{x + 1} = x - 2$.

Solution

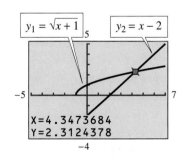

Figure 10.7

$y_1 = \sqrt{x + 1}$ $y_2 = x - 2$

X=4.3473684
Y=2.3124378

Figure 10.7 shows the graphs of $y_1 = \sqrt{x + 1}$ and $y_2 = x - 2$.

The one point of intersection represents the one real number solution of the equation. The solution is approximately 4.3.

$$\sqrt{x + 1} = x - 2$$
$$\left(\sqrt{x + 1}\right)^2 = (x - 2)^2 \qquad \text{Square both sides of the equation.}$$
$$x + 1 = x^2 - 4x + 4 \qquad \text{Square the binomial.}$$
$$0 = x^2 - 5x + 3 \qquad \text{Write the equation in standard form.}$$

Use the Quadratic Formula to solve for *x*.

$$x = \frac{5 \pm \sqrt{(-5)^2 - 4(1)(3)}}{2(1)} = \frac{5 \pm \sqrt{13}}{2}$$

$$x \approx 4.30 \quad \text{or} \quad x \approx 0.70$$

The solution 4.30 is consistent with the solution that we estimated from the graph. However, when we check the value 0.70, we find that it is an extraneous solution. ■

Quadratic Forms

Some equations are not quadratic equations but have the same structure or form as a quadratic equation, that is, the variable factor in the first term is the square of the variable factor in the second term. Such equations can often be converted to an equation that is quadratic by making an appropriate substitution.

The variable used to make the substitution is sometimes called a *dummy variable*. We let the dummy variable represent the variable factor in the middle term.

After making the appropriate substitution, we can use our routine methods for solving the resulting quadratic equation for the dummy variable. We then replace the dummy variable with the expression it represents and solve the resulting equations for the original variable.

EXAMPLE 3   *Solving Equations That Are in Quadratic Form*

Solve each equation.

(a) $x^4 - 6x^2 + 8 = 0$ (b) $16(z - 1)^2 - 8(z - 1) - 2 = 0$

(c) $28x^{-2} + x^{-1} - 2 = 0$

Solution

(a) In Figure 10.8 the graph of $y_1 = x^4 - 6x^2 + 8$ suggests two positive solutions and two negative solutions.

Figure 10.8

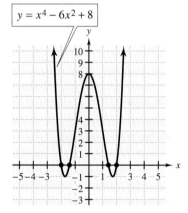

$$x^4 - 6x^2 + 8 = 0$$
$$u^2 - 6u + 8 = 0 \qquad \text{Let } u = x^2. \text{ Note that } u^2 = x^4.$$
$$(u - 2)(u - 4) = 0 \qquad \text{Factor.}$$
$$u - 2 = 0 \quad \text{or} \quad u - 4 = 0 \qquad \text{Zero Factor Property}$$
$$u = 2 \quad \text{or} \qquad u = 4 \qquad \text{Solve each case for } u.$$

Now replace u with x^2.

$$x^2 = 2 \qquad\qquad \text{or} \quad x^2 = 4$$
$$x = \pm\sqrt{2} \approx \pm 1.41 \quad \text{or} \quad x = \pm 2 \qquad \text{Square Root Property}$$

As indicated by the graph, the equation has two positive solutions and two negative solutions.

(b) Figure 10.9 shows the graph of $y_1 = 16(z - 1)^2 - 8(z - 1) - 2$. From the graph, we anticipate two positive solutions between 0 and 2.

Figure 10.9

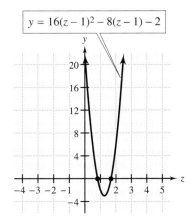

$$y = 16(z - 1)^2 - 8(z - 1) - 2$$

$$16(z - 1)^2 - 8(z - 1) - 2 = 0$$

$$16u^2 - 8u - 2 = 0 \qquad \text{Let } u = z - 1. \text{ Then } u^2 = (z - 1)^2.$$

Use the Quadratic Formula to solve for u.

$$u = \frac{8 \pm \sqrt{(-8)^2 - 4(16)(-2)}}{2(16)}$$

$$u = \frac{8 \pm \sqrt{64 + 128}}{32}$$

$$u = \frac{8 \pm \sqrt{192}}{32}$$

$$u \approx 0.68 \quad \text{or} \quad u \approx -0.18$$

To solve for z, replace u with the expression it represents.

$$z - 1 \approx 0.68 \quad \text{or} \quad z - 1 \approx -0.18$$

$$z \approx 1.68 \quad \text{or} \qquad z \approx 0.82$$

(c) $28x^{-2} + x^{-1} - 2 = 0$

$$28u^2 + u - 2 = 0 \qquad \text{Let } u = x^{-1}. \text{ Then } u^2 = x^{-2}.$$

$$(4u - 1)(7u + 2) = 0 \qquad \text{Factor.}$$

$$4u - 1 = 0 \quad \text{or} \quad 7u + 2 = 0 \qquad \text{Zero Factor Property}$$

$$4u = 1 \quad \text{or} \qquad 7u = -2 \qquad \text{Solve each case for } u.$$

$$u = \frac{1}{4} \quad \text{or} \qquad u = -\frac{2}{7}$$

To solve for x, replace u with the expression it represents.

$$x^{-1} = \frac{1}{4} \quad \text{or} \quad x^{-1} = -\frac{2}{7} \qquad \text{If the reciprocal of } x \text{ is } \tfrac{1}{4} \text{ or } -\tfrac{2}{7}, \text{ then } x \text{ is}$$

$$\qquad \qquad \qquad \qquad \qquad \qquad \qquad \text{the reciprocal of } \tfrac{1}{4} \text{ or of } -\tfrac{2}{7}.$$

$$x = 4 \quad \text{or} \qquad x = -\frac{7}{2}$$

An equation involving rational exponents may be quadratic in form if one of the exponents is twice another exponent. Recognizing such patterns is essential to detecting quadratic forms.

EXAMPLE 4   *Solving Equations with Rational Exponents*

Solve the equation $x^{2/3} - 2x^{1/3} - 24 = 0$.

Solution

$$x^{2/3} - 2x^{1/3} - 24 = 0$$

$$u^2 - 2u - 24 = 0 \qquad \text{Let } u = x^{1/3}. \text{ Then } u^2 = (x^{1/3})^2 = x^{2/3}.$$

$$(u - 6)(u + 4) = 0 \qquad \text{Factor.}$$

$$u - 6 = 0 \quad \text{or} \quad u + 4 = 0 \qquad \text{Zero Factor Property}$$

$$u = 6 \quad \text{or} \qquad u = -4 \qquad \text{Solve each case for } u.$$

Replace u with $x^{1/3}$.

$$x^{1/3} = 6 \qquad \text{or} \qquad x^{1/3} = -4$$

$$(x^{1/3})^3 = 6^3 \qquad \text{or} \quad (x^{1/3})^3 = (-4)^3 \qquad \text{Raise both sides to the third power.}$$

$$x = 216 \quad \text{or} \qquad x = -64 \qquad \blacksquare$$

Complex Solutions

We are now prepared to deal with a full range of possibilities when solving fractional equations, radical equations, and equations that are quadratic in form. Whenever quadratic equations emerge in the process of solving an equation, the solutions for both the dummy variable and the original variable may be imaginary numbers.

EXAMPLE 5   *Equations with Complex Solutions*

Solve the equation $x^4 + x^2 = 12$.

Solution

$$x^4 + x^2 = 12$$

$$x^4 + x^2 - 12 = 0$$

Figure 10.10 shows the graph of $y_1 = x^4 + x^2 - 12$. From the graph we anticipate two real number solutions. An algebraic method of solving is needed to determine any imaginary number solutions.

$$x^4 + x^2 = 12$$

$$u^2 + u = 12 \qquad \text{Let } u = x^2. \text{ Then } u^2 = x^4.$$

$$u^2 + u - 12 = 0 \qquad \text{Write the equation in standard form.}$$

$$(u + 4)(u - 3) = 0 \qquad \text{Factor.}$$

$$u + 4 = 0 \quad \text{or} \quad u - 3 = 0 \qquad \text{Zero Factor Property}$$

$$u = -4 \quad \text{or} \qquad u = 3 \qquad \text{Solve each case for } u.$$

Figure 10.10

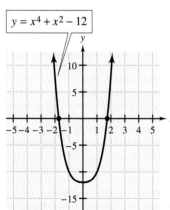

$y = x^4 + x^2 - 12$

The solutions for u are real numbers. To solve for x, replace u with the expression it represents.

$$x^2 = -4 \quad \text{or} \quad x^2 = 3$$
$$x = \pm 2i \quad \text{or} \quad x = \pm\sqrt{3} \approx \pm 1.73$$

Two of the solutions for x are real (irrational) numbers (as the graph suggested), and two solutions are complex conjugates. ■

In this section we have seen that the ability to solve quadratic equations benefits us well beyond quadratic equations themselves. Solving a quadratic equation may be one part of solving a higher degree equation; an equation involving rational expressions, radicals, or rational exponents; or any other equation that is quadratic in form.

10.4 *Quick Reference*

Fractional Equations	■ To solve a fractional equation, clear the fractions by multiplying both sides of the equation by the LCD. This step may lead to a quadratic equation. ■ Make sure that any solutions you obtain are not restricted values.
Radical Equations	■ To solve a radical equation, raise both sides of the equation to a power equal to the index of the radical. The result may be a quadratic equation. ■ Make sure that any solutions you obtain are not extraneous.
Quadratic Forms	■ To solve an equation that is quadratic in form, where the variable factor in the first term is the square of the variable factor in the second term, use a dummy variable as a temporary substitution for the variable factor in the second term. The resulting quadratic equation can then be solved for the dummy variable. ■ After obtaining solutions for the dummy variable, replace the dummy variable with the expression it represents. Then, solve for the original variable. ■ Check for restricted values and extraneous solutions.
Complex Solutions	■ When solving an equation that is quadratic in form, solutions for a dummy variable and for an original variable may be real numbers or imaginary numbers.

10.4   Exercises

1. To estimate the number of solutions of the equation

$$x + \frac{1}{x-1} - \frac{5-4x}{x-1} = 0,$$

we can produce the graph of

$$f(x) = x + \frac{1}{x-1} - \frac{5-4x}{x-1}$$

and observe that there is one *x*-intercept. (See figure.)

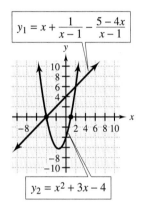

$$y_1 = x + \frac{1}{x-1} - \frac{5-4x}{x-1}$$

$$y_2 = x^2 + 3x - 4$$

Now clear the fractions and show that the result is the quadratic equation $x^2 + 3x - 4 = 0$. From the figure we can see that the graph of $g(x) = x^2 + 3x - 4$ has two intercepts. Explain why the graphs indicate only one solution for the original equation but two solutions for the resulting quadratic equation.

2. Other than to find algebraic mistakes, why is it essential to check solutions for equations with rational expressions?

In Exercises 3–14, solve the given equation.

3. $\dfrac{3}{x} = 2 + \dfrac{4}{x^2}$

4. $x + \dfrac{5}{x} = 3$

5. $\dfrac{t-3}{t} = \dfrac{2}{t-3}$

6. $\dfrac{y}{y+1} = \dfrac{y+1}{3}$

7. $x + 2 = \dfrac{6}{1-x}$

8. $x + 1 = \dfrac{2}{2x+3}$

9. $\dfrac{x}{x+3} + \dfrac{3}{x} = 4$

10. $\dfrac{2}{x-2} + \dfrac{4}{x+3} = 1$

11. $\dfrac{x}{x-2} - \dfrac{5}{x+2} = \dfrac{12}{x^2-4}$

12. $\dfrac{x}{x-5} - \dfrac{16}{x^2-25} = \dfrac{1}{x+5}$

13. $\dfrac{x+3}{x-2} - \dfrac{5}{2x^2-3x-2} = \dfrac{2}{2x+1}$

14. $\dfrac{x}{x-3} - \dfrac{1}{2x-1} = \dfrac{5x}{2x^2-7x+3}$

15. To estimate the number of solutions of the equation $\sqrt{x^2-5} + 2\sqrt{x} = 0$, we can produce the graph of $f(x) = \sqrt{x^2-5} + 2\sqrt{x}$ and observe that there are no *x*-intercepts. (See figure.)

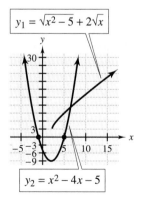

$$y_1 = \sqrt{x^2-5} + 2\sqrt{x}$$

$$y_2 = x^2 - 4x - 5$$

Now isolate the radicals, square both sides, and show that the result is $x^2 - 4x - 5 = 0$. From the figure we can see that the graph of the equation $g(x) = x^2 - 4x - 5$ has two *x*-intercepts. Explain why the graphs indicate that the original equation has no solution but that the resulting quadratic equation has two solutions.

 16. Other than to find algebraic mistakes, why is it essential to check your solutions for radical equations?

In Exercises 17–26, solve the radical equation.

17. $5 = 3\sqrt{x} + \dfrac{2}{\sqrt{x}}$

18. $\sqrt{x + 1} - \dfrac{2}{\sqrt{x + 1}} = 1$

19. $2y + 1 = \sqrt{11y - 1}$

20. $2 + x = \sqrt{2x + 3}$

21. $x - 3 = \sqrt{x - 1}$

22. $2x + 5 = \sqrt{4x + 14}$

23. $\sqrt{3x - 2} = \sqrt{x + 2} - 1$

24. $\sqrt{3x + 4} = \sqrt{x - 2} - 3$

25. $\sqrt{x}\,\sqrt{x + 3} = 1$

26. $\sqrt{2x}\,\sqrt{x - 3} = 6$

 27. Explain the role of a dummy variable in solving equations that are quadratic in form. After you have solved for the dummy variable, how do you determine the solution(s) of the original equation?

 28. Suppose an equation is quadratic in form. The two variable terms of the equation are $x^{1/6}$ and $x^{1/3}$ and you want to use a substitution to solve the equation. Which of the two expressions should your dummy variable represent? Why?

In Exercises 29–48, solve by using an appropriate substitution to rewrite the given equation in quadratic form.

29. $b^4 - 7b^2 + 12 = 0$

30. $x^4 - 14x^2 + 45 = 0$

31. $y^4 + 2y^2 = 63$

32. $x^4 + 17x^2 + 16 = 0$

33. $x - 6x^{1/2} + 8 = 0$

34. $c + 21 = 10c^{1/2}$

35. $y^{-2} + 2 = y^{-1}$

36. $1 - x^{-1} + x^{-2} = 0$

37. $y^{2/3} + 12 = 7y^{1/3}$

38. $3x^{1/10} + x^{1/5} = 4$

39. $(x^2 - 5x)^2 - 8(x^2 - 5x) = 84$

40. $(x^2 - x)^2 - 26(x^2 - x) + 120 = 0$

41. $(x + 1)^2 + 2(x + 1) = 2$

42. $(y - 2)^2 - 2(y - 2) = 4$

43. $(x^2 - 3)^2 - 5(x^2 - 3) - 6 = 0$

44. $(x^2 + 1)^2 - 2(x^2 + 1) - 15 = 0$

45. $3(\sqrt{x} - 2)^2 - 7(\sqrt{x} - 2) + 2 = 0$

46. $(\sqrt{x} - 8)^2 + 9(\sqrt{x} - 8) + 20 = 0$

47. $x - \sqrt{x} = 6$

48. $x + 3\sqrt{x} - 10 = 0$

In Exercises 49–78, determine all complex number solutions.

49. $x^4 - 7x^2 + 12 = 0$

50. $x^4 - 10x^2 + 16 = 0$

51. $\dfrac{x}{x + 2} + \dfrac{4}{x} = 5$

52. $\dfrac{4}{x} + \dfrac{5x}{x + 3} = 2$

53. $2x + 5 = 2\sqrt{x + 3}$

54. $2x = 1 + \sqrt{x + 7}$

55. $a^4 + a^2 = 20$

56. $x^4 + 5x^2 + 6 = 0$

57. $x - 13x^{1/2} + 42 = 0$

58. $b = 4b^{1/2} + 21$

59. $\dfrac{x + 3}{x - 1} = \dfrac{5}{x}$

60. $x + 2 = \dfrac{6}{1 - x}$

61. $\sqrt{3x + 12} - \sqrt{x + 8} = 2$

62. $\sqrt{3x - 2} = \sqrt{x + 3} + 1$

63. $a^{2/3} = 2a^{1/3} + 15$

64. $2x^{1/6} + x^{1/3} = 8$

65. $\sqrt{3x} = \sqrt{2 - 7x}$

66. $2x = \sqrt{3}\sqrt{2x + 1}$

67. $\dfrac{x + 1}{3 - x} + \dfrac{12}{x^2 - 3x} + \dfrac{5}{x} = 0$

68. $1 = \dfrac{3}{x + 4} - \dfrac{5x}{x^2 + 3x - 4}$

69. $(3 - y)^2 - 3(3 - y) = 6$

70. $(x + 1)^2 + (x + 1) - 3 = 0$

71. $\sqrt{2x - 3} = 4 - \sqrt{x + 2}$

72. $\sqrt{3x - 5} = 2 - \sqrt{x + 1}$

73. $(3 - x)^{-2} - 3(3 - x)^{-1} = 4$

74. $4(x + 1)^{-1} + (x + 1)^{-2} - 21 = 0$

75. $\dfrac{x}{4(x - 4)} = \dfrac{1}{x + 2} + \dfrac{6}{x^2 - 2x - 8}$

76. $\dfrac{x + 4}{2x + 1} + \dfrac{1}{x + 1} = \dfrac{-1}{2x^2 + 3x + 1}$

77. $(x^2 - x)^2 + 3(x^2 - x) + 2 = 0$

78. $(2x^2 - 7x)^2 - 2(2x^2 - 7x) - 8 = 0$

Exploring with Real Data

In Exercises 79–82, refer to the following background information and data.

In the 1980s the cost of the federally funded National School Lunch Program rose steadily, even though the number of students who were served by the program decreased. The accompanying figure shows the costs (in millions of dollars) for 1980, 1985, and 1990. (Source: Food and Nutrition Service, USDA.)

With t representing the number of years since 1980, the cost C (in millions of dollars) can be modeled by a function that is quadratic in form.

$$C(t) = 6.74(t + 5)^2 - 41.3(t + 5) + 2317$$

79. Use the model function to write an equation in which the cost $C(t)$ is 3000 (million dollars).

Figure for 79–82

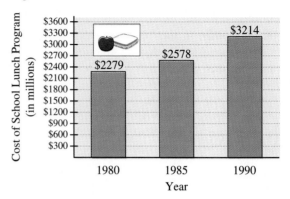

80. Use a dummy variable u for the expression $t + 5$ and solve this equation for u. Then, replace u with $t + 5$ and solve for t.

81. Select the positive solution for t and interpret the meaning of the solution.

82. The number of students who benefited from the school lunch program decreased by about 9% from 1980 to 1990. Calculate the corresponding percentage increase in costs during that period.

Challenge

In Exercises 83 and 84, rewrite the equation so that it is quadratic in form. Then make an appropriate substitution and solve.

83. $t = \sqrt{t + 3}$ (Hint: Consider adding 3 to both sides.)

84. $x^2 + 6x - \sqrt{x^2 + 6x - 2} = 22$

In Exercises 85–90, determine all complex number solutions.

85. $x^6 - 7x^3 = 8$

86. $x^6 + 26x^3 = 27$

87. $x^2 + 2x = |2x + 1|$

88. $|x - 3| = x^2 - 3x$

89. $y^{-2} + 1 = y^{-1}$

90. $2\sqrt{x} - \dfrac{1}{\sqrt{x}} = 2$

10.5 Applications

Rectangles ▪ *Ratio and Proportion* ▪ *Distance, Rate, Time* ▪ *Work Rate* ▪ *Investment*

In this section we illustrate how the ability to solve any quadratic equation can be of benefit in solving application problems. As we have seen, even if the initial equation used to describe the conditions of a problem is not quadratic, solving this equation may lead to another equation that is quadratic.

This section includes a sampling of applications in five different categories. Naturally, there are many other kinds of applications in which a quadratic equation may arise.

Rectangles

When a problem involves the *area* of a rectangle, we sometimes must multiply two variable expressions representing the width and length of the rectangle to represent its area. This can result in a quadratic equation.

EXAMPLE 1 *The Area of a Rectangular Play Area*

A day care operator plans to use 170 feet of fencing to enclose a rectangular play area of 2800 square feet. She plans to construct the play area so that it is centered against the back of her home, which is 50 feet in length. Determine the dimensions necessary to enclose the required area.

Solution

Let x = the distance from the corner of the house to the corner of the fence. (See Fig. 10.11.) Then the length of the rectangle is $50 + 2x$.

Figure 10.11

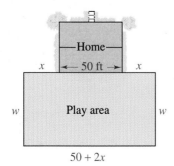

Let w = the width of the play area. Note that no fencing is needed along the back of the home.

$$x + w + (50 + 2x) + w + x = 170$$ Total fencing used is 170 feet.
$$2w + 50 + 4x = 170$$ Combine like terms.
$$2w = 120 - 4x$$ Solve for w.
$$w = 60 - 2x$$
$$L \cdot W = A$$ Formula for the area of a rectangle
$$(50 + 2x)(60 - 2x) = 2800$$ Required area is 2800 square feet.
$$3000 + 20x - 4x^2 = 2800$$ FOIL
$$-4x^2 + 20x + 200 = 0$$ Write the equation in standard form.
$$x^2 - 5x - 50 = 0$$ Divide both sides by -4.
$$(x - 10)(x + 5) = 0$$ Factor.
$$x - 10 = 0 \quad \text{or} \quad x + 5 = 0$$ Zero Factor Property
$$x = 10 \quad \text{or} \quad x = -5$$ Solve each case for x.

Because the distance is not negative, the only applicable solution is 10. The fence extends 10 feet on each side of the house, which means that the length of the play area is 70 feet. The width is $60 - 2(10) = 40$ feet.

Ratio and Proportion

According to early Greek mathematicians, the most perfect rectangles were those whose width W and length L satisfied the *Golden Ratio*.

$$\frac{W}{L} = \frac{L}{W + L}$$

We can use this formula in application problems in which a rectangle is to have dimensions that satisfy the Golden Ratio.

EXAMPLE 2   *Rectangles That Satisfy the Golden Ratio*

If a rectangle has a perimeter of 20 inches and satisfies the Golden Ratio, what are the dimensions?

Solution

Let L = the length of the rectangle, W = the width of the rectangle, and P = the perimeter, which is given as 20 inches.

$$2L + 2W = P$$ Formula for the perimeter of a rectangle
$$2L + 2W = 20$$ The required perimeter is 20 inches.
$$L + W = 10$$ Divide both sides by 2.
$$W = 10 - L$$ Solve for W.

The Golden Ratio formula is $\dfrac{W}{L} = \dfrac{L}{W + L}$.

$$\frac{10 - L}{L} = \frac{L}{10}$$ Replace *W* with 10 − *L*.

$$\frac{10L}{1} \cdot \frac{10 - L}{L} = \frac{10L}{1} \cdot \frac{L}{10}$$ Multiply by the LCD 10*L* to clear the fractions.

$$100 - 10L = L^2$$ Simplify.

$$L^2 + 10L - 100 = 0$$ Write the equation in standard form.

We use the Quadratic Formula to solve for *L*.

$$L = \frac{-10 \pm \sqrt{10^2 - 4(1)(-100)}}{2(1)}$$

$$L = \frac{-10 \pm \sqrt{100 + 400}}{2}$$

$$L = \frac{-10 \pm \sqrt{500}}{2}$$

$$L \approx 6.18 \quad \text{or} \quad L \approx -16.18$$

Because length is positive, the solution −16.18 is disqualified. The length is 6.18 inches, and the width is 10 − 6.18 = 3.82 inches. ■

Distance, Rate, Time

Recall that distance *D*, rate *R*, and time *T* are related by the formula *D* = *RT*.

EXAMPLE 3   *Determining Hiking Rates*

A hiker traveled 9.2 miles along the Appalachian Trail from Woody's Gap up to Blood Mountain. From Blood Mountain, he hiked 2.1 miles down the mountain to Neel's Gap. His rate down the mountain was 1 mph faster than his rate up the mountain, and the total hike took 6 hours. What were his hiking rates for the two portions of the trip?

Solution   Let *x* = rate up the mountain and *x* + 1 = rate down the mountain.

	D (miles)	R (mph)	$T = \dfrac{D}{R}$ (hours)
Up	9.2	x	$\dfrac{9.2}{x}$
Down	2.1	$x + 1$	$\dfrac{2.1}{x + 1}$

The chart helps us to write the equation.

$$\frac{9.2}{x} + \frac{2.1}{x + 1} = 6$$ The total hiking time was 6 hours.

$$\frac{x(x + 1)}{1} \cdot \frac{9.2}{x} + \frac{x(x + 1)}{1} \cdot \frac{2.1}{x + 1} = 6x(x + 1)$$ Multiply by the LCD $x(x + 1)$.

$$9.2(x + 1) + 2.1x = 6x(x + 1)$$

$$9.2x + 9.2 + 2.1x = 6x^2 + 6x$$ Distributive Property

$$11.3x + 9.2 = 6x^2 + 6x$$ Combine like terms.

$$6x^2 - 5.3x - 9.2 = 0$$ Write the equation in standard form.

We use the Quadratic Formula to solve for x.

$$x = \frac{5.3 \pm \sqrt{(-5.3)^2 - 4(6)(-9.2)}}{2(6)}$$

$$x = \frac{5.3 \pm \sqrt{28.09 + 220.8}}{12}$$

$$x = \frac{5.3 \pm \sqrt{248.89}}{12}$$

$$x \approx 1.76 \quad \text{or} \quad x \approx -0.87$$

Because rates are positive, the hiker's rate up the mountain is 1.76 mph and his rate down the mountain is $1.76 + 1 = 2.76$ mph.   ■

When an unknown quantity is represented by the square root of a variable expression, the resulting radical equation might lead to a quadratic equation.

EXAMPLE 4   *A Boating and Walking Trip*

From a dock, it is 3 miles directly across a lake to a marina. An angler travels by boat at 5 mph from the dock to a pier on the opposite shore. Then he walks at 4 mph from the pier to a fishing spot that is 6 miles down the shoreline from the marina. (See Fig. 10.12.) If he made the trip in 1.5 hours, how far did he walk?

Figure 10.12

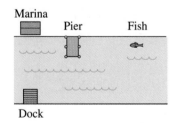

Figure 10.13

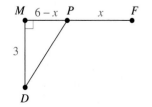

Solution

In Fig. 10.13, we identify locations with letters.

D: Dock

M: Marina

P: Pier

F: Fishing spot

Let $x =$ the distance the angler walked from P to F. The distance from the marina M to the fishing spot F is 6 miles. Therefore, the distance from the marina M to the pier P is $6 - x$.

We use the Pythagorean Theorem on the right triangle *DMP* to determine the distance *DP* traveled by boat.

$$(DP)^2 = (DM)^2 + (MP)^2$$
$$(DP)^2 = 3^2 + (6 - x)^2$$
$$DP = \sqrt{9 + (6 - x)^2}$$

	Distance (miles)	Rate (mph)	Time (hours)
Boat	$\sqrt{9 + (6 - x)^2}$	5	$\dfrac{\sqrt{9 + (6 - x)^2}}{5}$
Walking	x	4	$\dfrac{x}{4}$

$$\frac{\sqrt{9 + (6 - x)^2}}{5} + \frac{x}{4} = 1.5 \qquad \text{The total travel time was 1.5 hours.}$$

$$\frac{20}{1} \cdot \frac{\sqrt{9 + (6 - x)^2}}{5} + \frac{20}{1} \cdot \frac{x}{4} = 20 \cdot 1.5 \qquad \text{Clear fractions by multiplying by the LCD 20.}$$

$$4\sqrt{9 + (6 - x)^2} + 5x = 30 \qquad \text{Simplify.}$$

$$4\sqrt{9 + (6 - x)^2} = 30 - 5x \qquad \text{Isolate the radical.}$$

$$\left[4\sqrt{9 + (6 - x)^2}\right]^2 = (30 - 5x)^2 \qquad \text{Square both sides.}$$

$$16\left[9 + (6 - x)^2\right] = (30 - 5x)^2 \qquad (ab)^2 = a^2 b^2$$

$$16(9 + 36 - 12x + x^2) = (30 - 5x)^2 \qquad \text{Square the binomial } 6 - x.$$

$$16(45 - 12x + x^2) = 900 - 300x + 25x^2 \qquad \text{Square } 30 - 5x.$$

$$720 - 192x + 16x^2 = 900 - 300x + 25x^2 \qquad \text{Distributive Property}$$

$$9x^2 - 108x + 180 = 0 \qquad \text{Write the equation in standard form.}$$

$$x^2 - 12x + 20 = 0 \qquad \text{Divide both sides by 9.}$$

$$(x - 2)(x - 10) = 0 \qquad \text{Factor.}$$

$$x - 2 = 0 \quad \text{or} \quad x - 10 = 0 \qquad \text{Zero Factor Property}$$

$$x = 2 \quad \text{or} \qquad x = 10 \qquad \text{Solve each case for } x.$$

By substitution, we find that 10 is an extraneous solution. For $x = 2$, the distance $6 - x$ would be 4, which is an acceptable value. Therefore, the angler walked 2 miles. ■

Work Rate

If T is the time it takes to complete a task, then $1/T$ is the work rate. If two or more persons, groups, or machines work at the task, then the combined work rate is the sum of the individual work rates.

EXAMPLE 5   *Combined Work Rates*

Working together, two details of Air Force reservists can load a C-130 aircraft with relief supplies for hurricane victims in 6 hours. Working alone, the first detail can load the plane in 2 hours less time than the second detail can when working alone. How long would it take each detail, working alone, to load the plane?

Solution   Let t = number of hours for the second detail, working alone, to load the plane. Then $t - 2$ = number of hours for the first detail, working alone, to load the plane.

	Time Alone (hours)	Work Rate (loads/hour)
First Detail	$t - 2$	$\dfrac{1}{t-2}$
Second Detail	t	$\dfrac{1}{t}$
Together	6	$\dfrac{1}{6}$

$$\frac{1}{t-2} + \frac{1}{t} = \frac{1}{6}$$
Sum of the work rates equals the combined work rate.

$$\frac{6t(t-2)}{1} \cdot \frac{1}{t-2} + \frac{6t(t-2)}{1} \cdot \frac{1}{t} = \frac{6t(t-2)}{1} \cdot \frac{1}{6}$$
Multiply by the LCD $6t(t-2)$.

$$6t + 6(t - 2) = t(t - 2)$$ Simplify.

$$6t + 6t - 12 = t^2 - 2t$$ Distributive Property

$$t^2 - 14t + 12 = 0$$ Write the equation in standard form.

To solve for t, use the Quadratic Formula.

$$t = \frac{14 \pm \sqrt{(-14)^2 - 4(1)(12)}}{2(1)}$$

$$t = \frac{14 \pm \sqrt{196 - 48}}{2}$$

$$t = \frac{14 \pm \sqrt{148}}{2}$$

$$t \approx 13.08 \quad \text{or} \quad t \approx 0.92$$

The solution 0.92 makes the first detail's time negative. The only applicable solution is 13.08. Therefore, it would take 13.08 hours for the second detail to load the plane and 11.08 hours for the first detail to load the plane.   ■

Investment

In previous investment problems, we have used this relationship.

Number of shares · Price per share = Total investment

If the focus of an investment application is on price per share, then we can rewrite the relationship this way.

$$\text{Price per share} = \frac{\text{Total investment}}{\text{Number of shares}}$$

EXAMPLE 6 *An Investment Application Problem*

Six years ago a woman invested $42,000 in stock. Due to stock splits and dividend reinvestment, she now has 1100 more shares of the stock than her original purchase, and the market value is $73,600. However, the price per share is $3 per share less than the price per share of the original purchase. How many shares did she purchase originally and how many does she own now?

Solution   Let x = number of shares initially purchased and $x + 1100$ = number of shares currently owned. The price per share is the market value divided by the number of shares.

$$\frac{42,000}{x} = \text{Price per share of initial purchase}$$

$$\frac{73,600}{x + 1100} = \text{Current price per share}$$

The difference between the initial price per share and the current price per share is $3:

$$\frac{42,000}{x} - \frac{73,600}{x + 1100} = 3.$$

Clear fractions by multiplying by the LCD $x(x + 1100)$.

$$\frac{x(x + 1100)}{1} \cdot \frac{42,000}{x} - \frac{x(x + 1100)}{1} \cdot \frac{73,600}{x + 1100} = 3x(x + 1100)$$

$$42,000(x + 1100) - 73,600x = 3x(x + 1100)$$

$$42,000x + 46,200,000 - 73,600x = 3x^2 + 3300x$$

$$46,200,000 - 31,600x = 3x^2 + 3300x$$

$$3x^2 + 34,900x - 46,200,000 = 0$$

To solve for x, use the Quadratic Formula.

$$x = \frac{-34,900 \pm \sqrt{34,900^2 - 4(3)(-46,200,000)}}{2(3)}$$

$$x = \frac{-34,900 \pm \sqrt{1,772,410,000}}{6}$$

$$x = 1200 \quad \text{or} \quad x = -12,833.33$$

Because the number of shares must be positive, the only applicable solution is 1200. Thus, the woman bought 1200 shares initially, and she now owns 2300 shares. ■

10.5   *Exercises*

Rectangles

1. In an application problem, suppose the dimensions of a rectangle are to be found. If the dimensions are represented by first-degree expressions, and the area is given, what kind of equation will you need to solve? Why?

2. Suppose a rectangular garden of length L is surrounded by a walk of uniform width x. Consider the larger rectangle that includes the garden and the walk. (See figure.) Do you represent the length of this rectangle with $L + x$ or $L + 2x$? Explain.

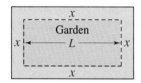

3. The dimensions of a rectangular front yard are 60 feet by 80 feet. If you begin mowing around the outside edge, how wide a border around the lawn must you mow to complete three-fourths of the job?

4. The dimensions of a rectangular hay field are 60 yards by 100 yards. A farmer is going to cut the hay around the outside edge. How wide a border around the field must the farmer cut to complete two-thirds of the job?

5. A plot of land next to a river is to be enclosed with 300 feet of fencing to form a rectangular area of 11,500 square feet. The river will serve as the fourth side. An 8-foot-wide gate is to be installed in the fence that is parallel to the river. (See figure.) How far should the fence extend from the river if the minimum distance is 70 feet?

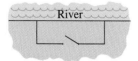

6. A woman bought a 21-square-yard rectangular piece of carpeting for a room that is 12 feet by 18 feet. She chose the dimensions of the carpet so that there would be a uniformly wide strip of uncovered floor around the outside of the carpet. How wide will this strip be?

Triangles

7. When using the Pythagorean Theorem to solve a right triangle problem, how do you identify the side of the triangle that is the hypotenuse?

8. If a and b are the lengths of the legs of a right triangle and c is the length of its hypotenuse, then $c^2 = a^2 + b^2$. Then, according to the Square Root Property, $c = \pm \sqrt{a^2 + b^2}$. Would this formula for c apply in this application problem? Why?

9. A plot of land is in the shape of a right triangle with the longest side 160 yards in length. What are the lengths of the perpendicular sides if one side is 30 yards longer than the other?

10. When a painter leans a 24-foot ladder against a house, the distance from the base of the ladder to the house is 8 feet less than the distance from the top of the ladder to the ground. How far up the house is the top of the ladder?

11. The area of a triangular mainsail is 440 square feet. If the height of the sail is 10 feet less than the base, what is the height?

12. A car and a truck left Luckenback, Texas, at noon. The car traveled due north and the truck traveled due east. In 2 hours time, they were 155 miles apart, and the car had traveled 30 miles more than the truck. How far was each vehicle from Luckenback at 2:00 P.M.?

Ratio and Proportion

13. We say that the dimensions of a rectangle satisfy the Golden Ratio if the length L and width W satisfy the equation

$$\frac{W}{L} = \frac{L}{W + L}.$$

Is this equation a ratio? If not, what should it be called? Why?

14. Explain how the corresponding sides of similar triangles are related.

 15. A rectangle satisfying the Golden Ratio has a perimeter of 56 inches. What are the dimensions?

 16. A rectangular book has a front cover that satisfies the Golden Ratio and has a perimeter of 31 inches. What are the dimensions of the front cover?

 17. In the accompanying figure, triangles *ABC* and *DEF* are similar triangles. Determine the lengths *AC* and *DE*.

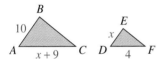

 18. In the accompanying figure, triangles *ABC* and *DEF* are similar triangles. Determine the lengths *AB* and *DF*.

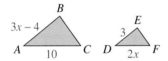

Distance, Rate, Time

 19. Suppose an application problem involves distance *D*, rate *R*, and time *T*. If you are given the total time required to travel two legs of a trip, how will you probably need to modify the formula $D = RT$, and what kind of equation is likely to model the conditions of the problem?

 20. Suppose you have correctly solved an application problem involving distance, rate, and time, and your answer is that the traveler's rate was 3 mph. If *x* represents the rate, explain why neither of the following equations could have modeled the conditions of the problem.

(i) $\dfrac{x}{x + 1} + \dfrac{5}{x - 3} = 2$

(ii) $2\sqrt{1 - x} = 4x - 1$

21. A boater drove a motorboat upstream a distance of 46 miles and then immediately drove back to the starting point. The total time of the trip was 3.5 hours. If the rate of the current was 4 mph, how fast would the boat have traveled in still water?

22. Two campers left camp at 8:00 A.M. to paddle their canoe upstream 10 miles to a fishing cove. They fished for 3.75 hours and then paddled back to the campsite and arrived at 6:00 P.M. If the rate of the current was 3 mph, how fast can the campers paddle in still water?

23. A family drove 45 miles and then hiked 1 mile to their campsite. The trip took 1 hour and 15 minutes. If they drove 55 mph faster than they hiked, how fast did they drive?

24. Two track team members ran a 200-meter relay in 29 seconds. (Each person runs 100 meters.) One sprinter runs 0.55 meters per second faster than the other. What is the speed of each runner?

Work Rate

25. Suppose a tank is half full when both the inlet and outlet valves are opened. Under what circumstances will the tank eventually be full?

26. In typical work rate problems it is assumed that the combined work rate of two people is the sum of their individual work rates. Realistically, is it possible that two people working together might take longer to complete a job than it would take for each person working alone? Why?

27. Working together, two roommates can paint their apartment in 6 hours. Working alone, one person can complete the job 2 hours sooner than the other person working alone. How long would it take each person, working alone?

28. Two landscape workers can complete a certain landscaping job in a total of 20 hours. Working alone, one can do this same job in 2 hours less time than it takes the other. How long would it take the faster worker to do the job if the slower worker were unable to work?

29. There are two fill pipes and a drain pipe to a backyard swimming pool. It takes 9 hours to fill the pool with the first fill pipe. It takes 1 hour longer to drain the pool with the drain pipe than it takes to fill the pool with the second fill pipe. If all three pipes are open at once, the pool can be filled in 8 hours. How long would it take to drain the pool with the two fill pipes closed?

30. Both the inlet and outlet pipes to a cider barrel were mistakenly left open. Nevertheless, the barrel filled in 12 hours. With only the outlet pipe open, a full barrel of cider can be emptied in 1

hour more than the time it takes to fill an empty barrel with only the inlet pipe open. How long would it take a full barrel of cider to drain with the outlet pipe open and the inlet pipe closed?

Investment

31. The value V of a stock holding can be found by multiplying the number N of shares by the price P per share. Write a formula for N in terms of V and P. If this formula is needed in an application problem, what type of equation would likely be used to model the conditions of the problem?

32. At a college recital, there were 40 more students in attendance than nonstudents. If n represents the number of nonstudents who attended and if the box office receipts from the sale of student tickets totaled $180, write an expression for the price of a student ticket. What additional information would be needed to write an expression for the price of a nonstudent ticket?

33. Ten years ago, a man had less than 400 shares of stock worth $16,095. Today, he has 405 more shares of stock than he had originally, and the investment is worth $34,875. The price per share today is $1.50 more than it was 10 years ago. How many shares of stock did the man have originally?

34. A woman invested $13,500 to buy more than 200 shares of a certain stock. After 5 years she had 22.23 more shares of stock, and her investment was worth $16,189.13. If each share increased in value by $4, how many shares did the woman purchase initially?

35. A community service organization sells citrus fruit each fall to raise money for the local library. In 1993 total fruit sales were $8568. In 1994 the organization sold 102 more boxes of fruit with total sales of $11,594. The selling price per box in 1994 was up $2 from the 1993 price. If fewer than 500 boxes were sold in 1993, how many boxes of fruit were sold each year and what was the price per box in 1994?

36. The community theater group performed a popular musical. The tickets for the Friday night performance were $2 more than the Thursday night tickets. The total receipts on Thursday night were $967.50, and the total receipts on Friday night were $2034.00. On Friday 65 more people attended the play than had on Thursday night. How many people attended on Friday night?

Miscellaneous

37. The owner of an apartment complex collects $24,000 per month in rent when all the apartments are occupied. Then, the owner raises the rent $80 per month, and ten apartments are vacated. However, the owner still collects $24,000 a month in rent. How many apartments are in the complex?

38. Find two consecutive odd integers such that the difference of their cubes is 866.

39. Find two consecutive integers such that the difference of their cubes is 169.

40. If a rectangular sign is 5 feet wide, how long must it be in order to satisfy the Golden Ratio?

41. A polygon of n sides has $\frac{1}{2}n(n-3)$ diagonals. If a polygon has 27 diagonals, how many sides does it have?

42. In geometry the altitude h to the hypotenuse of a right triangle is related to the two segments a and b shown in the figure by the equation

$$\frac{a}{h} = \frac{h}{b}.$$

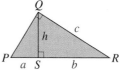

Suppose a home gardener uses the design shown in the figure to build a cold frame. The front segment b is 2.5 feet longer than the back segment a, and the height of the frame is 3 feet. How long is the front panel c?

43. The difference in the circumferences of two circles is 4π. If the sum of the areas of the circles is 164π, what is the radius of the smaller circle?

44. A rectangular table must be 2 feet longer than it is wide. The table must be designed so that its center can be reached from any point around the table. Assume that the distance from the center to any point around the table can be at most 3 feet. How wide must the table be?

45. A metal drum is designed to have a height of 4 feet and a total surface area of 24π square feet. Determine the diameter of the drum. (The total surface area S of a right circular cylinder is given by $S = 2\pi rh + 2\pi r^2$ where r is the radius and h is the height.)

46. When 1 is added to a number, the result is 2 more than the square root of the result. What is the number?

47. To construct a paved walk of uniform width around a rectangular swimming pool, a contractor uses 5.5 cubic yards of concrete. The pool is 30 feet long and 20 feet wide. If the walk is to be 3 inches thick, how wide can it be?

Challenge

48. A campus co-op bought a shipment of calculators for $10,000. The co-op first sold calculators to all students at a price required to cover the total cost. The remaining 40 calculators were then sold to nonstudents at the same price that the students paid, and the profit was distributed to the students. If all of the calculators had been sold to students, the price would have been $12.50 less than what they paid. What share of the profit did each student receive?

49. A fragile item is packed in a carton in the shape of a cube. This carton is then placed inside a larger carton, which is also in the shape of a cube and which has sides 2 feet longer than the sides of the smaller carton. The larger carton is then filled with 40 cubic feet of packing material around the smaller carton. What is the size of the smaller carton?

50. A spring is attached at point A [see Fig. (a)] and is connected to the rim of a wheel at point B so that the spring is tangent to the wheel. In this position the spring is 26 inches long. The wheel is turned clockwise so that the spring stretches and the end of the spring is now at point D. [See Fig. (b).] If the distance between points C and D is 8 inches, how far did the spring stretch from its original position? (Hint: A theorem from geometry relates AB, AC, and AD as $AB/AD = AC/AB$.)

(a)

(b)

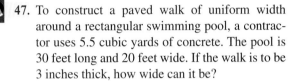
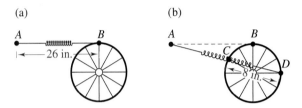

51. A ceremonial cannon, located on the bank of a river, is about to be fired. (See point C in the accompanying figure.) Observer 1 is directly across the river at point A, and Observer 2 is farther down the river at point B. Observer 1 is 2000 feet farther from Observer 2 than he is from the cannon. When the cannon is fired, Observer 2 hears the sound 3 seconds after she sees the flash. How wide is the river? (Assume that the speed of sound is 1100 feet per second.)

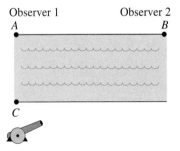

10.6 Quadratic Inequalities

Graphing and Algebraic Methods   ▪   *Special Cases*   ▪   *Applications*

Graphing and Algebraic Methods

The standard form of a quadratic equation is $ax^2 + bx + c = 0$. When the equal sign is replaced by an inequality symbol, we obtain a **quadratic inequality.**

> ### Definition of a Quadratic Inequality
>
> A **quadratic inequality** is an inequality that can be written in the form $ax^2 + bx + c > 0$, where a, b, and c are real numbers and $a \neq 0$. We call this the *standard form* of a quadratic inequality. The symbols $<$, $\leq$, and $\geq$ can also be used.

The **solution set** of a quadratic inequality is the set of all replacements for the variable that make the inequality true.

 EXPLORATION 1   *Using a Graph to Solve a Quadratic Inequality*

(a) Produce the graph of $y = x^2 + 2x - 8$ and estimate the x-intercepts.

(b) Verify the x-intercepts algebraically.

(c) For what x-values is $y < 0$?

(d) What is the solution set of the quadratic inequality $x^2 + 2x - 8 < 0$?

Discovery

Figure 10.14

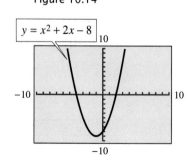

$y = x^2 + 2x - 8$

(a) Figure 10.14 shows the graph of $y = x^2 + 2x - 8$. The x-intercepts appear to be $(-4, 0)$ and $(2, 0)$.

(b) We can verify that the x-intercepts are $(-4, 0)$ and $(2, 0)$ by substitution, or we can algebraically determine the x-intercepts by solving the quadratic equation.

$$x^2 + 2x - 8 = 0$$
$$(x + 4)(x - 2) = 0 \qquad \text{Factor.}$$
$$x + 4 = 0 \quad \text{or} \quad x - 2 = 0 \qquad \text{Zero Factor Property}$$
$$x = -4 \quad \text{or} \quad x = 2 \qquad \text{Solve each case for } x.$$

(c) For $y < 0$, we look for that portion of the graph that is *below* the x-axis. From Fig. 10.14 we see that $y < 0$ in the interval between the two x-intercepts, that is, for all values of x such that $-4 < x < 2$.

Figure 10.15

(d) Solving $x^2 + 2x - 8 < 0$ is equivalent to solving $y < 0$. In interval notation, the solution set of the quadratic inequality is $(-4, 2)$. (See Fig. 10.15.)

Exploration 1 suggests a routine for solving quadratic inequalities. The process combines the accuracy of algebraic methods with the visual assistance of graphing methods.

> ### Solving a Quadratic Inequality with Graphing and Algebraic Methods
>
> 1. Write the inequality in standard form.
>
> 2. Produce the graph of the corresponding quadratic function and note the *x*-intercepts, if any.
>
> 3. Verify the estimated *x*-intercepts by substitution or determine them algebraically.
>
> 4. For inequalities involving $<$, the solutions are the interval(s) of *x*-values where the graph is *below* the *x*-axis; for inequalities involving $>$, the solutions are the interval(s) of *x*-values where the graph is *above* the *x*-axis. For inequalities involving $\leq$ or $\geq$, the *x*-intercepts also represent solutions.
>
> 5. The solution set includes all *x*-values determined in Step 4.

EXAMPLE 1   *Solving a Quadratic Inequality*

Solve the quadratic inequality $3x - x^2 \leq -10$.

Solution

$$-x^2 + 3x + 10 \leq 0 \qquad \text{Write the inequality in standard form.}$$

Figure 10.16 shows the graph of the quadratic function $y = -x^2 + 3x + 10$. The *x*-intercepts appear to be $(-2, 0)$ and $(5, 0)$. We can easily verify these *x*-intercepts by substitution.

Figure 10.16

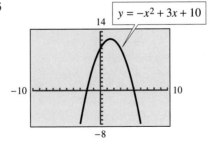

$y = -x^2 + 3x + 10$

Because the solutions of the inequality are all *x*-values for which $y \leq 0$, we note the two intervals where the graph is on or *below* the *x*-axis. One interval is to the *left* of $(-2, 0)$ and the other interval is to the *right* of $(5, 0)$. Note that the *x*-intercepts also represent solutions.

Figure 10.17

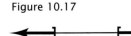

The solution set is the union of the two intervals of *x*-values: $(-\infty, -2] \cup [5, \infty)$. (See Fig. 10.17.)

When we use a graph to estimate the solutions of a quadratic inequality, our focus is on the *x*-intercepts. To determine the solution set accurately, we must verify these *x*-intercepts by substitution or determine the *x*-intercepts algebraically. We can use factoring, the Square Root Property, or the Quadratic Formula to determine the *x*-intercepts.

EXAMPLE 2   *Solving Quadratic Inequalities*

Solve the following inequalities.

(a) $1 > x(4 + x)$

(b) $12x < 4x^2 + 7$

Solution

(a)
$$1 > x(4 + x)$$
$$1 > 4x + x^2 \qquad \text{Distributive Property}$$
$$-x^2 - 4x + 1 > 0 \qquad \text{Write the inequality in standard form.}$$

Figure 10.18

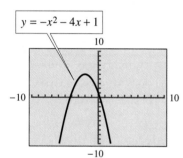

$y = -x^2 - 4x + 1$

Figure 10.18 shows the graph of $y = -x^2 - 4x + 1$. The estimated *x*-intercepts are $(-4, 0)$ and $(0, 0)$.

To determine the actual *x*-intercepts of the graph, we solve the associated quadratic equation $-x^2 - 4x + 1 = 0$ with the Quadratic Formula.

$$x = \frac{4 \pm \sqrt{(-4)^2 - 4(-1)(1)}}{2(-1)}$$

$$x = \frac{4 \pm \sqrt{16 + 4}}{-2}$$

$$x = \frac{4 \pm \sqrt{20}}{-2}$$

$$x \approx -4.24 \quad \text{or} \quad x \approx 0.24$$

Figure 10.19

$-4.24 \qquad 0.24$

We observe that the graph of the function is above the *x*-axis ($y > 0$) for *x*-values in the interval *between* the two *x*-intercepts. Therefore, the (approximate) solution set of the inequality is $(-4.24, 0.24)$. (See Fig. 10.19.)

(b)
$$12x < 4x^2 + 7$$
$$-4x^2 + 12x - 7 < 0 \qquad \text{Write the inequality in standard form.}$$

Figure 10.20 shows the graph of the quadratic function $y = -4x^2 + 12x - 7$. From the graph, we estimate the *x*-intercepts: $(1, 0)$ and $(2, 0)$.

Figure 10.20

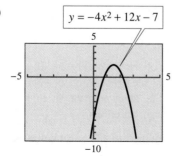

$y = -4x^2 + 12x - 7$

To determine the x-intercepts algebraically, we solve the associated quadratic equation $-4x^2 + 12x - 7 = 0$ with the Quadratic Formula.

$$x = \frac{-12 \pm \sqrt{12^2 - 4(-4)(-7)}}{2(-4)}$$

$$x = \frac{-12 \pm \sqrt{144 - 112}}{-8}$$

$$x = \frac{-12 \pm \sqrt{32}}{-8}$$

$$x \approx 0.79 \quad \text{or} \quad x \approx 2.21$$

Figure 10.21

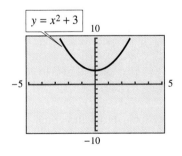

For $y < 0$, we note that the graph is *below* the x-axis to the *left* of $(0.79, 0)$ and to the *right* of $(2.21, 0)$. Therefore, an approximation of the solution set of the inequality is $(-\infty, 0.79) \cup (2.21, \infty)$. (See Fig. 10.21.) ■

Special Cases

In each of the previous examples, the solution set was an interval of x-values or the union of two intervals of x-values. The next example illustrates some special cases of quadratic inequalities that lead to different solution sets.

EXAMPLE 3   *Special Cases of Quadratic Inequalities*

Solve the inequalities.

(a) $x^2 + 3 \geq 0$ (b) $x^2 + 3 < 0$

(c) $x^2 + 6x + 9 \leq 0$ (d) $x^2 + 6x + 9 > 0$

Figure 10.22

Solution

(a) Figure 10.22 shows the graph of $y = x^2 + 3$. Because the entire graph lies *above* the x-axis, we can see that $x^2 + 3 \geq 0$ for all values of x. The solution set of the inequality is the set of all real numbers.

(b) For $x^2 + 3 < 0$, Fig. 10.22 shows that there is no part of the graph *below* the x-axis. Therefore, the solution set is empty.

(c) Figure 10.23 shows the graph of $y = x^2 + 6x + 9$. We see that no part of the graph is *below* the x-axis. However, there is one point, the x-intercept, for which $x^2 + 6x + 9 = 0$. This point is $(-3, 0)$, which easily can be verified by substitution. The only solution of the inequality is -3.

Figure 10.23

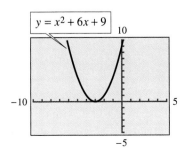

(d) The graph in Fig. 10.23 lies entirely *above* the x-axis except for one point, the x-intercept. Therefore, every point of the graph except $(-3, 0)$ represents a solution. In interval notation, the solution set is $(-\infty, -3) \cup (-3, \infty)$. ■

NOTE: When we use a graph to estimate the solutions of a quadratic inequality, we may find that the graph has no x-intercepts. However, as Example 3 shows, this does *not* imply that the inequality has no solution. ■

Applications

Applications involving quadratic inequalities are characterized by words and phrases such as the following.

more than	$>$	less than	$<$
at least	$\geq$	at most	$\leq$
no less than	$\geq$	no more than	$\leq$

EXAMPLE 4   *Minimum Area of a Rectangle*

If the perimeter of a rectangle is 400 feet, how long can the rectangle be so that the area is at least 8000 square feet?

Solution

Let L = the length of the rectangle and W = the width of the rectangle.

$$2L + 2W = 400 \qquad \text{The perimeter of the rectangle is 400 feet.}$$
$$2W = 400 - 2L \qquad \text{Solve for } W.$$
$$W = 200 - L$$

Figure 10.24 shows the rectangle with the dimensions labeled.

Figure 10.24

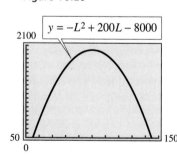

L

$200 - L$

The area of the rectangle is given by the formula $A = LW$.

$$A = LW = L(200 - L) = -L^2 + 200L$$
$$-L^2 + 200L \geq 8000 \qquad \text{The area must be at least 8000 square feet.}$$
$$-L^2 + 200L - 8000 \geq 0 \qquad \text{Write the inequality in standard form.}$$

Figure 10.25

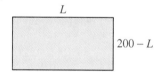

$y = -L^2 + 200L - 8000$

2100

50 ⌐ 150
0

Figure 10.25 shows the graph of $y = -L^2 + 200L - 8000$. From the graph we see that y is *at least* 0 for that portion of the graph between (and including) the x-intercepts.

Use the Quadratic Formula to verify that the solutions of $-L^2 + 200L - 8000 = 0$ are 55.28 and 144.72 (rounded to two decimal places). Therefore, the solution set of the inequality is approximately [55.28, 144.72].

The length of the rectangle can be any value between 55.28 feet and 144.72 feet to obtain an area of at least 8000 square feet.

NOTE: In Example 4 any length between 55.28 feet and 144.72 feet will produce a rectangle whose area is *at least* 8000 square feet. Later, we will determine the length that *maximizes* the area. Meanwhile, does the graph in Fig. 10.25 suggest what that optimum length is?

10.6 *Quick Reference*

Graphing and
Algebraic
Methods

- A **quadratic inequality** is an inequality that can be written in the form $ax^2 + bx + c > 0$ (or $<, \leq, \geq$) where a, b, and c are real numbers and $a \neq 0$. We call this the *standard form* of a quadratic inequality.

- The **solution set** of a quadratic inequality is the set of all replacements for the variable that make the inequality true.

- The following is a combined graphing and algebraic method for solving a quadratic inequality.

  1. Write the inequality in standard form.

  2. Produce the graph of the corresponding quadratic function and note the *x*-intercepts, if any.

  3. Verify the estimated *x*-intercepts by substitution or by determining them algebraically.

  4. For inequalities involving $<$, the solutions are the interval(s) of *x*-values where the graph is *below* the *x*-axis; for inequalities involving $>$, the solutions are the interval(s) of *x*-values where the graph is *above* the *x*-axis. For inequalities involving $\leq$ or $\geq$, the *x*-intercepts also represent solutions.

  5. The solution set includes all *x*-values determined in step 4.

Special Cases

- When we use a graph to estimate the solutions of a quadratic inequality, the graph may have only one *x*-intercept $(a, 0)$ or no *x*-intercepts. In such cases the solution set of the quadratic inequality may be one of the following special cases:

  1. the set of all real numbers,

  2. the empty set,

  3. the set containing just the one number a,

  4. the set of all real numbers except the number a.

10.6 *Exercises*

In Exercises 1–4, match the inequality in column B with the sentence in column A.

Column A	Column B
1. A number does not exceed 5.	(a) $x > 5$
2. A number is at least 5.	(b) $x \leq 5$
3. A number is more than 5.	(c) $x - 5 < 0$
4. A number is less than 5.	(d) $5 \leq x$

 5. Suppose you want to use a graph to estimate the solutions of the inequality $2x^2 + 7 < -3x$. What first step must you take if you want to produce a single graph rather than two graphs? What is the function whose graph you will produce?

6. Suppose you want to use a graph to estimate the solution set of a quadratic inequality, and you have already determined the *x*-intercepts. Describe the next step you will take in order to write the solution set.

In Exercises 7–10, for what values of x is the given expression positive?

7. $(3 - x)(1 + x)$ **8.** $2x^2 + 5x - 3$

9. $5x^2 - 8x$ **10.** $6 - x^2 + 3x$

In Exercises 11–14, for what values of x is the given expression negative?

11. $x^2 + 10 - 7x$ **12.** $12 + 4x - x^2$

13. $3 + x - 2x^2$ **14.** $x^2 - 3x - 6$

 15. If we use a graph to estimate the solution set of a quadratic inequality, we begin by estimating the x-intercepts. What methods can be used to determine the x-intercepts algebraically? Which of these methods will always work?

 16. Suppose you are using a graph to estimate the solutions of a quadratic inequality, and suppose the graph has two x-intercepts. Describe the possible intervals of x-values that are in the solution set.

In Exercises 17–22, use a graph to estimate the solution set of the given inequality.

17. $9 - x^2 < 0$ **18.** $x(x + 3) \geq 0$

19. $5 - 4x^2 \geq 8x$ **20.** $x^2 > 2(x + 3)$

21. $x(7 - x) \geq 8$

22. $4x - 2x^2 + 3 < 0$

In Exercises 23–30, solve the inequality.

23. $(x - 2)(x + 5) > 8$ **24.** $5 + 4x - x^2 \geq 0$

25. $x^2 \leq 5x$ **26.** $4x^2 < 7$

27. $x(12x + 5) \geq 2$ **28.** $x(4x + 9) > 9$

29. $2x(x - 2) < 7$ **30.** $x^2 + 2x \geq 7$

In Exercises 31–38, translate the given information into an inequality and then solve it.

31. The product of x and $x + 1$ is at least 6.

32. The product of $x + 3$ and $x - 1$ is less than 12.

33. The product of x and $x - 1$ is not more than 6.

34. The sum of $3x^2$ and $5x$ is less than 2.

35. Six less than the product of x and $6x - 5$ is negative.

36. The difference of 6 and $5x^2 + 7x$ is positive.

37. Seven more than the product of $2x$ and $2 - x$ is not positive.

38. The difference of $4x + 7$ and $2x^2$ is not negative.

In Exercises 39–42, for what values of x is the function nonnegative?

39. $g(x) = (4 - x)(2 + x)$

40. $f(x) = x^2 + 3 - 4x$

41. $g(x) = 4x^2 - 7x$ **42.** $h(x) = 8 - 5x^2$

In Exercises 43–46, for what values of x is the function nonpositive?

43. $f(x) = 16 - x^2$

44. $h(x) = 3(1 - x^2) - 8x$

45. $g(x) = x(x - 8) - 5$

46. $f(x) = 9x^2 + 6x - 8$

47. Suppose that you use a graph to solve a quadratic inequality and you find that the graph has no x-intercepts. List all of the possible solution sets of the inequality.

48. Suppose that you use a graph to solve a quadratic inequality and you find that the graph has only one x-intercept, $(3, 0)$. List all of the possible solution sets of the inequality.

In Exercises 49–74, solve the inequality.

49. $4x^2 + 23x + 15 \geq 0$ **50.** $x^2 + 7x + 6 < 0$

51. $5x - x^2 < 0$ **52.** $25 - x^2 \geq 0$

53. $x^2 + 3x + 6 < 0$ **54.** $3x - 2x^2 > 7$

55. $4x^2 + 20x + 25 \leq 0$

56. $6x^2 - 41x + 70 \leq 0$

57. $x^2 - 4x \geq 4$ **58.** $x^2 + 2x + 5 \leq 0$

59. $x^2 + 9 \leq 6x$ **60.** $x^2 + 16 \leq 8x$

61. $2x^2 < 4x + 5$ **62.** $2x^2 \geq 5x + 4$

63. $x^2 + 5 \geq 6x$ **64.** $5x - x^2 < 8$

65. $0.5x - 0.25x^2 \le 0.9$

66. $0.3x^2 - 0.7x + 0.9 \le 0$

67. $x(6x + 17) \le -7$ **68.** $3x(2 - 3x) \ge 1$

69. $x^2 + \dfrac{3}{4}x > \dfrac{5}{2}$ **70.** $\dfrac{x^2}{2} + x \ge 3$

71. $3x(x + 2) < 4$ **72.** $2x(3 - x) < 9$

73. $(2x + 1)(x + 2) \le (x - 3)(x + 2)$

74. $(2x + 3)(x + 1) > 2(2x + 3)(1 - x)$

In Exercises 75–78, use the discriminant to determine all values of k so that the solution set of the quadratic inequality is all real numbers.

75. $2x^2 - 3x + k > 0$

76. $kx^2 + 1 - 3x > 0$

77. $9 + kx - x^2 > 0$ **78.** $7x^2 - kx > 0$

In Exercises 79–82, use the discriminant to determine all values of k so that the real number solution set of the quadratic inequality is the empty set.

79. $x^2 + 5x + k < 0$

80. $kx^2 + 2x + 1 < 0$

81. $kx^2 + 8x - 5 > 0$

82. $5x^2 - 2x + k < 0$

In Exercises 83–86, use the discriminant to determine all values of k so that the real number solution set of the quadratic inequality is a set with only one element.

83. $x^2 + 12x + k \le 0$ **84.** $x^2 + 3x + k \ge 0$

85. $x^2 + kx + 9 \le 0$

86. $4x^2 + kx + 1 \le 0$

In Exercises 87–90, write a quadratic inequality that describes the conditions of the problem. Solve the inequality and answer the question.

87. The Woodstock Dairy Company wants to design a holding tank for chocolate milk. The tank is to be in the shape of a right circular cylinder 9.5 feet high. What should the radius of the tank be so that it holds at least 10,000 gallons? (1 gallon = 231 cubic inches.)

88. The Mathis Package Company wants to design a box that is 11 inches high with a base perimeter of 80 inches. The box's volume needs to be at least 2 cubic feet. What should be its length?

89. Consider a circle with radius r. For what values of r does the numerical value of the circumference of the circle exceed the numerical value of the area?

90. If an object is projected vertically upward, its height h (in feet) after t seconds is given by $h = 96t - 16t^2$. During what time interval will the object be at least 30 feet above the ground?

Exploring with Real Data

In Exercises 91–94, refer to the following background information and data.

The accompanying figure shows deaths (in thousands) due to heart disease in America from 1980 to 1990. (Source: National Institutes of Health.)

Figure for 91–94

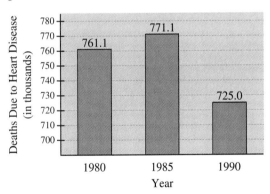

If H is the number (in thousands) of deaths due to heart disease, the data in the figure can be modeled by

$$H(t) = -1.12t^2 + 7.62t + 761.1$$

where t is the number of years since 1980.

91. Produce the graph of $y = H(t)$. According to the model and its graph, during what year did deaths due to heart disease fall below 750,000?

92. During what year of this 10-year period were deaths due to heart disease approximately 765,000?

93. According to the model and its graph, for what value of t will $H(t) = 0$? Interpret this result and indicate how realistic you think it is.

94. The data indicate that deaths due to heart disease reached a high point during the Reagan administration. What argument could you present that would put the Reagan years in a good light in connection with this issue?

Challenge

95. Produce the graph of $g(x) = x^2 - 2x - 15$.

    (a) By looking at the graph of g, explain how you would estimate the domain of
$$f(x) = \sqrt{x^2 - 2x - 15}.$$

    (b) By looking at the graph of g, what do you think the graph of $h(x) = |x^2 - 2x - 15|$ would look like? Why?

96. For $ax^2 > 0$ ($a \neq 0$), use the discriminant to prove that the solution set cannot be the set of all real numbers.

97. For nonzero real numbers a and b, use the discriminant to prove that the solution set of $ax^2 + bx > 0$ cannot be the set of all real numbers.

In Exercises 98 and 99, determine all values of k so that the solution set of the quadratic inequality is all real numbers.

    98. $x^2 + kx + 4 > 0$ **99.** $9x^2 + kx + 1 \geq 0$

In Exercises 100 and 101, determine all values of k so that the real number solution set of the quadratic inequality is the empty set.

100. $3x^2 + kx - 1 > 0$

101. $-2x^2 + kx - 2 > 0$

In Exercises 102 and 103, determine all values of k so that the real number solution set of the quadratic inequality is a set with only one element.

102. $kx^2 + 3x + (k - 1) \geq 0$

103. $2kx^2 + x + (k - 1) \leq 0$

10.7 Higher Degree and Rational Inequalities

Higher Degree Inequalities ▪ *Rational Inequalities* ▪ *Applications*

Higher Degree Inequalities

In general, any polynomial inequality can be solved in a manner similar to that used for solving quadratic inequalities. To solve a polynomial inequality such as $P(x) > 0$, we produce the graph of $y = P(x)$, estimate the x-intercepts, if any, and observe the interval(s) of x-values for which the graph lies above the x-axis. Similar procedures are used for $\leq$, $<$, and $\geq$ inequalities.

For quadratic inequalities it is always possible to determine the exact x-intercepts by solving the associated equation $P(x) = 0$. For polynomial inequalities of higher degree, this step will not be possible unless the polynomial can be factored.

This is one instance in which the power of the graphing method is evident. In real-data applications, higher degree polynomials are not often factorable, and graphing emerges as the best method for routine problem solving.

EXAMPLE 1 *Solving Higher Degree Polynomial Inequalities*

Solve the following inequalities.

(a) $x^3 + 2x^2 - 15x < 0$

(b) $2x^4 - 5x^3 + x - 1 \leq 0$

Solution

Figure 10.26

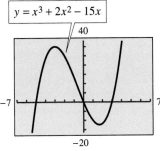

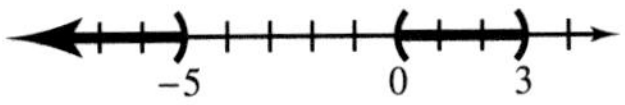

(a) Figure 10.26 shows the graph of the polynomial function $y = x^3 + 2x^2 - 15x$. From the graph, we estimate the x-intercepts to be $(-5, 0)$, $(0, 0)$, and $(3, 0)$.

Because $x^3 + 2x^2 - 15x$ is factorable, we can determine the x-intercepts by solving the associated equation.

$$x^3 + 2x^2 - 15x = 0$$
$$x(x^2 + 2x - 15) = 0$$
$$x(x + 5)(x - 3) = 0$$
$$x = 0 \quad \text{or} \quad x + 5 = 0 \quad \text{or} \quad x - 3 = 0$$
$$x = 0 \quad \text{or} \qquad x = -5 \quad \text{or} \qquad x = 3$$

The graph lies *below* the x-axis to the *left* of $(-5, 0)$ and *between* $(0, 0)$ and $(3, 0)$. Therefore, the solution set of the inequality is $(-\infty, -5) \cup (0, 3)$. (See Fig. 10.27.)

Figure 10.27

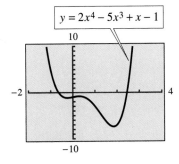

Figure 10.28

(b) Figure 10.28 shows the graph of the function $y = 2x^4 - 5x^3 + x - 1$.

Because $2x^4 - 5x^3 + x - 1$ is not factorable, we will not be able to determine the x-intercepts algebraically. This means we must be more careful in our estimates of the x-intercepts on the graph.

By zooming in on the two x-intercepts, and rounding to the nearest hundredth, we obtain the points $(-0.64, 0)$ and $(2.45, 0)$. Note that the x-intercepts represent solutions of the inequality.

Other solutions are represented by points that are *below* the x-axis, that is, *between* the two x-intercepts. The approximate solution set is $[-0.64, 2.45]$. (See Fig. 10.29.)

Figure 10.29

Rational Inequalities

We can use similar graphing procedures to solve inequalities with rational expressions. However, we must remember that any restricted values must be excluded from the solution set.

| EXAMPLE 2 | *Solving Rational Inequalities* |

Solve the following rational inequalities.

(a) $\dfrac{2}{x-3} \geq 1$ (b) $\dfrac{x}{x-16} < \dfrac{2}{6-x}$

Figure 10.30

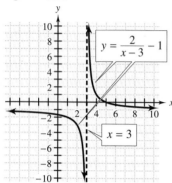

Solution

(a) $\dfrac{2}{x-3} \geq 1$ Note that 3 is a restricted value.

$\dfrac{2}{x-3} - 1 \geq 0$ We write the inequality with 0 on one side.

Figure 10.30 shows the graph of

$$y = \dfrac{2}{x-3} - 1.$$

From the graph we estimate the one x-intercept to be (5, 0). We can use substitution to verify this.

Figure 10.31

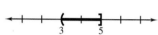

The dashed vertical line in Fig. 10.30 indicates that 3 is a restricted value. The graph lies *above* the x-axis in the interval from the vertical line $x = 3$ to the x-intercept (5, 0). The right endpoint of the interval 5 represents a solution, but the left endpoint 3 does not because it is a restricted value. Therefore, the solution set of the inequality is (3, 5]. (See Fig. 10.31.)

(b) $\dfrac{x}{x-16} < \dfrac{2}{6-x}$ Note that 16 and 6 are restricted values.

$\dfrac{x}{x-16} - \dfrac{2}{6-x} < 0$ Write the inequality with 0 on one side.

Figure 10.32

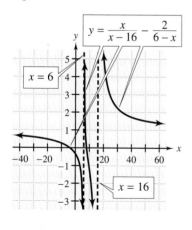

Figure 10.32 shows the graph of

$$y = \dfrac{x}{x-16} - \dfrac{2}{6-x}.$$

Note the two dashed vertical lines indicating the restricted values 6 and 16.

By tracing and zooming we estimate the x-intercepts to be (−4, 0) and (8, 0). We can determine these intercepts exactly by solving the associated equation

$$\dfrac{x}{x-16} = \dfrac{2}{6-x}.$$

Or we can simply verify them by substitution.

All points of the graph lie *below* the x-axis in the intervals

(i) from the x-intercept (−4, 0) to the vertical line at $x = 6$ and

(ii) from the x-intercept (8, 0) to the vertical line at $x = 16$.

Figure 10.33

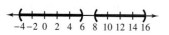

Therefore, the solution set of the inequality is (−4, 6) ∪ (8, 16). (See Fig. 10.33.) ∎

NOTE: In Example 2 we began each part by writing the inequality with 0 on one side. It is not essential to do this, but, instead of producing the graphs of both sides of the inequality, it is often easier to analyze just one graph with respect to the *x*-axis. ∎

Applications

A higher degree inequality or a rational inequality may be needed to describe the conditions in an application problem.

EXAMPLE 3 *Allowable Dimensions of a Box*

Figure 10.34

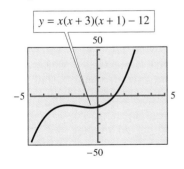

$x + 1$

$x + 3$

x

The length of a large cardboard carton must be 3 feet longer than the width. The height of the carton must be 1 foot greater than the width. (See Fig. 10.34.) The volume of the box must be no more than 12 cubic feet. Using only a graph, estimate the allowable widths of the box.

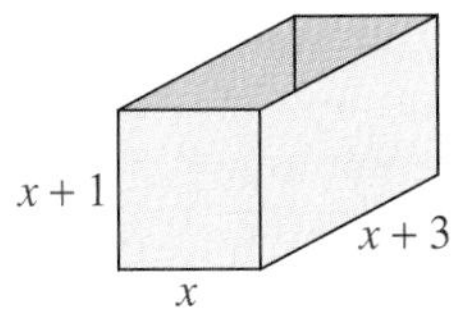

Solution   Let x = width of the box, $x + 3$ = length of the box, and $x + 1$ = height of the box.

The volume V is determined by the formula $V = LWH$.

$$x(x + 3)(x + 1) \leq 12 \qquad \text{The volume cannot exceed 12 cubic feet.}$$
$$x(x + 3)(x + 1) - 12 \leq 0 \qquad \text{Write the inequality with 0 on one side.}$$

Figure 10.35

$y = x(x + 3)(x + 1) - 12$

50

−5 ⊢⊢⊢⊢⊢⊢⊢⊢⊢⊢⊢ 5

−50

Figure 10.35 shows the graph of the function $y = x(x + 3)(x + 1) - 12$. From the graph we estimate the *x*-intercept to be (1.25, 0).

Because the inequality is of the form $y \leq 0$, we look for points of the graph that are *on* or *below* the *x*-axis. These are all points to the *left* of the *x*-intercept. However, because *x* represents the width of the box, *x* must be positive. Therefore, the valid interval of *x*-values is between the *y*-axis and the *x*-intercept.

The solution set is (0, 1.25]. The width of the box can be any positive number up to and including 1.25 feet. ∎

EXAMPLE 4 *Using a Rational Inequality to Calculate Election Returns*

An incumbent county commissioner can count on 1000 votes in the upcoming election. The challenger, who needs more than 50% of the vote to win, can count on 800 votes. Polls show that three-fourths of the remaining uncommitted votes are leaning toward the challenger while the other one-fourth are leaning toward the incumbent. For this reason, the challenger hopes that a great many of the currently uncommitted voters show up on election day. If the polls are completely accurate, at least how many of the uncommitted voters must vote to ensure the challenger's election?

Solution Let x = the number of uncommitted voters who do vote. Then $800 + \frac{3}{4}x$ = the number of votes the challenger receives, and $1000 + \frac{1}{4}x$ = the number of votes the incumbent receives. The total number of votes cast = $800 + \frac{3}{4}x + 1000 + \frac{1}{4}x = 1800 + x$. The percentage of votes cast for the challenger is given by

$$\frac{\text{Votes for challenger}}{\text{Total votes cast}} = \frac{800 + \frac{3}{4}x}{1800 + x}.$$

To win, the challenger's percentage must be more than 50%.

$$\frac{800 + \frac{3}{4}x}{1800 + x} > 0.50$$

$$\frac{800 + \frac{3}{4}x}{1800 + x} - 0.50 > 0$$

Figure 10.36

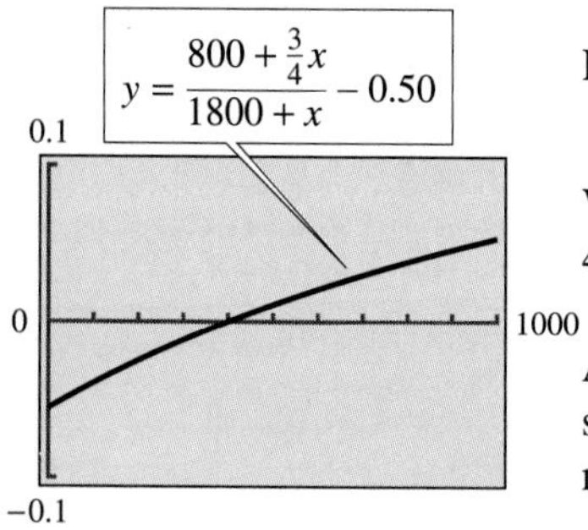

$$y = \frac{800 + \frac{3}{4}x}{1800 + x} - 0.50$$

Figure 10.36 shows the graph of $y = \dfrac{800 + \frac{3}{4}x}{1800 + x} - 0.50$.

We estimate the x-intercept to be (400, 0). This can be easily verified by storing 400 for x and evaluating the function.

All points of the graph lie *above* the x-axis from the x-intercept to the *right*. The solution set is (400, ∞). For the challenger to win, more than 400 of the uncommitted voters must vote.   ■

10.7 *Quick Reference*

Higher Degree Inequalities

- To solve a polynomial inequality $P(x) > 0$, produce the graph of $y = P(x)$, estimate the x-intercepts, if any, and observe the interval(s) of x-values for which the graph lies above the x-axis. Similar procedures are used for $\geq$, $<$, and $\leq$ inequalities.

- Unless $P(x)$ is factorable, it will not be possible to solve the equation $P(x) = 0$ in order to determine the x-intercepts algebraically.

Rational Inequalities

- The following is a summary of the graphing and algebraic methods for solving a rational inequality $R(x) > 0$.

  1. Produce the graph of the function $y = R(x)$ and estimate the x-intercepts, if any. Determine the x-intercepts by solving the associated rational equation $R(x) = 0$, or verify by substitution.

  2. Determine the interval(s) of x-values for which the graph is above the x-axis. Use the interval(s) to write the solution set of the inequality. *Exclude restricted values.*

  3. Similar procedures are used for $\geq$, $<$, and $\leq$ inequalities. Except for restricted values, the solution set includes endpoint x-values for $\geq$ and $\leq$ inequalities.

10.7 *Exercises*

 1. Suppose you use a graph to solve the inequality $P(x) < 0$ where $P(x)$ is a polynomial of degree greater than 2. From the graph you can estimate the x-intercepts. Under what conditions would it not be possible to determine the x-intercepts algebraically?

 2. Let $P(x) = x^3 - x^2 - 12x + k$, where k is a real number. Use a graph to estimate the solution set of $P(x) > 0$ when $k = 15$. What is the effect on the solution set if k is increased to 25?

In Exercises 3–16, solve the inequality. Determine all interval endpoints algebraically.

3. $(x - 3)(x + 2)(x - 1) > 0$

4. $3x(x - 3)(2x + 9) < 0$

5. $(x + 2)^2(x - 3)(x + 6) \leq 0$

6. $x^2(x + 4)(8 - x) \geq 0$

7. $x^3 - 3x^2 > 0$ 8. $2x^7 - x^6 \leq 0$

9. $x^5 + 6x^4 + 5x^3 < 0$

10. $x^6 - 7x^5 + 6x^4 \geq 0$

11. $x^4 - 13x^2 + 36 > 0$ 12. $x^4 - 3x^2 - 4 \geq 0$

13. $x^4 - 6x^2 + 5 \leq 0$

14. $x^4 + 2x^2 - 15 < 0$

15. $x^3 + 3x^2 - x - 3 > 0$

16. $9x^3 - 45x^2 - 16x + 80 \leq 0$

In Exercises 17–22, solve the given inequality. Verify all interval endpoints by substitution.

17. $x^3 - 4x^2 + x + 6 > 0$

18. $x^3 + 5x^2 + 2x - 8 < 0$

19. $x^3 + 6x^2 + 3x - 10 \leq 0$

20. $x^3 - 3x^2 - 6x + 8 \geq 0$

21. $x^4 + 6x^3 + 7x^2 - 6x - 8 < 0$

22. $x^4 + 5x^3 + 5x^2 - 5x - 6 \geq 0$

 23. Suppose that you are solving the rational inequality $R(x) \geq 0$. Explain how it is possible that every point of the graph of $y = R(x)$ is above the x-axis between the x-values a and b, and yet a and b are not solutions.

 24. Suppose that you are solving the rational inequality $R(x) \leq 0$, where m and n are restricted values. You produce the graph of $y = R(x)$ and find that there is one x-intercept $(a, 0)$ where $m < a < n$. From the following list, identify the two sets that could *not* be the solution set of the inequality and explain why.

(a) (m, n) (b) (a, n) (c) $(m, a]$

(d) $[m, a]$ (e) $\{a\}$ (f) $[a, n)$

In Exercises 25–34, solve the inequality.

25. $\dfrac{-2x}{x + 1} \geq 0$ 26. $\dfrac{x - 2}{x + 5} > 0$

27. $\dfrac{x - 6}{x + 1} < 3$ 28. $\dfrac{3}{4 - x} \geq 1$

29. $\dfrac{4 - 3x}{4x^2 + 1} > 0$ 30. $\dfrac{2x + 5}{(x - 1)^2} \leq 0$

31. $\dfrac{x - 2}{x + 4} > \dfrac{4}{x - 3}$ 32. $\dfrac{x + 3}{4 - x} < \dfrac{4}{x + 2}$

33. $\dfrac{x}{x + 3} \geq 2x$ 34. $\dfrac{x - 2}{x(3 + x)} \leq 2$

In Exercises 35 and 36, for what values of x is the expression nonnegative?

35. $x(3 - x)(3x + 4)$ 36. $\dfrac{x + 1}{(2x - 1)(x - 4)}$

In Exercises 37 and 38, for what values of x is the expression nonpositive?

37. $\dfrac{2x - 5}{x^2 + 9}$

38. $(4 - x)(x + 3)(x - 2)$

In Exercises 39–42, write the inequality and solve.

39. The quotient of $5 - 2x$ and $3x$ is at least $2x - 3$.

40. Three divided by $x - 4$ is not greater than 2.

41. The difference of $\dfrac{x}{x + 2}$ and $\dfrac{x - 1}{x + 1}$ is negative.

42. The product of $2x + 1$ and $x - 3$ divided by $x - 5$ is positive.

In Exercises 43 and 44, for what values of x is the function positive?

43. $f(x) = x^2(3 - x)(x + 5)$

44. $g(x) = \dfrac{x + 3}{x(x - 1)} - 1$

In Exercises 45 and 46, for what values of x is the function negative?

45. $h(x) = \dfrac{x^2 - 9x + 20}{x^2 - 9}$

46. $g(x) = (x + 1)(x - 3)^2(x + 5)$

 47. Suppose you are solving the rational inequality

$$\frac{x - 3}{x + 1} > 0.$$

Would it be correct to conclude that $x - 3$ and $x + 1$ must both be positive in order for the fraction to be positive? Why or why not?

 48. Suppose you are using a graph to solve a rational inequality. You notice that the displayed graph includes a vertical line at $x = 4$. Is the line part of the graph? Explain the significance of the line.

In Exercises 49–70, solve the inequality.

49. $\dfrac{1}{x} < \dfrac{-2}{3}$ **50.** $\dfrac{3}{x} \geq \dfrac{2}{5}$

51. $(x - 5)(3 - x)(2 + x) \geq 0$

52. $5x(2x - 1)(4 - x) \leq 0$

53. $\dfrac{x + 5}{x - 2} \leq \dfrac{x}{x - 2} - 3$

54. $\dfrac{x - 6}{x + 2} \geq \dfrac{5}{x + 2} - 2$

55. $(2x - 5)(x + 1)^2 > 0$

56. $x^3(x - 1)^2(5 - x) > 0$

57. $\dfrac{(x + 2)(7 - x)}{x - 5} < 0$

58. $\dfrac{x + 4}{(1 - 3x)(5 - x)} \geq 0$

59. $(x^2 + 1)(2x - 1) \leq 0$

60. $x^4 \leq 1$

61. $\dfrac{x + 2}{x + 3} - \dfrac{x - 1}{x - 2} \leq 1$

62. $\dfrac{1 - x}{x + 3} + \dfrac{x}{x - 1} \geq 2$

63. $x^4 - 5x^2 > 36$

64. $x^3 + 2x^2 < 16x + 32$

65. $\dfrac{x^2(2x + 1)(x - 3)}{(x + 1)^3(x - 1)} \leq 0$

66. $\dfrac{(3x - 2)^2(x + 4)(x + 1)}{(x - 3)(x + 2)^3} > 0$

67. $x^4 < 2x^5$

68. $x^5 + 9x^4 + 8x^3 \geq 0$

69. $\dfrac{8x^2 - 23x - 3}{2x^2 + 1} \geq 1$

70. $\dfrac{(2x - 3)^2}{x^2 + x + 5} \leq 1$

 71. For the inequality $\dfrac{x^2}{(x - 2)^2} \geq 0$, would it be correct to state that x^2 and $(x - 2)^2$ are nonnegative for *any* replacement for x; therefore, the solution set is the set of all real numbers? Why or why not?

 72. The inequality $x(x - 1)^2 < 0$ states that the product of x and $(x - 1)^2$ is negative. Why does the factor x determine the sign of the product while the factor $(x - 1)^2$ does not?

In Exercises 73–78, solve the given inequality by inspection.

73. $x(x - 5)^2 < 0$

74. $(x - 1)^4(x + 1)^2 > 0$

75. $\dfrac{x}{(x+2)^4} > 0$ 　　**76.** $\dfrac{x^4}{x-1} \le 0$

77. $\dfrac{x^2}{x+3} > 0$ 　　**78.** $\dfrac{-2x^4}{(x+1)^2} \le 0$

In Exercises 79–82, write an inequality that describes the conditions of the problem. Solve the inequality and answer the question.

79. The speed of the current in the Hillsborough River is 2 mph. At what speed would a canoeist have to paddle to go upstream 6 miles and back in less than 3 hours?

80. What are three consecutive odd integers whose product is at least 105?

81. The *successor* of an integer is one more than the integer. For example the successor of 8 is 9 and the successor of -5 is -4. What is the smallest positive integer for which the quotient of the integer and its successor is at least 2? Explain how a graph can be used to support your conclusion.

82. A builder plans to construct a home on a lot that is 100 feet wide. The covenants for the subdivision include the following provisions.

　(a) The front of a home must be at least 35 feet longer than its depth.

　(b) No home can be closer than 15 feet from any property line.

　(c) All homes must have at least 2556 square feet of ground-floor living space.

Show why the builder cannot construct the home.

Exploring with Real Data

In Exercises 83–86, refer to the following background information and data.

From 1985 through 1989, ticket sales to Broadway shows followed a roller-coaster pattern. The accompanying figure shows the sales (in thousands of tickets) for this period. (Source: *Variety.*)

The data in the figure can be modeled by the following third-degree polynomial function.

$$S(t) = -201.5t^3 + 4348.9t^2 - 30{,}254.5t + 74{,}919.5$$

In this function, $S(t)$ is the sales of tickets (in thousands) and t is the number of years since 1980.

Figure for 83–86

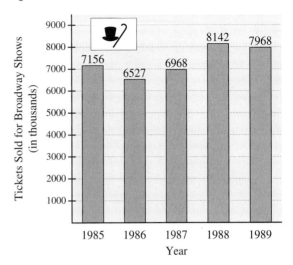

83. Write a third-degree inequality that describes the value of t for which at least 7.1 million tickets were sold. (Remember that $S(t)$ is in *thousands*.)

84. Write the inequality in Exercise 83 with 0 on the right side and produce the graph of the function defined by the expression on the left side.

85. Using the graph, estimate the years when at least 7.1 million tickets were sold. (Remember that t represents the number of years *since* 1980.)

86. What does the model predict for the years following 1990? Based on the shape of the graph, what is your opinion about the validity of the model before and after the specified years?

Challenge

87. Produce the graph of $y = (x-3)(x+2)(x-1)$. Explain how you would use the graph to determine the domain of the radical function $f(x) = \sqrt{(x-3)(x+2)(x-1)}$.

In Exercises 88–92, solve.

88. $x|x-4| \ge 0$ 　　**89.** $|x(x-4)| \ge 0$

90. $x^2 - 3x \le |x-2|$ 　　**91.** $x^3 - 1 < 0$

92. $\dfrac{x+1}{x-1} \le \dfrac{x+4}{2x+1} + \dfrac{6x}{2x^2 - x - 1}$

In Exercises 93–96, compare the solution sets of each pair of inequalities.

93. (a) $(2x - 1)^2(x + 4)^4 \geq 0$
    (b) $(2x - 1)^2(x + 4)^4 > 0$

94. (a) $(2x - 1)^2(x + 4)^4 \leq 0$
    (b) $(2x - 1)^2(x + 4)^4 < 0$

95. (a) $\dfrac{x}{(x - 3)^2} < 0$ (b) $\dfrac{x}{(x - 3)^2} \leq 0$

96. (a) $\dfrac{x}{(x - 3)^2} \geq 0$ (b) $\dfrac{x}{(x - 3)^2} > 0$

In Exercises 97–100, solve the inequality.

97. $\dfrac{2x^2 - 5x - 7}{x^2 - 1} \geq 0$

98. $(x + 1)(2 - 3x)^2 > 0$

99. $x^2(x + 6)(x - 2) \geq 0$

100. $3x^3(x + 2)^4 \geq 0$

10.8 Quadratic Functions

Properties of Graphs ▪ *Variations of the Graph of f(x) = x²* ▪ *Vertex and Intercepts* ▪ *Applications*

Properties of Graphs

In Section 10.1 we defined a quadratic function as a function of the form $f(x) = ax^2 + bx + c$, where a, b, and c are real numbers and $a \neq 0$. Throughout this chapter we have been using graphs of quadratic functions to assist us in solving quadratic equations and inequalities. In this section we consider quadratic functions and their graphs in greater detail.

We begin with the simple quadratic function $f(x) = ax^2$.

 **EXPLORATION 1**   *The Effect of a on the Graph of f(x) = ax²*

(a) Produce the graphs of $f(x) = ax^2$ for $a = \frac{1}{2}$, 1, 2, and 4.

(b) Produce the graphs of $f(x) = ax^2$ for $a = -\frac{1}{2}$, -1, -2, and -4.

(c) How does the sign of a affect the graph?

(d) What is the effect on the graph as $|a|$ increases?

(e) What are the domain and range of $f(x) = ax^2$?

Discovery

(a) Figure 10.37 (b) Figure 10.38

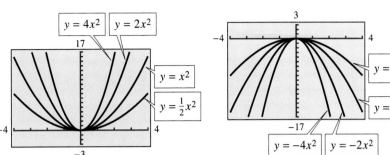

(c) If $a > 0$, the graph opens upward; if $a < 0$, the graph opens downward.

(d) As $|a|$ increases, the graph becomes narrower.

(e) Because the graph extends without bound to the left and right, the domain is the set of all real numbers. For $a > 0$ the range is $\{y \mid y \geq 0\}$; for $a < 0$, the range is $\{y \mid y \leq 0\}$.

The graph of a quadratic function is called a **parabola.** The parabola has a low point if the parabola opens upward and a high point if the parabola opens downward. This point is called the **vertex** of the parabola.

When the parabola opens upward ($a > 0$), the vertex corresponds to a minimum value of the function. When the parabola opens downward ($a < 0$), the vertex corresponds to a maximum value for the function.

Each graph in Exploration 1 is *symmetric with respect to the y-axis* or, in other words, *symmetric with respect to the line $x = 0$.* The line $x = 0$ is called the *axis of symmetry.* This means that if the graph is folded together along the line $x = 0$, the two sides of the parabola coincide.

Notice that the axis of symmetry contains the vertex.

Variations of the Graph of $f(x) = x^2$

We have seen that the absolute value of a affects the shape of a parabola and that the sign of a affects the direction in which it opens. But the vertex for $f(x) = ax^2$ is the origin for all values of a.

Now we consider how the vertex of a parabola can be shifted vertically and horizontally.

EXPLORATION 2   *Shifting the Vertex of a Parabola*

(a) Produce the graphs of $f(x) = x^2 + k$ for $k = 5$, -4, and -8. Compare them to the graph of $f(x) = x^2$.

(b) Produce the graphs of $f(x) = (x - h)^2$ for $h = -6$, 2, and 5. Compare them to the graph of $f(x) = x^2$.

(c) What is your conjecture about the effect of k and h on the vertex of the parabola?

Discovery

(a) Each graph is the same as the graph of $f(x) = x^2$ but shifted vertically. The axis of symmetry is the same for all of the graphs. (See Fig. 10.39.)

Figure 10.39

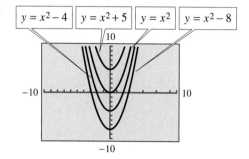

(b) Each graph is the same as the graph of $f(x) = x^2$ but shifted horizontally. The axis of symmetry is also shifted in each case. (See Fig. 10.40.)

Figure 10.40

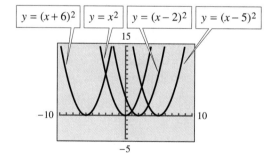

(c) The graph of $f(x) = x^2 + k$ is a vertical shift of the graph of $f(x) = x^2$. If $k > 0$, the graph is shifted upward k units; if $k < 0$, the graph is shifted downward $|k|$ units.

The graph of $f(x) = (x - h)^2$ is a horizontal shift of the graph of $f(x) = x^2$. If $h > 0$, the graph is shifted to the right h units; if $h < 0$, the graph is shifted to the left $|h|$ units.

A vertical or horizontal shift of $f(x) = x^2$ does not affect the domain of the function, but it may affect the vertex, the axis of symmetry, and the range.

	Vertex	*Axis of Symmetry*	*Range*
$f(x) = x^2 + k$	$(0, k)$	$x = 0$	$\{y : y \geq k\}$
$f(x) = (x - h)^2$	$(h, 0)$	$x = h$	$\{y : y \geq 0\}$

NOTE: Our conclusions about $f(x) = (x - h)^2$ are valid only if the function is written in that form. To analyze the function $f(x) = (x + h)^2$, we first write $f(x) = [x - (-h)]^2$. ■

In Example 1 we combine the conclusions drawn from our explorations.

EXAMPLE 1 *Properties of Graphs of Quadratic Functions*

Produce the graph of each of the following functions. Compare each graph to the graph of $y = x^2$. Describe the vertex, the axis of symmetry, and the range of the function.

(a) $f(x) = (x + 2)^2 - 3$ (b) $g(x) = (x - 4)^2 + 5$

Solution

(a) The graph of function f is the same as the graph of $y = x^2$ with the vertex shifted to the left 2 units and down 3 units. The vertex is $(-2, -3)$ and the axis of symmetry is the line $x = -2$. The range of f is $\{y \mid y \geq -3\}$. (See Fig. 10.41.)

(b) The graph of function g is the same as the graph of $y = x^2$ with the vertex shifted to the right 4 units and up 5 units. The vertex is $(4, 5)$, and the axis of symmetry is the line $x = 4$. The range of g is $\{y \mid y \geq 5\}$. (See Fig. 10.42.)

Figure 10.41

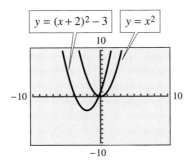

Figure 10.42

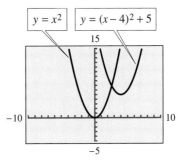

Example 1 prompts us to generalize as follows.

Properties of the Graph of $f(x) = (x - h)^2 + k$

The graph of $f(x) = (x - h)^2 + k$ looks like the graph of $y = x^2$ with the vertex shifted to the point (h, k). The graph of $f(x) = a(x - h)^2 + k$ looks like the graph of $y = ax^2$ with the vertex shifted to the point (h, k).

Vertex and Intercepts

By producing the graph of a quadratic function, we can estimate the coordinates of the vertex. Determining the exact coordinates requires an algebraic technique involving the method of completing the square and the general form $f(x) = a(x - h)^2 + k$.

EXAMPLE 2   *Determining the Vertex of a Parabola Algebraically*

Determine the vertex of the graph of $f(x) = x^2 - 8x + 10$.

Solution   You may want to begin by producing the graph of the function and tracing the graph to estimate the coordinates of the vertex.

By using the method of completing the square, we will write the function in the form $f(x) = (x - h)^2 + k$.

$$f(x) = x^2 - 8x + 10$$

$$f(x) - 10 = x^2 - 8x \qquad \text{Move the constant term to the left side.}$$

$$f(x) - 10 + 16 = x^2 - 8x + 16 \qquad \text{To complete the square, add } \left(\tfrac{8}{2}\right)^2 = 16 \text{ to both sides.}$$

$$f(x) + 6 = (x - 4)^2 \qquad \text{Factor and write in exponential form.}$$

$$f(x) = (x - 4)^2 + (-6) \qquad \text{Add } -6 \text{ to both sides to obtain the function.}$$

In this form we see that $h = 4$ and $k = -6$. Therefore, the vertex is $(4, -6)$. ■

The procedure illustrated in Example 2 is one we need to repeat each time we wish to determine the vertex of a parabola. By working through the same routine for the general quadratic function, we can derive formulas for determining the coordinates of the vertex.

$$f(x) = ax^2 + bx + c$$

$$f(x) - c = ax^2 + bx \qquad \text{Move the constant term to the left side.}$$

$$\frac{f(x)}{a} - \frac{c}{a} = x^2 + \frac{b}{a}x \qquad \text{Divide both sides by } a.$$

$$\frac{f(x)}{a} - \frac{c}{a} + \frac{b^2}{4a^2} = x^2 + \frac{b}{a}x + \frac{b^2}{4a^2} \qquad \text{Complete the square by adding } \left(\tfrac{1}{2} \cdot \tfrac{b}{a}\right)^2 = \frac{b^2}{4a^2} \text{ to both sides.}$$

$$\frac{f(x)}{a} - \frac{4ac - b^2}{4a^2} = \left(x + \frac{b}{2a}\right)^2 \qquad \text{Combine fractions and factor the right side.}$$

$$\frac{f(x)}{a} = \left(x + \frac{b}{2a}\right)^2 + \frac{4ac - b^2}{4a^2} \qquad \text{Add } \frac{4ac - b^2}{4a^2} \text{ to both sides.}$$

$$f(x) = a\left(x + \frac{b}{2a}\right)^2 + \frac{4ac - b^2}{4a} \qquad \text{Multiply both sides by } a \text{ to obtain } f(x).$$

With the function written in this form, we see that

$$h = -\frac{b}{2a} \quad \text{and} \quad k = \frac{4ac - b^2}{4a}.$$

The vertex is $\left(-\dfrac{b}{2a}, \ \dfrac{4ac - b^2}{4a} \right)$.

NOTE: The coordinates of the vertex can be found by using these two formulas. However, in practice, it is easier to determine the first coordinate by using the formula $-\dfrac{b}{2a}$. Then, the y-coordinate can be determined by evaluating $f\left(-\dfrac{b}{2a} \right)$. ∎

EXAMPLE 3 *Determining the Vertex Algebraically*

Determine the vertex of the graph of $f(x) = 2x^2 + 12x + 17$.

Solution The x-coordinate of the vertex is

$$x = -\frac{b}{2a} = -\frac{12}{2(2)} = -3.$$

Because $f(-3) = -1$, the vertex is $(-3, -1)$. ∎

In general, the y-intercept of a graph is a point whose x-coordinate is 0. For a quadratic function $f(x) = ax^2 + bx + c$, $f(0) = c$. Therefore, a parabola always has one y-intercept $(0, c)$.

To determine the x-intercepts, we solve the quadratic equation $ax^2 + bx + c = 0$. We know this equation has zero, one, or two real number solutions. Thus, a parabola has zero, one, or two x-intercepts.

The following summarizes the characteristics of the graph of a quadratic function $f(x) = ax^2 + bx + c$.

Figure 10.43

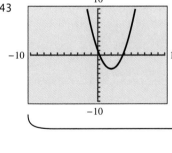
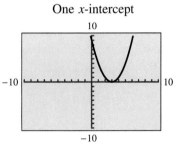

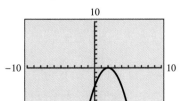
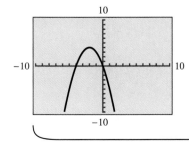
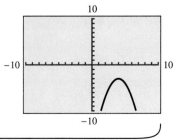

$b^2 - 4ac > 0$
Two x-intercepts

$b^2 - 4ac = 0$
One x-intercept

$b^2 - 4ac < 0$
No x-intercept

$a > 0$

$a < 0$

EXAMPLE 4 *Determining the Vertices and Intercepts of Parabolas*

Produce the graph of each of the following. Determine its vertex and intercepts algebraically.

(a) $f(x) = 6 - x - 2x^2$ (b) $g(x) = 4x^2 - 12x + 9$

(c) $h(x) = x^2 - 4x + 1$ (d) $s(x) = -2x^2 + 12x - 19$

Solution

Figure 10.44

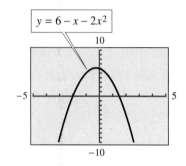

$y = 6 - x - 2x^2$

(a) Because $a = -2 < 0$, the graph opens downward. Also, $c = 6$, so the y-intercept is $(0, 6)$. The graph indicates two x-intercepts. (See Fig. 10.44.)

To determine the intercepts algebraically, solve the associated equation.

$$6 - x - 2x^2 = 0$$
$$(3 - 2x)(2 + x) = 0$$
$$3 - 2x = 0 \quad \text{or} \quad 2 + x = 0$$
$$3 = 2x \quad \text{or} \quad 2 + x = 0$$
$$x = \frac{3}{2} \quad \text{or} \quad x = -2$$

The x-intercepts are $(\frac{3}{2}, 0)$ and $(-2, 0)$.

Now we determine the coordinates of the vertex.

$$x = -\frac{b}{2a} = -\frac{-1}{2(-2)} = -\frac{1}{4} \quad \text{or} \quad -0.25$$

Because $f(-0.25) = 6.125$, the vertex is $(-0.25, 6.125)$.

Figure 10.45

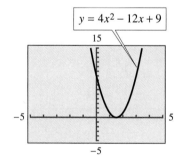

$y = 4x^2 - 12x + 9$

(b) Because $a = 4 > 0$, the graph opens upward. The y-intercept is $(0, 9)$. There appears to be only one x-intercept. (See Fig. 10.45.)

To determine the x-intercept exactly, solve the associated equation.

$$4x^2 - 12x + 9 = 0$$
$$(2x - 3)^2 = 0$$
$$2x - 3 = 0$$
$$2x = 3$$
$$x = \frac{3}{2}$$

The x-intercept is $(\frac{3}{2}, 0)$.

Now we determine the coordinates of the vertex.

$$x = -\frac{b}{2a} = -\frac{-12}{2(4)} = \frac{3}{2} = 1.5$$

Because $f(1.5) = 0$, the vertex is $(1.5, 0)$. (Note that the vertex and the x-intercept are the same point.)

(c) Because $a = 1 > 0$, the graph opens upward. The y-intercept is $(0, 1)$. There appear to be two x-intercepts. (See Fig. 10.46.)

Figure 10.46

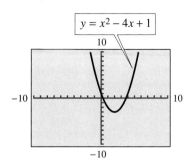

$$y = x^2 - 4x + 1$$

To determine the x-intercepts exactly, use the Quadratic Formula to solve the equation $x^2 - 4x + 1 = 0$.

$$x = \frac{4 \pm \sqrt{(-4)^2 - 4(1)(1)}}{2(1)} = \frac{4 \pm \sqrt{16 - 4}}{2} = \frac{4 \pm \sqrt{12}}{2}$$

$$x \approx 3.73 \quad \text{or} \quad x \approx 0.27$$

The (approximate) x-intercepts are $(3.73, 0)$ and $(0.27, 0)$. Now we determine the coordinates of the vertex.

$$x = -\frac{b}{2a} = -\frac{-4}{2(1)} = 2$$

Because $f(2) = -3$, the vertex is $(2, -3)$.

(d) Because $a = -2 < 0$, the parabola opens downward. The y-intercept is $(0, -19)$. There appear to be no x-intercepts. (See Fig. 10.47.)

Figure 10.47

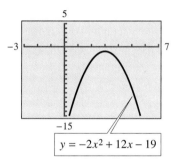

$$y = -2x^2 + 12x - 19$$

We confirm that there are no x-intercepts by evaluating the discriminant.

$$b^2 - 4ac = 12^2 - 4(-2)(-19) = 144 - 152 = -8$$

Because the discriminant is negative, there are no x-intercepts. Now we determine the coordinates of the vertex.

$$x = -\frac{b}{2a} = -\frac{12}{2(-2)} = 3$$

Because $f(3) = -1$, the vertex is $(3, -1)$.

Applications

Applications in which a quantity is to have an optimum (maximum or minimum) value can often be modeled with quadratic functions. The vertex of the graph is the point of interest because it is at that point that the function has its maximum or minimum value.

In Section 10.5 we saw an example of a day care operator who wished to enclose a rectangular play yard of a specified area. In Example 5, we see a similar problem, but this time the goal is to *maximize* the area.

EXAMPLE 5   *Maximizing a Rectangular Area*

A day care operator wants to use 200 feet of fencing to enclose a rectangular play area for her day care center. She plans to use an existing fence as one side. The existing fence extends from each back corner of her home to the property line. (See Fig. 10.48.) What dimensions should she use to maximize the rectangular area? What is the maximum area?

Figure 10.48

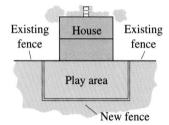

Solution

Let x = the width of the rectangle. Because the amount of fencing available is 200 feet, the length of the rectangle is $200 - 2x$. (See Fig. 10.49.) The area is given by the function $A(x) = x(200 - 2x) = -2x^2 + 200x$.

Figure 10.49

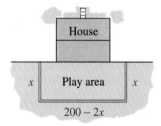

Because $a = -2 < 0$, the parabola opens downward. The maximum value of the function occurs at the vertex, which we estimate as (50, 5000). (See Fig. 10.50.)

Figure 10.50

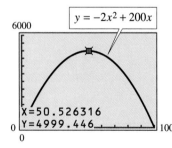

The x-coordinate of the vertex is determined as follows.

$$x = -\frac{b}{2a} = -\frac{200}{2(-2)} = 50$$

Because x represents the width of the play area, the maximum area is achieved when the width is 50 feet. The length is

$$200 - 2x = 200 - 2(50) = 100.$$

The area of the yard can be found either by evaluating the function for $x = 50$ or by using the formula $A = LW$.

$$A(50) = 5000$$
$$A = LW = 100 \cdot 50 = 5000$$

The maximum area of 5000 square feet is achieved when the width is 50 feet and the length is 100 feet.

10.8   *Quick Reference*

Properties of Graphs

- The graph of a quadratic function $f(x) = ax^2 + bx + c$ is called a **parabola.**

- If $a > 0$, the parabola opens upward; if $a < 0$, the parabola opens downward. As $|a|$ increases, the parabola becomes narrower; as $|a|$ decreases, the parabola becomes wider.

- The lowest point of a parabola (when $a > 0$) or the highest point (when $a < 0$) is called the **vertex.**

- The domain of a quadratic function is the set of all real numbers. If $V(h, k)$ is the vertex, then the range of the function is $\{y \mid y \geq k\}$ when a > 0 or $\{y \mid y \leq k\}$ when a < 0.

- The graph of a quadratic function is **symmetric** with respect to a vertical line containing the vertex. This line is called the **axis of symmetry.**

Variations of the Graph of $f(x) = x^2$

- For $f(x) = x^2 + k$, the vertex of the parabola is $V(0, k)$.

- For $f(x) = (x - h)^2$, the vertex of the parabola is $V(h, 0)$.

- The graph of $f(x) = a(x - h)^2 + k$ looks like the graph of $y = ax^2$ with the vertex shifted to $V(h, k)$.

Vertex and Intercepts

- To determine the vertex of the graph of a quadratic function, we can use the method of completing the square to rewrite the function in the form $f(x) = a(x - h)^2 + k$.

- In general, the x-coordinate of the vertex is $-\dfrac{b}{2a}$.

  The y-coordinate can be determined by evaluating $f\left(-\dfrac{b}{2a}\right)$.

- If $a > 0$, then the y-coordinate of the vertex represents the minimum value of the function; if $a < 0$, then the y-coordinate represents the maximum value of the function.

- The y-intercept of the graph of $f(x) = ax^2 + bx + c$ is $(0, c)$.

- To determine the x-intercepts, solve the quadratic equation $ax^2 + bx + c = 0$. The number of x-intercepts depends on the sign of the discriminant.

10.8   *Exercises*

 1. What do we call the graph of a quadratic function $f(x) = ax^2 + bx + c$? Describe how the value of a affects the shape and orientation of the graph.

2. What is the domain of any quadratic function? If you know the vertex $V(h, k)$ of the graph of a quadratic function, can you specify the range of the function? If so, state what the range is. If not, what additional information do you need?

In Exercises 3–6, for $f(x) = ax^2 + bx + c$, the value of a is given along with the vertex V of the graph of the function. What is the range of the function?

3. $a = -2$; $V(1, 3)$

4. $a = 1$; $V(-2, 1)$

5. $a = \dfrac{3}{5}$; $V(0, 2)$

6. $a = -3$; $V(-4, 0)$

7. If the vertex $V(h, k)$ of a parabola is known, explain how to write the equation of the vertical axis of symmetry.

8. If the equation of the axis of symmetry of a parabola is $x = a$, what can you conclude about the vertex of the parabola?

In Exercises 9–12, write the equation of the vertical axis of symmetry for a parabola with vertex V.

9. $V(0, -1)$ **10.** $V(-1, -5)$

11. $V(3, 0)$ **12.** $V(0, 4)$

In Exercises 13–18, match the function with its graph.

13. $f(x) = x^2 - 4$ **14.** $g(x) = x - 4$

15. $h(x) = -2x^2$ **16.** $f(x) = (x - 3)^2$

17. $g(x) = -\dfrac{1}{2}x - 2$

18. $h(x) = -\dfrac{1}{2}(x + 1)^2 - 2$

(a)

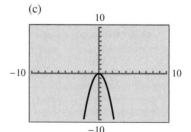

(b)

(c)

(d)

(e)

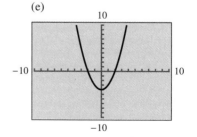

(f)

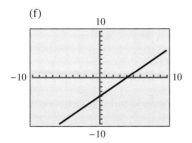

32. $g(x) = (x + 3)^2 - 4$

33. $h(x) = x^2 + 2x - 3$

34. $f(x) = x^2 - 6x + 5$

35. $g(x) = 10 - 6x - 3x^2$

36. $h(x) = -6 + 5x - 3x^2$

 19. Describe the graph of $f(x) = x^2 + k$ in comparison to the graph of $y = x^2$.

 20. Describe the graph of $f(x) = (x - h)^2$ in comparison to the graph of $y = x^2$.

In Exercises 21–28, compare the graphs of the given pair of functions to the graph of $y = x^2$. State whether the graphs are more or less narrow and describe any horizontal or vertical shifts of the vertices.

21. (a) $y = 3x^2$ (b) $y = \dfrac{1}{3}x^2$

22. (a) $y = -x^2$ (b) $y = -2x^2$

23. (a) $y = x^2 + 4$ (b) $y = x^2 - 4$

24. (a) $y = (x + 3)^2$ (b) $y = (x - 3)^2$

25. (a) $y = x^2 - 1$ (b) $y = (x - 1)^2$

26. (a) $y = \dfrac{1}{4}x^2 - 2$ (b) $y = -2x^2 + 4$

27. (a) $y = (x - 3)^2 - 2$ (b) $y = (x + 1)^2 + 4$

28. (a) $y = -2(x + 1)^2 - 1$ (b) $y = 3(x - 1)^2 + 2$

In Exercises 37–40, the range of a quadratic function is given along with the equation of the axis of symmetry of the graph of the function. What is the vertex of the graph?

37. $\{y \mid y \geq 3\}$; $x = 0$

38. $\{y \mid y \leq -2\}$; $x = -3$

39. $\{y \mid y \leq -1\}$; $x = 5$

40. $\{y \mid y \geq 0\}$; $x = -2$

In Exercises 41–46, for the given function, determine the vertex of its graph and the equation of the axis of symmetry.

41. $f(x) = x^2 - 6x + 9$

42. $g(x) = x^2 + 4x + 9$

43. $h(x) = 2x^2 + 4x - 7$

44. $g(x) = 3x^2 + 6x + 8$

45. $h(x) = 7 - 4x - x^2$

46. $f(x) = 5x - 2x^2$

 29. Given the function $f(x) = ax^2 + bx + c$, we can determine the vertex of the graph by using the method of completing the square. Describe a faster method for determining the x-coordinate of the vertex. How do we then determine the y-coordinate?

 30. How can we determine the minimum or maximum value of a quadratic function if we know the vertex of the graph of the function?

In Exercises 31–36, produce the graph of the given function and estimate the vertex. Then, determine the vertex algebraically and state the minimum or maximum value of the function.

31. $f(x) = (x - 3)^2 + 4$

 47. Use what you know about the domain of a quadratic function to explain why its graph always has a y-intercept. If you know what the function is, what is an easy way to determine the y-intercept?

 48. Describe how to use a graph to estimate the x-intercepts, if any, of a parabola. Then, describe how to determine the x-intercepts algebraically.

In Exercises 49–56, use a graph to estimate the y-intercepts and the x-intercepts, if any. Then, verify the intercepts algebraically.

49. $f(x) = x^2 - 3x - 28$

50. $h(x) = 2x^2 - 11x + 12$

51. $g(x) = 4x^2 - 5x + 9$

52. $f(x) = -3 + x - x^2$

53. $f(x) = 5 - 4x - 3x^2$

54. $h(x) = 2x^2 + 5x - 2$

55. $h(x) = 3x^2 + 5x$

56. $g(x) = 0.2x - 0.4x^2$

In Exercises 57–60, $y = f(x) = ax^2 + bx + c$, where a, b, and c are rational numbers. In each exercise the value of the discriminant is given for the associated equation $y = 0$. How many x-intercepts does the graph of $y = f(x)$ have? Are the x-coordinates of these points rational or irrational numbers?

57. -4 **58.** 0

59. 16 **60.** 20

In Exercises 61–66, $y = f(x) = ax^2 + bx + c$. In each exercise the value of a is given along with the vertex of the graph of $y = f(x)$. Determine the number of x-intercepts of the graph.

61. 2, $V(-2, 1)$ **62.** -1, $V(3, 2)$

63. $\dfrac{1}{2}$, $V(2, -4)$ **64.** -3, $V(-1, -2)$

65. 3, $V(2, 0)$ **66.** -4, $V(0, 1)$

In Exercises 67–72, information is given about the graph of a quadratic function. Determine the number of x-intercepts.

67. The vertex is $(-2, -5)$, and the graph opens upward.

68. The vertex is $(0, 1)$, and the graph opens upward.

69. The vertex is $(-3, 0)$.

70. The vertex is $(1, 2)$, and the y-intercept is $(0, 7)$.

71. The maximum value of the function occurs at $(-4, 3)$.

72. The minimum value of the function occurs at $(3, 0)$.

In Exercises 73–78, determine the following information for each function.

  (a) the domain and range of the given function

  (b) the x- and y-intercepts of its graph

  (c) the vertex and the equation of the axis of symmetry

73. $f(x) = x^2 - 6x + 8$

74. $g(x) = x^2 + 8x + 12$

75. $f(x) = 5 - 4x - x^2$

76. $g(x) = 12 - 4x - x^2$

77. $h(x) = 2x^2 + 4x + 9$

78. $g(x) = 2x - x^2 - 8$

In Exercises 79–86, you are given information about the location of the graph of $f(x) = (x - h)^2 + k$ relative to the location of the graph of $y = x^2$. Determine the values of h and k.

79. The graph of $f(x)$ is up 3 units.

80. The graph of $f(x)$ is down $\frac{1}{2}$ unit.

81. The graph of $f(x)$ is left $\frac{3}{5}$ unit.

82. The graph of $f(x)$ is right 1 unit.

83. The graph of $f(x)$ is up 2 units and left 4 units.

84. The graph of $f(x)$ is left 1 unit and down 4 units.

85. The graph of $f(x)$ is right 3 units and up 2 units.

86. The graph of $f(x)$ is right 2 units and down 3 units.

 87. If we know that the solutions of $ax^2 + bx + c = 0$ are m and n, what information do we have about the graph of $f(x) = ax^2 + bx + c$?

 88. If $(m, 0)$ and $(n, 0)$ are x-intercepts of the graph of $f(x) = ax^2 + bx + c$, what are two factors of the expression $ax^2 + bx + c$? How do you know?

In Exercises 89–94, determine the values of b and c in the quadratic function $f(x) = x^2 + bx + c$ so that the graph of f has the given x-intercept(s).

89. $(5, 0)$ and $(-3, 0)$ **90.** $(-2, 0)$ and $(-5, 0)$

91. $(0, 0)$ and $(-2, 0)$ **92.** $(-4, 0)$ and $(4, 0)$

93. $(4, 0)$ only

94. $(-1, 0)$ only

In Exercises 95–98, use the given information about the graph of the given quadratic function to write the specific function.

95. The graph of $f(x) = x^2 + bx + c$ has a y-intercept $(0, 3)$ and an axis of symmetry whose equation is $x = 2$.

96. The graph of $f(x) = 2x^2 + bx + c$ has the vertex $V(2, -3)$.

97. The graph of $f(x) = ax^2 + x + c$ has an x-intercept $(-1, 0)$, and the first coordinate of the vertex is $-\frac{1}{6}$.

98. The graph of $f(x) = ax^2 - 8x + c$ has the vertex $V(1, -9)$.

In Exercises 99–104, write a quadratic function that describes the conditions of the problem. Then, answer the question.

99. The sum of two numbers is 30. Determine the numbers such that the product is a maximum.

100. The difference of two numbers is 36. Determine the two numbers such that their product is a minimum.

 101. The Canton Package Co. wants to design a box that is 11 inches high with a perimeter around the base of 80 inches. What should be the length and width of the box to have the maximum volume? What is the maximum volume (in cubic feet)?

102. If an object is projected vertically upward, its height h (in feet) after t seconds is given by $h = 96t - 16t^2$. At what time will the object be at the maximum height above the ground? How high will it be?

103. An engineer wants to design a parabolic arch whose height is 60 feet. At ground level, the span of the arch is 100 feet. If one ground point of the arch is placed at the origin of a coordinate system and the other ground point is placed along the positive x-axis, determine the quadratic function that models the arch.

104. In Exercise 103 suppose the engineer places the vertex of the parabola on the positive y-axis and the ground points on the x-axis. Determine the quadratic function that models the arch and compare it to the function determined in Exercise 103.

Exploring with Real Data

In Exercises 105–108, refer to the following background information and data.

During the period 1988–1992, the net income of commercial banks decreased to a low in 1989–1990 and then increased from 1990–1992. (See figure.) (Source: Federal Deposit Insurance Corporation.)

Figure for 105–108

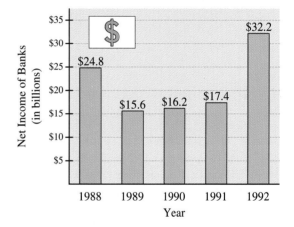

For the period 1988–1992, net income N can be modeled by

$$N(t) = 3.075t^2 - 59.65t + 305.2$$

where t is the number of years since 1980.

105. Find and interpret the vertex of the graph of the function.

106. Produce the graph of $y = N(t)$. How likely is the model to be valid beyond 1992? Why?

107. Why would the banking industry never want a model for N to have a graph with x-intercepts?

108. In the 1980s, believing that it would benefit the whole economy, the federal government drastically deregulated the banking industry. In response to dramatic declines in banking profits, regulations began to be put back into place in 1990. How do you think this graph might be explained by deregulation?

Challenge

In Exercises 109–112, the graph of the function $f(x) = ax^2 + bx + c$ is a parabola containing the given points. Determine $f(x)$.

109. $(-1, 0)$, $(1, -6)$, and $(2, -3)$

110. $(1, 6)$, $(-2, -12)$, and $(3, 8)$

111. $(1, 3)$, $(2, 7)$, and $(3, 13)$

112. $(0, 1)$, $\left(-\dfrac{1}{2}, 0\right)$, and $(1, 9)$

113. A commercial developer finds that the rental income I (in dollars) for a shopping mall is given by $I(x) = (700 + 25x)(30 - x)$, where x is the number of vacant stores.

(a) It seems reasonable to expect that the maximum income would be derived when there are no store vacancies, that is, when $x = 0$. Evaluate $I(0)$ to determine the income when no stores are vacant.

(b) Now determine the vertex of the graph of the income function. For how many store vacancies is the income maximized?

(c) How many stores are there in the shopping mall?

114. For a certain flight, a charter airline charges $96 per person. Each person is charged an additional $4 for each unsold seat. If the plane has 60 seats, find the number of unsold seats that produces the maximum revenue.

10 Chapter Review Exercises

Section 10.1

1. If you want to estimate the solutions of the equation $x^2 - 2 = 5x$ graphically, explain how to do so with (a) one graph and (b) with two graphs.

In Exercises 2 and 3, use the graphing method to solve.

2. $a^2 - 7a + 10 = 0$ **3.** $13 - 3z^2 = 0$

In Exercises 4 and 5, use the Zero Factor Property to solve.

4. $\left(4x - \dfrac{2}{7}\right)\left(6x - \dfrac{3}{5}\right) = 0$

5. $(d - 3)(d + 2) = 36$

In Exercises 6 and 7, use the factoring method to solve.

6. $16 + 9c^2 = 24c$ **7.** $x(x + 2) = 15$

In Exercises 8 and 9, solve by using the Square Root Property.

8. $25 - y^2 = 0$

9. $25(b + 4)^2 - 16 = 0$

In Exercises 10 and 11, determine the complex number solutions.

10. $a^2 + 36 = 0$

11. $(b - 2)^2 + 37 = 0$

12. The sum of the squares of two consecutive odd integers is 130. What are the integers?

Section 10.2

13. Suppose you use the method of completing the square to solve $2x^2 + 5x + 2 = 0$. Explain why you will eventually add $\frac{25}{16}$ to both sides of the equation.

In Exercises 14 and 15, solve by completing the square.

14. $x^2 + 10x - 1 = 0$ **15.** $3y(y - 5) = 12$

In Exercises 16 and 17, determine the imaginary number solutions by completing the square.

16. $x^2 - 6x + 10 = 0$

17. $3x^2 - 2x + 4 = 0$

18. The sum of the reciprocals of two consecutive integers is $\frac{7}{12}$. What are the integers?

Section 10.3

19. Which of the following equations can be solved with the Quadratic Formula?

    (i) $3x^2 = 12$

    (ii) $(x + 1)(x - 5) = 0$

    (iii) $x^2 + x + 1 = 0$ (iv) $(x + 3)^2 = 5$

    From this list, identify those equations that can be solved with an alternate method and state the method.

In Exercises 20–23, solve by using the Quadratic Formula.

20. $x(x + 3) = 7$

21. $5x^2 - 16 = 0$

22. $3 + \dfrac{4}{x^2} = \dfrac{7}{x}$

23. $4x^2 = 5x + 3$

In Exercises 24 and 25, determine the complex number solutions by using the Quadratic Formula.

24. $x^2 + 4x + 7 = 0$

25. $2x^2 + 9 = x$

In Exercises 26–29, use the discriminant to determine the number and the nature of the solutions of the equation.

26. $8x = x^2 + 16$

27. $4x = x^2 - 12$

28. $8x = 2x^2 - 7$

29. $6x^2 - 2x + 5 = 0$

30. The sum of the square of a number and 5 times the number is 7. What is the number?

Section 10.4

31. If $x > 0$, which of the following equations are quadratic in form?

    (a) $\dfrac{2}{x^2} + \dfrac{3}{x} - 5 = 0$ (b) $2x + \sqrt{2x} = 3$

    (c) $(x + 1)^2 + 2(x + 1) + 1 = 0$

    (d) $x^{1/4} - x^{1/2} + 3 = 0$

For the equations you selected, state the substitution you would use and rewrite the equation as a quadratic equation.

In Exercises 32–37, determine the real number solutions of the equation.

32. $\dfrac{3x}{x - 4} + \dfrac{3}{x + 3} = 4$

33. $\sqrt{3x - 4} = \sqrt{x + 1} - 2$

34. $a^4 - 5a^2 + 6 = 0$

35. $x - 3x^{1/2} = 10$

36. $x^{1/6} + x^{1/3} = 2$

37. $(x^2 + x)^2 + 72 = 18(x^2 + x)$

38. Use an appropriate substitution to write the equation $3 + 10x^{-1} + 8x^{-2} = 0$ as a quadratic equation. Then, determine the solutions of the original equation.

Section 10.5

39. Two tablecloths are to be cut from a rectangular piece of material. One tablecloth is a circle and the other is a rectangle whose length is 3 feet longer than the width. (See figure.) The total area of the tablecloths is 90 square feet. How many square yards of material are needed?

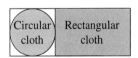

40. A fraternity sold tickets for its annual dinner-dance. Ten couples could not attend the dinner but did attend the dance. The fraternity collected $1680 for dinner tickets and $1300 for dance tickets. If the dinner tickets cost $15 more than the dance tickets, how many couples attended the dance?

41. For Hanukkah an aunt sent $6 to each child and a grandmother sent $60 to be divided equally among the children of the family. An uncle sent $56 to be divided among the children and three other relatives. A grandfather sent each child $5. If each child received a total of $30, how many children were in the family?

42. A boat traveled up a river and back, a total distance of 40 miles, in 6 hours. If the rate of the current was 5 mph, what would the rate of the boat have been in still water?

43. Working together, two people can wash all the windows in their home in 6 hours. It takes one person 5 hours longer than it takes the other to wash the windows alone. How long would it take the slower worker, working alone, to wash the windows?

Section 10.6

In Exercises 44–49, solve the inequality.

44. $2x^2 + 6 \geq 7x$

45. $x(x - 6) + 9 \leq 0$

46. $4x(x + 5) + 9 \leq 0$

47. $x^2 + 5 \geq 3x$

48. $2x - x^2 + 2 > 0$

49. $x - x^2 - 1 > 0$

 50. A dog owner has 250 feet of fencing to enclose a rectangular run area for his dogs. If he wants the area to be at least 3500 square feet, what can the length of the rectangle be?

Section 10.7

 51. Suppose that you are solving the inequality $(x + 5)(x - 3)^2 > 0$. The accompanying figure shows the graph of $y = (x + 5)(x - 3)^2$.

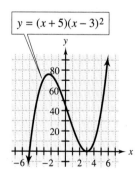

$y = (x + 5)(x - 3)^2$

From the following list, identify the correct solution set and explain why it is correct.

(i) $(-5, 3)$

(ii) $(-5, \infty)$

(iii) $(-5, 3) \cup (3, \infty)$

(iv) $(-5, 3]$

In Exercises 52–55, use a graph to estimate the x-intercepts. Then, determine the x-intercepts algebraically and write the solution set of the inequality.

52. $(x - 2)(x + 1)(x + 3) > 0$

53. $(x - 3)^2(x + 1) \geq 0$

54. $\dfrac{x + 3}{x - 4} < 0$

55. $\dfrac{3 - 4x}{3x - 4} \geq 1$

Section 10.8

 56. Explain how you can use a graph to estimate the domain and range of a quadratic function.

 57. Explain how you can use the graph of the function $f(x) = ax^2 + bx + c$ to estimate the solutions of the equation $ax^2 + bx + c = 0$.

In Exercises 58 and 59, determine the vertex and the intercepts of the graph of the given function.

58. $f(x) = x^2 - 5x + 9$

59. $g(x) = 3x^2 - 4x - 7$

In Exercises 60–63, determine whether the graph of the given function opens upward or downward. Then, find the vertex of the graph and the range of the function.

60. $f(x) = 17 - x^2$

61. $g(x) = 2 - \dfrac{4}{5}x - \dfrac{2}{3}x^2$

62. $h(x) = 2x^2 - 5x$

63. $f(x) = 4x^2 - 6x + 7$

64. A dog owner has 250 feet of fencing to enclose a rectangular run area for his dogs. If he wants the maximum possible area, what should the length of the rectangle be?

65. If an object is thrown vertically upward, its height (in feet) after t seconds is given by $h(t) = 64t - 16t^2$.

(a) In how many seconds will the object be at its maximum height?

(b) What is the maximum height reached by the object?

10 Chapter Test

1. Solve $3t^2 + 4t = 1$ by completing the square.

In Questions 2–8, solve the given equation.

2. $x(6x - 5) = 6$

3. $(2x - 1)^2 - 5 = 0$

4. $2 = \dfrac{2}{t} + \dfrac{1}{t^2}$

5. $(2y + 1)(y + 3) = 2$

6. $3x^2 - x + 2 = 0$

7. $9x^4 + 7x^2 = 2$

8. $\dfrac{2}{x + 7} + \dfrac{16}{x^2 + 6x - 7} = \dfrac{x}{1 - x}$

In Questions 9–11, write an equation to model the given conditions. Then, solve the equation and answer the question.

9. The area A of a triangle is 20 square feet. The base b is 3 feet longer than the height h. What are the dimensions of the triangle? $(A = \frac{1}{2}bh.)$

10. At a dairy a milk storage tank has an inlet pipe for receiving milk from the milking station. An outlet pipe takes the milk to the bottling room. The tank fills in 8 hours if both pipes are open. It takes 2 hours longer to empty the tank with the outlet pipe than to fill the tank with the inlet pipe. How long does it take to empty a full tank?

11. A gutter is formed from a 100-foot-long sheet of metal that is 12 inches wide. To make the gutter, the sides are folded up the same amount on each side. (See figure.) What should the height of the gutter be for maximum capacity?

 12. Suppose you use a graph to estimate the solutions of a quadratic inequality. Explain how the graph can have no x-intercepts, but the solution set of the inequality is the set of all real numbers.

In Questions 13–15, solve the given inequality.

13. $6 + x - x^2 \leq 0$

14. $\dfrac{2}{x - 3} < \dfrac{1}{x + 2}$

15. $t(2t + 1)(t - 5) \geq 0$

16. Suppose $f(x) = x^2 - 6x + 11$. Write the function in the form $f(x) = (x - h)^2 + k$. Then, give the coordinates of the vertex of the graph of f and write the equation of the axis of symmetry.

17. The figure shows the graph of $f(x) = ax^2 + bx + c$.

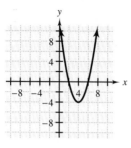

(a) What are the domain and range of the function?

(b) What do you know about the coefficient a?

(c) What are the estimated solutions of $ax^2 + bx + c = 0$?

18. To the nearest hundredth, determine the x-intercepts of the graph of $f(x) = x^2 + 6x - 20$.

11 Exponential and Logarithmic Functions

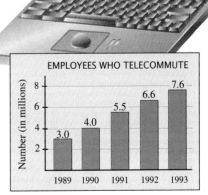

An increasing number of employees "commute without moving." *Telecommuting* refers to the use of computers and telephone technology for conducting work from home. The bar graph shows the trend in such activities during 1989–1993.

These data can be modeled by a **one-to-one** function. As we will see, such functions have **inverse functions** that can be evaluated, graphed, and analyzed like any other function. (For more on the telecommuting function and its inverse, see Exercises 93–96 at the end of Section 11.2.)

In Chapter 11 we discuss the algebra of functions and their inverses. In particular, we study the properties and graphs of two inverse functions called exponential and logarithmic functions. Finally, we learn how to solve equations and applications that involve exponential and logarithmic expressions.

EMPLOYEES WHO TELECOMMUTE

(Source: LINK Resources Corporation.)

11.1 Algebra of Functions

Review of Functions ▪ *Basic Operations on Functions* ▪ *Composition of Functions* ▪ *Applications*

Review of Functions

In Chapter 3 we defined a function, introduced function notation, and discussed methods for evaluating functions. We begin with a brief review of these topics.

> ### Definition of a Function
>
> A **function** is a set of ordered pairs in which no two different ordered pairs have the same first element.

If we produce the graph of a relation, we can use the Vertical Line Test to determine if it is a function. Because no two ordered pairs of a function can have the same first element, any vertical line will intersect the graph of a function at most once. For example, the graph in Fig. 11.1 is the graph of a function; the graph in Fig. 11.2 is not the graph of a function.

Figure 11.1

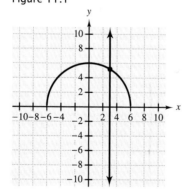

Figure 11.2

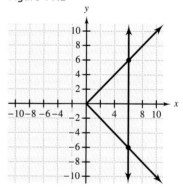

An equation in x and y describing a given relation also describes a function if each value of x corresponds to a unique value of y. For example, because every real number has only one absolute value, the equation $y = |x|$ describes a function. However, the equation $x = |y|$ does not describe a function because its solution set includes, for instance, the pairs $(4, 4)$ and $(4, -4)$.

Although the definition of *function* refers specifically to a set of ordered pairs, we usually call equations like $y = |x|$ functions because their solution sets are functions.

Recall that the **domain** of a function is the set of all first coordinates of the ordered pairs, and the **range** is the set of all second coordinates of the ordered pairs.

EXAMPLE 1 *Determining the Domain and Range of a Function*

Determine the domain and range of the function $f(x) = \sqrt{x - 3}$.

Solution

The square root is defined for all values of x for which the radicand is nonnegative.

$$x - 3 \geq 0$$
$$x \geq 3 \qquad \text{The domain is } \{x \mid x \geq 3\}.$$

Because the principal square root is always nonnegative, the range is $\{y \mid y \geq 0\}$.

Basic Operations on Functions

In previous chapters we have added, subtracted, multiplied, and divided algebraic expressions. Functions can also be combined with these operations.

Definitions of Basic Operations on Functions

If functions f and g have domains with at least one element in common, then the following operations are defined for each x in both the domain of f and the domain of g.

$(f + g)(x) = f(x) + g(x)$	Sum
$(f - g)(x) = f(x) - g(x)$	Difference
$(fg)(x) = f(x)g(x)$	Product
$\left(\dfrac{f}{g}\right)(x) = \dfrac{f(x)}{g(x)}, \quad \text{for } g(x) \neq 0$	Quotient

EXAMPLE 2 *Performing Basic Operations on Functions*

Let $f(x) = x^2 + 2$ and $g(x) = |x - 1|$. Evaluate each of the following.

(a) $(f + g)(1)$

(b) $(f - g)(-1)$

(c) $(fg)(2)$

(d) $\left(\dfrac{f}{g}\right)(0)$

Solution

(a) Both f and g are defined at 1.

$$(f + g)(1) = f(1) + g(1) \qquad \text{Definition of sum}$$
$$= [1^2 + 2] + |1 - 1| \qquad \text{Evaluate each function.}$$
$$= 3$$

(b) Both f and g are defined at -1.

$$(f - g)(-1) = f(-1) - g(-1) \qquad \text{Definition of difference}$$
$$= [(-1)^2 + 2] - |-1 - 1| \qquad \text{Evaluate each function.}$$
$$= [1 + 2] - |-2|$$
$$= 1$$

(c) Both f and g are defined at 2.

$$(fg)(2) = f(2)g(2) \qquad \text{Definition of product}$$
$$= [2^2 + 2]\,|2 - 1| \qquad \text{Evaluate each function.}$$
$$= [4 + 2]\,|1|$$
$$= 6$$

(d) Both f and g are defined at 0 and $g(0) \neq 0$.

$$\left(\frac{f}{g}\right)(0) = \frac{f(0)}{g(0)} \qquad \text{Definition of quotient}$$
$$= \frac{0^2 + 2}{|0 - 1|} \qquad \text{Evaluate each function.}$$
$$= \frac{2}{1}$$
$$= 2$$

EXAMPLE 3   *Performing Basic Operations on Functions*

Let $f(x) = 2x + 3$ and $g(x) = x - 5$. Determine each of the following.

(a) $(f + g)(x)$ (b) $(f - g)(x)$ (c) $(fg)(x)$ (d) $\left(\dfrac{f}{g}\right)(x)$

Solution   Note that the domain for both functions f and g is the set of all real numbers. Therefore, the operations can be performed for all values of x except as indicated in part (d).

(a) $(f + g)(x) = f(x) + g(x) = (2x + 3) + (x - 5) = 3x - 2$

(b) $(f - g)(x) = f(x) - g(x) = (2x + 3) - (x - 5) = x + 8$

(c) $(fg)(x) = f(x)g(x) = (2x + 3)(x - 5) = 2x^2 - 7x - 15$

(d) $\left(\dfrac{f}{g}\right)(x) = \dfrac{f(x)}{g(x)} = \dfrac{2x + 3}{x - 5} \quad (x \neq 5)$

Note that 5 is in the domain of both functions f and g, but 5 is not in the domain of f/g.

EXAMPLE 4   *Determining the Domain of a Combined Function*

Let $f(x) = \sqrt{x + 3}$ and $g(x) = \sqrt{2 - x}$. Determine $(f + g)(x)$. What is the domain of $(f + g)$?

Solution

$$(f + g)(x) = f(x) + g(x) = \sqrt{x + 3} + \sqrt{2 - x}$$

The domain of $(f + g)$ is the intersection of the domains of f and g. From the graphs of f and g in Fig. 11.3, we can estimate that the intersection of the domains of f and g is the interval of x-values between -3 and 2, inclusive.

Figure 11.3

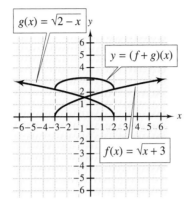

Figure 11.3 also shows the graph of $f + g$ whose domain appears to be the same interval. These estimates can be verified algebraically.

For function f, the domain is the interval of x-values such that $x + 3 \geq 0$, and so $x \geq -3$.

For function g, the domain is the interval of x-values such that $2 - x \geq 0$, and so $x \leq 2$.

Because the domain of f is $[-3, \infty)$ and the domain of g is $(-\infty, 2]$, the domain of $(f + g)$ is $[-3, \infty) \cap (-\infty, 2] = [-3, 2]$. ∎

EXAMPLE 5   *Determining the Domain of a Combined Function*

Let $f(x) = x^2 - 16$ and $g(x) = 2x - 8$. What is the domain of f/g?

Solution   Although the domain of both functions f and g is the set of all real numbers, f/g is defined only if $g(x) \neq 0$. Because $g(x) = 0$ if $x = 4$, the domain of f/g is the set of all real numbers *except* 4.

The process of determining f/g will also reveal the domain of the function.

$$
\begin{aligned}
\left(\frac{f}{g}\right)(x) &= \frac{f(x)}{g(x)} \\[4pt]
&= \frac{x^2 - 16}{2x - 8} \\[4pt]
&= \frac{(x + 4)(x - 4)}{2(x - 4)} \qquad \text{Factor.} \\[4pt]
&= \frac{x + 4}{2} \qquad\qquad \text{Divide out the common factor } x - 4.
\end{aligned}
$$

When we divide out the common factor $x - 4$, we are assuming that $x - 4$ is not 0, and so $x \neq 4$. That is why 4 must be excluded from the domain of f/g, even though the simplified expression for f/g appears to have no restricted values. ■

Composition of Functions

Another way to combine functions is to *compose* the functions.

Definition of Composition

The **composition** of functions f and g is a function $f \circ g$ where $(f \circ g)(x) = f(g(x))$ for all x in the domain of g for which $g(x)$ is in the domain of f.

NOTE: The notation $f \circ g$, used for the *composition* of f and g, does not mean the same thing as fg, which is used for the *product* of f and g. ■

EXAMPLE 6   *Composition of Two Functions*

Let $f(x) = x^2 - 1$ and $g(x) = x + 3$. Determine each of the following.

(a) $(f \circ g)(-2)$ (b) $(g \circ f)(-2)$

(c) $(f \circ g)(x)$ (d) $(g \circ f)(x)$

Solution

(a) Because $g(-2) = -2 + 3 = 1$,
$$(f \circ g)(-2) = f(g(-2)) = f(1) = 1^2 - 1 = 0.$$

(b) Because $f(-2) = (-2)^2 - 1 = 4 - 1 = 3$,
$$(g \circ f)(-2) = g(f(-2)) = g(3) = 3 + 3 = 6.$$

(c) We determine $(f \circ g)(x)$ by substituting $g(x)$ for all x in $f(x)$.
$$(f \circ g)(x) = f(g(x))$$
$$= f(x + 3)$$
$$= (x + 3)^2 - 1$$
$$= x^2 + 6x + 9 - 1$$
$$= x^2 + 6x + 8$$

(d) $(g \circ f)(x) = g(f(x)) = g(x^2 - 1) = (x^2 - 1) + 3 = x^2 + 2$ ■

In the following exploration, we compare the composite functions $(f \circ g)(x)$ and $(g \circ f)(x)$.

EXPLORATION 1

A Comparison of (f ∘ g)(x) and (g ∘ f)(x)

(a) Let $f(x) = x^2$ and $g(x) = \sqrt{x}$. Find the composite functions $(f \circ g)(x)$ and $(g \circ f)(x)$.

(b) Produce the graphs of $f \circ g$ and $g \circ f$.

(c) What conclusion can you make about $f \circ g$ and $g \circ f$?

Discovery

(a) Because g is defined for $x \geq 0$,

$$(f \circ g)(x) = f(g(x)) = f(\sqrt{x}) = [\sqrt{x}]^2 = x, \ x \geq 0.$$

For all real numbers x,

$$(g \circ f)(x) = g(f(x)) = g(x^2) = \sqrt{x^2} = |x|.$$

Note that the domain of $f \circ g$ is $[0, \infty)$ while the domain of $g \circ f$ is the set of all real numbers.

(b) Figure 11.4 shows the graph of $y = (f \circ g)(x)$, and Fig. 11.5 shows the graph of $y = (g \circ f)(x)$.

Figure 11.4

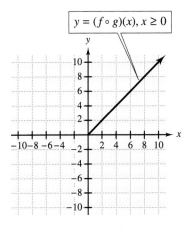

$y = (f \circ g)(x), \ x \geq 0$

Figure 11.5

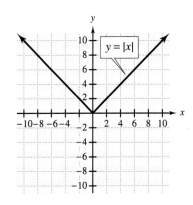

$y = |x|$

(c) Our observation that the domains of $f \circ g$ and $g \circ f$ are not the same is enough to conclude that the two composite functions are not equal. The graphs support this conclusion.

From Exploration 1 we observe that the operation of composition is not commutative. In general, $f \circ g \neq g \circ f$.

While $f \circ g$ is not generally equal to $g \circ f$, Example 7 illustrates some special cases in which the two composite functions are equal.

EXAMPLE 7   *Special Cases:* $(f \circ g)(x) = (g \circ f)(x) = x$

Show that $(f \circ g)(x) = (g \circ f)(x) = x$ for each pair of functions.

(a) $f(x) = 3x + 5$ and $g(x) = \dfrac{1}{3}(x - 5)$

(b) $f(x) = x^3$ and $g(x) = x^{1/3}$

Solution

(a) $(f \circ g)(x) = f(g(x))$

$$= f\left[\dfrac{1}{3}(x - 5)\right]$$

$$= 3\left[\dfrac{1}{3}(x - 5)\right] + 5$$

$$= x - 5 + 5$$

$$= x$$

$(g \circ f)(x) = g(f(x))$

$$= g(3x + 5)$$

$$= \dfrac{1}{3}[(3x + 5) - 5]$$

$$= \dfrac{1}{3}[3x]$$

$$= x$$

(b) $(f \circ g)(x) = f(g(x))$

$$= f\left(x^{1/3}\right)$$

$$= \left[x^{1/3}\right]^3$$

$$= x$$

$(g \circ f)(x) = g(f(x))$

$$= g(x^3)$$

$$= (x^3)^{1/3}$$

$$= x$$

Applications

In certain applications one variable may be a function of a second variable, which, in turn, is a function of a third variable. To describe how the first and third variables are related, a composite function can be used.

EXAMPLE 8   *Distance, Time, and Fuel Consumption*

The distance d (in miles) a race car has traveled after t minutes is given by $d(t) = 2.5t$. The number f of gallons of fuel remaining after d miles is given by $f(d) = 20 - 0.125d$.

(a) How far has the car traveled after 20 minutes?

(b) How much fuel remains after 20 minutes?

(c) Write a function to describe the amount of fuel remaining after t minutes.

Solution

(a) Distance and time are related by the function $d(t) = 2.5t$. For $t = 20$, $d(20) = 2.5(20) = 50$. The car traveled 50 miles.

(b) Remaining fuel and distance are related by the function $f(d) = 20 - 0.125d$. From part (a) if $t = 20$, $d = 50$. Thus, $f(50) = 20 - 0.125(50) = 13.75$. After 20 minutes, 13.75 gallons of fuel remain.

(c) Notice that remaining fuel f is given as a function of distance d, not of time t. In part (a) we used t to calculate d, and in part (b) we used d to calculate f.

We can establish a direct relationship between time t and remaining fuel f by composing the two functions. The amount of fuel f remaining after t minutes is given by $f(d(t))$ or $(f \circ d)(t)$.

$$(f \circ d)(t) = f(d(t))$$
$$= f(2.5t)$$
$$= 20 - 0.125[2.5t]$$
$$= 20 - 0.3125t$$

Now we can directly calculate the amount of fuel remaining after 20 minutes.

$$(f \circ d)(20) = 20 - 0.3125(20)$$
$$= 20 - 6.25$$
$$= 13.75$$

11.1 *Quick Reference*

Review of Functions

- A **function** is a set of ordered pairs in which no two different ordered pairs have the same first element.

- The **domain** of a function is the set of all first coordinates of the ordered pairs; the **range** is the set of all second coordinates.

- A relation is a function if

  1. the graph of the relation passes the Vertical Line Test;

  2. each element of the domain corresponds to a unique element of the range.

Basic Operations on Functions

- If functions f and g have domains with at least one element in common, then the following operations are defined for each x in both the domain of f and the domain of g.

  1. $(f + g)(x) = f(x) + g(x)$ Sum

  2. $(f - g)(x) = f(x) - g(x)$ Difference

  3. $(fg)(x) = f(x)g(x)$ Product

  4. $\left(\dfrac{f}{g}\right)(x) = \dfrac{f(x)}{g(x)}$, for $g(x) \neq 0$ Quotient

- The domain of $f + g$, $f - g$, and fg is the intersection of the domains of f and g. The domain of f/g is the intersection of the domains of f and g, excluding any x-values for which $g(x) = 0$.

Composition of Functions

- The **composition** of functions f and g is a function $f \circ g$ where $(f \circ g)(x) = f(g(x))$ for all x in the domain of g for which $g(x)$ is in the domain of f.

- In certain special cases, $(f \circ g)(x) = (g \circ f)(x)$, but in general, $f \circ g$ and $g \circ f$ are not equal.

11.1   *Exercises*

1. Suppose that $f(x) = x - 4$ and $g(x) = 3x^2$. Describe two methods for determining $(f + g)(1)$.

2. The set of all real numbers is the domain for the function $f(x) = x$ and for the function $g(x) = x + 1$. Is the domain of $(f/g)(x)$ also the set of all real numbers? Explain.

In Exercises 3–6, for the given functions f and g, determine each of the following.

   (a) $(f + g)(-1)$ (b) $(f - g)(1)$

   (c) $(fg)(0)$ (d) $\left(\dfrac{f}{g}\right)(0)$

3. $f(x) = |2x + 1|,\ g(x) = x^2$

4. $f(x) = \dfrac{1}{x + 2},\ g(x) = \dfrac{1}{(x + 2)^2}$

5. $f(x) = 2^x,\ g(x) = 2^{-x}$

6. $f(x) = 3^{x-1},\ g(x) = 3^{1-x}$

In Exercises 7–10, use the given functions f and g to evaluate the combined function in each part.

7. Let $f(x) = |x - 3|$ and $g(x) = 2x - 3$.

   (a) $(f + g)(2)$ (b) $(f - g)(-2)$

   (c) $(fg)(0)$ (d) $\left(\dfrac{f}{g}\right)(-1)$

8. Let $f(x) = 3x + 1$ and $g(x) = x - 6$.

   (a) $(f - g)(-1.2)$ (b) $(f + g)\left(-\dfrac{1}{2}\right)$

   (c) $\left(\dfrac{f}{g}\right)(2.5)$ (d) $(fg)\left(\dfrac{2}{3}\right)$

9. Let $f(x) = \sqrt{5 - x}$ and $g(x) = x - 5$.

   (a) $(f + g)(4)$ (b) $(f - g)(-4)$

   (c) $(fg)(5)$ (d) $\left(\dfrac{f}{g}\right)(1)$

10. Let $f(x) = x^2 - 2$ and $g(x) = \sqrt[3]{x}$.

   (a) $(f + g)(0)$ (b) $(f - g)(1)$

   (c) $\left(\dfrac{f}{g}\right)(27)$ (d) $(fg)(8)$

11. Suppose that $f(x) = x^2 + 3x$, $g(x) = x$, and $h(x) = x + 3$. Show that $(f/g)(x) = x + 3$. Is the domain of f/g the same as the domain of h? Explain.

12. Suppose that $A = \{x \mid x < 4\}$ is the domain of function f and $B = \{x \mid x > -3\}$ is the domain of function g. Use interval notation to express the domain of $f + g$, $f - g$, and fg.

In Exercises 13–18, functions f and g are given. Determine the simplified combined function in each part and state its domain.

13. Let $f(x) = x^2 - 9$ and $g(x) = x + 3$.

   (a) $(f - g)(x)$ (b) $(f + g)(x)$

   (c) $\left(\dfrac{f}{g}\right)(x)$ (d) $(fg)(x)$

14. Let $f(x) = x + 3$ and $g(x) = x^2 + 4x + 3$.

   (a) $(f + g)(x)$ (b) $(f - g)(x)$

   (c) $(fg)(x)$ (d) $\left(\dfrac{f}{g}\right)(x)$

15. Let $f(x) = x^2 + 4$ and $g(x) = x^2 - 4$.

   (a) $\left(\dfrac{f}{g}\right)(x)$ (b) $\left(\dfrac{g}{f}\right)(x)$

   (c) $\left(\dfrac{f}{f}\right)(x)$ (d) $\left(\dfrac{g}{g}\right)(x)$

16. Let $f(x) = x^2 - 5x + 6$ and $g(x) = x - 2$.

   (a) $\left(\dfrac{g}{f}\right)(x)$ (b) $\left(\dfrac{f}{g}\right)(x)$

   (c) $\left(\dfrac{g}{g}\right)(x)$ (d) $\left(\dfrac{f}{f}\right)(x)$

17. Let $f(x) = \dfrac{1}{x + 2}$ and $g(x) = \dfrac{1}{x - 3}$.

  (a) $(f - g)(x)$ (b) $(f + g)(x)$

  (c) $\left(\dfrac{f}{g}\right)(x)$ (d) $(fg)(x)$

18. Let $f(x) = \dfrac{x + 1}{x + 3}$ and $g(x) = \dfrac{x + 3}{2 - x}$.

  (a) $(f + g)(x)$ (b) $(f - g)(x)$

  (c) $\left(\dfrac{g}{f}\right)(x)$ (d) $\left(\dfrac{f}{g}\right)(x)$

In Exercises 19–22, let $f(x) = x^2 - 4$ and $g(x) = x - 1$. Determine the given combined function.

19. $(f - g)(2t)$ **20.** $(f + g)(-t)$

21. $\left(\dfrac{f}{g}\right)\left(\dfrac{t}{2}\right)$ **22.** $(fg)(t + 2)$

In Exercises 23–26, let $f(x) = x + 5$ and $g(x) = x^2 - 3$. Determine the given combined function.

23. $(f + g)(3t)$ **24.** $(f - g)(t + 1)$

25. $(fg)(-2t)$ **26.** $\left(\dfrac{f}{g}\right)\left(\dfrac{t}{3}\right)$

 27. Explain how you can use the graphs of functions f and g to estimate the domain of $f + g$.

28. Let $f(x) = \sqrt{x + 3}$ and $g(x) = \sqrt{4 - x}$.

  (a) Produce the graph of f and estimate the domain.

  (b) Produce the graph of g and estimate the domain.

  (c) Use the results of parts (a) and (b) to estimate the domain of $f + g$.

  (d) Verify the domain found in part (c) by producing the graph of $f + g$.

In Exercises 29–32, functions f and g are given. Determine the domain of the combined function given in each part.

29. Let $f(x) = \sqrt{2x - 3}$ and $g(x) = \sqrt{5 - x}$.

  (a) $(f + g)(x)$

  (b) $(f - g)(x)$

30. Let $f(x) = \sqrt{2 - x}$ and $g(x) = \sqrt{3 - x}$.

  (a) $(f - g)(x)$

  (b) $(f + g)(x)$

31. Let $f(x) = \sqrt{8 - x}$ and $g(x) = \sqrt{x - 1}$.

  (a) $\left(\dfrac{f}{g}\right)(x)$

  (b) $(fg)(x)$

32. Let $f(x) = \sqrt{2 - x}$ and $g(x) = \sqrt{x - 3}$.

  (a) $(fg)(x)$

  (b) $\left(\dfrac{g}{f}\right)(x)$

 33. For two given functions f and g, determining $(f \circ g)(5)$ requires two calculations. Describe these calculations in the order in which they are performed.

 34. Suppose $f = \{(1, 2), (2, 5), (4, -1)\}$ and $g = \{(1, 3), (2, 5)\}$. Explain why $(g \circ f)(1)$ is defined but $(f \circ g)(1)$ is not.

In Exercises 35–38, functions f and g are given. Evaluate the composite function given in each part.

35. Let $f(x) = x - 3$ and $g(x) = 2 - x^2$.

  (a) $(f \circ g)(1)$ (b) $(f \circ g)(0)$

  (c) $(g \circ f)(3)$ (d) $(g \circ f)(-1)$

36. Let $f(x) = 2x + 1$ and $g(x) = x^3$.

  (a) $(f \circ g)(-1)$ (b) $(f \circ g)(2)$

  (c) $(g \circ f)(0)$ (d) $(g \circ f)(-1)$

37. Let $f(x) = x^2$ and $g(x) = 2x - 3$.

  (a) $(g \circ f)(-3)$ (b) $(f \circ g)\left(-\dfrac{1}{2}\right)$

  (c) $(g \circ g)\left(\dfrac{3}{2}\right)$ (d) $(f \circ f)(-1)$

38. Let $f(x) = x^2 - 2x + 3$ and $g(x) = x + 4$.

   (a) $(f \circ g)(-5)$ (b) $(g \circ f)(3)$

   (c) $(f \circ f)(0)$ (d) $(g \circ g)(2)$

In Exercises 39 and 40, for the given functions f and g, determine each of the following.

   (a) $(f \circ g)(2)$

   (b) $(g \circ f)(-1)$

   (c) $(f \circ g)(-3)$

39. $f = \{(-2, 3), (2, -4), (1, 5), (-3, -2), (-1, 3)\}$
   $g = \{(-2, 5), (2, 1), (-3, 2), (3, -2)\}$

40. $f = \{(2, 5), (3, 6), (4, 7), (-2, 1), (-1, 2)\}$
   $g = \{(-5, -1), (-3, -2), (0, 4), (-1, 3), (2, 3)\}$

In Exercises 41–48, for the given functions f and g, determine $(f \circ g)(x)$ and $(g \circ f)(x)$.

41. $f(x) = 3 - x^2$, $g(x) = 2x + 1$

42. $f(x) = 3x - 2$, $g(x) = 4x - x^2$

43. $f(x) = x^2 - 3x + 5$, $g(x) = 3x - 2$

44. $f(x) = 2x - 5$, $g(x) = x^2 - 4x - 5$

45. $f(x) = x^3$, $g(x) = \sqrt[3]{x - 2}$

46. $f(x) = \sqrt[3]{2x - 3}$, $g(x) = -x^3$

47. $f(x) = 5$, $g(x) = 2x - 3$

48. $f(x) = x^2 - 3$, $g(x) = 7$

In Exercises 49–56, for the given functions f and g, determine $(f \circ f)(x)$ and $(g \circ g)(x)$.

49. $f(x) = 3x - 4$, $g(x) = x + 5$

50. $f(x) = x - 2$, $g(x) = 4 - 3x$

51. $f(x) = |x - 3|$, $g(x) = x^2 + 3$

52. $f(x) = x^2 - 2$, $g(x) = |x + 3|$

53. $f(x) = 2x^2 - 3x$, $g(x) = 2 - x$

54. $f(x) = x - 3$, $g(x) = 3x^2 - 2x$

55. $f(x) = 3$, $g(x) = 2$

56. $f(x) = -5$, $g(x) = -2$

In Exercises 57–60, show that $(f \circ g)(x) = x$ for each given pair of functions.

57. $f(x) = 2x + 3$, $g(x) = \dfrac{x - 3}{2}$

58. $f(x) = 2x - 5$, $g(x) = \dfrac{x + 5}{2}$

59. $f(x) = \dfrac{x}{3} + 2$, $g(x) = 3x - 6$

60. $f(x) = \dfrac{x}{4} - 2$, $g(x) = 4x + 8$

61. A study conducted at Tall Pines Golf Course showed that the number g of golfers on the course on any given day was related to the 9:00 A.M. temperature t (°F) by the function

$$g(t) = -\frac{1}{3}(t^2 - 140t + 4000).$$

The daily gross income h was given by $h(g) = 54g$.

   (a) How many golfers were on the golf course on May 15 if the 9:00 A.M. temperature was 76°F?

   (b) What was the gross income for May 15?

   (c) Write a function relating the daily gross income to the 9:00 A.M. temperature.

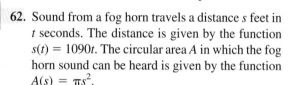

62. Sound from a fog horn travels a distance s feet in t seconds. The distance is given by the function $s(t) = 1090t$. The circular area A in which the fog horn sound can be heard is given by the function $A(s) = \pi s^2$.

   (a) How far from the fog horn can the sound be heard at 5 seconds?

   (b) In how much area can the sound be heard in 5 seconds?

   (c) Write a function to relate the area to the time.

63. On interstate trips, a driver averages 54 mph. The distance d (in miles) traveled in t hours is given by $d(t) = 54t$. Because the driver averages 25 miles per gallon, the number of gallons g used is given by $g(d) = d/25$. The cost per gallon is $1.03, so the total fuel cost c is given by $c(g) = 1.03g$.

   (a) Write a function describing the number of gallons used in t hours of travel.

(b) Write a function describing the total fuel cost in t hours of travel.

(c) Determine the total fuel cost of a 12-hour trip.

64. An oil tanker runs aground and springs a leak. The oil spreads out in a semicircular pattern from the shoreline. The distance r (in feet) from the tanker to the edge of the oil spill at time t (in minutes) is given by the function $r(t) = 20t$. If the area of the semicircle is given by $A(x) = \frac{1}{2}\pi x^2$, where x is the radius, write a function $A \circ r$ for the area covered by the oil at time t. What is the area of the oil spill after 5 minutes?

65. A party caterer charges \$5.00 per person for 100 people. For a gathering of less than 100 people, the caterer increases the price per person by \$0.05 for each person under 100 people. For example, 90 people is 10 less than 100, so the additional price per person is $10(0.05) = \$.50$. Then the total cost is $(90)(5.50) = \$495.00$.

(a) Write a function $f(x)$ to represent the number of people who attend the gathering.

(b) Write a function $c(x)$ to represent the per-person cost.

(c) Write a function $T(x) = (fc)(x)$ to represent the total cost of the gathering.

(d) Use the function in part (c) to determine the total cost for 84 people.

66. The distance between the bases of a baseball diamond (a square) is 90 feet. A ball is hit directly down the third base line at a speed of 60 feet per second. If x is the distance from home plate to the ball, then x is a function of time t in seconds. (See figure.) Let $f(x)$ represent the distance from the ball to first base.

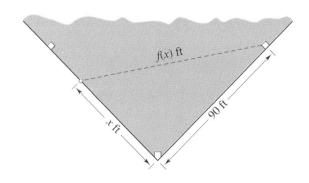

(a) Use the Pythagorean Theorem to write the function f.

(b) Write the function $x(t)$.

(c) Write the function $(f \circ x)(t)$. What does this function represent?

 Exploring with Real Data

In Exercises 67–70, refer to the following background information and data.

Federal funding for education steadily increased during the period 1985–1992. The accompanying figure shows federal spending for elementary, secondary, and vocational education compared to spending for higher education during that period. (Source: U.S. Office of Management and Budget.)

Figure for 67–70

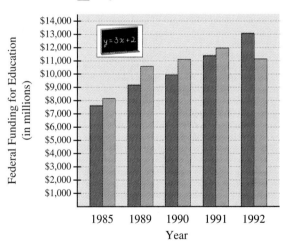

Funding f (in millions of dollars) for elementary, secondary, and vocational education can be modeled by the following function of t.

$$f(t) = 157t^2 - 1900t + 13{,}157$$

Funding h (in millions of dollars) for higher education can be modeled by the following function of t.

$$h(t) = 502t + 5875$$

In both models t is the number of years since 1980.

67. Use [X: 5, 12, 1] and [Y: 6000, 14000, 800] to produce the graphs of functions f and h in the same coordinate system.

68. What is the approximate first coordinate of the point of intersection of the two graphs? Letting t_0 represent this value, interpret the trend in federal funding for $t > t_0$.

69. Write the function $(f + h)(t)$. In the context of this exercise, what is the meaning of $f + h$?

70. During his 1988 campaign for the presidency, George Bush declared that he would be "The Education President." Based on the data for 1989–1992, what case can be made for or against this claim?

Challenge

Suppose $f(x) = 5\sqrt{x} - 7$. If $g(x) = \sqrt{x}$ and $h(x) = 5x - 7$, then

$$(h \circ g)(x) = h(g(x)) = h\left(\sqrt{x}\right) = 5\sqrt{x} - 7 = f(x).$$

Thus, $f(x) = (h \circ g)(x)$ where $g(x) = \sqrt{x}$ and $h(x) = 5x - 7$. In this way, we have written f as a composition of two simpler functions g and h.

In Exercises 71–78, write each function $f(x)$ as a composition of two simpler functions g and h. State specifically what the functions g and h are.

71. $f(x) = \sqrt{x^2 - 5x + 6}$

72. $f(x) = \dfrac{1}{x^2 + 4x}$

73. $f(x) = (3x + 2)^4$

74. $f(x) = \sqrt[3]{x^2 - 4x}$

75. $f(x) = 3x^{1/2} - 5$

76. $f(x) = (2x^2 - 5x)^{5/6}$

77. $f(x) = 3\sqrt{4x + 5} + 7$

78. $f(x) = |2x + 5| + 8$

79. Let $f(x) = 3$ and $g(x) = \dfrac{1}{x - 3}$. Determine $(f \circ g)(x)$ and $(g \circ f)(x)$.

80. The Associative Property of Addition states that

$$a + (b + c) = (a + b) + c.$$

To determine if there is a similar property for composition of functions, let $f(x) = x$, $g(x) = \sqrt{x}$ $(x > 0)$, and $h(x) = x^2$.

(a) Determine $f \circ (g \circ h)$.

(b) Determine $(f \circ g) \circ h$.

(c) What does this suggest about $f \circ (g \circ h)$ and $(f \circ g) \circ h$?

11.2 Inverse Functions

One-to-One Functions ▪ *Inverse Functions (Informal)* ▪ *Inverse Functions (Formal)* ▪ *Determining the Inverse*

One-to-One Functions

We have defined a function as a set of ordered pairs in which no two ordered pairs have the same *first* coordinate. If, in addition, we require that no two ordered pairs have the same *second* coordinate, then we have the special case of a **one-to-one function.**

> ### Definition of a One-to-One Function
>
> A function is a **one-to-one function** if no two different ordered pairs have the same second coordinate.

This means that for any value of y in the range of the function f, there is a *unique* value of x such that $f(x) = y$.

EXAMPLE 1 *One-to-One Functions*

Determine whether each function is a one-to-one function.

(a) $f = \{(2, 3), (4, 5), (6, 3)\}$

(b) $g = \{(1, 2), (2, 3), (3, 4), (4, 5), (5, 6)\}$

Solution

(a) Function f is not a one-to-one function because the ordered pairs (2, 3) and (6, 3) are different ordered pairs with the same second coordinate. A mapping diagram can help us to visualize the correspondence between elements of the domain and range. For a mapping diagram to represent a one-to-one function, only one arrow can point to any range element.

Figure 11.6

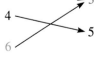

Figure 11.6 shows a mapping diagram of the given function f. Because two arrows point to the element 3 in the range of the function, f is *not* a one-to-one function.

Figure 11.7

1 ⟶ 2
2 ⟶ 3
3 ⟶ 4
4 ⟶ 5
5 ⟶ 6

(b) The mapping diagram in Fig. 11.7 shows that the given function g is a one-to-one function.

To determine if a graph represents a function, we use the Vertical Line Test. To determine if a graph of a function represents a *one-to-one* function, we use the Horizontal Line Test.

> ### The Horizontal Line Test for a One-to-One Function
>
> Since no two ordered pairs of a one-to-one function have the same second coordinate, a horizontal line cannot intersect the graph of the function at more than one point.

The graphs in Figs. 11.8 and 11.9 both pass the Vertical Line Test, so both represent functions.

Figure 11.8

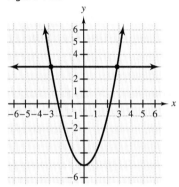

Figure 11.9

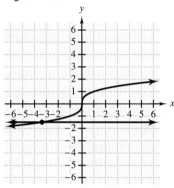

Using the Horizontal Line Test, we find that the graph in Fig. 11.8 does not represent a one-to-one function, but the graph in Fig. 11.9 does.

Inverse Functions (Informal)

If the corresponding *x*- and *y*-coordinates of a one-to-one function *f* are interchanged, the resulting set of ordered pairs is also a function. This new function is called the **inverse function** and is written f^{-1} (read "*f* inverse").

NOTE: In the notation f^{-1}, the -1 is not an exponent. We read f^{-1} as a single symbol that represents an inverse function. ∎

Consider the function

$$f = \{(-3, 5), (-1, -4), (3, -2), (6, 1)\}$$

To write the inverse function, interchange the *x*- and *y*-coordinates of each ordered pair.

$$f^{-1} = \{(5, -3), (-4, -1), (-2, 3), (1, 6)\}$$

Figure 11.10

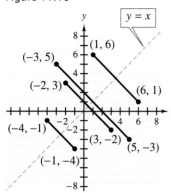

Note the domains and ranges of f and f^{-1}.

Domain of f: $\{-3, -1, 3, 6\}$ Range of f: $\{-4, -2, 1, 5\}$

Range of f^{-1}: $\{-3, -1, 3, 6\}$ Domain of f^{-1}: $\{-4, -2, 1, 5\}$

The domain of f is the same as the range of f^{-1}, and the range of f is the same as the domain of f^{-1}.

If we sketch the graphs of f and f^{-1} on the same coordinate system (see Fig. 11.10), we observe that the points lie the same distance on either side of the diagonal line whose equation is $y = x$. This line is called the **line of symmetry.**

EXPLORATION 1   *Recognizing the Inverse of a Function*

(a) Use the integer setting on your calculator to produce the graphs of the functions $y_1 = x + 12$, $y_2 = x - 12$, and $y_3 = x$ on the same coordinate axes.

(b) Trace to the point $(10, 22)$ on the graph of y_1. Is there a point $(22, 10)$ on the graph of y_2? Do the same for the points $(-7, 5)$ and $(-26, -14)$ on the graph of y_1. Are there points $(5, -7)$ and $(-14, -26)$ on the graph of y_2?

(c) Do the points of the graphs of y_1 and y_2 appear to be symmetric about the line $y_3 = x$?

(d) What is your conjecture about the relationship between the functions y_1 and y_2?

Discovery

Figure 11.11

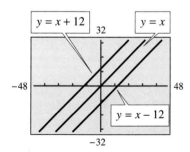

(a) Figure 11.11 shows the graphs of y_1, y_2, and y_3.

(b) For each of the given points on the graph of y_1, there is a corresponding point (with the coordinates reversed) on the graph of y_2.

(c) The line $y_3 = x$ is the line of symmetry for the graphs of y_1 and y_2.

(d) The functions y_1 and y_2 are inverse functions.

In Exploration 1 we observe that the function y_1 adds 12 to each value of x. The inverse function y_2 subtracts 12 from each value of x. Informally, we say that the inverse function f^{-1} reverses the function f.

Notice what happens when we compose $f(x) = x + 12$ and $f^{-1}(x) = x - 12$.

$$(f \circ f^{-1})(x) = f(f^{-1}(x)) = f(x - 12) = (x - 12) + 12 = x$$
$$(f^{-1} \circ f)(x) = f^{-1}(f(x)) = f^{-1}(x + 12) = (x + 12) - 12 = x$$

We can see that $(f \circ f^{-1})(x) = x$ and $(f^{-1} \circ f)(x) = x$. As we will see later, this important observation will be the basis for the formal definition of inverse functions.

To determine the inverse of a simple function, we can use our knowledge that the inverse function reverses the rule of the original function.

EXAMPLE 2   *Writing the Inverse of a Simple Function*

Write the inverse of the function $f(x) = 2x - 1$.

Solution

When we say that the inverse function f^{-1} reverses the rule of function f, we also mean that the changes are made *in reverse order*. Function f multiplies x by 2 and then subtracts 1. To write the inverse function f^{-1}, we begin by adding 1, and then we divide by 2.

$$f^{-1}(x) = \frac{x + 1}{2}$$ ∎

Note again the composition of the two functions in Example 2.

$$(f \circ f^{-1})(x) = f(f^{-1}(x)) = f\left(\frac{x+1}{2}\right) = 2\left(\frac{x+1}{2}\right) - 1 = x + 1 - 1 = x$$

$$(f^{-1} \circ f)(x) = f^{-1}(f(x)) = f^{-1}(2x - 1) = \frac{(2x - 1) + 1}{2} = \frac{2x}{2} = x$$

Inverse Functions (Formal)

The formal definition of an inverse function is expressed in terms of the composition of functions.

> **Definition of an Inverse Function**
>
> Functions f and g are **inverse functions** if $(f \circ g)(x) = x$ for each x in the domain of g and $(g \circ f)(x) = x$ for each x in the domain of f. We write $g = f^{-1}$ and $f = g^{-1}$.

EXAMPLE 3   *Verifying Inverse Functions*

Verify that the given pair of functions are inverses. Produce the graphs of each pair and observe that the graphs are symmetric with respect to the line $y = x$.

(a) $f(x) = \dfrac{x - 2}{3}$ and $g(x) = 3x + 2$

(b) $f(x) = \sqrt[3]{x + 2}$ and $g(x) = x^3 - 2$

Solution

(a) We apply the definition of inverse functions.

$$(f \circ g)(x) = f(g(x)) = f(3x + 2) = \frac{(3x + 2) - 2}{3} = \frac{3x}{3} = x$$

$$(g \circ f)(x) = g(f(x)) = g\left(\frac{x - 2}{3}\right) = 3 \cdot \frac{x - 2}{3} + 2 = (x - 2) + 2 = x$$

Because $(f \circ g)(x) = x$ and $(g \circ f)(x) = x$, f and g are inverse functions. Figure 11.12 shows their graphs and the line of symmetry.

Figure 11.12

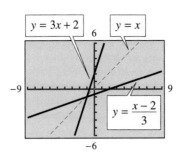

Figure 11.13

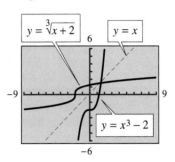

(b) $(f \circ g)(x) = f(x^3 - 2) = \sqrt[3]{(x^3 - 2) + 2} = \sqrt[3]{x^3} = x$
$(g \circ f)(x) = g(\sqrt[3]{x + 2}) = (\sqrt[3]{x + 2})^3 - 2 = (x + 2) - 2 = x$

Because $(f \circ g)(x) = x$ and $(g \circ f)(x) = x$, f and g are inverse functions. Figure 11.13 shows their graphs and the line of symmetry. ■

In Example 3 the graphs of the inverse functions would coincide if they were folded along the diagonal line shown in the figures. In general, the graphs of f and f^{-1} are symmetric with respect to the line of symmetry $y = x$.

Determining the Inverse

If a one-to-one function is defined by a rule, we can often determine a rule for the inverse function by interchanging the roles of x and y and solving for y.

EXAMPLE 4 *Determining an Inverse Function*
Let $f(x) = 2x - 3$. Determine f^{-1}.

Solution

$$f(x) = 2x - 3$$
$$y = 2x - 3 \qquad \text{Replace } f(x) \text{ with } y.$$
$$x = 2y - 3 \qquad \text{Interchange } x \text{ and } y.$$
$$x + 3 = 2y \qquad \text{Solve for } y.$$
$$\frac{x + 3}{2} = y$$
$$f^{-1}(x) = \frac{x + 3}{2} \qquad \text{Replace } y \text{ with } f^{-1}(x).$$

Figure 11.14

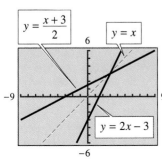

Figure 11.14 shows the graphs of f and f^{-1}. Notice that the graphs are symmetric with respect to the line $y = x$. Producing the graphs of f and f^{-1} gives you a visual check on your work. You can also use the definition of inverse functions to check that $(f \circ f^{-1})(x) = (f^{-1} \circ f)(x) = x$. ■

11.2   *Quick Reference*

One-to-One Functions	▪ A function is a **one-to-one function** if no two different ordered pairs have the same second coordinate.
	▪ The graph of a one-to-one function passes the Horizontal Line Test: A horizontal line intersects the graph in at most one point.

Inverse Functions (Informal)

▪ If the x- and y-coordinates of a one-to-one function f are interchanged, the resulting function is the **inverse function f^{-1}**.

▪ An inverse function f^{-1} reverses the rule of function f.

Inverse Functions (Formal)

▪ Functions f and g are defined to be **inverse functions** if $(f \circ g)(x) = x$ for each x in the domain of g and $(g \circ f)(x) = x$ for each x in the domain of f.

▪ We can use this definition to test whether two functions are inverse functions.

Determining the Inverse

▪ To determine the inverse of a function, perform the following steps.

   1. Replace the function notation $f(x)$ with y.

   2. Interchange the roles of x and y.

   3. Solve for y and then replace y with $f^{-1}(x)$.

▪ If f is a one-to-one function with inverse f^{-1}, the following are true about the two functions.

   1. $(f \circ f^{-1})(x) = x$ for all x in the domain of f^{-1} and $(f^{-1} \circ f)(x) = x$ for all x in the domain of f.

   2. The domain of f is the same as the range of f^{-1}.

   3. The range of f is the same as the domain of f^{-1}.

   4. The graphs of f and f^{-1} are symmetric with respect to the line $y = x$.

11.2   *Exercises*

1. Suppose that a function f is represented by a mapping diagram. What must be true about the diagram in order for f to be a one-to-one function?

2. Describe how the Horizontal Line Test is used to determine if a function is a one-to-one function.

In Exercises 3–6, determine whether the given function is a one-to-one function.

3. $\{(5, 2), (2, 5), (3, 2), (6, -4)\}$

4. $\{(5, 2), (3, 0), (1, -2), (7, 4)\}$

5. $\{(-3, -1), (-1, 1), (3, 5), (1, 3)\}$

6. $\{(-3, -1), (-1, -3), (3, -2), (2, -1)\}$

In Exercises 7–12, determine whether the given graph represents a one-to-one function.

7.
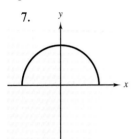

8.
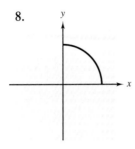

9.
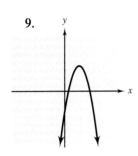

10.
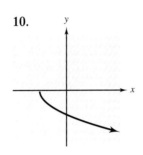

11.
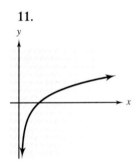

12.
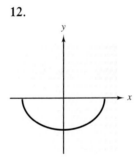

In Exercises 13–18, produce the graph of the given function. From the graph determine if the function is one-to-one.

13. $f(x) = x^3 - 3x^2 + 6x + 4$

14. $f(x) = x^3 - 2x^2 - 7x + 2$

15. $f(x) = |x - 7| - |x + 5|$

16. $f(x) = |x| + 3$

17. $f(x) = x^3 2^x$

18. $f(x) = x\sqrt{25 - x^2}$

In Exercises 19–22, determine f^{-1}. State the domain and range of f^{-1}.

19. $f = \{(1, 2), (3, 4), (5, 6), (7, -1)\}$

20. $f = \{(-1, 2), (-3, 4), (-5, -7), (5, 8)\}$

21. $f = \{(3, 2), (4, 3), (-5, 1), (-7, -2)\}$

22. $f = \{(5, -6), (7, -8), (-8, 5), (-6, 7)\}$

In Exercises 23–26, f is a one-to-one function. Suppose $f(3) = 4$, $f(5) = 7$, $f(-2) = -3$, and $f(-4) = -5$. Evaluate f^{-1} as indicated.

23. $f^{-1}(-3)$ **24.** $f^{-1}(-5)$

25. $f^{-1}(4)$ **26.** $f^{-1}(7)$

In Exercises 27–30, use the integer display to produce the graph of $f(x) = 2^x$. Then, trace the graph to evaluate f^{-1} as indicated.

27. $f^{-1}(64)$ **28.** $f^{-1}(8)$

29. $f^{-1}(0.0625)$ **30.** $f^{-1}(0.25)$

In Exercises 31–34, use the integer display to produce the graph of $f(x) = 2^{-x}$. Then, trace the graph to evaluate f^{-1} as indicated.

31. $f^{-1}(32)$ **32.** $f^{-1}(2)$

33. $f^{-1}(0.125)$ **34.** $f^{-1}(0.5)$

 35. Suppose that $f(x) = x$. If you enter x^{-1} on your calculator, will you obtain $f^{-1}(x)$? Explain.

 36. Suppose that you produce the graphs of functions f and g on your calculator and you notice that the graphs are mirror images about the line $y = x$. What is your conjecture about the two functions?

In Exercises 37–40, a verbal description of function f is given. Write a verbal description of the inverse function f^{-1}.

37. Function f adds 2 to x and then divides the result by 3.

38. Function f multiplies x by -2 and then subtracts 1 from the result.

39. Function f cubes 1 more than x.

40. Function f adds 4 to x and then takes half of the result.

In Exercises 41–48, translate the given algebraic rule for function f into a verbal description. Then, write a verbal description of f^{-1} and write f^{-1} as an algebraic rule.

41. $f(x) = \dfrac{1}{2}x$

42. $f(x) = x - 5$

43. $f(x) = 3(x - 1)$

44. $f(x) = 3x - 2$

45. $f(x) = \dfrac{1}{x + 4}$

46. $f(x) = \dfrac{1}{x} + 2$

47. $f(x) = x^3$

48. $f(x) = \sqrt[3]{x + 1}$

In Exercises 49–54, produce the graphs of each given pair of functions. Although the graphs are not conclusive, decide if it is reasonable to believe that the functions may be inverses.

49. $f(x) = 2x + 1$, $g(x) = \dfrac{x - 1}{2}$

50. $f(x) = 3x + 1$, $g(x) = 3x - 2$

51. $f(x) = x^2 + 1$, $g(x) = 1 - x^2$

52. $f(x) = |x - 1|$, $g(x) = x + 1$

53. $f(x) = x^3 - 2$, $g(x) = \sqrt[3]{x + 2}$

54. $f(x) = \sqrt[3]{x} - 1$, $g(x) = x^3 + 1$

In Exercises 55–58, use your calculator to produce the graph of the given function. Then, sketch the graph of the inverse function.

55. $f(x) = 2^x$

56. $f(x) = (0.7)^x$

57. $f(x) = x^3 - 1$

58. $f(x) = 0.2x^3 + 1$

 59. Suppose $f(x) = x^2$ and $g(x) = \sqrt{x}$, $x \geq 0$. In Exploration 1 of the previous section, we showed that $f \circ g$ and $g \circ f$ are not the same functions. What does this indicate about whether f and g are inverse functions?

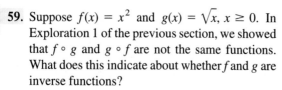 **60.** For the functions in Exercise 59, if we restrict the domain of f to $x \geq 0$, show that f and g are inverse functions.

In Exercises 61–66, determine whether the pairs are inverses by showing that $(f \circ g)(x) = (g \circ f)(x) = x$.

61. $f(x) = 2x$, $g(x) = -2x$

62. $f(x) = x + 2$, $g(x) = x - 2$

63. $f(x) = \sqrt[3]{x + 1}$, $g(x) = x^3 - 1$

64. $f(x) = \dfrac{1}{4}x^2$, $x \geq 0$; $g(x) = 2\sqrt{x}$, $x \geq 0$

65. $f(x) = \dfrac{3}{x}$, $g(x) = \dfrac{3}{x}$

66. $f(x) = x^{1/3}$, $g(x) = x^{-1/3}$

In Exercises 67–88, determine f^{-1} and verify that $(f \circ f^{-1})(x) = x$ for all x in the domain of f^{-1} and that $(f^{-1} \circ f)(x) = x$ for all x in the domain of f.

67. $f(x) = 3x + 5$

68. $f(x) = 5x - 7$

69. $f(x) = -2x$

70. $f(x) = 2(x - 1)$

71. $f(x) = \dfrac{x}{3} + 5$

72. $f(x) = \dfrac{2x - 1}{3}$

73. $f(x) = \dfrac{1}{x}$

74. $f(x) = \dfrac{1}{x} - 3$

75. $f(x) = \dfrac{1}{x + 2}$

76. $f(x) = \dfrac{3}{1 - x}$

77. $f(x) = \dfrac{x + 1}{x - 2}$

78. $f(x) = \dfrac{x + 3}{2x + 1}$

79. $f(x) = x^3 + 1$

80. $f(x) = -\dfrac{1}{8}x^3$

81. $f(x) = x^{-3}$

82. $f(x) = x^{-5} + 1$

83. $f(x) = \dfrac{\sqrt{x}}{2}$

84. $f(x) = \sqrt{3x}$

85. $f(x) = \sqrt[3]{2 - x}$

86. $f(x) = \sqrt[5]{3 - x}$

87. $f(x) = \sqrt[5]{x} + 3$

88. $f(x) = 1 - \sqrt[3]{x}$

89. From the following list, identify the pairs of functions that are inverse functions.

(i) $f(x) = \dfrac{1}{3}$, $g(x) = 3$

(ii) $f(x) = 5$, $g(x) = -5$

(iii) $f(x) = \dfrac{5}{x}$, $g(x) = \dfrac{5}{x}$

90. Suppose that the graph of function f lies entirely in the given quadrants. In what quadrants does the graph of function f^{-1} lie?

(a) I and II

(b) II and IV

(c) I and III

(d) II and III

 91. Explain why $y = \sqrt{9 - x^2}$, $x \geq 0$, has an inverse, but $y = \sqrt{9 - x^2}$ does not.

92. Let $f(x) = 2x - 1$. Determine $f^{-1}(x)$ by evaluating f at $f^{-1}(x)$ and then solving for $f^{-1}(x)$.

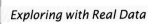 *Exploring with Real Data*

In Exercises 93–96, refer to the following background information and data.

A developing trend for American workers is *telecommuting*—using a computer or phone technology to work from home. The accompanying figure shows the number of employees (in millions) involved in telecommuting from 1989 through 1993. (Source: LINK Resources Corporation.)

Figure for 93–96

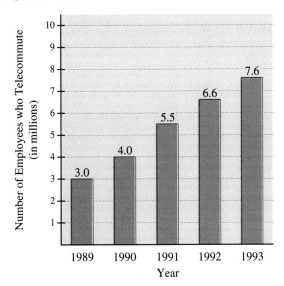

A function that models the data in the figure is

$$N(t) = 1.18t - 1.74,$$

where N is the number of employees (in millions) and t is the number of years since 1985.

93. Produce the graph of the model function and the graph of $y = x$ on the same coordinate system.

94. Trace the graph to estimate the point at which the graphs of function N and its inverse would intersect.

95. Determine $N^{-1}(t)$ and produce its graph. Do the graphs of $y_1 = N(t)$ and $y_2 = N^{-1}(t)$ intersect at the point you predicted in Exercise 94?

96. Solve the equation $N(t) = N^{-1}(t)$ algebraically and compare the solution for t to the first coordinate of the point of intersection in Exercise 95.

Challenge

In Exercises 97 and 98, produce the graph of the function. Then, sketch the graph of the inverse function.

97. $f(x) = x^2 - 4x, x \leq 2$

98. $f(x) = x^2 - 4x + 3, x \leq 2$

In Exercises 99–106, determine f^{-1} and verify that $(f \circ f^{-1})(x) = x$ for all x in the domain of f^{-1} and that $(f^{-1} \circ f)(x) = x$ for all x in the domain of f.

99. $f(x) = \sqrt{x}, x \geq 0$

100. $f(x) = x^2 + 1, x \geq 0$

101. $f(x) = -\sqrt{3 - x}, x \leq 3$

102. $f(x) = -x^2, x \leq 0$

103. $f(x) = x^2 + 1, x \leq 0$

104. $f(x) = |x - 4|, x \leq 4$

105. $f(x) = -|x + 5|, x \leq -5$

106. $f(x) = x^2 - 6x + 5, x \leq 3$

107. Let $f(x) = 3x$ and $g(x) = x - 3$. Find $f \circ g$, $(f \circ g)^{-1}, f^{-1}, g^{-1}$, and $g^{-1} \circ f^{-1}$. Note that $(f \circ g)^{-1} = g^{-1} \circ f^{-1}$, not $f^{-1} \circ g^{-1}$.

108. Suppose that $f(x) = \dfrac{3x - 1}{2x - 3}$.

(a) Show that $(f \circ f)(x) = x$.

(b) From part (a) what can we learn about the inverse of function f?

(c) Show that any nonconstant function of the form $f(x) = \dfrac{ax + b}{cx - a}$ is its own inverse.

Recall that a **relation** R is a set of ordered pairs. An **inverse relation** R^{-1} is defined as the set of all ordered pairs (y, x), where (x, y) is an element of R.

In Exercises 109–112, produce the graph of (a) the given relation and (b) the inverse relation. In each part state whether the graph represents a function.

109. $y = |x + 3|$

110. $y = x^2$

111. $y = \sqrt{16 - x^2}$

112. $y = 4 - x^2$

11.3 Exponential Functions

Definition ▪ Natural Base e ▪ Graphs of Exponential Functions ▪ Exponential Equations ▪ Applications

Definition

In our previous study we have worked with functions with *constant* exponents. A simple example is $f(x) = x^2$ whose graph is a parabola. In this section we study functions like $f(x) = 2^x$ and $f(x) = (\frac{2}{3})^x$ with *variables* as exponents.

Consider the expression 2^x. This expression is meaningful for any integer value of x. Moreover, in Section 8.2 we defined rational exponents: $b^{m/n} = \left(\sqrt[n]{b}\right)^m$ (if $\sqrt[n]{b}$ exists). Therefore, the expression 2^x is defined for all *rational* numbers.

It can also be shown that the value of 2^x can be defined when x is any *irrational* number. In short, an exponent can be any *real number*.

> ### Definition of an Exponential Function
>
> An **exponential function** can be written as $f(x) = b^x$, where $b > 0$, $b \neq 1$, and x is any real number.

Two observations about the definition are needed. First, the definition excludes $b = 1$. If $b = 1$, the function becomes 1^x, which has a value of 1 regardless of the value of x. This is simply the function $f(x) = 1$.

Also, the definition requires b to be positive. If the base were negative, there would be some values of x for which the function would not be defined in the real number system. For example, if $b = -4$, the function would be $f(x) = (-4)^x$. Then, for $x = \frac{1}{2}$, $(-4)^{1/2} = \sqrt{-4}$, which is not a real number. Therefore, in the real number system, the function is not defined at $x = \frac{1}{2}$.

EXAMPLE 1   *Evaluating an Exponential Function*

Evaluate $f(x) = 3^x$ at $x = 0, 1, 2, -1, -3,$ and $\frac{1}{2}$. (All but the last one can be done without a calculator, but use a calculator to verify your results.) Then, produce the graph of $f(x) = 3^x$ on your calculator. Trace to the given x-values and verify the corresponding y-values.

	Graph Point
Solution	

$$f(0) = 3^0 = 1 \qquad\qquad A$$
$$f(1) = 3^1 = 3 \qquad\qquad B$$
$$f(2) = 3^2 = 9 \qquad\qquad C$$

$$f(-1) = 3^{-1} = \frac{1}{3} = 0.3333 \ldots \qquad D$$

$$f(-3) = 3^{-3} = \frac{1}{3^3} = \frac{1}{27} = 0.037037 \ldots \qquad E$$

$$f\left(\frac{1}{2}\right) = 3^{1/2} = \sqrt{3} = 1.73205 \ldots \qquad F$$

Figure 11.15 shows the graph of $f(x) = 3^x$ and the six points.

Figure 11.15

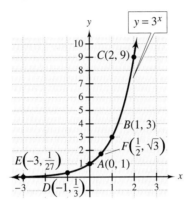

Natural base e

An irrational number approximately equal to 2.718281828 occurs frequently in applications in science, engineering, business, and other areas. Because it occurs so frequently, the letter e is used to represent it, just as the Greek letter π is used to represent the irrational number approximately equal to 3.14159.

The number e is often the base in an exponential function.

$$f(x) = e^x$$

This function is called the **natural exponential function,** but more often it is called simply *the* exponential function.

 BASE e Most calculators have a special key for the base e.

EXAMPLE 2   *Evaluating e^x*

Let $f(x) = e^x$. Use your calculator to evaluate the function at the indicated values. Round results to three decimal places.

(a) $f(0)$ (b) $f(1)$ (c) $f(-1)$ (d) $f(2)$

Solution

(a) $f(0) = e^0 = 1$ (b) $f(1) = e^1 = 2.718$

(c) $f(-1) = e^{-1} = 0.368$ (d) $f(2) = e^2 = 7.389$

Graphs of Exponential Functions

To investigate the behavior of exponential functions, we explore their graphs for $0 < b < 1$ and for $b > 1$.

EXPLORATION 1   *Graphs of Exponential Functions*

(a) For each of the following groups of functions, produce the graphs in the same coordinate system.

Group 1 ($b > 1$)	*Group* 2 ($0 < b < 1$)
$f(x) = 2^x$	$f(x) = \left(\dfrac{1}{2}\right)^x$
$g(x) = e^x$	$g(x) = \left(\dfrac{1}{3}\right)^x$
$h(x) = 7^x$	$h(x) = \left(\dfrac{1}{6}\right)^x$

(b) What are the domain and range for each group?

(c) What is the y-intercept for all the functions?

(d) Which functions, if any, are one-to-one functions?

(e) Describe the differences in the graphs for the two groups.

Discovery

(a) Figure 11.16 shows the graphs for Group 1; Fig. 11.17 shows the graphs for Group 2.

Figure 11.16

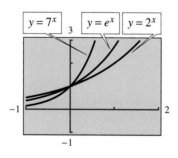

Figure 11.17

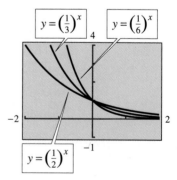

(b) The domain of each of the functions is the set of all real numbers. Because $b > 0$, b^x can never be 0. Therefore, the graph approaches but does not reach the x-axis. The range of each of the functions is $\{y \mid y > 0\}$.

(c) Because $b^0 = 1$, the y-intercept of each graph is (0, 1).

(d) All the graphs pass the Horizontal Line Test; therefore, all the functions are one-to-one functions.

(e) The graphs in Group 1 rise from left to right. As the base increases, the graph is steeper.

The graphs in Group 2 fall from left to right. As the base decreases, the graph is steeper.

Here is a summary of our findings about exponential functions.

> ### Properties of Exponential Functions and Their Graphs
>
> Let $f(x) = b^x$, $b > 0$ and $b \neq 1$.
>
> 1. The *y*-intercept is (0, 1); there is no *x*-intercept.
>
> 2. The function f is a one-to-one function.
>
> 3. The domain is the set of all real numbers; the range is $\{y \mid y > 0\}$.
>
> 4. The graph approaches but does not reach the *x*-axis.
>
> 5. For $b > 1$, the graph rises from left to right; for $0 < b < 1$, the graph falls from left to right.

Exponential Equations

Equations we have encountered to this point have always involved *constant* exponents. Equations which involve a *variable* in the exponent are called **exponential equations.**

The following are some examples of exponential equations.

$$2^x = 8 \qquad 2^{3x} = 16 \qquad 27^x = 81 \qquad \left(\frac{1}{3}\right)^x = 9$$

As we have done before, we can estimate the solutions of such equations by graphing.

EXAMPLE 3   *Solving an Exponential Equation by Graphing*

Use a graph to estimate the solution of $e^x = 2$.

Figure 11.18

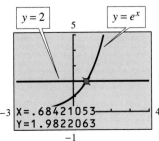

Solution   Produce the graphs of $y_1 = e^x$ and $y_2 = 2$. (See Fig. 11.18.) After one or two zoom-and-trace cycles, we estimate the solution to be $x \approx 0.68$.

We can verify by evaluating y_1 for $x = 0.68$ or by calculating $e^{0.68}$ directly. In either case, the result is close to 2, and the solution is approximately 0.68. ∎

To solve exponential equations algebraically requires new techniques.

We observed that an exponential function is a one-to-one function. Therefore, if $b^x = b^y$, then x must equal y.

For $b > 0$ and $b \neq 1$, if $b^x = b^y$, then $x = y$.

We can use this property to solve an exponential equation. If we can write each side of an exponential equation as an exponential function with the same base, then we can equate the exponents.

EXAMPLE 4   *Equating Exponents to Solve an Exponential Equation*

Use a graph to estimate the solution of $4^x = 8$. Then solve the equation algebraically.

Solution

Figure 11.19

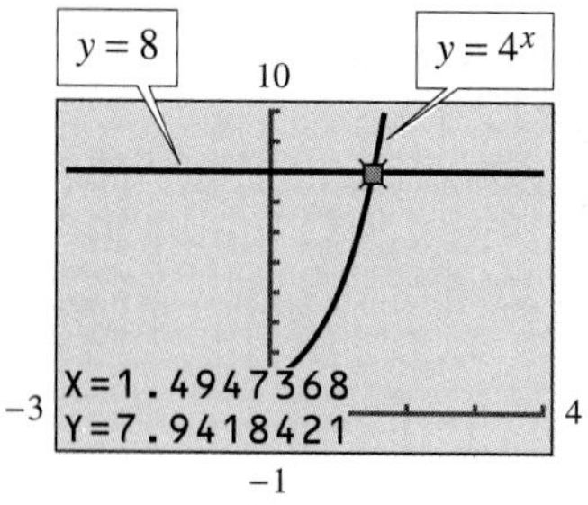

Figure 11.19 shows the graphs of $y_1 = 4^x$ and $y_2 = 8$. We estimate the solution to be about 1.5.

Now we solve algebraically by equating exponents.

$$4^x = 8$$
$$(2^2)^x = 2^3 \qquad \text{Write each side with a base of 2.}$$
$$2^{2x} = 2^3 \qquad \text{Power to a Power Rule for Exponents}$$
$$2x = 3 \qquad \text{Equate the exponents.}$$
$$x = \frac{3}{2}$$

As we will see in the next section, it is always possible to write exponential expressions with the same base. However, the method of equating exponents is recommended only for those special cases in which changing the base is easy to do.

EXAMPLE 5   *Solving Exponential Equations*

Solve each exponential equation.

(a) $4^{2x+1} = 32$ (b) $\left(\dfrac{2}{3}\right)^{1-x} = \left(\dfrac{9}{4}\right)^{2x}$ (c) $e^{x^2-1} = 1$

Solution

(a)
$$4^{2x+1} = 32$$
$$(2^2)^{2x+1} = 2^5 \qquad \text{Write both sides with base 2.}$$
$$2^{4x+2} = 2^5 \qquad \text{Power to a Power Rule for Exponents}$$
$$4x + 2 = 5 \qquad \text{Equate exponents.}$$
$$4x = 3 \qquad \text{Solve for } x.$$
$$x = \frac{3}{4}$$

(b) $\left(\dfrac{2}{3}\right)^{1-x} = \left(\dfrac{9}{4}\right)^{2x}$

$\left(\dfrac{2}{3}\right)^{1-x} = \left[\left(\dfrac{3}{2}\right)^2\right]^{2x}$ Write both sides with base $\frac{2}{3}$.

$\left(\dfrac{2}{3}\right)^{1-x} = \left[\left(\dfrac{2}{3}\right)^{-2}\right]^{2x}$ $\left(\dfrac{a}{b}\right)^{-n} = \left(\dfrac{b}{a}\right)^{n}$

$\left(\dfrac{2}{3}\right)^{1-x} = \left(\dfrac{2}{3}\right)^{-4x}$ Power to a Power Rule for Exponents

$1 - x = -4x$ Equate exponents.

$1 = -3x$ Solve for x.

$x = -\dfrac{1}{3}$

(c) $e^{x^2-1} = 1$

$e^{x^2-1} = e^0$ Write both sides with base e.

$x^2 - 1 = 0$ Equate exponents.

$(x + 1)(x - 1) = 0$ Factor.

$x + 1 = 0 \quad \text{or} \quad x - 1 = 0$ Zero Factor Property

$x = -1 \quad \text{or} \quad x = 1$ ■

Note that we would not be able to use this procedure to solve the equation $e^x = 2$ because we presently have no way of expressing 2 as a power of e. The techniques for solving this type of equation will be introduced later.

Applications

Exponential equations play a major role in describing the conditions of certain application problems. Sometimes the exponential equation is a formula that has been derived to relate the variables of the problem.

For example, the formula to compute the value A of an investment of P dollars at an interest rate r compounded n times per year for t years is

$$A = P\left(1 + \dfrac{r}{n}\right)^{nt}.$$

EXAMPLE 6   *Compound Interest*

If an investment earns 8% interest, compounded monthly, in how many years will the investment double?

Solution   If P is the original investment, then it must double to $2P$.

$$A = 2P$$
$$n = 12$$
$$r = 0.08$$

$$A = P\left(1 + \frac{r}{n}\right)^{nt}$$

$$2P = P\left(1 + \frac{0.08}{12}\right)^{12t} \qquad \text{Substitute for } A, \, n, \text{ and } r.$$

$$2 = \left(1 + \frac{0.08}{12}\right)^{12t} \qquad \text{Divide both sides by } P.$$

$$2 = (1.0067)^{12t}$$

Figure 11.20

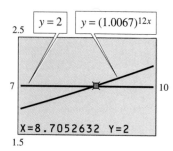

Because we cannot conveniently express both sides with the same base, we will use a graph (with x instead of t) to approximate the solution.

Figure 11.20 shows the graphs of $y_1 = 2$ and $y_2 = (1.0067)^{12x}$. The approximate solution for x is 8.7. It will take about 8.7 years for the investment to double. Note that the amount of the investment has no bearing on the result. ∎

11.3 *Quick Reference*

Definition	▪ An **exponential function** can be written $f(x) = b^x$ where $b > 0$, $b \neq 1$, and x is any real number.

Natural base e	▪ The irrational number 2.71828. . . is represented by the letter e. ▪ The function $f(x) = e^x$ is called the **natural exponential function.**

Graphs of Exponential Functions	▪ The following are properties of the graph of an exponential function $f(x) = b^x$, $b > 0$ and $b \neq 1$. 1. The y-intercept is $(0, 1)$; there is no x-intercept. 2. The function f is a one-to-one function. 3. The domain is the set of all real numbers; the range is $\{y \mid y > 0\}$. 4. The graph approaches but does not reach the x-axis. 5. For $b > 1$, the graph rises from left to right; for $0 < b < 1$, the graph falls from left to right.

Exponential Equations	▪ An equation in which the variable appears in the exponent is an **exponential equation.**

- Solutions of exponential equations can be estimated by graphing the two sides of the equation and determining the point of intersection.

- Because an exponential function is a one-to-one function, if $b^x = b^y$, then $x = y$. This rule can be used to solve an exponential equation as follows.

  1. Write each side of the equation as a single exponential expression with the same base.

  2. Use the rules of exponents to simplify both sides.

  3. Equate the exponents.

  4. Solve the resulting equation.

11.3 Exercises

1. For the exponential function $f(x) = b^x$, one restriction on b is that b cannot equal 1. Why is this restriction necessary?

2. For the exponential function $f(x) = b^x$, one restriction on b is that b must be positive. Why is this restriction necessary?

In Exercises 3–10, let $f(x) = 2^{3x-1}$ and $g(x) = \left(\frac{1}{3}\right)^{2x}$. Evaluate each of the following.

3. $f(2)$

4. $g(-1)$

5. $g(1)$

6. $f(-1)$

7. $g(0)$

8. $f\left(\dfrac{2}{3}\right)$

9. $f(0)$

10. $g\left(-\dfrac{3}{2}\right)$

In Exercises 11–18, let $g(x) = e^{-2x}$ and $h(x) = \left(\frac{1}{5}\right)^x$. Determine each of the following values to the nearest hundredth.

11. $g(2)$

12. $h(-2)$

13. $g(-3)$

14. $h(0)$

15. $h\left(-\dfrac{3}{2}\right)$

16. $g(-2)$

17. $h\left(\dfrac{1}{2}\right)$

18. $g(1)$

In Exercises 19–26, evaluate the given function as indicated. Round results to the nearest hundredth.

19. $f(x) = 4e^{3x}, \quad f(2)$

20. $g(x) = 100e^{-2x}, \quad g(3)$

21. $g(x) = \left(\sqrt{3}\right)^{2x}, \quad g(5)$

22. $f(x) = \left(\sqrt{5}\right)^{-3x}, \quad f(-3)$

23. $f(x) = (0.23)^{3x-1}, \quad f(-2)$

24. $g(x) = (1.45)^{1-2x}, \quad g(3)$

25. $g(x) = -3e^{2x} + 4, \quad g(-2)$

26. $f(x) = 5 + 3e^{-x+1}, \quad f(5)$

27. Explain the conditions under which the graph of $f(x) = b^x$

(a) rises from left to right.

(b) falls from left to right.

28. From the following list, identify the two functions that have the same graph and explain why.

(i) $f(x) = 3^x$ (ii) $g(x) = 3^{-x}$

(iii) $h(x) = \left(\frac{1}{3}\right)^x$

In Exercises 29–34, use a graph to estimate the range and domain of the given function.

29. $f(x) = -e^x - 2$ **30.** $g(x) = 2^{x-2} + 4$

31. $f(x) = 2 - 3^{-x}$ **32.** $h(x) = e^{x^2}$

33. $h(x) = e^{-|x|}$ **34.** $g(x) = e^x + e^{-x}$

35. If $f(x) = b^x$ with $b > 1$, as b increases, what is the effect on the graph of f?

36. If $f(x) = (1/b)^x$ with $b > 1$, as b increases, what is the effect on the graph of f?

37. Use your calculator to produce the graphs of the following three functions.

$$f(x) = 3^x \qquad g(x) = 3^x + 2 \qquad h(x) = 3^x - 4$$

In general, compare the graphs of b^x and $b^x + c$ when $c > 0$ and when $c < 0$.

38. Use your calculator to produce the graphs of the following three functions.

$$f(x) = 3^x \qquad g(x) = 3^{x+2} \qquad h(x) = 3^{x-4}$$

In general, compare the graphs of b^x and b^{x+k} when $k > 0$ and when $k < 0$.

In Exercises 39 and 40, compare the graphs of the given pair of functions to the graph of $h(x) = 2^x$.

39. (a) $f(x) = 2^x + 1$ (b) $g(x) = 2^x - 3$

40. (a) $f(x) = 2^{x+3}$ (b) $g(x) = 2^{x-4}$

In Exercises 41 and 42, compare the graphs of the given pair of functions to the graph of $h(x) = \left(\frac{1}{3}\right)^x$.

41. (a) $f(x) = \left(\frac{1}{3}\right)^x + 3$ (b) $g(x) = \left(\frac{1}{3}\right)^{x+3}$

42. (a) $f(x) = \left(\frac{1}{3}\right)^{-x}$ (b) $g(x) = 3^{-x}$

43. In each part compare the graphs of functions f and g to the graph of function h.

(a) $h(x) = 4^x$ (b) $h(x) = (0.5)^x$
    $f(x) = 2^x$ $f(x) = (0.3)^x$
    $g(x) = 6^x$ $g(x) = (0.9)^x$

44. The accompanying figure shows the graph of a function $f(x) = b^x$. Use this graph to assist you in matching each of the following functions to graph (A), (B), (C), or (D).

(a) $g(x) = b^x + 3$ (b) $f(x) = b^{-x}$

(c) $f(x) = 4 - b^x$ (d) $g(x) = b^{-x} - 5$

Figure for 44

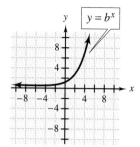

(A)

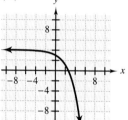

(B)

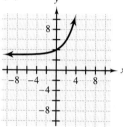

(C)

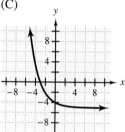

(D)

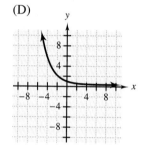

In Exercises 45–50, let $f(x) = e^x$ and $h(x) = (2.3)^x$. Use a graph to estimate the value of x (to the nearest hundredth) for which the function has the given value.

45. $h(x) = 0.76$ **46.** $f(x) = 0.59$

47. $h(x) = 7.51$ **48.** $f(x) = 4.85$

49. $f(x) = -3.45$ **50.** $h(x) = 0$

In Exercises 51–58, use a graph to estimate the solution of the given exponential equation.

51. $e^x = 4$ **52.** $2^{-x} = 7$

53. $(0.27)^x = 6$ **54.** $(1.56)^x = 8$

55. $e^x = -2x + 1$ **56.** $x + 4^x = -3$

57. $e^x = x$

58. $-4^x = 2 - x$

59. Why can we solve $4^x = 8$ without using a calculator but cannot use the same method to solve $4^x = 12$?

60. Explain why the graphs of $f(x) = 2^{2x}$ and $g(x) = 4^x$ are the same.

In Exercises 61–68, solve the exponential equation.

61. $5^{-x} = 125$

62. $2^{x-1} = 16$

63. $\left(\dfrac{2}{5}\right)^x = \dfrac{25}{4}$

64. $\left(\dfrac{3}{2}\right)^x = \dfrac{8}{27}$

65. $e^{3x+1} = e^{x-2}$

66. $e^{x+1} = e^{2x}$

67. $\dfrac{1}{9^x} = 27$

68. $25^x = \dfrac{1}{5}$

In Exercises 69–92, solve the exponential equation.

69. $49^x = 343$

70. $9^{-2x} = 27$

71. $16 = 8^{x-2}$

72. $e^{x+4} = \dfrac{1}{e^{3x}}$

73. $3 \cdot 8^x = 12$

74. $-3^x = -9$

75. $9 \cdot 3^x = \dfrac{1}{3}$

76. $4^x \cdot 16^{2-3x} = 8$

77. $2^{-x-3} = -4$

78. $-4^{2x} = 2$

79. $\dfrac{5^{x+1}}{5^{1-x}} = \dfrac{1}{25^x}$

80. $\dfrac{3^{x+1}}{27^x} = \dfrac{1}{9^x}$

81. $e^x + 4 = 3$

82. $7 - e^{3-2x} = 6$

83. $\left(\dfrac{1}{3}\right)^x 27^{x+1} = 9^{3-2x}$

84. $\left(\dfrac{1}{3}\right)^{x^2} = 27^{(2/3)x-1}$

85. $32^{1-x} = 8^{x^2-x}$

86. $125^{2-3x} = 25^{3x-5}$

87. $(0.25)^{2x+1} = 8^{2-x}$

88. $(0.2)^{x-3} = 25^{1-x}$

89. $e^{x^2-1} = 7^0$

90. $(e^{x+2})^x = (e^{x+2})^2$

91. $\left(\sqrt{3}\right)^{2-2x} = 9^{x^2}$

92. $\left(\sqrt{2}\right)^{4-2x} = \left(\sqrt[3]{2}\right)^{6x}$

93. Suppose that you want to place $1000 in a bank account to earn interest over the next year (365 days). You find two banks that pay the same interest rate of 6%, but one bank compounds daily while the other bank pays only simple interest. What would be the difference in the amount of interest you would earn at the two banks?

94. A woman places $2000 in an account that pays 7% compounded quarterly. Her sister places the same amount in another bank that pays 6% compounded daily (365 days). Which of the women will earn the most interest at the end of one year?

95. Two years ago, an investor put some money in a growth fund that paid 9% in dividend interest. Each quarter the interest was rolled back into the fund, and the account is now worth $5974.16. How much was the initial investment?

96. A man invested some money in two mutual funds. One fund paid 11% compounded daily (365 days), and the other fund paid 7% compounded quarterly. At the end of 1 year, the 11% fund had a balance of $3348.79 while the 7% fund balance was $3322.76. What was the total original amount of the two investments?

In Exercises 97 and 98, use the graphing method with the recommended scale to estimate the solution.

97. If $1500 is invested at 10% compounded daily (365 days) and $2000 is invested at 6% compounded quarterly, after approximately how many years will the two investments have the same value?

[X: 0, 10, 1]   [Y: 2500, 3500, 100]

98. When a child was born, a grandmother invested $3000 in a fund guaranteed to pay 7% compounded quarterly. The purpose was to be able to contribute at least $10,000 to the child's college education. Approximately how many years will it take to achieve this goal?

[X: 10, 20, 1]   [Y: 9000, 11000, 100]

Exploring with Real Data

In Exercises 99–102, refer to the following background information and data.

Not until the mid-1990s was any serious effort made to curtail the cost of health care in America. The accompanying figure shows the rapid increase in federal spending for Medicare during the period 1980–1992. (Source: U.S. Office of Management and Budget.)

Figure for 99–102

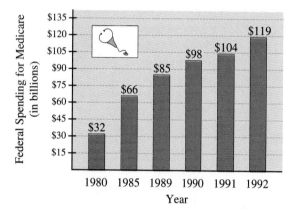

The data in the figure can be modeled by the exponential function $M(t) = 34.26(1.11)^t$, where M is federal spending (in billions) for Medicare and t is the number of years since 1980.

99. Use [X: 0, 30, 3] and [Y: 0, 400, 40] to produce the graph of the function. How does the appearance of the graph support the claim that Medicare costs were skyrocketing during the period 1980–1992?

100. Assuming that the model remains valid, trace the graph to estimate the cost of Medicare in the year 2000.

101. Assuming a federal budget of one trillion dollars, by what year would the cost of Medicare represent half the federal budget?

102. It has been estimated that 5% of Medicare costs are fraudulent. If so, by how much were American taxpayers cheated in 1992? Compare this figure to the approximately $11 billion of federal spending for higher education in 1992.

Challenge

In Exercises 103–108, use a graph to estimate the solution(s) of the given equation.

103. $6x \cdot 3^x = x + 1$ **104.** $5x^2 e^x - 1 = x$

105. $\dfrac{e^x - e^{-x}}{2} = 2.3$ **106.** $\dfrac{e^x + e^{-x}}{2} = 4.6$

107. $5xe^x = -1$ **108.** $7x^2 e^x = 1.8$

In Exercises 109 and 110, use a graph to estimate the solution set of the given inequality.

109. $2^x < 3^x$

110. $e^x < e^{-x}$

In Exercises 111–114, use a graph of the given function to estimate any local minimum or maximum.

111. xe^x **112.** $x^2 e^x$

113. $7x^2(0.5)^x$ **114.** $e^{|x|} + 3$

115. Compare the graphs of $f(x) = e^x$, $g(x) = e^{|x|}$, and $h(x) = e^{x^2}$.

11.4 Logarithmic Functions

Definition ▪ *Evaluation of Logarithms* ▪ *Logarithmic Equations* ▪
Graphs of Logarithmic Functions

Definition

In the previous section, we learned that the exponential function $f(x) = b^x$ is a one-to-one function. This means that an exponential function has an inverse.

Because the graph of a function and its inverse are symmetric with respect to the line $y = x$, we can sketch the graph of the inverse of a function by drawing the mirror image of the function on the other side of the line of symmetry.

 **EXPLORATION 1**   *The Inverse of an Exponential Function*

Consider the exponential functions $f(x) = 2^x$ and $g(x) = (\frac{1}{2})^x$.

(a) Use the fact that the graphs of a function and its inverse are symmetric about the line $y = x$ to sketch the graphs of $f^{-1}(x)$ and $g^{-1}(x)$.

(b) For f^{-1} and g^{-1}, estimate the intercepts of the graphs and the domain and range of the functions.

Discovery

(a) Figure 11.21 shows the graphs of f and f^{-1}; Fig. 11.22 shows the graphs of g and g^{-1}.

Figure 11.21   Figure 11.22

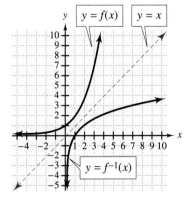

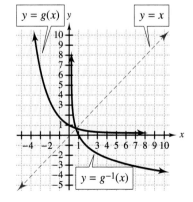

(b) The x-intercept for the graphs of both inverse functions is $(1, 0)$. The graphs approach but never reach the y-axis. Therefore, there is no y-intercept.

The graph of the inverse function lies entirely to the right of the y-axis, and so the domain is $\{x \mid x > 0\}$. The range of the inverse is the set of all real numbers.

In Section 11.2 we found inverse functions algebraically by interchanging the variables x and y and solving for y. Attempting this with an exponential function produces the following.

$$y = b^x$$
$$x = b^y$$

However, we cannot solve the resulting equation for y by using algebraic methods that we have studied so far. To develop new methods, we begin by defining the inverse of an exponential function. We call this function the **logarithmic function.**

Definition of Logarithmic Function

For $x > 0$, $b > 0$, and $b \neq 1$, the **logarithm of x with base b,** represented by $\log_b x$, is defined by the following:

$$\log_b x = y \text{ if and only if } x = b^y.$$

The function $g(x) = \log_b x$ is called the **logarithmic function with base b** and is the inverse of the function $f(x) = b^x$.

The notation $\log_b x$ is often read more simply as the "log base b of x." The number x is called the **argument** of the logarithm.

Natural and Common Logarithms

The logarithmic function with base e is called the **natural logarithmic function** and is written $y = \ln x$ rather than $y = \log_e x$.

The logarithmic function with base 10 is called the **common logarithmic function** and is written $y = \log x$ rather than $y = \log_{10} x$.

Evaluation of Logarithms

Expressions can be changed from exponential to logarithmic form or from logarithmic to exponential form.

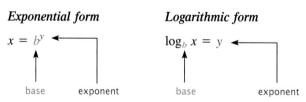

Observe that if $y = \log_b x$, then, in the exponential form $x = b^y$, $\log_b x$ is an exponent.

EXAMPLE 1   *Exponential Form to Logarithmic Form*

Write in logarithmic form.

(a) $3^4 = 81$ (b) $5^{-2} = 0.04$ (c) $10^3 = 1000$

Solution

Exponential form	**Logarithmic form**
(a) $3^4 = 81$	$\log_3 81 = 4$
(b) $5^{-2} = 0.04$	$\log_5 0.04 = -2$
(c) $10^3 = 1000$	$\log 1000 = 3$ Recall that log 1000 means $\log_{10} 1000$. ■

EXAMPLE 2   *Logarithmic Form to Exponential Form*

Write in exponential form.

(a) $\log_4 2 = \dfrac{1}{2}$ (b) $\ln e = 1$ (c) $\log_5 \dfrac{1}{5} = -1$

Solution

Logarithmic form	**Exponential form**
(a) $\log_4 2 = \dfrac{1}{2}$	$4^{1/2} = 2$
(b) $\ln e = 1$	$e^1 = e$ Recall that ln e means $\log_e e$.
(c) $\log_5 \dfrac{1}{5} = -1$	$5^{-1} = \dfrac{1}{5}$ ■

One method for evaluating a logarithmic expression begins with writing the expression in exponential form.

EXAMPLE 3   *Evaluating Logarithmic Expressions*

Evaluate the logarithms.

(a) $\log_6 6$ (b) $\log_{1/4} 2$ (c) $\log_5 125$

Solution

(a) Let $\log_6 6 = y$.

$6^y = 6$ Write in exponential form.

$6^y = 6^1$

$y = 1$ Equate exponents.

Therefore, $\log_6 6 = 1$.

(b) Let $\log_{1/4} 2 = y$.

$\left(\dfrac{1}{4}\right)^y = 2$ Write in exponential form.

$\left(\dfrac{1}{2^2}\right)^y = 2$ Write both sides with the same base.

$(2^{-2})^y = 2$ Definition of negative exponent

$2^{-2y} = 2^1$ Power to a Power Rule for Exponents

$-2y = 1$ Equate exponents.

$y = -\dfrac{1}{2}$

Therefore, $\log_{1/4} 2 = -\dfrac{1}{2}$.

(c) Let $\log_5 125 = y$.

$5^y = 125$ Write in exponential form.

$5^y = 5^3$ Write both sides with the same base.

$y = 3$ Equate exponents.

Therefore, $\log_5 125 = 3$.   ■

LOG

LN   Most calculators have keys for both common logarithms and natural logarithms.

EXAMPLE 4   *Evaluating Logarithms with a Calculator*

Use a calculator to evaluate the following logarithms. Round results to two decimal places.

(a) ln 8   (b) log 20

Solution

Figure 11.23

```
Ln 8
                    2.08
Log 20
                    1.30
```

■

Example 4 shows how easy it is to use the calculator to evaluate common logarithms and natural logarithms. However, most calculators do not have keys for evaluating logarithms with bases other than 10 or e.

It can be shown that a logarithm with any base b can be evaluated in terms of logarithms with any other base a.

Change of Base Formula

For $x > 0$ and for any positive bases a and b where $a \neq 1$ and $b \neq 1$,

$$\log_b x = \frac{\log_a x}{\log_a b}.$$

One use of this formula is for evaluating a logarithm with any base in terms of natural or common logarithms.

EXAMPLE 5 *Evaluating Logarithms with Bases Other Than 10 or e*

Evaluate.

(a) $\log_3 5$

(b) $\log_{2.3} 4.72$

Solution

(a) $\log_3 5 = \dfrac{\ln 5}{\ln 3} = 1.46$

(b) $\log_{2.3} 4.72 = \dfrac{\ln 4.72}{\ln 2.3} = 1.86$ ∎

Logarithmic Equations

We can often solve equations involving logarithms by converting logarithmic forms to exponential forms.

EXAMPLE 6 *Solving Logarithmic Equations*

Solve each equation for x.

(a) $\log_x 27 = 3$

(b) $\log_{2/3} x = 2$

(c) $\log x = -1$

(d) $\ln x = 0$

Solution

(a) $\log_x 27 = 3$

$\qquad\qquad x^3 = 27 \qquad$ Write in exponential form.

$\qquad\qquad x = 3 \qquad$ Odd Root Property

(b) $\log_{2/3} x = 2$

$\qquad\qquad \left(\dfrac{2}{3}\right)^2 = x \qquad$ Write in exponential form.

$\qquad\qquad x = \dfrac{4}{9}$

(c) $\log x = -1$

$\qquad\qquad 10^{-1} = x \qquad$ Common logarithm base is 10.

$\qquad\qquad x = \dfrac{1}{10}$

(d) $\ln x = 0$

$\qquad\qquad e^0 = x \qquad$ Natural logarithm base is e.

$\qquad\qquad x = 1$ ∎

Graphs of Logarithmic Functions

We can learn more about the behavior of the logarithmic functions by producing their graphs.

At the beginning of this section, we produced the graphs of the exponential functions $f(x) = 2^x$ and $g(x) = (\frac{1}{2})^x$. Then, we produced the graphs of f^{-1} and g^{-1} by drawing the mirror images of the graphs of f and g on the opposite side of the line $y = x$.

We now know that f^{-1} and g^{-1} are the logarithmic functions $y = \log_2 x$ and $y = \log_{1/2} x$, respectively. To produce the graphs of such functions on a calculator, we use the Change of Base Formula.

EXAMPLE 7 *Graphs of Logarithmic Functions*

Produce the graphs of the following logarithmic functions.

(a) $y = \log_3 x$ (b) $y = \log_{1/3} x$

Solution

(a) To produce the graph of $y = \log_3 x$, we use the Change of Base Formula:

$$y = \log_3 x = \frac{\log x}{\log 3}.$$

(See Fig. 11.24.) Because $b = 3$, $b > 1$, and the graph rises from left to right.

(b) To produce the graph of $y = \log_{1/3} x$, we use the Change of Base Formula:

$$y = \log_{1/3} x = \frac{\log x}{\log \frac{1}{3}}.$$

(See Fig. 11.25.) Because $b = \frac{1}{3}$, $0 < b < 1$, and the graph falls from left to right.

Figure 11.24

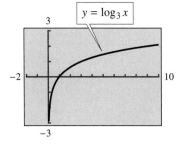

Figure 11.25

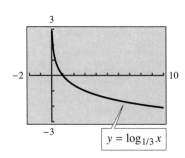

In Example 7 the choice of common logs to change the base was arbitrary. For $y = \log_3 x$, we could have used

$$y = \log_3 x = \frac{\ln x}{\ln 3}$$

and the graph would have been the same.

Here is a summary of properties of $f(x) = \log_b x$.

> ### Properties of $f(x) = \log_b x$ and Its Graph
>
> 1. The domain of f is $\{x \mid x > 0\}$.
> 2. The range of f is the set of all real numbers.
> 3. The x-intercept is $(1, 0)$.
> 4. The graph approaches but does not reach the y-axis; there is no y-intercept.
> 5. For $b > 1$, the graph rises from left to right.
> 6. For $0 < b < 1$, the graph falls from left to right.

11.4 *Quick Reference*

Definition

- For $x > 0$, $b > 0$, and $b \neq 1$, the **logarithm of x with base b,** represented by $\log_b x$, is defined by the following:

$$\log_b x = y \quad \text{if and only if} \quad x = b^y.$$

  The function $f(x) = \log_b x$ is called the **logarithmic function with base b** and is the inverse of the function $f(x) = b^x$.

- The logarithmic function with base e is called the **natural logarithmic function** and is written $y = \ln x$.

- The logarithmic function with base 10 is called the **common logarithmic function** and is written $y = \log x$.

Evaluation of Logarithms

- Exponential expressions can be changed to logarithmic form, and logarithmic expressions can be changed to exponential form.

$$x = b^y \longleftrightarrow \log_b x = y$$

- Logarithmic expressions can sometimes be evaluated without a calculator if they are first converted to exponential forms.

- In order to use a calculator to evaluate a logarithm with a base other than 10 or e, the Change of Base Formula is needed.

  For $x > 0$, and for any positive bases a and b where $a \neq 1$ and $b \neq 1$,

$$\log_b x = \frac{\log_a x}{\log_a b}.$$

Logarithmic Equations

- We can solve some (but not all) logarithmic equations by converting the logarithmic expressions to exponential expressions and equating exponents.

Graphs of Logarithmic Functions

▪ The following are properties of $f(x) = \log_b x$ and its graph.

1. The domain of f is $\{x \mid x > 0\}$.

2. The range of f is the set of all real numbers.

3. The x-intercept is $(1, 0)$.

4. The graph approaches but does not reach the y-axis; there is no y-intercept.

5. For $b > 1$, the graph rises from left to right.

6. For $0 < b < 1$, the graph falls from left to right.

▪ To use a calculator to produce graphs of logarithmic functions with bases other than 10 or e, use the Change of Base Formula.

11.4 Exercises

1. For the logarithmic function $y = \log_b x$, explain why b must be positive. Also explain why b cannot be 1.

2. From the following list, identify the logarithms that are not defined and explain why. (In each case, assume $x > 0$.)

(i) $\log_{-3} x$ (ii) $\log_0 x$ (iii) $\log_{0.27} x$

(iv) $\log_1 x$ (v) $\log_{8.4} x$

In Exercises 3–10, write the given logarithmic equation in exponential form.

3. $\log_7 49 = 2$

4. $\log_4 8 = \dfrac{3}{2}$

5. $\log 0.001 = -3$

6. $\log_{1/2} 8 = -3$

7. $\ln \sqrt[3]{e} = \dfrac{1}{3}$

8. $\log_8 \dfrac{1}{4} = -\dfrac{2}{3}$

9. $\log_m n = 2$

10. $\ln 3 = t$

In Exercises 11–18, write the given exponential equation in logarithmic form.

11. $4^3 = 64$

12. $27^{2/3} = 9$

13. $10^{-5} = 0.00001$

14. $8^{-2/3} = \dfrac{1}{4}$

15. $\left(\dfrac{1}{2}\right)^{-3} = 8$

16. $5^{-3} = \dfrac{1}{125}$

17. $P = e^{rt}$

18. $2^x = y$

In Exercises 19–26, for each function, determine f^{-1}.

19. $f(x) = e^x$

20. $f(x) = 3^x$

21. $f(x) = 4^{-x}$

22. $f(x) = \left(\dfrac{3}{4}\right)^x$

23. $f(x) = \log x$

24. $f(x) = \log_5 x$

25. $f(x) = \log_{1.5} x$

26. $f(x) = \ln x$

27. Explain why log 100 can be evaluated without a calculator but ln 100 cannot.

28. If b is neither 10 nor e, describe two ways to evaluate $\log_b x$.

In Exercises 29–52, without using a calculator, evaluate the given logarithms, if possible. Assume all variables represent positive numbers.

29. $\log_4 16$

30. $\log_3 81$

31. $\ln 1$

32. $\log_{11} 11$

33. $\log_9 243$

34. $\log_9 \dfrac{1}{27}$

35. $\log 1000$

36. $\ln e^5$

37. $\ln (-3)$

38. $\log_4 (-1)$

39. $\log_{1/9} 3$

40. $\log_{2/3} \left(\dfrac{9}{4} \right)$

41. $\log_2 8^{1/3}$

42. $\log_5 25^{-1/2}$

43. $\log_3 \sqrt{3^5}$

44. $\log \sqrt[3]{10^2}$

45. $\log 0$

46. $\log_7 (-7)$

47. $\log_{2/3} \left(\dfrac{3}{2} \right)$

48. $\log_{3/4} \left(\dfrac{4}{3} \right)$

49. $\log_b b^3$

50. $\log_b b$

51. $\log_a \sqrt{a}$

52. $\log_b 1$

In Exercises 53–66, evaluate the given logarithm. Round the results to two decimal places.

53. $\ln 4$

54. $\log 40$

55. $\log_4 7$

56. $\log_2 5$

57. $\ln \left(\dfrac{\sqrt{3}}{2} \right)$

58. $\dfrac{\ln \sqrt{5}}{\ln 3}$

59. $\log \left(7 - \sqrt{2} \right)$

60. $\log 5 + \log \sqrt{2}$

61. $\log_7 (\log 2)$

62. $\log \left(\dfrac{2 + \sqrt{5}}{4} \right)$

63. $\log_5 7^3$

64. $(\log_3 4)^2$

65. $\log \sqrt{5}$

66. $\sqrt{\log 5}$

 67. What is the first step in solving a logarithmic equation $y = \log_b x$?

 68. If $\log_b (x + 2) = y$, explain how you know that the solution of the equation must be greater than -2.

In Exercises 69–88, solve the given equation.

69. $\log_3 x = 5$

70. $\log_3 x = -2$

71. $\ln x = 1$

72. $\ln t = 0$

73. $\log_8 t = -\dfrac{2}{3}$

74. $\log_{27} x = \dfrac{4}{3}$

75. $\log_x 81 = 2$

76. $\log_x e^{17} = 17$

77. $\log_x 25 = \dfrac{2}{3}$

78. $\log_x 0.001 = -\dfrac{3}{2}$

79. $\log_x 1 = 0$

80. $\log_x x = 1$

81. $\log_2 \sqrt{x} = \dfrac{1}{2}$

82. $\log_2 \sqrt[3]{x} = \dfrac{2}{3}$

83. $\log_6 6^x = 3$

84. $\log_3 9^x = 2$

85. $\log_4 (x + 1) = 2$

86. $\log_5 (2x - 1) = 1$

87. $\log_5 x^2 = 4$

88. $\log_3 (x^2 - 1) = 1$

 89. Under what conditions will the graph of $\log_b x$

(a) rise from left to right?

(b) fall from left to right?

 90. Consider the graph of $y = \log_b x$. What is the x-intercept of the graph if

(a) $0 < b < 1$?

(b) $b > 1$?

In Exercises 91–96, use the graph of the given function to estimate the domain of the function.

91. $\ln (5 - 2x)$

92. $\log x^2$

93. $\ln |x - 3|$

94. $\ln (x^2 + 4)$

95. $\log (x^2 - x - 12)$

96. $\ln \left(\dfrac{x - 2}{x + 5} \right)$

In Exercises 97–104, let $f(x) = \ln x$ and $g(x) = \log x$. Use a graph to determine the value of x corresponding to the given functional value. Round results to the nearest hundredth.

97. $g(x) = 1.751$

98. $g(x) = 4.897$

99. $g(x) = -1.622$

100. $g(x) = -0.901$

101. $f(x) = 3.807$

102. $f(x) = 6.672$

103. $f(x) = -1.328$

104. $f(x) = -2.818$

 105. What is a function that you would enter in your calculator in order to produce the graph of $y = \log_3 x$?

106. If $f(x) = 7^x$ and $g(x) = \log_7 x$, what is $(f \circ g)(x)$? Explain how you know.

In Exercises 107–110, compare the graphs of each pair of functions to the graph of $f(x) = \ln x$.

107. (a) $g(x) = 3 + \ln x$ (b) $h(x) = \ln (3 + x)$

108. (a) $g(x) = \ln (x - 5)$ (b) $h(x) = -5 + \ln x$

109. (a) $g(x) = \ln (-x)$ (b) $h(x) = -\ln x$

110. (a) $g(x) = \log_2 x$ (b) $h(x) = \log_3 x$

111. The accompanying figure shows the graph of a function $f(x) = \log_b x$. Use this graph to assist you in matching each of the following functions to graph (A), (B), (C), or (D).

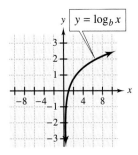

(a) $g(x) = \log_b (x + 3)$

(b) $h(x) = 3 + \log_b x$

(c) $h(x) = 1 - \log_b x$

(d) $g(x) = \log_b (x - 5)$

(A)

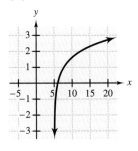

(B)

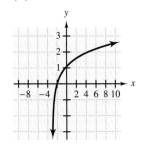

(C)

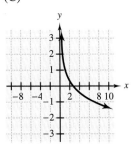

(D)

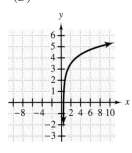

112. Produce the graphs of $y = \log x$ and $y = \ln x$ to see that the solution of $\log x = \ln x$ is 1. Using the Change of Base Formula to solve $\log x = \ln x$, we can write

$$\log x = \frac{\log x}{\log e}.$$

Explain why dividing both sides of this equation by $\log x$ leads to the wrong result.

113. An amplifier produces a power gain G (in decibels) given by $G = 10 \log (P_o/P_i)$, where P_o is the power output and P_i is the power input. To the nearest hundredth, what is the power gain when $P_o = 16$ and $P_i = 0.004$?

114. Use the formula in Exercise 113 to determine the power output for an amplifier that produces a 50-decibel power gain when the power input is 0.006.

115. The number d of cell divisions that occur in a given period of time is given by $d = \log_2 (n/n_0)$ where n_0 is the initial cell count and n is the final cell count. Approximately how many cell divisions occurred if the initial cell count was 50 and the final cell count was 100,000?

116. Use the formula in Exercise 115 to determine the final cell count if an initial cell count of 100 cells undergoes 12 cell divisions.

Challenge

117. Suppose $a > b$. If you multiply both sides of the inequality by $\ln \frac{1}{2}$, what is the result? Why?

118. Show that $e^{x - \ln x} = x^{-1} e^x$.

In Exercises 119–122, evaluate the given expression.

119. $e^{\ln 5 - \ln 2}$

120. $5^{\log_5 2 - \log_5 7}$

121. $e^{-\ln 2}$

122. $\ln e^{\ln e}$

In Exercises 123–128, for what values of x is the function positive?

123. $\ln |x|$

124. $\ln (2 - x)$

125. $1 - \ln x$

126. $\log \left(\dfrac{1}{x} \right)$

127. $\log \left(\dfrac{2x + 1}{x - 1} \right)$

128. $\ln (x^2 - 16)$

In Exercises 129 and 130, solve the given inequalities.

129. $\log_3 x < \log_4 x$

130. $\ln x > \log x$

In Exercises 131 and 132, show how to evaluate the given expression without a calculator.

131. $(\log_2 3)(\log_3 4)(\log_4 5)(\log_5 6)(\log_6 7)(\log_7 8)$

132. $(\log_2 4)(\log_4 6)(\log_6 8)$

11.5 Properties of Logarithms

Basic Properties ▪ *Product, Quotient, and Power Rules for Logarithms* ▪ *Combining the Properties*

Basic Properties

Because logarithms are defined in terms of exponents, the properties of logarithms are closely related to the properties of exponents.

The first group of properties follows directly from the definition of logarithm.

> ### Basic Properties of Logarithms
>
> 1. $\log_b 1 = 0$ 2. $\log_b b = 1$
> 3. $\log_b b^x = x$ 4. $b^{\log_b x} = x$

For each property an equivalent exponential form can be written. In the following table, the logarithmic properties in the left column are justified by the corresponding exponential equations in the right column.

Logarithmic Form	*Exponential Form*
$\log_b 1 = 0$	$b^0 = 1$
$\log_b b = 1$	$b^1 = b$
$\log_b b^x = x$	$b^x = b^x$
$\log_b x = \log_b x$	$b^{\log_b x} = x$

 EXAMPLE 1   *Using Properties of Logarithms to Evaluate Logarithmic Expressions*

Use the basic properties of logarithms to evaluate each expression.

(a) $\log_7 1$

(b) $\log_{3/4} \dfrac{3}{4}$

(c) $\log_8 8^5$

(d) $5^{\log_5 3}$

Solution

(a) $\log_7 1 = 0$

(b) $\log_{3/4} \dfrac{3}{4} = 1$

(c) $\log_8 8^5 = 5$

(d) $5^{\log_5 3} = 3$ ∎

Product, Quotient, and Power Rules for Logarithms

In the following exploration, we consider logarithms whose arguments are products, quotients, or exponential expressions.

EXPLORATION 1   *Logarithms of Products, Quotients, and Powers*

Use your calculator to evaluate each of the given pairs of expressions. For each part, propose an informal, general rule for the results that you obtained.

(a) $\ln (5 \cdot 2)$ $\qquad\qquad$ $\log (7 \cdot 3)$

$\quad\ \ \ln 5 + \ln 2$ $\qquad\qquad$ $\log 7 + \log 3$

(b) $\ln \dfrac{5}{2}$ $\qquad\qquad\quad$ $\log \dfrac{9}{4}$

$\quad\ \ \ln 5 - \ln 2$ $\qquad\qquad$ $\log 9 - \log 4$

(c) $\ln 2^3$ $\qquad\qquad\quad$ $\log 5^4$

$\quad\ \ 3 \ln 2$ $\qquad\qquad\quad$ $4 \log 5$

Discovery

(a) Figure 11.26 $\qquad\qquad\qquad$ Figure 11.27

```
ln (5*2)
          2.302585093
ln 5+ln 2
          2.302585093
```

```
log (7*3)
          1.322219295
log 7+log 3
          1.322219295
```

The logarithm of a product of two numbers is equal to the sum of the logarithms of the numbers.

(b) Figure 11.28 $\qquad\qquad\qquad$ Figure 11.29

```
ln (5/2)
           .9162907319
ln 5-ln 2
           .9162907319
```

```
log (9/4)
           .3521825181
log 9-log 4
           .3521825181
```

The logarithm of a quotient of two numbers is equal to the difference of the logarithms of the numbers.

(c) Figure 11.30

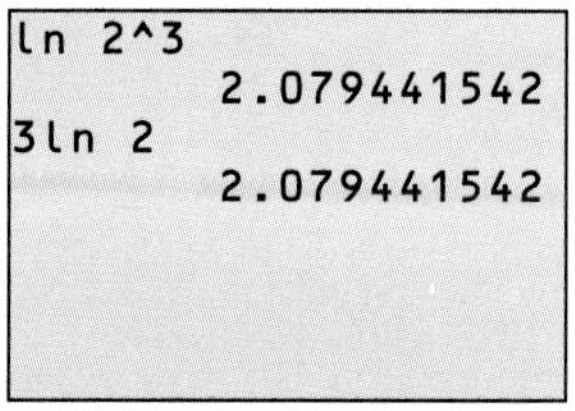

Figure 11.31

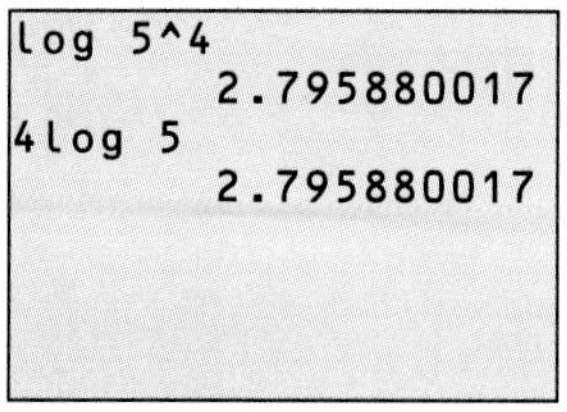

```
Ln 2^3
              2.079441542
3Ln 2
              2.079441542
```

```
log 5^4
              2.795880017
4log 5
              2.795880017
```

The logarithm of a number raised to a power is equal to the power times the logarithm of the number.

The informal generalizations stated in Exploration 1 are stated formally in the following summary.

Product, Quotient, and Power Rules for Logarithms

For $b > 0$, $b \neq 1$, positive real numbers M and N, and any real number c, the following are properties of logarithms.

1. $\log_b MN = \log_b M + \log_b N$ Product Rule for Logarithms

2. $\log_b \dfrac{M}{N} = \log_b M - \log_b N$ Quotient Rule for Logarithms

3. $\log_b M^c = c(\log_b M)$ Power Rule for Logarithms

The following demonstrates why the Product Rule for Logarithms is true.

Let $\log_b M = r$ and $\log_b N = s$. Then, $b^r = M$ and $b^s = N$. Therefore, the product $MN = b^r b^s = b^{r+s}$. But $b^{r+s} = MN$ implies that

$$\log_b MN = r + s = \log_b M + \log_b N.$$

Justifications for the Quotient and Power Rules for Logarithms are similar to that used for the Product Rule for Logarithms. The proofs are left as exercises.

NOTE: There is no property for the logarithm of a *sum* of two numbers or a *difference* of two numbers. In particular, $\log_b (x + y)$ is not equal to $\log_b x + \log_b y$, and $\log_b (x - y)$ is not equal to $\log_b x - \log_b y$. ■

EXAMPLE 2   *Rewriting Logarithms Whose Arguments are Products, Quotients, or Exponential Expressions*

Use the Product, Quotient, or Power Rule for Logarithms to write the given logarithm in an equivalent form. Assume all logarithms are defined.

(a) $\log_4 5(t + 2)$

(b) $\ln e(e - 1)$

(c) $\ln \dfrac{1}{y}$

(d) $\log_6 \dfrac{y + 4}{y}$

(e) $\log z^4$

(f) $\ln \dfrac{1}{x^3}$

Solution

(a) $\log_4 5(t + 2) = \log_4 5 + \log_4 (t + 2)$ Product Rule for Logarithms

(b) $\ln e(e - 1) = \ln e + \ln (e - 1)$ Product Rule for Logarithms

              $= 1 + \ln (e - 1)$

(c) $\ln \dfrac{1}{y} = \ln 1 - \ln y$ Quotient Rule for Logarithms

       $= 0 - \ln y$

       $= -\ln y$

(d) $\log_6 \dfrac{y + 4}{y} = \log_6 (y + 4) - \log_6 y$ Quotient Rule for Logarithms

(e) $\log z^4 = 4 \log z$ Power Rule for Logarithms

(f) $\ln \dfrac{1}{x^3} = \ln x^{-3} = -3 \ln x$ Power Rule for Logarithms   ■

EXAMPLE 3   *Writing a Logarithmic Expression as a Single Logarithm*

Write the given logarithmic expression as a single logarithm. Assume all logarithms are defined.

(a) $\ln 3 + \ln x$

(b) $\log (x + 1) + \log (x - 1)$

(c) $\ln (2x) - \ln 3$

(d) $\log (x^2 + 4) - \log (x + 2)$

(e) $\dfrac{1}{2} \ln t$

(f) $-2 \log_7 t$

Solution

(a) $\ln 3 + \ln x = \ln (3x)$ Product Rule for Logarithms

(b) $\log (x + 1) + \log (x - 1) = \log [(x + 1)(x - 1)]$

                         $= \log (x^2 - 1)$

(c) $\ln (2x) - \ln 3 = \ln \dfrac{2x}{3}$ Quotient Rule for Logarithms

(d) $\log (x^2 + 4) - \log (x + 2) = \log \dfrac{x^2 + 4}{x + 2}$

(e) $\dfrac{1}{2} \ln t = \ln t^{1/2} = \ln \sqrt{t}$ Power Rule for Logarithms

(f) $-2 \log_7 t = \log_7 t^{-2} = \log_7 \dfrac{1}{t^2}$ ■

Combining the Properties

When we need to expand a single logarithm into two or more logarithms or to write a logarithmic expression with a single logarithm, we will sometimes need to use more than one of the rules for logarithms.

EXAMPLE 4 *Expanding a Logarithmic Expression*

Use the properties of logarithms to expand each expression. Assume all logarithms are defined.

(a) $\log_5 3(x + 4)^2$ (b) $\ln \left(\dfrac{x + 2}{x + 1} \right)^2$ (c) $\log \sqrt{\dfrac{x^3}{x + 2}}$

Solution

(a) $\log_5 3(x + 4)^2 = \log_5 3 + \log_5 (x + 4)^2$ Product Rule for Logarithms

$\qquad\qquad\qquad\quad\; = \log_5 3 + 2 \log_5 (x + 4)$ Power Rule for Logarithms

(b) $\ln \left(\dfrac{x + 2}{x + 1} \right)^2 = 2 \ln \dfrac{x + 2}{x + 1}$ Power Rule for Logarithms

$\qquad\qquad\quad\;\; = 2[\ln (x + 2) - \ln (x + 1)]$ Quotient Rule for Logarithms

$\qquad\qquad\quad\;\; = 2 \ln (x + 2) - 2 \ln (x + 1)$ Distributive Property

(c) $\log \sqrt{\dfrac{x^3}{x + 2}} = \log \left(\dfrac{x^3}{x + 2} \right)^{1/2}$ Definition of rational exponent

$\qquad\qquad\quad = \dfrac{1}{2} \log \dfrac{x^3}{x + 2}$ Power Rule for Logarithms

$\qquad\qquad\quad = \dfrac{1}{2} [\log x^3 - \log (x + 2)]$ Quotient Rule for Logarithms

$\qquad\qquad\quad = \dfrac{1}{2} [3 \log x - \log (x + 2)]$ Power Rule for Logarithms

$\qquad\qquad\quad = \dfrac{3}{2} \log x - \dfrac{1}{2} \log (x + 2)$ Distributive Property   ■

EXAMPLE 5 *Combining Logarithms*

Use the properties of logarithms to combine each expression into a single logarithmic expression. Assume all logarithms are defined.

(a) $2 \ln t + 3 \ln (t + 2)$

(b) $4[\log_6 (y + 1) - \log_6 y]$

(c) $2 \log (2x + 1) + \log x - 4 \log (x + 3)$

Solution

(a) $2 \ln t + 3 \ln (t + 2) = \ln t^2 + \ln (t + 2)^3$ Power Rule for Logarithms

$\qquad\qquad\qquad\quad\;\; = \ln t^2(t + 2)^3$ Product Rule for Logarithms

(b) $4[\log_6 (y + 1) - \log_6 y] = 4 \log_6 \dfrac{y + 1}{y}$ Quotient Rule for Logarithms

$$= \log_6 \left(\dfrac{y + 1}{y}\right)^4$$ Power Rule for Logarithms

(c) $2 \log (2x + 1) + \log x - 4 \log (x + 3)$

$= \log (2x + 1)^2 + \log x - \log (x + 3)^4$ Power Rule for Logarithms

$= \log x(2x + 1)^2 - \log (x + 3)^4$ Product Rule for Logarithms

$= \log \dfrac{x(2x + 1)^2}{(x + 3)^4}$ Quotient Rule for Logarithms   ■

11.5   Quick Reference

Basic Properties

- Certain properties of logarithms follow directly from the definition of *logarithm* and can be used to evaluate a logarithmic expression.

  1. $\log_b 1 = 0$ 2. $\log_b b = 1$

  3. $\log_b b^x = x$ 4. $b^{\log_b x} = x$

Product, Quotient, and Power Rules for Logarithms

- For $b > 0$, $b \neq 1$, positive real numbers M and N, and any real number c, the following are properties of logarithms.

  1. $\log_b MN = \log_b M + \log_b N$ Product Rule for Logarithms

  2. $\log_b \dfrac{M}{N} = \log_b M - \log_b N$ Quotient Rule for Logarithms

  3. $\log_b M^c = c(\log_b M)$ Power Rule for Logarithms

Combining the Properties

- Logarithmic expressions may involve combinations of products, quotients, and powers. The Product, Quotient, and Power Rules for Logarithms can be used to expand a single logarithm into two or more logarithmic expressions.

- These same rules can be used to combine two or more logarithms into a single logarithmic expression.

11.5   Exercises

 1. Treating a logarithm as an exponent, we can evaluate $\log_4 16$ by asking, "What exponent on 4 results in 16?" Because the answer is 2, $\log_4 16 = 2$. What question would you ask in order to evaluate $\log_9 3$? What is the answer?

 2. If we write $\log x^2 = \log (x \cdot x) = \log x + \log x = 2 \log x$, what rule of logarithms are we using? What rule of logarithms would be more efficient in obtaining the same result?

In Exercises 3–18, evaluate the given expression.

3. $\log_5 5^7$

4. $\ln e^2$

5. $\log_{1.2} 1$

6. $\log_7 7$

7. $7^{\log_7 2}$

8. $3.5^{\log_{3.5} 10}$

9. $\log_9 \sqrt{3}$

10. $\log_4 \sqrt{8}$

11. $9^{-\log_3 2}$

12. $125^{-\log_5 4}$

13. $e^{2 \ln 3}$

14. $10^{3 \log 2}$

15. $\log_3 (\log_4 4)$

16. $\ln (\log 10)$

17. $\log_3 (\log_2 8)$

18. $\log_2 (\log_2 16)$

In Exercises 19–22, write the given logarithm as a sum or difference of logarithms. Assume all variables represent positive numbers.

19. $\log_7 3x$

20. $\log_8 5mn$

21. $\log_7 \dfrac{5}{y}$

22. $\log_6 \dfrac{y}{t}$

In Exercises 23–26, write the given logarithm with no exponents or radicals in the argument. Assume all variables represent positive numbers.

23. $\log x^2$

24. $\log_3 7^{-4}$

25. $\log_9 \sqrt{n}$

26. $\ln \sqrt[3]{x}$

In Exercises 27–34, use the rules of logarithms to expand the logarithm in column A. Then, letting $\log_3 2 = A$ and $\log_3 5 = B$, match the result with one of the forms in column B.

Column A   *Column B*

27. $\log_3 \left(\dfrac{5}{6}\right)^3$

(a) $A - \dfrac{1}{2}B + 1$

28. $\log_3 \dfrac{45}{8}$

(b) $-\dfrac{2}{3}A + \dfrac{1}{3}B - \dfrac{2}{3}$

29. $\log_3 \dfrac{6\sqrt{5}}{5}$

(c) $-3A + 3B - 3$

30. $\log_3 \dfrac{36\sqrt{2}}{125}$

(d) $-\dfrac{3}{4}A - \dfrac{3}{4}$

Column A   *Column B*

31. $\log_3 \dfrac{5\sqrt{10}}{9}$

(e) $\dfrac{1}{5}A - \dfrac{8}{5}B - 1$

32. $\log_3 \dfrac{\sqrt[3]{30}}{6}$

(f) $-3A + B + 2$

33. $\log_3 \dfrac{\sqrt[4]{96}}{12}$

(g) $\dfrac{1}{2}A + \dfrac{3}{2}B - 2$

34. $\log_3 \dfrac{\sqrt[5]{50}}{75}$

(h) $\dfrac{5}{2}A - 3B + 2$

 35. To evaluate $\log (AB)^n$, should you first apply the Product Rule for Logarithms or the Power Rule for Logarithms? Why?

 36. From the following list, identify two logarithms that cannot be expanded and explain why.

(i) $\log (AB)$

(ii) $\log (A - B)$

(iii) $\log (A + B)$

(iv) $\log \dfrac{A}{B}$

In Exercises 37–58, if possible, use the properties of logarithms to expand the given expression. Assume all variables represent positive numbers.

37. $\log_9 x^2 y^{-3}$

38. $\log_3 27(x + 5)^3$

39. $\log \dfrac{(x + 1)^2}{x + 2}$

40. $\log \dfrac{x^2 y^3}{z^4}$

41. $\ln \left(\dfrac{2x + 5}{x + 7}\right)^5$

42. $\log_3 \dfrac{(x^2 z)^3}{(y^2 z)^5}$

43. $\log \dfrac{\sqrt{10}}{x^2}$

44. $\ln \sqrt[4]{\dfrac{x^3 y}{z^2}}$

45. $\log_5 \sqrt[3]{xy^2}$

46. $\log_3 81x^2\sqrt{y}$

47. $\log_2 \dfrac{\sqrt{x + 1}}{32}$

48. $\log_8 \sqrt{x(x + 2)}$

49. $\log_2 \dfrac{x^2\sqrt{y}}{\sqrt[3]{x}}$

50. $\ln \dfrac{(x + 2)\sqrt{x}}{x^2}$

51. $\log (4x^2 + 4x + 1)$

52. $\log (x^2 + 3x + 2)$

53. $\ln 1000e^{0.08t}$

54. $\ln (e^2 - e)$

55. $\log_4 (x + 4)$

56. $\ln (e^2 + 1)$

57. $\ln Pe^{rt}$

58. $\ln P\left(1 + \dfrac{r}{t}\right)^{nt}$

59. Explain how to combine $2 \log 3 + \log 3$ into a single logarithm without the Product Rule for Logarithms. Can you combine $2 \log x + \log y$ into a single logarithm without the Product Rule for Logarithms if $x \neq y$? Explain.

60. Which rule of logarithms would you use to show that $-\log x = \log (1/x)$? Would you use the same rule to show that $\log (1/x) = -\log x$? Write the steps to show that both equations are true.

In Exercises 61–64, write the given logarithmic expression as a single logarithm. Assume all variables represent positive numbers.

61. $\log_4 3 + \log_4 5$

62. $\ln x^2 + \ln x^{-1}$

63. $\log_3 10 - \log_3 5$

64. $\log_7 u^5 - \log_7 u^2$

In Exercises 65–68, write the given logarithmic expression as a single logarithm with a coefficient of 1.

65. $3 \log_2 y$

66. $-\log_5 x$

67. $\dfrac{1}{2} \log x$

68. $\dfrac{2}{3} \ln t$

In Exercises 69–72, write the given expression as a single logarithm and evaluate the result. Use a calculator to verify that the single logarithm has the same value as the original expression.

69. $\log_4 60 - \log_4 15$

70. $\log_3 4 + \log_3 6 - \log_3 8$

71. $\log_5 50 + 2 \log_5 10 - \log_5 40$

72. $\dfrac{1}{3} \log_6 27 - \log_6 8 - \log_6 81$

In Exercises 73–90, use the rules of logarithms to combine the given expression into a single logarithmic expression with a coefficient of 1. Assume all variables represent positive numbers.

73. $\log_8 3 + \log_8 x - \log_8 y$

74. $\ln x - \ln 3 - \ln 2$

75. $2 \log_7 x - \log_7 y$

76. $\log_8 5 + 2 \log_8 t$

77. $\dfrac{1}{2} \log x - \dfrac{2}{3} \log y$

78. $\dfrac{3}{2} \log_6 r + \dfrac{1}{2} \log_6 s$

79. $3 \log y - 2 \log x - 4 \log z$

80. $\log y + 2 \log (x + 2) - 5 \log z$

81. $2 \log (x + 3) - 3 \log (x + 1)$

82. $\dfrac{1}{2} \log (t + 1) - 3 \log t$

83. $\dfrac{1}{2} [\log x - \log (x + 1)]$

84. $\dfrac{1}{3} [2 \log (x + 2) - \log (x - 5)]$

85. $3[\log (t + 2) - 2 \log t - 5 \log (t + 3)]$

86. $5[4 \log x - 2 \log (y + 3) - 3 \log z]$

87. $5 + 2 \log_3 x$

88. $1 + 4 \log_2 \sqrt{x}$

89. $rt + \ln P_0$

90. $0.06t + \ln 1000$

In Exercises 91–102, determine whether the given equation is always true. Assume all variables represent positive numbers.

91. $\log_2 x^x = x \log_2 x$

92. $\log 10^{x+1} = x + 1$

93. $\log_5 x^2 y = 2 \log_5 x + \log_5 y$

94. $\ln \dfrac{e}{x} = 1 - \ln x$

95. $\log_3 (x + 10) = \log_3 x + \log_3 10$

96. $\log x^2 = (\log x)^2$

97. $\dfrac{\log_8 3}{\log_8 5} = \log_8 \dfrac{3}{5}$

98. $\dfrac{\log_3 25}{\log_3 5} = 2$

99. $\dfrac{\log_3 81}{\log_3 9} = \log_3 9$

100. $\dfrac{\log_2 12}{\log_2 4} = \log_2 3$

101. $(\log 3)(\log 2) = \log 6$

102. $-2 = (\log_7 9)(\log_{1/3} 7)$

103. The formula $\log A = \log P + t \log (1 + r)$ relates the value A of an investment of P dollars compounded annually at an interest rate r for t years. Write the expression on the right side of the formula as a single logarithm.

104. If I_0 is the intensity of a small earthquake, then the Richter scale rating of a stronger earthquake of intensity I is given by $\log I/I_0$. Expand this expression into one with two logarithms.

105. Suppose that the sales (in thousands of units) of a certain new brand of cereal is given by $S(t) = 100 \log_2 (t + 2)$ where t is the number of months since the brand was introduced. How many units of the product were sold six months after its introduction? Show that an equivalent formula can be written with natural logarithms:

$$S(t) = \frac{\ln(t + 2)^{100}}{\ln 2}.$$

106. In the previous section, we used the formula $G = 10 \log P_o/P_i$, where G is the power gain of an amplifier with power input P_i and power output P_o. Show that $G = \log (P_o)^{10} - \log (P_i)^{10}$ is an equivalent formula.

Exploring with Real Data

In Exercises 107–110, refer to the following background information and data.

While the homicide rate in America increased during the period 1970–1990, the number of accidental deaths caused by firearms actually decreased. (See figure.) (Source: U.S. National Center for Health.)

Figure for 107–110

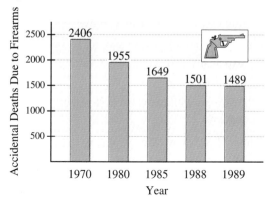

The data in the figure can be modeled by the logarithmic function $D(t) = 4447.074 - 869.79 \ln t$ where $D(t)$ is the number of accidental deaths and t is the number of years since 1960.

107. Use [X: 0, 30, 3] and [Y: 0, 5000, 500] to produce the graph of $y = D(t)$. Note how rapidly D decreases in the interval $0 < t < 10$. Is the model valid for this interval? Why?

108. The base of the logarithm in the function is e and $e > 1$. Why, then, does the graph fall from left to right?

109. If the model is assumed to be valid for the indefinite future, in what year would we anticipate accidental deaths caused by firearms to decline to 1300?

110. The Constitution guarantees the right of each state to maintain a militia. The National Rifle Association (NRA) interprets this to mean that every citizen has a right to bear arms. As a spokesperson for the NRA, how would you take credit for the data in the figure? As an opponent of the NRA, how would you argue that the NRA's position will reverse the trend shown in the figure?

Challenge

111. Let $\log_b M = r$ and $\log_b N = s$. Using steps similar to those used to justify the Product Rule for Logarithms, show that the Quotient Rule for Logarithms is true.

112. Let $\log_b M = r$. Show that the Power Rule for Logarithms is true.

113. Show that $2 + \log_3 x = \log_3 9x$.

114. Show that

$$\log_2 x + \log_3 x = \frac{(\log 2 + \log 3) \log x}{(\log 2)(\log 3)}.$$

115. Show that $\log_2 (xy) = \dfrac{\ln x + \ln y}{\ln 2}$.

116. Suppose $a^2 + b^2 = 7ab$. Show that the following is true.

$$\log (a + b) - \log 3 = \frac{1}{2} (\log a + \log b)$$

117. Let $f(x) = \ln x$. For $h > 0$, show that the following is true.

$$\frac{f(x + h) - f(x)}{h} = \ln \left(1 + \frac{h}{x} \right)^{1/h}$$

118. Show that

$$\log \left[\frac{x}{1 + \sqrt{1 + x^2}} \right] = \log \left[\frac{\sqrt{1 + x^2} - 1}{x} \right].$$

(Hint: Rationalize the denominator.)

11.6 Logarithmic and Exponential Equations

Definitions and Rules ▪ *Exponential Equations* ▪ *Logarithmic Equations* ▪ *Applications*

Definitions and Rules

An equation in which a variable appears in the argument of a logarithm is called a **logarithmic equation.** An equation in which a variable appears as an exponent is called an **exponential equation.**

Because both exponential and logarithmic functions are one-to-one functions, the following properties can be used to solve exponential and logarithmic equations.

1. If $x = y$, then $b^x = b^y$.   Equating exponential expressions

2. If $b^x = b^y$, then $x = y$.   Equating exponents

3. If $x = y$, then $\log_b x = \log_b y$.   Equating logarithmic expressions

4. If $\log_b x = \log_b y$, then $x = y$.   Equating arguments of logarithms

Exponential Equations

We can use a graph to estimate the solution of an exponential equation.

When an equation can be written as an exponential equation with both sides having the same base, we can equate the exponents and solve the resulting equation algebraically.

If both sides of an equation cannot be written with the same base, other algebraic techniques are needed to solve the equation.

Consider the equation $3^{2x} = 5$. Figure 11.32 shows the graphs of $y_1 = 3^{2x}$ and $y_2 = 5$ on the same axes. The point of intersection represents the solution of the equation. The solution is approximately 0.73.

Figure 11.32

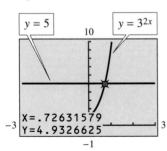

To solve $3^{2x} = 5$ algebraically, we choose a logarithm with a convenient base, usually ln or log, and equate the logarithms of both sides.

$$\ln 3^{2x} = \ln 5$$

$$2x \cdot \ln 3 = \ln 5 \qquad \text{Power Rule for Logarithms}$$

$$x = \frac{\ln 5}{2 \ln 3} \qquad \text{Divide both sides by 2 ln 3.}$$

$$x \approx 0.73$$

EXAMPLE 1   *Solving Exponential Equations*

Solve each equation.

(a) $4^{2x-1} = \dfrac{1}{16}$ (b) $7^{x+1} = 14$ (c) $e^x = 2$

Solution

(a) $4^{2x-1} = \dfrac{1}{16}$

$$[2^2]^{2x-1} = \frac{1}{2^4} \qquad \text{Write both sides with base 2.}$$

$$2^{4x-2} = 2^{-4} \qquad \begin{array}{l}\text{Power to a Power Rule for Exponents and definition of a}\\ \text{negative exponent}\end{array}$$

$$4x - 2 = -4 \qquad \text{Equate exponents.}$$

$$4x = -2$$

$$x = -\frac{1}{2}$$

(b) $7^{x+1} = 14$

$$\log 7^{x+1} = \log 14 \qquad \begin{array}{l}\text{Equate the logarithms of both}\\ \text{sides.}\end{array}$$

$$(x + 1) \log 7 = \log 14 \qquad \text{Power Rule for Logarithms}$$

$$x \log 7 + \log 7 = \log 14 \qquad \text{Distributive Property}$$

$$x \log 7 = \log 14 - \log 7 \qquad \text{Isolate the variable term.}$$

$$x = \frac{\log 14 - \log 7}{\log 7}$$

$$x \approx 0.36$$

(c) Because the equation involves the number e, we choose to use natural logarithms to solve.

$$e^x = 2$$

$$\ln e^x = \ln 2 \qquad \text{Equate the logarithms of both sides.}$$

$$x \ln e = \ln 2 \qquad \text{Power Rule for Logarithms}$$

$$x \cdot 1 = \ln 2 \qquad \text{ln } e = 1$$

$$x = \ln 2$$

$$x \approx 0.69$$

Logarithmic Equations

We can also use graphs to estimate the solutions of logarithmic equations.

Consider the equation $\ln(x - 3) + \ln(x + 1) = \ln 5$.

Figure 11.33 shows the graphs of $y_1 = \ln(x - 3) + \ln(x + 1)$ and $y_2 = \ln 5$.

The solution is approximately 4. (In fact, it is easy to verify by substitution that the solution is *exactly* 4.)

Figure 11.33

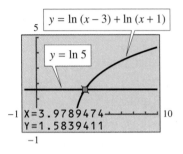

To solve the equation algebraically, we write the left side as a single logarithm and then equate the arguments of the logarithms.

$$\ln(x - 3) + \ln(x + 1) = \ln 5$$
$$\ln(x - 3)(x + 1) = \ln 5 \qquad \text{Product Rule for Logarithms}$$
$$(x - 3)(x + 1) = 5 \qquad \text{Equate the arguments of the logarithms.}$$
$$x^2 - 2x - 3 = 5 \qquad \text{FOIL}$$
$$x^2 - 2x - 8 = 0 \qquad \text{Write the quadratic equation in standard form.}$$
$$(x - 4)(x + 2) = 0 \qquad \text{Factor.}$$
$$x - 4 = 0 \quad \text{or} \quad x + 2 = 0 \qquad \text{Zero Factor Property}$$
$$x = 4 \quad \text{or} \qquad x = -2$$

The graph in Fig. 11.33 suggests that the equation has only one solution, but the algebraic approach produces two solutions.

Replacing x with -2 gives $\ln(-5) + \ln(-1) = \ln 5$. But logarithmic functions are defined only for positive numbers. Therefore, -2 is an extraneous solution. The only solution of the equation is 4.

EXAMPLE 2   *Solving Logarithmic Equations*

Use a graph to estimate the solution(s) of the given equation. Then, solve algebraically.

(a) $\ln(1 - 2x) = \ln 5$

(b) $\log_3 x + \log_3 (x - 2) = 1$

(c) $\ln(4x - 5) - \ln(x - 2) = \ln(2x + 1)$

Solution

Figure 11.34

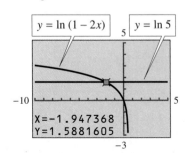

(a) Figure 11.34 shows the graphs of $y_1 = \ln(1 - 2x)$ and $y_2 = \ln 5$. We estimate the one solution to be -1.95.

$$\ln(1 - 2x) = \ln 5$$
$$1 - 2x = 5 \qquad \text{Equate arguments.}$$
$$-2x = 4$$
$$x = -2$$

The solution is easily verified by substitution.

(b) To produce the graph of $y_1 = \log_3 x + \log_3 (x - 2)$, we use the Change of Base Formula.

$$\log_3 x + \log_3 (x - 2) = \frac{\log x}{\log 3} + \frac{\log (x - 2)}{\log 3}$$

The graphs of $y_1 = \dfrac{\log x}{\log 3} + \dfrac{\log (x - 2)}{\log 3}$ and $y_2 = 1$ are shown in Fig. 11.35. We estimate the solution to be 3.05.

Figure 11.35

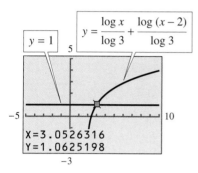

$$\log_3 x + \log_3 (x - 2) = 1$$

$\log_3 x(x - 2) = 1$	Product Rule for Logarithms
$x(x - 2) = 3^1$	Exponential form
$x^2 - 2x = 3$	Distributive Property
$x^2 - 2x - 3 = 0$	Quadratic Equation in standard form
$(x - 3)(x + 1) = 0$	Factor.
$x - 3 = 0$ or $x + 1 = 0$	Zero Factor Property
$x = 3$ or $x = -1$	

The value -1 is not in the domain of $\log_3 (x - 2)$, so it is an extraneous solution. The only solution is $x = 3$.

(c) We produce the graphs of $y_1 = \ln (4x - 5) - \ln (x - 2)$ and $y_2 = \ln (2x + 1)$. (See Fig. 11.36.) The estimated solution is 3.

Figure 11.36

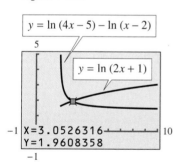

$$\ln (4x - 5) - \ln (x - 2) = \ln (2x + 1)$$

$\ln \dfrac{4x - 5}{x - 2} = \ln (2x + 1)$	Quotient Rule for Logarithms
$\dfrac{4x - 5}{x - 2} = 2x + 1$	Equate arguments.
$4x - 5 = (2x + 1)(x - 2)$	Clear the fraction.
$4x - 5 = 2x^2 - 3x - 2$	FOIL
$0 = 2x^2 - 7x + 3$	Standard form
$0 = (2x - 1)(x - 3)$	Factor.
$2x - 1 = 0$ or $x - 3 = 0$	Zero Factor Property
$2x = 1$ or $x = 3$	
$x = \dfrac{1}{2}$ or $x = 3$	

Because $\frac{1}{2}$ is not in the domain, it is an extraneous solution. Therefore, 3 is the only solution. ∎

Applications

When interest is *compounded continuously,* the amount A on deposit after t years is given by $A = Pe^{rt}$ where r is the interest rate and P is the initial investment.

EXAMPLE 3   *Continuously Compounded Interest*

At age 24, a person has dreams of becoming a millionaire. If this person's current net worth is $150,000, what growth rate is required to reach the $1,000,000 goal by age 50?

Solution

$$A = 1,000,000 \qquad \text{Desired net worth at age 50}$$

$$P = 150,000 \qquad \text{Current net worth}$$

$$t = 26 \qquad \text{Number of years until the age is 50}$$

$$A = Pe^{rt} \qquad \text{Model for continuous compounding}$$

$$1,000,000 = 150,000e^{26r} \qquad \text{Substitute } A, P, \text{ and } t.$$

$$6.667 \approx e^{26r} \qquad \text{Divide both sides by 150,000.}$$

$$\ln 6.667 \approx 26r \qquad \text{Convert to logarithmic form.}$$

$$r \approx \frac{\ln 6.667}{26}$$

$$r \approx 0.073$$

For the person to be a millionaire by age 50, the current assets must be invested at an interest rate of 7.3%, compounded continuously.   ■

If a quantity increases (growth) or decreases (decay) over a period of time, the formula $P = P_0 e^{kt}$ can sometimes be used to describe the amount P of the quantity after a certain time t. The initial amount of the quantity (at time $t = 0$) is P_0. The constant k is called the *growth* or *decay constant*. A positive value of k indicates growth and a negative value of k indicates decay.

Growth or decay that can be approximated by the model $P = P_0 e^{kt}$ is called *exponential growth* or *decay*.

EXAMPLE 4   *Exponential Decay of Toxic Waste*

Containers of toxic waste totaling 500 pounds are buried in a landfill. After 30 years, 400 pounds of toxic waste still remain. If the decay is exponential, how much will remain after 100 years?

Solution

$$P_0 = 500 \qquad \text{Initial amount of toxic waste}$$

$$P = 500e^{kt} \qquad \text{Exponential decay model with } P_0 = 500$$

$$400 = 500e^{30k} \qquad \text{When } t = 30, P = 400.$$

$$0.8 = e^{30k} \qquad \text{Divide both sides by 500.}$$

$$\ln 0.8 = 30k \qquad \text{Convert to logarithmic form.}$$

$$k = \frac{\ln 0.8}{30}$$

$$k \approx -0.0074 \qquad \text{Decay constant is negative.}$$

Now we use the formula again, this time with the known decay constant.

$$P \approx 500e^{-0.0074t}$$

$$P \approx 500e^{-0.0074(100)} \qquad \text{Time } t \text{ is 100 years.}$$

$$P \approx 500e^{-0.74}$$

$$P \approx 238.56$$

After 100 years, approximately 238.56 pounds of toxic waste will remain in the landfill. ∎

EXAMPLE 5 *Exponential Growth of Bacteria*

The number of bacteria in a culture dish increased from 100 to 175 in 15 minutes. If the growth is exponential, how many bacteria are present after 1.5 hours?

Solution

$P_0 = 100$	The initial number of bacteria
$P = 100e^{kt}$	Exponential growth model with $P_0 = 100$
$175 = 100e^{0.25k}$	When $t = 0.25$ (hours), $P = 175$.
$1.75 = e^{0.25k}$	Divide both sides by 100.
$\ln 1.75 = 0.25k$	Convert to logarithmic form.
$k = \dfrac{\ln 1.75}{0.25}$	
$k \approx 2.24$	Growth constant is positive.
$P \approx 100e^{2.24t}$	Substitute k in the growth model.
$P \approx 100e^{2.24(1.5)}$	Time is 1.5 hours.
$P \approx 100e^{3.36}$	
$P \approx 2878.92$	

After 1.5 hours, the bacteria population is about 2879. ∎

11.6 *Quick Reference*

Definitions and Rules

- An equation in which a variable appears in the argument of a logarithm is a **logarithmic equation.** An equation in which a variable appears as an exponent is an **exponential equation.**

- The following properties can be used to solve exponential and logarithmic equations.

  1. If $x = y$, then $b^x = b^y$.

  2. If $b^x = b^y$, then $x = y$.

  3. If $x = y$, then $\log_b x = \log_b y$.

  4. If $\log_b x = \log_b y$, then $x = y$.

Exponential Equations	▪ To solve an exponential equation that can be written with both sides having the same base, equate the exponents and solve the resulting equation.
	▪ To solve an exponential equation for which the bases are different, take the logarithm (with a convenient base) of both sides, simplify, and solve the resulting equation.

Logarithmic Equations	▪ To solve a logarithmic equation algebraically, follow these steps.
	1. Write one side of the equation as a single logarithm and the other as a number or as a single logarithm. The logarithms on each side must have the same base.
	2. If one side of the equation is a number, convert the equation to an exponential equation and solve.
	3. If both sides involve a logarithmic expression with the same base, equate the arguments and solve the resulting equation.
	4. Check all solutions in the original equation. Using the rules of logarithms can introduce extraneous solutions.

11.6 Exercises

 1. To solve the equation $b^x = 3$, under what conditions is it not necessary to write the equation in logarithmic form?

 2. To solve the equation $e^x = 5$, why is it not possible to use the method of equating exponents?

In Exercises 3–6, solve the given equation.

3. $3^{5x} = \dfrac{1}{27}$

4. $5^{2x} = \dfrac{1}{25}$

5. $7^{1-x} = 49$

6. $\left(\dfrac{1}{9}\right)^{x-3} = 27$

In Exercises 7–10, solve for x. Express the result to the nearest hundredth.

7. $3^x = 5$

8. $e^{3x} = 5$

9. $3^{x-3} = 2^{x+2}$

10. $4^{2x-1} = 10^{1-x}$

In Exercises 11–28, solve the exponential equation. Express the result to the nearest hundredth.

11. $e^{20k} = 0.6$

12. $e^{0.25t} = 10$

13. $4e^{2x-1} = 5$

14. $\dfrac{2}{3}e^{-x/2} = 10$

15. $\left(\dfrac{5}{3}\right)^{-x} = 20$

16. $5 \cdot 4^{1-3x} = 12$

17. $3 + e^x = 7$

18. $2e^x - 5 = 19$

19. $5^x + 7 = 2$

20. $15 - 2^x = 12$

21. $600(1 + 0.02)^{4t} = 1000$

22. $750\left(1 + \dfrac{0.06}{12}\right)^{12t} = 1200$

23. $5^{-x^2} = 0.2$

24. $3^{x^2} = 5$

25. $5^{|x+1|} = 2$

26. $3^{|x|} = 0.5$

27. $|2 - 3^x| = 1$

28. $|3e^x + 1| = 2$

 29. To solve the equation $\log_5 x = \log 8$, can you use the method of equating the arguments of the logarithms? Explain.

 30. Solving the equation $\log x + \log(x + 3) = \log 18$ leads to two values for x: 3 and -6. Assuming that all steps of the solving process are correct, what is the solution of the equation? Why?

In Exercises 31–44, solve for x.

31. $\log_5 x = \log_5 (2x - 3)$

32. $\log (3 - x) = \log (2x)$

33. $\log_5 (x - 1) = 2$

34. $7 + \log_2 (x + 1) = 10$

35. $\log_3 \sqrt[3]{3x - 5} = 1$

36. $1 + \log_2 \sqrt{x} = 5$

37. $\dfrac{1}{3} \log_8 (x - 7) = \log_8 3$

38. $\dfrac{1}{2} \ln (3x - 1) = \ln 5$

39. $\log_2 (x^2 - 7x) = 3$

40. $\log_6 x(x + 1) = 1$

41. $\log_4 (5 - 3x)^3 = 6$

42. $\log_2 (x^2 + 4x)^2 = 10$

43. $\dfrac{1}{2} \log_9 (6 - x) = \log_9 x$

44. $\dfrac{1}{2} \log_2 (x - 1) = \log_2 (x - 3)$

In Exercises 45–62, solve the given equation.

45. $2 - \log_9 x^2 = 1$

46. $1 + \log_4 (x + 1)^2 = 4$

47. $\log_4 6x - \log_4 (x - 5) = 2$

48. $\log (3t - 1) = 1 + \log (5t - 2)$

49. $\log x + \log (7 - x) = 1$

50. $2 \log_3 x = \log_3 (5x - 4)$

51. $1 - \log (x - 5) = \log \left(\dfrac{x}{5}\right)$

52. $\log t = 2 - \log (t + 21)$

53. $\log_5 (2x - 3) - \log_5 x = 1$

54. $\log_2 (4x + 5) - \log_2 (x - 7) = 2$

55. $\log_5 x(x + 2) = 0$

56. $\log_5 x(x + 3) = 1$

57. $\log x + \log (4 + x) = \log 5$

58. $\ln (2x + 1) = \ln 14 - \ln (x + 2)$

59. $\log (x + 6) - 2 \log x = \log 12$

60. $\ln 3 = \ln (6 - 7x) - 2 \ln x$

61. $\left|2 - \log_2 (x - 1)\right| = 3$

62. $\log_6 \left|2x + 1\right| = 1$

 63. The following is one of the basic exponent properties.

> For $b > 0$ and $b \neq 1$, if $b^x = b^y$, then $x = y$.

Using an example, explain why $b \neq 1$ is a necessary condition.

 64. Use what you know about the graphs of the functions $f(x) = e^x$ and $g(x) = \ln x$ to explain why the equation $e^x = \ln x$ has no solution.

A logarithmic or exponential equation may be quadratic in form.

Example:

$$(\log_7 x)^2 + \log_7 x^2 - 3 = 0$$

$$(\log_7 x)^2 + 2 \log_7 x - 3 = 0$$

Letting $u = \log_7 x$, we have $u^2 + 2u - 3 = 0$. Now we can use the methods discussed in Section 9.4 to continue the solving process.

In Exercises 65–72, use the method of the preceding example to solve the given equation.

65. $(\log x)^2 + \log x^3 - 4 = 0$

66. $(\ln x)^2 - \ln x - 6 = 0$

67. $\log_4 x - 3\sqrt{\log_4 x} + 2 = 0$

68. $\log_3 x - 4\sqrt{\log_3 x} + 3 = 0$

69. $e^{2x} + 6 = 5e^x$ **70.** $e^{2x} - 4e^x + 3 = 0$

71. $5^{2x} = 5^x + 42$ **72.** $7^{2x} + 7^x - 2 = 0$

In Exercises 73–80, solve for the indicated variable.

73. $R = \log I$ for I

74. $P = \log \left(\dfrac{10P_I}{P_0}\right)$ for P_0

75. $L = 10 \log \left(\dfrac{I}{I_0}\right)$ for I

76. $\ln M = \ln Q - \ln (1 - Q)$ for Q

77. $P = P_0 e^{rt}$ for r

78. $A = A_0 e^{-kt}$ for k

79. $2 = \left(1 + \dfrac{r}{n}\right)^{nt}$ for t

80. $A = P(1 + r)^t$ for t

Compound Interest

If P dollars is invested at an interest rate r for a period of t years and if the interest is compounded continuously, then the value A of the investment is given by $A = Pe^{rt}$.

81. At a certain bank where interest is compounded continuously, investments double in value in 14 years. What is the annual interest rate offered by the bank?

82. How long will it take $5000 invested at an annual interest rate of 5.5% (compounded continuously) to grow to $20,000?

Exponential Growth

During certain periods of time, populations of people or other life forms may grow exponentially. Therefore, models describing such growth may be exponential equations.

83. It is projected that t years from 1980, the population P (in millions) of a certain small country will be $P(t) = 14e^{0.03t}$.

   (a) What was the population in 1980?

   (b) What will the population be in the year 2000?

   (c) What will the population be in the year 2010?

84. In an industrial city the number of toxic particles in the atmosphere is found to be increasing over time according to the formula $P(t) = 400 \cdot 2^{0.4t}$ where P is measured in standard units. Letting $t = 0$ for the year 1988, estimate the number of toxic particles in the year 2000.

Exponential Decay

Certain radioactive substances decay according to the formula $P(t) = P_0 e^{rt}$, where P_0 is the original amount

of the substance, $P(t)$ is the amount of substance remaining after time t, and r is the rate of decay.

The *half-life* of a substance is the time t_h it takes for half of the original amount of the substance to remain: $P(t_h) = \frac{1}{2}P_0$.

85. The half-life of carbon-14 is 5730 years. In how many years will only 30% of an original amount of carbon-14 remain?

86. What is the half-life of a substance when only 30% of the substance remains after five days?

Depreciation

Accountants refer to the decrease in the value of an item over time as *depreciation*. Depreciation is sometimes modeled with an exponential equation.

87. A bulldozer depreciates so that the value after t years is given by the function $V(t) = V_0 e^{-0.19t}$ where V_0 is the initial cost. After 7 years the bulldozer is worth $22,480.57. What was the original cost?

88. A front-end loader depreciates so that the value after t years is given by the function $V(t) = V_0 e^{-0.27t}$ where V_0 is the initial cost. After 6 years the front-end loader is worth $9894.94. What was the original cost?

Law of Cooling

The temperature T of an object after an elapsed period of time t is given by the formula $T = T_m + D_0 e^{kt}$, where T_m is the constant temperature of the surrounding medium, D_0 is the difference between the initial temperature of the object and the temperature of the surrounding medium, and k is a cooling constant that depends on the composition of the object.

89. The coroner arrived at the murder scene at 8:00 A.M. The body was in a 60°F room and the body's temperature was 88°F. After 45 minutes, the body's temperature was 85°F. If you assume that a normal body temperature is 98.6°F, at what time did the murder occur?

90. A cake is removed from a 350°F oven and placed in a room whose temperature is 60°F. How long will it take for the cake to cool to 75°F? (If t is in minutes, $k = -0.1$.)

Richter Scale

A Richter scale reading is given by $R = \log I$ where I is the number of times more intense an earthquake is than a very small quake.

91. The 1988 earthquake in Armenia registered 6.9 on the Richter scale. What would be the Richter scale reading for an earthquake that is 10 times as intense?

92. The Richter scale reading for the 1964 Alaskan earthquake was 8.5. For the 1985 Mexico City earthquake, the Richter scale reading was 7.8. How much more intense was the Alaskan earthquake?

Decibel Scale

A decibel reading is given by $D = 10 \log I$ where I is the number of times more intense a sound is than a very faint sound.

93. If an average conversation registers 60 decibels and an airplane registers 140 decibels, how much more intense is the sound of an airplane than the sound of a conversation?

94. If a whisper registers 30 decibels and the sound of a rocket is 10^{15} times more intense, what would be the decibel reading for a rocket launch?

Exploring with Real Data

In Exercises 95–98, refer to the following background information and data.

Enrollment in higher education increased dramatically during the period 1970–1980, but the rate of increase was lower from 1980 to 1990. (See figure.) (Source: Council for Aid to Education.)

The data in the figure can be modeled by the logarithmic equation $N(t) = 6672.87 + 2107.35 \ln t$, where $N(t)$ is the enrollment (in thousands) and t is the number of years since 1970.

95. Why is it not possible to use this model to estimate the enrollment in 1970?

96. According to the model, what was the enrollment in 1975?

97. Produce the graph of $y = N(t)$. Trace the graph to compare the accuracy of the model for the years 1986 and 1988.

Figure for 95–98

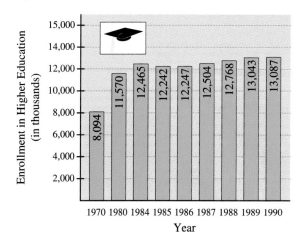

98. If enrollments in higher education began to decrease at some time after 1990, would the model function continue to be valid? Why or why not?

Challenge

In Exercises 99–102, $f(x) = \log x$ and $g(x) = \ln x$. Solve the given equation.

99. $[f(x)]^2 = f(x^2)$

100. $[g(x)]^3 = g(x^3)$

101. $(g \circ f)(x) = 1$

102. $(f \circ g)(x) = 1$

In Exercises 103–106, solve the given equation.

103. $2 \log_3 x + \log_3 (10 - x^2) = 2$

104. $\log_3 x^2 + \log_3 (10 - x^2) = 2$

105. $\log_4 |2x + 2| - \log_4 |3x + 1| = \dfrac{1}{2}$

106. $3^{x^2} = 7^{x+2}$

107. The following is a proof of the Change of Base Formula. Fill in the blanks with a rule or definition that justifies each step.

$x = b^{\log_b x}$ (a) _____

$\log_a x = \log_a b^{\log_b x}$ (b) _____

$\log_a x = (\log_b x)(\log_a b)$ (c) _____

$\dfrac{\log_a x}{\log_a b} = \log_b x$ (d) _____

108. Consider the following equation.

$$\ln (3 - 4x) - \ln (x - 2) = \ln (2x + 3)$$

Produce the graph of the left side of the equation. What is your conclusion about the solution set of the equation? Support your conclusion by determining the domains of the logarithmic expressions.

In Exercises 109–112, use graphs to estimate the solution(s) of the given equations.

109. $e^x + x = -2$ **110.** $e^x - x = -\ln x$

111. $2^x = x^2$ **112.** $3^x = x^3$

11   Chapter Review Exercises

Section 11.1

In Exercises 1–8, let $f(x) = x^2 - 3x$, $g(x) = 3x + 4$, and $h(x) = \sqrt{x + 1}$. Determine each of the following.

1. $(f + g)(x)$ **2.** $(f - g)(2.4)$

3. $\left(\dfrac{f}{g}\right)(3.7)$ **4.** $(fg)(x)$

5. $(f \circ g)(x)$ **6.** $(g \circ h)(x)$

7. $(f \circ h)(3.56)$ **8.** $(h \circ f)(4.23)$

 9. Let $f(x) = \sqrt{x - 1}$ and $g(x) = x + 1$. Since $g(-1)$ is defined, can we conclude that -1 is in the domain of $f \circ g$? Explain.

In Exercises 10 and 11, for the given functions f and g, determine $(f \circ g)(x)$ and $(g \circ f)(x)$.

10. $f(x) = 8$, $g(x) = 2x^2 - 3x + 4$

11. $f(x) = \dfrac{1}{x + 2}$, $g(x) = |2x + 5|$

12. For a 50-mile ride, a cyclist's time t (in minutes) increases as the temperature x increases. Suppose the following function describes the relationship between t and x, where $70 < x < 90$.

$$t(x) = 2x + 160$$

If the cyclist's rate is $r = 50/t$, write the function $(r \circ t)(x)$. What is your interpretation of this function?

Section 11.2

 13. Explain the difference between the Vertical Line Test and the Horizontal Line Test.

In Exercises 14 and 15, produce the graph of the function. From the graph determine if the function is one-to-one.

14. $f(x) = x^3 - 2x^2 + 3x - 4$

15. $f(x) = x^3 + x^2 - 4x - 3$

In Exercises 16 and 17, determine f^{-1}. Then, verify that $(f \circ f^{-1})(x) = x$ for all x in the domain of f^{-1}.

16. $f(x) = 5x + 8$

17. $f(x) = x^3 - 5$

 18. What is the equation of the line of symmetry for the graphs of inverse functions? Describe the graphs of f and f^{-1} if the equation $f(x) = f^{-1}(x)$ has no solution.

Section 11.3

19. For the exponential function $f(x) = b^x$, what restrictions, if any, are placed on b and x?

20. For the graph of $f(x) = b^x$, state each of the following.

(a) *x*-intercept   (b) *y*-intercept

(c) domain   (d) range

In Exercises 21 and 22, use a graph to estimate the solution(s) of the given equation.

21. $\pi^x = 5$ **22.** $(0.57)^x = 6$

In Exercises 23–26, solve the given exponential equation algebraically. Verify the solution with your calculator.

23. $4^x = 2048$ **24.** $16^{2-3x} = 8^{5-6x}$

25. $\left(\dfrac{1}{27}\right)^{3x} = 9^{4x-5}$ **26.** $\left(\sqrt{5}\right)^{6-2x} = 25^{x^2}$

 27. Suppose that you want to model the growth behavior of a certain quantity over time *t*, and you decide to use the function $f(t) = b^t$. Would the base *b* be greater than 1 or would it be between 0 and 1? Why?

28. Five years ago, some money was invested in an account at 7.5% interest compounded monthly. Today the account is worth $7629.80. How much was originally invested?

Section 11.4

In Exercises 29 and 30, write the given exponential equation in logarithmic form.

29. $8^{2/3} = 4$ **30.** $49^{-1/2} = \dfrac{1}{7}$

In Exercises 31 and 32, write the given logarithmic equation in exponential form.

31. $\log_{25} \dfrac{1}{5} = -\dfrac{1}{2}$ **32.** $\log_{1/2} 16 = -4$

 33. For the definition of the logarithmic function, explain why $b = 1$ is excluded.

34. What are the bases of the following logarithms?

(a) $\log_3 x$ (b) $\ln x$ (c) $\log x$

In Exercises 35–38, evaluate the logarithms without a calculator.

35. $\ln e^{3.23}$ **36.** $\log_{27} \dfrac{1}{9}$

37. $\log_{1/4} 32$ **38.** $\log 100^{2.5}$

In Exercises 39 and 40, use your calculator to solve for *x*. Round the results to two decimal places.

39. $\ln x = 5.2$ **40.** $\log x = 5.2$

In Exercises 41–44, solve for *x*.

41. $\log_{343} x = \dfrac{2}{3}$ **42.** $\log_x 676 = 2$

43. $\log_4 32 = x$ **44.** $\log_x 128 = \dfrac{7}{3}$

45. For the graph of $f(x) = \log_b x$, state each of the following.

(a) *x*-intercept   (b) *y*-intercept

(c) domain   (d) range

 46. Suppose $g(t) = \log_b t$, where $0 < b < 1$. In the 20th century, and for appropriate values of time *t*, would *g* be more likely to model the change in the cost of living or the incidence of polio? Why?

Section 11.5

In Exercises 47–50, use the basic properties of logarithms to simplify the given expression.

47. $\log_{21} 21^{41}$ **48.** $21^{\log_{21} 43}$

49. $\log_{21} 21$ **50.** $\log_b 1$

 51. Describe how you can use the fact that $\log 7 \approx 0.845$ to estimate $\log 49$ without a calculator.

52. Let $\log_2 3 = C$ and $\log_2 5 = D$. Write

$$\log_2 \dfrac{5\sqrt[3]{60}}{24}$$

as an expression involving *C* and *D*.

In Exercises 53 and 54, use the properties of logarithms to expand the given expression. Write your answer in terms of logarithms of x, y, and z. Assume x, y, and z represent positive numbers.

53. $\log \dfrac{\sqrt[4]{x^2 y^3}}{x^3 z^4}$

54. $\ln \sqrt[6]{\dfrac{x^5 y^4}{z^3}}$

In Exercises 55 and 56, use the properties of logarithms to combine the given expression into a single logarithmic expression. Assume x, y, and z represent positive numbers.

55. $\dfrac{1}{4}\left[3 \log (x + 4) - \log y - \log 2z\right]$

56. $3 \log (x + 2) - 5 \log (x + 1)$

57. Write $\log_2 24 + 2 \log_2 5 - \log_2 75$ as a single logarithm. Then evaluate it.

58. Write $\log_3 (x + 5)$ in terms of natural logarithms.

Section 11.6

59. Explain why $3^x = \frac{1}{9}$ can be solved by equating exponents, but $3^x = 4$ cannot. For the latter equation, what method should you use to solve it?

60. If you wished to solve $\log_2 x = 4$ by equating arguments of logarithms, you could replace 4 with $\log_2 16$. What is an easier method for solving the equation?

In Exercises 61–66, solve for x. Round your solutions to two decimal places.

61. $3^x = 7$

62. $e^{-x} = 0.78$

63. $\log_4 (x + 3) + \log_4 (x - 3) = 2$

64. $2^{x-1} = 3^{x+2}$

65. $\ln (x - 4) - \ln (3x - 10) = \ln \dfrac{1}{x}$

66. $\log_2 (x + 1) + \log_2 (3x - 5)$
$= \log_2 (5x - 3) + 2$

67. If $a = 6.9e^{-2.3b}$, find b when $a = 0.45$.

68. Suppose $7000 is invested in a savings fund in which interest is compounded continuously at the rate of 12% per year. How long will it take for the money to double in value?

69. Suppose the number N of bacteria present in a certain culture after t minutes is $N = ke^{0.04t}$ where k is a constant. If 5500 bacteria are present after 11 minutes, how many bacteria were present initially?

70. An amount of $800 is deposited in a savings account and earns interest for 8 years at 5.25% compounded monthly. If there were no withdrawals or additional deposits, how much is in the account at the end of the 8 years?

11 Chapter Test

For Questions 1–4, let $f(x) = x^2 - 4$, $g(x) = x + 2$, and $h(x) = \sqrt{x}$.

1. Find $(f + g)(x)$.

2. Find $\left(\dfrac{f}{g}\right)(x)$.

3. Find $(f \circ g)(x)$.

4. Find $(h \circ g)(x)$.

5. Verify that the functions $f(x) = \dfrac{x - 5}{3}$ and $g(x) = 3x + 5$ are inverse functions.

6. For the function $f(x) = \sqrt[3]{x + 2}$, determine f^{-1}.

7. The graph of a function is given in the figure. Sketch the graph of the inverse function.

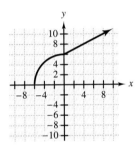

For Questions 8 and 9, evaluate the expression.

8. $\log_{25} 5$

9. $\log_2 4^{3/4}$

For Questions 10–13, solve the equation.

10. $5^{2-3x} = 25$

11. $\left(\dfrac{1}{16}\right)^{x-1} = 8$

12. $8^x = 3^{x+1}$

13. $e^{2x} = 10$

 14. Explain how to evaluate $\log_5 13$ with your calculator.

15. Write the expression $\log\left(A^2/\sqrt{B}\right)$, where A and B represent positive numbers, in terms of $\log A$ and $\log B$.

For Questions 16 and 17, write the given expression as a single logarithmic expression.

16. $\log (x^2 - x - 2) - \log (x - 2)$ $(x > 2)$

17. $2 \log_6 x + \log_6 (x - 1) - 1$ $(x > 1)$

For Questions 18–21, solve the given equation.

18. $\log (2x + 1) = \log 5$ **19.** $\log x + \log (x + 3) = 1$

20. $\ln (2t + 1) - \ln t = \ln 3$ **21.** $2 + \ln x = 0$

22. At the birth of a child, \$5000 is invested in a college fund. What rate of interest, compounded continuously, is required for the investment to be worth \$23,000 eighteen years later?

23. The demand (in thousands) for a certain brand of soap t weeks after a major advertising campaign is given by $D(t) = 2 + 5(0.89)^t$.

   (a) What was the demand when the ad campaign began?

   (b) Using a graph of $y = D(t)$, state your opinion about the effectiveness of the ad campaign.

   (c) After approximately how many weeks had the demand for this brand of soap fallen to half the initial demand?

24. The Richter scale rating of an earthquake is given by the formula $R = \log I$, where I is the number of times more intense the earthquake is than a very small quake. The San Francisco earthquake that occurred during the World Series in 1989 registered 7.1 on the Richter scale. How much more intense was the 1906 San Francisco earthquake, which registered 8.3 on the Richter scale?

25. The temperature T of an object after an elapsed period of time t is given by the formula $T = T_m + D_0 e^{kt}$, where T_m is the constant temperature of the surrounding medium, D_0 is the difference between the initial temperature of the object and the temperature of the surrounding medium, and k is a cooling constant.

A can of soda had been left in a hot car where the soda reached a temperature of 90°F. If the can was placed in a cooler of ice at 32°F, how long did it take to chill the beverage to 40°F? (With time measured in hours, the cooling constant is $k = -1.32$.)

9-11   Cumulative Test

1. Simplify.

    (a) $\sqrt{y^8}$

    (b) $\sqrt{y^{10}}$

2. Evaluate, if possible.

    (a) $(-27)^{2/3}$

    (b) $(-27)^{-2/3}$

    (c) $(-16)^{3/4}$

    (d) $-16^{-3/4}$

3. Simplify and write the result with positive exponents. Assume all variables represent positive numbers.

    (a) $\dfrac{a^{-2}b^{1/3}}{a^{1/2}b}$

    (b) $(x^{-1}y^2)^{-1/2}$

    (c) $\left(\dfrac{x^{-1}}{y^2}\right)^{-3}$

4. Simplify. Assume all variables represent positive numbers.

    (a) $\sqrt[3]{\dfrac{x^6}{8y^3}}$

    (b) $\sqrt[5]{3x}\,\sqrt[5]{2x^2y}$

    (c) $\dfrac{\sqrt{x^3y}}{\sqrt{xy^3}}$

5. Simplify. Assume all variables represent positive numbers.

    (a) $\sqrt{20x^9}$

    (b) $\sqrt[14]{y^{21}}$

6. Perform the indicated operations. Assume all variables represent positive numbers.

    (a) $\sqrt{12x^3} - x\sqrt{3x}$

    (b) $\left(\sqrt{a} + \sqrt{2}\right)^2$

7. What is the product of $3 - \sqrt{5}$ and its conjugate?

8. Rationalize the denominator. Assume all variables represent positive numbers.

    (a) $\dfrac{5}{\sqrt[3]{2}}$

    (b) $\dfrac{\sqrt{x}}{\sqrt{2x} + \sqrt{y}}$

9. Solve.

    (a) $\sqrt{2x + 1} = x\sqrt{3}$

    (b) $(2x - 3)^{2/3} = 9$

10. Simplify.

    (a) i^{19}

    (b) $(3 - i)^2$

    (c) $\dfrac{i}{2 + i}$

11. In addition to the method of completing the square, name three methods for solving a quadratic equation. Which of these three methods works for all quadratic equations?

12. Solve $(3x - 1)(2x + 3) = x(x - 2)$. Round your solution(s) to the nearest hundredth.

13. Six less than a number is 7 times the reciprocal of the number. What is the number?

14. If $ax^2 + bx + c = 0$ has two imaginary solutions, what is the relationship between b^2 and $4ac$? Why?

15. Solve $a + a^{1/2} = 12$.

16. A person jogged 2 miles and then walked back to the starting point. He jogged 3 mph faster than he walked. If the total time was 39 minutes, how fast did he jog?

17. Suppose $p(x) = x^2 + 2x - 24$. Solve each of the following.

    (a) $p(x) > 0$ (b) $p(x) < -25$ (c) $p(x) > -26$

18. Solve.

    (a) $x^3 - 3x^2 + x - 1 \leq 0$ (b) $\dfrac{x}{x - 3} > 0$

19. If $f(x) = ax^2 - 6x + c$, tell what you know about a or c for each of the following descriptions of the graph.

    (a) The y-intercept is 3.

    (b) The parabola opens downward.

    (c) The x-coordinate of the vertex is -1.

20. Suppose that the daily profit P from producing x items is given by $P(x) = 8x - x^2$.

    (a) How many items should be produced each day to maximize the profit?

    (b) What will the maximum daily profit be?

21. For $f(x) = x^2$ and $g(x) = 2x$, evaluate each of the following.

    (a) $(f + g)(0)$ (b) $(f - g)(2)$ (c) $(fg)(-2)$

22. For $f(x) = |x - 1|$ and $g(x) = 2x + 1$, evaluate each of the following, if possible.

    (a) $(f \circ g)\left(-\dfrac{1}{2}\right)$ (b) $\left(\dfrac{f}{g}\right)\left(-\dfrac{1}{2}\right)$

23. If $f(x) = \dfrac{x}{x - 3}$, determine f^{-1}. Then show that $(f \circ f^{-1})(x) = x$ for all x in the domain of f^{-1} and that $(f^{-1} \circ f)(x) = x$ for all x in the domain of f.

24. What properties do the graphs of $f(x) = 5x$ and $g(x) = \left(\dfrac{1}{5}\right)x$ have in common? Describe how the graphs are different.

25. Solve each of the following equations.

    (a) $\log_2 32 = x$ (b) $\log_3 (x - 1) = 2$ (c) $\log_x 7 = 2$

26. (a) Expand $\ln \dfrac{x^2 y}{\sqrt{z}}$, where $x, y, z > 0$.

    (b) Write $3 \log (x - 1) - \log (x + 1)$, where $x > 1$, as a single logarithm.

27. Solve each of the following equations. Round your results to the nearest hundredth.

    (a) $\log_3 x(x - 6) = 3$ (b) $3^{x+1} = 5$ (c) $4e^{2x} = 1$

28. For continuously compounded interest, the value A of an investment of P dollars after t years is given by $A = Pe^{rt}$, where r is the interest rate. Approximately what interest rate is needed for a \$1200 investment to be worth \$2187 after 10 years?

Answers to Odd-Numbered Exercises

Chapter 1

Section 1.1 (Page 9)

1. The decimal name of a rational number is a terminating or repeating decimal.

3. Terminating   **5.** Repeating   **7.** False   **9.** True

11. False   **13.** True   **15.** True   **17.** {0, 1, 2, 3, 4, 5, 6}

19. Ø   **21.** Negative integers

23. The number can be written as p/q such that $q = 1$

25. False   **27.** False   **29.** True   **31.** False

33. The number 1.75 is a terminating decimal, and the digit 5 is followed by zeros. The number $1.\overline{75}$ is a repeating decimal, and the block of digits 75 repeats without end.

35. Rational   **37.** Irrational   **39.** Rational

41. Irrational

43. (a) $498, \sqrt{16}$ (b) $0, 498, \sqrt{16}$ (c) $-4, 0, 498,$ $-17, \sqrt{16}$ (d) $-4, 3\frac{5}{8}, 0, \frac{3}{5}, 498, 0.25, -17, 0.\overline{63}, \sqrt{16}$
(e) $\sqrt{7}, \pi$ **45.** $2x + 3$ **47.** $x^2 - 9$ **49.** $2L + 2W = 50$

51. $2t + 7 = 10$ **53.** Transitive Property   **55.** Trichotomy Property   **57.**
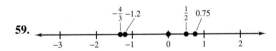

59.

61.

63. If $a < b$, then a is to the left of b on the number line. If $a > b$, then a is to the right of b.

65. <   **67.** <   **69.** >   **71.** >   **73.** >

75. The first inequality states that -4.1 is to the left of -4. The second inequality states that -4 is to the right of -4.1. Both descriptions are correct.

77. $10 > x$ **79.** $-14.2 < -14$

81. $\pi < 3.1416 < \frac{22}{7} < \sqrt{10}$ **83.** $\sqrt{2} < \frac{\pi}{2} < \frac{11}{7} < \sqrt{3}$

85. $3 < y$ **87.** $-2 \le x \le 3$ **89.** $y \le -1$

91. $a < 0$ **93.** $-2 < x < 2$

95. The graph of $x \ge 2$ includes 2. The graph of $x > 2$ does not.

97. $(-\infty, -4)$

99. $[-5, \infty)$

101. $(-3, -1)$

103. $[2, 5]$

105. $(-3, 7]$

107. $\{x \mid -3 \le x < 7\}$ **109.** $\{x \mid x \le 0\}$

111. The symbols $|7|$ and $|-7|$ represent the distances from 0 to 7 and from 0 to -7, respectively. Because these distances are the same, $|7| = |-7|$.

113. $-5, 5$ **115.** $-6, 2$ **117.** 6   **119.** -7 **121.** 0

123. $\frac{3}{5}$ **125.** 0   **127.** 1   **129.** 21

131. Reading $-a$ as "negative a" suggests that $-a$ represents a negative number. However, if $a < 0$, then $-a$ is a positive number.

133. 6   **135.** 0   **137.** $\frac{1}{2}$ **139.** 3.15 gallons

141. 9.6 ounces   **143.** >   **145.** >   **147.** >   **149.** =

151. <   **153.** (a) All x (b) 0   **155.** (a) $x \ge 0$
(b) $0 \le x \le 1$ **157.** $\frac{44}{125}$ **159.** $\frac{5}{3}$

Section 1.2 (page 18)

1. Add the absolute values and retain the common sign.

3. 3   **5.** -4 **7.** -10 **9.** -8 **11.** 0   **13.** 2   **15.** 8

17. 8   **19.** -7 **21.** 20   **23.** -55 **25.** 8   **27.** 0

29. -22 **31.** -1 **33.** -11 **35.** -17 **37.** -6

39. 1   **41.** -9 **43.** 0   **45.** -6.8 **47.** 2.8

49. -51.57 **51.** -10.31 **53.** -613 **55.** -4

57. -14 **59.** 8   **61.** $\frac{4}{3}$ **63.** $-\frac{6}{5}$ **65.** $-\frac{1}{15}$ **67.** $-\frac{1}{16}$

69. $-\frac{17}{3}$ **71.** -7 **73.** -12 **75.** 7   **77.** 13   **79.** >

81. >   **83.** 9   **85.** -4 **87.** 2   **89.** 0   **91.** $-\frac{27}{5}$

93. -45 **95.** 3   **97.** -18.3 **99.** $-\frac{2}{21}$ **101.** -7

103. 0   **105.** -0.222 **107.** $-\frac{27}{8}$ **109.** 2-yard gain

111. \$8.41   **113.** 1.86°   **115.** 41%

117. The total number of ATM card users is not given.

119. Positive   **121.** Negative   **123.** 81

Section 1.3 (page 26)

1. Change the minus sign to a plus sign and change 8 to -8: $-5 + (-8)$.

3. -11 5. -10 7. 13   9. 6   11. -13 13. 52

15. 18   17. -24 19. -3 21. 48   23. -20

25. -3 27. -10 29. -5 31. 13   33. -8

35. -14 37. 0   39. 1   41. -10 43. 13   45. $-\frac{5}{3}$

47. $\frac{2}{5}$ 49. $-\frac{1}{10}$ 51. $-\frac{14}{3}$ 53. $\frac{17}{16}$

55. The opposite of the difference of a and b is the difference of b and a.

57. -34 59. 33   61. 74.23   63. 5   65. 6

67. AB: 6
    BC: 10
    AC: 16

69. AB: 4.25
    BC: 4.25
    AC: 8.50

71. 4   73. -2 75. True   77. True   79. True

81. -5 83. -41.13 85. 43   87. 5   89. $-\frac{22}{15}$

91. -21 93. -32 95. $-\frac{19}{4}$ 97. 32   99. 59.76

101. -45 103. 0   105. -8 107. -80 109. -15

111. 14,775 feet   113. $410.99

115. 11,931, -2175, -1259, -1695, 4142, 11,908

117. 2.73 miles   119. Negative   121. Negative

123. This is not a valid rule. For example, if $a = 2$ and $b = -2$, the relation is false.

125. Because both addends are nonnegative, $|a| + |b|$ is nonnegative. However, $|a| - |b|$ is negative if, for example, $a = 1$ and $b = 2$.

127. Of the nine numbers, five are odd and four are even. Regardless of the signs, the result will be an odd number.

Section 1.4 (page 33)

1. Determine the product of the absolute values. The answer is positive.

3. -28 5. -10 7. 12   9. 12   11. -100

13. -44 15. -7 17. -24 19. 0   21. 6.21

23. -4.83 25. -32.4 27. 1525.92   29. 18.27

31. -3.24 33. -24 35. -30 37. $-\frac{1}{5}$ 39. $\frac{2}{5}$

41. $\frac{1}{4}$ 43. $\frac{10}{3}$ 45. -6 47. $\frac{85}{2}$

49. The result is positive because the number of negative factors is even.

51. -24 53. 70   55. 12   57. 48   59. -36 61. 0

63. $<$ 65. $<$ 67. -3 69. 6   71. 18   73. -99

75. $-\frac{11}{3}$ 77. 28.512   79. -20 81. -40 83. $\frac{4}{21}$

85. -163.0209 87. 12   89. 72   91. $-\frac{1}{3}$ 93. 96

95. 0   97. $\frac{1}{5}$ 99. -140 101. 2, 3   103. 4, -3

105. -2, -6 107. -8, 1   109. $51.95   111. 75

113. (a) The second factor is decreasing by 1.

(b) In column A the results are decreasing by 3. In column B the results are increasing by 3.

(c) -3, -6, -9
    3, 6, 9

(d) The product of two numbers with unlike signs is negative, and the product of two numbers with like signs is positive.

115. Choice (i) is easier to calculate. Because all the factors are -1 and there is an even number of negative signs, the result is 1.

117. $288.75 billion   119. $33,000   121. Negative

123. Positive   125. Negative

Section 1.5 (page 41)

1. Divide the absolute values. The result is positive.

3. -2 5. -3 7. 8   9. 8   11. 16   13. -3

15. -4 17. -7 19. Undefined   21. 0   23. $\frac{1}{2}$

25. 8   27. -4 29. -1 31. A nonzero number divided by itself is 1.   33. 2.19   35. -9.61

37. -0.51 39. -3.14 41. 4   43. $-\frac{11}{9}$ 45. 0

47. $\frac{1}{2}$ 49. $-\frac{3}{5}$ 51. $\frac{4}{3}$ 53. $-\frac{5}{6}$ 55. $\frac{15}{2}$ 57. $-\frac{3}{32}$

59. (a) -5 (b) $\frac{1}{5}$ 61. (a) $\frac{3}{4}$ (b) $-\frac{4}{3}$ 63. (a) 0

(b) No reciprocal   65. (a) -4.5 (b) $\frac{2}{9}$ 67. $>$

69. $<$ 71. 14   73. -40 75. 0.46   77. $-\frac{9}{4}$

79. -1 81. -28 83. $\frac{35}{2}$ 85. -3 87. 3   89. -2

91. 0   93. -19.34 95. $-\frac{1}{3}$ 97. 4   99. -9

101. -36 103. Undefined   105. Undefined

107. Division by zero is not defined.   109. $\frac{5}{8}$ 111. $-\frac{1}{6}$

113. $\frac{4}{7}$ 115. -4.875 117. -144.7

119. $-$28.94 million   121. Negative   123. Undefined

125. Positive   127. $\frac{5}{6}$ 129. $\frac{2}{3}$

Section 1.6 (page 50)

1. In 3^4, 3 is the base. In $\sqrt{3}$, 3 is the radicand.

3. 7^6 5. $-(-5)^3$ 7. a^3 9. $(2b)^4$

11. $\left(\frac{7}{9}\right) \cdot \left(\frac{7}{9}\right) \cdot \left(\frac{7}{9}\right) \cdot \left(\frac{7}{9}\right)$ 13. $2 \cdot x \cdot x$ 15. $5 \cdot 5 \cdot y \cdot y \cdot y$

17. $-4 \cdot x \cdot y \cdot y \cdot y$ 19. 25   21. -8 23. -49

25. 32   27. 7   29. $\frac{4}{9}$ 31. 1   33. 248,832

35. $-177,147$ 37. 6057.68

39. Because $6^2 = 36$ and $(-6)^2 = 36$, the two square roots of 36 are -6 and 6. The radical symbol refers to the principal (positive) square root. Thus, $\sqrt{36} = 6$.

41. $2, -2$ **43.** $6, -6$ **45.** Not a real number

47. $\frac{3}{4}, -\frac{3}{4}$ **49.** $23, -23$ **51.** 2 **53.** -6 **55.** $\frac{2}{3}$

57. 26 **59.** 2.9 **61.** 23.11 **63.** 252.74 **65.** 0.09

67. For $(-6)^2$, multiply $(-6)(-6)$ to obtain 36. For -6^2, multiply $-1(6)(6)$ to obtain -36.

69. -6 **71.** 10 **73.** 1300 **75.** 10 **77.** -3 **79.** 2

81. 16 **83.** $\frac{5}{16}$ **85.** -1 **87.** 14 **89.** -6

91. $-2 \cdot 7 - 4 = -18$ **93.** $\dfrac{-3 + 7}{2} - (-4) = 6$

95. $[3 - (-2)]^2 = 25$

97. **(a)** $-3(4 - 5)$ **(b)** $(-3 \cdot 4) - 5$ **(c)** $-(3 \cdot 4 - 5)$

99. **(a)** $(8 + 12) \div 4 - 1$ **(b)** $8 + 12 \div (4 - 1)$

(c) $8 + (12 \div 4 - 1)$ **101.** -36

103. 24 **105.** 7 **107.** 7 **109.** 0 **111.** 5 **113.** 25

115. 17 **117.** 0.22

119. Although $\sqrt{9} = 3$, $\sqrt{-9}$ is not a real number.

121.

x	-4	-1	0	1	4	9	16	25
$\sqrt{x}$	✕	✕	0	1	2	3	4	5
x^2	16	1	0	1	16	81	256	625

123. 1978, 1.05; 1979, 1.10; 1980, 1.16; 1981, 1.22

125. $(1.05)^N$ **127.** **(a)** No values of x **(b)** $x = 0$

(c) $x \neq 0$ **129.** **(a)** Positive **(b)** Either **(c)** Positive

(d) Negative **(e)** Negative **(f)** Negative **(g)** Positive

(h) Negative **131.** $3^3 + 3$

Section 1.7 *(page 60)*

1. When $-9 - 4x + x^2$ is written as $-9 + (-4)x + x^2$, the Commutative Property of Addition can be applied to write the expression as $x^2 + (-4)x + (-9) = x^2 - 4x - 9$.

3. Associative Property of Addition

5. Additive Identity Property

7. Distributive Property

9. Property of Multiplicative Inverses

11. Associative Property of Multiplication

13. Multiplicative Identity Property

15. Multiplication Property of 0

17. Distributive Property

19. Associative Property of Multiplication

21. Commutative Property of Multiplication

23. (i) **25.** (b) **27.** (c) **29.** (f) **31.** (d)

33. $x + (4 + 3)$ **35.** $3(x + 6)$ **37.** $-x + 8$

39. $x + 6$ **41.** $1x$ **43.** $3(4z + 5)$ **45.** 0

47. Suppose 0 had a reciprocal, r. Then, according to the Property of Multiplicative Inverses, $0 \cdot r = 1$. But the Multiplication Property of 0 states that $0 \cdot r = 0$. Therefore, 0 does not have a reciprocal.

49. $\frac{1}{5}$; Property of Multiplicative Inverses

51. 1; Multiplicative Identity Property

53. -1; Distributive Property

55. $[7 \cdot \frac{1}{7}]$; Associative Property of Multiplication

57. **(a)** Commutative Property of Addition
(b) Associative Property of Addition
(c) Property of Additive Inverses
(d) Additive Identity Property

59. **(a)** Multiplication Property of -1
(b) Distributive Property
(c) Commutative Property of Addition

61. **(a)** Associative Property of Addition

63. **(a)** Associative Property of Multiplication
(b) Property of Multiplicative Inverses
(c) Multiplicative Identity Property

65. $x + (3 + 2) = x + 5$ **67.** $y + (1 - 6) = y + (-5)$

69. $3x + (-5 + 7) = 3x + 2$ **71.** $(-5 \cdot 3)x = -15x$

73. $[-\frac{3}{5}(-\frac{5}{3})]a = 1a = a$ **75.** $[-\frac{5}{6}(-\frac{9}{10})]x = \frac{3}{4}x$

77. $20 + 15b$ **79.** $-3x - 12$ **81.** $-15 + 6x$

83. $xy + xz$ **85.** $\frac{3}{4}y + 1$ **87.** $5(x + y)$

89. $7(3x - 2y)$ **91.** $\frac{7}{4}(x + 3)$ **93.** $3(x + 4)$

95. $5(y + 1)$ **97.** $3(2 - x)$ **99.** $-x + 4$

101. $-2x - 5y + 3$ **103.** $12x$ **105.** $5x^3$ **107.** $12xy$

109. $\frac{8}{3}y$ **111.** $x + 8$ **113.** 3

115. $(\frac{3}{7} + \frac{4}{7}) + (\frac{2}{9} + \frac{5}{9} + \frac{2}{9}) = 2$

117. $(2 \cdot 5)[-87(-1)] = 870$

119. **(a)** Definition of subtraction
(b) Distributive Property
(c) Commutative Property of Multiplication
(d) Multiplication Property of -1
(e) Associative Property of Multiplication
(f) Commutative Property of Multiplication
(g) Multiplication Property of -1
(h) Definition of subtraction

121. $7(100 + 8) = 700 + 56 = 756$

123. $15(100 - 2) = 1500 - 30 = 1470$

125. $18(1 + \frac{1}{2}) = 18 + 9 = 27$

127. If n is the number, $9n = (10 - 1)n = 10n - n$.

129. $-\frac{5}{7}, \frac{7}{5}$ **131.** 0, none   **133.** $\frac{1}{3x}$ **135.** $\frac{3}{2x}$

137. $-x - 3$ **139.** $3x - 4$ **141.** True

143. Sometimes; True if $a = b$ or $a = -b$, otherwise false

145. Sometimes; True if a and b have the same sign, otherwise false

147. False   **149.** No

151. False if, for example, $a = 0$, $b = 16$, $c = 9$

Chapter Review Exercises   *(page 64)*

1. False   **3.** True   **5.** False   **7.** False

9. The first interval includes 5 but not -3, and the second interval includes -3 but not 5.

11. (a)

(b)

(c)

13. Addends are numbers that are added. In $4 + 6$, 4 and 6 are addends.

15. -3 **17.** $\frac{1}{12}$ **19.** 3   **21.** 11   **23.** 11°

25. In $6 - 2$, 6 is the minuend and 2 is the subtrahend.

27. 20   **29.** $-\frac{23}{20}$ **31.** -10 **33.** 4   **35.** 10,350 feet

37. The factors are the numbers that are multiplied.

39. -20 **41.** $\frac{5}{2}$ **43.** $-\frac{1}{6}$ **45.** 0   **47.** -4.5

49. 3000 gallons less

51. For $10 \div 2$, 10 is the dividend and 2 is the divisor.

53. -5 **55.** 0   **57.** $\frac{4}{81}$ **59.** $-\frac{1}{2}$

61. For -5^2, multiply $-1(5)(5)$ to obtain -25. For $(-5)^2$, multiply $(-5)(-5)$ to obtain 25.

63. $(-3)^5 = -243$ **65.** 16   **67.** 0   **69.** 1   **71.** 7

73. 25   **75.** -24 **77.** 5

79. The Multiplication Property of -1 states that $-1x = -x$.

81. Commutative Property of Addition

83. Associative Property of Multiplication

85. Additive Identity Property

87. $7 \cdot (c + 5)$ **89.** $5(x + 8)$ **91.** $-5(-\frac{1}{5})$

Chapter Test   *(page 67)*

1. *(1.1)* $-2, 0, 6, \sqrt{9}$

2. *(1.1)* $\pi, \sqrt{7}$

3. *(1.1)* $-3.7, -2, -0.\overline{14}, 0, \frac{2}{3}, 6, \sqrt{9}$

4. *(1.1)* $\{-2, -1, 0, 1, 2, 3, 4, 5\}$

5. *(1.1)*

6. *(1.1)*

7. *(1.3)* -4

8. *(1.4)* $-\frac{1}{15}$

9. *(1.6)* $\frac{5}{37}$

10. *(1.6)* -7

11. *(1.6)* -1

12. *(1.3)* 4

13. *(1.2)* $-\frac{5}{4}$

14. *(1.3)* -9.2

15. *(1.4)* 0

16. *(1.5)* 8

17. *(1.5)* **(a)** 0

   (b) Division by zero is not defined.

18. *(1.7)* $3x - 12$

19. *(1.7)* $-x + 4$

20. *(1.7)* $3 + (t - 3r)$

21. *(1.3)* 4

22. *(1.5)* $-\frac{1}{2}$

23. *(1.6)* 1.40

24. *(1.6)* -5

25. *(1.3)* $-3.49°$

Chapter 2

Section 2.1   *(page 73)*

1. For $x = 2$, the denominator is 0. Division by 0 is undefined.

3. 8   **5.** 4   **7.** -13 **9.** -3 **11.** 3   **13.** Undefined

15. -1 **17.** 3   **19.** -2 **21.** -5 **23.** 10

25. Undefined   **27.** 0   **29.** $\frac{5}{4}$ **31.** 2   **33.** $-\frac{2}{3}$ **35.** 2

37. $-\frac{2}{5}$ **39.** -9.41 **41.** -13.20 **43.** 19.02   **45.** 0

47. 7

49. For $t = 5$, $4 - t = -1$. Because $\sqrt{-1}$ is not a real number, your calculator should respond with an error message.

51. 4   **53.** $\frac{9}{5}$ **55.** -4 or 14

57. When we evaluate an expression, we perform all indicated operations to determine the value of the expression. To verify that a number is a solution of an equation, we substitute the number for the variable in the equation and then evaluate the left- and right-hand expressions to determine if they are equal.

59. 2 **61.** 4 **63.** $\frac{5}{2}$ **65.** Yes **67.** No **69.** Yes

71. No **73.** $n - 5 = 3$ **75.** $8n = 11$

77. $2n = n - 3$ **79.** No **81.** Yes **83.** Yes

85. Yes **87.** 84.9 feet

89. No. The ladder, building, and ground form a triangle whose sides are 6, 22, and 24. Because $6^2 + 22^2 \neq 24^2$, the building is not vertical with the ground.

91. 10°C **93.** 66.67 km/hr **95.** $1300

97. **(a)** 84 feet **(b)** 156 feet **(c)** 0 feet

99. $t = 5, 10, 12, 13$

101. From the model, helmet sales in 1980 would be 7.72 million. Because this sales figure is about 19 times higher than the sales figure for 1985, the model appears not to be valid prior to 1985.

103. Suppose n is the line number in the given table. Then, for $a^2 + b^2 = c^2$, $a = 2n + 1$, $b = n(a + 1)$, and $c = b + 1$. The next two lines are: $9^2 + 40^2 = 41^2$, $11^2 + 60^2 = 61^2$.

Section 2.2 *(page 81)*

1. Suppose A is an algebraic expression that has no explicitly written constant term. Because $A = A + 0$, the constant term of the expression is 0.

3. Terms: $2x, -y, 5$

Coefficients: $2, -1, 5$

5. Terms: $3a^3, -4b^2, 5c, -6d$

Coefficients: $3, -4, 5, -6$

7. Terms: $7y^2, -2(3x - 4), \frac{2}{3}$

Coefficients: $7, -2, \frac{2}{3}$

9. Terms: $\dfrac{x + 4}{5}, -5x, 2y^3$

Coefficients: $\frac{1}{5}, -5, 2$

11. The terms of an expression are separated by plus or minus signs that are not inside grouping symbols. The expression $2x + 3$ contains one plus sign that separates the terms $2x$ and 3. The plus sign in the expression $2(x + 3)$ is inside parentheses and is not considered when identifying the terms of $2(x + 3)$. Therefore, the expression has only one term.

13. $-x + 3$ **15.** $3x - 3y - 5$ **17.** $6x - 7y$ **19.** 4

21. $-x + y$ **23.** $17x^2 + 3x$ **25.** $10ac^2 - 13ac$

27. $2.75x^3 - 4.40x^2$

29. By the Multiplication Property of -1, $-(2x - 5) = -1(2x - 5)$. Similarly, $3 - (4a + 1) = 3 + [-(4a + 1)] = 3 + [-1(4a + 1)]$. In both cases, distributing -1 changes the signs of the terms inside the grouping symbols.

31. $-a + b$ **33.** $10x - 15$ **35.** $-2a + 3b - 7d$

37. $3x - 6y - z + 2$ **39.** (d) **41.** (f) **43.** (a)

45. $-x - 1$ **47.** $x - 12$ **49.** $-7a - 2b$ **51.** -8

53. $-3x + 17$ **55.** $x + 2$ **57.** $3x + 5$ **59.** $8t + 13$

61. $2x - 4$ **63.** $c - 125$ **65.** $-14x - 37$ **67.** 4

69. $x^2y^3 + 3x^3y^2$ **71.** 0 **73.** $\dfrac{2n}{12} = \dfrac{n}{6}$

75. $3n + n = 4n$ **77.** (b) **79.** (b) **81.** $3L$

83. $25q + 10d$ **85.** $x + 7$ **87.** $3n + 3$ **89.** $0.08d$

91. $\sqrt{a^2 + (a - 2)^2}$ **93.** $1.05s$

95. The phrase is ambiguous because we can read it as "(the product of three and a number) increased by seven" or as "the product of three and (a number increased by seven)." The corresponding translations are $3x + 7$ and $3(x + 7)$.

97. $\dfrac{d}{240,000,000}$ **99.** $11.76 billion **101.** (a)

103. (a) **105.** (b) **107.** (a) **109.** (b) **111.** 0

113. $x^2 - xy$

Section 2.3 *(page 91)*

1.

	Base	Exponent
-3^2	3	2
$(-3)^2$	-3	2
-2^3	2	3
$(-2)^3$	-2	3

Parentheses rank ahead of exponents in the Order of Operations.

3. **(a)** b^3 **(b)** $-b^3$ **(c)** $-b^3$ **(d)** b^3 **(e)** $-b^3$

Parts (a) and (d) are equivalent. Parts (b), (c), and (e) are equivalent.

5. The expansion is $2 \cdot 2 \cdot 2 \cdot 2 \cdot 2 \cdot 2 \cdot 2 \cdot 2$, which has a total of 8 factors of 2 and simplifies to 2^8. In effect, when we add exponents according to the Product Rule for Exponents, we are determining the total number of factors.

7. 2^{11} **9.** t^5 **11.** $-5x^6$ **13.** $18t^{10}$ **15.** $(6 + t)^9$

17. $3y^7$ **19.** $-6x^8$ **21.** $-6x^4y^3$ **23.** **(a)** Not possible

(b) $64y^9$ **25.** **(a)** $-2n$ **(b)** $-18n^3$ **27.** **(a)** $2y^3$

(b) $-20y^9$ **29.** **(a)** $8x^5$ **(b)** $8x^5$

31. When we multiply, we add the exponents: $x^3 \cdot x^3 = x^6$. When we combine terms, we do nothing with the exponents: $x^3 + x^3 = 2x^3$. When we raise a power to a power, we multiply the exponents: $(x^3)^3 = x^9$.

33. Raise 3 and 5 to the third power, then multiply the results: $3^3 \cdot 5^3$.

35. x^{20} **37.** x^{12} **39.** $(x-3)^{12}$ **41.** x^5y^5 **43.** $-x^6y^3$

45. $-8x^6y^{12}$ **47.** $\dfrac{a^6}{b^6}$ **49.** $\dfrac{16x^{12}}{y^{10}}$ **51.** $\dfrac{49x^6}{25y^{10}}$ **53.** 8

55. x^6 **57.** x^{11} **59.** $(x+1)^2$ **61.** $-\dfrac{x}{2}$ **63.** $\dfrac{3z^3}{4}$

65. a^2b **67.** a^4b^3 **69.** $\dfrac{x^6y^4}{2}$ **71.** $(7-2)^2$

73. $(8 \cdot 5)^2$ **75.** False **77.** True **79.** True
81. $-20x^8$ **83.** $27s^9t^{24}$ **85.** $-t^{11}$ **87.** $-2a^{10}$

89. x^4y^6 **91.** $-\dfrac{x^8}{4}$ **93.** $2x^6$ **95.** $\dfrac{2x^4}{5}$ **97.** a^4b^2

99. $x^6 = (x^2)^3 = 5^3 = 125$ **101.** c^6 **103.** $3x^5$

105. 4 **107.** 6 **109.** For example, let $c = 1$ and $d = 1$. Then $(1 + 1)^2 = 4$, but $1^2 + 1^2 = 2$. **115.** 2^{15}

117. 3^{16} **119.** 2^5 **121.** $>$ **123.** $<$ **125.** $=$
127. y^{3n+2} **129.** a^{12n^2} **131.** $-32a^{33}b^{50}$ **133.** $3x^9y^5$

135. $\dfrac{4x^2y^7}{5}$ **137.** x^3y^5 **139.** $\dfrac{-675x^{16}y^{21}}{z^{19}}$

Section 2.4 (page 100)

1. If 0^0 were defined, then according to the Quotient Rule for Exponents, $\dfrac{0^5}{0^5} = 0^{5-5} = 0^0$. But $0^5 = 0$ and division by 0 is undefined.

3. 1 **5.** 3 **7.** -1 **9.** $\frac{1}{25}$ **11.** $-\frac{1}{8}$ **13.** 15 **15.** 4
17. $\frac{1}{8}$ **19.** $\frac{3}{2}$ **21.** $\frac{7}{12}$
23. If b were 0, b^{-n} would equal $0^{-n} = \dfrac{1}{0^n} = \dfrac{1}{0}$, which is undefined.

25. $\dfrac{1}{x^7}$ **27.** $\dfrac{4}{x^4}$ **29.** $\dfrac{x^3}{32}$ **31.** $-\dfrac{1}{x^3y^4}$ **33.** $\dfrac{1}{y^2}$

35. $-\dfrac{k^6}{5}$ **37.** $\dfrac{1}{(2+y)^3}$ **39.** $-\dfrac{3b}{2a}$ **41.** $>$ **43.** $=$

45. $=$ **47.** $>$ **49.** $-1 - 4^0,\ -|8|^0,\ 0,\ (-10)^0,\ 4^0 + x^0$

51. 1 **53.** $-\dfrac{12}{x^3}$ **55.** $\dfrac{14y}{x^7}$ **57.** $\dfrac{1}{5^3}$ **59.** $\dfrac{1}{z^7}$ **61.** $\dfrac{1}{y^5}$

63. $\dfrac{3}{2z^3}$ **65.** $-v^5$ **67.** $\dfrac{y^{10}}{x^5}$ **69.** $-\dfrac{1}{64x}$ **71.** $\dfrac{y^2}{36x^{10}}$

73. $\dfrac{x^{39}}{y^{48}}$ **75.** x^{-8} **77.** x^4 **79.** $\dfrac{1}{x^{30}}$ **81.** $\dfrac{a^{20}}{16}$ **83.** $\dfrac{a^{24}}{b^{30}}$

85. $-\dfrac{8y^{12}}{x^6}$ **87.** $-\dfrac{k^5}{32}$ **89.** $\dfrac{-8000}{x^9y^{12}}$ **91.** $(4 - 3)^{-9}$

93. $(2^5 + 1)^0$ **95.** $2(xy)^{-1}$ **97.** $(x^{-1} + y)^{-1}$ **99.** $\dfrac{x^{12}}{16}$

101. $\dfrac{-4}{x^5}$ **103.** $-\dfrac{12}{y^2}$ **105.** y^8 **107.** $\dfrac{1}{a^{20}b^2}$

109. $-\dfrac{x^2}{2y^7}$ **111.** $-\dfrac{x^6y^{15}}{8}$ **113.** $\dfrac{1}{x^8}$ **115.** $\dfrac{b^2}{a^2}$

117. $3x^2y^4$ **119.** $\frac{1}{25}$ **121.** $\dfrac{3}{x^3}$ **125.** $26 \cdot 10^{-2}$

133. $m = 2,\ n = -6$ **135.** $n = 2$

137. $n = -5,\ m = 16$ **139.** $-\dfrac{27b}{a^{13}}$ **141.** $\dfrac{288y^8}{25x^{11}}$

143. $\dfrac{2c^{21}}{3a^{15}b^{12}}$ **145.** $\dfrac{3z^4}{5y^7}$

Section 2.5 (page 108)

1. The number n is such that $1 \le n < 10$, and p is an integer.

3. $1.25 \cdot 10^5$ **5.** $2.5704 \cdot 10^3$ **7.** $-5.376 \cdot 10^8$
9. $4.567 \cdot 10^{11}$ **11.** $5 \cdot 10^0$ **13.** $-4 \cdot 10^{-1}$
15. $6.45 \cdot 10^{-6}$ **17.** $3.749 \cdot 10^{-7}$ **19.** 1,340,000
21. -0.0000004214 **23.** 0.000001 **25.** 30,000
27. $-5,720,000,000$ **29.** 0.00089 **31.** $4.3 \cdot 10^{52}$
33. $-6.7 \cdot 10^{-24}$ **35.** $1.2 \cdot 10^{-11}$ **37.** $5.7 \cdot 10^{12}$
39. $>$ **41.** $>$ **43.** 7153 **45.** 2.34 **47.** 6000
49. 200 **51.** 5 **53.** $3.65 \cdot 10^{11}$ **55.** $5.93 \cdot 10^{-5}$
57. $8.13 \cdot 10^{12}$ **59.** $1.06 \cdot 10^{15}$ **61.** $1.20 \cdot 10^{-6}$
63. $3.88 \cdot 10^{-10}$ **65.** $2.24 \cdot 10^{13}$
67. 0.000000000053 meters **69.** 5,000,000
71. \$2,280,000 or \$$2.28 \cdot 10^6$ **73.** $3.456 \cdot 10^9$
75. $1.234 \cdot 10^8$ **77.** $9.75 \cdot 10^{-12}$ **79.** 8.5962 E4
81. 2.048 E-3 **83.** 1.45519152 E-11
85. $n = 6$ **87.** $n = 13$ **89.** 1.234 E3
91. 43.7124217 E-15 **93.** 931.3225746 E18

Chapter Review Exercises (page 110)

1. Evaluate $3^2 + 3 + 1$ or store 3 for x and evaluate $x^2 + x + 1$.

3. 23 **5.** 6.24 **7.** 6 **9.** (a) $2x - 3 = 7$ (b) No
11. (a) Yes (b) No
13. The exponents on x and y are not the same
15. Terms: $3x,\ -2y,\ -z$; Coefficients: 3, -2, -1
17. No **19.** $-2a - 3b + 4c$ **21.** $-9x^3 + 16x^2$
23. $-15n - 20$ **25.** $-2x + 6$
27. In the expression $x^2 \cdot x^3$, there is a total of 5 factors of x. This total is found by adding the exponents according to the Product Rule for Exponents.

29. -3^{12} **31.** 4^4 **33.** $\dfrac{-8}{125}$ **35.** $\dfrac{8h^9}{k^6}$ **37.** $-35x^{12}y^9$

39. Any nonzero number divided by itself is 1.   **41.** 1

43. $-\frac{1}{27}$ **45.** $\frac{1}{8}$ **47.** $-\dfrac{k}{h}$ **49.** $\dfrac{1}{r^9}$ **51.** $\dfrac{1}{y^2}$ **53.** $\dfrac{d^7}{c^{14}}$

55. $\dfrac{h^{20}k^{12}}{16}$ **57.** $\dfrac{-6z^7}{x^3y^{10}}$ **59.** 12   **61.** $\dfrac{b^{11}}{a^{22}}$

63. The number p is the number of places to move the decimal in 2.3 to obtain 23,000.

65. $2.45 \cdot 10^{10}$ **67.** 0.000000683   **69.** $-1.6 \cdot 10^{87}$

71. (a) $6.9 \cdot 10^4$; $4.9 \cdot 10^6$

   (b) 71 people per square mile

Chapter Test   *(page 113)*

1. *(2.1)* -2

2. *(2.1)* Terms: $5x^2$, $-4x$, 2; Coefficients: 5, -4, 2

3. *(2.1)* Terms are the addends of an expression. Factors are quantities that are multiplied.

4. *(2.3)* x^4y^6

5. *(2.4)* t^2

6. *(2.4)* $\dfrac{4b^8}{a^{10}}$

7. *(2.4)* $\dfrac{1}{a^4b^6}$

8. *(2.4)* $\dfrac{4b^2}{5a^3}$

9. *(2.4)* c^2d^6

10. *(2.4)* $-\frac{1}{32}$

11. *(2.4)* 3

12. *(2.4)* $-\frac{1}{8}$

13. *(2.5)* 27,000,000

14. *(2.5)* 0.0000456

15. *(2.5)* $5.7 \cdot 10^{-7}$

16. *(2.1)* -4

17. *(2.1)* 5

18. *(2.1)* 20.216

19. *(2.2)* $3x + 5y$

20. *(2.2)* $4s - t$

21. *(2.2)* $-4x + 2$

22. *(2.1)* Yes

23. *(2.1)* $\frac{1}{2}L - 3$

24. *(2.1)* **(a)** $n + 3n = 2n - 2$

   (b) Yes

25. *(2.5)* 102,600,000; $1.026 \cdot 10^8$

Cumulative Test: Chapters 1–2   *(page 115)*

1. *(1.1)* $(-2, 3]$

2. *(1.1)* **(a)** False

   (1.1) **(b)** True

   (1.1) **(c)** True

   (1.3) **(d)** True

   (1.1) **(e)** False

3. *(1.3)* 3

4. *(1.2)* $-\frac{5}{21}$

5. *(1.3)* -10

6. *(1.3)* $-\frac{16}{15}$

7. *(1.4)* -12

8. *(1.4)* $-\frac{9}{4}$

9. *(1.5)* Undefined

10. *(1.5)* 2

11. *(1.6)* -25

12. *(1.6)* 5

13. *(1.6)* 4

14. *(1.6)* Not a real number

15. *(1.6)* 5

16. *(1.6)* 2

17. *(1.7)* $3(3x + 4)$

18. *(1.7)* $-x$

19. *(1.7)* $(2 + 5)x$

20. *(1.7)* $1y$

21. *(2.1)* 0.5

22. *(2.1)* $-\frac{50}{49}$

23. *(2.1)* 6

24. *(2.1)* **(a)** No

   (b) Yes

25. *(2.2)* $-x + 6$

26. *(2.2)* 0

27. *(2.4)* 1

28. *(2.4)* $\dfrac{1}{x}$

29. *(2.4)* $\dfrac{1}{8x^6}$

30. *(2.4)* x^2y^2

31. *(2.4)* $\dfrac{2z^2}{3x^3}$

32. *(2.4)* $108x^4$

33. *(2.5)* 0.000027

34. *(2.5)* $2.5 \cdot 10^4$

Chapter 3

Section 3.1 *(page 127)*

1. Starting at the origin, move 2 units to the left along the x-axis and then 4 units vertically upward. The destination is the point that represents $A(-2, 4)$.

3.

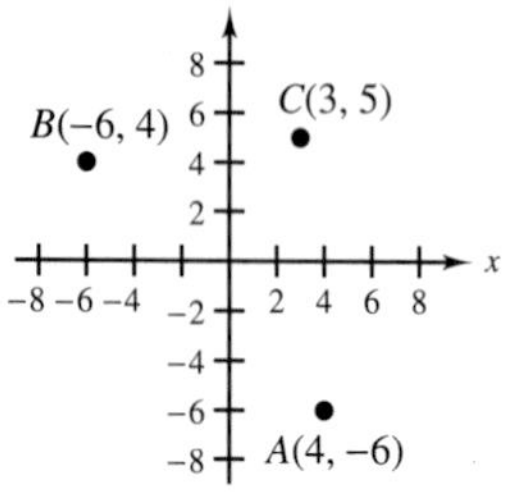

5.

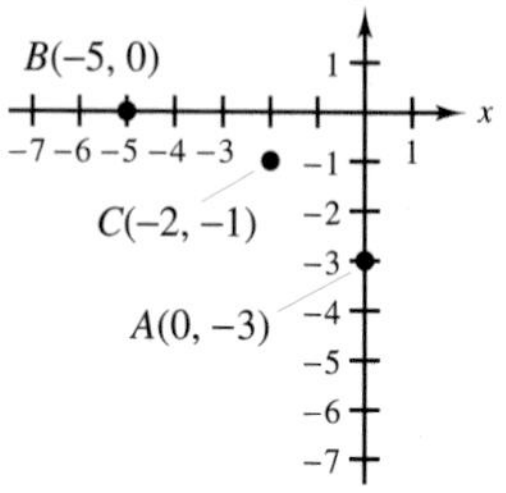

7. (6, 2)

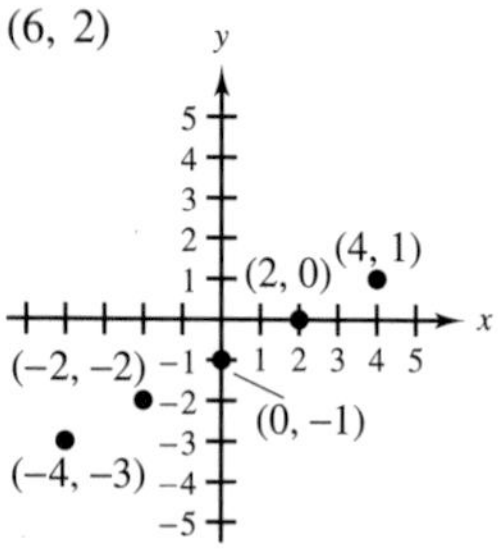

9. (2, 5)

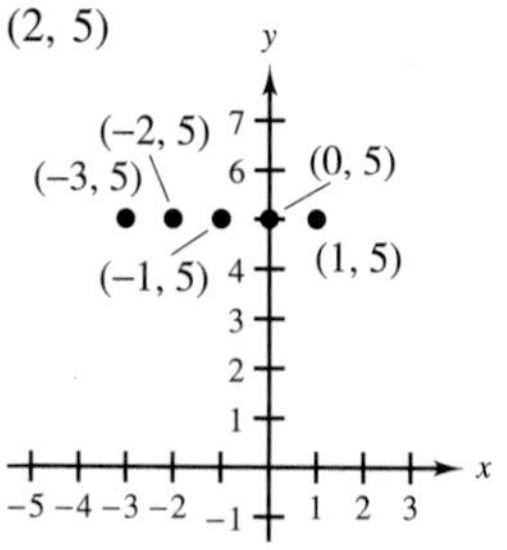

11. (6, 3)

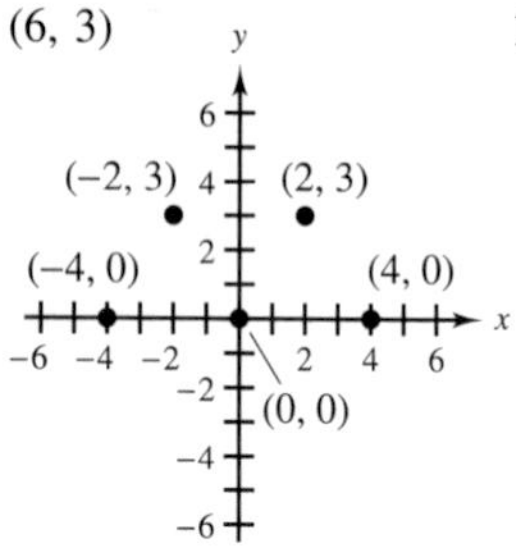

13. III

15. I **17.** IV **19.** $(-6, 4)$ **21.** (0, 0) **23.** (5, 0)

25. The ordered pair (5, 5) is the same as the ordered pair (5, 5). The ordered pair (x, y) is the same as the ordered pair (y, x) only if $x = y$. In general (x, y) is not the same as (y, x) because the order of the coordinates is significant.

27.

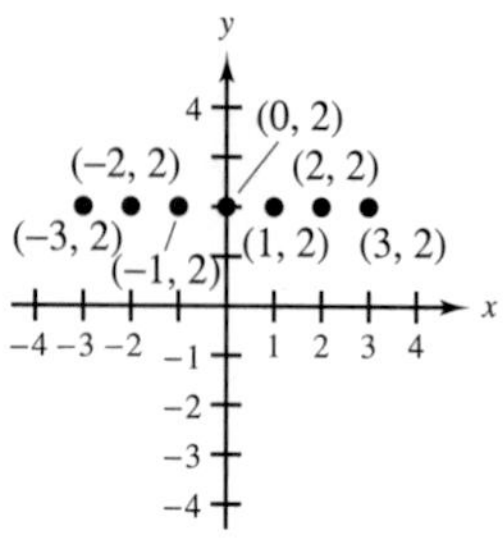

29.

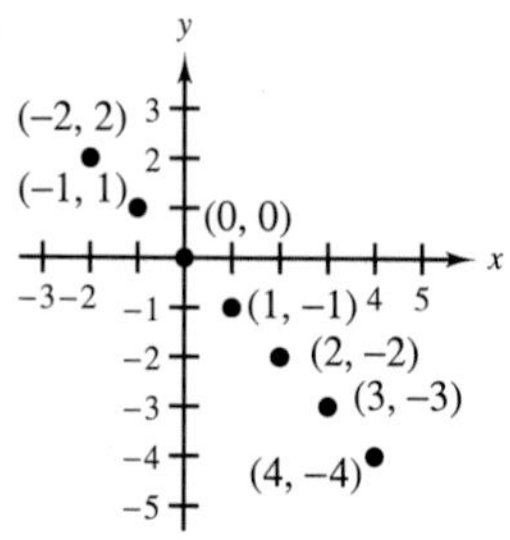

31.

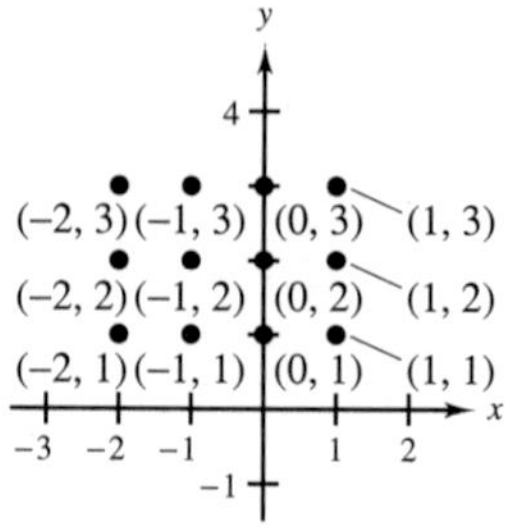

33. I, IV, or x-axis **35.** III, IV, or y-axis

37. II, III, or x-axis **39.** II, III, or x-axis

41. I, II, or y-axis

43. $x < 0, y = 0$ **45.** $x = 0, y > 0$ **47.** IV

49. Point Q is the same distance from the x-axis as point P, but on the opposite side of the y-axis.

51. **(a)** xmin = 0, ymin = −4

(b) xmax = 10, ymax = 12 **(c)** x-axis: 6; y-axis: 5

53. [X: 100, 500, 50], [Y: 100, 500, 50]

55. [X: −12, 0, 1], [Y: −20, 20, 2]

57. Find the square of the difference between the x-coordinates and the square of the difference between the y-coordinates. Add these quantities and take the square root of the result.

59. 17 **61.** 16 **63.** 10 **65.** 5 **67.** 10.20

69. Collinear **71.** Noncollinear **73.** (8, 5); 40

75. $(4, -2)$ **77.** (1, 5) **79.** (0, 1.75) **85.** Yes

87. No **89.** No **91.** 21.6 inches

93. (1960, 52), (1970, 57), (1980, 59), (1990, 60)

95. While the population grew in both regions, the rate of growth was higher in the South.

97.

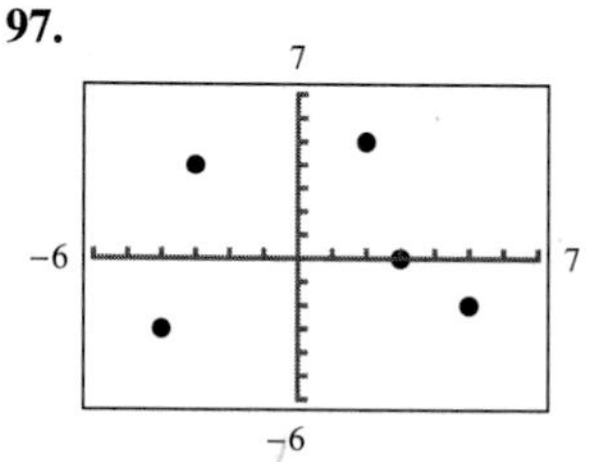

99.

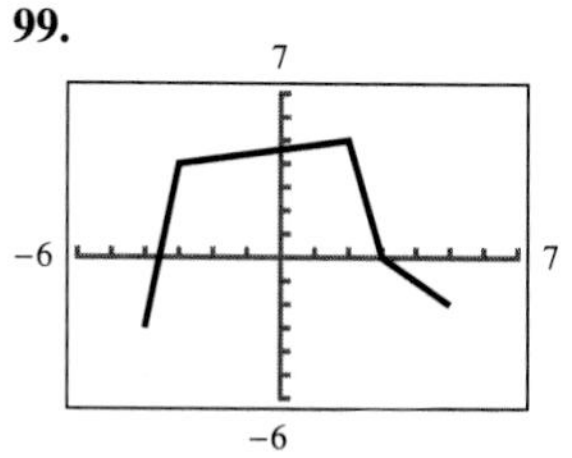

Section 3.2 *(page 138)*

1. A relation is a set of ordered pairs. A function is a relation in which each first coordinate is paired with exactly one second coordinate. Every function is a relation, but not every relation is a function.

3. **(a)** $C =$ {(Eisenhower, 2.1), (Nixon, 6.1), (Kennedy, 6.2), (Reagan, 5.2), (Carter, 4.7), (Bush, 4.2), (Clinton, 5.0)}

(b) Yes

5. $\{(-3, 9), (-2, 4), (-1, 1), (0, 0), (1, 1), (2, 4), (3, 9)\}$
Domain: $\{-3, -2, -1, 0, 1, 2, 3\}$
Range: $\{0, 1, 4, 9\}$

7. $\{(1, 4), (2, 8), (3, 12), (4, 16), (5, 20)\}$
Domain: $\{1, 2, 3, 4, 5\}$
Range: $\{4, 8, 12, 16, 20\}$

9. $\{(10, 70), (20, 140), (30, 210), (40, 280)\}$
Domain: $\{10, 20, 30, 40\}$
Range: $\{70, 140, 210, 280\}$

11.
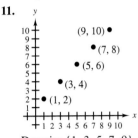
Domain: $\{1, 3, 5, 7, 9\}$
Range: $\{2, 4, 6, 8, 10\}$
Function

13.
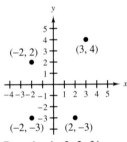
Domain: $\{3, 4, 5\}$
Range: $\{1\}$
Function

15.
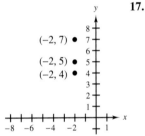
Domain: $\{-2\}$
Range: $\{4, 5, 7\}$
Not a function

17.
Domain: $\{-2, 2, 3\}$
Range: $\{-3, 2, 4\}$
Not a function

19.
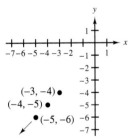
Domain: $\{\ldots, -5, -4, -3\}$
Range: $\{\ldots, -6, -5, -4\}$
Function

21. The domain of a relation can be estimated from a graph by determining the *x*-coordinates of the points of the graph.

23. For a graph to represent a function, any vertical line must intersect the graph in no more than one point. If a vertical line intersects a graph in more than one point, there are two ordered pairs with the same first coordinate. This violates the definition of a function.

25. Domain: $[-5, 5]$
Range: $[-3, 3]$
Not a function

27. Domain: $\{-5, -4, -3, -2, -1, 0, 1, 2, 3\}$
Range: $\{3\}$
Function

29. Domain: $[-4, 6]$
Range: $[-2, 6]$
Function

31. Domain: The set of all real numbers
Range: $(-\infty, -3] \cup [3, \infty)$
Not a function

33. Domain: $[-5, 4]$
Range: $[-5, 5]$
Not a function

35. Domain: The set of all real numbers
Range: $[-3, \infty)$
Function

37. Yes   **39.** No

41. Yes
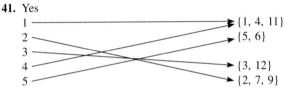

43. Yes   **45.** Yes   **47.** Yes   **49.** Yes   **51.** No
53. Yes   **55.** No

Section 3.3   *(page 146)*

1. The function $y = \frac{3}{4}x - 1$ is entered on the function screen with the following keystrokes: (3 / 4) x − 1.

3.

	x	y
(a)	18	5
(b)	33	15
(c)	−15	−17

5. Enter the function $y = 9x + 5$ on the function screen and display the graph. Then trace to the point whose -coordinate is 23 and read the corresponding *x*-coordinate.

7.

x	y
−15	−10
−9	−8
0	−5
15	0
27	4
39	8

9.

x	y
11	2
17	20
1	−12
−9	−6
−19	−20

11.

x	y
−5	13
−15	7
−30	−8
15	−7
20	−12

13.

x	y
14	2
−22	8
−31	11
−34	14
41	−1

15. (a) $x = -3.45$

(b) y-axis: $(0, -3)$; x-axis: $(-3, 0)$ and $(1, 0)$

17. (a) $x = 1.45$

(b) y-axis: $(0, 8)$; x-axis: $(-4, 0)$ and $(2, 0)$

19. $y = -x + 5$ **21.** -22 and 10 **23.** $-4, 2,$ and 10

25. If the main points of interest are those with integer coordinates, the integer setting would be appropriate. Otherwise, we would probably use the default setting.

27. (b) and (d) **29.** $-4.19, 1.19$ **31.** $-1.45, 3.45$

33. The domain is estimated by observing the x-coordinates of the displayed points; the range is estimated by observing the y-coordinates of the displayed points.

35. Domain: The set of all real numbers
Range: $[-5, \infty)$

37. Domain: The set of all real numbers
Range: $[-9, \infty)$

39. Domain: $[-8, \infty)$
Range: $[10, \infty)$

41. Domain: $[-4, \infty)$
Range: $(-\infty, 12]$

43. Domain: The set of all real numbers
Range: $(-\infty, 12]$

45. 4

47. (a) 0.00056239 (b) 0.071812 (c) 0.0011111

49. (a) The y-coordinate is not displayed. There is no point of the graph for which $x = -26$. Replacing x with -26 in the function results in a negative radicand. Thus, -26 is not in the domain of the function.

(b) $(-25, 5)$

51.

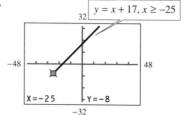

53.

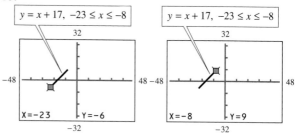

55.

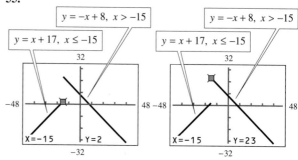

57.

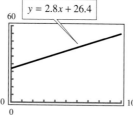

59. -0.1; Although the average of the differences is quite small, we cannot use this fact to conclude that the equation is an accurate model.

61. (a) $x \le 3$ (b) $x \ge 3$ **63.** (a) $-3 \le x \le 3$
(b) $x \le -3$ or $x \ge 3$ **65.** (a) $x = 0$ (b) $x \ge 0$

Section 3.4 *(page 156)*

1. To evaluate $f(3)$, we replace x in the expression $2x + 1$ with 3 and perform the indicated operations:
$f(3) = 2(3) + 1 = 7.$

3. 3   **5.** 2   **7.** 2   **9.** −1   **11.** 1   **13.** 3 + *t*   **15.** *a*

17. −13   **19.** −11   **21.** 9   **23.** 9

25. (a) $f(7) = -49,\ g(7) = 49,\ h(7) = 49$

(b) $f(-7) = -49,\ g(-7) = 49,\ h(-7) = 49$

(c) $f(x)$: B, A; $g(x)$: B; $h(x)$: A, B

(d) Functions *g* and *h* have the same graph because $x^2 = (-x)^2$.

(e)

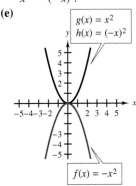

27. 2   **29.** 18   **31.** −3   **33.** −5

35. (a) $B(0) = 5,\ C(0) = 5$

(b) $B(3) = \sqrt{34},\ C(3) = 8$

(c) $B(-6) = \sqrt{61},\ C(-6) = -1$

(d) The two functions do not have the same graph because $B(x) \neq C(x)$ for all *x*.

37. 1   **39.** 1   **41.** (a) Yes   (b) No

(c)

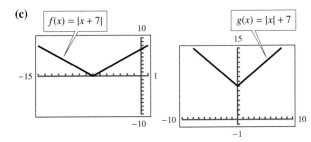

43. 1   **45.** 7

47. (a) $h(5) = -2,\ k(5) = -\frac{3}{4}$

(b) $h(14) = -8,\ k(14) = -\frac{6}{13}$

(c) The parentheses in function *k* instruct us to perform the subtraction before performing the division.

49. −2   **51.** 5   **53.** −59   **55.** 26

57. (a) $G(1) = G(-4) = G(2.3) = -3$

(b) The results in part (a) suggest that $G(x) = -3$ for all *x*.

59. Store the value for which the function is to be evaluated. Then enter the expression for the function on the home screen. The calculator returns the value of the function for the stored value of the variable.

61. −32.9   **63.** 2   **65.** −4.44   **67.** −4165.54

69. −30.13   **71.** 0.60

73. (a) −1   (b) 2 X − 3 (X − √ (X − 1))

75. (a) 8.15   (b) 2.59   **77.** (a) 9.49   (b) 3.77

79. (a) 3 − 5 * ABS (X − 3)   (b) −27

81. (a) 11,441   (b) 7736   **83.** (a) 0.67   (b) 2

(c) −1.11   (d) 0.45   (e) −3.32   **85.** $10a - 7$

87. $6z + 20$ **89.** $t^2 - 5t + 6$

91. $9t^2 - 12t - 8$ **93.** $\sqrt{a^2 + 5}$ **95.** $\sqrt{6 - 9b^2}$

97. $|-s^3 - 2|$ **99.** $|8r^3 + 4|$

101. (a) 1   (b) $2h + 1$ (c) $2h$ (d) 2

103. (a) −25   (b) $4h - 25$ (c) $4h$ (d) 4

105.

```
300 ┌──────────────────────────┐
    │ F(x) = 100/3 x + 120 (0 ≤ x ≤ 3) │
    │                          │
    │                          │
    │                          │
  0 └──────────────────────────┘ 8
  0   F(x) = −25x + 295 (3 ≤ x ≤ 7)
```

$F(x) = \frac{100}{3}x + 120\ (0 \le x \le 3)$

$F(x) = -25x + 295\ (3 \le x \le 7)$

107. 1997; Beyond 1992, we cannot rely on the model to reflect economic conditions, government actions, and other factors that could influence the number of bank failures.

109. (a) 3, 30, 300, 3000, 30,000

(b) As *x* becomes smaller, $3/x$ becomes larger.

(c) The result is an error message because division by 0 is undefined.

Section 3.5   *(page 166)*

1. (a) Trace to the point whose *x*-coordinate is 0 and read the corresponding *y*-coordinate.

(b) Evaluate the function for $x = 0$.

3. If the graph of a relation has two *y*-intercepts, the relation contains two ordered pairs with the same first coordinate, 0. The graph would not pass the Vertical Line Test. Thus, the relation is not a function.

5. (a) 10   (b) (4, 0); (0, −4)   **7.** (a) 9

(b) (24, 0); (0, −8)   **9.** (a) −15   (b) 10   (c) 50

11. (a) 27   (b) 5 or −3   (c) None   **13.** (a) 2

(b) 18   (c) 0   **15.** (a) 11   (b) 14 or −14   (c) None

17. Evaluate the function for the given value of *x*.

19. None; 0   **21.** 9; None   **23.** None; 6

25. (a) 5.56   (b) (3.69, 0); (0, −5.90)

27. (a) 1.37   (b) (0.17, 0); (0, 0.5)

29. (a) 6.65   (b) 3.26   (c) −4.74

31. (a) −4.14   (b) 2.21   (c) −0.74

33. None; −3.21   **35.** 0; None

37. A point can be the highest point in its neighborhood as well as the highest point of the graph.

39. (a) (0.82, −1.09)   **(b)** (−0.82, 1.09)

41. (a) (2.63, −9.71)   **(b)** (−0.63, 7.71)

43. (a)

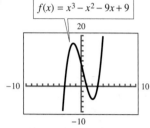

$f(x) = x^3 - x^2 - 9x + 9$

(b) Local maximum: (−1.43, 16.90)

Local minimum: (2.10, −5.05)

45. Produce a complete graph of the function and note the number of times the graph crosses the *x*-axis.

47. 1, 3

49. If the absolute maximum is *B* and the absolute minimum is *A*, then the range is [*A*, *B*].

51.

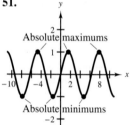

53.

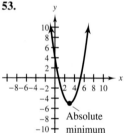

55.

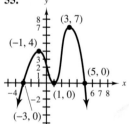

57. The graph is above the *x*-axis at *x* = 1 and below the *x*-axis at *x* = 3. Therefore, the graph must cross the *x*-axis somewhere between *x* = 1 and *x* = 3.

59. (a) Line
(b) 27.30
(c) 11.55
(d) The results in parts (b) and (c) are the total cost, including tax, of items whose prices are $26 and $11, respectively.
(e) 1.05
(f) The value of the expression in part (e) is the same as the coefficient of *x* in *C*(*x*) = 1.05*x*.

61. (a) 2 *x*-intercepts, 1 *y*-intercept
(b) 1 local minimum, 2 local maximums
(c) Domain: (−∞, ∞); Range: (−∞, 5.65]
(d) (−1.37, 0)

63. If *y* = −2, then |*x* + 2| = −2. But this is not possible because the absolute value of any expression is nonnegative.

65. (a) *x*-intercepts are (−3, 0) and (7, 0); *y*-intercept is (0, 3)
(b) There is no absolute minimum; absolute maximum is 5
(c) Domain: (−∞, ∞); Range: (−∞, 5]

67. (a) *x*-intercept is (21, 0); *y*-intercept is (0, 3)
(b) There is no absolute minimum; absolute maximum is 5
(c) Domain: [−4, ∞); Range: (−∞, 5]

69. For function *f*, 0 ≤ *x* ≤ 1 and so x^2 ≤ 1. Therefore, *f*(*x*) ≤ 1 and the absolute maximum of the function is 1. For function *g*, *x* represents any real number and so there is no limit to how large x^2 can be. Therefore, function *g* does not have an absolute maximum.

71. (a) *g*   **(b)** *f*   **(c)** *f* and *g*   **73. (a)** *g*   **(b)** *f*
(c) *f* and *g*   **75.** 1960; 800   **77.** 2294

79. (a)

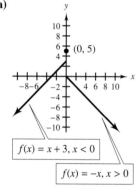

(0, 5)

$f(x) = x + 3, x < 0$

$f(x) = -x, x > 0$

(b) 5   **(c)** The range is the set containing 5 and all real numbers less than 3.

Chapter Review Exercises   *(page 171)*

1. (−4, 3)   **3.** (−2, −3)   **5.** (3, −4)   **7.** II   **9.** II
11. IV   **13.** II   **15. (a)** 9.49   **(b)** (−6.5, −0.5)
17. 36
19. Domain: {−4, 0, 3, 5}
Range: {−6, 1, 5, 7}
Function
21. Domain: [−7, 4]
Range: [−5, 6]
Not a function
23. −20   **25.** 9.00   **27.** 5   **29.** 13   **31.** 20

33. 1.64

35. ($\sqrt{}$ (X $\wedge$ 2 $-$ 3) $-$ 4) $\div$ ABS (X $\wedge$ (-1) -2)

(Calculators with a *negative* key do not require parentheses around negative exponents.)

37. **(a)** *x*-intercepts: $(-2, 0)$ and $(20, 0)$; *y*-intercept: $(0, 8)$

(b) 17   **(c)** None

39. **(a)** Local maximum: 22.5; Local minimum: -22.5

(b) Domain: $(-\infty, \infty)$; Range: $(-\infty, \infty)$

(c) $(0, 0)$

Chapter Test   *(page 173)*

1. *(2.1)* $(-3, 6)$

2. *(2.1)* $(0, 3)$

3. *(2.1)* $(3, 0)$

4. *(2.1)* $(6, -3)$

5. *(2.1)* I, IV, or *x*-axis

6. *(2.1)* II

7. *(2.1)* $(2.5, 1)$

8. *(2.1)* 5.1

9. *(2.1)* 76.2 miles; No

10. *(2.1)* No. The distance from the tornado to the home plus the distance from the home to the radar site is not equal to the distance from the tornado to the radar site.

11. *(2.2)* The graph represents a function because it passes the Vertical Line Test.

12. *(2.2)* The graph does not represent a function because it does not pass the Vertical Line Test.

13. *(2.2)* Domain: $[-4, 4]$; Range: $[-5, 5]$

14. *(2.2)* Domain: $\{1, 2, 3, 5\}$; Range: $\{-1, 3, 7\}$

15. *(2.4)* $\frac{1}{4} = 0.25$

16. *(2.4)* 3

17. *(2.4)* $3t$

18. *(2.3)* $[-30.25, \infty)$

19. *(2.3)* $(-1, 0)$ and $(10, 0)$

20. *(2.3)* -5

21. *(2.3)* $(-\infty, 4]$

22. *(2.3)* $(-4, 6)$ and $(4, 2)$

23. *(2.3)* $(0, 2)$

24. *(2.4)* **(a)** Local maximum

(b) Absolute minimum

(c) *y*-intercept

(d) *x*-intercept

(e) local minimum

(f) Absolute maximum

25. *(2.5)* 800; The maximum profit that can be earned

26. *(2.5)* 2.55 or 7.45; Ticket prices at which the profit is 500

Chapter 4

Section 4.1   *(page 183)*

1. An equation is a statement that two algebraic expressions have the same value.

3. Yes   **5.** No   **7.** Yes   **9.** No   **11.** Yes   **13.** Yes

15. No   **17.** Yes   **19.** No

21. A solution of an equation is any replacement for the variable that makes the equation true. The solution set of an equation is the set of all the solutions.

33. 5   **35.** -5 **37.** $-3, 0, 2$ **39.** -11

41. The graphs of the right and left sides of an inconsistent equation are parallel lines. The solution set is empty. An example is $x + 3 = x$.

43. 3   **45.** 3   **47.** No solution   **49.** 2   **51.** 2

53. All real numbers   **55.** -2 **57.** 0   **59.** No solution

61. -18 **63.** All real numbers   **65.** 6   **67.** 6

69. **(a)** The number *a* is the solution of the equation.

(b) The number *b* is the value of both the left and right sides when the variable is replaced with *a*.

71. 11; $(11, 25)$ **73.** -5; $(-5, 3)$ **75.** 15; $(15, -3.75)$

77. -12; $(-12, -7)$

79. The graphs of the left and right sides of the equation are parallel.

81. No, the equation has two variables.

83. From the line graph, when $y = 244$, $x = 8$ (1988).

85. 120   **87.** -8.5

89. If $x > 0$, then the equation is $2x + 3 = 7$. If $x < 0$, then the equation is $-2x + 3 = 7$. Both are linear equations in one variable.

91. If $x = -\dfrac{B}{A}$ $(A \neq 0)$, then

$$
\begin{aligned}
Ax + B &= A\left(-\frac{B}{A}\right) + B \\
&= -B + B \\
&= 0
\end{aligned}
$$

Therefore, $-\dfrac{B}{A}$ is a solution of $Ax + B = 0$, and so a linear equation in one variable is not inconsistent.

Section 4.2   *(page 194)*

1. Equivalent equations are equations that have exactly the same solution sets.

3. Substitution Property   **5.** Symmetric Property   **7.** 4

9. 8   **11.** 0   **13.** 3   **15.** -17 **17.** 6   **19.** -12

21. 35

23. The Addition Property has been used correctly, but a better first step would be to isolate the variable term by subtracting 8 from both sides.

25. 0   **27.** -9 **29.** $-\frac{2}{15}$ **31.** 36   **33.** 6   **35.** -5

37. (a) $\dfrac{3x}{5} = \dfrac{3x}{5 \cdot 1} = \dfrac{3}{5} \cdot \dfrac{x}{1} = \dfrac{3}{5}x$

(b) If, for example, $x = 15$, then

$$\frac{3}{5x} = \frac{3}{5 \cdot 15} = \frac{1}{5 \cdot 5} = \frac{1}{25},$$

but

$$\frac{3}{5}x = \frac{3}{5} \cdot 15 = 3 \cdot 3 = 9.$$

Because they do not have the same value for every replacement for x, the two expressions are not equivalent.

39. 3   **41.** 2   **43.** -12 **45.** 2   **47.** -6 **49.** $\frac{19}{9}$

51. All real numbers   **53.** -8 **55.** 17

57. No solution   **59.** 6   **61.** $\frac{11}{9}$

63. The Multiplication Property of Equations permits us to multiply both *sides* of the equation by 15. In other words, $15(\frac{1}{3} - 2x) = 15(\frac{2}{5})$. By the Distributive Property, this equation is equivalent to $15 \cdot \frac{1}{3} - 15 \cdot 2x = 15 \cdot \frac{2}{5}$.

65. $-\frac{1}{24}$ **67.** 1   **69.** $-\frac{10}{3}$ **71.** $\frac{40}{3}$ **73.** 5   **75.** 4

77. 52.86   **79.** -100

81. A conditional equation has at least one solution. An inconsistent equation has no solution.

83. Equivalent   **85.** Equivalent   **87.** Not equivalent

89. Conditional; $\{6\}$ **91.** Identity; The set of all real numbers   **93.** Inconsistent; The empty set

95. Conditional; $\{15\}$

97. The solution set of $2x = 8$ is $\{4\}$. Multiplying both sides by 0 results in $0 = 0$, which is true for all real numbers. Because the two equations are not equivalent, the Multiplication Property of Equations excludes multiplying both sides by 0.

99. -15 **101.** 8   **103.** 3   **105.** 1   **107.** 3   **109.** 5

111. 16   **113.** -22 **115.** $\frac{5}{2}$ **117.** -5 **119.** -4

121. All real numbers   **123.** $\frac{1}{3}$ **125.** No solution

127. 10   **129.** $\frac{21}{2}$ **131.** $\frac{23}{5}$ **133.** 3   **135.** 5

137. 1989

139. As x increases, $26x$ increases, and so a larger number is subtracted from 778. Thus, y decreases as x increases.

141. $\dfrac{c}{ab}$, $ab \neq 0$ **143.** $\dfrac{b - c}{a}$, $a \neq 0$

145. $\dfrac{d - b}{a - c}$, $a \neq c$ **147.** $7 \cdot 10^{12}$ **149.** $-\frac{26}{77}$

151. Not equivalent   **153.** Not equivalent

Section 4.3   *(page 203)*

1. To solve a formula for a given variable, treat all other variables as constants and use normal equation-solving procedures to isolate the given variable.

3. $m = \dfrac{F}{a}$ **5.** $r = \dfrac{I}{Pt}$ **7.** $t = \dfrac{s}{v}$ **9.** $b = \dfrac{2A}{h}$

11. $a = \dfrac{2A - bh}{h}$ **13.** $\pi = \dfrac{A}{r^2}$ **15.** $m = \dfrac{E}{c^2}$

17. $v = at + w$ **19.** $r = \dfrac{A - P}{Pt}$ **21.** $R = \dfrac{PV}{nT}$

23. $R_1 = 2R_T - R_2$ **25.** $y = -\dfrac{ax + c}{b}$

27. $v = \dfrac{2s - gt^2}{2t}$ **29.** $n = \dfrac{A - a + d}{d}$

31. $a = \dfrac{2S - n(n - 1)d}{2n}$

33. In the last step, we isolate y by dividing both sides by 3.

$$\frac{-2x + 6}{3} = \frac{1}{3}(-2x + 6)$$

$$= \frac{1}{3}(-2x) + \frac{1}{3}(6) = -\frac{2}{3}x + 2$$

Thus, the two methods give equivalent results.

35. $y = -\frac{2}{5}x + 2$ **37.** $y = x + 5$ **39.** $y = -\frac{2}{3}x + 3$

41. $y = \frac{3}{4}x - \frac{11}{4}$ **43.** $y = -\frac{1}{2}x$ **45.** $y = \frac{4}{3}x + 40$

47. $y = \frac{1}{2}x + 7$ **49.** $y = \frac{6}{5}x + 12$ **51.** $y = \frac{3}{4}x + \frac{17}{4}$

53. (a) 144.71 square inches   **(b)** $h = \dfrac{A - 2\pi r^2}{2\pi r}$
(c) 22 centimeters

55. (a) 127 miles   **(b)** 159 miles   **(c)** 239 miles

57. 7%   **59.** Yes

61. Because the diameter d is 6 feet, the circumference C of the tablecloth is $C = \pi d = 6\pi = 18.85$ feet. The 6 yards of fringe on hand is 18 feet, which is not quite enough.

63. (a) 1.30 inches   **(b)** 5.31 square inches

65. 555 feet   **67.** 63.66 meters

69. $S = 2WH + 2LH + 2WL$

71. (a) The radius of the ball is 1.44 inches. Therefore, the circumference of the ball (and the length of the metal band) is $C = 2\pi r = 2\pi(1.44) = 9.05$ inches. When we add 1 foot to the length of the band, its length is $9.05 + 12$ or 21.05 inches. Then the radius of the band is

$$r = \frac{C}{2\pi} \quad \text{or} \quad \frac{21.05}{2\pi} = 3.35 \text{ inches.}$$

Therefore, the difference between the radius of the ball and the radius of the band is 1.91 inches.

(b) The radius of the earth is 4000 miles or 21,120,000 feet. Then the circumference of the earth (and the length of the metal band) is $C = 2\pi r = 2\pi(21,120,000) = 132,700,873.7$ feet. When we add 1 foot to the length of the band, its length is 132,700,874.7 feet. Then the radius of the band is

$$r = \frac{C}{2\pi} = \frac{132,700,874.7}{2\pi} = 21,120,000.16 \text{ feet.}$$

Therefore, the difference between the radius of the earth and the radius of the band is 0.16 feet or 1.91 inches.

(c) If r_0 and C_0 represent the initial radius and circumference of the metal band, respectively, and if r_f represents the radius of the band after it has been lengthened by 1 foot, then

$$r_f - r_0 = \frac{C_0 + 1}{2\pi} - \frac{C_0}{2\pi} = \frac{1}{2\pi}$$
$$= 0.16 \text{ feet (or } 1.91 \text{ inches).}$$

Therefore, the result is the same regardless of the initial radius of the band.

73. For $f(x) = 128x + 382$,
$f(5) - f(3) = 1022 - 766 = 256$. From the actual data, the number of additional stores was $980 - 745 = 235$.

75. 1979; The equation is clearly not a valid model prior to 1983.

77. $x = \dfrac{ac + bc}{b - a}$ **79.** $x = \dfrac{ac + a}{1 - c}$ **81.** $r = \dfrac{S - a}{S}$

83. $V_2 = \dfrac{P_1 V_1 T_2}{P_2 T_1}$ **85.** $I = \dfrac{E}{R + r}$ **87.** $P = \dfrac{A}{1 + rt}$

Section 4.4 *(page 214)*

1. $T - n$
The sum of the numbers is $n + (T - n) = T$.

3. 13 **5.** 36 **7.** 442 **9.** 26 **11.** 40 ounces

13. $400, $800 **15.** 4 **17.** 2

19. Consecutive even integers and consecutive odd integers are 2 units apart on the number line. Depending on

whether x represents an even integer or an odd integer, the representations are the same in both cases.

21. 47, 48 **23.** −1, 1 **25.** 109 **27.** 38

29. If $x = $ smallest integer, then

$$\frac{x + (x + 2)}{2} = x + 1$$

is true for all x. Thus the condition holds for any three consecutive integers.

31. The sum of even integers is always even.

33. 49

35. The sum of the lengths of the pieces must equal the total length of the item. Therefore, $a + b + c = L$.

37. 8, 10, 22 **39.** 18, 22, 66

41. The sum of the measures of the angles of a triangle is $180°$. Therefore, $d + e + f = 180$.

43. 40°, 60°, 80° **45.** Yes: 30°, 60°, 90° **47.** 80°

49. The perimeter of a rectangle is the sum of the lengths of the four sides. The area is the product of the length and width.

51. 20 feet, 12 feet **53.** 13, 17

55. No, the book is 14 inches high. **57.** 6 **59.** 9 inches

61. No, the window is 45 inches high.

63. The value of the quarters is the number of quarters times the value of each one. The value of n quarters is $25n$ (in cents) or $0.25n$ (in dollars).

65. 21 **67.** 19 **69.** 12 **71.** 58 **73.** $40,000

75. Yes, the angles are 26°, 50°, 104°. **77.** 17 **79.** 18

81. 182 (Blaine), 219 (Cleveland) **83.** 111, 113

85. 30 yards, 25 yards, 50 yards **87.** 86 **89.** 56.4°

91. 8 acres **93.** 121, 123, 125 **95.** 50°, 40°, 90°

97. No, the frame is only 8 inches wide. **99.** 23

101. 17 **103.** Approx. 8.6 seconds

105. George's instructions translate into the expression $3(x - 1) - 2(x + 1) + 5$, which simplifies to x. Because $3(x - 1) - 2(x + 1) + 5 = x$ for all x, George's trick works for any number that is selected.

Section 4.5 *(page 227)*

1. In mathematics, *of* means *times*; $0.15(30)$ **3.** $205.33

5. $350, lost money **7.** $2000

9. In (i), x represents the wholesale price; in (ii), x represents the retail price.

11. 20% **13.** $25 **15.** $299

17. The costs in *a*, *c*, and *d* are variable costs; the costs in *b* and *e* are fixed costs.

19. 7.6 hours   **21.** 200

23. The value is the principal P plus 7% of the principal: (iii).

25. $571.43   **27.** $1100   **29.** $20,000

31. The interest owed is the amount borrowed times the interest rate divided by 100: (i).

33. $2500 at 9%, $4000 at 6%   **35.** $4000

37. The distance d is in miles, and the time t is in hours.

39. 4 mph   **41.** bus, 56 mph; car, 62 mph

43. 2:24 P.M., 168 miles

45. The total cost is the number of items times the unit value: (iii).

47. 10 pounds   **49.** 21 pounds   **51.** 3 pounds

53. The concentration is the amount of liquid substance divided by the total amount of liquid and water. The amount of substance is the total amount of liquid solution times the concentration of the substance.

55. 50 gallons of 3%, 100 gallons of 4.5%

57. 2 ounces of each   **59.** 64 ounces   **61.** $1475

63. English, $33; Biology, $37; Mathematics, $49

65. 4 pounds   **67.** 20°, 160°

69. 1.54 cups of 12%, 0.46 cups of 25%

71. 12:00 noon   **73.** $82,000

75. If x = wholesale cost, then $1.25x$ = retail value. The buyer's offer is $0.75(1.25x) = 0.9375x$. Because this is less than the amount you paid for the inventory, you would lose money.

Section 4.6   *(page 240)*

1. The interval $[a, b]$ contains a, but the interval $(a, b]$ does not.

3. $(-\infty, 2)$
$\{x \mid x < 2\}$

5. $[3, \infty)$
$\{x \mid x \geq 3\}$

7. $(-3, 2]$
$\{x \mid -3 < x \leq 2\}$

9. In both cases, we graph the left and right sides. For an equation, the solution is the x-coordinate of the point of intersection. For an inequality, the solution is the set of x-values for which the graph of one side is above (or below) the graph of the other side.

11. $[-14, \infty)$ **13.** $(-6, \infty)$ **15.** $(6, \infty)$ **17.** $(-\infty, 8)$

19. $[-5, \infty)$ **21.** $(-\infty, \infty)$ **23.** $(-\infty, 4]$ **25.** $\varnothing$

27. $(-\infty, 5)$ **29.** $(-\infty, 0]$ **31.** $(-\infty, -7)$

33. $(-\infty, -1]$

35. When we multiply both sides of $-x < 3$ by -1, we reverse the inequality symbol and obtain $x > -3$.

37. $<$ **39.** $>$ **41.** $>$ **43.** $<$ **45.** $c < 0$

47. Any real number   **49.** $c > 0$

51. The methods are the same, except we must reverse the inequality symbol if we multiply or divide both sides of an inequality by a negative number.

53. $(-\infty, 4]$ **55.** $(-5, \infty)$ **57.** $(-\infty, -8]$

59. $(-\infty, 15)$

61. (a) The graphs are parallel (or coincide for $<$ or $>$).
(b) The graphs are parallel (or coincide for $\leq$ or $\geq$).

63. $(-\infty, 6)$

65. $[5, \infty)$

67. $(-\infty, \infty)$

69. $[3, \infty)$

71. $(-\infty, 8)$

73. $(-\infty, \frac{2}{5}]$ **75.** $\varnothing$

77. $[0, \infty)$ **79.** $\varnothing$

81. $(-\infty, \frac{5}{2})$

83. $(-\infty, -8]$

85. $(-\infty, \frac{35}{4})$

87. $(-\frac{19}{2}, \infty)$

89. $(-\infty, \frac{5}{29}]$

91. $(-\infty, \frac{-5}{11})$

93. $[0.23, \infty)$

95. $(-\infty, 0.27]$

97. The sum of the lengths of any two sides must be greater than the length of the third side. Three true inequalities are $a + b > c$, $a + c > b$, and $b + c > a$.

99. $\left(\frac{10}{c}, \infty\right)$ **101.** $(-\infty, b - a)$

103. No. For example, $1 < 2$ but $\frac{1}{1} > \frac{1}{2}$.

Section 4.7 *(page 248)*

1. The solution set is the intersection of the solution sets of the individual inequalities.

3. Yes; No **5.** Yes; Yes **7.** Yes; Yes **9.** Yes; No

11. A conjunction is connected with *and*. A disjunction is connected with *or*.

13. 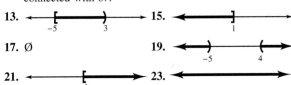 **15.**

17. Ø **19.**

21. **23.**

25. **(a)** The graph of y_2 intersects or is above the graph of y_1 at and to the right of the point $(-4, 8)$. The graph of y_2 intersects or is below the graph of y_3 at and to the left of the point $(8, 20)$. Points that satisfy both conditions have x-coordinates in the interval $[-4, 8]$.
(b) The graph of y_2 intersects or is below the graph of y_1 at and to the left of the point $(-4, 8)$. The graph of y_2 intersects or is above the graph of y_3 at and to the right of the point $(8, 20)$. Points that satisfy at least one condition have x-coordinates in the interval $(-\infty, -4] \cup [8, \infty)$.

27. $[\frac{-10}{3}, \frac{10}{3}]$ **29.** $(-5, 7)$

31. If we multiply each component of the inequality by -1, the inequality symbols are reversed, and we obtain $-2 < x < 3$.

33. **35.**

37. **39.**

41. **43.**

45.

47. The solution set of a disjunction is the union of intervals, while the solution set of a conjunction is the intersection of intervals. Thus, for example, the union of $(-\infty, 1)$ and $[1, \infty)$ is the set of all real numbers, but the intersection of those intervals is the empty set.

49. $[-5, 7]$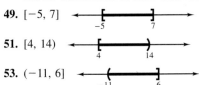

51. $[4, 14)$

53. $(-11, 6]$

55. $(-\infty, -4] \cup [3, \infty)$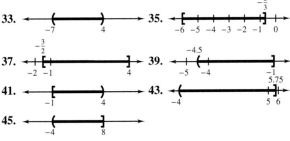

57. $(-\infty, -7) \cup (-1, \infty)$

59. $(-\infty, -1.25) \cup [1, \infty)$

61. $(-9, -6]$ **63.** $(-\infty, -1) \cup (5, \infty)$

65. $(-\infty, -3] \cup (4, \infty)$ **67.** Ø **69.** $[4, \infty)$

71. $[0, 2]$ **73.** $(-\infty, 0) \cup (6, \infty)$ **75.** $(-5, 1)$

77. **(a)** The only solution is the number c: $\{c\}$.
(b) No number can be greater than c and also less than c: Ø.

79. Ø **81.** $[3, \infty)$ **83.** $(2, \infty)$ **85.** $\{4\}$ **87.** $(-2, \infty)$

89. Ø **91.** $\{10\}$ **93.** $(-\infty, \infty)$

95. 1,605,000; The number of male students in the 25–34 age group

97. The union of F and M is the set of all students; The intersection of F and M is the set of students who are both male and female, that is, the empty set.

99. $(1 + c, 6 + c)$ **101.** $\left(\frac{5}{k}, \frac{-2}{k} \right]$ **103.** $k \geq 3$

105. $(6, 7]$ **107.** Ø **109.** $(0, 7)$

111. **(a)** Enter and graph the following functions:
$$y_1 = 0.5x - 15$$
$$y_2 = x - 5$$
$$y_3 = -x + 21$$
(b) The points of intersection are the points for which $y_1 = y_2$ and $y_2 = y_3$.
(c) The solution set is represented by that portion of the graph of y_2 that lies between the graphs of y_1 and y_3. The solution set is $(-20, 13)$.

113. $[-7, 3]$ **115.** $(-\infty, \infty)$

Section 4.8 *(page 259)*

1. The number represented by x is 5 units from 2.

3. $-12, 12$ **5.** 5 **7.** $-25, 5$ **9.** Ø **11.** $-9, 21$

13. $-2.5, 7.5$ **15.** 0.4 **17.** $-14, 6$

19. **(a)** $(-\infty, -14) \cup (8, \infty)$ **(b)** $(-14, 8)$

21. $(-12, 12)$ **23.** $(-3, 15)$ **25.** $(-\infty, \infty)$ **27.** Ø

29. The inequality $|x| \geq 5$ is equivalent to the disjunction $x \leq -5$ or $x \geq 5$ and has the solution set $(-\infty, -5] \cup [5, \infty)$. The inequality $|x| \leq 5$ is equivalent to the conjunction $-5 \leq x \leq 5$ and has the solution set $[-5, 5]$.

31. $(-4, 5)$

33. -7

35. $[-3, 11]$

37. $\emptyset$

39. $(-\infty, -3] \cup [\frac{5}{3}, \infty)$

41. $(-\infty, 5) \cup (5, \infty)$

43. $(-\infty, \infty)$

45. $(-\infty, -\frac{11}{4}) \cup (\frac{5}{4}, \infty)$

47. $|x - 5| < 7$ **49.** $|x + 7| \geq 10$ **51.** $|x + 1| < 3$

53. $|x| \geq 2$ **55.** $2, 6$ **57.** $-2.5, 5.5$ **59.** $-0.48, 4.39$

61. $-11, 9$ **63.** $\emptyset$ **65.** $-1, 11$ **67.** $\emptyset$ **69.** $\frac{4}{3}$

71. $-3, 2$ **73.** $\emptyset$ **75.** $-\frac{7}{3}, \frac{17}{3}$ **77.** $-\frac{1}{12}, \frac{17}{12}$ **79.** 1

81. 4.5 **83.** 10 **85.** $\frac{5}{3}, 5$ **87.** $(-\infty, \infty)$

89. Divide both sides by -3 to obtain $|x - 2| \leq -4$.

91. $(-7, 1)$ **93.** $(-\infty, -2] \cup [14, \infty)$

95. $(-\infty, -5) \cup (2, \infty)$ **97.** $(\frac{1}{3}, \frac{7}{3})$ **99.** $(-\infty, \infty)$

101. $(-\infty, -\frac{1}{3}] \cup [\frac{11}{3}, \infty)$ **103.** $\frac{5}{3}$ **105.** $(-\infty, 1) \cup (\frac{7}{3}, \infty)$

107. $\emptyset$ **109.** $[-11.62, 6.62]$ **111.** $(-1, 2)$

113. $(-\infty, -\frac{5}{2}) \cup (\frac{3}{2}, \infty)$ **115.** $[-\frac{1}{4}, \frac{7}{4}]$ **117.** $(-\infty, \infty)$

119. $(-\infty, 0] \cup [4, \infty)$ **121.** $(-\infty, -\frac{3}{2}]$ **123.** $(-\infty, -2)$

125. $(-\infty, -1)$ **127.** $[5, 15]$

129. (a) By definition of absolute value, $|x - 2| = x - 2$ only if $x - 2 \geq 0$ or $x \geq 2$.
(b) By definition of absolute value, $|x - 2| = -(x - 2)$ only if $x - 2 \leq 0$ or $x \leq 2$.
(c) The expressions $x - 2$ and $2 - x$ represent numbers that are opposites, and opposites have the same absolute value. Therefore the equation is true for all real numbers.

131. $x = \dfrac{-1 + k}{2}, \dfrac{-1 - k}{2}$ **133.** $\emptyset$ **135.** $x \geq 0$

137. (a) If c and d are positive, then $c + d$ is positive. Therefore, $|c| + |d| = c + d = |c + d|$.
(b) If c and d are negative, then $c + d$ is negative. Therefore, $|c| + |d| = -c + (-d) = -(c + d) = |c + d|$.
(c) Suppose $c > 0$ and $d < 0$ and $|c| > |d|$. Then $c + d > 0$. Therefore, $|c| + |d| = c + (-d) \geq c + d = |c + d|$. Therefore, $|c| + |d| \geq |c + d|$. Similar reasoning can be used for other cases.
(d) Parts (a)–(c) suggest that (ii) is correct for all numbers c and d.

Section 4.9   *(page 266)*

1. $x < 5$; $(-\infty, 5)$ **3.** $t \geq 70$; $[70, \infty)$ **5.** $s + 3$

7. $p - 200$ **9.** $-5 < x < 7$; $(-5, 7)$

11. Yes, must score at least 17

13. Third place, need at least 4 points

15. 25 feet   **17.** No   **19.** 61.5 feet

21. No, first number at most -2 **23.** 9   **25.** 32

27. If x is the number of people using the machine, the double inequality $28 \leq x \leq 51$ means the same thing as the conjunction $x \geq 28$ and $x \leq 51$.

29. 767, 800   **31.** 20 hours

33. $-17.78°C \leq$ temperature $\leq -12.22°C$

35. 201 feet $\leq$ length $\leq$ 301 feet

37. Length cannot be negative.

39. 2.6 miles $\leq$ distance $\leq$ 2.8 miles

41. $29 \leq$ toothbrushes ≤ 35

43. $3 \leq$ goals ≤ 6 **45.** $80 \leq$ seeds ≤ 100

47. By taking the absolute value, we guarantee that the distance is not negative.

49. $|L - 12| \leq 0.24$; $-0.24 \leq L - 12 \leq 0.24$

51. $|d - 3| \leq \frac{1}{16}$; $-\frac{1}{16} \leq d - 3 \leq \frac{1}{16}$ **53.** $44 \leq S \leq 56$

55. $98.1°F \leq t \leq 99.1°F$ **57.** 1989

59. \$1128.53 million

Chapter Review Exercises   *(page 269)*

1. No

5. Graph each side. The solution is the x-coordinate of the point of intersection. Store the solution and evaluate each side to check that both sides have the same value.

7. 6   **9.** All real numbers   **11.** -30 **13.** 55   **15.** 0

17. -7 **19.** Conditional; 1.4   **21.** Identity; All real numbers   **23.** -5 **25.** $H = \dfrac{V}{LW}$ **27.** $c = \dfrac{6xy}{ay + dx}$

29. $I = \dfrac{E}{R}$ **31.** $y = \frac{3}{4}x - \frac{25}{4}$ **33.** $r = \dfrac{A - P}{Pt}$; 6.5%

35. 2 feet   **37.** 32 feet, 33 feet, 34 feet   **39.** 225

41. \$155   **43.** 9 hours   **45.** 70 mph, 64 mph

47. 15°, 75°   **49.** $[2, \infty)$ **51.** $(-\infty, 4)$

53.

55. $x > 0$ **57.** $(4, \infty)$ **59.** $(-\infty, \infty)$

61. $(-\infty, 6)$

63. $[3, \infty)$

65. $(-2.5, -0.25]$

67. $[11.5, \infty)$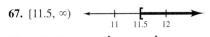

69. $(-7, 4)$

71. $(-5, 2]$

73. $(-\infty, \infty)$ **75.** $[-6, 2]$ **77.** $(-\infty, \infty)$

79. $(-\infty, -2] \cup [7, \infty)$ **81.** $(-\infty, -2) \cup (3, \infty)$

83. $(-\infty, -1)$ **85.** $-8, 8$ **87.** $-4.5, 7.5$ **89.** $1.5, 2.5$

91. $4, \frac{8}{3}$ **93.** $(-\frac{19}{3}, 3)$ **95.** $(-\infty, 2] \cup [8, \infty)$

97. $(-\infty, -7] \cup [7, \infty)$ **99.** $-5 \leq x - 3 \leq 5$

101. $(-\infty, 2)$ **103.** $(-3.5, \infty)$ **105.** 60 meters

107. 6 or 12   **109.** 0 to 600 miles

111. 131 pounds to 141 pounds

113. $|x - 1| < x$ implies $x > 0.5$ so x is positive.

Chapter Test   *(page 273)*

1. *(4.2)* **(a)** Addition Property of Equations
   (b) Property of Additive Inverses
   (c) Multiplication Property of Equations
   (d) Substitution

2. *(4.1)* Conditional; 3

3. *(4.1)* Identity; All real numbers

4. *(4.1)* Inconsistent; Ø

5. *(4.3)* $t = \dfrac{A - P}{Pr}$

6. *(4.2)* $y = -\frac{2}{3}x + 4$

7. *(4.5)* $2300

8. *(4.4)* 62.5 meters by 37.5 meters

9. *(4.5)* 1.2 gallons

10. *(4.5)* 42 mph, 54 mph

11. *(4.6)* $[-5, 2)$

12. *(4.6)* $x > 4$

13. *(4.6)* $x < -3$

14. *(4.6)* $x \leq -7.5$

15. *(4.6)* $x < 1.5$

16. *(4.7)* $[-2, 2]$

17. *(4.7)* $(-2, 3)$

18. *(4.7)* $[-3, \infty)$

19. *(4.7)* $(-\infty, \infty)$

20. *(4.8)* $-4, 0$

21. *(4.8)* $(-\infty, -1) \cup (3, \infty)$

22. *(4.8)* $[-2, 3]$

23. *(4.8)* Ø

24. *(4.9)* $12 \leq m \leq 42$ minutes

25. *(4.9)* At most 1.35 inches

Cumulative Test: Chapters 3–4   *(page 275)*

1. *(3.1)* **(a)** II
   (b) x-axis
   (c) y-axis
   (d) IV

2. *(3.2)* (i), (ii), and (iv)

3. *(3.1)* **(a)** 17.20   **(b)** $(1, -2)$

4. *(3.2)* For the graph to represent a function, no vertical line can be drawn that intersects the graph at more than one point.

5. *(3.3)* $-18, 2$

6. *(3.4)* **(a)** -19 **(b)** -5.39

7. *(3.4)* If $x = 1$, the denominator of the function has a value of 0. You will receive an error message because division by 0 is not defined.

8. *(3.5)* **(a)** $(0, 0), (2.2, 0), (-2.2, 0)$
   (b) Local maximum $(-1.3, 4.3)$
        Local minimum $(1.3, -4.3)$

9. *(4.1)* The solution is 4. When x is replaced with 4, the value of each side of the equation is 17.

10. *(4.2)* **(a)** $\frac{6}{5}$ **(b)** $\frac{49}{26}$

11. *(4.3)* $b = \dfrac{2A}{h}$

12. *(4.3)* 72

13. *(4.4)* $22°, 78°, 80°$

14. *(4.5)* 1:45 P.M.

15. *(4.9)* 11, 12, 13

16. *(4.5)* $3000 at 5%, $5000 at 6.5%

17. *(4.6)* **(a)** $(-\frac{2}{3}, \infty)$

   (b) $(-\infty, -1]$

18. *(4.7)* **(a)** $(-\infty, -1) \cup [3, \infty)$

   (b) $(3, 6)$

19. *(4.8)* **(a)** $(-5, 11)$ **(b)** $3, -5$
   (c) $(-\infty, -5] \cup [-1, \infty)$

20. *(4.9)* 15 inches

Chapter 5

Section 5.1   *(page 286)*

1. Its graph is a straight line.   **3.** Yes   **5.** No   **7.** No

9. Yes   **11.** No   **13.** No   **15.** 5   **17.** 5   **19.** -0.5

21. $a = 11, b = -3, c = 6$ **23.** $a = 3, b = 3, c = 3$

25. $a = -5, b = -1, c = 16$

27. For $\{-2, 2\}$, there are two solutions, -2 and 2. For $(-2, 2)$ the solutions are all numbers in the open interval from -2 to 2. For $(-2, 2)$, one solution is the ordered pair $(-2, 2)$.

29.

x	10	-6	7
y	-6	10	-3

31.

x	-3	0	4
y	3	3	3

33.

x	0	3	-5
$f(x)$	1	10	-14
(x, y)	$(0, 1)$	$(3, 10)$	$(-5, -14)$

35. There is a one-to-one correspondence between the points of the graph and the solutions of the equation.

37. (a) $2x - 3y - 6 = 0$ (b) $c = -1.5, d = -2/3$

39.

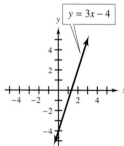

41.

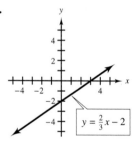

43.

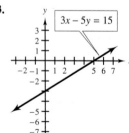

45.

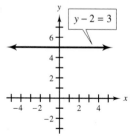

47.

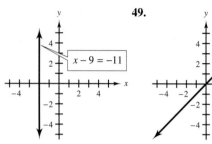

49.

51.

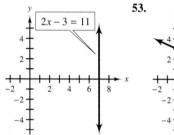

53.

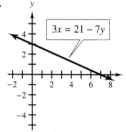

55. For the x-intercept, trace to the point where the graph crosses the x-axis and read the x-coordinate. For the y-intercept, trace to the point where the graph crosses the y-axis and read the y-coordinate.

57. $(0, 12), (4, 0)$ **59.** $(0, -10), (30, 0)$

61. $(0, -9), (12, 0)$ **63.** (a) $2x - 3y = 6$

65. $5x + 3y - 9 = 0$ and $6x + 2y - 6 = 0$

67. $y = -\frac{2}{5}x + \frac{14}{5}; (0, \frac{14}{5})$ **69.** $y = -4x - 10; (0, -10)$

71. $y = -\frac{4}{9}x + 12; (0, 12)$ **73.** $(0, 0)$

75. $(0, -3), (4, 0)$ **77.** $(0, 4), (3, 0)$ **79.** $(0, 15)$

81. If the x- and y-intercepts are the same, the line contains the origin. Therefore, $b = 0$.

83. (a) $y = x + 4$ (b) $y = x - 3$

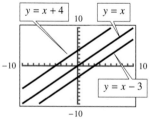

85. $a = -2, c = 8$ **87.** $a = 5, b = -3$

89. The graph of $x = $ constant is a vertical line, and the graph of $y = $ constant is a horizontal line.

91. Vertical   **93.** Neither   **95.** Horizontal   **97.** Neither

99. $x = -y$ **101.** $y = \frac{1}{2}x$ **103.** $x + y = 5$

105.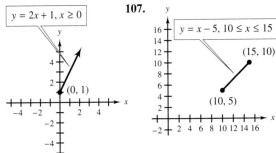

107.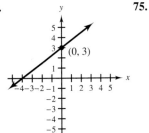

109. $y = 1.5x + 8$; Domain: $\{0, 1, 2, 3, 4, 5\}$
Cost $= 8, 9.50, 11, 12.50, 14, 15.50$

111. $y = 135 + 75x, x \geq 0$
Sells 3 to 5 cars

113. 55.58 loaves   **115.** No bread was consumed in 1910.

117. The x-intercept is $(g, 0)$ and the y-intercept is $(0, h)$.

119.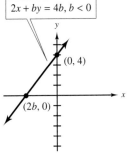

Section 5.2   *(page 298)*

1. The geometric interpretation is rise divided by run.

3. Negative   **5.** 0   **7.** Positive   **9.** Undefined   **11.** 2

13. -1 **15.** 0   **17.** Undefined

19. The algebraic definition is $m = \dfrac{y_2 - y_1}{x_2 - x_1}$.

21. 5   **23.** $\frac{7}{2}$ **25.** Undefined   **27.** -2.825 **29.** $-\frac{4}{5}$

31. 0   **33.** $\frac{5}{2}$

35. Procedures (i) and (iii) are correct. Procedure (ii) refers to a rise and run that are both negative. Thus the slope is $+3$.

37. $y = 3$ **39.** $x = 2$ **41.** $y = 4$

43. Method (ii) is the correct use of the slope formula. In method (i), the coordinates are not subtracted in the same order in the numerator and denominator.

45. $-\frac{3}{5}$ **47.** 0   **49.** $\frac{3}{4}$

51. No. For example, if $y_2 - y_1 = 8$ and $x_2 - x_1 = 10$, the slope is $\frac{4}{5}$. We only know that the ratio of rise to run is $\frac{4}{5}$.

53. $y = -2x + 17$; $m = -2$ **55.** $y = 2x$; $m = 2$

57. $y = 15$; $m = 0$ **59.** $y = -\frac{1}{3}x + 2$; $m = -\frac{1}{3}$

61. Undefined   **63.** $y = \frac{5}{3}x - \frac{7}{3}$; $m = \frac{5}{3}$

65. $y = \frac{9}{8}x - 18$; $m = \frac{9}{8}$ **69.** (i), (iii)   **71.** 0

73.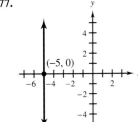

75.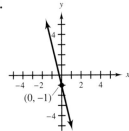

77.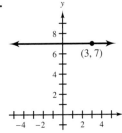

79.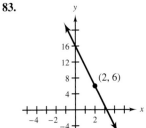

81.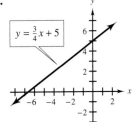

83.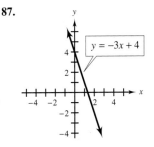

85.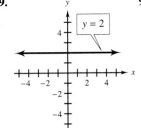

87.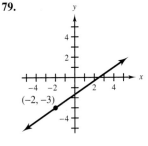

89.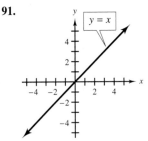

91.

93.

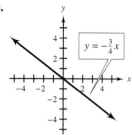

95.

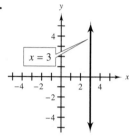

97. The graph retains the same slope, but it is shifted upward.

99. $y = -\frac{2}{5}x + 2$

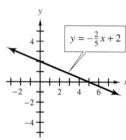

101. $x = 7$

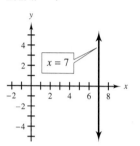

103. $y = 2x - 8$

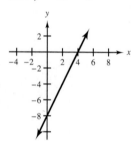

105. $y = 7$

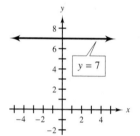

107. $y = 2x$

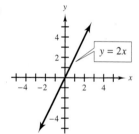

109. The student has probably entered $y = 2x + 3$.

111. 25.73%

113. Negative. The percentage of men decreases with increasing percentages of women.

115. The range settings for X must be made small or the range settings for Y must be large. The ratio of Y to X must be large.

117. [X: -5, 5, 1], [Y: -100, 500, 100], for example

119. [X: -900, 100, 100], [Y: -10, 100, 10], for example

121. [X: -1000, 3000, 200], [Y: -10, 50, 10], for example

Section 5.3 *(page 310)*

1. Perpendicular lines are lines that intersect to form a right angle.

3. Neither   **5.** Parallel   **7.** Perpendicular

9. (a) -3 **(b)** $\frac{1}{3}$ **11.** Parallel   **13.** Neither   **15.** Perpendicular   **17.** The slope of a vertical line is undefined.

19. Parallel   **21.** Perpendicular   **23.** Neither   **25.** Perpendicular   **27.** Parallel   **29.** Neither   **31.** Undefined

33. 0

35. $-\frac{2}{3}$

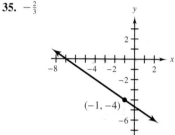

37. Undefined

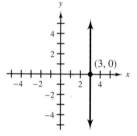

39. (a) 10.5   **(b)** $-\frac{14}{3}$ **43.** Collinear

45. The slope of a line represents the rate at which y changes with respect to a change in x.

47. $\frac{1}{2}$ **49.** -4 **51.** $\frac{3}{2}$ **53.** $-\frac{7}{3}$ **55.** y increased by 6

57. x decreased by 4   **59.** y decreased by 10

61. x increased by 4

63. (a) $c = 0.1m + 29$ **(b)** 0.1   **65.** 1.25

67. There exists a constant k such that $A = kB$.

69. Yes   **71.** Yes   **73.** No   **75.** Increases, decreases

77. 2820 pounds   **79.** 247.5 square feet   **81.** $156.25

83. 11.17 cents per year   **85.** $-$1.01 per year   **87.** 6

89. 1   **91.** 0

93. Show that the slopes of the opposite sides are equal. This proves that the opposite sides are parallel.

Section 5.4 *(page 321)*

1. To use the slope-intercept form, we must know the slope and the y-intercept.

3. $y = -4x - 3$ **5.** $y = \frac{1}{2}x + 3$ **7.** $y = -5x + 1$

9. $y = -\frac{3}{4}x - 5$ **11.** $y = 7x - \frac{2}{3}$ **13.** $y = \frac{3}{4}x + 2$

15. $x = 4$ **17.** $y = -\frac{5}{2}x - 3$

19. The slope of a vertical line is undefined.

21. $y = 2x + 1$ **23.** $y = \frac{2}{5}x - 8$ **25.** $y = -\frac{2}{3}x - \frac{19}{3}$

27. $y = -5$ **29.** $x = -3$ **31.** $x - y = 3$

33. $5x - 3y = 10$ **35.** $2x + 8y = -7$

37. Use the slope formula to determine the slope.

39. $y = -2x + 10$ **41.** $y = -2x + 3$ **43.** $y = x + 1$

45. $y = -\frac{2}{3}x - \frac{5}{3}$ **47.** $x = -4$ **49.** $y = \frac{1}{2}x - \frac{5}{2}$

51. $y = 5$

53. A line perpendicular to a horizontal line is a vertical line. Thus the equation is $x = $ constant.

55. $y = -4$ **57.** $y = 5$ **59.** $y = -\frac{2}{3}x - \frac{13}{3}$

61. $x = -2$ **63.** $y = 3x + 2$ **65.** $x = -2$

67. $y = 4x + 2$

69. $x = -2$ **71.** Because $m_1 m_2 = -1$, $m_2 = -\dfrac{1}{m_1}$.

73. $L_1: y = -\frac{4}{3}x + \frac{11}{3}$
$L_2: y = \frac{3}{4}x - \frac{1}{2}$

75. $y = -x - 7$ **77. (a)** $y = -x + 3$

(b) $y = -\frac{3}{4}x - 3$ **79.** 6   **81.** 4   **83.** $-\frac{1}{6}$ **85.** 1

87. (a) $v = -200t + 1500$ **(b)** \$700

89. 153.3 million in 1999; 6.8 million per year

91. $E = 1.24t + 6.82$, $t = $ number of years since 1987; \$22.94 in 2000

93. $D = 16.25t + 130$, $t = $ number of years since 1988; \$308.75 million in 1999

95. $y = 0.5x - 7.5$ **97.** $y = -1.25x - 428.25$

99. $PQ: y = \frac{1}{3}x + 2$; $PR: y = \frac{1}{3}x + 2$
The points are collinear because the equations are the same.

103. $(3, 0), (0, 2)$ **105.** $(\frac{5}{2}, 0), (0, -\frac{4}{3})$

107. $bx - ay = 0$

Section 5.5 *(page 333)*

1. The solution set of $x + 2y \le 6$ contains ordered pairs that satisfy $x + 2y = 6$, but the solution set of $x + 2y < 6$ does not.

3. No, yes, no   **5.** Yes, yes, no   **7.** No, yes, no

9. In the first we are assuming that $x + 3 > 7$ is an inequality in one variable. In the second we are assuming that $x + 3 > 7$ is an inequality in two variables.

11. Below line   **13.** Above line   **15.** Below line

17. (a) Choose any point not on the boundary line. If the point represents a solution, shade the half-plane containing the point. Otherwise shade the other half-plane.

(b) Solve the inequality for *y*. For $<$ or $\le$, shade below the line. For $>$ or $\ge$, shade above the line.

19.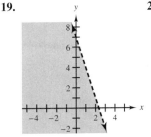

21.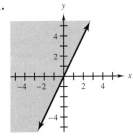

23.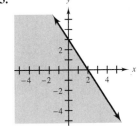

25.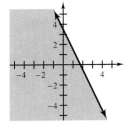

27.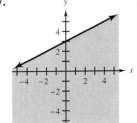

29.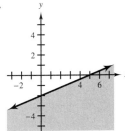

31.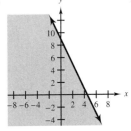

33.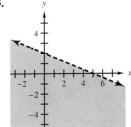

35.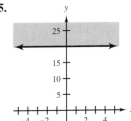

37.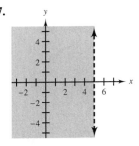

39. $y \le 3$ **41.** $y < 0.5x + 1$ **43.** $x \le 2$

45. $x \geq y + 1$

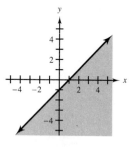

47. $y \geq 3x - 4$

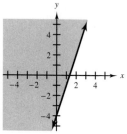

49. $x + 2y > 10$

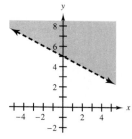

51. x = number of 4 × 4 pallets,
y = number of 3 × 5 pallets
$16x + 15y \leq 6000,\ x \geq 0,\ y \geq 0$

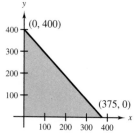

53.

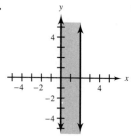

55.

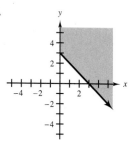

57.

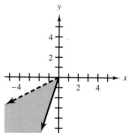

59.

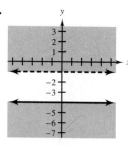

61.

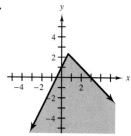

63.

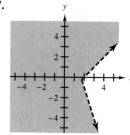

65.

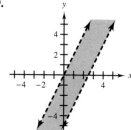

67.

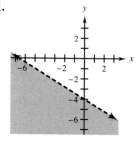

69.

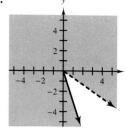

71.

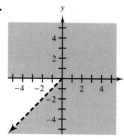

73.

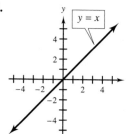

75.

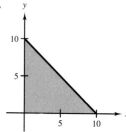

77. $y < 2$ and $y \geq -3$ **79.** $x \leq -2$ or $x > 1$
81. $y \geq x + 3$ or $y \leq x - 2$

83. c = number of cats, d = number of dogs,
$20 \leq c + d \leq 100,\ c \geq 0,\ d \geq 0$

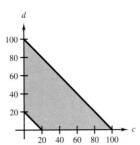

101. x = score of one team, y = score of the other team
$|x - y| \leq 5,\ x \geq 0,\ y \geq 0$

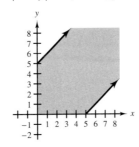

85.

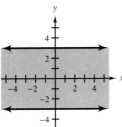

87.

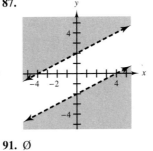

103. Entire plane

105.

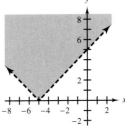

89.

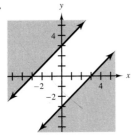

91. Ø

107.

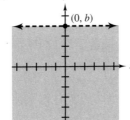

109.

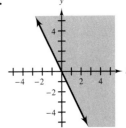

93.

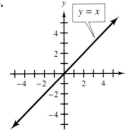

95.

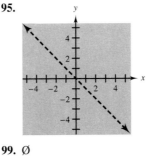

111.

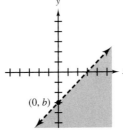

113.

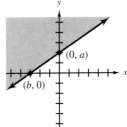

115. No, yes   **117.** No, yes

97.

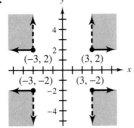

99. Ø

Chapter Review Exercises *(page 337)*

1. No   **3.** Yes   **7.** $a = 9,\ b = -7,\ c = 12$

9.

x	3	7	-7
y	9	15	-6

11. $(0, -7)$, $(3, 0)$ **13.** $(0, 20)$; No x-intercept

15. $y = \frac{1}{2}x - \frac{5}{2}$; $(0, -\frac{5}{2})$ **17.** $y = 2x - 8$; $(0, -8)$

19.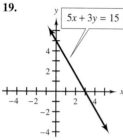

$5x + 3y = 15$

21.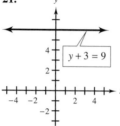

$y + 3 = 9$

23. (ii)   **25.** $\frac{5}{7}$ **27.** $-\frac{10}{9}$ **29.** $y = \frac{3}{4}x - \frac{15}{4}$; $\frac{3}{4}$

31.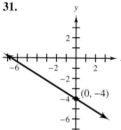

$(0, -4)$

33.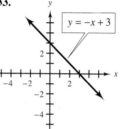

$y = -x + 3$

35.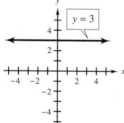

$y = 3$

37. Parallel

39. Perpendicular   **41.** Perpendicular   **43.** Yes   **45.** -3

47. y is decreased by 15   **49.** $1.25   **51.** No   **53.** Yes

55. 112,500   **57.** $y = -3x - 1$ **59.** $y = -x - 2$

61. $y = \frac{5}{4}x - \frac{9}{4}$ **63.** $y = 6$ **65.** $y = -\frac{2}{3}x + \frac{17}{3}$

67. $y = -0.4x + 2$ **69.** 1997   **71.** Yes, no, yes

73. No, yes, no

75.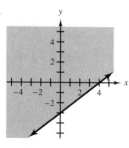

77.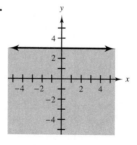

79.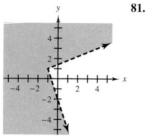

81.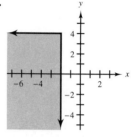

83.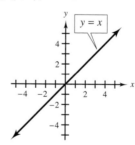

85.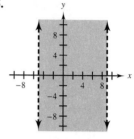

87. x = number of grandstand tickets sold, y = number of bleacher tickets sold, $6x + 4y < 11,400$, $x \geq 0$, $y \geq 0$

89. x = number of girls, y = number of boys, $x + y \leq 400$ and $x \geq 100$, $y \geq 0$

Chapter Test   *(page 341)*

1. *(4.1)* **(a)** $(-1, 0)$
    (b) $(0, 2)$
    (c) $(-2, -2)$
    (d) $(\frac{1}{2}, 3)$

2. *(4.2)* $(0, 0)$; $m = 1$

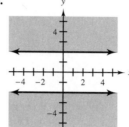

$y = x$

3. *(4.2)* $(0, -2)$, $(-2, 0)$; $m = -1$

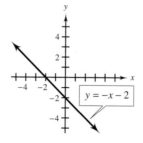

$y = -x - 2$

4. *(4.2)* $(0, -3)$, $(\frac{9}{2}, 0)$; $m = \frac{2}{3}$

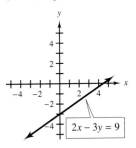

$2x - 3y = 9$

5. *(4.2)* $(5, 0)$; slope undefined

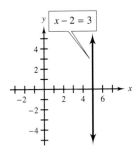

$x - 2 = 3$

6. *(4.2)* $m < 0$, $b < 0$

7. *(4.4)* $y = \frac{2}{3}x - \frac{13}{3}$

8. *(4.4)* $y = 2x + 5$

9. *(4.4)* $x = 2$

10. *(4.4)* $y = 3$

11. *(4.4)* $y = 2x + 2$

12. *(4.4)* $y = 0.75x + 3$

13. *(4.5)*

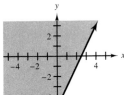

14. *(4.5)*

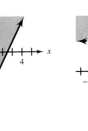

15. *(4.5)*

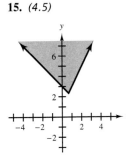

16. *(4.5)*

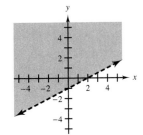

17. *(4.5)*

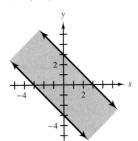

18. *(4.5)*

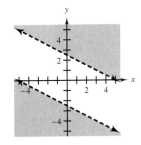

20. *(4.2)*

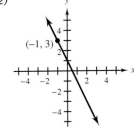

$(-1, 3)$

21. *(4.3)* Perpendicular

22. *(4.3)* Neither

23. *(4.3)* $v = 32t$; 64 feet per second

24. *(4.3)* $-\$0.45$

25. *(4.3)* $T = 2.5t + 88$, $t =$ number of hours since 6:00 A.M.; At 3:00 P.M. $t = 9$ and $T = 110.5$

Chapter 6

Section 6.1   *(page 352)*

1. Substitute the numbers into each equation. Both equations must be satisfied.

3. No, yes, no   **5.** No, yes, yes   **7.** $a = 1$, $b = 2$

9. $a = 4$, $c = 7$

11. Two lines cannot intersect at exactly two points.

13. 0   **15.** 1   **17.** No solution   **19.** 1

21. Infinitely many solutions

23. **(a)** (i) The graphs do not coincide.
(ii) The graphs coincide.
(b) Yes, the lines could be parallel.

25. $(0, 6)$ **27.** $(0, 0)$ **29.** $(3, 4)$ **31.** Ø   **33.** $(5, -2)$

35. $(11, 17)$ **37.** Infinitely many solutions

39. $(27, -18)$ **41.** $(-3, 4)$ **43.** $(-4, 7)$

45. **(a)** Yes, either equation can be solved for either variable.
(b) Solving the second equation for y is easier.

47. $(-22, -27)$ **49.** Dependent   **51.** Inconsistent

53. Dependent   **55.** $(2, -3)$ **57.** Inconsistent

59. The system is inconsistent and there are no solutions.

61. $(0, 5)$ **63.** $(80, 80)$ **65.** -2 **67.** 5   **69.** -2

71. -4 **73.** 10, 6   **75.** 20, 12   **77.** $67°, 23°$

79. $112°, 68°$ **81.** 27 feet, 10 feet

83. 10 inches, 30 inches   **85.** 78 T-shirts, 43 sweatshirts

87. $x \approx 90$, $y \approx 1250$

89. In 1990 the expenditure for health care and steel were both the same, $1.249 billion.

91. $\left(\dfrac{c}{a + mb}, \dfrac{cm}{a + mb}\right)$ **93.** $c \neq -2$ **95.** $a \neq 0.2$

97. $m_1 = m_2$, $b_1 \neq b_2$ **99.** Any b_1 and b_2

101. No, either infinitely many solutions or no solution

Section 6.2   *(page 362)*

1. **(a)** Multiply the second equation by -3 and add the equations.
(b) Multiply the first equation by 4 and add the equations.

3. $(6, 4)$ **5.** $(-10, 7)$ **7.** $(-2, 5)$ **9.** $(2, \tfrac{2}{3})$

11. No, it means there are infinitely many solutions.

13. $(10, -7)$ **15.** $(5, -3)$ **17.** $(\tfrac{1}{2}, \tfrac{2}{3})$ **19.** $(5, 4)$

21. The lines intersect at $(3, 2)$. Many lines can be drawn through this point.

23. $(1, 1)$ **25.** $(0, 2)$ **27.** $(-2, 1)$ **29.** $(0, 1)$

31. Both variables are eliminated, and you obtain an identity such as $0 = 0$.

33. $(-3, -4)$ **35.** Ø   **37.** Infinitely many solutions

39. Ø   **41.** $(4, -2)$ **43.** Infinitely many solutions

45. $(\tfrac{41}{7}, \tfrac{23}{7})$ **47.** Ø, system inconsistent

49. Infinitely many solutions, equations dependent

51. $(3, -3)$ **53.** $(1, 2)$

55. Infinitely many solutions, equations dependent

57. $a = 3$, $b = 2$ **59.** $a = 1$, $b = 3$ **61.** -4

63. -6 **65.** $c = 4$ **67.** $a = 3$, $b = -3$

69. Hamburger, $1.90; milk shake, $1.20   **71.** 27 dimes

73. 43 nickels, 21 dimes   **75.** $4250 at 7%, $1750 at 9%

77. $\left(\dfrac{b + c}{2}, \dfrac{b - c}{2}\right)$ **79.** $\left(\dfrac{c + 3}{a + 1}, \dfrac{3 - ac}{a + 1}\right)$, $a \neq -1$

81. $\left(\dfrac{1}{a}, 1\right)$, $a \neq 0$ **83.** $(1, 0)$, $a \neq -b$

85. [X: -300, 300, 50], [Y: $-100{,}000$, 100,000, 10,000], for example   **87.** $(94.75, 34{,}215.25)$; Interpret 94.75 as 1994.   **89.** $(2, 4)$ **91.** $(-5, 2.5)$

93. The system has no solution because one solution leads to the equation $1/x = 0$.

95. $b \neq -4$ **97.** $c = 5$ **99.** $a \neq 1$ **101.** $c \neq 10$

Section 6.3   *(page 374)*

1. No, not all coefficients can be zero.

3. Yes, no   **5.** No, yes   **7.** $a = 3$, $b = 0$, $c = -2$

9. $b = 3$, $c = 0$, $d = 1$

11. For an equation in two variables, the graph is a line in a plane. For an equation in three variables, the graph is a plane in 3-dimensional space.

13. **(a)** At least two planes are parallel or the third plane is parallel to the line of intersection of the other two planes.
(b) The three planes coincide or intersect in a common line.

15. $(-1, 3, -2)$ **17.** $(7, 5, 4)$ **19.** $(3, -2, 1)$

21. $(4, -3, -1)$ **23.** $(1, 3, -2)$ **25.** $(3, \tfrac{1}{2}, -\tfrac{1}{3})$

27. $(1, -\tfrac{1}{3}, \tfrac{1}{2})$ **29.** $(\tfrac{1}{2}, \tfrac{1}{3}, -\tfrac{2}{3})$

31. **(a)** After eliminating one variable, the resulting system of two equations in two variables is inconsistent.
(b) After eliminating one variable, the two equations in two variables in the resulting system are dependent.

33. Ø, inconsistent system

35. Solution set is the set of all solutions to $x + 2y - z = 3$. Dependent equations

37. $(3, -2, 1)$ **39.** $(1, 2, 3)$ **41.** $(5, -4, -1)$

43. $(5, -4, -3)$ **45.** $(-7, -4, 3)$ **47.** Ø

49. Solution set is the set of all solutions to $4x - 3y + 6z = 12$.

51. $(-3, -6, 4)$ **53.** $23°, 64°, 93°$

55. 27 yards, 52 yards, 21 yards

57. 27 nickels, 56 dimes, 72 quarters

59. T-shirt, $9; sweatshirt, $15; tank top, $4

61. $(-3, 2, 5)$ **63.** $(-1, -3, 2)$ **65.** $(1, 2, -1, 3)$

67. $(1, -1, 0.5)$

69. $\left(\dfrac{a - b + c}{2}, a - \dfrac{a - b + c}{2}, c - \dfrac{a - b + c}{2}\right)$

Section 6.4   *(page 384)*

1. A 2×4 matrix is an array of numbers with two rows and four columns.

3. 2×4 **5.** 2×2 **7.** 3×4

9. **(a)** A coefficient matrix is a matrix whose elements are the coefficients of the variables of the equations in a system of equations.

(b) A constant matrix is a matrix whose elements are the constants of the equations in a system of equations.

(c) An augmented matrix is a matrix whose elements are the coefficients and constants of the equations in a system of equations.

11. $\begin{bmatrix} 3 & 2 & | & 6 \\ 1 & -4 & | & 9 \end{bmatrix}$

13. $\begin{bmatrix} 1 & 2 & -3 & | & 5 \\ 1 & 0 & 2 & | & 15 \\ 0 & 2 & -1 & | & 6 \end{bmatrix}$

15. $\begin{aligned} 2x + y &= 1 \\ 3x - 2y &= 12 \end{aligned}$ **17.** $\begin{aligned} x + y - z &= 2 \\ 2x - 3y + z &= 5 \\ 3x + 2y - 4z &= 3 \end{aligned}$

19. The goal is to have 1's along the main diagonal of the matrix and 0's elsewhere.

21. (a) $\frac{1}{3}R_1$ (b) $-2R_1 + R_2 \to R_2$ (c) $-\frac{1}{7}R_2$
(d) $-2R_2 + R_1 \to R_1$ (e) $(2, 1)$

23. (a) $-2R_1 + R_2 \to R_2$ (b) $-1R_1 + R_3 \to R_3$
(c) $-\frac{1}{3}R_2$ (d) $-1R_3$ (e) $R_3 + R_1 \to R_1$
(f) $R_3 + R_2 \to R_2$ (g) $-1R_2 + R_1 \to R_1$
(h) $(1, 2, -1)$ **25.** (a) 2 (b) 2 (c) -56
(d) 2 (e) 1 (f) -3 (g) 1 (h) $(-3, 8)$

27. (a) 3 (b) -3 (c) 3 (d) -2 (e) 1 (f) -2
(g) -6 (h) -2 (i) 1 (j) 0 (k) 1 (l) 0
(m) 0 (n) 0 (o) $(1, 0, -2)$

29. The equations must be written in standard form.

31. $(3, -4)$ **33.** $(\frac{1}{2}, -\frac{1}{4})$ **35.** Inconsistent

37. Dependent **39.** $(4, 2, -3)$ **41.** $(1, -3, 2)$

43. $\begin{bmatrix} 2 & -5 \\ 5 & 2 \end{bmatrix}\begin{bmatrix} -5 \\ 2 \end{bmatrix}$ **45.** $\begin{bmatrix} 3 & -4 & 1 \\ 1 & 1 & 3 \\ 2 & 3 & -1 \end{bmatrix}\begin{bmatrix} 2 \\ 1 \\ 0 \end{bmatrix}$ **47.** $(-3, 1)$

49. $(3, -2, -2)$ **51.** $(0, -1, 2)$ **53.** $(1, -2, -1)$

55. $\begin{bmatrix} 60{,}332 & 72{,}517 & 89{,}917 \\ 10{,}221 & 10{,}380 & 10{,}680 \\ 30{,}381 & 52{,}170 & 63{,}725 \\ 633 & 576 & 663 \end{bmatrix}$; 4×3 **57.** Second row

59. (a) $z = -3, x = 4, y = -5$ (b) Yes

61. $(-2, -3, 1, 2)$ **63.** $(1, -1, -1, 2)$

65. Unique for $a \neq \frac{1}{3}$, no solution for $a = \frac{1}{3}$

Section 6.5 *(page 395)*

1. 12 pounds of assorted creams **3.** 5 pounds apples, 3 pounds grapes **5.** 42 cakes, 48 pies, 50 dozen cookies

7. 237 patrons, 219 others **9.** 14,261 box seats

11. 1450 orchestra seats, 1550 loge seats, 2500 balcony seats **13.** \$6366 at 6.5%, \$4634 at 9% **15.** \$3800 at

6%, \$3200 at 7% **17.** \$5400 in stock fund, \$4100 in bond fund, \$7000 in mutual fund **19.** 50 liters of 34% solution, 20 liters of 55% solution **21.** 200 gallons of 3.5% solution **23.** 6 liters of 10% solution, 4 liters of 20% solution, 8 liters of 50% solution **25.** canoe, 5 mph; current, 2 mph **27.** airplane, 450 mph; wind, 30 mph

29. canoe, 4 mph; current, 2 mph **31.** $a = -2, b = 4$

33. $a = 2, b = -3, c = -5$
$y = 2x^2 - 3x - 5$

35. $a = -4, b = 1, c = 7$
$y = -4x^2 + x + 7$

37. Father pays \$34, son pays \$17 **39.** 39 five-dollar bills, 31 ten-dollar bills **41.** 12 heavy duty drills, 19 home-owner's drills **43.** 32 gallons high gloss enamel, 43 gallons flat latex **45.** Orchid corsages, 37; carnation cor-sages, 58 **47.** 140 **49.** 8-mile trail, 22; 6-mile trail, 10; 5.5-mile trail, 8 **51.** A, 5 credit hours; B, 8 credit hours; C, 4 credit hours **53.** First set, 15 volumes; second set, 14 volumes; third set, 6 volumes

55. $(3, 12.97), (5, 13.13), (7, 17.00)$

57. $a = 0.46, b = -3.63, c = 19.69$
$S(t) = 0.46t^2 - 3.63t + 19.69$

59. 3.94 ounces **61.** Foreman, 5 mph; inspector, 7 mph

Section 6.6 *(page 404)*

1. The graph of each linear inequality is a half-plane. The graph of the solution set of the system is the intersection of the half-planes.

3. **5.**

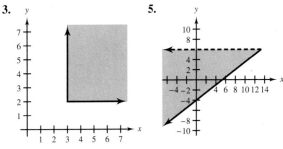

7. **9.**

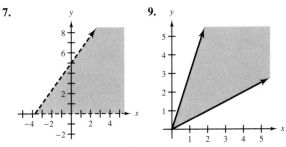

11.

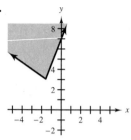

13. ∅   **15.** $y \leq 4$, $x \leq 5$, $x \geq 0$, $y \geq 0$

17. $y \leq -8x + 8$, $y > -\frac{1}{3}x - 2$

19. The graph is a square and its interior. The square is centered at the origin. The sides are parallel to the coordinate axes, and the length of each side is 10.

21.

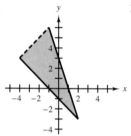

23.

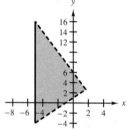

25.

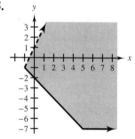

27. The number of people cannot be negative: $x \geq 0$ and $y \geq 0$.

29.

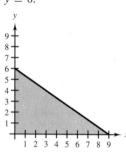

31.

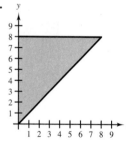

33.

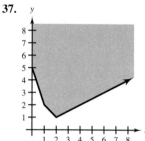

35.

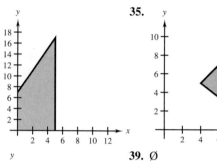

37.

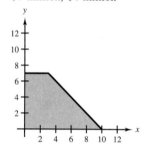

39. ∅

41. x = number of millions of dollars in short-term notes, y = number of millions of dollars in mortgages
$x + y \leq 10$, $y \leq 7$, $x \geq 0$, $y \geq 0$
$7 million, $4 million

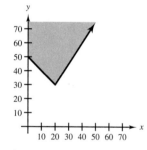

43. x = number of tuna salads, y = number of chicken salads
$x \leq \frac{2}{3}y$, $x + y \geq 50$, $x \geq 0$, $y \geq 0$
From 0 to 40 tuna salads can be prepared.

45. (a) $x =$ number of barns built,

$y =$ number of tool buildings built

$15x + 5y \leq 90, 7x + 6y \leq 84, x \geq 0, y \geq 0$

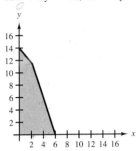

(b) Yes **(c)** No

47. $3 million in short-term notes, $7 million in mortgages

Chapter Review Exercises *(page 407)*

1. The pair represents a solution to both equations.

3. Yes **5.** $(3, 1)$ **7.** $\varnothing$ **9.** $(-3, 2)$ **11.** $(-1, -3)$

13. Parallel **15.** One point **17.** Coincident

19. Parallel **21.** One point **23.** $(0.6, 1.6)$ **25.** $\varnothing$

27. $(2.8, -1.6)$ **29.** $(0.75, -0.25)$ **31.** $\left(\frac{6}{5}, -\frac{2}{3}\right)$

33. Return to the original system and use the addition method again, but this time eliminate x.

35. $(2, -4)$ **37.** Infinitely many solutions **39.** $(8, -6)$

41. $\varnothing$ **43.** -12 **45.** -3

47. The solution set is represented by a plane drawn in a three-dimensional coordinate space.

49. No **51.** $(-2, 3, 1)$ **53.** $(2, -3, -5)$ **55.** $\varnothing$

57. $(-8, 1, 3)$ **59.** $(1, 2, -5)$ **61.** $(0, 2, 0)$

63. A coefficient matrix consists of the coefficients of the variables of a system of equations. An augmented matrix is the combination of a coefficient matrix and a constant matrix.

65. $\begin{bmatrix} 4 & -3 & | & 12 \\ 1 & -2 & | & -3 \end{bmatrix}$

67. $x + 2y - 3z = 4$
$3x + y - 2z = 5$
$x + y \quad\;\; = 0$

69. $(2, -1)$ **71.** $(-22, -27)$ **73.** $\left(-1, -\frac{5}{3}, -\frac{5}{6}\right)$

75. $(1, 2, 3)$ **77.** $(-10, 9)$ **79.** $(3, -1, -2)$

81. $(-4.375, -1.75, -5.5)$

83. 33 fives, 27 twenties

85. 15 sledge hammers, 27 claw hammers **87.** 435

89.

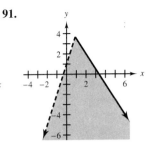

91.

93.

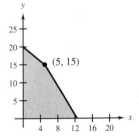

95.

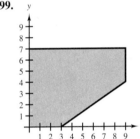

97.

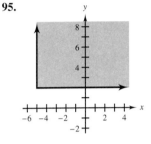

99.

101. $x =$ number of bags of mortar,

$y =$ number of bags of lime

$x + y \leq 20, 80x + 40y \leq 1000, x \geq 0, y \geq 0$

Chapter Test *(page 411)*

1. *(6.1)* $(2, -1)$

2. *(6.1)* $(-12, 11)$

3. *(6.1)* $(5, -3)$

4. *(6.2)* Infinitely many solutions

5. *(6.2)* $(10, -2)$

6. *(6.3)* $(2, -1, 1)$

7. *(6.4)* $(4, -2)$

8. *(6.4)* $(1, 3, -2)$

9. *(6.4)* $(-3, 0.5, 1)$

10. *(6.1)* Dependent

11. *(6.1)* Inconsistent

12. *(6.1)* Unique solution

13. *(6.1)* **(a)** The graphs coincide.
       (b) The graphs intersect at one point.
       (c) The graphs are parallel.

14. *(6.7)*

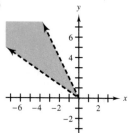

15. *(6.7)*

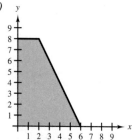

16. *(6.6)* Wind speed, 25 mph; airspeed of plane, 175 mph

17. *(6.6)* $31\frac{2}{3}°$, $58\frac{1}{3}°$

18. *(6.6)* 40 liters of the 15% solution, 60 liters of the 40% solution

19. *(6.6)* Shrimp, 45.5 pounds; crab, 43 pounds; lobster, 48 pounds

20. *(6.7)* p = number of pines, d = number of dogwoods
       $p + d \le 20$, $p \ge 2d$, $p \ge 0$, $d \ge 0$

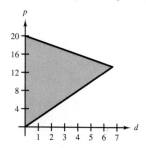

Cumulative Test: Chapters 5-6   *(page 413)*

1. *(5.1)* $a = \frac{1}{3}$, $b = 2$

2. *(5.1)* $(0, -10)$ $(-2, 0)$

3. *(5.2)* **(a)** Negative
       (b) Undefined
       (c) Zero
       (d) Positive

4. *(5.2)* $y = \frac{1}{2}x + 2$

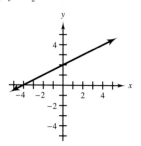

5. *(5.3)* The rate of change of y with respect to x is the slope of the line: $\frac{2}{3}$.

6. *(5.3)* $\frac{7}{5}$

7. *(5.4)* $x - y = 3$

8. *(5.4)* $y = -\frac{2}{3}x - \frac{10}{3}$

9. *(5.5)* **(a)** The boundary line is the dashed line
          $y = x - 7$.
       (b) Either test a point or write the inequality as $y > x - 7$ and shade above the line.

10. *(5.5)*

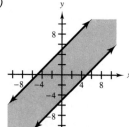

11. *(6.1)* **(a)** Empty set
        (b) One point
        (c) All points on the line

12. *(6.1)* $(-2, 6)$

13. *(6.2)* **(a)** Dependent
        (b) $(-1, 5)$
        (c) Inconsistent

14. *(6.2)* 200 mathematics books, 220 English books

15. *(6.3)* $(-1, 0, 4)$

16. *(6.3)* 40°, 70°, 140°

17. *(6.4)* **(a)** 11
        (b) 3
        (c) 2

18. *(6.4)* $(3, -2, 1)$

19. *(6.6)* 4 mph

20. *(6.6)* 150

21. *(6.7)*

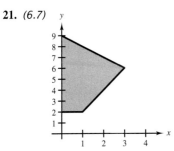

22. *(6.7)* $2L + 2W \le 40$, $W \le \frac{1}{2}L$, $L \ge 0$, $W \ge 0$

Chapter 7

Section 7.1 *(page 422)*

1. (a) The variable is in the denominator.

(b) The exponent is negative.

3. Yes   **5.** Yes   **7.** 5, 0   **9.** −4, 3   **11.** 1, 2

13. The degree of a term is the value of the exponent on the variable (or sum of exponents if more than one variable). The degree of a polynomial is the degree of the term with the highest degree.

15. 1   **17.** 5   **19.** 5, trinomial   **21.** 3, neither

23. 5, binomial   **25.** 2, $-3y^2 + 6y + 6$, $6 + 6y - 3y^2$

27. 9, $4x^9 - 7x^8 - 5x^6 + x^3$, $x^3 - 5x^6 - 7x^8 + 4x^9$

29. 12, $n^{12} + 4n^8 - 2n^6 - 3n^4$, $-3n^4 - 2n^6 + 4n^8 + n^{12}$

31. After you enter the function, you can store the value of x and evaluate the function for each of the 10 values.

33. 5, 0   **35.** −1, −11   **37.** −130.05, 68.33   **39.** −4

41. 15   **43.** −24   **45.** $7x + 1$ **47.** $4a$ **49.** $3xy^2$

51. $-7x + 6$ **53.** $4x^2 - x + 14$ **55.** $-2t^2 - 3t + 1$

57. $-4x^2 + 5x + 4$ **59.** $n^2 - m^2 + 7mn$

61. $-2xy + 5y^2$ **63.** $-2a^2 + 5a^2b$ **65.** $-2x^3 - 8x$

67. $-4x^2 + 5x - 8$ **69.** $x^2 - 12$ **71.** $2x^2 + 2x + 2$

73. (ii) The minuend is $3x + 4$ and the subtrahend is $x - 2$.

75. $3x^2 - 2x + 7$ **77.** −6   **79.** $-9x^2 - 3x - 9$

81. $-x + 7$ **83.** $x - 3$ **85.** $\frac{1}{2}y^2 - 1$

87. $3.5x^2 - 1.45x - 2.72$ **89.** $x^2 - 5x + 2$

91. $-a^3 - 13a^2 + 11a - 11$ **93.** $-x - 1$ **95.** −2

97. $-3y + 3$ **99.** $-2x^2 - 6xy + 2y^2 + 4xy^2$

101. $-x^{2a} - 2x^a - 5$ **103.** $x^2 + 4x - 2$

105. $3x^2 - 11x - 16$ **107. (a)** 1600   **(b)** 1000

(c) 232   **(d)** 0   **(e)** 0   **(f)** $40 \le t \le 100$

109. (a) \$2012   **(b)** \$9710   **(c)** $T(x) = x^2 + 17x + 22$

(d) \$11,722   **(f)** \$20,800   **(g)** $P(x) = x^2 - 9x - 22$

(h) \$9078   **111.** $0.04t^2 + 0.49t + 34$ **113.** Equal

115. $a = -1$, $b = 1$, $c = 4$

117. $a = 0$, $b = 6$, $c = -9$

119. When we simplify the left side, we obtain $(a + b)x^2 - 2 = 4x^2 - 2$, which means that $a + b = 4$. There are infinitely many values of a and b such that $a + b = 4$.

121. $-a^2 - 5a - 3$, $-4a^2 + 10a - 3$

123. $2ab^2 - 4ab$ **125.** 2, 3, 5

Section 7.2 *(page 432)*

1. Statement (ii) is false. For example, the product $(x + 2)(x - 2)$ is $x^2 - 4$, which is a binomial.

3. $6x^3 - 9x^2$ **5.** $-8x^5 - 20x^3$

7. $-x^3y^2 + x^2y^3 + 2xy^2$ **9.** $x^{n+1} + 5x^n$

11. $x^{2n+2} - 4x^{n+2} - 3x^2$ **13.** $x^2 - 2x - 24$

15. $y^2 - 9y + 14$ **17.** $-y^2 - 2y + 15$

19. $20x^2 - 9x - 18$ **21.** $9x^2 + 29x + 6$

23. $-10y^2 + 3y + 4$ **25.** $x^2 - 13xy + 36y^2$

27. $a^2 + 3ab - 10b^2$ **29.** $12x^2 - 28xy + 15y^2$

31. $6x^4 - 13x^2 + 6$ **33.** $3x^2 - \frac{3}{4}$

35. $x^3 - 2x^2 - 3x + 10$ **37.** $6x^3 - 7x^2 - 5x + 3$

39. $x^3 + y^3$ **41.** $4x^4 - 4x^3 + 7x^2 - 11x + 4$

43. $x^4 - 2x^2 + 1$ **45.** $x^3 - 8$

47. $2x^4 - 8x^3 + 19x^2 - 22x + 15$

49. Yes, but recognizing the special product saves the step of determining the inner and outer products.

51. $x^2 + 6x + 9$ **53.** $4 - 4y + y^2$

55. $49x^2 - 42x + 9$ **57.** $9a^2 - 6ab + b^2$

59. $4x^2 + 20xy + 25y^2$ **61.** $x^2 - 16$ **63.** $49 - y^2$

65. $4x^2 - 9$ **67.** $16x^2 - 25y^2$ **69.** $81n^2 - 16m^2$

71. The Associative Property of Multiplication guarantees that either grouping is correct. Method (i) is easier because it can be done in one step.

73. $\frac{4}{9}x^2 - 9$ **75.** $a^6 - b^6$

77. $16 + 4p^2q - 4pq^2 - p^3q^3$ **79.** $\frac{1}{4}x^2 + x + 1$

81. $x^4y^4 + 4x^2y^2 + 4$ **83.** $12x^5 - 26x^4 - 10x^3$

85. $-19x - 18$ **87.** $2b^3 - 5b^2 - 13b + 30$

89. $-20x - 50$ **91.** −40   **93.** $x^2 + x - 18$

95. $12x$ **97.** $-6x$ **99.** $32x^5 - 48x^4 + 18x^3$

101. $5y^2 + 20y - 29$ **103.** $a = 4$, $b = -2$

105. $a = 3$, $b = 2$ **107.** $2x^2 + 3x - 9$

109. $300w - 2w^2$ **111.** $9x^2 - 30x + 25$

113. $5x^2 + 9x + 4$ **115.** $x^3 + x^2 - 10x - 16$

117. \$2.31941 billion   **119.** \$2.349227 billion

121. $x^{2n} + 6x^n + 8$ **123.** $x^{2n} - 4x^n + 4$

125. $x^{2n} - 16$ **127.** $x^{2n+3} - x^n$

131. The Associative Property of Multiplication guarantees that the two expressions are equivalent. The grouping in (i) leads to the product of a binomial and a trinomial. The grouping in (ii) is easier because it leads to the product of two binomials.

133. $[(2x + 1)(2x - 1)](4x^2 + 1) = (4x^2 - 1)(4x^2 + 1)$
$$= 16x^4 - 1$$

135. $[(x + y) + 2]^2 = (x + y)^2 + 4(x + y) + 4$
$$= x^2 + 2xy + y^2 + 4x + 4y + 4$$

137. $[(x^2 + 5x) + 3][(x^2 + 5x) - 3]$
$$= (x^2 + 5x)^2 - 9 = x^4 + 10x^3 + 25x^2 - 9$$

139. $4y^2 - 4xy + 12y + x^2 - 6x + 9$

141. $x^2 - 6x + 9 - 25y^2$

97. $4a^2b^2(x - 4)(b + 3a)$ **99.** $5x^3(b - 2)^3(3x + 1)$

101. $A = P(1 + r)$ **103.** $A = 2\pi r(h + r)$

105. $T(t) =$
$$(1380.65 + 61.44t)(21.31t^2 + 312.68t + 5511.35)$$

107. Both the number of travelers and their expenditures increased.

109. Use the Distributive Property to multiply $x^{-6}(3x^2 + 2)$.

111. $2x - 5$ **113.** $3y - x^2$ **115.** $2x^{-4}(x^7 - 5)$

117. $5x^{-4}y^{-2}(y^9 - 3x^7)$ **119.** $a^n(a^n + 2)$

121. $4c^n(2c^{n+3} - c^2 + 3)$ **123.** $(a + 3)^n[(a + 3)^n - 1]$

125. $(y + 1)(x + 1)(x + y)$

127. $2n(2n + 1)$; $2n$ is even and $2n + 1$ is odd.

Section 7.3   *(page 441)*

1. In $2x + 3y$, $2x$ and $3y$ are terms. In $(2x)(3y)$, they are factors.

3. 12   **5.** 4   **7.** x^3 **9.** mn^3 **11.** $2x - 1$ **13.** 5

15. $-xy(y - 2x)$ **17.** $3(x + 4)$ **19.** $5(x + 2y - 6)$

21. $4x(3x - 1)$ **23.** Prime   **25.** $x^2(x - 1)$

27. $3x(x^2 - 3x + 4)$ **29.** $xy^2(x^4y^2 - x^3y + 1)$

31. $(a + by)(3x - 2y)$ **33.** $(2x + 3)(4x^2 - 7y^2)$

35. $(x + 5)(x - 4)$

37. All are correct because the product in each case is $4x - 12$.

39. $1(x - 3)$, $-1(3 - x)$

41. $1(6 + 3x - x^2)$, $-1(x^2 - 3x - 6)$

43. $-2y^2(2y + 1)$, $2y^2(-2y - 1)$

45. $(y - 3)(y - 2)$ **47.** $(3 - z)(5 + z)$

49. $(x - 4)(2x + 5)$

51. The answer is obtained from multiplying the two factors: $(2x - 5)(x + 3) = 2x^2 + 6x - 5x - 15$.

53. $(b + y)(a + x)$ **55.** $(d - 4)(c + 3)$ **57.** Prime

59. $(x^2 + y)(a + 3)$ **61.** $(3x - 4y)(a + 2b)$

63. $(x^2 + 3)(x + 2)$ **65.** $3(y^2 + 4)(y + 3)$

67. $(a^2 + 1)(b + y)$ **69.** $(x^2 + 3)(x - 5)$

71. $(3x^2 + 4)(2x - 3)$

73. It is not necessary to start over. Just factor the GCF from the first factor: $x^2(x + 1)(x^4 + 1)$.

75. $3(2x^2 - 5)(3x + 1)$ **77.** $x(3y^2 + 7)(y - 2)$

79. $4y - 5$ **81.** $3x^2 + 2x + 6$ **83.** $\frac{1}{3}(x + 15)$

85. $\frac{1}{12}(2x^2 + 3x + 36)$

87. Choose the smallest exponent with each variable.

89. $d^{15}(d^5 - 1)$ **91.** $11x^{48}(8 + x^3)$

93. $6ab^2(4a^4b - b^2 + 2a^2)$

95. $12x^2y^4z^3(4x - 5y^2z^5 + 6x^2yz^4)$

Section 7.4   *(page 447)*

1. A difference of two squares consists of two perfect square terms separated by a minus sign.

3. $(x + 3)(x - 3)$ **5.** $(1 + 2x)(1 - 2x)$

7. $(4x + 5y)(4x - 5y)$ **9.** $(\frac{1}{2}x + 1)(\frac{1}{2}x - 1)$

11. $(a^3 + b^8)(a^3 - b^8)$ **13.** $(4x^n + 5)(4x^n - 5)$

15. A perfect square trinomial is a trinomial whose first and last terms are perfect squares and whose middle term is twice the product of the square roots of the first and last terms.

17. $(a - 2)^2$ **19.** $(2x + 3)^2$ **21.** $(x - 7y)^2$

23. $(x + \frac{1}{2})^2$ **25.** $(x^2 - 5)^2$ **27.** $(x^n - 7)^2$

29. A difference of two cubes consists of two perfect cube terms separated by a minus sign.

31. $(x + 1)(x^2 - x + 1)$ **33.** $(3y - 2)(9y^2 + 6y + 4)$

35. $(5 - 2x)(25 + 10x + 4x^2)$

37. $(3x + y)(9x^2 - 3xy + y^2)$ **39.** $(x^2 + 1)(x^4 - x^2 + 1)$

41. $(x^a - 3)(x^{2a} + 3x^a + 9)$ **43.** $(x + 1)^2$

45. $(0.5x + 0.7)(0.5x - 0.7)$ **47.** $(x - 1)(x^2 + x + 1)$

49. $(c - 10)^2$ **51.** $2(x + 3)(x^2 - 3x + 9)$ **53.** Prime

55. $(6 - 7y)^2$ **57.** $4(x + 4)(x - 4)$

59. $(5x - 4y)^2$ **61.** $(x^2 + 4)(x + 2)(x - 2)$

63. $(2 + 5x^2)(4 - 10x^2 + 25x^4)$

65. $(0.2a - 0.1)^2$ **67.** $(x^n + y^{2m})(x^n - y^{2m})$ **69.** $(y^4 - 3)^2$

71. $(6x - y)(36x^2 + 6xy + y^2)$ **73.** $(x - 2)^2(x^2 + 2x + 4)^2$

75. $(x^6 + 2y^{10})(x^6 - 2y^{10})$ **77.** $(x^3 - 2)^2$

79. $2(7 + y^3)(7 - y^3)$ **81.** $x^2(5xy - 1)(25x^2y^2 + 5xy + 1)$

83. $x^3(7x^2 - 5)^2$ **85.** Prime

87. $(4 + 3xy^3)(16 - 12xy^3 + 9x^2y^6)$ **89.** $(x + 2)^2(x - 2)^2$

91. $(4x^2 + 5y)(4x^2 - 5y)$ **93.** $(a^7 + b^3)(a^{14} - a^7b^3 + b^6)$

95. $(8x^2 + 5y)^2$ **97.** $(4a^2 + 9d^2)(2a + 3d)(2a - 3d)$

99. $2x(x + 2)(x^2 - 2x + 4)$

101. $(5x^2 - 3y^4)(25x^4 + 15x^2y^4 + 9y^8)$

103. $4(a + 2b)(x + y)(x - y)$

105. $(a + b)(x + y)(x - y)$　**107.** $-2(a + 1)(a - 1)$

109. $4(x - 2)^2$　**111.** $9(2x - 1)^2$　**113.** $(x + 2)^2(x - 2)^2$

115. Because $x^6 - y^6 = (x^3)^2 - (y^3)^2$, the expression is a difference of two squares. Because $x^6 - y^6 = (x^2)^3 - (y^2)^3$, the expression is also a difference of two cubes.

117. $(x + 2)(x - 2)(x^2 - 2x + 4)(x^2 + 2x + 4)$

119. $(a + b)(a - b)(a^2 + ab + b^2)(a^2 - ab + b^2)$

121. 6　**123.** 1　**125.** 16

127. $(x^2 + 4x + 4) + 1 = (x + 2)^2 + (1)^2$

129. $(x^2 - 2x + 1) + (9) = (x - 1)^2 + (3)^2$

131. $A = \pi R^2 - \pi r^2; A = \pi(R + r)(R - r)$

133. $\frac{1}{2}mv_1{}^2 - \frac{1}{2}mv_2{}^2; \frac{1}{2}m(v_1 + v_2)(v_1 - v_2)$

135. $S(t) = -32.5t^2 + 234.5t + 141$

137. No, the tennis shoe sales could have risen.

139. $(a + b + 2)(a + b - 2)$

141. $(4 + x - y)(4 - x + y)$　**143.** $12y$

145. $(3 + 2x + 3y)[(3 + 2x)^2 - 3y(3 + 2x) + 9y^2]$

147. $(3y - 4)(21y^2 + 24y + 16)$

149. $(x + 1)(x - 1)(x + 2)(x^2 - 2x + 4)$

151. $(x - 2)(x + 2)^2(x^2 - 2x + 4)$

153. $(a + b)(x - y)(x^2 + xy + y^2)$

155. $(y + 3x - 2)(y - 3x + 2)$　**157.** $x^n(x^n + 2)(x^n - 2)$

159. $2(5x^{3a-2} + y^{5b+3})(5x^{3a-2} - y^{5b+3})$

161. $5x^4(x^{3n+2} + 2)(x^{3n+2} - 2)$

163. $(x^a + y^{2a})(x^{2a} - x^ay^{2a} + y^{4a})$

Section 7.5　*(page 458)*

1. The Commutative Property of Multiplication allows the factors to be written in either order.

3. $(x + 1)(x + 2)$　**5.** $(x - 3)(x - 6)$　**7.** $(x + 2)(x + 7)$

9. Prime　**11.** $(x - 8y)(x + 2y)$　**13.** $(2x + 1)(x + 2)$

15. $(2 + x)(6 - x)$　**17.** $(3x + 1)(x - 3)$

19. $(x - 5)(4x + 3)$　**21.** Prime　**23.** $(c + 2)(c - 1)$

25. $(6x - 5)(x + 3)$　**27.** $(x + 8)(x - 3)$

29. $(4 + x)(6 - x)$　**31.** $(b - 5)(b + 3)$

33. $(3x + 2)(x + 1)$　**35.** $(3x - 5)(2x - 3)$

37. $(x + 5)^2$　**39.** $(2 - 5x)(1 + 3x)$

41. $(9 - 10x)(2 + x)$　**43.** $(a - 6)(a - 4)$

45. $(3x + 2)(3x - 5)$　**47.** $(x - 7)(x - 9)$

49. $9(x - 2)(x + 1)$　**51.** $(2x + 3)(6x - 1)$

53. $(4x - 7y)(2x + 3y)$　**55.** $2(x + 3)(x - 2)$

57. $4(x + 3)^2$

59. There are fewer combinations to try because 17 and 23 have fewer factors than 24 and -18.

61. $8xy(x - 5y)(x + 2y)$　**63.** $-3x(2x - 5)(2x + 3)$

65. $2y(2x - 3)(7x + 3)$　**67.** $(ab - 5)(ab + 2)$

69. $xy(x + 6)^2$　**71.** $(xy - 8)(xy + 6)$　**73.** $2(3x - y)^2$

75. $(6m - n)(m + n)$　**77.** $2xy^2(4x - y)(2x - y)$

79. $(5x + 3)(x - 2)$　**81.** $(5 + 9y)(5 - 9y)$

83. $(x^2 - 3)(2x + 1)$　**85.** $(x - 2)(3x + 5y)$

87. $(5x + 1)(25x^2 - 5x + 1)$　**89.** $(x + 2)(x + 4)$

91. Prime　**93.** $(2x + 5)^2$　**95.** $(a - b)(x - 3)$

97. $25(y - 1)(y^2 + y + 1)$

99. $(y^2 + 4x^2)(y + 2x)(y - 2x)$　**101.** $a^2b^4(b^3 + 3ab - 1)$

103. $(3x + 1)(7x - 2)$　**105.** $(4 - a)^2$

107. $2(5 + 3x^2)(5 - 3x^2)$　**109.** $(y + 5)(2 - 3x)$

111. $4x^2(3x - 2)(2x + 3)$

113. $(c^{10} + d^4)(c^5 + d^2)(c^5 - d^2)$　**115.** Prime

117. $(a + 2)(a - 2)(x - y)(x^2 + xy + y^2)$

119. $a^5b^3(b - a^2b^5 + a)$　**121.** $(m - 8)(m - 2)$

123. $4(x^2 + 4y^2)(x + 2y)(x - 2y)$

125. $(a + b - 2)(a^2 + 2ab + b^2 + 2a + 2b + 4)$

127. $5(2x - 3)^2$　**129.** $2a^4(3a - 5)(5a - 9)$

131. $(c - 1)(c^2 + c + 1)(c + 2)(c - 2)$

133. $(x^2 + 11y^2)(x + 3y)(x - 3y)$

135. $x(x - 1)(x - 3)(x + 14)$

137. $(x + 5)(x - 5)(x + 3)(x - 3)(x + 1)(x - 1)$

139. $(xy + 1)(2xy - 3)(xy + 3)$

141. No, the factor $x^2 - 4$ can be factored into $(x + 2)(x - 2)$.　**143.** $(x^2 + 30)(x^2 - 2)$

145. $(x^3 + 5)(x^3 + 12)$　**147.** $(x + 3)(x - 3)(x + 2)(x - 2)$

149. $(5x^4 - 7)(x^4 + 2)$

151. $(2x + 1)(2x - 1)(x + 3)(x - 3)$

153. $(3x + 2)(3x - 2)(x + 1)(x - 1)$

155. $(x^n + 8)(x^n + 3)$　**157.** $(y^{2n} - 11)(y^n + 3)(y^n - 3)$

159. $(5x^n - 3)(x^n + 1)$　**161.** $(x + 3)(x + 6)$

163. $2(x - 3)(2x - 7)$

165. $(2x^2 + 4x - 3)(x^2 + 2x + 2)$　**167.** $x(x + 2)(x + 1)$

169. $(x + 8)(x - 3)$　**171.** $f(t) = 52(3.55t + 56.49)$

173. For both functions the value is $3675.88. This differs from data by $15.08.

175. $-7, 7, -11, 11$　**177.** $-1, 1, -4, 4, -11, 11$

179. $-8, 8, -16, 16$　**181.** $-5, 5, -13, 13$　**183.** 2

185. $x^{2n}(3x^n + 2)(x^n + 1)$

187. $3x^2 + 12x + 20 = x^2 + (x + 2)^2 + (x + 4)^2$

Section 7.6   *(page 467)*

1. Produce the graph and estimate the x-intercepts.

3. (a) $6, -2$ **(b)** No solution   **(c)** 2 **(d)** $8, -4$

5. The equation $3 = 0$ has no solution. The other two equations lead to solutions 1 and -2.

7. $-3, 4$ **9.** $0, 6$ **11.** 5 **13.** $-5, -1, 3$

15. Assuming that the equation has only one variable, the largest exponent on the variable is 1 for a first-degree equation and 2 for a second-degree equation.

17. $-7, 0$ **19.** $-3, 3$ **21.** $2, 3$ **23.** $-1, 2.5$

25. The right side of the equation is not 0.

27. $0, 5$ **29.** $-3, 7$ **31.** $-6, 1$ **33.** $5, 7$ **35.** $1.5, 4$

37. $-2, \frac{4}{9}$ **39.** $-6, 5$ **41.** $2, 30$ **43.** $-5, 1$

45. $-4, 2$ **47.** $2.5, 2$ **49.** $-3, 4$

51. $-4, 1.5$ **53.** $0, 1.25$

55. The value of c for which $p(c) = 0$ is -1. The quadratic factor cannot be factored. The graph has no x-intercepts.

57. $-6, 3, 5$ **59.** $-3, 0, 3$ **61.** $-1, 0, \frac{2}{3}$ **63.** $-1, 1, 5$

65. $-2, 1.5, 2$ **67.** $-3, -2, 2, 3$ **69.** $2, \frac{5}{2}, \frac{5}{3}$

71. $-3, 2, 3$ **73.** $b = 5; -0.5$ **75.** $c = 6; -3.5$

77. $x^2 + x - 12 = 0$ **79.** $x^2 - 8x + 16 = 0$

81. $2x^2 - x = 0$ **83.** $-c, 2c$ **85.** $-\dfrac{3y}{2}, \dfrac{y}{2}$

87. $-3a, a$ **89.** $-0.65, 4.65, 1, 3$ **91.** $-0.5, 2$

93. $-0.5, 0, 0.83, 1.33$ **95.** $-6, -3, -2, 1$

97. $-9, -7, -3, -1$ **99.** $-1, 1, 2, 4$ **101.** $-9, 8$

103. $3, 4, 5; -3, -4, -5$ **105.** $0, 2, 3$ **107.** $-2, 5$

109. 16 **111.** 22 meters, 27 meters   **113.** 2 feet

115. 8 inches, 15 inches, 17 inches

117. $[X: 0, 25, 1]; [Y: 0, 3000, 500]$

119. The model is more reasonably accurate for 1990. For 1970, the model indicates a negative value for V.

121. $x^2 - (a + b)x + ab = 0$

123. $x = 1, y =$ any real number   **125.** $-2, 1$

Section 7.7   *(page 475)*

1. Dividend: $x + 7$ Divisor: x
Quotient: 1 Remainder: 7

3. $\dfrac{3x^2}{2}$ **5.** $2x^2 + 4x$ **7.** $2x^3 + 3x + \frac{1}{2}$

9. $8a^3 + 6a^2 - 3a + 4$ **11.** $5x^3 - 3x^2 + \dfrac{3}{2} - \dfrac{2}{x}$

13. $5c^3d^5 + 2c^2d^3 - \dfrac{3d}{c}$ **15.** $9x^2 + 7x - 6$ **17.** 8

19. $4x^{2n} - 16$ **21.** $x^n - 1$

23. Stop dividing when the degree of the remainder is less than the degree of the divisor.

25. $Q(x) = 2$
$R(x) = 7$

27. $Q(x) = x + 5$
$R(x) = 0$

29. $Q(x) = x + 2$
$R(x) = -1$

31. $Q(x) = 3x + 14$
$R(x) = 44$

33. $Q(x) = 3x + 3$
$R(x) = 5$

35. $Q(a) = a^2 - a + 2$
$R(a) = -2$

37. $Q(a) = 2a^2 + 3a + 2$
$R(a) = 7$

39. $Q(x) = 3x^2 + 4x + 1$
$R(x) = 0$

41. Like terms are not lined up in columns. The missing term $0x^2$ should have been included in the dividend.

43. $Q(x) = 1$
$R(x) = 4$

45. $Q(x) = 1$
$R(x) = x - 4$

47. $Q(a) = a^2 + 2a$
$R(a) = a - 3$

49. $Q = x + 6y$
$R = 15y^2$

51. $Q = 4x^2 - 4xy + 7y^2$
$R = -5y^3$

53. $Q(x) = 2x - 13$
$R(x) = 21$

55. $Q(x) = 9x + 15$
$R(x) = 16$

57. $2x^2 + 5x - 1$ **59.** $x + 2$

61. The first expression is the dividend and the second expression is the divisor, so we mean divide $x + 2$ by $x^2 + 5x - 3$.

63. $Q(x) = x + 3$
$R(x) = 11x - 3$

65. $-x + 8$ **67.** $9x^5 + 12x^4 + 3x^3 - 12x^2$

69. $x + 8; 11$ **71. (a)** $x^2 - 8x - 1$ **(b)** 2 feet

73. For $N(t) \div S(t)$, the quotient is $-0.2t + 10.1$.

75. $Q(t)$ is remaining relatively constant. Because the slope is close to 0, the graph is nearly horizontal.

77. -8 **79.** 2

81. $Q(x) = x^{3n} - x^n + 5$
$R(x) = -14$

83. 2

Section 7.8   *(page 481)*

1. Write the divisor as $x - (-3)$. Then the divider is -3.

3. $Q(x) = 3x + 2$
$R(x) = 0$

5. $Q(x) = x + 4$
$R(x) = 15$

7. $Q(x) = 6x + 4$
$R(x) = 3$

9. $Q(x) = 3x - 6$
$R(x) = 0$

11. $Q(x) = 2x^2 + x - 5$
   $R(x) = 12$

13. $Q(x) = 2x^3 + x^2 + 3x + 3$
   $R(x) = 10$

15. $Q(x) = x^3 - 2x^2 + 5x - 10$
   $R(x) = 6$

17. $Q(x) = x^2 + 4x + 5$
   $R(x) = 4$

19. $Q(x) = x^4 - 2x^3 + 2x^2 - 4x + 9$
   $R(x) = -13$

21. $3x^6 + x^2 + x - 3 =$
   $(3x^5 - 3x^4 + 3x^3 - 3x^2 + 4x - 3)(x + 1) + 0$

23. $x^5 + 32 = (x^4 - 2x^3 + 4x^2 - 8x + 16)(x + 2) + 0$

25. $x^4 + 16 = (x^3 - 2x^2 + 4x - 8)(x + 2) + 32$

27. $a = 1; b = -2; c = 0; d = 1; e = -2$

29. Dividend: $x^3 + 3x^2 - 2x - 4$; divisor: $x + 1$

31. $Q(x) = x^6 + 2x^5 + 2$

33. $Q(x) = x^3 - x^2 + 2x - 2$
   $R(x) = 4$

35. No, the divisor cannot be second degree. It must be in the form $x - c$.

37. -3 **39.** -27 **41.** 12 **43.** 4

Section 7.9   *(page 488)*

1. The remainder is 5 according to the Remainder Theorem.

3. -2 **5.** -15 **7.** 9 **9.** -331 **11.** -1.125

13. We say that a is a root or solution of the equation because when x is replaced with a, the equation is true.

27. $(x^2 - 3x + 4)(x + 3)$ **29.** $(2x^2 + 3x + 6)(x + 2)$

31. $(2x^2 - 4x - 6)(x - \frac{1}{2})$ **33.** $(x^3 - x + 2)(x - 4)$

35. Divide the polynomial by $x - 1$ and factor the resulting second-degree polynomial.

37. $(x + 3)^2(x + 2)$ **39.** $(x + 5)(x - 4)(x - 1)$

41. $(x - 1)(2x - 1)(x + 2)$ **43.** $(x - 3)(x + 2)(4x - 3)$

45. $(x + 2)(4x - 3)(2x - 1)$

47. $(3x - 2)(x - 3)(x + 1)$

49. $(x + 1)(x - 2)(x - 3)$

51. $(x - 3)(x^2 - x + 2)$ **53.** $(x + 1)(x - 2)(2x - 3)$

55. $(x + 1)(3x - 2)(2x - 3)$

57. $(x + 3)(x - 1)(3x + 2)$

59. $(x + 2)(x + 1)(x - 3)(x + 3)$

61. $(x + 2)(x - 1)(x - 2)(x^2 + 1)$

63. Three factors are $(x - a)$, $(x - b)$, and $(x - c)$. This does not imply that $P(x)$ is of degree 3 because one or more factors might be repeated.

65. $x^2 - x - 6$ **67.** $x^3 + x^2 - 12x$ **69.** $x - 3$

71. A solution is c because $x - c$ is a factor if and only if $P(c) = 0$.

73. $-3, -2, 1, 3$ **75.** $-3, -1, 1, 3$

77. Evaluate the polynomial at -1. The remainder is 1.

79. $L = x + 3, W = x + 1$

81. $(x + 3)(x - 1)(x - 3)(x - 5)$

83. $t^2 - 8.30t - 87.68$

85. $-6.09, 14.39$; 1974, 1994
   The positive value of t would indicate the number of years since 1980 when the circulation would be 0.

87. -1 **89.** $1, 6$ **91.** $-2, 2$

93. Positive odd integer n

95. No, $c^4 + c^2 + 1$ is never 0 for any real number c.

Chapter Review Exercises   *(page 491)*

1. (a) Binomial   **(b)** 2   **(c)** $-2x^2 + 3x$ **(d)** -2

3. (a) Trinomial   **(b)** 5   **(c)** $9x^5 + x^2 - 3$ **(d)** 1

5. 6   **7.** 2   **9.** -55 **11.** $2x^2 - 8x + 13$

13. $3x^2 - 2x + 7$ **15.** $-8x^8$ **17.** $6x^5 - 8x^3$

19. $10a^2b - 2a^2b^3 + 6a^2b^2$ **21.** $x^2 - x - 12$

23. $-6x^2 + 7x + 24$ **25.** $2x^3 + x^2 - 5x + 2$

27. $-x^2 - 12x - 19$ **29.** $8x^2 - 22x + 15$

31. A difference of two squares consists of two perfect square terms separated by a minus sign.

33. $9x^2 - 4$ **35.** $9y^2 - 12y + 4$

37. $9 - 6nm^2 + n^2m^4$ **39.** $8y^3 + 12y^2 + 6y + 1$

41. $x^2 - x - 6$ **43.** $x^3 - x$

45. Factor out the common factor $-x^2y^2$.

47. $x^{25}(1 + x^5)(1 - x^5)$ **49.** $x^3y^3(x^3 - x^4y - y^2)$

51. $(x^2 - 2)(x + 3)$ **53.** $x^n(2x^4 - 5x^2 + 6)$

57. The constant term is not a perfect square.

59. $(7x + 4y)(7x - 4y)$ **61.** $(4 + a^2)(2 + a)(2 - a)$

63. $(3b^2 + 2a)(9b^4 - 6ab^2 + 4a^2)$ **65.** $(c + 8)^2$

69. $(x - 6)(x + 1)$ **71.** $2(x + 7)(x - 1)$

73. Because the numbers 6 and 24 have several factors, they lead to a large number of possible factorizations to try.

75. $(3x + 5)(x + 3)$ **77.** $(6x - 5)(2x + 3)$

79. $(x + 4)(x - 2)$ **81.** $(5x - 4)(25x^2 + 20x + 16)$

83. No, but you can factor out the 2 to obtain $[2(x + 2y)]^2 = 4(x + 2y)^2$.

85. $(3x - 5)(x + 3)$ **87.** $(x^3 + 3y^2)(x^6 - 3x^3y^2 + 9y^4)$

89. $x^3y^3(y - x)$ **91.** $x(x + 3)(6x + 7)$

93. In $a + b = 0$, we know that a and b are opposites, but we do not know what the numbers are. In $ab = 0$, either a or b must be 0.

95. $0, 8$ **97.** $-7, 6$ **99.** $-6, 4$ **101.** $-3, -2, 2$

103. $-3, \frac{2}{3}, -2, -\frac{1}{3}$ **105.** $-9, 7$

107. If the degree of P is 1 or greater, the degree of the quotient is 1 less than the degree of P.

109. $4x^2y^3$

111. $Q(x) = 3x^2 - 4x - 5$ **113.** $Q = x^2 + 3y^2$
$R(x) = -5$ $R = 8y^3$

115. $P(x) = 2x^3 + x^2 + 4$

117. Writing the divisor as $x - (-7)$, we see that the divider is -7.

119. $Q(x) = 2x^2 - x - 1$
$R(x) = 4$

121. $Q(x) = x^4 + 2x^3 + 4x^2 + 8x + 16$
$R(x) = 0$

123. $a = 1, b = -2, c = -1, d = 7$

125. The value of $P(2)$ is the remainder when $P(x)$ is divided by $x - 2$. Therefore, $P(2) = 3$.

127. 3

129. The graph of P has at least three x-intercepts: $(-4, 0)$, $(2, 0)$, and $(5, 0)$. If $P(x)$ is divided by $x - c$ and the remainder is zero, then $P(c) = 0$ and $(c, 0)$ is an x-intercept.

131. $(x + 1)(x - 3)(x - 2)$

Chapter Test   *(page 495)*

1. *(7.1)* -16

2. *(7.1)* **(a)** $(p + q)(x) = 3x^4 + x^3 - 5x^2 - 2x - 5$
$(p + q)(1) = -8$
$p(1) + q(1) = -7 + (-1) = -8$
(b) $(p - q)(x) = 3x^4 - 9x^3 - 5x^2 + 14x - 9$
$(p - q)(-1) = -16$
$p(-1) - q(-1) = -11 - 5 = -16$

3. *(7.2)* $-2x^5 + 10x^4 - 6x^3$

4. *(7.2)* $6t^2 + 11st + 4s^2$

5. *(7.2)* $x^3 - 11x + 6$

6. *(7.2)* $x^4 + 6x^2y + 9y^2$

7. *(7.2)* $25x^2 - 49z^2$

8. *(7.2)* $-6a^5b^6$

9. *(7.2)* $P(x) = x^5 + x^3 - x^2 - 1$; No portion of the graph is in Quadrant II.

10. *(7.7)* $P(x) = 2x^3 + 5x^2 - 1$
$2x^3 + 5x^2 - 1 = (2x^2 - x + 3)(x + 3) - 10$

11. *(7.7)* $Q(x) = 3x^2 - 8x + 21$
$R(x) = -49$

12. *(7.7)* $Q(x) = -5x^5y^2 + 7x^4y^6$
$R(x) = 0$

13. *(7.7)* $Q(x) = x + 3$
$R(x) = -2x + 4$

14. *(7.7)* $Q(x) = x^4 - 2x^3 + 4x^2 - 8x + 16$
$R(x) = -33$

15. *(7.8)* The number k is the remainder after dividing $P(x) = ax^3 + bx^2 + cx + d$ by $x + 2$.

17. *(7.3)* $6r^3s^4(3s - 2r^2)$

18. *(7.3)* $(y - 1)(x + 3)$

19. *(7.4)* $3(3 + 2x)(3 - 2x)$

20. *(7.4)* $(2y - 5x)^2$

21. *(7.4)* $(1 - 2x)(1 + 2x + 4x^2)$

22. *(7.5)* $(3x + 4)(2x + 1)$

23. *(7.9)* $(x + 3)(x - 2)(x - 4)$

24. *(7.9)* $(x - 2)(x - 7)(x + 5)$

25. *(7.6)* $0, \frac{4}{3}$

26. *(7.6)* $-3, 0.4$

27. *(7.6)* $-2, 0.5, 2$

28. *(7.6)* $-1, 0.5, 1, 2.5$

30. *(7.6)* 7 feet, 10 feet

31. *(7.6)* 100 feet, 120 feet

Chapter 8

Section 8.1   *(page 504)*

1. Any number that makes the denominator zero is excluded from the domain.

3. $x \neq 5$ **5.** $0.5, 0$, undefined   **7.** $-0.875, 0$, undefined

9. $\{x \mid x \neq 4\}$ **11.** Set of all real numbers

13. $\{x \mid x \neq -2\}$ **15.** $\{x \mid x \neq -3, 3\}$

17. $\{x \mid x \neq -4, 2\}$ **19.** Set of all real numbers

21. $x + 2$ **23.** $2x - 3$ **25.** $x^2 - 25$

27. **(a)** \$2 million   **(b)** No, the function is not defined for $x = 100$.   **(c)** 90%; \$10 million

29. $2x, 3, x, 3$ **31.** $2, x + 3, 7, x + 3$

33. The first expression has a factor $(x + 3)$ that is common to the numerator and denominator, so it can be divided out. There is no factor common to the numerator and denominator of the second expression.

35. $\dfrac{1}{8x}$ **37.** $\dfrac{5cd^2}{6}$ **39.** 1 **41.** $\dfrac{2x - 1}{3}$ **43.** 1

45. $x - 9$ **47.** Cannot be simplified   **49.** False

51. False   **53.** True   **55.** $\dfrac{1}{1 + 5x}$ **57.** $\dfrac{x + 4}{x + 3}$

59. $\dfrac{a(a + 1)}{6}$ **61.** $\dfrac{1}{2x + 1}$ **63.** $\dfrac{x - 5}{x - 8}$ **65.** 1

67. $\dfrac{a}{b+3}$ **69.** $\dfrac{1}{x-y}$ **71.** $\dfrac{c-2d}{2c+d}$ **73.** -2 **75.** -5

77. $3-x$

79. (a) Neither (b) 1 (c) -1

81. (a) -1 (b) 1 (c) Neither

83. -1 **85.** $-\dfrac{1}{3}$ **87.** $-(a+7)$ **89.** $\dfrac{-1}{5w+z}$

91. $-\dfrac{x+4}{x+3}$ **93.** $x-3$ **95.** $\dfrac{x^2-2x+4}{x-2}$

97. $\dfrac{3x(3x-2)}{2(2x+3)}$ **99.** $12x^2y$ **101.** 25 **103.** $(x+5)^2$

105. $5(x-3)$ **107.** $-3(x+4)$ **109.** $9(x-5)$

111. $-12t$ **113.** $-1(1+x)(1-x)$

115. $\dfrac{(26)(25)(15.75t^2-48.25t+445.5)}{-643t^2+5441t+43702}$

117. Attendance: 15.4% **119.** $-\dfrac{1}{x^3}$ **121.** x^5
Salaries: 44.8%

123. n even, $\{x \mid x \ne -1, x \ne 1\}$; n odd, $\{x \mid x \ne 1\}$

125. Set of all real numbers **127.** $(3x-1)^n$

129. $x^n - 3$ **131.** $\dfrac{(x+1)^2-(x-2)^2}{x(x+4)-(x+1)^2} = 3$

Section 8.2 *(page 513)*

1. Both methods are correct. It is generally easier to divide out common factors before multiplying, as in part (ii).

3. $\dfrac{1}{5a}$ **5.** $\dfrac{a-4}{3}$ **7.** 1 **9.** $\dfrac{c-6}{20c}$ **11.** $\frac{1}{3}$ **13.** $\dfrac{y+1}{y+8}$

15. $R(x) = 1$ only if $x \ne -2$ and $x \ne 2$. The graph is a horizontal line with holes at $x = -2$ and $x = 2$ because R is undefined for these x-values.

17. $3-x$ **19.** $2x^2-x-1$

21. The opposite of $x+3$ is $-(x+3) = -x-3$. The reciprocal of $x+3$ is $\dfrac{1}{x+3}$.

23. $\dfrac{x^3}{6y^2}$ **25.** $\dfrac{6}{x}$ **27.** $\dfrac{-x(x-2)}{x+2}$ **29.** $\dfrac{x-1}{8x}$

31. $\dfrac{x-6}{x-2}$ **33.** $\dfrac{1}{2x(3x-2)}$ **35.** $6x^2-11x+4$

37. x^3-8 **39.** The quotient is $\dfrac{\frac{A}{B}}{C}$. **41.** $\dfrac{-20x^3}{3y}$

43. $\dfrac{2(x+3)}{x-4}$ **45.** $\dfrac{x+2}{3x(x-1)}$ **47.** $\frac{1}{2}$ **49.** $\dfrac{a+9}{a-9}$

51. $\dfrac{x+1}{x+4}$ **53.** $\dfrac{x}{(2x-3)(2x-5)}$ **55.** 1 **57.** $x+5$

59. ERA $= \dfrac{9R}{I}$ **61.** $\dfrac{12}{x}$ **63.** $\dfrac{x+5}{x-5}$ **65.** $\dfrac{15}{m}$ **67.** $\frac{1}{3}$

69. $-\frac{9}{10}$ **71.** 4 **73.** $3x^2(x-1)$ **75.** 5 **77.** $\dfrac{x-3}{x(x+1)}$

79. Exclude values for which Q, R, or S is 0.

81. $\dfrac{x-9}{48x}$ **83.** $\dfrac{x-4}{2x}$ **85.** $\dfrac{x-3}{7x(x-2)}$ **87.** \$1081.61

89. \$14.34 **91.** $\dfrac{x-2}{x+3}$ **93.** $\dfrac{2x-3}{x+1}$ **95.** $\dfrac{(x+3)^2}{x}$

97. $\dfrac{(x^n+2)(x^m+2)}{2(x^m-1)(x^n-2)}$

99. As x approaches 0, the y-values become larger. By selecting x-values closer and closer to 0, we can make the y-value as large as we want.

Section 8.3 *(page 525)*

1. Add the numerators and retain the common denominator.

3. $\dfrac{8}{a}$ **5.** $\dfrac{-3x-9}{x}$ **7.** 6 **9.** -6 **11.** $\dfrac{3}{x-4}$

13. $x+7$ **15.** $\dfrac{1}{x+2}$ **17.** $\dfrac{9-2x}{2x+3}$ **19.** $\dfrac{1}{x-2}$

21. $x+6$ **23.** $3t-3$ **25.** $2x^2-x$

27. Multiply the numerator and denominator of the second fraction by -1 to obtain

$$\dfrac{15}{x} + \dfrac{-7}{x}.$$

29. 5; $a+5$ **31.** $8x$; $11x+9$ **33.** $\dfrac{9x+3}{7x-8}$

35. $x+4$ **37.** $\dfrac{a^2+b^2}{a-b}$ **39.** $30x^3y^2$ **41.** $6x(x-4)$

43. $-30t(t+5)$ **45.** $(3x-2)(x+1)(4x+1)$

47. $3(x+1)(x+2)(x-2)$

49. Multiply the numerator and denominator by x.

51. $\dfrac{5b+7a^2}{a^3b^2}$ **53.** $\dfrac{x^2-3}{x}$ **55.** $\dfrac{11x-12}{x(x-4)}$

57. $\dfrac{12}{(x-5)(x+7)}$ **59.** $\dfrac{t^2+2t+3}{t+1}$ **61.** $\dfrac{2t+5}{18(t-5)}$

63. $\dfrac{2t-1}{t(t+1)^2}$ **65.** $\dfrac{18x+27}{(x+5)(x+2)(x-2)}$ **67.** $\dfrac{4x+45}{(x+9)^2}$

69. $\dfrac{13t^2+2t-2}{3t^2(2t-1)(t+1)}$ **71.** $\dfrac{1}{2x+1}$ **73.** $\dfrac{x+5}{4(x-9)}$

75. $\dfrac{8x-49}{(x+7)(x+8)(x-8)}$ **77.** $\dfrac{x-12}{x-6}$

79. $\dfrac{x-6}{(x+3)(x-3)}$ **81.** $\dfrac{x+9}{(x+3)(x-1)}$

83. $(x + y)^{-1} = \dfrac{1}{x + y}$ and $x^{-1} + y^{-1} = \dfrac{1}{x} + \dfrac{1}{y}$
The expressions are not equivalent.

85. $\dfrac{y - x}{xy}$ **87.** $\dfrac{1}{x - y}$ **89.** $\dfrac{x^2 + x + 9}{(x + 3)(x - 2)}$ **91. (b)** 8

93. 20,498,000 people; $14,205,000,000

95. $A(5) = 395.91$; $A(10) = 559.75$; $A(15) = 693.29$

97. $\dfrac{-1}{x(x + h)}$ **99.** $\dfrac{x^n - 3}{x^n}$ **101.** $\dfrac{1 - x^2}{1 + x^2}$

Section 8.4 (page 535)

1. Each is the quotient of rational expressions.

3. The complex fraction can be written
$$\frac{w}{w + 1} \div \frac{w^2}{w - 1}.$$

5. $\dfrac{x^2}{3}$ **7.** $\dfrac{1}{3x}$ **9.** $\dfrac{x - 5}{2}$ **11.** $-\dfrac{x + 2}{x - 3}$

13. Add or subtract the fractions in the numerator and denominator. Then multiply the numerator by the reciprocal of the denominator.

15. $\dfrac{4(t + 6)}{3(t + 2)}$ **17.** $\dfrac{1}{y}$ **19.** $\dfrac{2x - 1}{x}$ **21.** -1 **23.** $\dfrac{1}{x}$

25. xy **27.** $\dfrac{-x - 15}{2x^2 - 17}$ **29.** $-\dfrac{2x^2 + 3x - 2}{2x^2 - 5x - 5}$ **31.** $\dfrac{1}{2}$

33. $\dfrac{x - 2}{2x - 1}$ **35.** $\dfrac{x + 2y}{x - 2y}$ **37.** $x + 2$

39. Multiply the numerator and denominator by the LCD.

41. $\dfrac{x + y}{x - 3y}$ **43.** $a - b$ **45.** $\dfrac{x - 1}{(x - 2)(x + 1)}$

47. $\dfrac{a + b}{a^3 b^3}$

49. When we simplify a complex fraction by multiplying the numerator and denominator by the LCD, we are multiplying the fraction by 1, which does not change the value of the expression. Multiplying the given expression by the LCD would change the value of the expression.

51. $\dfrac{1}{a + b}$ **53.** $\dfrac{x^3 + x}{x^4 + 1}$ **55.** $\dfrac{x - 1}{3x - 4}$ **57.** $\dfrac{x^3 + y^4}{x^2 y^3 (x - y)}$

59. $\dfrac{x}{3x + 1}$ **61.** $\dfrac{x + 1}{x^2 + x + 5}$

63. (a) $\dfrac{\dfrac{x}{x + 1}}{\dfrac{x + 1}{x + 2}}$ **(b)** $\dfrac{x^2 + 2x}{x^2 + 2x + 1}$

65. V **67.** 16.2% **69.** $\dfrac{-2}{(2x + 2h - 1)(2x - 1)}$

71. $\dfrac{-3}{(x + h - 2)(x - 2)}$ **73.** $\dfrac{3(a + 1)}{a}$ **75.** $\dfrac{-1}{x(x + h)}$

77. $\dfrac{14x^2 - 9x - 3}{7x - 2}$ **79.** $\dfrac{8}{5}$

Section 8.5 (page 545)

1. We call -3 and 0 restricted values because each one makes a denominator 0.

3. $-3, 0, 4$ **5.** $-3, 0, 3$

7. No, because 2 is a restricted value. **9.** $-3, 1$

11. $-3, -1$ **13.** $-3, -2, 2$ **15.** 20 **17.** 4

19. 7 **21.** No solution **23.** $\dfrac{2}{3}$ **25.** 8

27. The graph does not intersect the x-axis. This means the equation has no solution.

29. $-1.5, 5$ **31.** $-4, 5$ **33.** 3 **35.** $-4, 1$

37. In the first equation, 9 is an extraneous solution. The second equation leads to the equation $-4 = 4$, which is false. Thus, the second equation has no solution.

39. 4 **41.** -5 **43.** No solution **45.** $\dfrac{9}{8}, -\dfrac{1}{2}$

47. $-6, 0.5$ **49.** $-8, 0.5$ **51.** -2 **53.** $-\dfrac{121}{32}$

55. $-\dfrac{10}{7}$ **57.** 2 **59.** No solution **61.** No solution

63. To solve the equation, clear the fractions by multiplying both sides by the LCD. To perform the addition, rewrite each expression with the LCD as the denominator.

65. $\dfrac{x^2 - 5x - 9}{(x - 8)(x - 3)}$ **67.** -1 **69.** $\dfrac{y}{8}$

71. $\dfrac{-2t^2 - 7t + 3}{t + 3}$ **73.** $0, 6$ **75.** $\dfrac{4}{x + 2}$ **77.** $R = \dfrac{E}{I}$

79. $T_1 = \dfrac{P_1 V_1 T_2}{P_2 V_2}$ **81.** $r = \dfrac{I}{Pt}$ **83.** $m_2 = \dfrac{F \mu r^2}{m_1}$

85. $a = \dfrac{S(1 - r)}{1 - r^n}$ **87.** $a = \dfrac{bx}{b - y}$ **89.** $y = -2x - 2$

91. $y = -\dfrac{4}{5}(x + 4)$

93.

t	Year	$A(t)$	Actual Data
1	1987	1.0	1.0
2	1988	1.2	1.3
3	1989	1.5	1.5
4	1990	1.7	1.7
5	1991	1.9	1.9
6	1992	2.0	2.0

95. 0.61; The model does not seem valid beyond 1992 because there is no reason to believe that the amount spent on children's books will start to decline.

97. 7 **99.** $-3, -2, 3$ **101.** $y = m(x - x_1) + y_1$

103. $\dfrac{2a - 6b}{3}$ **105.** $\dfrac{5c}{2}$, $5c$ **107.** -1.25, 0

Section 8.6   *(page 557)*

1. In 1 hour, 1/5 of the job is completed. In 3 hours, 3/5 of the job is completed. In 5 hours, 5/5 (all) of the job is completed. If a job takes h hours, $1/h$ of the job is completed in 1 hour.

3. 15 minutes   **5.** 6 hours

7. Part (ii) illustrates an inverse variation.   **9.** 15

11. 6   **13.** 7.2 hours   **15.** $19.50

17. If we multiply both sides by the LCD, which is bd, we obtain $ad = bc$.

19. 154 miles   **21.** 0.72 minutes

23. In (i) we are asked to find the sum of two fractions. Write the fractions with the LCD and add. In (ii) we are asked to solve an equation. Clear the fractions by multiplying both sides of the equation by the LCD.

25. 3, 0.5   **27.** 10 and 6 or $\frac{2}{3}$ and $-\frac{10}{3}$

29. If the distance d is constant, then an increase in speed r corresponds to a decrease in time t.

31. Plane, 150 mph; train, 60 mph   **33.** 12 mph

35. Less than 1% of the bulbs should be defective. Multiply 2000 by 1% to obtain 20. Fewer than 20 bulbs should be defective.

37. 9   **39.** 1200

41. The following are not inverse relations:
(a) As humidity increases, drying time increases.
(b) As the number of skiers increases, greater demand for equipment results in increased prices.
(d) As accidents increase, so do insurance premiums. Parts (c) and (e) are inverse relations.

43. 60 feet   **45.** 66 minutes   **47.** 60 mph

49. $30,400, $45,600   **51.** 5, 7; -7, -5 **53.** 6 hours

55. 10 mph   **57.** 1190 parkas; 6.25°   **59.** 12   **61.** 83

63. 12 pounds of pecans and 15 pounds of cashews, or 15 pounds of pecans and 18 pounds of cashews

65. 6, 8, 10

67. Square, 0.6 inches by 0.6 inches
Rectangle, 0.6 inches by 5.6 inches

Chapter Review Exercises   *(page 562)*

1. The two numbers are $a_1 = 5$ and $a_2 = -3$. We call these numbers restricted values.

3. $\{x \mid x \ne 2.5\}$

5. In the first expression, 3 is a factor in both the numera-

tor and denominator, so it can be divided out. In the second expression, 3 is not a factor of the numerator or denominator.

7. $\dfrac{x + 4}{2x - 3}$ **9.** $\dfrac{b + 4}{b + 2}$ **11.** $5(x - 4)$ **13.** $\dfrac{8}{x^2}$

15. $\dfrac{x + 1}{x - 2}$ **17.** (a) The reciprocal of their product is $\dfrac{1}{xy}$.

(b) The product of their reciprocals is $\dfrac{1}{x} \cdot \dfrac{1}{y} = \dfrac{1}{xy}$.

(c) The two expressions are equivalent.

19. $\dfrac{6}{x + 2}$ **21.** $\dfrac{x - 1}{x - 4}$ **23.** $\dfrac{2}{x - 2}$

25. $9(x + 7)(x - 7)$

27. If the denominators were not the same, the terms would not have a common factor, and the Distributive Property would not apply.

29. $\dfrac{-4x + 5}{5x - 4}$ **31.** $\dfrac{10x - 37}{(x + 4)(x - 4)(x - 3)}$ **33.** $\dfrac{5x + 4}{x(x + 2)}$

35. $\dfrac{6x + 27}{(x + 8)(x - 6)(x + 5)}$

37. Multiply the numerator by the reciprocal of the denominator or multiply the numerator and denominator by the LCD.

39. $\dfrac{a - 3}{12}$ **43.** $\dfrac{x^2 + 16}{x}$

45. An apparent solution does not check if it makes a denominator zero. It is called an extraneous solution.

47. 0.5, -0.4 **49.** No solution   **51.** 2

53. No solution   **55.** 1 solution   **57.** 24 hours

59. 15 hours   **61.** 54 mph

Chapter Test   *(page 565)*

1. *(8.1)* When $x = 3$, the denominator is zero.

2. *(8.1)* $\{x \mid x \ne 0, -3\}$

3. *(8.1)* $\dfrac{2(x + 2)}{x(x - 2)}$

4. *(8.1)* $\dfrac{x + 5}{x - 9}$

5. *(8.1)* $-\dfrac{x + 2y}{x + y}$

6. *(8.2)* $\dfrac{t + 2}{t + 3}$

7. *(8.2)* $\dfrac{1}{2(y - 5)}$

8. *(8.3)* $3x(x - 2)$

9. *(8.3)* The expression can be written

$$\frac{r}{m+n} - \frac{r-1}{m+n}.$$

Because the denominators are the same, the subtraction is easily performed. The result is

$$\frac{1}{m+n}.$$

10. *(8.3)* 1

11. *(8.3)* $\dfrac{3-5t}{t^2}$

12. *(8.3)* $\dfrac{2x^2 - 16x - 25}{(2x-1)(x+5)(4x+3)}$

13. *(8.4)* All are complex fractions. The last two are equivalent.

14. *(8.4)* $\dfrac{6}{x-3y}$

15. *(8.4)* $\dfrac{x-3}{x+2}$

16. *(8.4)* $\dfrac{b^2}{b^2 - a^2}$

17. *(8.5)* 12

18. *(8.5)* $-2, 5$

19. *(8.5)* No solution

20. *(8.5)* -3.2

21. *(8.5)* In (a) the apparent solution is 1, but it is an extraneous solution. In (b) the equation leads to $-2 = 3$, which has no solution.

22. *(8.5)* $B = \dfrac{A}{2A-1}$

23. *(8.6)* 40%

24. *(8.6)* 2.4 hours

25. *(8.6)* 5 mph

Cumulative Test: Chapters 7–8   *(page 567)*

1. *(7.1)* $x^3 + x^2 + x - 1$

2. *(7.2)* $-6x + 18$

3. *(7.3)* **(a)** $x^3 y^3 (x-y)$ **(b)** $(x+3y)(a-7b)$

4. *(7.4)* **(a)** $(2x+5)(2x-3)$ **(b)** $2(x-3)^2$
       (c) $(y+2)(y^2 - 2y + 4)$

5. *(7.5)* **(a)** $(z+9)(z-7)$ **(b)** $(a-9b)(a+5b)$
       (c) $3(4-x)(1+x)$

6. *(7.6)* After factoring the left side, set each factor equal to 0. Then solve each case for x.

7. *(7.6)* **(a)** $0, 7$ **(b)** $-1, 1, -5, 5$

8. *(7.6)* 8 or -3

9. *(7.7)* $Q(c) = 2c^3 + c^2 + 3c - 2$
      $R(c) = -2$

10. *(7.8)* $x^5 - 3x^3 - 17x^2 + x - 13$
      $= (x^4 + 3x^3 + 6x^2 + x + 4)(x-3) + (-1)$

11. *(7.9)* Divide $P(x)$ by $x - c$. If the remainder is 0, then $x - c$ is a factor.

12. *(7.9)* -3

13. *(8.1)* $\dfrac{x+3}{4x+3}$

14. *(8.2)* $\dfrac{a+4}{a-4}$

15. *(8.3)* $\dfrac{y+12}{(y+10)(y+4)}$

16. *(8.4)* $\dfrac{x}{3}$

17. *(8.5)* -4

18. *(8.6)* $3\frac{3}{7}$ hours

19. *(8.6)* 55 mph

Chapter 9

Section 9.1   *(page 573)*

1. The number 9 has two square roots, -3 and 3, because squaring either 3 or -3 results in 9. The expression $\sqrt{9}$ means the principal or positive square root of 9, which is 3.

3. $6, -6$ **5.** Not a real number   **7.** 3   **9.** $5, -5$

11. -3 **13.** Not a real number   **15.** 4   **17.** -4

19. $\frac{2}{3}$ **21.** Not a real number   **23.** -4 **25.** 20   **27.** 4

29. 4   **31.** 2   **33.** $\frac{1}{27}$ **35.** -4

37. The expression x^3 could be either positive or negative depending on the value of x. The expression x^2 is nonnegative for all values of x.

39. 2, rational   **41.** -3.16, irrational   **43.** 0.25, rational

45. 3, rational   **47.** 2.29, irrational   **49.** Not a real number

51. 5, rational   **53.** 2   **55.** 21   **57.** $2\sqrt{x}$

59. $\sqrt[5]{a} - 9$ **61.** $\sqrt{c} + b$ **63.** $\sqrt[4]{xy}$ **65.** x^5 **67.** $4y^3$

69. x^7 **71.** $-y^5$ **73.** x^7 **75.** $-x^2$ **77.** $(3x)^3$ or $27x^3$

79. $(x+3)^5$ **81.** $|x^3|$ **83.** $-x^4$ **85.** $5|y^7|$ **87.** x^3

89. $|a|$ **91.** $|x+2|$ **93.** $|x^2 - 9|$ **95.** $-2|x-1|$

97. $|3x^3 - 5x|$ **99.** $1 - x$ **101.** $|x-7|$ **103.** $<$

105. $>$ **107.** $>$ **109.** $\sqrt{x^4}$ **111.** $\sqrt[8]{y^{24}}$ **113.** $-3\sqrt[4]{16}$

115. $4\sqrt[5]{x^{10}}$ **117.** 26.4 inches   **119.** 1986, 1990

121. 5.92%   **123.** $|t^n|$ **125.** t^2

127. (a) If we did not exclude $n = 1$, then $\sqrt[1]{b} = a$
would imply that $a^1 = b$, which would mean that
$\sqrt[1]{b} = b$. Thus nothing is gained by including
$n = 1$ in the definition.

(b) For $b = 1$, $\sqrt[0]{b} = a$ implies that $a^0 = b$ or
$a^0 = 1$. But this statement is true for all $a \neq 0$.
Thus, $\sqrt[0]{b}$ would not have a unique value if $b = 1$.
For $b \neq 1$, $\sqrt[0]{b} = a$ implies that $a^0 = b$, which
implies that $a^0 \neq 1$. Because this statement is
false, $\sqrt[0]{b}$ has no value.

Section 9.2   (page 581)

1. Choice (iii) is false because $16^{1/2}$ is the principal square
root 4.

3. 7   **5.** -4 **7.** $\frac{3}{4}$ **9.** Not a real number   **11.** $\frac{2}{3}$

13. -3 **15.** -2 **17.** Not a real number   **19.** -5

21. 1.97   **23.** Not a real number   **25.** -1.41

27. The numerator is the exponent and the denominator is
the index: $9^{2/5} = \left(\sqrt[5]{9}\right)^2$.

29. $\sqrt[4]{10}$ **31.** $3\sqrt[3]{x^2}$ **33.** $\sqrt[3]{(3x)^2}$ **35.** $\sqrt{x + 2}$

37. $\sqrt[3]{(2x + 3)^2}$ **39.** $3\sqrt{x} + \sqrt{2y}$ **41.** $20^{1/5}$ **43.** $y^{2/3}$

45. $(3x^3)^{1/4}$ **47.** $(x^2 + 4)^{1/2}$ **49.** $t^{1/2} - 5^{1/2}$

51. $(2x - 1)^{2/3}$ **53.** $3x^{1/2} + (3y)^{1/2}$ **55.** -2.46

57. -3.01 **59.** 29.93   **61.** -33.76 **63.** 2.25   **65.** $\frac{1}{3}$

67. 3   **69.** 9   **71.** $\frac{1}{9}$ **73.** 8   **75.** Not a real number

77. -8 **79.** $\frac{1}{4}$ **81.** $\frac{9}{4}$ **83.** $\frac{1}{8}$ **85.** $-\frac{1}{16}$

87. Not a real number   **89.** $-\frac{1}{32}$ **91.** 11   **93.** $\frac{5}{6}$ **95.** $\frac{7}{2}$

97. 8   **99.** -49 **101.** 2   **103.** $\frac{1}{3}$ **105.** 6.35

107. 0.72   **109.** 0.53   **111.** 9   **113.** 16   **115.** -8

117. $-\frac{1}{2}$ **119.** $\frac{2}{3}$

121. (a) The graphs intersect at $(0, 0)$ and $(1, 1)$.
If $x = 0$, then $x^{1/n} = 0^{1/n} = 0$ for all $n > 1$. If $x = 1$, then
$x^{1/n} = 1^{1/n} = 1$ for all $n > 1$.   (b) $0 < x < 1$ (c) $x > 1$

123. (a) 2 hours   (b) 4.2 additional hours
(c) 0; Yes. Even if a student does not study for the
final, it is unlikely that the exam grade would be
0. The model indicates that a perfect score can be
earned with about 10 hours of study. This seems
realistic.

125. 10 mph   **127.** 3.24 mph   **129.** $\frac{1}{2}$ **131.** $\frac{1}{2}$

133. The function is not defined if x is an even number.

Section 9.3   (page 586)

1. Choice (i) is true according to the Product Rule and
Power to a Power Rule for Exponents. Choice (ii) can

be shown to be untrue for $a = 3$ and $b = 4$, for
example.

3. 36   **5.** 16   **7.** Not a real number   **9.** $-\frac{5}{3}$ **11.** 6

13. $\frac{2}{3}$ **15.** $\frac{16}{49}$ **17.** 65   **19.** 7   **21.** 2   **23.** 6   **25.** $\frac{1}{3}$

27. $\frac{1}{2}$

29. Choice (i) is false because the exponents were multiplied when they should have been added. Choice (ii) is
true because the Product Rule for Exponents was
applied properly.

31. x **33.** $t^{12/7}$ **35.** $a^{1/4}b^{3/2}$ **37.** $z^{1/6}$ **39.** $a^{1/3}b^{4/3}$

41. $\frac{1}{a}$ **43.** y **45.** a^9b^6 **47.** $\frac{1}{3}$ **49.** $\frac{3}{2}$ **51.** 8   **53.** a^3

55. $a^2|b|$ **57.** $|x + y|$ **59.** $\frac{b^{1/6}}{a^2}$ **61.** $\frac{a^4}{b^6}$ **63.** $\frac{-9x^{3/2}}{y}$

65. $\frac{b^{10/3}}{a^3}$ **67.** $\frac{x^2}{y^3}$ **69.** $\frac{2y^2}{x^3z}$ **71.** ab^4 **73.** $\frac{a^4c^{7/2}}{b^7}$

75. $\frac{y^2}{3x^4}$ **77.** $a^2 + 1$ **79.** $x^2 - x^{5/12}$ **81.** $x - y$

83. $x + 3 - 2(3x)^{1/2}$ **85.** $A(t) = 2.96t^{0.11}$

87. No, 3.6 is not three times 2.96.   **89.** $x^{1/n}$ **91.** $x^{n/6}$

93. $\frac{x^n}{y^{2n}}$ **95.** a^3b^5 **97.** x^{n+1} **99.** $x^{-1/2}(3 + 5x^2)$

101. $5a^{-1/2}(a - 2)$ **103.** $(t^{1/5} + 5)(t^{1/5} - 1)$

105. $(2x^{1/6} + 3)(x^{1/6} - 5)$

Section 9.4   (page 594)

1. In the real number system, every number has a cube
root, but only nonnegative numbers have square roots.

3. $\{x \mid x \geq 2\}$ **5.** $\{x \mid x \geq \frac{3}{2}\}$ **7.** $\{x \mid x \leq 2\}$

9. Set of all real numbers   **11.** Set of all real numbers

13. $\{x \mid x \geq -3\}$ **15.** Set of all real numbers

17. The expression x^6 has two square roots, x^3 and $-x^3$.
But $\sqrt{x^6}$ means the principal square root and, because
x^3 could be negative, absolute value symbols are necessary. However, x^9 has only one cube root, x^3, and so
absolute value symbols are not used.

19. $5y$ **21.** $7w^4t^2$ **23.** $\frac{9x^2}{y^3}$ **25.** $2t^2$ **27.** $3x^3y^5$

29. $\frac{a^3}{5b}$ **31.** $\sqrt{9x^6y^8}$ **33.** $-\sqrt{4t^{10}}$ **35.** $\sqrt[3]{64x^3}$

37. $\sqrt{49x^8}$

39. The Product Rule for Radicals applies to products of
real numbers. The factors $\sqrt{-3}$ and $\sqrt{-12}$ are not real
numbers.

41. $\sqrt{6}$ **43.** 20   **45.** $\sqrt[3]{x^2 + x - 6}$ **47.** $9t^4$

49. $2x + 1$ **51.** $\sqrt[4]{8xy^2}$

53. Product Rule for Radicals does not apply.

55. $\dfrac{1}{y}$

57. The rule does not apply because the index is not the same on each radical.

59. 4   **61.** $\dfrac{1}{t^3}$ **63.** 3

65. Quotient Rule for Radicals does not apply.

67. x **69.** $\dfrac{1}{5a}$ **71.** $\dfrac{7a^3b^2}{2}$ **73.** $\sqrt{x}$ **75.** $\sqrt[3]{2x}$

77. $\sqrt[5]{x^2y^3}$ **79.** $\sqrt{a}$

81. Factor the trinomial to obtain $\sqrt{(x+5)^2} = |x+5|$. This step is necessary because there is no rule for simplifying a radical whose radicand is a sum. Although $x^2 + 10x + 21$ can be factored, the result is not a binomial squared. Therefore, the radical cannot be simplified.

83. $-2x^4$ **85.** $5|x^5|y^4$ **87.** $|xy^3|z^2$ **89.** $|x|(y-3)^2$

91. $|2y-1|$ **93.** $x^2|x+1|$ **95.** $3 \cdot 10^{-4}$ **97.** 60

99. $\sqrt[12]{x^{11}}$ **101.** $\dfrac{1}{\sqrt[12]{x^5}}$ **103.** $\sqrt[12]{7}$ **105.** $\sqrt[6]{200}$

107. (a) 1.57   (b) The result is the same. $\sqrt[4]{6}$
(c) $\sqrt{\sqrt{6}} = (6^{1/2})^{1/2} = 6^{1/4} = \sqrt[4]{6}$

109. $x+3$ **111.** 16.35%   **113.** 15.16%

115. $\{x \mid x \le -4 \text{ or } x \ge 3\}$ **117.** $\sqrt[8]{x}$ **119.** -1

121. 2   **123.** 2   **125.** Sometimes true; $n \ge 0$

127. Never true   **129.** $x \ge 2;\ x \le 2$

Section 9.5   (page 605)

1. The radicand 32 has a perfect square factor 16. The radicand 30 does not have a perfect square factor.

3. $2\sqrt{3}$ **5.** $5\sqrt{2}$ **7.** $3\sqrt[3]{2}$ **9.** $-2\sqrt[3]{6}$ **11.** $2\sqrt[4]{2}$

13. $2\sqrt[5]{8}$ **15.** $a^2b\sqrt{ab}$ **17.** $a^3b^2\sqrt[3]{b^2}$ **19.** $x^4y^2\sqrt[4]{y}$

21. $2\sqrt{15t}$ **23.** $5x^{12}\sqrt{x}$ **25.** $10a^3b^7\sqrt{5a}$

27. $-7x^3y^3\sqrt{y}$ **29.** $2w^3t^4\sqrt[4]{7t}$ **31.** $3x^3y^3\sqrt[3]{3y^2}$

33. $-2xy\sqrt[5]{x^2y^4}$ **35.** $\sqrt{16a^4b^3}$ **37.** $\sqrt[3]{3x^2y^3}\sqrt{3x^2y^3}$

39. $\dfrac{\sqrt{36x^9y^8}}{\sqrt{3x^2y^7}}$ **41.** $\sqrt[3]{16x^4y^3}$

43. Factor the radicand to obtain $4x^2 + 16 = 4(x^2 + 4)$. Then $\sqrt{4x^2 + 16} = \sqrt{4(x^2 + 4)} = \sqrt{4}\sqrt{x^2 + 4} = 2\sqrt{x^2 + 4}$.

45. $2\sqrt{3x + 2y}$ **47.** $2\sqrt{x^2 + 4}$ **49.** $3x^3y^3\sqrt{x + y}$

51. $3|y|\sqrt{1 + y^2}$ **53.** $r^2|t|\sqrt{1 + r^2t^2}$ **55.** $|y - 5|$

57. $n^2|3n + 1|$

59. The radical can be simplified if the monomial Q is a perfect square or if P is divisible by Q.

61. $\dfrac{2\sqrt{7}}{3}$ **63.** $2x^5y\sqrt{5x}$ **65.** $\dfrac{-\sqrt[3]{2a}}{5}$ **67.** -20

69. $3\sqrt[3]{6}$ **71.** $7x\sqrt{2}$ **73.** $x^4\sqrt{6x}$ **75.** $6x^2y^2\sqrt{2y}$

77. $5xy^2\sqrt[3]{x^2}$ **79.** $3xy\sqrt[4]{2xy}$

81. For $\sqrt[6]{x^4}$ the index and the exponent have a common factor. This is not so for $\sqrt[6]{x^5}$.

83. $x\sqrt[3]{x}$ **85.** $xy^2\sqrt{2xy}$ **87.** $\sqrt[4]{3x^2y^3}$ **89.** $\sqrt[7]{4x^6y^5}$

91. $x^2y\sqrt{2xy}$ **93.** $\dfrac{\sqrt{5}}{7}$ **95.** $\dfrac{a^2\sqrt{a}}{b^5}$ **97.** $3x^2d\sqrt{xd}$

99. $\dfrac{2}{x^4}$ **101.** 16   **103.** 36   **105.** 4   **109.** t^n **111.** t^2

113. x^ay^{2b} **115.** $\dfrac{x^m}{8^n}$

117. Because $\sqrt{\sqrt{(a+b)^n}} = \left(((a+b)^n)^{1/2}\right)^{1/2} = (a+b)^{n/4}$, the least positive value of n is 4.

Section 9.6   (page 612)

1. The Distributive Property is used to add or subtract radical expressions if they are like terms.
$$3\sqrt{5} + 4\sqrt{5} = (3 + 4)\sqrt{5} = 7\sqrt{5}$$

3. $6\sqrt{13}$ **5.** $-\sqrt{7}$ **7.** $16\sqrt{x}$ **9.** $-3\sqrt[3]{y}$

11. $-4\sqrt{2x} + \sqrt{3}$

13. (a) The radicands are different.   (b) The indices are different.   (c) The indices are different.   (d) The radicands are different.

15. $\sqrt{3}$ **17.** $8\sqrt[3]{2}$ **19.** $\sqrt[4]{2}$ **21.** $62\sqrt{2}$ **23.** $\sqrt{3}$

25. $2\sqrt{3} + 10\sqrt{2}$ **27.** $\sqrt{2x}$ **29.** $13x\sqrt{3}$ **31.** $11a^2\sqrt{5b}$

33. $\dfrac{5x + 2}{x^2}$ **35.** $6x\sqrt{2x} - 6x\sqrt{2}$ **37.** $4x\sqrt{xy} - 7y\sqrt{xy}$

39. 0   **41.** $-x\sqrt[3]{2}$

43. The Commutative Property can be used to write the factors in a different order, and the Associative Property can be used to group the factors as desired.
$$\left(3\sqrt{2}\right)\left(5\sqrt{7}\right) = (3 \cdot 5)\left(\sqrt{2} \cdot \sqrt{7}\right) = 15\sqrt{14}$$

45. $5\sqrt{21}$ **47.** 19   **49.** -72 **51.** $12x$ **53.** $30x\sqrt{2}$

55. $3x\sqrt[4]{x}$ **57.** $24 - 4\sqrt{21}$ **59.** $20\sqrt{21} + 35\sqrt{7}$

61. $5\sqrt{7} - 5\sqrt{3}$ **63.** $\sqrt[3]{3y} - 3\sqrt[3]{y}$ **65.** 8   **67.** $\sqrt{7}$

69. $\sqrt{6t}$ **71.** $x + 2\sqrt{3}$ **73.** $2\sqrt{2} - \sqrt{3}$

75. We are using the Product to a Power Rule for Exponents. No, the quantity being squared is a sum, not a product.

77. $\sqrt{6} - 2\sqrt{2} + 3 - 2\sqrt{3}$

79. $\sqrt{21x} + 7x - 3\sqrt{3} - 3\sqrt{7x}$ **81.** $3\sqrt{5} - 8\sqrt{10}$

83. $8 - 4\sqrt{3}$ **85.** $15 + 10\sqrt{2}$ **87.** $\sqrt{5} - 2$

89. $\sqrt{x} - \sqrt{y}$

91. To form the conjugate of a binomial, we change the sign of the second term. To form the opposite of a binomial, we change the signs of both terms. For the binomial $3 - \sqrt{5}$, the conjugate is $3 + \sqrt{5}$ and the opposite is $-3 + \sqrt{5}$.

93. -1 **95.** 3   **97.** $100 - 50t^2$ **99.** $\sqrt{x} - \sqrt{2}$

101. $\sqrt{x} + \sqrt{y}$ **103.** $\sqrt{3} + 2$

105. $P = 4\sqrt{x} + 2\sqrt{y};\ A = x + \sqrt{xy}$

107. (a) $d = \sqrt{x^2 - 2500}$ (b) $h = \sqrt{2600 - 20x}$

109. $-7 - 4\sqrt{3}$ **111.** $\dfrac{7 - \sqrt{21}}{4}$ **117.** 1.63

119. $2x^{2n}\sqrt{x}$ **121.** $x^{n+1} + x^{2n}$ **123.** $2\sqrt{x}$ **125.** $-\sqrt{2}$

127. $\dfrac{\sqrt{5} + 2}{3}$

Section 9.7   *(page 620)*

1. Both methods are correct.   **3.** $\sqrt{5}$ **5.** $-\dfrac{\sqrt{15}}{3}$

7. $\dfrac{3\sqrt{10}}{2}$ **9.** $\dfrac{\sqrt{21} + 2}{2}$ **11.** $\dfrac{7\sqrt[3]{9}}{3}$ **13.** $\dfrac{9\sqrt[4]{2}}{2}$

15. Either gives the correct result. Multiplying by just $\sqrt{3}$ is more efficient.

17. $\dfrac{2\sqrt{3x}}{x}$ **19.** $x\sqrt{x}$ **21.** $\dfrac{a\sqrt{2a}}{4}$ **23.** $\dfrac{2\sqrt{x+2}}{x+2}$

25. $\dfrac{x - 3\sqrt{x}}{x}$ **27.** $6x\sqrt{3y}$ **29.** $\dfrac{x^2\sqrt{5xy}}{10y}$ **31.** $2\sqrt[3]{b}$

33. $5\sqrt[5]{x^3}$ **35.** 5   **37.** $\sqrt{14}$ **39.** $\sqrt{3}$

41. The operation cannot be performed because the denominators are not the same. It is not necessary to rationalize the denominators in order to add. It is only necessary for the fractions to have a common denominator.

43. $\dfrac{16\sqrt{5}}{5}$ **45.** $\dfrac{2 - 3\sqrt{2}}{2}$ **47.** $\dfrac{2\sqrt{x+1}}{x+1}$ **49.** $6 - \sqrt{3}$

51. $\dfrac{5 - 3\sqrt{3}}{7}$ **53.** $\dfrac{-4 + \sqrt{2}}{2}$ **55.** $\dfrac{x\sqrt{5x} + 5}{3}$

57. $\dfrac{1 - 3x}{2}$ **59.** $3 + \sqrt{6}$ **61.** $4 - \sqrt{15}$

63. $\dfrac{y + 9\sqrt{y} + 20}{y - 16}$ **65.** $\dfrac{x - 2\sqrt{x}}{x - 4}$ **67.** $\dfrac{\sqrt{x+1} + 2}{x - 3}$

69. $\dfrac{x - 2\sqrt{xy} + y}{x - y}$ **71.** $x - 25$ **73.** -7

75. $\sqrt{7} - \sqrt{2}$ **77.** $\dfrac{-1}{2(\sqrt{5} - \sqrt{7})}$ **79.** $\dfrac{x - 9}{x - 6\sqrt{x} + 9}$

81. $\dfrac{1}{\sqrt{x+2} + \sqrt{x}}$

83. The product of the conjugates is 0, and division by 0 is undefined. An easy method is to combine like terms.

85. (a) 5.35 inches   (b) $r = \dfrac{\sqrt{3V\pi h}}{\pi h}$

87. $h = \dfrac{2A(\sqrt{x} - \sqrt{y})}{x - y}$ **89.** $P(t) = t^2 + 43$

91. $t = 7.55$

The model projects that everyone would wear a seat belt by 1995, which is not realistic.

93. $A^3 - B^3 = (A - B)(A^2 + AB + B^2)$

Multiply the numerator and denominator by $\sqrt[3]{x^2} + \sqrt[3]{xy} + \sqrt[3]{y^2}$.

95. $\dfrac{5(\sqrt[3]{x^2} + \sqrt[3]{3x} + \sqrt[3]{9})}{x - 3}$ **97.** $\dfrac{\sqrt{x} + 3}{x + 3}$ **99.** $\dfrac{\sqrt[4]{x^3}}{x}$

101. $\dfrac{3\sqrt[n]{x^3}}{x}$ **103.** $\sqrt{5}$

Section 9.8   *(page 631)*

1. The Odd Root Property allows us to write $A^n = B$ in the form $A = \sqrt[n]{B}$, which allows us to determine A.

3. $\pm\sqrt{6}$ **5.** No real number solution   **7.** 4.5, 0.5

9. $\dfrac{1 \pm \sqrt{3}}{3}$ **11.** ± 2 **13.** 5   **15.** 0.5

17. (a) 1   (b) ± 1 (c) 1   (d) ± 1 (e) 1   (f) ± 1
(g) For even n, the solutions of $x^n = 1$ are ± 1. For odd n, the solution is 1.

19. No, the first equation has no solution, and the second has one solution, 25. Because an extraneous solution is possible, the proposed solutions must be checked.

21. 18   **23.** 64   **25.** 8   **27.** 64

29. The radical in each equation can be eliminated by squaring once. The first two resulting equations are second-degree equations.

31. 3   **33.** 1   **35.** 3   **37.** No real number solution

39. 15   **41.** 4, -1 **43.** 9   **45.** 0.5, 2   **47.** 3

49. $-3, 2$ **51.** No solution   **53.** No solution

55. No solution   **57.** 4   **59.** 3

61. The first equation must be squared twice. In the second equation, isolate the radical and square once.

63. 4   **65.** 3, 7   **67.** No solution   **69.** 3   **71.** 6

73. 2, 3   **75.** 2, 6   **77.** 4

79. Raise both sides to the b/a power in order to isolate x.

81. ± 27 **83.** $\frac{1}{81}$ **85.** $\frac{25}{8}, \frac{23}{8}$ **87.** 6   **89.** $r = \sqrt{\dfrac{A}{\pi}}$

91. $r = \sqrt[3]{\dfrac{3V}{4\pi}}$ **93.** Yes, $r = 5.499$ feet   **95.** 2.32 inches

97. 4   **99.** 24.8 cm   **101.** $\sqrt{5} \approx 2.24$ **103.** 25.04 feet

105. $x = 1{,}000{,}001$; 31.62 feet   **107.** 82.92 feet

109. The domain of $\sqrt{x-3}$ is $\{x \mid x \ge 3\}$, and the domain of $\sqrt{2-x}$ is $\{x \mid x \le 2\}$. Because the intersection of these domains is empty, the equation has no solution.

111. $t = 10.25$

The model predicts that the price will be $5.50 in 1997.

113. The graph suggests a steady but gradual increase.

115. (a) $\{x \mid x \ge 0\}$ **(b)** $\{x \mid x \le 0\}$

(c) All real numbers   **117.** 8   **119.** 3   **121.** 1, 3

123. 19.4475   **125.** $\pm 1, \pm \sqrt{5}$

Section 9.9 *(page 644)*

1. The number is i and the properties are $i^2 = -1$ and $i = \sqrt{-1}$.

3. $3i$ **5.** $3i\sqrt{2}$ **7.** $5 + 6i$ **9.** $-2 - i\sqrt{3}$

11. The real parts are equal and the imaginary parts are equal.

13. $a = 2, b = -1$ **15.** $a = -1, b = 5$ **17. (a)** False

(b) True   **(c)** False   **19.** $1 + i$ **21.** $15 + 10i$

23. $6 + 21i$ **25.** 14   **27.** $13i$ **29.** $1 - 4i$

31. $a = -2, b = -2$ **33.** $a = -1, b = -2$

35. The Product Rule for Radicals was stated for radicals defined in the real number system. The rule does not apply to this product because $\sqrt{-3}$ is not a real number.

37. -10 **39.** $28 + 21i$ **41.** $-6 - 17i$ **43.** $8 + 6i$

45. -4 **47.** $-9 - 3\sqrt{2}$ **49.** $15 + 5i$

51. $a = 0, b = -5$ **53.** $a = 3, b = -4$ **55.** $-i$

57. 1   **59.** -1 **61.** i

63. In part (a) the result is $a^2 - b^2$, but in part (b) the result is $a^2 + b^2$.

65. $-7i, 49$ **67.** $2 - 3i, 13$ **69.** $\sqrt{5} + 4i, 21$

71. $a = -5, b = 1$ **73.** $\frac{3}{2}i$ **75.** $2 - 3i$ **77.** $\dfrac{1 - 7i}{5}$

79. $\dfrac{-1 + 43i}{74}$ **81.** $\dfrac{6 - 3i\sqrt{3}}{7}$

83. $\dfrac{\sqrt{15} + 3\sqrt{3} + 3i - 3i\sqrt{5}}{12}$ **85.** $a = -5, b = 2$

87. $a = 0, b = 4$ **89.** $8 - 3i$ **91.** 28   **93.** $\dfrac{2 - i}{3}$

95. $9 + 6i$ **97.** $11 - 60i$ **99.** $-i$ **101.** $\dfrac{-7 + 24i}{25}$

103. $7i$ **105.** $5 - 3i\sqrt{3}$ **107.** 5   **109.** $\dfrac{1 + i\sqrt{3}}{4}$

115. $x^2 + 36$ **117.** $x^2 + 25$ **119.** $(x + 3i)(x - 3i)$

121. $(2x + i)(2x - i)$

123. (a) -1 **(b)** i **(c)** $\dfrac{2 - 3i}{13}$ **(d)** $\dfrac{-3 + 4i}{25}$

125. A complex number has 3 cube roots.

127. $\dfrac{-23 + 37i}{26}$ **129.** $\dfrac{18 - 4i}{5}$

Chapter Review Exercises   *(page 647)*

1. The number x^2 has two square roots, x and $-x$, but $\sqrt{x^2}$ is the principal (nonnegative) square root, which means that absolute value symbols are required. The number x^3 has only one real cube root and absolute value symbols are not used.

3. -8 **5.** -4 **7.** x^2 **9.** $x - 4$

11. (a) $\sqrt{10} - a$ **(b)** $\sqrt{10 - a}$

13. There is not a real number whose square is -4.

15. 64   **17.** $\frac{1}{9}$ **19.** $-\frac{1}{3}$ **21.** $4\sqrt[7]{x^6}, \sqrt[7]{(4x)^6}$ **23.** 16

25. The initial result is always $|x|$. However, if x is nonnegative, then $|x| = x$.

27. $f^{2/7}$ **29.** $a^{1/2}$ **31.** $\dfrac{x^{3/2}}{2}$ **33.** $(x^8 y^{12})^{1/4} = x^2 y^3$

35. 16.55 feet

37. The square root function is defined only if the radicand is nonnegative. The cube root function is defined for all real numbers.

39. $\{x \mid x \le 5\}$ **41.** $\sqrt{42x}$ **43.** x **45.** $3a^2 b^3$

49. (a) If the index is n, the radicand must contain no factor that is a perfect nth power.

(b) The index must be reduced to its smallest value.

(c) The radicand must contain no fractions.

(d) There must be no radicals in the denominator of a fraction.

51. $x^2\sqrt{x}$ **53.** $c^8 d^{10}\sqrt{c}$ **55.** $\sqrt[5]{x^4 y^3}$

57. (b) Requires more than one operation.

   (a) $\sqrt{5}$ **(c)** $\dfrac{\sqrt{15}}{3}$

59. Because the radical factors are different, the terms are not like terms and the Distributive Property does not apply.

61. $13\sqrt{5x}$ **63.** $45x$ **65.** $x - 2$

67. (a) 45   **(b)** 4   **(c)** $14 + 6\sqrt{5}$
   Parts (a) and (b) are rational numbers.

69. The radicand would still not be a perfect cube.

71. $\sqrt{6}$ **73.** $\dfrac{3\sqrt{x}}{x^2}$ **75.** $\dfrac{x + 4\sqrt{x} - 21}{x - 9}$ **77.** $2\sqrt{3}$

79. $\sqrt{x}$

81. When an algebraic method is correctly used to solve an equation, an apparent solution may not satisfy the equation. Such a number is an extraneous solution.

83. 3   **85.** 1   **87.** 15.20 feet   **89.** 4 and 9

91. The square root of a negative number is not defined.
   $\sqrt{-4}\sqrt{-9} = (2i)(3i) = 6i^2 = -6$

93. $-2 + 2i$ **95.** $33 + 4i$ **97.** $4 - 2i$ **99.** i

101. $a = 0, b = -6$

Chapter Test   *(page 651)*

1. *(9.1)* $\{x \mid x \le 3\}$

2. *(9.2)* $\dfrac{-2y^4}{x^2}$

3. *(9.2)* $-25^{1/2} = -\sqrt{25} = -5$
   $(-25)^{1/2} = \sqrt{-25}$ is not a real number.

4. *(9.4)* $\sqrt{\dfrac{20}{49}} = \dfrac{2\sqrt{5}}{7}$

5. *(9.3)* $\sqrt[12]{x^7}$

6. *(9.3)* $\sqrt[6]{x^5}$

7. *(9.6)* $\sqrt{3}$

8. *(9.6)* $26 - 5\sqrt{5}$

9. *(9.5)* The two conditions are that b is divisible by a or that a is a perfect square.

10. *(9.7)* $\dfrac{10\sqrt{2} - 9}{7}$

12. *(9.5)* $7x^5 y^8 \sqrt{2x}$

13. *(9.3)* $\frac{125}{64}$

14. *(9.8)* $-\frac{7}{3}, 3$

15. *(9.8)* Extraneous solutions can occur.

16. *(9.8)* 9

17. *(9.8)* 1

18. *(9.9)* $-1 + 6i$

19. *(9.9)* $5 + 12i$

20. *(9.9)* $\dfrac{1 - i}{2}$

21. *(9.9)* $i\sqrt{7}, i\sqrt{3}$

22. *(9.9)* **(i)** True   **(ii)** False   **(iii)** False   **(iv)** True

23. *(9.2)* $s = V^{1/3}$
   5.85 inches

24. *(9.9)* $\sqrt{3} - 3 + i\sqrt{3} + 3i$

25. *(9.8)* **(a)** $B < 0$ **(b)** $a > 0$ or n is odd

Chapter 10

Section 10.1   *(page 658)*

1. The x-intercepts correspond to the solutions of the equation.

3. $0.5, -3$ **5.** $\frac{1}{3}, -3$ **7.** 3   **9.** 5   **11.** No real number solution   **13.** No real number solution

15. Write the equation in standard form and factor the polynomial. Set each factor equal to zero and solve.

17. $-6, -1$ **19.** $0, 1$ **21.** $-\frac{4}{3}, 2$ **23.** $1, 3$

25. $-1, 0.4$

27. If $A^2 = B$ and $B > 0$, then $A = \pm\sqrt{B}$. The equation $x^2 = 6$ has two solutions, $\pm\sqrt{6}$. If $A^2 = B$ and $B < 0$, then there are no real number solutions. Thus, the other equation $x^2 = -6$ has no real number solution.

29. $0.2, 1.4$ **31.** $-4.44, -5.56$ **33.** $0, 3$ **35.** $-3, 3$

37. $\pm 4i$ **39.** $\pm 4.90i$ **41.** $3 \pm 2i$ **43.** $1 \pm 5i$

45. $-2.5 \pm 1.22i$ **47.** $-1 \pm 1.5i$ **49.** $\pm\frac{2}{3}i$

51. $-0.5, 2.5$ **53.** $\pm 2i$

55. Because there are no x-intercepts, the equation has no real number solutions. There are two imaginary number solutions.

57. $0, 3$ **59.** $\pm 6.71i$ **61.** $-\frac{1}{6}$ **63.** $\pm 1.2i$ **65.** ± 3.16

67. $-5, 7$ **69.** $-1.25 \pm 1.75i$ **71.** $3.95, -5.29$

73. $\pm 2.45i$ **75.** $1, 4$ **77.** 1.5 **79.** $x^2 - 8x + 15 = 0$

81. $x^2 - 6x + 9 = 0$ **83.** $3x^2 - 13x + 12 = 0$

85. $x^2 + 16 = 0$ **87.** $-8, -9$

89. The two integers are even, not odd.   **91.** 18, 20

93. 5 minutes   **95.** 4 feet   **97.** $(-0.69, 101.88)$

99. 1990: 33.4 pounds
   The model projects zero consumption in 1994. The model does not appear to be reasonable for years not in the 1970–1985 period.

101. $\pm 0.8c$ **103.** $-2 \pm i\sqrt{y}$ **105.** $-a \pm 3$

107. $\dfrac{-1 \pm 4i}{a}$ **109.** $1 \pm i$ **111.** 4.65, -0.65

113. $-6, -\frac{4}{3}$ **115.** All real numbers

Section 10.2 *(page 666)*

1. The expression A is a first degree binomial.

3. 9   **5.** $\frac{49}{4}$ **7.** $\frac{4}{9}$

9. Both methods are correct. The factoring method can be used because $x^2 - 8x + 12$ is factorable. The method of completing the square can be used for any quadratic equation.

11. $-3, -1$ **13.** $-5, 3$ **15.** $-3, 4$ **17.** $-3, 2$

19. 0.76, 5.24   **21.** $-6, 4$ **23.** $-0.83, 4.83$

25. Divide both sides of the equation by 3 to make the leading coefficient 1.

27. $-1, -\frac{1}{3}$ **29.** $-5, 1.5$ **31.** $-1.5, 3.5$ **33.** 2.5

35. $-0.28, 1.78$ **37.** $-3.64, 0.14$ **39.** 9   **41.** $2 \pm 2i$

43. $2 \pm 1.41i$ **45.** $-1.5 \pm 1.66i$ **47.** $-0.25 \pm 1.20i$

51. 2, 8   **53.** $\pm\frac{7}{6}$ **55.** $-2, 5$ **57.** $-4.32, -1.68$

59. $-0.5 \pm 2.96i$ **61.** 1.15, 7.85   **63.** $-1.90, 0.40$

65. $0.5 \pm 2.96i$ **67.** $-2.11, 0.36$ **69.** $\frac{1}{3}, -\frac{2}{3}$

71. $-1, 2$ **73.** $0.25 \pm 1.48i$ **75.** $a = 4; -1.25$

77. $b = 7; -1.5$ **79.** $c = -48; -8$ **81.** 5

83. 5.54   **85.** 31 feet   **87.** 2 hours and 17 minutes

89. $(-1.12, 0), (18.50, 0)$

91. The x-intercepts represent 1978 and 1998. At these points there are no convenience stores. The model is not valid at these points.

93. $-5 \pm \sqrt{25 - c}$ **95.** $c \pm ci\sqrt{3}$

97. $\dfrac{-1 \pm \sqrt{1 - a^2}}{a}$ **99.** 0.76, -5.24

101. 1.41, -7.07

103. (a) $k \geq 2, k \leq -2$ (b) $-2 < k < 2$

105. (a) $k = 0$ (b) $k \neq 0$ **107.** $-2i \pm 2$ **109.** $-7i, i$

111. (a) $x^2 + 8x$ (b) 16   (c) $x^2 + 8x + 16$
    (d) $(x + 4)^2$ (e) $(x + 4)^2 = x^2 + 8x + 16$

Section 10.3 *(page 675)*

1. The equation must be in standard form.

3. 3, $-9, -6$ **5.** 1, $-1, 2$

7. The Quadratic Formula is derived by completing the square and is a generalized result of that method.

9. $-1, 2$ **11.** 4   **13.** $-4, 0.5$ **15.** 0, 5

17. $-0.29, 2.54$ **19.** $1 \pm 2.83i$ **21.** $-3.45, 1.45$

23. $-0.25 \pm 1.39i$

25. $(-B + \sqrt{(B \wedge 2 - 4 A C)}) / (2 A);$
    $(-B - \sqrt{(B \wedge 2 - 4 A C)}) / (2 A)$

It will divide the radical quantity by $2a$ but not $-b$.

27. 0, $\frac{2}{3}$ **29.** $\pm\frac{3}{7}$ **31.** 7   **33.** 0.17, 1.43   **35.** $-2, 4$

37. $-5, 1$ **39.** $-0.2 \pm 1.17i$ **41.** $1.5 \pm 2.40i$

43. 0.76, 5.24   **45.** $-2.5 \pm 1.66i$ **47.** $0.67 \pm 1.49i$

49. $-0.19, 2.69$ **51.** $-1.30, 2.30$

53. In the Quadratic Formula, the discriminant is $b^2 - 4ac$. Its value indicates the number and type of solutions of a quadratic equation.

55. 2 rational   **57.** 2 imaginary

59. 1 rational (double root)

61. 2 irrational   **63.** $k < 16$ **65.** $k < 4$

67. $k > 9$ **69.** $k > \frac{9}{4}$ **71.** ± 6 **73.** 0, 8

75. (a) The discriminant is $a^2 + 20$, which is positive for all values of a. Therefore, the solutions are always real numbers.
    (b) The discriminant is $a^2 + 4k$. Because k is positive, the discriminant is always positive and the solutions are always real numbers.

77. $(x - 20)(x + 24)$ **79.** $(x - 18)(x + 25)$ **85.** $\frac{2}{3}, -\frac{1}{3}$

87. $-1, 1$ **89.** 5, -1 **91.** Yes   **93.** 0

95. The discriminant is $(k - 2)^2$. For $k = 2$, the discriminant is zero, and there is only one real solution. For all other values of k, the discriminant is positive, and there are two real solutions.

97. 1983–1984

99. After the year 2000, beef consumption would be negative. The model might be valid for a few years after 1990, but not for the year 2000 and beyond.

101. $-b \pm \sqrt{b^2 - 3}$ **103.** $\dfrac{y \pm \sqrt{y^2 + 8}}{2}$

105. $-1 \pm \sqrt{9 + x}$ **107.** $\dfrac{3 \pm \sqrt{-11 + 4x}}{2}$

109. $16x^2 - 16x + 1 = 0$ **111.** $x^2 + 4x + 13 = 0$

113. $0.46, -2.19$ **115.** $1.41 \pm i$

117. $1.41 + i, -1.41 + i$ **119.** $2i, -3i$

Section 10.4 *(page 686)*

1. Multiplying both sides of the original equation by $x - 1$ introduces an extraneous solution.

3. $0.75 \pm 1.20i$ **5.** 6.65, 1.35   **7.** $-0.5 \pm 1.94i$

9. $-3.79, 0.79$ **11.** $-0.56, 3.56$ **13.** -2

15. Squaring both sides of the original equation introduces extraneous solutions.

17. $1, \frac{4}{9}$ **19.** 1.39, 0.36   **21.** 5   **23.** 0.82   **25.** 0.30

27. A dummy variable is used to represent a variable expression in an equation. After solving for the dummy variable, replace it with the variable expression it represents and solve the resulting equation(s) for the original variable.

29. $\pm 2, \pm 1.73$ **31.** $\pm 3i, \pm 2.65$ **33.** 4, 16

35. $0.25 \pm 0.66i$ **37.** 64, 27   **39.** 7, -2, 3, 2

41. $-0.27, -3.73$ **43.** $\pm 3, \pm 1.41$ **45.** $16, \frac{49}{9}$ **47.** 9

49. $\pm 2, \pm 1.73$ **51.** $0.85, -2.35$ **53.** -1.13

55. $\pm 2, \pm 2.24i$ **57.** 49, 36   **59.** $1 \pm 2i$ **61.** 8

63. $-27, 125$ **65.** 0.26   **67.** 1   **69.** $-1.37, 4.37$

71. 3.04   **73.** 2.75, 4   **75.** No solution

77. $0.5 \pm 1.32i, 0.5 \pm 0.87i$

79. $6.74(t + 5)^2 - 41.3(t + 5) + 2317 = 3000$

81. The cost of the program in 1988 was $3000 million.

83. 2.30   **85.** $2, -1, -1 \pm 1.73i, 0.5 \pm 0.87i$

87. $1, -3.73$ **89.** $0.5 \pm 0.87i$

Section 10.5   *(page 696)*

1. The area of a rectangle is its length times its width. If the length and width are represented by first-degree expressions, then the product is a second-degree expression. Thus, the equation to be solved will be a quadratic equation.

3. 16.97 feet   **5.** 90.38 feet

7. The hypotenuse is opposite the right angle and is the longest side.

9. 97.14 yards, 127.14 yards   **11.** 25.08 feet

13. No, each side of the equation is a ratio, but the equation itself is called a proportion. A proportion states that two ratios are equal.

15. 17.3 inches, 10.70 inches   **17.** *AC*: 12.26, *DE*: 3.26

19. Use the formula $T = D/R$. The equation will probably involve rational expressions.

21. 26.88 mph   **23.** 57.16 mph

25. If the rate of flow through the inlet pipe is greater than the rate of flow through the outlet pipe, then the tank will fill.

27. 11.08 hours, 13.08 hours   **29.** 9 hours

31. $N = V/P$. An equation with rational expressions is likely.   **33.** 370 shares

35. 1993: 459 boxes at $18.67
    1994: 561 boxes at $20.67

37. 60   **39.** 7, 8 or $-7, -8$ **41.** 9   **43.** 8   **45.** 4 feet

47. 4.96 feet   **49.** 1.52 feet   **51.** 1108.32 feet

Section 10.6   *(page 705)*

1. (b)   **3.** (a)

5. Write the inequality in standard form. The function is $f(x) = 2x^2 + 3x + 7$.

7. $(-1, 3)$ **9.** $(-\infty, 0) \cup (1.6, \infty)$ **11.** $(2, 5)$

13. $(-\infty, -1) \cup (1.5, \infty)$

15. The Quadratic Formula always works. Other methods include factoring and the Square Root Property.

17. $(-\infty, -3) \cup (3, \infty)$ **19.** $[-2.5, 0.5]$

21. $[1.44, 5.56]$ **23.** $(-\infty, -6) \cup (3, \infty)$ **25.** $[0, 5]$

27. $(-\infty, -0.67] \cup [0.25, \infty)$ **29.** $(-1.12, 3.12)$

31. $(-\infty, -3] \cup [2, \infty)$ **33.** $[-2, 3]$ **35.** $(-0.67, 1.5)$

37. $(-\infty, -1.12] \cup [3.12, \infty)$ **39.** $[-2, 4]$

41. $(-\infty, 0] \cup [1.75, \infty)$ **43.** $(-\infty, -4] \cup [4, \infty)$

45. $[-0.58, 8.58]$

47. The set of all real numbers or the empty set

49. $(-\infty, -5] \cup [-0.75, \infty)$ **51.** $(-\infty, 0) \cup (5, \infty)$

53. $\varnothing$ **55.** $\{-2.5\}$ **57.** $(-\infty, -0.83] \cup [4.83, \infty)$

59. $\{3\}$ **61.** $(-0.87, 2.87)$ **63.** $(-\infty, 1] \cup [5, \infty)$

65. $(-\infty, \infty)$ **67.** $[-2.33, -0.5]$

69. $(-\infty, -2) \cup (1.25, \infty)$ **71.** $(-2.53, 0.53)$

73. $[-4, -2]$ **75.** $k > 1.125$ **77.** Not possible

79. $k \geq 6.25$ **81.** $k \leq -3.2$ **83.** 36   **85.** ± 6

87. At least 6.69 feet   **89.** Between 0 and 2   **91.** 1988

93. $t \approx 30$
    This means that in the year 2010 there will be no deaths due to heart disease. Unfortunately, it is not realistic to believe the number of deaths will be reduced to zero.

95. (a) The domain of f is the set of all x for which the radicand is nonnegative. Therefore, we estimate the intervals for which the graph of g is on or above the x-axis: $(-\infty, -3] \cup [5, \infty)$.
    (b) Because $h(x)$ is nonnegative for all x, the portion of the graph of g that is below the x-axis is inverted to the other side of the x-axis.

97. The discriminant is $b^2 > 0$, which means the graph has two x-intercepts. The x-values of the x-intercepts must be excluded from the solution set of the inequality.

99. $[-6, 6]$ **101.** $[-4, 4]$ **103.** 1.11

Section 10.7   *(page 713)*

1. If the polynomial cannot be factored, we cannot determine the x-intercepts with algebraic techniques presented to this point.

3. $(-2, 1) \cup (3, \infty)$ 5. $[-6, 3]$ 7. $(3, \infty)$

9. $(-\infty, -5) \cup (-1, 0)$

11. $(-\infty, -3) \cup (-2, 2) \cup (3, \infty)$

13. $[-2.24, -1] \cup [1, 2.24]$ 15. $(-3, -1) \cup (1, \infty)$

17. $(-1, 2) \cup (3, \infty)$ 19. $(-\infty, -5] \cup [-2, 1]$

21. $(-4, -2) \cup (-1, 1)$

23. If a and b are restricted values, they are not included in the solution set.

25. $(-1, 0]$ 27. $(-\infty, -4.5) \cup (-1, \infty)$

29. $(-\infty, 1.33)$ 31. $(-\infty, -4) \cup (-1, 3) \cup (10, \infty)$

33. $(-\infty, -3) \cup [-2.5, 0]$ 35. $(-\infty, -1.33] \cup [0, 3]$

37. $(-\infty, 2.5]$ 39. $(-\infty, -0.5] \cup (0, 1.67]$

41. $(-2, -1)$ 43. $(-5, 0) \cup (0, 3)$

45. $(-3, 3) \cup (4, 5)$

47. No, both expressions could be negative.

49. $(-1.5, 0)$ 51. $(-\infty, -2] \cup [3, 5]$ 53. $[0.33, 2)$

55. $(2.5, \infty)$ 57. $(-2, 5) \cup (7, \infty)$ 59. $(-\infty, 0.5]$

61. $(-\infty, -4.19] \cup (3, 1.19] \cup (2, \infty)$

63. $(-\infty, -3) \cup (3, \infty)$ 65. $(-1, -0.5] \cup (1, 3]$

67. $(0.5, \infty)$ 69. $(-\infty, -0.17] \cup [4, \infty)$

71. Both are nonnegative for all x. However, the quotient is not defined for $x = 2$. The solution set is the set of all real numbers except 2.

73. $(-\infty, 0)$ 75. $(0, \infty)$ 77. $(-3, 0) \cup (0, \infty)$

79. More than 4.83 mph

81. There is no such positive integer. The graph of

$$\frac{x}{x+1} - 2$$

is never above the x axis.

83. $-201.5t^3 + 4348.9t^2 - 30{,}254.5t + 74{,}919.5 \geq 7{,}100$

85. 1987–1989

87. Because the radicand must be nonnegative, the domain of f is represented by portions of the graph that are on or above the x-axis.

89. $(-\infty, \infty)$ 91. $(-\infty, 1)$ 93. (a) All real numbers

(b) All real numbers except 0.5 and -4 95. (a) $(-\infty, 0)$

(b) $(-\infty, 0]$ 97. $(-\infty, -1) \cup (-1, 1) \cup [3.5, \infty)$

99. $(-\infty, -6] \cup \{0\} \cup [2, \infty)$

Section 10.8   *(page 726)*

1. The graph is a parabola. For positive a the graph opens upward and for negative a the graph opens downward. The graph becomes narrower as $|a|$ increases.

3. $\{y \mid y \leq 3\}$ 5. $\{y \mid y \geq 2\}$ 7. The vertical axis of symmetry is $x = h$.   9. $x = 0$ 11. $x = 3$ 13. (e)

15. (c)   17. (b)

19. The graph is shifted vertically k units.

21. (a) Narrower, same vertex   (b) Wider, same vertex

23. (a) Shifted up 4 units   (b) Shifted down 4 units

25. (a) Shifted down 1 unit   (b) Shifted right 1 unit

27. (a) Shifted right 3 units and down 2 units
   (b) Shifted left 1 unit and up 4 units

29. The x-coordinate is $-\dfrac{b}{2a}$.

To determine the y-coordinate evaluate the function at

$$-\frac{b}{2a}.$$

31. $(3, 4)$; Minimum: 4   33. $(-1, -4)$; Minimum: -4

35. $(-1, 13)$; Maximum: 13   37. $(0, 3)$ 39. $(5, -1)$

41. $(3, 0), x = 3$ 43. $(-1, -9), x = -1$

45. $(-2, 11), x = -2$

47. The domain is the set of all real numbers. Thus, the function is defined for $x = 0$. The y-intercept is $(0, c)$ where c is the constant term of the function.

49. $(0, -28), (7, 0), (-4, 0)$ 51. $(0, 9)$, no x-intercept

53. $(0, 5), (-2.12, 0), (0.79, 0)$ 55. $(0, 0), (-1.67, 0)$

57. None   59. Two, rational   61. 0   63. 2   65. 1

67. 2   69. 1   71. 2

73. (a) Domain: all real numbers
   Range: $\{y \mid y \geq -1\}$
   (b) $(0, 8), (4, 0), (2, 0)$
   (c) $(3, -1), x = 3$

75. (a) Domain: all real numbers
   Range: $\{y \mid y \leq 9\}$
   (b) $(0, 5), (-5, 0), (1, 0)$
   (c) $(-2, 9), x = -2$

77. (a) Domain: all real numbers
   Range: $\{y \mid y \geq 7\}$
   (b) $(0, 9)$ (c) $(-1, 7), x = -1$

79. $h = 0, k = 3$ 81. $h = -\frac{3}{5}, k = 0$

83. $h = -4, k = 2$ 85. $h = 3, k = 2$

87. The x-intercepts are $(m, 0)$ and $(n, 0)$.

89. $b = -2, c = -15$ 91. $b = 2, c = 0$

93. $b = -8, c = 16$ 95. $f(x) = x^2 - 4x + 3$

97. $f(x) = 3x^2 + x - 2$ 99. 15, 15

101. 20 inches, 20 inches; 2.55 cubic feet

103. $y = -0.024x^2 + 2.4x$

105. Vertex: (9.70, 15.92); In 1989, bank income was at a minimum.

107. An x-intercept would correspond to no income.

109. $y = 2x^2 - 3x - 5$ **111.** $y = x^2 + x + 1$

113. (a) $21,000   **(b)** Vertex: (1, 21025); The income is maximized if one store is vacant.   **(c)** 30

Chapter Review Exercises   *(page 730)*

1. (a) For one graph write the equation in standard form, graph, and estimate the x-coordinates of the x-intercepts.
  (b) For two graphs, graph the left and right sides and estimate the x-coordinates of the points of intersection.

3. ±2.08   **5.** −6, 7   **7.** −5, 3   **9.** −4.8, −3.2

11. $2 \pm 6.08i$

13. First you must divide both sides by 2 to obtain a leading coefficient of 1 and an x-coefficient of $\frac{5}{2}$. Then the square of half the coefficient of x is $\frac{25}{16}$.

15. −0.70, 5.70   **17.** $0.33 \pm 1.11i$

19. All can be solved with the Quadratic Formula and all can be solved by completing the square. Parts (i) and (iv) can be solved with the Square Root Property. Parts (i) and (ii) can be solved by factoring.

21. ±1.79   **23.** −0.44, 1.69   **25.** $0.25 \pm 2.11i$

27. 2, rational   **29.** 2, imaginary

31. All are quadratic in form.

(a) $u = \dfrac{1}{x}$; $2u^2 + 3u - 5 = 0$

(b) $u = \sqrt{2x}$; $u^2 + u - 3 = 0$

(c) $u = (x + 1)$; $u^2 + 2u + 1 = 0$

(d) $u = x^{1/4}$; $u - u^2 + 3 = 0$

33. No solution   **35.** 25   **37.** −4, −3, 2, 3

39. 10.95 square yards   **41.** 5   **43.** 15 hours

45. {3}   **47.** $(-\infty, \infty)$ **49.** Ø

51. The solutions are represented by those points of the graph that are above the x-axis. The solutions are all numbers greater than −5 except 3. Thus, (iii) is correct.

53. $[-1, \infty)$ **55.** [1, 1.33)

57. Trace to the x-intercepts. The estimated solutions of the equation are the x-coordinates.

59. Intercepts: (0, −7), (2.33, 0), (−1, 0)
  Vertex: (0.67, −8.33)

61. Downward
  Vertex: (−0.6, 2.24)
  Range: $\{y \mid y \le 2.24\}$

63. Upward
  Vertex: (0.75, 4.75)
  Range: $\{y \mid y \ge 4.75\}$

65. (a) 2 seconds   **(b)** 64 feet

Chapter Test   *(page 733)*

1. *(10.2)* −1.55, 0.22

2. *(10.1)* −0.67, 1.5

3. *(10.1)* −0.62, 1.62

4. *(10.3)* −0.37, 1.37

5. *(10.3)* −0.15, −3.35

6. *(10.3)* $0.17 \pm 0.80i$

7. *(10.3)* $\pm i, \pm 0.47$

8. *(10.1)* −2

9. *(10.5)* $h = 5$ feet, $b = 8$ feet

10. *(10.5)* 5.12 hours

11. *(10.5)* 3 inches

12. *(10.6)* If there are no x-intercepts, then the graph is either entirely above or entirely below the x-axis. If, for example, the graph is entirely above the x-axis and the inequality symbol is > or ≥, then the solution set is the set of all real numbers.

13. *(10.6)* $(-\infty, -2] \cup [3, \infty)$

14. *(10.7)* $(-\infty, -7) \cup (-2, 3)$

15. *(10.7)* $[-0.5, 0] \cup [5, \infty)$

16. *(10.8)* $f(x) = (x - 3)^2 + 2$
  Vertex: (3, 2)
  Axis of symmetry: $x = 3$

17. *(10.8)* **(a)** Domain: all real numbers
  Range: $\{y \mid y \ge -4\}$
  (b) $a > 0$ **(c)** 2, 6

18. *(10.8)* (2.39, 0), (−8.39, 0)

Chapter 11

Section 11.1   *(page 744)*

1. Evaluate f and g at 1 and add the result or algebraically add f and g and then evaluate the resulting function at 1.

3. (a) 2   **(b)** 2   **(c)** 0   **(d)** Undefined   **5. (a)** 2.5
(b) 1.5   **(c)** 1   **(d)** 1   **7. (a)** 2   **(b)** 12   **(c)** −9
(d) −0.8   **9. (a)** 0   **(b)** 12   **(c)** 0   **(d)** −0.5

11. No, 0 is in the domain of h, but 0 is not in the domain of $\dfrac{f}{g}$ because $g(0) = 0$.

13. (a) $x^2 - x - 12$; All real numbers   **(b)** $x^2 + x - 6$;
All real numbers   **(c)** $x - 3$; $\{x \mid x \neq -3\}$
(d) $x^3 + 3x^2 - 9x - 27$; All real numbers

15. (a) $\dfrac{x^2 + 4}{x^2 - 4}$; $\{x \mid x \neq -2, x \neq 2\}$

(b) $\dfrac{x^2 - 4}{x^2 + 4}$; All real numbers   **(c)** 1; All real numbers

(d) 1; $\{x \mid x \neq -2, x \neq 2\}$

17. (a) $\dfrac{-5}{(x + 2)(x - 3)}$; $\{x \mid x \neq -2, x \neq 3\}$

(b) $\dfrac{2x - 1}{(x + 2)(x - 3)}$; $\{x \mid x \neq -2, x \neq 3\}$

(c) $\dfrac{x - 3}{x + 2}$; $\{x \mid x \neq -2, x \neq 3\}$ **(d)** $\dfrac{1}{(x + 2)(x - 3)}$;

$\{x \mid x \neq -2, x \neq 3\}$ **19.** $4t^2 - 2t - 3$ **21.** $\dfrac{t^2 - 16}{2t - 4}$

23. $9t^2 + 3t + 2$ **25.** $-8t^3 + 20t^2 + 6t - 15$

27. Produce the graphs of f and g and note the intersection
of the x-values for which f and g are defined.

29. (a) $\{x \mid 1.5 \leq x \leq 5\}$ **(b)** $\{x \mid 1.5 \leq x \leq 5\}$
31. (a) $\{x \mid 1 < x \leq 8\}$ **(b)** $\{x \mid 1 \leq x \leq 8\}$
33. First evaluate $g(5)$ and then evaluate $f(g(5))$.
35. (a) -2 **(b)** -1 **(c)** 2 **(d)** -14 **37. (a)** 15
(b) 16 **(c)** -3 **(d)** 1 **39. (a)** 5 **(b)** -2 **(c)** -4
41. $-4x^2 - 4x + 2$; $-2x^2 + 7$
43. $9x^2 - 21x + 15$; $3x^2 - 9x + 13$
45. $x - 2$; $\sqrt[3]{x^3 - 2}$ **47.** 5; 7   **49.** $9x - 16$; $x + 10$
51. $\|x - 3| - 3|$; $x^4 + 6x^2 + 12$
53. $8x^4 - 24x^3 + 12x^2 + 9x$; x **55.** 3; 2   **61. (a)** 288
(b) \$15,552   **(c)** $h(g(t)) = -18(t^2 - 140t + 4000)$

63. (a) $g(d(t)) = \dfrac{54t}{25}$ **(b)** $c(g(d(t))) = 1.03\left(\dfrac{54t}{25}\right)$

(c) \$26.70   **65. (a)** $f(x) = x$ **(b)** $c(x) = 10 - 0.05x$
(c) $T(x) = 10x - 0.05x^2$ **(d)** \$487.20
67.

$y = 502t + 5875$

$y = 157t^2 - 1900t + 13{,}157$

69. The total federal spending on education is
   $(f + h)(t) = 157t^2 - 1398t + 19{,}032.$

71. $f = g \circ h$ where $g(x) = \sqrt{x}$ and $h(x) = x^2 - 5x + 6$
73. $f = g \circ h$ where $g(x) = x^4$ and $h(x) = 3x + 2$
75. $f = g \circ h$ where $g(x) = 3x - 5$ and $h(x) = \sqrt{x}$
77. $f = g \circ h$ where $g(x) = 3x + 7$ and $h(x) = \sqrt{4x + 5}$
79. 3; Not defined

Section 11.2 *(page 754)*

1. Only one arrow can point to any range element.
3. Not one to one   **5.** One to one   **7.** Not one to one
9. Not one to one   **11.** One to one   **13.** One to one
15. Not one to one   **17.** Not one to one
19. $f^{-1} = \{(2, 1), (4, 3), (6, 5), (-1, 7)\}$
   Domain $= \{2, 4, 6, -1\}$
   Range $= \{1, 3, 5, 7\}$
21. $f^{-1} = \{(2, 3), (3, 4), (1, -5), (-2, -7)\}$
   Domain $= \{2, 3, 1, -2\}$
   Range $= \{3, 4, -5, -7\}$
23. -2 **25.** 3   **27.** 6   **29.** -4 **31.** -5 **33.** 3
35. No, the X^{-1} key returns the reciprocal of x, not the
   inverse function. In this case $f(x) = f^{-1}(x) = x$.
37. Multiply x by 3 and then subtract 2.
39. Take the cube root of x and then subtract 1.
41. $f(x)$: half of x; $f^{-1}(x)$: twice x; $2x$
43. $f(x)$: subtract 1 from x and then multiply by 3;
   $f^{-1}(x)$: divide x by 3 and then add 1; $\dfrac{x}{3} + 1$
45. $f(x)$: add 4 to x and then take the reciprocal; $f^{-1}(x)$:
   take the reciprocal of x and then subtract 4; $\dfrac{1}{x} - 4$
47. $f(x)$: cube x; $f^{-1}(x)$: take the cube root of x; $\sqrt[3]{x}$
49. Yes   **51.** No   **53.** Yes
55.

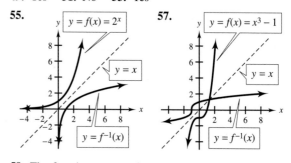

57.

59. The functions are not inverses.   **61.** No   **63.** Yes

65. Yes   **67.** $\dfrac{x - 5}{3}$ **69.** $-0.5x$ **71.** $3(x - 5)$ **73.** $\dfrac{1}{x}$

75. $\dfrac{1}{x} - 2$ **77.** $\dfrac{2x + 1}{x - 1}$ **79.** $\sqrt[3]{x - 1}$ **81.** $x^{-1/3}$

83. $4x^2, x \geq 0$ **85.** $2 - x^3$ **87.** $(x - 3)^5$ **89.** (iii)

91. The first is a one-to-one function, but the second is not.

93.

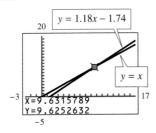

95. $\dfrac{t + 1.74}{1.18}$; The graphs intersect as predicted.

97.

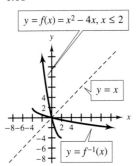

99. $x^2, x \geq 0$ **101.** $3 - x^2, x \leq 0$

103. $-\sqrt{x - 1}, x \geq 1$

105. $x - 5, x \leq 0$

107. $3x - 9, \dfrac{x + 9}{3}, \dfrac{1}{3}x, x + 3, \dfrac{x + 9}{3}$

109. **(a)** Function   **(b)** Not a function

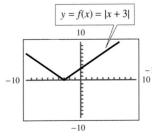 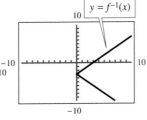

111. **(a)** Function   **(b)** Not a function

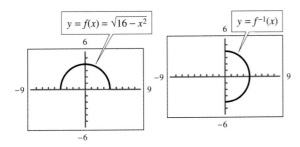

Section 11.3 *(page 765)*

1. For $b = 1$ the function is the constant function $f(x) = 1$.

3. 32   **5.** $\frac{1}{9}$ **7.** 1   **9.** $\frac{1}{2}$ **11.** 0.02   **13.** 403.43

15. 11.18   **17.** 0.45   **19.** 1613.72   **21.** 243

23. 29,370.08   **25.** 3.95

27. **(a)** The graph rises from left to right if $b > 1$.

(b) The graph falls from left to right if $0 < b < 1$.

29. Domain: All real numbers
$$\{y \mid y < -2\}$$

31. Domain: All real numbers
$$\{y \mid y < 2\}$$

33. Domain: All real numbers
$$\{y \mid 0 < y \leq 1\}$$

35. The graph becomes steeper.

37. For $c > 0$ the graph is shifted up c units, and for $c < 0$ the graph is shifted down $|c|$ units.

39. **(a)** Up 1 unit   **(b)** Down 3 units

41. **(a)** Up 3 units   **(b)** Left 3 units

43. **(a)** f is flatter and g is steeper.

(b) f is steeper and g is flatter.   **45.** -0.33

47. 2.42   **49.** None   **51.** 1.39   **53.** -1.37 **55.** 0

57. No solution

59. Both sides of the equation $4^x = 8$ can be written with the same base, and so the exponents can be equated. The left and right sides of $4^x = 12$ cannot be written with the same base.

61. -3 **63.** -2 **65.** -1.5 **67.** -1.5 **69.** 1.5

71. $\frac{10}{3}$ **73.** $\frac{2}{3}$ **75.** -3 **77.** No solution   **79.** 0

81. No solution   **83.** 0.5   **85.** $1, -\frac{5}{3}$ **87.** -8

89. ± 1 **91.** $-1, 0.5$ **93.** \$1.83   **95.** \$5000

97. 7.12 years

99. The graph rises sharply, which indicates a rapid increase in Medicare costs.

101. 2005   **103.** 0.16, -2.19 **105.** 1.57   **107.** -0.26, -2.54 **109.** $x > 0$ **111.** Minimum: $(-1, -0.37)$

113. Minimum: $(0, 0)$; maximum: $(2.89, 7.89)$

115.

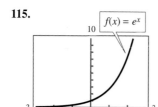

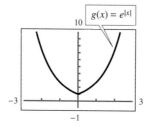

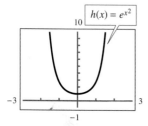

119. 2.5   **121.** 0.5   **123.** $(-\infty, -1) \cup (1, \infty)$
125. $(0, e)$ **127.** $(-\infty, -2) \cup (1, \infty)$ **129.** $(0, 1)$
131. 3

Section 11.5   (page 784)

1. What exponent on 9 results in 3? The answer is 0.5.
3. 7   **5.** 0   **7.** 2   **9.** 0.25   **11.** 0.25   **13.** 9   **15.** 0
17. 1   **19.** $\log_7 3 + \log_7 x$ **21.** $\log_7 5 - \log_7 y$
23. $2 \log x$ **25.** $\frac{1}{2} \log_9 n$ **27.** (c)   **29.** (a)   **31.** (g)
33. (d)
35. Begin with the Power Rule for Logarithms to obtain
$n \log (AB)$. In the Order of Operations, exponents
have priority over products.
37. $2 \log_9 x - 3 \log_9 y$
39. $2 \log (x + 1) - \log (x + 2)$
41. $5 \ln (2x + 5) - 5 \ln (x + 7)$
43. $\frac{1}{2} - 2 \log x$ **45.** $\frac{1}{3} \log_5 x + \frac{2}{3} \log_5 y$
47. $-5 + \frac{1}{2} \log_2 (x + 1)$ **49.** $\frac{5}{3} \log_2 x + \frac{1}{2} \log_2 y$
51. $2 \log (2x + 1)$ **53.** $0.08t + 3 \ln 10$
55. Cannot be expanded
57. $rt + \ln P$
59. Use the Distributive Property to obtain
$(2 + 1) \log 3 = 3 \log 3$. For $2 \log x + \log y$, $x \neq y$,
the Distributive Property does not apply.
61. $\log_4 15$ **63.** $\log_3 2$ **65.** $\log_2 y^3$ **67.** $\log \sqrt{x}$
69. 1   **71.** 3   **73.** $\log_8 \dfrac{3x}{y}$ **75.** $\log_7 \dfrac{x^2}{y}$ **77.** $\log \dfrac{\sqrt{x}}{\sqrt[3]{y^2}}$
79. $\log \dfrac{y^3}{x^2 z^4}$ **81.** $\log \dfrac{(x + 3)^2}{(x + 1)^3}$ **83.** $\log \sqrt{\dfrac{x}{x + 1}}$
85. $\log \left[\dfrac{t + 2}{t^2(t + 3)^5} \right]^3$ **87.** $\log_3 243x^2$ **89.** $\ln P_0 e^{rt}$
91. True   **93.** True   **95.** False   **97.** False   **99.** True
101. False   **103.** $\log P(1 + r)^t$ **105.** 300
107. The interval $0 < t < 10$ represents the years 1960–
1970. Although the model includes this period, the
actual data does not. Therefore, the model function
should be regarded as valid only for 1970–1990.
109. 1997

Section 11.4   (page 776)

1. The exponential function b^x is not defined for $b \leq 0$ or
for $b = 1$.
3. $7^2 = 49$ **5.** $10^{-3} = 0.001$ **7.** $e^{1/3} = \sqrt[3]{e}$
9. $m^2 = n$ **11.** $\log_4 64 = 3$ **13.** $\log 0.00001 = -5$
15. $\log_{1/2} 8 = -3$ **17.** $\ln P = rt$ **19.** $\ln x$
21. $\log_{1/4} x$ **23.** 10^x **25.** 1.5^x
27. If $x = \log 100$, then $10^x = 100 = 10^2$. Thus
$x = \log 100 = 2$. If $x = \ln 100$, then $e^x = 100$,
but 100 cannot be easily written as a power of e.
29. 2   **31.** 0   **33.** 2.5   **35.** 3   **37.** Not defined
39. -0.5 **41.** 1   **43.** 2.5   **45.** Not defined   **47.** -1
49. 3   **51.** 0.5   **53.** 1.39   **55.** 1.40   **57.** -0.14
59. 0.75   **61.** -0.62 **63.** 3.63   **65.** 0.35   **67.** Write
the equation in exponential form.   **69.** 243   **71.** e
73. 0.25   **75.** 9   **77.** 125   **79.** All $x > 0$, $x \neq 1$
81. 2   **83.** 3   **85.** 15   **87.** ± 25
89. (a) The graph rises from left to right if $b > 1$.
   (b) The graph falls from left to right if $0 < b < 1$.
91. $\{x \mid x < 2.5\}$ **93.** $\{x \mid x \neq 3\}$
95. $\{x \mid x < -3 \text{ or } x > 4\}$ **97.** 56.36   **99.** 0.02
101. 45.02   **103.** 0.27
105. Using the Change of Base Formula, we can enter
either $\dfrac{\ln x}{\ln 3}$ or $\dfrac{\log x}{\log 3}$.
107. (a) Up 3   (b) Left 3   **109.** (a) Reflected in y-axis
(b) Reflected in x-axis   **111.** (a) (B)   (b) (D)   (c) (C)
(d) (A)   **113.** 36.02   **115.** 11
117. Because $\ln \frac{1}{2} < 0$, reverse the inequality symbol to
obtain $a(\ln \frac{1}{2}) < b(\ln \frac{1}{2})$.

Section 11.6   (page 794)

1. If b can be written as a power of 3, then we can equate
exponents instead of writing the equation in logarithmic
form.
3. -0.6 **5.** -1 **7.** 1.46   **9.** 11.55   **11.** -0.03

13. 0.61   **15.** −5.86   **17.** 1.39   **19.** No solution

21. 6.45   **23.** ±1   **25.** −0.57, −1.43   **27.** 0, 1

29. No, the base must be the same to equate the arguments.

31. 3   **33.** 26   **35.** $\frac{32}{3}$ **37.** 34   **39.** 8, −1   **41.** $-\frac{11}{3}$

43. 2   **45.** ±3   **47.** 8   **49.** 2, 5   **51.** 10

53. No solution   **55.** −2.41, 0.41   **57.** 1   **59.** 0.75

61. 33, 1.5

63. If b were permitted to be 1, then $1^3 = 1^8$ but $3 \neq 8$.

65. 10, 10^{-4} **67.** 4, 256   **69.** ln 3, ln 2   **71.** $\dfrac{\ln 7}{\ln 5}$

73. $I = 10^R$ **75.** $I = I_0 \sqrt[10]{10^L}$ **77.** $r = \dfrac{1}{t} \ln \left(\dfrac{P}{P_0} \right)$

79. $t = \dfrac{\ln 2}{n \ln \left(1 + \dfrac{r}{n} \right)}$ **81.** 4.95%   **83.** (a) 14 million

(b) 25.51 million   (c) 34.43 million   **85.** 9953 years

87. $85,000   **89.** 5:53 A.M.   **91.** 7.9

93. 10^8 times as intense

95. For 1970, $t = 0$ and the natural logarithm is not defined at 0.

97. The model is more accurate for 1988 than for 1986.

99. 100, 1   **101.** 10^e **103.** 1, 3   **105.** 0, −0.5

107. (a) Basic Property of Logarithms   (b) Equate Logarithmic Expressions   (c) Power Rule for Logarithms
(d) Divide both sides of an equation.   **109.** −2.12

111. −0.77, 2, 4

Chapter Review Exercises   *(page 798)*

1. $x^2 + 4$ **3.** 0.17   **5.** $9x^2 + 15x + 4$ **7.** −1.85

9. No, $g(-1) = 0$, but f is not defined at 0.

11. $\dfrac{1}{|2x + 5| + 2}$, $\left| \dfrac{2}{x + 2} + 5 \right|$

13. Use the Vertical Line Test to determine if a relation is a function. Use the Horizontal Line Test to determine if a function is one to one.

15. Not one to one   **17.** $\sqrt[3]{x + 5}$ **19.** The exponent x can be any real number, but $b > 0$ and $b \neq 1$.   **21.** 1.41

23. 5.5   **25.** $\frac{10}{17}$ **27.** For growth, the function should be increasing. Thus $b > 1$.   **29.** $\log_8 4 = \frac{2}{3}$

31. $25^{-(1/2)} = \frac{1}{5}$

33. The function $y = 1^x$ is not one to one. Thus, it does not have an inverse.

35. 3.23   **37.** −2.5   **39.** 181.27   **41.** 49   **43.** 2.5

45. (a) (1, 0)   (b) None   (c) $\{x \mid x > 0\}$

(d) All real numbers   **47.** 41   **49.** 1

51. Because log 49 = 2 log 7, we estimate log 49 as $2(0.845) = 1.69$.

53. $\frac{3}{4} \log y - \frac{5}{2} \log x - 4 \log z$

55. $\log \sqrt[4]{\dfrac{(x + 4)^3}{2yz}}$ **57.** $\log_2 \dfrac{(24)(25)}{75} = \log_2 8 = 3$

59. In the first equation, both sides can be written as powers of 3. In the second equation, 4 cannot be easily written as a power of 3. The equation can be solved with logarithms.

61. 1.77   **63.** 5   **65.** 5   **67.** 1.19   **69.** 3542

Chapter Test   *(page 801)*

1. *(11.1)* $x^2 + x - 2$

2. *(11.1)* $x - 2, x \neq -2$

3. *(11.1)* $x^2 + 4x$

4. *(11.1)* $\sqrt{x + 2}$

5. *(11.2)* $(f \circ g)(x) = (g \circ f)(x) = x$

6. *(11.2)* $x^3 - 2$

7. *(11.2)*

8. *(11.4)* 0.5

9. *(11.4)* 1.5

10. *(11.3)* 0

11. *(11.3)* 0.25

12. *(11.6)* 1.12

13. *(11.6)* 1.15

14. *(11.4)* Use the Change of Base Formula to obtain $\log_5 13 = \dfrac{\ln 13}{\ln 5}$.

15. *(11.5)* $2 \log A - \frac{1}{2} \log B$

16. *(11.5)* $\log (x + 1)$

17. *(11.5)* $\log_6 \left[\dfrac{x^2(x - 1)}{6} \right]$

18. *(11.6)* 2

19. *(11.6)* 2

20. *(11.6)* 1

21. *(11.6)* $e^{-2} \approx 0.14$

22. *(11.6)* 8.48%

23. *(11.5)* **(a)** 7000 **(b)** As t (time) increases, demand (D) decreases. The ad campaign is ineffective. **(c)** 10.3 weeks

24. *(11.6)* 15.85 times as intense

25. *(11.6)* 1.5 hours

Cumulative Test: Chapters 9–11
(page 803)

1. *(9.1)* **(a)** y^4 **(b)** $|y^5|$

2. *(9.2)* **(a)** 9 **(b)** $\frac{1}{9}$ **(c)** Not a real number **(d)** $-\frac{1}{8}$

3. *(9.3)* **(a)** $\dfrac{1}{a^{5/2}b^{2/3}}$ **(b)** $\dfrac{x^{1/2}}{y}$ **(c)** $x^3 y^6$

4. *(9.4)* **(a)** $\dfrac{x^2}{2y}$ **(b)** $\sqrt[5]{6x^3 y}$ **(c)** $\dfrac{x}{y}$

5. *(9.5)* **(a)** $2x^4\sqrt{5x}$ **(b)** $y\sqrt{y}$

6. *(9.6)* **(a)** $x\sqrt{3x}$ **(b)** $a + 2\sqrt{2a} + 2$

7. *(9.6)* 4

8. *(9.7)* **(a)** $\dfrac{5\sqrt[3]{4}}{2}$ **(b)** $\dfrac{x\sqrt{2} - \sqrt{xy}}{2x - y}$

9. *(9.8)* **(a)** 1 **(b)** -12, 15

10. *(9.9)* **(a)** $-i$ **(b)** $8 - 6i$ **(c)** $\dfrac{1 + 2i}{5}$

11. *(10.3)* Graphing can be used to estimate the solutions. Algebraic methods include factoring, using the Square Root Property, and using the Quadratic Formula. The Quadratic Formula always works.

12. *(10.3)* -2.09, 0.29

13. *(10.5)* -1, 7

14. *(10.3)* If the equation has complex solutions, then the discriminant is negative. Thus, $b^2 - 4ac < 0$ and so $b^2 < 4ac$.

15. *(10.4)* 9

16. *(10.5)* 8 mph

17. *(10.6)* **(a)** $(-\infty, -6) \cup (4, \infty)$ **(b)** Ø **(c)** $(-\infty, \infty)$

18. *(10.7)* **(a)** $(-\infty, 2.77]$ **(b)** $(-\infty, 0) \cup (3, \infty)$

19. *(10.8)* **(a)** $c = 3$ **(b)** $a < 0$ **(c)** $a = -3$

20. *(10.8)* **(a)** 4 **(b)** 16

21. *(11.1)* **(a)** 1 **(b)** 0 **(c)** 1

22. *(11.1)* **(a)** 1 **(b)** Not defined

23. *(11.2)* $f^{-1}(x) = \dfrac{3x}{x - 1}$

24. *(11.3)* Both have a y-intercept at $(0, 1)$ and no x-intercept. The graph of f rises and the graph of g falls from left to right.

25. *(11.4)* **(a)** 5 **(b)** 10 **(c)** $\sqrt{7}$

26. *(11.5)* **(a)** $2 \ln x + \ln y - \frac{1}{2} \ln z$ **(b)** $\log \dfrac{(x - 1)^3}{x + 1}$

27. *(11.6)* **(a)** $9, -3$ **(b)** 0.46 **(c)** -0.69

28. *(11.6)* **(a)** 6%

Index of Applications

Index

Geometry Formulas

Plane Figures   Area (A), Perimeter (P), and Circumference (C)

Square
$A = s^2$
$P = 4s$

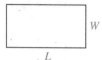

Rectangle
$A = LW$
$P = 2L + 2W$

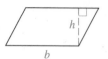

Parallelogram
$A = bh$

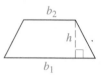

Trapezoid
$A = \frac{1}{2}(b_1 + b_2)h$

Circle
$A = \pi r^2$
$C = 2\pi r$
Diameter $= 2r$

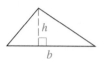

Triangle
$A = \frac{1}{2}bh$
Sum of measures of
three angles is 180°.

Right Triangle (one angle 90°)
$A = \frac{1}{2}ab$
Pythagorean Theorem:
$a^2 + b^2 = c^2$

Isoceles Triangle
Two equal sides
Two equal angles

Equilateral Triangle
All sides equal
All angles equal

Solid Figures   Volume (V) and Surface Area (S)

Cube
$V = s^3$

Rectangular Solid
$V = LWH$

Right Circular Cylinder
$V = \pi r^2 h$
$S = 2\pi rh + 2\pi r^2$

Right Circular Cone
$V = \frac{1}{3}\pi r^2 h$
$S = \pi r^2 + \pi r\sqrt{r^2 + h^2}$

Sphere
$V = \frac{4}{3}\pi r^3$
$S = 4\pi r^2$